# ESSENTIALS OF
# ORGANIC CHEMISTRY

**WCB** **Wm. C. Brown Publishers**

Dubuque, IA  Bogota  Boston  Buenos Aires  Caracas  Chicago  Guilford, CT
London  Madrid  Mexico City  Seoul  Singapore  Sydney  Taipei  Tokyo  Toronto

# ESSENTIALS OF
# ORGANIC
# CHEMISTRY

## ROBERT J. BOXER

### Georgia Southern University

## Project Team

Developmental Editor   *Brittany J. Rossman*
Marketing Manager   *Patrick E. Reidy*
Publishing Services Coordinator   *Julie Avery Kennedy*

## Wm. C. Brown Publishers

*President and Chief Executive Officer*   Beverly Kolz
*Vice President, Director of Editorial*   Kevin Kane
*Vice President, Sales and Market Expansion*   Virginia S. Moffat
*Vice President, Director of Production*   Colleen S. Yonda
*Director of Marketing*   Craig S. Marty
*National Sales Manager*   Douglas J. DiNardo
*Advertising Manager*   Janelle Keeffer
*Production Editorial Manager*   Renée Menne
*Publishing Services Manager*   Karen J. Slaght
*Royalty/Permissions Manager*   Connie Allendorf

*Production services by*   Textbook Writers Associates, Inc.
*Cover design and icons by*   Linda Dana Willis

Library of Congress Catalog Card Number: 95-76541

ISBN 0-697-14060-1

Printed in the United States of America

2460 Kerper Boulevard, Dubuque, IA 52001

10   9   8   7   6   5   4   3   2   1

**To my family**

To my wife, Riette, my children, Deborah and Mark,
and to Mark's bride, Betsy,

with love

# Brief Contents

# Contents

Optional sections are indicated in light purple type.

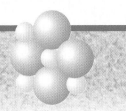

Optional sections are indicated in light purple type.

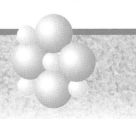

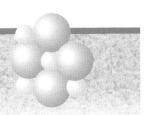

Optional sections are indicated in light purple type.

# Preface

## Audience

I wrote this textbook primarily to serve biology students and a variety of health science related majors. By no means, though, is the text limited to this audience. Students studying agronomy, forestry, materials engineering, geology, animal science, foods and nutrition, technology, and science education will benefit by the use of this textbook.

One of my goals in writing this textbook was to produce a flexible book designed for use in a two-quarter or a one-semester course sequence. The student should have previously taken the minimum of two quarters (or one semester) of general chemistry for science majors. The level of coverage is designed to provide the student the organic chemistry needed either in furthering his/her education or in practicing his/her profession.

## Philosophy

My main purpose in writing this textbook is to provide the student with a concisely written, detailed, yet easily understandable textbook. In a more specific sense, I know how organic chemistry is perceived by students and I want to show them that organic chemistry is not merely a head-aching agony of memorization, but contains within it laws and principles that can make the subject a great deal simpler and fun to learn and use. Also, in introducing the student to examples of the use of organic chemicals in society, they can see that without organic chemistry our existence would be very bleak indeed. Imaging being without soap, antibiotics, or gasoline! With this in mind, I have included the essential topics that I believe are most important to the student.

## Organization and Content

The text is organized in such a way as to proceed from the simple to the more complex. That is, Chapter 1 and some of Chapter 2 represent a review of general inorganic chemistry. Chapters 3 and 4 cover the relatively nonreactive alkanes; and Chapters 5 and 6 the more chemically complex alkenes. Chapters 7–11 cover the remainder of the hydrocarbons. Families with one functional group that contain elements other than carbon and hydrogen are studied in Chapters 12–25. Biomolecules which contain two or more functional groups are covered in Chapters 26–28.

I used unifying "threads" to weave material together, not only from chapter to chapter, but also from section to section. These threads bring continuity to the text by emphasizing:

    a. structure and geometry

    b. nomenclature

    c. natural occurrence and uses

    d. physical properties

    e. synthesis and reactions

    f. mechanisms

The textbook is flexible in that an instructor who wishes to provide more limited coverage can opt to omit certain chapters and sections. Each individual instructor can add or subtract material based on individual preference and time restraints. A suggested one-semester course sequence is indicated in black type in the expanded table of contents. The material in purple type is material that may more easily be covered in a two-quarter course or other course which needs more detail than the typical one-semester course.

For the instructor who prefers in-depth coverage, I introduced synthesis early in the text with reaction mechanisms in Chapter 4. I believe this enables the student to get over the "shock" of reasoning out syntheses and learning reaction mechanisms early on, so as to gain confidence and excel as he/she meets these concepts later. For a shorter course, refer to the table of contents for more detailed, precise recommendations, but in general, only cover the first three sections of Chapter 13 (Organic Halogen Compounds), the first nine sections of Chapter 27 (Carbohydrates), and the first eight sections in Chapter 28 (Amino Acids and Proteins). Methods of synthesis and reactions are covered at the end of each chapter and can be omitted. Chapter 12 (Nucleophilic Substitution and Elimination) can be deleted in a limited coverage curriculum.

An optional chapter covering spectroscopy is provided in Chapter 29 (Spectroscopy: Instrumental Methods of Structure Determination). Along with the standard in-chapter exercises and the end-of-chapter problems, two spectroscopy-related problems are provided at the very end of the end-of-chapter problems in the synthesis chapters starting with Chapter 13 (Organic Halogen Compounds).

**Pedagogical Devices**

As a teacher, I always count on the *teaching* aspect of any text to help me communicate the key points of content. As I tested this manuscript on my own two-quarter organic students, I also tested various pedagogical features. The acid test for each feature is that it must help the student. The most helpful features are included in this finished product.

1. *Complete table of contents.* A detailed summary of the contents of each chapter is provided. Topics are divided and subdivided so as to provide the student with quick access to a given topic, and to help students quickly identify key concepts in each chapter.

2. *In-chapter examples, solved problems, and exercises.* When a given concept is introduced, at least one example is given as a part of the text. Often, a step-by-step solution of a problem related to the concept is provided. Following the solved problem, the student has the opportunity to test his/her mastery of the material just covered by solving an exercise.

3. *Listing and glossary of key terms.* The key terms are in italics and are colored magenta. Each such term is listed at the end of each chapter, along with the section or sections in which the term is to be found, and each term is defined in a glossary located at the end of this textbook. Again, in the glossary, the section location of each key term is specified.

4. *Chapter cross-references.* Often a topic is discussed that has been previously introduced. In such an event, in the narrative of the textbook, the student is directed to the chapter, section, or page in which the concept was first discussed. Far more rare is a situation where the student is directed to a chapter, section, or page in a higher numbered chapter to explain a concept more fully.

5. *Chapter accomplishments.* At the end of each chapter, a number of questions are asked of the student encompassing material in each section in the chapter. These questions are not difficult, and are designed to establish if the student has minimal mastery of the material. Beside each question is a box the student can check off when he/she is able to answer the particular question. No answers are supplied for these questions.

6. *End-of-chapter problem sets.* These represent a number of more challenging problems than the chapter accomplishments to attempt to stimulate the student to think.

7. *Answers to chapter exercises and end-of-chapter problem sets.* The answers to all the odd-numbered exercises and problems are found at the end of the textbook. Each problem has been checked by two Ph.D. organic chemists and a great deal of care has been taken to assure that they are accurate.

8. *Boxed readings.* In most chapters there are one or more boxed readings, which are mini-essays that relate the material to the reality of our lives. Topics are of agricultural, biochemical, historical, medical, and industrial interest.

9. *Ilustrations.* Each chapter is profusely illustrated with figures, tables, and chemical formulas. All illustrations are annotated to ensure clarity.

10. *Marginal notes.* There are located throughout the textbook a number of notes in blue which are written in the margins. Each note either includes more information about a concept, reviews a concept, or serves as a cross-reference.

11. *Biological applications.* At appropriate points where the biological significance of a particular topic or concept is illuminated, an icon of the bombardier beetle (see back cover for explanation) is present to highlight the section.

12. *"Envision the Reaction" boxes.* Each box contains a step-by-step description of the progress of a reaction. Students are encouraged to use a mechanistic approach to understand a reaction rather than rote memorization.

13. *Color-coding scheme.* Because it is difficult for students to understand the chemical changes that occur in complex reactions, we have color-coded reactions so that chemical groups being added or removed in a reaction can be quickly and clearly identified.

14. *Colored backgrounds.* In various sections there appear numbered lists of instructions or explanations for students, such as nomenclature and bonding rules. These lists are set off against a light orange background for easy identification.

15. *Appendixes.* The textbook contains five appendixes. General inorganic chemistry problems with some relationship to organic chemistry are provided in Appendix 1. Appendix 2 covers the less-often-used common system of nomenclature for alkanes. A comparison of common and IUPAC nomenclature of alkanes is discussed in Appendix 3. Appendix 4 treats the common nomenclature of alkenes, and Appendix 5 compares common and IUPAC nomenclature for alkenes.

## Supplementary Materials

1. *Instructor's Manual/Test Item File.* A list of transparencies, answers to the even-numbered end-of-chapter questions, and the printed Test Item File are featured in the Instructor's Manual/Test Item File.

2. *Student Study Guide/Student Solution Manual.* Authored by Jerome Maas of Oakton Community College, this helpful tool guides students through the study of organic chemistry by offering a summary of each chapter's material, a student self-quiz, and answers to the even-numbered questions.

3. *Laboratory Manual.* By Murray Zanger and James McKee of the Philadelphia College of Pharmacy and Science, this includes thirteen small-scale organic experiments appropriate for the one-semester or two-quarter organic chemistry course.

4. *Instructor's Manual for Lab Manual.* Accompanies the above laboratory manual.

5. *Transparencies.* A collection of 100 key full-color illustrations and figures is available to adopting professors.

6. *Customized Transparency Service.* For those adopters interested in receiving acetates of text figures not included in the standard transparency package, a select number of acetates will be custom-made upon request.

7. *MicroTest.* This computerized classroom management system/service includes a database of test questions, reproducible student self-quizzes, and a grade-recording program. Disks are available for IBM, Windows, and Macintosh computers, and require no programming experience.

8. *How to Study Science.* Written by Fred Drewes of Suffolk County Community College, this excellent workbook offers students helpful suggestions for meeting the considerable challenges of a science course. It offers tips on how to take notes and how to overcome science anxiety. The book's unique design helps to stir critical thinking skills, while facilitating carefully notetaking on the part of the student.

9. *Exploring Chemistry Videotapes.* Narrated by Ken Hughes of the University of Wisconsin-Oshkosh, the tapes provide six hours of laboratory demonstrations. Many of the demonstrations are of high interest experiements, too expensive or dangerous to be performed in the introductory laboratory.

Words cannot express how deeply grateful I am to my wife, Riette, whose patience with this project and support of this project enabled it to come to fruition. Staying home on a Saturday night while one's husband works on a manuscript can get old. I wish to thank the staff of Wm. C. Brown Publishers for all the work they put in on this textbook. A special thank you goes to my Developmental Editors, Elizabeth Sievers, John Berns, and Brittany Rossman for "holding my hand" so effectively over the years. I would also like to thank Molly Kelchen, Editorial Secretary, Julie Kennedy, Publishing Services Coordinator at Wm. C. Brown, and Marty Tenney, Project Manager of Textbook Writers Associates, for all the work they did to ensure a quality book was produced.

I would like to thank my colleague Robert Nelson for reading every word of the manuscript and providing me with the necessary energy of activation to satisfy some of the reviewers. Also, my thanks to Bill Ponder for the innumerable pesky nit-picking questions that he so graciously answered, and to Michael Hurst for his answers to my amino acid and protein questions. Last, but not least, thanks to Norman Schmidt for the infrared spectra and Jeffrey Orvis for the nuclear magnetic resonance spectra.

Simple thanks are not sufficient to convey my gratitude to the reviewers who read the manuscript in its many stages. They include:

**Acknowledgments**

**James R. Ames**
University of Michigan-Flint

**Barbara C. Andrews**
Catawba College

**Charles R. Bacon**
Ferris State University

**Sukhamaya (Sam) Bain**
Xavier University of Louisiana

**Satinder Bains**
Arkansas State University-Beebe Branch

**Ronald J. Baumgarten**
University of Illinois-Chicago

**Barrett Benson**
Bloomsburg University

**C. Larry Bering**
Clarion University of Pennsylvania

**Thomas Berke**
Brookdale Community College

**K. D. Berlin**
Oklahoma State University

**Dale R. Buck**
Cape Fear Community College

**Charles K. Buller**
Tabor College

**G. Lynn Carlson**
University of Wisconsin-Parkside

**Dana Chatellier**
University of Delaware

**Sheldon I. Clare**
University of Pittsburgh-Johnstown

**Brian E. Cox**
Cochise College

**Jerry Easdon**
College of the Ozarks

**Jeffrey E. Elbert**
South Dakota State University

**David K. Erwin**
Rose-Hulman Institute of Technology

**Sharbil J. Firsan**
Oklahoma State University

**Richard Fish**
California State University, Sacramento

**Kenneth A. French**
Blinn College

**Joanna H. Fribush**
North Adams State College

**Joseph J. Friederichs**
Dawson Community College

**John Fulkrod**
University of Minnesota-Duluth

**Ana M. Gaillat**
Greenfield Community College

**Roy Garvey**
North Dakota State University

**Theresa Gioannini**
Baruch College

**Helen S. Hauer**
Delaware Technical & Community College

**Byron L. Hawbecker**
Ohio Northern University

**Allan D. Headley**
Texas Tech University

**Christine K. F. Hermann**
Radford University

**Richard Hoffman**
Illinois Central College

**Philip Hogan**
Lewis University

**Tamera S. Jahnke**
Southwest Missouri State University

**T. G. Jackson**
University of South Alabama

**Dennis N. Kevill**
Northern Illinois University

**Angela Glisan King**
Wake Forest University

**Marilyn J. Kouba**
Harold Washington College (retired)

**Julia A. H. Lainton**
Clemson University

**Joseph Landesberg**
Adelphi University

**Michael Looney**
Schreiner College

**Fulgentius N. Lugemwa**
Murray State University

**Jerome Maas**
Oakton Community College

**William L. "Hank" Mancini**
Paradise Valley Community College

**Guy Mattson**
University of Central Florida

**Larry D. Martin**
Morningside College

**William A. Meena**
Rock Valley College

**Robert D. Miess**
Mayville State University

**Nathan C. Miller**
University of South Alabama

**Aaron P. Monte**
University of Wisconsin-La Crosse

# CHAPTER 1

# An Introduction to Organic Chemistry

Humans used organic compounds thousands of years ago. Ethyl alcohol was produced by fermentation of honey and grapes. Sucrose (from sugar cane or berries) satisfied the human sweet tooth, while a variety of plant parts (containing organic compounds) served as drugs to combat disease.

By 1750, a sufficient number of substances were isolated from plant and animal products to stimulate the curiosity of scientists. Thus, between 1750 and 1820, a very large number of compounds were obtained, such as morphine from the opium poppy, citric acid from lemon, uric acid and urea from human urine, cholesterol from animal fats, and hippuric acid from horse urine.

Since these compounds were associated with living and once-living organisms, the science of organic chemistry gradually evolved. The term *organic* referred to the fact that these compounds were obtained from the *organs* of living and once-living things.

The modern definition of organic chemistry is based on the experimental work of Lavoisier (in the late 1700s), which showed that the known organic compounds always contained carbon and hydrogen and sometimes contained nitrogen and oxygen. Organic chemistry is the chemistry of *most* of the compounds of carbon. You should realize that not all compounds of carbon are organic. The metal carbonates ($K_2CO_3$), bicarbonates ($NaHCO_3$), and cyanides (LiCN) are considered inorganic, along with gases such as carbon dioxide ($CO_2$), carbon monoxide (CO), and hydrogen cyanide (HCN).

## 1.1 WHERE IT ALL BEGAN

During the 1700s, most chemists held that organic compounds could be obtained *only* from a living or once-living source. They believed there was a *vital force* present only in living things that enabled them to synthesize these compounds. This was the *vital force theory*.

Friedrich Wohler discredited the vital force theory in 1828 when he evaporated an aqueous solution of ammonium cyanate (an inorganic compound obtained from the reaction of silver cyanate and ammonium chloride) for several hours, producing urea, an organic compound.

## 1.2 WOHLER AND THE VITAL FORCE THEORY

This theory held back organic chemistry for many years because it discouraged chemists from attempting any laboratory synthesis of an organic compound from inorganic sources.

$$NH_4OCN \xrightarrow[\text{several hours}]{\text{evaporation}} NH_2CONH_2$$

Ammonium cyanate (inorganic)      Urea (organic)

For the first time, the white crystals of urea were produced outside the human body. Vitalism died slowly. It took 25 more years before the last diehard chemists were convinced.

# 1.3
# THE ATOM

You might ask, Why do we consider the probability of finding an electron? During the early 1900s, the electron was believed to be a tiny hard sphere of negative charge revolving around the nucleus of an atom the same way a planet revolves around our Sun. By the 1930s, the theory of quantum mechanics established that the electron is a cloud of negative charge in space surrounding the nucleus, with the shape of the cloud depending on the energy of the electron. The density of matter in the electron cloud of a particular shape (the orbital) varies and is a function of the probability of locating the electron.

Before we begin a more detailed discussion of organic chemistry, we should review some basic concepts, starting with the structure of the atom. Atoms are fundamental, very small particles of an element, composed of an extremely dense nucleus packed with positively charged protons and neutral neutrons. The number of protons in the nucleus of an atom is called the *atomic number* of the atom. Surrounding the nucleus are a number of negatively charged electrons equal to the number of protons. These electrons (one or two) of equal energy are found most probably in a section of space called an *orbital*.

Let us begin our study of orbitals by considering $s$ and $p$ orbitals (Fig. 1.1a, b). The greatest electron density in the spherical $s$ orbital (Fig. 1.1a) is at the center of the sphere. The electron density gradually decreases as the surface of the sphere is approached. A somewhat similar situation exists for each of the dumbbell-shaped $p$ orbitals in that there is a variation of electron density within the orbital with the greatest electron densities in two regions close to but on opposite sides of the nucleus (Fig. 1.1b). The electron population in $s$ and $p$ orbitals is detailed in Table 1.1.

Since each orbital contains a maximum of two electrons, the electron configuration for the first 10 elements is as follows. Hydrogen, with an atomic number of 1, contains only one electron (designated by the superscript 1) and has the configuration $1s^1$. Helium, with two electrons, completes the 1s subshell (the first main shell) with a $1s^2$ electron configuration.

Because no orbital can hold more than two electrons, and since the first main shell contains only one subshell, the third electron of lithium must begin to fill the second main shell, as you can see in Fig. 1.2.

For the following elements let us consider the filling of the $p$ orbitals. The $2p$ orbital is filled in accordance with *Hund's rule*, which states that each orbital in a $p$ subshell is half-filled with electrons before any $p$ orbital is completely filled. Thus consider the electron configuration of boron and carbon (Fig. 1.3a, b). Note how in carbon, instead of having two electrons in the same orbital, Hund's rule requires that they occupy separate $p$ orbitals until each $2p$ orbital is half-filled.

The outer shell electrons are called *valence electrons*. The valence electrons are available for transfer or sharing with an atom or atoms of other elements and determine the chemical reactivity of those elements.

**FIGURE 1.1** Geometry of an $s$ orbital (a) and $p$ orbital (b). The three $p$ orbitals, each of which is perpendicular to the others, make up a $p$ subshell.

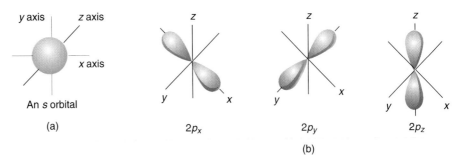

| TABLE 1.1 | Electron Population in the First and Second Main Shells | | |
|---|---|---|---|
| Number of Main Shell | Name of Subshell | Number of Orbitals in the Subshell | Maximum Number of Electrons in the Subshell |
| 1 | 1s | 1 | 2 |
| 2 | 2s | 1 | 2 |
| 2 | 2p | 3 | 6 |

---

### HISTORICAL BOXED READING 1.1

Philippus Aureolus Theophrastus Bombastus von Hohenheim (1493–1540), or Paracelsus, as he preferred to be called, was trained as an alchemist and a physician.

The main objective of alchemy was the conversion of common (base) metals into gold. Paracelsus rebelled at this and insisted that alchemy should instead be used to prepare medicines for healing of the sick. He believed that the human being was a chemical factory to be dosed with neutralizing chemicals when things went wrong. Thus he used opium to alleviate pain, sugar to sweeten his less pleasant tasting preparations, and ethyl alcohol to dull the senses.

His main arsenal of drugs consisted of a number of salts of iron, lead, copper, antimony, and mercury, along with sulfur and a number of arsenic compounds. A large part of his skill as a physician was in realizing that high doses of many of these compounds would do more harm than good.

All in all, unlike many physicians of that day, Paracelsus cured more than he killed. He simply could not accept the old cures: gout by playing a flute, sciatica by blowing a trumpet, jaundice with liverwort, or pneumonia with feverwort. He relied on his arsenal of compounds instead.

Paracelsus had a Dr. Jekyll and Mr. Hyde kind of personality. On the one hand, he believed in the importance of observation and experience in treating patients, utterly rejected the useless medical dogma of his day, and was a good chemical experimentalist. On the other hand, however, he was boastful, far too critical of his peers, and an alcoholic. Other physicians and apothecaries either loved him or hated him; there was no in-between.

Paracelsus was the catalyst that made the science of medicinal chemistry throb with life.

---

Determine the number of valence electrons in the nitrogen atom.

**SOLVED PROBLEM 1.1**

**SOLUTION**

Write the electron configuration of nitrogen, atomic number 7.

$$N \qquad 1s^2 2s^2 2p^3$$

---

**FIGURE 1.2** The outer-level electron configuration (valence electron configuration) is repeated for all elements within the same group of the periodic table.

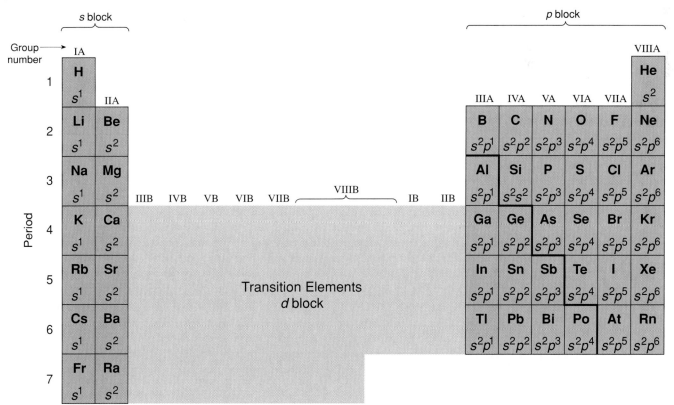

The second main shell is the outer-most shell in the nitrogen atom. Since this shell contains five electrons ($2s^2 2p^3$), the number of valence electrons in the nitrogen atom is 5.

## EXERCISE 1.1

Determine the number of valence electrons in the hydrogen atom, the carbon atom, and the neon atom.

## 1.4
## THE IONIC BOND: IONIC COMPOUNDS

**FIGURE 1.3** The electron configuration of boron (a). The electron configuration of carbon (b).

$$\underset{2s}{\underline{\uparrow\downarrow}} \quad \underset{2p_x}{\underline{\uparrow}} \quad \underset{2p_y}{\underline{\phantom{\uparrow}}} \quad \underset{2p_z}{\underline{\phantom{\uparrow}}}$$

(a)        (Boron $1s^2 2s^2 2p^1$)

$$\underset{2s}{\underline{\uparrow\downarrow}} \quad \underset{2p_x}{\underline{\uparrow}} \quad \underset{2p_y}{\underline{\uparrow}} \quad \underset{2p_z}{\underline{\phantom{\uparrow}}}$$

(b)        (Carbon $1s^2 2s^2 2p^2$)

Although the preceding orbital diagram indicates that the fifth electron of boron occupies the $2p_x$ orbital, for our purposes the three $2p$ orbitals are of equal energy, and the fifth electron of boron could just as well occupy the $2p_y$ or the $2p_z$ orbital.

A chemical bond can form between two elements in one of two ways: either valence electrons of one element are transferred to the other element to form an *ionic bond*, or valence electrons from one element are shared with valence electrons from another element to form a *covalent bond*.

The driving force in the formation of an ionic bond is the formation of a *noble gas electron configuration* ($ns^2 np^6$) for each element participating in bond formation. Therefore, an element with one or two valence electrons (a metal in Group IA or IIA) readily gives up the electrons to produce a positively charged ion, or *cation*. In the same way, an element with six or seven valence electrons (a nonmetal in Group VIA or VIIA) accepts the electrons donated by the metal to form a negatively charged ion, or *anion*. The electrostatic forces of attraction between oppositely charged ions is an ionic bond.

Consider an element from Group IA combining with an element from Group VIIA to form an ionic bond.

$$\text{Na } (1s^2 2s^2 2p^6 3s^1) + \text{Cl } (1s^2 2s^2 2p^6 3s^2 3p^5)$$

The $3s$ electron of a sodium atom is donated to fill the $3p$ orbital of the chlorine atom to form a $\text{Na}^+$ ion and a $\text{Cl}^-$ ion, which combine to give sodium chloride, an ionic compound:

$$\text{Na}^+ (1s^2 2s^2 2p^6) + \text{Cl}^- (1s^2 2s^2 2p^6 3s^2 3p^6) \longrightarrow \text{NaCl}$$

Any given ionic compound, like sodium chloride, does not exist as a discrete molecule but as a network of ions. Using sodium chloride as an example, six sodium ions are displayed around each chloride ion, and six chloride ions are displayed around each sodium ion (Fig. 1.4).

## 1.5
## THE COVALENT BOND

Note that both the resulting chloride ion (a negatively charged ion, or anion) and the sodium ion (a positively charged ion, or cation) have a *stable octet* of electrons ($ns^2 np^6$).

As a rule, ionic compounds are high-melting solids, usually soluble in water. The high melting points are an indication of the strength of the ionic bond.

Expressing it another way, a monopolar covalent bond is a bond where the electrons are shared equally between the two atoms. Examples of molecules, each of which contains a nonpolar covalent bond, are the fluorine and hydrogen molecules.

H : H      F : F

The driving force in the formation of a covalent bond is similar to that of an ionic bond, i.e., the formation of a noble gas electron configuration for each element participating in bond formation. However, covalent bond formation occurs by a different route, most often where one element shares an electron pair with another element. Multiple covalent bonds exist, but for now let us confine our attention to the single-electron-pair covalent bond.

Consider the combination of two atoms of hydrogen to give a molecule of hydrogen. Each hydrogen atom furnishes an electron to be shared by both atoms in the molecule, and this provides each atom in the resulting molecule with a stable duet of electrons.

There are three kinds of covalent bonds: a *nonpolar covalent bond*, a *polar covalent bond*, and a *coordinate covalent bond* or *dative bond*. A bond where the electron density is placed fairly equally between the two atoms of the bond is called a *nonpolar covalent bond*.

On the other hand, when the electrons of the bond are closer to one atom than to the other, the bond is called a *polar covalent bond*. You might ask, Why are the electrons of the bond closer to one atom than to the other? The reason is that the two atoms differ in *electronegativity*.

The electronegativity of an element is the tendency of a bonded atom to attract electrons toward itself. In general terms, the electronegativity of an atom increases across a period of the periodic table from left to right and increases from the bottom to the top of a group in the periodic table (refer to Fig. 1.5).

**FIGURE 1.4** (a) Each Cl⁻ ion in this distance-exaggerated crystal lattice of NaCl is surrounded by six Na⁺ ions. Although not shown in this structure, each Na⁺ ion is surrounded by six Cl⁻ ions. (b) The NaCl crystal lattice (an orderly arrangement of ions in a crystal).

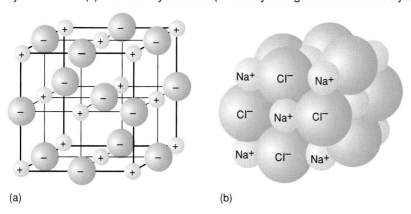

(a)                                    (b)

Consider a molecule of hydrogen chloride.

$$\overset{\delta^+\ \ \delta^-}{H\ :\ Cl}$$

Since chlorine is the more electronegative element, it bears the partial negative charge. The greater the electronegativity difference between the two bonding atoms, the more polar is the bond. For example, the carbon-chlorine bond (electronegativity difference is $3.0 - 2.5 = 0.5$) is a more polar bond than the carbon-bromine bond (electronegativity difference is $2.8 - 2.5 = 0.3$) (see Fig. 1.5).

A bond is considered covalent when the electronegativity difference between bonded atoms is 1.7 or less. However, a bond is classified as ionic when the electronegativity difference is 1.7 or greater. A bond with an electronegativity difference of 1.7 is considered to be about 50% covalent and 50% ionic. The point is that the degree of covalent character in a bond can range from 100% for a bond such as that in $H_2$ to essentially zero for a bond such as that in CsF. Since the electronegativity difference between cesium and fluorine is $4.0 - 0.8 = 3.2$, we classify the bond as ionic (Fig. 1.5).

In both the polar covalent bond and the nonpolar covalent bond, each participating atom donates one electron to form the electron-pair covalent bond. Chemical species exist, however, in which one atom supplies both electrons. Such a bond is called a

Each lowercase Greek delta ($\delta$) designates that the given element bears a partial positive or negative charge.

Bond polarity also can be estimated using the electronegativity trends across a period and down a group in the periodic table. For example, one would predict that the carbon-oxygen bond is more polar than the nitrogen-oxygen bond because nitrogen lies to the right of carbon in the periodic table. One important bond in organic chemistry, the carbon-carbon bond, is nonpolar. The carbon-hydrogen bond, another important bond, is essentially nonpolar. Although the two atoms participating in the bond are different, you can note from Figure 1.5 that the difference in electronegativity between the atoms is small ($2.5 - 2.1 = 0.4$), and therefore the electron pair is close to midway between the two participating atoms.

**FIGURE 1.5** A scale of electronegativities according to Linus Pauling.

| H | | | | | | | | | | | | | | | | | |
|---|---|---|---|---|---|---|---|---|---|---|---|---|---|---|---|---|---|
| 2.1 | | | | | | | | | | | | | | | | | |

| Li | Be | | | | | | | | | | | B | C | N | O | F |
|----|----|---|---|---|---|---|---|---|---|---|---|----|----|----|----|----|
| 1.0 | 1.5 | | | | | | | | | | | 2.0 | 2.5 | 3.0 | 3.5 | 4.0 |

| Na | Mg | | | | | | | | | | | Al | Si | P | S | Cl |
|----|----|---|---|---|---|---|---|---|---|---|---|----|----|----|----|----|
| 0.9 | 1.2 | | | | | | | | | | | 1.5 | 1.8 | 2.1 | 2.5 | 3.0 |

| K | Ca | Sc | Ti | V | Cr | Mn | Fe | Co | Ni | Cu | Zn | Ga | Ge | As | Se | Br |
|---|----|----|----|---|----|----|----|----|----|----|----|----|----|----|----|----|
| 0.8 | 1.0 | 1.3 | 1.5 | 1.6 | 1.6 | 1.5 | 1.8 | 1.9 | 1.9 | 1.9 | 1.6 | 1.6 | 1.8 | 2.0 | 2.4 | 2.8 |

| Rb | Sr | Y | Zr | Nb | Mo | Tc | Ru | Rh | Pd | Ag | Cd | In | Sn | Sb | Te | I |
|----|----|---|----|----|----|----|----|----|----|----|----|----|----|----|----|---|
| 0.8 | 1.0 | 1.2 | 1.3 | 1.6 | 1.8 | 1.9 | 2.2 | 2.2 | 2.2 | 1.9 | 1.7 | 1.7 | 1.8 | 1.9 | 2.1 | 2.5 |

| Cs | Ba | La | Hf | Ta | W | Re | Os | Ir | Pt | Au | Hg | Ti | Pb | Bi | Po | At |
|----|----|----|----|----|---|----|----|----|----|----|----|----|----|----|----|----|
| 0.8 | 1.0 | 1.1 | 1.3 | 1.5 | 1.7 | 1.9 | 2.2 | 2.2 | 2.2 | 2.4 | 1.9 | 1.8 | 1.9 | 1.9 | 2.0 | 2.2 |

| Fr | Ra | Ac | Th | Pa |
|----|----|----|----|----|
| 0.8 | 1.0 | 1.1 | 1.3 | 1.4 |

*coordinate covalent bond* or *dative bond*. For example, when water reacts with a proton to form the hydronium ion, a coordinate covalent bond forms because the oxygen of the water provides both electrons to produce the bond. Once formed, a coordinate covalent bond cannot be distinguished from any other polar covalent bond.

$$ \overset{H}{\underset{..}{..O..}}\overset{H}{} \ + \ H^+ \longrightarrow \ \overset{H}{\underset{H}{:\overset{+}{O}}}\text{—}H $$

The vast majority of organic compounds are covalent. On the other hand, some inorganic compounds are ionic, and some are covalent.

## 1.6
## INTERMOLECULAR ATTRACTIVE FORCES

Covalent substances exist as discrete molecules with relatively weak intermolecular forces (Sec. 1.6). Covalent substances do exist in all states of matter. Covalent solids melt at low temperatures and covalent liquids boil at low temperatures, reflecting the relatively weak intermolecular forces compared with the stronger interionic forces.

Physical properties of covalent substances are determined by the relative strength of *intermolecular attractive forces*, i.e., forces between molecules. We will be discussing the effect of intermolecular forces on the physical properties of a considerable number of organic compounds in this textbook.

For our purposes, the three most important intermolecular forces we need to discuss are dipole-dipole attractions, hydrogen bonds, and London forces. Polar covalent molecules have oppositely charged ends. These partially charged molecules, called *dipoles*, tend to orient themselves so as to place a positively charged end of one dipole next to the negatively charged end of another dipole to produce *dipole-dipole attractions*. For example, consider hydrogen iodide:

$$ \overset{\delta^+ \ \ \delta^-}{H\text{—}I} \ \ \overset{\delta^+ \ \ \delta^-}{H\text{—}I} \ \ \overset{\delta^+ \ \ \delta^-}{H\text{—}I} $$

Hydrogen iodide dipoles

These electrostatic interactions tend to play an important role in determining the physical properties of polar covalent compounds.

One kind of dipole-dipole attraction needs to be studied separately because it is particularly strong. This interaction occurs when a partially positively charged hydrogen atom of one dipole interacts with a small, highly electronegative, partially negatively charged fluorine, oxygen, or nitrogen atom of another dipole. Since these species are small, the positive end of one dipole can approach very close to the negative end of a nearby dipole. This leads to a large force of attraction between neighboring dipoles that is called a *hydrogen bond*. The high boiling point of water (100°C) compared with hydrogen sulfide ($-62$°C) is due to hydrogen bonding.

$$ \overset{H}{\underset{}{}}\overset{}{\underset{}{}}\overset{H}{} $$

Hydrogen bond

Permanent dipole-dipole attractions cannot exist in nonpolar covalent compounds such as elemental bromine ($Br_2$). However, since nonpolar substances such as bromine can be liquefied, some intermolecular attractions must be present. These attractions, called *London forces*, are weaker than dipole-dipole attractions. Consider a bromine molecule. Over a period of time, the electrons are shared equally by each bromine atom, and the molecule is nonpolar. At a given instant, however, more electrons can be on one side of the molecule than on the other. We then say that the molecule becomes an *instantaneous dipole* because it exists for less than a second. However, during its short life the molecule can repel the electrons in a nearby molecule of bromine, creating another instantaneous dipole. These dipoles orient in the usual way to produce the instantaneous dipole-dipole attractions we know as London forces. These forces are weak because of the fleeting existence of the dipoles. London forces are short range and tend to increase with increasing molecular weight.

FIGURE 1.6 Various carbon atom combinations.

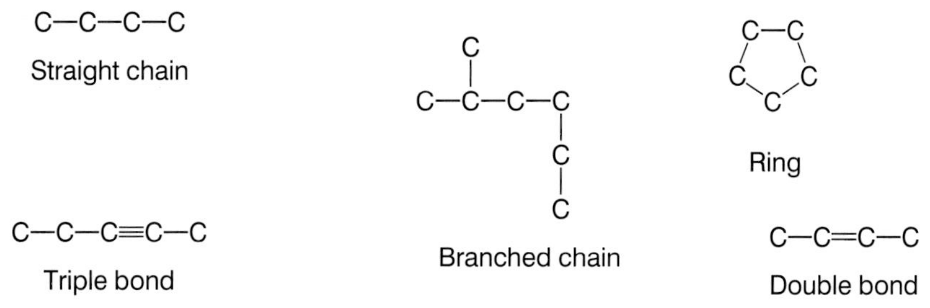

C—C—C—C
Straight chain

```
      C
      |
C—C—C—C
      |
      C
      |
      C
```
Branched chain

```
 C—C
C     C
 C
```
Ring

C—C—C≡C—C
Triple bond

C—C=C—C
Double bond

Only London forces are present in nonpolar covalent substances; both London forces and dipole-dipole attractions exist in polar covalent substances. The combination of London forces and dipole-dipole attractions found in polar covalent molecules is called *van der Waal's forces*. London forces play a minor role because they are so weak.

Organic chemicals play a vital role in the health professions, biological processes, and chemical industries. Almost all drugs prescribed by physicians and dispensed by pharmacists are organic, from lifesaving penicillin to pain-killing codeine. For the biochemist or biologist, organic compounds are involved in all the body functions of living organisms. Some of the major industries dependent to a large degree on organic compounds are petroleum, food, clothing, pharmaceuticals, and construction.

## 1.7
## THE EXTRAORDINARY BEHAVIOR OF THE CARBON ATOM

Although well over 8 million organic compounds exist, all containing the element carbon, less than 350,000 inorganic compounds containing the other 108 elements are known. What is so special about the carbon atom that it can be a part of this huge number of different substances? First, carbon can combine with itself to form straight chains, branched chains, or rings, and each combination can contain a varying number of carbon atoms. Second, the carbon atom can form multiple covalent bonds (double and triple bonds) with neighboring carbon atoms (Fig. 1.6).

## 1.8
## MOLECULAR FORMULA AND TYPES OF STRUCTURAL FORMULAS

Chemical formulas play a vital role in organic chemistry. A *molecular formula* tells us what kinds of atoms and the number of each kind of atom in a molecule. Consider methane, a major constituent of natural gas. The molecular formula of methane is $CH_4$. This means that one molecule of methane contains one carbon atom and four hydrogen atoms. In addition, organic chemists make use of the *structural formula (structure)* of a molecule. A structural formula shows the way in which atoms in a molecule are linked to one another. For example, the structural formula of methane shows that the carbon atom is bonded to four hydrogen atoms by means of four electron-pair covalent bonds.

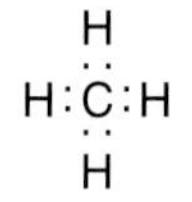

```
    H
    ··
H : C : H
    ··
    H
```

Structural formula of methane

Let us begin our study of structural formulas used in organic chemistry with the *electron-dot structure*. Each electron-pair covalent bond and each unshared pair of electrons on a particular atom are designated by a pair of dots. Consider the compound methyl alcohol ($CH_3OH$):

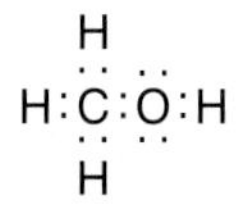

```
    H    ··
    ··   ··
H : C : O : H
    ··
    H
```

In the *line-bond structure*, each electron pair is represented by a line. Note that the unshared electrons on oxygen are not shown but are assumed to be present.

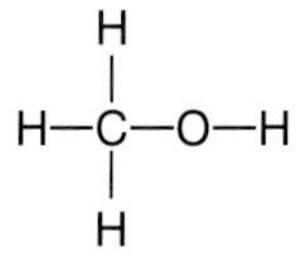

```
    H
    |
H — C — O — H
    |
    H
```

Line-bond formulas are useful in structure writing but take time to write and take up considerable space when written. Thus we often use an abbreviated structural formula called the *condensed formula*. In the *semicondensed* or condensed formula, most or all the bonds are not shown.

$$CH_3{-\!}OH \qquad CH_3OH$$

Semicondensed     Condensed

Consider the abbreviations of the line-bond structure of butane.

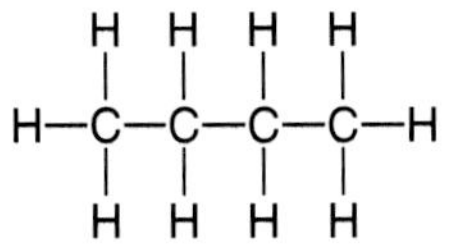

$$CH_3{-\!}CH_2{-\!}CH_2{-\!}CH_3 \qquad CH_3CH_2CH_2CH_3 \text{ or } CH_3(CH_2)_2CH_3$$

Butane: line-bond     Butane: semicondensed     Butane: condensed

See also Fig. 1.7 and Table 1.2.

Note in the condensed structure of butane that the $CH_2CH_2$ grouping is further condensed to $(CH_2)_2$.

---

**FIGURE 1.7** Line-bond, semicondensed, and condensed structures for isobutane.

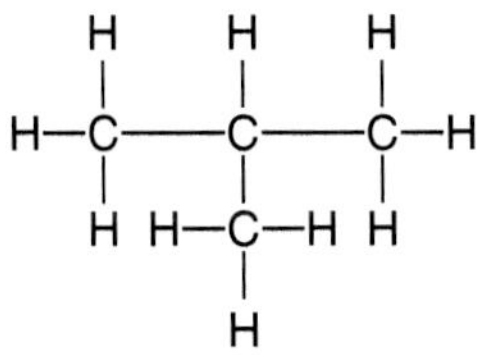

$(CH_3)_3CH$

Isobutane: condensed

Isobutane: line-bond

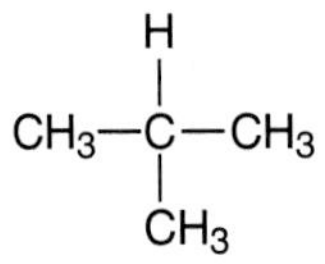

Isobutane: semicondensed

Note that the three $CH_3$ groups surrounding the central carbon in the line-bond structure are set off by parentheses in the condensed structural formula. For now, all you need to know is how to convert one type of structure into another structure. We will study the rules of structure writing given a molecular formula in the next section.

Note in the line-bond structural formulas in Table 1.2 that unshared electrons on nitrogen, oxygen, and the halogens are assumed to be present.

| TABLE 1.2 | A Comparison of Structural Formulas | | | |
|---|---|---|---|---|
| **Name** | **Electron Dot** | **Line-Bond** | **Semicondensed** | **Condensed** |
| Methylamine | 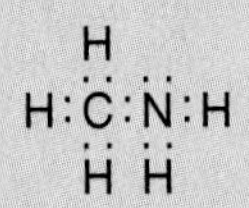 | 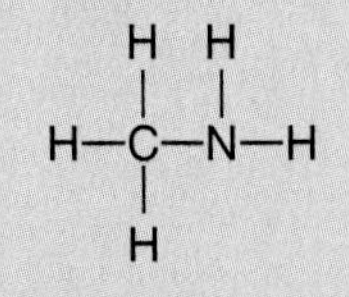 | $CH_3{-\!}NH_2$ | $CH_3NH_2$ |
| Methyl ether | 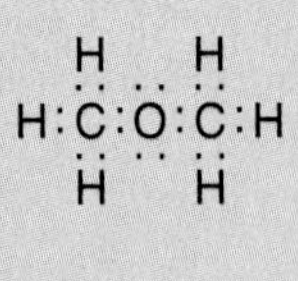 | 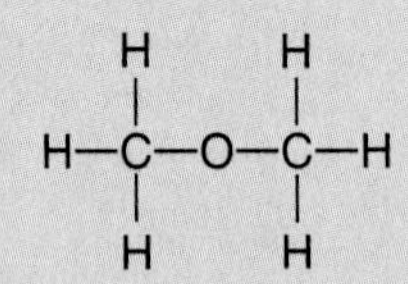 | $CH_3{-\!}O{-\!}CH_3$ | $CH_3OCH_3$ |
| Acetone | 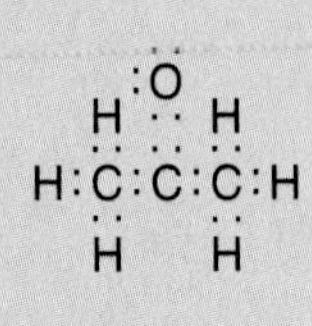 | 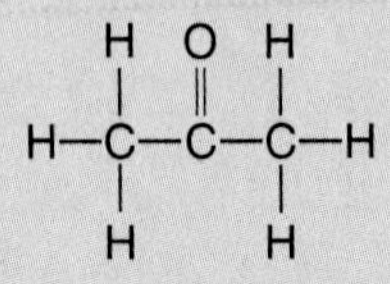 | 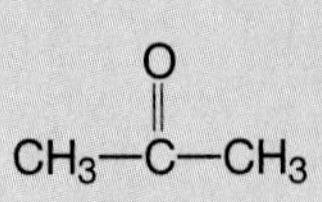 | $CH_3COCH_3$ |
| Propylene | 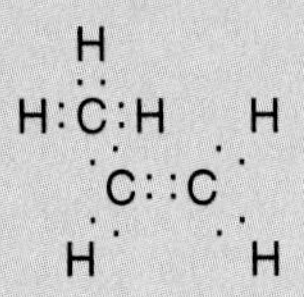 | 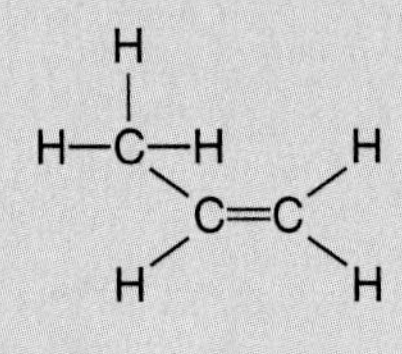 | $CH_3{-\!}CH{=}CH_2$ | $CH_3CH{=}CH_2$ |

Draw a semicondensed structural formula and a condensed structural formula for each of the following line-bond structures:

**SOLVED PROBLEM 1.2**

a.

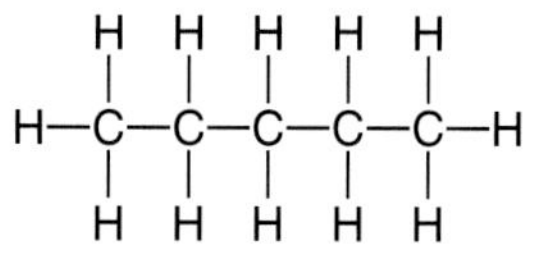

$$H-\underset{\underset{H}{|}}{\overset{\overset{H}{|}}{C}}-\underset{\underset{H}{|}}{\overset{\overset{H}{|}}{C}}-\underset{\underset{H}{|}}{\overset{\overset{H}{|}}{C}}-\underset{\underset{H}{|}}{\overset{\overset{H}{|}}{C}}-\underset{\underset{H}{|}}{\overset{\overset{H}{|}}{C}}-H$$

b.

$$H-\underset{\underset{H}{|}}{\overset{\overset{H}{|}}{C}}-\underset{\underset{\underset{\underset{H}{|}}{C-H}}{|}}{\overset{\overset{H}{|}}{C}}-\underset{\underset{H}{|}}{\overset{\overset{H}{|}}{C}}-\underset{\underset{H}{|}}{\overset{\overset{H}{|}}{C}}-H$$

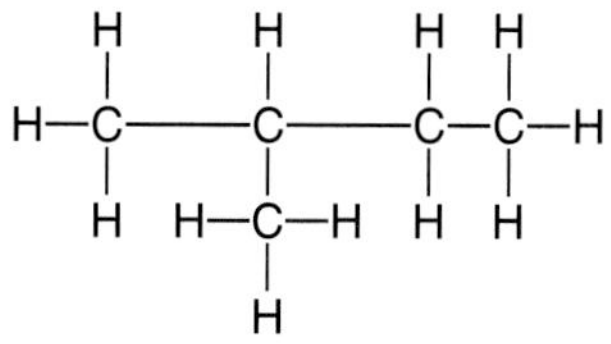

**SOLUTION**

a.  Draw each end carbon with the bonded hydrogens as $CH_3$ and each internal carbon with the bonded hydrogens as $CH_2$. Thus, for the semicondensed structure, we get $CH_3-CH_2-CH_2-CH_2-CH_3$. To draw the condensed structural formula, enclose the $CH_2$ groups in parentheses to give $CH_3(CH_2)_3CH_3$.

b.  In a similar way as in Solved Problem 1.2a, draw each end carbon and the bottom carbon with the bonded hydrogens as $CH_3$. Thus the semicondensed structure is

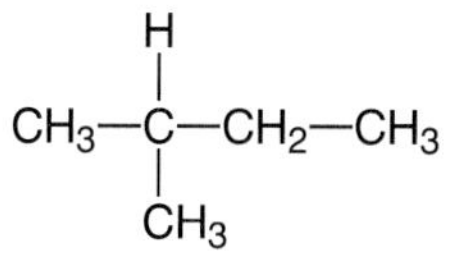

$$CH_3-\underset{\underset{CH_3}{|}}{\overset{\overset{H}{|}}{C}}-CH_2-CH_3$$

To draw a condensed structural formula, we place the two $CH_3$ groups on the left side of the structure in parentheses and get $(CH_3)_2CHCH_2CH_3$.

Draw a semicondensed and a condensed structural formula for each of the following line-bond structures:

**EXERCISE 1.2**

a.

$$H-\underset{\underset{H}{|}}{\overset{\overset{H}{|}}{C}}-\underset{\underset{H}{|}}{\overset{\overset{H}{|}}{C}}-\underset{\underset{H}{|}}{\overset{\overset{H}{|}}{C}}-\underset{\underset{H}{|}}{\overset{\overset{H}{|}}{C}}-\underset{\underset{H}{|}}{\overset{\overset{H}{|}}{C}}-\underset{\underset{H}{|}}{\overset{\overset{H}{|}}{C}}-H$$

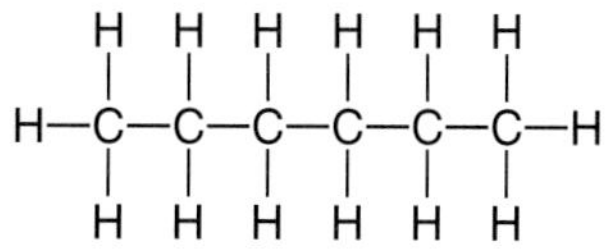

b.

$$H-\underset{\underset{H}{|}}{\overset{\overset{H}{|}}{C}}-\underset{\underset{\underset{\underset{H}{|}}{C-H}}{|}}{\overset{\overset{H}{|}}{C}}-\underset{\underset{H}{|}}{\overset{\overset{H}{|}}{C}}-\underset{\underset{H}{|}}{\overset{\overset{H}{|}}{C}}-\underset{\underset{H}{|}}{\overset{\overset{H}{|}}{C}}-H$$

It is important to be able to write or draw one or more structural formulas from a given molecular formula. The following rules will be helpful in formulating uncharged organic structures:

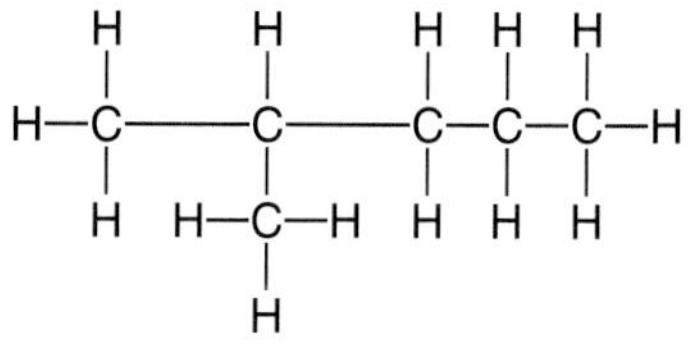

**1.9
RULES OF BONDING FOR STRUCTURE WRITING OF ORGANIC COMPOUNDS**

1.  To be neutral in a structure, carbon must share four bonds, with no electrons unshared.
2.  To be neutral in a structure, hydrogen must share one bond, with no electrons unshared.

3. To be neutral in a structure, nitrogen must share three bonds, with two electrons unshared as one electron pair.

4. To be neutral in a structure, oxygen must share two bonds, with four electrons unshared as two electron pairs.

5. To be neutral in a structure, a halogen must share one bond, with six electrons unshared as three electron pairs.

**SOLVED PROBLEM 1.3**

Write as many structural formulas consistent with the preceding rules as possible for each of the following molecular formulas:

a. $C_3H_6$ (line-bond)      b. $C_3H_9N$ (condensed)

**SOLUTION**

a.  Carbon is used to make the skeleton of the compound: C—C—C. Then hydrogen can be filled in.

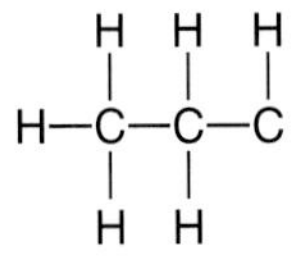

H—C—C—C   with H's    This structure is incorrect because the carbon atom on the right shares only two bonds.

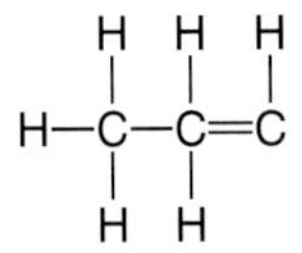

H—C—C=C   with H's    This structure is also incorrect because the middle carbon atom shares five bonds, while the carbon atom on the right shares only three bonds.

A double bond is placed between adjacent carbon atoms. If a hydrogen atom is moved from the middle carbon to the carbon on the right, then the criterion that each carbon has four bonds (a stable octet of electrons) is satisfied.

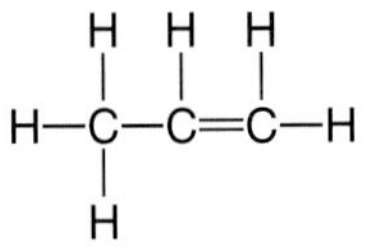

H—C—C=C—H   Correct structure

Another correct structure is cyclic.

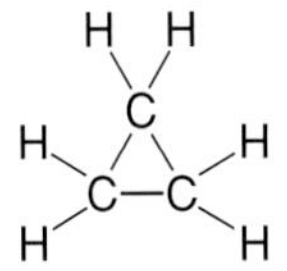

b.  This time form a skeleton with carbon and nitrogen atoms arranged in a variety of ways and then fill in with hydrogens. Remember, to be neutral, each nitrogen shares only three bonds.

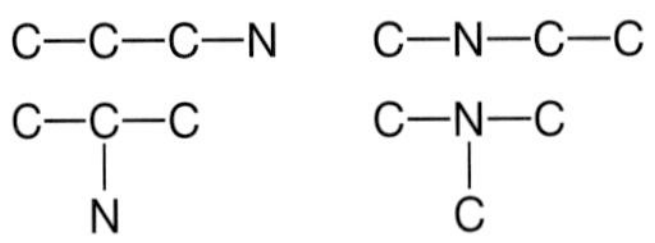

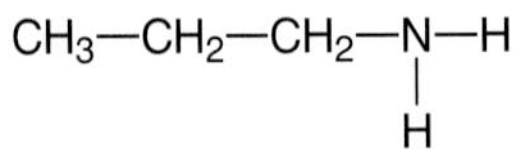

C—C—C—N      C—N—C—C

C—C—C        C—N—C
   |              |
   N              C

$CH_3$—$CH_2$—$CH_2$—N—H   or   $CH_3CH_2CH_2NH_2$ 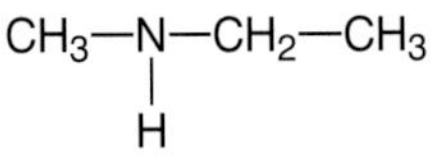
                     |
                     H

$CH_3$—N—$CH_2$—$CH_3$   or   $CH_3NH(CH_2CH_3)$
        |
        H

$$CH_3-\overset{\overset{\displaystyle H}{|}}{\underset{\underset{\displaystyle NH_2}{|}}{C}}-CH_3 \quad \text{or} \quad CH_3CH(NH_2)CH_3$$

$$CH_3-\overset{}{\underset{\underset{\displaystyle CH_3}{|}}{N}}-CH_3 \quad \text{or} \quad (CH_3)_3N$$

It is important to note that all four of the preceding structures are correct.

---

Write as many line-bond structural formulas as you can using the rules of bonding for each of the following compounds:

a. $HNO_2$

b. $C_3H_7Cl$

c. $N_2H_4$

**EXERCISE 1.3**

---

A question that usually comes up is, How do you know if a structural formula represents a neutral substance or a charged species? A simple answer to this question is to calculate the charge (known as the *formal charge*) on each atom in the structure. If the sum of the formal charges within a structure is zero, the structure is neutral; if the sum is not zero, the structure is an ion. The formal charge on each atom in a structure is calculated using the following expression:

$$\text{Formal charge} = O - \frac{S}{2} - U$$

where $O$ represents the number of outer valence electrons the atom would have if it was by itself ($C = 4, N = 5, O = 6, F = 7$), $S$ equals the number of shared electrons with the atom in the structure, and $U$ represents the number of unshared electrons held by the atom in the structure. To determine whether the structure as a whole is neutral or an ion, use the following formula:

$$\text{Charge on the species} = \Sigma \, FC_J \cdot n(J)$$

The charge on the species equals the sum of the formal charge on each atom $FC_J$ times the number of atoms $n(J)$ with that charge. $J$ represents the element involved. In other words, the charge on the species equals the sum of the formal charge on each atom times the number of atoms with that charge. Let us consider the ammonium ion.

$$H-\overset{\overset{\displaystyle H}{|}}{\underset{\underset{\displaystyle H}{|}}{\overset{+}{N}}}-H$$

Using the formulas, we get

$$\text{Formal charge on nitrogen} = O - \frac{S}{2} - U = 5 - \frac{8}{2} - 0 = +1$$

$$\text{Formal charge on each hydrogen} = 1 - \frac{2}{2} - 0 = 0$$

$$\text{Charge (ionic) on the species} = \Sigma \, FC_J \cdot n(J)$$

$$\text{Charge (ionic) on the species} = +1 \cdot 1 \, (N) + 0 \cdot 4 \, (H) = +1$$

## 1.10
## CHARGED SPECIES IN ORGANIC CHEMISTRY

Since nitrogen in the ammonium ion is bonded by four covalent bonds rather than the usual three, it seems reasonable to assume that nitrogen does carry a formal charge. We can reason out the magnitude and sign of the charge as follows. The neutral uncombined nitrogen atom contains five electrons (the number of valence electrons). Now, in determining the formal charge carried by a given atom, we stipulate that the atom, when combined, "possesses" one-half its shared electrons and all its unshared electrons. Thus combined nitrogen has $8/2 = 4$ electrons. Since neutral nitrogen has five and combined nitrogen only four, nitrogen bears a formal charge of $+1$. Using a similar procedure for hydrogen, you can demonstrate that each hydrogen bears a formal charge of zero. Thus the formal charge on nitrogen ($+1$) is carried to the ammonium ion as a whole.

**SOLVED PROBLEM 1.4**

Calculate the formal charge on each atom in each of the following structures, and determine if each structure represents an ion.

a. 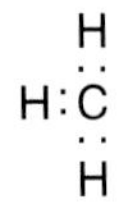
```
       H
      ..
   H:C
      ..
       H
```

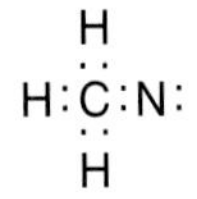

b.
```
       H
      ..
   H:C:N:
      ..
       H
```

c.
```
        ..
   H:O:H
      ..
       H
```

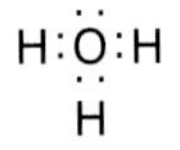

**SOLUTION**

a.    Formal charge on carbon $= 4 - \dfrac{6}{2} - 0 = +1$

Formal charge on each hydrogen $= 1 - \dfrac{2}{2} - 0 = 0$

Since each hydrogen is neutral (bears a formal charge of zero = no formal charge) and since carbon carries a $+1$ charge, the charge on the entire ion is $+1$. Using the formula for ionic charge, we have

b.              Ionic charge $= +1 \cdot 1\ (C) + 0 \cdot 3\ (H) = +1$

Formal charge on carbon $= 4 - \dfrac{8}{2} - 0 = 0$

Formal charge on each hydrogen $= 1 - \dfrac{2}{2} - 0 = 0$

Formal charge on nitrogen $= 5 - \dfrac{2}{2} - 2 = +2$

The charge on the entire ion is $+2$, which is carried by the formal charge on nitrogen. Using the formula for ionic charge, we get

c.              Ionic charge $= 0 \cdot 1\ (C) + 0 \cdot 3\ (H) + (+2) \cdot 1\ (N) = +2$

Formal charge on each hydrogen $= 1 - \dfrac{2}{2} - 0 = 0$

Formal charge on oxygen $= 6 - \dfrac{6}{2} - 2 = +1$

This ion has a $+1$ charge that is carried by the formal charge on oxygen. With the formula for ionic charge, we obtain

$$\text{Ionic charge} = +1 \cdot 1\ (O) + 0 \cdot 3\ (H) = +1$$

**EXERCISE 1.4**

a. Calculate the formal charge on each atom in $H-C\equiv C:$

b. What is the charge on the entire species, if any?

It is important to note that two or more atoms in a structure can bear a formal charge, and yet the entire structure can be neutral.

**SOLVED PROBLEM 1.5**

Show that $H_3N-BH_3$ is a neutral substance.

**SOLUTION**

$$\text{Formal charge on nitrogen} = 5 - \frac{8}{2} - 0 = +1$$

$$\text{Formal charge on boron} = 3 - \frac{8}{2} - 0 = -1$$

$$\text{Formal charge on each hydrogen} = 1 - \frac{2}{2} - 0 = 0$$

Thus

$$\text{Ionic charge} = +1 \cdot 1 \text{ (N)} + (-1) \cdot 1 \text{ (B)} + 0 \cdot 6 \text{ (H)} = 0$$

This substance is neutral.

---

Show that $CH_3NO_2$ is a neutral substance.

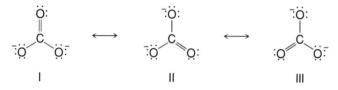

**EXERCISE 1.5**

---

*Resonance* is an extension of the structural theory that can explain molecule or ion stability as well as the formation of certain reaction products that are isolated. Putting it another way, resonance comes into play when one or more pairs of unshared or pi electrons (Sec. 2.4) can exist throughout the molecule. We say that these electrons are *delocalized*. Consider the carbonate ion, $CO_3^{2-}$.

## 1.11 RESONANCE

Note that resonance is discussed in much greater depth in Section 7.2.

Note that the three structures do not flip from one structure to another, but the correct structure is a blend of structures I, II, and III with the pi (Sec. 2.4) and unshared electrons delocalized and spread around the structure. *Contributing structures* or *resonance structures* can be quickly recognized by the use of one or more *double-headed arrows*, as is the case with the carbonate ion.

Looking at resonance from another point of view, each carbon-oxygen bond is neither a single bond nor a double bond but somewhere in between. This state of affairs is shown by means of a structural formula that we designate as a *resonance hybrid*.

resonance hybrid of
carbonate ion

Since there are over 8 million organic compounds, a system of classification for these compounds is absolutely necessary to simplify the study of their chemistry. Fortunately, these compounds can be classified into a limited number of families, each of which contains one or more *functional groups*. A functional group is an atom or group of atoms (other than hydrogen) that determines the chemical properties of the molecule. The functional group is a part of a given molecule. However, whether the molecule is small and simple or large and complex, the presence of the same functional group in two molecules means that both molecules will react in the same way.

Table 1.3 summarizes the functional groups we will study, listing the name of each functional group, a general structural formula in which each line bond with no attachment is assumed to be bonded to either a carbon or hydrogen atom, and an example of a typical compound containing the functional group.

## 1.12 FUNCTIONAL GROUPS

| TABLE 1.3 | Important Functional Groups | | |
|---|---|---|---|
| **Functional Group Family Name** | **Functional Group Structure** | **Example** | |
| Alkane (paraffin) | 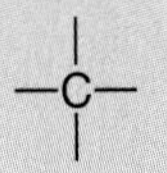 $-\overset{\mid}{\underset{\mid}{C}}-$ | 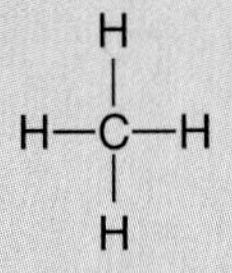 $H-\overset{\overset{\displaystyle H}{\mid}}{\underset{\underset{\displaystyle H}{\mid}}{C}}-H$ | Methane, major component of natural gas |
| Alkene (olefin) | 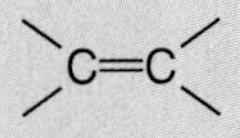 $\overset{\diagup}{\underset{\diagup}{C}}=\overset{\diagdown}{\underset{\diagdown}{C}}$ | 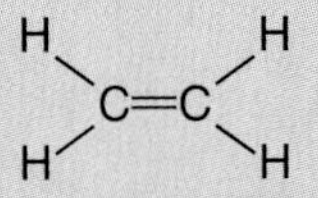 $\overset{H}{\underset{H}{}}C=C\overset{H}{\underset{H}{}}$ | Ethylene, raw material for the production of polyethylene |
| Alkyne | $-C\equiv C-$ | $H-C\equiv C-H$ | Acetylene, used in the oxyacetylene torch to generate a hot flame |
| Halide X = F, Cl, Br, and I | 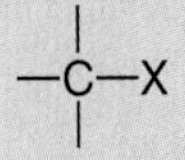 $-\overset{\mid}{\underset{\mid}{C}}-X$ | $CCl_4$ | Carbon tetrachloride, nonpolar solvent |
| Aromatic hydrocarbon | 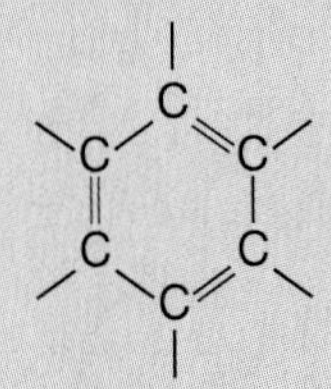 | 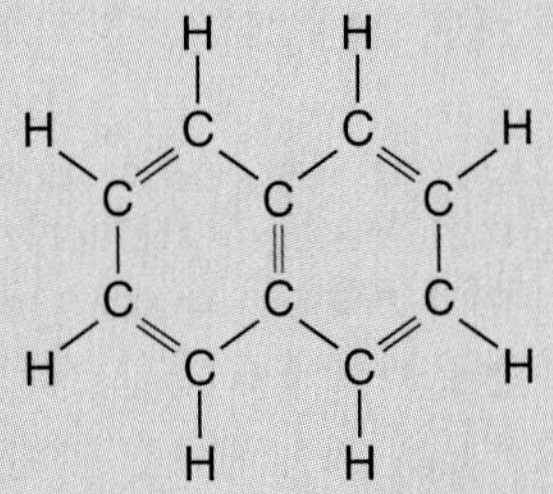 | Naphthalene, moth repellant |
| Alcohol | 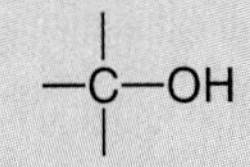 $-\overset{\mid}{\underset{\mid}{C}}-OH$ | 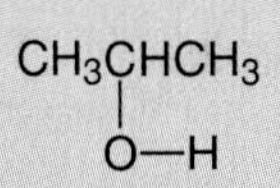 $CH_3\overset{}{\underset{\underset{\displaystyle O-H}{\mid}}{C}}HCH_3$ | Isopropyl (rubbing) alcohol, furnishes the "sting" of an after shave lotion |
| Mercaptan (thiol) | 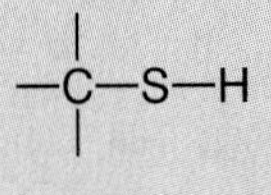 $-\overset{\mid}{\underset{\mid}{C}}-S-H$ | $(CH_3)_2CHCH_2CH_2SH$ | Isopentyl mercaptan, essence of skunk |
| Disulfide | 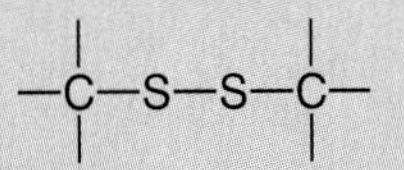 $-\overset{\mid}{\underset{\mid}{C}}-S-S-\overset{\mid}{\underset{\mid}{C}}-$ | $CH_3-S-S-CH_3$ | Dimethyl disulfide |
| Phenol | 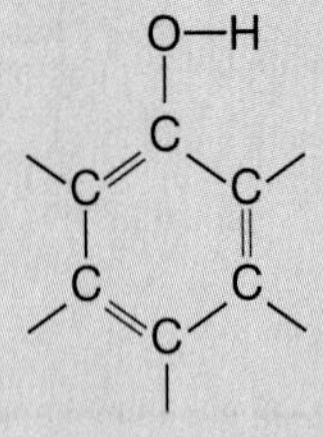 | 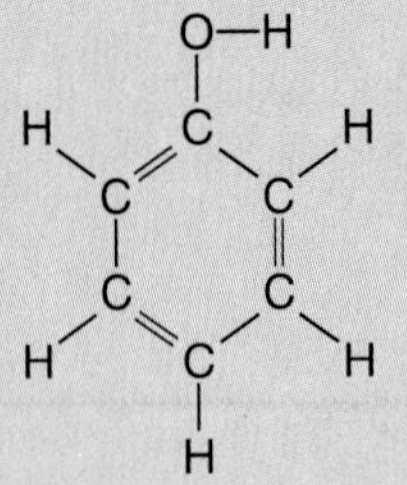 | Phenol, an antiseptic |
| Ether | 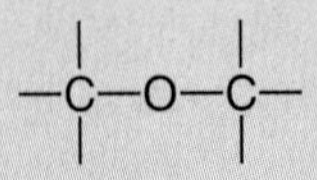 $-\overset{\mid}{\underset{\mid}{C}}-O-\overset{\mid}{\underset{\mid}{C}}-$ | $CH_3CH_2OCH_2CH_3$ | Diethyl ether, the first inhalation anesthetic |
| Aldehyde | 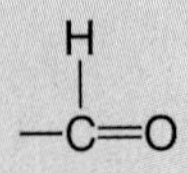 $-\overset{\overset{\displaystyle H}{\mid}}{C}=O$ | 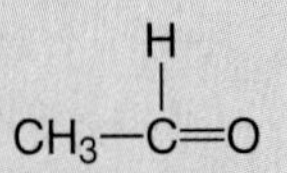 $CH_3-\overset{\overset{\displaystyle H}{\mid}}{C}=O$ | Acetaldehyde |
| Ketone | 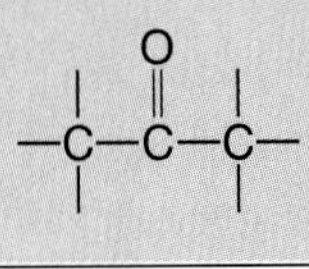 $-\overset{\mid}{\underset{\mid}{C}}-\overset{\overset{\displaystyle O}{\parallel}}{C}-\overset{\mid}{\underset{\mid}{C}}-$ | 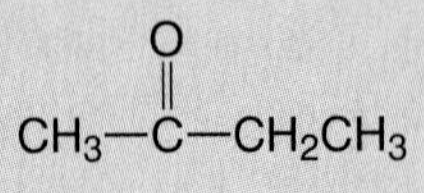 $CH_3-\overset{\overset{\displaystyle O}{\parallel}}{C}-CH_2CH_3$ | Methyl ethyl ketone, solvent used in paints |

| TABLE 1.3 | Important Functional Groups *(Continued)* | | |
|---|---|---|---|
| **Functional Group Family Name** | **Functional Group Structure** | **Example** | |
| Carboxylic acid | $\overset{\displaystyle O}{\underset{}{-C-O-H}}$ | $CH_3-\overset{\displaystyle O}{C}-OH$ | Acetic acid, component of vinegar |
| Cyanide (nitrile) | $-\overset{\mid}{\underset{\mid}{C}}-C\equiv N$ | $CH_3C\equiv N$ | Methyl cyanide, not nearly as deadly as $H-C\equiv N$ |
| Acid chloride | $-\overset{\mid}{\underset{\mid}{C}}-\overset{\displaystyle O}{C}-Cl$ | $CH_3-\overset{\displaystyle O}{C}-Cl$ | Acetyl chloride |
| Acid anhydride | $-\overset{\mid}{\underset{\mid}{C}}-\overset{\displaystyle O}{C}-O-\overset{\displaystyle O}{C}-\overset{\mid}{\underset{\mid}{C}}-$ | $CH_3-\overset{\displaystyle O}{C}-O-\overset{\displaystyle O}{C}-CH_3$ | Acetic anhydride, used in the manufacture of aspirin |
| Ester | $-\overset{\displaystyle O}{C}-O-\overset{\mid}{\underset{\mid}{C}}-$ | $CH_3-\overset{\displaystyle O}{C}-OCH_2(CH_2)_6CH_3$ | *n*-Octyl acetate, odor of oranges |
| Amide | $-\overset{\displaystyle O}{C}-\overset{H}{\underset{}{N}}-H$ | $CH_3-\overset{\displaystyle O}{C}-\overset{H}{\underset{}{N}}-H$ | Acetamide |
| Amine | $-\overset{\mid}{\underset{\mid}{C}}-\overset{H}{\underset{}{N}}-H$ | $CH_3NH_2$ | Methylamine |

## 1.13 STRUCTURAL ISOMERS

Two or more compounds that have the same molecular formula but different structural formulas and therefore different physical properties are called *isomers*. For example, the molecular formula $C_4H_{10}O$ could represent diethyl ether or its isomer *n*-butyl alcohol. The presence of two or more compounds with the same molecular formula but different structural arrangements is known as *isomerism*.

There are two different kinds of structural isomers. Two or more compounds in which each belongs to a different family of compounds are known as *functional-group isomers*. Diethyl ether and *n*-butyl alcohol are functional-group isomers. Each isomer belongs to a different family of compounds. Note that the oxygen in diethyl ether is linked to two carbons, whereas the oxygen in *n*-butyl alcohol is bonded to one carbon and one hydrogen.

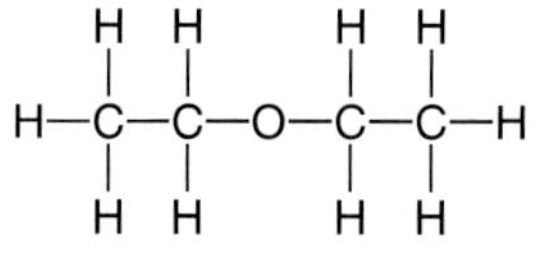

*Diethyl ether*
(a member of the ether family)

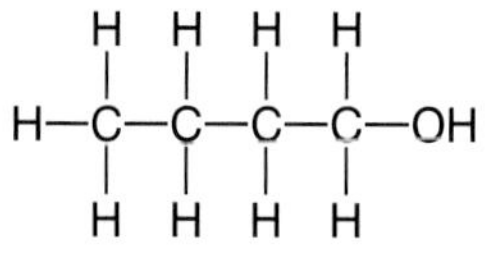

*n-Butyl alcohol*
(a member of the alcohol family)

Another example of functional-group isomerism is represented by propionic acid and methyl acetate (Fig. 1.8).

In both examples of functional-group isomerism, the atoms are arranged in different ways. This leads to different structural formulas and therefore different physical and chemical properties. The differences in physical properties (Table 1.4) are due to significant differences in intermolecular forces between, say, molecules of *n*-butyl alcohol as opposed to those forces between molecules of diethyl ether. These differences are explored in greater depth later in this textbook.

**FIGURE 1.8** Two functional-group isomers. Note that the carbon-oxygen double bond is attached to the end carbon in propionic acid, while it is linked to the middle carbon in methyl acetate.

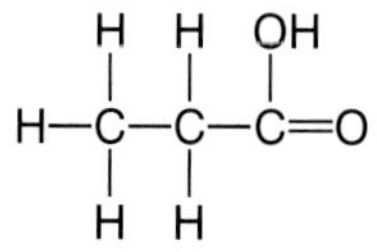

Propionic acid
(a carboxylic acid)

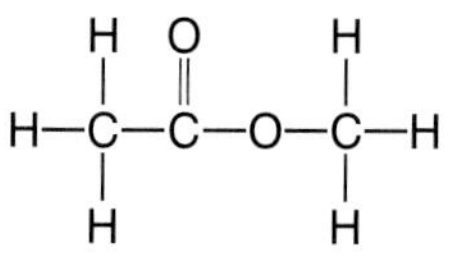

Methyl acetate
(an ester)

| TABLE 1.4 | A Comparison of the Physical Properties of Two Functional-Group Isomers | |
|---|---|---|
| | **Diethyl Ether** | **n-Butyl Alcohol** |
| Molecular formula | $C_4H_{10}O$ | $C_4H_{10}O$ |
| Boiling point (°C) | 35 | 117 |
| Water solubility (room temperature) | Slightly soluble | Soluble |
| Specific gravity | 0.71 | 0.81 |

A second kind of isomerism involves two or more compounds that belong to the same chemical family, where the same functional group occupies a different position in each molecule. This is called *positional isomerism*. For example, compound (a) (*n*-butyl alcohol) is an isomer of compound (b) (*sec*-butyl alcohol) because in (a) the hydroxyl group is bonded to the end carbon, whereas in (b) the hydroxyl group is bonded to the carbon next to the end carbon.

*n*-butyl alcohol
(a)

*sec*-butyl alcohol
(b)

Note in Table 1.5 that the physical properties shown by positional isomers are not as significantly different as those properties of functional-group isomers (see Table 1.4). This is because positional isomers contain the same functional group, so their intermolecular forces are similar. Thus each isomer will exhibit somewhat similar physical properties. On the other hand, since each functional-group isomer contains a different functional group, the physical properties of the isomers are considerably different. These same relationships hold for the chemical properties of isomers. Functional-group isomers show different chemical properties. For example, diethyl ether does not react with sodium, whereas *n*-butyl alcohol reacts with sodium to give the corresponding alkoxide and hydrogen.

$$CH_3CH_2OCH_2CH_3 + Na \longrightarrow \text{no reaction}$$

$$2\,CH_3CH_2CH_2CH_2OH + 2\,Na \longrightarrow 2\,CH_3CH_2CH_2CH_2O^-Na^+ + H_2$$

On the other hand, since positional isomers contain the same functional group, these isomers display similar chemical properties. For example, each isomer reacts with sodium in a similar way but at different rates.

$$2\,CH_3CH_2CH_2CH_2OH + 2\,Na \longrightarrow 2\,CH_3CH_2CH_2CH_2O^-Na^+ + H_2 \text{ (faster)}$$

$$2\,CH_3\underset{\underset{OH}{|}}{C}HCH_2\,CH_3 + 2\,Na \longrightarrow 2\,CH_3\underset{\underset{O^-Na^+}{|}}{C}HCH_2CH_3 + H_2 \text{ (slower)}$$

| TABLE 1.5 | A Comparison of the Physical Properties of Two Positional Isomers | |
|---|---|---|
| | **n-Butyl Alcohol** | **sec-Butyl Alcohol** |
| Molecular formula | $C_4H_{10}O$ | $C_4H_{10}O$ |
| Boiling point (°C) | 117 | 100 |
| Water solubility (room temperature) | Soluble | Very soluble |
| Specific gravity | 0.81 | 0.81 |

Now let us consider another illustration of positional isomers. Isomer (c) has one chlorine bonded to each carbon, while isomer (d) has no chlorine bonded to one carbon and two chlorines bonded to the other.

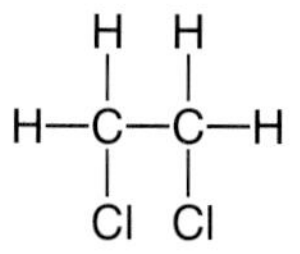

$$
\begin{array}{cc}
\underset{\underset{Cl}{|}}{\overset{\overset{H}{|}}{H-C}} \; \underset{\underset{Cl}{|}}{\overset{\overset{H}{|}}{C-H}} & \underset{\underset{Cl}{|}}{\overset{\overset{H}{|}}{Cl-C}} \; \underset{\underset{H}{|}}{\overset{\overset{H}{|}}{C-H}} \\
(c) & (d)
\end{array}
$$

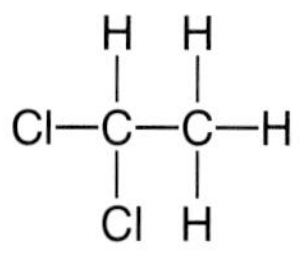

If two structures have the same atomic surroundings, the structures represent the same compound. Thus both (e) and (f) are identical to compound (b).

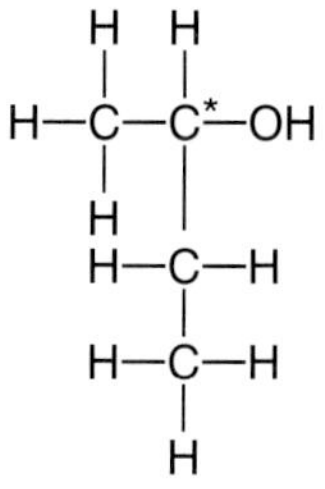

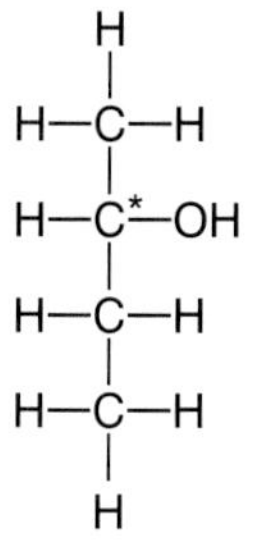

$$
\begin{array}{cc}
(e) & (f)
\end{array}
$$

Never mind that the structures are bent or twisted in a different way. Simply check the atomic environment of each starred carbon in (e) and (f), and you will see that it is identical to (b). That is, each starred carbon has bonded to it one H, one OH, one $CH_3$, and one $C_2H_5$ group. This same similarity of environment holds for the other three carbons.

Carbon-carbon single bonds can be freely rotated. This will be discussed in greater detail in Section 2.3.

---

**SOLVED PROBLEM 1.6**

Draw two functional-group isomers and two positional isomers that have the condensed formula $C_3H_6O$.

**SOLUTION**

Form a skeleton with various combinations of carbon and oxygen, and fill in with hydrogen. Two positional isomers are as follows:

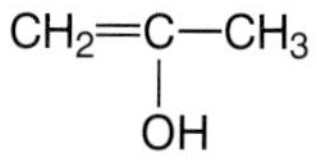

$$
\begin{array}{cc}
\underset{\underset{OH}{|}}{CH_2{=}C{-}CH_3} & CH_2{=}CHCH_2OH \\
(g) & (h)
\end{array}
$$

Note that the hydroxyl group in (g) is bonded to the middle carbon, whereas the hydroxyl group in (h) is bonded to an end carbon. Both compounds (g) and (h) are alcohols. Although compound (g) happens to be unstable, structure (g) is a correct response to this question because the rules of bonding are correctly followed in writing the structure. Two functional-group isomers are shown as follows:

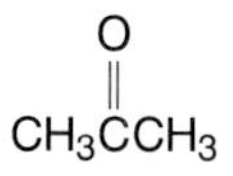

$$
\begin{array}{cc}
\overset{\overset{O}{\|}}{CH_3CCH_3} & \overset{\overset{O}{\|}}{CH_3CH_2C{-}H} \\
(i) & (j)
\end{array}
$$

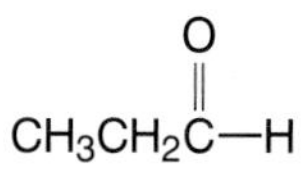

Compound (i) is an example of a ketone, and compound (j) is an aldehyde.

---

**EXERCISE 1.6**

Draw a structural line-bond formula for another positional isomer of compounds (g) and (h) in Solved Problem 1.6.

**SOLVED PROBLEM 1.7**

Which of the following pairs of compounds are functional-group isomers, positional isomers, unrelated, or identical?

a.  $CH_3CH{=}CHCH_3$    and    $CH_3CH_2CH{=}CH_2$

b.  $CH_3CH_2CH_2CH_2CH_3$    and    $CH_3(CH_2)_4CH_3$

c.

$$CH_3{-}\overset{\overset{\displaystyle H}{|}}{\underset{\underset{\displaystyle CH_2CH_2CH_3}{|}}{C}}{-}H \quad \text{and} \quad CH_3CH_2CH_2{-}\overset{\overset{\displaystyle CH_3}{|}}{\underset{\underset{\displaystyle H}{|}}{C}}{-}H$$

**SOLUTION**

a.  Notice that the molecular formula of each compound is $C_4H_8$. Therefore, the compounds can be isomers or identical. Observe that each compound contains a carbon-carbon double bond. Since in one compound the double bond links the middle two carbons while in the other compound the double bond links an end carbon to an internal carbon, these compounds represent positional isomers.

b.  Note that $(CH_2)_4$ corresponds to $CH_2CH_2CH_2CH_2$. With this fact in mind, a quick check should reveal that one compound contains five carbon atoms, while the other compound contains six carbon atoms. Therefore, these compounds are unrelated.

c.  The molecular formula of each structure is the same; thus the structures are either isomers or identical. The atom environment around each carbon of the pair of structures is the same irrespective of the arrangement of the carbons in the chain. The two structures, therefore, are identical and represent the same compound.

**EXERCISE 1.7**

Which of the following compounds are functional-group isomers, and which are positional isomers?

a.  $CH_3CH_2OCH(CH_3)_2$    and    $HOCH_2CH_2CH_2CH_2CH_3$

b.

$$CH_3{-}\overset{\overset{\displaystyle OH}{|}}{\underset{\underset{\displaystyle NO_2}{|}}{C}}{-}CH_3 \quad \text{and} \quad CH_2{-}\overset{\overset{\displaystyle NO_2}{|}}{\underset{\underset{\displaystyle OH\ \ H}{|}}{C}}{-}CH_3$$

**1.14 ACIDITY**

Many reactions that occur in organic chemistry involve the reaction of an acid with a base. Thus a detailed discussion of acids and bases is reasonable at this point.

Let us begin with the *Brønsted* definition. Brønsted defined an acid as a proton donor and a base as a proton acceptor. This definition broadened the scope of acidity and basicity in that a base is not confined to being a compound with one or more hydroxide groups present. For example, a chloride ion is a Brønsted base because it can accept a proton (from hydronium ion) to produce HCl (and $H_2O$). Thus

$$Cl^- + H_3O^+ \rightleftharpoons HCl + H_2O$$
$$\text{Base}\quad \text{Acid}\quad \text{Conjugate}\quad \text{Conjugate}$$
$$\text{acid}\qquad \text{base}$$

Here, chloride ion (Brønsted base) accepts a proton from hydronium ion (Brønsted acid) to produce hydrogen chloride (conjugate acid) and water (conjugate base). Because chloride ion is a weak base, the equilibrium lies far to the left, and essentially, the reaction does not occur. The reverse reaction, on the other hand, goes to completion.

$$HCl + H_2O \longrightarrow H_3O^+ + Cl^-$$
$$\text{Acid}\quad \text{Base}\quad \text{Conjugate}\quad \text{Conjugate}$$
$$\text{acid}\qquad \text{base}$$

| TABLE 1.6 | Relative Strengths of Some Brønsted Acid-Base Pairs | |
|---|---|---|
| $pK_a$ | Acid | Base |
| ~ −6 | $HNO_3$, HCl    strongest | $NO_3^-$, $Cl^-$    weakest |
| 4.8 | $HC_2H_3O_2$ | $C_2H_3O_2^-$ |
| 9.3 | HCN | $^-CN$ |
| 9.9 | $C_6H_5OH$ (phenol) | $C_6H_5O^-$ (phenoxide ion) |
| 15.7 | $H_2O$ | $^-OH$ |
| ~17 | ROH | $RO^-$ |
| 25 | $HC{\equiv}CH$ | $HC{\equiv}C^-$ |
| 35 | $NH_3$ | $^-NH_2$ |
| 44 | $CH_2{=}CH_2$ | $CH_2{=}CH^-$ |
| 50 | $CH_3CH_3$ weakest | $CH_3CH_2^-$ strongest |

$pKa = -\log Ka$. Refer to Section 20.3.

An important relationship to remember is that the stronger acid produces a weaker conjugate base and the weaker acid produces a stronger conjugate base. In the same way, the stronger the base, the weaker its conjugate acid. Note that the equilibrium tends to favor formation of the weaker acid. Refer to Table 1.6 for $pK_a$ values of selected acids. It should be helpful to realize that the weaker the acid, the greater is the $pK_a$ value.

Some other Brønsted acid-base reactions are as follows:

$$^-NH_2 \quad + \quad H_2O \quad \rightleftharpoons \quad NH_3 \quad + \quad ^-OH$$

Stronger base    Stronger acid    Weaker    Weaker
($pK_a = 15.7$)    conjugate acid    conjugate base
($pK_a = 35$)

$$HC_2H_3O_2 \quad + \quad ^-OH \quad \rightleftharpoons \quad C_2H_3O_2^- \quad + \quad H_2O$$

Stronger acid    Stronger base    Weaker    Weaker
($pK_a = 4.8$)    conjugate base    conjugate acid
($pK_a = 15.7$)

**SOLVED PROBLEM 1.8**

Which is the stronger base, $NO_3^-$ or $C_2H_3O_2^-$?

**SOLUTION**

Note in Table 1.6 that $HC_2H_3O_2$ ($pK_a = 4.8$) is a weaker acid than $HNO_3$ ($pK_a \sim -6$). Thus $C_2H_3O_2^-$, the conjugate base of $HC_2H_3O_2$, must be a stronger base than $NO_3^-$, the conjugate base of $HNO_3$.

**SOLVED PROBLEM 1.9**

Indicate if each of the following sets of reagents reacts (i.e., if the reaction essentially proceeds to completion). If so, complete the equation; if not, write "No reaction." Explain your answers.

a.   $CH_3O^- + HCN \longrightarrow$

b.   $NO_3^- + H_2O \longrightarrow$

**SOLUTION**

a.   This reaction essentially proceeds to completion. Methoxide ion ($CH_3O^-$) is a base and should abstract a proton from hydrocyanic acid (HCN) as follows:

$$CH_3O^- + HCN \longrightarrow CH_3OH + {}^-CN$$

Since $CH_3OH$ ($pK_a \approx 17$) (see Table 1.6) is a weaker acid than HCN ($pK_a = 9.3$), and since the equilibrium shifts in favor of the weaker acid, this reaction essentially proceeds to completion.

b. No reaction. If the reaction were to proceed, the base $NO_3^-$ would abstract a proton from acidic $H_2O$ as follows:

$$NO_3^- + H_2O \longrightarrow HNO_3 + {}^-OH$$

Since $HNO_3$ ($pK_a \approx -6$) is a stronger acid than $H_2O$ ($pK_a = 15.7$), the equilibrium remains in favor of the reactants, and the reaction does not occur.

---

**EXERCISE 1.8**

Which is the stronger base? Explain your answers.

a. $^-CN$ or $RO^-$

b. $^-OH$ or $C_2H_3O_2{}^-$

c. $HC{\equiv}C^-$ or $Cl^-$

---

**EXERCISE 1.9**

Indicate if each of the following sets of reagents reacts (i.e., if the reaction essentially proceeds to completion). If they do, complete the equation; if not, write "No reaction." Explain your answers.

a. $RC{\equiv}CH + {}^-NH_2 \longrightarrow$

b. $HC_2H_3O_2 + CH_3O^- \longrightarrow$

c. $C_6H_5OH + Cl^- \longrightarrow$

---

The *Lewis definition* of an acid broadens the scope of acidity beyond that of the Brønsted concept. A Lewis acid is an electron-pair acceptor; a Lewis base is an electron-pair donor. With this definition, the Lewis system classifies substances as acids even if they contain no hydrogen atoms. For example, boron trifluoride is a Lewis acid, yet the molecule does not contain a hydrogen atom and cannot be classified as an Brønsted acid.

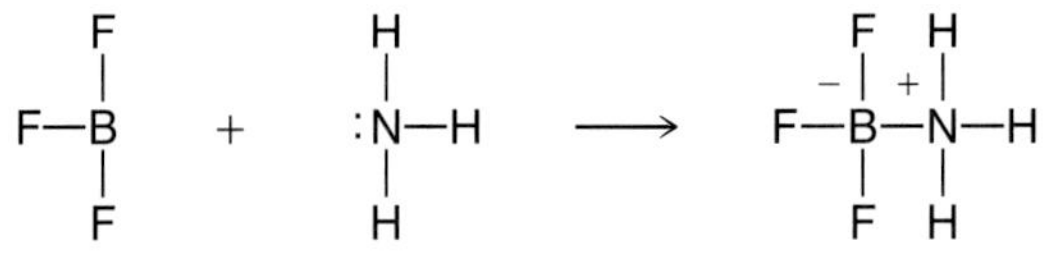

Acid (boron trifluoride)    Base (ammonia)    Adduct

The Lewis base (gaseous ammonia) donates a pair of electrons to the Lewis acid (gaseous boron trifluoride) to produce a covalent bond between nitrogen and boron and yield an adduct. This adduct, a high-melting solid, has properties similar to a salt.

---

## ▶ CHAPTER ACCOMPLISHMENTS

### 1.1 Where It All Began

☐ Define the term *organic* as it was used 200 years ago.
☐ Identify the naturally occurring source of cholesterol.
☐ Furnish a modern definition of organic chemistry.
☐ Identify the two elements that are almost always present in organic compounds.

### 1.2 Wohler and the Vital Force Theory

☐ Explain the vital force theory.
☐ Identify the scientist who disproved this theory.

### 1.3 The Atom

☐ List the particles in a typical atom.
☐ Give the maximum number of electrons found in an orbital.
☐ Distinguish between *s* and *p* orbitals in terms of shape.
☐ Write the electron configuration of the neon atom.
☐ Define valence electrons.

### 1.4 The Ionic Bond: Ionic Compounds

☐ Differentiate between
    a. an anion and a cation.
    b. a covalent bond and an ionic bond.

☐ Explain the significance of a stable octet of electrons.

## 1.5 The Covalent Bond

☐ Distinguish between a polar covalent bond, a nonpolar covalent bond, and a coordinate covalent bond.

☐ Supply the electronegativity difference between two bonded atoms that represents a bond with about 50% ionic character and 50% covalent character.

## 1.6 Intermolecular Attractive Forces

☐ Explain why the boiling point of water is high for its molecular weight.

☐ Name the weakest intermolecular forces of attraction.

☐ Name each of the intermolecular attractions that when combined represent van der Waal's forces.

## 1.7 The Extraordinary Behavior of the Carbon Atom

☐ Explain why fewer than 350,000 inorganic compounds are known, while over 8 million organic compounds have been identified.

☐ List some of the uses of organic chemicals today.

☐ Estimate the number of organic compounds known today.

## 1.8 Molecular Formula and Types of Structural Formulas

☐ Draw an electron-dot structural formula, a line-bond structural formula, a semicondensed formula, and a condensed formula for methylamine.

☐ Write two condensed structural formulas for butane.

## 1.9 Rules of Bonding for Structure Writing of Organic Compounds

☐ List the rules of bonding for drawing the structure of an organic compound.

☐ Identify the element that, to be neutral in a structure, shares three bonds and contains one unshared electron pair.

## 1.10 Charged Species in Organic Chemistry

☐ Determine that each hydrogen in an organic compound such as methane ($CH_4$) bears a formal charge of zero.

☐ Show by means of a calculation that methane is a neutral substance.

## 1.11 Resonance

☐ Draw a line-bond structural formula for a resonance structure of the carbonate ion.

☐ Draw a line-bond structural formula for the resonance hybrid that represents the "true" structure of the carbonate ion.

☐ Provide an alternate name for a resonance structure.

## 1.12 Functional Groups

☐ Provide a family name for each of the following compounds:

  a. $CH_3CH_3$
  b. $CH_3CH_2Br$
  c. $CH_3CH{=}CH_2$
  d. $CH_3C{\equiv}CH$
  e. $CH_3C{\equiv}N$

☐ Structurally distinguish between

  a. an alcohol and a mercaptan.
  b. an aldehyde and a ketone.
  c. a carboxylic acid and an ester.
  d. an amide and an amine.

## 1.13 Structural Isomers

☐ Explain why

$$CH_3CHCH_3 \quad \text{and} \quad CH_3CH_2CH_2{-}Cl$$
$$\underset{Cl}{|}$$

cannot be classified as functional-group isomers.

☐ Draw a line-bond structural formula for a pair of functional-group isomers.

## 1.14 Acidity

☐ Name the concept of acidity that is most general in scope; narrowest in scope.

☐ Recite the relationship between the $pK_a$ of an acid and its acid strength.

☐ State the relationship, in terms of acid and base strength, between a Brønsted acid and its conjugate base.

---

## ▶ KEY TERMS

The section in which each key term is first mentioned is given in parentheses. If the term is briefly introduced, the sections in which the term is discussed in depth are also provided.

*organic* (1.1)
*organ* (1.1)
*vital force theory* (1.2)

*atomic number* (1.3)
*orbital* (1.3)
*Hund's rule* (1.3)
*valence electron* (1.3)
*ionic bond* (1.4)
*covalent bond* (1.4)
*noble gas electron configuration* (1.4)
*cation* (1.4)
*anion* (1.4)

*stable octet* (1.4)
*nonpolar covalent bond* (1.5)
*polar covalent bond* (1.5)
*electronegativity* (1.5)
*coordinate covalent bond* or *dative bond* (1.5)
*intermolecular attractive forces* (1.6)
*dipole* (1.6)
*dipole-dipole attraction* (1.6)

*hydrogen bond* (1.6, 14.2)
*London forces* (1.6, 3.3)
*instantaneous dipole* (1.6)
*van der Waal's forces* (1.6)
*molecular formula* (1.8)
*structural formula (structure)* (1.8)
*electron-dot structure* (1.8)
*line-bond structure* (1.8)

*condensed formula* (1.8)
*semicondensed formula* (1.8)
*formal charge* (1.10)
*resonance* (1.11, 7.4)
*delocalized* (1.11)
*contributing structure or resonance*
   *structure* (1.11, 7.2)
*double-headed arrow* (1.11, 7.2)

*resonance hybrid* (1.11, 7.2)
*functional group* (1.12)
*isomer* (1.13)
*isomerism* (1.13)
*functional-group isomers* (1.13)
*positional isomerism* (1.13)
*Brønsted concept of acidity* (1.14)
*Lewis concept of acidity* (1.14)

## ▶ PROBLEMS

1. Sucrose (table sugar) has been isolated from sugar cane and sugar beets for many years. Explain how the isolation of sucrose justified the vital force theory.

2. Define or explain each of the following.
   a. ionic bond
   b. covalent bond
   c. atomic orbital
   d. electronegativity
   e. Hund's rule
   f. polar covalent bond
   g. nonpolar covalent bond
   h. molecular formula
   i. coordinate covalent bond
   j. vital force theory
   k. isomerism
   l. anion
   m. cation
   n. dipole-dipole attractions
   o. hydrogen bond
   p. van der Waal's forces
   q. London forces

3. Write an electronic structure for
   a. the carbon atom.
   b. the nitrogen atom.
   c. the oxygen atom.

4. Using an orbital diagram, show how the electronic structure of the nitrogen atom represents an example of Hund's rule.

5. Determine the number of valence electrons in each of the following atoms:
   a. P
   b. Mg
   c. F
   d. Li

6. Which of the following covalent substances are polar?
   a. HI
   b. $I_2$

7. Without looking at Fig. 1.5, determine which of the following pairs of atoms has the greater electronegativity.
   a. C or Si
   b. C or F
   c. Br or I

8. Which of the following compounds would you classify as ionic?
   a. HBr
   b. CO
   c. CsCl
   d. CaO

9. For each bond, label the atom with the greater electronegativity $\delta^-$ and the atom with the lesser electronegativity $\delta^+$.
   a. C—O
   b. H—S
   c. B—I
   d. C—N
   e. C—S

10. List each of the bonds in Problem 9 in order of increasing ionic character. Can any bond in Problem 9 be classified as ionic? Why or why not?

11. Classify each of the following compounds as organic or inorganic.
   a. $CH_3CH_2OH$
   b. $CO_2$
   c. $K_2CO_3$
   d. $CH_3COOH$
   e. $NH_4Cl$

12. Classify each of the following compounds as ionic or covalent.
   a. Compound A melts at 150°C, is insoluble in water, and does not conduct electricity in the melt.
   b. Compound B melts at 801°C, is water-soluble, and conducts an electric current both in solution and in the melt.

13. List the intermolecular forces present in
   a. HI
   b. $H_2$
   c. $CH_3CH = CHCH_3$

14. Why does carbon form many more compounds than all the other elements put together?

15. Draw a semicondensed and condensed structural formula for each of the following line-bond structures.
   a.
   $$\begin{array}{ccccc} & H & H & H & \\ & | & | & | & \\ H - & C & - C & - C & - H \\ & | & | & | & \\ & H & Cl & H & \end{array}$$

b.

$$H-\overset{\overset{\displaystyle H}{|}}{\underset{\underset{\displaystyle H}{|}}{C}}-\overset{\overset{\displaystyle H}{|}}{\underset{\underset{\displaystyle H}{|}}{C}}-\overset{\overset{\displaystyle H}{|}}{C}=\overset{\overset{\displaystyle H}{|}}{C}-H$$

16. Draw a line-bond structural formula for each of the following semicondensed structures.
    a. $CH_3-CH_2-CH_2-NH_2$
    b. $Cl-CH_2-CH_2-CH_2-Cl$
    c.

$$CH_3-\overset{\overset{\displaystyle H}{|}}{\underset{\underset{\displaystyle CH_3}{|}}{C}}-CH_2-CH_3$$

17. Draw a line-bond structural formula for each of the following condensed structures.
    a. $(CH_3)_2CHCH_2CH_2CH_2CH_3$
    b. $CH_3(CH_2)_6CH_3$

18. Draw a condensed structural formula for each of the compounds in Problem 16.

19. Draw as many line-bond structural formulas as you can for each of the following compounds. Use the rules of bonding.
    a. $N_2O_4$        e. $C_5HN$        i. CHON
    b. $C_2H_4O_2$     f. $C_3H_4$
    c. $C_4H_6$        g. $C_3H_3N$
    d. $C_2H_3N$       h. $CH_3NO$

20. Draw as many semicondensed and condensed structural formulas as you can for each of the following compounds. Use the rules of bonding, and do not include cyclic isomers.
    a. $C_2H_4O_2$     e. $C_3H_4$
    b. $C_3H_8$        f. $C_3H_6$
    c. $C_2H_6O$       g. $C_2H_5Br$
    d. $C_2H_7N$       h. $C_4H_8O$

21. Draw a line-bond structural formula for ethane $(C_2H_6)$. Since the bonds present are carbon-carbon and carbon-hydrogen, would you expect this covalent compound to be polar or nonpolar? Explain.

22. Calculate the formal charge, if any, on each atom of the following chemical species.
    a.

$$H-\overset{\overset{\displaystyle H}{|}}{C}=\overset{\overset{\displaystyle H}{|}}{C}:$$

    b.

$$H-\overset{\overset{\displaystyle H}{|}}{\underset{\underset{\displaystyle H}{|}}{C}}-\overset{\cdot\cdot}{C}:$$

    c.

$$CH_3-\overset{\overset{\displaystyle \cdot\overset{\cdot\cdot}{O}\cdot}{\|}}{C}-\overset{\cdot\cdot}{N}:$$

    d.

$$H-\overset{\overset{\displaystyle H}{|}}{C}-\overset{\overset{\displaystyle H}{|}}{\underset{\underset{\displaystyle O}{}}{C}}-H$$
$$\underset{\underset{\displaystyle H}{|}}{\overset{\cdot\cdot}{O}}$$

23. Determine whether each species in Problem 22 is neutral or an ion.

24. Given that the cyanide ion $(:C\equiv N:^-)$ is negatively charged, which atom carries the formal charge?

25. True or false: Hydrogen can carry a formal charge in one or more covalent compounds. Explain.

26. Draw a line-bond structure for nitric acid $(HNO_3)$; include unshared electron pairs. Then calculate the formal charge on each atom and show that nitric acid is a neutral substance.

27. Show that carbon monoxide (CO) is a neutral substance.

28. Name the functional group in each of the following compounds:
    a. $CH_3CH_2CH_2NH_2$
    b.

$$CH_3\overset{\overset{\displaystyle}{}}{\underset{\underset{\displaystyle O}{\|}}{C}}-Cl$$

    c. $CH_3CH_2CH_2SH$
    d. $CH_3C\equiv CCH_2CH_3$
    e.

$$CH_3CH_2CH_2\overset{\overset{\displaystyle}{}}{\underset{\underset{\displaystyle O}{\|}}{C}}-OH$$

29. Each of the following well-known organic compounds contains one or more functional groups. Identify the group(s).
    a.

Cholesterol

    b.

Amphetamine

    c.

D-Glucose, a sugar

d.

Benadryl, an antihistamine

e.

Morphine, a narcotic painkiller

f.

$$H_2N - C - NH_2$$

Urea, the main constituent
of human urine

g.          $CH_3(CH_2)_6COOH$

Caprylic acid, "essence" of goat

h.

$$CH_3-C-OCH_2CH_2-\overset{CH_3}{\underset{H}{C}}-CH_3$$

Isoamyl acetate,
banana flavor

30. Which of the following pairs of compounds are functional-group isomers, positional isomers, unrelated, or identical?

a. $CH_3OC_2H_5$   and   $C_2H_5\overset{|}{O}$
                               $CH_3$

b.
                                        $CH_3$
                                         $|$
   $CH_3(CH_2)_6CH_3$   and   $CH_3-C-(CH_2)_4CH_3$
                                         $|$
                                         $H$

c.
                                    $H$
                                    $|$
   $CH_3CH_2CH_2$   and   $CH_3-C-CH_3$
         $|$                        $|$
         $OH$                       $OH$

d.
   $O$                        $O$
   $\|$                       $\|$
   $H-C-OC_2H_5$   and   $CH_3-C-OCH_3$

e.
   $O$
   $\|$
   $CH_3CCH_2CH_3$   and   $CH_2=C-CH_2CH_3$
                                $|$
                                $OH$

f.
        $H$                    $H$ $H$
        $|$                    $|$ $|$
   $CH_3-C-OH$   and   $H-C-C-H$
        $|$                    $|$ $|$
        $OH$                   $OH$ $OH$

g. $CH_3CH_2Cl$   and   $ClCH_2CH_3$

h. $CH_3-CH_2$   and   $CH_3(CH_2)_2CH_3$
            $|$
            $CH_2-CH_3$

i. $CH_3-\overset{|}{\underset{CH_3}{N}}-CH_3$   and   $CH_3CH_2CH_2NH_2$

j. $CH_3CH=CH_2$   and   $CH_2=CHCH_3$

31. The compound with a molecular formula $C_4H_{10}O$ has seven isomers. Draw a line-bond and a condensed structural formula for
   a. each of the isomers containing the alcohol functional group.
   b. each of the isomers containing the ether functional group.

32. Draw a line-bond and condensed structural formula for all ketones with the molecular formula $C_5H_{10}O$.

33. Four additional isomers with the molecular formula $C_3H_6O$ (Solved Problem 1.6) can be isolated. Draw a line-bond structural formula for each isomer. *Hint*: Three of these isomers are cyclic.

34. Draw a line-bond structural formula for each of the five isomers of $C_3H_5I_3$.

35. Which of the following compounds are identical and which are unrelated?
   a. $CH_3CH_2COOH$   and   $CH_3COOH$
   b.
      $O$                        $O$
      $\|$                       $\|$
      $CH_3CH_2CCH_3$   and   $CH_3CCH_2CH_3$
   c.
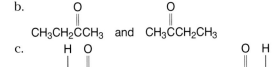

36. Define or explain
   a. Brønsted base
   b. Lewis acid

37. Which is the stronger acid (see Table 1.6.)?
   a. $HC_2H_3O_2$ or $CH_3CH_2OH$
   b. $NH_3$ or $CH_3CH_2CH_3$

38. Write a formula for the conjugate base of
   a. $CH_4$
   b. $HCO_3^-$

39. Write a formula for the conjugate acid of
   a. $^-OCH_3$
   b. $Br^-$

40. Indicate if each of the following sets of reactants reacts (i.e., if the reaction essentially proceeds to completion); if they do, complete the equation; if not, write "No reaction." Explain your answers. Use Solved Problem 1.9 as a guide.

 a. $HCN + {}^-OH \longrightarrow$

 b. $^-CH_3 + H_2O \longrightarrow$

 c. $NH_3 + {}^-OCH_3 \longrightarrow$

 d. $CH_3CH_3 + NO_3^- \longrightarrow$

41. Calculate the formal charge on boron and nitrogen in a. and aluminum and oxygen in b.

 a.

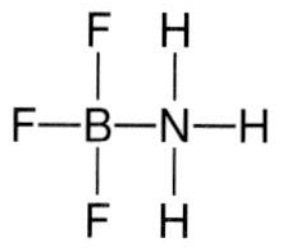

$$\begin{array}{ccc} & F & H \\ & | & | \\ F\!-\!&B\!-\!N&\!-\!H \\ & | & | \\ & F & H \end{array}$$

 b.

$$\begin{array}{ccc} & Cl & C_2H_5 \\ & | & | \\ Cl\!-\!&Al\!-\!O&\!-\!C_2H_5 \\ & | & \\ & Cl & \end{array}$$

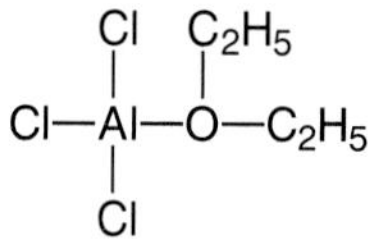

# CHAPTER 2

# Bonding in Hydrocarbons

We begin our study of functional groups by first examining a number of families of compounds, each of which contains only the elements carbon and hydrogen, called *hydrocarbons*.

There are a number of different hydrocarbon families with varying chemical structures. The following paragraphs list some of the more commonly found classes of hydrocarbons that we will study in depth in succeeding chapters (see Table 2.1).

1. **Saturated acyclic** ($C_nH_{2n+2}$): A hydrocarbon containing a straight or branched chain with no carbon-carbon multiple bonds (see Chaps. 3 and 4). The formula $C_nH_{2n+2}$ is a *generic formula* for this family. Examples of its use are given on page 27.

2. **Saturated cyclic** ($C_nH_{2n}$): A hydrocarbon in the form of a ring with no carbon-carbon multiple bonds, a cycloalkane (see Chaps. 3 and 4). See Table 2.1 for a typical structural formula.

3. **Unsaturated** ($C_nH_{2n}$, alkene: $C_nH_{2n-2}$, alkyne): A hydrocarbon that contains one or more double or triple bonds or a combination of the two (see Chaps. 5 through 8). See Table 2.1 for a typical structural formula for each hydrocarbon family.

4. **Aromatic** ($C_nH_{2n-6}$): A cyclic hydrocarbon with alternating single and double bonds (see Chaps. 9 and 10). See Table 2.1 for a typical structural formula.

Each of the families listed in Table 2.1, except the *aromatic hydrocarbons*, represents a series of compounds called a *homologous series*. Consider the molecular formula calculation for successive alkanes: Using the generic formula $C_nH_{2n+2}$, where $n$ is the number of carbon atoms in a given molecule, we have

$$
\begin{array}{llll}
C_1H_{2(1)+2} = CH_4 & n = 1 & \text{methane} \\
C_2H_{2(2)+2} = C_2H_6 & n = 2 & \text{ethane} \\
C_3H_{2(3)+2} = C_3H_8 & n = 3 & \text{propane} \\
C_4H_{2(4)+2} = C_4H_{10} & n = 4 & \text{butane}
\end{array}
$$

Note that the difference between successive formulas is, as expected, $CH_2$. Thus the alkanes represent a homologous series of compounds.

## 2.1 HYDROCARBONS

Table 1.3 contains a number of functional group families.

Alkanes and cycloalkanes are *saturated* when they contain only carbon-carbon single bonds. Alkenes and alkynes are *unsaturated hydrocarbons* because each compound contains one or more carbon-carbon double or triple bonds. Hydrocarbons are classified as *acyclic* if the carbon chains are straight or branched, and *cyclic* if the carbons link together to form a ring.

Successive members of a homologous series differ in molecular formula by $CH_2$ (a carbon and two hydrogens).

**FIGURE 2.1** The hybridization process depicted with methane.

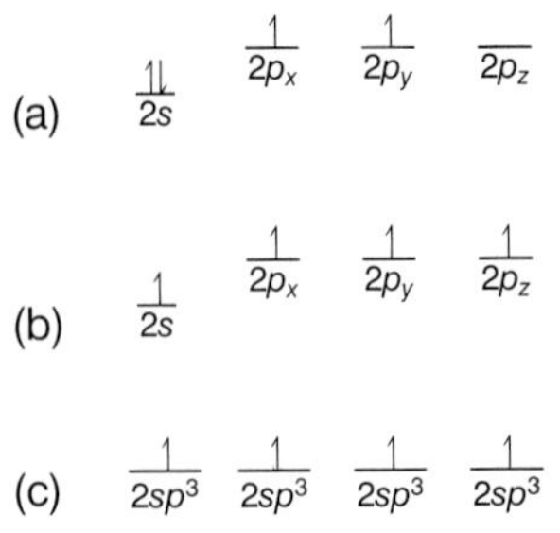

(a) $\quad \underset{2s}{\underline{\text{11}}} \quad \underset{2p_x}{\overline{\underline{1}}} \quad \underset{2p_y}{\overline{\underline{1}}} \quad \underset{2p_z}{\overline{\phantom{1}}}$

(b) $\quad \underset{2s}{\underline{1}} \quad \underset{2p_x}{\overline{\underline{1}}} \quad \underset{2p_y}{\overline{\underline{1}}} \quad \underset{2p_z}{\overline{\underline{1}}}$

(c) $\quad \underset{2sp^3}{\overline{\underline{1}}} \quad \underset{2sp^3}{\overline{\underline{1}}} \quad \underset{2sp^3}{\overline{\underline{1}}} \quad \underset{2sp^3}{\overline{\underline{1}}}$

We know from experiment that methane, for example, has four carbon-hydrogen bonds of equal energy. Although the orbital diagram in Fig. 2.1b shows the presence of four unpaired electrons to produce four bonds, these bonds would have different energies if formed. One bond would be produced from a lower-energy $2s$ electron, while the other three bonds would be formed from three different higher-energy $2p$ electrons.

Note that in the hybridization process the number of orbitals and electrons is not changed; i.e., we start with four of each and finish with four of each.

| TABLE 2.1 | Families of Hydrocarbons | | | |
|---|---|---|---|---|
| Family | Generic Formula | Saturated/ Unsaturated | Typical Molecular Formula | Typical Structural Formula |
| Alkane (paraffin) | $C_nH_{2n+2}$ | Saturated | $C_4H_{10}$ | $CH_3CH_2CH_2CH_3$ |
| Alkene (olefin) | $C_nH_{2n}$ | Unsaturated | $C_3H_6$ | $CH_3CH{=}CH_2$ |
| Cycloalkane | $C_nH_{2n}$ | Saturated | $C_3H_6$ | 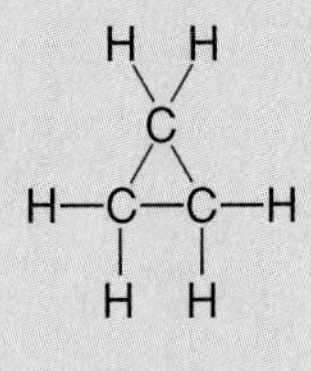 (structural formula of cyclopropane) |
| Alkyne | $C_nH_{2n-2}$ | Unsaturated | $C_3H_4$ | $CH_3C{\equiv}CH$ |
| Aromatic hydrocarbon | $C_nH_{2n-6}$ | Unsaturated* | $C_6H_6$ | 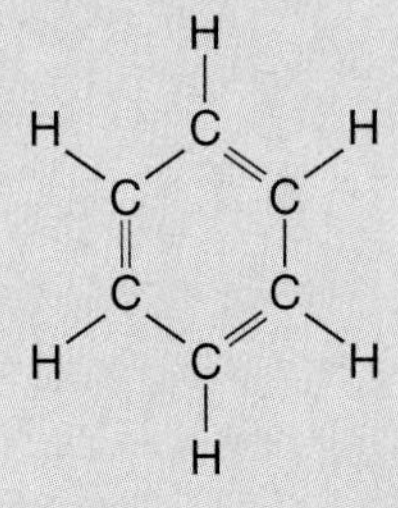 (structural formula of benzene) |

* Although aromatic hydrocarbons contain a number of carbon-carbon double bonds, these compounds react differently than typical unsaturated compounds.

## 2.2 HYBRIDIZATION IN METHANE—AND ALL ALKANES

From now on, for convenience these $2sp^3$ hybrid orbitals will be designated $sp^3$.

Consider the electron configuration of the carbon atom: $1s^2 2s^2 2p^2$. The inner ($1s$) shell of electrons lies too close to the positively charged nucleus of the atom and is unimportant because chemical activity is due to the electrons in the outermost subshells. Since carbon contains two unpaired electrons (refer to the orbital diagram of carbon, Fig. 2.1a), one could expect carbon to form the compound $CH_2$. However, we know that carbon almost always forms four bonds. Somehow we must devise an electron configuration for carbon with four unpaired electrons to produce four covalent bonds. This is done (as put forward by Pauling) by promoting one $2s$ electron to a higher-energy $2p$ orbital to produce the electron distribution in Fig. 2.1b.

To produce four unpaired electrons of equal energy and thus form four bonds of equal energy, Pauling additionally suggested to mix (hybridize) the one $2s$ orbital and three $2p$ orbitals into four $2sp^3$ *hybrid orbitals* of equal energy, where each orbital has $\frac{1}{4}s$ and $\frac{3}{4}p$ character (Fig. 2.1c). Thus, we now have a conceptual or theoretical model of the carbon atom that is consistent with experimental evidence.

These $2sp^3$ orbitals (each containing an electron) are directed to the corners of a regular tetrahedron (Fig. 2.2). In this way, the hydrogen atoms in methane are as far apart from each other as possible, resulting in maximum stability due to minimal electron-pair repulsions between orbitals. These repulsions, known as *London repulsions*, exist because electron pairs on adjacent carbon-hydrogen bonds are too close together and repel each other due to similar electric charges (see Fig. 2.2). Each of the $2sp^3$—$1s$ bonds (abbreviated $sp^3$—$s$ for convenience) is called a *sigma ($\sigma$) bond*. We say the orbitals constituting the sigma bond overlap end to end (Fig. 2.2).

Methane can be represented in a variety of ways. You are familiar with both the line-bond structural formula and the condensed formula (Sec. 1.8). The *ball-and-stick model* represents the molecule as *tetrahedral*, where a large ball represents an atom of carbon, a smaller ball signifies an atom of hydrogen, and a stick depicts a carbon-hydrogen

**FIGURE 2.2** Orbital direction in two methane molecules with one molecule containing inscribed tetrahedron.

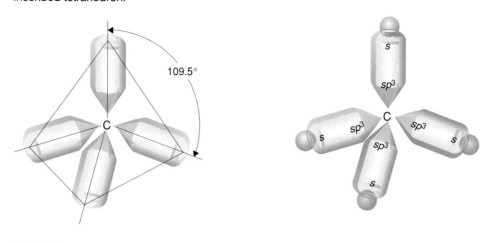

**FIGURE 2.3** Methane.

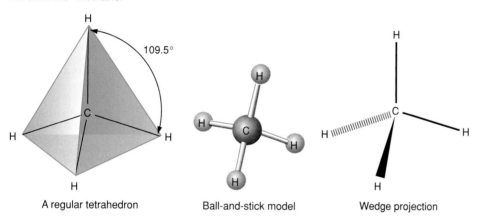

A regular tetrahedron          Ball-and-stick model          Wedge projection

H—C—H bond angle = 109.5 degrees

bond. In a similar manner, the *wedge projection* of methane shows the wedged carbon-hydrogen bond projecting toward the observer and the dashed carbon-hydrogen projecting away from the observer. The remaining two carbon-hydrogen bonds of the wedge projection are on the plane of the paper (Fig. 2.3).

The orbital description of ethane is similar to that of methane with the addition of a carbon-carbon ($sp^3$—$sp^3$) sigma ($\sigma$) bond (Fig. 2.4).

Ethane molecules consist of two tetrahedra placed head to head and can be represented by means of a line-bond structural formula, condensed formula, ball-and-stick model, and wedge projection (Fig. 2.5). The carbon-carbon ($sp^3$—$sp^3$,$\sigma$) bond in ethane can be rotated freely (see Sec. 3.1 for an explanation). This means that a number of different spatial arrangements are possible for ethane and higher homologues in the alkane family. Each arrangement is called a *conformational isomer* or *conformation*. Conformations can be described using the sawhorse diagram, Newman projection, and wedge projection (Fig. 2.6). In the *sawhorse diagram*, each intersection of three lines represents a carbon. The *Newman projection* consists of a carbon atom at the intersection of the three lines nearest you. The large circle farther away is the second carbon of ethane. If you view the ball-and-stick model of ethane from one end (see Fig. 2.5), the eclipsed Newman projection shown in Fig. 2.6 can be observed. In the wedge projection, the solid lines represent bonds and hydrogen atoms in the plane, the wedges represent bonds bond hydrogen atoms projecting toward you, and the dashed lines

## 2.3
## ETHANE

Each student should have a set of molecular models. Molecular models are invaluable in clarifying for the student spatial relationships in various chemical species.

**FIGURE 2.4**  (a) Orbital description of ethane; (b) orbital description of ethane with back-to-back inscribed tetrahedra.

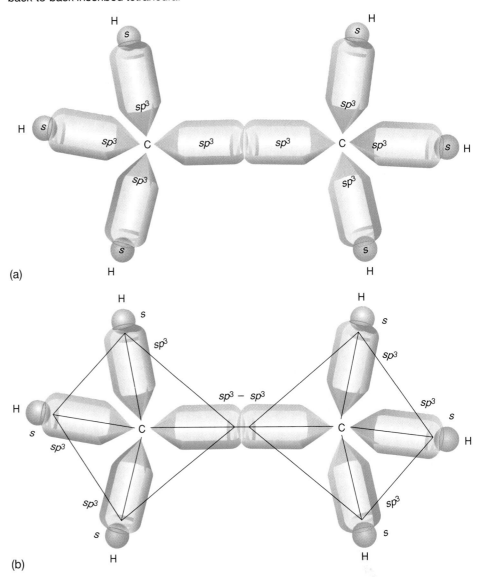

(a)

(b)

represent bonds and hydrogen atoms pointing away from you. *Eclipsed* and *staggered conformations* are discussed in greater depth in Sec. 3.1.

## 2.4 BONDING IN ETHYLENE—AND OTHER ALKENES

Let us begin our creation of an orbital model for alkenes by considering the orbital diagram of carbon, just as we did previously for alkanes.

$$\underset{2s}{\underline{\text{1\hspace{-0.2em}l}}} \quad \overset{\underline{1}}{2p} \quad \overset{\underline{1}}{2p} \quad \overset{\phantom{1}}{\overline{2p}}$$

(orbital diagram of carbon)

Again, as with the alkanes, one $2s$ electron is promoted to a higher-energy $2p$ orbital to produce the following orbital diagram of carbon:

$$\underset{2s}{\underline{1}} \quad \overset{\underline{1}}{2p} \quad \overset{\underline{1}}{2p} \quad \overset{\underline{1}}{2p}$$

We know from experiment that each carbon in ethylene, for example, is bonded to only three other atoms. Yet the rule of eight is followed, and each carbon in ethylene—and

**FIGURE 2.5** Ethane.

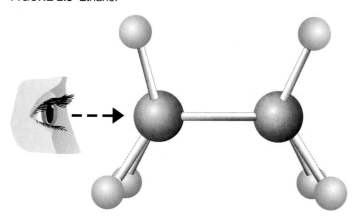

Ball and stick model

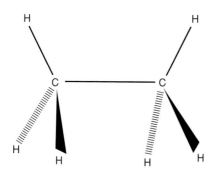

Wedge projection

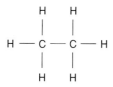

Line–bond structural formula

$CH_3CH_3$

Condensed structural formula

other alkenes—shares eight electrons. To reconcile theory with experiment, Pauling proposed to hybridize the $2s$ orbital and two $2p$ orbitals into three $2sp^2$ (abbreviated $sp^2$) hybrid orbitals of equal energy, where each orbital has $\frac{1}{3}s$ and $\frac{2}{3}p$ character.

$$\underline{\quad}_{2sp^2}^{\frac{1}{\phantom{2}}} \qquad \underline{\quad}_{2sp^2}^{\frac{1}{\phantom{2}}} \qquad \underline{\quad}_{2sp^2}^{\frac{1}{\phantom{2}}} \qquad \underline{\quad}_{2p}^{\frac{1}{\phantom{2}}}$$

These $sp^2$ orbitals (each containing an electron) are directed to the corners of an equilateral triangle and are located on a plane. Finally, the one unhybridized $2p$ orbital remains and is perpendicular to the $sp^2$ orbitals (Fig. 2.7).

In this way, the orbitals are as far away from each other as possible, resulting in maximum stability due to minimal electron-pair repulsion between orbitals (London repulsions).

**FIGURE 2.6** Conformations of ethane.

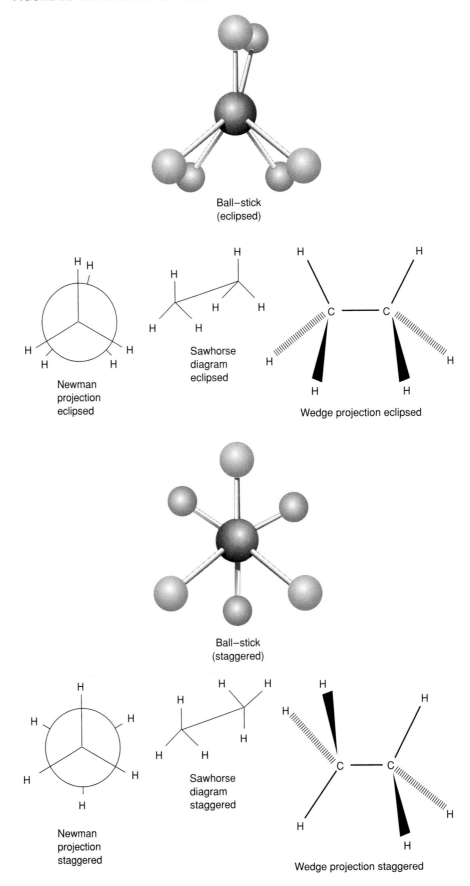

Ball–stick
(eclipsed)

Newman
projection
eclipsed

Sawhorse
diagram
eclipsed

Wedge projection eclipsed

Ball–stick
(staggered)

Newman
projection
staggered

Sawhorse
diagram
staggered

Wedge projection staggered

**FIGURE 2.7** Oblique and top view of the orbitals projecting from a trigonal carbon atom following hybridization.

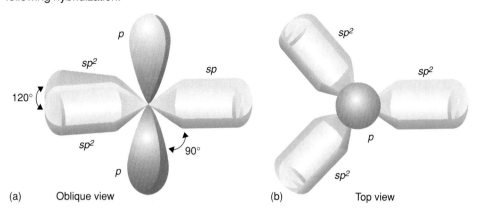

(a)  Oblique view  (b)  Top view

**FIGURE 2.8** Bonding in ethylene: orbital overlap. The sigma bond is a bond where the greatest electron density is in the same plane as the atoms participating in the bond, between the atoms concerned. The $p$—$p$, $\pi$ bond, on the other hand, is a bond where the greatest electron density exists above and below the plane of the carbon atoms.

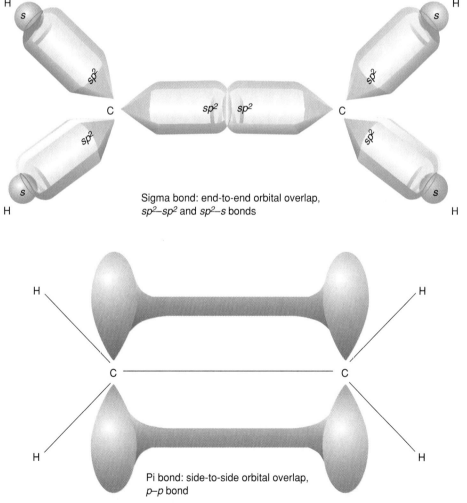

Sigma bond: end-to-end orbital overlap,
$sp^2$–$sp^2$ and $sp^2$–$s$ bonds

Pi bond: side-to-side orbital overlap,
$p$–$p$ bond

Since the simplest alkene contains two carbons (ethylene, $CH_2{=}CH_2$), let us bring two $sp^2$-hybridized carbons together. This results in the formation of an $sp^2$—$sp^2$,$\sigma$ bond. This bond is strong due to a significant amount of end-to-end overlap of the $sp^2$ orbitals. Once the $sp^2$—$sp^2$, $\sigma$ bond is formed, the $2p$ orbitals are in position to overlap. This overlap takes place side to side, and the bond formed is called a *pi* ($\pi$) *bond*. To be more precise, this is a $2p$—$2p,\pi$ (abbreviated $p$—$p,\pi$) bond. To sum up, the double bond consists of a $sp^2$—$sp^2$, $\sigma$ bond and a $p$—$p,\pi$ bond (Fig. 2.8). The amount of orbital overlap is far less in the pi bond than in the sigma bond. Therefore, the pi bond is weaker and is broken in most chemical reactions, while the stronger sigma bond is left intact.

In addition to the double bond, each carbon in ethylene forms two $sp^2$—$s,\sigma$ bonds with hydrogen in the usual manner (Fig. 2.8). The two carbons participating in the double bond and the four atoms bonded to them, in ethylene and other alkenes, lie in a plane. The H—C—H and H—C—C bond angles in ethene are approximately 120 degrees to maximize the distance between $sp^2$ orbitals and thereby minimize London repulsions (Fig. 2.9). It is interesting to note that the carbon-carbon double bond is both shorter and stronger than the carbon-carbon single bond. For example, in ethene, the length of the carbon-carbon double bond is 1.34 Å, and the strength of the bond is 152 kcal/mol, while these data for the carbon-carbon single bond in ethane are 1.54 Å and 88 kcal/mol, respectively. Both the shorter bond length and the greater bond strength of the carbon-carbon double bond in ethylene are due to the additional bond in ethylene drawing the carbons together and requiring more energy to break two bonds.

---

## SOLVED PROBLEM 2.1

**FIGURE 2.9** Bonding in ethylene; bond angles.

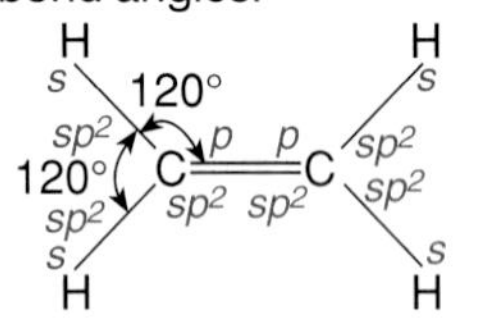

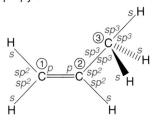

**FIGURE 2.10** Bonding in propylene.

Draw a line-bond structural formula for propylene ($CH_3$—$CH$=$CH_2$) similar to that of ethylene in Fig. 2.9. Do each of the following for propylene:

a.  Locate the $sp^2$—$sp^3$ hybridized bond.

b.  Locate the $p$—$p$ bond and the $sp^2$—$sp^2$ hybridized bond.

c.  Locate the $sp^2$—$s$ bonds.

d.  Locate the $sp^3$—$s$ bonds.

### SOLUTION

a.  Since $sp^2$ bonding is found on a *trigonal carbon* (a carbon bonded to three other atoms) and $sp^3$ bonding on a *tetrahedral carbon* (a carbon bonded to four other atoms), the $sp^2$—$sp^3$ bond must be located between C-2 and C-3.

b.  The $p$—$p$ bond and the $sp^2$—$sp^2$ bond are located between two trigonal carbons and thus are located between C-1 and C-2.

c.  The $sp^2$ orbitals are projected from trigonal carbons, and $s$ orbitals from hydrogen. Therefore, two $sp^2$—$s$ bonds project from C-1 to hydrogen and one from C-2 to hydrogen.

d.  The three $sp^3$—$s$ bonds project from C-3 to hydrogen, since $sp^3$ bonding is associated with a tetrahedral carbon. All the bonding in the propylene molecule is summarized in Fig. 2.10.

---

## EXERCISE 2.1

Draw a line-bond structural formula for 1-pentene ($CH_3CH_2CH_2CH$=$CH_2$) as in Fig. 2.9. Locate and determine the number of each of the following bonds in the molecule.

a.  $sp^2$—$s$

b.  $sp^2$—$sp^2$

c.  $p$—$p$

d.  $sp^3$—$s$

e.  $sp^2$—$sp^3$ or $sp^3$—$sp^2$

f.  $sp^3$—$sp^3$

To sum up, introducing a second bond between carbon atoms in a molecule both shortens and strengthens the resulting bond.

Let us create an orbital model for alkynes in the same manner that we did for alkenes, i.e., by examining the orbital diagram for carbon.

$$\frac{\underline{\uparrow}}{2p} \quad \frac{\underline{\uparrow}}{2p} \quad \frac{\phantom{\uparrow}}{2p}$$

$$\frac{\underline{\uparrow\downarrow}}{2s} \quad \text{(orbital diagram of carbon)}$$

Again, as with the alkenes, one $2s$ electron is promoted to a higher-energy $2p$ orbital to produce the following orbital diagram of carbon:

$$\frac{\underline{\uparrow}}{2s} \qquad \frac{\underline{\uparrow}}{2p} \quad \frac{\underline{\uparrow}}{2p} \quad \frac{\underline{\uparrow}}{2p}$$

We know from experiment that each carbon in, for example, acetylene ($C_2H_2$) is bonded to only two other atoms. Yet the rule of eight is followed, and each carbon in acetylene—and other alkynes—must share eight electrons. To reconcile the theoretical model with experiment, Pauling proposed to hybridize the $2s$ orbital and one $2p$ orbital into two $2sp$ (abbreviated $sp$) hybrid orbitals of equal energy, where each orbital has $\frac{1}{2}s$ and $\frac{1}{2}p$ character. Note that in the hybridization process the number of orbitals and electrons is not changed; i.e., we start with two of each and finish with two of each.

$$\frac{\underline{\uparrow}}{2sp} \qquad \frac{\underline{\uparrow}}{2sp} \qquad \frac{\underline{\uparrow}}{2p} \quad \frac{\underline{\uparrow}}{2p}$$

These two $sp$ orbitals (each containing an electron) are directed to both ends of a straight line and are said to be *linear*. In this way, the orbitals are as far apart from each other as possible, resulting in maximum stability due to minimal electron-pair repulsion between orbitals (London repulsions). Finally, the two unhybridized $2p$ orbitals remain and are both mutually perpendicular and perpendicular to the $sp$ orbitals (Fig. 2.11).

Since the simplest alkyne contains two carbons (acetylene), let us bring two $sp$-hybridized carbons together. This results in the formation of a strong $sp$—$sp,\sigma$ bond with end-to-end overlap. Once the $sp$—$sp,\sigma$ bond is formed, the two remaining $2p$ orbitals in each carbon atom overlap side to side to form two weaker $p$—$p,\pi$ bonds, each of which is perpendicular to the other and to the plane of the $sp$—$sp$ bond (Fig. 2.12).

The two carbons participating in the triple bond and the two atoms bonded to them in acetylene and other alkynes lie on a straight line (are linear). The carbon-carbon triple bond is the shortest bond of the three classes of hydrocarbons considered thus far. Since both alkanes and alkynes have a sigma bond available for bonding but only the alkynes

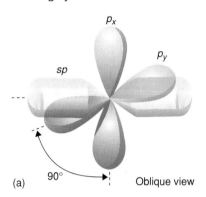
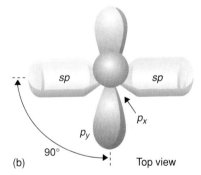
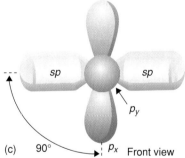

**FIGURE 2.11** Selected views of the orbitals projecting from a carbon atom bonded to two atoms following hybridization.

For the sake of clarity, one $p$ orbital is arbitrarily designated $p_x$ and the other $p_y$. Thus, in the top view of Fig. 2.11b we are looking down the top of the $p_x$ orbital to show more clearly the 90 degree bond angle between the $p_y$ and $sp$ orbitals. In the same way, in Fig. 2.11c, we are looking in front of the $p_x$ orbital to observe more clearly the 90 degree bond angle between the $p_x$ and $sp$ orbitals.

**FIGURE 2.12** Bonding in acetylene. Note the $p$ bonds are perpendicular to each other.

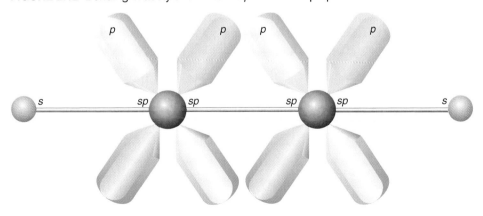

To sum up, the triple bond consists of an $sp$—$sp$, $\sigma$ bond and two $p$—$p$, $\pi$ bonds. In addition to the triple bond, each carbon in acetylene forms one $sp$—$s$, $\sigma$ bond with hydrogen in the usual manner.

have an additional two pi bonds available, alkynes have the shortest bond distance between the functional group carbons due to greater orbital overlap resulting from the presence of the additional two pi bonds in the alkyne.

| SOLVED PROBLEM 2.2 | Draw a line-bond structural formula for propyne (CH₃C≡CH). Then do each of the following for propyne. |
|---|---|

Draw a line-bond structural formula for propyne ($CH_3C{\equiv}CH$). Then do each of the following for propyne.

   a.  Locate the *sp—sp* hybridized bond.

   b.  Locate the *p—p* bonds.

   c.  Locate the *sp—sp³* hybridized bond.

   d.  Locate the *sp—s* bond.

   e.  Locate the *sp³—s* bonds.

**FIGURE 2.13**  Line-bond structural formula of propyne.

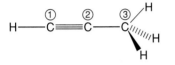

**SOLUTION**

   a.  Since *sp* bonding is found between the carbons of the triple bond, the only *sp—sp* bond is located between C-1 and C-2 (Fig. 2.13).

   b.  The two *p—p* bonds also are located between the carbons of the triple bond and therefore are found between C-1 and C-2.

   c.  Since *sp* bonding is found on a triple-bonded carbon and *sp³* bonding is found on a tetrahedral carbon, the *sp—sp³* bond must be located between C-2 and C-3.

**FIGURE 2.14**  Bonding in propyne.

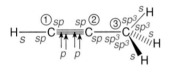

   d.  The *sp* orbital is projected from a triple-bonded carbon, and *s* orbitals from hydrogen. Therefore, the only *sp—s* bond projects from C-1 to hydrogen.

   e.  The three *sp³—s* bonds project from C-3 to hydrogen because *sp³* bonding is associated with a tetrahedral carbon. All the bonding in the propyne molecule is summarized in Fig. 2.14.

| EXERCISE 2.2 | Draw a line-bond structural formula for 1-butyne (CH₃CH₂C≡CH) as in Fig. 2.13 (neglect spacial geometry). |
|---|---|

Draw a line-bond structural formula for 1-butyne ($CH_3CH_2C{\equiv}CH$) as in Fig. 2.13 (neglect spacial geometry). Locate and determine the number of each of the following bonds in the molecule.

   a.  *sp—s*

   b.  *sp—sp*

   c.  *p—p*

   d.  *sp³—s*

   e.  *sp—sp³*

   f.  *sp³—sp³*

*Note that in the geometric condensed structure (Secs. 3.6 and 3.11), one carbon atom, by convention, is assumed to be present at the intersection of two lines. The presence of one hydrogen on each carbon also is understood.*

## 2.6
## BONDING IN BENZENE

In 1858, Professor Friedrich Kekulé proposed a hexagonal structure for benzene ($C_6H_6$) with alternating single and double bonds to form a *conjugated system* as

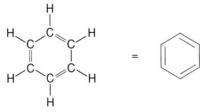

Geometric condensed    Geometric condensed
structural formula      structural formula

Benzene is a planar molecule with the shape of a regular hexagon. Each carbon in the benzene ring is $sp^2$ hybridized to the single hydrogen bonded to it and to each of the neighboring carbons. Thus three sigma bonds join each carbon on the ring: two $sp^2$—$sp^2$,$\sigma$ bonds, each to a neighboring carbon, and an $sp^2$—$s$,$\sigma$ bond to hydrogen (Fig. 2.15).

**FIGURE 2.16** Pi bonding in benzene.

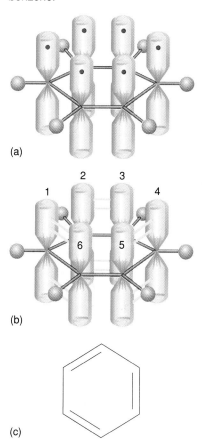

(a)

(b)

(c)

Note that in (b), $p$ orbital overlap is shown between carbons 2 and 3, 4 and 5, along with 6 and 1. You need to realize that $p$ orbital overlap also can occur between carbons 1 and 2, 3 and 4, along with 5 and 6. We will discuss this phenomenon of complete pi electron delocalization around the benzene ring when we study resonance in benzene (Sec. 9.6).

**FIGURE 2.15** Sigma bonding in benzene.

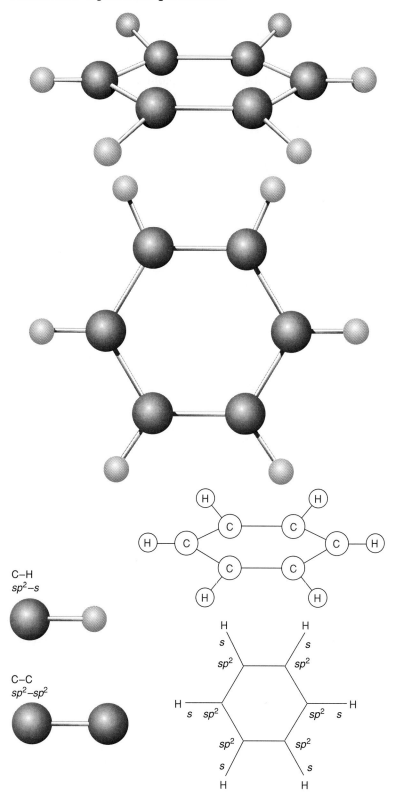

C–H
$sp^2$–$s$

C–C
$sp^2$–$sp^2$

In addition, each carbon has a *p* orbital, each orbital containing one electron (Fig. 2.16a). Side-to-side overlap of neighboring *p* orbitals takes place as shown in Fig. 2.16b to give three *p—p*,π bonds. Thus each conjugated double bond consists of an *sp²—sp²*,σ bond and a *p—p*,π bond. This orbital description is translated into a standard *geometric condensed structure* of benzene (Fig. 2.16c) often called a *Kekulé structural formula* in honor of Friedrich Kekulé.

## 2.7
## A SUMMARY OF HYBRIDIZATION, GEOMETRY, AND BOND ANGLES IN THE VARIOUS CLASSES OF HYDROCARBONS

*Hybridization* takes place when *s* and *p* atomic orbitals of carbon are mixed to produce hybrid atomic orbitals, which are used by the carbon atom because overlapping with orbitals of other atoms is more extensive. Thus the bonds produced are more stable. The kinds of hybridized atomic orbitals used by carbon, the geometry of hydrocarbons, and the appropriate bond angles in selected classes of hydrocarbons are summarized in Table 2.2.

| TABLE 2.2 | Hybridization, Geometry, and Bond Angles in Selected Classes of Hydrocarbons | | |
|---|---|---|---|
| **Class of Hydrocarbon** | **Hybrid Orbitals Used** | **Geometry** | **Bond Angle (degrees)** |
| Alkane | $sp^3$ | Tetrahedral | 109.5 (H—C—H) |
| Alkene | $sp^2$ | Planar | 120 (H—C—H) |
| Alkyne | $sp$ | Linear | 180 (H—C—C) |
| Aromatic | $sp^2$ | Planar | 120 (H—C—C) |

## 2.8
## CARBON-CARBON BOND STRENGTHS AND BOND LENGTHS IN HYDROCARBONS

The strength of a bond is the amount of energy necessary to break the bond *homolytically*; i.e., each atom participating in the bond has one electron after the break.

$$A—B \longrightarrow A\cdot + \cdot B$$

Since the carbon-carbon sigma bond has a bond strength of 88 kcal/mol in an alkane, we can calculate the bond strength of a pi bond in an alkene if we assume that the bond strength of an $sp^2—sp^2$,σ bond in the alkene is equal to the bond strength of an $sp^3—sp^3$,σ bond in a corresponding alkane. The data necessary for the calculation are as follows:

$$-\overset{|}{\underset{|}{C}}-\overset{|}{\underset{|}{C}}- \longrightarrow -\overset{|}{\underset{|}{C}}\cdot \ \cdot\overset{|}{\underset{|}{C}}- \qquad \Delta H = 88 \text{ kcal/mol}$$

We also know that

$$\overset{}{\underset{}{}}C{=}C\overset{}{\underset{}{}} \longrightarrow \overset{}{\underset{}{}}\ddot{C} \ \ddot{C}\overset{}{\underset{}{}} \qquad \Delta H = 152 \text{ kcal/mol}$$

Thus we can calculate the bond strength of the pi bond as

$$\Delta H \text{ pi bond} = 152 - 88 = 64 \text{ kcal/mol}$$

From this inexact calculation (different researchers have obtained different values for some bond energies), we can see that the pi bond of the double bond is weaker than the sigma bond. A similar calculation will demonstrate that both pi bonds in an alkyne are much weaker (show lower bond energies) than the *sp—sp*, σ bond. Thus, for an alkyne, we have

$$-C{\equiv}C- \longrightarrow -\dot{C} \ \cdot\dot{C}- \qquad \Delta H = 200 \text{ kcal/mol}$$

$$\Delta H \text{ sigma bond } (sp—sp) = 88 \text{ kcal/mol}$$
$$\Delta H \text{ two pi bonds } (p—p) = \underline{112 \text{ kcal/mol}} \ (56 \times 2)$$
$$\text{Total} = 200 \text{ kcal/mol}$$

This makes sense experimentally because in the addition reactions of an alkene or alkyne, it is the pi bond(s) that is(are) broken, while the sigma bond most often is left intact (see Secs. 5.6, 6.2, 7.4, and 8.7).

In general, the greater the number of bonds between two carbons, the shorter the bond distance and the stronger the bond (Table 2.3). In general also, the greater the percentage of $s$ character in a bond, the shorter the bond distance and the stronger the bond (see Table 2.3).

## 2.9 A SUMMARY OF CARBON-CARBON BOND LENGTHS AND BOND STRENGTHS IN VARIOUS CLASSES OF HYDROCARBONS

| TABLE 2.3 | A Comparison of Bond Lengths and Bond Strengths in Various Classes of Hydrocarbons | | |
|---|---|---|---|
| **Functional Group** | **Carbon-Carbon Bond Length (Å) (Number of Bonds Between the Two Carbons)** | **Bond Strength (kcal/mol)** | **Percent s Character** |
| Alkane —C—C— | 1.54 (1) | 88 | 25 |
| Alkene C=C | 1.34 (2) | 152 | 33 |
| Alkyne —C≡C— | 1.20 (3) | 200 | 50 |
| Aromatic hydrocarbon | 1.39 (1.5) (due to resonance, p. 172) | ~115 (based on bond length) | — |

## ▶ CHAPTER ACCOMPLISHMENTS

### 2.1 Hydrocarbons
☐ Define the term *hydrocarbon*.
☐ Supply the generic formula for
   a. a saturated acyclic hydrocarbon.
   b. a saturated cyclic hydrocarbon.
   c. an unsaturated acyclic hydrocarbon containing one double bond.
☐ Provide a family name for each class of hydrocarbon.

### 2.2 Hybridization in Methane—And All Alkanes
☐ Explain why the orbital diagram of a carbon atom shown on page 4 contains no completely filled $p$ orbital.
☐ Draw a wedge projection of methane.
☐ Name the hybridized orbitals projecting from a carbon atom in methane—and all alkanes.
☐ Give the geometry of methane—and all alkanes.

### 2.3 Ethane
☐ Draw a sawhorse diagram of the eclipsed conformation of an ethane molecule.
☐ Supply the hybrid orbitals used in the carbon-carbon bond in ethane.

### 2.4 Bonding in Ethylene—And Other Alkenes
☐ Differentiate between a trigonal carbon and a tetrahedral carbon.
☐ Explain why the carbon-carbon double bond is shorter than the carbon-carbon single bond.

### 2.5 Bonding in Acetylene—And All Alkynes
☐ Supply the geometry around the carbon-carbon triple bond.
☐ Explain why a carbon-carbon triple bond is shorter than a carbon-carbon double bond.

**2.6  Bonding in Benzene**
☐ Draw a geometric condensed structure for benzene.
☐ Provide another name for a geometric condensed structural formula of benzene that honors a scientist.

**2.7  A Summary of Hybridization, Geometry, and Bond Angles in the Various Classes of Hydrocarbons**
☐ Summarize the hybrid orbitals used, geometry, and bond angle present for each class of hydrocarbon.

**2.8  Carbon-Carbon Bond Strengths and Bond Lengths in Hydrocarbons**
☐ Compare the bond strength of a carbon-carbon double bond in an alkene with that of a pi bond in an alkene.

☐ Compare the bond strength of a carbon-carbon triple bond in an alkyne with that of the pi bonds in an alkyne.

**2.9  A Summary of Carbon-Carbon Bond Lengths and Bond Strengths in Various Classes of Hydrocarbons**
☐ Select the class of hydrocarbon that contains the greatest percent *s* character in its hybridized orbitals; the least percent *s* character.

## ▶ KEY TERMS

hydrocarbon (2.1)
saturated hydrocarbon (2.1)
unsaturated hydrocarbon (2.1, 5.1, 7.1, 8.1)
generic formula (2.1)
acyclic hydrocarbon (2.1, 3.2)
cyclic hydrocarbon (2.1, 3.7)
aromatic hydrocarbon (2.1, 9.1)
homologous series (2.1)
hybrid orbital (2.2)

London forces of repulsion (2.2)
sigma (σ) bond (2.2)
ball-and-stick model (2.2, 2.3)
tetrahedral geometry (2.2)
wedge projection (2.2, 2.3)
conformational isomer or conformation (2.3, 3.1)
sawhorse diagram (2.3)
Newman projection (2.3)
eclipsed conformation (2.3)

staggered conformation (2.3)
pi (π) bond (2.4)
trigonal carbon (2.4, 5.1)
tetrahedral carbon (2.4)
conjugated system of bonds (2.6)
geometric condensed structure (2.6, 3.6, 3.11)
Kekulé structural formula (2.6)
hybridization (2.7, 2.2, 2.4, 2.5, 2.6)
homolytic dissociation (2.8, 4.3)

## ▶ PROBLEMS

1. Which of the following hydrocarbons is saturated?
   a. $CH_3(CH_2)_{14}CH_3$
   b. $CH_3—CH=CH—CH_3$
   c. $CH_3—C\equiv C—H$

2. Name the family to which each compound in Problem 1 can be assigned.

3. Which of the following compounds contains one or more pi bonds?
   a. $CH_3CH_2CH_3$
   b. $CH_3CH_2CH_2CH=CH_2$
   c. $CH_3C\equiv CH$

4. Calculate the molecular formula for each of the following. Use the generic formula given in Table 2.1.
   a. A saturated cyclic hydrocarbon containing seven carbon atoms.
   b. An unsaturated hydrocarbon, with one double bond, containing seven carbon atoms.

5. Why does methane take the shape of a regular tetrahedron as opposed to remaining planar?

6. Without looking at Figs. 2.3 and 2.6, draw each of the following. A set of molecular models will be useful here.
   a. A wedge projection of methane

   b. A Newman projection of ethane
   c. A sawhorse diagram of ethane

7. Draw a structural formula for propane ($CH_3CH_2CH_3$) of the type given in Fig. 2.9, and determine the number of each of the following bonds in the molecule. Neglect spatial geometry.
   a. $sp^3—sp^3$
   b. $sp^3—s$

8. Why is an $sp^2—sp^2,\sigma$ bond stronger than a $p—p,\pi$ bond?

9. Draw a line-bond structure for 1-butene ($CH_2= CHCH_2CH_3$) and for 2-butene ($CH_3CH=CHCH_3$) similar to that of ethylene in Fig. 2.9. Fill in the following table by listing the number of bonds of each type in each structure. Neglect spatial geometry.

| Bond Type | 1-Butene | 2-Butene |
|-----------|----------|----------|
| $p—p$ | | |
| $sp^2—s$ | | |
| $sp^3—s$ | | |
| $sp^2—sp^2$ | | |
| $sp^2—sp^3$ | | |
| $sp^3—sp^3$ | | |

10. Designate each of the following bonds as sigma or pi.
    a. $sp^2$—$s$
    b. $sp^2$—$sp^2$
    c. $s$—$s$
    d. $p$—$p$

11. Describe the bonds that make up the double bond. How do they differ?

12. Draw a line-bond structural formula for 2-butyne ($CH_3C{\equiv}CCH_3$) as in Fig. 2.13 (neglect spacial geometry). Locate and determine the number of each of the following bonds in the molecule.
    a. $sp$—$sp$
    b. $p$—$p$
    c. $sp^3$—$s$
    d. $sp$—$sp^3$

13. Designate each of the following bonds as sigma or pi.
    a. $sp$—$sp^3$
    b. $p$—$p$
    c. $sp$—$sp$

14. Why does a carbon-carbon triple bond show a greater bond strength than a carbon-carbon single bond?

15. Draw a line-bond structural formula for vinylacetylene ($CH_2{=}CH$—$C{\equiv}CH$) as in Fig. 2.13 (neglect spacial geometry). Locate and determine the number of each of the following bonds in the molecule.
    a. $sp$—$sp$
    b. $sp^2$—$sp^2$
    c. $sp^2$—$sp$
    d. $p$—$p$
    e. $sp$—$s$
    f. $sp^2$—$s$

16. Which of the following represents a homolytic bond-breaking process?
    a. $CH_3CH_2Cl \longrightarrow CH_3CH_2{\cdot} + {\cdot}Cl$
    b. $CH_3CH_2Cl \longrightarrow CH_3CH_2{}^+ + {}^-Cl$

# CHAPTER 3

# Alkanes and Cycloalkanes (I)

## CONFORMATION, NOMENCLATURE, PHYSICAL PROPERTIES, AND NATURAL SOURCES

In this chapter we will learn more about the conformation of alkanes, learn and use rules for the systematic nomenclature of alkanes and cycloalkanes (and note the relative lack of rules in nonsystematic nomenclature), discuss the physical properties of alkanes and cycloalkanes, and finally consider the various sources and uses of these compounds.

## 3.1 MORE ON THE CONFORMATIONS OF ALKANES

Conformational isomers (see Sec. 2.3) of an alkane cannot be isolated at room temperature because of the low energy barrier between the isomers, whereas most functional-group and positional isomers are isolable. For example, for ethane, the staggered conformation is the most stable due to greater hydrogen separation on adjacent carbons. On the other hand, the eclipsed conformation, with adjacent hydrogens relatively close together, is the least stable. This instability exists because electron pairs on adjacent C—H bonds are too close together. This crowding results in destabilizing London repulsions between electron pairs. As you can observe in Fig. 3.1, the staggered conformation of ethane is 3 kcal/mol more stable than the eclipsed conformation. This means that at any given split second in time, the vast majority of ethane molecules are in the staggered conformation, some molecules are in conformations close to staggered, and very few ethane molecules assume the eclipsed conformation.

Therefore, ethane (or any other organic molecule with at least one carbon-carbon single bond) represents a mixture of conformers, each of which is an ethane molecule.

**SOLVED PROBLEM 3.1**

Actually, as can be observed from a molecular model (Fig. 3.2), the propane molecule is zigzag in shape and far from flat due to the tetrahedral geometry around each carbon atom. Draw a Newman projection, sawhorse diagram, and wedge projection for the eclipsed conformation of propane.

### SOLUTION

Let us begin by replacing a hydrogen on the front Newman carbon of ethane with a $CH_3$ (Fig. 3.3). Now, in the Newman projection, let us align the $CH_3$ with a hydrogen looking from the front of the molecule to the rear so that the $CH_3$ group and hydrogens bonded to the front carbon are directly in front of the hydrogens bonded to the rear carbon. Actually, the rear hydrogens are placed slightly to one side so that they can be seen. The unique perspective of the sawhorse diagram shows the $CH_3$ and two hydrogens bonded to the front carbon to be directly in front of the hydrogens bonded to the rear carbon. In

There is a 3 kcal/mol energy barrier to carbon-carbon bond rotation. However, this barrier is a small one, and for convenience, we can assume free rotation around any carbon-carbon single bond. This free rotation means very rapid conversion of one conformation to another with no possibility of isolating any conformer at room temperature.

**FIGURE 3.1** Energy difference between two conformations of the compound ethane.

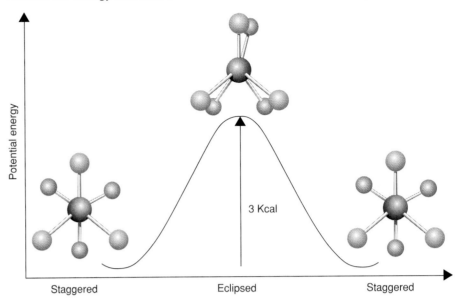

**FIGURE 3.2** Molecular model of propane.

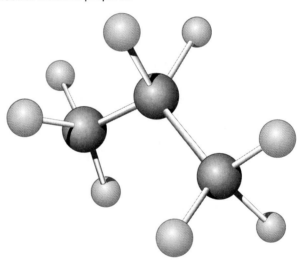

**FIGURE 3.3** Newman projection, sawhorse diagram, and wedge projection of an eclipsed conformation of propane.

Newman projection
(eclipsed conformation of propane)

Sawhorse diagram
(eclipsed conformation of propane)

Wedge projection
(eclipsed conformation of propane)

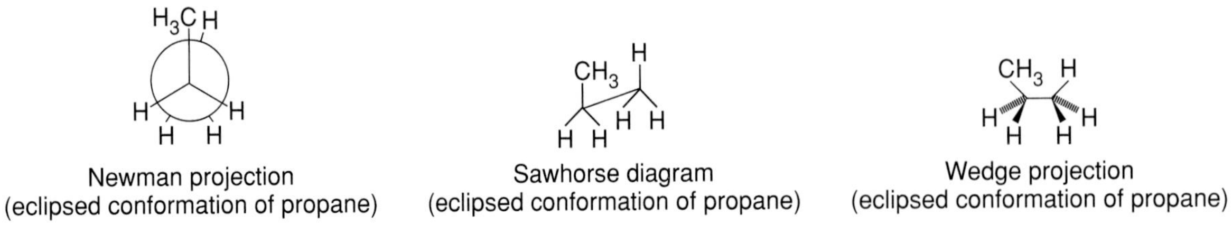

## HISTORICAL BOXED READING 3.1

The first oil well was drilled well over a thousand years ago by the Chinese,* and petroleum was first separated into its components (gasoline, kerosene, heating oils, and lubricating oils) at about the same time by the Arabs and Persians. George Washington observed a burning oil spring and thought that oil would eventually become an important industrial commodity.

In the 18th and first half of the 19th centuries, whale oil was used for lamps. By 1855, almost all the whales were killed, and the problem of fueling the lamps of the world remained. Thus the stage was set for petroleum to come into its own. In the meantime, petroleum was more of a bother than anything else in the United States. The reason: Many a well drilled for salt (a vital commodity before the days of the refrigerator) also found oil as a contaminant.

The breakthrough came in 1852 when kerosene (coal oil) was found to be a fine fuel for lamps. Since kerosene could be obtained easily from petroleum by distillation, petroleum as a commerical entity was on its way, and conditions were right for the oil well drilled by Edwin L. Drake, which was completed in 1859.

*Colliers Encyclopedia, Vol. 18, page 628 (1987).

the wedge projection, the wedged $CH_3$ and H groups jut out. This indicates an eclipsed conformation, since the $CH_3$ and H can be projected to be one directly behind the other. Thus it follows that the other two hydrogens bonded to the carbon bearing the $CH_3$ also are directly in front of the two hydrogens bonded to the other carbon.

There are two staggered conformations of butane. Using the two middle carbons in the molecule as the Newman-projection carbons, draw a Newman projection for each staggered conformation. Which staggered conformation is less stable? Why?

**EXERCISE 3.1**

Now that we can draw various kinds of structural formulas for a given alkane, let us consider identifing an alkane by means of a name. Before we begin to actually name compounds, a few terms must be explained, i.e., *parent compound* and *alkyl group*. Before you do anything else in this section, you need to memorize Table 3.1.

Let us examine Table 3.1 and look at two memory aids to help you. First, note that the -*ane* suffix is used in each name. Second, the combining forms *pent-* (five), *hex-* (six), *hept-* (seven), *oct-* (eight), *non-* (nine), and *dec-* (ten) represent the number of carbon atoms in a given compound.

The name of an *alkyl group* is derived from its parent alkane by dropping the -*ane* suffix of the alkane and substituting the suffix -*yl*. Structurally, the alkyl group is created by removing one hydrogen from an alkane.

## 3.2 NOMENCLATURE OF ALKANES

### The Alkyl Group

Alkyl groups do not exist! They are artificially created structural units to serve as an aid in organic nomenclature.

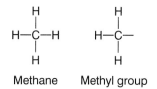

Methane        Methyl group

| TABLE 3.1 | Formulas and Names of Some Unbranched Alkanes | | |
|---|---|---|---|
| Formula | Name | Formula | Name |
| $CH_4$ | *meth*ane | $CH_3(CH_2)_4CH_3$ | *hex*ane |
| $CH_3CH_3$ | *eth*ane | $CH_3(CH_2)_5CH_3$ | *hept*ane |
| $CH_3CH_2CH_3$ | *prop*ane | $CH_3(CH_2)_6CH_3$ | *oct*ane |
| $CH_3CH_2CH_2CH_3$ | *but*ane | $CH_3(CH_2)_7CH_3$ | *non*ane |
| $CH_3(CH_2)_3CH_3$ | *pent*ane | $CH_3(CH_2)_8CH_3$ | *dec*ane |

**FIGURE 3.4** Equivalent representations of the methyl group.

H—C—H  H  |  Methane
(Methane and Methyl group structures shown)

| Methane | Methyl group | Methyl group | Methyl group | Methyl group |

No matter which of the four methane hydrogens is removed, the resulting structure is the same methyl group because all four hydrogens have the same chemical environment in the molecule and thus are said to be structurally *equivalent* (Fig. 3.4).

In ethane, all six hydrogen atoms are equivalent. Thus only one possible alkyl group can be derived from ethane.

(Ethane and Ethyl group structures shown)

Ethane    Ethyl group

The prefix *n*- represents a straight-chain alkane with one of the hydrogens removed that is bonded to either of the end carbons. Thus, in a similar way, we have the *n*-pentyl group derived from pentane and the *n*-hexyl group derived from hexane.

CH₃CH₂CH₂CH₂CH₃

Pentane

CH₃CH₂CH₂CH₂CH₂—

*n*-Pentyl group or pentyl

CH₃CH₂CH₂CH₂CH₂CH₃

Hexane

CH₃CH₂CH₂CH₂CH₂CH₂—

*n*-Hexyl group or hexyl

Propane contains two sets of hydrogens. One set of six equivalent hydrogen atoms is circled; the other set of two equivalent hydrogens is boxed. Therefore, two different alkyl groups can be derived from propane (Fig. 3.5).

When any hydrogen bonded to an end carbon is removed from propane, the *n*-propyl group is formed. In the same way, when either of the two hydrogens bonded to the central carbon is taken away, the isopropyl group is produced (see Figs. 3.5b and 3.6). There are two isomeric butanes: butane and isobutane.

**FIGURE 3.5** The *n*-propyl and isopropyl groups are derived from propane.

**FIGURE 3.6** The *iso*- group is formulated when a methyl is bonded to a carbon next to the end carbon.

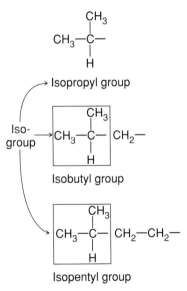

→ Isopropyl group

Iso-group → Isobutyl group

→ Isopentyl group

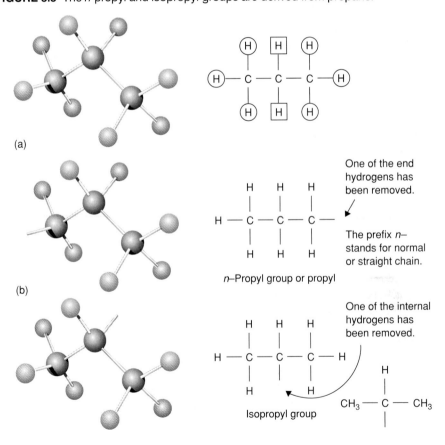

(a)

(b)

One of the end hydrogens has been removed.

The prefix *n*– stands for normal or straight chain.

*n*–Propyl group or propyl

One of the internal hydrogens has been removed.

Isopropyl group

$$CH_3CH_2CH_2CH_3$$

Butane

$$\begin{array}{c} CH_3 \\ | \\ CH_3-C-CH_3 \\ | \\ H \end{array}$$

Isobutane

Both butane and isobutane contain two sets of equivalent hydrogens. One set of six hydrogen atoms in butane is circled; the other set of four is boxed. Removal of any one of the six hydrogens bonded to either end carbon produces an *n*-butyl or butyl group. In a similar manner, loss of any of the four hydrogens bonded to the middle two carbons results in the formation of the secondary butyl (*sec*-butyl) group (Fig. 3.7).

In isobutane, one set of nine hydrogens is circled; the other set of one is boxed (Fig. 3.8). When any of the nine circled equivalent hydrogens are removed, the isobutyl group is formed. When the boxed hydrogen is withdrawn, the tertiary butyl (*tert*-butyl) group is created.

A more convenient way to identify equivalent hydrogens (most of the time) involves use of the terms *primary, secondary,* and *tertiary*:

*Primary* (abbreviated 1°)—a hydrogen or other atom or group bonded to a carbon which, in turn, is bonded to one other carbon. Each propane and each butane molecule contains six primary hydrogens (circled on pages 46 and 47).

*Secondary* (abbreviated 2° or *sec*-)—a hydrogen (or other atom or group) bonded to a carbon which, in turn, is bonded to two other carbons. Each propane molecule contains two secondary hydrogens (boxed on page 46), while every butane molecule contains four secondary hydrogens (boxed on page 47).

*Tertiary* (abbreviated 3° or *tert*-)—a hydrogen (or other atom or group) bonded to a carbon which, in turn, is bonded to three other carbons. Each isobutane molecule contains one tertiary hydrogen (boxed on page 49).

The prefix *iso*- represents a methyl branch on a carbon adjacent to the end carbon.

It is interesting to note that although the middle carbon of the isopropyl group is secondary, the end carbon of the isobutyl group and higher homologues are primary.

$$\begin{array}{c} CH_3 \\ | \\ CH_3-C- \\ | \\ H \end{array}$$

Isopropyl (2°)

$$\begin{array}{cc} CH_3 & H \\ | & | \\ CH_3-C-&C- \\ | & | \\ H & H \end{array}$$

Isobutyl (1°)

$$\begin{array}{ccc} CH_3 & H & H \\ | & | & | \\ CH_3-C-&C-&C- \\ | & | & | \\ H & H & H \end{array}$$

Isopentyl (1°)

**FIGURE 3.7** The *n*-butyl and *sec*-butyl groups are derived from butane.

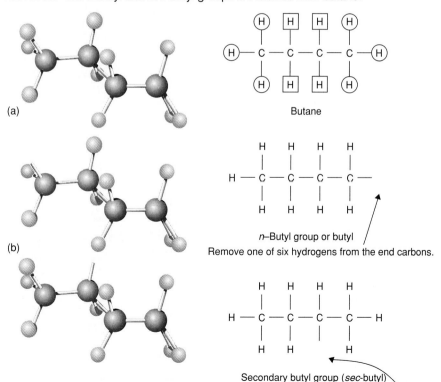

(a)

Butane

(b)

*n*–Butyl group or butyl
Remove one of six hydrogens from the end carbons.

Secondary butyl group (*sec*-butyl)
Remove one of four hydrogens from the middle carbons.

The first five alkanes (containing four carbons or less) and the alkyl groups derived from them are summarized in Table 3.2. You need to memorize this table, too.

| TABLE 3.2 | Summary of the First Five Alkanes (Containing Four Carbons or Less) and Their Derived Alkyl Groups |
|---|---|
| **Alkane** | **Derived Alkyl Group(s)** |

| Alkane | Derived Alkyl Group(s) |
|---|---|
| 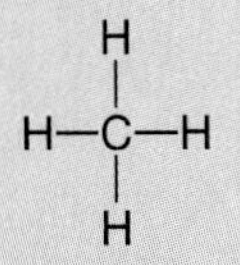 $CH_4$  —  Methane | 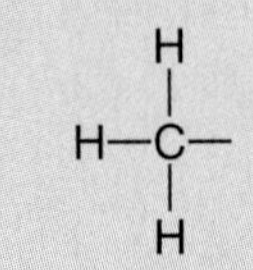 $CH_3-$  —  Methyl |
| 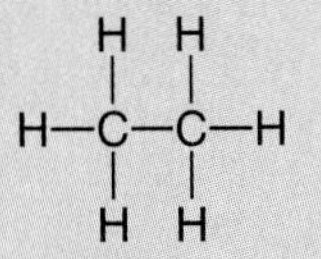 $CH_3CH_3$  —  Ethane | 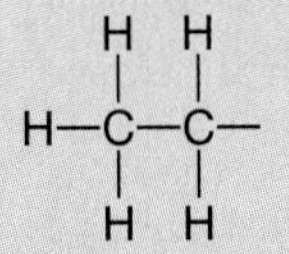 $CH_3CH_2-$  —  Ethyl |
| 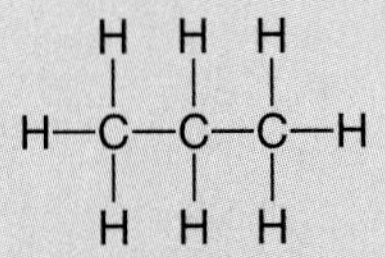 $CH_3CH_2CH_3$  —  Propane | 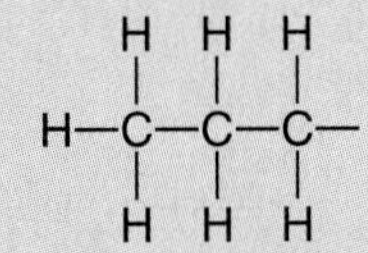 $CH_3CH_2CH_2-$  —  *n*-Propyl or propyl |
| 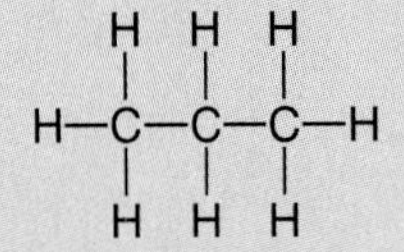 $CH_3CH_2CH_3$ | 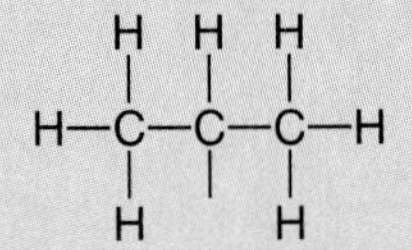 $CH_3CHCH_3$  —  Isopropyl |
| 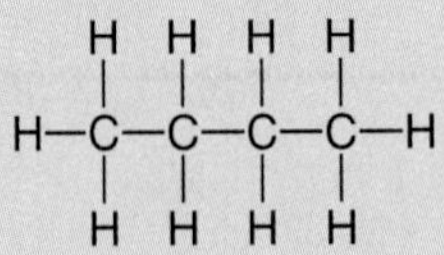 $CH_3CH_2CH_2CH_3$  —  Butane | 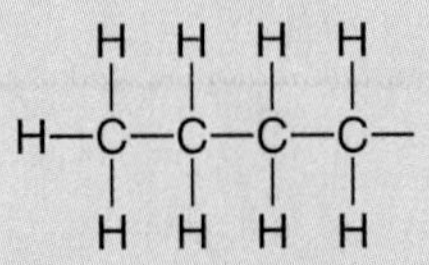 $CH_3CH_2CH_2CH_2-$  —  *n*-Butyl or butyl |
| 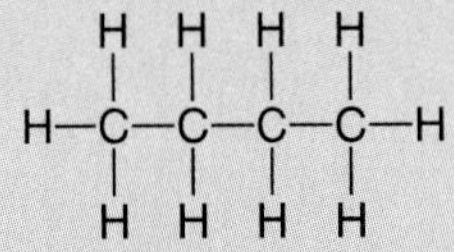 $CH_3CH_2CH_2CH_3$ | 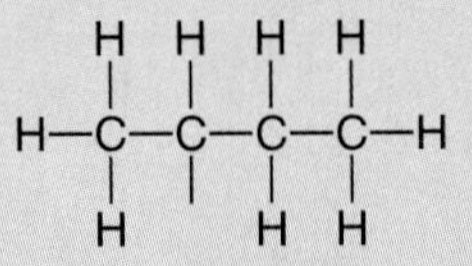 $CH_3CHCH_2CH_3$  —  *sec*-Butyl |

| TABLE 3.2 | Summary of the First Five Alkanes (Continued) |
|---|---|
| **Alkane** | **Derived Alkyl Group(s)** |

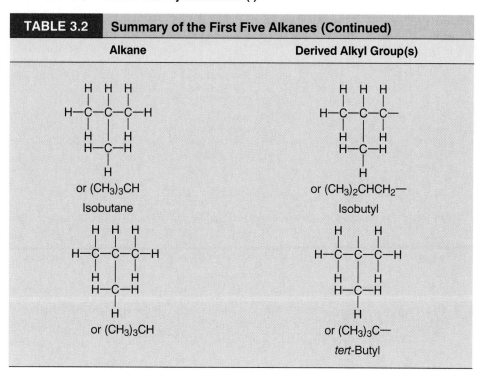

or $(CH_3)_3CH$

Isobutane

or $(CH_3)_2CHCH_2—$

Isobutyl

or $(CH_3)_3CH$

or $(CH_3)_3C—$

*tert*-Butyl

**FIGURE 3.8** The isobutyl and *tert*-butyl groups are derived from isobutane.

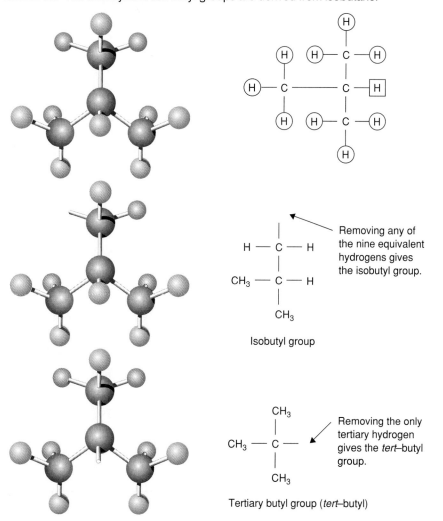

Removing any of the nine equivalent hydrogens gives the isobutyl group.

Isobutyl group

Removing the only tertiary hydrogen gives the *tert*–butyl group.

Tertiary butyl group (*tert*–butyl)

**SOLVED PROBLEM 3.2**

Identify each of the following alkyl groups as primary, secondary, or tertiary.

a. $CH_3CH_2—$

b. 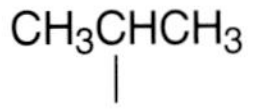 $CH_3CHCH_3$
  |

c. $(CH_3)_2CHCH_2—$

**SOLUTION:**

a.

$$\begin{array}{cc} H & H \\ | & | \\ H—C—C— \\ | & | \\ H & H \end{array}$$

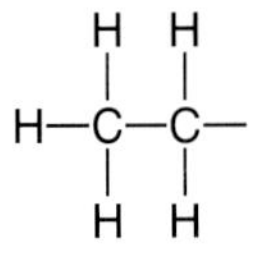

Since the hydrogen removed is primary, that carbon of the ethyl group is primary, and all remaining five hydrogens are primary.

b.

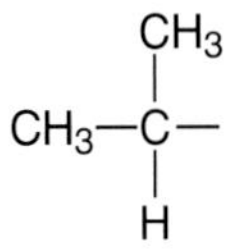

$$\begin{array}{c} CH_3 \\ | \\ CH_3—C— \\ | \\ H \end{array}$$

The middle carbon of the isopropyl group is secondary.

c.

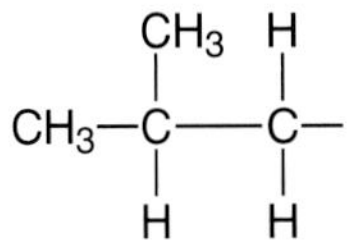

$$\begin{array}{cc} CH_3 & H \\ | & | \\ CH_3—C——C— \\ | & | \\ H & H \end{array}$$

The carbon with the dash or "dangling bond" in the isobutyl group is primary.

**EXERCISE 3.2**

Putting it another way, you are asked to identify the carbon with the dash or "dangling bond" as primary, secondary, or tertiary.

Identify each of the following alkyl groups as primary, secondary, or tertiary, and name each group.

a. $CH_3CH_2—$

b. $CH_3CH_2CH_2—$

c.

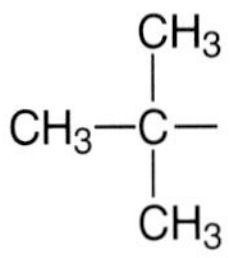

$$\begin{array}{c} CH_3 \\ | \\ CH_3—C— \\ | \\ CH_3 \end{array}$$

d.

$$\begin{array}{c} CH_3 \\ | \\ H—C— \\ | \\ CH_2CH_3 \end{array}$$

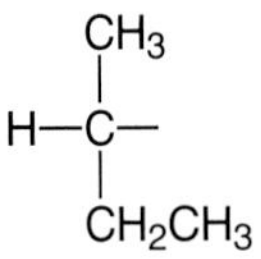

## The IUPAC System of Nomenclature

$$\begin{array}{c} {}^1C—C^2 \\ | \\ {}_3C—C_4 \\ | \\ {}^5C \end{array}$$

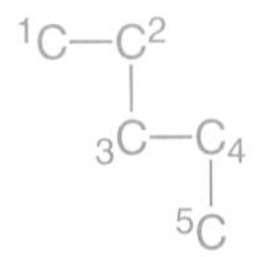

Pentane

(There are 5 carbons; thus the fragment is *pent-*, and since the compound is an alkane, add the suffix *-ane* to get pentane.)

Alkyl and other groups ($NO_2$ = nitro, Cl = chloro, Br = bromo, I = iodo) linked to the main carbon chain are called *substituent groups.*

As recently as a century ago, organic nomenclature was a mess. A compound could have seven or eight different names, none of which was systematic. To eliminate this chaos, the *International Union of Pure and Applied Chemistry (IUPAC)*, a body of respected chemists, met in 1892 in Geneva, Switzerland, to create a system of nomenclature that would give each organic compound a specific and deducible name.

The IUPAC system is based on a series of rules that apply to all classes of compounds (not only alkanes). These rules are as follows:

**Rule 1.** Number and name the longest continuous chain of carbon atoms, called the main chain (in the *parent compound*), as shown in Table 3.1 (note the *-ane* suffix). Since there is free rotation about the carbon-carbon bonds in alkanes, the shape of the carbon chain as depicted on paper has no bearing on the name of the compound.

**Rule 2.** Number the carbon atoms in this longest continuous chain (main chain) from left to right or from right to left so that the *substituent group* first met has the lowest possible number. To name a compound with one substituent, number the carbon of the main chain bonded to the group followed by a dash, add the name of the substituent group, and then add the name of the parent compound.

Name the following compound.

$$CH_3CH_2CH_2CHCH_2CH_3$$
$$|$$
$$CH_2CH_3$$

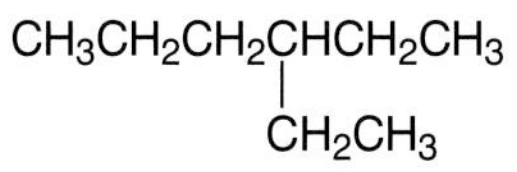

**SOLUTION**

*Step 1.* Name and number the longest continuous chain, which contains six carbons (rule 1). Thus the parent compound is hexane (see Table 3.1).

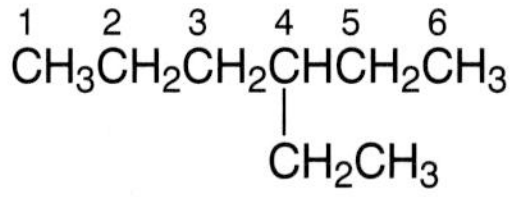

$$\overset{1}{C}H_3\overset{2}{C}H_2\overset{3}{C}H_2\overset{4}{C}H\overset{5}{C}H_2\overset{6}{C}H_3$$
$$|$$
$$CH_2CH_3$$

*Step 2.* We must number from right to left in this case because this will give the ethyl group the lower number—the number 3 (rule 2). Should we number from left to right, we would get the number 4.

$$\overset{1}{C}H_3\overset{2}{C}H_2\overset{3}{C}H_2\overset{4}{C}H\overset{5}{C}H_2\overset{6}{C}H_3 \quad \text{or} \quad \overset{6}{C}H_3\overset{5}{C}H_2\overset{4}{C}H_2\overset{3}{C}H\overset{2}{C}H_2\overset{1}{C}H_3$$
$$CH_2CH_3 \qquad\qquad\qquad CH_2CH_3$$

*Step 3.* Write the name:

3-ethylhexane (*not* 4-ethylhexane)

Name each of the following compounds.

a. 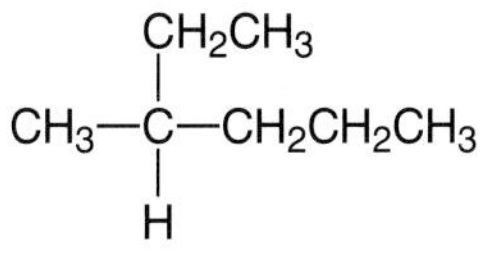
$$CH_2CH_3$$
$$|$$
$$CH_3-C-CH_2CH_2CH_3$$
$$|$$
$$H$$

b.   $BrCH_2CH_2CH_2CH_2CH_2CH_2CH_3$

**Rule 3.**  If the same substituent group is found bonded two or more times on the main chain, precede each substituent group name by the number(s) of the carbons on the main chain where the group appears. Since the group appears more than once, precede the group name with the prefix *di-*, *tri-*, or *tetra-*, showing that the group appears two, three, or four times. Separate the last number from the prefix by a dash; do not separate the prefix from the group. Finally, follow the name of the group with the name of the parent compound written as one word.

One number for each appearance, separated by commas.

Name the following compound.

$$\overset{2}{C}H_3$$
$$|$$
$$\overset{1}{C}H_3-\overset{}{C}-\overset{3}{C}H_2\overset{4}{C}H_2\overset{5}{C}H\overset{6}{C}H_2\overset{7}{C}H_3$$
$$| \qquad\qquad |$$
$$CH_3 \qquad\quad CH_3$$

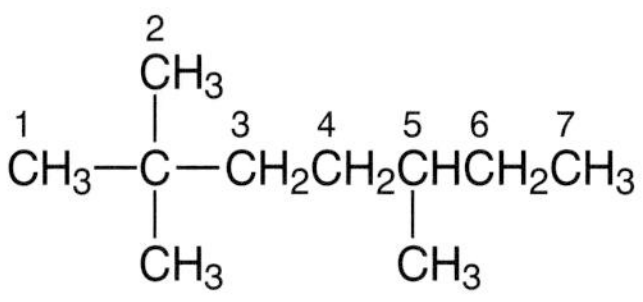

**SOLUTION**

*Step 1.* The longest continuous chain contains seven carbons. Thus the parent compound is heptane (rule 1).

*Step 2.* We must number from left to right in this case because this will give the first group encountered the lower number—the number 2 (rule 2).

*Step 3.* Write the name:

2,2,5-trimethylheptane (rule 3)

*not* 2,5-trimethylheptane

---

| | |
|---|---|
| **EXERCISE 3.4** | Name each of the following compounds. |

a.  $CH_3CH(Br)CH_2CH_2CH(Br)CH_2CH_3$

b.  $CH_3CCl_3$

**Rule 4.**  If several different substituent groups are bonded to one or more carbons of the main chain, list these groups in alphabetical order. When determining alphabetical order, ignore Greek prefixes such as *di-* and *tri-*, along with prefixes such as *n-*, *sec-*, and *tert-*, but include the *iso-* prefix. For example:

$ClCH_2CBr_2CH_3$

2,2-di*b*romo-1-*c*hloropropane

Note that the *b* and *c* are italicized in the name only to emphasize alphabetical order.

---

| | |
|---|---|
| **SOLVED PROBLEM 3.5** | Name the following compound. |

$$
\begin{array}{ccccc}
 & H & NO_2 & CH_3 & \\
 & 4| & 3| & 2| & 1 \\
5 & & & & \\
CH_3- & C- & C- & C- & CH_3 \\
 & | & | & | & \\
 & Cl & H & CH_3 & \\
\end{array}
$$

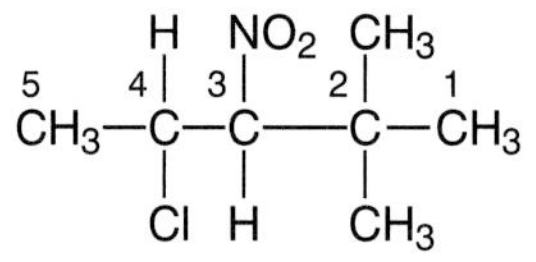

**SOLUTION**
*Step 1.*  The longest continous chain contains five carbons. Thus the parent compound is pentane (rule 1).
*Step 2.*  Number from right to left because two substituent methyl groups are on the second carbon from the right, as opposed to one chloro group on the second carbon from the left. That is, we want to give the low numbers to as many groups as possible (rule 2).
*Step 3.*  Write the name:

4-*c*hloro-2,2-di*m*ethyl-3-*n*itropentane (rules 3 and 4)

Note that the *c*, *m*, and *n* are italicized in the name only to emphasize alphabetical order.

---

| | |
|---|---|
| **EXERCISE 3.5** | Name each of the following compounds. |

a. 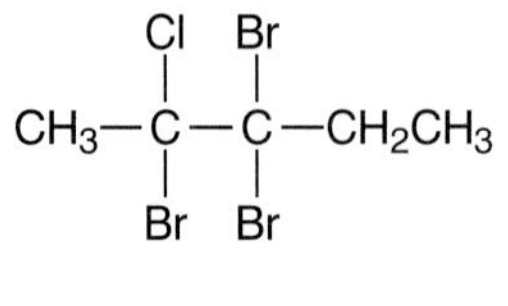
$$
\begin{array}{ccc}
Cl & Br & \\
| & | & \\
CH_3-C-C-CH_2CH_3 \\
| & | & \\
Br & Br & \\
\end{array}
$$

b. 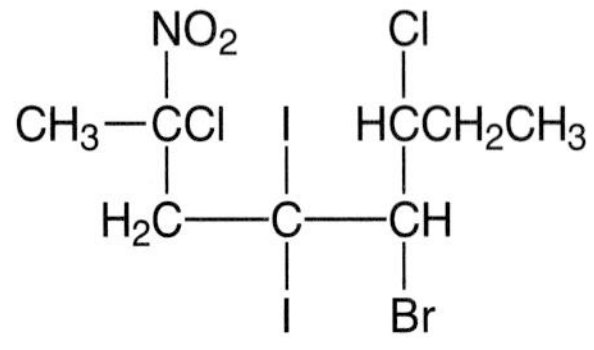
$$
\begin{array}{ccc}
NO_2 & & Cl \\
| & & | \\
CH_3-CCl & | & HCCH_2CH_3 \\
 & | & | \\
H_2C- & C- & CH \\
 & | & | \\
 & I & Br \\
\end{array}
$$

**Rule 5.**  If two (or more) carbon chains in the same molecule have the same number of carbon atoms, use the chain that has the larger number of smaller substituent groups.

Name the following compound.

$$\begin{array}{c} \overset{5}{C}H_3\overset{4}{C}H{\longrightarrow}\overset{3}{C}HCH_2CH_3 \\ \underset{CH_3}{\overset{2}{|}} \quad \underset{\underset{CH_3}{\overset{1}{|}}}{HC{\longrightarrow}CH_3} \end{array}$$

**SOLUTION**

*Step 1.* There are several continuous chains of five carbons each. Thus the parent compound is pentane (rule 1).

$$\begin{array}{ccc} \overset{3}{C}H_3\overset{4}{C}H{\longrightarrow}\overset{5}{C}HCH_2CH_3 & & \overset{1}{C}H_3\overset{2}{C}H{\longrightarrow}\overset{3}{C}HCH_2CH_3 \\ \underset{CH_3}{|} \quad \underset{\underset{CH_3}{|}}{HC{\longrightarrow}CH_3} & \text{and} & \underset{CH_3}{|} \quad \underset{\underset{CH_3}{|}}{HC{\longrightarrow}CH_3} \end{array}$$

*Step 2.* Select the chain that has two methyl and one ethyl group (larger number of smaller groups), as opposed to those which have the isopropyl group (smaller number of larger groups) (rule 5).

*Step 3.* Number from right to left or from left to right as appropriate. (In this case it makes no difference) (rule 2).

*Step 4.* Name the compound.

3-ethyl-2,4-di*m*ethylpentane (rules 3 and 4)

Name each of the following compounds.

a. $(CH_3)_3CCH_2CH_3$ with $CH_3$ above and $CH_2CH_3$ below

$$\text{a.} \quad \overset{\overset{\displaystyle CH_3}{|}}{\underset{\underset{\displaystyle CH_2CH_3}{|}}{(CH_3)_3C{-}CH_2CH_3}}$$

$$\text{b.} \quad CH_3CH_2CH_2\overset{\overset{\displaystyle CH_3}{|}}{\underset{\underset{\displaystyle CH_3}{|}}{C}}{\longrightarrow}\overset{\overset{\displaystyle H}{|}}{\underset{\underset{\displaystyle CH_2CH_3}{|}}{C}}CH(CH_3)_2$$

For a discussion of the common system of nomenclature and trivial nomenclature, refer to Appendices 2 and 3.

**The Common System of Nomenclature and Trivial Nomenclature**

Most of the substances you deal with in everyday life are water soluble, e.g., soap, salt, etc. The alkanes, on the other hand, are very different.

Since alkanes are nonpolar molecules, they are insoluble in water and soluble in nonpolar solvents, such as other alkanes, benzene and carbon tetrachloride. In addition, their boiling and melting points are low because only London forces are present between molecules.

As the carbon content of the alkanes increases, the melting and boiling points also increase. The presence of more atoms in a molecule produces greater intermolecular forces (London forces). These forces require more energy to be overcome, so the melting and boiling points of the series of compounds gradually increase (Table 3.3).

**3.3
PHYSICAL PROPERTIES
OF THE ALKANES**

| TABLE 3.3 | Physical Properties of the Alkanes | | |
|---|---|---|---|
| IUPAC Name of Hydrocarbon | Boiling Point (°C) | Melting Point (°C) | Density (g/ml) at 20°C |
| Methane | −162 | −183 | 0.67* |
| Ethane | − 89 | −183 | 1.25* |
| Propane | − 42 | −190 | 1.66* |
| Butane | 0 | −138 | 2.24* |
| Pentane | 36 | −130 | 0.626 |
| Hexane | 69 | − 95 | 0.660 |
| Heptane | 98 | − 91 | 0.684 |
| Octane | 126 | − 57 | 0.703 |
| Nonane | 151 | − 51 | 0.718 |
| Decane | 174 | − 30 | 0.730 |
| Water | 100 | 0 | 1.00 |

*Gas at 20°C; values are for each gas at 1 atm pressure and 20°C. Densities are in grams per liter, not grams per milliliter.

Note: Alkanes are colorless in the gaseous ($C_1$ to $C_4$) and liquid ($C_5$ to $C_{17}$) states and white in the solid state ($C_{18}$ to $C_\infty$). Densities listed for methane, ethane, propane, and butane are in grams per liter. These compounds are gases, and density is more conveniently listed in units of grams per liter rather than grams per milliliter.

A straight-chain alkane has a higher boiling point than a branched isomer.

| $CH_3(CH_2)_3CH_3$ | $CH_3\overset{\overset{\displaystyle CH_3}{|}}{\underset{\underset{\displaystyle H}{|}}{C}}CH_2CH_3$ | $CH_3-\overset{\overset{\displaystyle CH_3}{|}}{\underset{\underset{\displaystyle CH_3}{|}}{C}}-CH_3$ |
|---|---|---|
| *n*-pentane bp = 36°C (1 atm) | isopentane bp = 28°C (1 atm) | neopentane bp = 10°C (1 atm) |

The more branched the isomer, the more spherical the compound and the less surface area is exposed to other molecules. This causes a decrease in intermolecular forces and thus a decreased boiling point. All alkanes are less dense than water.

**SOLVED PROBLEM 3.7**

Predict the compound in each of the following pairs with the lower boiling point.

a. Heptane and octane

b. Heptane and 2-methylhexane

**SOLUTION**

a. Since an octane molecule contains more carbons (is larger) than a heptane molecule, we would predict that greater London forces exist in a sample of octane. These forces require more energy to be overcome, so octane should boil at a higher temperature than heptane.

b. Both compounds are isomers. Thus the lower boiling point can be determined on the basis of branching. Because 2-methylhexane is branched, molecules of the compound are spherical, and less surface area is exposed to other molecules than molecules of heptane that are not branched. This adds up to decreased London forces between molecules of 2-methylhexane, and therefore, 2-methylhexane should boil at a lower temperature than heptane.

Without looking at Table 3.3, predict the compound in each of the following pairs with the lower boiling point. Explain each prediction.

a. Ethane and hexane

b. 2-Methylhexane and 2,2-dimethylpentane

c. 3,3-Dimethylhexane and octane

Methane, a colorless, odorless gas, is found in coal mines, where it is known as "fire damp." Since methane-air mixtures are explosive, miners need to be careful about sparks in mines. Methane also is produced by the action of bacteria on vegetation at the bottom of marshes and swamps, where it is known as "marsh gas." Large amounts of methane gas are also trapped together with ethane and a little propane in deposits underground. This mixture is called *natural gas* and is an important fuel used to heat our homes.

Black, crude, liquid petroleum is a complex mixture of hydrocarbons. Petroleum was formed as a result of the decomposition of single-celled marine organisms subjected to heat and pressure deep under the surface of the earth over a long period of time. Deposits of natural gas (mostly methane) accompany the deposits of petroleum. Crude petroleum can be *refined*, i.e., separated into its constituents by distillation, to give the fractions listed in Table 3.4.

Paraffin wax is used in ointments to raise the melting point so that when applied on the skin, the medication will not turn into a liquefied, dripping mess. Paraffin wax is obtained as a white solid from certain types of petroleum, such as Pennsylvania crude, by cooling the lubricating oils fraction and consists of a mixture of normal alkanes having 26 to 30 carbons.

Mineral oil is a viscous, colorless, odorless liquid obtained by distilling petroleum and collecting the fraction that boils at 300°C. Solid paraffins and other impurities are removed by cooling. Mineral oil is used as a laxative and a lubricant.

Vaseline (petrolatum, petroleum jelly) has been a mother's helper for many years when used as an ointment. It soothes diaper rash, cleanses the face, prevents chapping, and serves as a moisturizer. Vaseline is prepared from the highest-boiling fraction of petroleum. It is a colloidal dispersion of branched-chain solid hydrocarbons in a mixture of high-boiling liquid alkanes.

## 3.4 NATURAL AND COMMERCIAL SOURCES OF ALKANES

Many alkanes are found throughout the plant and animal kingdoms. Alkanes have been detected in earthworms (*Lumbricus terrestris*); beef hearts; mormon crickets (*Anabrus simplex*); human liver, urine, and sweat; foxglove plant leaves (*Digitalis purpurea*); and many other representatives of the plant and animal kingdoms.

The function of these alkanes depends on the source. In the stable fly, *Stomoxys calcitrans*, a mixture of 15-methyl- and 15,19-dimethyltriacontanes (Table 3.5) was found to have the highest activity as a mating stimulant. These substances are part of a chemical

## 3.5 ALKANES IN BIOLOGICAL SYSTEMS

| TABLE 3.4 | Name, Carbon Content, and Boiling Point Range for Selected Petroleum Fractions | | |
|---|---|---|---|
| **Petroleum Fraction** | **Carbon Content** | **Boiling Point Range (°C)** | **State of Matter** |
| Gasoline | $C_6$–$C_{12}$ | 85–200 | Liquid |
| Kerosene | $C_{10}$–$C_{15}$ | 185–300 | Liquid |
| Heating oils | $C_{14}$–$C_{18}$ | 285–400 | Liquid |
| Lubricating oils, greases, paraffin wax | $C_{18}$– | 400+ | Solid |

## BIOCHEMICAL BOXED READING 3.1

The palmetto beetle eats palmetto leaves, which contain a wax. The animal converts the compounds in the wax to a series of hydrocarbons ($C_{17}$ to $C_{23}$). The beetle secretes these hydrocarbons as a glue from its feet to anchor itself on the palmetto leaf. This serves as a mechanism for protection. Between the animal's hard exposed shell and its ability to cling to the smooth leaf with the glue, it is almost impossible for a predator to dislodge and harm the beetle.

message system employed by animals to find food, choose a mate, or evade enemies. These chemical messengers are known as *pheromones*. Even though as little as 200 μg $(1 \mu g = 10^{-6} g)$ of the sex pheromone of the housefly was isolated, this is more than enough to attract male houseflies. Male gypsy moths are summoned to the female by as little as 1 ng $(1 \times 10^{-9} g)$ of sex pheromone. One-nanogram to one-picogram $(1 \times 10^{-12} g)$ sex messages are not at all unusual in the insect world.

In most animals or plants, as many as four homologous series of alkanes are present. Typical are the hydrocarbons found in the cuticular lipids of the mormon cricket, which are of four types: *n*-alkanes, 3-methylalkanes, other internally branched monomethylalkanes, and internally branched dimethylalkanes. The main *n*-alkane present is *n*-nonacosane (the *ei*- prefix is sometimes dropped), while they range from $C_{23}$ to $C_{33}$. The three 3-methylalkanes detected had total carbon numbers of 28, 30, and 32, respectively. A typical internally branched monomethylalkane isolated was 9-methylhentriacontane; in addition, a total of 27 compounds were detected, ranging from $C_{28}$ to $C_{38}$. The internal dimethylalkanes found had their methyl groups separated by three methylene ($-CH_2-$) groups, e.g., 15,19-dimethylpentatriacontane.

*Paraffins* (a synonym for alkanes; see page 69) also have been isolated from human liver, urine, and sweat. The presence of alkanes in human beings usually is considered to be caused by the ingestion of these compounds in the diet. When *n*-nonacosane is the principal *n*-alkane found in the liver, a vegetable diet is indicated.

Hydrocarbons isolated from living materials almost always contain more than 10 carbon atoms. Thus a supplement to alkane nomenclature is in order and is found in Table 3.5.

| TABLE 3.5 | IUPAC Nomenclature of Straight Chain Alkanes Found in Living Organisms |
|---|---|
| **Name** | **Formula** |
| Decane | $C_{10}H_{22}$ |
| Hendecane (undecane) | $C_{11}H_{24}$ |
| Dodecane | $C_{12}H_{26}$ |
| Tridecane | $C_{13}H_{28}$ |
| Tetradecane | $C_{14}H_{30}$ |
| Pentadecane | $C_{15}H_{32}$ |
| Hexadecane | $C_{16}H_{34}$ |
| Heptadecane | $C_{17}H_{36}$ |
| Octadecane | $C_{18}H_{38}$ |
| Nonadecane | $C_{19}H_{40}$ |
| Eicosane | $C_{20}H_{42}$ |
| Heneicosane (hencosane, uncosane) | $C_{21}H_{44}$ |
| Doeicosane (docosane) | $C_{22}H_{46}$ |
| Triacontane | $C_{30}H_{62}$ |
| Tetracontane | $C_{40}H_{82}$ |

Name each of the following alkanes.

a. $CH_3(CH_2)_{24}CH_3$

b. $CH_3(CH_2)_{36}CH_3$

**SOLUTION**

a. This is an alkane that contains 26 carbons. Thus we must put together the prefix *hexa-* (6) and the stem *eicosane* (20) (see Table 3.5), and we come up with *hexaeicosane*, which is shortened to *hexacosane*.

b. This is an alkane that contains 38 carbons. Thus we must put together the prefix *octa-* (8) and the stem *triacontane* (30) (see Table 3.5), and we come up with *octatriacontane*.

Name each of the following compounds.

a.  $CH_3(CH_2)_{25}CH_3$

b. 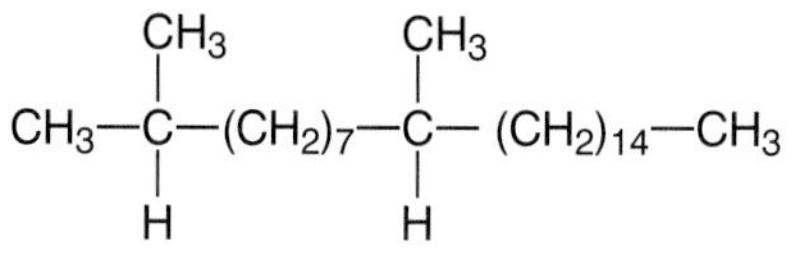  
$$
\begin{array}{ccc}
& CH_3 & CH_3 \\
& | & | \\
CH_3-C-(CH_2)_7-C-(CH_2)_{14}-CH_3 \\
& | & | \\
& H & H
\end{array}
$$

## 3.6 INTRODUCTION TO CYCLOALKANES

Cycloalkanes represent a homologous series of cyclic saturated compounds with a ring size of three-carbon skeleton or greater. These alicyclic (*aliphatic cyclic*) compounds have the generic formula $C_nH_{2n}$ or $(CH_2)_n$ and can be considered to be formed by removing one hydrogen from each end of the alkane with the same number of carbon atoms and linking one end carbon with the other. This process does *not* represent a chemical reaction; the yield arrows are employed merely for the sake of clarity.

*Aliphatic compounds* are open-chain compounds; i.e., they contain straight or branched chains but are not cyclic.

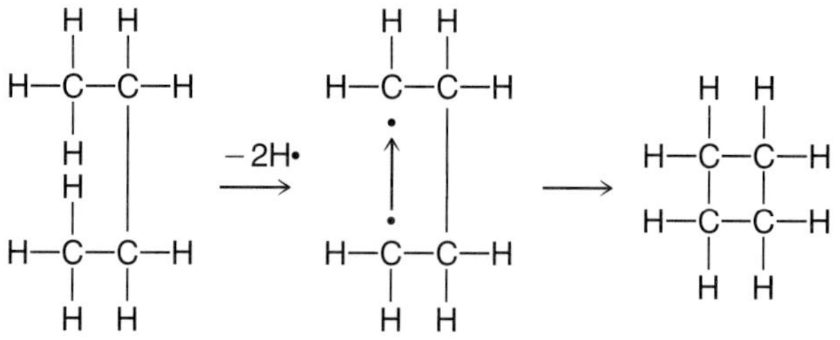

Cycloalkane structures can be abbreviated by representing a methylene group ($-CH_2-$) as the intersection of two lines. These are called *geometric condensed structures* (Fig. 3.9) (see Sec. 2.6).

## 3.7 NOMENCLATURE OF CYCLOALKANES

Use the prefix *cyclo-* before the alkane stem (Fig. 3.10a). When only one substituent group is bonded to a ring carbon, that ring carbon is assigned the number 1, and that number is not used in the name (Fig. 3.10b). If two or more substituents are present, the usual rules are used (Fig. 3.10c and d). However, the group with the highest alphabetical priority is assigned the lower number.

When a substituent group contains more carbons than the ring bonded to it, the compound is named as a derivative of the corresponding alkane, and the ring is considered to be a cycloalkyl group. On the other hand, when the ring contains a greater number of carbons than the alkyl group, the compound is named as an alkylcycloalkane (Fig. 3.11).

Note the mnemonic that both *c*arbon and *c*orner start with the letter *C*.

**FIGURE 3.9** Selected geometric condensed structures.

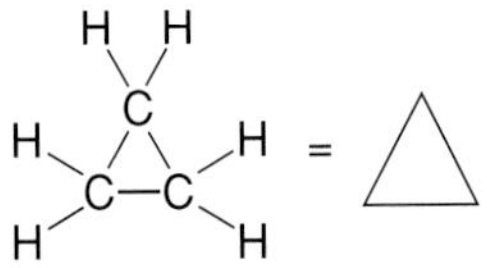

(a)

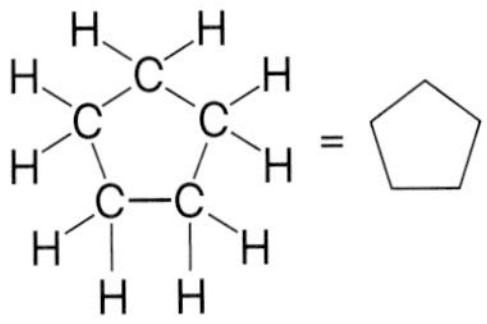

(b)

**FIGURE 3.10** Examples of the IUPAC nomenclature of selected cycloalkanes. We number in the direction of the nearest substituent. Refer to step 2 of the solution for Solved Problem 3.9b.

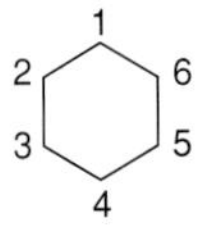

Cyclohexane
(a)

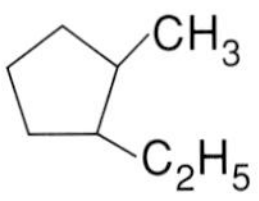

1-*ethyl*-2-*methyl*cyclopentane
(not 1-ethyl-5-methylcyclopentane)
(c)

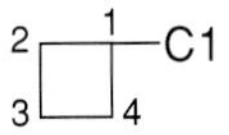

Chlorocyclobutane
(b)

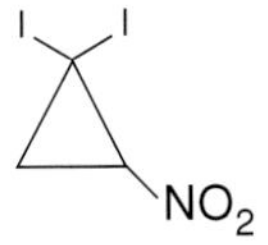

1,1-di*iodo*-2-*nitro*cyclopropane
(d)

**FIGURE 3.11** A cycloalkylalkane (a) versus an alkylcycloalkane (b).

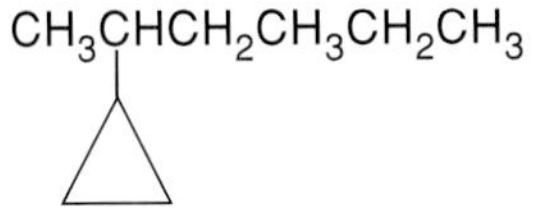

$CH_3CHCH_2CH_3CH_2CH_3$

$CH_2CH_2CH_3$

2-cyclopropylhexane
(a)

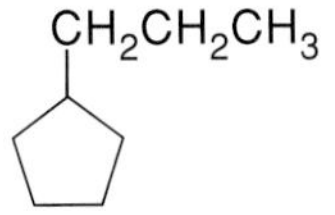

*n*-propylcyclopentane
(b)

## SOLVED PROBLEM 3.9

Name the following compounds.

a. 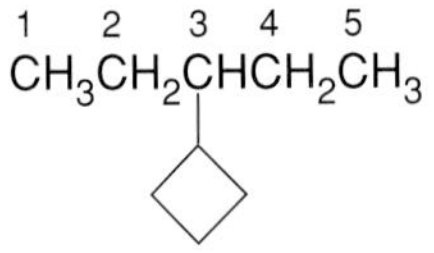
$$\overset{1}{C}H_3\overset{2}{C}H_2\overset{3}{C}H\overset{4}{C}H_2\overset{5}{C}H_3$$

b. 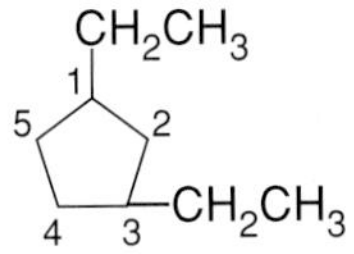
$CH_2CH_3$
$CH_2CH_3$

**SOLUTION**

a. *Step 1.* Since the alkyl group contains more carbons (five) than the ring (four), this compound is named as a cycloalkylalkane.

*Step 2.* Locate the cyclobutyl group on the parent pentane chain. The cyclobutyl group is bonded to carbon number 3 on the chain.

*Step 3.* Name the compound *3-cyclobutylpentane*.

b. *Step 1.* Since the ring contains more carbons (five) than the alkyl groups (two carbons per group), this compound is named as a substituted cycloalkane.

*Step 2.* Number the carbons in the ring in such a way that the carbons bonded to the substituent groups are assigned the lowest possible numbers.

*Step 3.* Name the compound: *1,3-diethylcyclopentane*.

## EXERCISE 3.9

Name the following compounds.

a. 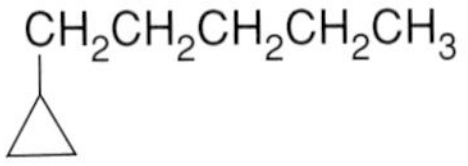
$CH_2CH_2CH_2CH_2CH_3$

b. 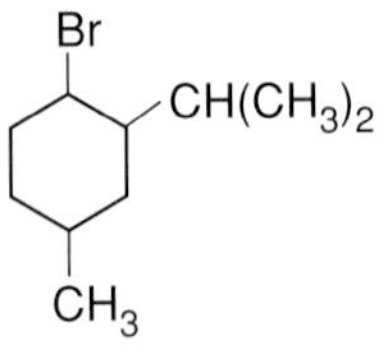
Br
$CH(CH_3)_2$
$CH_3$

| TABLE 3.6 | A Comparison of Physical Properties of Selected Cycloalkanes and Alkanes | | |
|---|---|---|---|
| **Name** | **Melting Point (°C)** | **Boiling Point (°C, at 1 atm)** | **Density (g/ml) at 20°C** |
| Cyclopentane | −94 | 49 | 0.751 |
| Pentane | −130 | 36 | 0.626 |
| Cyclooctane | 14.3 | 149 | 0.835 |
| Octane | −57 | 126 | 0.703 |

Although cycloalkanes as well as acyclic (no rings) alkanes are nonpolar, their physical properties are somewhat different (Table 3.6).

Cycloalkanes have substantially higher melting points than the corresponding alkanes, while densities are roughly 0.13 g/ml greater (Table 3.6). These properties can be explained by the fact that the cycloalkanes are more symmetrical molecules than the alkanes. Therefore, they pack into a crystal lattice more efficiently. This means that more energy is needed to break up the lattice, resulting in higher melting points. Also, since the cycloalkane molecules are more symmetrical, more of them can be placed in a given volume, resulting in a greater density.

Boiling points of cycloalkanes average 10°C higher than those of the corresponding alkanes. This property is explained by the fact that the symmetrical cycloalkanes stack up better in the liquid state due to less carbon-carbon free rotation in the more rigid cyclic structure. Therefore, intermolecular London forces are greater in the cycloalkanes, and more energy is needed to separate the molecules, resulting in higher boiling points.

## 3.8 PHYSICAL PROPERTIES OF CYCLOALKANES

**FIGURE 3.12** Bond angles in selected cycloalkanes.

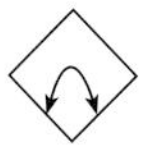

cyclobutane
(90° bond angles)

(a)

cyclopropane
(60° bond angles)

(b)

## 3.9 RING SHAPE OF CYCLOALKANES

The ring shape of a cycloalkane varies with the lower members of this homologous series. Cyclopropane is a flat molecule (with four $sp^3$ hybridized orbitals surrounding each carbon nucleus) and to be most stable would have tetrahedral bond angles of 109.5 degrees. Since the C—C—C bond angles of cyclopropane (60 degrees) (Fig. 3.12b) show a deviation of 49.5 degrees from this optimal bond angle of 109.5 degrees (109.5 − 60 = 49.5 degrees), the ring is very strained and, therefore, highly reactive (see Sec. 4.5). The C—C—C bonds in planar cyclobutane are less strained than those in cyclopropane (109.5 − 90 = 19.5 degrees) (Fig. 3.12a), so cyclobutane is less reactive than cyclopropane.

Cyclopentane and higher homologues have more stable structures because these compounds are not planar but puckered. These puckered structures result in bond angles of 105 degrees in cyclopentane (Fig. 3.13a) and 109.5 degrees in cyclohexane (Fig. 3.13b).

Cyclohexane is an alkane of particular importance because many naturally occurring substances contain one or more cyclohexane rings. Therefore, the conformations of cyclohexane and its derivatives have been studied in great depth. Cyclohexane exists as

## 3.10 CONFORMATIONS OF CYCLOHEXANE AND DERIVATIVES OF CYCLOHEXANE

**FIGURE 3.13** Cyclopentane and cyclohexane are nonplanar molecules. Note the puckering in cyclohexane is more pronounced than in cyclopentane.

(a) Cyclopentane

(b) Cyclohexane

**FIGURE 3.14** Conformations of cyclohexane.

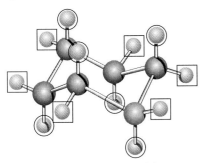

(a)  Chair conformation of cyclohexane

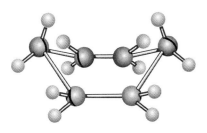

(b)  Boat conformation of cyclohexane

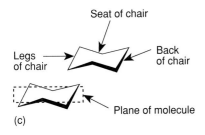

(c)

an equilibrium mixture of two conformational isomers: the more stable *chair* form (Fig. 3.14a) and the less stable *boat* form (Fig. 3.14b).

The chair conformation (named because it is shaped somewhat like a chair; see Fig. 3.14c) contains two different kinds of hydrogen. *Axial hydrogens* (circled in Fig. 3.14a) are oriented in a direction perpendicular to the plane of the molecule. *Equatorial hydrogens* (boxed) are directed on the plane of the molecule. The greater stability of the chair form is due to the fact that hydrogens on adjacent carbons are staggered, whereas in the boat form they are eclipsed (Fig. 3.15).

Since the chair conformation is considerably more stable, let us examine the structure of the chair conformation of cyclohexane in more depth. Note that three axial bonds project upward, while three axial bonds project downward, and all six axial bonds are parallel and alternate in the direction of projection (Fig. 3.16a). On the other hand, the six equatorial bonds can be divided into three groups of two bonds per group (labeled 1, 2, and 3) that are parallel (Fig. 3.16b). In addition, each equatorial bond is parallel to two ring bonds. For example, in Fig. 3.16c, the equatorial bond projecting from C-4 is parallel to the ring bond between C-2 and C-3, along with the ring bond between C-5 and C-6.

By a process known as *ring flipping*, one chair cyclohexane conformation is converted to and is in equilibrium with another chair conformation. One end of the molecule flips up, while the opposite side of the molecule flips down at the same time. This process is very fast at room temperature and proceeds by way of a boat form. The complete equilibrium system is as follows:

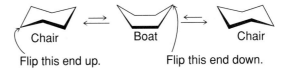

More than 99% of the cyclohexane molecules are in the more stable chair conformation at any given time, with less than 1% in the less stable boat conformation. Note that an axial hydrogen ($H_b$) in ring *a* of Fig. 3.17 is converted, by means of a ring flip, to an equatorial hydrogen in ring *b*, while an equatorial hydrogen ($H_c$) in ring *a* is converted, in the same way, to an axial hydrogen in ring *b*. Both chair conformations of

**FIGURE 3.15**  A guide to the greater stability of the chair conformation of cyclohexane.

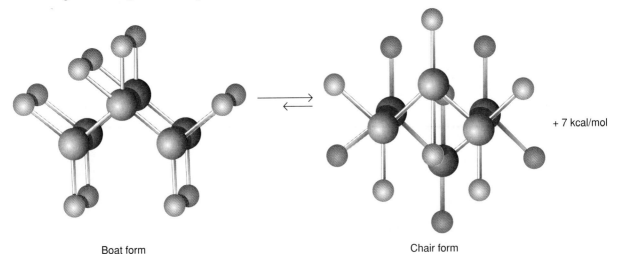

Boat form

Note the unstable eclipsed conformation of hydrogens on two adjacent carbons in the Newman projection of adjacent carbons in the boat form of cyclohexane.

Chair form

Note the stable staggered conformation of hydrogens on two adjacent carbons in the Newman projection of adjacent carbons in the chair form of cyclohexane.

cyclohexane contain the same energy and are therefore of equal stability. Thus each conformation is present in the same concentration in the equilibrium mixture at any given time.

When an alkyl or other group replaces a hydrogen on the cyclohexane ring, the group can be axial or equatorial. *The preferred conformation is the one in which a group larger than hydrogen is in the equatorial position.* The equatorial conformation is favored because of crowding between, for example, an axial methyl group on C-1 and axial hydrogens on C-3 and C-5. This crowding, in turn, leads to London repulsions between the methyl group and the hydrogens, destabilizing the molecule. These destabilizing interactions are known as *1,3-diaxial interactions*. This crowding is relieved when the methyl (or larger) group is placed in an equatorial position. Thus about 95% of the methylcyclohexane molecules contain the methyl group in an equatorial position, with only 5% of the methyl groups in the axial position.

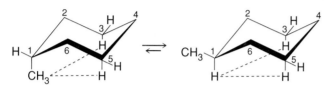

| Crowding is more important between an axial methyl and hydrogens on C-3 and C-5. Thus London repulsive forces are significant, and methylcyclohexane would have only 5% of its molecules in the conformation with the methyl in an axial position | Crowding is less important between an axial hydrogen and hydrogens on C-3 and C-5. Thus London repulsions are less important, and methylcyclohexane would have 95% of its molecules in the conformation with the methyl in an equatorial position. |

Observe that a ring flip inverts the conformation of a substituent. That is, an axial methyl and equatorial hydrogen on ring *a* below is converted by a ring flip to an equatorial methyl and axial hydrogen on ring *b*.

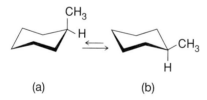

(a)                            (b)

If a group larger than the methyl group is placed in an axial position on the ring, crowding and London repulsive forces are increased. Therefore, over 99% of the molecules of *tert*-butylcyclohexane are found with the *tert*-butyl group in an equatorial position as opposed to less than 1% with the *tert*-butyl group in the axial position. In other words, the larger and bulkier the group on the ring, the greater is the percentage of molecules with the group in an equatorial position present in an equilibrium mixture.

**FIGURE 3.16** Projection of axial and equatorial bonds in cyclohexane.

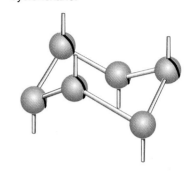

(a)

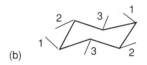

(b)

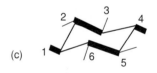

(c)

**FIGURE 3.17** The conversion of an axial hydrogen to an equatorial hydrogen (and vice versa) by a ring flip.

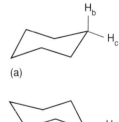

(a)

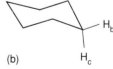

(b)

Consider each of the following bonds projecting from the cyclohexane ring. Classify each as axial or equatorial and explain your choice.

**SOLVED PROBLEM 3.10**

a.                    b.

**SOLUTION**

a. The bond is equatorial. The bond is on the plane of the molecule, and it is parallel to the two ring bonds designated 1 and 2 below.

b. The bond is axial. The bond is represented by a vertical line and is perpendicular to the plane of the molecule.

a.    b.   ← Plane of the molecule

---

a. Draw structures representing the chair-chair equilibrium for isopropylcyclohexane.

b. Estimate the percentage composition of the chair conformers where the isopropyl group is found in an axial position versus an equatorial position.

---

**3.11
GEOMETRIC
CONDENSED
STRUCTURES**

One type of structural formula we have studied is the geometric condensed structure for the cycloalkanes (see Sec. 3.6). For example, see Fig. 3.18. Here, the intersection of two lines represents a methylene ($-CH_2-$) group. A wide variety of other compounds can be condensed in this way. To accomplish this, a straight line extending from a structure represents a methyl group at the end of the line, while the intersection of three lines describes the following group:

$$-\overset{|}{\underset{|}{C}}-H$$

Two examples of the use of geometric condensed structures along with the corresponding line-bond structures are given below.

**FIGURE 3.18** Some selected geometric condensed structures of cycloalkanes.

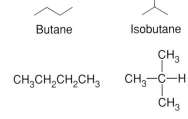

Butane        Isobutane

Cyclopropane

$CH_3CH_2CH_2CH_3$    $CH_3-\overset{CH_3}{\underset{CH_3}{\overset{|}{\underset{|}{C}}}}-H$

Cyclohexane

---

Draw a geometric condensed structure for the following.

a.   $CH_3-\overset{CH_3}{\underset{H}{\overset{|}{\underset{|}{C}}}}-CH_2-CH_3$

b.   $CH_3-\overset{CH_3}{\underset{\underset{CH_3}{|}}{\overset{|}{CH}}}-CH-CH_2-CH_3$

**SOLUTION**

a. Since the longest carbon chain contains four carbons, draw a W-style geometric condensed formula with three direction changes to represent a four-carbon chain. Now draw a straight line from the structure at carbon-2 to represent the methyl group.

b. Since a five-carbon chain is present, draw a W-style geometric condensed formula with four direction changes to represent a five-carbon chain. The two methyl branches are represented by drawing two straight lines from the structure at carbon-2 and carbon-3.

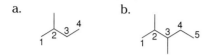

a.                          b.

---

Draw a geometric condensed structural formula for the following compounds.

**EXERCISE 3.11**

a. 2-Methylhexane

b. Isohexane

c. 2,2-Dimethylpropane

---

## ▶ CHAPTER ACCOMPLISHMENTS

### 3.1 More on the Conformations of Alkanes
☐ Explain why the staggered conformation of ethane is the most stable.
☐ Explain why we can assume that free rotation exists around any carbon-carbon acyclic single bond.

### 3.2 Nomenclature of Alkanes
☐ Explain how the ethyl group is formulated from ethane.
☐ Explain what characteristic equivalent hydrogens have in common.
☐ Determine the number of sets of equivalent hydrogens in butane and the number of hydrogens in each set. Locate each set using a line-bond structural formula of butane.
☐ Determine the number of primary, secondary, and tertiary hydrogens in butane.

### 3.3 Physical Properties of the Alkanes
☐ Identify the only intermolecular forces present in alkanes.
☐ List each of the physical properties of the alkanes that result from the presence of only the weakest type of intermolecular forces.
☐ Explain why alkanes are insoluble in water and soluble in carbon tetrachloride.

### 3.4 Natural and Commercial Sources of Alkanes
☐ Explain why the presence of sparks in a mine represents a deadly danger to the working miners.
☐ List the main constituents of natural gas.

### 3.5 Alkanes in Biological Systems
☐ Calculate the number of picograms in 1 gram.
☐ Define a pheromone.
☐ Supply a condensed formula for nonadecane; for docosane.

### 3.6 Introduction to Cycloalkanes
☐ Draw a geometric condensed structure for cyclopentane.
☐ Provide a molecular formula for cyclooctane.

### 3.7 Nomenclature of Cycloalkanes
☐ Explain why a number is not assigned to the bromo group in bromocyclopentane.
☐ Explain the following paradoxical nomenclature in that the first compound is named as a cycloalkane while the second is is named as an alkane.
   chlorocyclopropane
   1-cyclopropylpentane

### 3.8 Physical Properties of Cycloalkanes
☐ Explain why cyclopentane has a greater density than pentane.
☐ Predict why the density of cyclobutane is given more conveniently in grams per liter rather than grams per milliliter.

### 3.9 Ring Shape of Cycloalkanes
☐ Identify a cycloalkane that is flat (planar).
☐ Explain why cyclopropane is more reactive than cyclopentane.

### 3.10  Conformations of Cyclohexane and Derivatives of Cyclohexane

☐ Distinguish between (a) the chair conformation and boat conformation of cyclohexane and (b) an axial hydrogen and an equatorial hydrogen—both bonded to a cyclohexane carbon in the chair conformation.

☐ Illustrate a ring flip.

### 3.11  Geometric Condensed Structures

☐ Draw a geometric condensed formula for isobutane; for butane.

---

▶ **KEY TERMS**

*alkyl group* (3.2)
*equivalent* (3.2)
*primary* (3.2)
*secondary* (3.2)
*tertiary* (3.2)
*International Union of Pure and Applied Chemistry* (3.2)
*main chain* (3.2)

*parent compound* (3.2)
*substituent group* (3.2)
*natural gas* (3.4)
*refined* (3.4)
*pheromone* (3.5)
*paraffin* (3.5, 4.2)
*aliphatic compound* (3.6)

*chair conformation* (3.10)
*boat conformation* (3.10)
*axial bond* (3.10)
*equatorial bond* (3.10)
*ring flip* (3.10)
*1,3-diaxial interaction* (3.10, 5.3)
*quaternary carbon* (Problem 11)

---

▶ **PROBLEMS**

1. Draw a Newman projection, wedge projection, and sawhorse diagram for each of the following.
   a. The most stable conformation of 1-bromoethane
   b. The least stable conformation of 1-bromoethane
   c. A conformation of 1-bromoethane midway in stability between the conformational isomers in parts a and b.

2. Draw a sawhorse diagram for the staggered conformation of propane.

3. How many conformations are possible for an ethane molecule between the staggered and eclipsed conformations?

4. Draw a structural formula for each of the following alkyl groups.
   a. methyl
   b. isoheptyl
   c. *n*-octyl

5. Consider isohexane

$$(CH_3)_2CHCH_2CH_2CH_3$$

   a. Draw a structural formula for each alkyl group that can be derived from isohexane.
   b. Classify each alkyl group as primary, secondary, or tertiary.

6. Determine the number of primary, secondary, and tertiary hydrogens in the following.
   a. $CH_3CH_2CH_2CH_2CH_3$
   b. Each of the alkanes from which the alkyl groups in Exercise 3.2 are derived.

7. Is the methyl group primary? Explain.

8. Identify each of the following groups as primary, secondary, or tertiary, and name each group.

   a. $(CH_3)_2CHCH_2CH_2CH_2—$, which is the same alkyl group as

$$CH_3—\underset{\underset{H}{|}}{\overset{\overset{CH_3}{|}}{C}}—CH_2—CH_2—CH_2—$$

   b.

$$CH_3—\underset{\underset{CH_2CH_3}{|}}{\overset{\overset{CH_3}{|}}{C}}—$$

   c.

$$CH_3—\underset{\underset{CH_3}{\underset{|}{CH_2}}}{\overset{\overset{H}{|}}{C}}—$$
   Compare this group with the group in Exercise 3.2d.

9. Determine the number of primary, secondary, and tertiary hydrogens in each of the alkanes from which the alkyl groups in Problem 8a and b are derived.

10. Draw a structural formula for every monochloro derivative of the following compounds.
    a. Isopentane
    b. $(CH_3)_4C$ (neopentane)
    c. *n*-pentane

11. Tabulate the number of primary, secondary, and tertiary hydrogens in the following.
    a. Each of the hydrocarbons in Problem 10.
    b. $CH_3CH_2CHCH_2CH_2CH_3$
       with $CH_3$ branch

c. $(CH_3)_2\underset{\underset{H}{|}}{C}-\underset{\underset{H}{|}}{C}(CH_3)_2$

d. Which hydrocarbon in Problem 10 contains a quaternary carbon? A *quaternary carbon* is a carbon that is bonded to four carbons.

12. Name each of the following compounds by the IUPAC method.

a. $CH_3CH_2C(CH_3)_2CH_2CH_2CH_3$

b. $CH_3(CH_2)_7CH_3$

c.

$CH_3-\underset{\underset{CH_3}{|}}{\overset{\overset{H}{|}}{C}}-\underset{\underset{CH_3}{|}}{\overset{\overset{H}{|}}{C}}-CH_3$

d. $CCl_4$

e.

$CH_3-CH_2$
$\quad\quad\quad CH_2$
$\quad\quad HC-CH_2-CH_3$
$CH_3-\underset{\underset{H}{|}}{C}-CH_3$

f.

$CH_3-CH_2-\underset{\underset{CH_2CH_3}{|}}{\overset{\overset{H}{|}}{C}}-CH_2-CH_3$

g. $CH_2ClCHBrCH_3$

h.

$CH_3-\underset{\underset{NO_2}{|}}{\overset{\overset{CH_2CH_3}{|}}{C}}-CH_2\underset{\underset{Br}{|}}{C}HCH(CH_3)_2$

13. Draw a structural formula for each of the following compounds.

a. 3-Iodopentane

b. 2-Methyloctane

14. Draw a structural formula for each of the following compounds.

a. 1,1,2-Triiodopentane

b. 3,4,4,5-Tetrachlorononane

15. Draw a structural formula for each of the following compounds.

a. 1,1-Dibromo-2,4-dichloro-5-isopropyl-3-methyldecane

b. 3-Bromo-3,4-dimethylhexane

16. Each of the following IUPAC names is incorrect. Draw a structural formula for each incorrect name. Then write the correct name, and explain why the first name is incorrect.

a. 4-Methylhexane

b. 3-Chlorocyclopropane

c. 2-Nitro-3-bromopentane

d. 1,4-Dimethyltriacontane

e. 3-*n*-Propylbutane

17. Arrange the following series of compounds in order of increasing boiling point.

a. $(CH_3)_3CCH_2CH_3$, $CH_3(CH_2)_4CH_3$, $(CH_3)_2CHCH_2CH_2CH_3$

b. Ethane, *n*-heptane, *n*-butane

c. *n*-Hexane, isobutane, isohexane, ethane, neopentane

18. Predict the compound of each of the following pairs with the higher melting point.

a. Propane and nonane

b. Nonane and pentadecane

19. Define or explain each of the following.

a. Eclipsed conformation

b. Alkyl group

20. Name each of the following compounds.

a. $CH_3(CH_2)_{17}CH_3$

b. $CH_3CH(Br)CH_2(CH_2)_{23}CH_3$

c. $CH_3CH_2CH_2C(CH_3)_2(CH_2)_{15}CH_3$

d. $CH_3(CH_2)_{36}CH_3$

21. Draw a structural formula for each of the following compounds.

a. Tricosane

b. Hexatriacontane

c. Hentetracontane

22. Draw a structural formula for each of the following compounds.

a. Hendecane

b. Octacosane

c. 5,7,9-Trimethylnonatriacontane

23. How do human beings accumulate small quantities of alkanes in their bodies?

24. Which hydrocarbon when found in relatively large amounts in human beings indicates a vegetarian diet?

25. Draw a structural formula for each of the following compounds.

a. 1,1,3-Trimethylcyclopentane

b. 1,4-Dicyclobutylheptane

c. 1-*sec*-Butyl-4-isobutylcyclohexane

26. Name each of the following compounds.

a.

b.

c.

d.

e.

27. Draw a structural formula for each of the following compounds.
    a. 1-Ethyl-2-isopropylcyclohexane
    b. 1-Chloro-3,3-dimethylcyclobutane
    c. 2-Cyclopentyloctane

28. Tabulate the number of primary, secondary, and tertiary hydrogens in the following.
    a. The alicyclic hydrocarbons in Problem 27a and c.
    b. 1,1,2-Trimethylcyclohexane
    c. Which compound in Problem 28 contains a quaternary carbon? A *quaternary carbon* is a carbon that is bonded to four other carbons.

29. Each of the following IUPAC names is incorrect. Draw a structural formula for each incorrect name. Then write the correct name, and explain why the first name is incorrect.
    a. 2-Methyl-1-ethylcycloheptane
    b. 2-Butyl-1-chlorocyclopentane
    c. *n*-Pentylcyclopropane

30. Explain the difference in the value of each physical property between hexane and cyclohexane in the following table.

|  | Density (g/ml) at 20°C | mp (°C) at 1 atm | bp (°C) at 1 atm |
| --- | --- | --- | --- |
| Hexane | 0.66 | −94.3 | 69.0 |
| Cyclohexane | 0.78 | 6.5 | 81.4 |

31. Draw a structural formula for methylcyclohexane in which the methyl group is (a) in an axial position and (b) in an equatorial position.

32. Explain why in a methylcyclohexane molecule the preferred conformation is that in which the methyl group is in the equatorial position?

33. a. Draw a structural formula for 1,2-dimethylcyclohexane in which each group is axial.
    b. Draw a structural formula for 1,3-diethylcyclohexane in which each group is equatorial.

34. Classify each of the following cyclohexane bonds projecting from the cyclohexane ring as axial or equatorial.

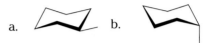

a.            b.

35. Draw a geometric condensed formula for the following compounds.
    a. Pentane
    b. Isopentane
    c. Cyclopropylcyclopentane

36. Draw line-bond and condensed structural formulas corresponding to each of the following compounds.

    a.            b.

    c.            d.   —CH₂(CH₂)₅CH₃

37. Draw a line-bond structural formula and name each of the following compounds using the IUPAC method of nomenclature.

    a.            b.

# CHAPTER 4

# Alkanes and Cycloalkanes (II)

## SYNTHESIS AND CHEMICAL PROPERTIES

Most alkanes needed by chemists are isolated from petroleum and plants. A few low-molecular-weight alkanes are obtained from natural gas. However, on occasion a chemist will need an alkane that cannot be obtained conveniently from these natural sources. In that event, one of the following synthetic methods is used in the laboratory.

As a result of the reaction, one of the two carbon-carbon bonds of the double bond is broken, and two carbon-hydrogen bonds are formed: one bond and hydrogen on each carbon of the double bond. In general,

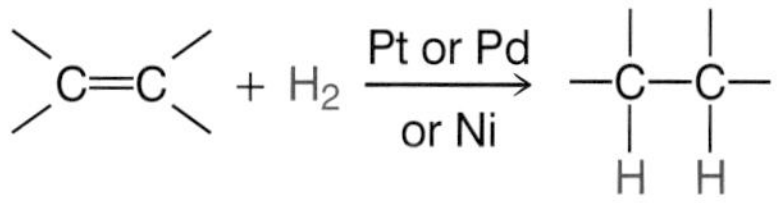

$$>C=C< \ + \ H_2 \ \xrightarrow[\text{or Ni}]{\text{Pt or Pd}} \ -\overset{|}{\underset{H}{C}}-\overset{|}{\underset{H}{C}}-$$

Example:

$$\overset{H}{\underset{CH_3}{}}C=C\overset{H}{\underset{H}{}} \ + \ H_2 \ \xrightarrow{\text{Ni}} \ CH_3-\overset{\overset{H}{|}}{\underset{\underset{H}{|}}{C}}-\overset{\overset{H}{|}}{\underset{\underset{H}{|}}{C}}-H$$

When an alkyl halide is treated with magnesium turnings in water-free, or anhydrous, diethyl ether (see Table 1.3), an important synthetic reagent known as the *Grignard reagent* forms. In general,

$$R-X + Mg \xrightarrow{\text{Anhydrous diethyl ether}} R-Mg-X$$

alkyl halide                  Grignard reagent

where R = any alkyl group
X = Cl, Br, or I

Example:

$$H-\overset{\overset{H}{|}}{\underset{\underset{H}{|}}{C}}-Br \ + \ Mg \xrightarrow{\text{Anhydrous diethyl ether}} H-\overset{\overset{H}{|}}{\underset{\underset{H}{|}}{C}}-Mg-Br$$

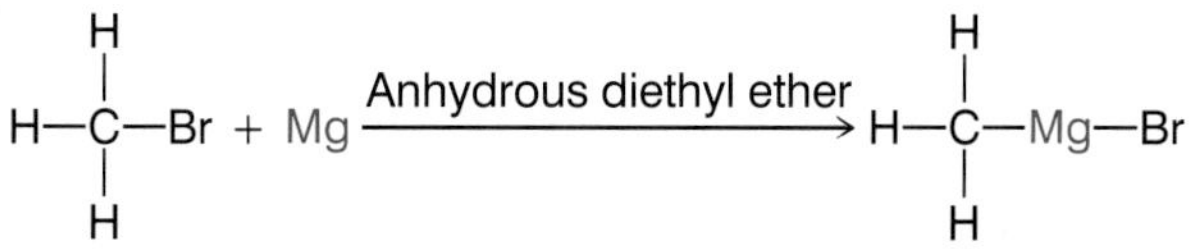

Methylmagnesium
bromide

## 4.1 SYNTHESIS OF ALKANES

### Reduction of an Alkene

*Reduction* is the addition of one or more hydrogen atoms to a species.

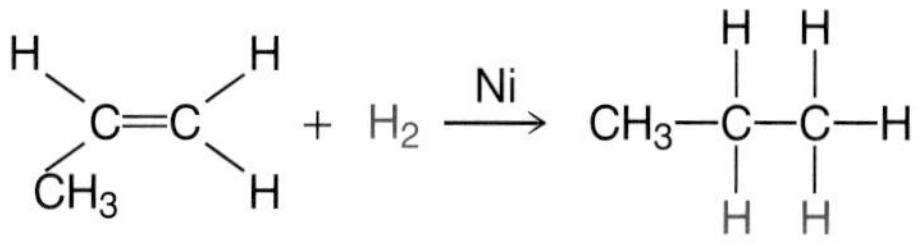

Finely divided platinum, palladium, or nickel can serve as a catalyst. Yields are excellent at room temperature.

### Reduction of an Alkyl Halide via the Grignard Reagent

This reagent is named after Victor Grignard, a Nobel Prize–winning organic chemist who discovered the reagent.

Anhydrous diethyl ether is used as a solvent for three reasons. First, diethyl ether in particular and ethers in general do not react with either organic halides or the Grignard reagents when produced. Second, both the organic halides and the corresponding Grignard reagents are diethyl ether–soluble. Finally, diethyl ether helps stabilize the Grignard reagent once it forms by combining with it.

As a result of the reaction, the magnesium is placed between the carbon of the alkyl group bearing the halogen and the halogen. The diethyl ether must be anhydrous (free of water); if not, the Grignard reagent cannot form. The Grignard reagent acts as a Lewis acid and the diethyl ether as a Lewis base (Sec. 1.14) to form a coordination complex.

$$R—Mg—X \quad + \quad 2\ C_2H_5\ddot{O}C_2H_5 \longrightarrow$$

$$
\begin{array}{c}
H_5C_2 \\
\diagdown \\
\overset{+}{O}{-}C_2H_5 \\
Mg^{2-}{-}X \\
\overset{+}{O}{-}C_2H_5 \\
\diagup \\
H_5C_2
\end{array}
$$

Lewis acid          Lewis base

Coordination complex

The Grignard reagent does not have to be isolated. Hydrolysis (treatment with water) of the Grignard reaction mixture gives the alkane. In general,

$$RMgX + H_2O \longrightarrow R—H + X—Mg—OH$$

Example:

$$CH_3MgBr + H_2O \longrightarrow CH_3—H + Br—Mg—OH$$

As a result of this hydrolysis, the Mg—X moiety is replaced by a hydrogen from the water to produce the alkane. In the laboratory, a dilute solution of hydrochloric acid usually is used rather than water. This produces a water-soluble salt, $MgCl_2$ or $MgClX$, rather than an insoluble magnesium hydroxyhalide ($HOMgX$), thus facilitating isolation of the water-insoluble product. For example:

$$
\begin{array}{c}
CH_3 \\
| \\
CH_3—C—CH_2MgBr + HCl(aq) \\
| \\
H
\end{array}
\longrightarrow
\begin{array}{c}
CH_3 \\
| \\
CH_3—C—CH_3 + MgBrCl \\
| \\
H
\end{array}
$$

## The Corey-House Synthesis

The *Corey-House synthesis* furnishes the chemist with a means of preparing an alkane with a carbon content greater than that of the starting material. In general,

$$\textit{Step 1:}\quad R—X + 2\ Li \xrightarrow{\text{Anhydrous diethyl ether}} R—Li + LiX$$

Alkyl lithium

where R = any alkyl group
X = Cl, Br, or I

$$\textit{Step 2:}\quad 2\ RLi + CuI \longrightarrow R_2CuLi + LiI$$

Lithium dialkylcuprate

$$\textit{Step 3:}\quad R_2CuLi + R'X \longrightarrow R—R' + LiX + RCu$$

where R′ = any primary alkyl group or methyl
X = Cl, Br, or I

Example:

$$(CH_3)_3CCl + 2\ Li \longrightarrow (CH_3)_3CLi + LiCl$$

$$2\ (CH_3)_3CLi + CuI \longrightarrow [(CH_3)_3C]_2CuLi + LiI$$

Lithium di-*tert*-butylcuprate

$$[(CH_3)_3C]_2CuLi + CH_3CH_2Cl \longrightarrow (CH_3)_3CCH_2CH_3 + (CH_3)_3CCu + LiCl$$

The result of step 1 is to substitute a lithium for a halogen, producing an alkyl lithium, RLi. In step 2, 2 mol of alkyl lithium combines with copper(I) from CuI to form a lithium dialkyl cuprate, $R_2CuLi$, while step 3 represents a link-up of the alkyl group in the cuprate (R) with the alkyl group of the halide (R′) to produce a hydrocarbon R—R′. The sole limitation of the Corey-House synthesis is that good yields are obtained only if the alkyl halide in step 3 is primary or methyl.

*n*-Heptane can be detected in the wood terpentine taken from the western pine *Pinus jeffreyi*. Synthesize *n*-heptane from *n*-butyl bromide and *n*-propyl bromide.

### SOLUTION

Since the number of carbons in the product compound is greater than the number of carbons in either reactant, a reaction increasing the number of carbons in the product is necessary. Thus the Corey-House synthesis is employed.

*Note*: In a synthesis problem, inorganic and organic by-products are not usually shown.

$$CH_3CH_2CH_2Br \xrightarrow{Li} CH_3CH_2CH_2Li \xrightarrow{CuI} (CH_3CH_2CH_2)_2CuLi \xrightarrow{CH_3CH_2CH_2CH_2Br} \textit{n-heptane}$$

Prepare each of the following compounds using any needed inorganic reagents.

a. Pentane from $CH_3CH_2Br$ and $CH_3CH_2CH_2Br$

b. Isobutane from $CH_3CH(I)CH_3$ and $CH_3Cl$. *Hint*: In step 3 of the synthesis, use care in selecting which of the preceding alkyl groups is R′. Why?

## 4.2 CHEMICAL REACTIONS OF ALKANES

The chemistry of the alkanes is relatively simple because the alkanes are chemically unreactive. Alkanes do not react with aqueous solutions of strong acids ($H_2SO_4$ or HCl), strong bases (NaOH or KOH), oxidizing agents ($KMnO_4$ or $Na_2Cr_2O_7$ + $H_2SO_4$), or reducing agents ($FeSO_4$ or $SnCl_2$). Thus alkanes are known as *paraffins* because of their relative chemical inertness (L. *parum*, "little"; *affinis*, "affinity"). However, there are several reactions that they will undergo. We will study combustion and halogenation.

### Combustion

When treated with excess oxygen (in air), alkanes burn to produce carbon dioxide, water, and heat. Only a small amount of heat is necessary to start the reactions. Once started, the reactions continue because of the heat produced. In general,

$$\text{alkane} + O_2 \longrightarrow CO_2 + H_2O + \text{heat}$$

Example:

$$2\,CH_3CH_3 + 7\,O_2 \longrightarrow 4\,CO_2 + 6\,H_2O + \text{heat}$$

Combustion reactions of alkanes are highly *exothermic*. An exothermic reaction is a reaction that liberates heat to the surroundings. Thus alkanes make excellent fuels.

Incomplete combustion, due to a lack of oxygen, would produce carbon and/or deadly carbon monoxide. This is why an automobile engine, which burns gasoline as fuel, should not be run in a closed-in garage. In general,

$$\text{Alkane} + O_2 \longrightarrow C + CO + H_2O$$

Example:

$$C_2H_6 + 2\,O_2 \longrightarrow C + CO + 3\,H_2O$$

Alkanes react with certain halogens to produce alkyl halides when heated or exposed to ultraviolet light. In general,

$$RH + X_2 \xrightarrow[\text{or } \Delta]{h\nu} RX + HX$$

where $X$ = Cl or Br
$\Delta$ = heat

Although the preceding equation is balanced, some equations in this textbook will *not* be balanced, particularly when two or more organic products are produced.

For example, natural gas, a mixture of over 90% methane with some ethane and very little propane, is used as a heating fuel along with heating oil (see Sec. 3.4). On the other hand, gasoline and jet fuel (kerosene fraction of petroleum) are sources of power.

### Halogenation

The symbols $h\nu$ represent light energy ($E = h\nu$), where $h$ is Planck's constant (erg · s) and $\nu$ is the frequency of light ($s^{-1}$).

Example:

$$CH_3CH_3 + Br_2 \xrightarrow{\Delta} CH_3CH_2Br + HBr$$
$$\text{ethyl bromide}$$

Since the net effect of the reaction is to replace a hydrogen with a halogen, these are called *substitution reactions*. Alkanes react with fluorine to give considerable heat, which breaks carbon-carbon bonds, while iodine is relatively unreactive with alkanes.

Both methane and ethane have one set of equivalent hydrogens. Thus only one monosubstituted alkyl halide (see Table 1.3) product is formed.

$$CH_4 + Br_2 \xrightarrow{h_\nu} CH_3Br + HBr$$
$$\text{Methyl bromide (common)}$$
$$\text{Bromomethane (IUPAC)}$$

But propane with two types of hydrogens (primary and secondary) gives two monosubstituted products.

$$CH_3CH_2CH_3 + Cl_2 \xrightarrow{\Delta} \underset{\underset{Cl}{|}}{\overset{\overset{H}{|}}{CH_3CCH_3}} + CH_3CH_2CH_2Cl + HCl$$

*n*-Propyl chloride (common)
1-Chloropropane (IUPAC)

Isopropyl chloride (common)
2-Chloropropane (IUPAC)

| | |
|---|---|
| Note that the names directly below the formulas here are IUPAC and those beneath the IUPAC names are common (see App. 2). | Treatment of an alkane with excess halogen gives a mixture of polyhalogenated products. |

$$CH_4 + n\,Cl_2 \xrightarrow{\Delta} CH_3Cl + CH_2Cl_2$$

Chloromethane              Dichloromethane
Methyl chloride            Methylene chloride

$$+ CHCl_3 + CCl_4 + n\,HCl$$

Trichloromethane      Tetrachloromethane
Chloroform            Carbon tetrachloride

To name an alkyl halide by the common system, name the alkyl group and add the word *halide*. This method also works for the methylene group, —$CH_2$—. Chloroform is a trivial name that must be memorized (see App. 2).

This mixture can be separated into its four individual components by fractional distillation when large quantities of reactants are used in an industrial synthesis.

## 4.3
## REACTION MECHANISM

A reaction mechanism is a step-by-step description of the chemical species involved during the course of a reaction. Each reaction mechanism is based on experimental data and is useful to predict the reaction conditions needed to maximize the yield of product.

**Reactive Chemical Species.**   Before we begin our study of reaction mechanisms, we need to examine a number of reactive chemical species often produced during the course of a chemical reaction. You will recall that the loss of a hydrogen from a methane (or any alkane) molecule gives us an artificially created methyl group, $CH_3$— (or a comparable alkyl group) that is used as an aid in organic nomenclature. Structural formulas of some chemical species derived from methane that do exist are compared with the structure we assign to the methyl group in Table 4.1.

The methyl *carbocation* and all other carbocations (see Table 4.1) are species in which carbon bearing the positive charge shares only six electrons. The negatively charged species is the methyl *carbanion*. The methyl carbanion and other carbanions contain a negatively charged carbon that has two unshared electrons. The neutral methyl *free radical* and other free radicals contain a carbon that has one unshared electron. Each of these species exists but is very unstable. The imaginary methyl group and other alkyl groups can be distinguished from the other real species by the fact that

| TABLE 4.1 | A Comparison of One Carbon Chemical Species | |
|---|---|---|
| **Structure** | **Name** | **State in Nature** |
| H<br>\|<br>H—C$^+$<br>\|<br>H | Methyl carbocation | It exists but is unstable |
| H<br>\|<br>H—C:$^-$<br>\|<br>H | Methyl carbanion | It exists but is unstable |
| H<br>\|<br>H—C·<br>\|<br>H | Methyl free radical | It exists but is unstable |
| H<br>\|<br>H—C—<br>\|<br>H | Methyl alkyl group | It does not exist |

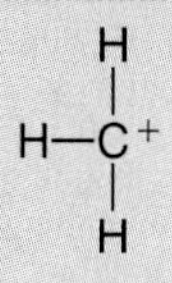

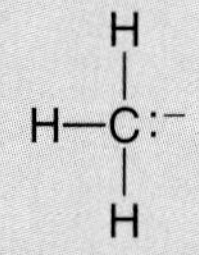

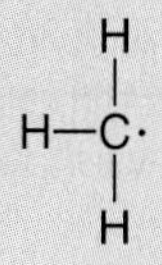

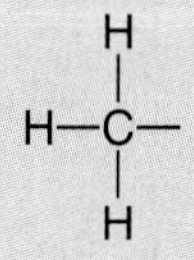

Let us consider a few more examples of some of these highly reactive species:

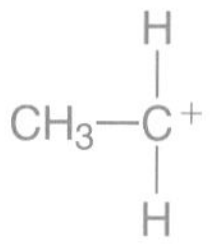

$$CH_3—\overset{H}{\underset{H}{C^+}}$$

Ethyl carbocation

$$CH_3—\overset{CH_3}{\underset{H}{C\cdot}}$$

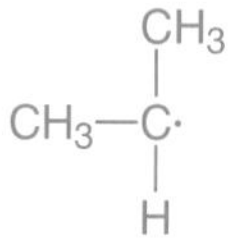

Isopropyl free radical

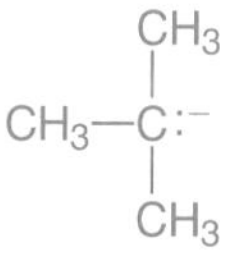

$$CH_3—\overset{CH_3}{\underset{CH_3}{C:^-}}$$

*tert*-Butyl carbanion

a dash (dangling bond) is placed adjacent to the carbon in the alkyl group that shares six electrons. We will discuss these real species in more depth further along in this chapter and later in this textbook.

To name a reactive species, simply name the alkyl group and then add the word *carbocation, carbanion,* or *free radical.*

**EXERCISE 4.2**

Name each of the following chemical species.

a. 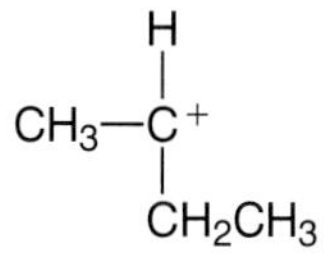 $CH_3—\overset{H}{\underset{CH_2CH_3}{C^+}}$   b. $CH_3—\overset{CH_3}{\underset{CH_2CH_3}{C\cdot}}$   c. $CH_3CH_2CH_2\overset{..}{C}H_2^-$

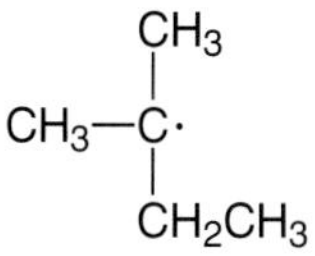

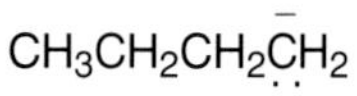

**EXERCISE 4.3**

Draw a structural formula for each of the following chemical species.

a. Isobutyl carbocation

b. *sec*-Butyl free radical

c. *n*-Pentyl carbanion

**Use of the Curved Arrow Symbol.**  *Curved arrows* are used throughout this textbook to show the movement of one or more electrons to explain the various steps in a reaction mechanism. If one electron is moved, an arrow with a half barb is employed; if two electrons are moved, an arrow with a full barb is used. Let us first use the curved arrow symbol to show the homolytic breaking of a bond, where each participating atom of the bond acquires an electron after the break. Consider a bond A : B breaking as follows:

$$A \overset{\frown}{\phantom{i}} B \longrightarrow A\cdot + \cdot B$$

For *homolytic bond formation*, we have the opposite process, where both A and B each supply an electron to produce the bond.

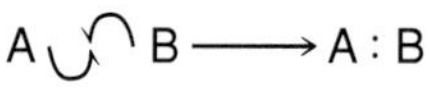

$$A \overset{\smile}{\phantom{i}} B \longrightarrow A : B$$

Note page 72 for use of the curved arrow symbols to explain the mechanism of the halogenation of an alkane.

The curved arrow symbol is also used to demonstrate *heterolytic bond breaking*, in which one or the other atom participating in the bond acquires both electrons. For bond breaking, we have:

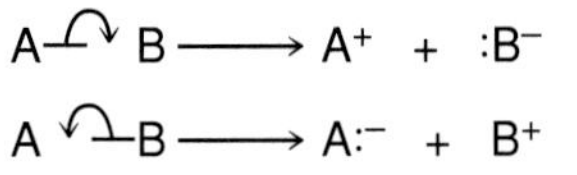

$$A \overset{\frown}{\phantom{a}} B \longrightarrow A^+ \; + \; :B^-$$

$$A \overset{\frown}{\phantom{a}} B \longrightarrow A:^- \; + \; B^+$$

Note that when both electrons of a bond are moved together, we employ only one curved arrow to show the movement.

For *heterolytic bond formation*, we have

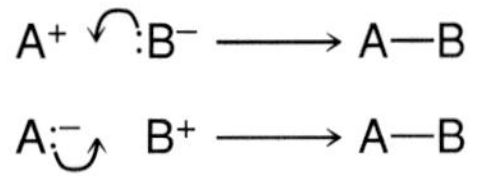

$$A^+ \overset{\frown}{\phantom{a}} :B^- \longrightarrow A{-}B$$

$$A:\overset{\smile}{\phantom{a}} \; B^+ \longrightarrow A{-}B$$

To show bond formation sometimes, observe that the curved arrow is placed between the two atoms concerned to emphasize that the two electrons are shared by each participating atom. Under other circumstances, the curved arrow is shown directly from A to B

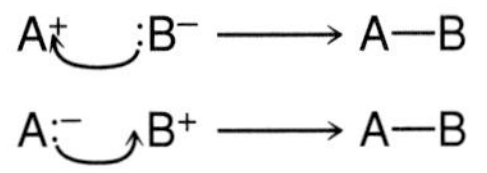

$$A^+ \overset{\smile}{\phantom{aa}} :B^- \longrightarrow A{-}B$$

$$A:^- \overset{\smile}{\phantom{aa}} B^+ \longrightarrow A{-}B$$

We will use the mechanism of halogenation of an alkane to illustrate curved arrow use.

<table>
<tr><td>Envision the Reaction</td></tr>
</table>

## Halogenation of Alkanes

Initiation:    $Br \overset{\frown}{:} Br \xrightarrow{\Delta} Br\cdot \; + \; \cdot Br$

Propagation:

Ethyl free radical

Termination:    $CH_3CH_2\cdot \overset{\frown}{\phantom{a}} Br \longrightarrow CH_3CH_2Br$

$CH_3CH_2\cdot \overset{\frown}{\phantom{a}} CH_2CH_3 \longrightarrow CH_3CH_2CH_2CH_3$

$Br\cdot \overset{\frown}{\phantom{a}} Br \longrightarrow Br_2$

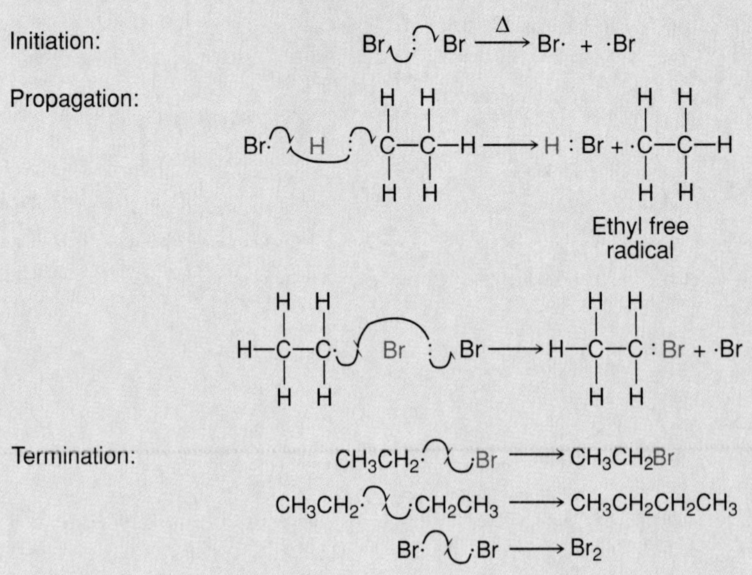

The first step in the mechanism of the halogenation of alkanes (ethane) involves the absorption of light or heat by the halogen, which dissociates it into two reactive halogen (bromine) atoms, each with one unpaired electron. This first step is called the *initiation step*. An atom or group of atoms which has one (or more) unpaired electron(s) is called a *free radical*.

The halogen atom produced is so reactive that it attacks an alkane molecule, abstracting a hydrogen atom, and forming a highly reactive alkyl free radical and a stable hydrogen halide. The free radical R· collides with another $X_2$ molecule to produce a stable molecule of product and a highly reactive halogen atom. This resulting halogen

Free radicals are neutral and can be inorganic (Cl ·) or organic (R ·) (see Table 4.1).

atom can collide and react with another molecule of alkane, R—H. This is characteristic of a *chain reaction*. These two steps are known as *propagation steps*. The final three steps of the mechanism are *termination steps*. The progress of an alkane bromination reaction can be followed by means of a color change. In the bromination of an alkane, the red-orange color of bromine is slowly dissipated when substitution takes place:

It is the reactivity of the unpaired electron species (free radicals) that causes the reaction to proceed.

$$C_2H_6 \quad + \quad Br_2 \quad \xrightarrow{h\nu} \quad C_2H_5Br \quad + \quad HBr$$

Colorless    Red-orange          Colorless    Colorless

A *multiline arrow* (a series of straight lines ending in an arrowhead) is employed to show the movement of an atom by itself or an atom with a pair of electrons. For example,

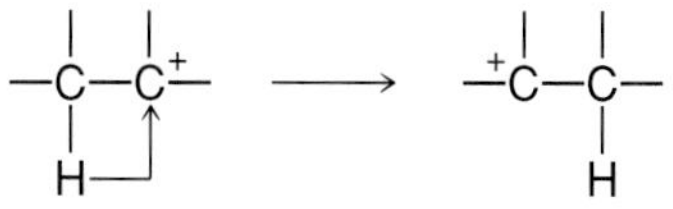

**Use of the Multiline Arrow Symbol**

Use of the multiline arrow is shown in Sec. 8.7.

---

Most cycloalkanes are isolated by the chemical industry from petroleum. In the laboratory, cycloalkanes can be produced by hydrogenation of cycloalkenes (cyclobutene and higher homologues), use of the Corey-House procedure, and Grignard synthesis from cycloalkyl halides (cyclobutyl halides and higher homologues). The Corey-House procedure is further limited in that R′ must be methyl or primary.

Examples:

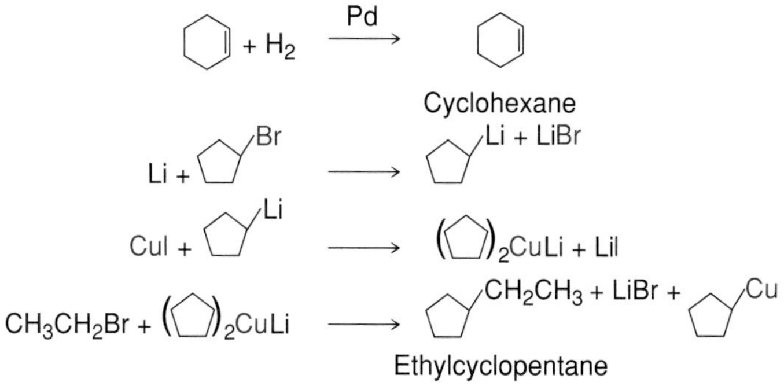

**CYCLOALKANES**

**4.4
CYCLOALKANES:
METHODS OF
PREPARATION**

Cyclopropane can be prepared by a ring-closing reaction. The two bromines are abstracted by the zinc, and the end carbons are bonded to each other.

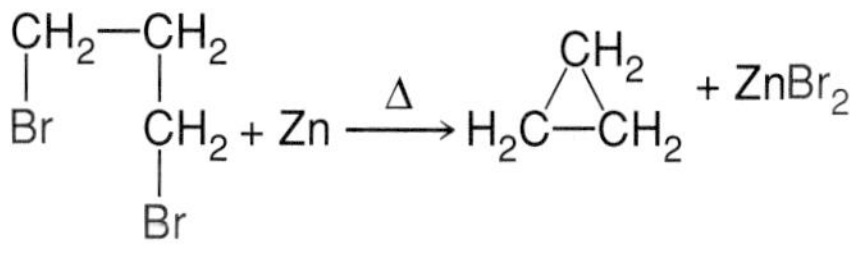

Cyclopropane and cylobutane are very reactive molecules because of the geometry of each of the molecules. Cyclopropane is a planar (flat) molecule (with four $sp^3$ hybridized orbitals surrounding each carbon nucleus) and to be most stable would have tetrahedral bond angles of 109.5 degrees. Since the C—C—C bond angle of cyclopropane (60 degrees) shows a deviation of 49.5 degrees from this optimal bond angle of 109.5 degrees (109.5 − 60 = 49.5 degrees), the ring is very strained, and there is a strong tendency for reactions to induce ring opening to relieve the strain, thus producing a stable tetrahedral molecule (Fig. 4.1).

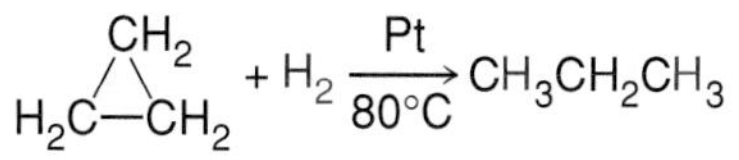

**4.5
RING SHAPE OF
CYCLOALKANES:
A GUIDE TO
CHEMICAL REACTIVITY**

**FIGURE 4.1** Bond angles of cyclobutane and cyclopropane.

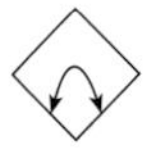

cyclobutane
(90° bond angles)

cyclopropane
(60° bond angles)

**FIGURE 4.2** Note, the puckering in cyclohexane is more pronounced than in cyclopentane.

Cyclopentane

Cyclohexane

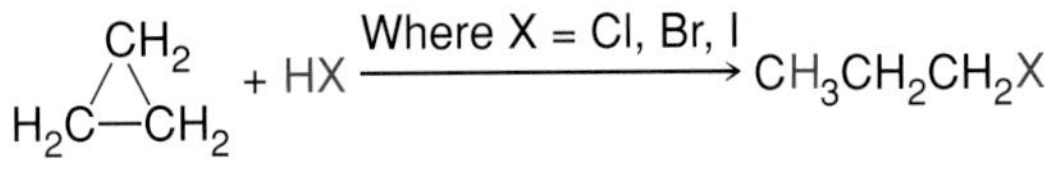

$$\underset{H_2C-CH_2}{\overset{CH_2}{\triangle}} + HX \xrightarrow{\text{Where X = Cl, Br, I}} CH_3CH_2CH_2X$$

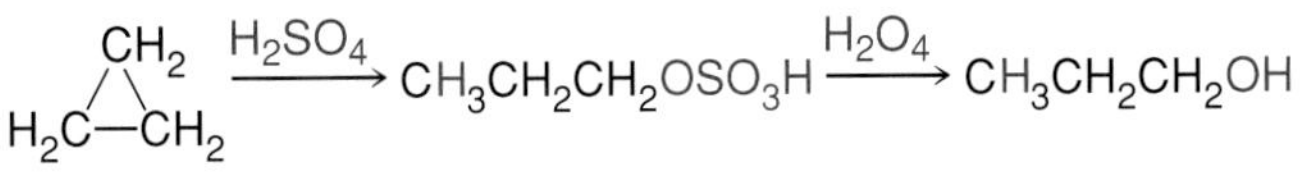

$$\underset{H_2C-CH_2}{\overset{CH_2}{\triangle}} \xrightarrow{H_2SO_4} CH_3CH_2CH_2OSO_3H \xrightarrow{H_2O_4} CH_3CH_2CH_2OH$$

Since the bonds in cyclobutane are less strained than those in cyclopropane, more rigorous conditions are necessary to effect ring opening.

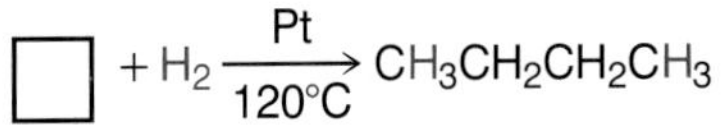

$$\square + H_2 \xrightarrow[120°C]{Pt} CH_3CH_2CH_2CH_3$$

Under the same reaction conditions, cyclobutane reacts with the same reagents as cyclopropane—only much slower.

Cyclopentane and higher homologues have more stable structures because these compounds are not planar but puckered. These puckered structures result in bond angles of 105 degrees in cyclopentane and 109.5 degrees in cyclohexane (Fig. 4.2).

The reaction of cyclopentane and higher cycloalkanes with $Cl_2$ and $Br_2$ is analogous to alkanes because of their strain-free, nonplanar conformations.

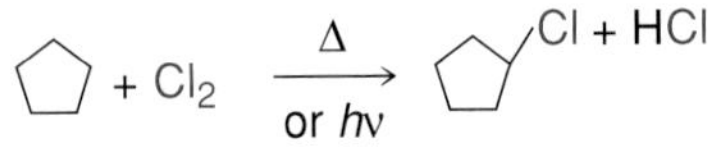

$$\pentagon + Cl_2 \xrightarrow[\text{or } h\nu]{\Delta} \overset{Cl + HCl}{\pentagon}$$

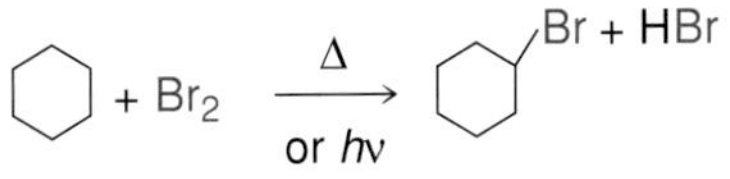

$$\hexagon + Br_2 \xrightarrow[\text{or } h\nu]{\Delta} \overset{Br + HBr}{\hexagon}$$

Cyclopropane was used as an anesthetic in the 1930s. However, its use was soon discontinued when it was realized that a wide range of mixtures of air and cyclopropane would explode.

In general,

$$\text{Cycloalkane} + O_2 \xrightarrow{\Delta} CO_2 + H_2O$$

Example:

$$\triangle + \tfrac{9}{2} O_2 \xrightarrow{\Delta} 3\ CO_2 + 3\ H_2O$$

# ▶ CHAPTER ACCOMPLISHMENTS

### 4.1 Synthesis of Alkanes
☐ Write a general equation illustrating each of the following methods of preparing an alkane.
  a. Hydrogenation of an alkene
  b. Hydrolysis of the Grignard reagent
  c. Corey-House synthesis
☐ Explain why diethyl ether used as a solvent to prepare a Grignard reagent from an alkyl halide must be dry.

### 4.2 Chemical Reactions of Alkanes
☐ Explain why alkanes are also known as paraffins.
☐ Explain why running an automobile in a closed-in garage is dangerous.
☐ Classify a chemical reaction that liberates heat.

☐ Write a general equation for the reaction of an alkane with 1 mol of halogen.
☐ Explain why monochlorination of ethane produces only one isomer of $C_2H_5Cl$, while monochlorination of propane produces two isomers of $C_3H_7Cl$.

### 4.3 Reaction Mechanism
☐ Differentiate structurally between a carbocation, a free radical, and a carbanion.
☐ Use a curved arrow to show the breaking of bond A : B
  a. homolytically.
  b. heterolytically.
☐ Write the equation for the initiation step of the bromination of ethane in the presence of light.

### 4.4 Cycloalkanes: Methods of Preparation
☐ Write an equation showing the reaction of lithium with cyclopentyl bromide.

☐ Identify the natural source of most isolated cycloalkanes.

### 4.5 Ring Shape of Cycloalkanes: A Guide to Chemical Reactivity
☐ Explain why cyclobutane is a more reactive compound than cyclohexane.

☐ Write an equation showing
  a. cyclopentane reacting with an excess of oxygen.
  b. cyclopropane reacting with hydrogen in the presence of platinum at 80°C.
  c. cyclopentane heated with chlorine.

---

### ▶ KEY TERMS

*Corey-House synthesis* (4.1)
*Grignard reagent* (4.1, 14.4, 15.1)
*reduction* (4.1)
*exothermic* (4.2)
*substitution reaction* (4.2)
*carbocation* (4.3, 6.2)

*carbanion* (4.3)
*free radical* (4.3)
*curved arrow symbol* (4.3)
*homolytic bond formation* (4.3)
*heterolytic bond breaking* (4.3)
*heterolytic bond formation* (4.3)

*initiation step* (4.3)
*chain reaction* (4.3)
*propagation step* (4.3)
*termination step* (4.3)
*multiline arrow* (4.3, 8.7)
*endothermic* (Problem 3)

---

### ▶ PROBLEMS

1. Draw a structural formula for the following.
   a. Ethyl carbanion
   b. Isopropyl free radical
   c. Neopentyl carbocation (Refer to Problem 3.10 for a structural formula of neopentane.)

2. Prepare each of the following compounds.
   a. Nonadecane from $CH_2{=}CH(CH_2)_{16}CH_3$ (one-step synthesis)
   b. Butane from $CH_3CH_2CH_2CH_2Br$ (two-step synthesis)
   c. Butane from $C_2H_6$ (multistep synthesis)
      *Note:* Use any inorganic reagents necessary. You need not write inorganic products or organic by-products.

3. Natural gas would be useless as a fuel if the reaction of an alkane with oxygen should be endothermic. Explain. An *endothermic* reaction is a reaction that absorbs energy from the surroundings.

4. Draw a structural formula for every dichloroderivative of:
   a. isobutane.
   b. $(CH_3)_4C$ (neopentane).
   c. *n*-butane.

5. Define or explain each of the following.
   a. Carbocation
   b. Exothermic reaction
   c. Free radical

6. Classify each of the following dissociations as homolytic or heterolytic.
   a. $CH_3{-}O{-}O{-}CH_3 \longrightarrow 2\ CH_3{-}O\cdot$
   b. $I{-}I \longrightarrow 2\ I\cdot$
   c. $H{-}I \longrightarrow H^+ + I^-$

7. In the chlorination of methane, very small amounts of ethane can be formed. Explain using the mechanism for the chlorination of methane.

8. Complete each of the following equations by drawing a structural formula for each organic product and writing a molecular formula or formula unit for each inorganic product formed. Do not balance the equations.

   a. $CH_3CH_2CH_2CH_3 + Cl_2 \xrightarrow{h\nu}$
   b. $C_2H_5MgBr + HCl(aq) \longrightarrow$
   c. $CH_3CH{=}CHCH_3 + H_2 \xrightarrow{Ni}$
   d. $CH_3CH_2Br + Mg \xrightarrow{\text{Anhydrous diethyl ether}}$
   e. $C_2H_5Li + CuI \longrightarrow$
   f. $C_2H_5Br + (CH_3)_2CuLi \longrightarrow$
   g. $CH_3(CH_2)_{10}CH_3 + O_2 \xrightarrow{\text{Limited}}$
   h. $CH_3(CH_2)_{10}CH_3 + O_2 \longrightarrow$
   i. $\underset{\underset{I}{|}}{CH_3CHCH_3} + Li \xrightarrow{\text{Anhydrous diethyl ether}}$

9. Prepare each of the following compounds.
   a. Cyclopentane from cyclopentene (one-step synthesis)

b. Cyclopentane from bromocyclopentane (two-step synthesis)

c. Methylcyclohexane from methyl bromide and cyclohexyl bromide (bromocyclohexane) (multistep synthesis)

   *Note:* Use any inorganic reagents necessary. You need not write inorganic products or organic by-products.

10. Cyclopentylcyclohexane cannot be produced in good yield using the Corey-House synthesis. Why?

11. Complete each of the following equations by drawing a structural formula for each organic product and writing a molecular formula or formula unit for each inorganic product formed. Do not balance the equations.

   a. Cyclopentane + $O_2 \longrightarrow$

   b. Cyclopentene + $H_2 \xrightarrow{\text{Ni}}$

   c. Cyclobutane + $H_2 \xrightarrow[\Delta]{\text{Pt}}$

   d. Cyclohexyl bromide + Li $\xrightarrow{\text{Anhydrous diethyl ether}}$

   e. $BrCH_2CH_2CH_2Br$ + Zn $\xrightarrow{\Delta}$

12. Why is cyclopropane far more reactive than cyclopentane?

13. A chemist reacted cyclopropane with bromine in the presence of ultraviolet light. She was surprised when she isolated two bromine-containing products; one monobromo product and one dibromo product rather than the sole product she expected.

   a. Draw a structural formula for the expected product. *Hint:* Keep in mind the ring-opening tendency of highly strained rings.

   b. Draw a structural formula for the unexpected product. *Hint:* Hydrogen bromide is formed as a by-product.

14. Use of methylchloroform, a dry cleaning fluid, is being phased out leading to a total ban by the year 2005 because it causes deterioration of the ozone layer in the Earth's upper atmosphere.

   a. Draw a structural formula for methyl chloroform.

   b. Name the compound using the IUPAC system.

15. Name each of the following compounds using the common method of nomenclature.

   a. $CH_3CHICH_2CH_3$

   b. $CH_2Br_2$

   c. $(CH_3)_3CCl$

# CHAPTER 5

# Alkenes (I)

## STRUCTURE, NOMENCLATURE, GEOMETRIC ISOMERISM, PHYSICAL PROPERTIES, AND REPRESENTATIVE REACTIONS

Let us now turn our attention from the saturated hydrocarbons we studied in Chaps. 3 and 4 to a class of unsaturated hydrocarbons that contain one or more double bonds. These compounds are called *alkenes* or *olefins*. The most important feature of the typical alkene molecule is the presence of the carbon-carbon double bond:

$$\text{\Large $\diagup$}C{=}C\text{\Large $\diagdown$}$$

The alkene family is unsaturated in that each carbon of the double bond is bonded to only three atoms, as opposed to a carbon in an alkane, which is bonded to the maximum (saturation) number of four atoms (Fig. 5.1).

## 5.1 INTRODUCTION, STRUCTURE, AND CLASSIFICATION: HYDROGEN DEFICIT

They have a generic formula of $C_nH_{2n}$ for a *monoene* with one double bond, $C_nH_{2n-2}$ for a *diene* that contains two double bonds, etc. *Trienes* have three double bonds; *tetraenes*, four.

---

**FIGURE 5.1** A carbon bonded to three atoms is called a trigonal carbon.

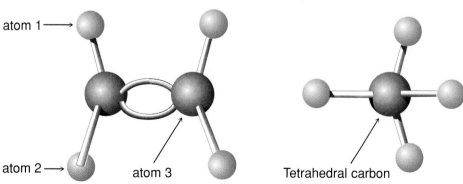

atom 1 →
atom 2 →
atom 3
Tetrahedral carbon

Structures of some simple alkenes are as follows:

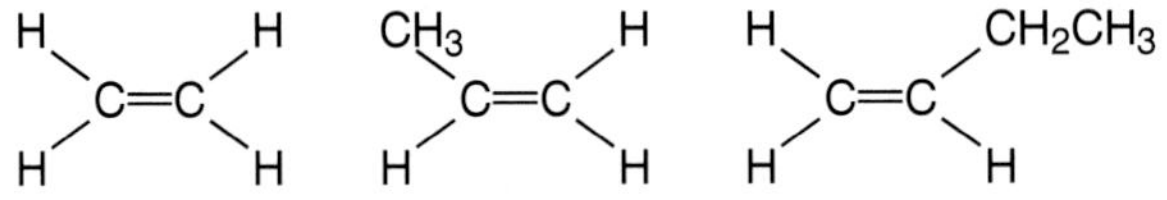

| $C_2H_4$ | $C_3H_6$ | $C_4H_8$ |
| Ethene | Propene | 1-Butene |

There are three classes of dienes. *Isolated dienes* have at least one methylene group ($-CH_2-$) between the two double bonds:

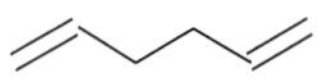

1,5-Hexadiene
(an isolated diene) 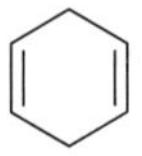                                 1,4-Cyclohexadiene
                                                    (an isolated diene)

*Conjugated dienes* are characterized by alternate double and single bonds:

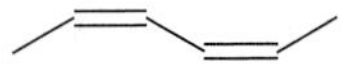

(*Z,Z*)-2,4-Hexadiene
(a conjugated diene)                                 1,3-Cyclohexadiene
                                                    (a conjugated diene)

*Cumulated dienes* contain two double bonds bonded to the same carbon atom. Examples of this class of compound are often unstable and will not be further discussed in this textbook.

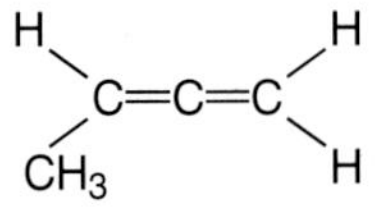

1,2-Butadiene

Some compounds with more than two double bonds are as follows:

$$CH_2{=}CHCH_2CH{=}CHCH_2CH{=}CH_2 \qquad CH_2{=}CH-CH{=}CH-CH{=}CH_2$$

1,4,7-Octatriene
(an isolated triene)                                1,3,5-Hexatriene
                                                    (a conjugated triene)

The generic formula for a monoene is $C_nH_{2n}$.

Note that these polyenes also can be classified as isolated or conjugated in the same way as dienes.

---

## SOLVED PROBLEM 5.1

Note that $C_3H_6$ can represent an alkene ($CH_3CH{=}CH_2$) or a cyclo-alkane, e.g., cyclopropane. In general terms, the loss of two hydrogens (a *two-hydrogen deficit*) from the formula of the corresponding alkane results in the formation of a double bond or a ring.

a.  A compound, $C_5H_8$, is known to be acyclic and to contain one or more double bonds. Based on the formula of the compound, deduce the number of double bonds present.

b.  Let us assume a hydrocarbon, $C_6H_{10}$, can contain one or more double bonds and/or rings. List the possible double-bond/ring combinations possible.

### SOLUTION

a.  An alkane containing five carbons has the formula $C_5H_{12}$. Subtracting $C_5H_8$ from $C_5H_{12}$ gives a four-hydrogen deficit. Since one double bond results from a two-hydrogen deficit, then the structure must contain two double bonds.

b.  With a four-hydrogen deficit ($C_6H_{14} - C_6H_{10}$), the following structural combinations are possible:

One ring and one double bond
Two rings
Two double bonds

**EXERCISE 5.1**

Deduce possible combinations of rings and double bonds present in the following molecular formulas.

a. $C_6H_{12}$

b. $C_7H_{12}$

## 5.2 NOMENCLATURE

All the rules we learned with the alkanes and cycloalkanes regarding IUPAC nomenclature (see Secs. 3.2 and 3.7) also hold with alkenes and cycloalkenes, with just a few changes and additions as follows:

1. The suffix *-ene* is used rather than *-ane*.
2. Number the longest chain containing the carbon-carbon double bond from the end closest to the double bond, with the lower of the two numbers of the carbons of the double bond locating the double bond.
3. In cyclic alkenes, the double-bonded carbons are assumed to be carbons 1 and 2.

   Some examples of alkene nomenclature:

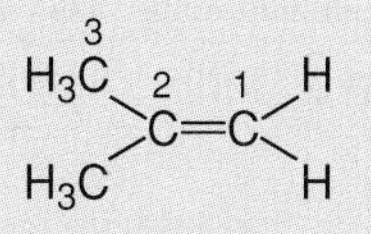

2-Methylpropene

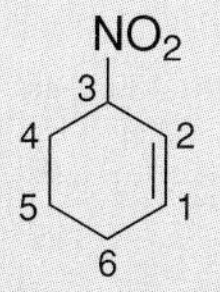

3-Nitrocyclohexene
(The double bond is located on C-1 in such a way as to give any substituent group the lowest possible number)

   Other examples of alkene nomenclature can be seen in Fig. 5.2.

4. Nomenclature for di-, tri- and tetraalkenes is essentially the same as for the monoalkenes. Here, we drop the *-ane* (alkane suffix) and add *-adiene*, *-atriene*, or *-atetraene.* As with monoenes, each double bond is located using the lower of the two numbers of the carbons of the double bond. Two or more numbers are separated by a comma or commas. For example,

$$CH_2{=}CH{-}CH_2{-}CH_2{-}CH_2{-}CH{=}CH_2$$

1,6-Heptadiene

**SOLVED PROBLEM 5.2**

Name the following compounds.

a. $\overset{4}{C}H_3\overset{3}{C}H_2{-}\overset{2}{C}({=}\overset{1}{C}H_2){-}CH_2CH_3$

b. $Cl\overset{5}{C}H_2\overset{4}{C}H_2\overset{3}{C}H{=}\overset{2}{C}H\overset{1}{C}H_3$

c. $Br\overset{1}{C}H_2\overset{2}{C}H{=}\overset{3}{C}H\overset{4}{C}H_3$

**FIGURE 5.2** Some other examples of IUPAC nomenclature of alkenes.

(a) $\overset{1}{C}H_3\overset{2}{C}H{=}\overset{3}{C}H\overset{4}{C}H_2\overset{5}{C}H_3$

2-Pentene

(b) 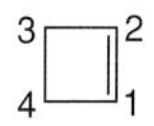

Cyclobutene

(c) $CH_3(CH_2)_{14}\underset{3}{C}H{=}\underset{2}{C}H\underset{1}{C}H_3$

2-Octadecene

(d) $\underset{1}{C}H_3\underset{2}{C}H_2\underset{3}{C}H{=}\underset{4}{C}H(CH_2)_{25}CH_3$

3-Triacontene

**SOLUTION**

a. *Step 1.* Although the longest chain contains five carbon atoms, it does not include the double bond. Thus the four-carbon parent compound containing the double bond is used.

   *Step 2.* Write the name:

2-ethyl-1-butene

b. *Step 1.* We must number here from right to left because the carbon-carbon double bond functional group must be assigned the lower number—the number 2—irrespective of the chlorine being assigned the number 5.

   *Step 2.* Write the name:

5-chloro-2-pentene

c. *Step 1.* When the carbon-carbon double bond is assigned the same number when numbering left to right or right to left, number so that the substituent group on the first branch is assigned the lower number.

   *Step 2.* Write the name:

1-bromo-2-butene

---

**EXERCISE 5.2**

Name each of the following compounds using the IUPAC system of nomenclature.

a. $(CH_3)_2C{=}CHCH_2CH_2CH_3$

b.
$$CH_2{=}C\begin{array}{l} CH_2CH_2CH_3 \\ \\ CH_2CH_2CH_3 \end{array}$$

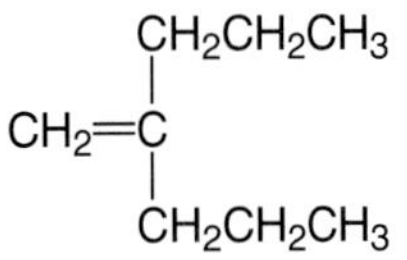

c.
$$CH_3{-}\underset{\underset{CH_3}{|}}{\overset{\overset{CH_3}{|}}{C}}{-}CH_2{-}CH_2{-}CH{=}CH_2$$

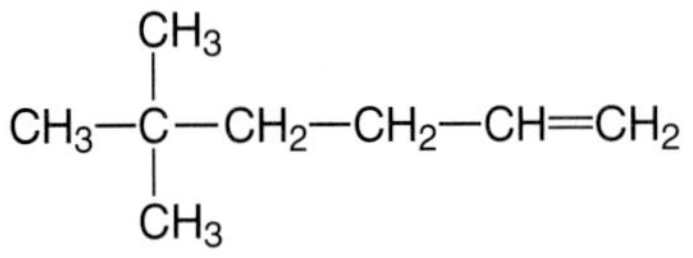

d. $CH_2{=}CH(CH_2)_{23}CH_3$

---

5. The alkene functional group often is found with other functional groups:

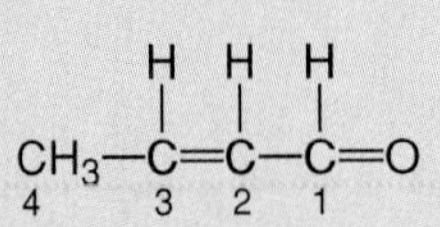

$$CH_3\underset{4}{-}\overset{\overset{H}{|}}{\underset{3}{C}}{=}\overset{\overset{H}{|}}{\underset{2}{C}}{-}\overset{\overset{H}{|}}{\underset{1}{C}}{=}O$$

2-Buten-1-al

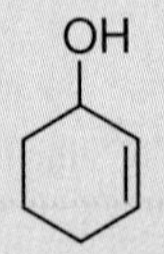

2-Cyclohexen-1-ol

Notice that both the aldehyde group (*-al*) and the alcohol functional group (*-ol*) dominate the carbon-carbon double bond functional group (*-ene*) and thus show the lower number.

---

**SOLVED PROBLEM 5.3**

Name the following compounds, and classify each compound as conjugated or isolated.

a. 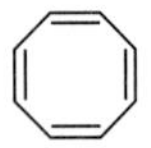     b.

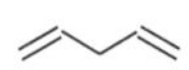

SOLUTION

a. 1,3,5,7-cyclooctatetraene. This is a conjugated tetraene because the compound contains alternating double and single bonds.

b. 1,4-pentadiene. This is an isolated diene because a methylene group is located between the two double bonds.

---

Name and classify, as isolated or conjugated, each of the following compounds.

a. $CH_2=CH-CH_2-CH=CH-CH_2-CH=CH_2$ (neglect geometric isomerism; see Sec. 5.3)

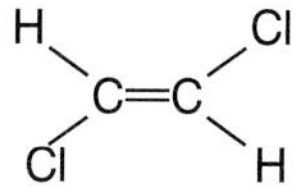

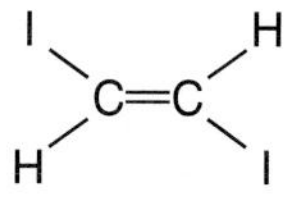

b. (do not name)

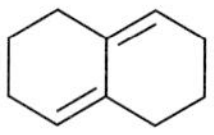

c.

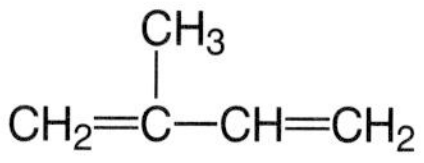

$$CH_3$$
$$|$$
$$CH_2=C-CH=CH_2$$

---

See Appendices 4 and 5 for a discussion of the common system of nomenclature.

## 5.3 GEOMETRIC ISOMERISM: AN EXAMPLE OF STEREOISOMERISM

### The Cis-Trans System

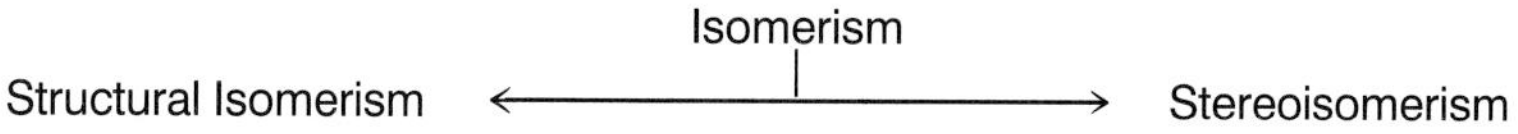

Isomerism

Structural Isomerism ⟵————————⟶ Stereoisomerism

*Stereoisomers* differ from structural isomers (see Sec. 1.13) in that structural isomers have different structural formulas, while stereoisomers have the same general structural formula but different orientations in space. *Geometric* or *cis-trans isomerism* occurs because unlike a carbon-carbon single bond, a carbon-carbon double bond cannot be rotated freely. This results in the formation of two possible isomers for an alkene with the general structures as shown in Fig. 5.3.

---

Name the following compounds.

**FIGURE 5.3** A generalized example of *cis-trans* isomerism.

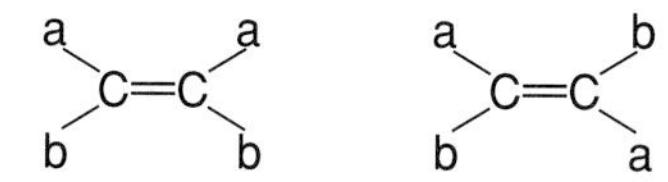

(a) *cis*-isomer          (b) 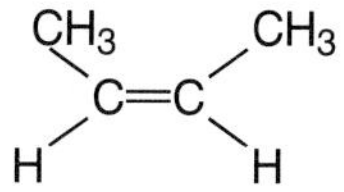 *trans*-isomer

The prefix *cis-* (Latin: *on this side*) refers to a structure that has two identical atoms (groups) on the same side of the molecule; *trans-* (Latin: *across*) refers to a structure that has two identical atoms (groups) on opposite sides of the molecule.

SOLUTION

a. *trans*-1,2-Dichloroethene

b. Since the two methyl groups in the main chain are positioned on the same side of the molecule, the molecule is named *cis*-2-butene.

*Note:* You need to be alert to the fact that the structures of the compounds given require the cis-trans method of specifying configuration to answer this question even though the question gives no hint of this.

---

Name the following compounds.

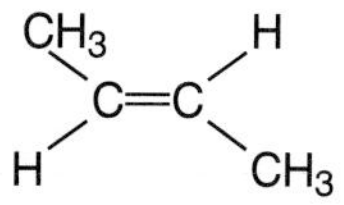

It is important to note that if, for example, you are asked to draw the structural formula for *trans*-1,2-dichloroethene, you should show the two chlorine atoms and the two hydrogen atoms on opposite sides of the molecule, as in Solved Problem 5.4a, or for *cis*-2-butene on the same side of the molecule, as in Solved Problem 5.4b. Some incorrect methods of writing the structure of a geometric isomer are as follows:

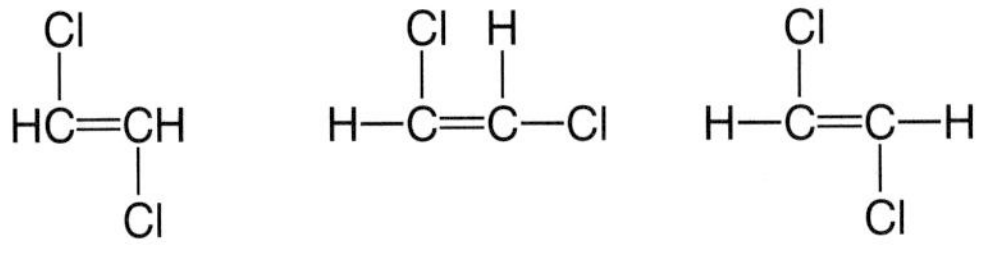

Geometric isomerism can be ruled out only in an alkene where identical atoms (groups) are on a double-bonded carbon. For example, 1,1-dichloroethene cannot show cis-trans isomerism because the two chlorine atoms are bonded to the same carbon of the double bond:

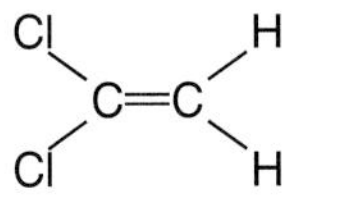

1,1-Dichloroethene

**SOLVED PROBLEM 5.5**

Which of these compounds cannot show cis-trans isomerism?

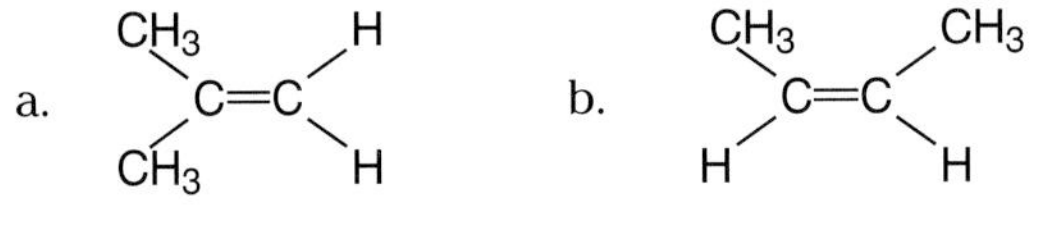

**SOLUTION**

a.  The carbon on the left has two methyl groups bonded to it. Thus cis-trans isomerism is not possible.

b.  Because each carbon of the double bond has a bonded hydrogen atom and methyl group [two different atoms (groups)], cis-trans isomerism exists.

**EXERCISE 5.5**

Determine if the following compounds can exist as cis-trans isomers. (If not, explain why not.)

a.  Propylene

b.  2,3-Dimethyl-2-butene

c.  3-Hexene

d.  1-Butene

e.  Other general examples of compounds that can form cis-trans isomers are

Since two identical atoms (groups) are on different carbon atoms, the *cis*- or *trans*- prefix can be used. For example,

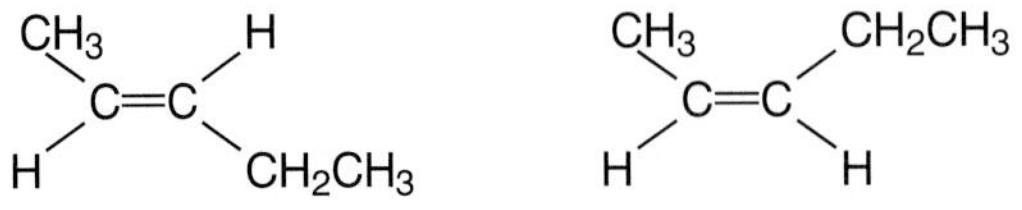

*trans*-2-Pentene                          *cis*-2-Pentene

In *trans*-2-pentene, two hydrogens are on opposite sides of the molecule, while in *cis*-2-pentene, the two hydrogens are on the same side of the molecule. Geometric isomers also can be formed in the following general formula of an alkene:

$$\begin{array}{c} a \\ \diagdown \\ \phantom{x}\;\;C{=}C \\ \diagup \\ b \end{array}\begin{array}{c} d \\ \diagup \\ \\ \diagdown \\ e \end{array}$$

However, the *cis*- or *trans*- prefix cannot be used here because there are not two identical groups (atoms) on different carbons of the double bond. Instead, all four atoms and/or groups are different. To solve this problem, the ambiguous cis-trans method of determining configuration is replaced with the *(Z)-(E) (zusammen-entgegen) method*, which makes use of a system based on priority rankings. The priority rules apply to the four atoms (groups) attached to the doubly bonded carbons.

**The Zusammen-Entgegen System**

   Priority rankings are established as follows:

1. When the two atoms directly attached to a carbon of the double bond are different, the atom of direct attachment to the carbon of the double bond with the higher atomic number shows the higher priority.
   Example:

$$\begin{array}{c}\diagdown\\C{=}C{-}NH_2\\\diagup\end{array}\quad\text{versus}\quad\begin{array}{c}\diagdown\\C{=}C{-}CH_3\\\diagup\end{array}$$

   The —$NH_2$ group shows a higher priority than the —$CH_3$ group because the atomic number of nitrogen (7) is higher than that of carbon (6).

2. If the two atoms directly attached to the carbon of the double bond are the same, then the atomic numbers of the next atom(s) further away from the double bond determine the priority.
   Example:

$$\begin{array}{c}\diagdown\\C{=}C{-}CH_3\\\diagup\end{array}\quad\text{versus}\quad\begin{array}{c}\diagdown\\C{=}C{-}CH_2CH_3\\\diagup\end{array}$$

The —$CH_2CH_3$ group shows the higher priority because the carbon of the methyl group is bonded to three hydrogens, each with an atomic number of 1, while the methylene carbon of the ethyl group is bonded to two hydrogens and a carbon, with an atomic number of 6.

   Once group priorities are established, the following rules are used: When the two groups of higher priority, each on different carbons of the double bond, are on the same side of the molecule, the name of the molecule is prefixed with (Z)- (from the German *zusammen*, "together"). When the groups of higher priority are on opposite sides of the molecule, the designation (E)- (from the German *entgegen*, "across") is used. In general, we have

$$\begin{array}{cc}\underset{2}{\overset{1}{\diagdown}}\;\;C{=}C\;\;\underset{2}{\overset{1}{\diagup}} & \underset{2}{\overset{1}{\diagdown}}\;\;C{=}C\;\;\underset{1}{\overset{2}{\diagup}}\end{array}$$

(Z)-Isomer      (E)-Isomer

where 1 is an atom (or group) with a higher priority and 2 is an atom (or group) with a lower priority.

---

Name the following alkene.

$$\begin{array}{c} Br \\ \diagdown \\ \phantom{x}\;\;\underset{Cl}{\overset{}{C}}\underset{1}{=}\underset{2}{C}\underset{H}{} \end{array}\begin{array}{c} CH_3 \\ \diagup \end{array}$$

**SOLUTION**

Since there are four different atoms attached to the carbons of the double bond, the (*Z*)-(*E*) method of specifying configuration is used.

*Step 1.* On carbon 1, bromine has a higher priority than chlorine because the atomic number of bromine (35) is higher than that of chlorine (17).

*Step 2.* On carbon 2, the carbon of the methyl group has a higher priority than hydrogen because the atomic number of carbon (6) is higher than that of hydrogen (1).

*Step 3.* Since the bromine and the carbon of the methyl group are on the same side of the molecule, the compound is zusammen, (*Z*)-.

*Step 4.* Thus, the name of the compound is

(Z)-1-Bromo-1-chloropropene

*Note:* The prefix *Z* is placed in parentheses in nomenclature.

---

<table>
<tr><td>

**EXERCISE 5.6**

</td><td>

Name the following compounds using the (*Z*)-(*E*) method of specifying configuration.

a.    b.

</td></tr>
</table>

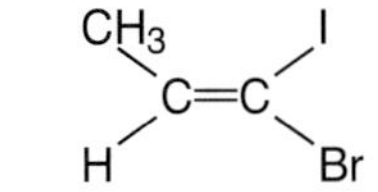

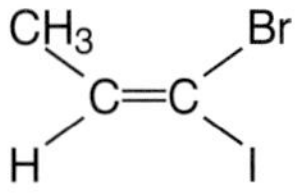

---

## Cis-Trans Isomerism in Cycloalkanes

Just as alkenes show cis-trans isomerism, cycloalkanes also can exist as cis-trans isomers. Cis-trans isomerism occurs in certain cycloalkanes because of restricted rotation around carbon-carbon single bonds in the rigid cyclic structure. In order to exist as cis-trans isomers, each carbon of two adjacent carbons in the cycloalkane ring must be bonded to two different atoms and/or groups, but at least one atom or group on each of the two carbons must be identical. Some examples are as follows:

**FIGURE 5.4** For convenience, we disregard the ring puckering in drawing the structures of *cis*- and *trans*-cyclopentane and cyclohexane derivatives.

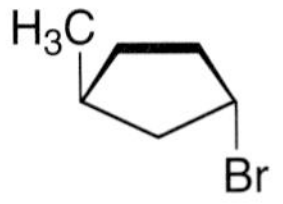

cis-1,2-Dichlorocyclobutane        trans-1,2-Dichlorocyclobutane

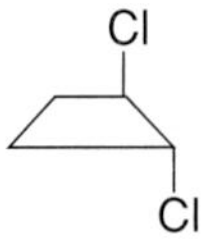

*trans*-1-Bromo-3-methylcyclopentane

Further examples can be seen in Fig. 5.4. Cis and trans disubstituted cyclohexanes can be represented more accurately using the chair conformation of cyclohexane:

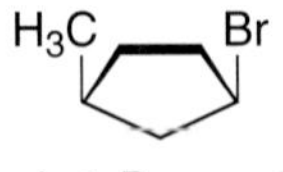

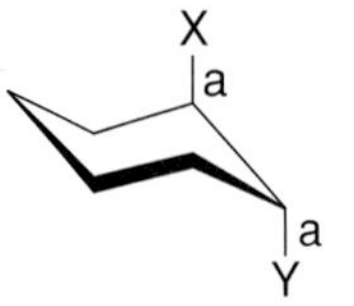

*cis*-1-Bromo-3-methylcyclopentane

1,2-Diaxial      1,2-Equatorial, axial      1,2-Diequatorial<br>(trans)           (cis)                  (trans)

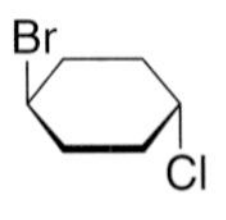

*trans*-1-Bromo-4-chlorocyclohexane

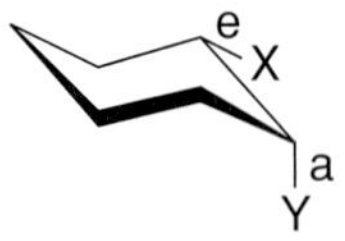

In most cases, the 1,2-diaxialcyclohexane conformation is less stable than the 1,2-diequatorial conformation. This is due to destabilizing 1,3-diaxial interactions between th/e two axial hydrogens and X (or Y):

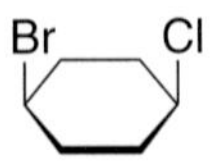

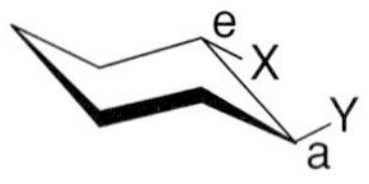

*cis*-1-Bromo-4-chlorocyclohexane

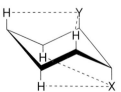

These destabilizing interactions are present because X and Y in axial positions are so close to the two axial hydrogens on the same side of the cyclohexane ring that repulsive London forces come into play (see Sec. 3.10). For example, *trans*-1,2-dimethyl-cyclohexane consists almost entirely of the diequatorial conformation (Fig. 5.5). Notice that the substituents in a cis conformation are oriented to the same side of the molecule, while in a trans conformation, the substituents are directed to opposite sides of the molecule. Steric orientations of 1,3-disubstituted and 1,4-disubstituted stereoisomers are as follows:

1,3-Diequatorial
(*cis*)

1,3-Axial-equatorial
(*trans*)

1,3-Diaxial
(*cis*)
Less stable

1,4-Diequatorial
(*trans*)

1,4-Axial-equatorial
(*cis*)

1,4-Diaxial
(*trans*)
Less stable

**FIGURE 5.5** Diaxial and diequatorial conformations of *trans*-1,2-dimethylcyclohexane.

1,2-Diaxial methyl groups
(*trans*)

1,2-Diequatorial methyl groups
(*trans*)

---

Draw two conformations for each of the following compounds. (Which conformation is more stable?)

a. *trans*-1,2-Diethylcyclohexane

b. Ethylcyclohexane

**SOLVED PROBLEM 5.7**

### SOLUTION

a. *Trans*-1,2-Disubstituted cyclohexanes can exist in the diaxial or diequatorial conformation. The diequatorial conformation is more stable.

b. Any monosubstituted cyclohexane can exist in either an axial or an equatorial conformation. The equatorial conformation is more stable.

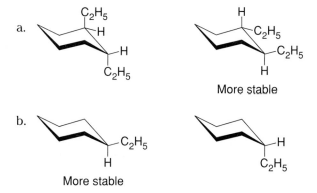

a.

More stable

b.

More stable

| EXERCISE 5.7 |
|---|

Identify the following conformations as cis or trans and as a,a (diaxial); a,e; or e,e (diequatorial).

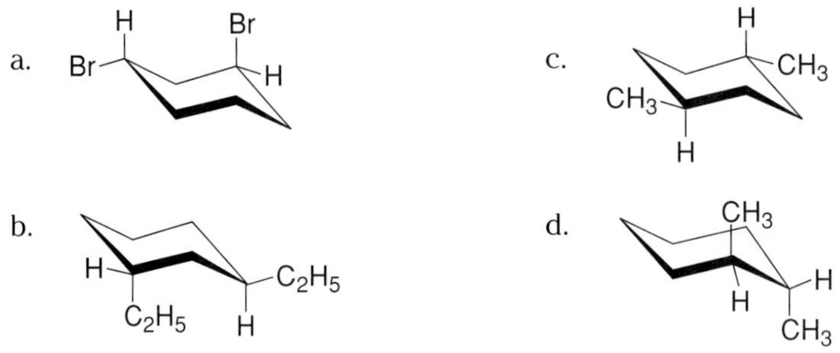

a.

c.

b.

d.

## 5.4 PHYSICAL PROPERTIES

The physical properties of alkenes are similar to those of the alkanes (see Sec. 3.3). The alkenes, like the alkanes, are nonpolar molecules, insoluble in polar water and soluble in nonpolar pentane and carbon tetrachloride.

Boiling points are comparable; they increase with increasing molecular weight and decrease with increasing branching for both alkanes and alkenes. Like alkanes, alkenes are less dense than water. As the double bond is located further inside the carbon chain, the boiling point and density values usually increase.

For example, consider the data in Table 5.1.

As the double bond is located further within the carbon chain, the molecules become more symmetrical. Thus, more molecules of a given alkene can be placed in a set volume, resulting in a higher density. The boiling point of an internally double-bonded alkene is higher than that of a corresponding 1-alkene because the internally bonded alkenes stack up better in the liquid state than the 1-alkene, since the internally bonded alkenes are more symmetrical. Therefore, intermolecular London forces are greater between the internally double-bonded alkene molecules, and more energy is necessary to separate the molecules, resulting in a higher boiling point.

| SOLVED PROBLEM 5.8 |
|---|

Predict which compound of the following pairs boils at a higher temperature. Explain your prediction.

a.   1-Octene or *cis*-3-octene

b.   1-Hexene or 2-methyl-1-pentene

**SOLUTION**

a.   Both alkene isomers are unbranched. Thus the higher-boiling alkene can be determined on the basis of the location of the carbon-carbon double bond. Since both

| TABLE 5.1 | A Comparison of Boiling Point and Density Data for Selected Isomeric Alkenes | |
|---|---|---|
| **Alkene** | **Boiling Point (°C)** | **Density (g/ml)** |
| 1-Pentene | 30.0 | 0.641 |
| *cis*-2-Pentene | 36.9 | 0.656 |
| *trans*-2-Pentene | 36.4 | 0.648 |
| 1-Hexene | 63.7 | 0.673 |
| *cis*-2-Hexene | 68.8 | 0.687 |
| *trans*-2-Hexene | 67.9 | 0.678 |
| 1,4-Pentadiene | 26.0 | 0.661 |

alkenes show limited free rotation due to a carbon-carbon double bond, the more symmetrical alkene, *cis*-3-octene, stacks up better in the liquid state. This results in greater intermolecular London forces, which require more energy to effect separation of molecules, and thus *cis*-3-octene boils at a higher temperature.

b. Both isomers are 1-alkenes. Thus the higher boiling point can be determined on the basis of branching. Like alkanes, the unbranched alkene, 1-hexene, is a less spherical compound than 2-methyl-1-pentene. This causes an increase in London intermolecular forces and thus an increased boiling point.

---

**EXERCISE 5.8**

Predict which compound of the following pairs boils at a higher temperature. Explain your prediction.

a. 1-Heptene or *cis*-3-heptene

b. 1-Heptene or 2-methyl-1-hexene

c. 1-Pentene or 1-hexene (Use Table 5.1 for help.)

d. 1-Heptene or 1-octene

e. *cis*-2-Heptene or *trans*-2-heptene (*Hint:* Which is the more symmetrical alkene?)

---

## 5.5 ALKENES IN BIOLOGICAL SYSTEMS

A significant number of insect pheromones contain the double bond along with one or more other functional groups in a long chain. For example, gyplure is the sex attractant of the gypsy moth, while citronellal is an alarm pheromone for the smaller yellow ant, *Acanthomyops claviger* (Fig. 5.6).

Some pheromones contain the carbon-carbon double bond as the only functional group. For example, the staphylinid beetle (*Aleochara curtula*) secretes 1-undecene and *cis*-4-tridecene, two major components of a combined defensive secretion and mating stimulant, while *cis*-9-tricosene represents the sexual lure of the female common housefly (Fig. 5.7a).

Dienes also act as pheromones. For example, (*E,Z*)-10,12-hexadecadien-1-ol represents the sexual allure of the female silkworm moth (*Bombyx mori*) (Fig. 5.7b).

Ethylene (ethene) has been shown to be an important plant hormone. A *hormone* is a substance, formed in very small quantities (5 μg/kg of plant material is typical), that is produced in one part of a plant and has a specific effect on the cells in another part of the plant. Ethylene (ethene) regulates plant growth and fruit ripening and, in some cases, controls the sex of flowers of certain plants.

## 5.6 REPRESENTATIVE REACTIONS OF THE ALKENES

### Addition

The double bond consists of a sigma bond (*sp2*—*sp2*) and a pi bond (*p*—*p*). Orbital overlap in the pi bond is less than that in the sigma bond. Therefore, the pi bond is a

---

**FIGURE 5.6** Selected pheromones that contain the carbon-carbon double bond with at least one additional functional group.

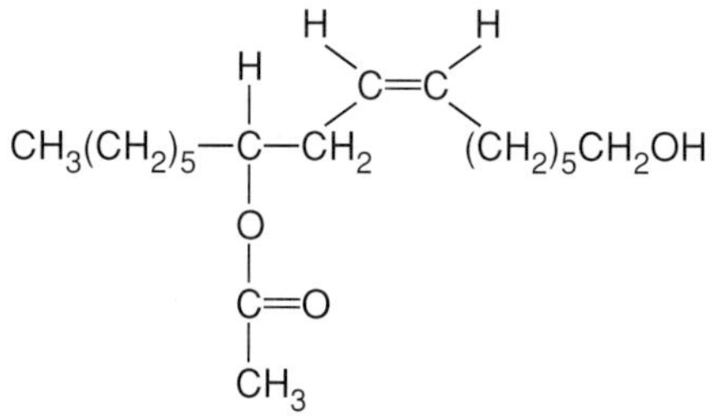

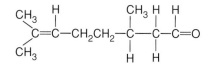

(a) 10-Acetoxy-*cis*-7-hexadecen-1-ol (gyplure)

(b) 3,7-Dimethyl-6-octenal (citronellal)

Note, the numbering (-6-octenal) locates the aldehyde group on carbon 1 and results in the greater priority of the aldehyde group over the alkene.

**FIGURE 5.7** Selected pheromones that contain the carbon-carbon double bond.

1-Undecene

cis-4-Tridecene

cis-9-Tricosene

(a)

(E,Z)-10,12-Hexadecadien-1-ol

Note, the numbering (-1-ol) locates the alcohol on carbon 1 and results in the greater priority of the alcohol group over the alkene.

(b)

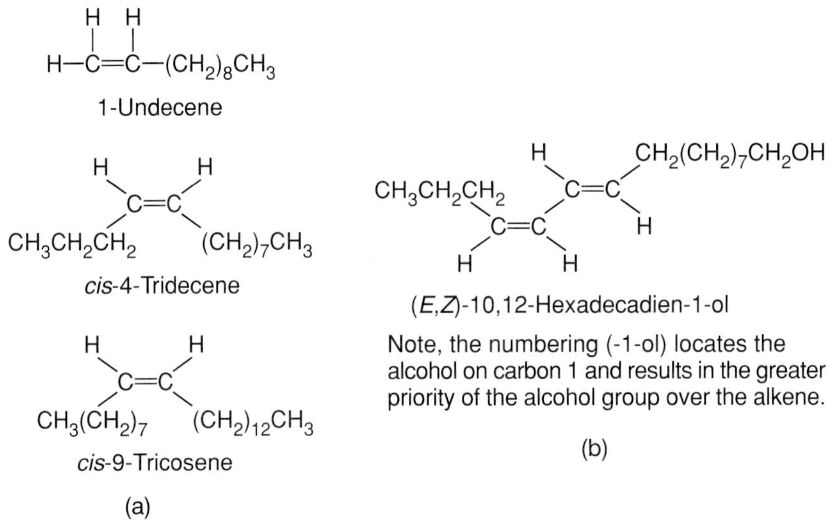

weaker bond than the sigma bond, and therefore it follows that the weaker pi bond is broken during a chemical reaction. Thus,

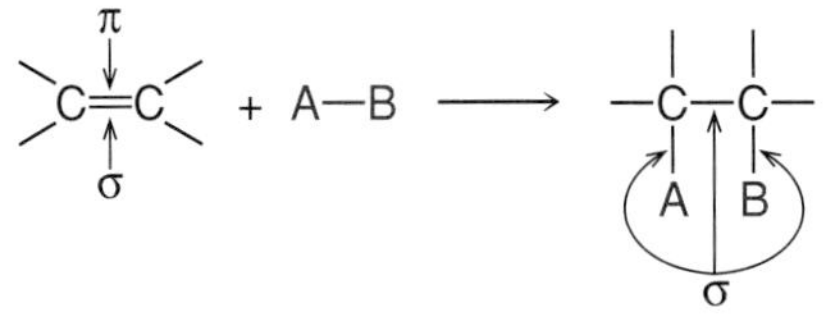

Since the units of A—B are added across the double bond, this type of reaction is known as an *addition reaction*. A significant number of reactions of the alkenes are addition reactions.

**Hydrogenation**   In the presence of a catalytic amount of platinum, paladium, or nickel, an alkene reacts with hydrogen to produce an alkane (see Sec. 4.1). In general,

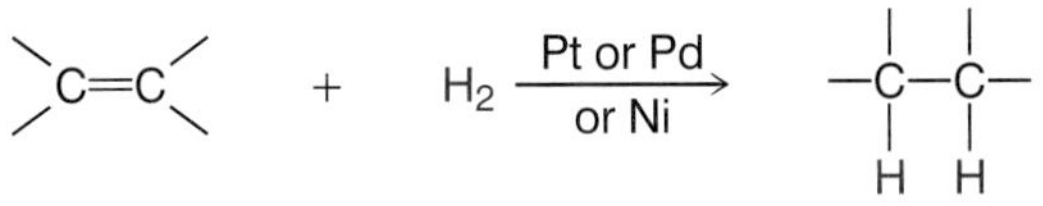

Example:

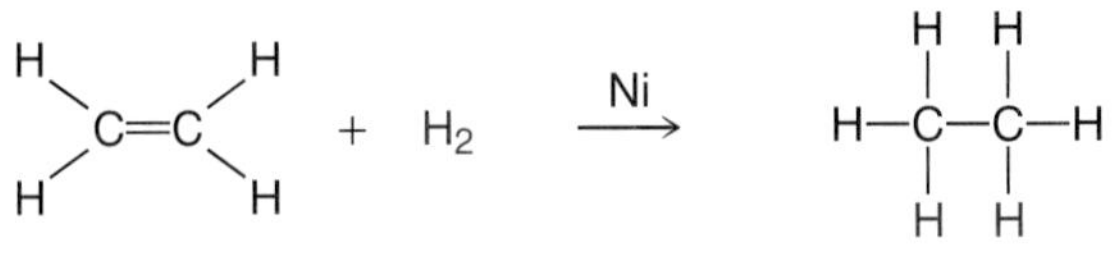

The net effect of the reaction is the breaking of one bond of the carbon-carbon double bond, followed by the substitution of a bond and a hydrogen on each carbon of the double bond.

**Addition of Sulfuric Acid—Hydration**   When an alkene is mixed with cold, concentrated sulfuric acid, the mixture warms and a vigorous chemical reaction takes place. In effect, a hydrogen adds across one carbon of the double bond, while the moiety $OSO_3H$ adds across the other carbon of the double bond. In general,

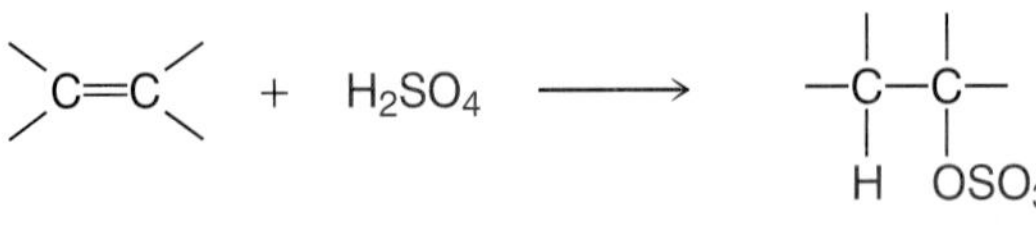

An alkyl hydrogen sulfate

The ripening of many familiar fruits such as apples and bananas is controlled by ethylene production within the fruit. This ethylene production leads to techniques in the storage of fruit that decrease the concentration of ethylene around the fruit and therefore increase the storage life of the fruit. For example, since ethylene production is directly proportional to temperature, fruits such as strawberries, cantaloupes, oranges, tomatoes, and others are chilled to prolong shelf life.

Another storage technique is to remove the ethylene around the fruit by flushing the fruit with a gas such as carbon dioxide or nitrogen. Bananas cannot tolerate cold, so ripening is delayed by storing in an atmosphere of 3% to 5% carbon dioxide. Also, storage in polyethylene bags is a useful technique because the carbon dioxide concentration increases and oxygen decreases inside. This slows down the metabolic processes in the fruit and thus decreases ethylene production. In time, however, when the ethylene concentration increases, these bags must be punctured to permit the ethylene to escape.

An additional method is to store the fruit in bags filled with an ethylene absorber. A chemical ethylene absorber is made by preparing a saturated solution of potassium permanganate (see Sec. 6.2) that is absorbed on alumina (aluminum oxide). This technique is effective in decreasing the yellowing of stored cucumbers and lengthening their storage time. Another advantage of using a permanganate absorber is that the permanganate also reacts with other polluting substances in air such as hydrogen sulfide, sulfur dioxide, and ozone. Finally, bananas and about 50% of the apples harvested are stored in an atmosphere with a decreased concentration of oxygen (5% to 10%). Apples can be stored for as long as 6 months using this technique. It is interesting to note that none of these techniques is effective in storing peaches.

Sometimes, rather than decreasing the concentration of ethylene around the fruit, ethylene is deliberately added to fruit. For example, tomatoes often are harvested at the green stage. After being shipped, the fruit is deliberately sprayed with ethylene to bring on the characteristic red-ripe color, although flavor and firmness are sacrificed because these shipped green tomatoes are not ripened on the vine.

Example:

$$CH_2{=}CH_2 \; + \; H_2SO_4 \; \longrightarrow \; \underset{\underset{\displaystyle H \quad OSO_3H}{|\qquad|}}{CH_2{-}CH_2}$$

Ethyl hydrogen sulfate

Alkyl hydrogen sulfates are hydrolyzed by heat to form alcohols and regenerate sulfuric acid. Note that the symbol for heat is the capital Greek letter delta ($\Delta$). Here we have an OH substituting for an $OSO_3H$ in the organic structure. In general,

$$\underset{\underset{\displaystyle H \quad OSO_3H}{|\quad|}}{-\overset{|}{C}-\overset{|}{C}-} \; + \; H_2O \; \xrightarrow{\Delta} \; \underset{\underset{\displaystyle H \quad OH}{|\quad|}}{-\overset{|}{C}-\overset{|}{C}-} \; + \; H_2SO_4$$

Example:

$$CH_3CH_2OSO_3H \; + \; H_2O \; \xrightarrow{\Delta} \; CH_3CH_2OH \; + \; H_2SO_4$$
Ethanol
(ethyl alcohol)

If both reactions are combined, the net effect is to add the units of water (H and OH) across the carbon-carbon double bond. In general,

$$\overset{\diagdown}{\underset{\diagup}{}}C{=}C\overset{\diagup}{\underset{\diagdown}{}} \quad \xrightarrow[\text{2. } H_2O, \, \Delta]{\text{1. } H_2SO_4} \quad \underset{\underset{\displaystyle H \quad OH}{|\quad|}}{-\overset{|}{C}-\overset{|}{C}-}$$

Example:

$$\underset{H}{\overset{H}{\diagdown}}C{=}C\underset{H}{\overset{H}{\diagup}} \quad \xrightarrow[\text{2. } H_2O, \, \Delta]{\text{1. } H_2SO_4} \quad H{-}\underset{\underset{\displaystyle H \quad OH}{|\quad|}}{\overset{\overset{\displaystyle H \quad H}{|\quad|}}{C}{-}C}{-}H$$

A number of different alkenes under special experimental conditions can link themselves to form giant molecules consisting of many repeating units of the original alkene. The original alkene is known as a *monomer*, while the giant molecule is called a *polymer*. In general,

**Polymerization**

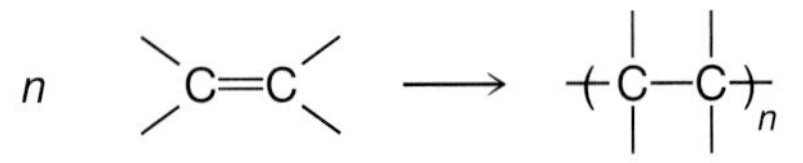

$$n \quad \diagdown C = C \diagdown \quad \longrightarrow \quad \left( C - C \right)_n$$

where $n$ is a large number.

Example:

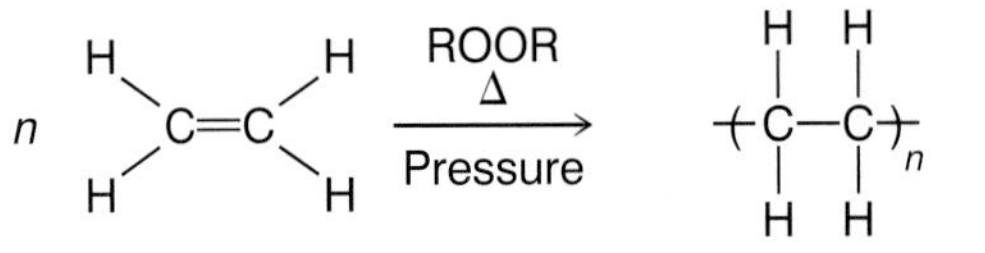

$$n \quad \underset{H}{\overset{H}{\diagdown}} C = C \underset{H}{\overset{H}{\diagup}} \quad \xrightarrow[\text{Pressure}]{\overset{\text{ROOR}}{\Delta}} \quad \left( \overset{H}{\underset{H}{\overset{|}{C}}} - \overset{H}{\underset{H}{\overset{|}{C}}} \right)_n$$

Ethylene                                    Polyethylene

You can draw an expanded structural formula for polyethylene as follows:

$$-CH_2CH_2CH_2CH_2CH_2CH_2CH_2CH_2CH_2CH_2-$$

Just picture several hundred to several thousand methylene units extending from each dash in the structure, and that would essentially complete the structure of polyethylene. Polyethylene is heavily used in packaging, cable insulation, and bottles.

To formulate a convenient abbreviated structural formula for a given polymer from the corresponding monomer, polypropylene, for example, first break one bond of the carbon-carbon double bond contained in propylene:

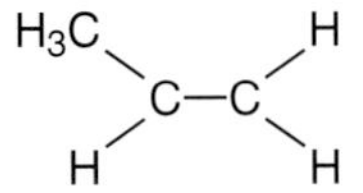

$$\underset{H}{\overset{H_3C}{\diagdown}} C - C \underset{H}{\overset{H}{\diagup}}$$

Second, extend a bond to the left from the trigonal carbon on the left and a bond to the right from the trigonal carbon on the right:

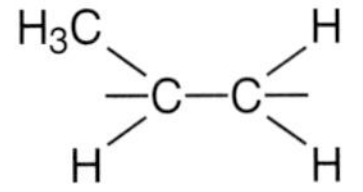

$$\underset{H}{\overset{H_3C}{\diagdown}} {=} C - C {=} \underset{H}{\overset{H}{\diagup}}$$

Third, place a curved line around each extended bond and incorporate the subscript $n$ adjacent to the curved line on the right to produce the final structure of polypropylene:

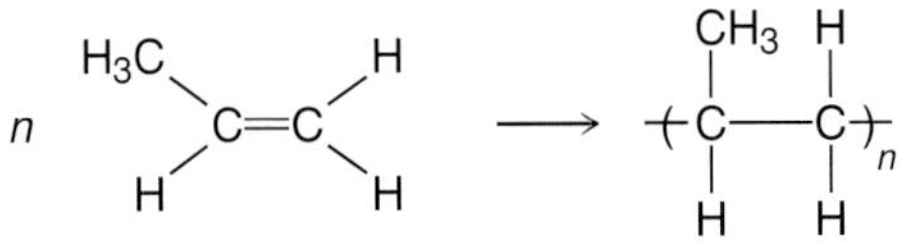

$$n \quad \underset{H}{\overset{H_3C}{\diagdown}} C = C \underset{H}{\overset{H}{\diagup}} \quad \longrightarrow \quad \left( \overset{CH_3}{\underset{H}{\overset{|}{C}}} - \overset{H}{\underset{H}{\overset{|}{C}}} \right)_n$$

Propylene          Polypropylene

Polypropylene can be found in carpet fibers and in packaging material. In Chap. 6, we will discuss the structure and uses of some other frequently used polymers.

**SOLVED PROBLEM 5.9**

Complete the following equations by drawing a structural formula for each organic product produced.

a.   $CH_3(H)C{=}C(H)CH_3 + H_2 \xrightarrow{\text{Ni}}$

    Ignore cis-trans isomerism.

b.   $n \, H_2C{=}CHCl \xrightarrow{\text{Polymerization}}$

**SOLUTION**

a.   The first step in any addition reaction involving an alkene is to break the weaker pi bond of the double bond to get

$$CH_3(H)C-C(H)CH_3$$

Finally, we add a bond and a hydrogen to each trigonal carbon to formulate the product

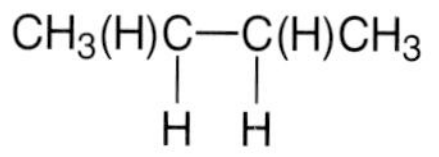

$$CH_3(H)C—C(H)CH_3$$
$$\quad\ \ |\quad\ \ |$$
$$\quad\ \ H\quad H$$

b.  As the first step to formulate a polymer from the corresponding monomer, we break one bond of the double bond to get

$$H_2C—CHCl$$

The second step is to extend a bond to the left from the left trigonal carbon and to the right from the right trigonal carbon to produce

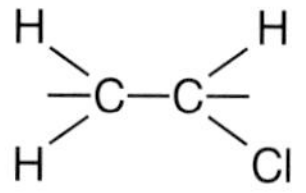

Finally, place a curved line around each extended bond and incorporate the *n* as a subscript adjacent to the curved line on the right to give

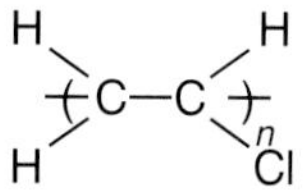

---

| | |
|---|---|
| Complete the following equations by drawing a structural formula for each organic product. | **EXERCISE 5.9** |

a.  cyclopentene + $H_2$ $\xrightarrow{Ni}$

b.  $n\ CH_2=CCl_2$ $\xrightarrow{Polymerization}$

c.  $CH_3(H)C=C(H)CH_3$ $\xrightarrow[\text{2. } H_2O,\ \Delta]{\text{1. } H_2SO_4}$

 Ignore cis-trans stereoisomerism.

d.  $CH_3(H)C=C(H)CH_3 + H_2SO_4 \longrightarrow$

 Ignore cis-trans stereoisomerism.

e.  $CH_3CH_2CH_2OSO_3H + H_2O \xrightarrow{\Delta}$

---

## ▶ CHAPTER ACCOMPLISHMENTS

**5.1 Introduction, Structure, and Classification: Hydrogen Deficit**

☐ Draw a structural formula for (a) a five-carbon isolated diene and (b) a six-carbon conjugated triene.

☐ Explain the significance of a hydrogen deficit in a molecule.

☐ Determine the molecular formula of a seven-carbon acyclic hydrocarbon containing three double bonds.

**5.2 Nomenclature**

☐ Draw a geometric condensed structural formula of (a) propene and (b) 2-methylpropene.

☐ Explain why 2-buten-4-al is an incorrect IUPAC name.

☐ Draw a geometric condensed structural formula for 1,5-hexadiene.

**5.3 Geometric Isomerism: An Example of Stereoisomerism**

☐ Explain why alkenes can show geometric isomerism but acyclic hydrocarbons cannot.

□ Draw a general formula for (a) a *cis*-alkene and (b) a *trans*-alkene.

□ Supply the structural formula of an alkene in which the cis-trans method of stating configuration fails and the zusammen-entgegen method must be used instead.

□ Draw a chair conformation of (a) *cis*-1,2-dimethylcyclohexane, (b) *trans*-1,2-dimethylcyclohexane (the more stable conformation), and (c) *trans*-1,2-dimethylcyclohexane (the less stable conformation).

### 5.4 Physical Properties

□ Explain why an internally double-bonded alkene is denser and has a higher boiling point than the corresponding 1-alkene.

□ Predict the solubility of an alkene in water.

### 5.5 Alkenes in Biological Systems

□ Draw a structural formula for the pheromone *cis*-9-tricosene secreted by the female brown housefly.

□ Name an important plant hormone.

### 5.6 Representative Reactions of the Alkenes

□ Write a general equation to show the addition of a reagent, A—B, to an alkene:

$$\ce{>C=C<}$$

□ List three metal catalysts used in the hydrogenation of an alkene.

□ Complete the following equation:

$$H_2C=CH_2 \quad + \quad \xrightarrow[\text{2. } H_2O]{\text{1. } H_2SO_4} \text{(organic product only)}$$

□ Write an equation showing the polymerization of ethylene.

---

► **KEY TERMS**

---

*alkene* (5.1)
*olefin* (5.1)
*monoene* (5.1)
*diene* (5.1)
*triene* (5.1)
*tetraene* (5.1)

*isolated diene* (5.1)
*conjugated diene* (5.1)
*cumulated diene* (5.1)
*hydrogen deficit* (5.1)
*stereoisomers* (5.3)
*geometric or cis-trans isomerism* (5.3)

*(Z)-(E) zusammen-entgegen method of specifying configuration* (5.3)
*hormone* (5.5)
*addition reaction* (5.6, 6.2)
*monomer* (5.6, 6.3)
*polymer* (5.6, 6.3)

---

► **PROBLEMS**

---

1. Classify each of the following compounds as isolated or conjugated.

   a.           b.

   c. $$\ce{CH3\underset{CH3}{C}=\overset{H}{C}-CH2-\overset{H}{C}=\underset{CH3}{C}CH3}$$

   d. $$\ce{H-C#C-\overset{H}{C}=CH2}$$

   e. $$\ce{H2C-\overset{H}{C}=\overset{H}{C}-\overset{H}{C}=CH2}$$

   f. $$\ce{H2C=\overset{H}{C}CH2CH2\overset{H}{C}=O}$$

   g. ⟋⌒⌒⟍

2. A compound, $C_6H_6$, can contain one or more double bonds and/or rings.

   a. List the possible double-bond/ring combinations possible.

   b. Draw a structural formula for each combination.

3. Deduce a molecular formula for a hydrocarbon that contains seven carbons with

   a. two double bonds and one ring.

   b. one double bond and two rings.

   c. three double bonds.

4. Deduce a molecular formula for a hydrocarbon that contains four carbons with

   a. two double bonds.

   b. one ring.

5. Draw a structural formula for a compound with each structure combination in Solved Problem 5.1b.

6. a. Name each of the following compounds by the IUPAC system. Omit geometric isomers.

   (1) $$\ce{CH3CH2\underset{CH3CH2}{C}=\overset{CH2CH3}{C}CH2CH3}$$

(2) 

$$\underset{CH_3}{\overset{CH_3}{>}}C=C\underset{H}{\overset{CH_3}{<}}$$

(3)

$$\underset{F}{\overset{Cl}{>}}C=C\underset{I}{\overset{CH_3}{<}}$$

(4)

$$\underset{H}{\overset{CH_3}{>}}C=C\underset{H}{\overset{CH_2CH_3}{<}}$$

(5)

(6) $CH_3$

(7)

(8)

(9)

$$\underset{H}{\overset{CH_3CH_2}{>}}C=C\underset{CH_2}{\overset{H}{<}}$$
$$CH_2$$
$$CH$$
$$CH_3CH_2CH_2CH_2CH_2CH_2\ CH_3$$

b. A peach aphid secretes an alarm pheromone when threatened. This alarm pheromone is 3-methylene-7,11-dimethyl-1,6,10-dodecatriene. Given that the methylene group is —CH₂—, and that the methylene carbon is attached to carbon 3 by means of a double bond

$$\underset{-C-}{\overset{CH_2}{\|}}$$

draw a structure formula for this alarm pheromone.

7. Draw a structural formula for each of the following compounds. (Omit geometric isomers.)
   a. 2,3-Dimethylcyclohexene
   b. Vinyl bromide (App. 5)
   c. 3-Hexadecene
   d. 2,3-Dimethyl-1-hexene
   e. 3-Methylcyclopentene
   f. Allyl iodide (App. 5)
   g. 1-Docosene

8. Each of the following IUPAC names is incorrect. Write a structure and name correctly.
   a. 2-Vinylbutane (App. 5)
   b. 2-Propene
   c. 5-Methylcyclopentene

d. 2-Propyl-2-butene

9. Label each hydrogen in the following compounds as primary, secondary, tertiary, vinylic, or allylic (App. 5).
   a. Ethylene
   b. 2-Butene
   c. 1-Methylcyclopentene

10. Which of the following compounds can show geometric isomerism?
   a.

   b. $NO_2$ H, Br

   c. 

   $$\underset{CH_3}{\overset{CH_3}{>}}C=C\underset{H}{\overset{CH_3}{<}}$$

   d. 

   $$\underset{Br}{\overset{Cl}{>}}C=C\underset{H}{\overset{CH_2CH_3}{<}}$$

   e. 

   $$\underset{H}{\overset{CH_3}{>}}C=C\underset{CH_2CH_3}{\overset{NO_2}{<}}$$

11. Name each of the compounds that shows geometric isomerism in Problem 10.

12. Draw a structural formula as in Exercise 5.4 for each of the following compounds.
   a. *trans*-3-Hexene
   b. *cis*-3-Hexene
   c. *trans*-3,4-Dichloro-3-hexene

13. Determine which compounds in Problem 6a exhibit geometric isomerism, and then name each as a geometric isomer.

14. Draw a structural formula for
   a. (*Z*)-2-chloro-1-fluoropropene.
   b. (*E*)-3-methyl-2-pentene.
   c. (*Z*)-2-bromo-2-butene.
   d. (*Z*)-1,2-dibromopropene.
   c. (*E*)-1,3-dichloropropene.
   f. (*Z*)-1-chloro-2-methyl-1-butene.

15. Draw a structural formula of each of the following compounds.
   a. (*Z,E*)-2,4-Hexadiene
   b. (*E,E*)-2,4-Hexadiene
   c. 1,4-Cyclooctadiene

16. Disubstituted cycloalkanes, along with alkenes, show cis-trans isomerism.
   a. Why?
   b. Name each of the following.

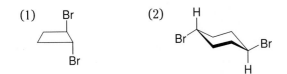

(1)    Br         (2)

c. Draw a structural formula for each of the following.
   (1) *trans*-1,2-Dibromocyclopentane
   (2) *cis*-1-Chloro-2-methylcyclopropane

17. Draw a chair conformational structure(s) for each of the following.
    a. *trans*-1,2-Dichlorocyclohexane
    b. *trans*-1,3-Dichlorocyclohexane
    c. *cis*-1,4-Dichlorocyclohexane

18. Identify each of the following conformations as cis or trans and as a,a (diaxial), a,e, or e,e (diequatorial).
    a.
        Cl          Cl

    b.
               Br
       Br

    c.
               CH$_3$
            CH$_3$

19. Which compound of the following pairs of compounds boils at a higher temperature? Explain your answers.
    a. *cis*-2-Hexene or *cis*-3-hexene
    b. 1-Hexene or 1-octene
    c. *cis*-2-Hexene or 2-methyl-2-pentene

20. Which bond of the double bond in an alkene is broken when the alkene undergoes an addition reaction? Why?

21. Draw a structural formula for the compound that would produce each of the following products after undergoing the indicated reaction.

    a. ? + H$_2$ $\xrightarrow{\text{Ni}}$ 3-methylpentane
       Draw a structural formula for each possible reactant in Problem 21a that correctly answers the question.

    b. ? + H$_2$O $\xrightarrow{\Delta}$ 2-propanol CH$_3$CH(OH)CH$_3$

    c. ? $\xrightarrow{\text{Polymerization}}$ $\left(\begin{matrix} \text{C} \equiv \text{N} & \text{H} \\ | & | \\ \text{C} & \text{C} \\ | & | \\ \text{H} & \text{H} \end{matrix}\right)_n$

# CHAPTER 6

# Alkenes (II)

## SYNTHESIS AND CHEMICAL PROPERTIES OF MONOENES

Like alkanes, a number of alkenes are produced commercially from petroleum. In the laboratory, one of two synthetic methods is used. In general,

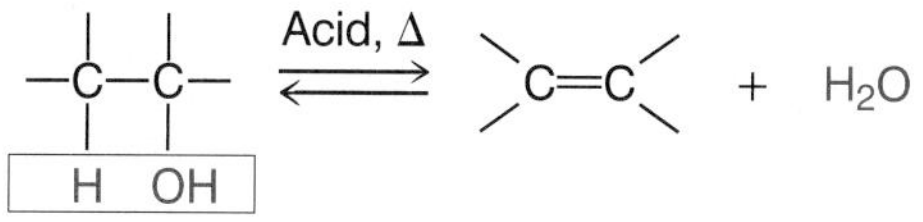

$$\overset{\displaystyle |\quad |}{\underset{\boxed{H\quad OH}}{-C-C-}} \xrightarrow{\text{Acid, } \Delta} \overset{\displaystyle \diagdown}{\diagup}C=C\overset{\displaystyle \diagup}{\diagdown} \;+\; H_2O$$

## 6.1
## SYNTHESIS OF ALKENES
## Dehydration of an Alcohol

Note that an H and an OH bonded to adjacent carbons are eliminated as water. Therefore, this type of reaction is known, in general, as an *elimination reaction*. Since water is eliminated, this is a *dehydration* reaction.

Either concentrated sulfuric acid or concentrated phosphoric acid can be used as a catalyst. The temperature required for this reaction is a function of the structure of the alcohol. Note that the reversed arrows used in the general equation indicate that the reaction is reversible. The use of heat and concentrated acid favors the formation of an alkene from alcohol. This occurs because the alkene is considerably more volatile than the corresponding alcohol. The concentration of the lower-boiling alkene is decreased in the reaction mixture as the alkene is distilled. This stress in the equilibrium is relieved, according to *Le Châtelier's principle*, by the formation of more alkene. Example:

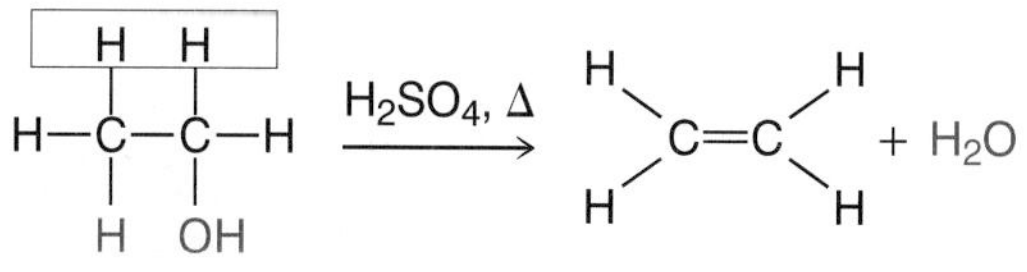

$$\overset{\displaystyle \boxed{H\quad H}}{\underset{\displaystyle H\quad OH}{H-\overset{|}{C}-\overset{|}{C}-H}} \xrightarrow{H_2SO_4,\ \Delta} \overset{H}{\underset{H}{\diagup}}C=C\overset{H}{\underset{H}{\diagdown}} + H_2O$$

Ethanol                    Ethene
(ethyl alcohol)          (ethylene)

Note, in the preceding example, that only one possible alkene can form. Now let us consider an example where more than one alkene can be produced.

**FIGURE 6.1** Use of Saytzeff's rule to predict the preferred product of an elimination reaction.

Carbon bears three hydrogens — minor product is obtained.

Carbon bears only two hydrogens — major product is obtained.

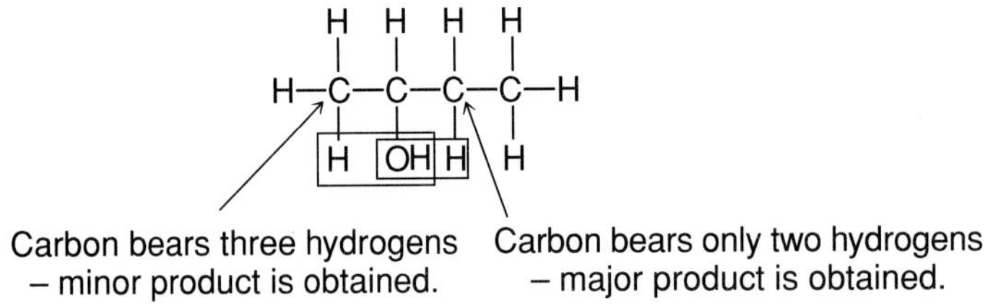

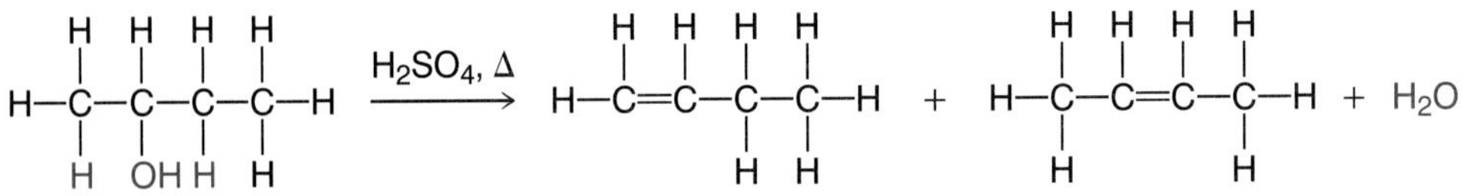

2-Butanol
(sec-butyl alcohol)

1-Butene
(minor product)

2-Butene
(major product)

An obvious question to ask is, Why is 2-butene the preferred product? *Saytzeff's rule* states that in an elimination reaction, the major product will be that which involves loss of hydrogen from the carbon that possesses the smallest number of hydrogens. Consider 2-butanol again (Fig. 6.1).

Certain dehydration reactions produce both an expected and an unexpected product. For example, consider the dehydration of 1-butanol (*n*-butyl alcohol).

$$CH_3CH_2CH_2CH_2OH \xrightarrow[H_2SO_4, \Delta]{} CH_3CH_2CH{=}CH_2 + CH_3CH{=}CHCH_3 + H_2O$$

1-Butene          2-Butene

The product 1-butene is expected, whereas 2-butene represents an unexpected product obtained due to a structural *rearrangement*. In this case, the carbon-carbon double bond in 2-butene is present in an unexpected place—between carbons 2 and 3. Other rearrangements involve the movement of an atom or a group of atoms (see Sec. 10.1). The preceding reaction will be discussed in depth in Section 12.5.

**Dehydrohalogenation of Alkyl Halides**

In general,

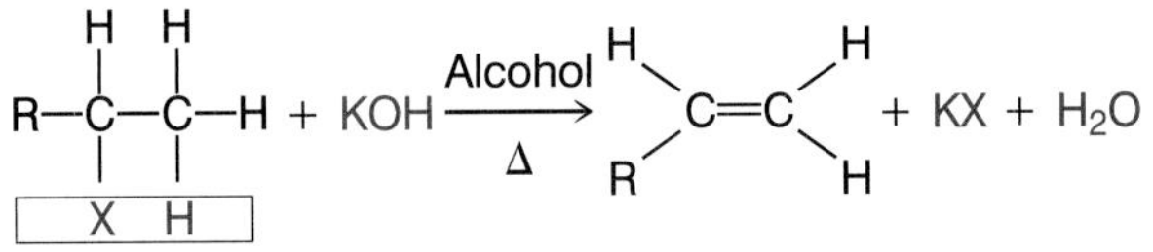

where X = Cl, Br, or I.

The loss of HX (hydrogen halide) is called *dehydrohalogenation*

Example:

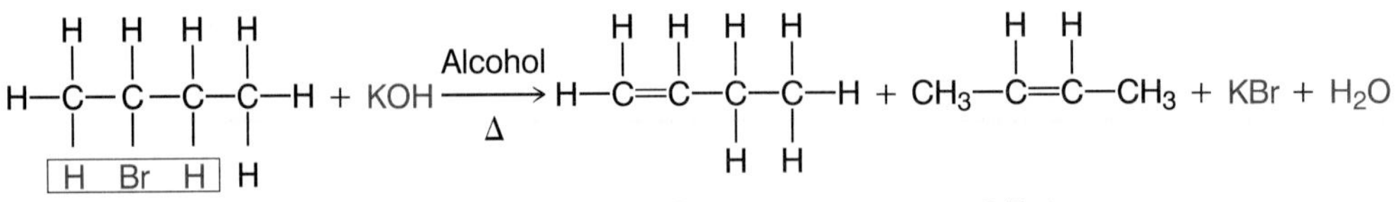

2-Bromobutane
(sec-butyl bromide)

1-Butene
(minor product)

2-Butene
(major product)

Saytzeff's rule also accounts for the product distribution in the dehydrohalogenation process.

**SOLVED PROBLEM 6.1**

Complete the following equation by writing a structural formula for each organic product and a molecular formula or formula unit for each inorganic product. Use Saytzeff's rule to indicate the major organic product and minor organic product.

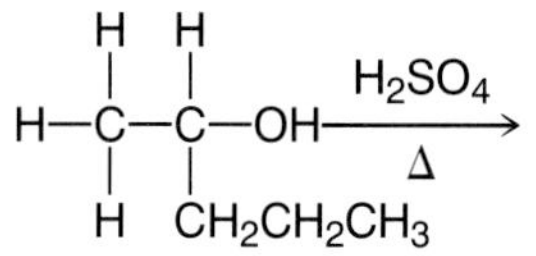

$$H-\overset{\overset{\displaystyle H}{|}}{\underset{\underset{\displaystyle H}{|}}{C}}-\overset{\overset{\displaystyle H}{|}}{\underset{\underset{\displaystyle CH_2CH_2CH_3}{|}}{C}}-OH \xrightarrow[\Delta]{H_2SO_4}$$

**SOLUTION**

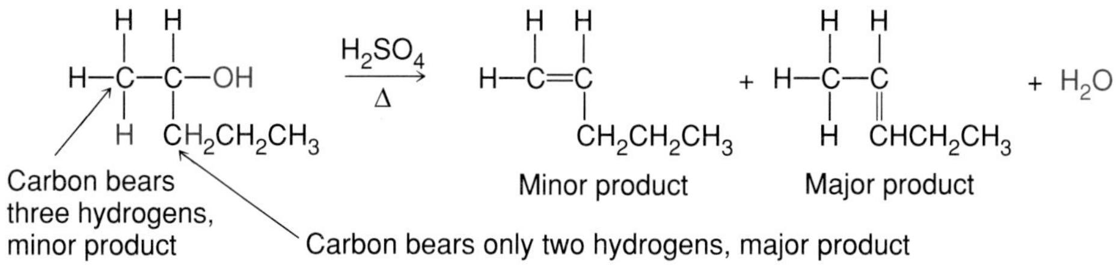

$$H-\overset{\overset{\displaystyle H}{|}}{\underset{\underset{\displaystyle H}{|}}{C}}-\overset{\overset{\displaystyle H}{|}}{\underset{\underset{\displaystyle CH_2CH_2CH_3}{|}}{C}}-OH \xrightarrow[\Delta]{H_2SO_4} \quad H-\overset{\overset{\displaystyle H}{|}}{C}=\overset{\overset{\displaystyle H}{|}}{\underset{\underset{\displaystyle CH_2CH_2CH_3}{}}{C}} \quad + \quad H-\overset{\overset{\displaystyle H}{|}}{\underset{\underset{\displaystyle H}{|}}{C}}-\overset{\overset{\displaystyle H}{|}}{\underset{\underset{\displaystyle CHCH_2CH_3}{||}}{C}} \quad + \quad H_2O$$

Carbon bears three hydrogens, minor product

Carbon bears only two hydrogens, major product

Minor product        Major product

---

Complete the following equations by writing a structural formula for each organic product and a molecular formula or formula unit for each inorganic product. Use Saytzeff's rule to indicate the major and minor organic products.

**EXERCISE 6.1**

a.   $CH_3CH(OH)CH_2CH_2CH_2CH_3 \xrightarrow[\Delta]{H_2SO_4}$

b.   $CH_3\overset{\overset{\displaystyle CH_3}{|}}{\underset{\underset{\displaystyle H}{|}}{C}}-\overset{\overset{\displaystyle CH_3}{|}}{\underset{\underset{\displaystyle Br}{|}}{C}}CH_2CH_2CH_3 + KOH \xrightarrow[\Delta]{Alcohol}$

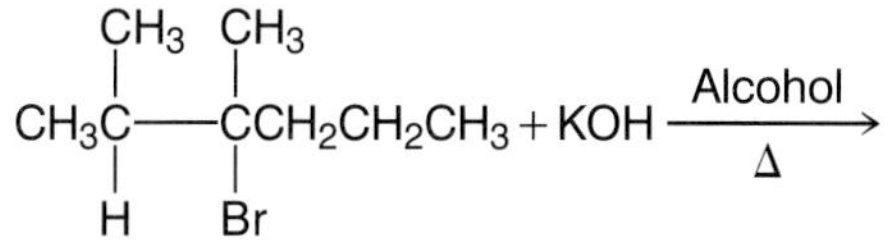

c.   

$$\underset{\underset{\displaystyle CH_3}{|}}{\overset{\overset{\displaystyle CH_3}{|}}{\bigcirc}}\overset{-OH}{\underset{-H}{}} \xrightarrow[\Delta]{H_3PO_4}$$

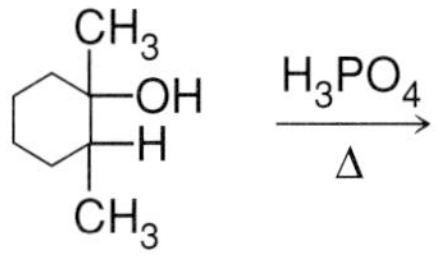

Assume that no rearrangement occurs.

d.   

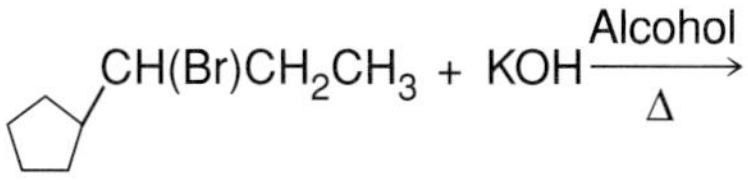

$$\bigcirc\!\!-CH(Br)CH_2CH_3 + KOH \xrightarrow[\Delta]{Alcohol}$$

---

Although Saytzeff's rule is effective in predicting major and minor products in dehydration and dehydrohalogenation processes, the rule is merely a mnemonic device. To actually understand why 2-butene is formed in greater yield than 1-butene when 2-bromobutane is reacted with hot alcoholic potassium hydroxide or when 2-butanol is heated in the presence of concentrated sulfuric acid, we need to examine alkene structure in more depth.

Let us focus on a method of describing alkenes based on the number of alkyl groups bonded to one or both carbons of the double bond. For example, an alkene with one such alkyl group is said to be *monosubstituted*; with two groups, *disubstituted*; with three groups, *trisubstituted*; and with four such groups, *tetrasubstituted* (Fig. 6.2). In general, the larger the number of alkyl groups bonded to one or more carbons of the double bond, the more stable the resulting alkene. For example, a tetrasubstituted alkene, 2,3-dimethyl-2-butene, is more stable than 1-pentene, a monosubstituted alkene (Fig. 6.3).

This greater stability of alkyl-substituted alkenes can be rationalized in the following way: Let us begin with the statement that an $sp^2$—$sp^3$ bond is stronger than an $sp^3$—$sp^3$ bond. This makes sense because an $sp^2$ orbital has more $s$ character (33.3%) than an $sp^3$

**Stability of Alkenes**

**FIGURE 6.2** Classes of substituted alkenes.

$$RCH{=}CH_2$$
Monosubstituted alkene

$$R_2C{=}CH_2 \text{ or } RCH{=}CHR$$
Disubstituted alkene

$$R_2C{=}CHR$$
Trisubstituted alkene

$$R_2C{=}CR_2$$
Tetrasubstituted alkene

**FIGURE 6.3** A typical tetrasubstituted alkene and monosubstituted alkene.

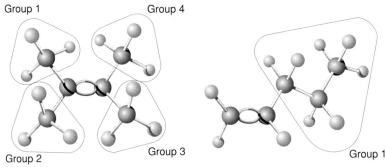

Group 1            Group 4

Group 2            Group 3                                    Group 1

2, 3–Dimethyl–2–butene,              1– Pentene, a monosubstituted alkene
a tetrasubstituted alkene

**FIGURE 6.4** Bond hybridization in 2-butene versus 1-butene.

$$CH_2{=}CH{-}CH_2{-}CH_3 \qquad CH_3{-}CH{=}CH{-}CH_3$$
$$sp^2{-}sp^3 \; sp^3{-}sp^3 \qquad\qquad sp^3{-}sp^2 \;\; sp^2{-}sp^3$$

1-Butene                          2-Butene
(monosubstituted-less stable)    (disubstituted-more stable)

orbital (25%), and therefore the $sp^2$ orbital is positioned closer to the carbon nucleus than the $sp^3$ orbital. Thus the bond formed with an $sp^3$ orbital is less under the influence of the carbon nucleus and therefore weaker, whereas a bond formed with an $sp^2$ orbital is more under the influence of the carbon nucleus and therefore stronger. It can be stated further that the compound with the larger number of stronger bonds is more stable. This also makes sense, because the larger the number of stronger bonds, the more energy is necessary to break these bonds, resulting in a more stable compound.

Now let us consider 1-butene, a monosubstituted alkene, and 2-butene, a disubstituted alkene. The monosubstituted alkene contains only one $sp^2$—$sp^3$ stronger bond along with one weaker $sp^3$—$sp^3$ bond and is less stable; the disubstituted alkene contains two $sp^2$—$sp^3$ stronger bonds and no $sp^3$—$sp^3$ weaker bonds and is more stable (Fig. 6.4). Thus, when 2-butanol is dehydrated or 2-bromobutane is dehydrohalogenated, 2-butene is the major product and 1-butene is the minor product because 2-butene, a disubstituted alkene, is a more stable alkene than 1-butene, a monosubstituted alkene.

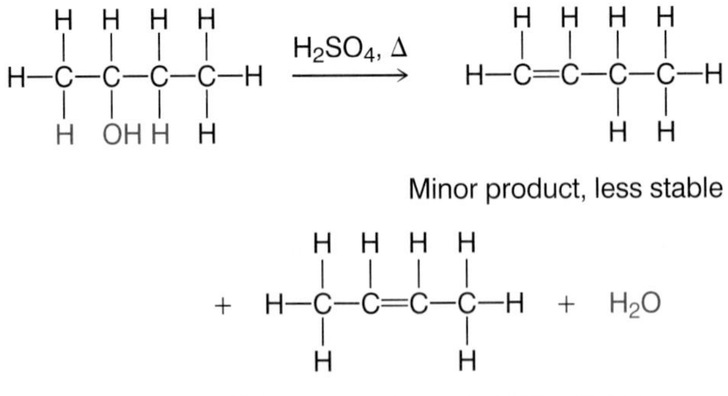

Minor product, less stable alkene

Major product, more stable alkene

You can determine the structure of a major or minor product in a dehydration or dehydrohalogenation reaction using Saytzeff's rule. Then use the principle of alkene stability as a check, or vice versa. Both methods are reliable.

Rank the following alkenes in order of increasing stability:

$$CH_3CH{=}CH_2, (CH_3)_2C{=}CH_2, (CH_3)_2C{=}CHCH_3, CH_2{=}CH_2.$$

**SOLUTION**

Since ethylene ($CH_2{=}CH_2$) contains no alkyl groups bonded to the carbons of the double bond, it is the least stable alkene. Propene ($CH_3CH{=}CH_2$) with one alkyl group is somewhat more stable, followed by isobutylene [$(CH_3)_2C{=}CH_2$] with two alkyl groups and 2-methyl-2-butene [$(CH_3)_2C{=}CHCH_3$] with three alkyl groups. Thus, in order of increasing stability, we have

$$CH_2{=}CH_2 < CH_3CH{=}CH_2 < (CH_3)_2C{=}CH_2 < (CH_3)_2C{=}CHCH_3$$

Point out each alkyl group bonded to a double-bonded carbon in each of the alkenes in Solved Problem 6.2 and thus prove in your own mind that you see why, for example, 2-methyl-2-butene contains three alkyl groups. Name the alkyl group present in each compound in Solved Problem 6.2.

Refer to Section 5.6 for an introduction to addition reactions of the alkenes.

## 6.2 ADDITION REACTIONS OF THE ALKENES

### Hydrogenation

This catalytic reaction was discussed in Sections 4.1 and 5.6. Note that the addition of hydrogen is *syn-* in that both hydrogens add to the same side of the molecule in a *cis* manner.

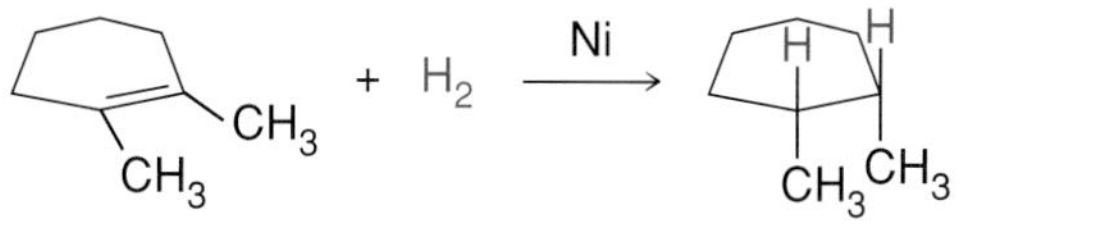

1,2-Dimethylcyclohexene          *cis*-1,2-Dimethylcyclohexane

In general,

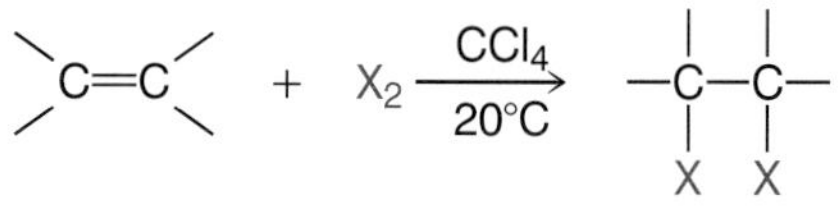

Vicinal dihalide

### Halogenation—At Room Temperature

where $X_2 = Cl_2$ or $Br_2$.

*Note:* A *vicinal dihalide (vic-)* is a halide where the two halogens are bonded to two adjacent carbon atoms in the molecule.

Example:

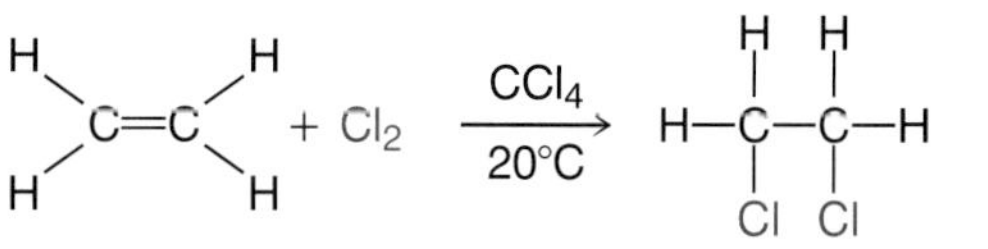

Ethene          1,2-Dichloroethane
(ethylene)

 Fluorine reacts explosively with alkenes. The heat liberated causes both sigma bonds and pi bonds to break, and the reaction has little utility. Iodine reacts with most alkenes to give poor yields of diiodide compounds.

 The bromination reaction is used as a qualitative test for a carbon-carbon double bond. A positive test for an alkene is the immediate disappearance of the red-orange bromine color in $CCl_4$ to form a colorless or pale yellow vicinal dibromide as the bromine adds across the double bond. Alkanes decolorize (react with) bromine at a much slower rate to evolve hydrogen bromide, but in the presence of light, heat, or both, some alkanes do react easily and rapidly with bromine to produce alkyl bromides and hydrogen bromide (see Sec. 4.2).

Before we discuss the mechanism of halogenation, we need to define a type of reagent called an *electrophilic reagent*, or an *electrophile*. An electrophilic reagent is an electron pair–seeking reagent that is electron-deficient, a Lewis acid (see Sec. 1.14). This reagent usually is positively charged but not necessarily so. Some examples of an electrophile are $Ag^+$, $H^+$, and $AlCl_3$.

**Envision the Reaction**

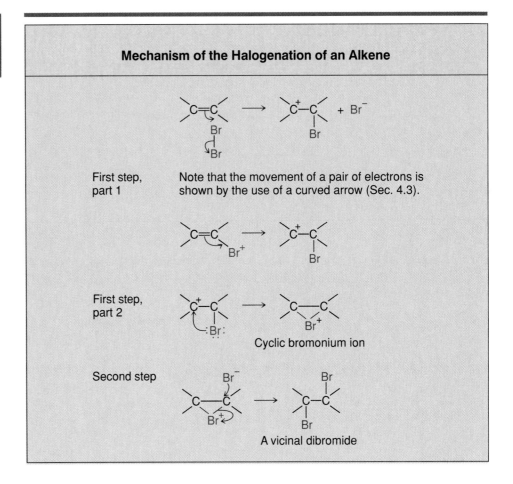

### Mechanism of the Halogenation of an Alkene

First step, part 1 — Note that the movement of a pair of electrons is shown by the use of a curved arrow (Sec. 4.3).

First step, part 2 — Cyclic bromonium ion

Second step — A vicinal dibromide

The mechanism of halogenation is a two-step process. Think of the first step of this addition mechanism as subdivided into two parts. First, an unstable species with a positive charge on carbon (called a *carbocation*; see Sec. 4.3) is formed from an attack of the bromine molecule on the pi electrons of the double bond. This resulting carbocation also can be thought of as resulting from an attack of the species $Br^+$ on the pi electrons of the double bond. Thus we call $Br^+$ an *electrophile* because it is an electron-poor reagent that forms a bond with the pi bond of the electron-rich double-bonded carbon. In the second part of this first step, the pair of unshared electrons on the bromine attack the positively charged carbon to produce a cyclic *bromonium ion*

In the second step, a bromide ion adds from the backside of the cyclic bromonium ion to give the dibromide. The addition is *anti-* in that the groups add to opposite sides of the double bond in a *trans* manner. The mechanism of the addition of chlorine to an alkene is similar to that of bromination.

Let us examine the bromination of 2-methylpropene. Since the product of the reaction, 1,2-dibromo-2-methylpropane, is a saturated acyclic molecule, free rotation exists around the bond between C-1 and C-2. Therefore, it is impossible to prove, using this example, that *anti-* addition takes place.

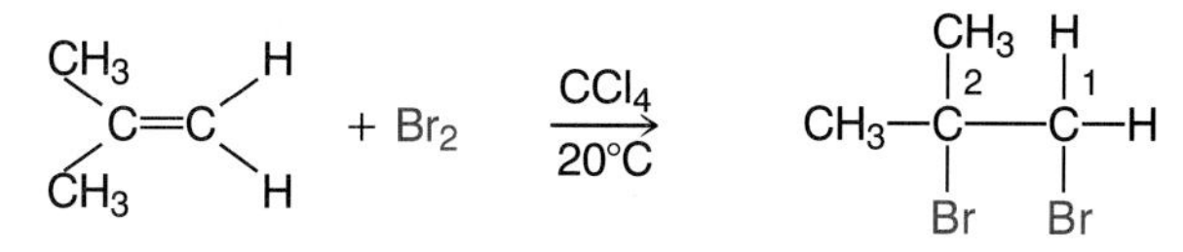

2-Methylpropene
(isobutylene)                          1,2-Dibromo-2-methylpropane

We can prove that addition is *anti-* by brominating a cyclic alkene such as cyclopentene. Because there is limited free rotation around carbon-carbon single bonds in a cycloalkane ring, the addition could produce either *cis*-1,2-dibromocyclopentane or *trans*-1,2-dibromocyclopentane. Since only the *trans* isomer is isolated, addition of bromine (or chlorine) must be *anti-*.

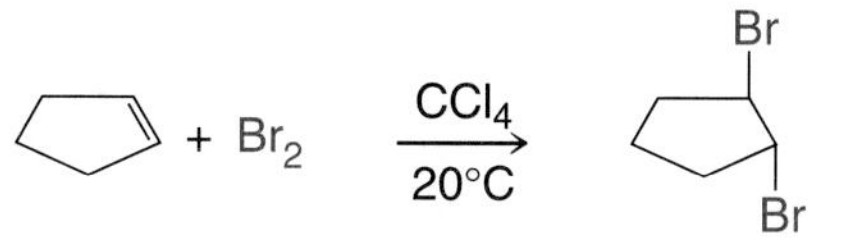

Cyclopentene        *trans*-1,2-Dibromocyclopentane

Treating an alkene with a dilute, neutral solution of potassium permanganate produces a glycol (a vicinal dihydroxyalkane, or α-diol). In general,

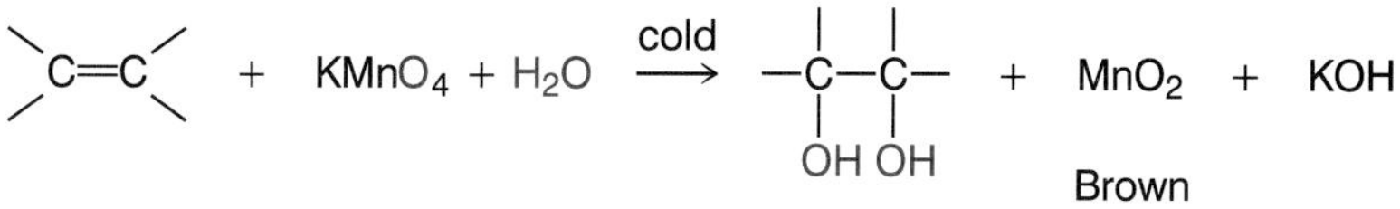

**Hydroxylation**

Example:

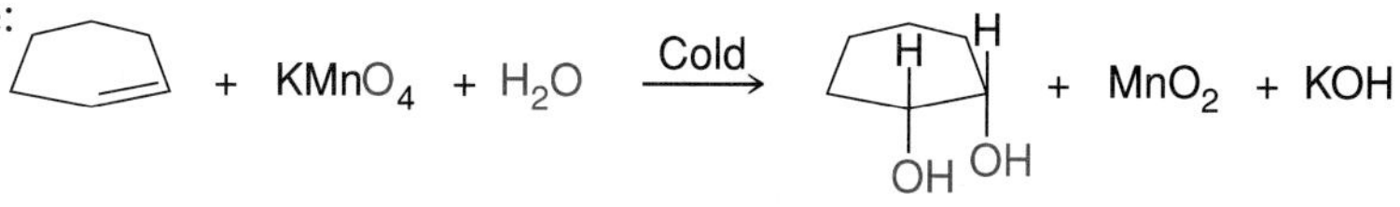

Cyclohexene                        *cis*-1,2-Cyclohexanediol

It can be determined by analysis that hydroxylation with permanganate in the cold occurs by a *syn-* addition; that is, the two OH units add to the same side of the alkene in a *cis* manner. When cyclohexene is reacted with permanganate in the cold, only the *cis*-diol is isolated.

Permanganate hydroxylation of an alkene serves as another test for an alkene. A positive test is the loss of the purple permanganate color and the formation of a brown precipitate of manganese dioxide.

---

Complete each of the following equations by drawing a structural formula for each organic product formed.

**SOLVED PROBLEM 6.3**

a.  1-Pentene + H$_2$ $\xrightarrow{\text{Pd}}$

b.  Cyclohexene + Cl$_2$ $\xrightarrow[20°C]{\text{CCl}_4}$

**SOLUTION**

a.  This is an example of a hydrogenation reaction. We know that addition of hydrogen across the double bond occurs in a *syn-* manner. However, since free rotation of carbon-carbon single bonds occurs, the stereochemical orientation of the product molecule is moot, so we obtain

CH$_3$CH$_2$CH$_2$CH$_2$CH$_3$
Pentane

b.  This is an example of a halogenation reaction at room temperature. We know that addition of halogen across the double bond takes place in an *anti-* manner to give a *trans* product. Since the product is cyclic, free rotation of carbon-carbon single bonds is limited, and the stereochemical orientation of the product is important. Thus we produce

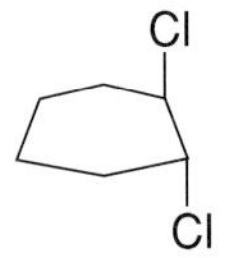

*trans*-1,2-Dichlorocyclohexane

---

**EXERCISE 6.3**

Complete each of the following equations by drawing a structural formula for each organic product and writing a molecular formula or formula unit for each inorganic product formed. Be sure to note the stereochemical orientation of the organic product when appropriate.

a.  Cyclopentene + $KMnO_4$ + $H_2O$ $\xrightarrow{\text{cold}}$

b.  1-Hexene + $Br_2$ $\xrightarrow[20°C]{CCl_4}$

c.  1-Butene + $H_2$ $\xrightarrow{Pd}$

d.  1,2-Diethylcyclohexene + $H_2$ $\xrightarrow{Ni}$

---

**SOLVED PROBLEM 6.4**

Using a test tube reaction, show how you would distinguish between 2-hexene and hexane. Indicate what reagent(s) should be used and what color change(s) take place.

**SOLUTION**
React each compound with a solution of bromine in carbon tetrachloride. The 2-hexene instantaneously decolorizes the bromine solution, hexane slowly decolorizes bromine in carbon tetrachoride, and fumes of hydrogen bromide are observed.

React each compound with a cold, neutral solution of potassium permanganate. 2-Hexene decolorizes the potassium permanganate solution to give a brown precipitate of manganese dioxide. Hexane does not react.

---

**EXERCISE 6.4**

Using a test tube reaction, show how you would distinguish between (a) 2-hexene and cyclohexane and (b) cyclohexane and cyclohexene. Indicate what reagent(s) should be used and what color change(s) take place.

---

**Addition of Hydrogen Halides**

In general,

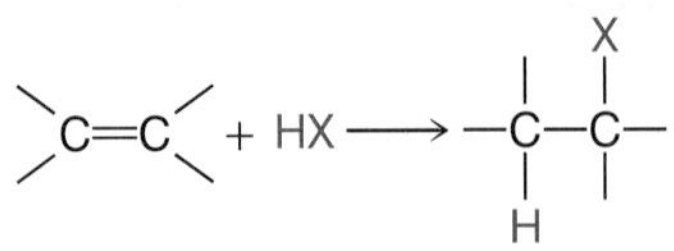

where X = Cl, Br, or I.

Example:

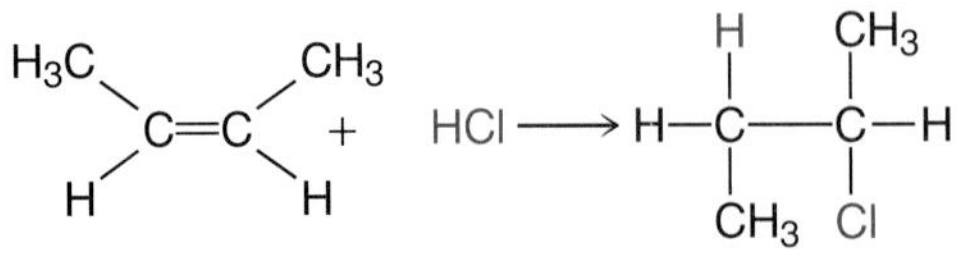

---

## Mechanism of Hydrogen Halide Addition to an Alkene

**First step**

(Slow)

**Second step**

**Envision the Reaction**

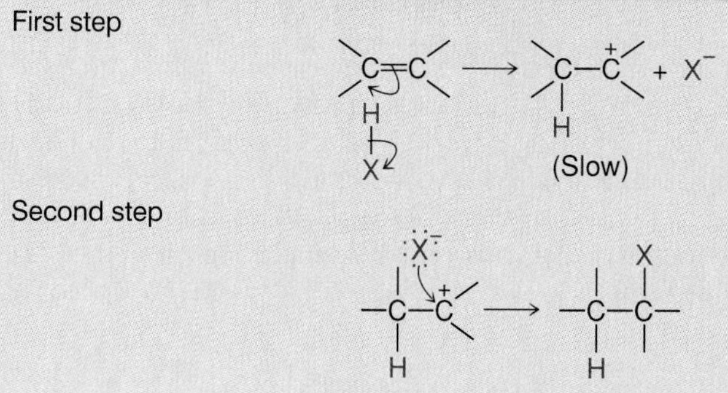

The mechanism of hydrogen halide addition to an alkene is well understood. The first step is the formation of a carbocation derived from the alkene. In effect, an electrophile, $H^+$, combines with the pi electrons of the alkene. Formation of the carbocation is the rate-determining step, or slowest step, in the reaction mechanism. The second step of the mechanism is the addition of halide ion to the carbocation formed in the first step.

With hydrogenation, halogenation, and hydroxylation in the cold, it makes no difference if the alkene is symmetrical or unsymmetrical; the product is the same irrespective of which group adds to which carbon. This occurs because the group added across each carbon of the double bond is the same (the reagent is symmetrical as $H_2$ or $Br_2$).

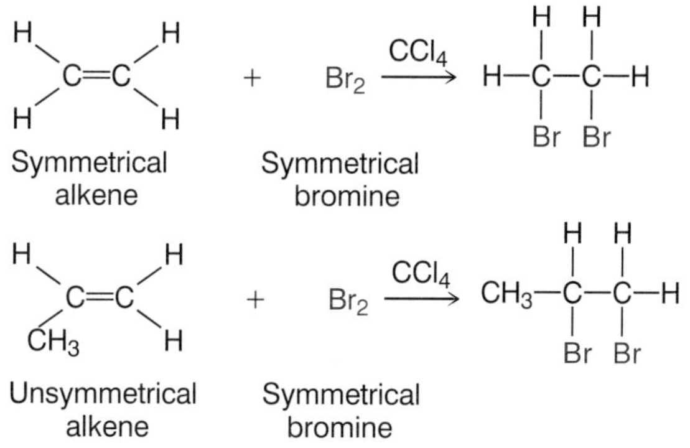

**The Complications of an Unsymmetrical Reagent— Markownikoff's Rule**

Symmetrical alkene     Symmetrical bromine

Unsymmetrical alkene     Symmetrical bromine

This same state of affairs holds for an unsymmetrical reagent reacting with a symmetrical alkene. It makes no difference which group adds to which carbon of the double bond; the product is the same.

---

Demonstrate in your own mind that the addition of hydrogen bromide (an unsymmetrical reagent) to ethene (a symmetrical alkene) produces a product that is the same irrespective of which group adds to which carbon of the double bond.

**EXERCISE 6.5**

---

However, when an unsymmetrical reagent is added to an unsymmetrical alkene, if the groups of the reactant add in one way, a particular product is obtained; if the groups add in another way, a different product is obtained. For example,

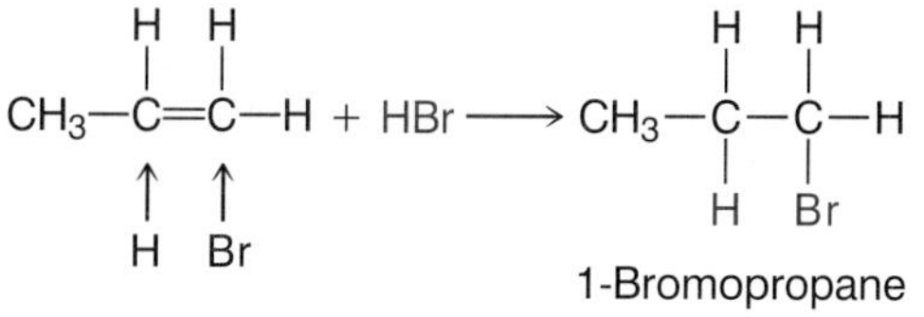

1-Bromopropane

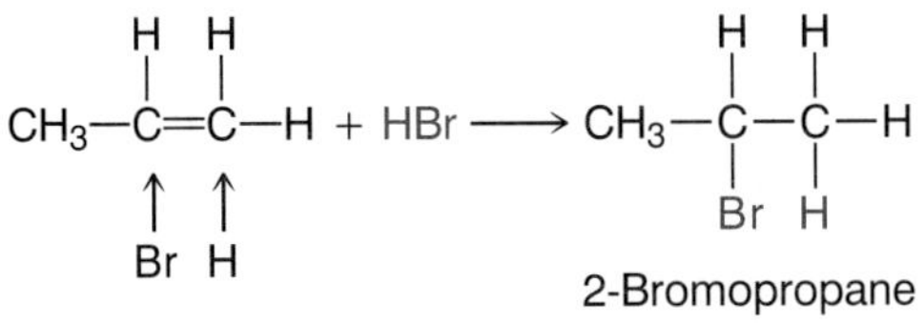

$$CH_3-\underset{\underset{Br}{\uparrow}}{\overset{\overset{H}{|}}{C}}=\underset{\underset{H}{\uparrow}}{\overset{\overset{H}{|}}{C}}-H + HBr \longrightarrow CH_3-\overset{\overset{H}{|}}{\underset{\underset{Br}{|}}{C}}-\overset{\overset{H}{|}}{\underset{\underset{H}{|}}{C}}-H$$

2-Bromopropane

Experimentally, only 2-bromopropane forms when propene reacts with hydrogen bromide in the absence of air or peroxides. No 1-bromopropane is isolated. Vladimir Markownikoff, director of the Chemical Institute of the University of Moscow, studied the orientation of unsymmetrical reagents added to unsymmetrical alkenes and established the following principle called *Markownikoff's rule*: An acid (HX) adds to an asymmetrical alkene so as to place the proton on the carbon of the double bond with the greater number of hydrogens. In general,

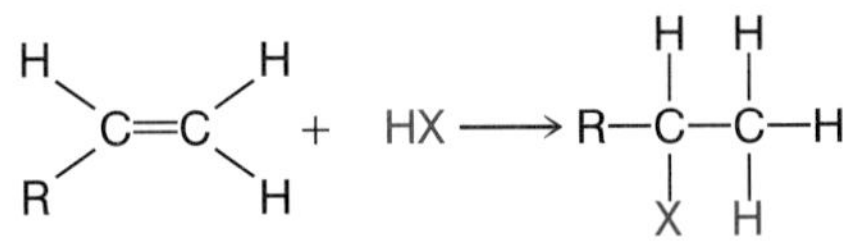

$$\underset{R}{\overset{H}{\diagdown}}C=C\underset{H}{\overset{H}{\diagup}} + HX \longrightarrow R-\overset{\overset{H}{|}}{\underset{\underset{X}{|}}{C}}-\overset{\overset{H}{|}}{\underset{\underset{H}{|}}{C}}-H$$

where X = Cl, Br, or I.

Example:

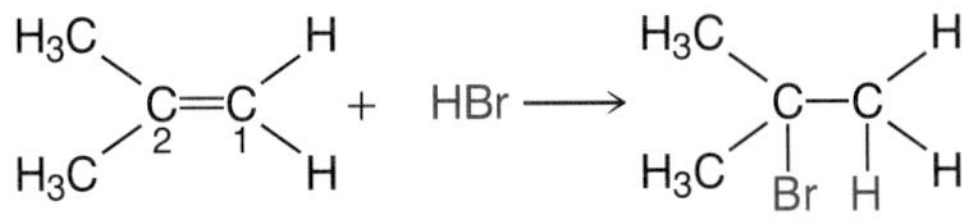

$$\underset{H_3C}{\overset{H_3C}{\diagdown}}\overset{2}{C}=\overset{1}{C}\underset{H}{\overset{H}{\diagup}} + HBr \longrightarrow \underset{H_3C}{\overset{H_3C}{\diagdown}}\overset{}{\underset{\underset{Br}{}}{C}}-\overset{}{\underset{\underset{H}{}}{C}}\underset{H}{\overset{H}{\diagup}}$$

Note that C-1 has two bonded hydrogens, while C-2 has no bonded hydrogens. Thus the proton of HBr adds to C-1.

**The Basis for Markownikoff's Rule**

Why does Markownikoff's rule work? In other words, why does the proton add to the carbon of the double bond with the greater number of hydrogens?

To answer this question, we need to look at the mechanism that could account for each possible product formed. Consider propene reacting with HCl to give two possible products.

**Envision the Reaction**

### Mechanism of the Addition of Hydrogen Chloride to Propene to Explain Markownikoff's Rule

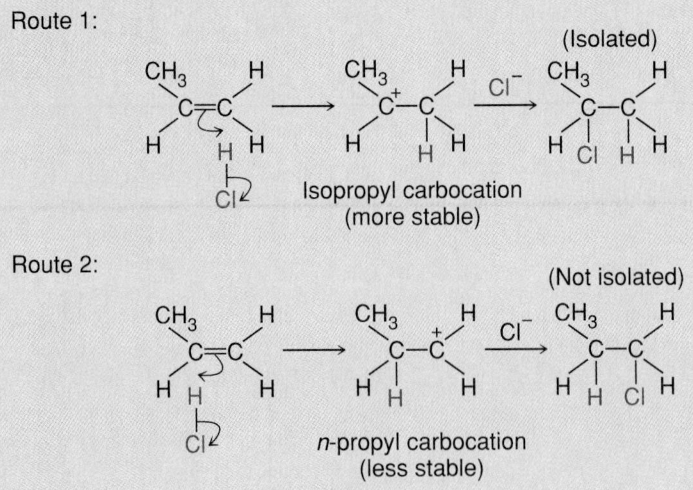

Route 1:

(Isolated)

Isopropyl carbocation
(more stable)

Route 2:

(Not isolated)

n-propyl carbocation
(less stable)

Since route 1 leads to Markownikoff addition, route 1 is the preferred route. The reason lies in the fact that the isopropyl carbocation, which is formed much faster than the n-propyl carbocation, is much more stable than the n-propyl carbocation. Thus a secondary carbocation is much more stable than a primary carbocation.

In general, the stability of carbocations is as follows:

$$3° > 2° > 1° > CH_3$$

Most stable ⟶ Least stable

To understand the reason for this order of stability, let us begin by classifying these carbocations. Carbocations are classified according to the number of alkyl groups bonded to the carbon with the positive charge. A primary carbocation has one bonded alkyl group; a secondary, two; and a tertiary, three. Thus

$$R{-}\overset{+}{C}H_2 \qquad\qquad R{-}\overset{\overset{\displaystyle H}{|}}{\underset{}{\overset{+}{C}}}{-}R \qquad\qquad R{-}\overset{\overset{\displaystyle R}{|}}{\underset{}{\overset{+}{C}}}{-}R$$

1° Carbocation (primary carbocation)      2° Carbocation (secondary carbocation)      3° Carbocation (tertiary carbocation)

Each alkyl group has an electron-releasing effect called a positive *inductive effect* ($+I$) that shifts electron density across a sigma bond toward the positive charge, delocalizing the positive charge and making the carbocation more stable. Since three alkyl groups are on the tertiary carbocation, that positively charged carbon atom is neutralized to the greatest extent and is the most stable. Thus the isopropyl carbocation with two electron-releasing alkyl groups is more stable than the n-propyl carbocation with only one electron-releasing alkyl group.

In some cases, two addition products are formed and isolated:

$$CH_3\overset{\overset{\displaystyle H}{|}}{C}{=}\overset{\overset{\displaystyle H}{|}}{C}CH_2CH_3 \ +\ HI \longrightarrow CH_3\overset{\overset{\displaystyle H}{|}}{\underset{\underset{\displaystyle I}{|}}{C}}{-}\overset{\overset{\displaystyle H}{|}}{\underset{\underset{\displaystyle H}{|}}{C}}CH_2CH_3 \ +\ CH_3\overset{\overset{\displaystyle H}{|}}{\underset{\underset{\displaystyle H}{|}}{C}}{-}\overset{\overset{\displaystyle H}{|}}{\underset{\underset{\displaystyle I}{|}}{C}}CH_2CH_3$$

Two products form because both carbocations are secondary and are nearly equal in stability. Since the methyl group ($CH_3{-}$) is more efficient at stabilization of a carbocation than the ethyl group ($CH_3CH_2{-}$), slightly more 2-iodopentane is produced.

---

Write a detailed, step-by-step mechanism of the reaction of 2-pentene with hydrogen iodide. Note that the secondary carbocation precursor to 2-iodopentane is stabilized by a methyl group.

**EXERCISE 6.6**

---

Refer to Section 5.6 for an introductory discussion of the addition of sulfuric acid to an alkene—*hydration*. Note that Markownikoff's rule is used to predict the product(s) here. In general,

$$\overset{}{\underset{}{>}}C{=}C\overset{}{\underset{}{<}} \ +\ H_2SO_4 \longrightarrow {-}\overset{|}{\underset{\underset{\displaystyle H}{|}}{C}}{-}\overset{|}{\underset{\underset{\displaystyle OSO_3H}{|}}{C}}{-}$$

An alkyl hydrogen sulfate

**Addition of Sulfuric Acid—Hydration**

When an alkane is treated with sulfuric acid, no reaction occurs at room temperature, and a two-phase system (two layers) is formed. On the other hand, alkenes easily react with sulfuric acid, liberating heat and forming a homogeneous solution. Thus a sulfuric acid test tube reaction can be used as another test to distinguish an alkene from an alkane.

Example:

$$\overset{\overset{\displaystyle H}{\diagdown}}{\underset{\underset{\displaystyle CH_3}{\diagup}}{}}C{=}C\overset{\overset{\displaystyle H}{\diagup}}{\underset{\underset{\displaystyle H}{\diagdown}}{}} \ +\ H_2SO_4 \longrightarrow CH_3\overset{\overset{\displaystyle H}{|}}{\underset{\underset{\displaystyle HO_3SO}{|}}{C}}{-}\overset{\overset{\displaystyle H}{|}}{\underset{\underset{\displaystyle H}{|}}{C}}{-}H$$

Isopropyl hydrogen sulfate

The mechanism involves the typical two-step electrophilic addition.

**Envision the Reaction**

---

**Mechanism of the Addition of Sulfuric Acid to Propene**

First step:

$$CH_3-\overset{\overset{\displaystyle H}{|}}{C}=\overset{\overset{\displaystyle H}{|}}{C}-H \longrightarrow CH_3-\overset{\overset{\displaystyle H}{|}}{\overset{+}{C}}-\overset{\overset{\displaystyle H}{|}}{\underset{\underset{\displaystyle H}{|}}{C}}-H + {}^-OSO_3H$$

$$H-OSO_3H$$

Isopropyl carbocation

Second step:

$$CH_3-\overset{\overset{\displaystyle H}{|}}{\overset{+}{C}}-\overset{\overset{\displaystyle H}{|}}{\underset{\underset{\displaystyle H}{|}}{C}}-H \longrightarrow CH_3-\overset{\overset{\displaystyle H}{|}}{\underset{\underset{\displaystyle HO_3SO}{|}}{C}}-\overset{\overset{\displaystyle H}{|}}{\underset{\underset{\displaystyle H}{|}}{C}}-H$$

$${}^-OSO_3H$$

---

Sulfuric acid is regenerated when an alkyl hydrogen sulfate is hydrolyzed by heat to produce the corresponding alcohol. In general,

$$-\overset{|}{\underset{\underset{\displaystyle H}{|}}{C}}-\overset{|}{\underset{\underset{\displaystyle OSO_3H}{|}}{C}}- \quad + \quad H_2O \quad \overset{\Delta}{\longrightarrow} \quad -\overset{|}{\underset{\underset{\displaystyle H}{|}}{C}}-\overset{|}{\underset{\underset{\displaystyle OH}{|}}{C}}- \quad + \quad H_2SO_4$$

Example:

$$CH_3-\overset{\overset{\displaystyle H}{|}}{\underset{\underset{\displaystyle HO_3SO}{|}}{C}}-CH_3 \quad + \quad H_2O \quad \overset{\Delta}{\longrightarrow} \quad CH_3-\overset{\overset{\displaystyle H}{|}}{\underset{\underset{\displaystyle OH}{|}}{C}}-CH_3 \quad + \quad H_2SO_4$$

2-Propanol
(isopropyl alcohol)

## Addition of Hypohalous Acids

Note that Markownikoff's rule can be used to predict the product(s) of halohydrin addition. This reaction represents a *Markownikoff-type addition*, with the positively charged halogen substituting for $H^+$ and attacking the carbon of the double bond with the greater number of hydrogens.

When bromine (or chlorine) in water is added to an alkene, a halohydrin is produced. In general,

$$\overset{\diagup}{\underset{\diagdown}{C}}=\overset{\diagup}{\underset{\diagdown}{C}} + X_2 \quad \overset{H_2O}{\longrightarrow} \quad -\overset{|}{\underset{\underset{\displaystyle X}{|}}{C}}-\overset{\overset{\displaystyle OH}{|}}{\underset{|}{C}}-$$

where $X_2 = Cl_2$ or $Br_2$.

Example:

$$\overset{H}{\underset{CH_3}{\diagup}}C=C\overset{\diagup H}{\diagdown H} \quad + \quad Br_2 \quad \overset{H_2O}{\longrightarrow} \quad CH_3-\overset{\overset{\displaystyle H}{|}}{\underset{\underset{\displaystyle OH}{|}}{C}}-\overset{\overset{\displaystyle H}{|}}{\underset{\underset{\displaystyle Br}{|}}{C}}-H$$

Propylene bromohydrin

Mechanistically, the reaction starts by formation of the bromonium ion. This occurs when a pair of unshared electrons on bromine attack the positively charged carbon of the carbocation. Since water is usually present in a much higher concentration than $Br^-$, the bromonium ion is preferentially attacked by water from the backside in the second step of the mechanism. Loss of a proton to form HBr completes the mechanism.

## Mechanism of the Addition of Hypobromous Acid to Propene

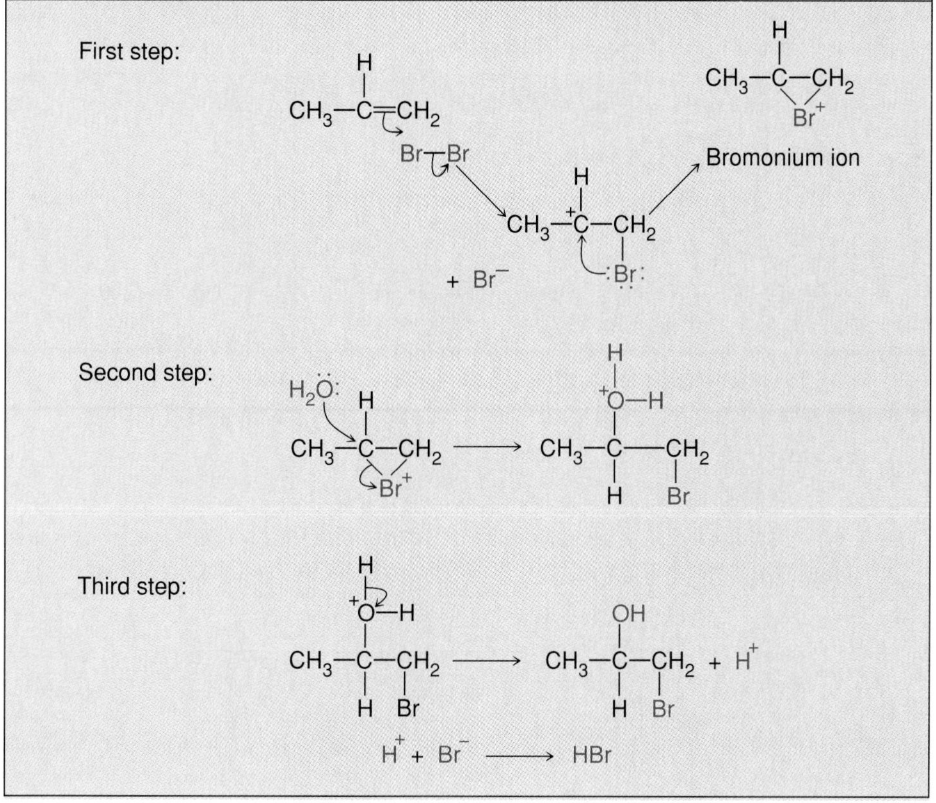

**Envision the Reaction**

**First step:**

CH₃—C=CH₂

Br—Br

+ Br⁻

Bromonium ion

**Second step:**

**Third step:**

H⁺ + Br⁻ ⟶ HBr

**Hydroboration-Oxidation**

Reacting an alkene with diborane ($B_2H_6$) followed by oxidation of the intermediate borane yields an alcohol. This reaction, discovered by H. C. Brown, professor of chemistry at Purdue University (Nobel prize 1979) in 1956, is important because the units of water (H and OH) add in an anti-Markownikoff orientation. Thus this reaction is an important tool in organic synthesis. Diborane ($B_2H_6$) can be represented chemically as $BH_3$ for convenience. Note that $BH_3$ is electron-deficient and is acting as a Lewis acid with the alkene as a Lewis base (see Sect. 1.14). In general,

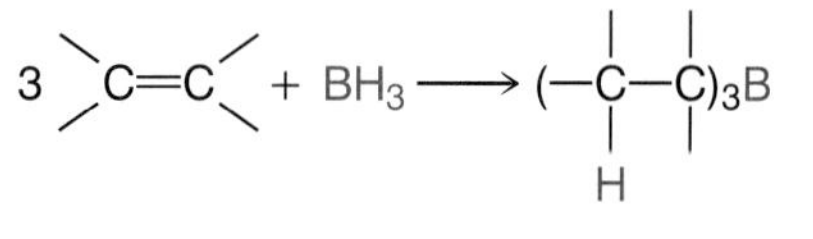

A trialkylborane

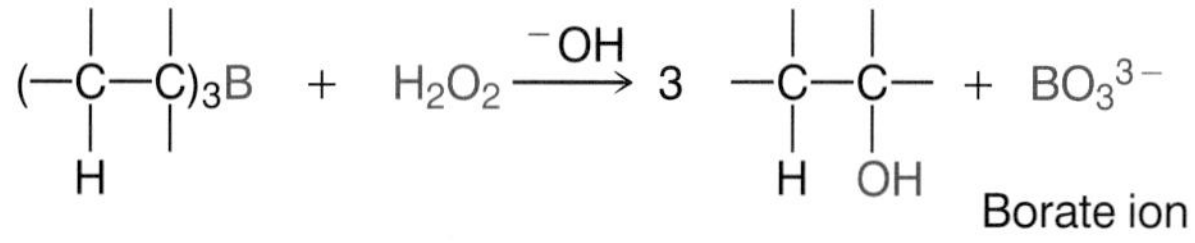

Borate ion

Examples:

$$3\ CH_3CH_2CH{=}CH_2 + BH_3 \longrightarrow (CH_3CH_2CH_2CH_2)_3B$$

Tri($n$-butyl)borane

$$(CH_3CH_2CH_2CH_2)_3B + H_2O_2 \xrightarrow{^-OH} 3\ CH_3CH_2CH_2CH_2OH + BO_3{}^{3-}$$

1-Butanol
($n$-butyl alcohol)

A trialkylborane is formed in a series of steps.

$$CH_3CH_2CH{=}CH_2 + BH_3 \longrightarrow CH_3CH_2CH_2CH_2BH_2$$

$n$-Butylborane

$$CH_3CH_2CH_2CH_2BH_2 + CH_3CH_2CH{=}CH_2 \longrightarrow (CH_3CH_2CH_2CH_2)_2BH$$

Di($n$-butyl)borane

$$(CH_3CH_2CH_2CH_2)_2BH + CH_3CH_2CH{=}CH_2 \longrightarrow (CH_3CH_2CH_2CH_2)_3B$$

Tri($n$-butyl)borane

---

## SOLVED PROBLEM 6.5

Complete each of the following equations by drawing a structural formula for each organic product and writing a molecular formula or formula unit for each inorganic product.

a.  $(CH_3)_2\ \overset{2}{C}{=}\overset{1}{C}H_2 \xrightarrow[\text{2. }H_2O_2,\ ^-OH]{\text{1. }BH_3}$

b.  $(CH_3)_2\ \overset{2}{C}{=}\overset{1}{C}H_2 \xrightarrow[\text{2. }H_2O,\ \Delta]{\text{1. }H_2SO_4}$

**SOLUTION**

a.  Since the units of water (H and OH) add across the double bond in an anti-Markownikoff orientation, H adds to the carbon of the double bond bonded to the smaller number of hydrogens (C-2) to give

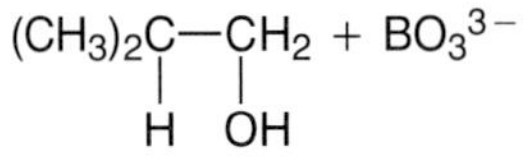

$$(CH_3)_2\underset{\underset{H}{|}}{C}{-}\underset{\underset{OH}{|}}{C}H_2 + BO_3{}^{3-}$$

b.  Here, the units of water add across the double bond in a Markownikoff orientation. Thus H adds to the carbon of the double bond bonded to the larger number of hydrogens (C-1) to give

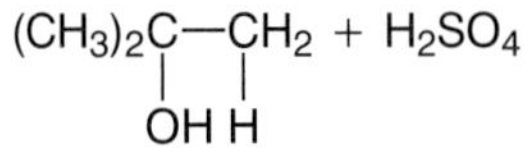

$$(CH_3)_2\underset{\underset{OH}{|}}{C}{-}\underset{\underset{H}{|}}{C}H_2 + H_2SO_4$$

---

## EXERCISE 6.7

Complete each of the following reactions by drawing a structural formula for each organic product and writing a molecular formula or formula unit for each inorganic product.

a.  2-Methyl-2-butene + $Cl_2 \xrightarrow[20°C]{CCl_4}$

b.  2-Methyl-2-butene + HCl $\longrightarrow$

c.  2-Pentene $\xrightarrow[\text{2. }H_2O,\ \Delta]{\text{1. }H_2SO_4}$

| TABLE 6.1 | Stereochemical Orientations of Some Selected Addition Reactions of Alkenes |
| --- | --- |

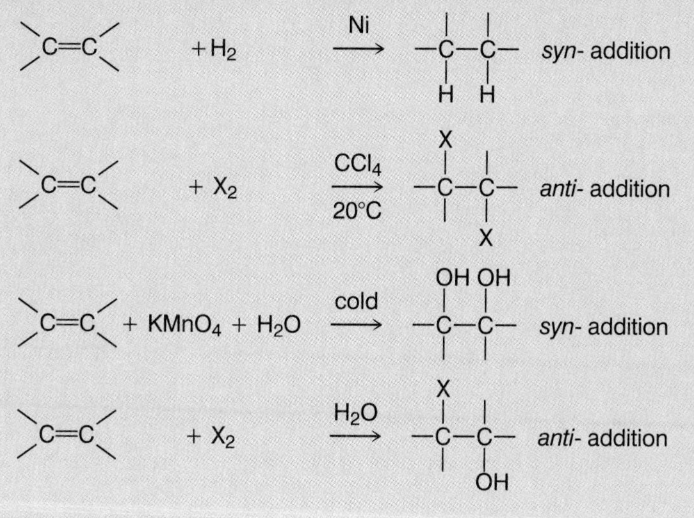

$$\text{C=C} + H_2 \xrightarrow{\text{Ni}} \underset{\underset{H}{|}}{-C}-\underset{\underset{H}{|}}{C}- \quad syn\text{- addition}$$

$$\text{C=C} + X_2 \xrightarrow[20°C]{\text{CCl}_4} \underset{|}{\overset{X}{\underset{|}{-C}}}-\underset{\underset{X}{|}}{C}- \quad anti\text{- addition}$$

$$\text{C=C} + KMnO_4 + H_2O \xrightarrow{\text{cold}} \underset{|}{\overset{OH}{\underset{|}{-C}}}-\underset{|}{\overset{OH}{\underset{|}{C}}}- \quad syn\text{- addition}$$

$$\text{C=C} + X_2 \xrightarrow{H_2O} \underset{|}{\overset{X}{\underset{|}{-C}}}-\underset{\underset{OH}{|}}{C}- \quad anti\text{- addition}$$

1. 1-Butene + $H_2 \xrightarrow{\text{Ni}}$

2. Propene $\xrightarrow[\text{2. } H_2O_2,\ ^-OH]{\text{1. } BH_3}$

Let us use Table 6.1 to summarize the stereochemical orientations of some selected reactions. It is important to remember that if a straight- or branched-chain alkene is reacted, the stereochemical orientation of the product is academic because of free rotation around carbon-carbon single acyclic bonds. Only when the product is cyclic does the stereochemical orientation become important.

Moreover, note that in almost all addition reactions of the alkene functional group where an unsymmetrical alkene is reacting with an unsymmetrical reagent, Markownikoff's rule is used to predict the product(s); only in hydroboration-oxidation does position orientation of the units of the reagent proceed in an anti-Markownikoff manner. These data are summarized in Table 6.2.

| TABLE 6.2 | Position Orientation of the Units of an Unsymmetrical Reagent(s) When Added to an Unsymmetrical Alkene |
| --- | --- |

| Reagent(s) | Position Orientation |
| --- | --- |
| HCl, HBr, HI | Markownikoff |
| $H_2SO_4$ then $H_2O$ | Markownikoff |
| $Cl_2$ and $H_2O$ or HOCl | Markownikoff-type |
| $Br_2$ and $H_2O$ or HOBr | Markownikoff-type |
| $BH_3$ then $H_2O_2$, $^-OH$ | Anti-Markownikoff |

## 6.3
## OTHER REACTIONS OF ALKENES

### Reaction with *N*-Bromosuccinimide

---

**FIGURE 6.5** The allyl group and allylic hydrogens.

$$CH_2{=}CH{-}CH_2{-}$$
Allyl group

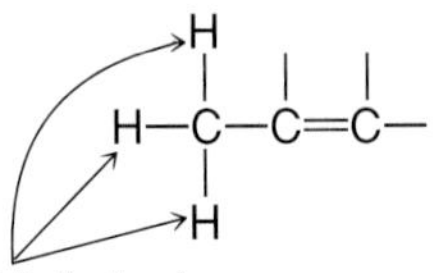

Allylic hydrogens

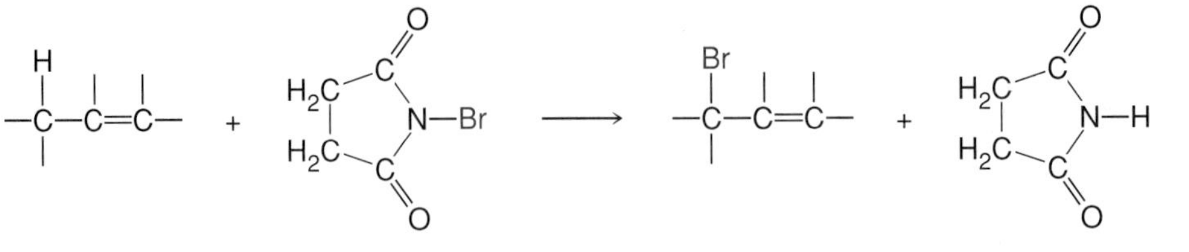

Treating an alkene (with a least one allylic hydrogen) at room temperature with *N*-bromosuccinimide gives an alkene brominated in an allylic position. An *allylic hydrogen* is a hydrogen bonded to a carbon which, in turn, is bonded to a carbon with the double bond (Fig. 6.5). Note the structure of the allyl group. The reaction shows a substitution of bromine for hydrogen and is not an addition. In general,

*N*-Bromosuccinimide (NBS)                                    Succinimide

Example:

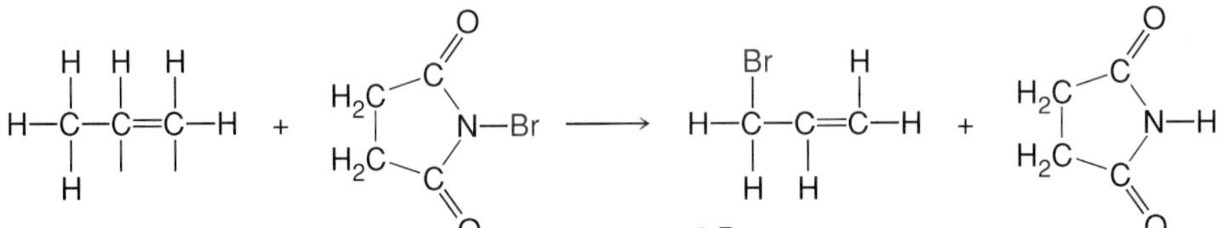

*3-Bromopropene*

This same reaction occurs if the alkene is treated with bromine in the presence of high temperatures or ultraviolet light. In each case (use of NBS or bromine in the presence of heat or light), the reaction occurs by means of a free-radical mechanism similar to that in Section 4.3. However, the mechanism of the reaction of an alkene with NBS is more complex, and we will not study it.

### Ozonolysis: Cleavage of the Carbon-Carbon Double Bond

Alkenes, when treated with the strong oxidizing agent ozone and when the resulting mixture is hydrolyzed with Zn and $H_2O$, give one or more aldehydes, ketones, or both. In general,

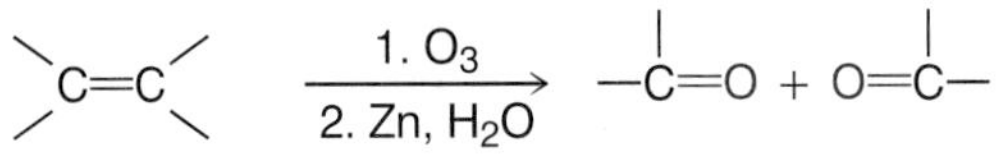

$$\diagdown C{=}C\diagup \quad \xrightarrow[\text{2. Zn, }H_2O]{\text{1. }O_3} \quad {-}C{=}O + O{=}C{-}$$

Example:

$$\xrightarrow[\text{2. Zn, }H_2O]{\text{1. }O_3}\ 2\ CH_3{-}\overset{\displaystyle CH_3}{C}{=}O$$

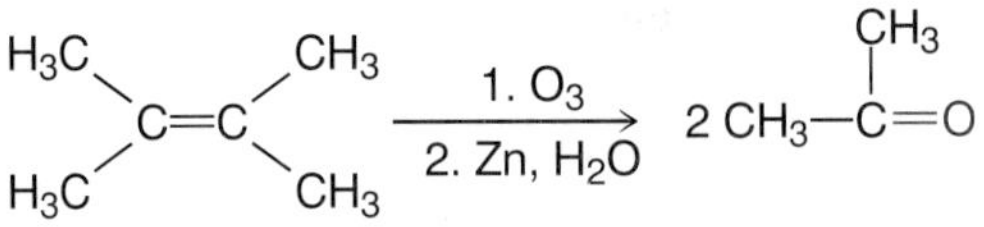

To predict the product of an ozonolysis, simply break both bonds of the carbon-carbon double bond and add to each carbon the $={}$ O grouping.

---

**SOLVED PROBLEM 6.6**

Predict the product(s) of the following ozonolysis:

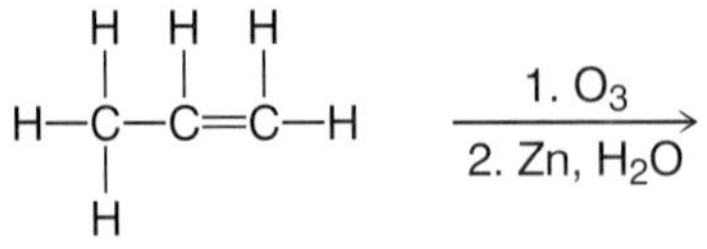

$$\xrightarrow[\text{2. Zn, }H_2O]{\text{1. }O_3}$$

**SOLUTION**

*Step 1.* Break both bonds of the carbon-carbon double bond:

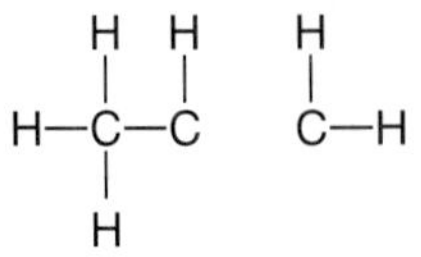

*Step 2.* Add to each carbon of the double bond the grouping $=O$, thus

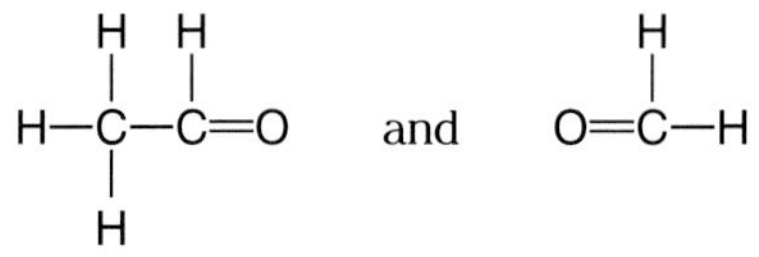

and

---

Complete each of the following equations by drawing a structural formula for each organic compound produced.

a.   $CH_3CH\!=\!CHCH_2CH_2CH_3 \xrightarrow[\text{2. Zn, H}_2\text{O}]{\text{1. O}_3}$

b.

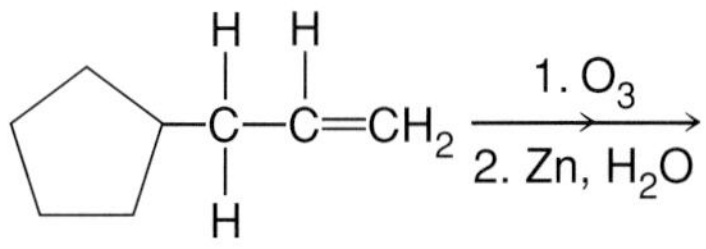

c.

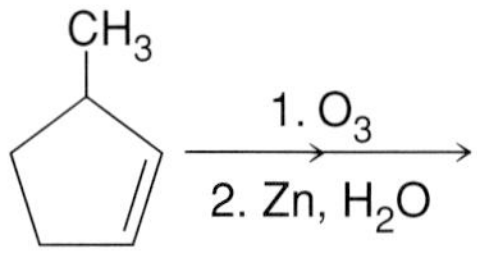

---

Ozonolysis frequently is used to determine the structure of an alkene by examining the ozonolysis products.

---

One or more alkenes gave

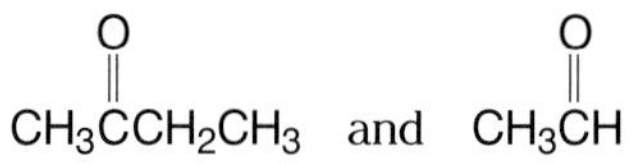

$CH_3CCH_2CH_3$   and   $CH_3CH$

on ozonolysis. Identify the parent alkene(s).

**SOLUTION**
Reverse the process of Solved Problem 6.6.

*Step 1.* Break the two carbon-oxygen double bonds:

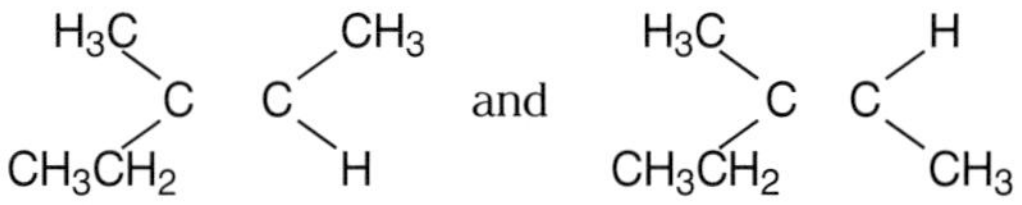

*Step 2.* Place a double bond between each electron-deficient carbon as follows:

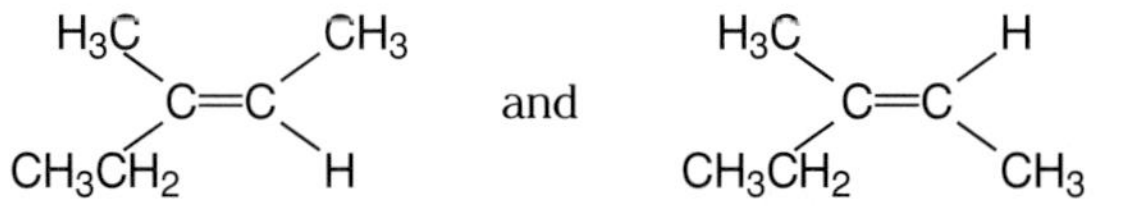

(*E*)-3-Methyl-2-pentene          (*Z*)-3-Methyl-2-pentene

**EXERCISE 6.9**

Provide the structure of one or more alkenes that on ozonolysis would give each of the following product(s).

a.  $CH_3CH_2CH_2\overset{\displaystyle H}{\underset{\displaystyle |}{C}}=O$ (2 mol)

b.  $O=\overset{\displaystyle H}{\underset{\displaystyle |}{C}}CH_2CH_2CH_2CH_2\overset{\displaystyle H}{\underset{\displaystyle |}{C}}=O$

c.  $CH_2=O + CH_3CH_2CH_2CH_2-\overset{\displaystyle O}{\overset{\displaystyle ||}{C}}-CH_3$

---

**Envision the Reaction**

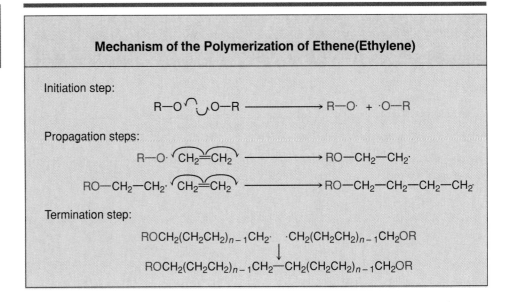

**Mechanism of the Polymerization of Ethene(Ethylene)**

Initiation step:

$$R-O \frown O-R \longrightarrow R-O\cdot + \cdot O-R$$

Propagation steps:

$$R-O\cdot \,\,CH_2{=}CH_2 \longrightarrow RO-CH_2-CH_2\cdot$$

$$RO-CH_2-CH_2\cdot \,\,CH_2{=}CH_2 \longrightarrow RO-CH_2-CH_2-CH_2-CH_2\cdot$$

Termination step:

$$ROCH_2(CH_2CH_2)_{n-1}CH_2\cdot \quad \cdot CH_2(CH_2CH_2)_{n-1}CH_2OR$$
$$\downarrow$$
$$ROCH_2(CH_2CH_2)_{n-1}CH_2-CH_2(CH_2CH_2)_{n-1}CH_2OR$$

**Polymerization**

For an introduction to polymerization, refer to Section 5.6. Polyethylene is produced by the use of a *promoter* such as an organic peroxide (ROOR). Since it is not recovered at the end of the reaction, the organic peroxide is not a true catalyst. The oxygen-oxygen bond is relatively weak and, in the first step of the mechanism, called an *initiation step*, undergoes a homolytic bond breaking to form two alkoxy free radicals (RO·). Remember, a *free radical* is a neutral reactive species that contains one unpaired electron.

This highly reactive alkoxy free radical attacks the pi electrons of ethylene to produce an alkoxyalkyl free radical, which in turn reacts with another molecule of ethylene to give an alkoxyalkyl free radical with a higher formula weight. This larger free radical can continue to react with ethylene molecules, one by one, until the supply of ethylene molecules is exhausted, at which time hundreds or thousands of ethylene molecules have been incorporated into the structure to produce a long carbon chain. Each of these chain-lengthening steps is called a *propagation step*.

When the supply of ethylene is exhausted, the reaction undergoes a *termination step*, in which two long chain alkoxyalkyl free radicals couple. Note the similarity between this reaction mechanism and that of the free-radical chlorination of an alkane.

Regardless of the *-ene* suffix in polyethylene, the polymer is alkane-like in terms of both structure (note the many methylene groups throughout the structure) and chemical properties (polyethylene is a relatively inert substance). Expressing it in another way, the two ether groups (— OR) represent a very small contribution to the structure of the molecule as a whole—so much so that the ether groups are not shown in the abbreviated structure for polyethylene.

Ethylene dibromide (EDB), or 1,2-dibromoethane, came on the scene about half a century ago with the introduction of tetraethyl lead into gasoline to prevent engine "knock." EDB is prepared commercially by bubbling gaseous ethylene (obtained from petroleum) into bromine. EDB is a colorless liquid that boils at a fairly low temperature (131°C) and is insoluble in water.

The use of EDB as an antiknock agent has been severely curtailed for the last decade or so for two reasons. First, EDB is not compatible with catalytic converters and antismog devices in automobiles. Second, the overwhelming majority of motorists use unleaded gasoline that does not require EDB. Ethylene dibromide was a vital part of the antiknock process. It would react with lead(II) oxide, a product of combustion of leaded gasoline, to produce volatile lead tetrabromide that is expelled with the exhaust. Since lead(II) oxide is easily reduced to metallic lead, which causes pits to form on the cylinder walls, elimination of lead as the monoxide is desirable.

Since 1948, however, ethylene dibromide also has been heavily used by farmers as an insecticide and fumigant. It controls the Khapra beetle, rootworms, and the infamous Mediterranean fruit fly.

In the early 1980s, scientists discovered that EDB produced cancer, very quickly, in mice and rats. There was no proof that EDB caused cancer in humans, so the Environmen-

tal Protection Agency (EPA) permitted the continued use of EDB as a pesticide.

The EPA began to change its mind during the summer of 1983 when EDB was found in the groundwater in Florida and other states. The compound diffused into the groundwater from the topsoil when it was injected into the citrus grove soil to control rootworms. To add insult to injury, higher than expected levels were found in cake mixes on grocery shelves because EDB often is used as a fumigant for grain, both in the field and in the storehouse.

Several years ago, the EPA suggested new guidelines for EDB: no more than 30 parts per billion (ppb) in food ready to eat and no more than 900 ppb in grain that has just been harvested.

Manufacturers of EDB insisted that the EPA has overreacted to possible health problems caused by EDB. And they have a point, since studies have shown that EDB decomposes when the air temperature is above 65°F. Thus grain stored in silos and cake mixes heated in an oven should lose just about all the EDB present.

The EPA would rather be safe than sorry, so EDB was removed from the commercial market as a grain fumigant in 1984. Recently, the decision to ban EDB was proven to be wise because new studies indicate that the rate of decomposition of EDB at 72°F is very slow; only half the EDB decomposed after 400 days.

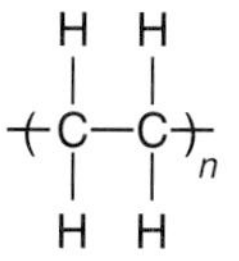

Abbreviated structure of polyethylene

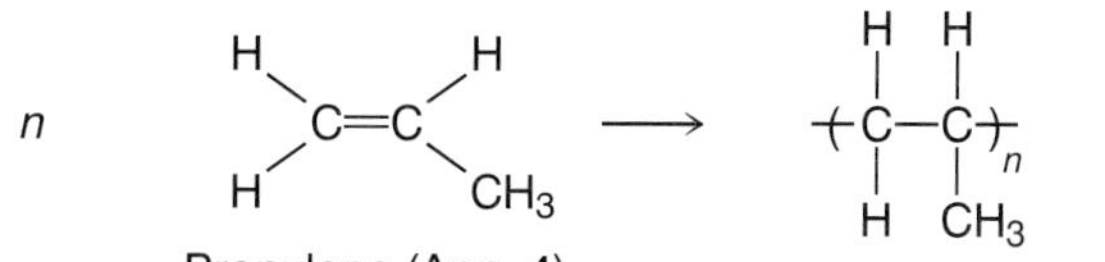

Propylene (App. 4) (propene) → Polypropylene

Polypropylene is used in plastics and fibers.

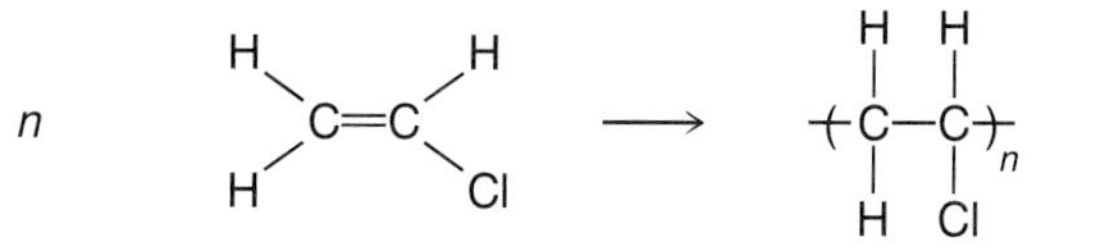

Vinyl chloride (1-chloroethene) → Polyvinyl chloride

Note the vinyl group:

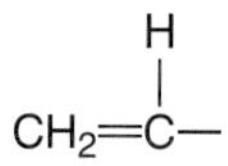

These polymers are sometimes called *vinyl polymers* when the monomer is represented as $CH_2{=}CHG$ or $CH_2{=}CG_2$, where G is any group. Polyvinyl chloride is used in the manufacture of water pipe and phonograph records.

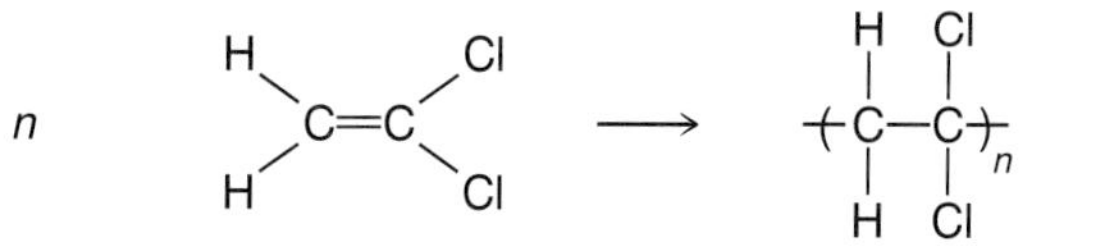

Vinylidine chloride 1,1-dichloroethene → Saran

Saran is used as a transparent packaging material to preserve the freshness of food.

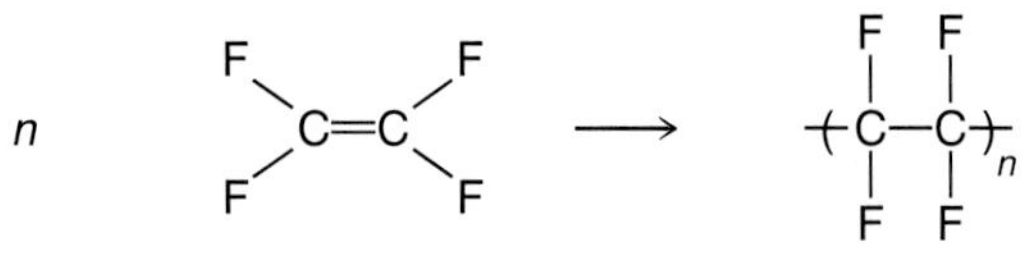

$$n \quad \begin{array}{c} F \\ \diagdown \\ F \end{array} C = C \begin{array}{c} F \\ \diagup \\ F \end{array} \longrightarrow \left( \begin{array}{c} F \quad F \\ | \quad | \\ C - C \\ | \quad | \\ F \quad F \end{array} \right)_n$$

Tetrafluoroethylene                          Teflon

Teflon is coated on pots and pans to make them resistant to food sticking on the cooking surface. These alkene polymers, especially Teflon, are unreactive to most reagents and are superb insulators.

---

**SOLVED PROBLEM 6.8**

Draw an expanded line-bond structural formula for polyvinyl chloride showing five repeating units.

**SOLUTION**

Since polymerization proceeds in a head-to-tail structural mode, we can draw the following:

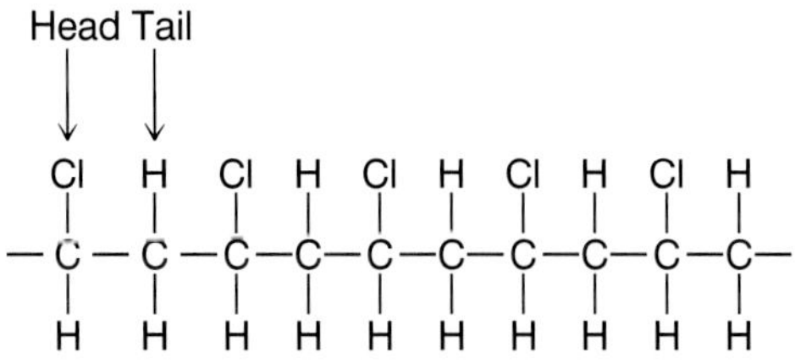

Head  Tail
↓      ↓

Cl   H   Cl   H   Cl   H   Cl   H   Cl   H
|    |    |    |    |    |    |    |    |    |
—C — C — C — C — C — C — C — C — C — C—
|    |    |    |    |    |    |    |    |    |
H    H   H    H   H    H   H    H   H    H

---

**EXERCISE 6.10**

Polyacrylonitrile (Orlon) is a vinyl polymer that is used mainly as a fiber in clothing. The monomer of Orlon is

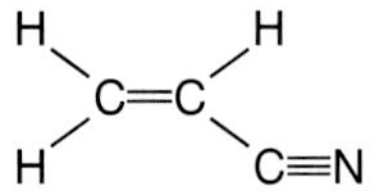

$$\begin{array}{c} H \\ \diagdown \\ H \end{array} C = C \begin{array}{c} H \\ \diagup \\ C \equiv N \end{array}$$

Acrylonitrile

Draw an expanded line-bond structural formula for Orlon showing five repeating units.

---

**6.4
HOW TO DO A
MULTISTEP ORGANIC
SYNTHESIS**

Exercise 6.11 and Problems 20 through 22 ask you to do a number of syntheses, some of which are multistep. Before you tackle these synthesis problems, you should know the methods of preparation and the reactions of both alkanes and alkenes. In working a synthesis, you are not to write the structure of any by-products formed. Simply write the structure of the organic product as in Solved Problem 6.9. It is sometimes helpful to work the problem backwards—from final product to starting material.

---

**SOLVED PROBLEM 6.9**

Prepare

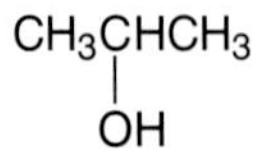

$$\begin{array}{c} CH_3CHCH_3 \\ | \\ OH \end{array}$$

2-Propanol

from

$$CH_3CH_2CH_2OH$$

1-Propanol

## SOLUTION

Note the carbon contents of the starting material and the product. Since there is no change in the number of carbons, a reaction such as ozonolysis or the Corey-House synthesis cannot be used. Since the product alcohol is derived from a secondary carbocation, Markownikoff's rule can be used to prepare it from propene.

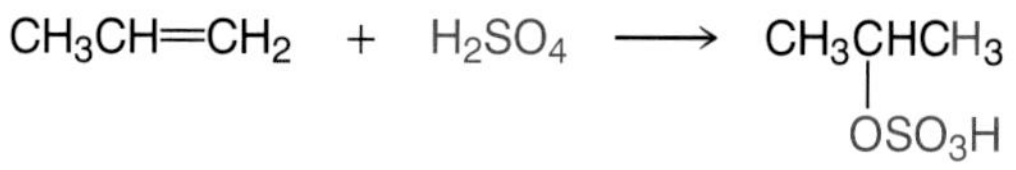

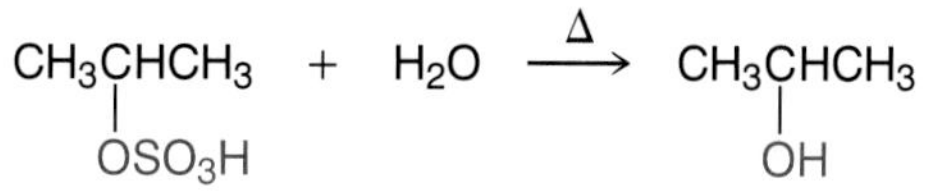

$$CH_3CH{=}CH_2 \;+\; H_2SO_4 \longrightarrow CH_3\underset{\overset{|}{O}SO_3H}{C}HCH_3$$

$$CH_3\underset{\overset{|}{O}SO_3H}{C}HCH_3 \;+\; H_2O \xrightarrow{\;\Delta\;} CH_3\underset{\overset{|}{O}H}{C}HCH_3$$

All that remains is to prepare propene from *n*-propyl alcohol.

$$CH_3CH_2CH_2OH \xrightarrow[\Delta]{H_3PO_4} CH_3CH{=}CH_2$$

Putting it all together we get

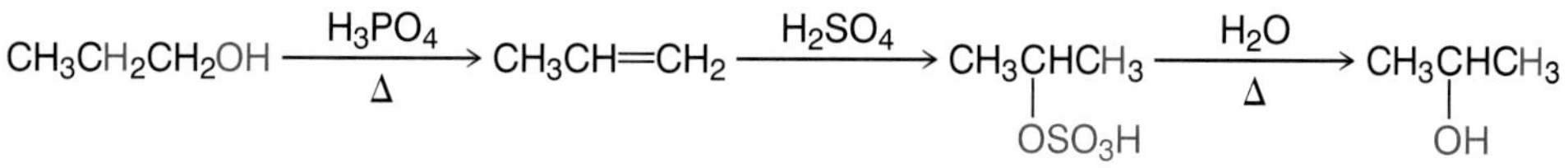

$$CH_3CH_2CH_2OH \xrightarrow[\Delta]{H_3PO_4} CH_3CH{=}CH_2 \xrightarrow{H_2SO_4} CH_3\underset{\overset{|}{O}SO_3H}{C}HCH_3 \xrightarrow[\Delta]{H_2O} CH_3\underset{\overset{|}{O}H}{C}HCH_3$$

---

**SOLVED PROBLEM 6.10**

Prepare

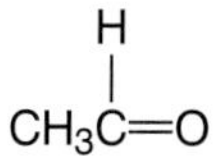

$$CH_3\overset{\overset{\textstyle H}{|}}{C}{=}O$$

Acetaldehyde

from

$$CH_3\underset{\overset{|}{O}H}{C}HCH_3$$

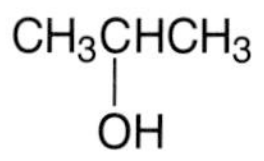

2-Propanol

## SOLUTION

Note that the product molecule has only two carbons, while the reactant molecule has three carbons. This must mean that an ozonolysis reaction is part of the synthesis.

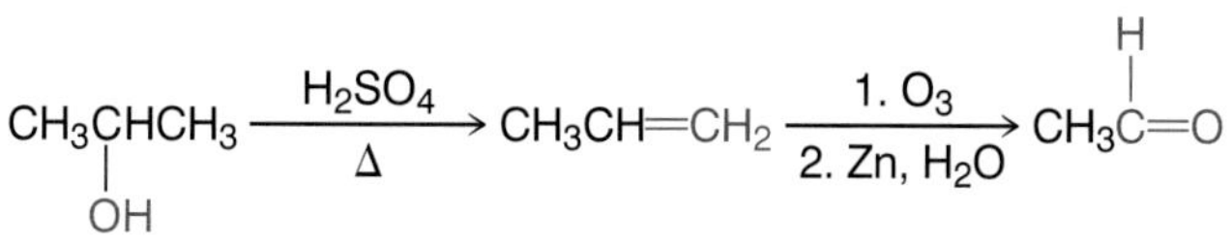

$$CH_3\underset{\overset{|}{O}H}{C}HCH_3 \xrightarrow[\Delta]{H_2SO_4} CH_3CH{=}CH_2 \xrightarrow[\text{2. Zn, H}_2O]{\text{1. O}_3} CH_3\overset{\overset{\textstyle H}{|}}{C}{=}O$$

Note, that

$$\overset{\textstyle CH_2}{\underset{\textstyle O}{\|}}$$

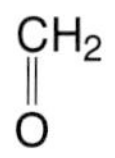

named formaldehyde, also is formed as a by-product, which, in this case, is of no interest to us.

| **EXERCISE 6.11** | Prepare each of the following compounds starting with $CH_3CH(OH)CH_3$ (isopropyl alcohol). (You may use any inorganic reagents necessary.) |

a. $CH_3CH_2CH_3$

b. $CH_3CH(Br)CH_3$

c. $CH_3CH(OH)CH_2Br$

d. $CH_3CH(Li)CH_3$

e. $CH_3CH_2CH_2OH$

f. $CH_3CH(OSO_3H)CH_3$

## ▶ CHAPTER ACCOMPLISHMENTS

### 6.1 Synthesis of Alkenes
☐ Explain why a mixture of the 2-butenes (*cis*-2-butene and *trans*-2-butene) is a major product and 1-butene a minor product when 2-butanol is dehydrated.

☐ List two inorganic acid catalysts used in the dehydration of an alcohol.

☐ Draw a general equation to represent the dehydrohalogenation of an alkyl halide to produce an alkene.

☐ Show that propene is a monosubstituted alkene by drawing a structural formula for propene and circling the only alkyl group present.

### 6.2 Addition Reactions of the Alkenes
☐ Explain why the reagent bromine in carbon tetrachloride can be used as a qualitative test to distinguish an alkene from an alkane.

☐ Draw a structural formula for the bromonium ion derived from propene.

☐ Provide the mode of addition (*syn* or *anti*) for each of the following alkene addition reactions.
   a. hydrogenation
   b. bromination
   c. hydroxylation
   d. bromohydrin formation

☐ State Markownikoff's rule.

☐ List the alkene addition reagents that show a Markownikoff or Markownikoff-type position orientation.

☐ Explain why the isopropyl carbocation is more stable than the *n*-propyl carbocation.

☐ Write a detailed, step-by-step mechanism for the reaction of HCl with propene.

☐ Supply a formula for diborane as represented chemically in this textbook.

### 6.3 Other Reactions of Alkenes
☐ Identify the two allylic hydrogens in the allyl group.

☐ Draw a structural formula for succinimide.

☐ Identify the three allylic hydrogens located in the propylene molecule.

☐ Draw an expanded line-bond structural formula for polyvinyl chloride showing five repeating units.

☐ Explain the value of coating a pot or pan with Teflon.

☐ Name an alkene reaction in which the product or products contain a larger number of carbons than the alkene used.

### 6.4 How to Do a Multistep Organic Synthesis
☐ Explain why 2-propanol cannot be prepared from propene using a hydroboration-oxidation reaction.

☐ Name an alkene reaction in which the product or products contain a smaller number of carbons than the alkene used.

## ▶ KEY TERMS

*elimination reaction* (6.1)
*dehydration* (6.1, 7.3, 12.5, 15.2)
*Le Châtelier's principle* (6.1, 23.6)
*Saytzeff's rule* (6.1)
*rearrangement* (6.1, Problem 6.22, 10.1, 12.5)
*dehydrohalogenation* (6.1, 7.3, 8.6, 8.6)
*monosubstituted alkene* (6.1)

*disubstituted alkene* (6.1)
*trisubstituted alkene* (6.1)
*tetrasubstituted alkene* (6.1)
*vicinal dihalide* (*vic-*) (6.2)
*electrophilic reagent or electrophile* (6.2)
*bromonium ion* (6.2)

*Markownikoff's rule* (6.2, 8.7)
*inductive effect* (6.2, 10.2)
*hydration* (6.2)
*Markownikoff-type addition* (6.2)
*allylic hydrogen* (6.3)
*promoter* (6.3, 19.3)
*vinyl polymer* (6.3)

## PROBLEMS

1. Which compound, of the following pairs of compounds, is more stable? Explain your answers.
   a. 1-pentene or 2-pentene
   b. 1-hexene or 2-methyl-2-pentene
   c. 2-methyl-2-pentene or 2,3-dimethyl-2-butene

2. Complete each of the following equations by supplying a structural formula for each organic product formed. Indicate which product is the principal product when appropriate by the application of Saytzeff's rule, and then use the principle of alkene stability as a check.

   a. $CH_3CHCH_2CH_3 + KOH \xrightarrow[\Delta]{alcohol}$
      |
      $CH_2Br$

   b. $CH_3CCH_2CH_2CH_3 + KOH \xrightarrow[\Delta]{alcohol}$
      with $C_2H_5$ above C and $Br$ below

   c. 1-Bromo-1-cyclohexylethane + KOH $\xrightarrow[\Delta]{alcohol}$

3. Draw a structural formula for each organic product, and indicate whether each product is major or minor (when applicable). Use the principle of alkene stability, and then employ Saytzeff's rule as a check.

   a. 2-Iodo-2-methylpentane + KOH $\xrightarrow[\Delta]{alcohol}$

   b. 1-Bromo-1,2-dimethylcyclopentane + KOH $\xrightarrow[\Delta]{alcohol}$

   c. $CH_3CH_2CH_2OH \xrightarrow[\Delta]{H_2SO_4}$

4. Calculate the formal charge on each atom of the following species, and demonstrate that the structure is an ion with a +1 charge.

   H H
   |  |
   H—C—C—H
   |
   ·Br·

5. Draw the structural formula of an alkene that contains three ethyl groups bonded to the carbons of the double bond. Name the alkene.

6. Why is it simple, experimentally, to demonstrate the *anti*-mode of addition of $Br_2$ to cyclohexene but impossible to demonstrate it with 2-pentene? Explain.

7. Since EDB is prepared by treating bromine with ethylene, draw a structural formula for ethylene dibromide.

8. Select the most stable carbocation from each of the following pairs.

   a. $CH_3$—C—H with $CH_2^+$ below   or   $CH_3$—$\overset{+}{C}$—$CH_2CH_3$ with H above

   b. cyclopentane ring with $\overset{+}{CH}$   or   $(CH_3)_3C^+$

   c. cyclopentane ring with $CH_2$—$\overset{+}{CH}CH_3$   or   $CH_3CH_2CH_2\overset{+}{CH}_2$

9. Name a carbocation that is neither primary, secondary, nor tertiary.

10. Using one or more test tube reactions, show how you would distinguish between
    a. 1-pentene and pentane.
    b. 1-pentene and cyclopentane.

    Indicate what reagent(s) should be used and what color changes, etc. would be expected.

11. Supply the electrophile in each of the following reagents reacting with an alkene.
    a. $Br_2$
    b. HBr
    c. $H_2SO_4$

12. Write a detailed, step-by-step mechanism for the reaction of 2-methylpropene with
    a. $Br_2$.
    b. HCl.
    c. $Br_2 + H_2O$.

13. Draw a structural formula for
    a. triethylborane.
    b. ozone.
    c. *N*-bromosuccinimide.
    d. the allyl group (App. 5).
    e. the vinyl group (App. 5).

14. Provide the structure of an alkene that on ozonolysis would give each of the following product(s).

    a. $O=CCH_2—CH—C=O$ with H, $CH_3$, H below

    b. $H—CCH_2CH_3$ (2 mol) with O double-bonded above C

    c. $CH_2=O + CH_3CH_2CH=O$

    d. two cyclohexane rings attached to a central C=O group   +   $CH_2O$

15. Classify each of the following dissociations (bond-breaking representations) as homolytic or heterolytic.
    a. $H{-}Br \longrightarrow H^+ + Br^-$

    b. $CH_3\overset{\overset{O}{\|}}{C}{-}O{-}O{-}\overset{\overset{O}{\|}}{C}CH_3 \longrightarrow 2\ CH_3\overset{\overset{O}{\|}}{C}{-}O\cdot$

    c. $HOBr \longrightarrow H^+ + {}^-OBr$

16. Draw the appropriate structural formula and write the molecular formula, formula unit to replace the question mark(s) in each of the following equations.

    a. $? + H_2 \xrightarrow{Pd}$ 2-methylpropane

    b. Cyclopentene $+ ? \xrightarrow{\ ?\ }$ trans-1,2-dibromocyclopentane

    c. $? + Br_2 \xrightarrow{H_2O}$ propylene bromohydrin

    d. 3 Propylene $+ BH_3 \longrightarrow$ ?

17. Draw an expanded line-bond structural formula for each of the following polymers showing five repeating units.
    a. Teflon

    b. polystyrene $\displaystyle \underset{\substack{| \\ H}}{\overset{\substack{H \\ |}}{C}}{-}\underset{\substack{| \\ H}}{\overset{\substack{C_6H_5 \\ |}}{C}}\Big)_n$

    c. polyvinyl acetate $\displaystyle \underset{\substack{| \\ H}}{\overset{\substack{H \\ |}}{C}}{-}\underset{\substack{| \\ O}}{\overset{\substack{H \\ |}}{C}}\Big)_n$
    $\overset{|}{\underset{|}{C}}{=}O$
    $CH_3$

18. Lucite is a vinyl polymer that is used as an unbreakable substitute for glass in automobiles. The monomer of Lucite is

    $\displaystyle \overset{H}{\underset{H}{\phantom{}}}\!\!\!\diagdown C{=}C\diagup\overset{CH_3}{\underset{C{-}O{-}CH_3}{\phantom{}}}$
    $\overset{\|}{O}$

    **Methyl methacrylate**

    Draw an expanded line-bond structural formula for polymethyl methacrylate showing five repeating units.

19. Draw a structural formula for each monomer of the polymers in Problem 17.

20. Prepare each of the following compounds starting with $C_2H_5Br$. You may use any inorganic reagents necessary.
    a. $CH_3CH_3$
    b. $H_2C{=}O$
    c. $C_2H_5OH$
    d. $\underset{\substack{| \\ OH}}{CH_2}{-}\underset{\substack{| \\ Br}}{CH_2}$
    e. $\underset{\substack{| \\ OH}}{CH_2}{-}\underset{\substack{| \\ OH}}{CH_2}$

f. $C_2H_5Li$
g. polyethylene
h. $CH_3CH_2OSO_3H$
i. $CH_3CH_2Cl$

21. Prepare each of the following. You may use any inorganic reagents necessary.
    a. $CH_3CH_2CH_2OH$ from $CH_3CH(OH)CH_3$
    b. $CH_3CH_2CH(Br)CH_3$ from $CH_3CH_2CH_2CH_2Cl$
    c. $CH_3CH_2CH_2CH_3$ from $CH_3CH_2Br$
    d. $CH_3CH_2CH(OH)CH_2Br$ from $CH_3CH_2CH_2CH_2Br$

22. Prepare each of the following. (You may use any inorganic reagents necessary.) Assume that no rearrangements occur. A *rearrangement* is a reaction that can result in a product with an unexpected pattern of carbon branching or a product with a functional group in an unexpected location on the carbon chain (see Sec. 6.1). These unexpected products are due to the shifting of a hydrogen or a methyl from one carbon to a neighboring carbon. We will study rearrangements in Chaps. 10 and 12.
    a. 2-triacontanol from 1-triacontanol

    $$CH_3(CH_2)_{27}CH_2CH_2OH \longrightarrow CH_3(CH_2)_{27}CH(OH)CH_3$$

    b. 2-chlorotriacontane from 1-triacontanol
    c. 1-bromo-2-hydroxytriacontane from 1-triacontanol

23. Complete each of the following equations by writing a structural formula for each organic product and a molecular formula or formula unit for each inorganic product formed.

    a. $CH_3CH_2CH_2CH_2CH_2OH \xrightarrow{\Delta,\ H_3PO_4}$

    b. $\underset{\substack{CH_3CH_2}}{\overset{\substack{H_3C}}{\phantom{}}}\!\!\!\diagdown C{=}CH_2 + H_2 \xrightarrow{Ni}$

    c. $CH_3CH_2CH_2I + Mg \xrightarrow[\text{diethyl ether}]{\text{anhydrous}}$

    d. $CH_3{-}\underset{\substack{| \\ H}}{\overset{\substack{H \\ |}}{C}}{=}\underset{\substack{| \\ H}}{\overset{\substack{H \\ |}}{C}}{-}CH_2CH_3 + H_2 \xrightarrow{Pd}$

    e. Cyclopentene $+ KMnO_4 + H_2O \xrightarrow{cold}$

    f. $CH_3CH_2\overset{\substack{H \\ |}}{C}{=}\overset{\substack{H \\ |}}{C}CH_3 + Cl_2 \xrightarrow[20°C]{CCl_4}$

    g. $CH_3CH_2\overset{\substack{H \\ |}}{C}{=}\overset{\substack{H \\ |}}{C}CH_3 + HCl \longrightarrow$

    h. $\overset{OSO_3H}{\diagup\!\!\!\bigcirc\!\!\!\text{-cyclopentane}} + H_2O \xrightarrow{\Delta}$

    i. $\underset{\substack{H_3C}}{\overset{\substack{H_3C}}{\phantom{}}}\!\!\!\diagup CH{-}\overset{\substack{H \\ |}}{C}{=}CH_2 + NBS \longrightarrow$

j.   Cyclohexene + $Cl_2$ $\xrightarrow{H_2O}$

k.   $CH_2{=}CHCH_2CH_3$ + $H_2SO_4$ $\longrightarrow$

l.   1-Methylcyclohexene $\xrightarrow[\text{2. } H_2O_2,\ ^-OH]{\text{1. } BH_3}$

m. $n$ $CH_3CH{=}CH_2$ $\xrightarrow{\text{polymerization}}$

24. Write a detailed, step-by-step mechanism for the following reaction.

$$CH_3CH{=}CH_2 + Br_2 \xrightarrow{\Delta} BrCH_2\,CH{=}CH_2 + HBr$$

# CHAPTER 7

# Alkenes (III)

## RESONANCE, REACTIONS OF DIENES, AND TERPENES

In some ways, the chemical properties of the dienes and monoenes are similar; in other ways, they are quite different. This chapter explores the similarities and differences in some depth.

Before considering the synthesis and reactions of the dienes, we need to look at a few properties. Isolated dienes behave like the monoenes. They are stable and can be isolated from a reaction mixture. Conjugated dienes (see Sec. 5.1) are more stable than the corresponding isolated dienes. This stability difference can be determined by examining the heat of hydrogenation of selected monoenes and dienes.

## 7.1 STABILITY OF THE DIENES

$$\text{C=C} + H_2 \xrightarrow{\text{Pd}} -\underset{H}{\overset{|}{C}}-\underset{H}{\overset{|}{C}}- + \text{heat}$$

The heat liberated by a monoene depends on the structure of the alkene but can be well represented by 30.1 kcal/mol. Since 1,4-pentadiene contains two double bonds, we would predict the heat of hydrogenation of the diene to be twice the value of the heat of hydrogenation of a monoene such as 1-pentene ($30.1 \times 2 = 60.2$) because the product of hydrogenation of both compounds is the same—pentane.

Furthermore, it can be verified by experiments (Table 7.1) that the isolated diene 1,4-pentadiene has an experimental heat of hydrogenation of 60.8 kcal/mol, whereas the conjugated diene 1,3-pentadiene has an experimental value of 54.1 kcal/mol. The difference between the two values ($60.2 - 54.1 = 6.1$ kcal/mol) is a measure of the stability

*The hydrogenation of an alkene is always an exothermic reaction.*

*The product n-pentane contains the same internal energy regardless of which starting diene was employed.*

*Note that the internal energy of a molecule is the sum of the potential energy (energy of position) and the kinetic energy (energy of motion) of that molecule.*

| TABLE 7.1 | Heats of Hydrogenation of Selected Alkenes | |
|---|---|---|
| Compound | Experimental Heat of Hydrogenation (kcal/mol of alkene) | Predicted Heat of Hydrogenation (kcal/mol of alkene) |
| 1-Pentene (1 mol $H_2$) | 30.1 | 30.1 |
| 1,4-Pentadiene (2 mol $H_2$) | 60.8 | 60.2 |
| 1,3-Pentadiene (2 mol $H_2$) | 54.1 | 60.2 |

Note that any difference in $\Delta H$ (the heat of hydrogenation) is directly proportional to the internal energy differences in the starting dienes $(e_{1,4\text{-pentadiene}} - e_{n\text{-pentane}})$ and $(e_{1,3\text{-pentadiene}} - e_{n\text{-pentane}})$. Since $e_{n\text{-pentane}}$ is the same value no matter what the diene in this case, $\Delta H$ is directly proportional to $e_{\text{diene}}$.

**FIGURE 7.1** Plot of internal energy for some selected dienes.

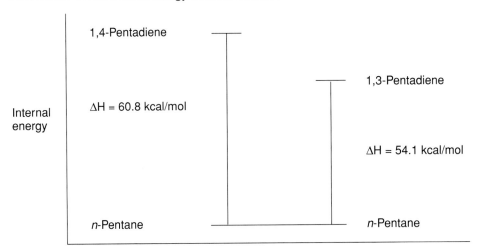

Both conjugated and isolated dienes are stable molecules, but conjugated diene molecules are more stable.

of a conjugated diene over an isolated diene. Since 1,3-pentadiene contains less internal energy, it is the more stable molecule (Fig. 7.1).

To understand why 1,3-pentadiene is the more stable diene, let us refer to Fig. 7.1. Note that when the internal energy of a molecule is high, the increased movement of electrons within the molecule would lead to bond rupture and fragmentation of the molecule, which translates into low stability. On the other hand, a molecule that has a low internal energy is less likely to undergo bond rupture and is more stable. Owing to this significant difference in stability, conjugated dienes show a number of differences in chemical behavior with respect to isolated dienes.

## 7.2 RESONANCE

Resonance is an extension of the structural theory that can explain molecule or ion stability as well as the formation of certain unexpected reaction products that are isolated. Five basic tenets of the resonance theory are as follows:

1. *Resonance occurs when pi electrons of a double bond or unshared electrons on an atom can shift so as to create a structurally reasonable alternative. The position of atoms cannot change, only that of pi and unshared electrons.* For example, consider the two resonance structures for nitromethane ($CH_3NO_2$):

Note that the only reasonable line-bond structure that can be written for nitromethane must have a positive formal charge on nitrogen and a negative formal charge on oxygen.

2. *These resulting imaginary structures are known as contributing structures or resonance structures. In general, the greater the number of contributing structures, the more stable the molecule or ion.* Since the contributing structures of nitromethane exhibit similar bonding, they are said to be equivalent.

3. *The true structure lies somewhere between the contributing structures and is known as the resonance hybrid*

In the case of nitromethane, the resonance hybrid shows that each nitrogen-oxygen bond is halfway between a single and double bond in character. Since the two contributing structures are equivalent, each structure contributes equally to the structure of the hybrid.

4. *Not all contributing structures are equivalent.* For example, consider formaldehyde:

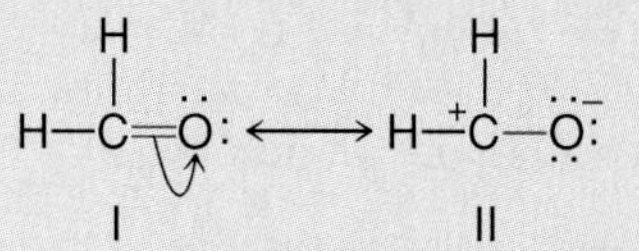

Structure I is a *major contributor* to the resonance hybrid, while structure II is a *minor contributor*

5. *Characteristics of a major resonance contributor include the following*:

a. The major contributing structure is the most stable.

b. A major contributor is particularly stable when each period 2 element (of the periodic table) in the structure contains an octet of electrons.

c. Each major contributing structure contains the largest number of covalent bonds.

d. Each major contributor contains a negative charge on a more electronegative atom or a positive charge on a less electronegative atom.

e. Each major contributor has a minimum amount of separation of opposite charges.

Let us examine some other examples of resonance involving positively charged species:

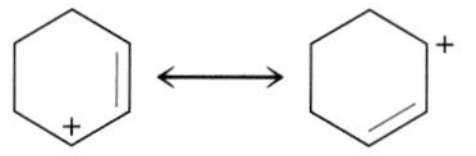

Structures III and IV are equivalent in that both exhibit similar bonding.

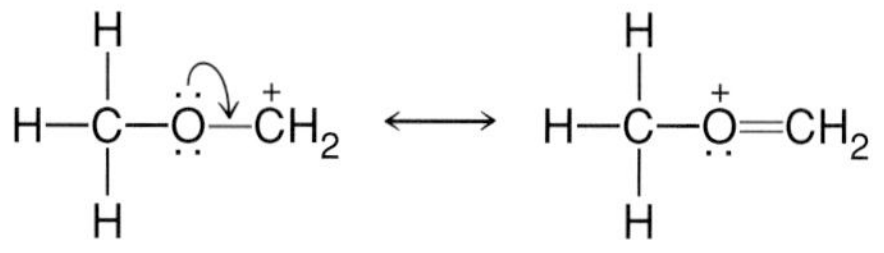

At first glance you might predict that structure V is the major contributor because the positive charge is on the less electronegative element, carbon. In actuality, structure VI is the major contributor because each period 2 element in the structure contains an octet of electrons, and this type of contributing structure is particularly stable.

Now, there are some instances in which contributing structures cannot be formed by delocalization of either pi or unshared electrons. Consider the following chemical species:

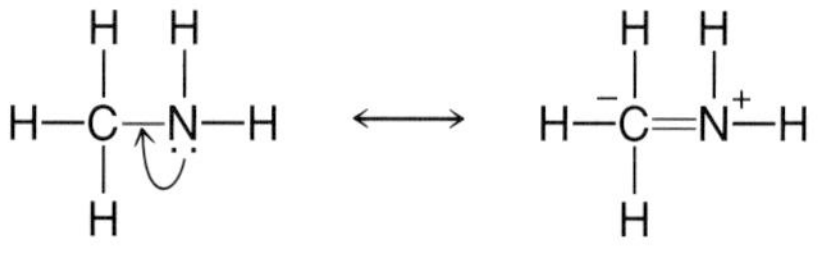

Structure VIII is not a contributing resonance structure of methylamine because carbon shares 10 electrons. No resonance structure can be written if the structure represents a violation of the structural theory. Thus the compound methylamine is not resonance stabilized.

Resonance is indicated by a double-headed arrow. We use curved arrows to show the movement of pi and unshared electrons. Although the double-headed arrow may suggest it, these structures are not in equilibrium. Instead, each structure represents an aspect of (a contribution to) the true structure. Let us now discuss the distinction between an example of resonance and an example of an equilibrium. Resonance involves the movement only of pi and unshared electrons (i.e., no movement of atoms takes place), whereas in an equilibrium one or more atoms are moved along with two or more electrons. The symbolism for each is different: for resonance, a double-headed arrow ($\longleftrightarrow$); for an equilibrium, two arrows with oppositely positioned heads ($\rightleftarrows$). An example of an equilibrium is

$$NH_3 + H_2O \rightleftarrows {}^{+}NH_4 + {}^{-}OH$$

---

**SOLVED PROBLEM 7.1**

Which of the following represents an equilibrium, and which represents an example of resonance?

a.  $CH_3COOH$ to $CH_3COO^- + H^+$

b.  $CH_2{=}CH_2$ to ${}^{+}CH_2{-}CH_2^{-}$

**SOLUTION**

a.  This is an equilibrium. This cannot be an example of resonance because an atom (H) is moved. In resonance, only pi and unshared electrons are moved (delocalized). Using the symbolism for an equilibrium, we have

$$CH_3COOH \rightleftharpoons CH_3COO^- + H^+$$

b.  This is an example of resonance because only pi electrons are delocalized. Using the symbolism for resonance, we have

$$CH_2{=}CH_2 \longleftrightarrow \overset{+}{C}H_2{-}\overset{-}{C}H_2$$

---

**EXERCISE 7.1**

Indicate which of the following is an example of resonance. Explain.

a.  $CH_3{-}\underset{\overset{\|}{O}}{C}{-}H$ to $CH_2{=}\underset{\overset{|}{OH}}{C}{-}H$

b.  $CH_2{=}CH{-}CH{=}CH_2$ to $CH_2{-}CH{=}CH{-}\overset{-}{\underset{\cdot\cdot}{C}}H_2$

c.  $CH_3CH_2CH{=}CH_2$ to $CH_3CH_2\overset{+}{C}H{-}\overset{-}{\underset{\cdot\cdot}{C}}H_2$

---

**SOLVED PROBLEM 7.2**

Draw resonance contributing structures for each of the following. Select the major contributing structure (if any), and explain why it is the major contributor.

a.  $CH_3\overset{-}{\underset{\overset{\|}{O}}{C}}CH_2$

c.  $^-{:}CH_2\overset{+}{C}HCH{=}CH_2$

b.  $CH_3\overset{\cdot\cdot}{\underset{\overset{\|}{O}}{C}}\overset{\cdot\cdot}{O}{:}^-$

d.  $CH_3\underset{\overset{\|}{O}}{C}NH_2$

**SOLUTION**

a.  First, expand the condensed formula given. Then, using curved arrows to show electron delocalization, formulate the second contributing structure. Notice that more than one pair of electrons can be delocalized in one contributing structure to produce another contributing structure.

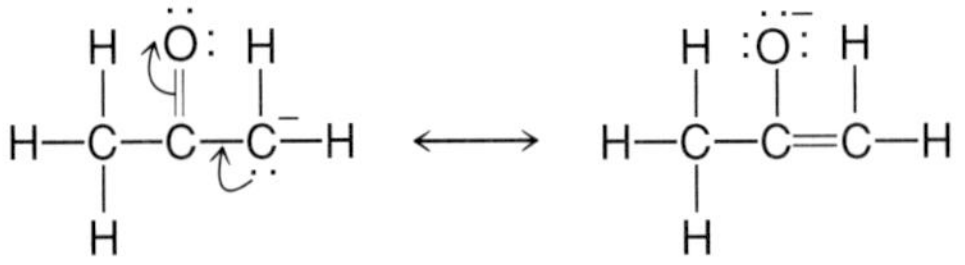

The distinguishing feature between structures 1 and 2 is that in structure 1 the negative charge is located on a carbon atom, whereas in structure 2 it is positioned on an oxygen atom. Since oxygen is the more electronegative atom, structure 2 is the major contributor to the resonance hybrid.

b.  Again, expand the condensed formula given. Then, using curved arrows to show electron delocalization, formulate the second contributing structure.

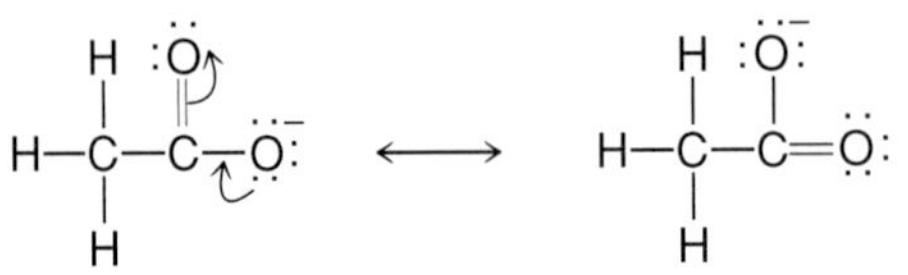

Note that the number (seven) and kind of covalent bonds (three carbon-hydrogen, one carbon-oxygen:single, one carbon-oxygen:double) in structures 3 and 4 are the same, along with the location of a negative charge in both structures 3 and 4 on the oxygen of the carbon-oxygen single bond. Thus the structures are equivalent, and each structure contributes equally to the resonance hybrid.

c.  Expand the condensed formula given. Then, using a curved arrow to show electron delocalization, formulate the second contributing structure.

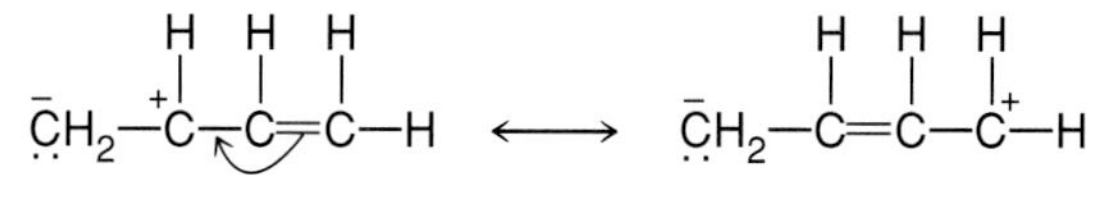

5                                    6

Structure 5 is the major contributor because there is a minimum amount of separation of opposite charges in the structure.

d.  Expand the condensed structure given. Then, using curved arrows to show electron delocalization, formulate the second contributing structure.

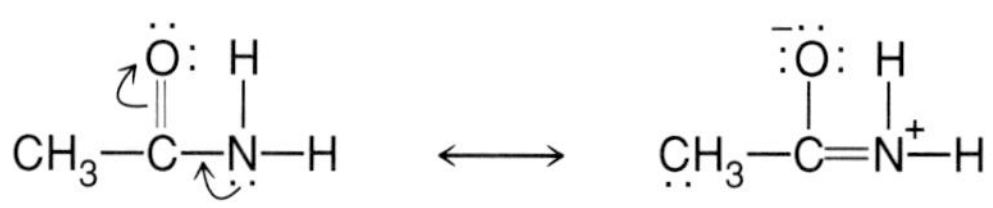

7                                    8

Structure 7 is the major contributor because it is a neutral molecule. The positive and negative charges present in structure 8 are destabilizing.

---

**EXERCISE 7.2**

Designate which of the following contributing structures is the more important contributor (if any). Explain your answer.

a.

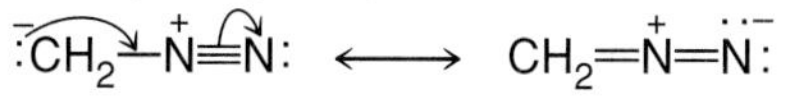

b.

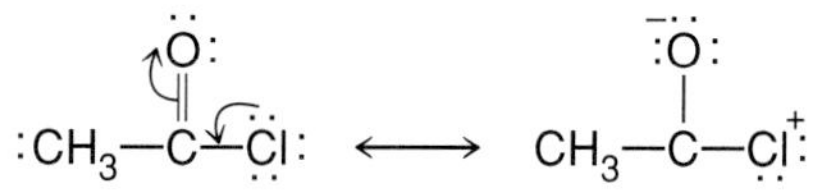

---

In this section, since we have examined a variety of chemical species containing one or more charges, you should refer to Section 1.10 for a review of formal charge to enable you predict when an atom in a structure bears a charge.

---

**SOLVED PROBLEM 7.3**

Consider the following species:

$$^-\!:CH_2\!-\!\overset{+}{N}\!\equiv\!N:$$

a.  Show that carbon bears a negative charge.
b.  Show that the middle nitrogen bears a positive charge.
c.  Show that the end nitrogen is neutral.
d.  Show that the species is neutral.

**SOLUTION**

a.  For carbon, we have

$$4 - 6/2 - 2 = -1$$

b.  For the middle nitrogen, we have

$$5 - 8/2 - 0 = +1$$

c.  For the end nitrogen, we have

$$5 - 6/2 - 2 = 0$$

d.  Given that the two hydrogens are neutral in the species (formal charge on hydrogen $= 1 - 2/2 - 0$), we have

Charge on the species $= -1 \cdot 1(C) + 1 \cdot 1(N) + 0 \cdot 1(N) + 0 \cdot 2(H) = 0$
Middle        End<br>nitrogen   nitrogen

---

**EXERCISE 7.3**

Show that the magnitude and sign of the charge on each of two carbons in structure 6 in Solved Problem 7.2 are correct.

---

Now let us explain the greater stability of 1,3-pentadiene over 1,4-pentadiene, a difference of 6.1 kcal/mol by the use of resonance theory. This difference in stability is associated with resonance and is therefore known as *resonance energy*

Three contributing structures involving both double bonds can be written for 1,3-pentadiene.

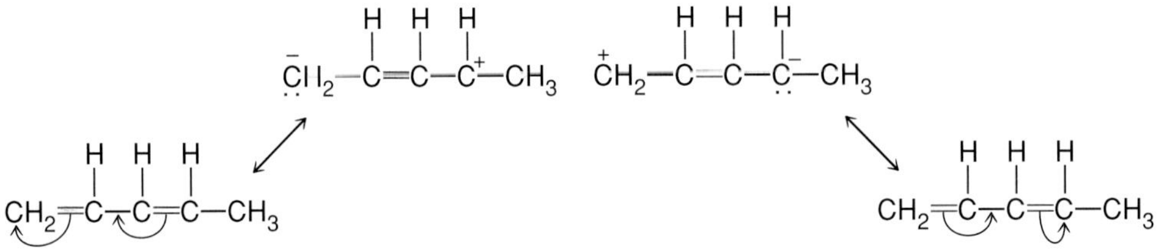

No comparable contributing structures can be written for 1,4-pentadiene because the double bonds are separated by a methylene group.

## 7.3
## SYNTHESIS OF DIENES

### Dehydration of Diols

In general,

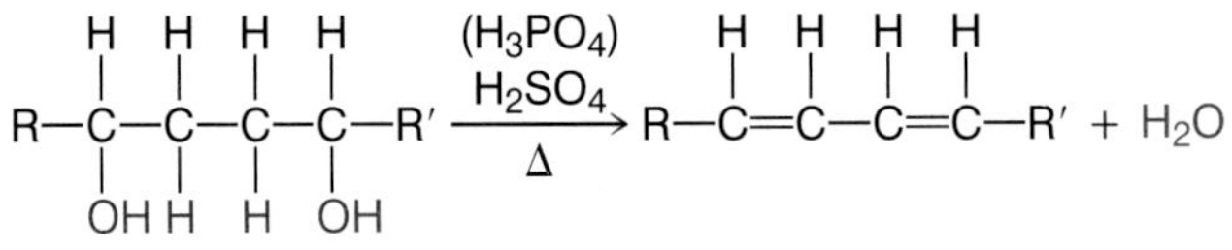

Example:

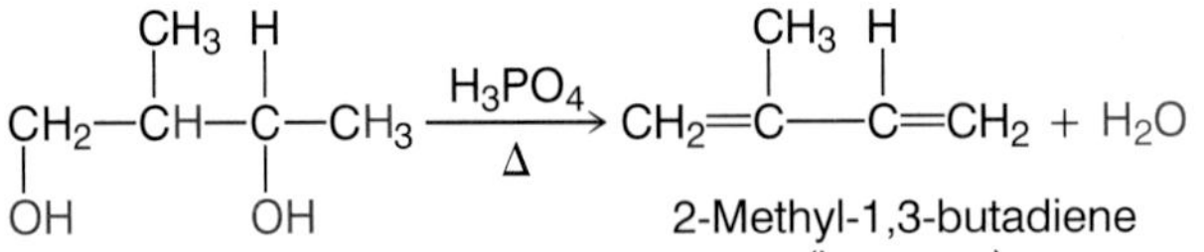

2-Methyl-1,3-butadiene
(isoprene)

### Dehydrohalogenation of Dihalides

In general,

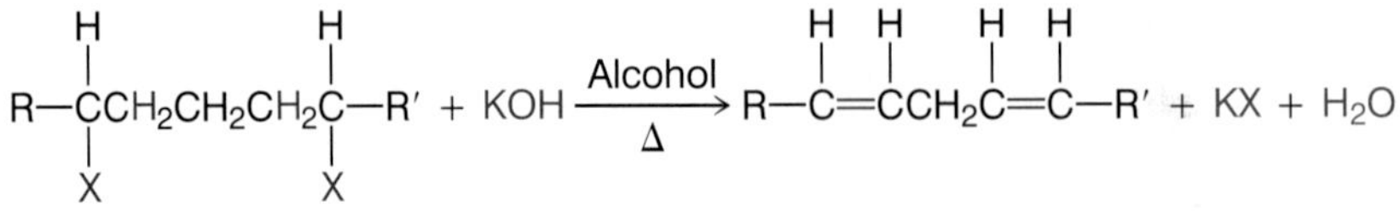

where X = Cl, Br, or I.
Example:

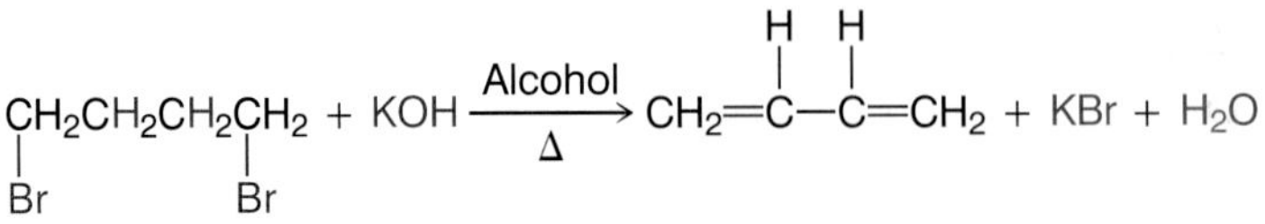

In general,

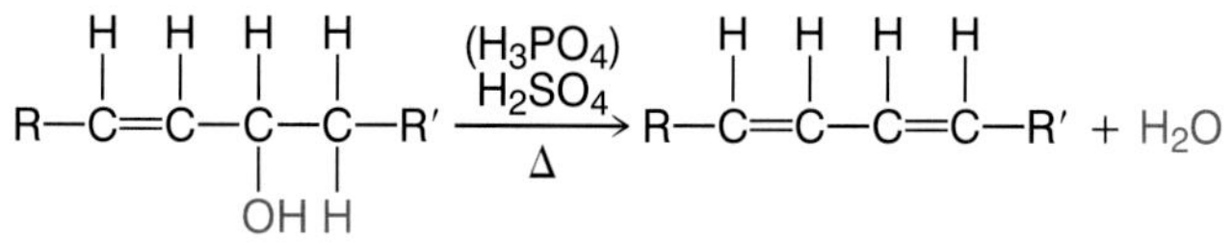

$$R-\overset{H}{\underset{}{C}}=\overset{H}{\underset{}{C}}-\overset{H}{\underset{OH}{\underset{|}{C}}}-\overset{H}{\underset{H}{\underset{|}{C}}}-R' \xrightarrow[\Delta]{\underset{H_2SO_4}{(H_3PO_4)}} R-\overset{H}{\underset{}{C}}=\overset{H}{\underset{}{C}}-\overset{H}{\underset{}{C}}=\overset{H}{\underset{}{C}}-R' + H_2O$$

Example:

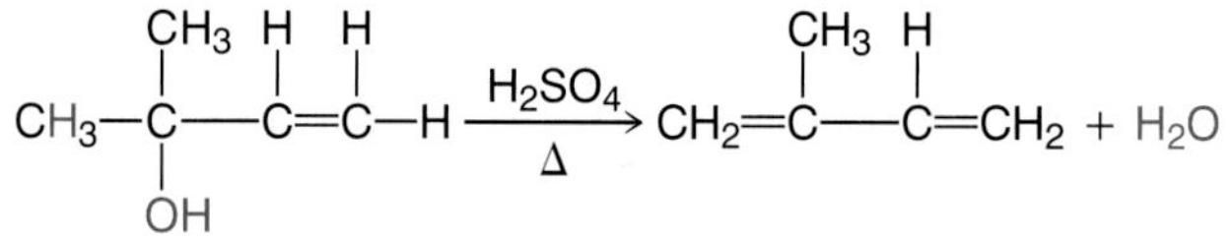

$$CH_3-\overset{CH_3}{\underset{OH}{\underset{|}{C}}}-\overset{H}{\underset{}{C}}=\overset{H}{\underset{}{C}}-H \xrightarrow[\Delta]{H_2SO_4} CH_2=\overset{CH_3}{\underset{}{C}}-\overset{H}{\underset{}{C}}=CH_2 + H_2O$$

In general,

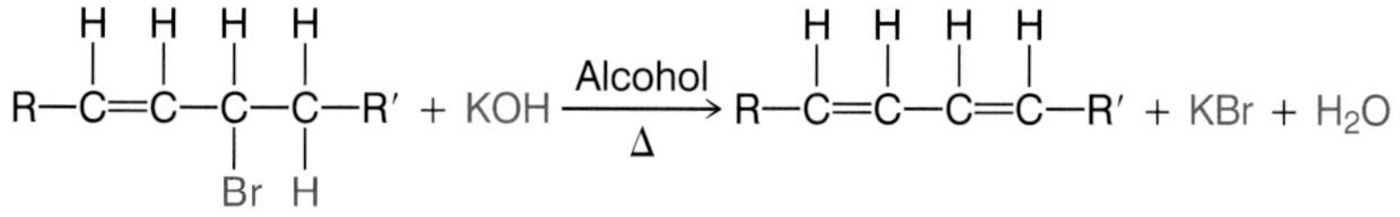

$$R-\overset{H}{\underset{}{C}}=\overset{H}{\underset{}{C}}-\overset{H}{\underset{Br}{\underset{|}{C}}}-\overset{H}{\underset{H}{\underset{|}{C}}}-R' + KOH \xrightarrow[\Delta]{Alcohol} R-\overset{H}{\underset{}{C}}=\overset{H}{\underset{}{C}}-\overset{H}{\underset{}{C}}=\overset{H}{\underset{}{C}}-R' + KBr + H_2O$$

Example:

+ KOH $\xrightarrow[\Delta]{Alcohol}$ + KBr + H$_2$O

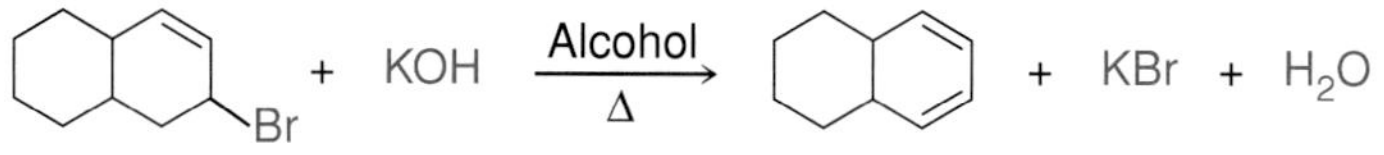

Note that only conjugated dienes
are prepared by the last two
methods.

---

Prepare 1,3-butadiene from *n*-butyl bromide (1-bromobutane) and NBS. You may use
any inorganic reagents necessary.

**SOLUTION**
Since 1,3-butadiene can be prepared by the elimination of HX from an allyl halide, we
can begin the synthesis as follows:

$$CH_2=CHCH(Br)CH_3 + KOH \xrightarrow[\Delta]{Alcohol} CH_2=CHCH=CH_2$$
3-Bromo-1-butene

To prepare 3-bromo-1-butene, react 1-butene with NBS.

$$CH_2=CHCH_2CH_3 + NBS \longrightarrow CH_2=CHCH(Br)CH_3$$
1-Butene

All that remains is to prepare 1-butene from *n*-butyl bromide.

$$BrCH_2CH_2CH_2CH_3 + KOH \xrightarrow[\Delta]{Alcohol} CH_2=CHCH_2CH_3$$
*n*-Butyl bromide

Putting it all together, we get

$$BrCH_2CH_2CH_2CH_3 \xrightarrow[\underset{\Delta}{KOH}]{Alcohol} CH_2=CHCH_2CH_3 \xrightarrow{NBS} CH_2=CHCH(Br)CH_3 \xrightarrow[\underset{\Delta}{KOH}]{Alcohol} CH_2=CHCH=CH_2$$

Prepare each of the following compounds. (You may use any inorganic reagents necessary.)

a.   1,3-cyclohexadiene from bromocyclohexane and NBS

b.   1,3-butadiene from 3-buten-2-ol [$CH_2$=CHCH(OH)$CH_3$]

## 7.4
## ADDITION REACTIONS OF THE DIENES

Note that Markownikoff's rule predicts the product when an isolated diene reacts with an unsymmetrical reagent such as HBr.

Isolated dienes, when treated with reagents ($X_2$, HX, $H_2SO_4$, $X_2$, and $H_2O$, $H_2$ and Pt, or Pd or Ni), react in the same way as monoenes except that 2 mol of reagent is used rather than 1 mol. For example,

$$CH_2=\overset{\overset{\displaystyle H}{|}}{C}-\overset{\overset{\displaystyle H}{|}}{\underset{\underset{\displaystyle H}{|}}{C}}-\overset{\overset{\displaystyle H}{|}}{C}=CH_2 + HBr \longrightarrow CH_2-\overset{\overset{\displaystyle H}{|}}{\underset{\underset{\displaystyle H}{|}}{C}}-CH_2-\overset{\overset{\displaystyle H}{|}}{\underset{\underset{\displaystyle Br}{|}}{C}}=CH_2$$

**4-Bromo-1-pentene**

A second mole of HBr produces the expected product, 2,4-dibromopentane.

$$CH_2-\overset{\overset{\displaystyle H}{|}}{\underset{\underset{\displaystyle Br}{|}}{C}}-CH_2-\overset{\overset{\displaystyle H}{|}}{C}=CH_2 + HBr \longrightarrow CH_2-\overset{\overset{\displaystyle H}{|}}{\underset{\underset{\displaystyle Br}{|}}{C}}-CH_2-\overset{\overset{\displaystyle H}{|}}{\underset{\underset{\displaystyle H}{|}}{\underset{Br}{C}}}-CH_2$$

**2,4-Dibromopentane**

Both the monobromo and dibromo compounds can be isolated. Since fragments of the molecule add to adjacent carbon atoms, this is known as *1,2-addition*

Conjugated dienes, on the other hand, when treated with 1 mol of reagent, form a mixture of two products. Thus,

**FIGURE 7.2** Line-bond and molecular orbital representations of the resonance structures of an allylic carbocation derived from the addition of a proton to 1,3-butadiene.

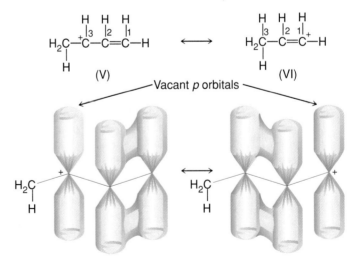

$$\underset{\substack{1 \quad 2 \quad 3 \quad 4 \quad 5}}{CH_2=\overset{\overset{\textstyle H}{|}}{C}-\overset{\overset{\textstyle H}{|}}{C}=CH-CH_3} + Br_2 \xrightarrow[20°C]{CCl_4} \underset{\substack{\quad 5 \quad 4 \quad 3 \quad 2 \quad 1}}{CH_2-\overset{\overset{\textstyle H}{|}}{\underset{\underset{\textstyle Br}{|}}{C}}-\overset{\overset{\textstyle H}{|}}{\underset{\underset{\textstyle Br}{|}}{C}}=\overset{\overset{\textstyle H}{|}}{C}-CH_3} + \underset{\substack{1 \quad 2 \quad 3 \quad 4 \quad 5}}{CH_2-\overset{\overset{\textstyle H}{|}}{C}=\overset{\overset{\textstyle H}{|}}{C}-\overset{\overset{\textstyle H}{|}}{\underset{\underset{\textstyle Br}{|}}{C}}-CH_3}$$

4,5-Dibromo-2-pentene  1,4-Dibromo-2-pentene

The product, 4,5-dibromo-2-pentene, results from 1,2-addition and is expected. However, 1,4-dibromo-2-pentene is not expected. The new carbon-carbon double bond is located between carbons 2 and 3, and fragments of $Br_2$ are added to the 1 and 4 carbons. This is known as *1,4-addition*

Resonance theory can account for 1,4-addition when we examine the mechanism of electrophilic addition to a conjugated diene.

---

### Mechanism of 1,2-Addition and 1,4-Addition of HCl (an Unsymmetrical Reagent) to 1,3-Butadiene (a Conjugated Diene)

First step:

$$\overset{+}{CH_2}=\overset{\overset{\textstyle H}{|}}{C}-\overset{\overset{\textstyle H}{|}}{C}=\overset{\overset{\textstyle H}{|}}{C}-H \longleftrightarrow CH_2-\overset{\overset{\textstyle H}{|}}{\overset{+}{C}}-\overset{\overset{\textstyle H}{|}}{C}=\overset{\overset{\textstyle H}{|}}{C}-H + Cl^-$$

with H—Cl adding below.

Allylic carbocation

Second step:

$$H-\overset{\overset{\textstyle H}{|}}{\underset{\underset{\textstyle H}{|}}{C}}-\overset{\overset{\textstyle H}{|}}{\overset{+}{C}}-\overset{\overset{\textstyle H}{|}}{C}=\overset{\overset{\textstyle H}{|}}{C}-H + Cl^- \longrightarrow H-\overset{\overset{\textstyle H}{|}}{\underset{\underset{\textstyle H}{|}}{C}}-\overset{\overset{\textstyle H}{|}}{\underset{\underset{\textstyle Cl}{|}}{C}}-\overset{\overset{\textstyle H}{|}}{C}=\overset{\overset{\textstyle H}{|}}{C}-H$$

Third step: See Fig. 7.2.
Fourth step:

$$H-\overset{\overset{\textstyle H}{|}}{\underset{\underset{\textstyle H}{|}}{C}}-\overset{\overset{\textstyle H}{|}}{C}=\overset{\overset{\textstyle H}{|}}{C}-\overset{\overset{\textstyle H}{|}}{\overset{+}{C}}-H + Cl^- \longrightarrow H-\overset{\overset{\textstyle H}{|}}{\underset{\underset{\textstyle H}{|}}{C}}-\overset{\overset{\textstyle H}{|}}{C}=\overset{\overset{\textstyle H}{|}}{C}-\overset{\overset{\textstyle H}{|}}{\underset{\underset{\textstyle Cl}{|}}{C}}-H$$

See Fig. 7.3.

---

**Envision the Reaction**

**FIGURE 7.3** Line-bond and molecular orbital representations of a resonance hybrid derived from an allylic carbocation resulting from the addition of a proton to 1,3-butadiene.

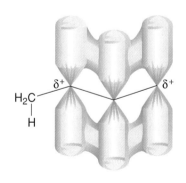

Note the hybrid structure of the allylic carbocation (Fig. 7.3), where the positive charge is delocalized over two carbon atoms (1 and 3), stabilizing the carbocation.

---

The first step is the usual Markownikoff addition of $H^+$ (from HCl) to a typical conjugated diene such as 1,3-butadiene. The allylic carbocation formed can add $Cl^-$ to form the expected 1,2-addition product. An *allylic carbocation* is one in which the carbon bearing the positive charge is next to a carbon with a double bond. Unexpected 1,4-addition can make sense when we realize that an allylic carbocation is a candidate for resonance stabilization (see Fig. 7.2). If chloride ion is added to the contributing structure on the right (VI) in Fig. 7.2, the 1,4-addition product results.

We can now expand the order of carbocation stability, for our purposes,* as follows:

allyl > $3^0$ > $2^0$ > $1^0$ > methyl

The greater stability of the allyl carbocation is due to resonance, i.e., the delocalization of the positive charge between carbons 1 and 3 of the allylic carbocation.

---

*In actuality, allyl carbocations are about as stable as secondary carbocations. This textbook, for pedagogical reasons, is taking a simplistic view that allyl carbocations are more stable than tertiary carbocations.

**SOLVED PROBLEM 7.5**

Which of the following is the more stable carbocation?

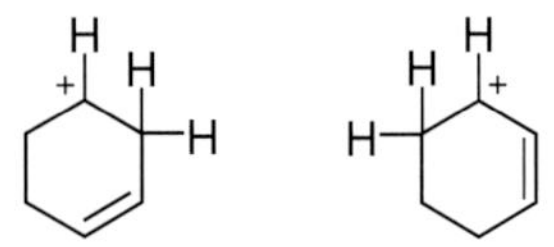

9                    10

**SOLUTION**

Carbocation 10 is more stable. Although both carbocations are secondary, only carbocation 10 is allylic and is thus stabilized by resonance.

**SOLVED PROBLEM 7.6**

Draw a contributing structure for the following allylic carbocation.

$$CH_3CH_2CH{=}CHCH_2^+$$
        5    4    3    2  1

**SOLUTION**

Use the curved arrow to delocalize the pi electrons from between C-2 and C-3 to between C-1 and C-2 of the ion as follows:

$$CH_3{-}CH_2CH{=}CH{-}CH_2^+ \longleftrightarrow CH_3{-}CH_2\overset{+}{C}H{-}CH{=}CH_2$$
      5      4      3     2     1              5      4     3      2     1

**EXERCISE 7.5**

Draw a contributing structure of allylic carbocation 10.

**EXERCISE 7.6**

Draw a contributing structure for each of the following allylic carbocations.

a.   $CH_2{=}CH\overset{+}{C}H_2$

b.

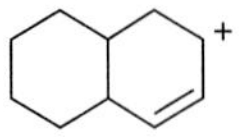

**SOLVED PROBLEM 7.7**

Draw a structural formula for a resonance hybrid based on the following contributing structures.

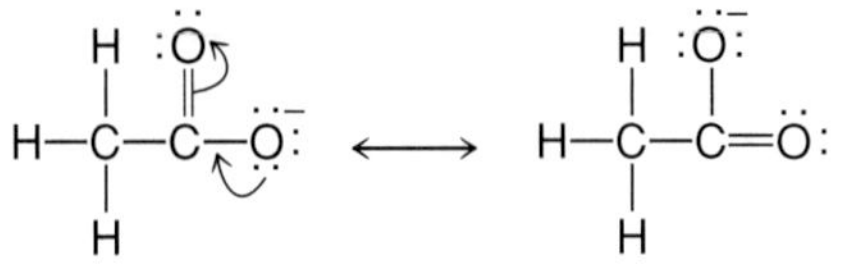

11                    12

**SOLUTION**

Note that the bonding in these structures is the same, and therefore, structures 11 and 12 are equivalent. This means that the negative charge is equally distributed between the two oxygen atoms, and we show the overall electron delocalization by means of dashed lines. The solid line and the dashed line between the carbon and oxygen atoms represents 1.5 bonds. Electron delocalization is confirmed by the presence of partial negative charges on the oxygen atoms (structure 13).

$$
\begin{array}{c}
\quad\ \ \text{H}\quad\ \text{O}^{\delta-} \\
\quad\ \ |\quad\ \ \| \\
\text{H}-\text{C}-\text{C}\!\!=\!\!\!\text{O}^{\delta-} \\
\quad\ \ | \\
\quad\ \ \text{H}
\end{array}
$$

13

---

**EXERCISE 7.7**

Draw the structural formula of the resonance hybrid based on each of the following pairs of contributing structures.

a.   III and IV (page 123)

b.   5 and 6 (page 125)

---

**SOLVED PROBLEM 7.8**

Complete the following equations by writing a structural formula for each organic product produced.

a. 1,3-butadiene + HBr (1 mol) $\longrightarrow$

b. 1,5-hexadiene + HBr (2 mol) $\longrightarrow$

**SOLUTION**

a.   Since 1,3-butadiene is a conjugated diene, two organic products are produced: one product by means of 1,2-addition and the other as a result of 1,4-addition.

$$
\begin{array}{cc}
\text{CH}_2-\text{CH}-\text{CH}=\text{CH}_2 & \text{CH}_2-\text{CH}=\text{CH}-\text{CH}_2 \\
\ \ |\quad\ \ | & \ \ |\qquad\qquad\ \ | \\
\ \ \text{H}\quad\ \text{Br} & \ \ \text{H}\qquad\qquad\ \text{Br}
\end{array}
$$

3-Bromo-1-butene     1-Bromo-2-butene<br>
(1,2-addition)       (1,4-addition)

b.   Since 1,5-hexadiene is an isolated diene, the only product formed is produced as a result of 1,2-addition.

$$
\begin{array}{c}
\text{CH}_2-\text{CH}-\text{CH}_2-\text{CH}_2-\text{CH}-\text{CH}_2 \\
\ \ |\quad\ \ |\qquad\qquad\qquad\quad |\quad\ \ | \\
\ \ \text{H}\quad\ \text{Br}\qquad\qquad\qquad\ \ \text{Br}\quad\ \text{H}
\end{array}
$$

---

**EXERCISE 7.8**

Complete the following equations by writing a structural formula for each organic product produced.

a.   1,3-butadiene + $H_2SO_4$ (1 mol) $\longrightarrow$

b.   1,6-heptadiene + $H_2$ (1 mol) $\xrightarrow{\text{Pt}}$

c.   1,3-hexadiene + $H_2$ (2 mol) $\xrightarrow{\text{Pt}}$

d.   1,6-heptadiene + $H_2$ (2 mol) $\xrightarrow{\text{Pt}}$

---

To sum up, 1,2- and 1,4-addition occur when 1 mol of a conjugated diene is reacted with reagents such as $X_2$, HX, $H_2SO_4$, $X_2$ and $H_2O$, and $H_2$ (with a metal catalyst). When a second mole of reagent is added to the resulting alkene, 1,2-addition occurs in the usual manner.

How can the presence of 1,4-addition products be explained? To answer this question, we must realize that the 1,4-addition products must be a consequence of the additional stability of conjugated dienes (in the case of 1,3-pentadiene, 6.1 kcal/mol) compared with isolated 1,4-pentadiene. Thus this 6.1 kcal of energy is known as resonance energy (see Sec. 7.2) with good reason.

**7.5
HYDROGENATION
OF CONJUGATED
DIENES—IN
GREATER DETAIL**

Hydrogenation of conjugated dienes produces both 1,2- and 1,4-addition products with the addition of 1 mol of hydrogen. In general,

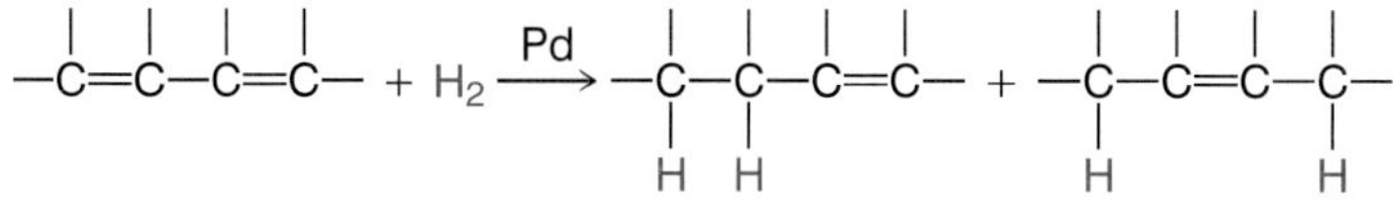

Example:

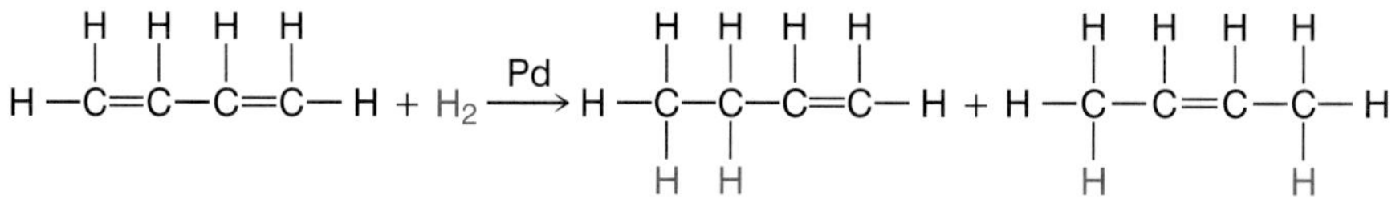

| 1,3-Butadiene | 1-Butene (5% yield) | 2-Butenes (93.2% yield) |

When 1,3-butadiene was reacted with 1 mol of hydrogen, three monoenes were isolated as follows: a mixture of two disubstituted alkenes, *cis*-2-butene and *trans*-2-butene, in 93.2% yield, and a monosubstituted alkene, 1-butene, in 5% yield. Unreacted 1,3-butadiene and saturated butane also were isolated and accounted for the remaining 1.8% yield to obtain the expected 100% recovery from the hydrogenation.

Thus the decreasing yields of monoenes produced can be predicted in terms of the stability of each alkene produced, just as in Section 6.1. This means that the most stable alkenes produced (disubstituted 2-butenes) should be produced in greater yield (93.2%), while the least stable alkene produced (monosubstituted 1-butene) would be expected to be produced with the smaller yield (5%).

---

**SOLVED PROBLEM 7.9**

A mixture of *cis*-2-butene and *trans*-2-butene was isolated in the reaction of 1,3-butadiene with 1 mol hydrogen in the presence of a palladium catalyst (the preceding reaction) in 93.2% yield. Predict which alkene of the mixture would be isolated in greater yield.

**SOLUTION**
Both *cis*-2-butene and *trans*-2-butene are disubstituted alkenes. Thus we must look for another factor that would account for a difference in stability between the two compounds. Consider the structural formula of each compound:

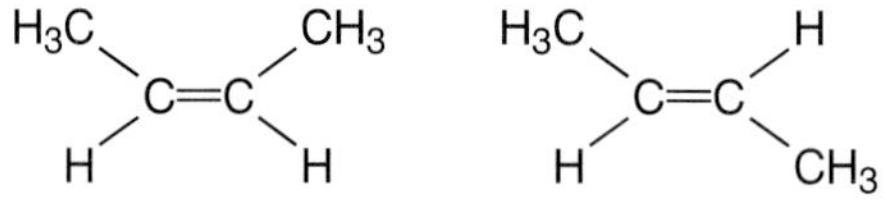

cis-2-Butene                 trans-2-Butene

Note that use of a set of molecular models would show that the two methyl groups in a *cis*-2-butene molecule are closer than in *trans*-2-butene. This means that the carbon-hydrogen bonds in the two methyl groups in *cis*-2-butene exhibit significant London electron-pair repulsions, destabilizing the molecule. Since a *trans*-2-butene molecule is free of these destabilizing repulsions, *trans*-2-butene is the more stable molecule and should be produced in greater yield. Experimentally, *trans*-2-butene was isolated in 67.5% yield; *cis*-2-butene, in 25.7% yield.

---

**EXERCISE 7.9**

Complete each of the following equations, and then determine the major and minor products based on alkene stability.

a.  2,3-dimethyl-1,3-butadiene + $H_2$ (1 mol) $\xrightarrow{\text{Pd}}$

b.  2-methyl-1,3-cyclohexadiene + $H_2$ (1 mol) $\xrightarrow{\text{Pt}}$

Conjugated dienes react with ozone in the same way that monoenes react (see Sec. 6.3). The only difference is that 2 mol of ozone is used rather than 1 mol. For example:

$$CH_2=CH-CH=CHCH_2CH_3 \xrightarrow[\text{2. Zn, H}_2\text{O}]{\text{1. O}_3 \text{ (excess)}} \overset{O}{\overset{\|}{C}}H_2 + \overset{O}{\overset{\|}{C}}H-\overset{O}{\overset{\|}{C}}H + \overset{O}{\overset{\|}{C}}HCH_2CH_3$$

**7.6**
**OZONOLYSIS OF**
**CONJUGATED DIENES**

---

Predict the products of the following ozonolysis.

**SOLVED PROBLEM 7.10**

$$CH_2=CH-CH_2-CH=CH_2 \xrightarrow[\text{2. Zn, H}_2\text{O}]{\text{1. O}_3 \text{ (excess)}}$$

**SOLUTION**

*Step 1.* Break both bonds of each carbon-carbon double bond:

$$CH_2 \qquad CH-CH_2-CH \qquad CH_2$$

*Step 2.* Add to each carbon of both double bonds the grouping $=O$; thus

$$CH_2=O \quad \text{and} \quad O=CH-CH_2-CH=O \quad \text{and} \quad CH_2=O$$

---

Complete the following equations by drawing a structural formula for each organic compound produced.

**EXERCISE 7.10**

a. 2-methyl-1,3-butadiene $\xrightarrow[\text{2. Zn, H}_2\text{O}]{\text{1. O}_3 \text{ (excess)}}$

b. 1,3-cyclopentadiene $\xrightarrow[\text{2. Zn, H}_2\text{O}]{\text{1. O}_3 \text{ (excess)}}$

---

The Diels-Alder reaction, an important example of 1,4-addition to a conjugated diene, was developed in 1928 by Otto Diels and Kurt Alder. The reaction consists of heating a conjugated diene with an alkene (often containing more than one functional group) called a *dienophile* to give an *adduct*. The delocalization of pi electrons is used to predict the product.

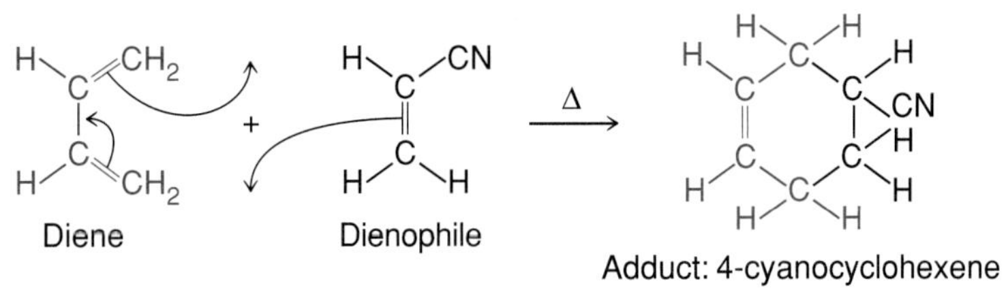

**7.7**
**THE DIELS-ALDER**
**REACTION**

Diene     Dienophile     Adduct: 4-cyanocyclohexene

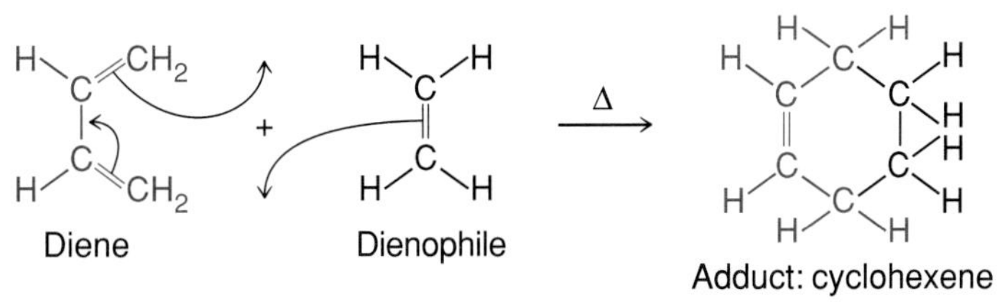

Diene     Dienophile     Adduct: cyclohexene

Using geometric condensed structures for the last reaction, we get

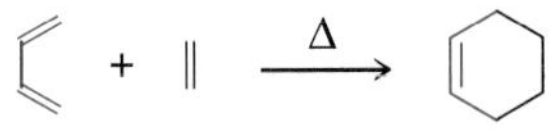

## SOLVED PROBLEM 7.11

Deduce the product of the following Diels-Alder reaction.

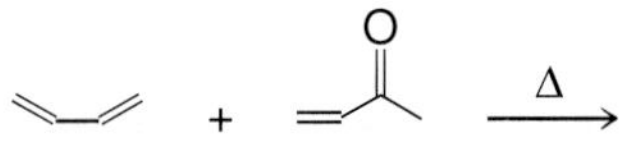

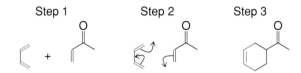

1,3 Butadiene (the diene) Methyl vinyl ketone (the dienophile)

### SOLUTION

*Step 1.* Draw a geometric condensed structure for each reagent in such a way that the double bonds are in position so that electron delocalization to form the cyclic adduct is readily accomplished.

*Step 2.* Show the electron delocalization.

*Step 3.* Draw the product (the adduct).

Step 1
Step 2
Step 3

## EXERCISE 7.11

Complete each of the following equations by drawing a structural formula for the Diels-Alder adduct produced.

a.

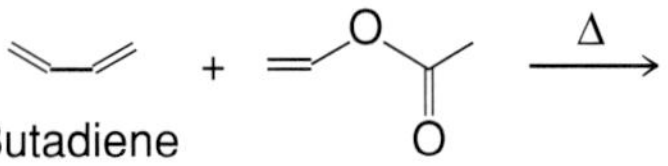

1,3-Butadiene Vinyl acetate

b.

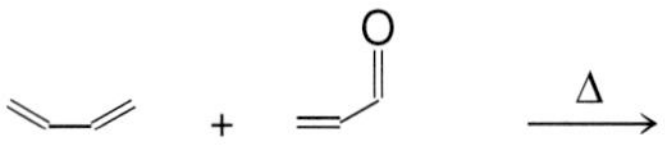

1,3-Butadiene Acrolein

Here is a somewhat more complex example of a Diels-Alder reaction, but the same electron shifting principles hold to predict the product.

CH2
HC=CH
HC=CH
H—C=O
Δ
H
H
CHO
H

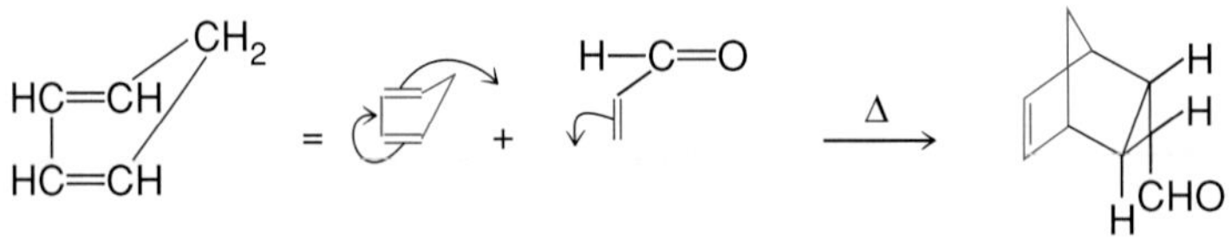

At this point, we need to discuss two stereochemical characteristics of the Diels-Alder reaction. First, notice that the product of the preceding reaction could be written as one of two possible isomers.

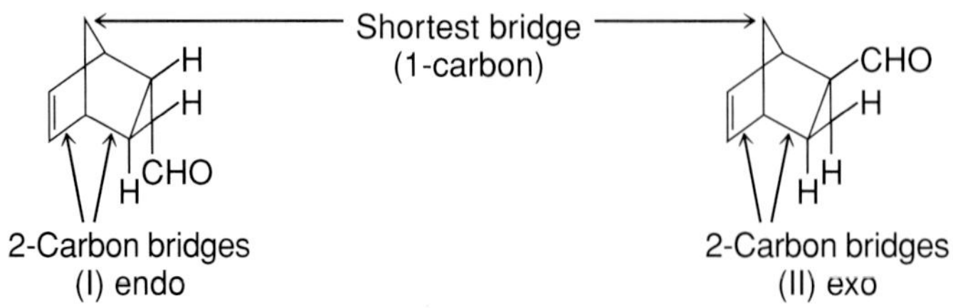

Shortest bridge
(1-carbon)
H
H
CHO
H
2-Carbon bridges
(I) endo
CHO
H
H
H
2-Carbon bridges
(II) exo

Isomer I has the aldehyde group

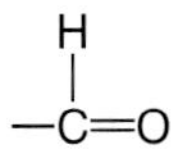

$$\begin{array}{c} H \\ | \\ -C=O \end{array}$$

on the opposite side of the molecule to that of the shortest bridge (a one-carbon bridge). The other two bridges in both isomers are longer (two-carbon bridges). Isomer II, on the other hand, has the aldehyde group on the same side of the molecule as the shortest bridge. Thus isomer I is called the *endo* isomer; isomer II, *exo*. A molecule with three bridges is called a *bridged bicyclic ring*.

Now for the first stereochemical principle: The Diels-Alder reaction yields endo product rather than exo. Expressing it another way, the reaction proceeds via an endo mode.

**SOLVED PROBLEM 7.12**

For each structural formula of a bridged bicyclic ring, determine if the group Cl is exo or endo.

a. 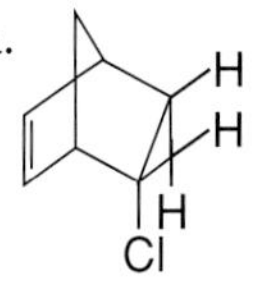      b.

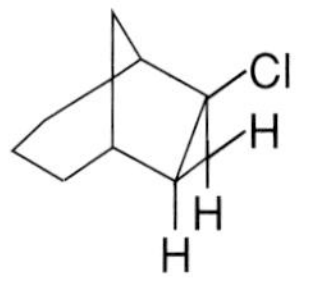

**SOLUTION**

a.  The Cl has taken the endo configuration, because the group is on the opposite side of the shortest bridge—a one-carbon bridge in this case.

b.  The Cl has assumed the exo configuration, because the group is located on the same side of the shortest bridge.

**EXERCISE 7.12**

Draw a structural formula for the adduct resulting from each of the following Diels-Alder reactions.

a.

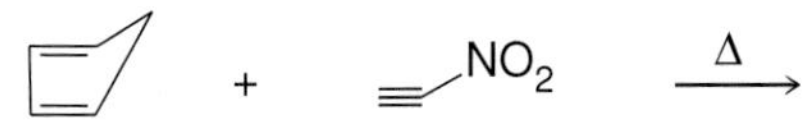

1,3-Cyclopentadiene     Nitroacetylene

b.

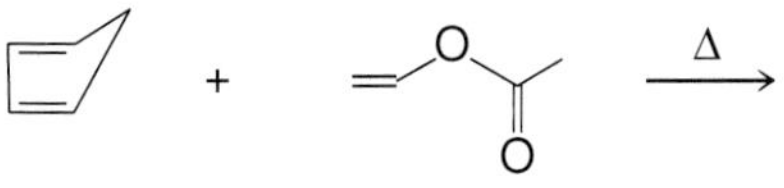

Vinyl acetate

c.

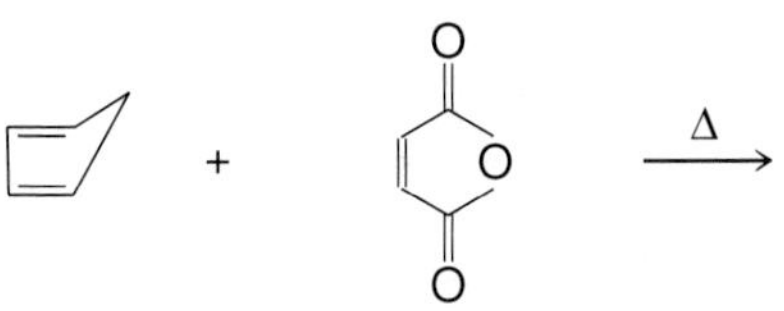

Maleic anhydride
(the dienophile)

The second stereochemical principle can be illustrated with the following equation:

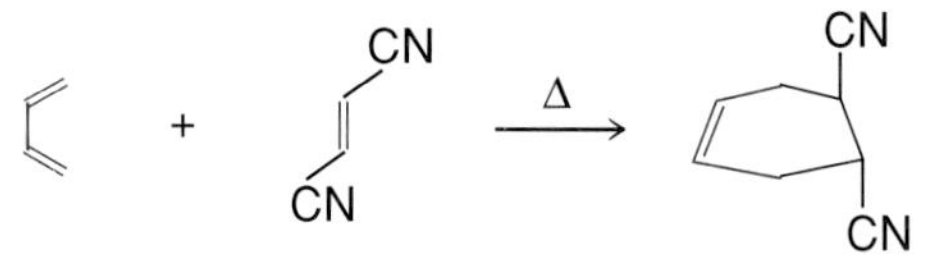

The principle states that substituent groups on the dienophile retain their configuration during the course of the reaction. In other words, since the cyano (—CN) groups in the dienophile in the preceeding equation are trans, they remain trans in the adduct.

---

**SOLVED PROBLEM 7.13**

Complete the following equation by drawing a structural formula for each organic product produced.

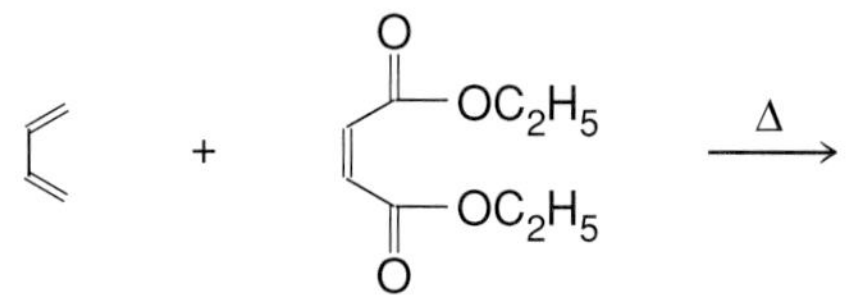

*cis*-Diethyl butenoate

**SOLUTION**

*Step 1.* Draw a geometric condensed structure for each reagent in such a way that the double bonds are in position so that electron delocalization to form the cyclic adduct is readily accomplished.

*Step 2.* Show the electron delocalization.

*Step 3.* Draw the product (the adduct). According to the second stereochemical principle of the Diels-Alder reaction, since the dienophile shows a cis configuration, the adduct also must be cis.

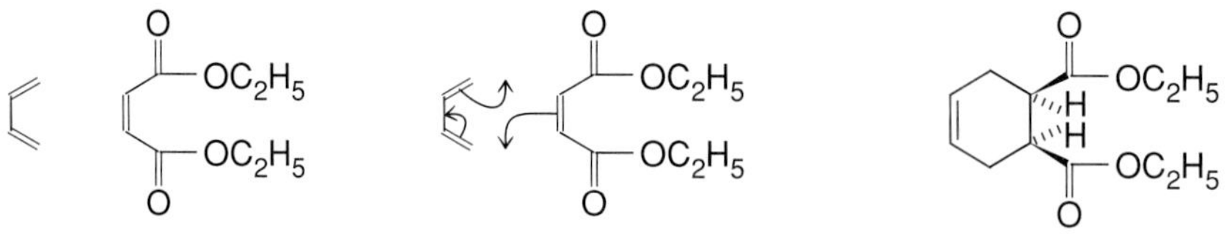

| Step 1 | Step 2 | Step 3 |
|---|---|---|

1,2-Diethyl
*cis*-4-cyclohexene-1,2-dicarboxylate

---

**EXERCISE 7.13**

Complete the following equations by drawing a structural formula for each adduct.

a.   1,3-butadiene + *cis*-1,2-dicyanoethene $\xrightarrow{\Delta}$

b.   1,3-cyclopentadiene + *cis*-1,2-dicyanoethene $\xrightarrow{\Delta}$

**7.8
POLYMERS OF
CONJUGATED DIENES**

When isoprene (2-methyl-1,3-butadiene) is polymerized to give *cis*-polyisoprene by 1,4-addition, the resulting compound is natural rubber.

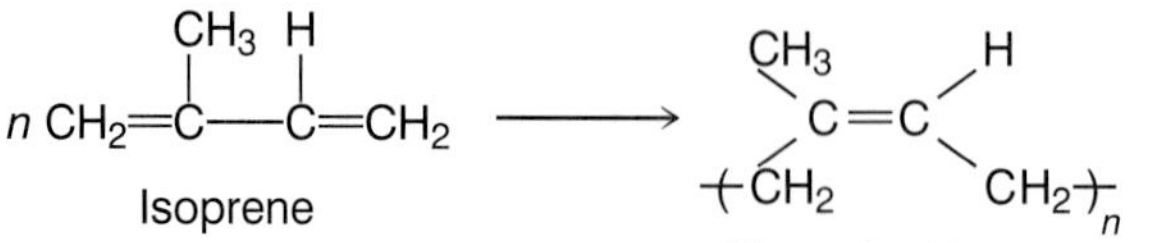

Isoprene                                          Natural rubber

| TABLE 7.2 | Terpene Classification | |
|---|---|---|
| **Number of Carbon Atoms** | **Number of Isoprene Skeletal Units** | **Class of Terpene** |
| 10 | 2 | Monoterpene |
| 15 | 3 | Sesquiterpene |
| 20 | 4 | Diterpene |
| 30 | 6 | Triterpene |
| 40 | 8 | Tetraterpene |

Natural rubber worked well when used in tires but could not be used in contact with organic solvents because a variety of organic solvents, including gasoline, would dissolve the rubber. This problem was solved with the development of neoprene in 1932. The monomer of neoprene is chloroprene.

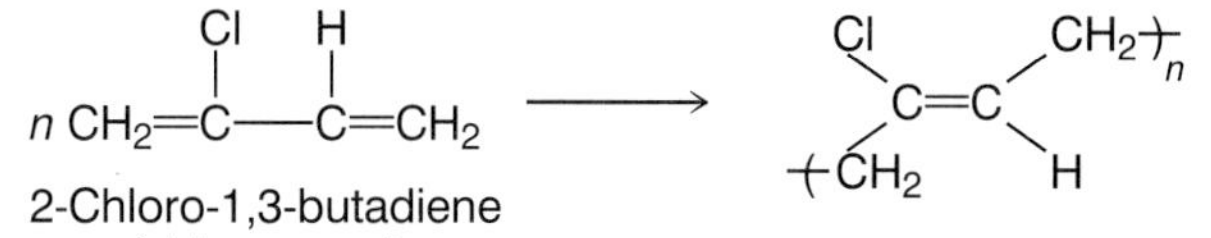

$$n \ CH_2{=}C{-}C{=}CH_2 \longrightarrow$$

2-Chloro-1,3-butadiene
(chloroprene)

*trans*-Poly-2-chloro-1,3-butadiene
(neoprene)

Neoprene does not dissolve in the presence of organic solvents and therefore is used in gasoline pump hoses.

## 7.9 TERPENES

Isoprene is a building block for a number of compounds extracted in many instances from plants. These compounds (along with others isolated from plants and animals) are called *natural products* because they are found in nature and are the basis of the pleasant odors emanating from the various plants. These *essential oils* have been used heavily in the perfume industry. As a number of compounds constituting these oils were isolated, it was found that many compounds had the formula $(C_5H_8)_x$, where $x$ can vary from 1 to 8.

These compounds are known as *terpenes* and, in a structural sense, are *oligomers* of isoprene formed by head-to-tail polymerizations. An oligomer is a substance formed from 2 to 10 monomer units; relatively few replicating monomeric structural units are involved. Note in Table 7.2 that the most complex terpene (tetraterpene) that we study contains only 8 isoprene skeletal units. Terpenes are not biosynthesized (made in the plant) from isoprene but are produced from a biologically important compound called *acetyl coenzyme A*

Terpene molecules can be cyclic or open chain, are carbon-carbon saturated or unsaturated, and may contain a wide variety of functional groups such as aldehydes, alcohols, ketones, and esters (Fig. 7.4).

**FIGURE 7.4** Some selected terpenes. Note: Ends of isoprene skeletal units are marked by dashed lines. Observe the head-to-tail orientations.

Limonene
(oil of lemon)

Vitamin A
(Retinol)

α-Phellandrene
(eucalyptus)

Menthol
(mint)

Myrcene
(bayberry)

## BIOCHEMICAL BOXED READING 7.1

Cockroaches are found almost everywhere on earth. And wherever they are found, they are considered to be bothersome insects. They breed very quickly. In just 1 year, one pair of roaches can produce over 15,000 offspring.

The common-sense way to eliminate cockroaches in the home is to keep all food and trash enclosed in containers. If this does not work, the average person will most likely try an insecticide. Favorites are chlordane, dieldrin, or malathion, which are sprayed in and around the cracks in the walls or floor of the infested room. These insecticides do a good job in destroying the cockroaches, but these chemicals are toxic, and therefore, spraying around food is undesirable.

This is where bay leaves come in. According to an old wives' tale, bay or laurel leaves from the sweet bay tree can repel cockroaches. The tree (*Laurus nobilis*) will grow up to 30 ft high in Mexico and California. The dried leaves of the plant give off a pleasant odor.

In a fascinating paper by Varma and Meloan that appeared in an issue of *American Laboratory*, Professor Meloan* takes the reader through the step-by-step process of proving the repellent effect of bay leaves on cockroaches and isolating and identifying the pure compounds (present in very low con-

centration) responsible for the repellent effect. In order of decreasing repellent effect, these compounds are cineole, geraniol, piperidine, and phenylhydrazine.

Since cineole and geraniol (two monoterpenes) will cause no problems for people if eaten, then a possible way to eliminate cockroaches from the kitchen would be to spray the kitchen area with a mixture of the repellent compounds. Varma and Meloan suggest that a time-release coating consisting of the repellents be produced for kitchen shelf paper and packaging use.

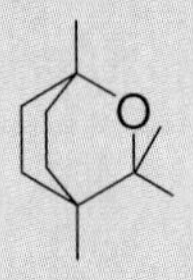

$$(CH_3)_2C{=}CHCH_2CH_2C{=}CHCH_2OH$$
$$CH_3$$

Cineole                                     Geraniol

*M. Varma and C. E. Meloan, A natural cockroach repellent in bay leaves, *American Laboratory* 13:64–69, October 1981.

---

### SOLVED PROBLEM 7.14

First, classify the terpene α-phellandrene. Then, subdivide the structural formula of this terpene into isoprene skeletal units using dashed lines.

**SOLUTION**

Since the molecule (see Fig. 7.4) contains 10 carbon atoms, according to Table 7.2, it can be classified as a monoterpene. Now let us examine the structure of isoprene (drawn W-style). To help us break down a terpene into isoprene skeletal units, we use a skeletal structure of isoprene as follows:

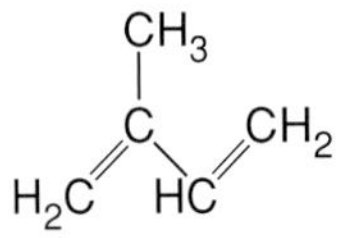

Isoprene 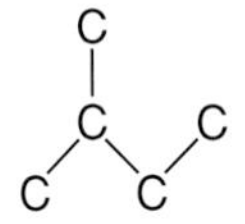                          An isoprene skeletal unit

We pick an isoprene skeletal unit by selecting 4 connecting carbons with a carbon branch on the second carbon atom of the skeletal unit. Note the arrows in Fig. 7.5 that point to the carbons that make up an isoprene skeletal unit in α-phellandrene.

---

### EXERCISE 7.14

Classify the following terpenes. Then subdivide the structural formula of each terpene into isoprene skeletal units using dashed lines.

a.  menthol

b.  myrcene

**FIGURE 7.5** Selecting an isoprene skeletal unit.

Isoprene
skeletal
unit

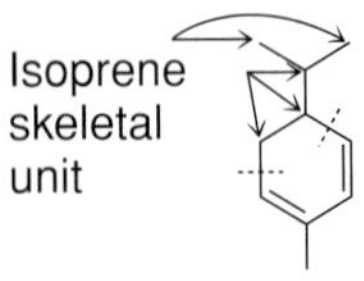

α-Phellandrene

Note that for a significant number of the terpenes, a trivial name is used rather than an IUPAC name. These trivial names are derived from the source of the terpene rather than from a set of rules (see App. 2).

## ► CHAPTER ACCOMPLISHMENTS

### 7.1 Stability of the Dienes

☐ Complete the following statement: The sum of the potential energy and kinetic energy of a molecule is called the _____ of the molecule.

☐ Relate the high internal energy of a molecule to low stability.

☐ State which diene is more stable: 1,3-pentadiene or 1,4-pentadiene.

### 7.2 Resonance

☐ Supply the term given to the true structure of a species stabilized by resonance.

☐ Explain why structure 1 (page 124) is a minor contributing structure, while structure 2 is a major contributor.

☐ Explain why the following is not a contributing structure of methylamine:

$$\overset{-}{C}H_3=\overset{+}{N}H_2$$

☐ Show by calculation that the formal charge on carbon, in the preceding structure of methylamine, is −1 and on nitrogen, +1.

### 7.3 Synthesis of Dienes

☐ Prepare 1,3-butadiene from n-butyl bromide.

☐ Name the class of diene that is produced when an allyl alcohol is dehydrated.

☐ Supply an IUPAC name for isoprene.

### 7.4 Addition Reactions of the Dienes

☐ Draw a structural formula of the 1,2-addition and the 1,4-addition products resulting from the reaction of 1,3-pentadiene with 1 mol of $Cl_2$ in the presence of $CCl_4$ at 20°C.

☐ Write a detailed, step-by-step mechanism for the following reaction:

$$CH_2{=}CH{-}CH{=}CH_2 + HCl \longrightarrow \underset{\overset{|}{H}}{CH_2}{-}CH{=}CH{-}\underset{\overset{|}{Cl}}{CH_2}$$

☐ Explain why the primary carbocation

$$^+CH_2CH{=}CH_2$$

is more stable than the following tertiary carbocation:

$$\underset{\overset{|}{CH_3}}{\overset{\overset{CH_3}{|}}{^+C}}{-}CH_3$$

### 7.5 Hydrogenation of Conjugated Dienes—In Greater Detail

☐ Explain why trans-2-butene is more stable than cis-2-butene.

☐ Predict the major and minor products for the reaction of 1 mol of hydrogen with 1,3-butadiene in the presence of palladium.

### 7.6 Ozonolysis of Conjugated Dienes

☐ Complete the following equation:

$$CH_2{=}CH{-}CH{=}CHCH_2CH_3 \xrightarrow[\text{2. Zn, }H_2O]{\text{1. }O_3\text{ (excess)}}$$

### 7.7 The Diels-Alder Reaction

☐ Give a general name to the product of a Diels-Alder reaction.

☐ Show the electron delocalizations needed to formulate the product of butadiene and ethylene.

☐ Differentiate between an exo configuration and an endo configuration using the following examples:

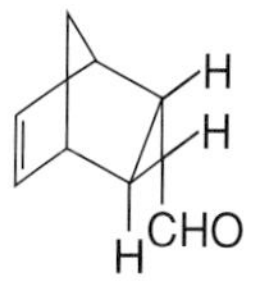

I 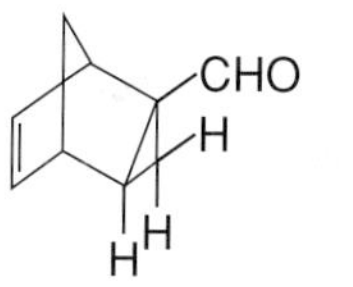       II

### 7.8 Polymers of Conjugated Dienes

☐ Draw an abbreviated structure for the polymer natural rubber.

### 7.9 Terpenes

☐ Define or explain:
  a. natural product.
  b. terpene.

☐ Draw a structural formula for isoprene.

☐ Distinguish between a monoterpene and a sesquiterpene.

☐ Subdivide the structural formula of limonene into isoprene skeletal units using dashed lines.

## ► KEY TERMS

resonance hybrid (7.2)
major contributor (7.2)
minor contributor (7.2)
resonance energy (7.2, 9.6)
1,2-addition (7.4)
1,4-addition (7.4)

*allylic carbocation* (7.4)
*dienophile* (7.7)
*adduct* (7.7)
*endo* (7.7)
*exo* (7.7)
*bridged bicyclic ring* (7.7)
*natural product* (7.9)
*essential oils* (7.9)
*terpene* (7.9)
*oligomer* (7.9)
*acetyl coenzyme A* (7.9)

## PROBLEMS

1. Which of the following dienes is more stable? Explain.
   a. 2,4-heptadiene or 1,4-heptadiene
   b. 1,3-cyclohexadiene or 1,4-cyclohexadiene
2. Give an IUPAC name for each of the following compounds.

   a. 

   b. 

   c. 

   d. 

   e. 

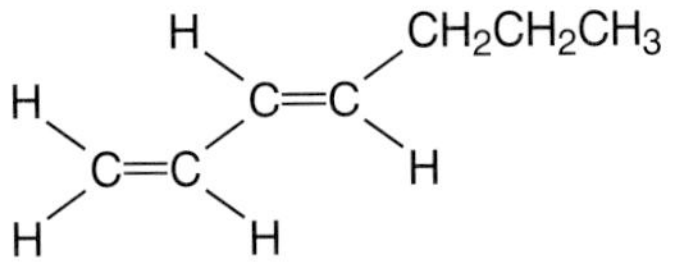

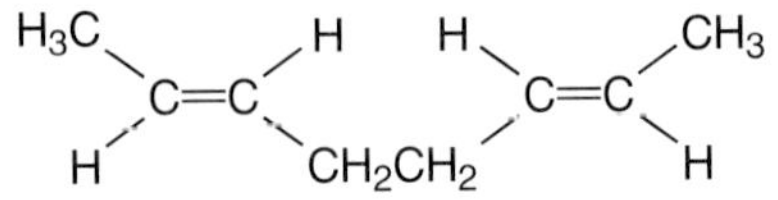

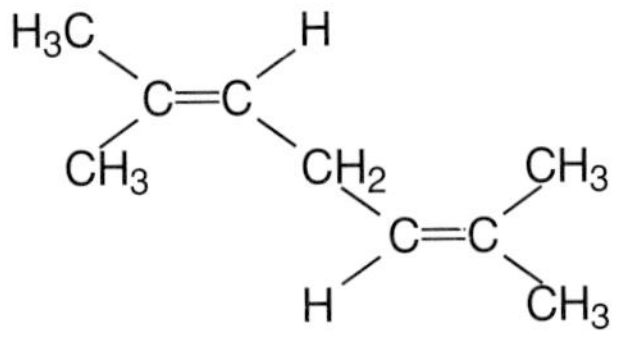

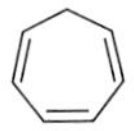

3. Draw one or more contributing structures for each of the following.

   a. 

   b. 

   c. 

   d. 

   e. 

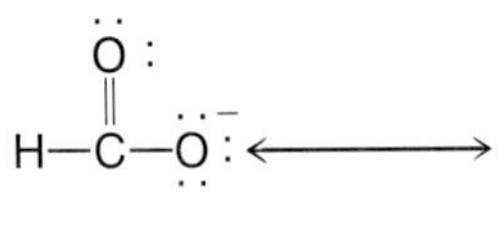

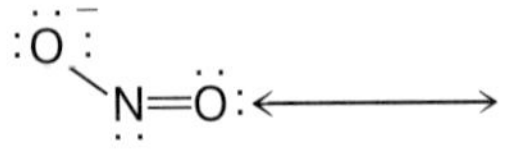

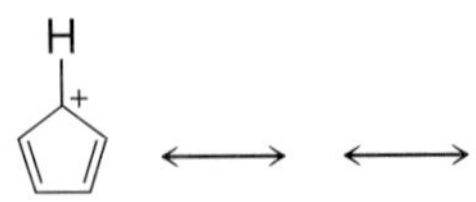

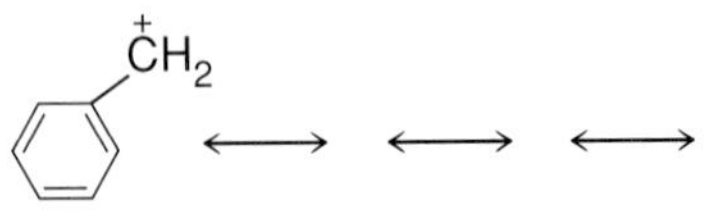

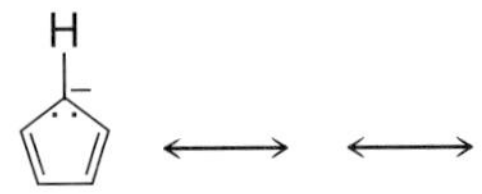

4. Do as many formal charge calculations as necessary to show that the species in Problem 3d has a positive charge and that the species in Problem 3e has a negative charge.
5. Draw the structural formula of each resonance hybrid formulated from each set of contributing structures in Problem 3.
6. Draw a line-bond formula for each of the following. Then draw at least one more contributing structure for each, and select the major contributor to the resonance hybrid for each, if any. Explain your choices.
   a. HCOOH
   b. $CH_3CH_2COO^-$
   c. $CH_3COCl$
   d. $CH_2{=}\overset{+}{C}H$
7. Ethyl alcohol ($CH_3CH_2OH$) cannot show resonance. Explain.
8. Draw a contributing structure for each of the following compounds. Select the major contributing structure, and explain why it is the major contributor.
   a. $^{-}:\ddot{O}-\overset{+}{N}\equiv N:$
   b. $CH_2{=}CH_2$
9. Show that the magnitude and sign of the charges on nitrogen and oxygen in structure 8 (page 125) are correct.
10. Calculate the resonance energy of 1,3-butadiene from the following data. (Show your reasoning.)

| Compound | Experimental Heat of Hydrogenation (kcal heat liberated/mol alkene) |
|---|---|
| 1-Butene (1 mol $H_2$) | 30.3 |
| 1,3-Butadiene (2 mol $H_2$) | 57.1 |

11. Prepare each of the following compounds. You may use any inorganic reagents necessary and NBS.
    a. butane from 1,3-butadiene
    b. 1,3-butadiene from $CH_3CH_2CH_2CH_2Cl$
    c. 1,4-dibromo-2-butene from 3-chloro-1-butene
    d.  from 1,4-cyclohexadiene

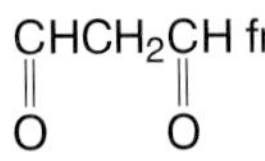

   e. 3-bromocyclohexene from bromocyclohexane

12. Which is the more stable carbocation? Explain.

   a. $CH_3CH_2\overset{+}{C}H_2$ or $CH_2{=}CH\overset{+}{C}H_2$

   b. $CH_3CH_2CH{=}CH\overset{+}{C}H_2$ or $CH_3CH_2\overset{+}{C}HCH{=}CH_2$

13. Write a detailed, step-by-step mechanism to explain why 1,3-cyclohexadiene, when reacted with 1 mol of concentrated sulfuric acid at 20°C, produces only one product.

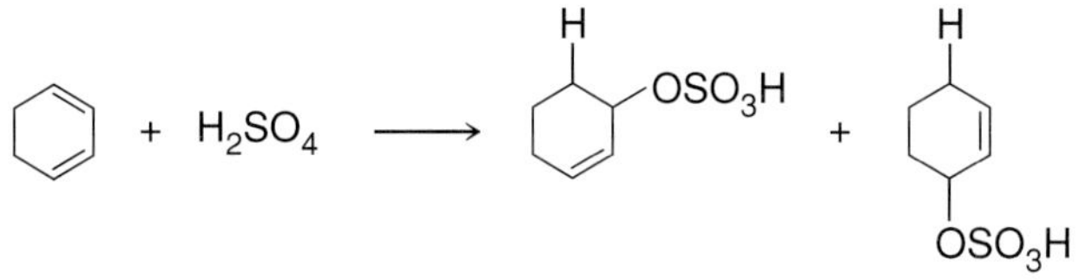

14. Each of the following compounds can be prepared by means of a Diels-Alder reaction. Select the appropriate diene and dienophile that will produce

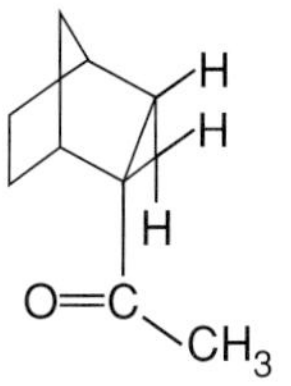

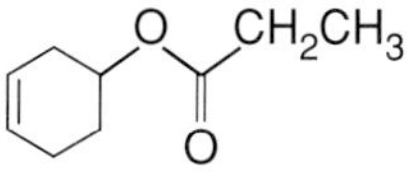

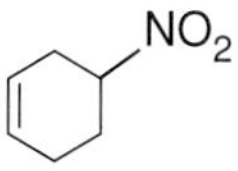

15. Complete each of the following equations by drawing a structural formula for each organic product.

   a. 1,3-cyclopentadiene $\xrightarrow[\text{2. Zn, H}_2\text{O}]{\text{1. O}_3\text{ (excess)}}$

   b. 1,3-cyclohexadiene + $Br_2$ (1 mol) $\xrightarrow[\text{20°C}]{\text{CCl}_4}$

   c. 1,3-butadiene + $Br_2$ (1 mol) $\xrightarrow{\text{H}_2\text{O}}$

   d. 1,3-butadiene + $\xrightarrow{\Delta}$

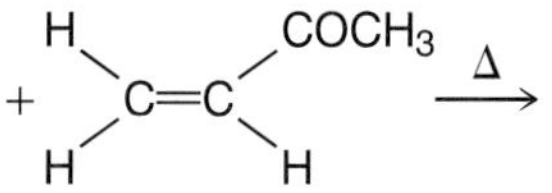

   e. 1,3-butadiene + $\xrightarrow{\Delta}$

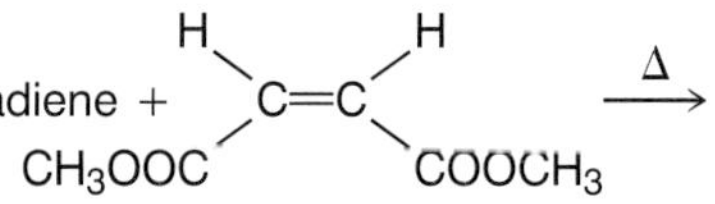

   f. 1,4-pentadiene + HBr (1 mol) $\longrightarrow$

   g. 1,4-pentadiene + HBr (2 mol) $\longrightarrow$

   h. 1,3-butadiene $\longrightarrow$ (cis polymer)

   i. 1,3-butadiene + HI (1 mol) $\longrightarrow$

16. The *trans* polymer produced when isoprene is polymerized is called *gutta percha* and is used in temporary dental fillings. Draw a structural formula for this polymer showing *n* structural units of isoprene in the polymer.

17. Pick out the isoprene skeletal units in each of the following using dashed lines.

   a.

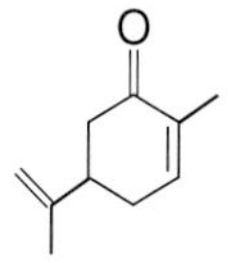

Carvone
(oil of spearmint)

   b. $(CH_3)_2C{=}CHCH_2CH_2C{=}CHCH_2OH$
                                 $CH_3$

Geraniol
(oil of roses)

   c.

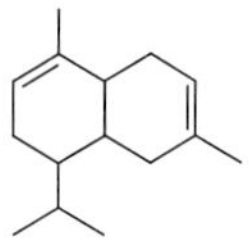

Cadinene
(oil of cade)

18. a. Prepare menthol from

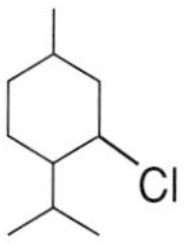

   b. Prepare geraniol from

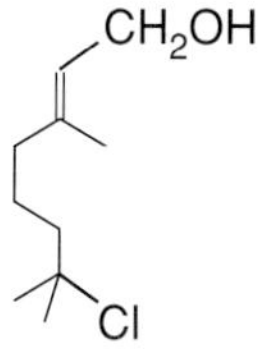

19. Using a test tube reaction, show how you would distinguish between
   a. 1,3-pentadiene and pentane.
   b. 1,3-pentadiene and cyclopentane.

20. Draw an expanded structural formula for
   a. carvone.
   b. cineole.
   c. limonene.
   d. vitamin A.

21. Draw a geometric condensed structure for geraniol.

22. Provide the class of each of the following terpenes.
   a. limonene
   b. cadinene
   c. vitamin A

# CHAPTER 8

# Alkynes

Previously, we discussed the bonding and geometry in acetylene and all alkynes (Sec. 2.5). Now, let us consider in more detail the structure of alkynes, because it is that unusual bonding structure and electron configuration in the region of the carbon-carbon triple bond which give alkynes their chemical characteristics.

The generic formula for alkynes is $C_nH_{2n-2}$ (see Table 2.1), the same generic formula as dienes. Like the alkene, the alkyne is unsaturated, in that each carbon of the triple bond is bonded to only two atoms. The most important feature of the alkyne is the carbon-carbon triple bond. Substituting in the generic formula, we get

**8.1 STRUCTURE**

$$n = 2: \quad C_2H_2 \qquad n = 3: \quad C_3H_4 \qquad n = 4: \quad C_4H_6 \qquad \text{etc.}$$

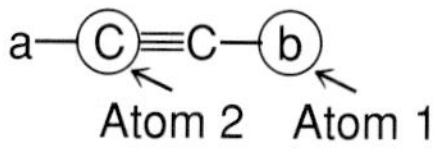

$$a-\boxed{C}\!\equiv\!C-\boxed{b}$$
$$\text{Atom 2} \quad \text{Atom 1}$$

Structurally, alkynes can be subdivided into two classes: terminal and internal. A *terminal alkyne* contains a hydrogen bonded to a carbon with the triple bond; an *internal alkyne* contains no such hydrogen.

| Terminal | Internal |
|----------|----------|
| $CH_3-C\!\equiv\!C-H$ | $CH_3-C\!\equiv\!C-C_2H_5$ |

The different classes of alkynes react differently, as we will see in Section 8.4.

The rules for naming alkynes using the IUPAC system of nomenclature are the same as those for naming alkenes—with one exception: the suffix *-yne* is used rather than *-ene* (page 79). Some examples of IUPAC nomenclature are shown in Fig. 8.1.

**8.2 NOMENCLATURE**

**FIGURE 8.1** Some examples of IUPAC nomenclature in alkynes.

$$CH_3-C\equiv CH$$
Propyne
(replace -ane of propane with -yne)

$$CH_3-C\equiv C-CH_3$$
2-Butyne

$$HC\equiv C-CH_2-C\equiv C-CH_3$$
1,4-Hexadiyne

Cyclooctyne

(a)

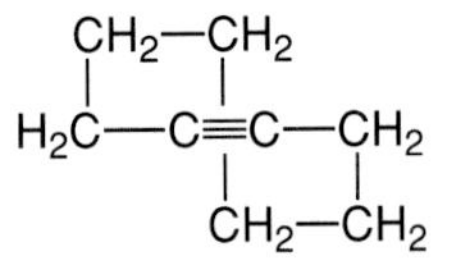

$$
\begin{array}{c}
CH_2-CH_2 \\
| \quad\quad | \\
H_2C-C\equiv C-CH_2 \\
| \quad\quad | \\
CH_2-CH_2
\end{array}
$$

(b)

*Note:* Cyclooctyne is drawn in part (a) as planar for simplicity. Actually, although the carbons of the triple bond and the atoms bonded to these carbons are linear, the remainder of the molecule is puckered and can be better represented as in part (b).

## SOLVED PROBLEM 8.1

Name the following compound by the IUPAC method.

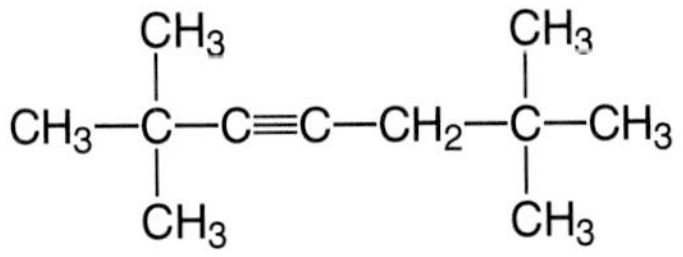

$$
\begin{array}{c}
\quad\quad CH_3 \quad\quad\quad\quad CH_3 \\
\quad\quad | \quad\quad\quad\quad\quad\quad | \\
CH_3-C-C\equiv C-CH_2-C-CH_3 \\
\quad\quad | \quad\quad\quad\quad\quad\quad | \\
\quad\quad CH_3 \quad\quad\quad\quad CH_3
\end{array}
$$

### SOLUTION

*Step 1.* The longest chain of carbons contains seven carbons. Thus the compound is a heptyne.

*Step 2.* Number the carbons from left to right so as to give the carbon of the triple bond the lowest possible number. Thus

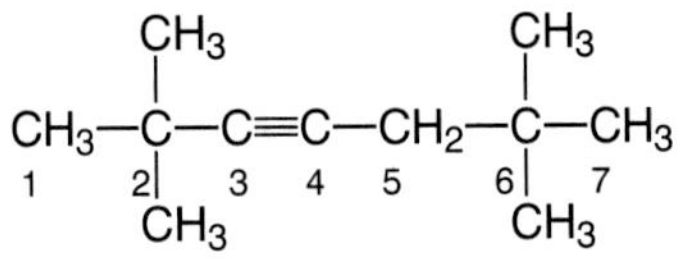

$$
\begin{array}{c}
\quad\quad CH_3 \quad\quad\quad\quad CH_3 \\
\quad\quad | \quad\quad\quad\quad\quad\quad | \\
CH_3-C-C\equiv C-CH_2-C-CH_3 \\
\;1\quad\; 2| \;\; 3\;\;4\;\;5 \quad\; 6|\;\;7 \\
\quad\quad CH_3 \quad\quad\quad\quad CH_3
\end{array}
$$

*Step 3.* Name the compound:

2,2,6,6-tetramethyl-3-heptyne

## EXERCISE 8.1

Name each of the following compounds by the IUPAC method:

a.  $CH_3C\equiv CCH_2CH_2CH_3$

b. 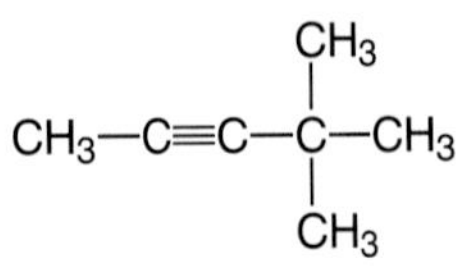
$$
\begin{array}{c}
\quad\quad\quad\quad CH_3 \\
\quad\quad\quad\quad | \\
CH_3-C\equiv C-C-CH_3 \\
\quad\quad\quad\quad | \\
\quad\quad\quad\quad CH_3
\end{array}
$$

c.  $HC\equiv CCH_2CH_2CH_3$

d.  $CH_3C\equiv CCH_2CH_2C\equiv CCH_2C\equiv CH$

A number of complex polyalkynes have been isolated from various plants. For example, 1,3,5-tridecatriene-7,9,11-triyne has been isolated from *Achillea ptarmica*, commonly known as sneezewort or white tansy.

$$CH_3-C{\equiv}C-C{\equiv}C-C{\equiv}C-\underset{|}{\overset{|}{C}}{=}\underset{|}{\overset{|}{C}}-\underset{|}{\overset{|}{C}}{=}\underset{|}{\overset{|}{C}}-\underset{|}{\overset{|}{C}}{=}CH_2$$

with H atoms on the four middle carbons

1,3,5-Tridecatriene-7,9,11-triyne

Note that the carbon-carbon double bond has a higher numbering priority than the carbon-carbon triple bond when both are in the same molecule. We can therefore formalize the priority system of functional-group nomenclature that has been developed.

$$C{=}C > C{\equiv}C$$

---

**SOLVED PROBLEM 8.2**

Name the following compounds by the IUPAC system of nomenclature.

a.  $CH_2{=}CHCH_2CH_2C{\equiv}CCH_2CH_3$

b.  $CH_3C{\equiv}C(CH_2)_{23}CH_3$

**SOLUTION**

a.  *Step 1:* Since the double bond has a higher numbering priority than the the triple bond, the compound is named as an *-en-yne* rather than an *-yn-ene*, and the first carbon of the double bond, from the end closest to the double bond, is assigned the lower of the two numbers of the carbons of the double bond locating the double bond.

   *Step 2:* Name the compound:

   1-octen-5-yne

b.  *Step 1:* This is an alkyne that contains 27 carbons. Thus we must put together the prefix *hepta-* (7) and the stem *eicosyne* (20) (see Table 3.5), and we come up with *heptaeicosyne*, which is shortened to *heptacosyne*.

   *Step 2:* Name the compound:

   2-heptacosyne

---

**EXERCISE 8.2**

Name each of the the following compounds:

a.  $HC{\equiv}C(CH_2)_4CH{=}CH_2$

b.  $CH_2{=}CHCH_2CH{=}CHCH_2C{\equiv}CC{\equiv}CH$

c.  $HC{\equiv}C(CH_2)_{12}C{\equiv}C(CH_2)_5C{\equiv}CH$

---

# 8.3 PHYSICAL PROPERTIES

The physical properties of alkynes are similar to those of the alkanes and alkenes in that compounds of all three families of hydrocarbons are water-insoluble, show densities of less than 1.0 g/ml, and boil at relatively low temperatures characteristic of nonpolar compounds (see Secs. 3.3 and 5.4).

# 8.4 ACIDITY OF ALKYNES

Terminal alkynes are more reactive than alkanes. To understand the reactivity of acetylene (for example), we need to look at what makes it more reactive than ethane, the corresponding alkane. One factor is the presence of the addition-prone triple bond; the other is that acetylene is more acidic than ethane.

For a detailed discussion of the Brønsted and Lewis concepts of acidity, refer to Section 1.14.

**FIGURE 8.2** Heterolytic dissociation of ethane to form the ethyl carbanion and acetylene to produce acetylide ion.

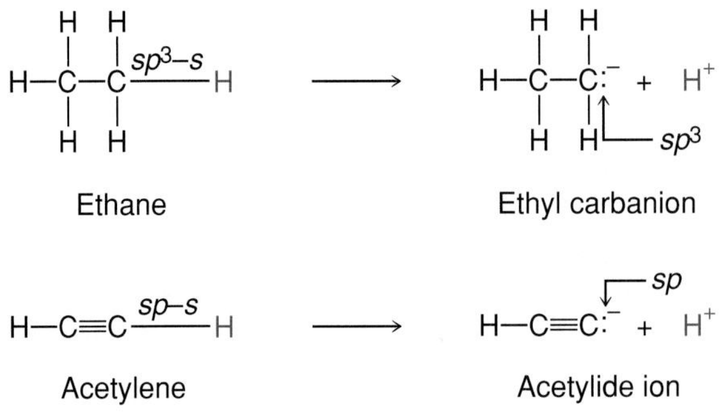

Ethane  Ethyl carbanion

Acetylene  Acetylide ion

## Why Is Acetylene More Acidic than Ethane?

To understand why acetylene is more acidic than ethane, we need to consider the hybridized orbitals in both compounds and in each corresponding conjugate base.

Consider Fig. 8.2. In the ethyl carbanion, the carbon with the unshared electron pair is $sp^3$ hybridized, whereas the acetylide ion is $sp$ hybridized. Since $sp^3$ orbitals have 75% $p$ character (25% $s$) and $sp$ orbitals have only 50% $p$ character (50% $s$), and since $p$ orbitals are farther from the carbon nucleus than $s$ orbitals, then the ethyl carbanion unshared electrons are farther from the carbon nucleus than the corresponding unshared electrons in the acetylide ion. Because the unshared electrons of the ethyl carbanion are farther from the carbon nucleus, they are less under the influence of the carbon and, therefore, more apt to be donated and act as a base than the corresponding unshared electrons in the acetylide ion. It follows from the Brønsted conjugate acid-base relationship that ethane is a weaker acid than acetylene.

An alternate explanation (the electronegativity explanation) of the greater acidity of an alkyne versus a corresponding alkane (or alkene) is as follows: A triple-bonded carbon in acetylene shows $sp$ hybridization, whereas a corresponding carbon in ethane is $sp^3$ hybridized. The electrons in the $sp$ orbital are closer to the carbon nucleus than the electrons of the $sp^3$ hybrid orbitals. Thus the $sp$ hybridized carbon in acetylene is more electronegative than the $sp^3$ hybridized carbon in ethane. Therefore, it follows that the carbon-hydrogen bond in acetylene is more polar. Thus the partial positive charge on the hydrogen in acetylene is greater than the partial positive charge on the corresponding carbon in ethane. This means that a proton is more easily removed from the acetylene molecule, making it the stronger acid. This electronegativity explanation is summmarized in Fig. 8.3.

*Let us designate this explanation as the Brønsted explanation of the greater acidity of an alkyne as opposed to the corresponding alkene or alkane.*

*Thus the ethyl carbanion is a stronger base than the acetylide ion.*

## Reactions of Terminal Alkynes as Acids

Remember (Sec. 1.14) that a Brønsted acid is a proton doner, and a Brønsted base is a proton acceptor. When a Brønsted acid reacts with a Brønsted base to produce a conjugate base of the acid and a conjugate acid of the base, reaction is favored by formation of a weaker conjugate acid and base. Thus Table 1.6 indicates that any base stronger than $R-C\equiv C:^-$ is strong enough to convert an alkyne to its conjugate base. Amide ion is the base of choice, though other stronger bases can be used as well. This

**FIGURE 8.3** Summary of the electronegativity explanation of the greater acidity of acetylene as opposed to ethane.

Carbon is more electronegative. Thus the carbon-hydrogen bond is more polar and a proton is more easily lost, resulting in a stronger acid.

Carbon is less electronegative. Thus the carbon-hydrogen bond is less polar and a proton is less easily lost, resulting in a weaker acid.

## HISTORICAL BOXED READING 8.1

Acetylene can be treated with arsenic trichloride in the presence of a catalytic amount of anhydrous aluminum chloride to produce a poison gas (actually a volatile liquid at 25°C) known as *lewisite*:

$$HC{\equiv}CH + AsCl_3 \xrightarrow{\text{AlCl}_3} ClCH{=}CHAsCl_2$$
β-Chlorovinyldichloroarsine
(lewisite)

In effect, Cl— and $AsCl_2$— groups add across the carbons of the triple bond to produce lewisite. Lewisite is a blistering agent that also significantly damages the lungs. Fortunately, the substance was developed too late to be used in World War I.

Today, lewisite is obsolete and is seldom used. There are far more lethal chemical weapons available such as sarin and tabun (nerve agents) that are stored by countries (even third-world countries) for possible use.

---

occurs because amide ion produces a conjugate acid ($NH_3$) that is weaker than the alkyne, and since the equilibrium lies in favor of the weaker acid, the equilibrium shifts to the right, and a good yield of alkynide is obtained. In general,

$$R{-}C{\equiv}CH + {}^-NH_2 \rightleftharpoons R{-}C{\equiv}C{:}^- + NH_3$$

| Stronger acid $pK_a \approx 25$ | Stronger base | Weaker base | Weaker acid $pK_a = 35$ |
|---|---|---|---|

Internal alkynes cannot react with base.

$$R{-}C{\equiv}C{-}R' + NaNH_2 \rightarrow$$
no reaction

Example:

$$CH_3C{\equiv}CH + NaNH_2 \rightleftharpoons CH_3C{\equiv}C^-Na^+ + NH_3$$

Terminal alkynes yield insoluble salts when treated with a silver(I) or copper(I) ammonia complex, whereas internal alkynes do not react. Formation of these salts serves as a qualitative test to distinguish a terminal alkyne from an internal alkyne. In general,

**A Qualitative Test for a Terminal Alkyne**

$$RC{\equiv}CH + {}^+Ag(NH_3)_2 + {}^-OH \longrightarrow RC{\equiv}CAg + H_2O + 2\,NH_3$$
Precipitate

$$RC{\equiv}CH + {}^+Cu(NH_3)_2 + {}^-OH \longrightarrow RC{\equiv}CCu + H_2O + 2\,NH_3$$
Precipitate

$$RC{\equiv}CR + {}^+Ag(NH_3)_2 + {}^-OH \longrightarrow \text{no reaction}$$

Examples:

$$CH_3CH_2C{\equiv}CH + {}^+Ag(NH_3)_2 + {}^-OH \longrightarrow CH_3CH_2C{\equiv}CAg + 2\,NH_3 + H_2O$$
Precipitate

$$HC{\equiv}CH + 2\,{}^+Cu(NH_3)_2 + 2\,{}^-OH \longrightarrow CuC{\equiv}CCu + 4\,NH_3 + 2\,H_2O$$
Precipitate

The copper and silver acetylides are unstable, sensitive to shock, and explode in the dry state. Thus it is important to avoid copper fittings or tubing when using an oxyacetylene torch or when acetylene is used in any other way (e.g., in an organic synthesis).

---

Use a test tube reaction to distinguish between 1-pentyne and 2-pentyne. Indicate what reagents should be used, and show what experimental results you would observe.

**SOLVED PROBLEM 8.3**

### SOLUTION

React both 1-pentyne and 2-pentyne with a silver ammonia complex [$^+Ag(NH_3)_2 + {}^-OH$]. Since 1-pentyne is a terminal alkyne, a reaction occurs and a precipitate is formed. No reaction takes place (a clear solution remains) in the test tube that contains the 2-pentyne because 2-pentyne is an internal alkyne.

**EXERCISE 8.3**

Use a test tube reaction to distinguish between each of the following pairs of compounds. Indicate what reagents should be used, and show what experimental results you would observe.

a. 1-hexyne and 2-hexyne

b. 1-hexyne and 1-hexene

c. 1-hexyne and cyclohexane

d. 2-hexyne and hexane

## 8.5 INDUSTRIAL PREPARATION OF ACETYLENE

Acetylene is the only alkyne produced in sufficiently large quantities to justify an industrial preparation. Acetylene is used in the oxyacetylene torch for welding and cutting steel and as a starting material for the synthesis of other organic compounds. An inexpensive method of preparation is available from calcium carbide as follows:

$$CaC_2 + 2\,H_2O \xrightarrow{20°C} HC\equiv CH + Ca(OH)_2$$

Calcium carbide

The hydrolysis of calcium carbide also can be used as a laboratory preparation of acetylene. The calcium carbide method has been replaced by the pyrolysis (heating in the absence of oxygen) of methane at high temperatures.

$$2\,CH_4 \xrightarrow{1500°C} C_2H_2 + 3\,H_2$$

There are significant differences between an industrial method of preparation and a laboratory method of preparation, as shown in Table 8.1.

| TABLE 8.1 | A Comparison of Industrial and Laboratory Methods of Preparation | |
|---|---|---|
| | **Industrial Preparation** | **Laboratory Preparation** |
| Amount of reactants | Usually huge quantities—sometimes tons—of reactants can be made to react at one time. | The upper limit of reagents used is about 1 lb. Much less can be used—usually as little as 1 g to as much as 100 g for each preparation. |
| Cost | Cost-effectiveness for the process is vital, since competition between chemical companies is significant. | Cost of the preparation is unimportant. Reagents can be expensive. |
| By-products | By-products usually present no problems in terms of product purification. They can be separated from the main product by a variety of techniques and the by-products sold or used in some other application. | By-products are undesirable. The presence of one or more by-products makes the isolation and purification of the main product more difficult. |
| Use of catalysts | A number of catalysts are used. | Catalysts are often used but are less significant than those used in industrial preparations. |
| Reaction conditions | Generally high temperatures and pressures are needed. | Conditions are usually milder than for industrial preparations. |

In general,

$$2\ NaNH_2 + \ \underset{\overset{|}{X}\ \overset{|}{X}}{\overset{\overset{H}{|}\ \overset{H}{|}}{-C-C-}} \ \xrightarrow[\Delta]{Mineral\ oil} \ -C{\equiv}C- \ + \ 2\ NH_3 + 2\ NaX$$

where X = Br or Cl. For example:

$$NaNH_2 + CH_2{=}CHBr \ \xrightarrow[Mineral\ oil]{\Delta} \ HC{\equiv}CH + NaBr + NH_3$$

Use of $NaNH_2$, a strong base, yields the alkyne.

**Mechanism of the Dehydrohalogenation of a *vic*-Dihalide**

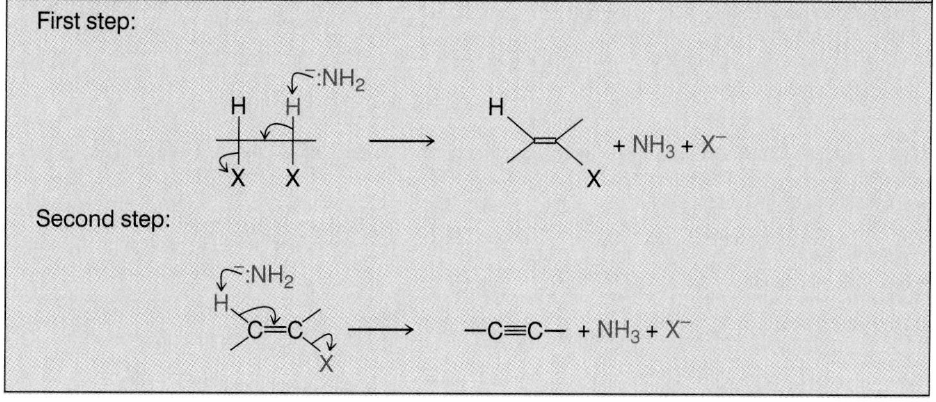

First step:

Second step:

This is a two-step process. The first step would work with hydroxide ion, but the second step is more difficult to accomplish and requires amide ion, a stronger base. Conditions can be arranged to stop the reaction after the first step is completed to form the vinyl halide, or the reaction can be continued to give the alkyne.

Other examples:

$$KOH + CH_2BrCH_2Br \ \xrightarrow[Alcohol]{\Delta} \ CH_2{=}CHBr + H_2O + KBr$$

Use of KOH, a weaker base, stops the reaction at the alkene stage.

$$2\ NaNH_2 + CH_3CH{=}CHCl + CH_3CCl{=}CH_2 \ \xrightarrow[Mineral\ oil]{\Delta} \ 2\ CH_3C{\equiv}CH + 2\ NaCl + 2\ NH_3$$

This mixture of two alkenes, in the presence of a strong base ($NaNH_2$), yields a single alkyne.

$$2\ NaNH_2 + CH_3CH(Br)CH_2Br \ \xrightarrow[Mineral\ oil]{\Delta} \ CH_3C{\equiv}CH + 2\ NaBr + 2\ NH_3$$

A vicinal dihalide, in the presence of 2 mol of $NaNH_2$, directly gives the corresponding alkyne.

**8.6
LABORATORY
PREPARATION
OF ALKYNES**

Dehydrohalogenation
of *vic*-Dihalides

**Envision the Reaction**

Steps 1 and 2 are known as E2 mechanisms (E stands for elimination; 2 for second order), which we will study in depth in Section 12.4.

## Dehydrohalogenation of *gem*-Dihalides

A *geminal (gem-) dihalide* is a dihalide that has both halogens bonded to the same carbon. The term *geminal* is derived from the Latin *gemini* meaning "twins." In general,

$$2\ NaNH_2 + \underset{\underset{X\ \ H}{|\ \ \ |}}{\overset{\overset{X\ \ H}{|\ \ \ |}}{-C-C-H}} \xrightarrow[\text{Mineral oil}]{\Delta} -C{\equiv}CH + 2\ NH_3 + 2\ NaX$$

where X = Br or Cl. For example:

$$2\ NaNH_2 + \underset{\underset{Cl}{|}}{\overset{\overset{Cl}{|}}{\bigcirc\!\!-C-CH_3}} \xrightarrow[\substack{\text{Mineral}\\\text{oil}}]{\Delta} \bigcirc\!\!-C{\equiv}CH + 2\ NH_3 + 2\ NaCl$$

The mechanism and the two-step process are the same here as with the dehydrohalogenation of a *vic*-dihalide (Sec. 8.6).

## From Acetylene and Other Terminal Alkynes (The Terminal Alkyne Synthesis)

This is also a two-step synthesis. In the first step, the terminal alkyne is treated with a strong base (see Sec. 8.4). In general,

$$R(H)C{\equiv}CH + NaNH_2 \longrightarrow R(H)C{\equiv}C^-Na^+ + NH_3$$
Sodium alkynide (ethynide)

This is an example of an $S_N2$ mechanism (substitution, nucleophilic, second order) that we will study in depth in Section 12.1.

The second step has the alkynide treated with a primary alkyl halide (or methyl) to give the alkyne.

$$R(H)C{\equiv}C^-Na^+ + XCH_2R' \longrightarrow R(H)C{\equiv}CCH_2R' + NaX$$

where X = Cl, Br, or I. For example:

Note that a multiline arrow is used to show the movement of an atom or group of atoms.

$$H{-}C{\equiv}C{-}H + NaNH_2 \longrightarrow H{-}C{\equiv}C^-\ Na^+ + NH_3$$

$$H{-}C{\equiv}C^-\boxed{Na^+ + Br}CH_2CH_3 \longrightarrow H{-}C{\equiv}C{-}CH_2CH_3 + NaBr$$

Since acetylene contains two terminal hydrogens, another alkyl group can be placed on the alkyne molecule.

$$H{-}C{\equiv}C{-}CH_2CH_3 + NaNH_2 \longrightarrow {}^+Na^-C{\equiv}CCH_2CH_3 + NH_3$$

$$CH_3CH_2CH_2I + {}^+Na^-C{\equiv}CCH_2CH_3 \longrightarrow CH_3CH_2CH_2C{\equiv}CCH_2CH_3 + NaI$$

You might ask why secondary and tertiary halides cannot be used in the second step of this synthesis. The answer to this question is discussed in Section 12.6 ("Nucleophilic Substitution and Elimination").

---

**SOLVED PROBLEM 8.4**

Complete the following equations by writing a structural formula for each organic product and a molecular formula or formula unit for each inorganic product. If no reaction occurs, write "no reaction" and explain why a reaction did not take place.

a.   $CH_3CH_2C{\equiv}CCH_3 + NaNH_2 \longrightarrow$

b.   $CH_3CH_2CH_2CHBr_2 + 2\ NaNH_2 \xrightarrow[\text{Mineral oil}]{\Delta}$

c.   $CH_3CH_2CH_2CHBr_2 + KOH \xrightarrow[\text{Alcohol}]{\Delta}$

**SOLUTION**

a.   No reaction. Since the alkyne is internal rather than terminal, no reaction can take place with NaNH$_2$ to produce the corresponding pentynide.

b.  Three clues are present that indicate that the product is an alkyne. First, 2 mol of base is used. Second, the base ($NaNH_2$) is strong enough to produce the alkyne. Third, the solvent employed is mineral oil. Thus we obtain

$$CH_3CH_2C\equiv CH + 2\,NaBr + 2\,NH_3$$

c.  Three clues are present that indicate that the product is an alkene. First, only 1 mol of base is used. Second, KOH is too weak a base to form the alkyne. Third, the solvent employed is alcohol. Thus we obtain

$$CH_3CH_2CH\!=\!CHBr + KBr + H_2O$$

---

Complete the following equations by writing a structural formula for each organic product and a molecular formula or formula unit for each inorganic product. If no reaction occurs, write "no reaction" and explain why a reaction did not take place.

**EXERCISE 8.4**

a.  $CH_3CH_2CH_2C\equiv CH + NaNH_2 \longrightarrow$

b.  $CH_3CH_2C(Br_2)CH_3 + KOH \xrightarrow[\text{Alcohol}]{\Delta}$

c.  $CH_3CH_2C(H)\!=\!CHBr + KOH \xrightarrow[\text{Alcohol}]{\Delta}$

d.  $CH_3CH_2CH_2CH(Cl)CH_2Cl + 2\,NaNH_2 \xrightarrow[\text{Mineral oil}]{\Delta}$

e.  Product Exercise 8.4a $+ CH_3CH_2Br \longrightarrow$

f.  $CH_3CH_2CH_2C\equiv CCH_3 + NaNH_2 \longrightarrow$

---

Alkynes react in a way similar to alkenes—that is, by addition. However, 1 or 2 mol of reagent can add across the triple bond—depending on the reagents, reactant ratios, and reaction conditions used.

## 8.7 CHEMICAL PROPERTIES OF ALKYNES

### Hydrogenation

When treated with hydrogen and a catalyst such as finely divided platinum, palladium, or nickel, an alkyne is reduced to an alkane. Although an alkene is formed first, it reacts with hydrogen so quickly that it cannot be isolated. In general,

$$RC\equiv CR' + 2\,H_2 \xrightarrow{Ni} RCH_2CH_2R'$$

$$RC\equiv CH + 2\,H_2 \xrightarrow{Pd} RCH_2CH_3$$

Examples:

$$CH_3C\equiv CCH_3 + 2\,H_2 \xrightarrow{Ni} CH_3CH_2CH_2CH_3$$

$$CH_3CH_2C\equiv CH + 2\,H_2 \xrightarrow{Pd} CH_3CH_2CH_2CH_3$$

This reaction can be halted at the alkene stage by use of *Lindlar's catalyst* to produce a *cis*-alkene and sodium (or lithium) in liquid ammonia to give a *trans*-alkene. In general,

Lindlar's catalyst is a mixture of palladium catalyst precipitated on calcium carbonate and deactivated by a lead salt and an aromatic amine.

$$R\!-\!C\equiv C\!-\!R' + H_2 \xrightarrow[\text{catalyst}]{\text{Lindlar's}} \underset{H}{\overset{R}{>}}C\!=\!C\underset{H}{\overset{R'}{<}}$$

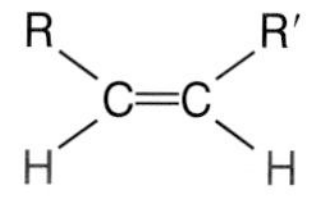

*cis*-Alkene

Example:

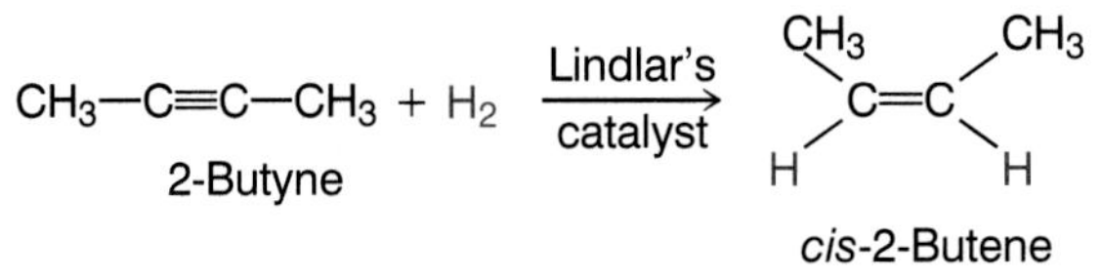
CH3—C≡C—CH3 + H2
2-Butyne
Lindlar's catalyst
CH3 CH3 C=C H H
cis-2-Butene

In general,

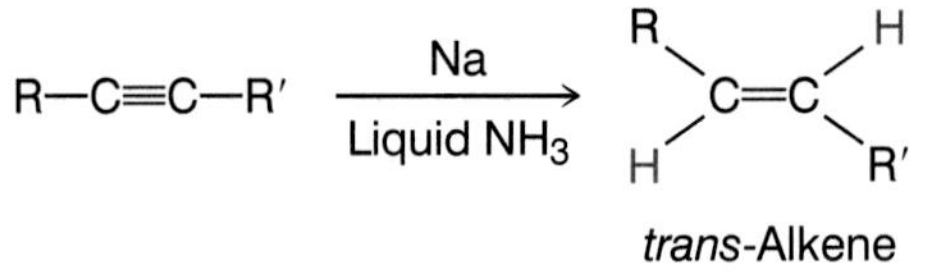
R—C≡C—R′
Na
Liquid NH3
R H C=C H R′
trans-Alkene

Example:

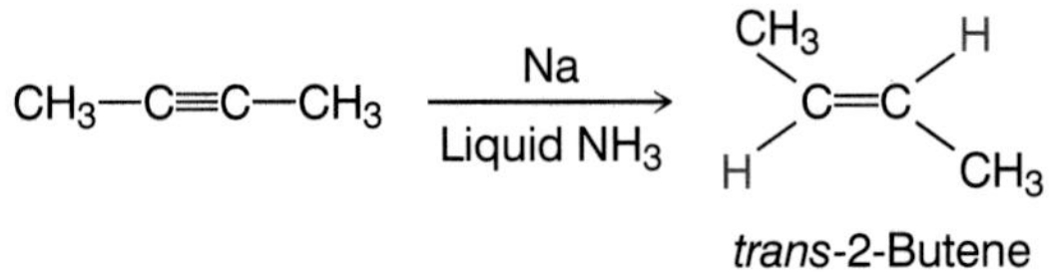
CH3—C≡C—CH3
Na
Liquid NH3
CH3 H C=C H CH3
trans-2-Butene

## Addition of Halogens

Bromine and chlorine react with alkynes. If 1 mol of halogen is added at a low temperature, a *trans*-dihaloalkene is isolated; if 2 mol of halogen is added at room temperature, the tetrahaloalkane is obtained. In general (1 mol of halogen at low temperature),

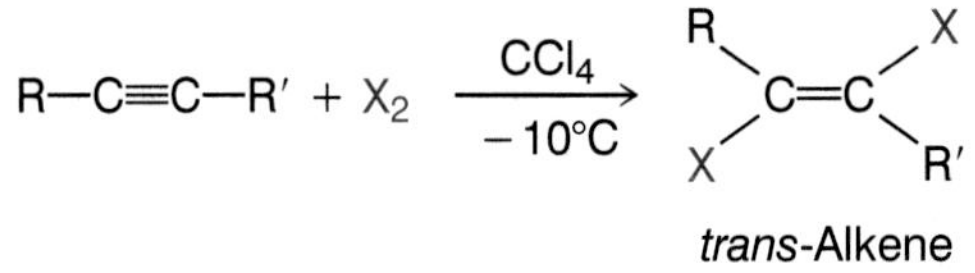
R—C≡C—R′ + X2
CCl4
– 10°C
R X C=C X R′
trans-Alkene

where for example:

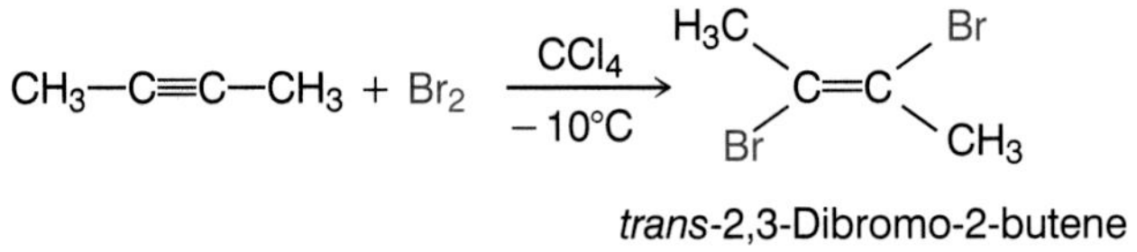
CH3—C≡C—CH3 + Br2
CCl4
– 10°C
H3C Br C=C Br CH3
trans-2,3-Dibromo-2-butene

In general (2 mol of halogen at room temperature),

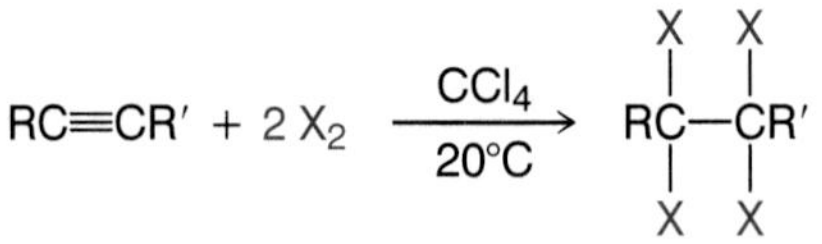
RC≡CR′ + 2 X2
CCl4
20°C
X X RC—CR′ X X

where $X_2 = Br_2$ or $Cl_2$. For example:

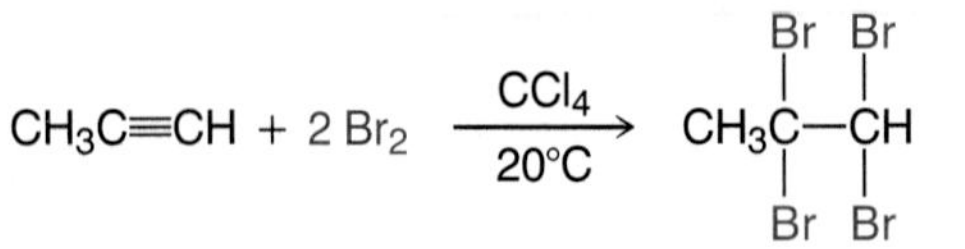
CH3C≡CH + 2 Br2
CCl4
20°C
Br Br CH3C—CH Br Br

**SOLVED PROBLEM 8.5**

Prepare *trans*-1,2-dichloropropene from *n*-propyl alcohol ($CH_3CH_2CH_2OH$). You may use any organic and inorganic reagents necessary.

**SOLUTION**

Since *trans*-1,2-dichloropropene can be prepared by the addition of 1 mol of chlorine to propyne, a reasonable last step of the synthesis is as follows:

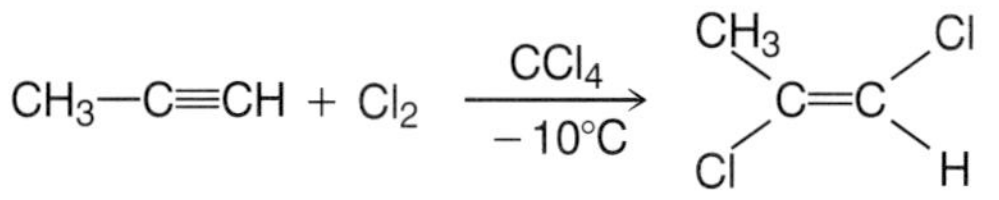

$$CH_3—C{\equiv}CH + Cl_2 \xrightarrow[-10°C]{CCl_4} \underset{Cl}{\overset{CH_3}{\underset{\diagdown}{\phantom{.}}}} C{=}C \underset{H}{\overset{Cl}{\phantom{.}}}$$

To prepare propyne, react 1,2-dichloropropane with $NaNH_2$.

$$CH_3CH(Cl)CH_2Cl + 2\,NaNH_2 \xrightarrow[\text{Mineral oil}]{\Delta} CH_3C{\equiv}CH$$

All that remains is to prepare 1,2-dichloropropane from *n*-propyl alcohol.

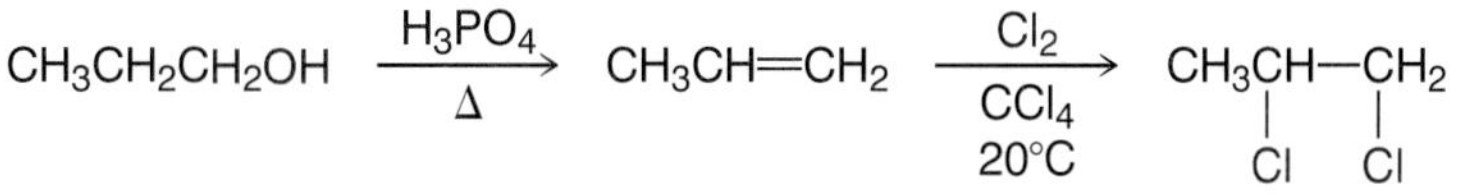

$$CH_3CH_2CH_2OH \xrightarrow[\Delta]{H_3PO_4} CH_3CH{=}CH_2 \xrightarrow[\substack{CCl_4\\20°C}]{Cl_2} \underset{\substack{|\;\;\;|\\Cl\;\;\;Cl}}{CH_3CH—CH_2}$$

Putting it all together, we get

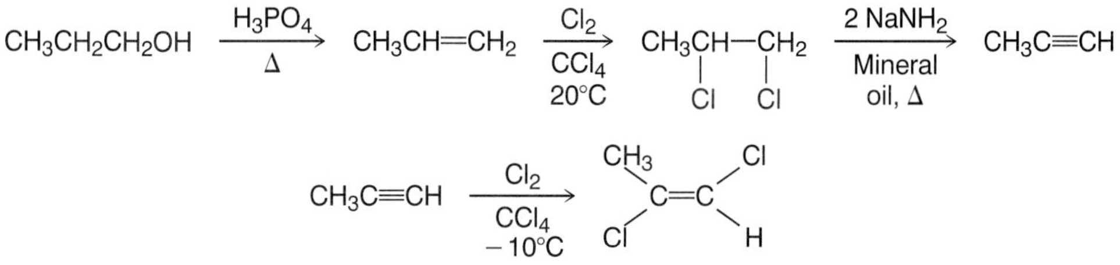

$$CH_3CH_2CH_2OH \xrightarrow[\Delta]{H_3PO_4} CH_3CH{=}CH_2 \xrightarrow[\substack{CCl_4\\20°C}]{Cl_2} \underset{\substack{|\;\;\;|\\Cl\;\;\;Cl}}{CH_3CH—CH_2} \xrightarrow[\substack{\text{Mineral}\\\text{oil, }\Delta}]{2\,NaNH_2} CH_3C{\equiv}CH$$

$$CH_3C{\equiv}CH \xrightarrow[\substack{CCl_4\\-10°C}]{Cl_2} \underset{Cl}{\overset{CH_3}{\underset{\diagdown}{\phantom{.}}}} C{=}C \underset{H}{\overset{Cl}{\phantom{.}}}$$

---

Prepare each of the following from *n*-propyl alcohol. You may use any inorganic and organic reagents necessary.

a.   1,1,2,2-tetrachloropropane

b.   *trans*-1,2-dibromopropene

c.   sodium propynide

**EXERCISE 8.5**

---

The reaction of $X_2$ with alkynes can stop at the alkene stage. So also can the reaction of HX with alkynes stop at the alkene stage. Markownikoff's rule is employed to predict the product(s) of the reaction. In general (1 mol of HX at low temperature),

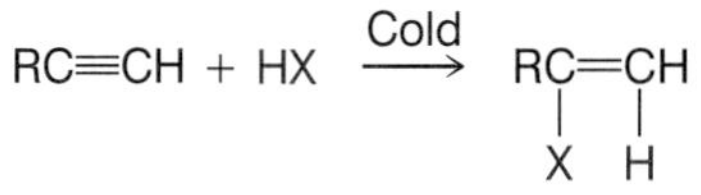

**Addition of a
Hydrogen Halide**

$$RC{\equiv}CH + HX \xrightarrow{\text{Cold}} \underset{\substack{|\;\;|\\X\;\;H}}{RC{=}CH}$$

where X = Br, Cl, or I. For example:

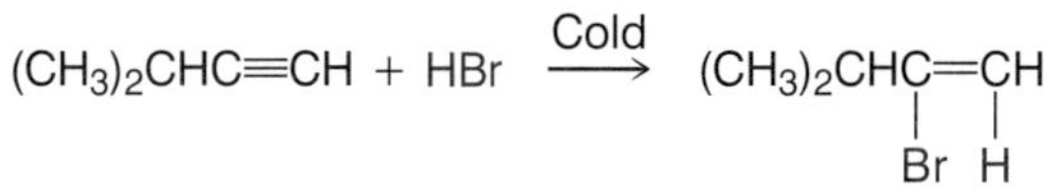

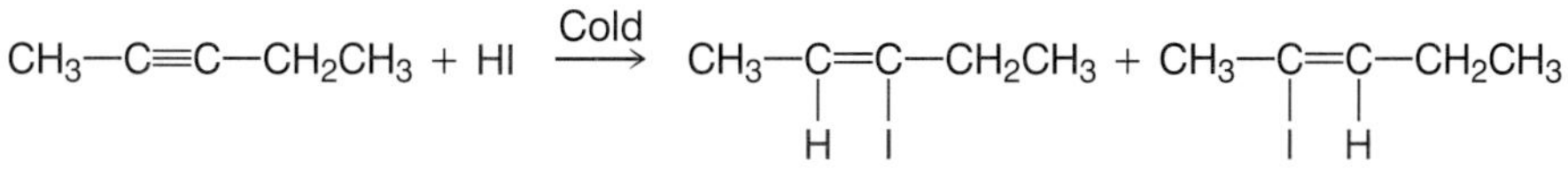

$$(CH_3)_2CHC{\equiv}CH + HBr \xrightarrow{\text{Cold}} \underset{\substack{|\;\;|\\Br\;\;H}}{(CH_3)_2CHC{=}CH}$$

$$CH_3—C{\equiv}C—CH_2CH_3 + HI \xrightarrow{\text{Cold}} \underset{\substack{|\;\;|\\H\;\;I}}{CH_3—C{=}C—CH_2CH_3} + \underset{\substack{|\;\;|\\I\;\;H}}{CH_3—C{=}C—CH_2CH_3}$$

Note that where each carbon of the triple bond has the same number of hydrogens, a mixture of products results. In general (2 mol of HX at room temperature),

$$RC\equiv CH + 2\,HX \xrightarrow{20°C} \underset{\underset{X}{|}}{\overset{\overset{X}{|}}{R}}C-\underset{\underset{H}{|}}{\overset{\overset{H}{|}}{C}}H$$

Example:

$$CH_3C\equiv CH + 2\,HBr \xrightarrow{20°C} CH_3\underset{\underset{Br}{|}}{\overset{\overset{Br}{|}}{C}}-\underset{\underset{H}{|}}{\overset{\overset{H}{|}}{C}}H$$

Adding a proton to an alkyne produces a vinyl carbocation as follows:

$$R-C\equiv CH + H^+ \longrightarrow R-\overset{+}{C}=CH_2$$

Remember, we have already looked at the relative stability of various classes of carbocations (Sec. 6.2). The stability of a vinyl carbocation is an extension of this. It is interesting to notice that the vinyl carbocation is less stable than a corresponding alkyl carbocation.

$$R-CH=CH_2 + H^+ \longrightarrow R-\overset{+}{C}H-CH_3$$

This is so because the vinyl carbocation contains only one alkyl group delocalizing the positive charge, whereas the saturated alkyl carbocation has two such groups, which translates into greater stability for the saturated alkyl carbocation. The order of carbocation stability can now be expanded:

$$\text{allyl} > 3° > 2° > 1° > \text{vinyl}$$

---

**SOLVED PROBLEM 8.6**

Synthesize 2-chloropropene [$CH_3-C(Cl)=CH_2$] from acetylene. You may use any organic and inorganic reagents necessary.

**SOLUTION**

Since the product molecule contains one more carbon atom than acetylene, the terminal alkyne synthesis (see Sec. 8.6) is used. Let us begin by preparing 2-chloropropene from propyne and hydrogen chloride.

$$CH_3-C\equiv CH + HCl \xrightarrow{\text{Cold}} CH_3-C(Cl)=CH_2$$

Now we prepare propyne from acetylene using the terminal alkyne synthesis.

$$HC\equiv CH \xrightarrow{\text{NaNH}_2} HC\equiv C^-Na^+ \xrightarrow{\text{CH}_3\text{I}} CH_3C\equiv CH$$

Putting it all together, we get

$$HC\equiv CH \xrightarrow{\text{NaNH}_2} HC\equiv C^-Na^+ \xrightarrow{\text{CH}_3\text{I}} CH_3C\equiv CH \xrightarrow[\text{Cold}]{\text{HCl}} CH_3-C(Cl)=CH_2$$

---

**EXERCISE 8.6**

Synthesize each of the following compounds from acetylene using the terminal alkyne synthesis. You may use any organic and inorganic reagents necessary.

a.  2-pentyne

b.  2-butyne

c.  1-eicosyne

In the presence of sulfuric acid and mercuric sulfate, water adds across the triple bond of an alkyne to give an unstable vinyl alcohol, which rearranges to produce a carbonyl compound (aldehyde or ketone). Markownikoff's rule is used to predict the product. In general,

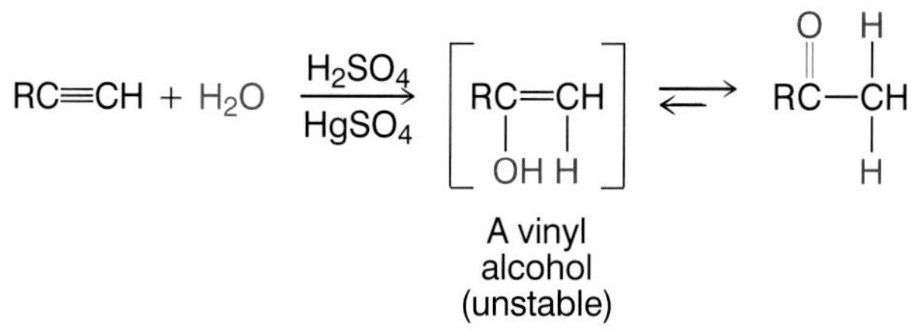

A vinyl
alcohol
(unstable)

Examples:   Acetylene produces the only aldehyde.

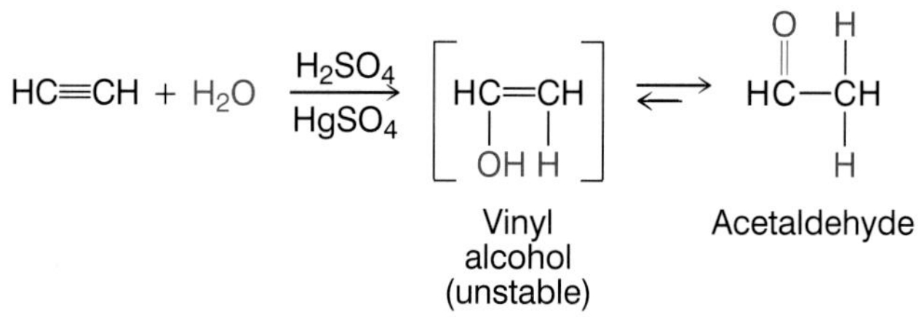

Vinyl          Acetaldehyde
alcohol
(unstable)

Methyl ethyl
ketone

Diethyl ketone     Methyl *n*-propyl
ketone

## Addition of Water—Tautomerism

Note that brackets around a chemical species signify that the species is unstable. A substituted vinyl alcohol is known as an *enol* The *-en(e)* segment of the suffix refers to the carbon-carbon double bond, while the *-ol* portion designates an —OH group (an alcohol) that is bonded to a carbon of the double bond. These enols are unstable and rapidly rearrange to the appropriate carbonyl compound (aldehyde or ketone).

A terminal alkyne produces a methyl ketone.

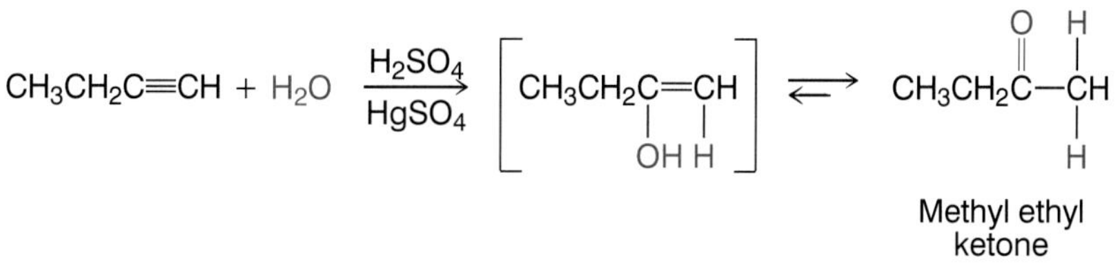

An internal alkyne produces a mixture of two, sometimes different, ketones.

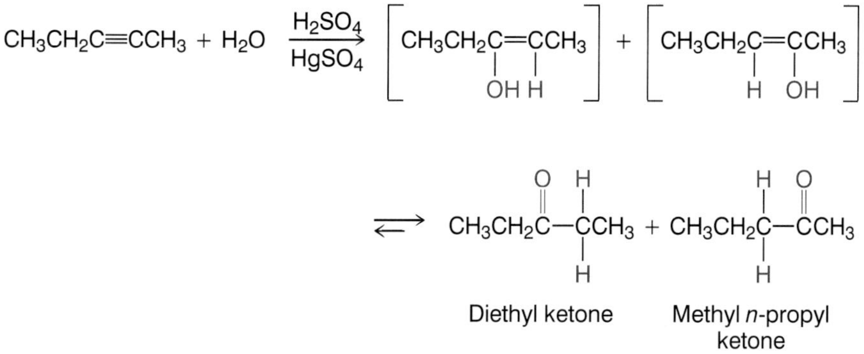

### Envision the Reaction

### Mechanism of the Reaction of an Enol to Its Tautomer (a Ketone)

The conversion of a vinyl alcohol to a ketone (or aldehyde) is catalyzed by acid and proceeds as follows:

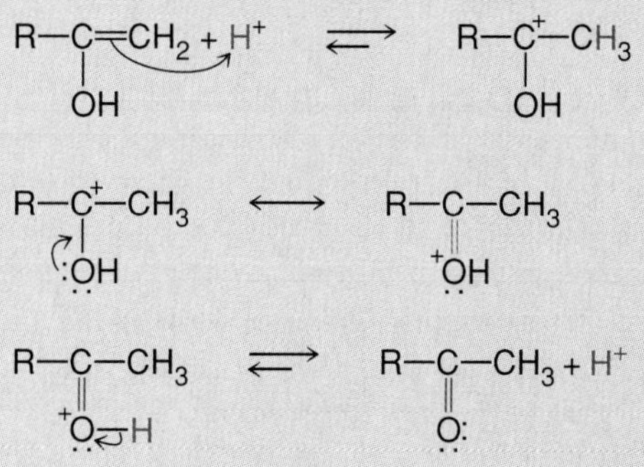

Acetaldehyde and ketones rapidly reach an equilibrium with the enol form. This equilibrium mixture is called *keto-enol tautomerism*, with the ketone (or acetaldehyde) and enol form referred to as *tautomers*.

It is important to realize that tautomers are not resonance structures of a given molecule but represent two different molecules. Since the conversion of a vinyl alcohol to an aldehyde or ketone involves the shift of a proton from the oxygen of the hydroxyl group in the enol form to a carbon adjacent to the carbon bonded to the carbonyl group in the keto form, resonance is ruled out. Note the following illustration of a tautomeric equilibrium. The movement of electrons is shown by a curved arrow, whereas the movement of an atom or group of atoms is represented by a multiline arrow.

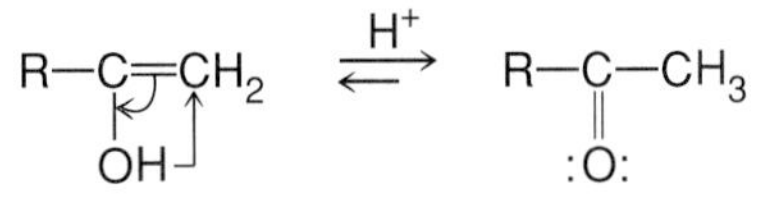

$$R—C{=}CH_2 \quad \underset{}{\overset{H^+}{\rightleftharpoons}} \quad R—C—CH_3$$
$$\overset{|}{OH} \qquad\qquad\qquad \overset{\|}{:O:}$$

On the other hand, observe in the two resonance structures of a generalized enol form shown below that only electrons have shifted; no atom has moved.

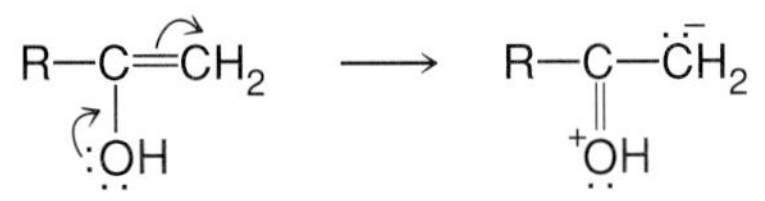

$$R—C{=}CH_2 \quad \longrightarrow \quad R—C—\overset{-}{C}H_2$$
$$\overset{|}{:OH} \qquad\qquad\qquad \overset{\|}{{}^+OH}$$

To formulate a carbonyl compound from the corresponding vinyl alcohol, use the following two steps:

*Step 1.* Move the hydroxyl hydrogen and its bond with oxygen to the other double-bonded carbon; at the same time, break a bond of the carbon-carbon double bond.[*]

*Step 2.* Add a pair of electrons to the carbon-oxygen single bond.

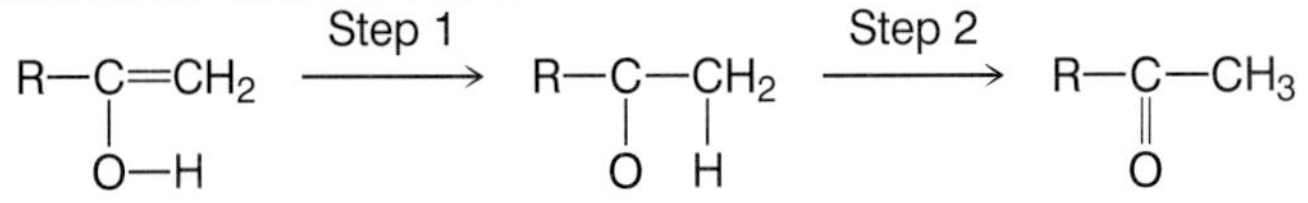

$$\underset{\overset{|}{O—H}}{R—C{=}CH_2} \quad\overset{Step\ 1}{\longrightarrow}\quad \underset{\overset{|}{O}\ \overset{|}{H}}{R—C—CH_2} \quad\overset{Step\ 2}{\longrightarrow}\quad \underset{\overset{\|}{O}}{R—C—CH_3}$$

$$I$$

For ketones that contain one carbonyl group, the equilibrium favors the keto form; there is little enol present. However, with certain compounds containing two carbonyl groups, there is a significant amount of enol form present. For example, ethyl acetoacetate contains 7% enol in the equilibrium mixture, whereas 2,4-pentanedione contains 72% enol in the equilibrium mixture.

$$\underset{\overset{\textstyle O}{\|}}{CH_3—C}—CH_2—\underset{\overset{\textstyle O}{\|}}{C}—OCH_2CH_3 \quad \rightleftharpoons \quad CH_3—\underset{\overset{\textstyle OH}{|}}{C}{=}CH—\underset{\overset{\textstyle O}{\|}}{C}—OCH_2CH_3$$

Ethyl acetoacetate<br>93% keto                                  7% enol

$$\underset{\overset{\textstyle O}{\|}}{CH_3—C}—CH_2—\underset{\overset{\textstyle O}{\|}}{C}—CH_3 \quad \rightleftharpoons \quad CH_3—\underset{\overset{\textstyle OH}{|}}{C}{=}CH—\underset{\overset{\textstyle O}{\|}}{C}—CH_3$$

2,4-pentanedione<br>28% keto                               72% enol

**FIGURE 8.4** Hydrogen bond stabilization of the enol form of ethyl acetoacetate.

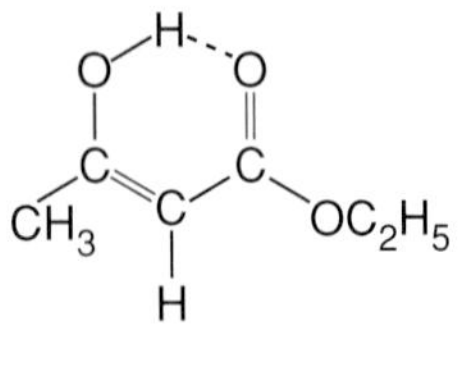

You might ask why the preceding compounds contain so much enol in the tautomeric mixture. The main reason is that an intramolecular hydrogen bond is present in the enol form of ethyl acetoacetate—between the hydroxyl group and the carbonyl of the ester functional group—stabilizing the enol by forming a quasi–six-membered ring. (Six-membered rings are very stable.) An *intramolecular hydrogen bond* is a bond that links a hydrogen atom with an oxygen or nitrogen atom in the same molecule. Figure 8.4

---

[*]Structure I does not exist. It is merely a means of conveniently enabling a student without a mechanistic background to formulate the structural formula of a carbonyl compound given the structural formula of the corresponding vinyl alcohol, and vice versa.

shows an intramolecular hydrogen bond (the dashed line) stabilizing the enol form of ethyl acetoacetate. The enol form of 2,4-pentanedione is stabilized in the same manner.

Draw a structural formula for an enol form of

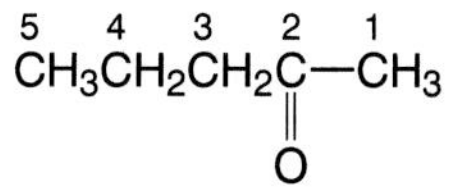

$$\overset{5}{C}H_3\overset{4}{C}H_2\overset{3}{C}H_2\overset{2}{\underset{\underset{O}{\|}}{C}}\text{—}\overset{1}{C}H_3$$

2-Pentanone

**SOLUTION**

We can formulate the enol form in a qualitative way by locating a carbon-carbon double bond between C-1 and C-2 (*-ene*) and an hydroxyl group on C-2 (*-ol*) to give

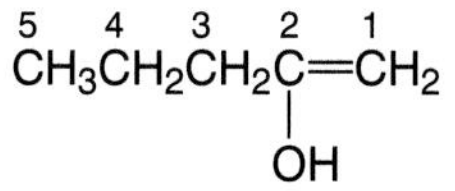

$$\overset{5}{C}H_3\overset{4}{C}H_2\overset{3}{C}H_2\overset{2}{\underset{\underset{OH}{|}}{C}}=\overset{1}{C}H_2$$

A more quantitative method of correctly drawing the structure of the enol is to formulate a mechanism.

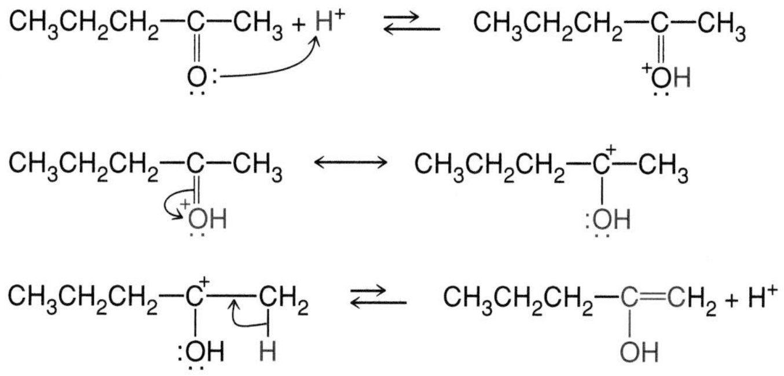

$$CH_3CH_2CH_2\text{—}\underset{\underset{\ddot{O}:}{\|}}{C}\text{—}CH_3 + H^+ \;\rightleftarrows\; CH_3CH_2CH_2\text{—}\underset{\underset{^+\ddot{O}H}{\|}}{C}\text{—}CH_3$$

$$CH_3CH_2CH_2\text{—}\underset{\underset{^+\ddot{O}H}{\|}}{C}\text{—}CH_3 \;\longleftrightarrow\; CH_3CH_2CH_2\text{—}\overset{+}{\underset{\underset{:\ddot{O}H}{|}}{C}}\text{—}CH_3$$

$$CH_3CH_2CH_2\text{—}\overset{+}{\underset{\underset{:\ddot{O}H\;\;H}{|}}{C}}\text{—}CH_2 \;\rightleftarrows\; CH_3CH_2CH_2\text{—}\underset{\underset{OH}{|}}{C}=CH_2 + H^+$$

Complete the following reactions by supplying a structural formula for each organic product. Draw the unstable enol form in brackets and the keto form for each product.

a. $CH_3C{\equiv}CCH_3 + H_2O \;\xrightarrow[\text{HgSO}_4]{\text{H}_2\text{SO}_4}\;$

b. $CH_3CH_2CH_2C{\equiv}CH + H_2O \;\xrightarrow[\text{HgSO}_4]{\text{H}_2\text{SO}_4}\;$

c. $CH_3CH_2CH_2CH_2C{\equiv}CCH_2CH_3 + H_2O \;\xrightarrow[\text{HgSO}_4]{\text{H}_2\text{SO}_4}\;$

**Ozonolysis**

Ozone ruptures all three bonds of the triple bond to produce one or more products. In general,

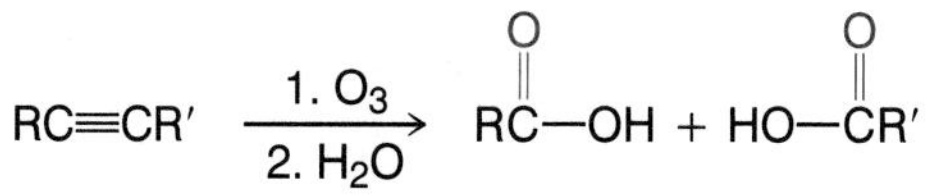

$$RC{\equiv}CR' \;\xrightarrow[\text{2. H}_2\text{O}]{\text{1. O}_3}\; \underset{\underset{\|}{O}}{R}C\text{—}OH + HO\text{—}\underset{\underset{\|}{O}}{C}R'$$

Example:

$$CH_3C{\equiv}CCH_2CH_3 \xrightarrow[\text{2. } H_2O]{\text{1. } O_3} \underset{}{CH_3\overset{O}{\overset{\|}{C}}{-}OH} + \underset{\text{Propionic acid}}{HO{-}\overset{O}{\overset{\|}{C}}CH_2CH_3}$$

Note that a terminal alkyne forms carbon dioxide as a product.

---

**SOLVED PROBLEM 8.8**

Predict the products of the following ozonolysis.

$$CH_3CH_2CH_2C{\equiv}CH \xrightarrow[\text{2. } H_2O]{\text{1. } O_3}$$

**SOLUTION**

*Step 1.* Break all three bonds of the carbon-carbon triple bond.

$$CH_3CH_2CH_2C \quad \text{and} \quad CH$$

*Step 2.* The electron-deficient terminal CH species is converted to carbon dioxide ($CO_2$).

*Step 3.* To the other electron-deficient carbon of the triple bond, add the groupings $=O$ and $-OH$; thus

$$CH_3CH_2CH_2\overset{O}{\overset{\|}{C}}{-}OH$$

*Step 4.* The products of the ozonolysis are

$$CH_3CH_2CH_2\overset{O}{\overset{\|}{C}}{-}OH \quad \text{and} \quad CO_2$$

---

**EXERCISE 8.8**

Predict the products of each of the following reactions.

a.   $CH_3CH_2CH_2CH_2C{\equiv}CH \xrightarrow[\text{2. } H_2O]{\text{1. } O_3}$

b.   cyclopentylacetylene $\xrightarrow[\text{2. } H_2O]{\text{1. } O_3}$

c.   $\underset{\underset{CH_3}{|}}{CH_3CHC}{\equiv}CCH_2CH_3 \xrightarrow[\text{2. } H_2O]{\text{1. } O_3}$

---

**SOLVED PROBLEM 8.9**

Ozonolysis can be used to identify an unknown alkyne. Consider the following example: An alkyne was reacted with ozone, followed by hydrolysis to give butyric acid and acetic acid.

$$\underset{\substack{\text{Butyric acid} \\ \text{II}}}{CH_3CH_2CH_2\overset{O}{\overset{\|}{C}}{-}OH} \qquad \underset{\substack{\text{Acetic acid} \\ \text{III}}}{HO{-}\overset{O}{\overset{\|}{C}}{-}CH_3}$$

Identify the parent alkyne.

**SOLUTION**

*Step 1.* Draw the structural formula of each carboxylic acid so that the two carboxyl groups are facing each other (refer to structures II and III). Then break the two carbon-oxygen double bonds and the two carbon-oxygen single bonds in the product carboxylic acids.

$$CH_3CH_2CH_2C \quad CCH_3$$

*Step 2.* Place a triple bond between each electron-deficient carbon, as follows, to give 2-hexyne:

$$CH_3CH_2CH_2C{\equiv}CCH_3$$

---

Provide the structure of the alkyne that on ozonolysis would give the following product(s).

**EXERCISE 8.9**

a.  $CH_3CH_2CH_2CH_2\overset{\displaystyle O}{\overset{\|}{C}}{-}OH \ + \ CH_3CH_2\overset{\displaystyle O}{\overset{\|}{C}}{-}OH$

b.  $CH_3CH_2\underset{\underset{\displaystyle CH_3}{|}}{CH}\overset{\displaystyle O}{\overset{\|}{C}}{-}OH \ + \ CO_2$

c.  $CO_2$

d.  $CH_3CH_2CH_2CH_2\overset{\displaystyle O}{\overset{\|}{C}}{-}OH$

---

## ▶ CHAPTER ACCOMPLISHMENTS

### 8.1 Structure

☐ Distinguish between a terminal alkyne and an internal alkyne.

☐ Formulate a molecular formula for an alkyne containing five carbons.

### 8.2 Nomenclature

☐ Draw a structural formula for 4-methyl-2-pentyne.

☐ Name each of the following according to the IUPAC system of nomenclature:

a. $CH_3CH_2C{\equiv}CCH_2CH_3$
b. $CH_3C{\equiv}CCH_3$
c. $HC{\equiv}CH$

☐ State which of the following functional groups has the higher priority using the IUPAC system of nomenclature: $C{=}C$ or $C{\equiv}C$.

### 8.3 Physical Properties

☐ Predict which of the following compounds would most likely be a liquid:

$$C_2H_2 \quad \text{or} \quad C_8H_{14}$$

### 8.4 Acidity of Alkynes

☐ Explain why acetylene is more acidic than ethane.

☐ Complete the following equation:

$$CH_3C{\equiv}CH \ + \ {}^-NH_2 \longrightarrow$$

☐ Suggest a test tube reaction that can distinguish between 1-butyne and 2-butyne. Describe what you do and see.

### 8.5 Industrial Preparation of Acetylene

☐ Name the alkyne that is produced in the largest quantities to justify an industrial preparation.

☐ Compare the typical industrial preparation with a laboratory preparation with respect to

a. amount of reactants used.
b. cost.
c. by-products produced.
d. use of catalysts.
e. reaction conditions employed.

### 8.6 Laboratory Preparation of Alkynes

☐ Write a detailed, step-by-step mechanism for the following reaction:

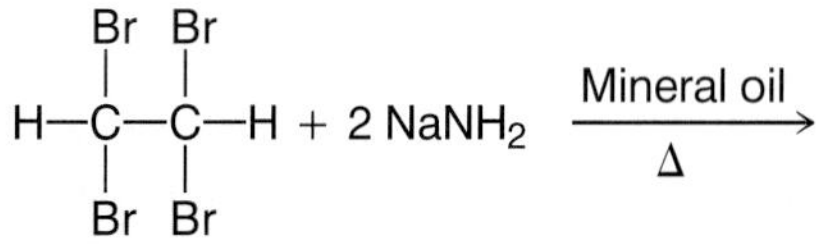

$$\underset{\underset{\displaystyle Br \ \ Br}{|\ \ \ |}}{\overset{\overset{\displaystyle Br \ \ Br}{|\ \ \ |}}{H{-}C{-}C{-}H}} \ + \ 2 \ NaNH_2 \ \xrightarrow[\Delta]{\text{Mineral oil}}$$

☐ Differentiate between the products produced when a *vic-* dihalide is treated with $NaNH_2$ versus $KOH$.

☐ Explain why 2-butyne cannot be used as a starting reagent in the terminal alkyne synthesis of 2-pentyne.

### 8.7 Chemical Properties of Alkynes

☐ Draw a structural formula for each product when 2-butyne is reacted with

  a. $H_2$ (2 mol).

  b. $H_2$ (1 mol—Lindlar's catalyst).

  c. $H_2$ (1 mol—Na, liquid $NH_3$).

  d. $Br_2$ (1 mol—low temperature).

  e. $Br_2$ (2 mol—room temperature).

  f. $HCl$ (1 mol—low temperature).

  g. $HCl$ (2 mol—room temperature).

  h. $H_2O$ ($H_2SO_4$, $HgSO_4$).

  i. $O_3$ then $H_2O$.

---

### ▶ KEY TERMS

*terminal alkyne* (8.1)            *Lindlar's catalyst* (8.7)          *tautomers* (8.7)

*internal alkyne* (8.1)            *enol* (8.7)                        *intramolecular hydrogen bond* (8.7)

*geminal (gem-) dihalide* (8.6)    *keto-enol tautomerism* (8.7)

---

### ▶ PROBLEMS

1. Why can't alkynes display geometric isomerism? *Hint:* Refer to Section 2.5.

2. For the sake of argument, let us assume that alkynes could show geometric isomerism. Draw a structural formula for an imaginary *cis*-2-butyne and an imaginary *trans*-2-butyne.

3. Given that an alkyne results from the loss of four hydrogens (a four-hydrogen deficit) from the corresponding alkane (refer to Sec. 5.1 for a review), and knowing that the compound $C_7H_8$ is acyclic and contains only one or more triple bonds, deduce the number of triple bonds present.

4. Let us assume that a hydrocarbon, $C_6H_8$, can contain one or more triple bonds and/or double bonds and/or rings. List the possible triple-bond–double-bond–ring combinations (refer to Sec. 5.1).

5. Draw a structure for and name using the IUPAC method each of the isomeric alkynes with the formula $C_5H_8$. Classify each alkyne as terminal or internal.

6. Name each of the following alkynes using the IUPAC method:

  a. $CH_3—C≡C—CH_2CH_2—C≡CH$

  b. $CH_2=CH—C≡CH$

  c. $CH_2=CH—C≡C—CH_3$

  d. ⌐⌐⌐—=—⌐⌐

  e. ⌐⌐⌐⌐—=

7. Draw a structural formula for a (an)

  a. isolated diyne.

  b. conjugated diyne.

8. Each of the following names is incorrect. Explain why the name is incorrect, and write the correct name of the compound.

  a. 3-methyl-4-hexyne

  b. *trans*-2-pentyne

  c. 2,4-pentadiyne

9. Propyne and 1-cyano-1,3-butadiene have been detected in interstellar space. Given that the cyano group is $—C≡N$, draw a structural formula for each compound.

10. Draw a structural formula for each of the following compounds that has been isolated from a different flowering plant.

  a. 1,7,9-heptadecatriene-11,13,15-triyne (isolated from *Centaurea cyanus*, commonly known as the cornflower)

  b. 1,3,5,11-tridecatetraene-7,9-diyne (isolated from various *Coreopsis* species, commonly known as tickweeds)

11. Draw a condensed formula for each of the following compounds.

  a. 2-pentyne

  b. 1,3-hexadiyne

12. Would you expect 2-pentyne to dissolve in water? Explain.

13. Explain why ethylene is less acidic than acetylene. Use both the Brønsted and electronegativity explanations.

14. How would you distinguish between each of the following using a test tube reaction? Write what reagents you would add and the experimental results you would observe.

  a. 1-butyne and 1-butene

  b. cyclononane and cyclononyne

  c. 1-chloropropane and 3-chloropropene

  d. hexane and 2-hexyne

15. Prepare each of the following from propyne. Use any organic and inorganic reagents necessary.

  a. (*E*)-1,2-dibromopropene

  b. propene

c. $CH_3\overset{O}{\underset{\|}{C}}-OH$

d. $CH_3CBr_2CH_3$

e. $CH_3\overset{O}{\underset{\|}{C}}CH_3$

f. propane

g. $CH_3CH_2CH_2OH$

h. 2-hexyne

i. (Z)-2-pentene

16. Prepare $CH_3CH_2C\equiv CH$ from each of the following. You may use any organic and inorganic reagents necessary.
   a. $CH_3CH_2CH=CH_2$
   b. $CaC_2$
   c. $CH_3CH_2CH_2CHBr_2$

17. Prepare each of the following compounds from 2-chloropentane [$CH_3CH_2CH_2CH(Cl)CH_3$]. You may use any organic and inorganic reagents necessary.
   a. *cis*-2-pentene
   b. *trans*-2-pentene
   c. 2,2,3,3-tetrabromopentane

18. Select the more stable carbocation from each of the following pairs.
   a. $CH_3\overset{+}{C}HCH_3$ or $CH_2=\overset{+}{C}-CH_3$
   b. $CH_3CH_2CH_2{}^+$ or $CH_3-CH=\overset{+}{C}-CH_3$

19. Complete each of the following equations by writing a structural formula for each organic product and a molecular formula or formula unit for each inorganic compound formed. If no reaction occurs, write "no reaction."
   a. $CH_3-C\equiv CH + HCl \xrightarrow{Cold}$
   b. $CH_3-C\equiv CH + 2 HCl \xrightarrow{20°C}$
   c. $CH_3-C\equiv C-CH_3 + NaNH_2 \longrightarrow$
   d. $(CH_3)_3C-C\equiv CH + {}^+Ag(NH_3)_2 + {}^-OH \longrightarrow$
   e. $CH_3CH=CHBr + KOH \xrightarrow[\Delta]{Alcohol}$
   f. $CH_3CH_2CH_2Br + CH_3\underset{\underset{CH_3}{|}}{CH}-C\equiv C^-Na^+ \longrightarrow$
   g. $CH_3-C\equiv C-C_2H_5 + 2 H_2 \xrightarrow{Pd}$
   h. $C_2H_5-C\equiv C-C_2H_5 \xrightarrow[Liquid\ NH_3]{Na}$
   i. $C_2H_5-C\equiv C-C_2H_5 + H_2 \xrightarrow[catalyst]{Lindlar's}$
   j. $HC\equiv CH + 2 Br_2 \xrightarrow[20°C]{CCl_4}$
   k. $CH_3CH_2-C\equiv C-CH_2CH_3 + Br_2 \xrightarrow[-10°C]{CCl_4}$
   l. $C_2H_5-C\equiv C-C_2H_5 + H_2O \xrightarrow[HgSO_4]{H_2SO_4}$
   m. $CH_3(CH_2)_{15}C\equiv CCH_2CH_3 \xrightarrow[2.\ H_2O]{1.\ O_3}$

20. Draw a structural formula for an alternate enol form of 2-pentanone (see Solved Problem 8.7).

21. Draw a structural formula for the keto tautomer of each of the following unstable enol tautomers. Draw a structural formula of the alkyne that would produce each unstable enol when treated with $H_2O$, $H_2SO_4$, and $HgSO_4$.
   a. $CH_3CH_2\underset{\underset{OH}{|}}{C}=CH_2$
   b. $CH_2=\underset{\underset{OH}{|}}{CH}$
   c. $CH_3CH_2CH_2CH_2CH_2CH_2\underset{\underset{OH}{|}}{C}=CH_2$

22. Draw a structural formula to illustrate hydrogen bond stabilization of the enol form of 2,4-pentanedione.

23. Write a detailed, step-by-step mechanism for the following reaction:

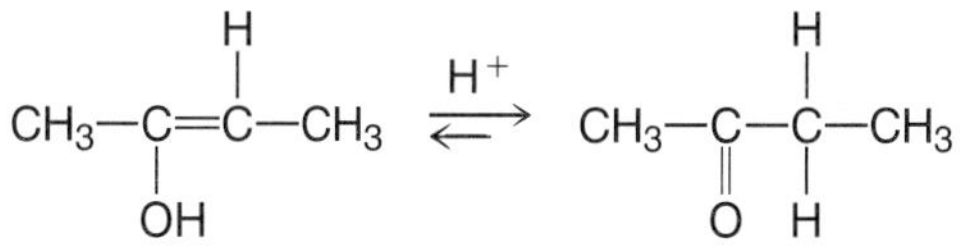

$$CH_3-\underset{\underset{OH}{|}}{\overset{\overset{H}{|}}{C}}=C-CH_3 \xrightleftharpoons{H^+} CH_3-\underset{\underset{O}{|}}{\overset{\overset{H}{|}}{C}}-\underset{\underset{H}{|}}{C}-CH_3$$

24. Draw a structure of one or more possible enol forms for each of the following tautomeric keto forms.
   a. $CH_3\overset{O}{\underset{\|}{C}}CH_2CH_3$
   b. $CH_3CH_2\underset{\underset{O}{\|}}{CH}$

25. Can formaldehyde ($H_2C=O$) exist as a component of a keto-enol tautomeric equilibrium? Explain.

26. Draw the structure of the alkyne that produces each of the following on reaction with ozone followed by water.
   a. $CH_3COOH$
   b. $CH_3CH_2COOH + CH_3COOH$
   c. $HOOCCH_2(CH_2)_5CH_2COOH$

27. Compounds $A, B, C,$ and $D$ have the formula $C_4H_6$. Each of these compounds decolorized bromine in carbon tetrachloride instantaneously. Compound $A$ gave a positive test with ${}^+Ag(NH_3)_2 + {}^-OH$, while compound $C$ showed both 1,2- and 1,4-addition products when treated with 1 mol of HCl. Compound $B$ gave a negative test with ${}^+Ag(NH_3)_2 + {}^-OH$ and reacted with ozone then water to give acetic acid ($CH_3COOH$). Compound $D$ reacted with hydrogen in the presence of a nickel catalyst to yield cyclobutane.
   a. Draw a structural formula for each isomer, and assign a name to each isomer. Write equations for the chemical reactions involved.
   b. What product would you expect to be produced when isomer $D$ is reacted with ozone then zinc and water?

CHAPTER 9

# Aromatic Compounds (I)

## STRUCTURE, NOMENCLATURE, AND AROMATICITY

Benzene and its derivatives were originally called *aromatic* because many of the compounds known at that time (early 19th century) had a fragrant odor. For example, vanillin was isolated from the vanilla bean, and methyl salicylate from oil of wintergreen obtained from the wintergreen plant (Fig. 9.1). Today, the term *aromatic* has a completely different meaning; instead of a number of compounds with pleasant odors, it refers to a family of compounds consisting of benzene and its derivatives.

Most of these pleasant-smelling compounds were isolated from plants.

Benzene was first isolated by Michael Faraday in 1825 from compressed illuminating gas, which was used in England as a source of light (Fig. 9.2). By the year 1858, chemists knew a good deal about benzene. The empirical formula of the compound was found to be CH, and when the molecular weight was determined to be 78 g/mol, the molecular formula of $C_6H_6$ was assigned to the compound. In addition, benzene was found to show the following chemical properties:

## 9.1
## THE STRUCTURE OF BENZENE

As we discussed in Section 2.6, benzene was assigned a hexagonal structure.

1. Benzene formed only one monosubstituted derivative, $C_6H_5X$. This means that all six hydrogens in benzene are equivalent. (For a more detailed discussion, refer to Sec. 9.6.)

2. Benzene reacted to yield three disubstituted derivatives, $C_6H_4XY$ or $C_6H_4X_2$. (For more detail, refer to page 165, "Disubstituted Benzenes.")

3. One mole of benzene reacted with ozone followed by zinc and water to produce 3 mol of $O=CH-HC=O$, or glyoxal.

**FIGURE 9.1** Two pleasant-smelling aromatic compounds.

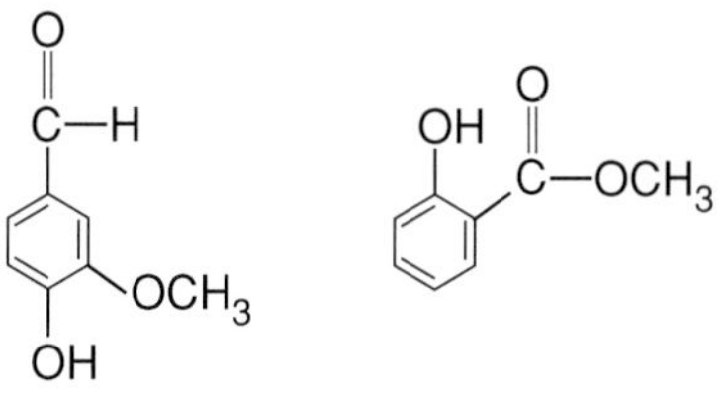

Vanillin

Methyl salicylate

**FIGURE 9.2** Various structural representations of benzene.

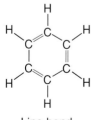

Line-bond
structural formula

Geometric
condensed
structural
formula

Resonance
hybrid
structural
formula

## HISTORICAL BOXED READING 9.1

For many years, only chemical data were available to support the hypothesis that benzene is a cyclic, conjugated six-carbon triene. Physical evidence from electron diffraction did indeed show that all the bonds were identical and that the molecule was planar. Invention of the scanning tunneling microscope in 1981 allowed us (during the 1980s) to visualize the actual hexagonal shape of benzene and confirm that this model was correct.

The microscope consists of a tungsten probe with a very fine needle-like tip (only a few atoms in width). The tip of this probe, in turn, is positioned about 5 Å (a knowable) above the sample that consists of a layer of gaseous benzene over a layer of carbon monoxide gas adsorbed on rhodium metal. The carbon monoxide and rhodium are needed to anchor the benzene so that a picture can be taken of the benzene molecules.

The probe moves back and forth across the sample, generating an electric current that is measured. This current is a function of the distance between the probe and the sample. Therefore, as the current changes, so does the distance between the probe and sample. A computer records these data and creates a topographic picture of the sample (Fig. 9.3).

Figure 9.4 is a photograph taken through the scanning tunneling microscope showing rows of hexagonal-shaped benzene molecules.

**FIGURE 9.3** A scanning tunneling microscope.

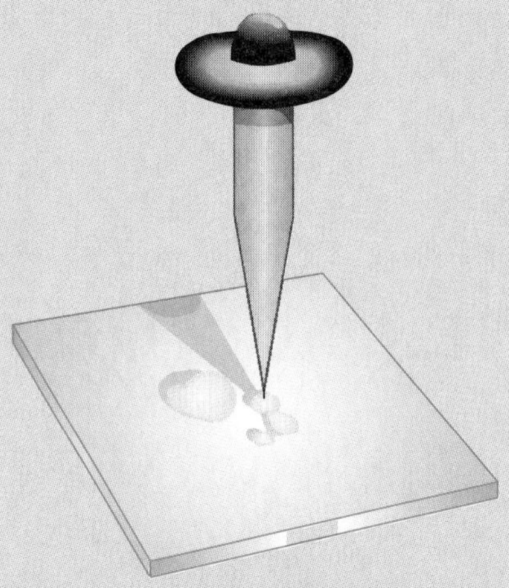

**FIGURE 9.4** A photograph of a number of benzene molecules anchored to a rhodium metal surface.

One evening, in the year 1858, after spending some time working on a textbook of organic chemistry at his apartment in Ghent (Belgium), Professor Friedrich Kekulé could not concentrate and dozed off. With the preceding experimental facts in mind, he dreamed that a number of atoms were moving with a snakelike motion. One of these molecular snakes, he dreamed, seized its own tail with its mouth. From that dream grew his hypothesis that benzene must be a cyclic molecule.

## 9.2 NOMENCLATURE

Historically, a significant number of aromatic compounds were known before the superior IUPAC system of nomenclature was introduced. Therefore, it seems reasonable that the use of nonsystematic names to designate aromatic compounds has carried over through the years. In fact, this carryover has been so extreme that several of the trivial names of the monosubstituted benzenes (along with benzene) have been accepted by the IUPAC as systematic names (Table 9.1).

**Monosubstituted Benzenes.**   These compounds are named as derivatives of benzene, such as nitrobenzene, the halobenzenes, and a number of the alkylbenzenes, or *arenes*. No number is needed to locate a substituent in a monosubstituted benzene because the six hydrogens in benzene are equivalent. Some examples of the nomenclature of selected monosubstituted derivatives are given in Table 9.2. Practicing scientists tend to use the name toluene rather than methylbenzene, phenol rather than hydroxybenzene, etc.

| TABLE 9.1 | Former Trivial Nomenclature of Selected Monosubstituted Benzene Derivatives |
|---|---|

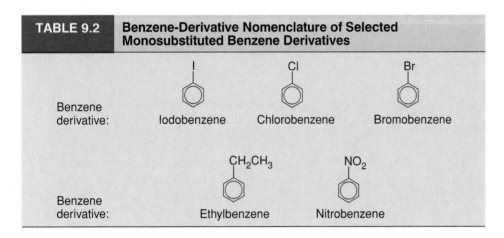

Former trivial: Toluene    Phenol    Aniline    Benzaldehyde

Former trivial: Styrene    Benzenesulfonic acid    Benzoic acid

Trivial names could be based on the natural source of the compound or on how the compound was prepared.

| TABLE 9.2 | Benzene-Derivative Nomenclature of Selected Monosubstituted Benzene Derivatives |
|---|---|

Benzene derivative: Iodobenzene    Chlorobenzene    Bromobenzene

Benzene derivative: Ethylbenzene    Nitrobenzene

Benzene-derivative and former trivial names are differentiated, although both are considered to be IUPAC today. Only commonly accepted names are given.

Just as a hydrogen is lost by an alkane to give an alkyl group, in the same manner a hydrogen can be lost by an aromatic compound to give an imaginary *aryl group*. Two important aryl groups used in nomenclature are shown in Table 9.3. Aryl groups can be used as substituent groups. Examples are found in Fig. 9.5.

**Disubstituted Benzenes.**   These derivatives are named using the prefixes *ortho-*, *meta-*, and *para-*. The prefix *ortho-* (abbreviated *o-*) designates that the two groups are bonded to carbons 1 and 2 of the ring; *meta-* (*m-*) indicates a 1,3-group placement, and *para-* (*p-*) signifies a 1,4-group placement on the carbons of the ring. In general, we have

| TABLE 9.3 | Summary of Selected Aryl Groups Derived From Benzene and Toluene |
|---|---|

Benzene    Phenyl    Phenylethane (or ethylbenzene)

Toluene    Benzyl    Benzyl bromide (common name)

with respect to group A the layout of *ortho-*, *meta-*, and *para-* substituents as seen in Fig. 9.6. When the compound is to be assigned a former trivial name, use the prefix *o-* , *m-* , or *p-* followed by the name of the group (when called for) and the name of the parent compound (Table 9.4).

| TABLE 9.4 | Former Trivial Nomenclature of Selected Disubstituted Benzene Derivatives |
|---|---|

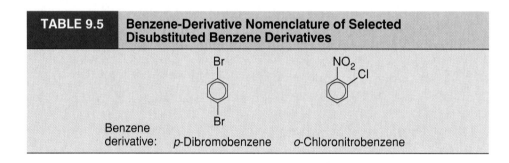

| | $CH_3$ / $CH_3$ | COOH / $NH_2$ |
|---|---|---|
| | | Note: The —COOH group has a higher nomenclature priority than the —$NH_2$ group. |
| Former trivial: | *m*-Xylene | *p*-Aminobenzoic acid |

When the two groups bonded to the benzene ring are the same, the prefix *di-* is used immediately following the prefix *o-*, *m-*, or *p-*. The name of the group and the stem *benzene* complete the nomenclature of a disubstituted benzene derivative. In benzene-derivative nomenclature, we list the substituent groups in alphabetical order. For example, look at Table 9.5.

| TABLE 9.5 | Benzene-Derivative Nomenclature of Selected Disubstituted Benzene Derivatives |
|---|---|

| | Br / Br | $NO_2$ / Cl |
|---|---|---|
| Benzene derivative: | *p*-Dibromobenzene | *o*-Chloronitrobenzene |

**FIGURE 9.5** Use of selected aryl groups in nomenclature.

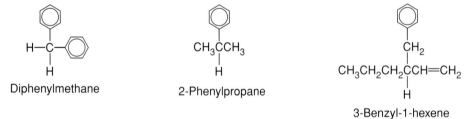

Diphenylmethane　　　　　2-Phenylpropane　　　　　3-Benzyl-1-hexene

**FIGURE 9.6** Meaning of prefixes *ortho-*, *meta-*, and *para-* in the nomenclature of disubstituted benzene derivatives.

Ortho→ 6　1　2 ←Ortho
Meta→ 5　　3 ←Meta
　　　　　4
　　　Para

---

**SOLVED PROBLEM 9.1**

Assign an IUPAC name to each of the following compounds.

a.　COOH / $C_2H_5$　　　　b.　Br / Cl

**SOLUTION**

a.　Since the two groups are located on the 1 and 3 carbons in the benzene ring, this represents a meta orientation. The parent compound is benzoic acid, so we use a former trivial name:

*m*-ethylbenzoic acid

b.  Since the two groups are located on adjacent carbons in the benzene ring, this represents an ortho orientation. Because no parent compound is present, we name the compound as a benzene derivative and list the substituents in alphabetical order to give

o-bromochlorobenzene

---

Assign an IUPAC name to each of the following compounds.

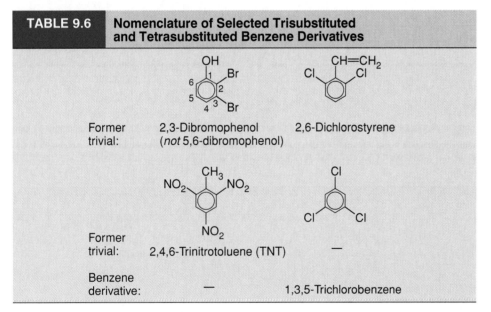

---

**Benzene Derivatives with Three or More Substituents.**  These compounds are named by assigning to each substituent the smallest possible carbon number on the ring. The numbering of substituents is used with both the former trivial and benzene-derivative variations of IUPAC nomenclature. If a group (such as —OH) is located on carbon 1 and the other substituents are different, the number 1 is not used with either variation of IUPAC nomenclature. However, if the substituents are the same, the number 1 is used when the compound is named as a benzene derivative. Examples are shown in Table 9.6.

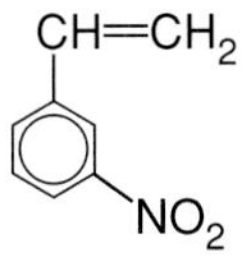

| TABLE 9.6 | Nomenclature of Selected Trisubstituted and Tetrasubstituted Benzene Derivatives |

| Former trivial: | 2,3-Dibromophenol (*not* 5,6-dibromophenol) | 2,6-Dichlorostyrene |
| Former trivial: | 2,4,6-Trinitrotoluene (TNT) | — |
| Benzene derivative: | — | 1,3,5-Trichlorobenzene |

Occasionally, if a group with the greater alphabetical priority is located on carbon 1 and the other substituents are different, the substituents are not listed in alphabetical order when the compound is named as a benzene derivative. For example,

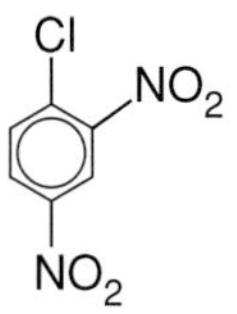

2,4-Dinitrochlorobenzene

**SOLVED PROBLEM 9.2**

Assign an IUPAC name to each of the following compounds.

a. 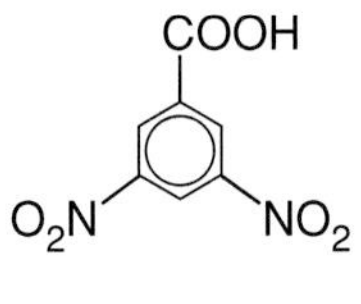
COOH
$O_2N$          $NO_2$

b. 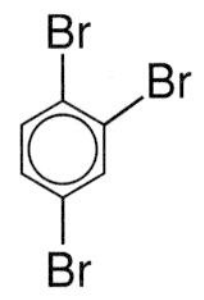
Br
Br
Br

**SOLUTION**

a.  The parent compound is benzoic acid. Thus the compound is named with a former trivial name to give

3,5-dinitrobenzoic acid

b.  A parent compound is not present, so the compound is named as a benzene derivative to give

1,2,4-tribromobenzene

**EXERCISE 9.2**

Assign an IUPAC name to each of the following compounds.

a. 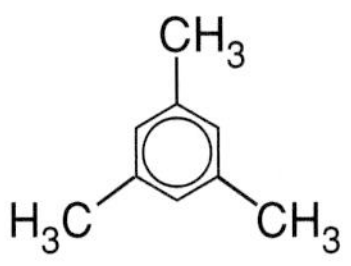
$CH_3$
$H_3C$          $CH_3$

b. 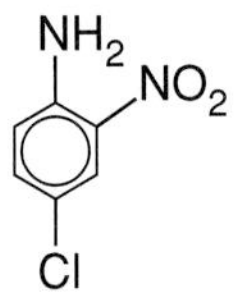
$NH_2$
$NO_2$
Cl

Polycyclic aromatic compounds contain two or more benzene rings fused together at two carbons. Two examples are shown in Fig. 9.7.

**SOLVED PROBLEM 9.3**

Furnish an IUPAC name for each of the following compounds.

a.

Br

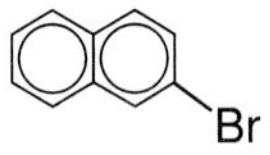

b. 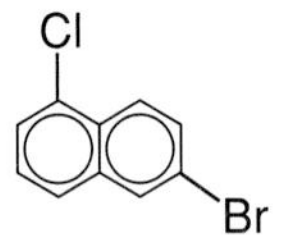
Cl
Br

c. 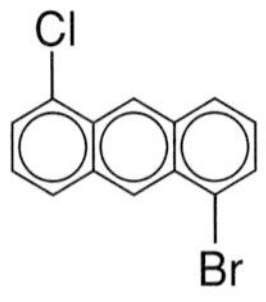
Cl
Br

**FIGURE 9.7** Structure and numbering pattern of two selected polycyclic aromatic hydrocarbons.

Naphthalene          Anthracene

**SOLUTION**

A substituted polycyclic aromatic compound is named as a derivative of the unsubstituted hydrocarbon. One or more substituents are located by means of the lowest possible numbers.

a.  This compound is a substituted naphthalene. From the structure and numbering pattern of naphthalene in Fig. 9.7, you might assign the name 5-bromonaphthalene. However, since positions 1, 4, 5, and 8 are equivalent, we can use the following equivalent numbering pattern to assign a lower number to the substituent group:

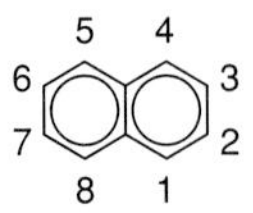

5    4
6          3
7          2
8    1

From this equivalent numbering pattern we get the correct name:

2-bromonaphthalene

b.  From the numbering pattern assigned to naphthalene in Fig. 9.7, we get

6-bromo-1-chloronaphthalene
(*not* 2-bromo-5-chloronaphthalene)

c.  From the numbering pattern assigned to anthracene in Fig. 9.7, we get

5-bromo-1-chloroanthracene
(*or* 1-bromo-5-chloroanthracene)
(*not* 4-bromo-8-chloroanthracene)

---

Furnish an IUPAC name for each of the following compounds.

**EXERCISE 9.3**

a.

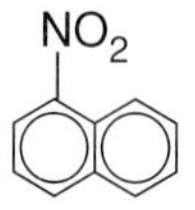

b.

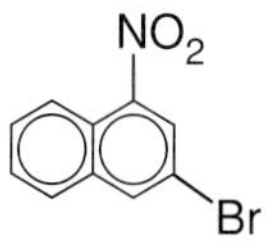

c.

---

Like the aliphatic hydrocarbons, aromatic hydrocarbons are low-boiling. The only intermolecular forces present are short-range London forces, just as with the alkanes (Table 9.7). Aromatic hydrocarbons, along with aliphatic hydrocarbons, are nonpolar and are therefore insoluble in water but soluble in nonpolar solvents such as pentane, carbon tetrachloride, and ether.

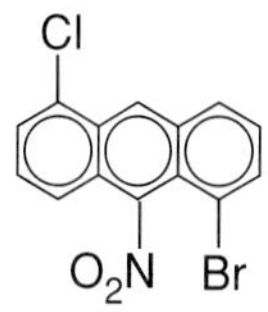

## 9.3
## PHYSICAL PROPERTIES

The two main industrial sources of aromatic hydrocarbons are coal tar and petroleum. Coke and coal tar are obtained by heating coal in the absence of air. If coal tar is distilled, a number of aromatic hydrocarbons are obtained, including benzene, toluene, xylenes (dimethylbenzenes), naphthalene, and anthracene. Many other hydrocarbons may be obtained as well. Aromatic compounds also may be obtained from petroleum by a process called *reforming* that converts alkanes and cycloalkanes into aromatic compounds.

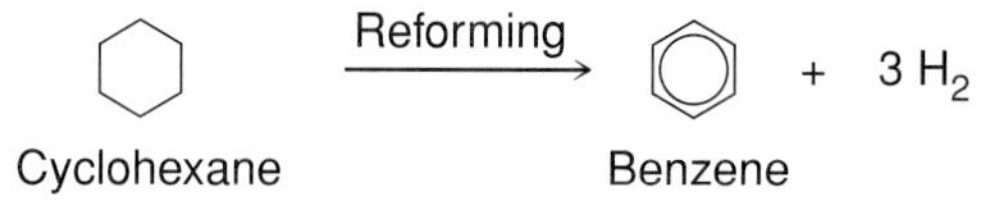

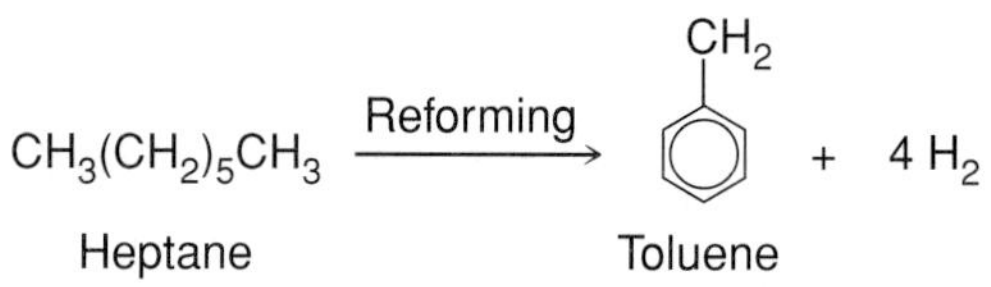

## 9.4
## SOURCES AND USES OF AROMATIC HYDROCARBONS

| TABLE 9.7 | A Comparison of the Boiling Points of Certain Members of Selected Classes of Hydrocarbons | |
| --- | --- | --- |
| **Compound** | **Molecular Weight (g/mol)** | **Boiling Point (°C)** |
| Benzene | 78 | 80 |
| Hexane | 86 | 69 |
| Cyclohexane | 84 | 81 |

## MEDICAL BOXED READING 9.2

Benzene is a colorless, low-boiling (80°C) liquid. The compound is often used as a nonpolar solvent and as a reagent in the synthesis of a wide variety of organic compounds. Thus it is a compound of great value, both in industry and in the laboratory.

$$\text{Benzene} \xrightarrow[\text{metabolism}]{\text{Human}} \text{Phenol}$$

Benzene → Phenol (OH)

One disadvantage of using benzene, however, is the toxicity of the compound. When a mixture of as little as 2% benzene vapor is inhaled by a person for 5 minutes, death is always the result—due to respiratory failure. Most fatalities of this type occur when workers clean large empty tanks that once contained liquid benzene.

Benzene poisoning also can be a long-term phenomenon when a person is exposed to a low concentration of benzene vapor day after day. Typical symptoms are headache, irritability, and dizziness. Today, benzene is an acknowledged carcinogen (a substance that produces cancer). The compound causes leukemia when exposure is long term.

About half the benzene absorbed is excreted by the lungs; the remaining benzene is metabolized to produce phenol.

Treatment of benzene poisoning consists of removing the patient from the source of benzene vapor. The urine should be monitored and phenol concentration determined to establish the degree of benzene poisoning.

The Environmental Protection Agency created a series of regulations in 1989 to cut industrial benzene emissions drastically—by 90%. However, these regulations do not include benzene in gasoline, cigarette smoke, and latex paints that the general public is exposed to. This could represent a deadly danger to people because of the pervasive presence of this toxic chemical in our society.

---

Benzene and toluene are used as solvents in organic reactions and as starting materials for many organic syntheses. Naphthalene has long been employed as an insecticide to keep clothing free of moths during the summer season.

A number of polynuclear aromatic hydrocarbons classified as carcinogens can be isolated from cigarette ash and smoke. One of the most potent of the known carcinogens is 1,2-benzopyrene. The benzo group is a $C_6H_4$ aromatic unit that is bonded to two adjacent carbons of the pyrene ring.

Pyrene            1,2-Benzopyrene

---

## BIOCHEMICAL BOXED READING 9.3

It seems illogical that 1,2-benzopyrene should be a carcinogen because the compound is nonpolar and therefore insoluble in blood and the aqueous environment of the tissues.

Recent experimental work indicated that the actual carcinogen is a more polar substance that is water soluble and can combine with DNA (deoxyribonucleic acid—a high-molecular-weight compound that contains genetic information) in cells. In the case of 1,2-benzopyrene, the polar compound, produced by enzymatic oxidation, is a diol epoxide derivative of 1,2-benzopyrene.

$$\text{1,2-Benzopyrene} \xrightarrow{\text{Enzymes}} \text{A diol epoxide}$$

1,2-Benzopyrene            A diol epoxide

Once the diol epoxide combines with cell DNA, the cell mutates. If the mutation does not kill the cell and is passed from mother cell to daughter cell, and if the mutation causes unchecked cell mitosis (reproduction), cancer is the grim result.

The benzene molecule contains three double bonds and is highly unsaturated. Therefore, the compound should react, by addition, with hydrogen bromide at room temperature and with potassium permanganate in the cold. However, it doesn't.

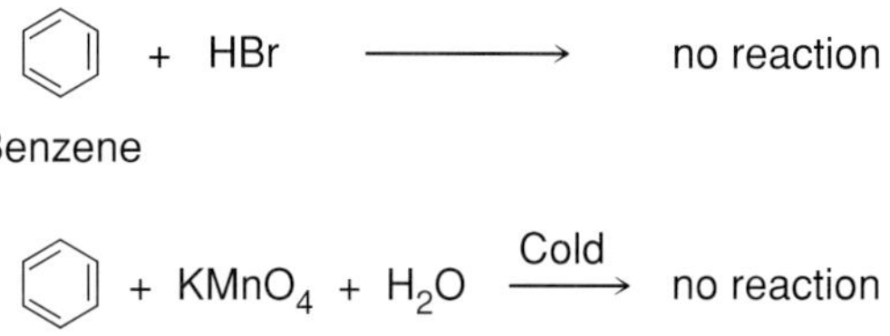

$$\text{Benzene} + \text{HBr} \longrightarrow \text{no reaction}$$

$$\text{Benzene} + \text{KMnO}_4 + \text{H}_2\text{O} \xrightarrow{\text{Cold}} \text{no reaction}$$

Instead, it reacts by substitution. Some typical substitution reactions of benzene are as follows.

Bromination of benzene:

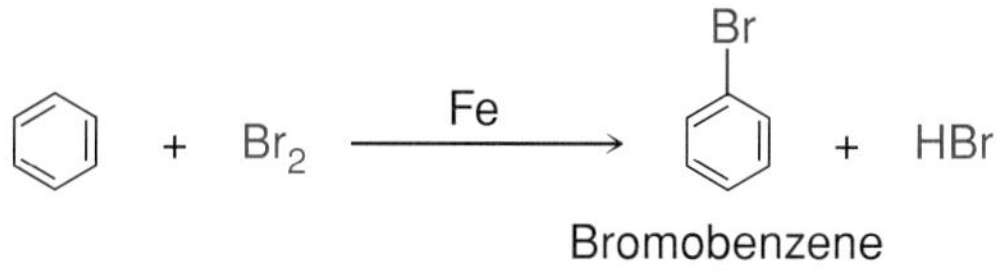

$$\text{Benzene} + \text{Br}_2 \xrightarrow{\text{Fe}} \text{Bromobenzene (Br)} + \text{HBr}$$

Sulfonation of benzene:

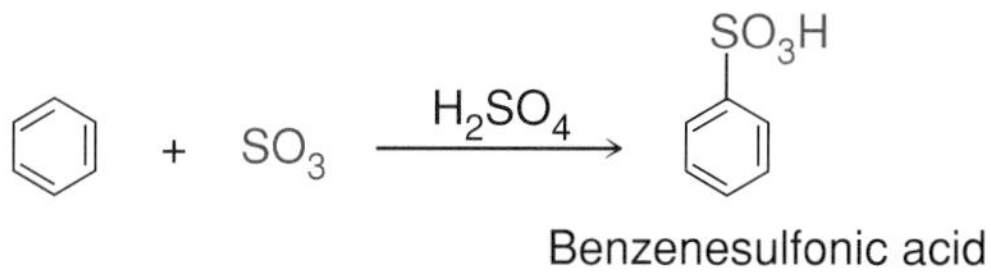

$$\text{Benzene} + \text{SO}_3 \xrightarrow{\text{H}_2\text{SO}_4} \text{Benzenesulfonic acid (SO}_3\text{H)}$$

Nitration of benzene:

$$\text{Benzene} + \text{HNO}_3 \xrightarrow{\text{H}_2\text{SO}_4} \text{Nitrobenzene (NO}_2\text{)} + \text{H}_2\text{O}$$

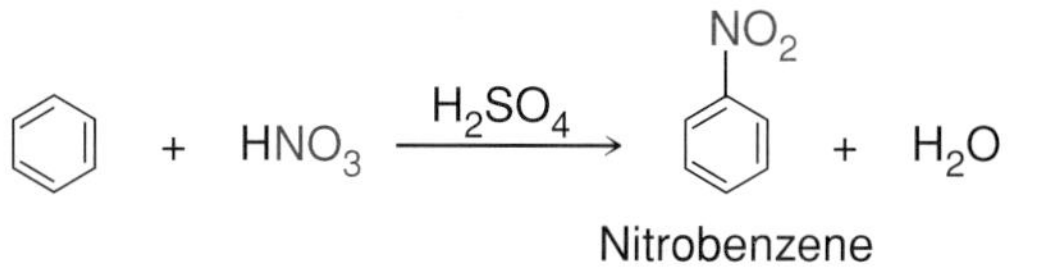

Why do benzene and its derivatives react by substitution rather than addition? The reason is that benzene and its derivatives are unusually stable compounds because of the pattern of conjugated double bonds within the molecule. Any addition to the molecule would result in the loss of one or more double bonds that would destroy this unique structural arrangement and decrease the stability of the resulting product molecule.

The stability of benzene can be demonstrated by examining the heat of hydrogenation of some carefully selected hydrocarbons in Table 9.8. Since benzene has three double bonds, we would expect a heat of hydrogenation of 28.6 kcal/mol × 3, or 85.8 kcal/mol; experimentally, however, only 49.8 kcal/mol is measured. The difference between the two (85.8 − 49.8) is 36 kcal/mol, and it represents the greater stability (less internal energy in benzene) of benzene compared with an imaginary cyclic triene with three

## 9.5
## THE STRANGE CHEMICAL BEHAVIOR OF BENZENE

*Use of an iron catalyst differentiates this bromination from that of an alkane.*

## 9.6
## THE STABILITY OF BENZENE— RESONANCE ENERGY

*Hydrogenation* is the breaking of a carbon-carbon π bond of a double bond with the formation of two carbon-hydrogen σ bonds. Refer to Section 5.6.

| TABLE 9.8 | Heats of Hydrogenation of Selected Cyclic Hydrocarbons | |
|---|---|---|
| Compound | Experimental Heat of Hydrogenation (kcal/mol) | Predicted Heat of Hydrogenation (kcal/mol) |
| Cyclohexene | 28.6 | 28.6 |
| 1,3-Cyclohexadiene | 55.4 | 57.2 |
| Benzene | 49.8 | 85.8 |

Note that each carbon-carbon single bond in structure I is a double bond in structure II, and vice versa. Thus each carbon-carbon link represents 1.5 bonds if we take an average of structures I and II. Structure III then shows each carbon-carbon link as 1.5 bonds. Each solid line of the hexagon in structure III represents 1 bond; the dashed circle inscribed within the hexagon represents 0.5 bond. As a matter of convenience, we usually represent the 0.5 bond with a solid circle as in structure IV. Since each carbon-carbon link in benzene represents 1.5 bonds, then the six hydrogens in the benzene molecule are chemically equivalent. Thus benzene produces only one monosubstituted derivative, $C_6H_5X$.

isolated double bonds. This energy (36 kcal/mol) is called the *resonance energy* of benzene. We can represent the stability of benzene by drawing the following equivalent resonance structures:

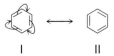

The "true," or hybrid, structure of benzene lies somewhere between the two imaginary contributing structures (I and II) and is represented in this textbook as

The presence of resonance in the benzene molecule indicates that each $p$ orbital (see Fig. 2.16) overlaps with both neighboring $p$ orbitals to give an extended molecular orbital around the molecule, as shown in Fig. 9.8. You can think of the delocalized pi electrons as being represented by two doughnut-shaped clouds, one above the plane of the molecule and the other below (Fig. 9.9).

We feel secure about resonance theory because it is well supported by experiment. Experimentally, we find that each carbon-carbon bond in benzene is 1.39 Å in length, between the experimentally determined value of a carbon-carbon single bond adjacent to a double bond in cyclohexene (1.46 Å, $sp^3$—$sp^2$,σ), and the experimentally determined value of a carbon-carbon double bond in cyclohexene (1.34 Å, $sp^2$—$sp^2$,σ bond plus a $p$—$p$,π bond) (Table 9.9). This fits because the hybrid structure of benzene would be expected to contain six carbon-carbon bonds of the same length, with each bond having half double-bond character (1.34 Å) and half single-bond character (1.46 Å). We can therefore predict a bond length of 1.40 Å for each carbon-carbon bond in the

**FIGURE 9.8** Delocalized pi bonding in benzene.

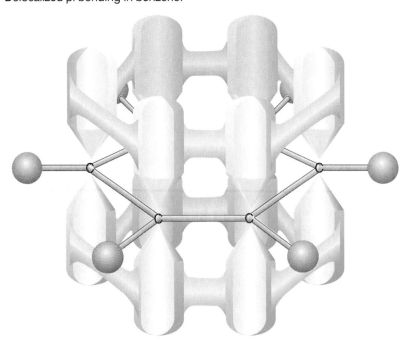

| TABLE 9.9 | A Comparison of the Experimentally Determined Bond Lengths of Selected Bonds in Cyclohexene | |
| --- | --- | --- |
| **Compound** | **Bond Type** | **Bond Length, Å** |
| Cyclohexene | $sp^2$—$sp^3$ | 1.46 |
| Cyclohexene | $sp^2$—$sp^2$ + $p$—$p$ | 1.34 |

| TABLE 9.10 | A Comparison of Bond Types and Bond Lengths of Selected Bonds in Cyclohexane and Cyclohexene | | |
| --- | --- | --- | --- |
| **Compound** | **Bond Type** | **Percent s Character in a Given Orbital** | **Bond Length, Å** |
| Cyclohexane | C—C $sp^3$—$sp^3$ | $sp^3$ 25.0 | 1.54 |
| Cyclohexene | C—C $sp^3$—$sp^2$ | $sp^2$ 33.3 | 1.46 |

benzene molecule $[(1.34 + 1.46)/2 = 1.40]$. This corresponds nicely with the experimentally determined bond length of 1.39 Å.

You might ask why the carbon-carbon single bond in cyclohexane is longer (1.54 Å) than a carbon-carbon single bond (1.46 Å) in cyclohexene. The answer is that the carbon-carbon single bond in cyclohexene is adjacent to the carbon-carbon double bond. Thus we have differing bond types representing the single bonds, as you can see in Table 9.10.

Since the carbon-carbon single bond in cyclohexene is adjacent to the double bond, the bond type is $sp^3$—$sp^2$ as opposed to the corresponding bond in cyclohexane, which is $sp^3$—$sp^3$. Thus the carbon-carbon single bond in cyclohexene shows more s character than in cyclohexane. This, in turn, leads to a stronger and shorter carbon-carbon single bond in cyclohexene because a 2s electron is closer to the carbon nucleus than a 2p electron, and orbital overlap is increased.

Figure 9.10 shows the resonance hybrid structures of some other aromatic compounds. Thus benzene (a *monocyclic aromatic*), every derivative of benzene, and *polycyclic aromatics* can be written with a solid circle within the hexagon(s) as in the

**FIGURE 9.9** Delocalized pi bonding in benzene represented as two doughnut-shaped electron clouds.

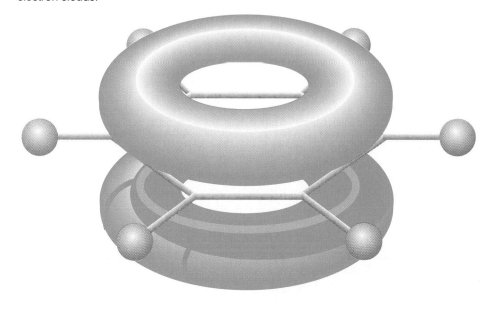

**FIGURE 9.10** Resonance hybrid structural formulas for some selected aromatic compounds.

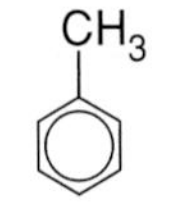

Toluene

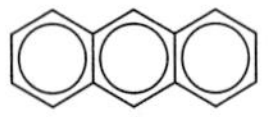

Anthracene
(a polycyclic aromatic)

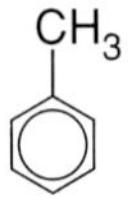

Chlorobenzene

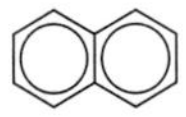

Naphthalene

figure to represent the resonance hybrid. In practice, benzene, its derivatives, and polycyclic aromatics are represented either by a Kekulé structural formula (a contributing or resonance structure) or a structure of the resonance hybrid, depending on the circumstances.

---

**SOLVED PROBLEM 9.4**

Draw the requested number of equivalent resonance structures for

a.  chlorobenzene (two).

b.  naphthalene (three).

**SOLUTION**

a.  We use Kekulé structures here in order to better follow the movement of pi electrons. Every two adjacent carbons in each monosubstituted benzene derivative is connected by means of a carbon-carbon single bond in one resonance structure and a carbon-carbon double bond in the other resonance structure. Thus, by delocalizing the pi electrons, we get

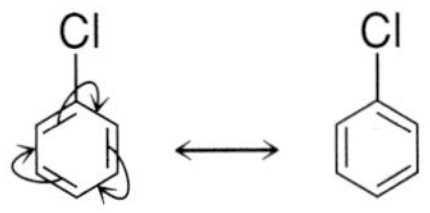

b.  Naphthalene, a bicyclic aromatic compound, shows more than two contributing structures as follows:

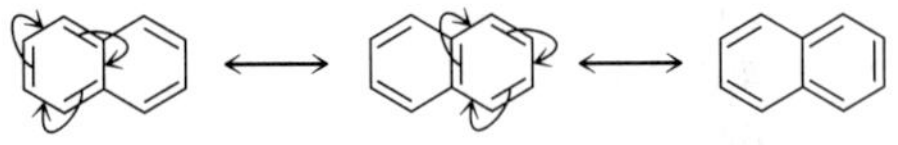

---

**EXERCISE 9.4**

Draw the requested number of equivalent resonance structures for

a.  toluene (two).

b.  anthracene (four).

---

**9.7
AROMATICITY**

Previously, we defined the term *aromatic* as representing a family of compounds consisting of benzene and its derivatives. Actually, based on much experimental data, the term *aromatic* is more general; that is, some substances (compounds or ions) exist that are not derivatives of benzene yet are aromatic. We say that any substance that is aromatic displays *aromatic character*, or *aromaticity*, and tends to undergo substitution reactions rather than addition reactions. How can you determine by looking at the structural formula of a species if that substance is aromatic or not? To be aromatic, the substance must satisfy three conditions:

1. The substance must be cyclic in structure.

2. The substance must be resonance-stabilized; that is, the species must be represented by at least two or more contributing structures.

3. The substance must conform to *Huckel's rule*. This rule states that to be aromatic, a species must contain $4n + 2$ pi electrons. The letter $n$ is an integer and can be 0, 1, 2, 3, etc. Substituting, we find that a species follows Huckel's rule when it contains 2, 6, 10, etc. pi electrons.

$$4(0) + 2 = 2$$
$$4(1) + 2 = 6$$
$$4(2) + 2 = 10$$

Note that the letter $n$ has no relationship to the number of rings or double bonds or unshared electron pairs in the species. It is simply an integer.[*]

---

Note which of the following species is aromatic and explain; if it is not aromatic, list which of the conditions of aromaticity is not satisfied.

**SOLVED PROBLEM 9.5**

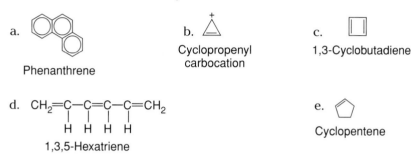

a. Phenanthrene

b. Cyclopropenyl carbocation

c. 1,3-Cyclobutadiene

d. $CH_2{=}C{-}C{=}C{-}C{=}CH_2$ with H H H H below
1,3,5-Hexatriene

e. Cyclopentene

## SOLUTION

a. Aromatic. Draw a Kekulé structure of phenanthrene to view more clearly the number of double bonds. Thus

The compound is cyclic (condition 1). Since the double bonds are conjugated, the pi electrons are delocalized, and the compound is resonance-stabilized (condition 2). There are 14 pi electrons (7 pi bonds) present in this compound. This compound does conform to Huckel's rule

$$4n + 2 = 14$$
$$4n = 12$$
$$n = 3 \qquad n \text{ is equal to the integer 3}$$

Since all three conditions are met, the compound is aromatic.

b. Aromatic. The ion is cyclic (condition 1). Since the double bond is conjugated with the positive charge, the pi electrons are delocalized, and the ion is resonance-stabilized (condition 2). There are two pi electrons (one pi bond) present in this ion. This ion does conform to Huckel's rule using the following calculation:

$$4n + 2 = 2$$
$$4n = 0$$
$$n = 0$$

Since all three conditions are met, the ion is aromatic.

---

[*]Huckel originally formulated the rule that bears his name only for monocyclic aromatic species. However, experience has demonstrated that the rule also can be extended to include a number of polycyclic aromatic species.

c. This compound is not aromatic because condition 3 is not fulfilled.

$$4n + 2 = 4$$
$$4n = 2$$
$$n = 0.5 \quad n \text{ is not an integer}$$

d. This compound is not aromatic because condition 1 is not complied with. The compound is acyclic (not cyclic).

e. This compound is not aromatic because condition 2 is not fulfilled; that is, the compound is not resonance-stabilized.

---

**EXERCISE 9.5**

Note which of the following species is aromatic and explain; if it is not aromatic, list which of the conditions of aromaticity is not satisfied.

a.

Cyclopentadienyl
carbanion

b.

Cyclopentadienyl
carbocation

c.

1,3,5,7-Cyclooctatetrene

---

There exist a number of cyclic compounds that contain one or more atoms other than carbon as part of the ring skeleton. These compounds are called *heterocyclic*, and some of them are aromatic. In order to be aromatic, the same three conditions that are used to test for aromaticity in the *homocyclic* or *carbocyclic species* we just discussed must be satisfied. A homocyclic species is a cyclic compound or ion that contains only carbon atoms in the ring skeleton.

---

**SOLVED PROBLEM 9.6**

Which of the following compounds are aromatic? Explain.

a.

Pyridine

b.

H
Pyrrole

c.

Thiophene

**SOLUTION**

Each of these compounds satisfies conditions 1 and 2 in that each compound is cyclic and resonance-stabilized. The problem here is deciding whether an unshared electron pair or pairs are to be considered as pi electrons or not. The rule for this (let us designate this rule as the *heterocyclic pi electron rule*) is as follows: If the heteroatom has a double bond, two unshared electrons on the heteroatom are *not* counted as pi electrons. If the heteroatom does not have a double bond, count two unshared electrons on the heteroatom as pi electrons.

a. Aromatic. Since nitrogen has a double bond, the unshared electrons located on nitrogen are *not* considered to be pi electrons according to the heterocyclic pi electron rule. Thus the pyridine molecule contains six pi electrons (three double bonds), and the compound is aromatic.

b. Aromatic. Since nitrogen does not have a double bond, the unshared electrons located on nitrogen are considered to be pi electrons. Thus the pyrrole molecule contains six pi electrons (two double bonds and one unshared electron pair on nitrogen), and the compound is aromatic.

c. Aromatic. Since sulfur does not have a double bond, two unshared electrons held by sulfur are considered to be pi electrons (the remaining two unshared electrons are sigma electrons). Thus the thiophene molecule contains six pi electrons (two double bonds and one unshared electron pair on sulfur), and the compound is aromatic.

Which of the following compounds is aromatic? Explain using the heterocyclic pi electron rule.

a.

Pyran

b.

Furan

c.

Pyrazole

These compounds we designate as aromatic are expected to react by substitution rather than addition and are very stable because of pi electron delocalization.

## ► CHAPTER ACCOMPLISHMENTS

Give the reason why benzene and its derivatives were originally called *aromatic compounds.*

### 9.1 The Structure of Benzene
☐ Explain the role a dream played in the discovery by Friedrich Kekulé that benzene must be a cyclic molecule.
☐ Give the empirical formula of benzene.

### 9.2 Nomenclature
☐ Draw a structural formula for an arene, and name that particular arene.
☐ Draw a structural formula for
  a. phenol.
  b. benzoic acid.
  c. aniline.
  d. bromobenzene.
  e. chlorobenzene.
  f. nitrobenzene.
  g. toluene.
  h. benzaldehyde.
  i. styrene.
  j. iodobenzene.
  k. benzenesulfonic acid.
  l. ethylbenzene.
  m. naphthalene.
  n. anthracene.
☐ Draw a structural formula for the
  a. phenyl group.
  b. benzyl group.
☐ Give the prefix that designates that two groups are bonded to carbons 1 and 3 in the benzene ring.
☐ Name the following compound:

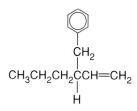

$$CH_3CH_2CH_2\overset{\overset{\displaystyle CH_2}{|}}{\underset{\underset{\displaystyle H}{|}}{C}}CH=CH_2$$

### 9.3 Physical Properties
☐ Name the only intermolecular forces present in aromatic hydrocarbons.
☐ Predict the solubility of an arene in water.

### 9.4 Sources and Uses of Aromatic Hydrocarbons
☐ Identify an aromatic hydrocarbon that has been used as an insecticide to keep clothing moth-free during the summer.
☐ Draw a structural formula for 1,2-benzopyrene, a potent carcinogen.

### 9.5 The Strange Chemical Behavior of Benzene
☐ Explain why benzene reacts by substitution rather than addition.
☐ Write an equation to represent the reaction of benzene with bromine in the presence of iron.

### 9.6 The Stability of Benzene—Resonance Energy
☐ Draw two equivalent resonance structures for benzene.
☐ Draw a resonance hybrid structure of
  a. benzene.
  b. chlorobenzene.
  c. anthracene.
☐ Explain why a carbon-carbon single bond in cyclohexane is longer than a carbon-carbon single bond in cyclohexene that is adjacent to the carbon-carbon double bond.
☐ Draw three equivalent resonance structures for naphthalene.

### 9.7 Aromaticity
☐ Give the three conditions that a structural formula must satisfy in order to classify the substance as aromatic.
☐ Supply the condition of aromaticity that is not met with
  a. 1,3,5,7-cyclooctatetrene.
  b. 1,3,5-hexatriene
☐ Distinguish between a heterocyclic compound and a carbocyclic compound.

☐ State the heterocyclic pi electron rule, and use the rule to predict whether thiophene is aromatic. Note that thiophene conforms to conditions 1 and 2.

## ▶ KEY TERMS

aromatic (Introduction)
arene (9.2)
aryl group (9.2)
ortho- (9.2)
meta- (9.2)

para- (9.2)
reforming (9.4)
monocyclic aromatic compound (9.6)
polycyclic aromatic compound (9.6)
aromatic character or aromaticity (9.7)

Huckel's rule (9.7)
heterocyclic compound (9.7)
homocyclic or carbocyclic species (9.7)
heterocyclic pi electron rule (9.7)

## ▶ PROBLEMS

1. Draw a structural formula for each of the three diiodobenzenes.

2. Supply an IUPAC name for each of the following.

a., 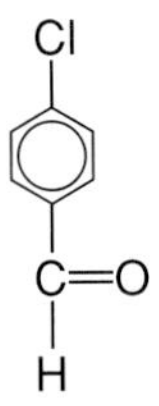 b., 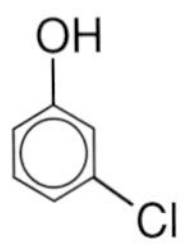 c., 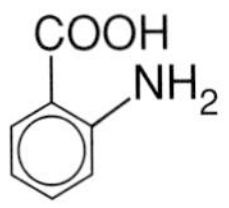 d., 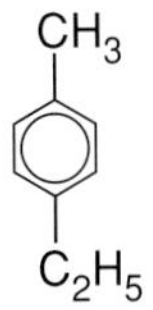 e., 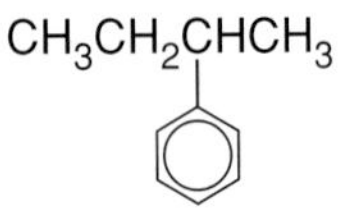 f., 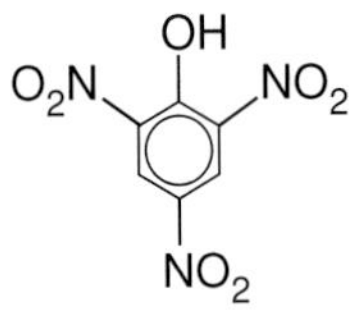 g. 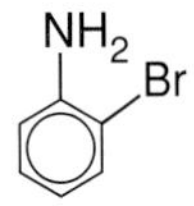 (structural formulas)

3. Draw a structural formula (resonance hybrid type) for each of the following compounds.
   a. *p*-nitroaniline
   b. anthracene
   c. benzyl chloride
   d. *o*-iodophenol
   e. 9,10-dibromoanthracene
   f. styrene
   g. triphenylmethane
   h. 1,3-dinitronaphthalene
   i. *p*-xylene
   j. *o*-toluenesulfonic acid

4. Draw a structural formula for each of the following compounds.
   a. *m*-bromoiodobenzene
   b. *m*-chlorobenzaldehyde
   c. *o*-chloroiodobenzene
   d. *o*-dinitrobenzene

5. Draw a structural formula for each of the following compounds.
   a. 2,4,6-trichlorophenol
   b. 3,5-dinitrobenzaldehyde
   c. pentachlorobenzene
   d. 2,4-dinitrofluorobenzene
   e. 2,3-dibromostyrene

6. Draw a structural formula for each of the following compounds.
   a. 2,3-dibromonaphthalene
   b. 1-bromo-8-nitronaphthalene
   c. 2-methylanthracene
   d. 9,10-dichloroanthracene

7. Define or explain each of the following.
   a. aromatic
   b. *ortho-*
   c. reforming
   d. *benzo-*

8. Using Table 9.8, calculate the resonance energy for 1,3-cyclohexadiene.

9. a. Draw two equivalent resonance structures and a resonance hybrid structure for
      (1) ethylbenzene.
      (2) benzaldehyde.
      (3) styrene.
   b. Draw a Kekulé structural formula for
      (1) toluene.
      (2) bromobenzene.
      (3) aniline.

10. Draw a contributing structure for the cyclopropenyl carbocation.

11. Which of the following species are aromatic? If not aromatic, indicate which condition of aromaticity is not met.

    a.

    b. $CH_2{=}CH{-}CH{=}CH_2$

    c.

    d.

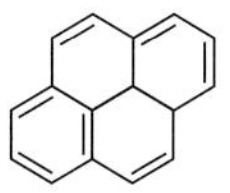

12. Which of the following compounds are aromatic? Explain using the heterocyclic pi electron rule.

    a.

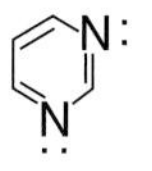

    Pyrimidine

    b.

    Imidazole

    c.

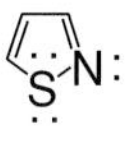

    Isoxazole

    d.

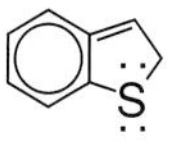

    Benzothiophene

    e.

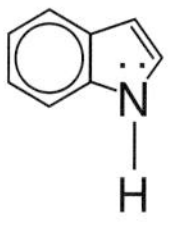

    Indole

13. Given the structural formula for benzothiophene in Problem 12d, draw a structural formula for benzofuran.

14. A dibromobenzene ($C_6H_4Br_2$) forms two mononitro derivatives. Draw a structural formula for the dibromobenzene.

15. Consider several aromatic isomers, each with the molecular formula $C_8H_{10}$.

    a. Draw a structural formula for two such isomers that form three monobromo derivatives.

    b. Draw a structural formula for each isomer that forms
       (1) two monobromo derivatives.
       (2) one monobromo derivative.

# CHAPTER 10

# Aromatic Compounds (II)

## ELECTROPHILIC AROMATIC SUBSTITUTION, RATE OF REACTION, ORIENTATION, OTHER REACTIONS, AND AROMATIC SYNTHESIS

This chapter considers the reactivity of benzene (and other aromatic compounds), electrophilic species that attack the aromatic ring, and some important factors that influence both the reactivity of aromatic compounds and the attack site in the aromatic system.

Benzene, like other aromatic compounds, is rich in delocalized electrons. We call an electron-rich reagent a *nucleophilic reagent*. Therefore, it seems reasonable to assume that a reaction would take place by attack of an electrophile (an electron-poor reagent) on the benzene ring, a nucleophile. Consider an electrophile $E^+$ reacting with benzene to give $C_6H_5E$ and $H^+$.

**10.1
SUBSTITUTION
REACTIONS OF
BENZENE:
ELECTROPHILIC
AROMATIC
SUBSTITUTION**

In the following pages we will look at a series of *electrophilic aromatic substitution* reactions.

**Envision the Reaction**

---

### General Mechanism for an Electrophilic Aromatic Reaction
### [the Reaction of $E^+$ (an Electrophile) with Benzene]

First step:

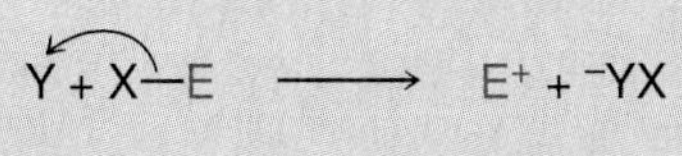

$$Y + X{-}E \longrightarrow E^+ + {}^-YX$$

Second step:

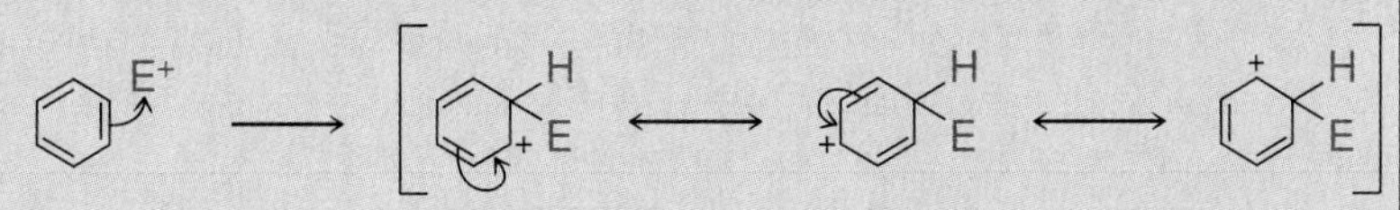

Resonance-stabilized arenium ion

Third step:

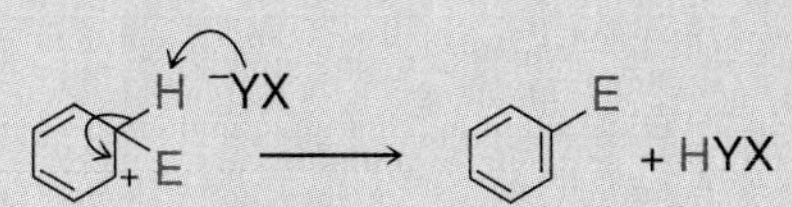

$$HYX \longrightarrow Y + HX$$

The mechanism of each of these electrophilic aromatic substitutions follows a set pattern. First, the electrophile $E^+$ is generated by some reagent (often a catalyst) Y that liberates $E^+$ from EX, where X is an atom or group of atoms bonded to E. Second, $E^+$ attacks the benzene ring to give an intermediate carbocation called an *arenium ion* that is stabilized by resonance. The resonance forms are set off by brackets. The final step in the mechanism involves the loss of a proton from the arenium ion to give the product. Sometimes, HYX decomposes to regenerate the catalyst Y and produce by-product HX. Note that the aromatic system of conjugated double bonds is regenerated in the product in order to maintain the stability of the aromatic ring.

## Halogenation of Benzene

When benzene is reacted with a mixture of chlorine (or bromine) and a catalytic amount of the corresponding iron(III) (ferric) halide (or metallic iron), the halobenzene is formed.

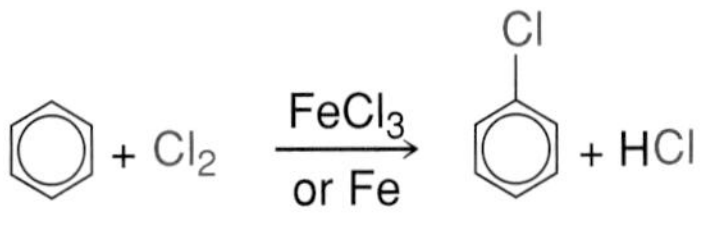

Chlorobenzene

---

**Envision the Reaction**

### Mechanism of the Chlorination of Benzene

First step:

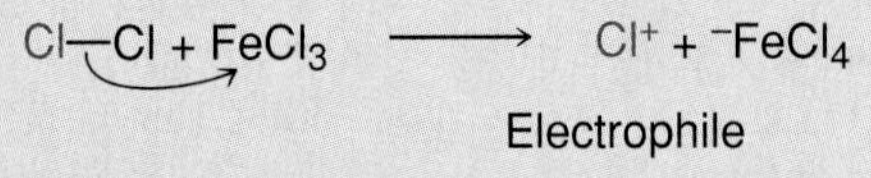

$$Cl-Cl + FeCl_3 \longrightarrow Cl^+ + {}^-FeCl_4$$

Electrophile

Second step:

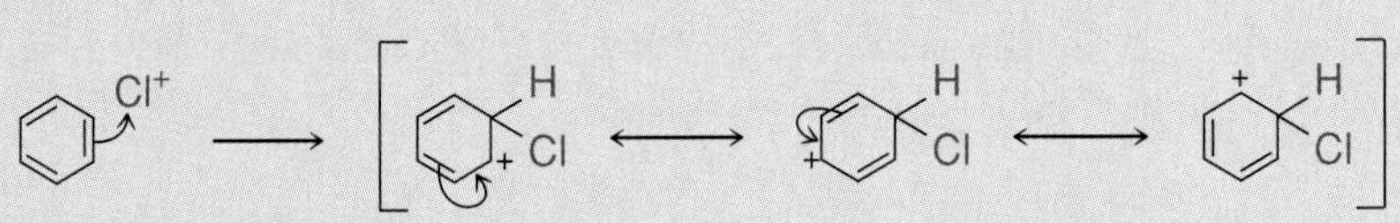

Third step:

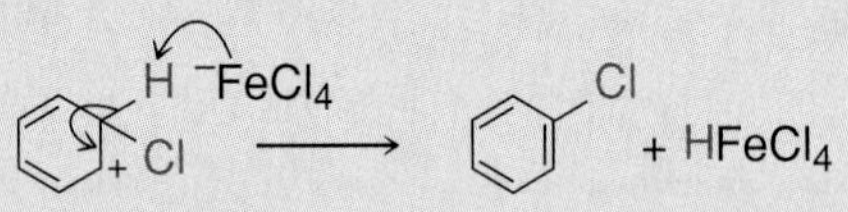

$$HFeCl_4 \longrightarrow HCl + FeCl_3$$

---

The mechanism is analogous to the model presented in Section 10.1 and is used for both the $Br_2$-$FeBr_3$ and the $Cl_2$-$FeCl_3$ systems. First, the $Cl^+$ electrophile is generated. Second, a pair of pi electrons from a double bond in the benzene ring (a nucleophile) is attacked by $Cl^+$ (an electrophile) to form an arenium ion that is resonance-stabilized. Third, $^-FeCl_4$ abstracts a proton from the resonance-stabilized arenium ion to give chlorobenzene. Finally, $HFeCl_4$ decomposes to regenerate the catalyst ($FeCl_3$) and by-product HCl.

Note that the —Cl group (or any other group) can be positioned on any carbon of the benzene ring to produce the same compound, chlorobenzene, because the carbons are equivalent. Putting it another way, there exists only one isomer of a monosubstituted benzene derivative with the formula $C_6H_5X$, where X is any group.

The reaction also takes place when metallic iron is substituted for a ferric halide. Metallic iron reacts with halogen to give the ferric halide, the catalyst.

$$2\,Fe + 3\,X_2 \longrightarrow 2\,FeX_3$$

## Friedel-Crafts Alkylation

In the presence of aluminum chloride (a catalyst), benzene reacts with an alkyl chloride to give an alkylbenzene (arene). In general,

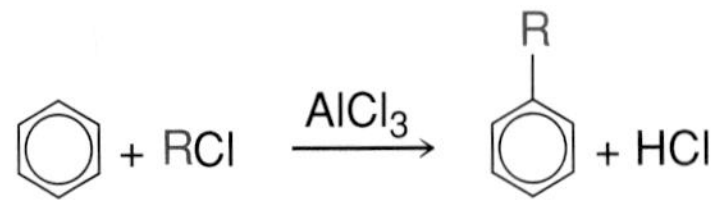

Example:

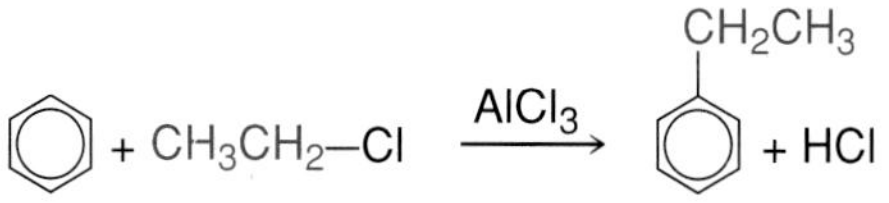

$$\text{benzene} + CH_3CH_2-Cl \xrightarrow{\text{AlCl}_3} \text{ethylbenzene} (CH_2CH_3) + HCl$$

The Friedel-Crafts reaction can give more than one product due to rearrangement within the alkyl group.

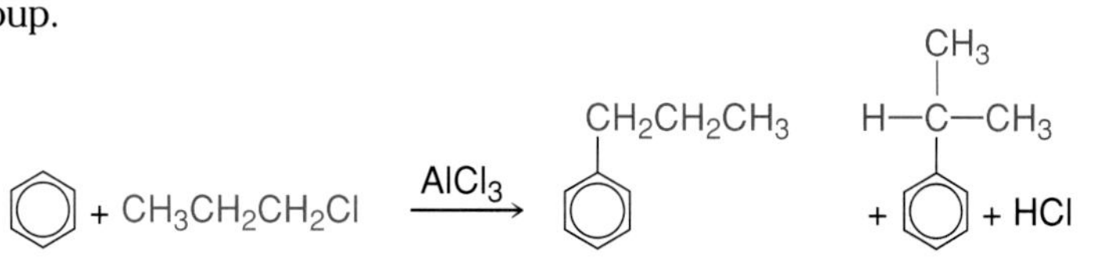

$$\text{benzene} + CH_3CH_2CH_2Cl \xrightarrow{\text{AlCl}_3}$$

| Major product *n*-propylbenzene ($CH_2CH_2CH_3$) | + Minor product isopropylbenzene ($H-\overset{CH_3}{\underset{CH_3}{C}}-$) + HCl |

### Mechanism of the Reaction of Benzene with *n*-Propyl Chloride to Produce an Expected Product (*n*-Propylbenzene) and an Unexpected Product (Isopropylbenzene): A Friedel-Crafts Alkylation

**Envision the Reaction**

First step:

$$CH_3CH_2CH_2-Cl + AlCl_3 \longrightarrow CH_3CH_2\overset{+}{C}H_2 + {}^-AlCl_4$$

Second step:

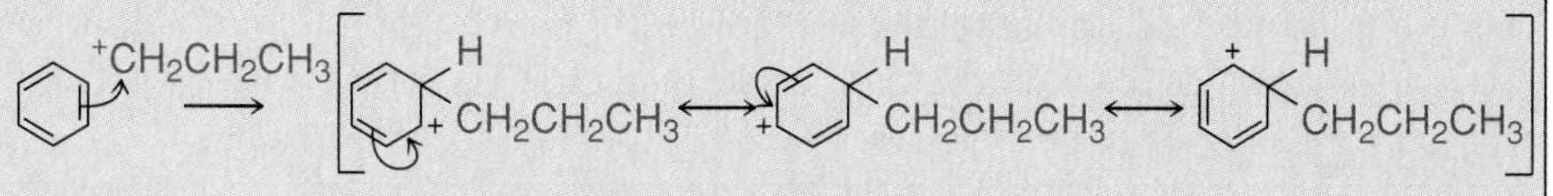

$$\overset{+}{C}H_2CH_2CH_3 \longrightarrow [ \text{Arenium ion} ]$$

Third step:

$$\overset{H}{\underset{+CH_2CH_2CH_3}{}} \,{}^-AlCl_4 \longrightarrow \text{C}_6H_5-CH_2CH_2CH_3 + HAlCl_4$$

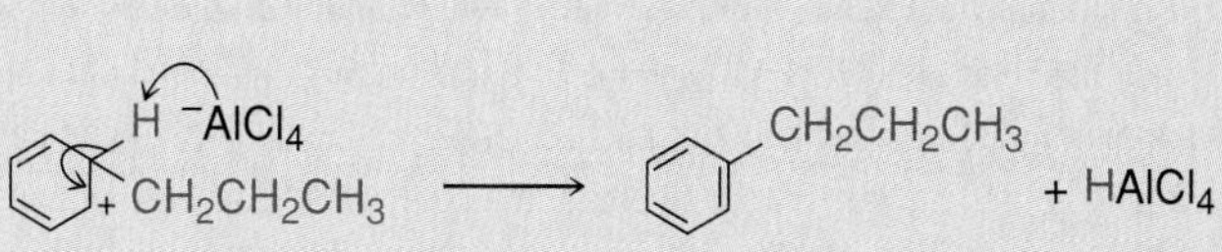

$$HAlCl_4 \rightarrow HCl + AlCl_3$$

Second R step:

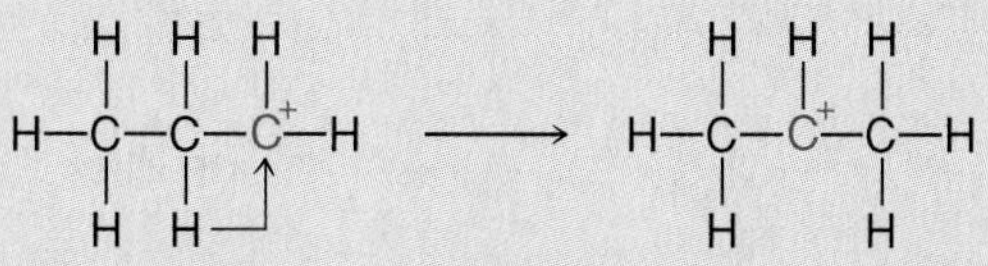

$$H-\overset{H}{\underset{H}{C}}-\overset{H}{\underset{H}{C}}-\overset{H}{\overset{+}{C}}-H \longrightarrow H-\overset{H}{\underset{H}{C}}-\overset{H}{\overset{+}{C}}-\overset{H}{\underset{H}{C}}-H$$

Third R step:

$$H-\overset{CH_3}{\underset{CH_3}{\overset{+}{C}}} \longrightarrow [ \quad +CH_2(CH_3)_2 \longleftrightarrow CH(CH_3)_2 \longleftrightarrow CH(CH_3)_2 \quad ]$$

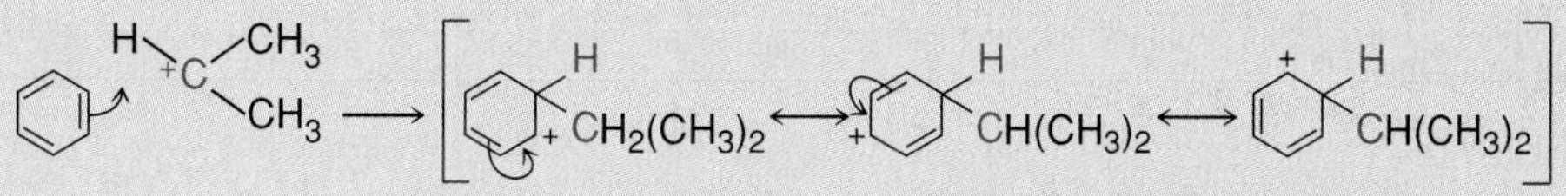

Fourth R step:

$$\overset{H}{\underset{+CH_2(CH_3)_2}{}} \,{}^-AlCl_4 \longrightarrow \overset{H}{\underset{CH_3}{C}}-CH_3 + HAlCl_4$$

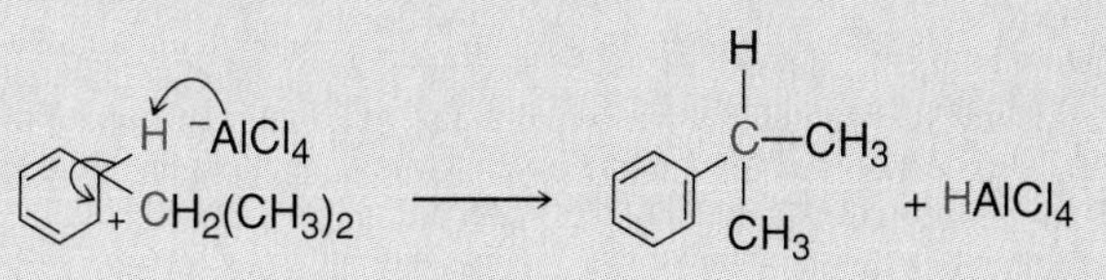

The first step of the mechanism involves abstraction of chlorine from the alkyl chloride, in this example, *n*-propyl chloride, to produce the *n*-propyl carbocation. Two pi electrons in a double bond in benzene, in turn, are attacked by the *n*-propyl carbocation (second step) to give the corresponding resonance-stabilized arenium ion, which in turn loses a proton (third step) to produce *n*-propylbenzene. Finally, HAlCl$_4$ decomposes to regenerate the aluminum chloride catalyst and produce hydrogen chloride, a by-product of the reaction.

To derive the major product (isopropylbenzene), the *n*-propyl carbocation (1°) rearranges (second R step; R for rearrangement) to form the more stable isopropyl carbocation (2°) by means of a hydride ion transfer (a rearrangement). Here, a hydride ion (hydrogen and its two electrons, H:$^-$) migrates from a terminal carbon to an adjacent carbon atom. The driving force for this rearrangement is the formation of a more stable secondary carbocation from a less stable primary carbocation. The pi electrons of a double bond in the benzene ring are attacked by the isopropyl carbocation in the usual way to give a resonance-stabilized arenium ion (third R step). The arenium ion, in turn, loses a proton to give isopropylbenzene (fourth R step). Finally, HAlCl$_4$ decomposes to regenerate the aluminum chloride catalyst as in the formation of *n*-propylbenzene.

Alkyl chlorides can be used in Friedel-Crafts alkylation, but chlorobenzene and vinyl chloride cannot be used because they are too unreactive. Vinyl chloride and chlorobenzene are unreactive because the carbon-chlorine bond contains some double-bond character due to resonance.

There exist a number of variations of Friedel-Crafts alkylation involving different reagents such as HF and CH$_3$CH$_2$CH$_2$OH with H$_3$PO$_4$. Each set of reagents must produce a carbocation that attacks the benzene ring.

This partial double-bond character present in both vinyl chloride and chlorobenzene makes it impossible for aluminum chloride to abstract chlorine from vinyl chloride or chlorobenzene to form the vinyl and phenyl carbocations, respectively.

## SOLVED PROBLEM 10.1

When *n*-propyl alcohol is reacted with H$_3$PO$_4$ (phosphoric acid) in the presence of benzene, both isopropylbenzene and *n*-propylbenzene are isolated as products. Write a detailed mechanism for the formation of both the *n*-propyl and isopropyl carbocations.

### SOLUTION
The first step in the mechanism is formation of the *n*-propyloxonium ion by attack of an unshared electron pair located on the oxygen atom in the alcohol on the phosphoric acid molecule.

*n*-Propyloxonium ion

This *n*-propyloxonium ion, in turn, loses a molecule of water to form the *n*-propyl carbocation.*

*For simplicity, the *n*-propyl carbocation is shown as being produced by means of an S$_N$1 reaction (see Sec. 12.2). In actuality, the carbon-oxygen bond is only partially broken, and rearrangement takes place by means of an S$_N$2 substitution (see Sec. 12.2) as follows:

The *n*-propyl carbocation is produced from the corresponding oxonium ion for two reasons: First, in the carbocation, the positive charge is carried by a less electronegative element, carbon. In the *oxonium ion*, the positive charge is carried by the more electronegative element, oxygen. Second (and more important), water, a stable molecule, is produced. The primary carbocation then rearranges to a more stable secondary carbocation in the usual manner.

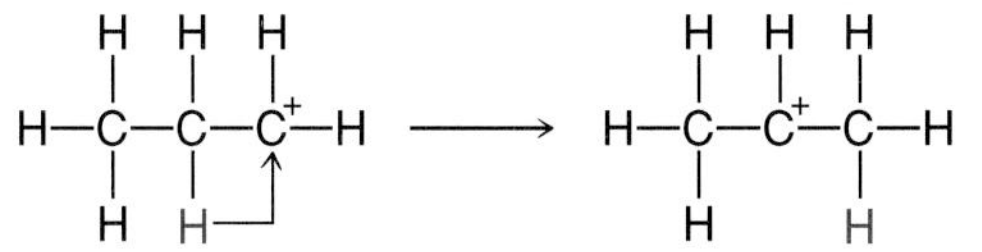

$$H-\underset{\underset{H}{|}}{\overset{\overset{H}{|}}{C}}-\underset{\underset{H}{|}}{\overset{\overset{H}{|}}{C}}-\overset{\overset{H}{|}}{\overset{+}{C}}-H \longrightarrow H-\underset{\underset{H}{|}}{\overset{\overset{H}{|}}{C}}-\overset{+}{\overset{\overset{H}{|}}{C}}-\underset{\underset{H}{|}}{\overset{\overset{H}{|}}{C}}-H$$

When the pi electrons of a double bond in benzene are attacked by the *n*-propyl carbocation in the usual way, *n*-propylbenzene is produced. On the other hand, when the isopropyl carbocation attacks the pi electrons of a double bond in benzene, isopropylbenzene is the resulting product.

**EXERCISE 10.1**

Write a detailed, step-by-step mechanism for each of the following reactions.

a. 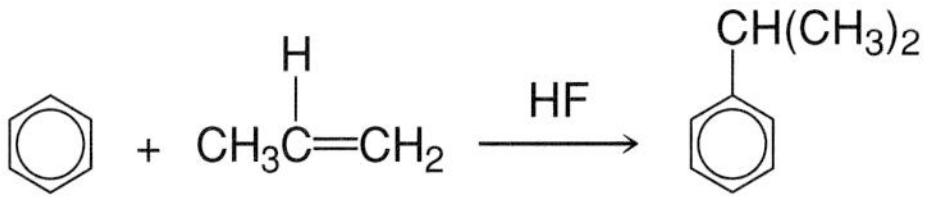 (benzene) $+ \; CH_3\overset{\overset{H}{|}}{C}{=}CH_2 \;\xrightarrow{\text{HF}}\;$ (benzene with $CH(CH_3)_2$)

b. 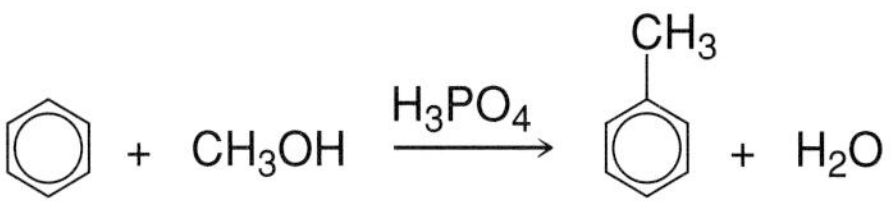 (benzene) $+ \; CH_3OH \;\xrightarrow{\text{H}_3\text{PO}_4}\;$ (benzene with $CH_3$) $+ \; H_2O$

**SOLVED PROBLEM 10.2**

Complete the following equations by drawing a structural formula for each organic product and writing a molecular formula or formula unit for each inorganic product formed.

a.  benzene $+ \; CH_3CH_2OH \xrightarrow{\text{H}_3\text{PO}_4}$

b.  benzene $+ \; CH_3CH_2CH{=}CH_2 \xrightarrow{\text{HF}}$

**SOLUTION**

a.  Mechanistically, reacting $CH_3CH_2OH$ with $H_3PO_4$ results in the formation of the ethyl carbocation. Thus the products are

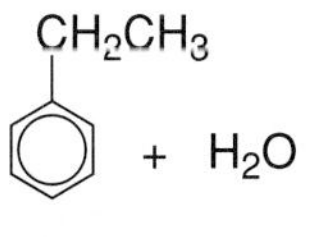

(benzene with $CH_2CH_3$) $+ \; H_2O$

b.  Mechanistically, reacting $CH_3CH_2CH{=}CH_2$ with HF results in the formation of the *sec*-butyl carbocation. Thus the product is

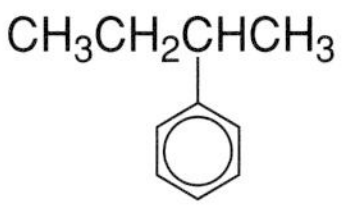

$CH_3CH_2CHCH_3$ (attached to benzene)

**EXERCISE 10.2**

Complete the following equations by drawing a structural formula for each organic product and writing a molecular formula or formula unit for each inorganic product formed.

a.  benzene + $CH_3CH(OH)CH_3 \xrightarrow{H_3PO_4}$

b.  benzene + $CH_3CH{=}CHCH_3 \xrightarrow{HF}$

## Friedel-Crafts Acylation

In this reaction, an acyl group ($R{-}\overset{\overset{\displaystyle O}{\|}}{C}{-}$) rather than an alkyl group ($R{-}$) is introduced into the benzene ring. In general,

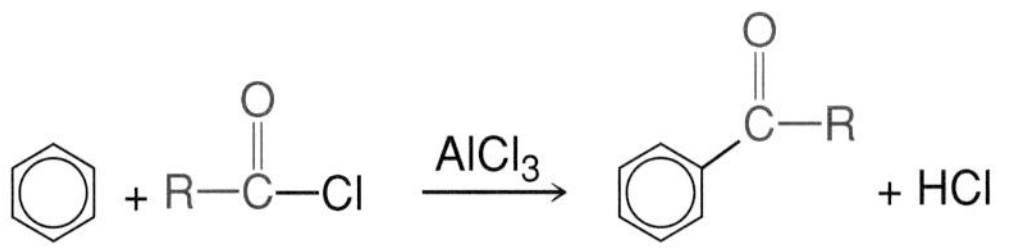

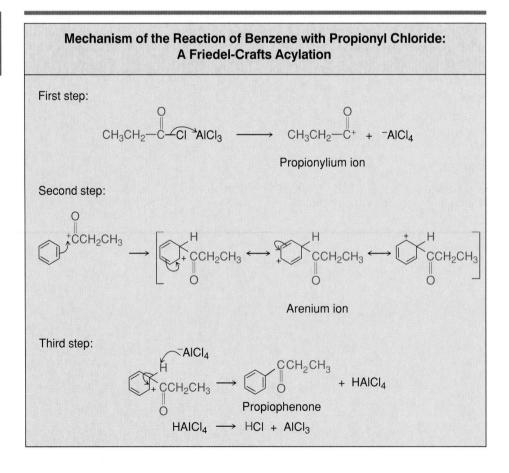

Example:

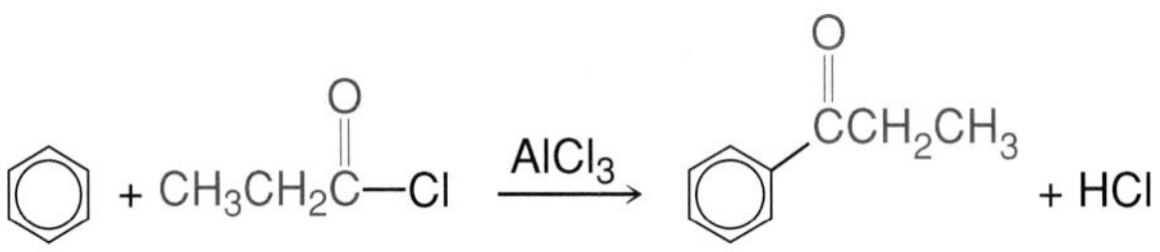

Propionyl chloride                    Propiophenone

**Envision the Reaction**

### Mechanism of the Reaction of Benzene with Propionyl Chloride: A Friedel-Crafts Acylation

First step:

$$CH_3CH_2{-}\overset{\overset{\displaystyle O}{\|}}{C}{-}Cl \; AlCl_3 \longrightarrow CH_3CH_2{-}\overset{\overset{\displaystyle O}{\|}}{C}{}^+ \; + \; {}^-AlCl_4$$

Propionylium ion

Second step:

Arenium ion

Third step:

Propiophenone

$$HAlCl_4 \longrightarrow HCl + AlCl_3$$

The mechanism of Friedel-Crafts acylation begins when the acylium ion (propionylium ion) is formed while propionyl chloride reacts with the aluminium chloride catalyst. Once generated, the propionylium ion attacks the pi electrons of a double bond in benzene to give a resonance-stabilized acylium ion (second step). The arenium ion, in turn, loses a proton to $^-AlCl_4$ to give propiophenone* and $HAlCl_4$ (third step). Finally, $HAlCl_4$ decomposes to regenerate the aluminum chloride catalyst and produce hydrogen chloride, a by-product of the reaction.

Friedel-Crafts acylation is the preferred means of introducing a primary alkyl group, three carbons or more in length, into a benzene ring without rearrangement, as opposed to Friedel-Crafts alkylation. It is the likelihood of rearrangement that makes alkylation less desirable as a synthetic method because the carbocation can undergo hydride migration while the charge on the acylium ion remains on the acyl carbon.

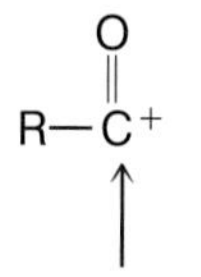

Acyl carbon

In order to reduce the aromatic ketone produced to the corresponding arene, the *Clemmensen reduction* is employed. In general,

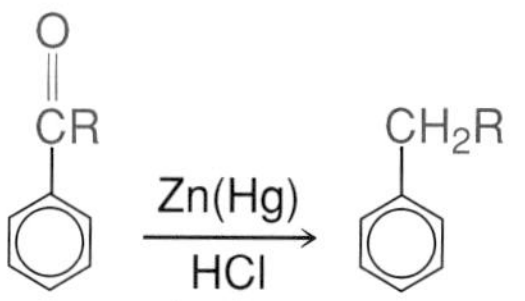

Yields are excellent in both Friedel-Crafts acylation and Clemmensen reduction.

The synthesis is as follows: Benzene is reacted with propionyl chloride in the presence of aluminum chloride to give propiophenone; the ketone, propiophenone, in turn, is reacted with a zinc amalgam in concentrated hydrochloric acid (a Clemmensen reduction) to produce *n*-propylbenzene (or propylbenzene).

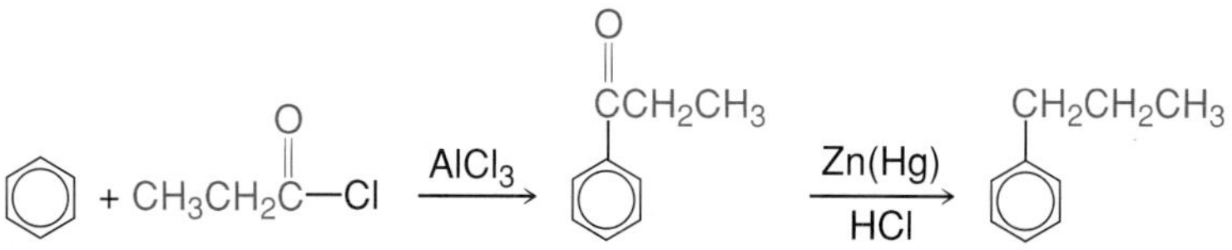

Propionyl chloride          Propiophenone

In contrast, the yield of *n*-propylbenzene is poor using the Friedel-Crafts alkylation reaction because a considerable amount of rearrangement takes place. Thus *n*-propylbenzene is the minor product and is produced in poor yield.

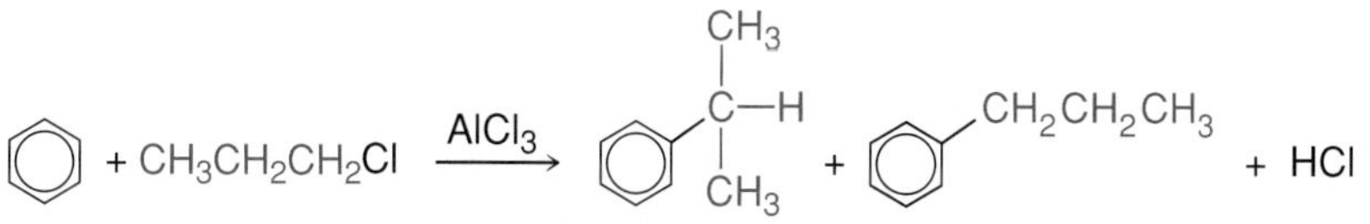

Major product     Minor product

The reaction of an aromatic compound with fuming sulfuric acid produces an aromatic sulfonic acid. Fuming sulfuric acid is a mixture of sulfuric acid and sulfur trioxide.                    **Sulfonation**

---

*For the sake of simplicity, this textbook assumes that the two mechanisms are similar. In actuality, the Friedel-Crafts acylation reaction is somewhat more complex than Friedel-Crafts alkylation, and the mechanism is beyond the scope of this textbook.

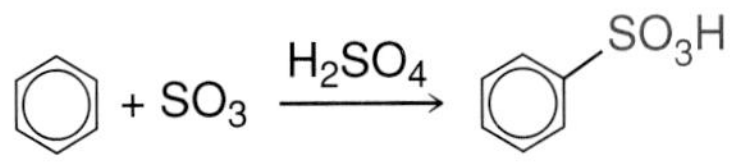

Benzenesulfonic acid

---

**Envision the Reaction**

| **Mechanism of the Reaction of Benzene with Fuming Sulfuric Acid: Sulfonation** |
|---|

First step:

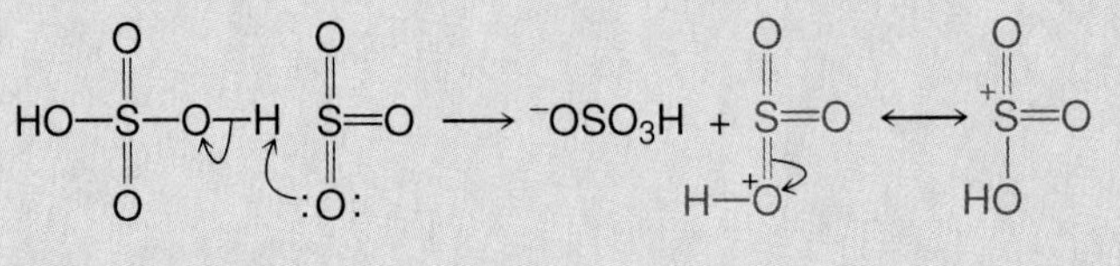

Second step:

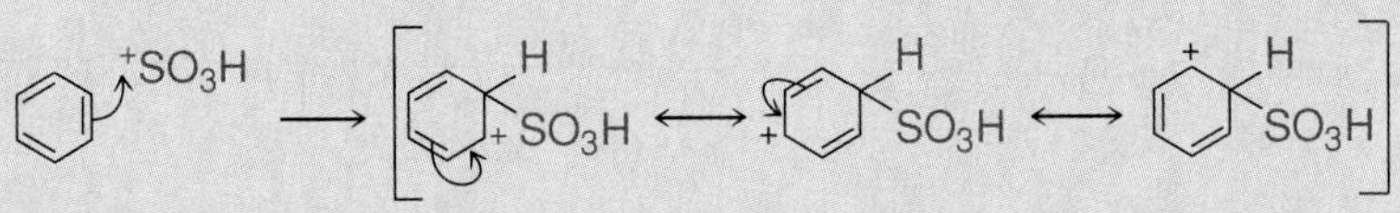

Third step:

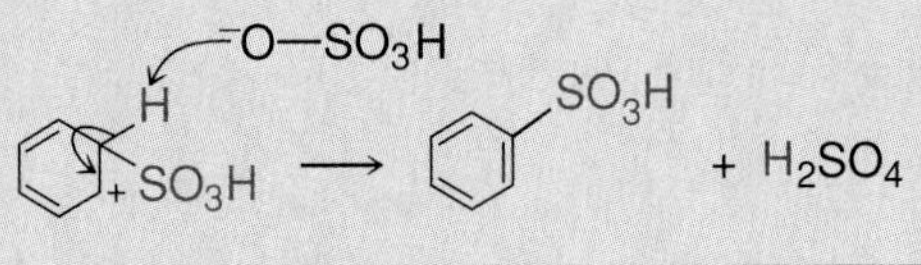

The electrophile is a protonated sulfur trioxide species ($^+$SO$_3$H) that is generated by a Brønsted acid-base reaction where sulfuric acid is the acid and sulfur trioxide is the base (first step). Note that the electrophile is a resonance-stabilized species. The $^+$SO$_3$H group then attacks the two pi electrons of a double bond of the aromatic ring to give a resonance-stabilized arenium ion (second step). The arenium ion, in turn, loses a proton to the bisulfate ion to give benzenesulfonic acid and regenerate sulfuric acid (third step).

**Nitration**

The benzene ring is nitrated by reacting an aromatic compound with a mixture of concentrated nitric and sulfuric acids. The concentrated sulfuric acid serves as a catalyst in the reaction.

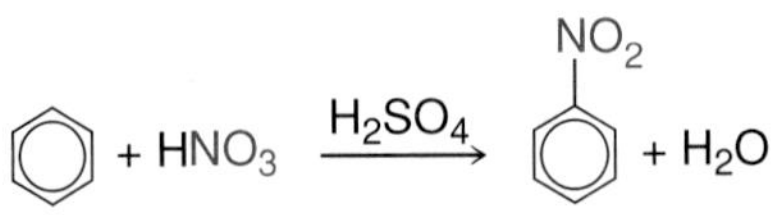

Nitrobenzene

The mechanism begins with a proton from sulfuric acid (the Brønsted acid) transferred to nitric acid (the Brønsted base) to form a protonated nitric acid species (first step). This species is unstable due to the fact that each of two adjacent atoms bears a positive charge. Since like charges repel, a molecule of water is lost by the species to give the

## Mechanism of the Reaction of Benzene with a Mixture of Nitric and Sulfuric Acids: Nitration

Envision the Reaction

First step:

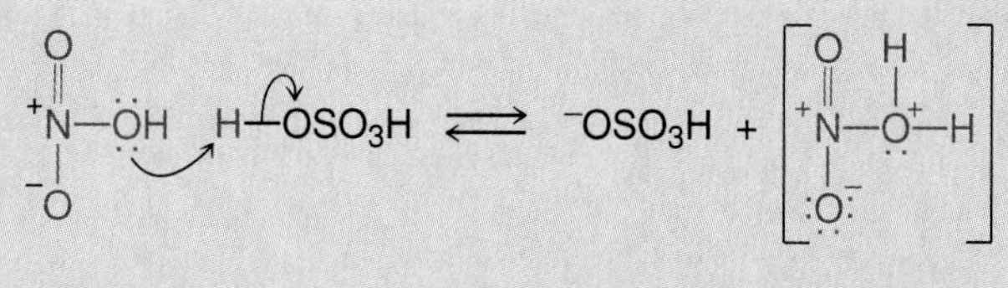

Second step:

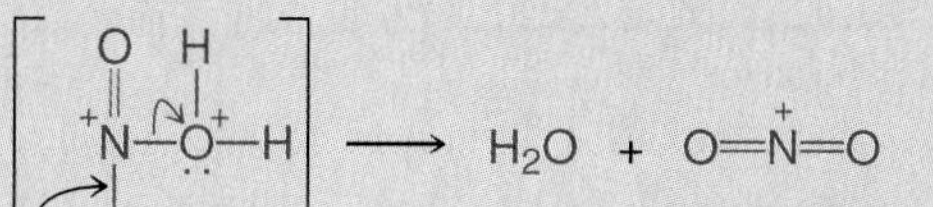

Nitronium ion

Third step:

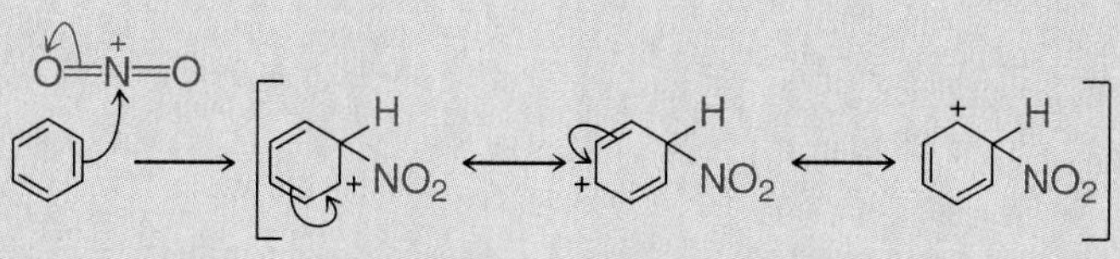

Fourth step:

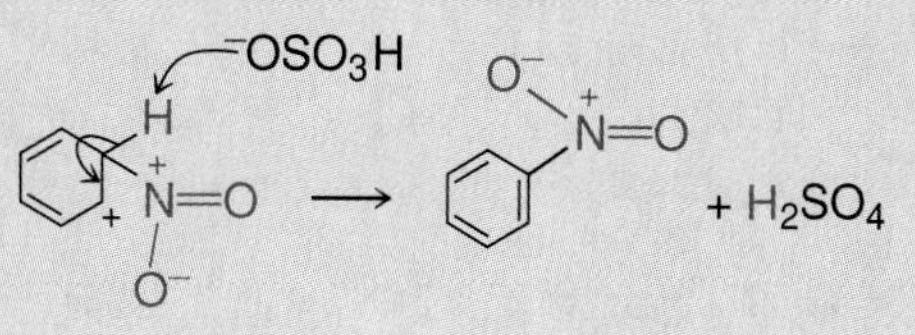

$+ H_2SO_4$

nitronium ion (second step). It is the nitronium ion that attacks two pi electrons of a double bond in benzene to give a resonance-stabilized arenium ion (third step). The arenium ion, in turn, loses a proton to bisulfate ion to produce nitrobenzene and regenerate sulfuric acid (fourth step).

Write an equation to show the regeneration of the catalyst (the last step of the mechanism) in

**SOLVED PROBLEM 10.3**

a.  Friedel-Crafts alkylation of benzene.

b.  the sulfonation of benzene.

**SOLUTION**

With each reaction, a proton is expelled when the product is produced. That proton reacts with an anion to regenerate the catalyst.

a.  $H^+ + {}^-AlCl_4 \longrightarrow HCl + AlCl_3$

b.  $H^+ + {}^-OSO_3H \longrightarrow H_2SO_4$

Write an equation to show the regeneration of the catalyst (the last step of the mechanism) in the

a. bromination of benzene.

b. nitration of benzene.

## 10.2 RATE OF AROMATIC SUBSTITUTION

In Section 10.1 we studied a number of reactions in which benzene reacted with an electrophile $E^+$ to produce a number of monosubstituted derivatives of benzene. At this time, a reasonable question to ask is, What happens in detail when we react an electrophile $E^+$ with a monosubstituted benzene derivative? We find that the substituent group on the benzene ring affects both the rate of reaction and the orientation (placement of $E^+$) in electrophilic aromatic substitution. We will first discuss the rate of reaction in some depth, and in Section 10.3 we will examine the orientation of $E^+$ in the ring with respect to the substituent group in electrophilic aromatic substitution.

The rate of reaction of a substituted aromatic compound is affected by the nature of the substituent on the benzene ring. Certain groups are *activating* (i.e., increase the rate of reaction compared with benzene as a standard), while other groups are *deactivating* (i.e., decrease the rate of reaction when compared with benzene).

The methyl group represents an example of an activating group. The methyl group and other activating groups donate electrons to the benzene ring and therefore make the benzene ring more negative and more likely to welcome a positively charged electrophile when compared with benzene.

On the other hand, the trifluoromethyl group ($-CF_3$) is a deactivating group. The trifluoromethyl group and other deactivating groups withdraw electrons from the benzene ring and therefore make the benzene ring more positive and less likely to be attacked by a positively charged electrophile when compared with benzene.

Trifluoromethylbenzene

Remember that the release of electrons by a substituent toward an organic group (e.g., an aromatic ring) or the attraction of electrons by a substituent away from an organic group (e.g., an aromatic ring) across a sigma bond is known as the *inductive effect* (see Sec. 6.2). The methyl group, for example, and other alkyl groups donate electrons and exhibit a $+ I$ effect, whereas the trifluromethyl group attracts electrons and shows a $- I$ effect. The inductive effect an atom or group of atoms exhibits is a function of the electronegativity of the atom or atoms concerned. For example, the chlorine in the carbon-chlorine bond exhibits a $- I$ effect because chlorine is more electronegative than carbon. Thus an electron shift occurs toward the chlorine, as indicated by the arrow:

$$\overrightarrow{\text{C—Cl}}$$

Since hydrogen is slightly less electronegative (more electropositive) than carbon, the hydrogens in a methyl group* release electrons, which, in turn, causes some electron release toward the carbon of the aromatic ring bonded to the methyl group. The methyl group, we say, exhibits a $+I$ effect.

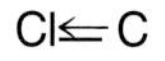

Cl⟸C     Electron shift from the less electronegative element (C) to the more electronegative element (Cl) across the C—Cl sigma bond. The chlorine shows a $-I$ effect.

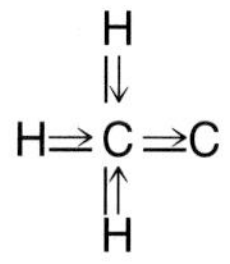

H⇒C⇒C     Electron shift from the less electronegative atoms of hydrogen to the more electronegative element (C), resulting in an electron shift toward the carbon on the right. The methyl group exhibits a $+I$ effect.

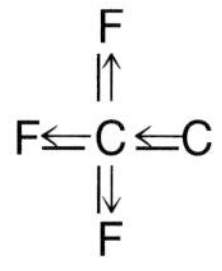

F⟸C⟸C     Electron shift from the less electronegative element (C) to the atoms of the more electronegative element (F), resulting in an electron shift away from the carbon on the right, producing a $-I$ effect.

Thus we can say that a reaction of toluene with electrophile $E^+$ proceeds at a faster rate than that of the corresponding reaction with benzene because the $+I$ effect of the methyl group increases, to a slight extent, the electron density in the ring and thus speeds up reaction with a positively charged electrophile. Using this same reasoning, the trifluoromethylbenzene reacts with $E^+$ at a slower rate than benzene due to the $-I$ effect of the trifluoromethyl group that decreases the electron density in the aromatic ring and thus slows down reaction with a positively charged electrophile.

In addition to the inductive effect, electrons can be moved into and out of an aromatic ring across a pi bond or bonds via the *resonance effect*. Therefore, the rate of aromatic substitution is also influenced by the resonance effect. Consider the resonance structures that can be drawn for phenol:

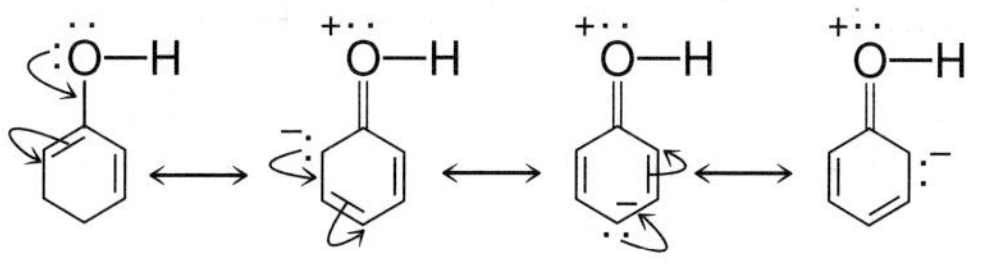

Electron density is increased in the aromatic ring by the resonance effect, as shown by the preceding structures. On the other hand, since oxygen is more electronegative than carbon, the inductive effect represents a movement of electrons toward oxygen. Thus what we have in phenol are two opposing forces at work: an activating resonance effect and a deactivating inductive effect. Experimentation has demonstrated that phenol reacts much faster with a given electrophile than benzene. Therefore, we can conclude that the activating resonance effect dominates the deactivating inductive effect because the —OH group is highly activating.

The nitro group in nitrobenzene is highly deactivating due to additive inductive and resonance effects. Since both nitrogen and oxygen are more electronegative than carbon, a $-I$ inductive movement toward the nitro group exists that is exacerbated by the $+1$ formal charge carried on nitrogen, making the nitro group deactivating. The deactivating effect is further increased by the resonance effect that also represents a withdrawal of electrons from the ring as follows:

---

*This explanation of the $+I$ effect of the methyl group is oversimplified for convenience. In actuality, the cause of the $+I$ effect of the methyl group is unknown.

| TABLE 10.1 | Summary of Inductive and Resonance Effects Regulating Reactivity in Electrophilic Aromatic Substitution: Activating Groups | | |
|---|---|---|---|
| Substituent Group | Inductive Effect | Resonance Effect | Overall Effect |
| —OH, | Deactivating | Activating | Strongly activating |
| —NH₂, —NHR —NR₂ | Deactivating | Activating | Strongly activating |
| —NHCOR | Deactivating | Activating | Moderately activating |
| —OR | Deactivating | Activating | Moderately activating |
| —R | Activating | None | Weakly activating |

Although the rate of reaction of any alkyl and the trifluoromethyl groups is regulated by the inductive effect alone, for most substituent groups the rate of reaction is regulated by a combination of the inductive and resonance effects. For a summary of inductive and resonance effects on the rate of substitution of monosubstituted benzene derivatives, refer to Tables 10.1 and 10.2. *Note that chemists are most interested in the overall effect on reactivity as a guide to setting experimental reaction conditions.*

Each of the halogens listed in Table 10.2 is weakly deactivating due to a strong electron-withdrawing inductive effect on the one hand negated to a degree by a weak electron-releasing resonance effect. In the same manner, the —OH, —NH₂, —NHR, and —NR₂ groups (Table 10.1) show a weak electron-withdrawing inductive effect and a strong electron-releasing resonance effect. Thus the overall effect is strongly activating.

**SOLVED PROBLEM 10.4**

Suppose chlorobenzene, phenol, benzene, toluene, and nitrobenzene are brominated. List the aromatic compounds in order of decreasing reactivity.

**SOLUTION**
From Table 10.1, since the overall effect of the —OH group is strongly activating, phenol shows the greatest reactivity, with toluene next in order due to the overall weakly activating effect of the methyl group. Benzene follows because it is the standard of reactivity. Chlorobenzene follows (Table 10.2) because the chloro group is weakly deactivating, whereas tail-end Charlie is the nitro group of nitrobenzene, which is strongly deactivating.

**SOLVED PROBLEM 10.5**

Consider a reaction between nitric acid and phenol. Would you need to heat these reagents to enable the reaction to occur? Explain.

**SOLUTION**
The reaction should proceed smoothly at room temperature because the —OH group is strongly activating in electrophilic aromatic substitution. Heating the reaction mixture could cause the reaction to proceed too fast, and an explosion could result.

| TABLE 10.2 | Summary of Inductive and Resonance Effects Regulating Reactivity in Electrophilic Aromatic Substitution: Deactivating Groups | | |
|---|---|---|---|
| Substituent Group | Inductive Effect | Resonance Effect | Overall Effect |
| —$NO_2$ | Deactivating | Deactivating | Strongly deactivating |
| —$CF_3$ | Deactivating | None | Strongly deactivating |
| —C≡N | Deactivating | Deactivating | Moderately deactivating |
| —$SO_3H$ | Deactivating | Deactivating | Moderately deactivating |
| —COOH | Deactivating | Deactivating | Moderately deactivating |
| —COOR | Deactivating | Deactivating | Moderately deactivating |
| $\overset{O}{\overset{\|}{—CHO, —C—R}}$ | Deactivating | Deactivating | Moderately deactivating |
| —Cl, —Br —I | Deactivating | Activating | Weakly deactivating |

**EXERCISE 10.4**

Arrange the following in decreasing order of reactivity to electrophilic aromatic substitution.

a.  chlorobenzene, benzaldehyde, nitrobenzene, aniline

b.  bromobenzene, $C_6H_5$—O—$CH_3$ (anisole), phenol, benzoic acid

**EXERCISE 10.5**

Consider a reaction between nitric acid (with sulfuric acid as a catalyst) and nitrobenzene. Would you need to heat these reagents to enable the reaction to occur? Explain.

## 10.3 ORIENTATION OF AN ELECTROPHILE IN AROMATIC SUBSTITUTION TO PRODUCE A DISUBSTITUTED BENZENE DERIVATIVE

Now that we have examined the relative rate of aromatic substitution to a monosubstituted benzene derivative, we need to consider where the entering group will substitute in the benzene ring with respect to the group already present on the ring.

Consider a group —X on the benzene ring. The electrophile $E^+$ could substitute in one of three locations as follows:

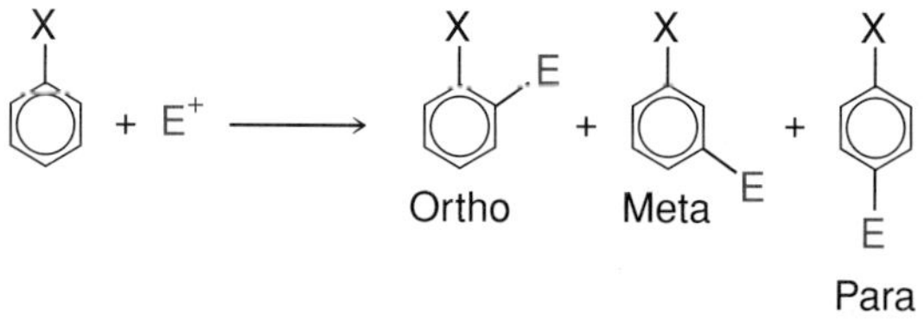

Experiment has demonstrated two patterns of substitution orientation: ortho-para and meta. Orientation, like rate of reaction, is affected by the group already located on the benzene ring. Table 10.3 summarizes the results of the orienting effect of a number of monosubstituted benzene substituents on an electrophile to form a disubstituted benzene derivative. When we react toluene with a mixture of nitric acid and a catalytic amount of sulfuric acid,* we obtain a mixture of mononitrotoluenes.

*In mixed-acid nitrations, the sulfuric acid is actually present in large amounts. It serves to generate the nitronium ion electrophile and reacts with the liberated water to prevent dilution of the nitric acid as the reaction proceeds.

| TABLE 10.3 | Ortho-Para and Meta Directing Substituents in Electrophilic Aromatic Substitution |
|---|---|
| **Substituents** | **Direction of Orientation** |
| —OH, —NH$_2$, —NHR, —NR$_2$, —NHCOR, —OR, —R, —Cl, —Br, —I | Ortho-para |
| —NO$_2$, —CF$_3$, —C≡N, —SO$_3$H, —COOH, —COOR, —CHO, —COR | Meta |

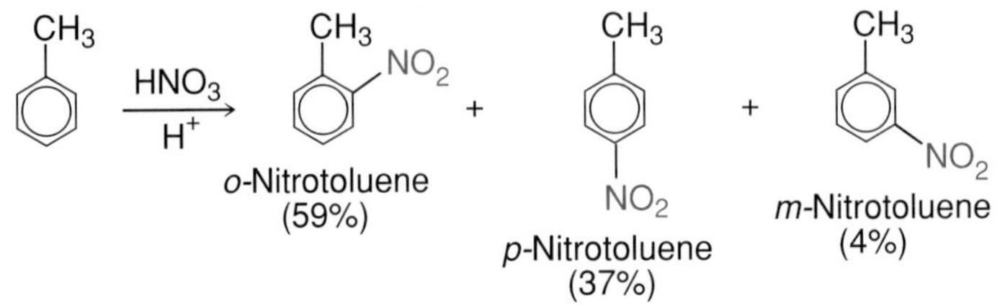

o-Nitrotoluene (59%) + p-Nitrotoluene (37%) + m-Nitrotoluene (4%)

Note that this reaction illustrates an ortho-para pattern of orientation. The total yield of ortho-para substitution is 96% (59% + 37%). Only a 4% yield of meta isomer is obtained.

On the other hand, when we react benzaldehyde with chlorine and a catalytic amount of iron, the meta isomer is produced in largest yield.

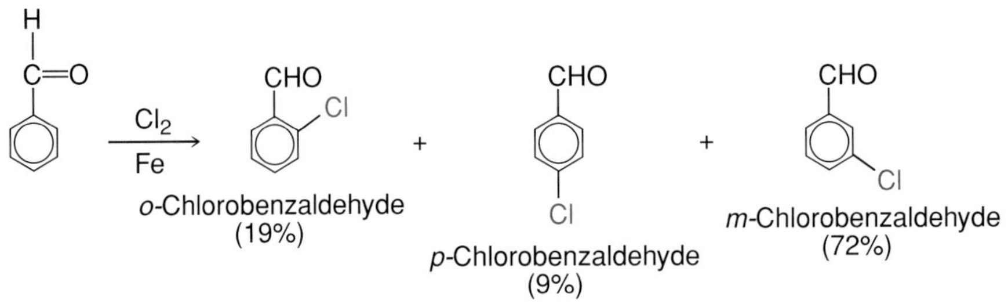

o-Chlorobenzaldehyde (19%) + p-Chlorobenzaldehyde (9%) + m-Chlorobenzaldehyde (72%)

Note that in the chlorination of benzaldehyde, the total yield of ortho-para substitution is only 28% (19% + 9%), whereas the meta isomer is produced in a substantial 72% yield. The important point to remember here is that the group already on the ring directs the orientation of a given electrophile. Thus we obtain a good yield of ortho and para isomers when an ortho-para directing group is on the benzene ring and a good yield of meta isomer when a meta directing group is on the ring (see Table 10.3).

As a rule, ortho-para directors are activating (with the halogens the only exception), whereas meta directors are deactivating without exception (see Tables 10.1, 10.2, and 10.3).

A question that should be asked is, How can you tell if a group located on a benzene ring orients ortho-para or meta based on the structure of the group? In other words, how can you predict the position on a ring to which an already existing group will direct an incoming group? Resonance theory provides an answer to this question. Consider aniline (containing the ortho-para director —NH$_2$) attacked by any electrophile E$^+$ in the ortho position. Note that the arenium ion formed is stabilized by four contributing structures.

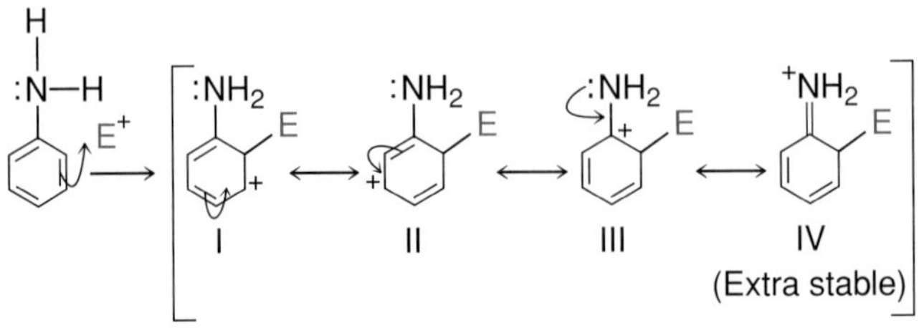

In a similar way, four contributing resonance structures can be drawn for the intermediate produced by para substitution.

Now suppose that $E^+$ attacks at the meta position.

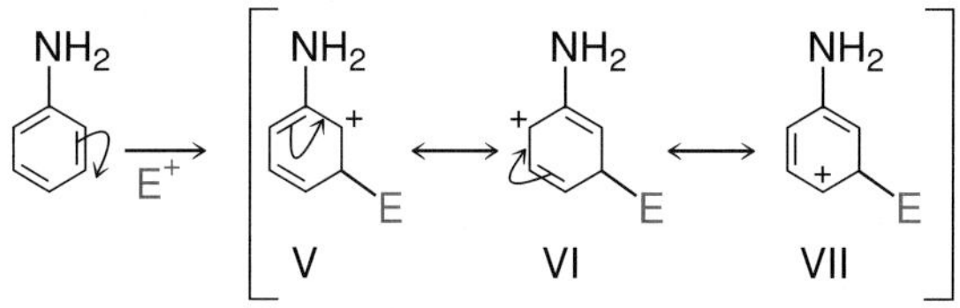

Note that only three resonance structures can be drawn for the arenium ion intermediate that results in meta substitution.

The amino group is ortho-para directing because the intermediate arenium ion leading to the ortho and para products is more stable than the intermediate arenium ion leading to meta substitution. Why? There are two reasons:

1. Only three resonance structures can be drawn for the arenium intermediate resulting in a meta product, as opposed to four resonance structures for the carbocation intermediate produced to give the ortho (and para) substitution products. Since the greater the number of contributing resonance structures drawn for a structure, the more stable the structure, the arenium ion intermediate produced by means of ortho-para substitution is more stable.

2. Structure IV is more stable than any of the other six structures (I, II, III, V, VI, and VII) because the nitrogen atom and each carbon in structure IV contain a completely filled octet of electrons. Each of the other structures contains a positively charged carbon with only a sextet of electrons.

Note that four resonance structures can be drawn for the ortho (and para) arenium ion intermediates because a positive charge exists on the carbon bonded to the mono-substituted group (structure III). The fourth resonance structure (IV) can then be drawn because a pair of electrons on the atom bonded to the ring (the nitrogen atom) interacts with the ring in structure III. This brings us to *Aromatic Orientation Rule 1*: Every atom (or group) with one or more unshared electron pairs on the atom bonded to the aromatic ring is an ortho-para director. A model for an ortho-para director is as follows, where Ar— represents an aryl group:

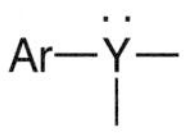

$$Ar-\overset{..}{\underset{|}{Y}}-$$

a. Show, using Aromatic Orientation Rule 1, why the —OH group in phenol would be expected to be an ortho-para director.

b. Draw the four resonance structures obtained when $E^+$ attacks phenol in the para position. Indicate the resonance structure that is particularly stable.

**SOLVED PROBLEM 10.6**

**SOLUTION**

a. Since the —OH group has two pairs of unshared electrons on the oxygen that is bonded to the benzene ring, we would predict that phenol would be an ortho-para director, as stated in Aromatic Orientation Rule 1.

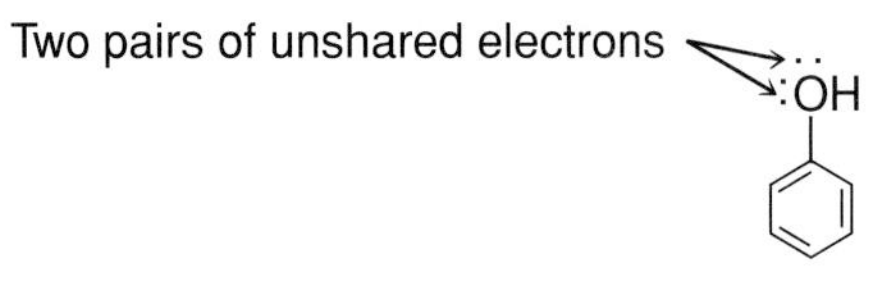

Two pairs of unshared electrons

:ÖH

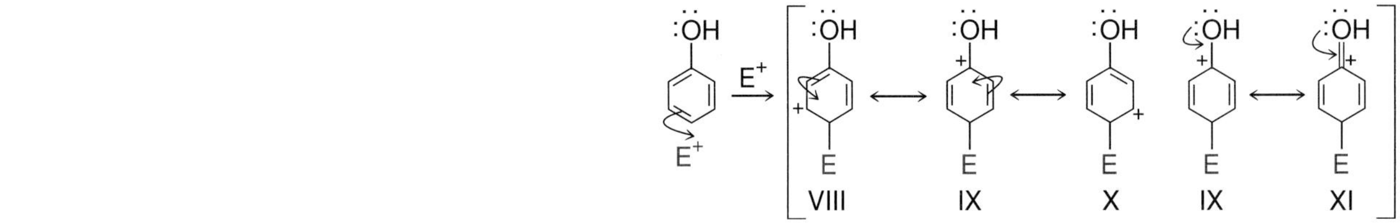

VIII IX X IX XI

Structure XI is the most stable because the oxygen carries a positive charge and every carbon contains a stable octet of electrons. In structures VIII, IX, and X, each carbon carrying a positive charge contains a less stable sextet of electrons.

**EXERCISE 10.6**

Draw the four resonance structures obtained when $E^+$ attacks aniline in the para position. Indicate the resonance structure that is particularly stable.

One general exception exists for Aromatic Substitution Rule 1. An alkyl group is an ortho-para director yet does not have one or more pairs of unshared electrons on the atom (carbon) bonded to a carbon of the benzene ring. To illustrate this phenomenon, let us brominate toluene and focus on ring attack by the $Br^+$ electrophile to form the intermediate arenium ion: first in the para (or ortho) position and then in the meta position. We will finally examine the arenium ion intermediates and explain why the methyl group is an ortho-para director.

XII XIII XIV

XV XVI XVII

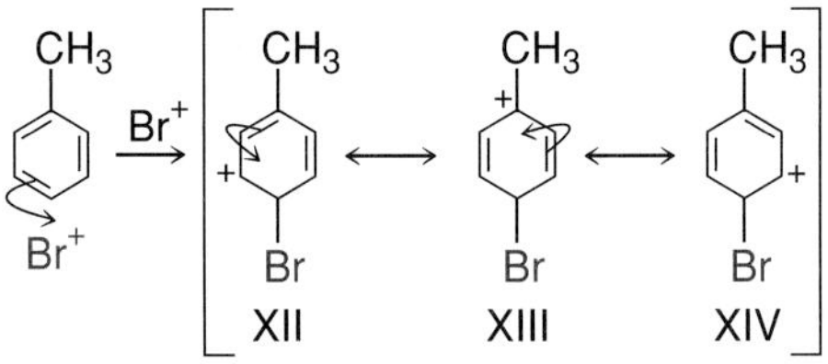

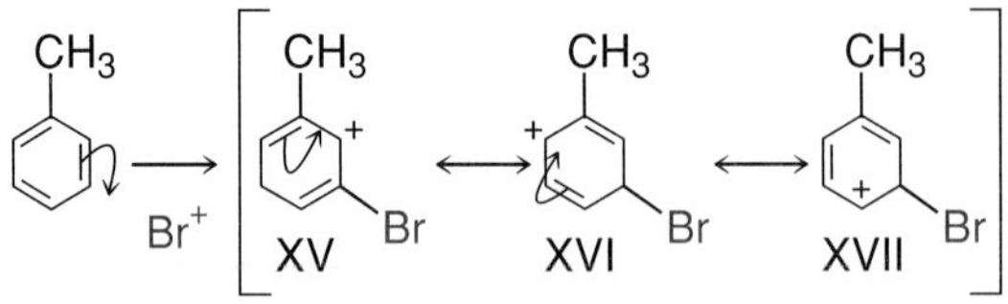

We can state that since the methyl group is known to be an ortho-para director by experimentation, the arenium ions formed leading to the ortho and para products are more stable than the arenium ion leading to formation of the meta isomer. This is so because the methyl group exhibits a $+I$ inductive effect, and since the positively charged carbon in structure XIII is bonded to the methyl group, then electrons are being released by the methyl group across the sigma bond and are delocalizing the positive charge on carbon and thus stabilizing structure XIII. It is interesting to note that no such positively charged carbon bonded to the methyl group exists in contributing structures XV, XVI, and XVII that would result in formation of the meta isomer.

Since we have explained the effect of ortho-para directing groups, let us now examine meta directing groups. Consider the nitration of benzaldehyde. As usual, we need to focus on the formation of the intermediate arenium ion when an electrophile $E^+$ (in this case, nitronium ion) attacks the benzene ring. First, consider benzaldehyde attacked at the ortho (or para) position by the nitronium ion.

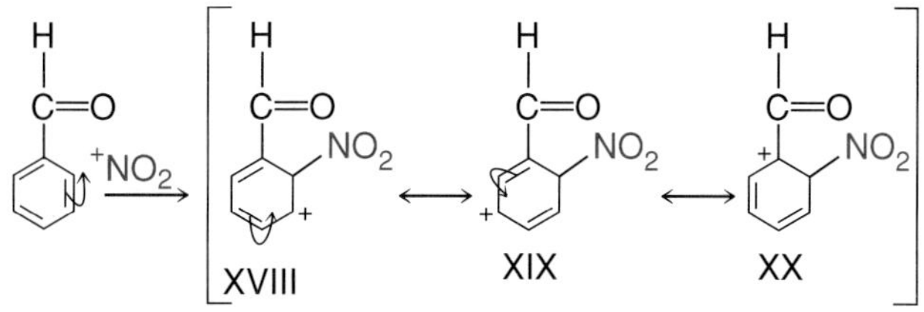

The resulting arenium ion can be shown as three resonance structures. A similar number of resonance structures can be generated for the intermediate arenium ion when the nitronium ion ($^+NO_2$) electrophile attacks meta as follows:

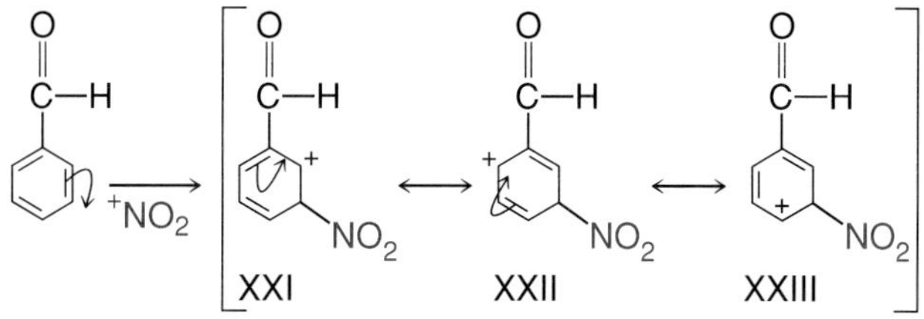

Since experiment has established that the formyl group (—CHO) is a meta director, we can state that the arenium ion produced by meta attack is more stable than the arenium ions formed as a result of ortho and para attack. This is so because structure XX is less stable than any of the other structures (XVIII, XIX, XXI, XXII, and XXIII). This is true, in turn, because only in structure XX is the positive charge carried by the ring carbon bonded to the carbon of the formyl group. Since the formyl group shows a −I inductive effect, electrons are stripped from the carbon bearing the positive charge, intensifying the positive charge and destabilizing the arenium ion. Thus ortho-para substitution is disfavored compared with meta substitution, and the meta isomer is produced in largest yield.

This brings us to *Aromatic Orientation Rule 2*: Every atom or group of atoms with no unshared electron pairs on the atom bonded to the aromatic ring is a meta director. A model for a meta director is as follows, where Ar— represents an aryl group:

$$Ar—A{=}B$$
$$|$$

One exception to this rule is any alkyl group. The carbon of a methyl group, for example, has no unshared electron pairs, yet the methyl group is an ortho-para director.

---

Predict and draw a structural formula for each major organic product or products formed from the following reactions.

**SOLVED PROBLEM 10.7**

a.  isopropylbenzene (cumene) + HNO₃ $\xrightarrow{\text{H}_2\text{SO}_4}$

b.  benzoic acid + SO₃ $\xrightarrow{\text{H}_2\text{SO}_4}$

**SOLUTION**

a.  Since the isopropyl group is an ortho-para director (see Table 10.3), the ortho and para substitution products are the major organic products.

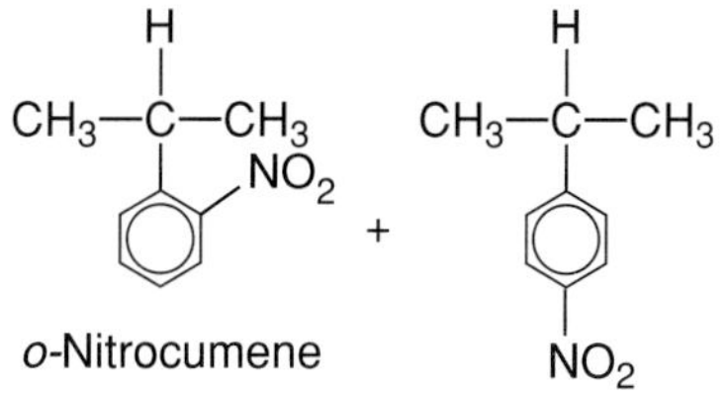

o-Nitrocumene

p-Nitrocumene

b.  Since the carboxyl group is a meta director (see Table 10.3), the meta substitution
    product is the major organic product.

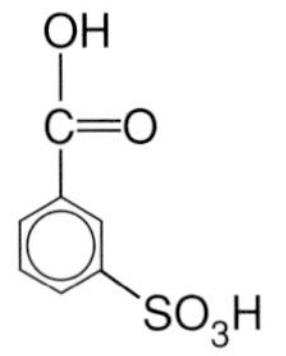

*m*-Sulfobenzoic acid

**SOLVED PROBLEM 10.8**

Without looking at Table 10.3, use an aromatic orientation rule to predict whether the
cyano group ($-C\equiv N$) in benzonitrile ($C_6H_5-CN$) is an ortho-para or meta director.
Then prove your prediction by the use of resonance structures.

**SOLUTION**

Since no unshared electron pairs surround the carbon of the cyano group, we can
predict that the cyano group is a meta director.

$$C_6H_5-C\equiv N$$

No unshared electron pairs

To prove this, we draw the necessary resonance structures of the arenium ions produced by both para (or ortho) and meta attack.

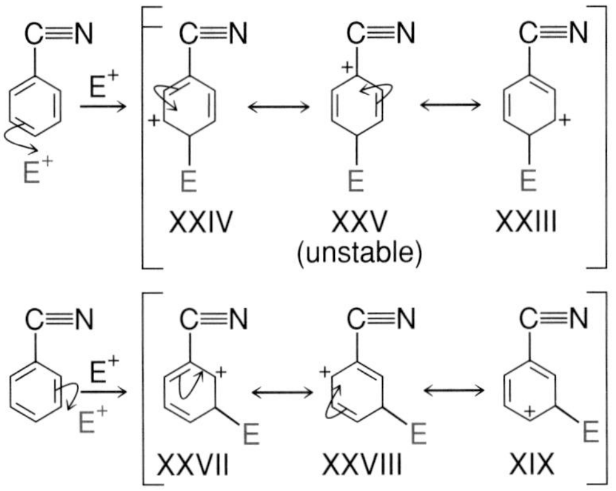

Note that structure XXV is unstable due to the $-$I inductive effect of the cyano group (see
Table 10.2). This electron withdrawal increases the charge on the ring carbon bearing
the positive charge bonded to the cyano group and thus destabilizes the arenium ion
leading to ortho-para substitution. Meta substitution is preferred because none of the
resonance structures of the arenium ion leading to meta substitution contains a positive
charge on the ring carbon bonded to the cyano group.

**EXERCISE 10.7**

Predict and draw a structural formula for each major organic product or products
formed from the following reactions.

a.  ethylbenzene + $HNO_3$ $\xrightarrow{H_2SO_4}$

b.  chlorobenzene + $CH_3-\overset{\overset{O}{\|}}{C}-Cl$ $\xrightarrow{AlCl_3}$

Note the fact that Friedel-Crafts reactions are not possible on aromatics more deactivating than the halobenzenes.

c.   nitrobenzene + $Cl_2$ $\xrightarrow{\text{FeCl}_3}$

d.   toluene + $SO_3$ $\xrightarrow{\text{H}_2\text{SO}_4}$

---

**EXERCISE 10.8**

Without looking at Table 10.3, use an aromatic substitution rule to predict whether the carbomethoxy group ($-COOCH_3$) in methyl benzoate ($C_6H_5-COOCH_3$) is an ortho-para or meta director. Then support your prediction with resonance structures.

---

## 10.4 ORIENTATION IN AROMATIC SUBSTITUTION TO PRODUCE A TRISUBSTITUTED BENZENE DERIVATIVE

Assuming two substitutents on the benzene ring, the same or different, in an ortho, meta, or para orientation, where on the ring will a third substituent attack? Let us consider a number of situations to predict one or more sites of substitution.

1. A good yield of a single product can be produced if the directing effects of the two substituents are additive. For example,

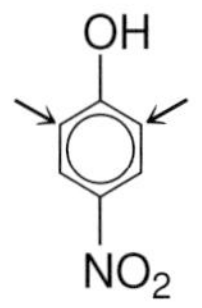

Points of attack are ortho to the ortho director $-OH$ and meta to the meta director $-NO_2$. Since both positions are equivalent, only one product is obtained.

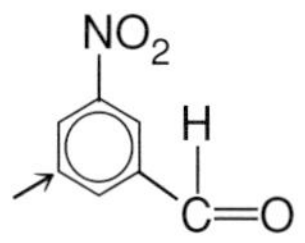

The single point of attack is meta to both meta directing groups. Thus one product is obtained. Since both groups are deactivating, reaction conditions need to be more severe.

2. A good yield of a single product can be obtained if the directing effects of the two groups are contrary to each other. The directing influence of the strongly activating group dominates that of a weakly activating group or deactivating group.

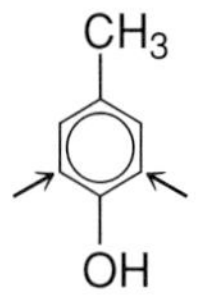

Points of attack are ortho to the dominant, strongly activating $-OH$ group and meta to the passive, weakly activating $-CH_3$ group. Since both positions are equivalent, only one product is obtained.

3. Two or more different products are formed if a moderately activating, weakly activating, or deactivating group is present with a deactivating group and if their directing influences are contrary to each other.

CH₃  No substitution occurs here due to steric hindrance

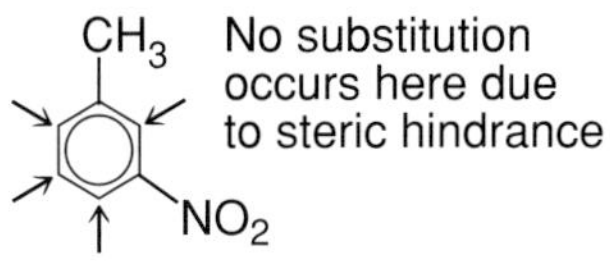

(Three products)

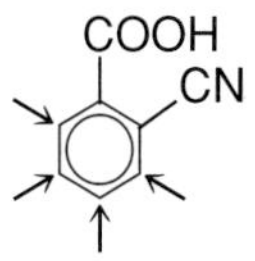

(Four products)

*Steric hindrance* is crowding around a reaction site due to the presence of bulky atoms or groups. This crowding makes it difficult for a reactant to approach the reaction site and thus decreases the rate of reaction.

---

Predict the product(s) formed from the nitration of *p*-chlorophenol.

**SOLVED PROBLEM 10.9**

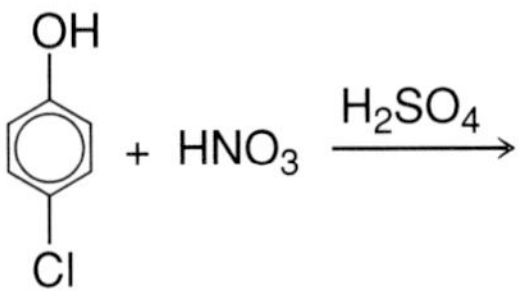

$$\text{(structure: } p\text{-chlorophenol)} + HNO_3 \xrightarrow{H_2SO_4}$$

## SOLUTION

Since both the —OH and —Cl groups are ortho-para directing, the directing effect of the —OH group is contrary to that of the —Cl group. Thus, the directing influences of the —OH group are dominant because it is strongly activating, whereas those of the —Cl group are passive because it is slightly deactivating. Therefore, the product formed is located ortho to the —OH group, and we get

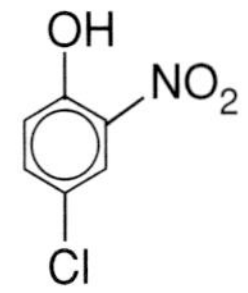

$$\text{(structure: 4-chloro-2-nitrophenol with OH, } NO_2, \text{ and Cl substituents)}$$

Note that the —NO$_2$ group cannot substitute para to the —OH group because the —Cl group is located in that position. In order to obtain a solution to Solved Problem 10.9, it is necessary that you know the directive (Table 10.3) and reactive (Tables 10.1 and 10.2) properties of the various groups located on the benzene ring.

---

### EXERCISE 10.9

Predict the product(s) formed from each of the following reactions, and explain your prediction.

a. nitration of $p$-hydroxybenzoic acid

b. bromination of $m$-nitrobenzenesulfonic acid

c. Friedel-Crafts alkylation (with $CH_3Cl$) of $p$-HOOC—C$_6$H$_4$—NHCOCH$_3$

d. sulfonation of $m$-cresol ($m$-methylphenol)

---

## OTHER REACTIONS OF BENZENE AND ITS DERIVATIVES

## 10.5 REDUCTION

Benzene and its derivatives can be hydrogenated in the presence of a rhodium-on-carbon catalyst to give cyclohexane or a cyclohexane derivative, respectively.

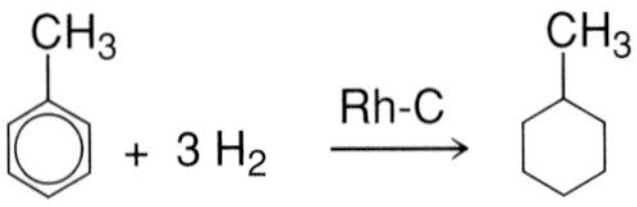

$$\text{(toluene)} + 3\,H_2 \xrightarrow{Rh-C} \text{(methylcyclohexane)}$$

Nitrobenzene and its derivatives are reduced by a metal-acid combination. This reaction produces the amine salt that is treated with sodium hydroxide to give amine. A frequently used metal-acid combination is tin and hydrochloric acid.

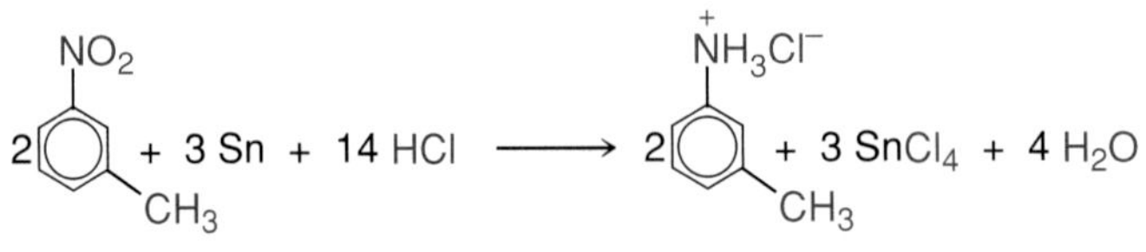

The rhodium-on-carbon catalyst is a stronger hydrogenation catalyst than finely divided platinum, palladium, or nickel. This is necessary in order to accomplish the reduction of the very stable aromatic benzene ring at room temperature and in excellent yield.

Note that the ionic nature of the organic salts is emphasized by showing the formula as Ar$\overset{+}{N}H_3X^-$, where X$^-$ is a halide ion. It is assumed that the student is aware of the ionic nature of a compound such as NaCl (Na$^+$Cl$^-$) (see Sec. 16.4).

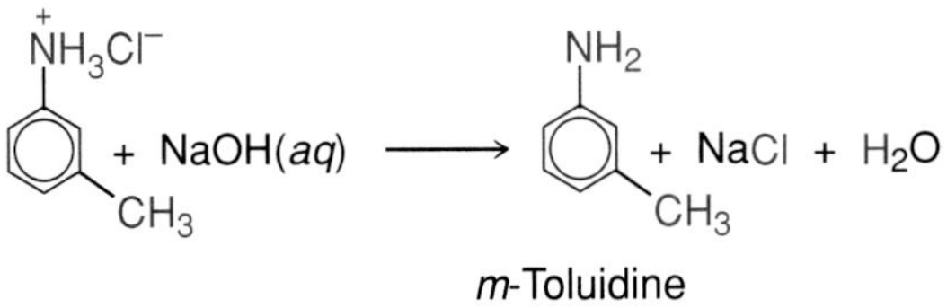

$$2\,\text{(}m\text{-Nitrotoluene, } NO_2, CH_3\text{)} + 3\,Sn + 14\,HCl \longrightarrow 2\,\text{(}\overset{+}{N}H_3Cl^-, CH_3\text{)} + 3\,SnCl_4 + 4\,H_2O$$

*m*-Nitrotoluene

$$\text{(}\overset{+}{N}H_3Cl^-, CH_3\text{)} + NaOH(aq) \longrightarrow \text{(}NH_2, CH_3\text{)} + NaCl + H_2O$$

*m*-Toluidine

## 10.6 SIDE-CHAIN HALOGENATION

Alkyl-substituted benzenes (arenes), when treated with chlorine or bromine in the presence of light or heat, undergo substitution where one or more hydrogens of the alkyl group are replaced by one (or more) halogen atoms. In general,

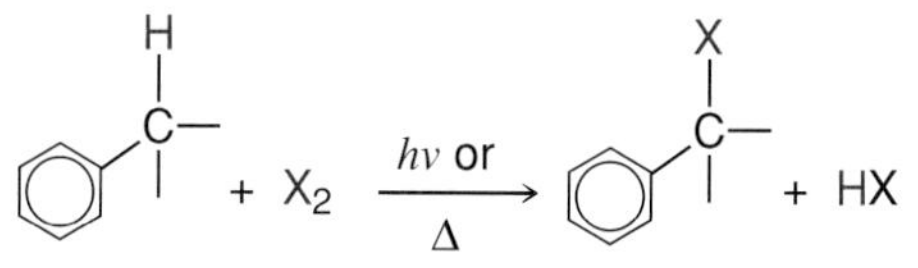

where $X_2 = Cl_2$ or $Br_2$. For example:

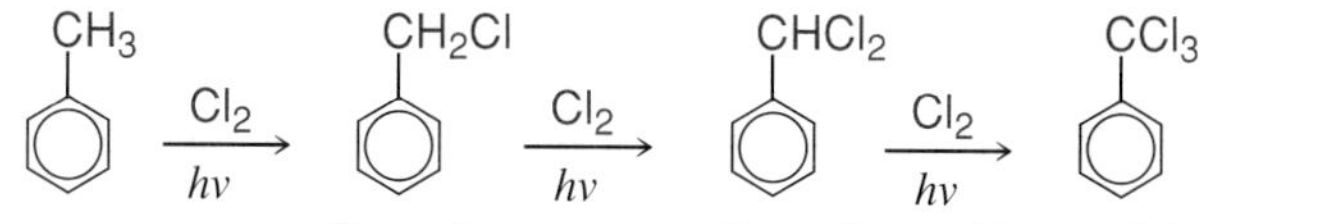

Benzyl chloride — Benzal chloride — Benzotrichloride

Reacting gaseous toluene with 1, 2, or 3 mol chlorine produces the corresponding chlorinated product and hydrogen chloride as a by-product. The products of these reactions, benzyl chloride, benzal chloride, and benzotrichloride, are the basis for further synthesis. For example:

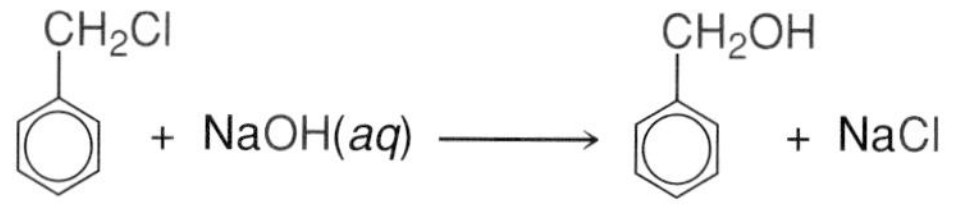

Benzyl chloride — Benzyl alcohol

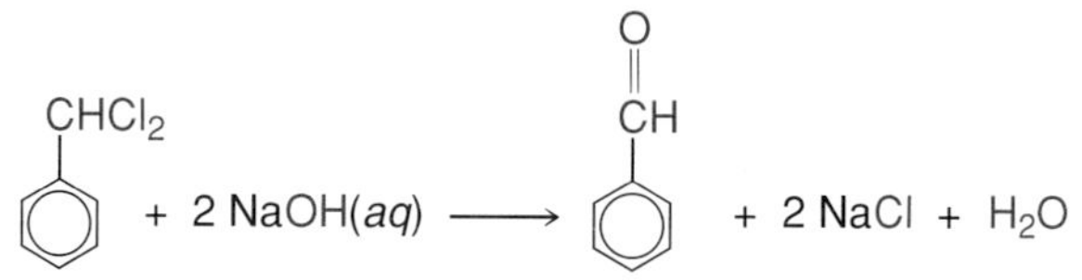

Benzal chloride — Benzaldehyde

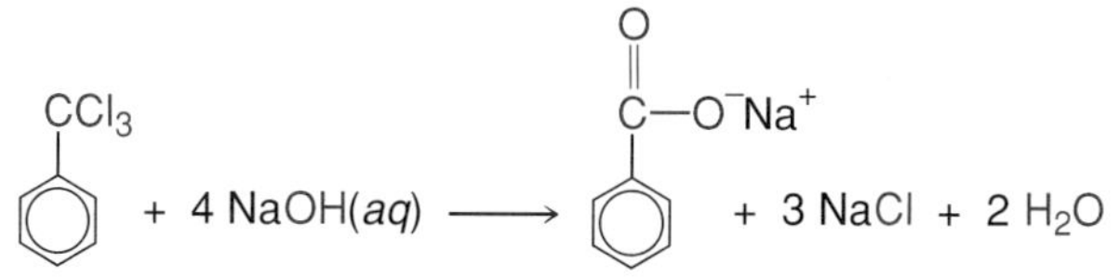

Benzotrichloride — Sodium benzoate

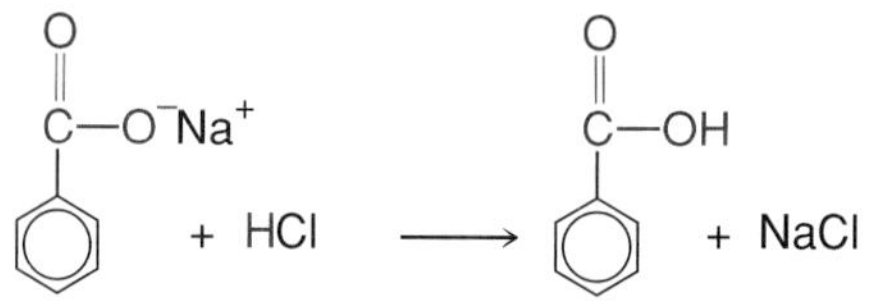

Benzoic acid

Notice that when an arene containing more than one aliphatic carbon (with more than one set of equivalent hydrogens) is reacted with bromine in the presence of sunlight (or is heated with bromine), only an α-substituted product is isolated in significant yield. In the same way, when an arene containing more than one aliphatic carbon (with more than one set of equivalent hydrogens) is reacted with chlorine in the presence of sunlight (or is heated with chlorine), a significant amount of α-substituted product is isolated along with small amounts of β-substituted and γ-substituted product. In this example, note that the *n*-propyl group contains three different sets of equivalent hydrogens.

Major product

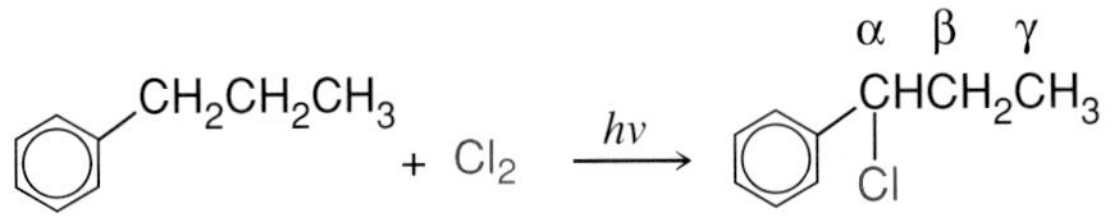

An obvious question to ask is, Why is the formation of the α-product preferred in both chlorination and bromination? For an explanation of the overwhelming amount of α-substituted product produced, refer to Problem 10.38.

Minor product Minor product

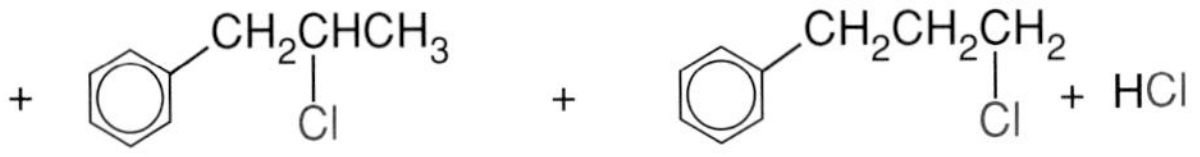

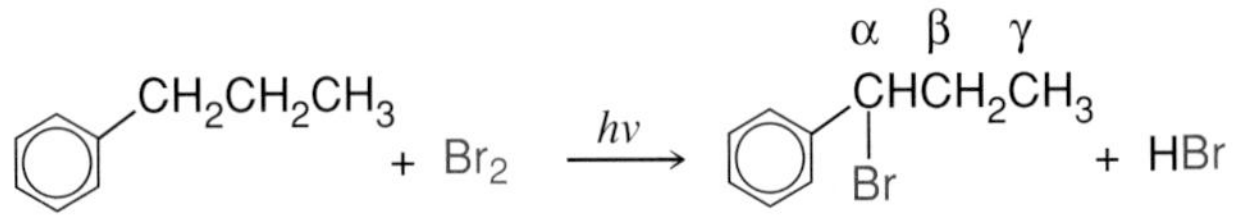

The reason we obtain by-products in chlorination and not in bromination is beyond the scope of this textbook. However, one conclusion we can reach is that bromination of an arene represents a better synthetic tool than chlorination when the alkyl group of the arene contains more than one equivalent set of hydrogens.

Note the differences between free-radical side-chain halogenation of an arene and electrophilic aromatic substitution of an arene in terms of both the difference in reagents used and the location of the substituted halogen.

Benzyl chloride

*o*-Chlorotoluene

*p*-Chlorotoluene

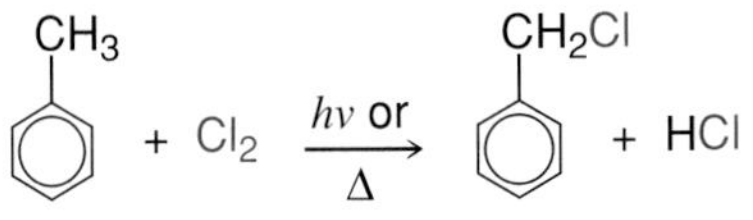

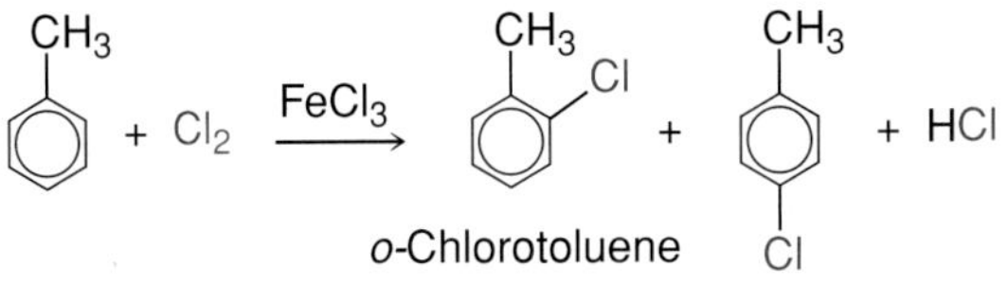

## 10.7 SIDE-CHAIN OXIDATION

An arene, when heated with a basic aqueous solution of potassium permanganate ($KMnO_4$), produces the potassium salt of a carboxylic acid. The salt, in turn, is reacted with hydrochloric acid to give a carboxylic acid. For example, the permanganate oxidation of toluene ultimately yields benzoic acid.

Benzoic acid

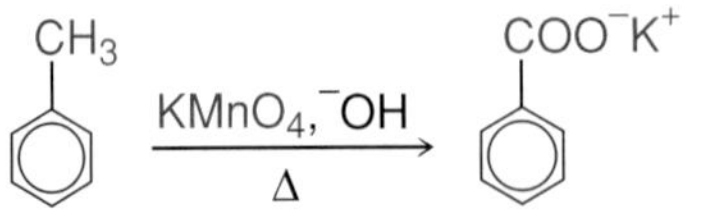

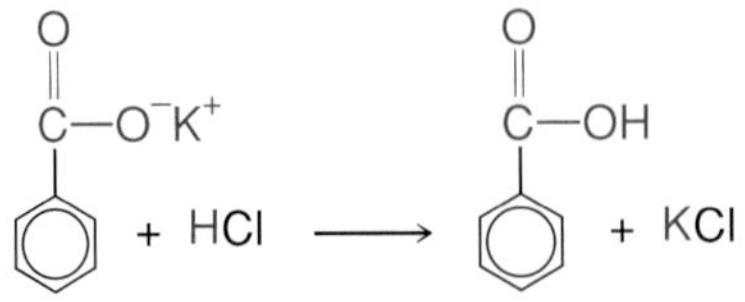

It is interesting to note that irrespective of the number of carbon atoms contained in the side chain of the arene, ultimately only benzoic acid is produced. The other carbon atoms of the alkyl group are converted to carbonate ion and finally to carbon dioxide and water.

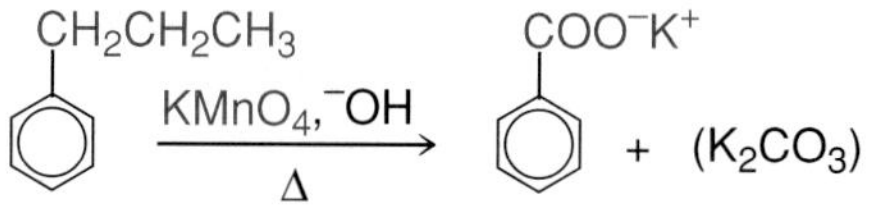

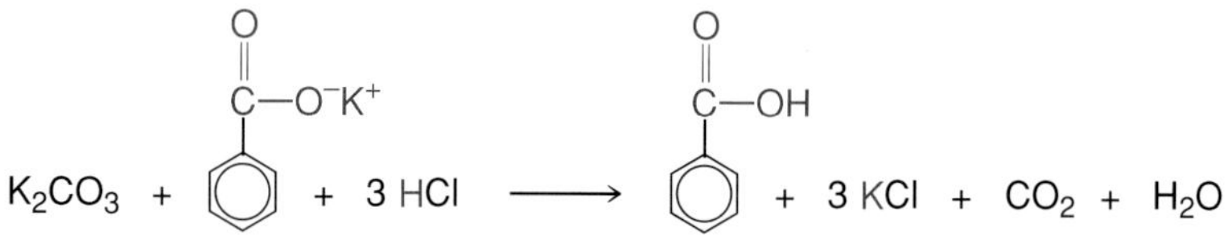

The alkyl group must contain at least one benzylic hydrogen in order to react. Since the isopropyl group in isopropylbenzene (cumene) contains one such hydrogen, reaction takes place.

The *tert*-butyl group in *tert*-butylbenzene does not contain a benzylic hydrogen, so reaction does not occur.

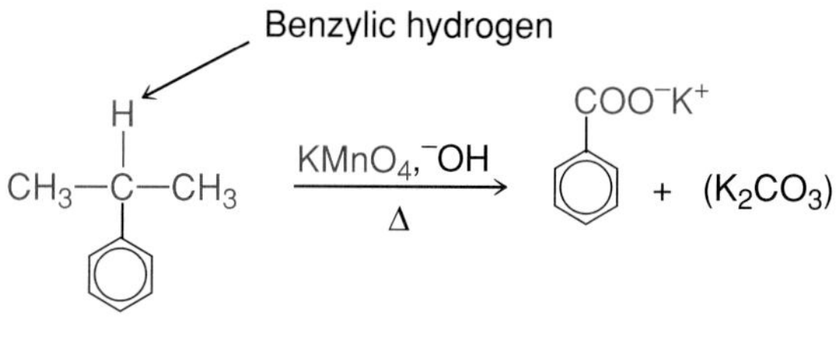

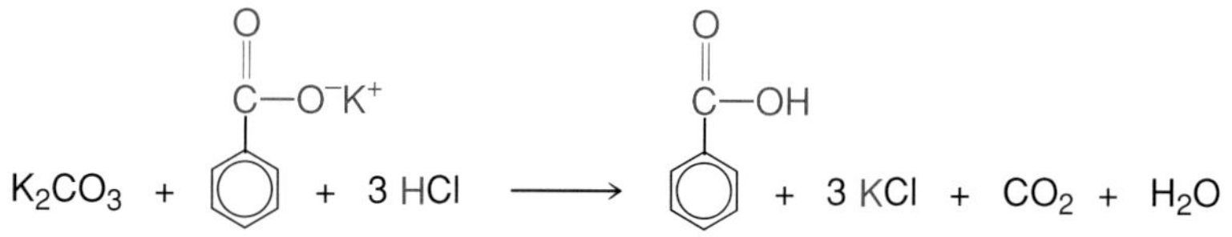

The oxidation of *p*-xylene gives terephthalic acid.

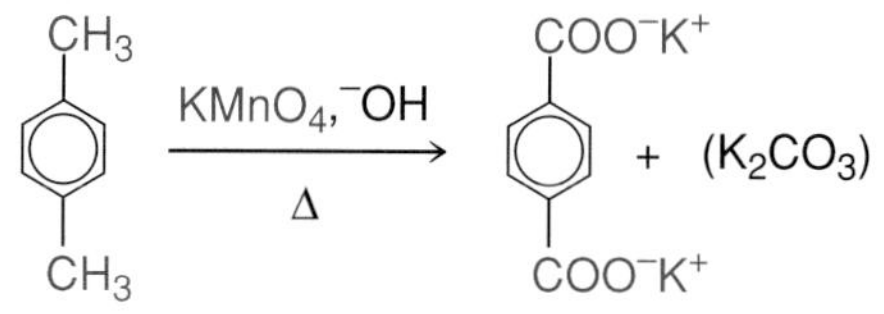

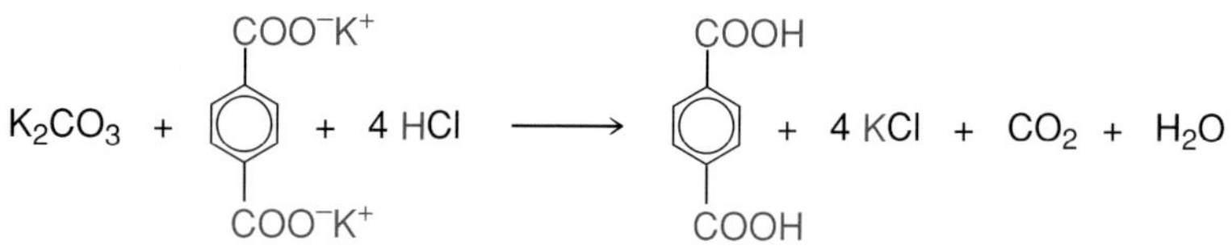

Terephthalic acid

Complete the following equations by drawing a structural formula for each organic product, and when applicable, write a molecular formula or formula unit for each inorganic product. (Do not balance each equation.)

**EXERCISE 10.10**

a.  propylbenzene + 3 H$_2$ $\xrightarrow{\text{Rh-C}}$

b.  *p*-chloronitrobenzene + Sn + HCl $\longrightarrow$

c.  product (b) + NaOH(*aq*) $\longrightarrow$

d.  *p*-chlorobenzyl chloride + NaOH(*aq*) $\longrightarrow$

e.  isopropylbenzene + Br$_2$ $\xrightarrow{h\nu}$

f.  isopropylbenzene + Cl$_2$ $\xrightarrow{h\nu}$

Chloramphenicol is the only naturally occurring antibiotic that contains a nitrophenyl group.

$$O_2N- \bigcirc -\underset{\underset{HO}{|}}{\overset{H}{\underset{|}{C}}}-\underset{\underset{H}{|}}{\overset{NH-\overset{\overset{O}{\|}}{C}-CHCl_2}{\underset{|}{C}}}-CH_2OH$$

Chloramphenicol

The compound was first found to be produced by a mold *Streptomyces venezuelae* during the late 1940s. The mold, in turn, was found in a soil sample taken from Venezuela. When tested, it showed great promise as an antibiotic. By 1948, the compound was synthesized in the laboratory, and an effective method of manufacture was worked out.

The compound was a superb antibiotic. It cured typhus, typhoid fever, meningitis, brucillosis, and bacteremia, along with other nasty infections. It took only 2 years of extensive use, however, for physicians to realize the drug had a dark side. Among other problems, it caused aplastic anemia. Thus, today, the drug is used with great caution in two ways. First, it is not used against any fly-by-night infection but only against nasty infections that are penicillin-resistant. Second, when used, the patient's blood must be monitored every 48 hours to make sure things are under control in that the white cell count does not fall below 4000 per cubic millimeter. The white cell count must be monitored because chloramphenicol destroys the ability of bone marrow to produce white cells. Should aplastic anemia develop, the only effective treatment is a bone marrow transplant, which would not be a procedure for a patient to anticipate with great joy.

g.  isopropylbenzene $\xrightarrow[\Delta]{KMnO_4, \ ^-OH}$

h.  product (g) + HCl $\longrightarrow$

# 10.8
# ORGANIC SYNTHESIS WITH AROMATIC COMPOUNDS

In order to effectively solve organic synthesis problems, you must have a good working knowledge of the organic reactions studied in the last two chapters. Let us learn about synthesis by solving two practice problems.

**SOLVED PROBLEM 10.10**

Prepare *p*-chloronitrobenzene from benzene and any necessary inorganic reagents.

**SOLUTION**

As usual, let us solve this synthesis problem by working backwards. The first question we must ask is, Do we prepare *p*-chloronitrobenzene by nitrating chlorobenzene or chlorinating nitrobenzene?

$$chlorobenzene + HNO_3 \xrightarrow{H_2SO_4}$$

or

$$nitrobenzene + Cl_2 \xrightarrow{FeCl_3}$$

Attempting this synthesis by reacting nitrobenzene with chlorine in the presence of ferric chloride (the second option) leads to disaster. Since the nitro group is a meta director, mostly *m*-chloronitrobenzene is produced; very little para isomer is obtained. We are left with the first option, and sure enough, because the chloro group is an ortho-para director, reacting chlorobenzene with nitric acid (with a sulfuric acid catalyst) produces a mixture of *o*- and *p*-nitrochlorobenzenes in good yield with very little meta isomer formed. (Note that by-products are not shown in the syntheses. Assume that when a mixture of ortho and para isomers is produced, these isomers can easily be separated in good yield.)

$$\underset{}{\overset{Cl}{\bigcirc}} \xrightarrow[H_2SO_4]{HNO_3} \underset{NO_2}{\overset{Cl}{\bigcirc}}$$

*p*-Chloronitrobenzene

Chlorobenzene is prepared by reacting benzene with chlorine in the presence of a ferric chloride ($FeCl_3$) catalyst.

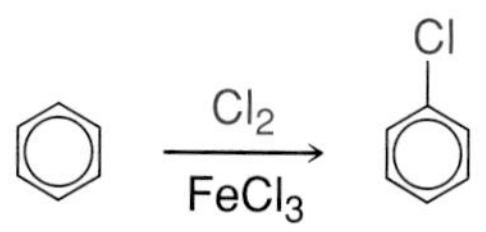

Putting the synthesis all together, we have

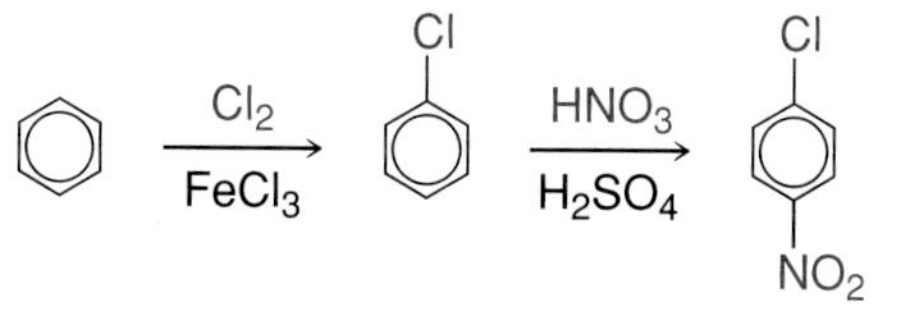

*p*-Chloronitrobenzene

If we choose the second option to prepare *p*-chloronitrobenzene, we would produce *m*-chloronitrobenzene as follows:

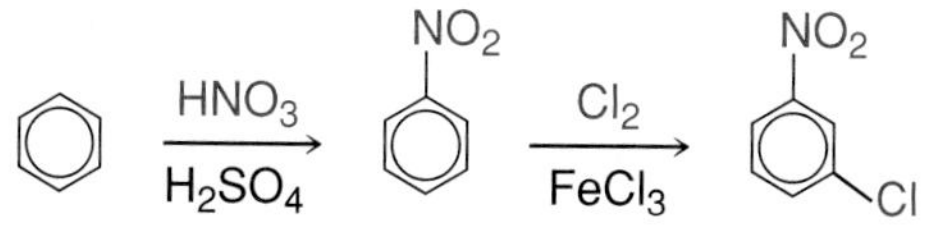

*m*-Chloronitrobenzene

It is important that you know the directing properties of the various groups in Table 10.3 in order to solve problems like Solved Problem 10.10.

---

Prepare *p*-bromobenzyl chloride from benzene and methyl chloride ($CH_3Cl$). You may use any inorganic reagents necessary.

**SOLVED PROBLEM
10.11**

**SOLUTION**
Let us solve this problem by working backwards. First, prepare *p*-bromobenzyl chloride by exposing a mixture of *p*-bromotoluene and chlorine to ultraviolet light.

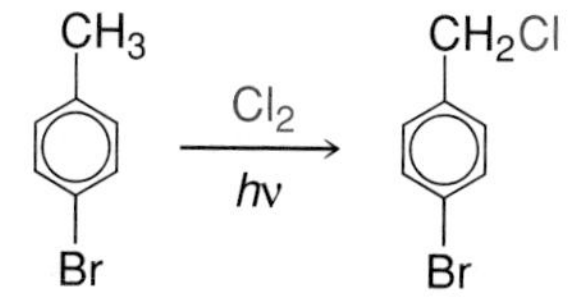

In turn, *p*-bromotoluene is made by reacting toluene (containing the methyl group, an ortho-para director) with bromine in the presence of ferric bromide ($FeBr_3$), a catalyst.

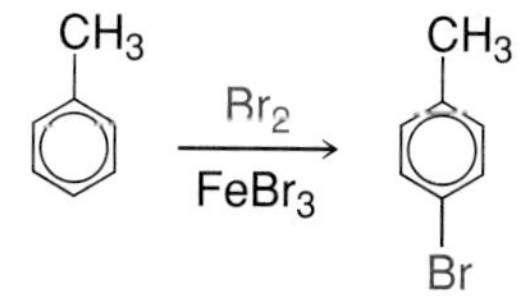

Finally, to prepare toluene, benzene is reacted with methyl chloride ($CH_3Cl$) and a catalyst, aluminum chloride ($AlCl_3$).

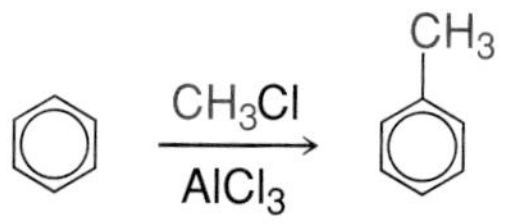

## HISTORICAL BOXED READING 10.2

2,4,6-Trinitrotoluene (TNT) is an excellent military explosive. First prepared from toluene by nitration in 1863 (J. Wilbrand), it has been used widely as an explosive in mines, grenades, bombs, and artillary shells since 1904.

The compound is a colorless crystalline substance that melts at 82°C. TNT is the military explosive of choice because it does not explode spontaneously below 240°C and is resis-

tant to shock. This means that it can be melted with steam and the liquid poured into shell, mine, or bomb casings with ease—without concern that the compound would explode spontaneously during the process.

1,3,5-Trinitrobenzene is a useful explosive, but reaction conditions are too extreme for an efficient preparation from benzene.

Putting these reactions all together, we get

---

### EXERCISE 10.11

Prepare each of the following compounds. You may use any inorganic reagents necessary.

a.  *o*-chlorobenzoic acid from benzene and methyl chloride

b.  *m*-bromonitrobenzene from benzene

c.  benzenesulfonic acid from benzene

d.  butyrophenone ($C_6H_5$—$\overset{\overset{\displaystyle O}{\|}}{C}$—$CH_2CH_2CH_3$) from benzene and any other organic reagent necessary

e.  *n*-butylbenzene from benzene and any other organic reagent necessary

---

## ► CHAPTER ACCOMPLISHMENTS

### 10.1  Substitution Reactions of Benzene: Electrophilic Aromatic Substitution

☐ Write a detailed step-by-step mechanism for each of the electrophilic aromatic substitution reactions given in Section 10.1.

☐ Name the electrophilic aromatic substitution reaction in which a rearrangement takes place.

☐ Write a formula for the hydride ion.

☐ Explain why vinyl chloride cannot be used in the Friedel-Crafts alkylation reaction.

### 10.2  Rate of Aromatic Substitution

☐ Explain the difference between a $+I$ effect and a $-I$ effect.

☐ State whether each of the following groups is either activating or deactivating.

   a. —OH
   b. —$SO_3H$
   c. —COOH
   d. —COOR
   e. —CHO
   f. —$NH_2$

☐ Draw the resonance structures that contribute to the fact that the nitro group in nitrobenzene is highly deactivating.

☐ Explain why the —OH group has been shown by experiment to be highly activating yet is deactivating by means of the inductive effect.

### 10.3  Orientation of an Electrophile in Aromatic Substitution to Produce a Disubstituted Benzene Derivative

☐ List the ortho-para directors that are deactivating.

☐ Draw the four resonance structures obtained when $E^+$ attacks aniline in the ortho position. Indicate the resonance structure that is particularly stable.

☐ Draw the three resonance structures obtained when $E^+$ attacks aniline in the meta position.

☐ State Aromatic Orientation Rule 1.

☐ Draw the three resonance structures obtained when $E^+$ attacks benzaldehyde in the ortho position. Indicate the resonance structure that is particularly unstable.

☐ Draw the three resonance structures obtained when $E^+$ attacks benzaldehyde in the meta position.

☐ State Aromatic Orientation Rule 2.

## 10.4 Orientation in Aromatic Substitution to Produce a Trisubstituted Benzene Derivative

☐ Predict the product(s) formed from the nitration of

a. *p*-nitrophenol.

b. *p*-methylphenol.

c. *m*-nitrotoluene.

## 10.5 Reduction

☐ Complete the following equations by supplying a structural formula for each organic product.

a. benzene $+ 3\,H_2 \xrightarrow{\text{Rh-C}}$

b. nitrobenzene $+ \text{Sn} + \text{HCl} \longrightarrow$

c. product (b) $+ \text{NaOH(aq)} \longrightarrow$

## 10.6 Side-Chain Halogenation

☐ Complete the following equations by supplying a structural formula for each organic product.

a. benzyl chloride $+ \text{NaOH}(aq) \longrightarrow$

b. propylbenzene $+ \text{Br}_2 \xrightarrow{hv}$

c. propylbenzene $+ \text{Cl}_2 \xrightarrow{hv}$

## 10.7 Side-Chain Oxidation

☐ Explain why potassium benzoate is the principal product obtained when both propylbenzene and toluene are heated with a basic aqueous solution of $\text{KMnO}_4$.

☐ Name an arene that does not react when heated with a basic aqueous solution of $\text{KMnO}_4$.

## 10.8 Organic Synthesis with Aromatic Compounds

☐ Prepare *m*-chloronitrobenzene from benzene and any necessary inorganic reagents.

☐ Prepare *p*-bromobenzyl chloride from benzene, methyl chloride, and any necessary inorganic reagents.

---

## ▶ KEY TERMS

*nucleophilic reagent* (10.1, 12.1)
*electrophilic aromatic substitution* (10.1)
*arenium ion* (10.1)

*oxonium ion* (10.1, 12.1, 12.2)
*Clemmensen reduction* (10.1)
*activating group* (10.2)
*deactivating group* (10.2)

*resonance effect* (10.2)
*Aromatic Orientation Rule 1* (10.3)
*Aromatic Orientation Rule 2* (10.3)
*steric hindrance* (10.4, 12.1)

---

## ▶ PROBLEMS

1. Draw stabilizing resonance structures for

2. Write a detailed, step-by-step mechanism for the following.

a.

b.

c.

3. An arenium ion is allylic. Explain.

4. Prepare *n*-penylbenzene from benzene. You may use any organic and inorganic reagents necessary.

5. Is HF (hydrogen fluoride) a catalyst in the reaction given in Exercise 10.1a? Explain.

6. Draw the structure of the carbocation obtained by a hydride ion rearrangement in each of the following, when possible. Indicate if rearrangement does not occur and explain.

a. $CH_3CH_2CH_2\overset{+}{C}H_2 \longrightarrow$

b. $CH_3-\underset{\underset{H}{|}}{\overset{\overset{CH_3}{|}}{C}}-\underset{\underset{H}{|}}{\overset{\overset{H}{|}}{C}}{}^+ \longrightarrow$

c. $(CH_3)_3C^+ \longrightarrow$

7. Explain why the following Friedel-Crafts reaction does not occur.

8. In the sulfonation of benzene, sulfur trioxide is a base. Explain.

9. What compound is the catalyst in the nitration of benzene? Explain.

10. Draw a structural formula of the electrophile(s) that reacts with benzene in each of the following electrophilic aromatic substitutions.
    a. a sulfonation
    b. a Friedel-Crafts alkylation where the organic halide is $CH_3CH_2CH_2Br$
    c. a nitration

11. Complete the following equations by writing a structural formula for each organic product and a molecular formula or formula unit for each inorganic product.
    a. benzene + $CH_3CH_2CH_2CH_2Cl$ $\xrightarrow{AlCl_3}$
    b. benzene + 1-butene $\xrightarrow{HF}$
    c. benzene + $Cl_2$ $\xrightarrow{FeCl_3}$

12. Which of the following resonance structures is the major contributor and thus more stable? Explain.

13. Consider the following possible reaction of an arenium ion. Why does it not occur?

14. Draw a resonance structure for the acylium ion.

    Which resonance structure is the major contributor and thus the more stable? Explain.

15. Draw resonance structures to illustrate the activating resonance effect of anisole ($C_6H_5$—$OCH_3$).

16. Draw resonance structures, using nitrobenzene (page 192) as a guide, to illustrate the deactivating resonance effect of the carboxyl group (—COOH) in benzoic acid.

17. Draw resonance structures to illustrate the activating resonance effect of the amino group (—$NH_2$) in aniline.

18. The amino group is deactivating by means of the inductive effect, yet the group has been shown by experiment to be highly activating. Explain.

19. Draw resonance structures to show that the group

    in benzamide ($C_6H_5$—$CONH_2$) is deactivating.

20. Which compound undergoes nitration at a faster rate?
    a. benzene or bromobenzene
    b. styrene or phenol
    c. toluene or nitrobenzene
    d. benzaldehyde or acetanilide ($C_6H_5NHCOCH_3$)

21. Which of the following compounds react with $Cl_2$ and $FeCl_3$ more rapidly than benzene and which react less rapidly? Explain.

22. State both qualitative aromatic orientation rules to predict the direction of orientation of an electrophile controlled by a group on the benzene ring.

23. Use the aromatic orientation rules to predict the direction of orientation for each of the following groups.

    a.
    b. —$OCH_2CH_3$
    c.
    d. —Br
    e.

24. Draw the three resonance structures obtained when $E^+$ attacks phenol in the meta position.

25. Draw resonance structures to show that
    a. the nitroso group (—$\ddot{N}$=O) in nitrosobenzene is an ortho-para director.
    b. the —C—$CH_3$ group in acetophenone is a meta director

26. Which resonance structure is more stable? Explain.

a. 
:OCH$_3$ / Br (structure 1)  or  :OCH$_3$ / Br (structure 2)

   1                           2

b. COOH / NO$_2$ (structure 3)  or  COOH / NO$_2$ (structure 4)

   3                              4

27. Draw an expanded structural formula for structure X to convince yourself that the carbon bearing a positive charge only has a sextet of electrons.

28. Consider the reaction of toluene with CH$_3$Cl and a catalytic amount of AlCl$_3$.
    a. Write an equation for the reaction.
    b. Do you think the organic products of the reaction would react with CH$_3$Cl (AlCl$_3$) at a slower or faster rate than toluene? Explain. This represents a real weakness of Freidel-Crafts alkylation.

29. Which of the reactions in Exercise 10.9 would you expect to be the slowest? Why?

30. Complete the following equations by writing a structural formula for each organic product formed and, when appropriate, a molecular formula or formula unit for each inorganic product formed.

a. C≡N (benzonitrile) + Br$_2$ $\xrightarrow{\text{FeBr}_3}$

b. OH / NO$_2$ (p-nitrophenol) + HNO$_3$ $\xrightarrow{\text{H}_2\text{SO}_4}$

c. CH$_3$ (toluene) + CH$_3$CH$_2$CH$_2$Cl $\xrightarrow{\text{AlCl}_3}$

d. CH$_3$ (toluene) + (CH$_3$)$_2$CHCl $\xrightarrow{\text{AlCl}_3}$

e. CH$_2$CH$_3$ (ethylbenzene) $\xrightarrow[\Delta]{\text{KMnO}_4, \ ^-\text{OH}}$

f. CH$_2$CH$_3$ (ethylbenzene) + Cl$_2$ $\xrightarrow{\Delta}$

g. NHCOCH$_3$ (benzene ring) + CH$_3$–C(=O)–Cl $\xrightarrow[\Delta]{\text{AlCl}_3}$

h. Cl / NO$_2$ (benzene ring) + Cl$_2$ $\xrightarrow{\text{FeCl}_3}$

i. CH$_2$CH$_3$ (ethylbenzene) + 3 H$_2$ $\xrightarrow{\text{Rh-C}}$

j. CH$_2$CH$_3$ (ethylbenzene) + SO$_3$ $\xrightarrow{\text{H}_2\text{SO}_4}$

k. Br / Br (dibromobenzene) + CH$_3$Cl $\xrightarrow{\text{AlCl}_3}$

l. CH$_3$ / NO$_2$ (o-nitrotoluene) + Sn + HCl $\longrightarrow$

m. product of part l + NaOH(aq) $\longrightarrow$

n. CCl$_3$ (naphthalene ring) + NaOH(aq) $\longrightarrow$

o. product of part n + HCl $\longrightarrow$

31. An alkene with three carbon-carbon isolated double bonds reacts with hydrogen and finely divided platinum under far milder conditions of temperature and pressure than an aromatic compound such as toluene. Explain.

32. An arene, C$_9$H$_{12}$, produced benzoic acid on reaction with basic potassium permanganate and then hydrochloric acid. Halogenating the arene with excess bromine followed by hydrolysis gave propiophenone (C$_6$H$_5$—COCH$_2$CH$_3$). Deduce a structural formula for the arene, and show your reasoning.

33. A disubstituted benzene derivative, C$_6$H$_4$Br$_2$, on nitration produced three isomers of C$_6$H$_3$Br$_2$NO$_2$. Draw a structural formula for the dibrombenzene. Show your reasoning.

34. a. 2,4,6-Trinitrotoluene (TNT) is a toxic high explosive used as a filler in artillery shells. Prepare TNT from benzene, CH$_3$Cl, and any inorganic reagents necessary.
    b. 1,3,5-Trinitrobenzene (TNB) is a better explosive than TNT in that TNB produces more shattering in an explosion than the same amount of TNT. Yet TNT is by far the more often used explosive. Explain.

35. Prepare each of the following from toluene. Use any inorganic reagents necessary.
    a. benzaldehyde
    b. benzoic acid
    c. *p*-toluidine (*p*-H₂N—C₆H₄—CH₃)
    d. *p*-chlorobenzyl bromide

36. Prepare each of the following from benzene. Use any organic and inorganic reagents necessary.

    a.

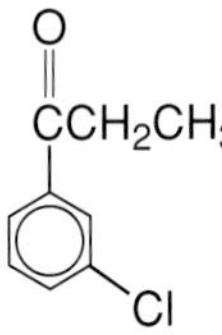

    c.

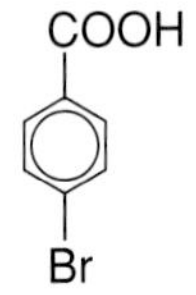

    b.

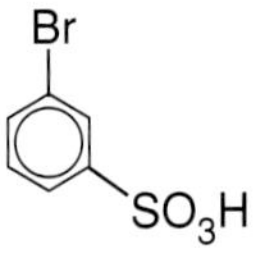

    d.

    e. 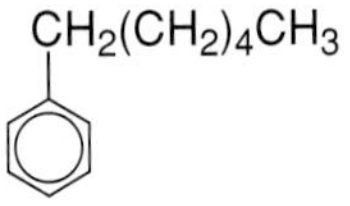 CH₂(CH₂)₄CH₃

37. Nitrobenzene does not undergo a Friedel-Crafts alkylation. As a matter of fact, nitrobenzene is sometimes used as a solvent in Friedel-Crafts alkylation reactions. Explain.

38. To explain why an overwhelming amount of α-product is produced in the side-chain chlorination of *n*-propylbenzene,
    a. write a detailed, step-by-step mechanism for the following reaction:

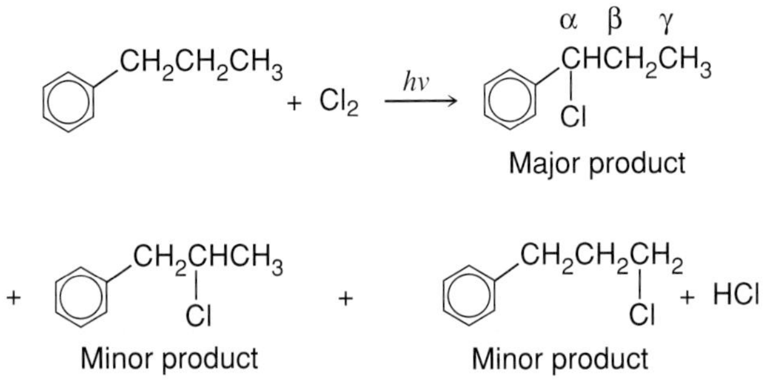

    Note that the mechanism of side-chain halogenation of an arene is the same as that for halogenation of an alkane.
    b. Why is C₆H₅—ĊHCH₂CH₃ a more stable species than C₆H₅—CH₂ĊHCH₃? Explain using appropriate resonance structures.

# CHAPTER 11

# Optical Isomerism

Let us summarize the various classes of isomers we have encountered up to this point in the textbook:

1. *Structural isomers:* Two or more compounds with the same molecular formula but different structural formulas and therefore different physical properties.

   a. *Functional-group isomers:* Two or more compounds with the same molecular formula that are represented by different functional groups. Consider the examples in Fig. 11.1.

   b. *Positional isomers:* Two or more compounds that belong to the same family (contain the same functional group) but where the functional group is either bonded to different nonequivalent carbons in the parent chain of the molecule or is located on different carbons of the parent chain. Consider the examples in Fig. 11.2.

2. *Stereoisomers:* Two or more compounds with the same molecular and structural formulas but which differ in how the atoms are oriented in space. In other words, stereoisomers emphasize the three-dimensional nature of organic structures.

   a. *Cis-trans isomers:* One stereoisomer (cis) has two identical (or different) groups on the same side of the molecule; the other (trans), on opposite sides of the molecule. Consider the examples in Fig. 11.3.

   b. *Optical isomers:* We will discuss another type of stereoisomerism at this point called *optical isomerism*. In this chapter we will learn to identify an *optical isomer*

## 11.1 INTRODUCTION

**FIGURE 11.1** Two functional group isomers.

$$CH_3CH_2OCH_2CH_3$$
Diethyl ether
(an ether)

$$CH_3CH_2CH_2CH_2OH$$
*n*-Butyl alcohol
(an alcohol)

*Stereoisomerism* is the study of stereoisomers. Cis-trans isomerism, previously studied in Section 5.3, is an example of stereoisomerism.

**FIGURE 11.2** Two sets of positional isomers.

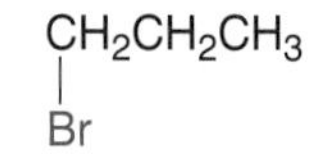

$$\overset{CH_2CH_2CH_3}{\underset{Br}{|}}$$
1-Bromopropane

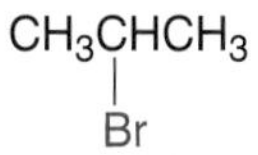

$$\overset{CH_3CHCH_3}{\underset{Br}{|}}$$
2-Bromopropane

$$CH_3CH_2C{\equiv}CH$$
1-Butyne

$$CH_3C{\equiv}CCH_3$$
2-Butyne

**FIGURE 11.3** Two sets of *cis-trans* isomers.

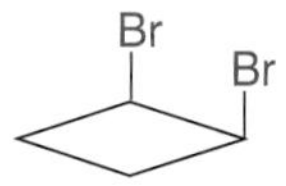

*cis*-1,2-Dibromocyclobutane 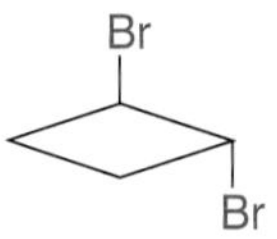      *trans*-1,2-Dibromocyclobutane

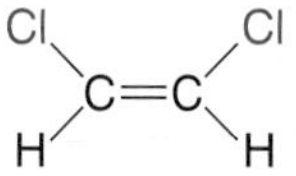

*cis*-1,2-Dichloroethene 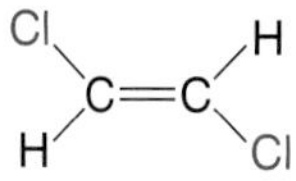      *trans*-1,2-Dichloroethene

**FIGURE 11.4** Handedness in humans and molecules.

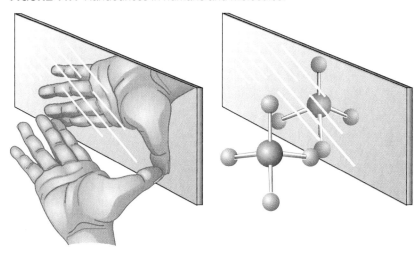

based on the structural formula of the molecule and develop the relationship between optical isomerism and light.

Optical isomerism is usually associated with an $sp^3$ hybridized, saturated, tetrahedral carbon. For our purposes, an $sp^2$ or $sp$ hybridized carbon cannot exhibit optical isomerism. Thus neither ethylene (a planar molecule) nor acetylene (a linear molecule) can show optical isomerism.

## 11.2
## THE TETRAHEDRAL CARBON ATOM AND CHIRALITY

**FIGURE 11.5** Nonsuperimposability of chiral mirror images.

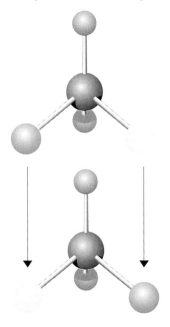

In order to exhibit optical isomerism, a molecule must be *chiral* (from the Greek meaning "hand"). That is, two chiral molecules are related to each other as the left hand is to the right hand. The two molecules, like your two hands, are mirror images of each other (Fig. 11.4). In addition, the molecules, like your hands, are not superimposable on their mirror images and therefore are not identical. Take your left hand, palm down, and place your right hand, palm down, on it. You can see the thumb of your left hand fits under the little finger of your right hand while the little finger of your left hand fits under the thumb of your right hand. Thus your hands are not superimposable images and are not identical.

The nonsuperimposablity of chiral molecules works in the same way (Fig. 11.5). Placing structure II under structure I in the figure shows that atom d of structure I corresponds to atom b of structure II, while atom b of structure I corresponds to atom d of structure II. Thus these chiral molecules are not identical and represent a pair of optical isomers.

Not every tetrahedral carbon-centered molecule is chiral. To be chiral, a molecule must lack a *plane of symmetry*. Consider two molecules:

$$
\begin{array}{cc}
\begin{array}{c} a \\ | \\ d-C-b \\ | \\ a \end{array}
&
\begin{array}{c} a \\ | \\ d-C-b \\ | \\ x \end{array}
\\
\text{III} & \text{IV} \\
\text{Achiral (not chiral)} & \text{Chiral}
\end{array}
$$

Molecule III is achiral because it contains the plane of symmetry through d—C—b (Fig. 11.6). A plane of symmetry exists in a molecule when an imaginary line drawn through the molecule creates two halves that are mirror images. Since molecule IV does not show a plane of symmetry, it is chiral.

**FIGURE 11.6** Superimposability of selected achiral mirror images.

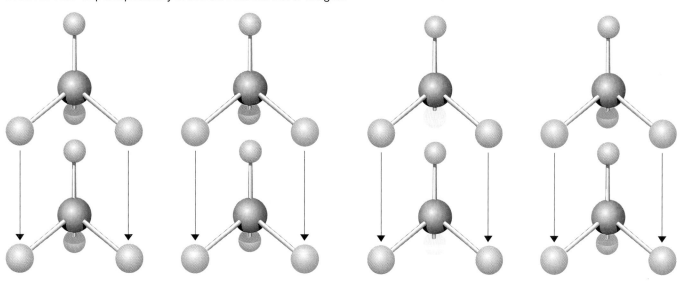

The asymmetrical carbon atom in a chiral molecule is known as a *chiral center*. A molecule with one chiral center is chiral when that chiral center is bonded to four different groups (Cabdx). Only this type of grouping (in this textbook*) represents a chiral molecule, which is not superimposable on its mirror image and which lacks a plane of symmetry. Thus $Ca_3x$, $Ca_2x_2$, $Ca_2xb$, and $Cax_3$ are *achiral*—not chiral (Fig. 11.6). These molecules exhibit superimposable mirror images and do not show optical isomerism.

Every molecule with one chiral center is chiral. However, we will learn later that some molecules with two or more chiral centers may be achiral. Two optical isomers, one of which is the nonsuperimposable mirror image of the other, are known as *enantiomers*. The chiral carbon atom(s) in an enantiomer are often designated by an asterisk.

$$CH_3-\overset{\overset{\displaystyle OH}{|}}{\underset{\underset{\displaystyle H}{|}}{C}}{}^{*}-CH_2CH_3 \qquad CH_3-\overset{\overset{\displaystyle OH}{|}}{\underset{\underset{\displaystyle H}{|}}{C}}{}^{*}-\overset{\overset{\displaystyle Br}{|}}{\underset{\underset{\displaystyle H}{|}}{C}}{}^{*}-C_2H_5 \qquad I-\overset{\overset{\displaystyle H}{|}}{\underset{\underset{\displaystyle Br}{|}}{C}}{}^{*}-Cl$$

**SOLVED PROBLEM 11.1**

Draw a structural formula for each enantiomer of bromochloroiodomethane as mirror images.

**SOLUTION**

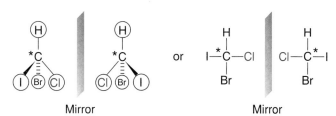

Mirror          Mirror

Since we are given that bromochloroiodomethane exists as a pair of enantiomers, let us place an asterisk next to the chiral carbon in each enantiomer.

---

*In actuality, there exist a number of chiral compounds that do not contain a chiral carbon atom. For example, substituted allenes (RCH=C=CHR) are chiral molecules.

**SOLVED PROBLEM 11.2**    Place an asterisk next to every chiral center in the following structural formulas. If no chiral center is found, write "no chiral center is present."

a.  glucose

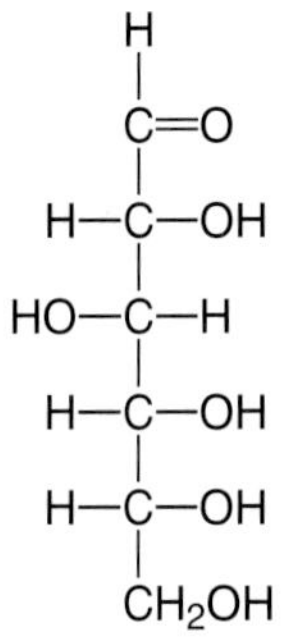

b.  lactic acid

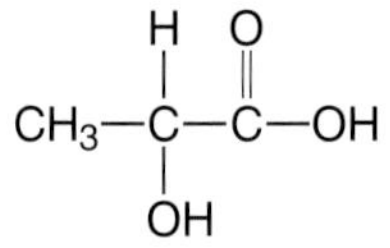

c.  ethylbenzene

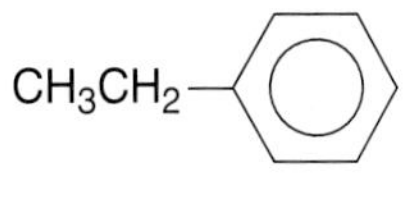

d.  1-bromo-2-chlorocyclohexane

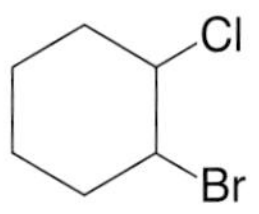

**SOLUTION**

a.

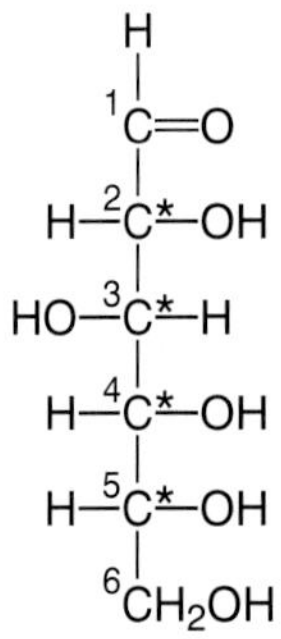

b.

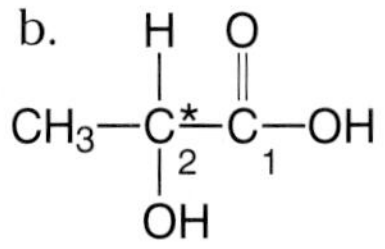

c.

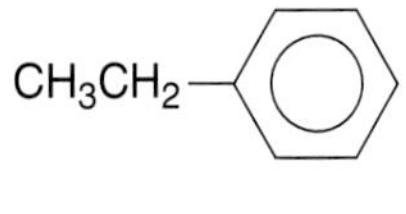

d.

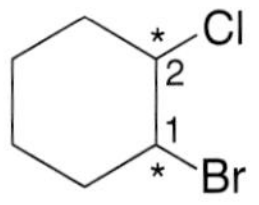

a.  Carbon 1 in glucose is not a chiral center because of the presence of the double bond attached to it. Carbons 2, 3, 4, and 5 are chiral centers because each carbon is bonded to four different groups. For example, carbon 2 is attached to the following different groups: —H, —OH, —CHO, —[CH(OH)]$_3$CH$_2$OH. Carbon 6 is achiral because of the bonding of two hydrogens to that carbon.

b.  Carbon 2 is chiral. Therefore, the lactic acid molecule is chiral.

c.  No chiral center is present. Each carbon of the phenyl group is not a chiral center because of the presence of a carbon-carbon double bond in a Kekulé structure. The two carbons of the ethyl group are not chiral centers because of the presence of two or more bonded hydrogens to each carbon. Therefore, the ethylbenzene molecule is achiral.

d.  Carbons 1 and 2 in 1-bromo-2-chlorocyclohexane are chiral centers. Carbon 1 is bonded to —Br, —H, —(CH$_2$)$_4$CH(Cl)—, and —CH(Cl)(CH$_2$)$_4$—, four different groups. In the same way, we have the following different groups bonded to carbon 2: —Cl, —H, —(CH$_2$)$_4$CH(Br)—, and —CH(Br)(CH$_2$)$_4$—.

**EXERCISE 11.1**

a.  Draw a line-bond structural formula for each enantiomer of the following as mirror images: lactic acid and *sec*-butyl chloride CH$_3$CH(Cl)CH$_2$CH$_3$

b.  Demonstrate in your own mind that ethylbenzene (Solved Problem 11.2c) is achiral by drawing a line-bond expanded structural formula. You should observe that no carbon in the structure is bonded to four different groups.

---

Place an asterisk next to every chiral center in the following. If no chiral center is found, write "no chiral center is present."

a.

$$
\begin{array}{c}
O \\
\| \\
C-H \\
| \\
H-C-OH \\
| \\
H-C-OH \\
| \\
CH_2OH
\end{array}
$$

Erythrose, a sugar

b.

$$
\begin{array}{c}
OH \quad H \\
| \quad\;\; | \\
CH_3-C-C-CH_3 \\
| \quad\;\; | \\
H \quad H
\end{array}
$$

sec-Butyl alcohol

c.  cumene (isopropylbenzene)

d. 1-bromo-2-chlorocyclopentane

---

Let us summarize what has been brought out thus far:

1.  A tetrahedral molecule displays optical isomerism when the molecule is chiral.

2.  To be chiral, a molecule must contain at least one chiral center. A chiral center is a carbon atom in the molecule that is bonded to four different groups (or atoms). A molecule of the Cabdx type does not show mirror-image superimposability (identical mirror images) and does not exhibit a plane of symmetry.

3.  Chiral molecules display optical isomerism. Optical isomers exist in pairs called *enantiomers*, one of which is the nonsuperimposable mirror image of the other.

Now that we see the relationship between chirality and structure, let us explore the experimental measurement of chirality. Since polarized light plays a vital part in this measurement, let us first discuss plane-polarized light.

## 11.3 PLANE-POLARIZED LIGHT

White light consists of waves with a variety of wavelengths that vibrate on all planes perpendicular to the plane of propagation, i.e., the direction the light comes from. When light is composed of waves with only one wavelength that vibrate on all planes, the light is said to be *monochromatic*. Monochromatic light can be produced either by passing white light through one or more filters or by means of a vapor lamp.

Now, when white light or monochromatic light travels through a sheet of *Polaroid*, *plane-polarized light* is produced that vibrates only on one plane perpendicular to the direction the light comes from (Fig. 11.7). Polaroid is made of a complex organic substance

---

**FIGURE 11.7** Generation of plane-polarized light.

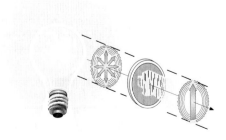

**FIGURE 11.8** A typical polarimeter.

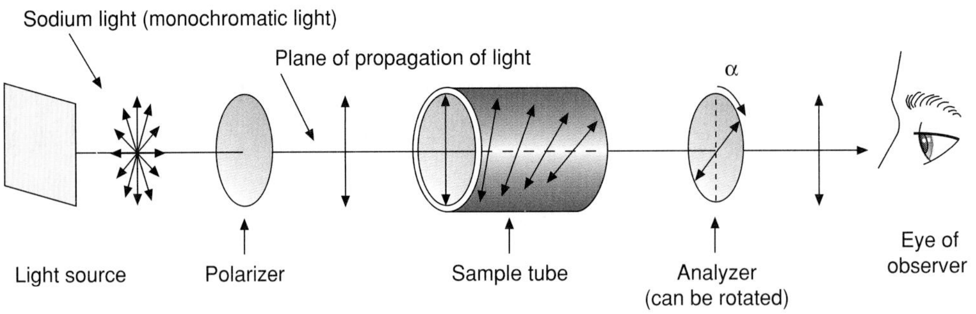

Sodium light (monochromatic light)

Plane of propagation of light

$\alpha$

Light source        Polarizer              Sample tube           Analyzer          Eye of
                                                                (can be rotated)    observer

that has the capacity to reduce glare by producing plane-polarized light when incorporated into the lenses of a pair of eyeglasses or a car window. In addition, when incorporated into a sheet of plastic, it makes a convenient source of polarized light for laboratory use. Molecules of a chiral compound rotate the plane of polarized light a set number of degrees as measured in an instrument called a *polarimeter*; molecules of an achiral compound do not.

In the polarimeter (Fig. 11.8), a beam of monochromatic light passes through a disk of Polaroid (called the *polarizer*), polarizing the light. This plane-polarized light passes through a tube containing the chiral substance that rotates the plane of polarized light. This rotated plane-polarized light passes, in turn, through another Polaroid disk known as the *analyzer*, which is rotated a set number of degrees so that the light is vibrating along the same plane it was before it reached the sample tube. The number of degrees the analyzer rotated is read on a scale by the observer and is called the *observed optical rotation* (abbreviated $\alpha$) of the sample. The sample is said to be *optically active*.

Besides being used to measure the magnitude of rotation (in degrees), the polarimeter is also employed to determine the direction of rotation of the sample. Should an optically active substance rotate the plane of polarized light to the right, it is *dextrorotary*; to the left, *levorotary*. The symbol $(+)$ is used for dextrorotary; $(-)$ for levorotary. For example, dextrorotary lactic acid is abbreviated as $(+)$-lactic acid, whereas levorotary camphene is written as $(-)$-camphene.

## 11.4 SPECIFIC ROTATION

The observed rotation of a given compound often varies from experiment to experiment because observed rotation depends on the concentration of the optically active substance, the length of the sample tube, and the wavelength of the light used. For example, a rotation of $-1.13$ degrees can be observed for a sample of $(-)$-lactic acid present in a 2.5-dm-long tube at a concentration of 0.2 g/ml. We define *specific rotation* as the rotation where the product of the tube length (in decimeters) and the concentration (in grams per milliliter), $(l \cdot C) = 1.0$ dm $\cdot$ g/ml. In this case, the product is only 0.5 dm $\cdot$ g/ml, so we calculate the specific rotation $[\alpha]$ as

$$[\alpha] = \alpha/(l \cdot C) = \frac{-1.13}{2.5 \text{ dm} \cdot 0.2 \text{ g/ml}} = -2.26 \text{ degrees} \cdot \text{ml}/(\text{dm} \cdot \text{g})$$

This intensive property is characteristic of the chemical substance measured, in this case $(-)$-lactic acid, and can help to identify it.

The observed rotation is usually taken at room temperature using a sodium lamp that emits light at 5896 Å ($1 \times 10^8$ Å = 1 cm), a wavelength known as the D line of sodium. Using the following relationship, the specific rotation can be calculated:

$$[\alpha] = \frac{\alpha}{l \cdot C}$$

where $[\alpha]$ is the specific rotation of the compound in degrees, $\alpha$ is the observed rotation in degrees, $l$ is the length of the sample tube in decimeters (1 dm = 10 cm), and $C$ is the concentration of the sample in grams per milliliter (or the density of a pure liquid). Let us assume in this textbook that every polarimetric determination is made using the sodium lamp at room temperature.

The specific rotation $[\alpha]$ of $(+)$-*sec*-butylamine is $+7.4$ degrees. Calculate the observed rotation of a sample of $(+)$-*sec*-butylamine (density $= 0.72$ g/ml) in a 2.0-dm sample tube.

**SOLUTION**
Substituting in the expression for $[\alpha]$, we get

$$+7.4 = \frac{\alpha}{(2.0)(0.72)}$$
$$\alpha = +10.7 \text{ degrees} \quad (+11 \text{ degrees to two significant figures})$$

Cholesterol, a water-insoluble lipid that is found throughout our bodies, was observed to have a rotation of -0.63 degrees when 1.0 g cholesterol was dissolved in 50 ml ethyl alcohol ($C_2H_5OH$) and the resulting solution placed in a polarimeter tube with a length of 10.0 cm. Calculate the specific rotation $[\alpha]$ of cholesterol.

## 11.5 PROPERTIES OF ENANTIOMERS: THE RACEMIC MIXTURE

Each enantiomer of a pair has the same physical properties as the other with one exception—specific rotation. (Note that the slight difference in density between enantiomers in Table 11.1 is due to experimental error.) The enantiomers show the same magnitude of rotation but differ in the direction of rotation. Consider the stereoisomeric forms of *sec*-butylamine [$CH_3CH(NH_2)CH_2CH_3$] in Table 11.1.

When equal amounts of two enantiomers are intimately mixed, the resulting mixture is called a *racemic mixture*, or *racemate*. The racemic mixture is not optically active because the optical rotation of every molecule of $(+)$-isomer is canceled by the optical rotation of a molecule of $(-)$-isomer to give a net optical rotation of zero. A racemic mixture is designated by the prefix $(\pm)$. Thus $(\pm)$-*sec*-butylamine represents a racemic mixture (racemate). The physical properties of a racemic mixture usually differ from those of the $(+)$ and $(-)$ enantiomers (see Table 11.1).

## 11.6 THE (R)/(S) METHOD OF WRITING CONFIGURATIONS (ARRANGEMENTS IN SPACE) OF MOLECULES

Consider wedge projections of the two enantiomers of *sec*-butylamine:

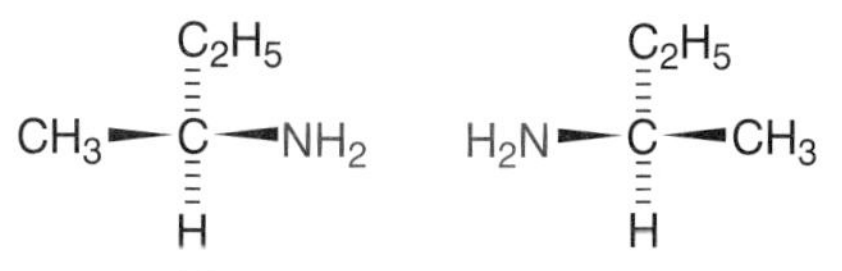

The solid triangles in structures V and VI mean that the bonded groups are oriented toward the reader from the plane of the paper; the dashed lines show that the bonded

| TABLE 11.1 | Physical Properties of the Stereoisomers of sec-Butylamine | | | |
|---|---|---|---|---|
| | Boiling Point (°C) (at 760 mmHg) | Density (g/ml) | Refractive Index (20°C) | [α] (degrees) |
| (+)-isomer | 63 | 0.724 | 1.344 | +7.4 |
| (−)-isomer | 63 | 0.721 | 1.344 | −7.4 |
| Racemic mixture | 63 | 0.725 | 1.393 | 0.0 |

groups are oriented away from the reader from the plane of the paper. These orientations reflect the tetrahedral geometry of the molecules.

A convenient method of identifying the configuration of structures V and VI is the rectus-sinister method. This method uses the same rules of priority ranking as are used in Section 5.3 to determine the ($Z$) (zusammen)–($E$) (entgegen) configuration of an alkene as follows:

1. When the four atoms directly attached to a chiral carbon are different, the atom with the highest atomic number is assigned the highest priority.

2. If the four atoms directly attached to the chiral carbon are the same, then the atomic number(s) of the next atom(s) farther away from the chiral carbon will decide priority.

Let us add one more rule.

3. If a carbon atom is attached to a group that contains a double or triple bond, the priority of that group is assigned so as to break the multiple bond into single-bond units. Thus:

to establish priorities.

To specify the configuration of a molecule, the group of lowest priority must be oriented away from the reader. Then, the remaining three groups are arranged in descending order of priority. If this arrangement occurs in a clockwise direction, the

---

**HISTORICAL BOXED READING 11.1**

Just as the concept of handedness exists for certain organic molecules, so does handedness play a role in the affairs of human beings. About 10% of all Americans are left-handed; i.e., they bat, throw a ball, eat, and write with the left hand. For years they have struggled with scissors, wrenches, and can openers designed for right-handed people and have twisted their bodies across the horrible college and university student chairs to be able to write.

If you think this isn't fair, take a look at the definition of left-handed in the *Webster's Encyclopedia of Dictionaries*, "using the left hand more easily than the right, awkward." Hardly complimentary.

It gets worse if you dig deeper. The Latin word for left is *sinister*. Again, from Webster, for sinister: "evil-looking, unlucky (left being regarded as the unlucky side)." This prejudice extends to almost all languages. In Italian, the word for left, *macino*, means "deceitful"; in German, *linkisch* is "awkward", *na levo* is "sneaky" in Russian, and to cap it all, *zurdo*, in Spanish, means "malicious."

You guessed it, to be right-handed is exactly the opposite—brave, noble, and powerful. A "right hand man" is a trusted assistant.

Why does this prejudice exist? It started with the Bible and got worse and worse with time. Episodes such as the following certainly contributed: When fighting knights decided to make peace, they would approach one another with their sword hand (right hand) extended in peace. We know this gesture as a hand shake. Since the right hand was extended, no weapon could be hidden in the more powerful right hand. However, if one of the knights was a lefty, he could certainly hide a dagger or short sword in his powerful left hand, and "zap."

Some of the forms of prejudice are hard to believe. Jack Fincher, in his book *Lefties*[*], reports that Bedouin tribes stipulate that women can only gather in the left side of a tent, while the men get to use the right side. Several years ago in isolated farming areas of Japan, if a married woman was found to be left-handed, this was grounds for divorce.

Left-handers of the world unite. The left-handed pencil sharpener will be right (oops) around the corner.

[*]Jack Fincher, *Lefties*, (New York, G. P. Putnam's Sons, 1980), p. 46.

compound has the ($R$) for *rectus* (Latin: "to the right") configuration; if counterclockwise, the compound has the ($S$) for *sinister* (Latin: "to the left") configuration. Consider compound V:

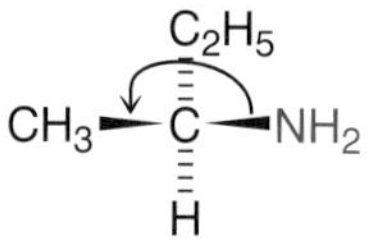

$$CH_3 \overset{C_2H_5}{\underset{H}{\overset{|}{\underset{|}{C}}}} NH_2$$

V

Since the atom of lowest priority, hydrogen, is directed away from you and the order of priority of the remaining groups is —$NH_2$ > —$C_2H_5$ > —$CH_3$, then the decreasing order of priority proceeds in a counterclockwise direction and compound V is ($S$)-*sec*-butylamine. We can show this process using a molecular model as follows:

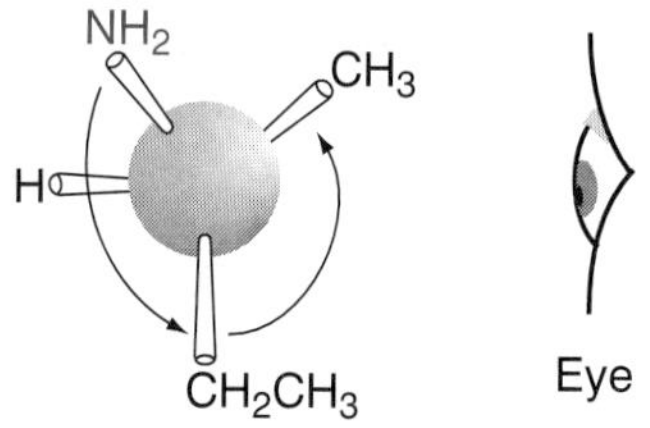

V          Eye

Position the molecular model in a way that places the group of lowest priority, —H, away from you. Observe that the decreasing order of priority of the remaining three groups proceeds in a counterclockwise direction, confirming that the structure is ($S$). Thus compound V is named ($S$)-*sec*-butylamine.

---

a. Show that compound VI is ($R$)-*sec*-butylamine.

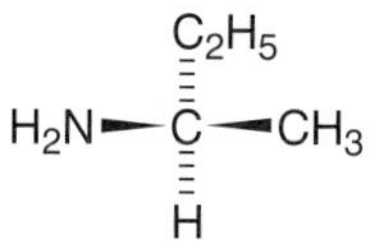

**SOLVED PROBLEM 11.4**

$$H_2N \overset{C_2H_5}{\underset{H}{\overset{|}{\underset{|}{C}}}} CH_3$$

VI

b. Show that compound VII is ($S$)-3-chloro-1-butene.

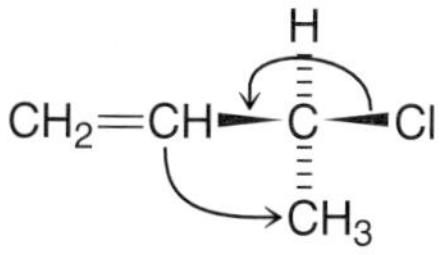

$$CH_2{=}CH \overset{H}{\underset{CH_3}{\overset{|}{\underset{|}{C}}}} Cl$$

VII

**SOLUTION**

a.

$$NH_2 \overset{C_2H_5}{\underset{H}{\overset{|}{\underset{|}{C}}}} CH_3$$

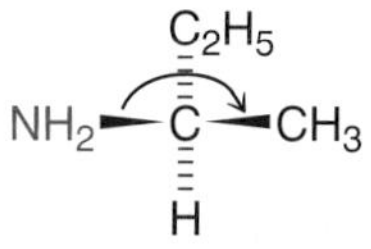

Again, the atom of lowest priority, hydrogen, is directed away from you. Since the order of priority of the remaining groups is —$NH_2$ > —$C_2H_5$ > —$CH_3$, then this order of priority proceeds in a clockwise fashion, and the compound is designated ($R$)-*sec*-butylamine. Rule 1 establishes the higher priority of —$NH_2$ over —$CH_2CH_3$ and —$CH_3$ because the atomic number of nitrogen (seven) is higher than that of carbon (six). Rule 2 confirms the higher priority of —$C_2H_5$ over —$CH_3$. Consider both groups:

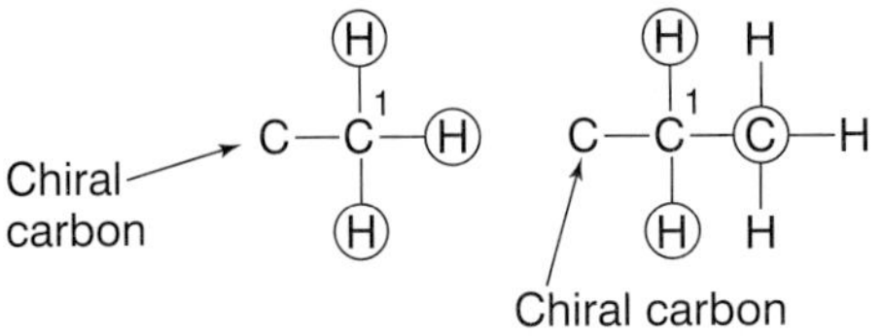

Chiral carbon

Chiral carbon

Since C-1 is directly bonded to the chiral carbon in both groups, we need to consider the atomic numbers of the next atoms further away. For methyl, we have three hydrogens (circled); for ethyl, two hydrogens and a carbon (circled). Since the atomic number of carbon (six) is higher than that of hydrogen (one), the ethyl group demonstrates the higher priority.

b.  Since the group of lowest priority, —H, is projected away from you, we can now consider the relative priorities of the three other groups. The chloro group shows the highest priority. The problem is to decide whether the vinyl group or the methyl group is next highest in priority because a carbon in each group is directly bonded to the chiral carbon. In terms of priority, we have

Since the carbon bonded to the methyl group is, in turn, attached to three hydrogens, while in terms of priority, the carbon of the vinyl group is attached to two carbons and a hydrogen, the vinyl group shows the higher priority. Thus we have, in order of decreasing priority, —Cl > —CH=CH$_2$ > —CH$_3$, and the compound is (S)-3-chloro-1-butene.

---

**EXERCISE 11.4**

Assign a configuration to each of the following compounds.

a. 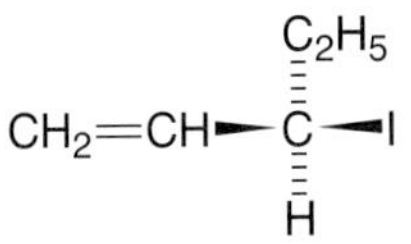  CH$_2$=CH—C—I

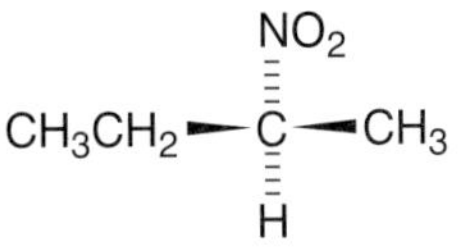

b.  CH$_3$CH$_2$—C—CH$_3$

c.  HC≡C—C—NH$_2$

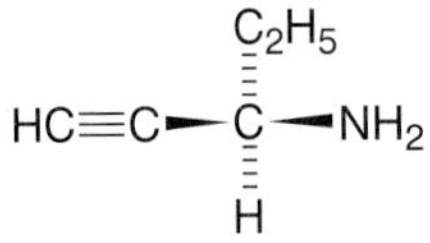

d. 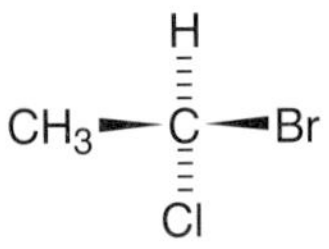  CH$_3$—C—Br

---

## 11.7
## ABSOLUTE
## CONFIGURATION

Note that there is no relationship between the configuration of a substance [(R) or (S)] and the direction of rotation [(+) or (−)] the substance displays. That is, a compound can show an (R) configuration and rotate the plane of polarized light either to the right (+) or to the left (−). In a similar manner, a compound with an (S) configuration can be either dextrorotary or levorotary.

The compound 2-chlorbutane exists in two absolute configurations (R) and (S). One enantiomer rotates the plane of polarized light to the right (+); the other, to the left (−). A question to be asked is which enantiomer rotates (+) and which rotates (−)? When we know which enantiomer rotates in which direction, we have established the *absolute configuration* of the enantiomers.

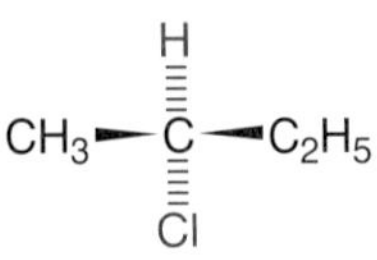

(S)-(+)-2-Chlorobutane
(absolute configuration)

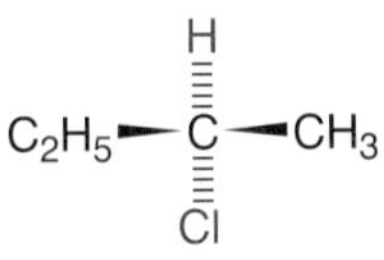

(R)-(−)-2-Chlorobutane
(absolute configuration)

We can also think of the absolute configuration of a molecule as knowing the actual three-dimensional orientation of each group bonded to a chiral carbon in a given (+) or (−) isomer.

The first absolute configuration determination of a chiral compound was made by J. M. Bijvoet in 1951. Using x-ray analysis, he determined the actual location in space of each atom of a salt of (+)-tartaric acid.

To determine the absolute configuration of (+)-2-chlorobutane experimentally, the salt of tartaric acid is used as a starting material, and (+)-2-chlorobutane is synthesized from the tartaric acid salt, with the configuration of each given product determined after each step. Knowing the absolute configuration of the tartaric acid salt and the configuration changes (if any) of each step, the absolute configuration of (+)-2-chlorobutane can be determined.

## 11.8 THE FISCHER PROJECTION

The *Fischer projection* is a simple and popular method of representing a three-dimensional tetrahedral carbon compound on a two-dimensional page. In the projection, the chiral center is represented by the intersection of a vertical and horizontal line. Groups on a horizontal line are directed toward you; groups on a vertical line are directed away from you. Thus

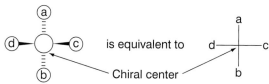

Swapping the position of any two groups of a Fischer projection gives its enantiomer.[*]

Thus, swapping —a and —d:

$$\begin{array}{c} d \\ a \!-\!\!\!+\!\!\!- c \\ b \end{array} \quad \text{is an enantiomer of} \quad \begin{array}{c} a \\ d \!-\!\!\!+\!\!\!- c \\ b \end{array}$$

VIII                                      IX

Other enantiomeric representations of compound VIII are

$$\begin{array}{c} d \\ c \!-\!\!\!+\!\!\!- a \\ b \end{array} \quad \text{and} \quad \begin{array}{c} d \\ b \!-\!\!\!+\!\!\!- c \\ a \end{array}$$

(Swap —a and —c)      (Swap —a and —b)

X                                      XI

Swapping any four groups (i.e., a double swap of any two groups) bonded to a chiral carbon gives the same molecule. Thus

$$\begin{array}{c} b \\ c \!-\!\!\!+\!\!\!- d \\ a \end{array}$$

IX

is the same molecule as

$$\begin{array}{c} a \\ d \!-\!\!\!+\!\!\!- c \\ b \end{array} \quad \text{(First swap —d and —c, then swap —a and —b)}$$

IX

---

[*]There are limits on the ways the Fischer projection can be used. In actuality, a Fischer projection can only be rotated 180 degrees in the plane of the paper to give the same stereoisomer. Rotating the stereoisomer 90 degrees forms the enantiomer. A technique has been introduced in the chemical literature in which atoms or groups incorporated in a Fischer projection are swapped. Of course, this swapping suggests that bonds are being broken. In actuality, however, the swapping technique represents a simple and convenient method—that always works—of using the Fischer projection to predict the configuration of a molecule without the need for molecular models or any other type of projection. This technique will be employed in this textbook.

**SOLVED PROBLEM 11.5**

Which of the following represent enantiomers, and which are identical?

a.

$$
\begin{array}{c}
\text{H} \\
\text{CH}_3-\!\!\!\overset{\displaystyle |}{\underset{\displaystyle |}{\text{C}}}\!\!\!-\text{Br} \\
\text{CH}_2\text{CH}_3
\end{array}
\quad\text{and}\quad
\begin{array}{c}
\text{CH}_2\text{CH}_3 \\
\text{CH}_3-\!\!\!\overset{\displaystyle |}{\underset{\displaystyle |}{\text{C}}}\!\!\!-\text{Br} \\
\text{H}
\end{array}
$$

XII              XIII

b.

$$
\begin{array}{c}
\text{CH}=\!\text{CH}_2 \\
\text{CH}_3-\!\!\!\overset{\displaystyle |}{\underset{\displaystyle |}{\text{C}}}\!\!\!-\text{OH} \\
\text{H}
\end{array}
\quad\text{and}\quad
\begin{array}{c}
\text{OH} \\
\text{CH}_2=\!\text{CH}-\!\!\!\overset{\displaystyle |}{\underset{\displaystyle |}{\text{C}}}\!\!\!-\text{CH}_3 \\
\text{H}
\end{array}
$$

XIV              XV

**SOLUTION**

a.  Enantiomers. By swapping the —H and —CH$_2$CH$_3$ groups in XII, we get XIII.

b.  Identical. First, swap the —CH$_3$ and —CH=CH$_2$ groups in structure a to produce structure b. Then swap the —OH and —CH$_3$ groups in structure b to produce structure c:

$$
\begin{array}{c}
\text{CH}=\!\text{CH}_2 \\
\text{CH}_3-\!\!\!\overset{\displaystyle |}{\underset{\displaystyle |}{\text{C}}}\!\!\!-\text{OH} \\
\text{H}
\end{array}
\;
\begin{array}{l}
\text{Swap —CH}_3 \\
\text{and —CH}=\!\text{CH}_2 \\
\text{to give}
\end{array}
\;
\begin{array}{c}
\text{CH}_3 \\
\text{CH}=\!\text{CH}-\!\!\!\overset{\displaystyle |}{\underset{\displaystyle |}{\text{C}}}\!\!\!-\text{OH} \\
\text{H}
\end{array}
\;
\begin{array}{l}
\text{Swap —CH}_3 \\
\text{and —OH} \\
\text{to give}
\end{array}
\;
\begin{array}{c}
\text{OH} \\
\text{CH}_2=\!\text{CH}-\!\!\!\overset{\displaystyle |}{\underset{\displaystyle |}{\text{C}}}\!\!\!-\text{CH}_3 \\
\text{H}
\end{array}
$$

a                    b                    c
(XIV)                                     (XV)

Now, let us show that structures a and c are identical; they certainly do not look identical. One way to demonstrate that structures a and c are identical is with the use of a set of molecular models. Make a model of structure a and one of structure c, and one will be superimposable on the other. Another technique is to determine the configuration of each structure. If the configurations are the same, the two structures are identical.

$$
\begin{array}{c}
\text{CH}=\!\text{CH}_2 \\
\text{CH}_3\curvearrowleft\!\!\!\overset{\displaystyle |}{\underset{\displaystyle |}{\text{C}}}\!\!\!-\text{OH} \\
\text{H}
\end{array}
\qquad
\begin{array}{c}
\text{OH} \\
\text{CH}_2=\!\text{CH}\!\!-\!\overset{\displaystyle |}{\underset{\displaystyle |}{\text{C}}}\curvearrowright\!\text{CH}_3 \\
\text{H}
\end{array}
$$

(a)                          (c)
(S)-3-Buten-2-ol             (S)-3-Buten-2-ol

Since the group of lowest priority is directed away from you in both structures a and c and the decreasing order of priority of the other groups is —OH > —CH=CH$_2$ > —CH$_3$, both structures a and c show the decreasing order of group priorities in a counterclockwise direction. Thus both structures exhibit the (S) configuration and are identical.

**EXERCISE 11.5**

Which of the following represent enantiomers, and which are identical?

a.

$$
\begin{array}{c}
\text{H} \\
\text{CH}_3-\!\!\!\overset{\displaystyle |}{\underset{\displaystyle |}{\text{C}}}\!\!\!-\text{CH}_2\text{CH}_3 \\
\text{I}
\end{array}
\quad\text{and}\quad
\begin{array}{c}
\text{CH}_2\text{CH}_3 \\
\text{CH}_3-\!\!\!\overset{\displaystyle |}{\underset{\displaystyle |}{\text{C}}}\!\!\!-\text{I} \\
\text{H}
\end{array}
$$

b.
$$CH=CH_2$$
$$Cl—C—CH_3 \quad \text{and} \quad CH_2=CH—C—Cl$$
$$H \qquad\qquad CH_3 \quad H$$

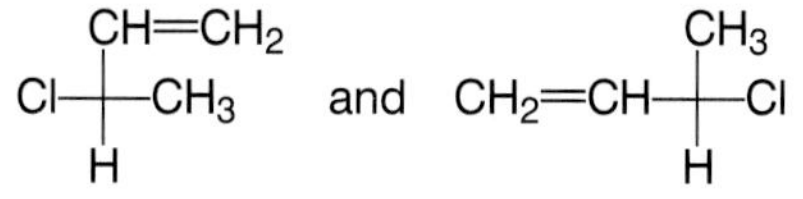

c.
$$C_2H_5$$
$$HO—C—CH_2OH \quad \text{and} \quad C_2H_5—C—OH$$
$$H \qquad\qquad CH_2OH \quad H$$

---

Designate the following compound as (R) or (S).

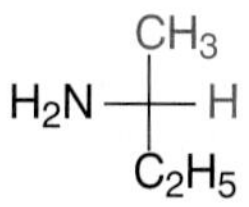

**SOLVED PROBLEM 11.6**

$$CH_3$$
$$H_2N—C—H$$
$$C_2H_5$$

Compound d

**SOLUTION**

The group with the lowest priority is —H. The problem is that the —H group is oriented toward the reader, and therefore, a determination of configuration cannot be made. This problem is solved by exchanging the —H group with either the —CH$_3$ or the —CH$_2$CH$_3$ group so as to cause the —H group to be attached to a vertical bond in the Fischer projection. This satisfies the condition for the determination of an (R) or (S) configuration in that the group of lowest priority must be oriented away from the reader and thus a determination of configuration can be made. Let us swap the —H group with the —CH$_3$ group to give the enantiomer of compound d:

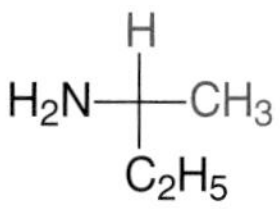

$$H$$
$$H_2N—C—CH_3$$
$$C_2H_5$$

Compound e

Now, the decreasing order of priority of the groups other than —H is —NH$_2$ > —C$_2$H$_5$ > —CH$_3$. Thus this compound (compound e) is (S):

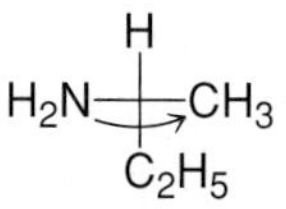

$$H$$
$$H_2N—C—CH_3$$
$$C_2H_5$$

Compound e
(S)-sec-Butylamine

However, we want the configuration of compound d. Since compound d was converted to compound e by a group swap, then compound d must have the opposite configuration of compound e and must be (R)-sec-butylamine.

---

Determine the configuration of and name each of the following compounds.

**EXERCISE 11.6**

a.
$$CH_3$$
$$H—C—Br$$
$$CH_2CH_3$$

c.
$$I$$
$$CH_3—C—Br$$
$$Cl$$

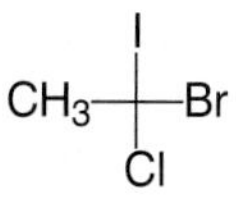

b.
$$CH=CH_2$$
$$H—C—CH_3$$
$$Cl$$

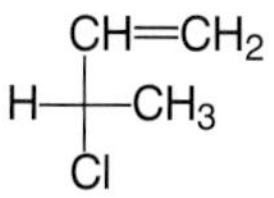

d.
$$CH_3$$
$$CH_3CH_2CH_2—C—Cl$$
$$H$$

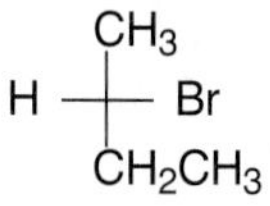

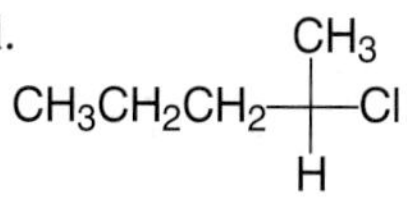

FIGURE 11.9 A pair of enantiomers.

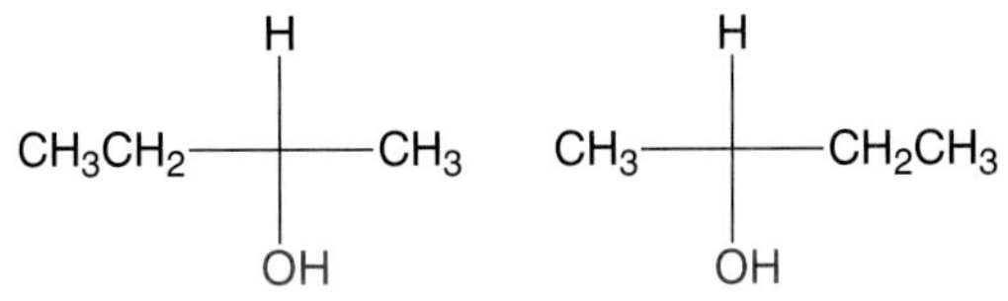

(R)-2-Butanol
$[\alpha] = -13.5°$

(S)-2-Butanol
$[\alpha] = +13.5°$

## 11.9 STEREOISOMERISM IN COMPOUNDS WITH ONE CHIRAL CENTER

To review, any compound containing one chiral center exists as a pair of enantiomers. When a 50:50 mixture of enantiomers is present, we call this a *racemic mixture*. Let us consider the stereoisomers of 2-butanol [$CH_3CH(OH)CH_2CH_3$] (Fig. 11.9). There exists a (+)-isomer, a (−)-isomer and a 50:50 racemic mixture of the two enantiomers. These represent three stereoisomeric forms of 2-butanol. Each enantiomer is optically active, whereas the racemic mixture is optically inactive. This loss of optical activity in the racemic mixture is due to the fact that one (+)-rotating molecule of 2-butanol ($[\alpha] = +13.5$ degrees) is neutralized by one (−)-rotating molecule of 2-butanol ($[\alpha] = -13.5$ degrees) to give an optically inactive mixture.

## 11.10 STEREOISOMERISM IN COMPOUNDS WITH TWO OR MORE DISSIMILAR CHIRAL CENTERS

Two dissimilar chiral centers are carbons that differ in at least one bonded group. For example, consider the following stereoisomer of 3-chloro-2-butanol:

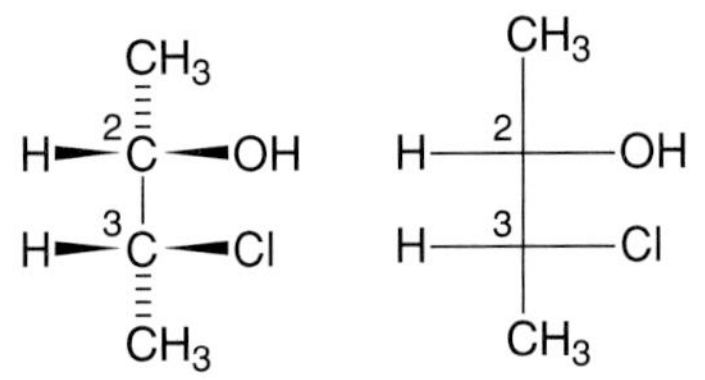

Observe that chiral carbon 2 is linked to —H, —CH₃, —OH, and —CH(Cl)CH₃, while chiral carbon 3 is bonded to the following four groups: —H, —CH₃, —Cl, and —CH(OH)CH₃. Two of the groups bonded to each chiral center are the same (—H, —CH₃), but since the other two groups bonded to each chiral carbon differ [—OH versus —Cl and —CH(Cl)CH₃ versus —CH(OH)CH₃], we say these chiral carbons are dissimilar.

The number of optically active stereoisomers for a molecule with two or more dissimilar chiral carbons can be calculated using the expression

$$\text{Number of stereoisomers} = 2^n$$

where $n$ represents the number of dissimilar chiral centers in the molecule. This expression is known as *van't Hoff's rule*. Substituting in van't Hoff's rule, we calculate that 3-chloro-2-butanol can exist as four possible stereoisomers ($2^2 = 4$).

Consider 2,3-dihydroxybutanioc acid, [$HOOC\overset{*}{C}H(OH)\overset{*}{C}H(OH)CH_3$], a structure with two dissimilar chiral carbons. Using van't Hoff's rule, we can calculate the number of stereoisomers to be $2^2 = 4$. Note that a structure, using a Fischer projection, for each optically active stereoisomer is as follows:

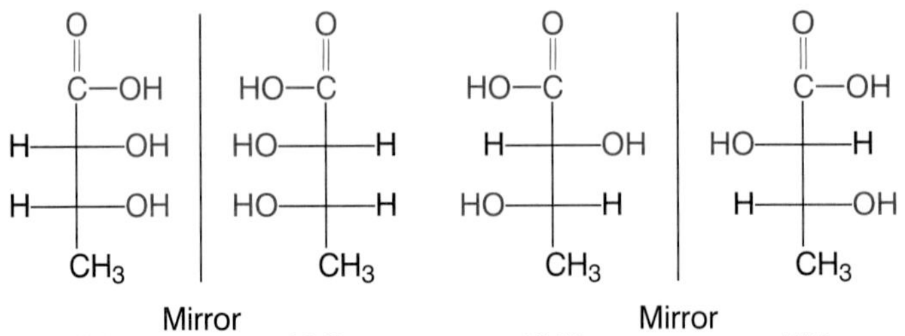

Structures XVI and XVII are enantiomers; structures XVIII and XIX are also enantiomers. Stereoisomers that are not enantiomers are *diastereomers*. Thus structures XVI and XVIII, XVI and XIX, XVII and XVIII, and XVII and XIX are diastereomers.

Now let us name and assign an $(R)$ or $(S)$ configuration to each chiral center in structures XVI to XIX. Since the lowest-priority groups on carbons 2 and 3 are oriented toward us, we must swap groups to get both hydrogens orienting away from us in order to make an $(R)$ or $(S)$ determination.

Priorities C-2 XVIa: —OH > —COOH > —CH(OH)CH$_3$

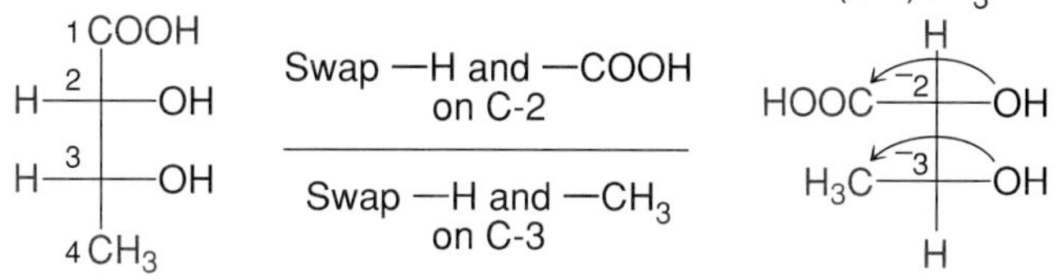

Swap —H and —COOH
on C-2
_____
Swap —H and —CH$_3$
on C-3

XVI                    XVIa

Priorities C-3 XVIa: —OH > —CH(OH)COOH > —CH$_3$

Since C-2 is $(S)$ and C-3 is $(S)$ in structure XVIa, C-2 must be $(R)$ and C-3 must be $(R)$ in structure XVI.

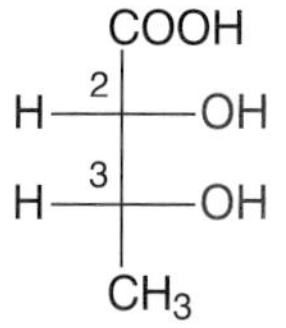

and for the mirror image XVII, we get

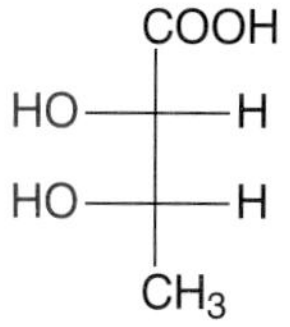

XVI
(2R,3R)-Dihydroxybutanoic acid

XVII
(2S,3S)-Dihydroxybutanoic acid

In the same way, a configuration is assigned to each chiral carbon in structures XVIII and XIX, and a name is given as follows:

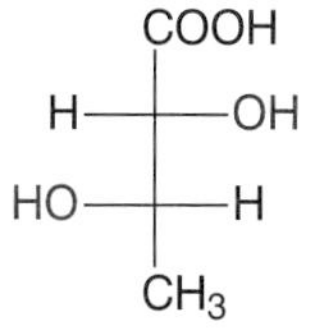

and for the mirror image XIX, we get

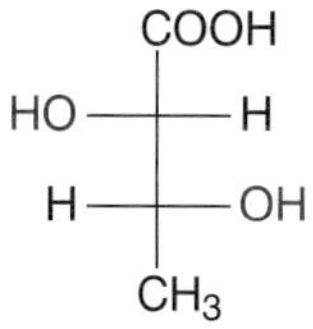

XVIII
(2R,3S)-Dihydroxybutanoic acid

XIX
(2S,3R)-Dihydroxybutanoic acid

## 11.11 STEREOISOMERISM IN COMPOUNDS WITH TWO SIMILAR CHIRAL CENTERS

Let us consider 2,3-butanediol, $\overset{*}{CH_3}\overset{}{CH}(OH)\overset{*}{CH}(OH)CH_3$, a structure that contains two similar chiral centers. To demonstrate the fact that 2,3-butanediol does contain two similar chiral carbons, note that carbons 2 and 3 are bonded to the same four groups as follows: —H, —OH, —CH$_3$, and —CH(OH)CH$_3$.

The number of optically active stereoisomers of compounds with two similar chiral carbons should be four using van't Hoff's rule ($2^2 = 4$). Experimentally, however, only two optically active stereoisomers are isolated. To understand why, examine 2,3-butanediol once more. Only one pair of nonsuperimposable optically active stereoisomers can be written:

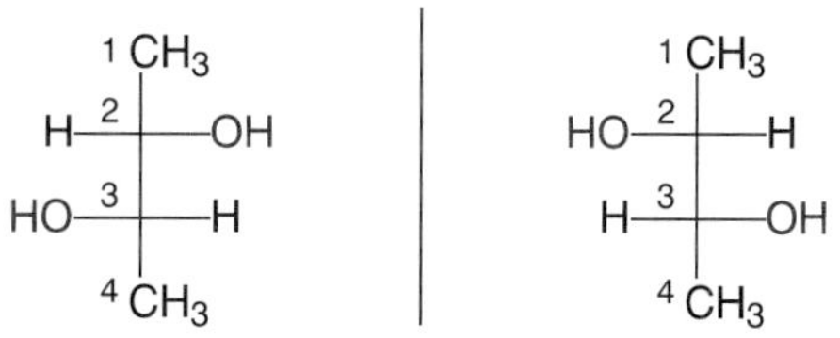

Mirror

XX
(2S,3S)-Butanediol

XXI
(2R,3R)-Butanediol

Since structures XX and XXI represent nonsuperimposable mirror images of each other, they are enantiomers, and an equal quantity of each enantiomer, when intimately mixed, produces a racemic mixture.

Now, consider structures XXII and XXIII:

XXII
(2R,3S)-Butanediol

XXIII
(2S,3R)-Butanediol

Although structure XXII is a mirror image of structure XXIII, they are superimposable and therefore represent the same molecule. Look at this phenomenon in another way: Rotating structure XXII 180 degrees in the plane gives structure XXIII. (A Fischer projection can be lifted out of the plane of the textbook, rotated 180 degrees, placed back into the plane of the book, and still keep its identity.) This compound, XXII or XXIII, is known as a *meso form*.

Any meso form is achiral as a whole and is optically inactive, even though the structure contains two chiral centers. A simple test to confirm meso achirality in a molecule is the presence of a plane of symmetry in the molecule, as shown in Fig. 11.10. Never mind that compound XXII is (2R,3S)-butanediol while compound XXIII is (2S,3R)-butanediol. Both structures represent the same molecule (the meso form) because of the presence of a plane of symmetry in the molecule. One final point: We can number a molecule with a plane of symmetry from either end and end up with the same name for the molecule. For example, consider structure XXII again:

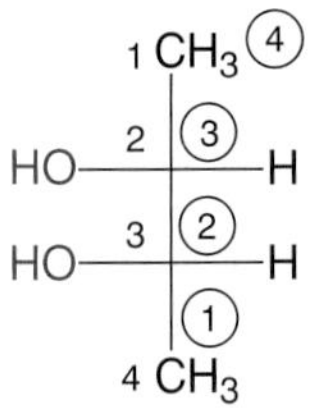

XXII

Numbering from top to bottom gives (2R,3S)-butanediol. Numbering from bottom to top gives (2S,3R)-butanediol. However, this is the name we formulated for compound XXIII, which we have already shown is the same molecule as compound XXII.

The meso form, like the racemic mixture, exhibits physical properties that differ from those of the enantiomers. Consider the stereoisomeric forms of tartaric acid [HOOCCH(OH)CH(OH)COOH] shown in Table 11.2.

**FIGURE 11.10** A plane of symmetry in *meso*-2,3,-butanediol.

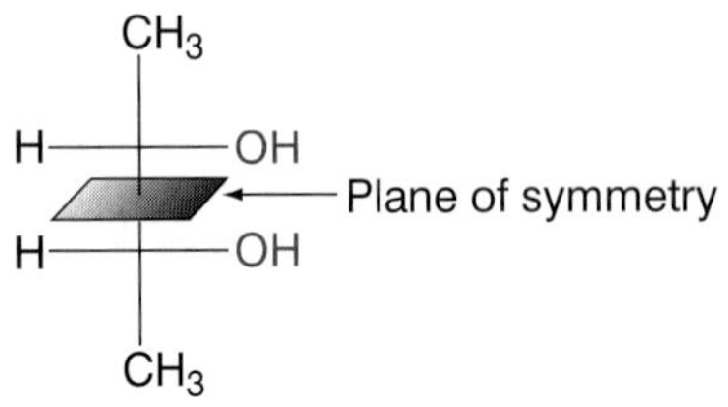

XXII or XXIII

| TABLE 11.2 | Physical Properties of the Stereoisomers of Tartaric Acid | | |
|---|---|---|---|
| | **Melting Point (°C)** | **Density (g/ml)** | **[α] (degrees)** |
| (+)-isomer | 171–174 | 1.7598 | + 12.7 |
| (−)-isomer | 171–174 | 1.7598 | − 12.7 |
| Racemic mixture | 206 | 1.788 | 0.0 |
| Meso form | 146–148 | 1.666 | 0.0 |

Draw a Fischer projection for each of the following compounds.

a. (2S,3R)-dibromopentane

b. (2R,3S)-dibromopentane

**SOLUTION**

The first step is to determine that two chiral centers exist in each structure. Then draw the Fischer projection of each compound with the groups of lowest priority on carbons 2 and 3 (in each case, —H) placed on a vertical line in the projection so that the groups orient away from you. Then position the arrows around both chiral compounds to aid in the placement of groups so as to produce the appropriate configuration. For example, in compound a, place a counterclockwise arrow around C-2. Thus the —Br group is placed to the left of C-2, the —CH(Br)CH₂CH₃ group is placed below C-2, and the —CH₃ group is placed to the right of C-2.

 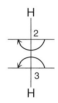

Finally, fill in the other groups so that the (2S,3R) configuration is obtained in compound a and the (2R,3S) in compound b.

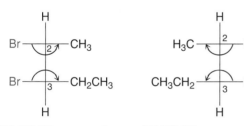

(2S,3R)-Dibromopentane    (2R,3S)-Dibromopentane
Priorities C-2 compounds a and b: —Br > —CH(Br)CH₂CH₃ > —CH₃
Priorities C-3 compounds a and b: —Br > —CH(Br)CH₃ > —CH₂CH₃

Draw a Fischer projection for each of the following compounds.

a. (2R,3R)-dibromopentane

b. (2S,3S)-dibromobutane

c. (2R,3S)-dichloropentane

Suppose we react an achiral alkene such as 1-butene with HX, where X is —Br, —Cl, —I, or —OSO₃H, in the laboratory. From the equation, we can see that a new chiral center is created in the product molecule to give a chiral compound:

**11.12**
**GENERATION OF A**
**NEW CHIRAL CENTER**
**IN ORGANIC SYNTHESIS**

**FIGURE 11.11** (S)—(−)—Dopa.

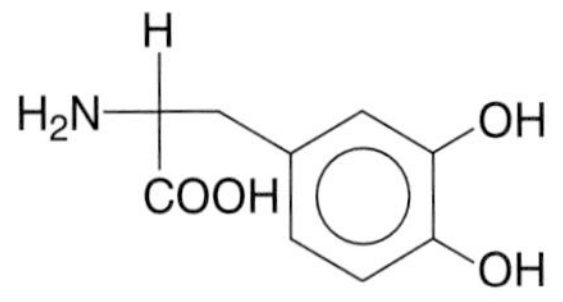

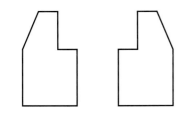

**FIGURE 11.11** (S)—(−)—Dopa.

**FIGURE 11.12** Enantiomeric crystals painstakingly separated by tweezers.

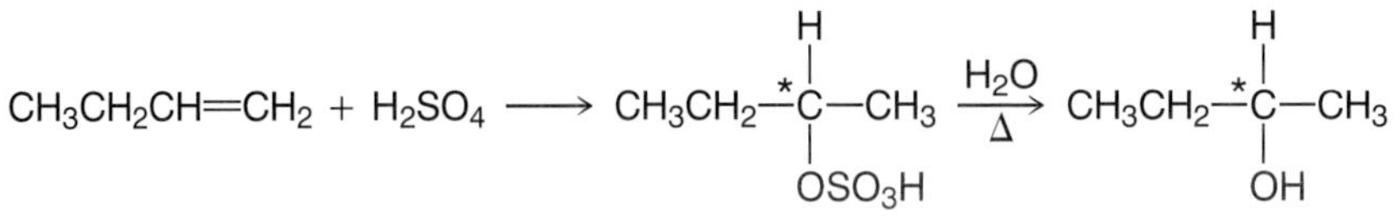

## 11.13
## RESOLUTION

## 11.14
## IMPORTANCE OF CHIRAL MOLECULES IN NATURE AND THE LABORATORY

$$CH_3CH_2CH{=}CH_2 + H_2SO_4 \longrightarrow CH_3CH_2{-}\overset{*}{\underset{OSO_3H}{\overset{H}{C}}}{-}CH_3 \xrightarrow[\Delta]{H_2O} CH_3CH_2{-}\overset{*}{\underset{OH}{\overset{H}{C}}}{-}CH_3$$

When a new chiral center is created by reacting an achiral reagent (such as 1-butene) with one or more reagents (first, $H_2SO_4$, then $H_2O$ and heat) in the laboratory, a racemic mixture of the product (*sec*-butyl alcohol) is always produced. Thus it often is necessary to separate the racemic mixture to isolate one or more enantiomers. For example, when a racemic mixture of 2-amino-3-(3,4-dihydroxyphenyl)propanoic acid (abbreviated Dopa; Fig. 11.11) is prepared, the (−)-isomer must be isolated because of its ability to ease the symptoms of Parkinson's disease, a progressive neuromuscular condition suffered by the elderly. The (+)-isomer, on the other hand, shows no biological effect on human beings and can be discarded.

The separation of a racemic mixture into pure (R) and (S) enantiomers is known as *resolution*. The first resolution of a racemic mixture was accomplished by Louis Pasteur in 1848. While working with a crystalline salt of tartaric acid, he observed that two different kinds of crystals formed, one kind the mirror image of the other (Fig. 11.12). Laboriously, using a microscope and tweezers, he separated one enantiomer from the other. This resolution method never became popular because it was so time consuming.

A currently used method of resolution is chemical in nature and involves reacting a racemic alcohol (± alcohol) with an optically active carboxylic acid to give diastereomeric esters that can be separated by fractional distillation. The individual alcohols are regenerated by treating the individual esters with base (Fig. 11.13).

Mother Nature is very picky when it comes to using chiral molecules. For example, our bodies convert (S)-(+)-valine, an amino acid, along with other amino acids, into proteins necessary for our growth and development. However, (R)-(−)-valine is not employed at all. This specificity of action is due to a series of biological catalysts called *enzymes*, each of which can easily distinguish between a (+)- and (−)-isomer.

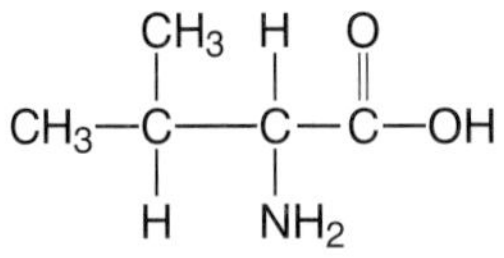

$$CH_3{-}\overset{CH_3}{\underset{H}{\overset{|}{C}}}{-}\overset{H}{\underset{NH_2}{\overset{|}{C}}}{-}\overset{O}{\overset{\|}{C}}{-}OH$$

Valine

$$(+)\text{-valine} \xrightarrow{\text{Enzyme 1}} A \xrightarrow{\text{Enzyme 2}} B \xrightarrow[\text{2. Other amino acids}]{\text{1. Several enzymes}} \text{protein}$$

$$(-)\text{-valine} \xrightarrow{\text{Enzyme 1}} \text{no reaction}$$

Laboratory-synthesized (+)-amphetamine excites the central nervous system in humans three to four times more than (−)-amphetamine. (−)-Ascorbic acid (vitamin C) prevents scurvy, while (+)-ascorbic acid does not. In general, two enantiomers react with chiral enzyme systems in the body at different rates.

Since the odor and taste of compounds are actually chemical properties, a pair of enantiomers or diastereomers often has different odors and tastes. For example, (R)-carvone, a terpene, has a caraway odor, while its enantiomer (S)-carvone has an odor of spearmint. One enantiomer of limonene (page 137) smells like lemons; the other

**FIGURE 11.13** Chemical resolution.

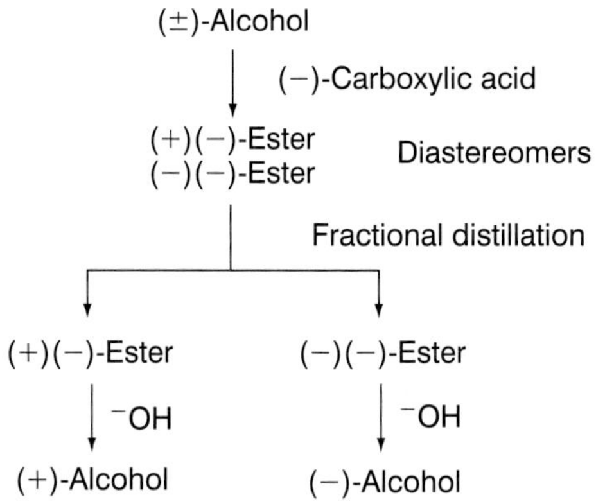

smells like oranges. ($-$)-Amino acids are tasteless or bitter; however, ($+$)-amino acids are sweet.

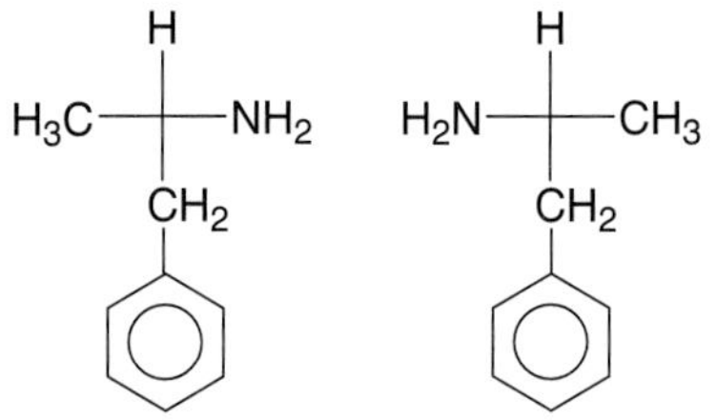

(R)-($-$)-Amphetamine      (S)-($+$)-Amphetamine

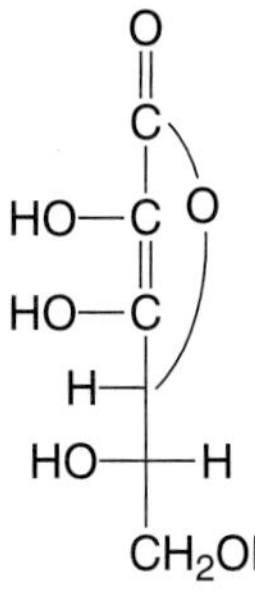

Ascorbic acid (vitamin C)

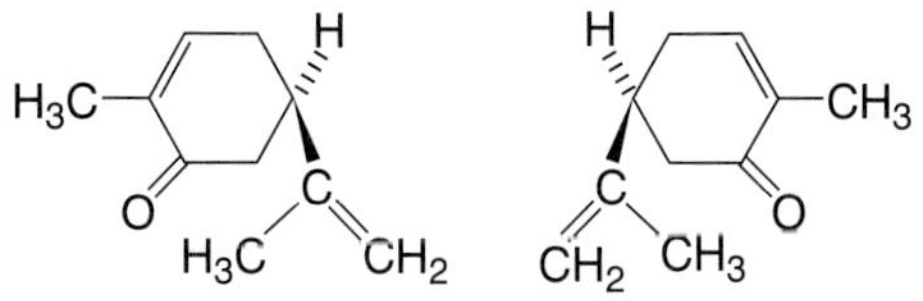

(R)-($-$)-Carvone      (S)-($+$)-Carvone

## ▶ CHAPTER ACCOMPLISHMENTS

### 11.1 Introduction

☐ Draw a structural formula for each of two functional-group isomers.

☐ Draw a structural formula for each of two positional isomers.

☐ Distinguish between structural isomers and stereoisomers.

### 11.2 The Tetrahedral Carbon Atom and Chirality

☐ Provide a structural condition for a chiral molecule.

☐ Provide a structural criterion for a carbon atom in a molecule to be designated as a chiral center.

☐ Draw a line-bond structural formula for each enantiomer of *sec*-butyl alcohol as mirror images.

☐ Determine the number of chiral carbons in:

    a. glucose.

    b. *sec*-butyl chloride.

    c. lactic acid.

    d. 3-bromo-2-pentanol [$CH_3CH(OH)CH(Br)CH_2CH_3$].

☐ Recognize the structural formulas of a pair of enantiomeric compounds.

## 11.3 Plane-Polarized Light

☐ Describe the operation of the polarimeter.

☐ Explain the difference between a dextrorotary substance and a levorotary substance.

## 11.4 Specific Rotation

☐ Write the equation relating specific rotation to observed rotation, length of the polarimeter tube, and concentration.

☐ State the advantage of calculating the specific rotation of a chiral substance as opposed to determining the observed rotation.

## 11.5 Properties of Enantiomers: The Racemic Mixture

☐ State which physical property differs from one enantiomer to another.

☐ Supply the name for a 50:50 mixture of two enantiomers.

## 11.6 The (*R*)/(*S*) Method of Writing Configurations (Arrangements in Space) of Molecules

☐ State in your own words the rules of priority ranking used to identify the configuration of a chiral substance as rectus or sinister.

☐ Specify the configuration of the following compound as (*R*) or (*S*).

$$H_2N \blacktriangleright \overset{\displaystyle C_2H_5}{\underset{\displaystyle H}{C}} \blacktriangleleft CH_3$$

## 11.7 Absolute Configuration

☐ Define the term *absolute configuration*.

☐ State what instrumental tool was used to determine the first absolute configuration of a chiral compound.

## 11.8 The Fischer Projection

☐ Supply the configuration of the compound in Chapter Accomplishment 11.6 when

    a. any two groups are swapped.

    b. any four groups are swapped.

☐ Supply the configuration of the following compound:

$$H_2N \overset{\displaystyle CH_3}{\underset{\displaystyle C_2H_5}{|\!\!-\!\!-\!\!|}} H$$

## 11.9 Stereoisomerism in Compounds with One Chiral Center

☐ List each stereochemical form of 2-butanol.

## 11.10 Stereoisomerism in Compounds with Two or More Dissimilar Chiral Centers

☐ Draw a Fischer projection for each stereochemical form of 2,3-dihydroxybutanoic acid. Label each enantiomeric pair of compounds. Label each diastereomeric pair of compounds.

☐ Name and assign an (*R*) or (*S*) configuration to each chiral center in the following compound.

$$\begin{array}{c} COOH \\ H \!-\!\!\!-\!\!\!-\!\!\!-\!\! OH \\ H \!-\!\!\!-\!\!\!-\!\!\!-\!\! OH \\ CH_3 \end{array}$$

## 11.11 Stereoisomerism in Compounds with Two Similar Chiral Centers

☐ Draw a Fischer projection for (2*S*,3*S*)-butanediol.

☐ Draw a Fischer projection for the meso form of 2,3-butanediol.

## 11.12 Generation of a New Chiral Center in Organic Synthesis

☐ Supply the stereochemical form of 2-butanol that is produced when an achiral reagent such as 2-butene is reacted first with $H_2SO_4$ and then with $H_2O$ and heat.

## 11.13 Resolution

☐ Define the term *resolution*.

☐ Name the scientist who accomplished the first resolution of a racemic mixture.

☐ Write a flowchart (see Fig. 11.13) describing the separation of a racemic mixture of 2-butanol using a chemical resolution.

## 11.14 Importance of Chiral Molecules in Nature and the Laboratory

☐ Supply an example of the specificity of an enzyme as a biological catalyst in reacting with one optical stereoisomer but not with its enantiomer.

☐ Supply an example in which one optical stereoisomer has a different odor than its enantiomer.

# KEY TERMS

stereoisomerism (11.1)
optical isomerism (11.1)
optical isomer (11.1)
chiral (11.2)
plane of symmetry (11.2, 11.11)
chiral center (11.2)
achiral (11.2)
enantiomers (11.2)
monochromatic light (11.3)
Polaroid (11.3)

plane-polarized light (11.3)
polarimeter (11.3)
polarizer (11.3)
analyzer (11.3)
observed optical rotation (α) (11.3)
optically active (11.3)
dextrorotary (11.3)
levorotary (11.3)
specific rotation (11.4)
racemic mixture or racemate (11.5)

rectus (R) (11.6)
sinister (S) (11.6)
absolute configuration (11.7)
Fischer projection (11.8)
van't Hoff's rule (11.10)
diastereomers (11.10)
meso form (11.11)
resolution (11.13)
enzyme (11.14)

# PROBLEMS

1. Draw a line-bond structural formula for each of the dichloropropanes. Indicate which are chiral.

2. Which of the following represent functional-group isomers, and which represent positional isomers.

   a. $CH_2CH_2-NH_2$  and  $CH_3-\overset{\overset{H}{|}}{\underset{\overset{|}{NH_2}}{C}}-NH_2$
      $\underset{NH_2}{|}$

   b. $CH_3CH_2CH_2CH_3$  and  $(CH_3)_3CH$

   c. $CH_3CH_2\overset{\overset{H}{|}}{C}=O$  and  $CH_3\overset{\overset{O}{||}}{C}CH_3$

3. Locate at least one chiral center in each of the following.

   a. ⬡$CH_2CH(NH_2)COOH$  b. $H_3C$  $CH_3$  $H_3C$  $O$

   c. [structure with CH₃, OH, H–C–CH₃, CH₃]  d. [structure: $\overset{O}{||}$C–H, HO–C–H, HO–C–H, H–C–OH, HO–C–H, CH₂OH]

   e. $CH_3CH(Cl)CH(Cl)CH(Cl)CH_3$

4. A solution of 10 g of heroin is dissolved in 100 ml of methyl alcohol. Knowing the specific rotation of heroin is −166 degrees, calculate the observed rotation of this solution if it is placed in a 2-dm tube.

5. Calculate the weight of cholesterol in 100 ml of ether if the solution was placed in a 1-dm tube and the observed rotation was determined to be −1.2 degrees. The specific rotation of cholesterol is −31.5 degrees.

6. a. Fructose, a constituent of candy, has a specific rotation of −92.00 degrees. Calculate the rotation observed when a 2.00% solution of fructose is measured in a 10.0-cm tube.
   b. Now calculate the observed rotation of the fructose solution in part (a) but use a 20.0-cm tube.
   c. Finally, calculate the observed rotation of the fructose solution under the same conditions as in part (a) but use a 4.00% solution.
   d. State the relationship between the observed rotation and the length of the polarimeter tube and the observed rotation and the concentration of the solution measured.

7. The specific rotation of morphine, a narcotic, is −132 degrees. Calculate the observed rotation of morphine when 10.0 g of morphine is dissolved in 25 ml of ethyl alcohol and the resulting solution placed in a 2-dm polarimeter tube.

8. Predict the specific rotation of toluene.

9. sec-Butyl chloride cannot be used as a solvent in a polarimeter procedure. Why?

10. Draw a wedge projection for each of the following.
    a. (R)-3-methyl-1-pentene
    b. (S)-2-iodopentane
    c. (R)-1-bromo-1-phenylethane
    d. (R)-4-bromo-2-pentyne

11. Designate each compound in Problem 15 as (R) or (S).

12. Designate each of the following as (R) or (S).

    a. $HOCH_2-\overset{\overset{H}{|}}{\underset{\overset{|}{CH_3}}{C}}-Cl$  b. $CH_3-\overset{\overset{C_2H_5}{|}}{\underset{\overset{|}{CH_2CH_2CH_3}}{C}}-CH(CH_3)_2$

    c. $HO-\overset{\overset{H}{|}}{\underset{\overset{|}{CH}}{C}}-\overset{\overset{H}{|}}{C}=CH_2$  $\underset{O}{}$  d. $CH_3-\overset{\overset{OH}{|}}{\underset{\overset{|}{H}}{C}}-Br$

13. Which of the following has the higher priority?

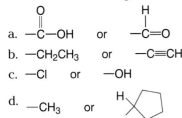

a. $-\overset{O}{\overset{\|}{C}}-OH$   or   $-\overset{H}{\overset{|}{C}}=O$

b. $-CH_2CH_3$   or   $-C\equiv CH$

c. $-Cl$   or   $-OH$

d. $-CH_3$   or   [cyclopentane with H]

14. List the following in order of decreasing priority.
    a. $-Cl, -I, -NH_2$
    b. $-H, -CH_3, -CH_2CH_2CH_3$
    c. $-CH_2CH_2CH_3, -CH(CH_3)_2, -CH_2CH_3$

15. Which of the following are identical or enantiomers?

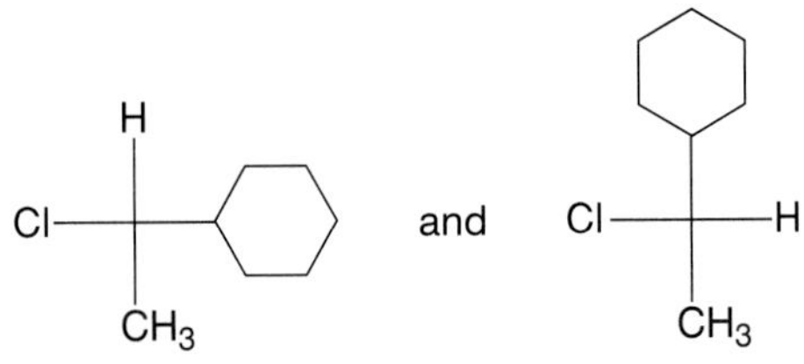

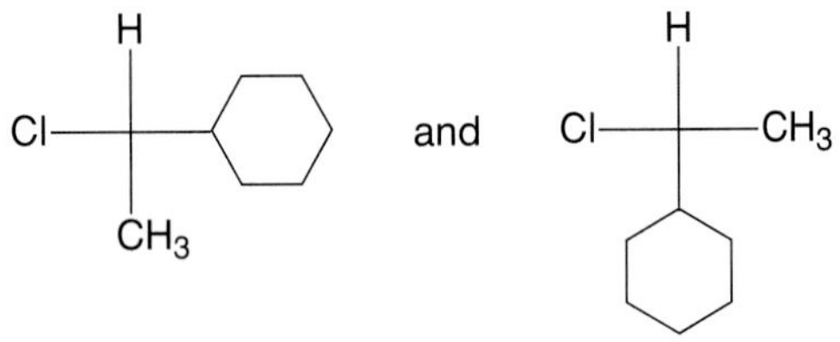

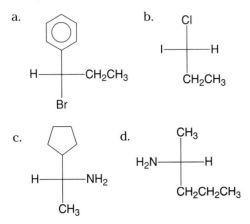

   c. Draw a Fischer projection and a wedge projection for each of the following.
   d. (R)-2-bromopentane
   e. (S)-3-chloro-1-butene
   f. (S)-1-bromo-1-chloro-1-iodoethane
   g. (R)-1-bromo-1-phenylethane
   h. (S)-1-iodo-2-methylbutane

16. Draw a Fischer projection for each of the following.
    a. (R)-2-iodobutane
    b. (S)-2-bromopentane
    c. (S)-3-iodo-1-pentene
    d. (R)-3-methylhexane

17. Designate each of the following as (R) or (S).

    a. [structure: benzene ring, H—C—CH₂CH₃, Br]
    b. [Cl, I—C—H, CH₂CH₃]
    c. [cyclopentane, H—C—NH₂, CH₃]
    d. [CH₃, H₂N—C—H, CH₂CH₂CH₃]

19. Determine the number of optically active stereoisomers (if any) possible for each of the following structures, and draw a Fischer projection for each.
    a. 3-methyl-1-pentyne
    b. isobutane
    c. 2-hexanol $CH_3CH(OH)CH_2CH_2CH_2CH_3$
    d. 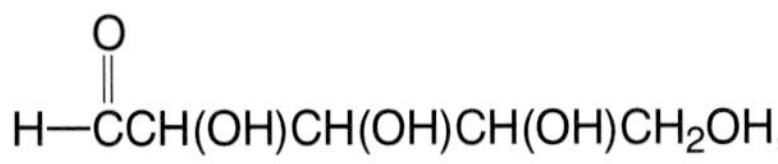 $H-\overset{O}{\overset{\|}{C}}CH(OH)CH(OH)CH(OH)CH_2OH$, a sugar

20. Draw a structural formula and determine the configuration at each chiral center for every stereoisomer of 3-chloro-2-butanol (page 224).

21. Draw a Fischer projection, as in Section 11.10, for each of the stereoisomers of 2,3-dichloropentane. Label enantiomers and diastereomers. Label each chiral center as (R) or (S), and name each stereoisomer.

22. Draw Fischer projections, as in Section 11.11, for 2,3-difluorobutane. Label enantiomers, diastereomers, and meso forms. Label each chiral carbon as (R) or (S), and name each stereoisomer.

23. Demonstrate that (2R,3S)-dibromobutane is a meso form.

24. Determine the configuration [(R) or (S)] at each chiral center in the following.

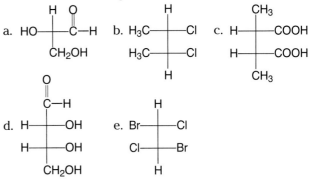

25. Define or explain each of the following.
    a. racemic mixture
    b. specific rotation
    c. (R)
    d. (−)
    e. meso form
    f. superimposable
    g. resolution
    h. chiral
    i. optically active

26. What is the difference between
    a. (R) and (+).
    b. chiral and achiral.
    c. optically active and optically inactive.

27. a. A compound (A), $C_{10}H_{14}$, produced benzoic acid when treated with hot $KMnO_4$. Compound A is optically active. What is its structural formula?
    b. A compound (B), an isomer of compound A, did not react when treated with hot $KMnO_4$. What is the structural formula of compound B?

c. A compound (C), an isomer of compound A, produced benzoic acid when treated with hot $KMnO_4$. When compound C was treated with bromine in the presence of light, only one monobromo substitution product was isolated, although four products could have formed. Deduce a structural formula for compound C.

d. A compound (D), an isomer of compound A, produced benzoic acid when treated with hot $KMnO_4$. When compound D was reacted with bromine in the presence of light, only one monobromo substitution product was isolated, although three products could have formed. Draw a structural formula for compound D.

28. How many stereoisomers can exist for each of the following?

a. 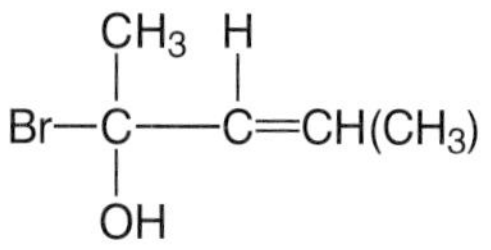 Br—C——C=CH(CH₃)
   with CH₃ and H above, OH below

b. ascorbic acid (page 229)
c. valine (page 228)

29. Draw a Fischer projection for each stereoisomer in Problem 28a and c.

30. Assign an (R) or (S) configuration to each chiral center in these molecules.

a. 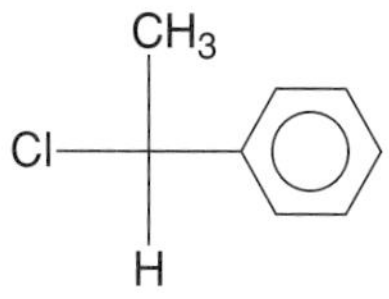 Cl—C—(phenyl), with CH₃ up and H down

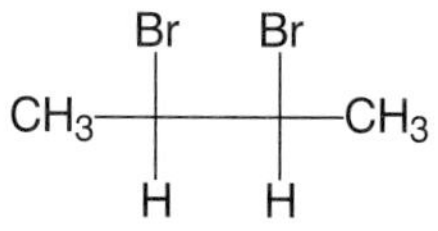

b. 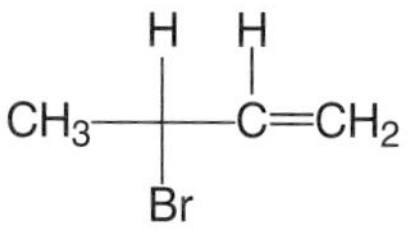 CH₃—C—C=CH₂, with H, H up and Br down

c. CH₃—C—C—CH₃, with Br, Br up and H, H down

d. 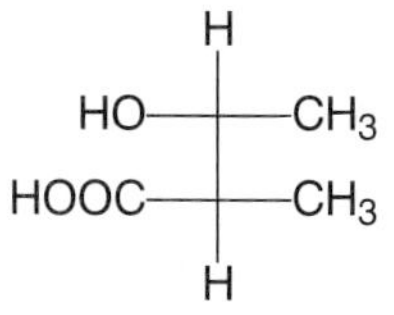 HO—CH₃; HOOC—CH₃, with H up and H down

31. Draw a structural formula for each organic product (include all possible stereoisomers).

a. 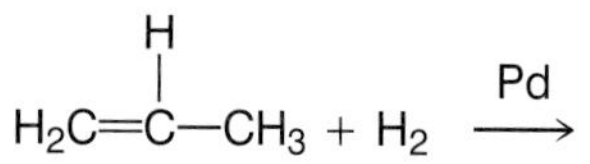 $H_2C=C-CH_3 + H_2 \xrightarrow{Pd}$ (with H on the central carbon)

b. $CH_3CH_2CH=CH_2 + HBr \longrightarrow$

c. 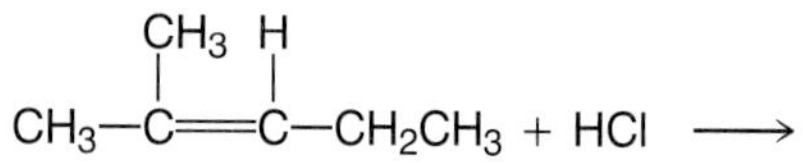 $CH_3-C=C-CH_2CH_3 + HCl \longrightarrow$ (with CH₃ and H on the double-bond carbons)

d. 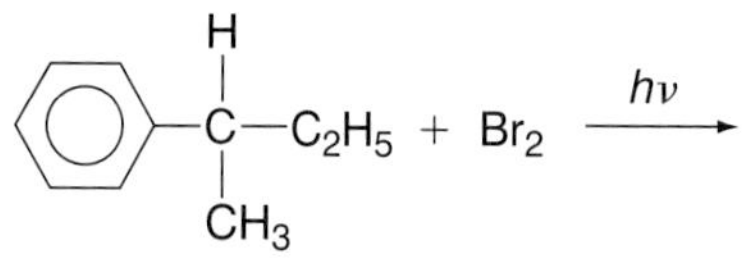 (phenyl)—C—C₂H₅ + Br₂ $\xrightarrow{h\nu}$ (with H on top and CH₃ on bottom)

32. Which of the reactions in Problem 31 yields a racemic product?

33. When a chiral alcohol, $C_5H_{11}OH$, was heated with acid, a compound, $C_5H_{10}$, was isolated. Treating $C_5H_{10}$ with ozone and then zinc and water gave acetone,

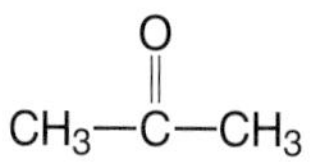

$$CH_3-\overset{\displaystyle O}{\overset{\displaystyle \|}{C}}-CH_3$$

and acetaldehyde,

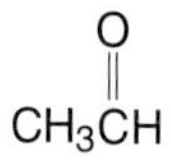

$$CH_3\overset{\displaystyle O}{\overset{\displaystyle \|}{C}}H$$

Draw a structural formula for the alcohol. Show your reasoning.

# CHAPTER 12

# Nucleophilic Substitution and Elimination

Chemists, often faced with compounds or functional groups that are not what they want them to be, must resort to substitution reactions—that is, they will replace one functional group with another. For example, suppose the compound on hand is an alcohol, ROH, and the chemist needs the corresponding alkyl halide, RX. He or she needs a means by which —OH can be replaced by —X. It is under such circumstances that we discover the utility of an important class of chemical reactions referred to as *nucleophilic substitution*. Competing elimination reactions are also discussed.

Since the carbon-halogen bond in organic halogen compounds is slightly polar, carbon is somewhat lacking in electrons, while more electronegative halogen is somewhat more electron-rich. You can also think of this state of affairs as the result of a $-I$ inductive effect produced by the halogen (see Secs. 6.2 and 10.2).

$$-\overset{|}{\underset{|}{C}}\overset{\delta+}{-}X^{\delta-}$$

The lowercase Greek deltas mean the presence of a partial positive and negative charge on carbon and halogen, respectively—far less charge than on an ionic halide such as sodium chloride ($Na^+ Cl^-$). It is this small positive charge on carbon that is open to attack by nucleophilic reagents as follows:

$$\underset{\underset{\text{Nucleophile}}{\uparrow}}{RX + :N^-} \longrightarrow RN + \underset{\underset{\text{Leaving group}}{\uparrow}}{X^-}$$

Substrate ↓

In this *nucleophilic substitution reaction* :N$^-$ is the *nucleophile*, —X is —Cl, —Br or —I, R—X is called the *substrate* (the reagent), while X$^-$ is known as the *leaving group*. A nucleophile or nucleophilic reagent is an electron-rich species that is attracted to a carbon nucleus. Nucleophiles are bases. A nucleophilic substitution is a reaction where one nucleophile is substituted for another in the substrate.

## 12.1 NUCLEOPHILIC SUBSTITUTION REACTIONS: ORGANIC HALOGEN COMPOUNDS

### Introduction to Nucleophilic Substitution

A nucleophile can be a neutral compound (ammonia), but more often it is an anion [hydroxide ion, iodide ion, or methoxide ion ($^-OCH_3$)].

As a rule, the best leaving groups are weak bases, and the poorest leaving groups are strong bases. Because halide ions are weak bases, they are superb leaving groups, and most organic halogen compounds are good reagents (substrates) for nucleophilic substitution.

**SOLVED PROBLEM 12.1**

Which of the following nucleophilic substitutions take place? Explain.

a.  $CH_3OH + Cl^- \longrightarrow CH_3Cl + {}^-OH$

b.  $CH_3NH_2 + {}^-OH \longrightarrow CH_3OH + {}^-NH_2$

c.  $CH_3Cl + {}^-OH \longrightarrow CH_3OH + Cl^-$

**SOLUTION**

a.  The reaction does not take place. Since $^-OH$ is a strong base, it represents a poor leaving group.

b.  Same as (a). The poor leaving group is $^-NH_2$.

c.  The reaction does take place because $Cl^-$ is a good leaving group, since it is a weak base.

**EXERCISE 12.1**

Which of the following nucleophilic substitutions take place? Explain. Assume that a base stronger than $^-CN$ is too poor a leaving group to result in reaction. You will need to refer to Table 1.6.

a.  $CH_3CH_2CH_2Br + {}^-OCH_3 \rightarrow$

b.  $CH_3CH_2CH_2OCH_3 + :H^- \rightarrow$

c.  $CH_3CH_2C \equiv N + HC \equiv C^- \rightarrow$

Refer to Section 13.1 for the classification and nomenclature of organic halogen compounds. Using ethyl iodide as a reagent, the nucleophilic substitutions shown in Fig. 12.1 take place.

**Envision the Reaction**

### Mechanism of the S$_N$2 Reaction of Ethyl Bromide with Hydroxide Ion

Bond forming          Bond breaking

Transition state

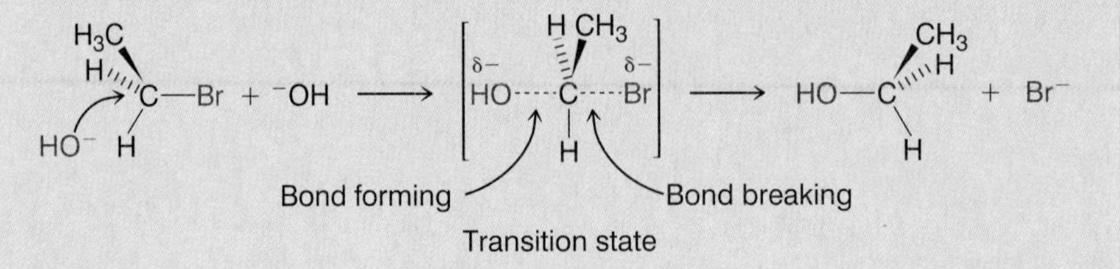

**The S$_N$2 Reaction (Substitution-Nucleophilic, Second Order)**

Consider the reaction of ethyl bromide with hydroxide ion as follows:

$$CH_3CH_2Br + {}^-OH \rightarrow CH_3CH_2OH + Br^- \qquad \text{(reaction 1)}$$

Kinetic studies have proved that the rate of reaction depends on the concentrations of both substrate and nucleophile. In other words, if the hydroxide ion concentration is held constant and the ethyl bromide concentration is doubled, the rate would double. In the same way, if the ethyl bromide concentration is kept constant and the hydroxide ion

**FIGURE 12.1** The reaction of ethyl iodide with selected nucleophiles.

$C_2H_5I$

| | | |
|---|---|---|
| + $^-NH_2$ <br> Amide ion | $\longrightarrow$ | $C_2H_5NH_2$ + $I^-$ <br> Ethylamine |
| + $NH_3$ <br> Ammonia | $\longrightarrow$ | $C_2H_5\overset{+}{N}H_3I^-$ (Sec. 25.1) <br> Ethylammonium iodide |
| + $^-CN$ <br> Cyanide ion | $\longrightarrow$ | $C_2H_5CN$ + $I^-$ (Sec. 21.1) <br> Propanenitrile |
| + $^-OC_2H_5$ <br> Ethoxide ion | $\longrightarrow$ | $C_2H_5OCH_5$ + $I^-$ (Sec. 17.4) <br> Diethyl ether |
| + $^-SH$ <br> Hydrogen sulfide ion | $\longrightarrow$ | $C_2H_5SH$ + $I^-$ (Sec. 14.5) <br> Ethanethiol |
| + $^-SCH_3$ <br> Methyl sulfide ion | $\longrightarrow$ | $C_2H_5SCH_3$ + $I^-$ <br> Methyl ethyl sulfide |
| + $H^-$ <br> Hydride ion | $\longrightarrow$ | $C_2H_6$ + $I^-$ <br> Ethane |
| + $^-OH$ <br> Hydroxide ion | $\longrightarrow$ | $C_2H_5OH$ + $I^-$ <br> Ethyl alcohol |
| + $^-C\equiv CH$ <br> Acetylide ion | $\longrightarrow$ | $C_2H_5-C\equiv CH$ + $I^-$ (Sec. 8.6) <br> 1-Butyne |
| + $^-\overset{\displaystyle O}{\overset{\|}{O}}CCH_3$ <br> Acetate ion | $\longrightarrow$ | $C_2H_5-\overset{\displaystyle O}{\overset{\|}{O}}CCH_3$ + $I^-$ <br> Ethyl acetate |

concentration is doubled, the rate is again doubled. Thus we can write the following expression called the *rate law* for the reaction:

$$\text{Rate} = k[CH_3CH_2Br][^-OH]$$

where $k$ is the *specific rate constant* for the reaction, and each set of brackets represents the concentration of a reactant in moles per liter. Since the rate of the reaction depends on the concentration of both reactants, we say the reaction follows *second-order kinetics* and is a *second-order reaction*. Another way of expressing the rate law is that the rate is first-order with respect to ethyl bromide and first-order with respect to hydroxide ion. Since the rate of reaction follows second-order kinetics overall, we call this type of reaction *substitution nucleophilic, second order* (abbreviated $S_N2$).

The rate of an $S_N2$ reaction also depends on the nature of the alkyl halide, as shown in Table 12.1. Data from Table 12.1 point out that the tertiary halide reacts slowly, while the methyl halide reacts far more rapidly.

| **TABLE 12.1** | **Relative Rate of $S_N2$ Substitution of Certain Alkyl Halides with Hydroxide Ion** | |
|---|---|---|
| | $RX + {}^-OH \longrightarrow ROH + X^-$ | |
| **Alkyl Halide** | **Class** | **Relative Rate of Reaction** |
| $CH_3Br$ | Methyl | Very fast |
| $CH_3CH_2Br$ | 1° | Fast |
| $(CH_3)_2CHBr$ | 2° | Slow |
| $(CH_3)_3CBr$ | 3° | Very slow |

Note: Refer to Section 13.1 for a discussion of the classes of alkyl halides.

Another interesting feature of the $S_N2$ reaction can be demonstrated by reacting a chiral organic halide with a nucleophile. Let us study the reaction of (R)-2-bromopentane with ethoxide ion. Consider

$$Br{-}\underset{CH_3}{\overset{H}{\underset{|}{\overset{|}{C}}}}{-}CH_2CH_3 \; + \; {}^-OC_2H_5 \longrightarrow CH_3CH_2{-}\underset{CH_3}{\overset{H}{\underset{|}{\overset{|}{C}}}}{-}OC_2H_5 \; + \; Br^-$$

(R)-2-Bromobutane                    (S)-2-Ethoxybutane

Note that the configuration of the reactant is (R); of the product, (S). This change of configuration is known as an *inversion of configuration* (or *Walden inversion*). This inversion of configuration is characteristic of $S_N2$ reactions and involves a reactant molecule of one configuration converted to a product molecule with the inverted molecular configuration.

Inversion of configuration is not the only stereochemical option a nucleophilic substitution reaction can show. For example:

$$Br{-}\underset{CH_3}{\overset{H}{\underset{|}{\overset{|}{C}}}}{-}CH_2CH_3 \; + \; {}^-OC_2H_5 \longrightarrow CH_3CH_2O{-}\underset{CH_3}{\overset{H}{\underset{|}{\overset{|}{C}}}}{-}CH_2CH_3 \; + \; Br^-$$

(R)-2-Bromobutane                    (R)-2-Ethoxybutane

In this stereochemical option, the configuration of the product molecule is the same as that of the reactant molecule. This is known as *retention of configuration*. In a third stereochemical option, the product molecule forms a racemic mixture.

$$Br{-}\underset{CH_3}{\overset{H}{\underset{|}{\overset{|}{C}}}}{-}CH_2CH_3 \; + \; {}^-OC_2H_5 \longrightarrow CH_3CH_2{-}\underset{CH_3}{\overset{H}{\underset{|}{\overset{|}{C}}}}{-}OC_2H_5 \; + \; C_2H_5O{-}\underset{CH_3}{\overset{H}{\underset{|}{\overset{|}{C}}}}{-}CH_2CH_3 \; + \; Br^-$$

(R)-2-Bromobutane                    (S)-2-Ethoxybutane          (R)-2-Ethoxybutane
                                        (50% yield)                (50% yield)

The Walden inversion is named for Paul Walden, who first reported this phenomenon in 1896.

Scientists used the knowledge gained from the $S_N2$ reaction to write a mechanism consistent with the experimental data obtained about the characteristics of the reaction. To explain the inversion of configuration and the observed rates of reaction listed in Table 12.1 for an $S_N2$ reaction, a backside attack of the nucleophile was suggested; i.e., the nucleophile attacks the central carbon from the side opposite to the location of the carbon-halogen bond. Since the reaction follows second-order kinetics, the rate-determining step, or slowest step (the only step of this mechanism), must involve both reactants. Finally, a transition state was proposed, the stability of which is inversely proportional to the amount of crowding around the carbon bonded to the functional group that is displaced.

Bond forming ⌣ ⌣ Bond breaking

Transition state

The *transition state* is an unstable species (note the brackets) because bonds are in the process of being formed and broken; in this example, as the nucleophile ($^-OH$) approaches the central carbon (the carbon bearing the halogen), the carbon-oxygen bond is being formed at the same time the carbon-bromine bond is being broken. Thus the $S_N2$ reaction is a one-step *concerted reaction mechanism*. As the carbon-oxygen bond is formed, the oxygen atom gradually loses negative charge until it bears a partial negative charge in the transition state. In the same way as the carbon-bromine bond breaks, the bromine gradually acquires negative charge until in the transition state the bromine also bears a partial negative charge.[*]

---

[*]The preceding diagram is somewhat oversimplified for convenience. In actuality, before the nucleophile attacks, the carbon-halogen bond is slightly polar. Thus carbon bears a partial positive charge, while

| TABLE 12.2 | Specific Rotation of Each Stereoisomer of 2-Chlorobutane and 2-Butanol | |
|---|---|---|
| **Name and Configuration of a Stereoisomer** | | **[α] (degrees)** |
| (R)-2-chlorobutane | | − 36 |
| (S)-2-chlorobutane | | + 36 |
| (R)-2-butanol | | − 13.5 |
| (S)-2-butanol | | + 13.5 |

Note in the preceding example illustrating the $S_N2$ mechanism that both the substrate, ethyl bromide, and the product, ethyl alcohol, are achiral substances. Thus no polarimetric measurements can be used to prove the change in configuration. When a chiral substance is used, polarimetric measurements prove the change in configuration. For example, polarometric laboratory work has provided us with data for the specific rotation of each of the stereoisomers of 2-chlorobutane and 2-butanol (Table 12.2) in order to understand the stereochemistry of the following reaction:

$$CH_3-\underset{\underset{Cl}{|}}{CH}-CH_2-CH_3 \xrightarrow{\ ^-OH\ } CH_3-\underset{\underset{OH}{|}}{CH}-CH_2-CH_3 + Cl^-$$

When (R)-2-chlorobutane with a specific rotation of − 36 degrees was reacted with sodium hydroxide under $S_N2$ conditions, polarimetric measurement of the product (2-butanol) resulted in a specific rotation of + 13.5 degrees. Therefore, (S)-2-butanol was produced, as can be observed in Table 12.2, and the reaction is proven to proceed by way of an $S_N2$ mechanism.

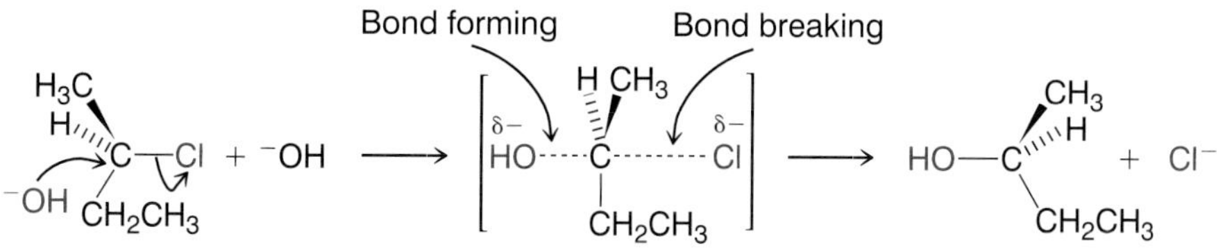

(R)-2-Chlorobutane           (S)-2-Butanol

The greater the crowding around the central carbon of the alkyl halide, the harder it is for the nucleophile to attack that carbon and the slower is the rate of reaction. This crowding is known as *steric hindrance* (see Sec. 10.4) and explains the relative rates of reaction in Table 12.1.

Figure 12.2 shows that due to steric hindrance, *tert*-butyl bromide provides little access to the nucleophile. On the other hand, methyl bromide shows far less crowding with far easier access of the nucleophile to the carbon bearing the halide.

Vinyl halides and unsubstituted aryl halides do not undergo $S_N2$ substitution reactions. This is due to two factors. The first is the partial double-bond character of the carbon-halogen bond, which leads to a stronger bond, as can be observed in the following resonance structures:

Vinyl bromide      The partial double-bond character of the carbon-bromine bond strengthens the bond and makes $S_N2$ substitution impossible.

---

bromine carries a partial negative charge. Thus, in the transition state, the partial positive charge on carbon is minimized, a neutral nucleophile acquires a partial positive charge (in the case of a negatively charged nucleophile, a partial negative charge results), and the partial negative charge on the halogen (the leaving group) is enhanced. These changes are difficult to illustrate using a diagram. Thus the preceding diagram does not show the polarity of the carbon-halogen bond.

**FIGURE 12.2** Crowding in $S_N2$ reactions involving selected alkyl halides.

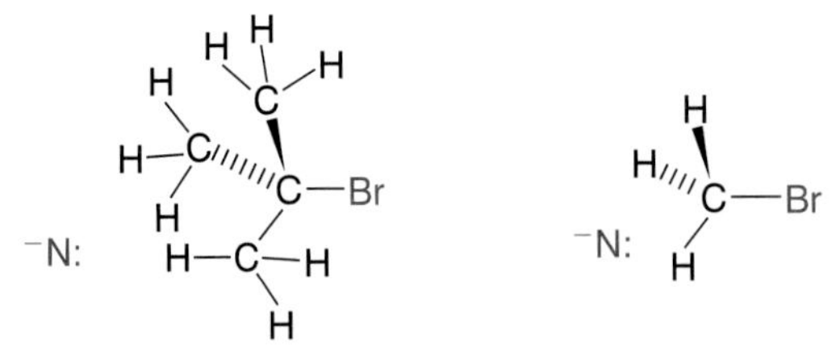

| Crowded central carbon | Less crowded carbon |
|---|---|
| (Difficult for nucleophile to attack the central carbon because of steric hindrance.) | (Less difficult for the nucleophile to attack the carbon because of less steric hindrance.) |

Second, electrons of the double bond and benzene ring, respectively, repel the attacking nucleophile.

---

**SOLVED PROBLEM 12.2**

Write a detailed $S_N2$ mechanism, using wedge projections, for the reaction of (R)-2-iodobutane with hydrogen sulfide ion, $HS^-$.

**SOLUTION**

First, we need to draw a wedge projection of (R)-2-iodobutane. Note that the chiral carbon contains four groups: —H, —CH₃, —CH₂CH₃, and —I. Since the group of lowest priority is —H, this is the group projecting away from us. The other three groups are then arranged in a clockwise order of decreasing priority to give an often-used wedge projection as follows:

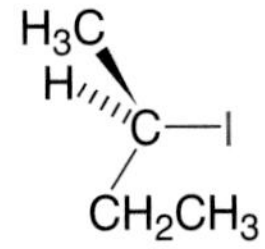

The remainder of the mechanism is shown in the usual manner.

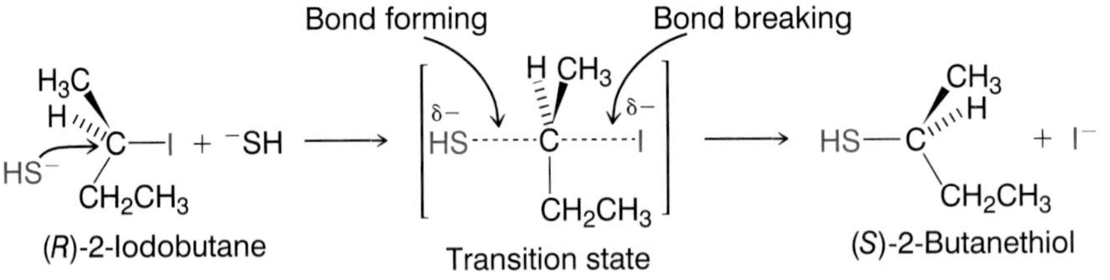

---

**SOLVED PROBLEM 12.3**

What is the difference between a transition state and an intermediate?

**SOLUTION**

A transition state is an unstable species characterized by partially formed and broken bonds (the C—X and C—Y bonds).

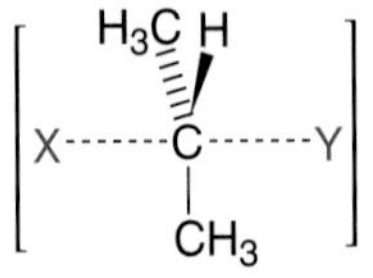

Transition state

An intermediate is somewhat more stable than a transition state because all bonds are completely formed and broken.

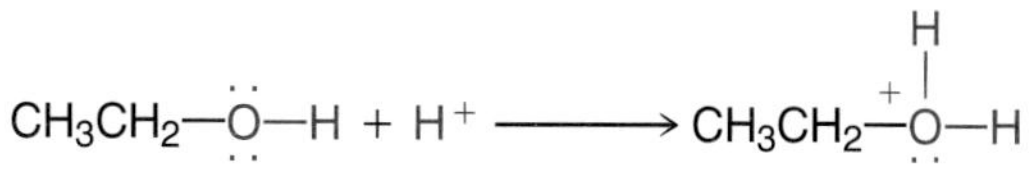

$$CH_3CH_2-\overset{..}{\underset{..}{O}}-H + H^+ \longrightarrow CH_3CH_2-\overset{H}{\underset{}{\overset{+|}{O}}}-H$$

Intermediate

---

List each of the following in order of decreasing reactivity in an $S_N2$ reaction.

a. 1-bromobutane

b. 2-bromo-2-methylbutane

c. 2-bromopentane

**SOLVED PROBLEM 12.4**

### SOLUTION

Let us draw a structural formula for each halide.

a. $CH_3CH_2CH_2CH_2Br$

b. 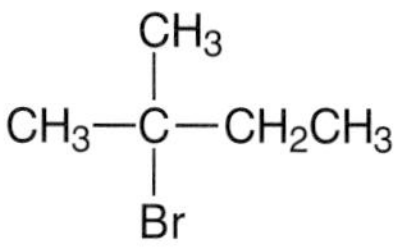 $CH_3-\overset{CH_3}{\underset{Br}{\overset{|}{\underset{|}{C}}}}-CH_2CH_3$

c. $CH_3CH(Br)CH_2CH_2CH_3$

Since 1-bromobutane is a primary halide, it reacts fastest. 2-Bromopentane is a second-ary halide and reacts at a slower rate than a primary halide. Finally, the tertiary halide 2-bromo-2-methylbutane reacts slower than the secondary halide. Thus the order of reactivity is (a) > (c) > (b). Steric hindrance is greatest for the carbon bearing the halogen in (b); least for (a).

---

Complete each of the following equations. Assume $S_N2$ reaction conditions.

**SOLVED PROBLEM 12.5**

a. 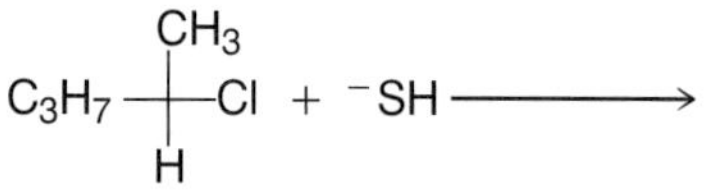 $C_3H_7-\overset{CH_3}{\underset{H}{\overset{|}{\underset{|}{C}}}}-Cl + \ ^-SH \longrightarrow$

b. 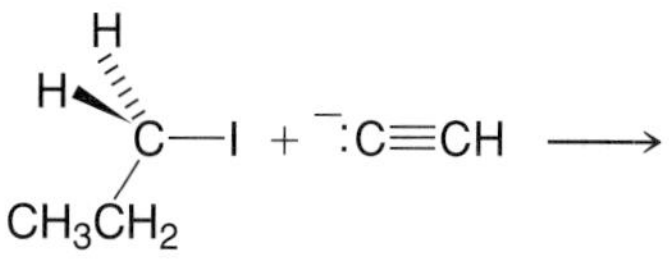 $\overset{H}{\underset{CH_3CH_2}{\overset{|}{\underset{}{H\cdots C}}}}-I + \ ^-:C\equiv CH \longrightarrow$

### SOLUTION

a. The $S_N2$ mechanism is operable here; thus products are as follows due to an inver-sion of configuration.

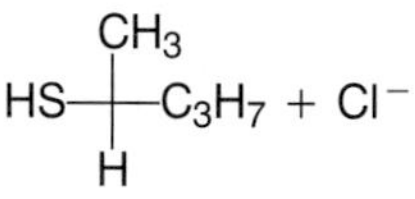

$$HS-\overset{CH_3}{\underset{H}{\overset{|}{\underset{|}{C_3H_7}}}} + Cl^-$$

b. Again, the $S_N2$ mechanism is operable. Note that the substrate and product are achiral.

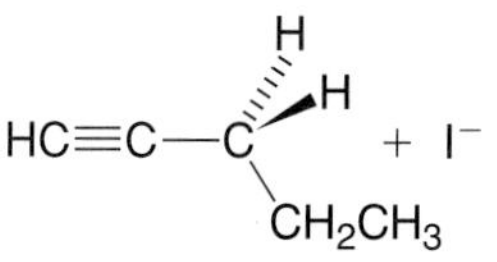

$$HC\equiv C-\overset{H}{\underset{CH_2CH_3}{\overset{\diagup H}{\underset{}{C}}}} + I^-$$

## EXERCISE 12.2

Design a one-step synthesis using an $S_N2$ reaction to prepare the following.

a. $CH_3CH_2CH_2SH$

b. The (*R*) isomer of $CH_3CH_2CH(OCH_3)CH_3$

c. $CH_3CH_2C{\equiv}CH$

## EXERCISE 12.3

Can polarimetric measurements prove the change in configuration for each $S_N2$ reaction in Exercise 12.2? Explain.

## EXERCISE 12.4

Write a detailed mechanism for the one-step syntheses in Exercises 12.2a and b. Use wedge projections, and label the transition state.

## EXERCISE 12.5

Criticize the following one-step synthesis.

$$CH_2{=}CHCl + {}^{-}CN \longrightarrow CH_2{=}CHCN + Cl^{-}$$

## Envision the Reaction

**Mechanism of the $S_N1$ Reaction of *tert*-Butyl Bromide with Water to Produce *tert*-Butyl Alcohol and HBr**

First step:

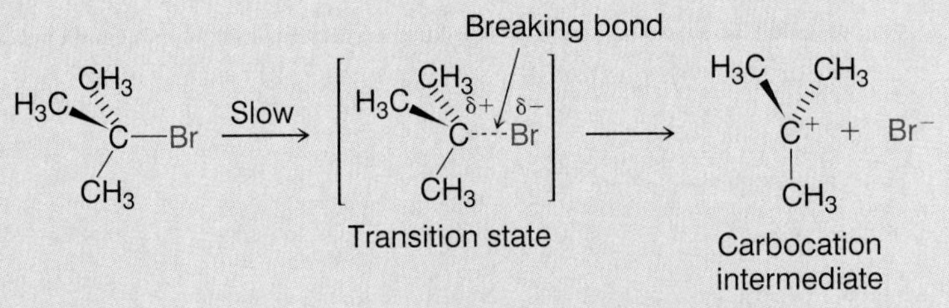


Second step:

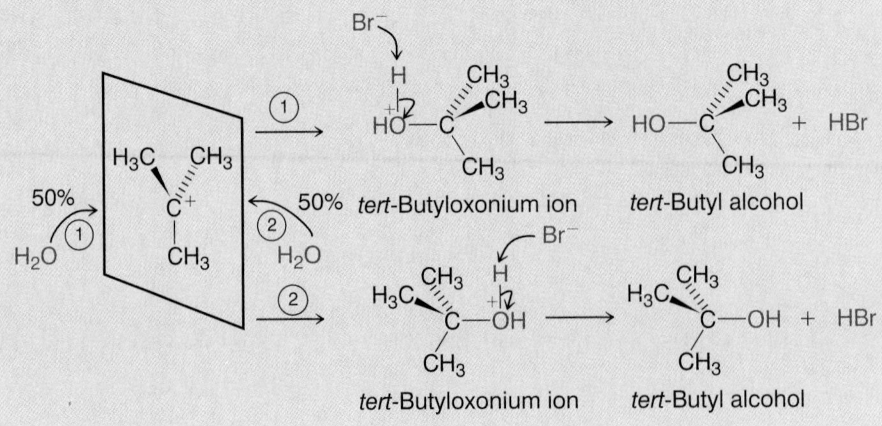


## The $S_N1$ Reaction (Substitution Nucleophilic, First Order)

Let us examine the reaction of *tert*-butyl bromide with water. You might ask, Why not use hydroxide ion, as in the case of reaction 1 (page 236), in which ethyl bromide reacts with

| TABLE 12.3 | Relative Rate of $S_N1$ Substitution of Certain Alkyl Halides with Water | |
|---|---|---|
| | $RX + H_2O \longrightarrow ROH + HX$ | |
| **Alkyl Halide** | **Class** | **Relative Rate of Reaction** |
| $CH_3Br$ | Methyl | Very slow |
| $CH_3CH_2Br$ | 1° | Slow |
| $(CH_3)_2CHBr$ | 2° | Fast |
| $(CH_3)_3CBr$ | 3° | Very fast |

Note: Refer to Section 13.1 for a discussion of the classes of alkyl halides.

hydroxide ion? This is so because elimination—rather than substitution—would predominate if we were to react *tert*-butyl bromide with hydroxide ion (see Sec. 12.4). The reaction of a tertiary halide with a nucleophile that also functions as a solvent is called *solvolysis*. Solvents most often used are water, ethyl alcohol ($CH_3CH_2OH$), and methyl alcohol ($CH_3OH$). As you would surmise because of the different nucleophile employed, this reaction proceeds by a pathway different from the $S_N2$.

$$(CH_3)_3CBr + H_2O \longrightarrow (CH_3)_3COH + HBr$$

Kinetic studies have demonstrated that the rate of reaction depends only on the concentration of alkyl halide. Thus, if the concentration of alkyl halide is kept constant, no matter what concentration of water is used, the rate of the reaction remains the same. Thus we can write for the rate law:

$$Rate = k[(CH_3)_3CBr]$$

where $k$ is the specific rate constant for the reaction, and the brackets represent the concentration of *tert*-butyl bromide in moles per liter. The rate of the reaction also depends on the nature of the alkyl halide, as shown in Table 12.3.

It is interesting to note that the rate of an $S_N1$ reaction increases progressively from a primary to a tertiary halide. The reverse order holds for the rate of an $S_N2$ reaction; i.e., the rate of reaction increases progressively from a tertiary halide to the corresponding methyl halide. Obviously, a factor other than crowding around the central carbon is responsible for $S_N1$ reactivity.

Stereochemical studies have been useful in formulating a mechanism for $S_N1$ reactions, just as for $S_N2$ reactions. Consider, then, a chiral tertiary organic halide reacting with ethyl alcohol:

(R)-2-Chloro-2-phenylbutane    (R)-2-Ethoxy-2-phenylbutane    (S)-2-Ethoxy-2-phenylbutane

Racemic mixture

The resulting ether is found to be optically inactive due to formation of a racemic mixture.[*]

Now let us formulate a mechanism of $S_N1$ substitution[†] consistent with the experimental data obtained. Since the reaction is first-order in organic halide concentration, only the organic halide participates in the rate-determining, or slow, step. The only process the halide could undergo by itself would be a dissociation to form the carbocation

It is unlikely that the halide would dissociate to form a carbanion because bromine is more electronegative than carbon.

*Although some $S_N1$ reactions produce a racemic mixture, most give partially racemized products—typical is 47% of one stereoisomer and 53% of the enantiomer. Refer to Problem 12.5.
†Both $S_N1$ reactions and E1 reactions (page 252) are equilibrium processes, but for pedagogical reasons, this text will not treat them as such.

intermediate and halide ion. Using *tert*-butyl bromide reacting with water as an example, we propose as the first, rate-determining step

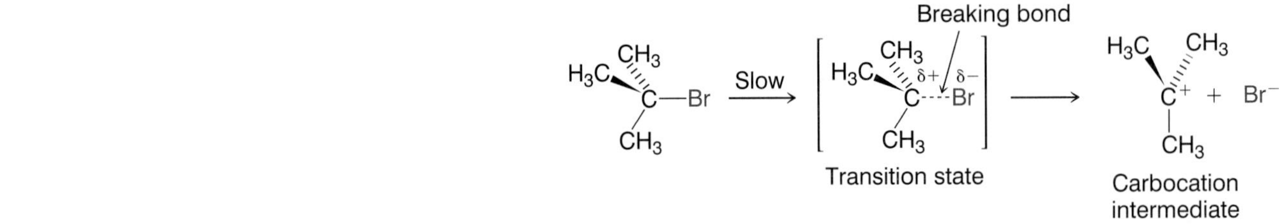

Formation of the carbocation explains the relative rates of reaction in Table 12.3. Generation of the relatively stable *tert*-butyl carbocation (see Sec. 6.2) results in a comparable fast rate-determining step and therefore a fast reaction, whereas formation of the fairly unstable methyl carbocation results in a slow rate-determining step and therefore a slow reaction.

Another reason to propose the formation of the carbocation is the formation of racemic product from an optically active reagent. Consider the rate-determining step when (*R*)-2-chloro-2-phenylbutane is reacted with ethyl alcohol in a solvolysis. The intermediate formed is planar and achiral, leading to the racemic product.

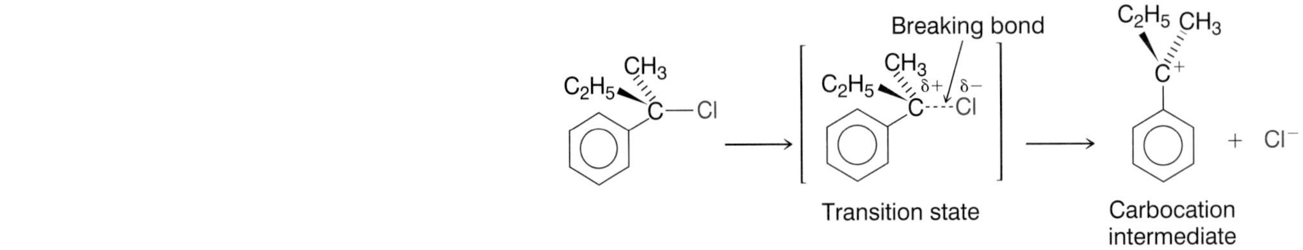

All carbocations are planar (flat) in structure so that the nucleophile can attack the carbocation equally from the left side or from the right side to give equal amounts of a pair of enantiomers—a racemic mixture. The final step of the mechanism involves a rapid attack on the positively charged carbocation by ethyl alcohol to give the oxonium ion, followed by the loss of a proton to give product. An oxonium ion is a positively charged species in which the positive charge is carried by an oxygen atom.

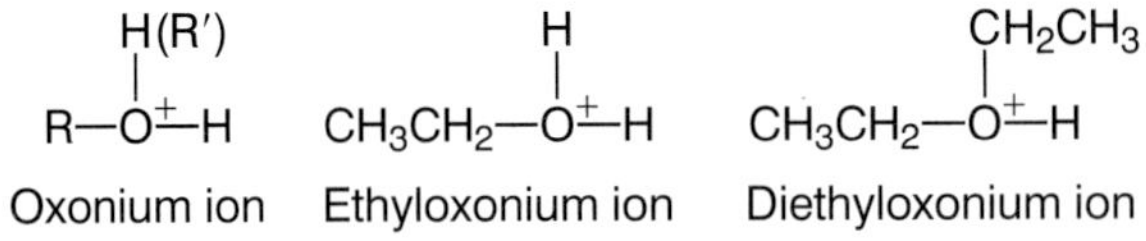

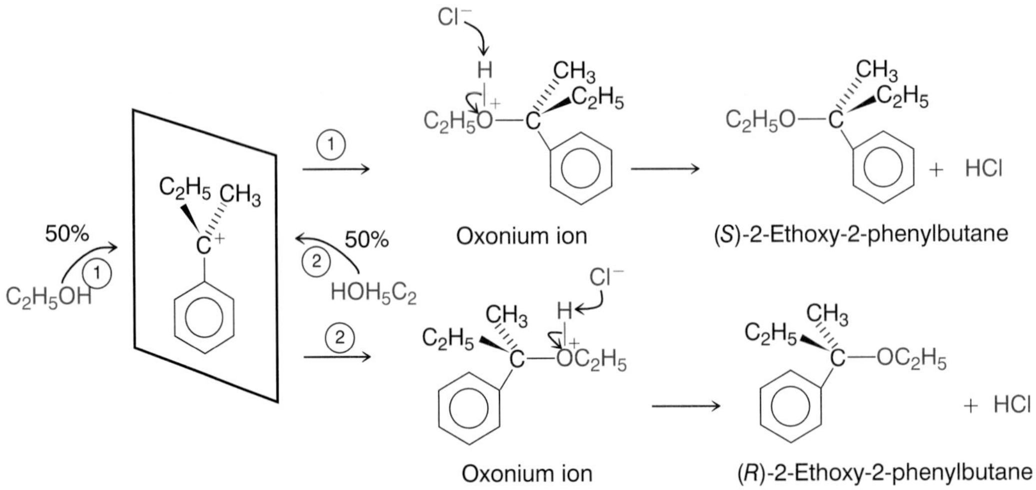

Let us summarize this two-step $S_N1$ mechanism using the reaction of *tert*-butyl bromide with water as an example. The first step of the mechanism is the dissociation of the organic compound to give a carbocation.

First step:

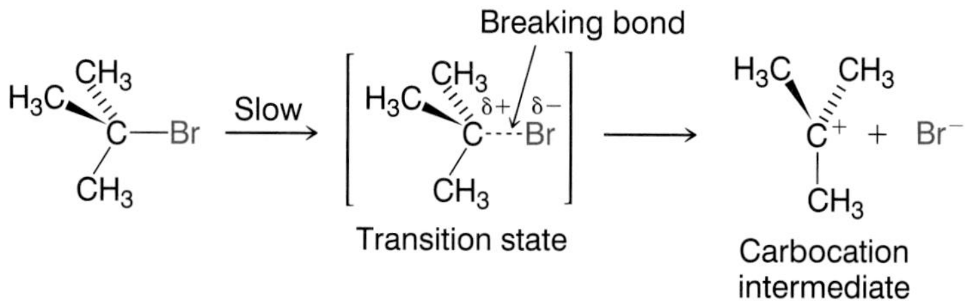

Finally, the nucleophile, water, attacks the intermediate carbocation to give the oxonium ion, which, in turn, loses a proton to give product.

Second step:

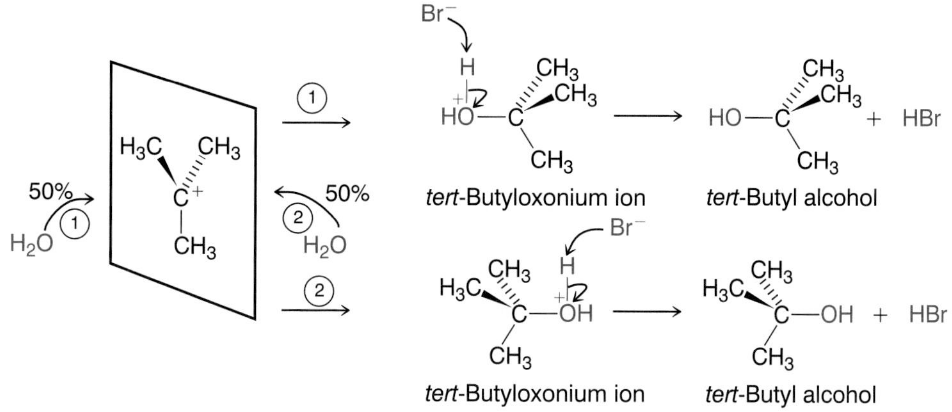

Note that no stereochemical considerations are relevant in this case because *tert*-butyl bromide is an achiral reagent and *tert*-butyl alcohol, an achiral compound, is the only product formed.

---

**SOLVED PROBLEM 12.6**

Consider the following $S_N1$ solvolysis reaction:

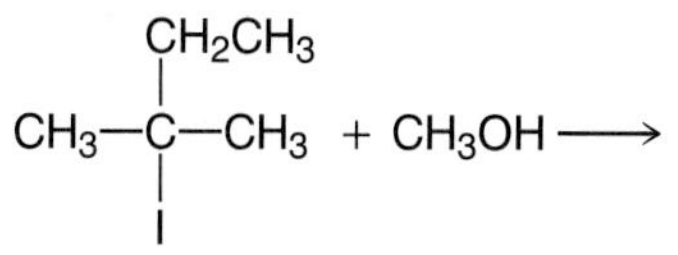

$$CH_3-\underset{\underset{I}{\overset{|}{C}}}{\overset{\overset{CH_2CH_3}{|}}{}}-CH_3 \ + \ CH_3OH \longrightarrow$$

*tert*-Pentyl iodide

a. Complete the equation.

b. Write a rate expression for the reaction.

c. What effect does doubling the concentration of *tert*-pentyl iodide have on the reaction rate? What effect does doubling the concentration of $CH_3OH$ have on the rate of reaction?

**SOLUTION**

a. 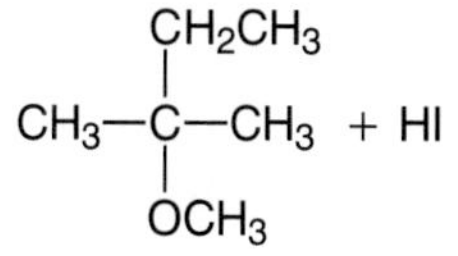 $CH_3-\underset{\underset{OCH_3}{\overset{|}{C}}}{\overset{\overset{CH_2CH_3}{|}}{}}-CH_3 \ + \ HI$

b. Since this is an $S_N1$ reaction, the rate depends only on the concentration of the substrate, the alkyl halide, and we can write

$$\text{Rate} \ = \ k[\text{\textit{tert}-pentyl iodide}]$$

c. Doubling the concentration of *tert*-pentyl iodide doubles the rate. Doubling the concentration of $CH_3OH$ does not affect the rate because $CH_3OH$ is not involved in the rate-determining, or slow, step of the mechanism.

## SOLVED PROBLEM 12.7

List the following carbocations in order of increasing stability.

a. $CH_3(CH_2)_3\overset{+}{C}H_2$

b. $(CH_3)_2\overset{+}{C}CH_2CH_3$

c. $CH_3\overset{+}{C}H(CH_2)_2CH_3$

### SOLUTION

These carbocations are stabilized by the inductive effect (see Sec. 6.2). The carbocation with the greatest number of alkyl groups bonded to the carbon with the positive charge most effectively delocalizes the positive charge and stabilizes the carbocation to the greatest extent. Thus a tertiary carbocation is more stable than a secondary carbocation, which, in turn, is more stable than a primary carbocation. Thus we have (a) < (c) < (b).

## EXERCISE 12.6

Draw a detailed, step-by-step mechanism for each of the following $S_N1$ reactions. Use wedge projections (page 244), and label the transition state and intermediate. Refer to Section 13.1 for the nomenclature of organic halogen compounds.

a. (*R*)-2-bromobutane + $CH_3OH \longrightarrow$

b. 2-iodopropane + $H_2O \longrightarrow$

## EXERCISE 12.7

Select the more favorable reaction pathway ($S_N1$ or $S_N2$), and then complete each of the following reactions.

a. $CH_3CH_2CH_2I + {}^-OCH_3$ (methoxide ion) $\longrightarrow$

b. $(CH_3CH_2)_3CI + CH_3OH \longrightarrow$

Vinyl compounds are also unreactive in $S_N1$ substitution. The presence of a double bond between carbon and bromine in structure II makes dissociation impossible in the $S_N1$ mechanism.

I                                        II

Allylic and benzylic systems substitute using the $S_N1$ mechanism (even though the allyl and benzyl groups are primary) when the nucleophile is poor (a weak base such as water, methyl alcohol, or ethyl alcohol) because allylic and benzylic carbocations are significantly stabilized by resonance. The poor nucleophile does not attack the halogen-bearing carbon until the positively charged carbocation has been formed. On the other hand, if the nucleophile is good (strong base), substitution occurs via the $S_N2$ mechanism. The good nucleophile attacks the carbon bearing the halogen before carbocation formation occurs. The driving force for this attack is the small amount of steric hindrance in the $S_N2$ transition state.

$$CH_2{=}\overset{\overset{\displaystyle H}{|}}{C}{-}CH_2{-}Cl + H_2O \xrightarrow{\ S_N1\ } CH_2{=}\overset{\overset{\displaystyle H}{|}}{C}{-}CH_2OH + HCl$$

Allyl chloride                                Allyl alcohol
Weak base

$$CH_2{=}\overset{\overset{\displaystyle H}{|}}{C}{-}CH_2{-}Cl + {}^-OH \xrightarrow{\quad S_N2 \quad} CH_2{=}\overset{\overset{\displaystyle H}{|}}{C}{-}CH_2{-}OH + Cl^-$$

Strong base

---

Draw resonance structures for the benzyl carbocation, and draw a structure of the resonance hybrid.

**SOLVED PROBLEM 12.8**

**SOLUTION**

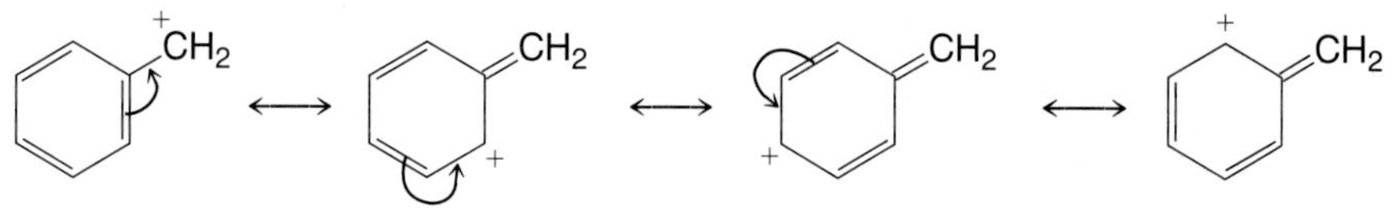

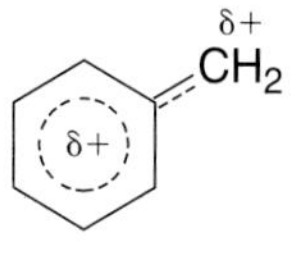

Resonance hybrid

---

Draw resonance structures for the methylbenzyl carbocation ($C_6H_5\overset{+}{C}HCH_3$), and draw a structure of the resonance hybrid.

**EXERCISE 12.8**

---

Reacting an alcohol with hydrochloric acid (and zinc chloride), hydrobromic acid, or hydriodic acid produces the corresponding halide. In general,

**12.2 NUCLEOPHILIC SUBSTITUTION REACTIONS: ALCOHOLS**

$$ROH + HX \longrightarrow RX + H_2O$$

Tertiary, allyl, and benzyl alcohols react by way of an $S_N1$ mechanism; primary alcohols are converted to the corresponding halide by means of an $S_N2$ mechanism. Secondary alcohols can react by means of either mechanism. Refer to Section 14.1 for the classification and nomenclature of alcohols.

The order of reactivity is

$$3° > 2° > 1°$$

irrespective of the mechanism. For example:

$$CH_3CH_2Cl\,l_2OH + HBr \xrightarrow{\quad \Delta \quad} CH_3CH_2CH_2Br + H_2O$$

In order to prepare a primary chloride from an alcohol, a catalyst, zinc chloride ($ZnCl_2$), is necessary.

$$CH_3CH_2CH_2OH + HCl \xrightarrow[\Delta]{\quad ZnCl_2 \quad} CH_3CH_2CH_2Cl + H_2O$$

A typical $S_N1$ mechanism proceeds as follows when *tert*-butyl alcohol is reacted with hydrobromic acid:

$$CH_3{-}\overset{\overset{\displaystyle CH_3}{|}}{\underset{\underset{\displaystyle CH_3}{|}}{C}}{-}OH + HBr \xrightarrow[\text{temperature}]{\text{Room}} CH_3{-}\overset{\overset{\displaystyle CH_3}{|}}{\underset{\underset{\displaystyle CH_3}{|}}{C}}{-}Br + H_2O$$

**Envision the Reaction**

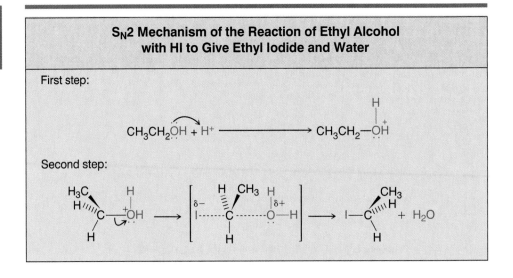

### $S_N1$ Mechanism of the Reaction of *tert*-Butyl Alcohol with HBr to Give *tert*-Butyl Bromide and Water

First step:

$$(CH_3)_3C\ddot{O}H + H^+ \longrightarrow (CH_3)_3C\overset{H}{\underset{+}{O}}H$$

Second step:

$$(CH_3)_3C\overset{H}{\underset{+}{O}}H \longrightarrow (CH_3)_3C^+ + H_2O$$

Third step:

$$(CH_3)_3C^+ + :\ddot{B}r:^- \longrightarrow (CH_3)_3CBr$$

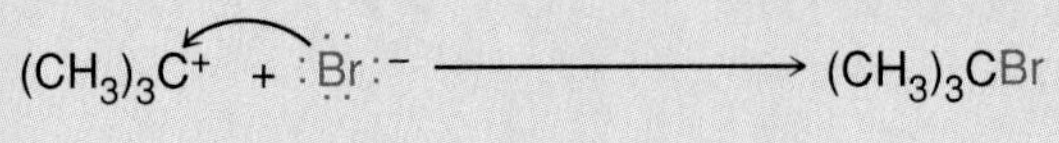

$$R\!-\!OH \longrightarrow R^+ + {}^-OH$$

Strong base
(poor leaving group)
(dissociation step does
not take place)

$$R\!-\!\overset{H}{\underset{+}{O}}\!-\!H \longrightarrow R^+ + H_2O$$

(reaction          Weak
step            base
takes       (excellent
place)       leaving
group)

Notice that an oxonium ion is first formed (see Sec. 12.1) before the corresponding carbocation is produced. This extra step in the mechanism is necessary because the hydroxide ion is too strong a base and thus a very poor leaving group for dissociation to occur. It is the oxonium ion that is readily converted in the second step of the mechanism to a carbocation because water is an excellent leaving group.

The following illustrates a typical $S_N2$ mechanism. Here, ethyl alcohol is reacting with hydriodic acid.

$$CH_3CH_2OH + HI \xrightarrow{\Delta} CH_3CH_2I + H_2O$$

**Envision the Reaction**

### $S_N2$ Mechanism of the Reaction of Ethyl Alcohol with HI to Give Ethyl Iodide and Water

First step:

$$CH_3CH_2\ddot{O}H + H^+ \longrightarrow CH_3CH_2\!-\!\overset{H}{\underset{\ddot{}}{O}}H$$

Second step:

$$\underset{H}{\overset{H_3C}{\underset{H^{////}}{C}}}\!-\!\overset{+}{O}H \longrightarrow \left[ \overset{\delta^-}{I}\cdots C\cdots\overset{\delta^+}{O}\!-\!H \right] \longrightarrow I\!-\!C\overset{CH_3}{\underset{H}{\overset{H}{\phantom{|}}}} + H_2O$$

### SOLVED PROBLEM 12.9

When 2-buten-1-ol is reacted with HBr, both the expected product, 1-bromo-2-butene, and an unexpected product, 3-bromo-1-butene, are produced. Write a detailed, step-by-step mechanism for the formation of the unexpected product as follows:

$$CH_3CH{=}CHCH_2OH + HBr \longrightarrow CH_3CH(Br)CH{=}CH_2 + H_2O$$

Assume that addition of HBr across the carbon-carbon double bond does not take place.

**SOLUTION**

Since the product produced is unexpected (we would expect $CH_3CH{=}CHCH_2Br$ to form), an $S_N1$ mechanism is indicated as follows:

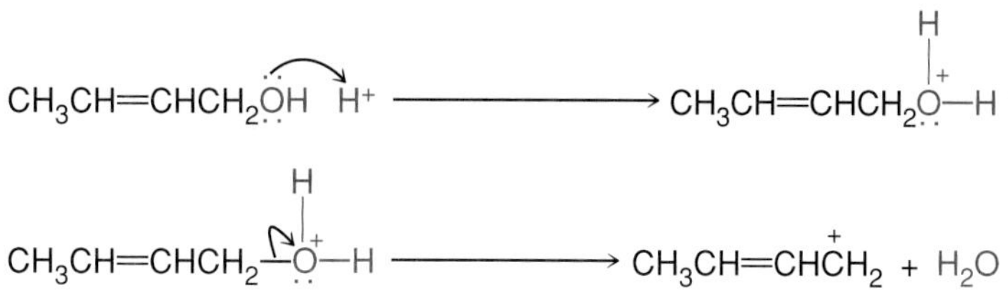

II

Structure II is an allylic carbocation and thus can be assigned a stabilizing resonance structure.

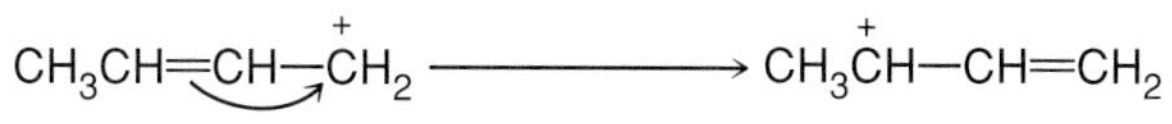

III

To complete the mechanism, structure III is attacked by bromide ion to give product.

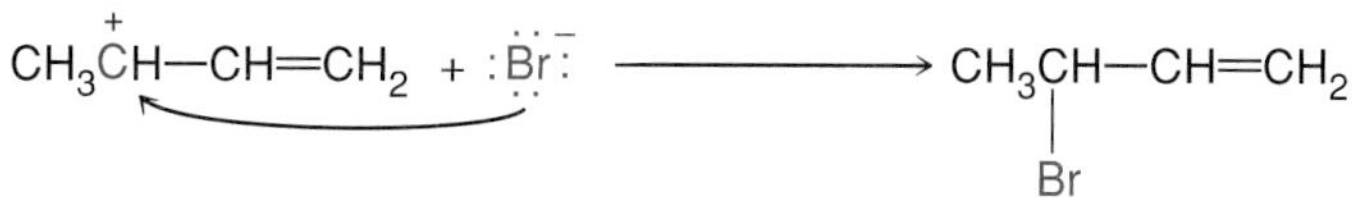

Solved Problem 12.9 illustrates an important characteristic of $S_N1$ reactions—the occurrence of rearrangements. Rearranged products can result from resonance considerations or when a less stable carbocation rearranges to a more stable carbocation. Since a carbocation is not formed in an $S_N2$ reaction, rearrangements cannot occur.

### EXERCISE 12.9

Write a detailed, step-by-step mechanism for each of the following reactions.

a. $CH_3CH_2CH(OH)CH_3 + HI \xrightarrow{\Delta} CH_3CH_2CH(I)CH_3 + H_2O \quad (S_N2)$
(*S*)-Isomer

b. $(CH_3)_3COH + HI \xrightarrow[\text{temperature}]{\text{Room}} (CH_3)_3CI + H_2O \quad (S_N1)$

## 12.3 A COMPARISON OF $S_N2$ AND $S_N1$ MECHANISMS

Table 12.4 represents a summary of the characteristics of the $S_N2$ and $S_N1$ reaction mechanisms.

| TABLE 12.4 | A Comparison of $S_N2$ and $S_N1$ Reaction Mechanisms | |
|---|---|---|
| | $S_N2$ | $S_N1$ |
| Order | Second order | First order |
| Number of steps | One-step concerted mechanism (halides) | Two-step mechanism (halides) |
| | Two-step mechanism (last step concerted) (alcohols) | Three-step mechanism (alcohols). |
| Rate and structure of substrate | Methyl > 1° > 2° > 3° (halides) | 3° > 2° > 1° > methyl (halides) (alcohols for $S_N1$ and $S_N2$)] |
| Rate and concentration of reactants | Depends on concentration of both organic compound and nucleophile | Depends on concentration of substrate only |
| Stereochemistry | Complete inversion of configuration | Racemization |
| Nucleophile | Good (strong base) | Poor (weak base) |
| Rearrangements | Never observed | Often observed |

## 12.4 ELIMINATION REACTIONS: ORGANIC HALOGEN COMPOUNDS

Consider treating an alkyl halide with a nucleophile (base). We find that two competing reactions can take place: substitution and elimination.

$$CH_3CH_2Cl + OH^- \underset{\text{Elimination}}{\overset{\text{Substitution}}{\rightleftarrows}} \begin{array}{l} CH_3CH_2OH + Cl^- \\ CH_2{=}CH_2 + Cl^- + H_2O \end{array}$$

In some instances it is desirable to maximize substitution; in other instances, elimination. Since substitution was already discussed in some depth, let us now turn our attention to the mechanism of elimination.

### The E2 Mechanism

The *E2 mechanism* is similar in character to the $S_N2$ mechanism, with second-order kinetics and a one-step concerted reaction process depending on the stability of the alkene formed. The more highly substituted the alkene, the more stable is the alkene and the faster it is formed. Thus the order of reactivity is 3° > 2° > 1°.

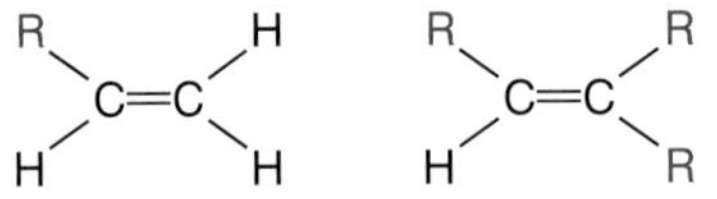

Less stable          More stable

(Refer to Sec. 6.1.)

### Envision the Reaction

**General E2 Mechanism of an Alkyl Halide Reacting with Hydroxide Ion to Form an Alkene, Water, and Halide Ion**

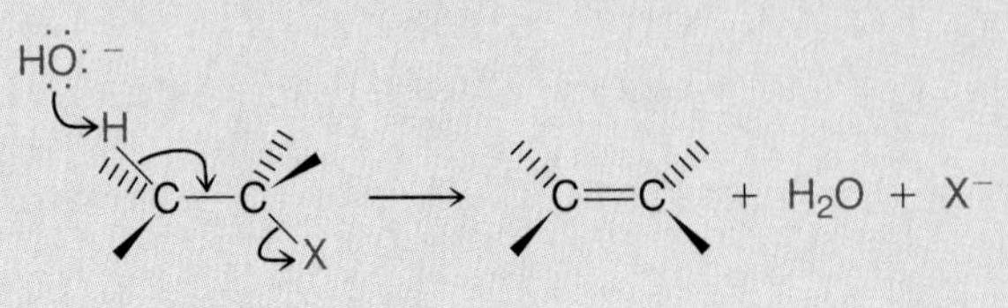

In this first representation of the E2 mechanism, note the hydrogen attacked by base; the halogen and the carbons bonded to these atoms must lie on the same plane. This is a necessary condition for elimination and is called *periplanar geometry*. Since the two groups, hydrogen and halogen, leave from opposite sides of the molecule, we call this *antiperiplanar geometry*. The usual curved arrows represent the flow of electron pairs. In addition, notice that the hydrogen attacked by base is bonded to a carbon that is next to the carbon bearing the halogen.

This next representation of the E2 mechanism depicts the bonds breaking and forming on the one hand (shown as dots) and shows a representation of the transition state of the reaction on the other. The four atoms concerned (the hydrogen attacked by base and the halogen and the carbons bonded to these atoms) must lie on the same plane to ensure maximum overlap of the *p* orbitals in the alkene and partial overlap in the transition state.

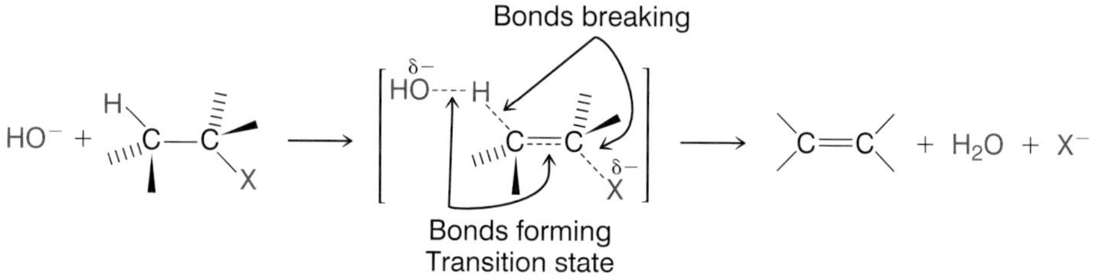

Bonds breaking

Bonds forming
Transition state

where X = Cl, Br, or I

---

Consider the elimination of (2*S*,3*R*)-2-bromo-3-methylpentane under E2 conditions using potassium hydroxide in ethyl alcohol. Determine the configuration of and name the alkene formed.

**SOLVED PROBLEM 12.10**

**SOLUTION**

First, we must decide on the location of the double bond to be formed. Placing the double bond between carbon 2 and carbon 3 is called for because this produces the alkene with the greatest stability. Next, we create a sawhorse projection in such a manner that H-3, C-3, C-2, and Br-2 are antiperiplanar. A set of molecular models is invaluable here.

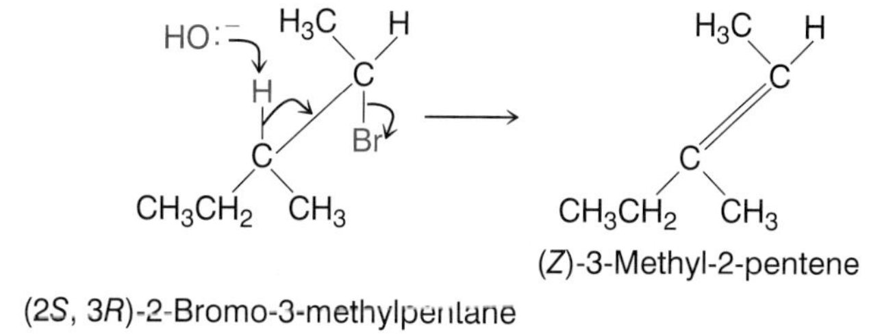

(Z)-3-Methyl-2-pentene

(2S, 3R)-2-Bromo-3-methylpentane

---

Consider the elimination of (2*R*,3*R*)-2-bromo-3-methylpentane under E2 conditions using potassium hydroxide in ethyl alcohol. Determine the configuration of and name the alkene formed.

**EXERCISE 12.10**

**Envision the Reaction**

---

**E1 General Mechanism of the Elimination
of an Alkyl Bromide to Form
an Alkene, Water, and Bromide Ion**

First step:

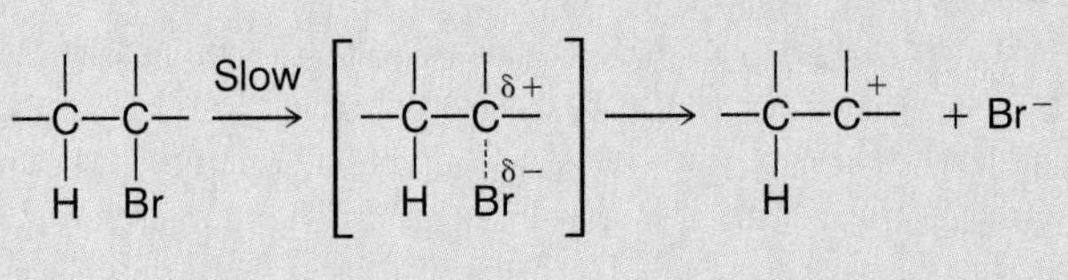

$$-\overset{|}{\underset{H}{C}}-\overset{|}{\underset{Br}{C}}- \xrightarrow{\text{Slow}} \left[ -\overset{|}{\underset{H}{C}}-\overset{|}{\underset{Br}{\overset{\delta+}{C}}}_{\delta-} \right] \longrightarrow -\overset{|}{\underset{H}{C}}-\overset{|}{\overset{+}{C}}- + Br^-$$

Second step:

$$-\overset{|}{\underset{\overset{|}{H}}{C}}-\overset{|}{\overset{+}{C}}- \xrightarrow{\text{Fast}} \overset{}{\underset{}{>}}C=C\overset{}{\underset{}{<}} + H_2O$$
$$\curvearrowleft OH^-$$

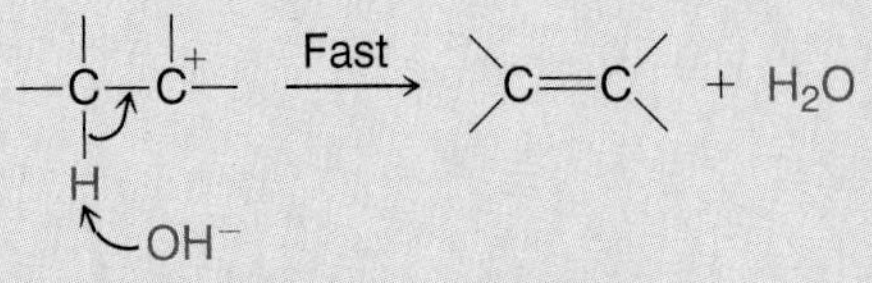

Note: The proton lost must be on
the carbon next to the carbon with
the positive charge.

---

**The E1 Mechanism**

The *E1 mechanism* is similar to the $S_N1$ in that both show first-order kinetics and both are two-step mechanisms. Mechanistically, the E1 mechanism is similar to the $S_N1$ mechanism in that the carbocation is a reaction intermediate. The order of reactivity is

$$3° > 2° > 1°$$

**12.5
ELIMINATION
REACTIONS:
ALCOHOLS**

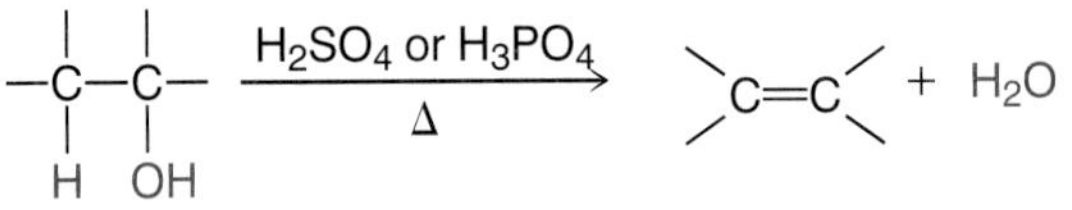

$$-\overset{|}{\underset{H}{C}}-\overset{|}{\underset{OH}{C}}- \xrightarrow[\Delta]{H_2SO_4 \text{ or } H_3PO_4} \overset{}{>}C=C< + H_2O$$

An alcohol can eliminate water to form an alkene by way of an E1 or an E2 mechanism. For previous discussions of this reaction, refer to Section 6.1. The mechanism followed depends on the structure of the alcohol used. Tertiary, substituted allyl, and substituted benzyl alcohols react by an E1 mechanism.

## E1 Mechanism of the Reaction of *tert*-Butyl Alcohol to Form 2-Methylpropene and Water

**Envision the Reaction**

First step:

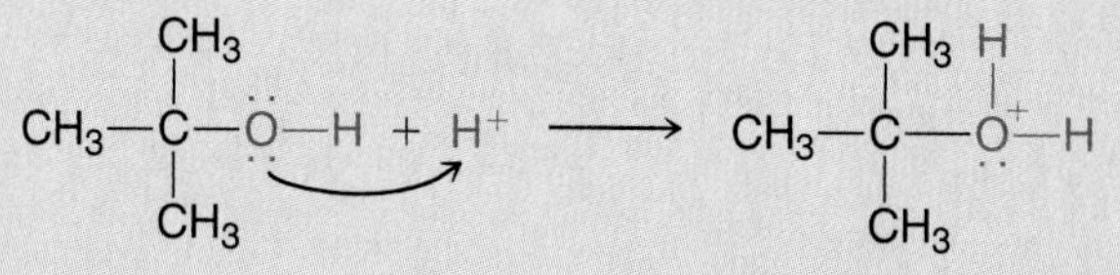

*tert*-Butyloxonium ion

Second step:

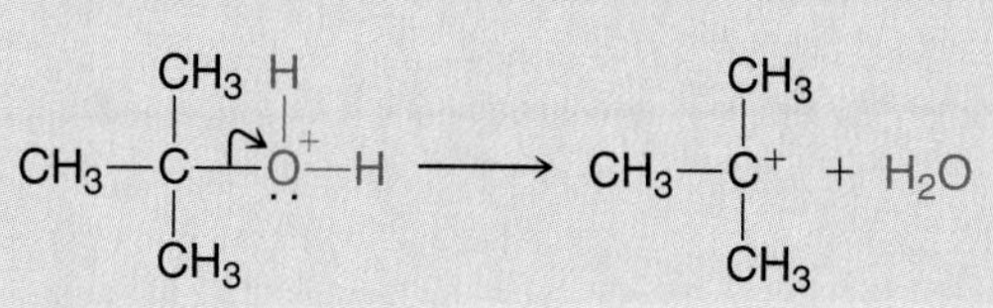

Third step:

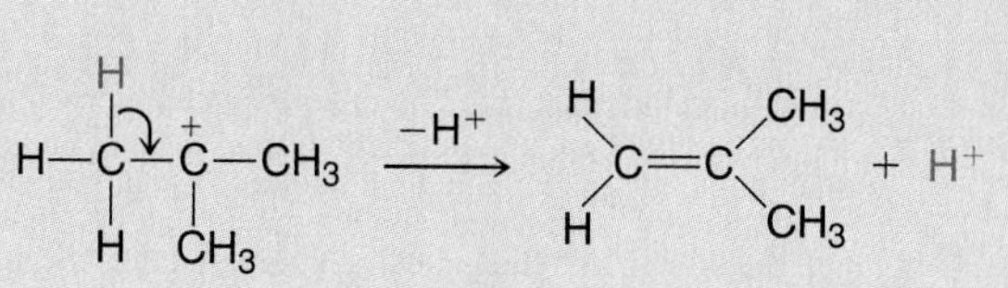

The first step in the E1 mechanism involves an acid-base reaction where a proton from the acid attacks the electron-rich oxygen of the alcohol (e.g., *tert*-butyl alcohol) to give an oxonium ion. The second step involves the loss of water to give the carbocation. The driving force of the second step is a combination of the shifting of a positive charge from the electronegative element, oxygen, to a more electropositive element, carbon, stabilizing the positive charge, and the formation of a very stable water molecule.

If the carbocation formed is relatively stable, such as a tertiary, substituted allyl, or substituted benzyl carbocation, the mechanism of the elimination is E1. A less stable carbocation such as a simple (not resonance-stabilized) primary carbocation would drive the corresponding alcohol into a E2 elimination in which the carbocation is not produced. The third step of the E1 mechanism involves the loss of a proton to produce the alkene.

Note that the proton lost must be bonded to the carbon next to the carbon bearing the positive charge. Since the proton is introduced into the reaction in the first step and is liberated from the reaction in the third step, the proton is a true catalyst because it is not consumed in the reaction. This dehydration reaction is a reversible process* because an alkene can react with sulfuric acid to form an alkyl hydrogen sulfate and ultimately an alcohol when the alkyl hydrogen sulfate reacts with water.

When *n*-propyl alcohol reacts with concentrated phosphoric acid, propene (propylene) is produced by means of an E2 mechanism.

---

*Each step in the E1 mechanism just discussed is reversible. However, for the sake of simplicity and convenience, each step is shown as going to completion.

**Envision the Reaction**

---

### E2 Mechanism of the Reaction of *n*-Propyl Alcohol in the Presence of $H_2SO_4$ to Give Propene and Water

First step:

$$CH_3CH_2CH_2\text{—}OH + H^+ \longrightarrow CH_3CH_2CH_2\overset{\overset{\displaystyle H}{|}}{\underset{\cdot\cdot}{O}}{}^+\!\!\text{—}H$$

Second step:

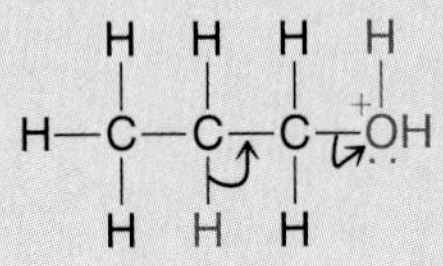

$$\underset{\underset{H}{|}}{\overset{\overset{H}{|}}{H\text{—}C}}\text{—}\underset{\underset{H}{|}}{\overset{\overset{H}{|}}{C}}\text{—}\underset{\underset{H}{|}}{\overset{\overset{H}{|}}{C}}\text{—}\overset{+}{\underset{\cdot\cdot}{O}}H \longrightarrow H_2O + CH_3(H)C{=}CH_2 + H^+$$

---

The first step results in the formation of the oxonium ion. This step is necessary to convert a poor leaving group ($^-OH$) into an excellent leaving group ($H_2O$). Water is a weak base and thus an excellent leaving group. The second step represents a concerted loss of a proton with formation of the double bond and loss of water. Remember, a carbocation does not form in an E2 mechanism.

Under certain circumstances, dehydration of an alcohol produces a totally unexpected product. For example, dehydration of *n*-butyl alcohol produces not only 1-butene (expected) but also 2-butene (unexpected).

$$CH_3CH_2CH_2CH_2OH \xrightarrow[\Delta]{H_2SO_4}$$

$$CH_3CH_2(H)C{=}CH_2 + CH_3(H)C{=}C(H)CH_3 + H_2O$$

1-Butene (expected)         2-Butene
(27.5% yield)           (unexpected)
                     (72.5% yield)

## Mechanism of the Dehydration of *n*-Butyl Alcohol to Produce 1-Butene (E2) and 2-Butene (E1)

**First step:**

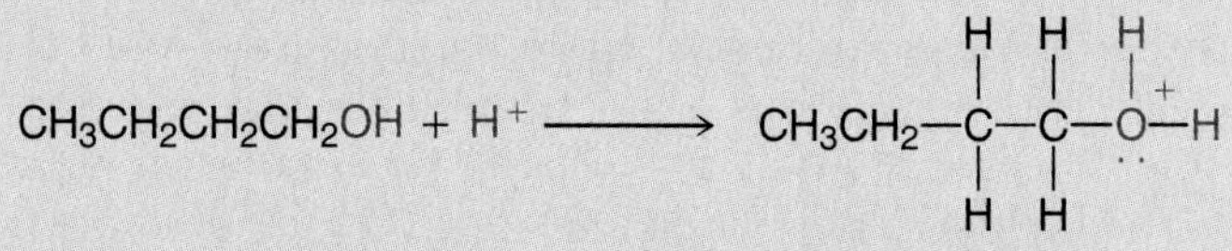

$$CH_3CH_2CH_2CH_2OH + H^+ \longrightarrow CH_3CH_2-\overset{\overset{H}{|}}{\underset{\underset{H}{|}}{C}}-\overset{\overset{H}{|}}{\underset{\underset{H}{|}}{C}}-\overset{+}{\underset{..}{O}}-H$$

*n*-Butyloxonium ion

**Second step:**

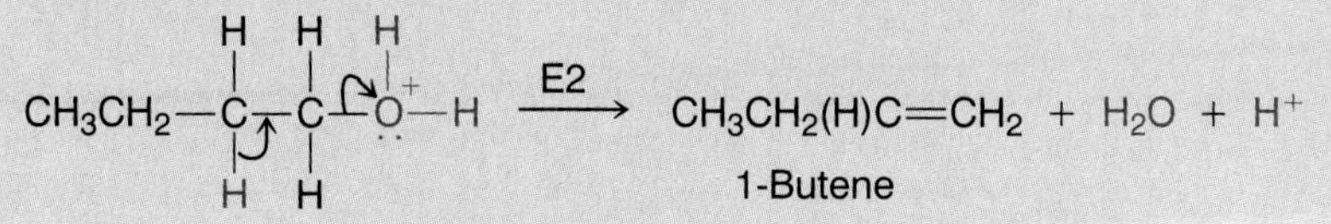

$$CH_3CH_2-\overset{\overset{H}{|}}{\underset{\underset{H}{|}}{C}}-\overset{\overset{H}{|}}{\underset{\underset{H}{|}}{C}}-\overset{+}{O}-H \xrightarrow{\text{E2}} CH_3CH_2(H)C{=}CH_2 + H_2O + H^+$$

1-Butene

**Third step:**

$$CH_3CH_2CH_2\overset{\overset{H}{|}}{\underset{\underset{H}{|}}{C}}-\overset{+}{\underset{..}{O}}-H \xrightarrow{-H_2O} CH_3CH_2CH_2\overset{+}{C}H_2 + H_2O$$

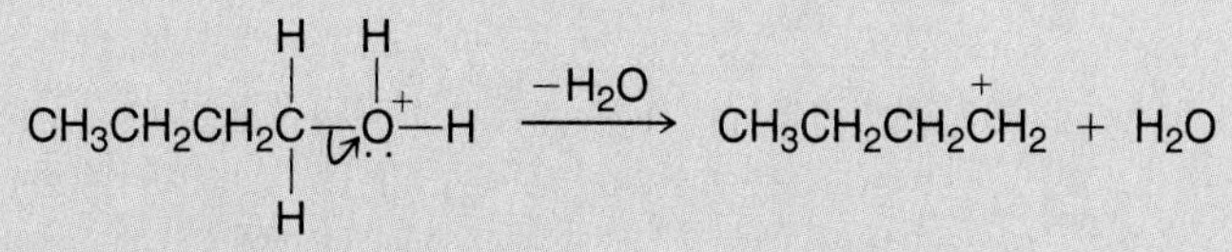

**Fourth R step (hydride ion transfer):**

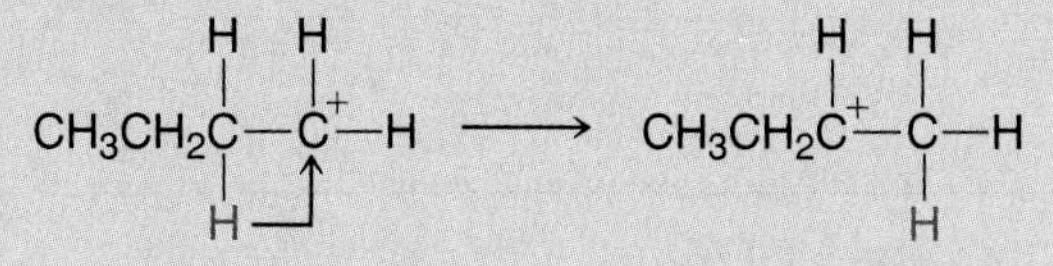

$$CH_3CH_2\overset{\overset{H}{|}}{\underset{\underset{H}{|}}{C}}-\overset{\overset{H}{|}}{\overset{+}{C}}-H \longrightarrow CH_3CH_2\overset{\overset{H}{|}}{\overset{+}{C}}-\overset{\overset{H}{|}}{\underset{\underset{H}{|}}{C}}-H$$

    *n*-Butyl carbocation         *sec*-Butyl carbocation
      (primary)               (secondary)

**Fifth R step:**

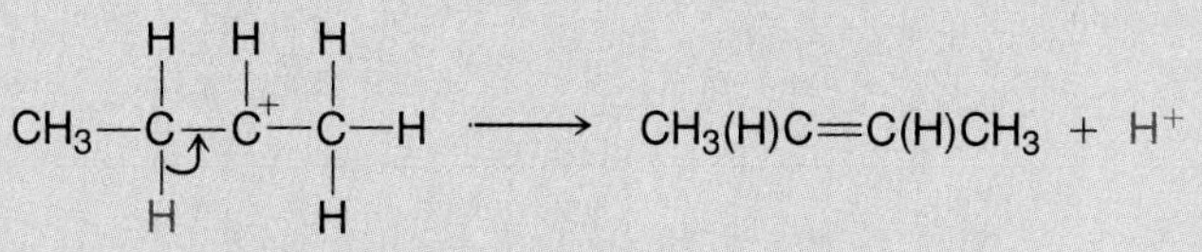

$$CH_3-\overset{\overset{H}{|}}{\underset{\underset{H}{|}}{C}}-\overset{\overset{H}{|}}{\overset{+}{C}}-\overset{\overset{H}{|}}{\underset{\underset{H}{|}}{C}}-H \longrightarrow CH_3(H)C{=}C(H)CH_3 + H^+$$

**Sixth R step:**

$$CH_3-\overset{\overset{H}{|}}{\underset{\underset{H}{|}}{C}}-\overset{\overset{H}{|}}{\overset{+}{C}}-\overset{\overset{H}{|}}{\underset{\underset{H}{|}}{C}}-H \longrightarrow CH_3CH_2(H)C{=}CH_2 + H^+$$

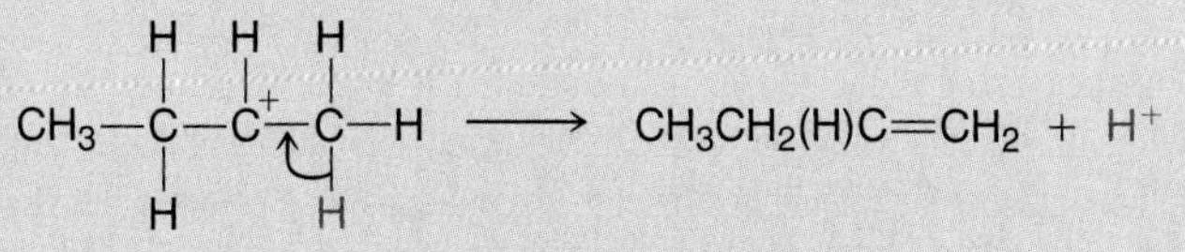

The mechanism begins in the usual manner with formation of the *n*-butyloxonium ion (first step), followed by concerted loss of a proton and water (E2) to produce 1-butene, as shown* in the second step. However, only some of the *n*-butyloxonium ion is converted to alkene. The remainder is transformed into the *n*-butyl carbocation with loss of water in the third step. This *n*-butyl carbocation is, in turn, converted to the *sec*-butyl carbocation (fourth R step; R stands for rearrangement) by means of a hydride ion

---

*Each step in the E2 mechanism just shown is reversible.

transfer (see Sec. 10.1).* Note that a hydride ion consists of a hydrogen nucleus along with a pair of electrons (H:⁻). The driving force for this transformation is the formation of a more stable secondary carbocation from a less stable primary carbocation. Finally, in the fifth R step, a proton is lost to give the more stable alkene. Very little 1-butene is formed (sixth R step) because 1-butene is a less stable alkene than 2-butene.

Since carbocations are produced, the formation of 2-butene proceeds by means of an E1 mechanism. So here is a reaction where butenes are formed from *n*-butyl alcohol by two different mechanisms at the same time. In actuality, 2-butene is isolated in this and other reactions as a mixture of two isomers: *cis*-2-butene and *trans*-2-butene, with the trans isomer predominating. The trans isomer is preferentially formed because it contains less internal energy than the cis isomer and is therefore more stable. In this reaction, the mixture of 2-butenes is isolated in 72.5% yield; 1-butene in 27.5% yield.

**SOLVED PROBLEM 12.11**

Predict the mechanism type for the following reaction.

$$CH_3CH_2CH_2CH_2CH_2OH \xrightarrow[\Delta]{H_2SO_4} CH_3CH_2CH{=}CHCH_3 + H_2O$$

**SOLUTION**
Since the —OH group of the alcohol is located on C-1, we would expect the double bond to form between C-1 and C-2. Because the double bond is located between C-2 and C-3 in the alkene, this is an example of a reaction that results in the formation of an unexpected alkene. Thus a hydride ion migration must have occurred. Therefore, a carbocation must form during the course of the mechanism, and the mechanism must be E1.

**EXERCISE 12.11**

Predict the mechanism(s) for each of the following reactions.

a.   $CH_3CH_2CH_2CH_2CH_2CH_2OH \xrightarrow[\Delta]{H_3PO_4} CH_3CH_2CH_2CH_2CH{=}CH_2$
$+ CH_3CH_2CH_2CH{=}CHCH_3 + H_2O$

b. 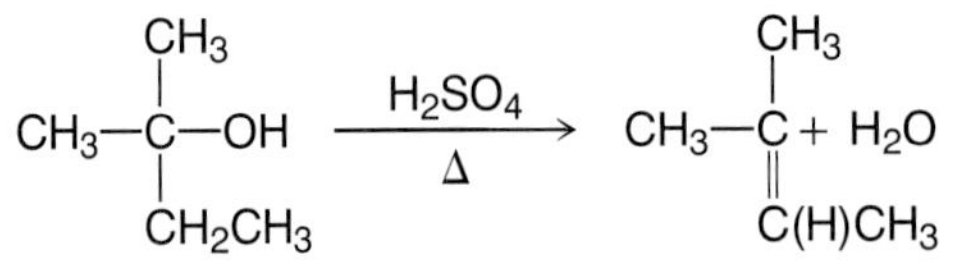
$$\begin{array}{c} CH_3 \\ | \\ CH_3{-}C{-}OH \\ | \\ CH_2CH_3 \end{array} \xrightarrow[\Delta]{H_2SO_4} \begin{array}{c} CH_3 \\ | \\ CH_3{-}C + H_2O \\ \| \\ C(H)CH_3 \end{array}$$

## 12.6
## SUBSTITUTION VERSUS ELIMINATION

In general, an alkyl halide (or other substrate with a good leaving group) that substitutes $S_N2$ eliminates E2. Unlike the $S_N2$ reaction, the E2 reaction shows tertiary substrates reacting faster than secondary substrates, which, in turn, react faster than primary substrates because the tertiary compound, being more branched, produces the more highly substituted and therefore more stable alkene.

In general, a tertiary compound substitutes $S_N1$ and eliminates E1 in solvolysis reactions. When a strong base is used (⁻OH or ⁻OCH$_3$), the E2 mechanism is favored.

*The fourth R step in the mechanism is irreversible.

Secondary substrates tend to undergo $S_N2$ reactions with weak bases, $S_N1$ reactions in solvolysis, and E2 reactions with strong bases. Primary systems give mainly $S_N2$ reactions regardless of conditions. Only hindered strong bases such as *tert*-butoxide ion $[(CH_3)_3CO^-]$ give an E2 reaction with an organic halide.

With a tertiary compound, substitution is favored by relatively low temperatures and a good nucleophile (a weak base). If a strong base is used, it should be in dilute solution. For example:

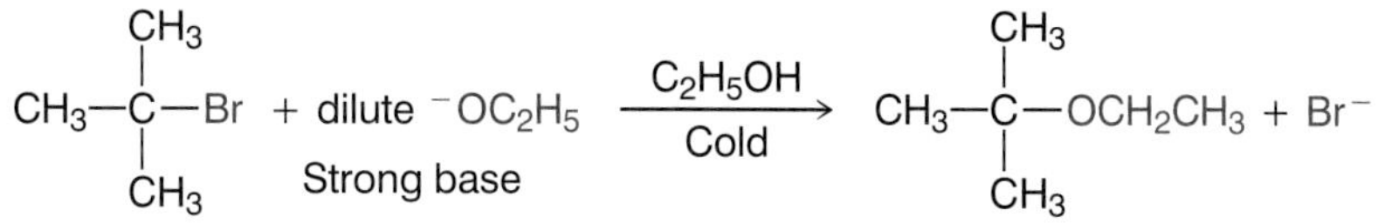

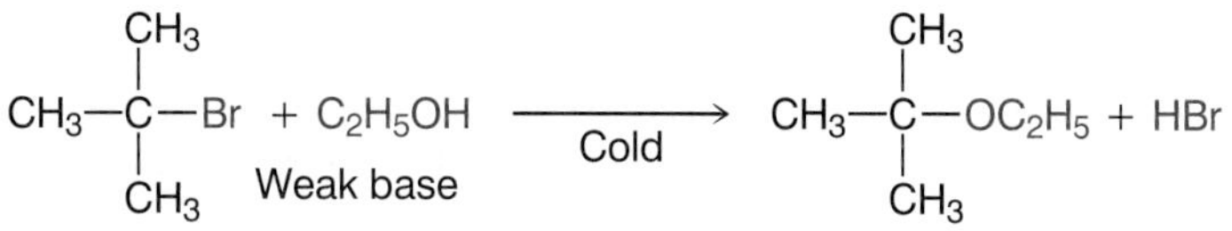

$$CH_3{-}\underset{\underset{CH_3}{|}}{\overset{\overset{CH_3}{|}}{C}}{-}Br \ + \ dilute \ {}^-OC_2H_5 \ \xrightarrow[\text{Cold}]{C_2H_5OH} \ CH_3{-}\underset{\underset{CH_3}{|}}{\overset{\overset{CH_3}{|}}{C}}{-}OCH_2CH_3 \ + \ Br^-$$

Strong base

$$CH_3{-}\underset{\underset{CH_3}{|}}{\overset{\overset{CH_3}{|}}{C}}{-}Br \ + \ C_2H_5OH \ \xrightarrow[\text{Cold}]{} \ CH_3{-}\underset{\underset{CH_3}{|}}{\overset{\overset{CH_3}{|}}{C}}{-}OC_2H_5 \ + \ HBr$$

Weak base

Elimination in a tertiary system, on the other hand, is favored by high temperatures and a strong base (poor nucleophile).

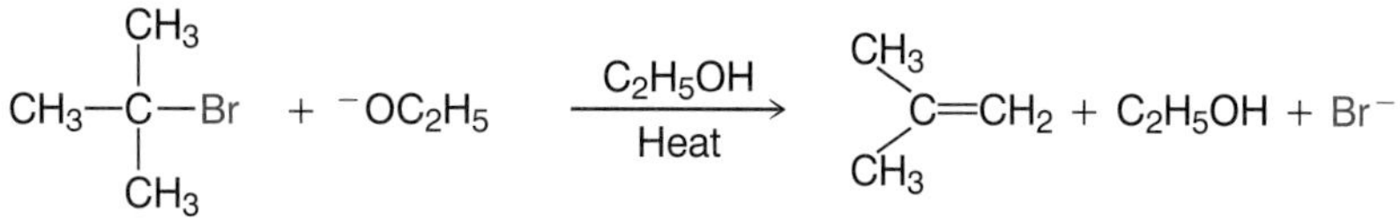

$$CH_3{-}\underset{\underset{CH_3}{|}}{\overset{\overset{CH_3}{|}}{C}}{-}Br \ + \ {}^-OC_2H_5 \ \xrightarrow[\text{Heat}]{C_2H_5OH} \ \underset{\underset{CH_3}{}}{\overset{\overset{CH_3}{}}{C}}{=}CH_2 \ + \ C_2H_5OH \ + \ Br^-$$

---

Classify each of the following reactions as $S_N1$, $S_N2$, E1, or E2. Draw a structural formula for the main organic product.

a. *tert*-pentyl iodide $+ \ CH_3O^- \ \xrightarrow{\Delta}$

b. *tert*-pentyl iodide $+ \ CH_3OH \ \xrightarrow{\text{Cold}}$

c. isopropyl bromide $+ \ I^- \ \longrightarrow$

d. *n*-propyl chloride $+ \ {}^-OH \longrightarrow$

**SOLUTION**

a. The combination of tertiary halide, strong base, and heat add up to an E2 mechanism to yield the alkene:

$$(CH_3)_2C{=}CH(CH_3)$$

b. The solvolysis with a weak base and low temperature indicates an $S_N1$ mechanism to form the ether:

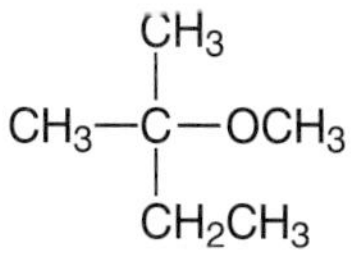

$$CH_3{-}\underset{\underset{CH_2CH_3}{|}}{\overset{\overset{CH_3}{|}}{C}}{-}OCH_3$$

c. Iodide ion is a weak base, so this secondary bromide should form the iodide by means of an $S_N2$ mechanism.

$$CH_3CH(I)CH_3$$

d. Although the base used is strong, primary halides tend to substitute using an $S_N2$ mechanism to give

$$CH_3CH_2CH_2OH$$

**EXERCISE 12.12**

Classify each of the following reaction mechanisms as proceeding by means of an $S_N1$, $S_N2$, E1, or E2 mechanism. Draw a structural formula for the main organic product.

a.  $CH_3CH_2CH_2I + (CH_3)_3CO^- \xrightarrow{\Delta}$

b.  (R)-sec-butyl bromide + $CH_3OH \xrightarrow{\text{Cold}}$

c.  tert-butyl iodide + $CH_3CH_2OH \xrightarrow{\Delta}$

d.  $CH_3I + {}^-OH \longrightarrow$

e.  tert-butyl bromide + $^-OCH_3 \xrightarrow{\Delta}$

In general, when alcohols are reacted with HX, low temperatures favor substitution, and higher temperatures favor elimination. Tertiary, allyl, and benzyl alcohols substitute $S_N1$ and, when practical, eliminate E1. Primary alcohols substitute $S_N2$ and eliminate E2. Secondary alcohols can substitute and eliminate using either reaction path. For example:

$$(CH_3)_3COH + HBr \xrightarrow{\text{Cold}} (CH_3)_3CBr + H_2O \quad (S_N1)$$

$$(CH_3)_3COH \xrightarrow[\Delta]{HBr} (CH_3)_2C{=}CH_2 + H_2O \quad (E1)$$

**SOLVED PROBLEM 12.13**

Classify each of the following reactions as proceeding by means of an $S_N1$, $S_N2$, E1, or E2 mechanism, if possible. Draw a structural formula for the main organic product.

a.
$$\underset{\underset{OH}{|}}{\overset{\overset{CH_3}{|}}{CH_3{-}C{-}CH_2{-}CH_3}} + HBr \xrightarrow{\text{Cold}}$$

b.  $\underset{\underset{OH}{|}}{CH_3CH_2CHCH_2CH_3} \xrightarrow[\Delta]{H_2SO_4}$

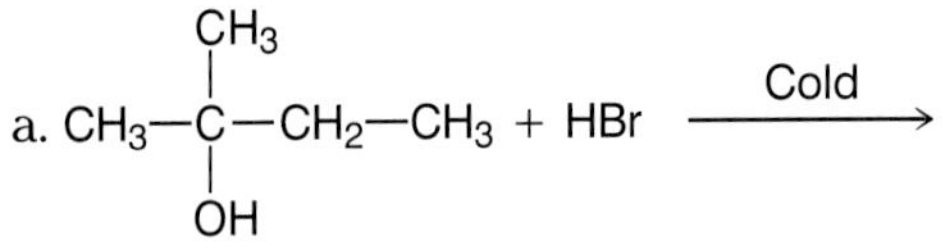

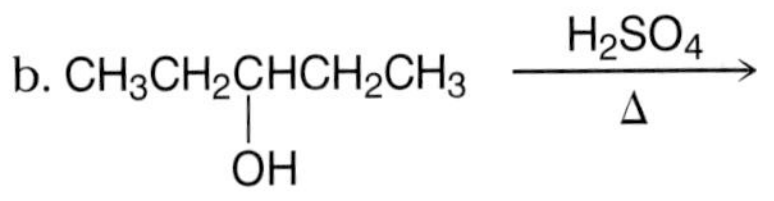

**SOLUTION**

a.  Since the reacting alcohol is tertiary, this suggests a first-order reaction. Because the reaction takes place in the cold, substitution rather than elimination should occur. Putting these two facts together, we can deduce that the mechanism should be $S_N1$. The product is 2-bromo-2-methylbutane:

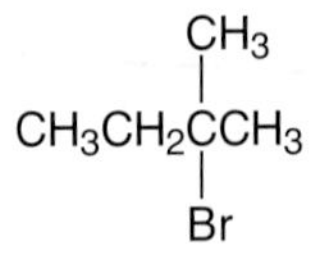

$$\underset{\underset{Br}{|}}{\overset{\overset{CH_3}{|}}{CH_3CH_2CCH_3}}$$

2-Bromo-2-methylbutane

b.  Because the alcohol is heated, an elimination is indicated. The reaction can proceed by either an E1 or E2 mechanism to form 2-pentene:

$$CH_3CH{=}CHCH_2CH_3$$

2-Pentene

Classify each of the following reactions as proceeding by means of an $S_N1$, $S_N2$, E1, or E2 mechanism, if possible. Draw a structural formula for the main organic product.

a.  $C_6H_5-CH_2CH_2OH \xrightarrow[\Delta]{H_2SO_4}$

b.  *n*-butyl alcohol + HI $\longrightarrow$
    (Assume a substitution reaction.)

---

## ▶ CHAPTER ACCOMPLISHMENTS

### 12.1 Nucleophilic Substitution Reactions: Organic Halogen Compounds

☐ Distinguish between a nucleophile and a substrate.
☐ Explain why halide ions are very good leaving groups.
☐ Complete each of the following equations.

  a. $C_2H_5I + {}^-OH \longrightarrow$
  b. $C_2H_5I + {}^-OCH_3 \longrightarrow$

☐ Write a rate law expression for the following reaction:

$$CH_3CH_2Br + {}^-OH \longrightarrow$$

☐ Write a detailed $S_N2$ mechanism for the reaction of ethyl bromide with hydroxide ion.
☐ Explain why the rate of an $S_N2$ reaction decreases as follows:

$$1° \text{ halide fastest} \longrightarrow 3° \text{ halide slowest}$$

☐ Explain the difference between a transition state and an intermediate.
☐ Write a detailed $S_N1$ mechanism for the reaction of *tert*-butyl bromide with water.
☐ Explain why the rate of an $S_N1$ reaction decreases as follows:

$$3° \text{ halide fastest} \longrightarrow 1° \text{ halide slowest}$$

☐ Can you explain the term *solvolysis*?
☐ Explain the formation of a racemic product when a chiral substrate undergoes an $S_N1$ reaction with a nucleophile.
☐ Write a rate law expression for the following reaction: $(CH_3)_3C(I) + H_2O \longrightarrow$

### 12.2 Nucleophilic Substitution Reactions: Alcohols
☐ Write a detailed, step-by-step mechanism for each of the following reactions.

  a. $(CH_3)_3COH + HBr \xrightarrow{\text{Room temperature}}$

  b. $CH_3CH_2OH + HI \xrightarrow{\Delta}$

☐ Write the mechanism of a substitution reaction that produces a rearranged product.

### 12.3 A Comparison of $S_N2$ and $S_N1$ Mechanisms
☐ Compare $S_N2$ and $S_N1$ reaction mechanisms with respect to
  a. order.
  b. number of steps (for an organic halide substrate).
  c. number of steps (for an alcohol substrate).
  d. rate and structure of substrate.
  e. rate and concentration of reactants.
  f. stereochemistry.
  g. nucleophile.
  h. rearrangements.

### 12.4 Elimination Reactions: Organic Halogen Compounds
☐ Write a detailed mechanism for each of the following reactions.
  a. $CH_3CH_2Cl + {}^-OH \longrightarrow$ (E2)
  b. $(CH_3)_3CBr + {}^-OH \longrightarrow$ (E1)

☐ Explain what is meant by the term *periplanar geometry*.

### 12.5 Elimination Reactions: Alcohols
☐ Write a detailed, step-by-step mechanism for each of the following reactions.

  a. $CH_3CH_2OH \xrightarrow{H_2SO_4}$ (E2)

  b. $(CH_3)_3COH \xrightarrow{H_2SO_4}$ (E1)

### 12.6 Substitution versus Elimination
☐ Compare substitution and elimination tendencies with respect to
  a. temperature.
  b. base strength (organic halides).
  c. class of substrate.

## ► KEY TERMS

nucleophilic substitution reaction (12.1)
nucleophile (12.1)
substrate (12.1)
leaving group (12.1)
rate law (12.1)
specific rate constant (12.1)
second-order kinetics (12.1)
second-order reaction (12.1)

substitution nucleophilic second-order
    reaction ($S_N2$ reaction) (12.1)
inversion of configuration (Walden
    inversion) (12.1)
retention of configuration (12.1)
transition state (12.1)
concerted reaction mechanism (12.1)

substitution nucleophilic first-order
    ($S_N1$ reaction) (12.1)
solvolysis (12.1)
E2 mechanism (12.4)
periplanar geometry (12.4)
antiperiplanar geometry (12.4)
E1 mechanism (12.4)

## ► PROBLEMS

1. Consider the $S_N2$ reaction

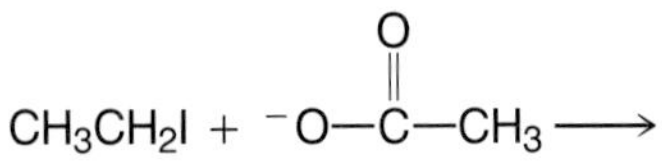

$$CH_3CH_2I + {}^-O-\overset{\overset{\displaystyle O}{\|}}{C}-CH_3 \longrightarrow$$

   a. Complete the equation.
   b. Write a rate-law expression for the reaction.
   c. Identify the substrate and the nucleophile.
   d. Write a detailed mechanism for the reaction (use wedge projections).

2. Which of the following pairs is the less effective leaving group? Why?
   a. $^-OH$ or $H_2O$?
   b. $^-NH_2$ or $NH_3$?
   c. $I^-$ or $Cl^-$?
   d. $H^-$ or $Br^-$?
   e. $^-OH$ or $^-OC_2H_5$?

3. Write a detailed $S_N2$ mechanism for

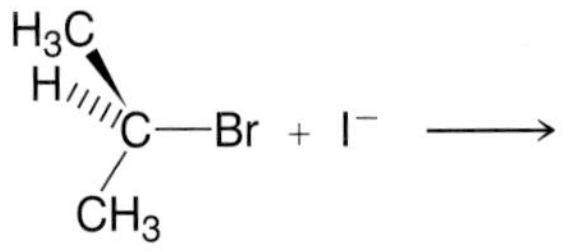

4. Write a detailed, step-by-step mechanism using wedge projections for the following reaction using the yield data given.
   Refer to Section 14.1 for a discussion of carbinol nomenclature.

5. When (R)-2-bromobutane was treated with dilute methoxide ion to give 2-methoxybutane via an $S_N1$ mechanism, a racemic mixture of the ethers was not formed. Instead, 53% (S)-2-methoxybutane and 47% (R)-2-methoxybutane formed. Explain why the racemic mixture did not form. (*Hint*: Examine the position of the leaving group.) Use wedge projections to draw the substrate, transition state, intermediate, and the products of the reaction.

6. Write a detailed, step-by-step $S_N1$ mechanism for

$$(CH_3)_2CHBr + AgNO_3 \xrightarrow{\text{Ethyl alcohol}}$$

   Note that ethyl alcohol in this reaction only functions as a solvent.

7. Why is the rate of reaction of a tertiary halide slower than that of a primary halide in an $S_N2$ pathway?

8. Why is the rate of reaction of a tertiary halide faster than that of a primary halide in an $S_N1$ pathway?

9. Consider the following reaction:

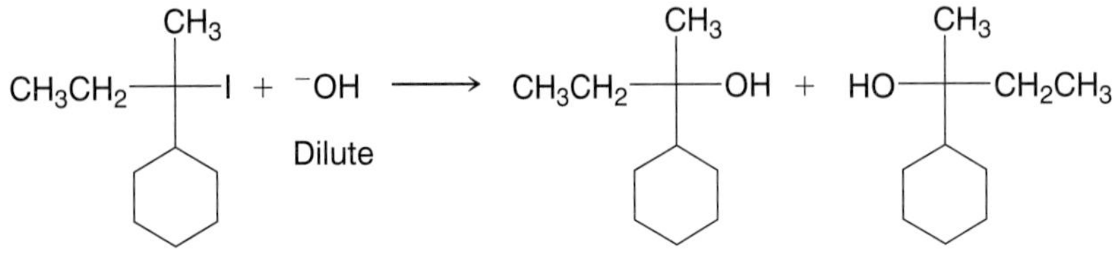

(R)-α-Iodo-sec-butylcyclohexane

(S)-Cyclohexylethylmethylcarbinol
(50%)

(R)-Cyclohexylethylmethylcarbinol
(50%)

a. This solvolysis proceeds by way of an $S_N1$ mechanism even though the $C_6H_5CH_2$— group is primary. Explain.

b. Write a rate expression for the reaction.

c. Write a detailed, step-by-step mechanism for the reaction.

d. Elimination does not occur under these reaction conditions. Why not?

e. Changing the nucleophile to $^-OCH_3$ changed the mechanism.

   (i) Name this mechanism.

   (ii) Why did the mechanism change?

   (iii) Write this mechanism in detail.

10. Neopentyl bromide, a primary alkyl bromide, does not substitute by means of an $S_N2$ mechanism. Explain.

11. a. When sodium acetylide is treated with isopropyl bromide, 3-methylbutyne is not obtained. What are the actual products of the reaction?

    b. On the other hand, when sodium acetylide is reacted with *n*-propyl bromide, 1-pentyne is produced. What generalization can you make from (a) and (b)?

12. Indicate whether substitution or elimination takes place, and complete each of the following equations by drawing a structural formula for each organic product and a molecular formula or formula unit for each inorganic product. Classify each reaction mechanism as $S_N1$, $S_N2$, E1, or E2. If no reaction occurs, write "no reaction." Note that the substances above and below the yield arrows are solvents.

a.

$$CH_3CH_2O-\underset{\underset{O}{\|}}{\overset{\overset{O}{\|}}{S}}-CH_3 \ + \ ^-OH \ (dilute) \ \xrightarrow{\text{Ethyl alcohol, water}}$$

b. $(CH_3)_3C(l) \ + \ ^-OH \ (con.) \ \xrightarrow[\text{Ethyl alcohol}]{\Delta}$

c. $CH_2{=}CHBr \ + \ ^-OCH_3 \ (con.) \ \xrightarrow{\text{Methyl alcohol}}$

d. $CH_3CH_2Br \ + \ ^-O-\overset{\overset{O}{\|}}{C}-CH_3 \ (dilute) \ \xrightarrow{\text{Acetic acid}}$

e.

$$(CH_3)_2CH-O-\underset{\underset{O}{\|}}{\overset{\overset{O}{\|}}{S}}-\bigcirc-CH_3 \ + \ H{:}^- \ \xrightarrow{\Delta}$$

f. $CH_3CH_2CH_2CH_2Br \ + \ I^- \ \xrightarrow{\text{Acetone}}$

g. $CH_3CH_2CH_2CH_2I \ + \ (CH_3)_3CO^- \ \xrightarrow{\textit{tert}\text{-Butyl alcohol}}$

Note that $CH_3SO_3^-$, methanesulfonate ion, and

$$CH_3-\bigcirc-SO_3^-$$

*p*-toluenesulfonate ion, are excellent leaving groups.

13. Consider the following transformation:

$$CH_3-\underset{\underset{CH_3}{|}}{\overset{\overset{CH_3}{|}}{C}}-CH_2OH \ \xrightarrow[\Delta]{H_2SO_4} \ \underset{H_3C}{\overset{H_3C}{>}}C{=}C\underset{CH_3}{\overset{H}{<}}$$

a. The reaction must proceed by means of an E1 mechanism even though neopentyl alcohol is primary and thus an E2 route would be expected. Explain.

b. Why isn't 3-methyl-1-butene a significant product?

c. Write a detailed, step-by-step mechanism for the transformation.

14. Given that the heat of hydrogenation of *cis*-2-pentene is 28.6 kcal/mol and that of *trans*-2-pentene is 27.6 kcal/mol, determine which compound is the more stable alkene. Show your reasoning.

# CHAPTER 13

# Organic Halogen Compounds

Other than the compounds derived from petroleum, there is probably no more important class of industrial organic intermediates than the organic halogen compounds. Because organic halogen compounds can participate in a variety of substitution and elimination reactions (see Chap. 12), they are of great interest to the synthetic chemist. These compounds are used as insecticides, solvents, and in a variety of ways by health professionals. Thus we will study the organic halogen compounds in this chapter.

One or more halogen atoms can be bonded to an alkyl, alkenyl, alkynyl, aromatic, or alicyclic group. Thus a large variety of organic halogen compounds is known.

The common name of an organic halogen compound is usually created by naming the alkyl, alkenyl, or cycloalkyl group and adding the term *halide* (Fig. 13.1). However, some common names must be learned because these names cannot be deduced from the structure of the molecules. For example:

| $CH_2Cl_2$ | $CHBr_3$ | $CCl_4$ |
|---|---|---|
| Methylene chloride | Bromoform | Carbon tetrachloride |

Common nomenclature is effective only for those molecules which do not exhibit a complex branching pattern.

Organic halogen compounds are named by the IUPAC method of nomenclature as a derivative of the corresponding hydrocarbon. Thus all rules of hydrocarbon nomenclature are used here. IUPAC nomenclature can be used no matter how complex the branching pattern of the organic halogen compound. Note that a primary organic halogen compound (1°) includes a halogen atom bonded to carbon, which, in turn, is bonded to one carbon, a secondary organic halogen compound (2°) contains a halogen atom bonded to carbon, which, in turn, is bonded to two carbons, while a tertiary organic halogen compound (3°) contains a halogen atom bonded to a carbon, which, in turn, is bonded to three other carbons (Fig. 13.2). Examples of IUPAC nomenclature for a number of organic halogen compounds are given in Fig. 13.3.

## 13.1 STRUCTURE, NOMENCLATURE, AND GEOMETRY

**FIGURE 13.1** Common names of selected halogen compounds.

$CH_3CH_2CH_2Cl$     $CH_2{=}CHCH_2Br$
*n*-Propyl chloride     Allyl bromide

Cyclopentyl iodide

**FIGURE 13.2** General representation of primary, secondary, and tertiary organic halogen compounds.

$$R—CH_2—Br$$

1° bromide

$$R—\overset{\displaystyle R}{\underset{\displaystyle H}{\overset{|}{\underset{|}{C}}}}—Cl \qquad R—\overset{\displaystyle R}{\underset{\displaystyle R}{\overset{|}{\underset{|}{C}}}}—I$$

2° chloride   3° iodide

$$R—\overset{\displaystyle R'}{\underset{\displaystyle H}{\overset{|}{\underset{|}{C}}}}—Cl \qquad R—\overset{\displaystyle R'}{\underset{\displaystyle R''}{\overset{|}{\underset{|}{C}}}}—I$$

2° chloride   3° iodide

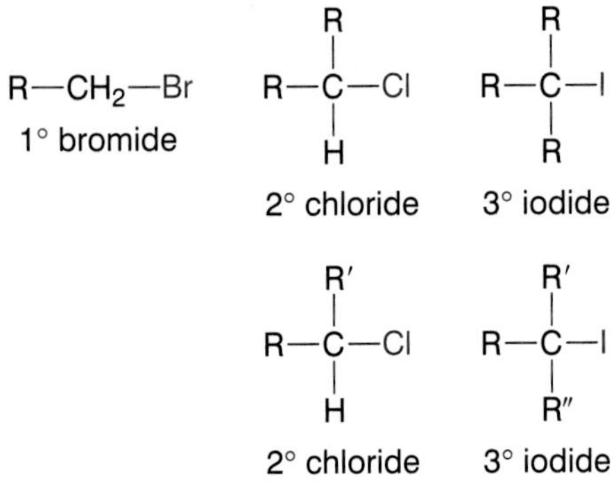

R—, R'—, and R"— represent organic groups (alkyl, alkenyl, alkynyl, cycloalkyl, aryl) that may differ from one another.

**FIGURE 13.3** Examples of IUPAC nomenclature of organic halogen compounds.

$CH_3CH_2CH_2Cl$
1-Chloropropane
(a primary chloride)

$CH_2$=CHBr
Bromoethene

CH≡Cl
Iodoethyne

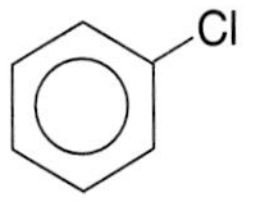

Chlorobenzene

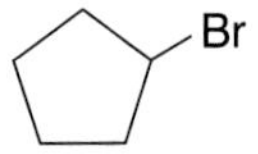

Bromocyclopentane
(a secondary bromide)

$(CH_3)_2CHCH_2(I)$
1-Iodo-2-methylpropane
(a primary iodide)

$(CH_3)_3C(I)$
2-Iodo-2-methylpropane
(a tertiary iodide)

---

**SOLVED PROBLEM 13.1**

Name the following compound using the IUPAC method of nomenclature.

$$\underset{\displaystyle I}{\overset{\displaystyle |}{CH_2CH_2\underset{\displaystyle H}{\overset{\displaystyle Cl}{\overset{|}{C}}}—\underset{\displaystyle Br}{\overset{\displaystyle Br}{\overset{|}{C}}}CH_2\underset{\displaystyle HCCH_3}{\overset{\displaystyle H}{\overset{|}{C}}}CH_2CH_3}$$
$$\underset{\displaystyle CH_3}{}$$

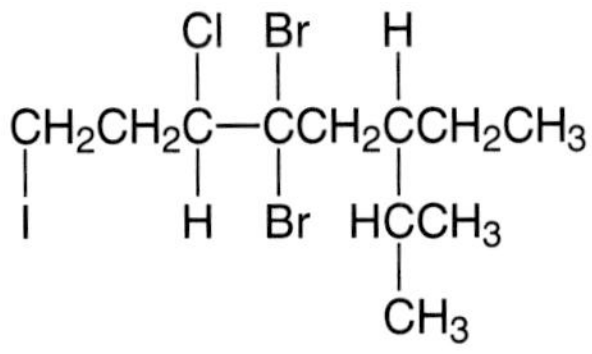

**SOLUTION**

*Step 1.* Establish the longest chain of carbon atoms—eight. Since there are two different eight-carbon chains, select the chain with the larger number of substituent groups but smaller groups.

*Step 2.* Number the chain from left to right because of the presence of the iodo group on C-1.

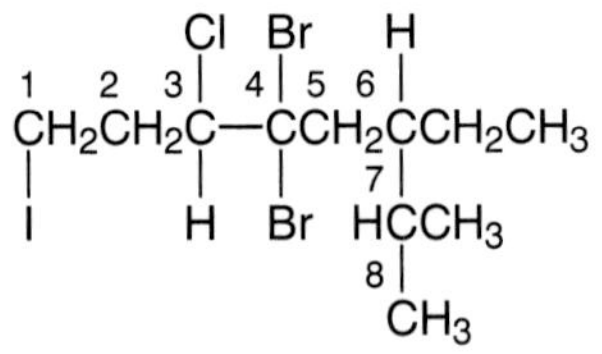

$$\overset{\text{1  2  3 | 4 | 5  6 |}}{CH_2CH_2C—CCH_2CCH_2CH_3}$$

*Step 3.* Name the compound: 4,4-dibromo-3-chloro-6-ethyl-1-iodo-7-methyloctane.

---

**EXERCISE 13.1**

Using the IUPAC method of nomenclature, name each of the following compounds. Supply a common name when appropriate.

a. 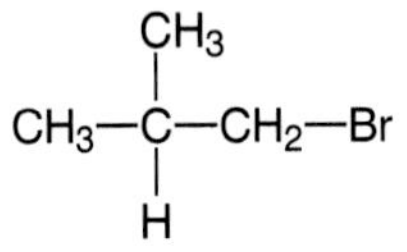 $CH_3—\overset{\displaystyle CH_3}{\underset{\displaystyle H}{\overset{|}{\underset{|}{C}}}}—CH_2—Br$

b. $CHCl_3$

c. $CBr_4$

d. $CH_3—CH_2—CH$=$CH—CH_2—Cl$

| TABLE 13.1 | Geometry and Carbon-Halogen Bond Hybridization of Organic Halogen Compounds | | |
|---|---|---|---|
| Class | Structure | Geometry | Hybridization of the Bonds Around the Halogen-Bonded Carbon |
| Alkyl halides | 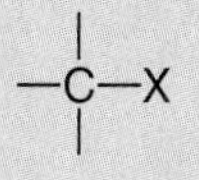 $-\overset{\displaystyle\vert}{\underset{\displaystyle\vert}{C}}-X$ | Tetrahedral | $sp^3$ |
| Cycloalkyl halides | 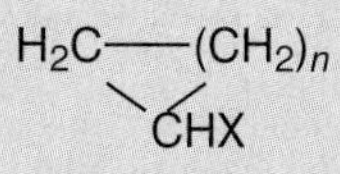 $H_2C\!-\!\!-\!(CH_2)_n$ $\diagdown$ $CHX$ | Tetrahedral* | $sp^3$ |
| Vinyl halides | 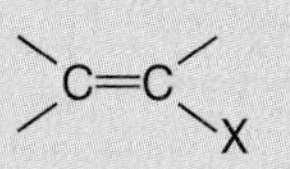 $\diagdown$ C$=$C $\diagup$ $\diagdown$X | Planar | $sp^2$ |
| Aryl halides | 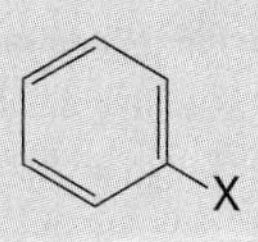 (benzene ring)—X | Planar | $sp^2$ |
| Alkynyl halides | $-C\equiv C-X$ | Linear | $sp$ |

*Planar for cyclopropyl and cyclobutyl halides ($sp^2$ hybridization).

e.  $CH_3-C\equiv C-CH_2-I$

f. 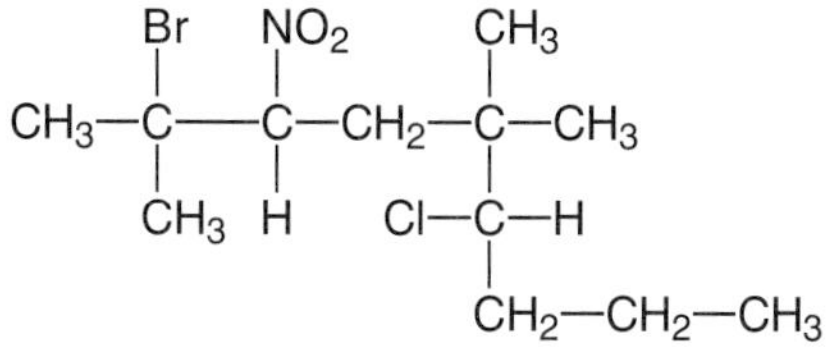
$$CH_3-\overset{\overset{\displaystyle Br}{\vert}}{\underset{\underset{\displaystyle CH_3}{\vert}}{C}}-\overset{\overset{\displaystyle NO_2}{\vert}}{\underset{\underset{\displaystyle H}{\vert}}{C}}-CH_2-\overset{\overset{\displaystyle CH_3}{\vert}}{\underset{\underset{\displaystyle \underset{\displaystyle CH_2-CH_2-CH_3}{Cl-C-H}}{\vert}}{C}}-CH_3$$

Alkyl halides and most alicyclic halides are tetrahedral in geometry (Table 13.1). Vinyl halides are planar around the carbon-carbon double bond (i.e., the two carbons of the carbon-carbon double bond and the atoms bonded to these carbons lie on the same plane), whereas alkynyl halides of the type R—C≡C—X, where X is a halogen, are linear around the carbon-carbon triple bond (i.e., the two carbons of the carbon-carbon triple bond and the atoms adjacent to these carbons lie on a straight line). Aryl halides are planar.

Indicate the carbon-halogen hybridization and the geometry of the following compound.

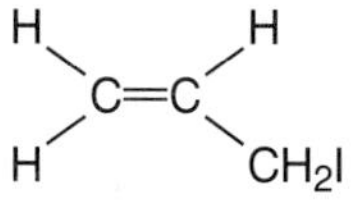

**SOLVED PROBLEM 13.2**

$$\underset{H}{\overset{H}{\diagdown}}C=C\underset{CH_2I}{\overset{H}{\diagup}}$$

**SOLUTION**

Hybridization is $sp^3$. The compound is planar about the carbon-carbon double bond and tetrahedral around the carbon bonded to iodine.

| **EXERCISE 13.2** |
|---|

Indicate the carbon-halogen hybridization and the geometry of each of the following compounds.

a. $CH_3CH_2Br$

b. bromobenzene

c. $CH_2\!=\!CHI$

d. $HC\!\equiv\!CBr$

## 13.2
## PHYSICAL PROPERTIES

Organic halogen compounds, like the hydrocarbons, show no intermolecular hydrogen bonding. However, unlike hydrocarbons, the carbon-halogen bond is slightly polar. This polarity is due to the electronegativity difference between carbon and halogen on the one hand and an innate tendency on the part of the halogen to be polarized, called *polarizability*, on the other hand. This phenomenon of polarizability exists because of the relatively large size of the halogen. As the size of an atom increases, the outermost electrons are held more loosely and are thus more easily moved. Consider an organobromine compound:

$$\overset{\displaystyle |}{\underset{\displaystyle |}{-C}}\overset{\delta^+\ \delta^-}{-Br}$$

Organobromine dipole

The presence of this dipole leads to dipole-dipole interactions between molecules. Thus we would expect an organohalogen compound to boil at a slightly higher temperature than a corresponding hydrocarbon. As predicted, the boiling point of an alkyl halide is a function of the carbon chain length and atomic weight of the halide and the degree of branching.

These conclusions can be verified by noting the increase in boiling point from ethyl chloride to ethyl iodide, from methyl chloride to carbon tetrachloride, and from isopropyl chloride to *n*-propyl chloride (Table 13.2). Most of the alkyl halides are liquids at room temperature. Low-molecular-weight halides such as methyl chloride, ethyl chloride, and ethyl bromide are gases, whereas high-molecular-weight halides such as iodoform ($CHI_3$) are solids. Liquid alkyl monochlorides are less dense than water, whereas monobromides and iodides are more dense than water. All polyhalogen compounds are more

| **TABLE 13.2** | **Physical Properties of the Organic Halogen Compounds** | | | |
|---|---|---|---|---|
| Formula | Name | Molecular Weight | Boiling Point (°C) | Density (g/ml) |
| $CH_3Cl$ | Methyl chloride | 51 | −24 | — |
| $CH_3CH_2Cl$ | Ethyl chloride | 65 | 13 | — |
| $CH_3CH_2Br$ | Ethyl bromide | 109 | 38 | 1.44 |
| $CH_3CH_2I$ | Ethyl iodide | 156 | 72 | 1.93 |
| $CH_2Cl_2$ | Methylene chloride | 85 | 43 | 1.34 |
| $CHCl_3$ | Chloroform | 120 | 61 | 1.49 |
| $CCl_4$ | Carbon tetrachloride | 154 | 77 | 1.60 |
| $CH_3CH_2CH_2Cl$ | *n*-Propyl chloride | 79 | 47 | 0.89 |
| $CH_3CH_2CH_2CH_2Cl$ | *n*-Butyl chloride | 93 | 79 | 0.88 |
| $CH_3CH(Cl)CH_3$ | Isopropyl chloride | 79 | 36 | 0.86 |
| | Chlorobenzene | 113 | 132 | 1.11 |
| $CH_3(CH_2)_4CH_2Cl$ | *n*-Hexyl chloride | 121 | 134 | 0.88 |

dense than water. Aromatic halides have boiling points in the same ballpark value as alkyl halides with the same carbon content. Note the boiling points in Table 13.2 of chlorobenzene and *n*-hexyl chloride.

All organic halogen compounds are too weakly polar to dissolve in water and are only soluble in such low-polarity solvents as ether and hexane.

**SOLVED PROBLEM 13.3**

Predict the substance with the higher boiling point. Explain your choice.

a.  *n*-propyl bromide or *n*-propyl iodide

b.  *n*-propyl bromide or isopropyl bromide

**SOLUTION**

a.  Since the molecular weight of *n*-propyl iodide is higher than that of *n*-propyl bromide, *n*-propyl iodide should boil at a higher temperature. The *n*-propyl iodide boils at a higher temperature because the larger molecular weight indicates a larger molecule than the *n*-propyl bromide with greater intermolecular London forces.

---

## HISTORICAL BOXED READING 13.1

Most insects of the world are harmless or useful to humans. The ladybug, for example, has a wonderful appetite for a wide variety of small insect pests. Bees are a source of honey and pollinate flowering plants. The fruit fly has served research biologists for many decades, and the scarab beetle removes animal waste that could become a source of disease. Insects represent a major source of food for many creatures: birds, fish, amphibians, reptiles, and mammals. On occasion, humans have been known to eat an insect or two. For example, chocolate-covered ants are considered to be a great delicacy by some people. Also, the ant was considered to represent a paragon of industry. The Bible puts it as follows: "Go to the ant, thou sluggard; consider her ways and be wise."

The total bug population in the world today is about a billion billion, $1 \times 10^{18}$. Only 0.1% of these are harmful to humans, but they pack quite a wallop. Diseases such as typhus, typhoid, malaria, sleeping sickness, yellow fever, plague, and dengue fever are transmitted by insects. Fire ants can injure or kill livestock with their sting, while the stinger of the wasp and bee is as annoying as ever to both humans and beasts. The southern pine beetle is devastating our forests along with the gypsy moth.

Of particular concern is the recent success of a wide variety of pests in eating or destroying our crops. From the army worm to the grasshopper and the Japanese beetle to the boll weevil, our farmers are finding it more and more difficult to eradicate the insects that harm our crops—both in the field and in storage.

And no wonder. These insects are superbly built for survival. Many species produce thousands of young after mating. Several pass from egg to larva to pupa and adult in just a few days or weeks. Thus, within 3 short summer months, the insect population can skyrocket.

Insect control has been attempted for thousands of years. The ancient Greeks and Romans used sulfur to eliminate pests from crops, while the Chinese employed arsenic sulfide. To-

bacco, pyrethrum, rotenone, and a number of synthetic insecticides were used at various times from 1600 to 1942 with varying degrees of success.

A gigantic breakthrough occurred in 1942 when dichlorodiphenyltrichloroethane (DDT for short) was found to do a marvelous job of killing insects. DDT was the first of an army of insecticides that went into action during World War II. The compound killed big chunks of the bug population cheaply and thus kept a number of infectious diseases under control and made it possible for farmers to have a bountiful harvest year after year.

Unfortunately, after years of spraying tons of DDT on the land, scientists found that insects were becoming immune, and more expensive insecticides had to be developed. Active use of DDT as an insecticide in the United States was terminated when the EPA banned DDT as possibly being carcinogenic to humans. To make matters worse, it was discovered that DDT remained in the environment to be ingested by animals and fish and, ultimately, to be found in increasing amounts in fat cells of human beings. DDT persists in the environment for so long a time because most microorganisms cannot degrade halogenated aromatic compounds.

One possible way to solve this problem involves the use of a white rot fungus, *Phanerochaete chrysosporium*, which, when starved for nitrogen, produces an enzyme that degrades DDT to carbon dioxide, chloride ion, and water.

Dichlorodiphenyltrichloroethane (DDT)

b.  Both compounds are isomers. Thus the higher-boiling compound can be deter-
    mined on the basis of branching. Because *n*-propyl bromide is unbranched, London
    forces between molecules are greater in comparison with the London forces be-
    tween branched isopropyl bromide molecules. Thus *n*-propyl bromide would be
    expected to boil at a higher temperature.

Predict the substance with the higher boiling point. Explain your choice.

a.  *n*-butyl bromide or *n*-hexyl bromide

b.  *n*-pentyl chloride or *tert*-pentyl chloride

c.  *n*-pentyl chloride or *n*-pentyl iodide

## 13.3
## ORGANIC HALOGEN COMPOUNDS AS DRUGS, CARCINOGENS, AND IRRITANTS

Compressed ethyl chloride, when sprayed on the skin from a can, instantly vaporizes to
cool the skin and acts as an anesthetic to ease muscle pain. Most baseball, basketball,
and football trainers have a can on hand.

Chloroform at one time was used as an anesthetic in place of ether in areas of the
world with high ambient temperatures. Ether (bp = 35°C) could not be stored in a bottle
because it would evaporate rapidly, so chloroform (bp = 61°C) was used in place of
ether because chloroform could be stored in a bottle without significant evaporation.
The use of chloroform as an anesthetic was discontinued when it was determined that
chloroform caused liver damage in humans. Today, it is considered to be carcinogenic,
along with carbon tetrachloride. Iodoform serves as a urinary antiseptic.

Many halogen compounds are irritating to human beings. Compounds such as
*p*-bromobenzyl bromide, 1-chloro-2-butene, and benzyl chloride are irritating to the
eyes, whereas 2.4-dinitrochlorobenzene causes dermatitis.

## 13.4
## SYNTHESIS

### Hydrohalogenation of an Alkene (see Sec. 6.2)

In general,

$$\diagdown C=C \diagup + HX \longrightarrow -\underset{H}{\overset{|}{C}}-\underset{X}{\overset{|}{C}}-$$

where X = Cl, Br, or I. For example:

(cyclopentene with CH₃) + HBr ⟶ (cyclopentane with CH₃, Br)

### Halogenation of an Alkane (see Sec. 4.2)

In general,

$$RH + X_2 \xrightarrow[\text{or }h\nu]{\Delta} RX + HX$$

where $X_2$ = $Cl_2$ or $Br_2$. For example:

$$CH_4 + Cl_2 \xrightarrow{h\nu} CH_3Cl + HCl$$

### Halogenation of an Alkene (see Sec. 6.2)

In general,

$$\diagdown C=C\diagup + X_2 \xrightarrow[20°C]{CCl_4} -\overset{|}{\underset{X}{C}}-\overset{X}{\underset{|}{C}}-$$

where $X_2$ = $Br_2$ or $Cl_2$. For example,

$$CH_3CH{=}CH_2 + Br_2 \xrightarrow[20°C]{CCl_4} CH_3CH(Br)CH_2Br$$

Notice the difference in reagents and reaction conditions between ring and side-chain halogenation. In general,

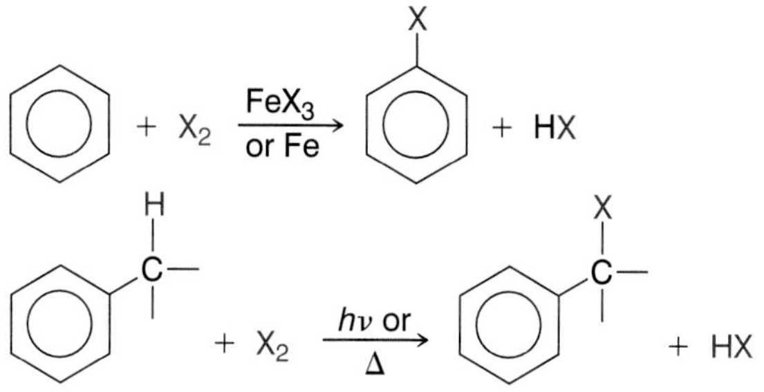

**Halogenation of an Aromatic Compound (see Secs. 9.5, 10.1, and 10.6)**

where $X_2$ = $Cl_2$ or $Br_2$. For example:

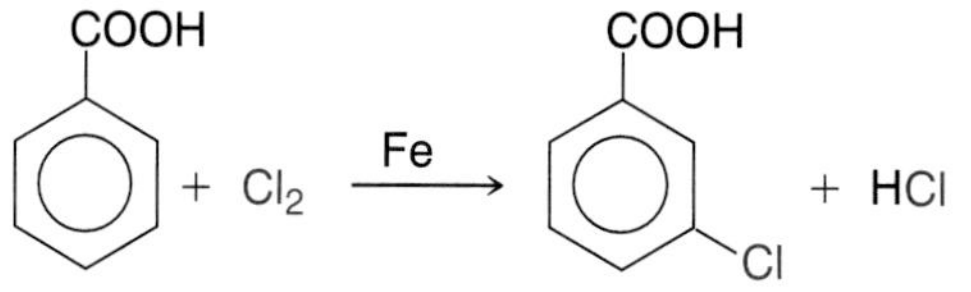

Benzoic acid          *m*-Chlorobenzoic acid

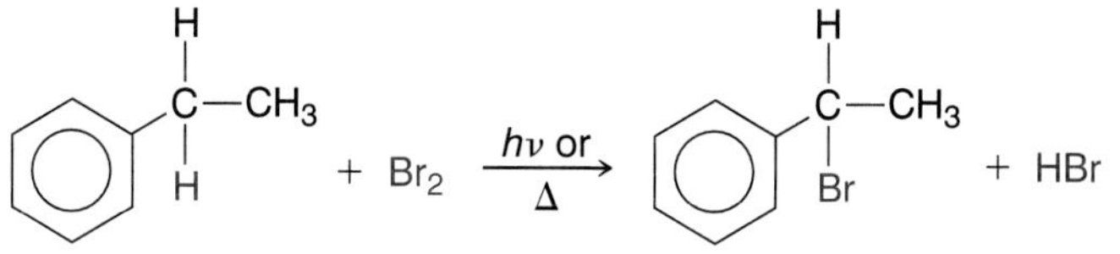

Ethylbenzene          1-Bromo-1-phenylethane

Reacting an alkene (with one or more allylic hydrogens) with *N*-bromosuccinimide (NBS) at room temperature yields an alkene brominated in an allylic position. In effect, a bromine atom substitutes for an allylic hydrogen. In general,

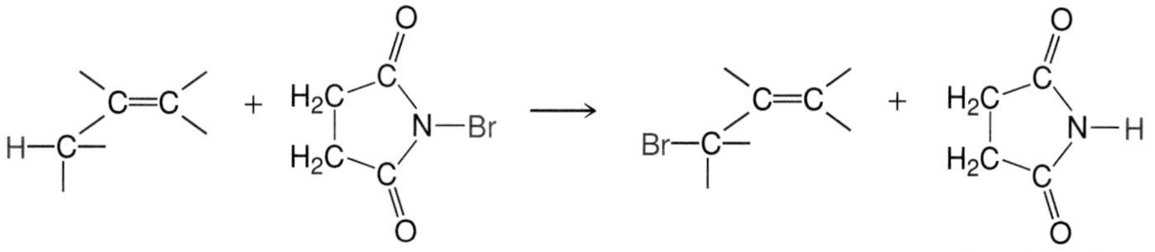

**Bromination with *N*-Bromosuccinimide (see Sec. 6.3)**

*N*-Bromosuccinimide
(NBS)

Succinimide

Example:

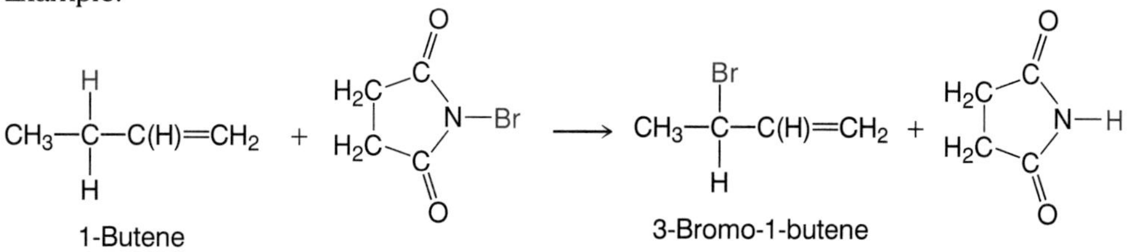

1-Butene          3-Bromo-1-butene

## Halogenation and Hydrohalogenation of Alkynes

A review of Section 8.7 will recall to you that 1 or 2 mol of halogen can be added to an alkyne depending on the temperature. In general,

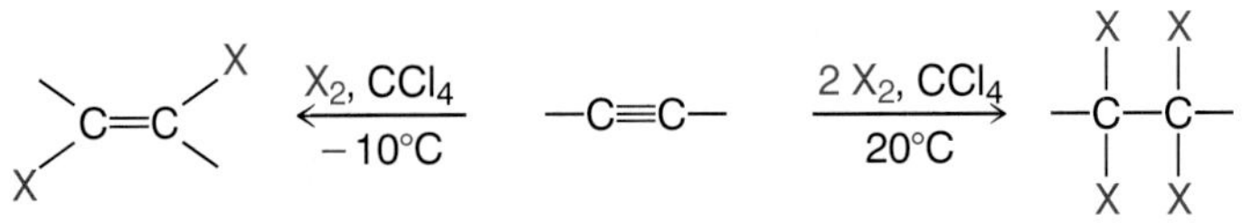

A *trans*-dihaloalkene                                    A tetrahaloalkane

where $X_2 = Cl_2$ or $Br_2$. For example:

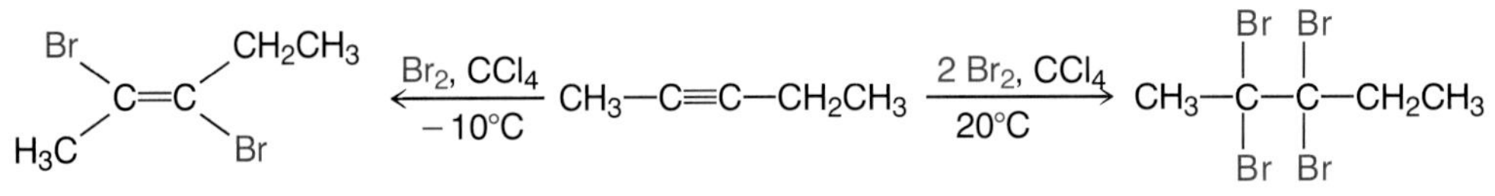

(*E*)-2,3-Dibromo-2-pentene          2-Pentyne          2,2,3,3-Tetrabromopentane

This same sort of situation exists for the addition of 1 or 2 mol of HX to an alkyne, as you can observe by reviewing Section 8.7. In general,

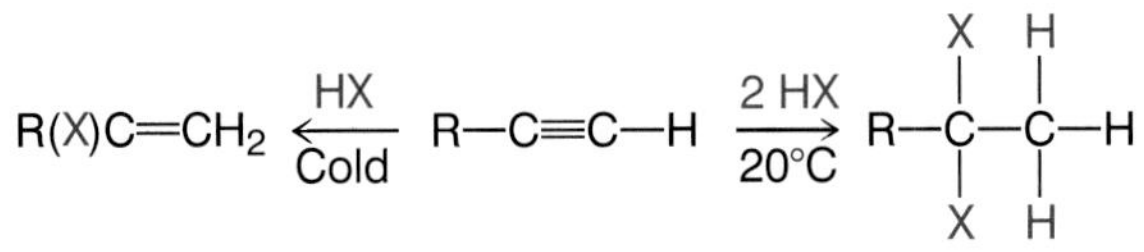

where X = Cl, Br, or I. For example:

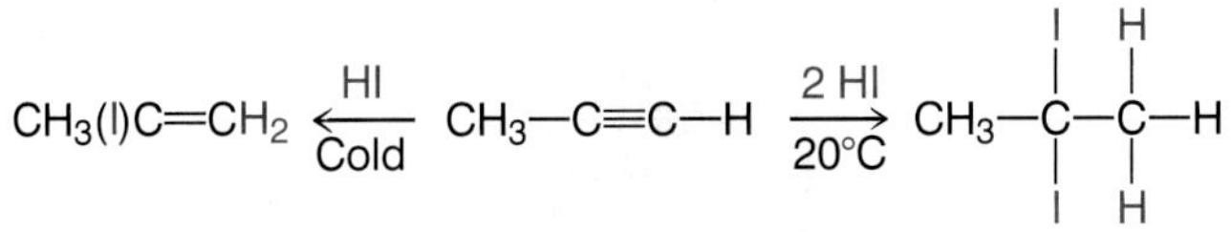

## Alkyl Halides from Alcohols (see Sec. 12.2)

In general,

$$R-OH + HX \longrightarrow R-X + H_2O$$

Example:

$$CH_3CH_2OH + HI \xrightarrow{\Delta} CH_3CH_2I + H_2O$$

(See also Sec. 15.2)

## 13.5 REACTIONS OF ORGANIC HALOGEN COMPOUNDS: FORMATION OF ORGANOMETALLIC COMPOUNDS

Besides producing a Grignard reagent from alkyl halides (see Sec. 4.1) and alicyclic halides (see Sec. 4.4) a Grignard reagent also can be prepared from an aryl, vinylic, allylic, or benzylic halide. Simply treat the halide with magnesium turnings in anhydrous diethyl ether. The reaction is exothermic, so no heat is needed; indeed, cooling and a reflux condenser should be available. For example:

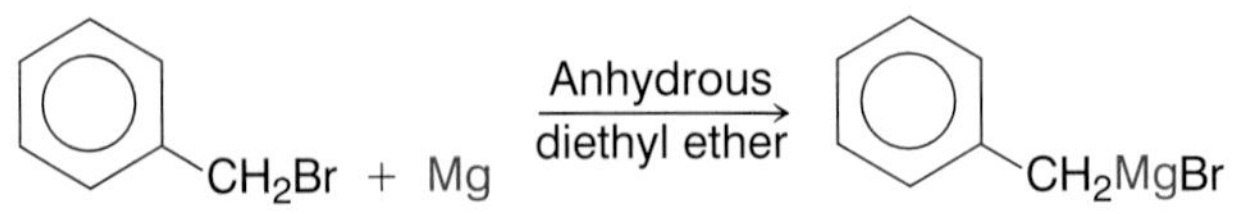

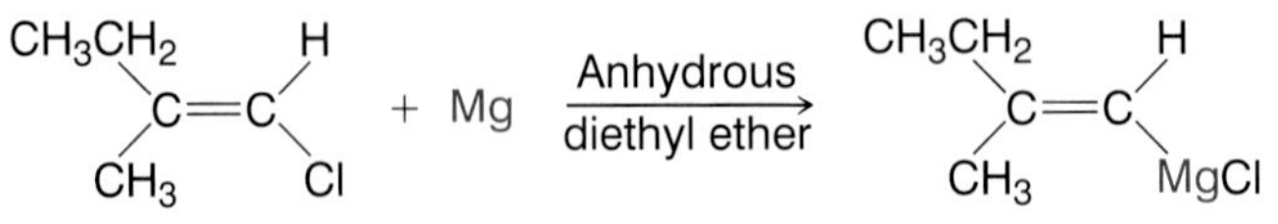

$$CH_3(H)C{=}C(H)CH_2I + Mg \xrightarrow[\text{diethyl ether}]{\text{Anhydrous}} CH_3(H)C{=}C(H)CH_2MgI$$

Refer to Section 4.1 for the formation of organolithium compounds as a part of the Corey-House synthesis. In general,

$$R\!-\!X + 2\,Li \xrightarrow[\text{diethyl ether}]{\text{Anhydrous}} R\!-\!Li + LiX$$

where R = any alkyl group
X = Cl, Br, or I

For example:

$$CH_3CH_2CH_2CH_2Br + 2\,Li \xrightarrow[\text{diethyl ether}]{\text{Anhydrous}} CH_3CH_2CH_2CH_2Li + LiBr$$

*n*-Butyl lithium

A detailed discussion of nucleophilic substitution is presented in Sections 12.1 and 12.3, while a detailed analysis of elimination is presented in Section 12.4.

$$RX + :N^- \longrightarrow RN + X^-$$

where X = Cl, Br, or I, and :N⁻ is a nucleophile.

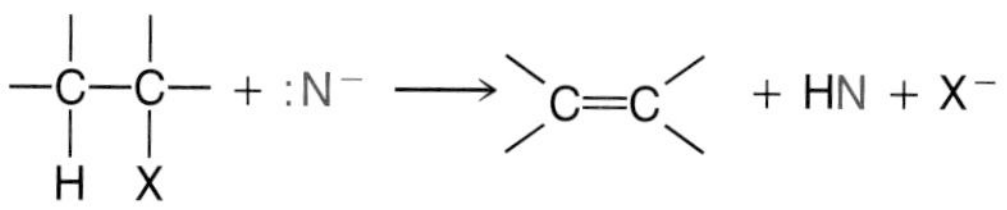

where X = Cl, Br, or I, and :N⁻ is a base.

Each of two nucleophilic substitution reactions is used as a qualitative test to classify organic halogen compounds, whereas one of the two nucleophilic substitution reactions can be used to identify the halogen in an organic halogen compound.

A qualitative test to characterize the class (see Sec. 13.1) of a given organic chloride or bromide is based on an $S_N2$ reaction. When an organobromide or organochloride is treated with a solution of sodium iodide in acetone, a precipitate of sodium bromide or sodium chloride is obtained. Primary compounds react instantaneously, secondary compounds react slowly (3 to 5 minutes), while tertiary, vinyl, and aryl halides do not react (a precipitate does not form). In general,

$$RX + NaI \xrightarrow{\text{Acetone}} RI + NaX \downarrow$$

where X = Cl or Br. For example:

$$CH_3CH_2CH_2Cl + NaI \xrightarrow{\text{Acetone}} CH_3CH_2CH_2I + NaCl \downarrow$$

Instantaneous reaction—a primary chloride

$$(CH_3)_2CHBr + NaI \xrightarrow{\text{Acetone}} (CH_3)_2CHI + NaBr \downarrow$$

Slow reaction—a secondary bromide

## 13.6
## REACTIONS OF ORGANIC HALOGEN COMPOUNDS: NUCLEOPHILIC SUBSTITUTION AND ELIMINATION

Substitution

Elimination

The Sodium Iodide Qualitative Test for Organic Halogen Compounds

**SOLVED PROBLEM 13.4**

Predict the rate of the following reaction.

$$CH_3Cl + NaI \xrightarrow{\text{Acetone}} CH_3I + NaCl \downarrow$$

**SOLUTION**

Methyl chloride cannot be classified as a primary halide. However, since the carbon bearing the halogen is not bonded to another carbon, as is the case with a primary halide, the compound should react even more rapidly than a primary halide. Since primary halides react instantaneously, so will methyl chloride.

---

| **EXERCISE 13.4** | Two organic halides, each with the formula $C_5H_{11}Br$, when reacted with NaI in acetone, did not produce a precipitate. Draw a structural formula for each organic halogen compound. One of the organic halides is tertiary, as expected; the other is primary. Explain why this particular primary halide did not react with NaI in acetone. |

---

**The Silver Nitrate Qualitative Test for Organic Halogen Compounds**

One test tube reaction used to identify the halogen in an organic halide is an $S_N1$ reaction. Warming a halogen compound with an alcoholic solution of silver nitrate produces a precipitate of silver halide (AgCl = white, AgBr = light yellow, AgI = dark yellow). In general,

$$R\text{—}X + AgNO_3 \xrightarrow{\text{Ethyl alcohol}} AgX \downarrow + R\text{—}NO_3$$

where X = Cl, Br, or I. Besides identifying the halogen, this test can be used to establish the class of the organic group. Tertiary, allylic, and benzylic halides react instantaneously, secondary halides react slowly (3 to 5 minutes), while vinyl, phenyl, methyl, and primary halides do not react (precipitates do not form). For example:

$$CH_2\text{=}CHCH_2\text{—}Cl + AgNO_3 \xrightarrow{\text{Ethyl alcohol}} AgCl \downarrow + CH_2\text{=}CHCH_2\text{—}NO_3$$

Instantaneous reaction

$$(CH_3)_2CH\text{—}I + AgNO_3 \xrightarrow{\hspace{2cm}} AgI \downarrow + (CH_3)_2CH\text{—}NO_3$$

Slow reaction

---

# ▶ CHAPTER ACCOMPLISHMENTS

## 13.1 Structure, Nomenclature, and Geometry

☐ Supply a common name for each of the following compounds.

   a. $CH_2Cl_2$
   b. $CHBr_3$
   c. $CCl_4$

☐ Draw a structural formula for:

   a. cyclopentyl iodide
   b. 1-chloropropane
   c. 2-iodo-2-methylpropane
   d. bromoethene

☐ Classify each of the following compounds as primary, secondary, tertiary, vinylic, allylic, or benzylic.

   a. cyclopentyl iodide
   b. 1-chloropropane
   c. 2-iodo-2-methylpropane
   d. $p$-ethylbenzyl iodide
   e. 3-chloropropene
   f. 1-bromo-1-butene

☐ Supply the geometry of the carbon atom in a molecule based on the following hybridized orbitals around the halogen-bonded carbon.

   a. $sp^2$

   b. $sp^3$
   c. $sp$

## 13.2 Physical Properties

☐ Draw a dipole-dipole interaction between two dipoles of methyl bromide.

☐ Predict which compound of the following pairs of compounds would boil at a higher temperature.

   a. $CH_3CH_2Cl$ or $CH_3CH_2I$
   b. $CH_3Cl$ or $CCl_4$

## 13.3 Organic Halogen Compounds as Drugs, Carcinogens, and Irritants

☐ Explain why, many years ago, chloroform was used in place of ether as an anesthetic in places with high normal temperatures.

☐ Draw a structural formula for $p$-bromobenzyl bromide.

## 13.4 Synthesis

☐ Write an equation illustrating the preparation of

   a. 2-iodobutane from 1-butene.
   b. ethyl bromide from ethane.
   c. 1,2-dibromopropane from propene.
   d. bromobenzene from benzene.
   e. 1-bromo-1-phenylethane from ethylbenzene.

f.  3-bromo-1-butene from 1-butene.

g.  (*E*)-2,3-dibromo-2-pentene from 2-pentyne.

h.  2,2-diiodopropane from propyne.

## 13.5  Reactions of Organic Halogen Compounds: Formation of Organometallic Compounds

☐ Complete the following equations by drawing a structural formula for each organic product and a molecular formula or formula unit for each inorganic product.

a.  $C_6H_5CH_2Br + Mg \xrightarrow{\text{Anhydrous diethyl ether}}$

b.  $CH_3CH_2CH_2CH_2Br + 2\ Li \xrightarrow{\text{Anhydrous diethyl ether}}$

## 13.6  Reactions of Organic Halogen Compounds: Nucleophilic Substitution and Elimination

☐ Complete each of the following equations by drawing a structural formula for each organic product and writing a molecular formula or formula unit for each inorganic product.

a.  $CH_2{=}CHCH_2{-}Cl + AgNO_3 \xrightarrow{\text{Ethyl alcohol}}$

b.  $CH_3CH_2CH_2Cl + NaI \xrightarrow{\text{Acetone}}$

☐ Draw a structural formula for an alkyl chloride, $C_3H_7Cl$, that produced a precipitate in 4 minutes when reacted with NaI in acetone.

☐ Draw a structural formula for an alkyl halide that contains four carbons and produced an instantaneous reaction with alcoholic $AgNO_3$ to give a white precipitate.

---

▶ **KEY TERM**

*polarizability* (13.2)

---

▶ **PROBLEMS**

1.  Name each of the following compounds using the IUPAC system and the common system, when applicable. Classify each monohalogen compound as primary, secondary, tertiary, allylic, benzylic, or vinylic.

    a.  $CHBr_3$

    b.  $CH_2{=}\overset{\overset{\textstyle H}{|}}{C}{-}Cl$

    c.

    d.

    e.  $(CH_3)_2CHCCl_2CHIBr$

    f.

    g.  $(CH_3)_2CHCH_2Br$

    h.  $(C_6H_5)_3CCl$

    i.  $CH_3\overset{\overset{\textstyle H}{|}}{\underset{\underset{\textstyle I}{|}}{C}}{-}\overset{\overset{\textstyle H}{|}}{\underset{\underset{\textstyle I}{|}}{C}}CH_2\overset{\overset{\textstyle H}{|}}{\underset{\underset{\textstyle CH_3}{|}}{C}}CH_2CH_3$

    j.  $H{-}C{\equiv}C{-}Br$  (do not classify)

    k.

2.  Give the carbon-halogen hybridization and the geometry for compounds (b), (d), and (j) in Problem 1.

3.  Draw a structural formula for each of the following compounds.

    a.  neopentyl chloride
    b.  2,4-dinitrobromobenzene
    c.  cyclohexyl bromide
    d.  *tert*-pentyl iodide
    e.  isopentyl chloride
    f.  *sec*-butyl iodide
    g.  2-bromo-3,3-dichloro-5-methyl-6-nitrononane
    h.  2-bromoeicosane

4.  Draw a structural formula to illustrate each of the following classes of halogen compounds.

    a.  primary
    b.  secondary

   c. tertiary

   d. vinylic

   e. allylic

   f. benzylic

5. Which of the following compounds is vinylic, allylic, tertiary, or benzylic?

   a. $CH_3CH{=}CHBr$

   b. $(CH_3)_3CCl$

   c.

   d. $(CH_3)_2C{=}C(CH_3)CH_2Br$

6. Draw a structural formula for each of the following components.

   a. 2,4-dichlorophenol

   b. *p*-chlorobenzyl iodide

   c. benzyl chloride

   d. (*Z*)-1-chloro-2-butene

   e. methylene iodide

   f. iodoform

7. Arrange the following compounds in order of decreasing boiling point.

   a. $CH_3(CH_2)_2CH_2Br$, $(CH_3)_3CBr$, and $CH_3CH(Br)CH_2CH_3$

   b. $CH_3CH_2CH_2I$, $CH_3CH_2CH_2Cl$, and $CH_3CH_2CH_2Br$

8. Prepare each of the following compounds from *n*-propyl chloride. Use any inorganic reagents necessary.

   a. propene

   b. propane (two methods)

   c. *n*-propyl alcohol, $CH_3CH_2CH_2OH$

   d. propyne

   e. 2-hexyne

   f. *cis*-2-hexene

   g. *trans*-2-hexene

   h. 1,2-dibromopropane

   i. 2,2-dibromopropane

   j. isopropyl alcohol, $CH_3CH(OH)CH_3$

   k. 2-chloropropene

   l. *n*-propylmagnesium bromide

   m. 3-bromopropene

9. a. Draw a Fischer projection for (*R*)- and (*S*)-2-bromohexane.

   b. Draw a wedge projection for (*R*)- and (*S*)-2-bromohexane.

10. Complete the following equations by writing a structural formula for each organic product and a molecular formula or formula unit for each inorganic product.

   a. $CH_3CH_2CH_2CH_2CH_2Br + Mg \xrightarrow[\text{diethyl ether}]{\text{Anhydrous}}$

   b. $CH_3CH(Cl)CH_2CH_3 + KOH \xrightarrow[\Delta]{\text{Ethyl alcohol}}$

   c. $CH_3CH_2CH_2C{\equiv}CH + 2\ HBr \xrightarrow{20°C}$

   d. $CH_3CH{=}CH_2 + NBS \longrightarrow$

   e. $CH_3CH_2CH{=}CH_2 + HI \longrightarrow$

   f. cyclopentene + NBS $\longrightarrow$

   g. toluene + $Br_2 \xrightarrow{\Delta}$

   h. benzaldehyde + $Cl_2 \xrightarrow{Fe}$

   i. 2-butyne + $Br_2 \xrightarrow[-10°C]{CCl_4}$

11. Treating an alkyl halide with alcoholic silver nitrate gave a white precipitate in 4 minutes. The compound must be

   a. *tert*-butyl chloride.

   b. isobutyl iodide.

   c. isopropyl chloride.

   d. *n*-butyl bromide.

12. Provide a simple test tube reaction to distinguish between each of the following. (Describe any color change, gas, or precipitate formation.)

   a. *tert*-butyl chloride and *tert*-butyl bromide

   b. *n*-propyl iodide and *n*-propyl bromide

   c. *tert*-butyl chloride and *n*-butyl chloride

   d. *n*-propyl bromide and isopropyl bromide

   e. 1-pentene and benzyl bromide

   f. *n*-propyl bromide and hexane

13. The coupled $^{13}C$ NMR spectrum for $C_4H_9Br$ is as follows:

$$\delta = 16\ (3H)\ \text{quartet}$$
$$\delta = 23\ (3H)\ \text{quartet}$$
$$\delta = 33\ (2H)\ \text{triplet}$$
$$\delta = 55\ (1H)\ \text{doublet}$$

Draw a structural formula for the compound.

14. Draw a structural formula for the compound showing each of the following proton NMR spectra.

   a. Molecular formula = $C_3H_3Br_5$

$$\delta = 4.3\ (1H)\ \text{triplet}$$
$$\delta = 5.9\ (2H)\ \text{doublet}$$

   b. Molecular formula = $C_3H_7I$

$$\delta = 1.0\ (3H)\ \text{triplet}$$
$$\delta = 2.0\ (2H)\ \text{hextet}$$
$$\delta = 3.2\ (2H)\ \text{triplet}$$

# CHAPTER 14

# Alcohols, Thiols, and Disulfides

An alcohol can be considered an organic derivative of water:

R—OH      H—OH

An alcohol     Water

where the organic (or R—) group can be alkyl, alkenyl, alkynyl, or cycloalkyl. An alcohol can contain an aryl group, but there must be present at least one aliphatic carbon separating the aromatic ring (or substituted aromatic ring) from the hydroxyl group (or groups). If a structure contains one or more hydroxyl groups bonded directly to an aryl group, the structure represents a *phenol*.

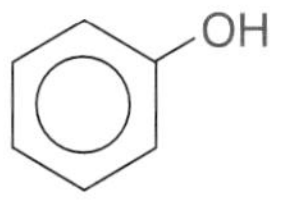

Phenol

The —OH group is bonded directly to a carbon of the aryl group.

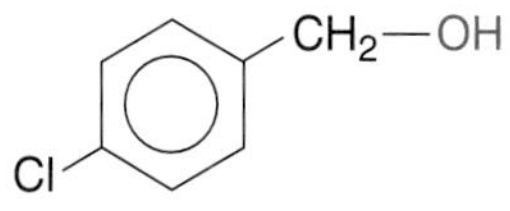

Alcohol

At least one methylene is located between the aromatic group and the hydroxyl group.

To name an alcohol using the common system, name the organic group and then add the word *alcohol*. For examples of common nomenclature, refer to Fig. 14.1.

Alcohols are classified primary (1°), secondary (2°), or tertiary (3°) in the same way as are organic halogen compounds (Fig. 14.2).

IUPAC nomenclature, useful no matter how complex the alcohol, is based on a few simple rules:

1. Name the longest continuous chain of carbons containing the hydroxyl group.
2. Replace the last letter of the corresponding alkane with the suffix *-ol*.
3. Number the carbons from right to left or left to right so as to give the carbon bearing the hydroxyl group the lowest possible number.
4. In any alicyclic compound, the hydroxyl group is assumed to be on carbon 1.

## 14.1 INTRODUCTION, STRUCTURE, AND NOMENCLATURE OF ALCOHOLS

Refer to Section 15.2 for a discussion of the Lucas test, a test to distinguish primary from secondary from tertiary alcohols.

**FIGURE 14.1** Examples of common nomenclature of alcohols.

CH$_2$OH

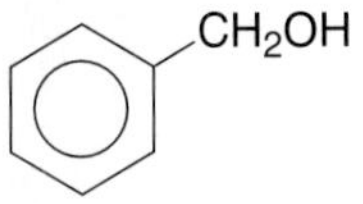

Benzyl alcohol
(a benzylic alcohol)
(a primary alcohol)

(CH$_3$)$_2$CHOH

Isopropyl alcohol
(a secondary alcohol)

CH$_3$
|
H$_3$C—C—OH$_3$
|
C$_2$H$_5$

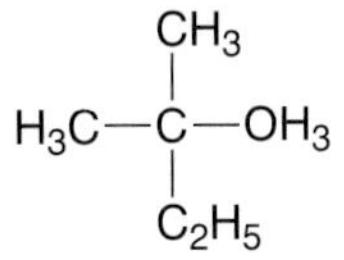

*tert*-Pentyl alcohol
(a tertiary alcohol)

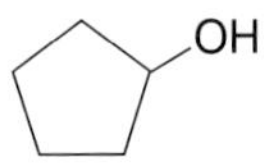

OH

Cyclopentyl alcohol
(a secondary alcohol)

H$_2$C=CH—CH$_2$OH

Allyl alcohol
(an allylic alcohol)
(a primary alcohol)

HC≡C—CH$_2$OH

Propargyl alcohol
(a primary alcohol)

Note that the propargyl group is HC≡CCH$_2$—

**FIGURE 14.2** General representations of primary, secondary, and tertiary alcohols.

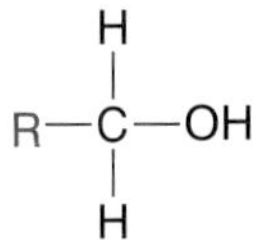

H
|
R—C—OH
|
H

A primary alcohol

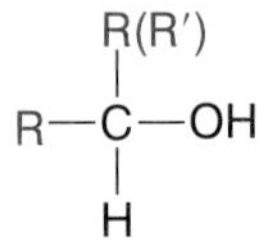

R(R′)
|
R—C—OH
|
H

A secondary alcohol

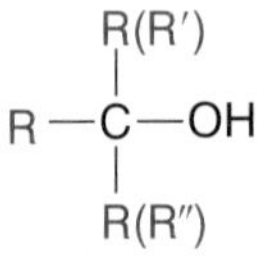

R(R′)
|
R—C—OH
|
R(R″)

A tertiary alcohol

Some examples of the use of these four rules are given in Fig. 14.3. Notice that in 2-buten-1-ol (the last *e* in butene is dropped) the hydroxyl group has priority over the double bond and thus is bonded to the carbon that carries the lower number. The hydroxyl group, of course, always has priority over substituent groups such as the nitro group, alkyl groups, and halo groups. We can now expand our priority system of functional-group nomenclature to include the hydroxyl and halo groups.

$$OH > C=C > C≡C > Cl(Br)(I)$$

**SOLVED PROBLEM 14.1**

Provide an IUPAC name for the following compound.

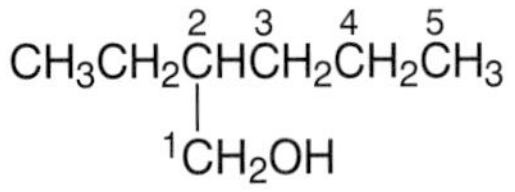

$$\overset{2}{C}H_3CH_2\overset{3}{C}HCH_2\overset{4}{C}H_2\overset{5}{C}H_3$$
$$|$$
$$^1CH_2OH$$

**SOLUTION**

Although the longest chain of carbon atoms is six in number in this compound, the compound is a pentanol because the longest chain containing the hydroxyl group is only five carbons in length. Note that we must specify in the name that the hydroxyl group is located on C-1. Thus we name the compound 2-ethyl-1-pentanol.

**FIGURE 14.3** Examples of IUPAC nomenclature of alcohols.

$$\overset{4}{C}H_3\overset{3}{C}H=\overset{2}{C}H\overset{1}{C}H_2OH$$

IUPAC: 2-Buten-1-ol

CH$_3$

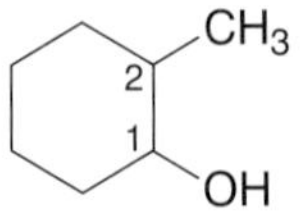

2-Methylcyclohexanol

CH$_3$CH(OH)CH$_2$CH$_3$

IUPAC: 2-Butanol
common: *sec*-Butyl alcohol

$$\overset{\gamma}{C}lCH_2\overset{\beta}{C}H_2\overset{\alpha}{C}H_2OH$$

3-Chloro-1-propanol
γ-Chloro-*n*-propyl alcohol

Supply an IUPAC name for each of the following compounds.

a.   H—C≡CCH$_2$CH$_2$CH$_2$OH

b.   CH$_3$CH$_2$CH(OH)CH$_2$CH$_3$

c.   (CH$_3$)$_2$CHCH$_2$CH$_2$OH

 

A sometimes useful method for naming alcohols is the *carbinol system* (a common system of nomenclature), particularly when two or more phenyl groups (and one or more cycloalkyl groups) and the hydroxyl group are bonded to the same carbon. Methyl alcohol is carbinol, and other alcohols are named as derivatives of carbinol. That is, one or more hydrogens bonded to the carbinol carbon are removed and replaced by the appropriate group or groups that are circled in the examples of the carbinol system of common nomenclature given in Fig. 14.4.

Alcohols can contain more than one —OH group. An alcohol that contains two hydroxyl groups per molecule is called *dihydric,* whereas an alcohol with three hydroxyl groups per molecule is said to be *trihydric*.

CH$_2$—CH$_2$            CH$_2$—CH—CH$_2$
|     |                     |     |     |
OH   OH                 OH  OH  OH

Ethylene glycol (common)      Glycerol (trivial)
1,2-Ethanediol (IUPAC)    1,2,3-Propanetriol (IUPAC)

Note that a *monohydric alcohol* contains one —OH group.

To name such an alcohol using the IUPAC system, assign a number to locate each hydroxyl group on the main chain of carbons, and separate the numbers by one or more commas. Then name the corresponding alkane and add the suffix *-diol, -triol*, etc. depending on the number of hydroxyl groups present. Note that ethylene glycol is a common name. To name a glycol using the common system, name the corresponding alkene, and add the word *glycol.* A common name cannot be written for 1,2,3-propanetriol. Unlike common and IUPAC nomenclature of alcohols, trivial nomenclature is not based on one or more rules but depends on the source of the compound or who discovered the compound. For example, since methyl alcohol was originally obtained by the destructive distillation of wood, it was named *wood alcohol.* In the same way, since ethyl alcohol was first produced from wheat or rye, it was named *grain alcohol.* Also, 1,2,3-propanetriol was named *glycerol* because it is a sweet viscous liquid

---

**FIGURE 14.4** Examples of the carbinol system of alcohol nomenclature.

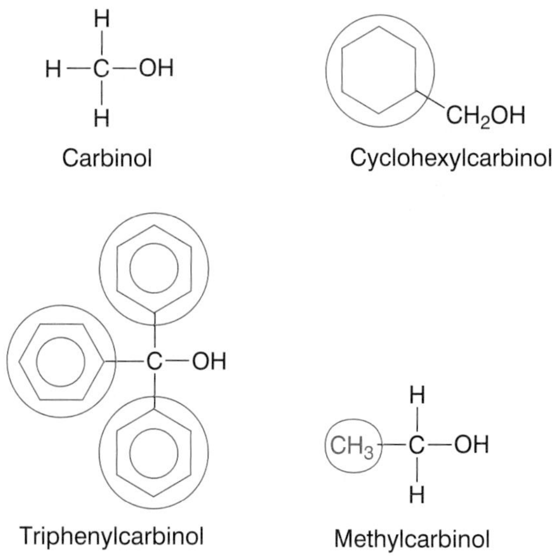

Carbinol

Cyclohexylcarbinol

Triphenylcarbinol

Methylcarbinol

| TABLE 14.1 | A Comparison of Common, Derived, and IUPAC Nomenclature for Some Selected Alcohols | | |
|---|---|---|---|
| Structure | Common | IUPAC | Derived |
| $CH_3OH$ | Methyl alcohol | Methanol | Carbinol |
| $CH_3CH_2OH$ | Ethyl alcohol | Ethanol | Methylcarbinol |
| $CH_3CH(OH)CH_3$ | Isopropyl alcohol | 2-Propanol | Dimethylcarbinol |
| $CH_3CH_2CH_2OH$ | n-Propyl alcohol | 1-Propanol | Ethylcarbinol |
| $(CH_3)_2CHCH_2OH$ | Isobutyl alcohol | 2-Methyl-1-propanol | Isopropylcarbinol |

(Greek: *glykys,* meaning "sweet"). Table 14.1 shows a comparison of common, derived, and IUPAC nomenclature for some simple alcohols.

## 14.2 PHYSICAL PROPERTIES OF ALCOHOLS

Alcohols boil at a significantly higher temperature than do alkanes of comparable molecular weight, as can be observed in Table 14.2. In each comparison of an alcohol with an alkane, the alcohol of about the same molecular weight boils at least 50°C higher than the corresponding hydrocarbon. The alcohols boil higher because of the presence of intermolecular hydrogen bonds (see Sec. 1.6).

The hydrogen bond is an intermolecular force that is created when a hydrogen atom on a highly electronegative element such as oxygen, nitrogen, or fluorine is attracted to a highly electronegative element of another molecule. The highly electronegative atom to which the hydrogen is chemically attached partially strips the electrons from the hydrogen, giving it a partial positive charge ($\delta^+$). This positive charge is attracted to the electron-rich nitrogen, oxygen, or fluorine in another molecule. This resulting intermolecular dipole-dipole interaction is the hydrogen bond.

Hydrogen bond

The association of alcohol molecules increases the effective molecular weight of the alcohol and thus increases the boiling point.

As the molecular weight of the alcohol increases and the alcohol becomes more alkane-like, water solubility of the alcohol decreases (Table 14.3). We regard monohydroxy alcohols (also known as *monohydric alcohols*) with five carbon atoms or more as water-insoluble. A compound is designated as soluble in water if 3.0 g or more dissolves in 100 ml of water.

Note that hydrocarbons cannot form hydrogen bonds because hydrogen is only bonded to carbon. Thus the hydrocarbon molecules cannot associate, and boiling points are low.

| TABLE 14.2 | A Comparison of the Boiling Point of Selected Alcohols with Comparable Alkanes | | |
|---|---|---|---|
| Name | Structure | Molecular Weight | Boiling Point (°C) |
| Ethanol | $CH_3CH_2OH$ | 46 | 78 |
| Propane | $CH_3CH_2CH_3$ | 44 | $-42$ |
| 1-Butanol | $CH_3(CH_2)_2CH_2OH$ | 74 | 118 |
| Pentane | $CH_3(CH_2)_3CH_3$ | 72 | 36 |
| Cyclopentanol | | 86 | 140 |
| Cyclohexane | | 84 | 81 |

| TABLE 14.3 | Physical Properties of Alcohols | | |
| --- | --- | --- | --- |
| Name | Formula | Boiling Point (°C) | Water Solubility (g/100 ml $H_2O$) |
| Methyl alcohol | $CH_3OH$ | 65 | ∞ |
| Ethyl alcohol | $CH_3CH_2OH$ | 78 | ∞ |
| n-Propyl alcohol | $CH_3CH_2CH_2OH$ | 97 | ∞ |
| n-Butyl alcohol | $CH_3CH_2CH_2CH_2OH$ | 118 | 8 |
| Isobutyl alcohol | $(CH_3)_2CHCH_2OH$ | 108 | 11 |
| sec-Butyl alcohol | $CH_3CH(OH)CH_2CH_3$ | 100 | 13 |
| tert-Butyl alcohol | $(CH_3)_3COH$ | 83 | ∞ |
| n-Pentyl alcohol | $CH_3(CH_2)_3CH_2OH$ | 138 | 2 |
| n-Hexyl alcohol | $CH_3(CH_2)_4CH_2OH$ | 158 | 0.7 |

Methyl alcohol is a good low-boiling (65°C) solvent often used in the laboratory. Unfortunately, it is toxic, and consumption of less than 30 ml can cause blindness and, ultimately, death. Unscrupulous people who made moonshine whiskey illegally sometimes would substitute methyl alcohol for the liquor (ethyl alcohol) with disastrous results. At the present time, methyl alcohol is made industrially by treating carbon monoxide with hydrogen at a high temperature and pressure in the presence of catalysts.

$$CO + H_2 \xrightarrow{\Delta} CH_3OH$$
$$\text{Pressure, catalysts}$$

Before this process, methanol was obtained industrially by the destructive (absence of air) distillation of wood. Thus methyl alcohol is often called *wood alcohol.*

Ethyl alcohol is known as *grain alcohol* because it has been prepared for thousands of years from various grains by fermentation. Actually, the starch in the wheat, rye, rice, or corn is hydrolyzed to sugars, which, in turn, are fermented by the enzymes in yeast to produce ethyl alcohol.

Ethyl alcohol can be obtained from the fermentation mixture by distillation. However, no matter how many times the mixture is redistilled, the resulting solution is not pure ethyl alcohol but a mixture of 95% ethanol and 5% water. This 95%-5% solution is known as an *azeotrope,* or a *constant-boiling mixture,* and it represents the maximum concentration of ethanol that can be obtained by distillation in the presence of water.

To prepare *water-free,* or *absolute, ethyl alcohol,* add benzene to a mixture of 95% ethanol and 5% water and distill. The mixture of benzene, ethanol, and water forms a different azeotrope that boils at 65°C. Distilling this azeotrope results in the removal of the

## 14.3
## THE MORE IMPORTANT ALCOHOLS

As with other families of organic compounds, increased branching within a series of isomeric alcohols results in a gradual decrease in boiling point. Note the isomeric butyl alcohols in Table 14.3.

## BIOCHEMICAL BOXED READING 14.1

Pure ethyl alcohol is a clear and colorless liquid with a pungent odor. Alcohol cannot be consumed pure because it is a powerful dehydrating agent that will cause discomfort. Thus the alcohol is diluted with water to form a solution containing 45% to 50% alcohol by volume. The percentage alcohol in a whiskey is one-half the term known as *proof.* Thus a whiskey containing 50% alcohol by volume is 100 proof. The term *proof* is short for *proof of spirit* and came into use about 150 years ago. A person who wished to buy whiskey from a farmer wanted to be sure it was whiskey—not water—he or she was paying for. Thus the liquid was poured on some gunpowder and the mess contacted with a flame. If it did not burn, too much water was present. If it

immediately blazed up, the whiskey contained too much alcohol. Whiskey of the right alcohol content (100 proof) would burn slowly with a blue flame. This represented a 50:50 alcohol-water mix, which was considered to be the perfect mix because it contained just enough alcohol to ignite the gunpowder and was called 100 proof—as it is today.

The body can break down (metabolize) ⅓ ounce of ethyl alcohol per hour. Drinking black coffee, exercising, and taking cold showers will not change this metabolic rate. In other words, drinking black coffee or taking amphetamines counteracts the depression of the drunk and keeps him or her awake, but it does not change his or her state of intoxication at all.

water and some ethanol along with the benzene. Left behind is absolute ethanol that is easily distilled at 78°C. Because traces of benzene remain in the alcohol, such a distilled product is not safe to drink. Ethyl alcohol (as whiskey, wine, and beer) is the only alcohol that human beings deliberately consume and have done so for thousands of years.

Ethylene glycol is a well-known automobile antifreeze. At one time methanol was used as an automobile antifreeze. Problems developed when a typical down-and-out alcoholic examined the radiator tag labeled "This radiator protected with methyl alcohol." The alcohol addict saw the word *alcohol,* and a bell rang in his or her head. He or she would drain the radiator, drink the alcohol-water mixture, and die. Although situations like this contributed to the elimination of methanol as an automobile antifreeze, the most pressing reason for its removal was that it is too volatile to remain in the cooling system of an automobile. This is why ethylene glycol-based antifreeze is called *permanent* antifreeze.

Glycerol plays an important role in the chemical makeup of fats and oils. In addition, the production of glycerol permits the arctic caterpillar, *Gynaephora groenlandica*, to live in temperatures far below 0°C. The glycerol acts an antifreeze, protecting the insect and permitting it to function in the severe cold.

## 14.4 REPRESENTATIVE REACTIONS OF THE ALCOHOLS

### Dehydration

Refer to Section 6.1 for a more detailed discussion. In the presence of sulfuric acid or phosphoric acid, an alcohol, when heated, loses water to form the corresponding alkene.

$$-\overset{|}{\underset{H}{C}}-\overset{|}{\underset{OH}{C}}- \;\; \underset{}{\overset{\text{Acid, }\Delta}{\rightleftharpoons}} \;\; -\overset{|}{C}=\overset{|}{C}- \;\; + \;\; H_2O$$

Notice that an H and an OH bonded to adjacent carbons are lost as water. Thus this reaction is known, in general, as an *elimination reaction*. Because water is eliminated, this reaction is known as a *dehydration reaction*. For example:

$$\text{Cyclopentanol} \;\; \overset{H_2SO_4}{\underset{\Delta}{\longrightarrow}} \;\; \text{Cyclopentene} \;\; + \;\; H_2O$$

Cyclopentanol
(cyclopentyl
alcohol)

Cyclopentene

---

## INDUSTRIAL BOXED READING 14.2

The typical *perfume* sold in a store is made up of three components. A mixture of as many as 200 essential oils represents the first component. A number of these oils are obtained from plants such as peppermint leaves (menthol is a main chemical constituent) or lemon peel (citral is a main chemical constituent). Essential oils from flowers such as rose, jasmine, and gardenia, from animals such as musk deer, and from trees such as sandalwood, cedar, and pine are blended in.

The second component is a solvent—most often ethyl alcohol is used. When the ethyl alcohol evaporates, it carries with it a few molecules of the essential oil mixture, so a delicate scent surrounds the person using the perfume.

The third component is a *fixative* that is present to slow the rate of evaporation of the essential oils and extend the pleasant odor for a longer time. Naturally occuring fixatives are present in civit (from the civit cat) and orrisroot (from the Spanish iris), among other sources.

The percentage of essential oils present in a fragrance defines how the fragrance is classified. If the fragrance contains 20% to 40% essential oils, it is called a *perfume*. A *cologne* typically contains about 4% essential oils, whereas a *splash* includes approximately 2% essential oils.

Menthol
(mint)

$$(CH_3)_2C=CHCH_2CH_2\underset{CH_3}{\overset{|}{C}}=CH\underset{O}{\overset{||}{C}}H$$

Citral

For a more detailed discussion of the oxidation of alcohols, refer to Section 15.2. Primary alcohols are oxidized to aldehydes and, finally, to carboxylic acids using an oxidizing agent such as a mixture of sodium dichromate and sulfuric acid ($Na_2Cr_2O_7 + H_2SO_4$). An *oxidizing agent* is a substance that undergoes *reduction* and thus produces an *oxidation*. In a similar way, a *reducing agent* is a substance that undergoes *oxidation* and thus produces a *reduction*. For our purposes, *oxidation* can be defined as either a gain of oxygen or a loss of hydrogen, while *reduction* is a gain of hydrogen or loss of oxygen. In general,

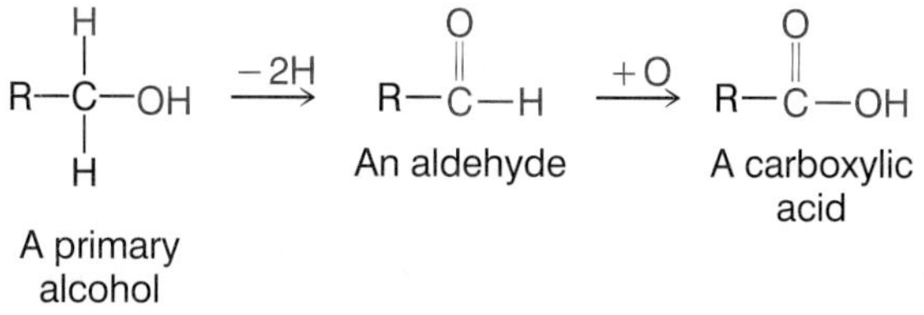

**Oxidation**

$$
\begin{array}{ccc}
\underset{\substack{| \\ \text{H} \\ \text{A primary} \\ \text{alcohol}}}{\overset{\text{H}}{\underset{|}{\text{R—C—OH}}}} & \xrightarrow{-2\text{H}} & \underset{\text{An aldehyde}}{\overset{\overset{\text{O}}{\|}}{\text{R—C—H}}} & \xrightarrow{+\text{O}} & \underset{\substack{\text{A carboxylic} \\ \text{acid}}}{\overset{\overset{\text{O}}{\|}}{\text{R—C—OH}}}
\end{array}
$$

When sodium dichromate and sulfuric acid are used as the oxidizing agent, usually the intermediate aldehyde product is not isolated, and the reaction proceeds all the way to the acid. For example:

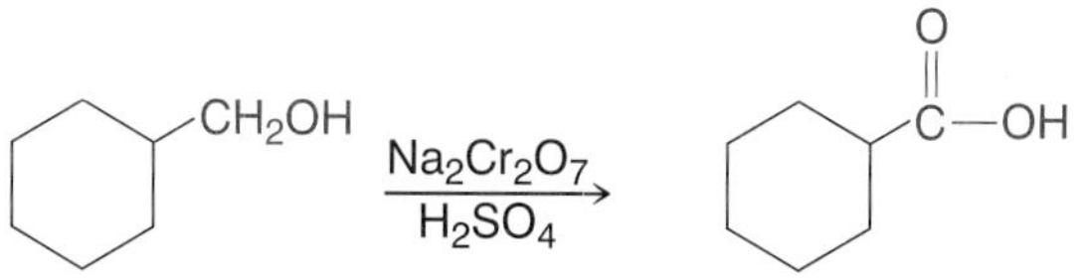

Cyclohexylcarbinol                    $\xrightarrow[\text{H}_2\text{SO}_4]{\text{Na}_2\text{Cr}_2\text{O}_7}$                    Cyclohexylcarboxylic acid

In a similar way, secondary alcohols are oxidized to ketones. In general,

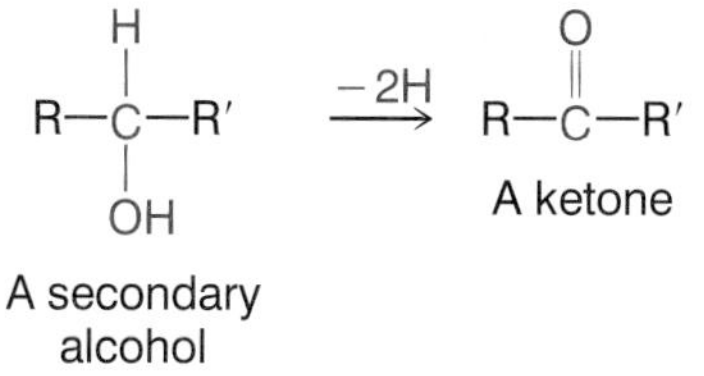

$$
\underset{\substack{\text{A secondary} \\ \text{alcohol}}}{\overset{\text{H}}{\underset{\underset{\text{OH}}{|}}{\text{R—C—R}'}}} \xrightarrow{-2\text{H}} \underset{\text{A ketone}}{\overset{\overset{\text{O}}{\|}}{\text{R—C—R}'}}
$$

Example:

$$
\underset{\underset{\text{OH}}{|}}{\text{CH}_3\text{CHCH}_2\text{CH}_3} \xrightarrow[\text{H}_2\text{SO}_4]{\text{Na}_2\text{Cr}_2\text{O}_7} \overset{\overset{\text{O}}{\|}}{\text{CH}_3\text{CCH}_2\text{CH}_3}
$$

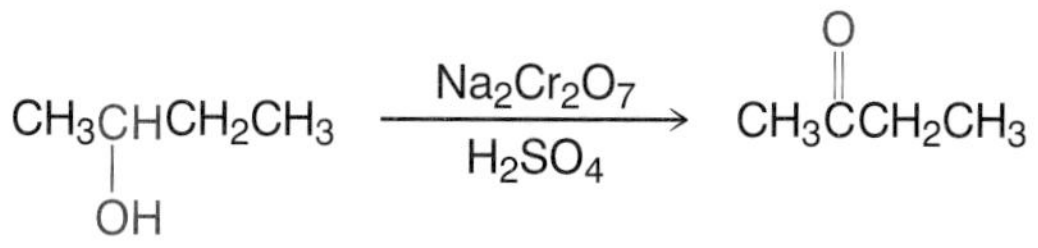

Tertiary alcohols do not react with the sodium dichromate–sulfuric acid mixture. Refer to Section 15.2 for a test tube reaction that can qualitatively distinguish primary and secondary alcohols from tertiary alcohols.

---

Complete the following equations by supplying a structural formula for every organic compound formed. If no reaction takes place, write "no reaction."

**SOLVED PROBLEM 14.2**

a.  $CH_3CH_2CH_2CH_2OH \xrightarrow[\Delta]{H_2SO_4}$

b.  2-methyl-2-pentanol $\xrightarrow[H_2SO_4]{Na_2Cr_2O_7}$

**SOLUTION**

a. This is a dehydration reaction. Since the hydroxyl group is located on C-1 and the neighboring hydrogen eliminated with it to produce water must be on C-2,* the carbon-carbon double bond produced must exist between C-1 and C-2, and the organic product of this reaction is

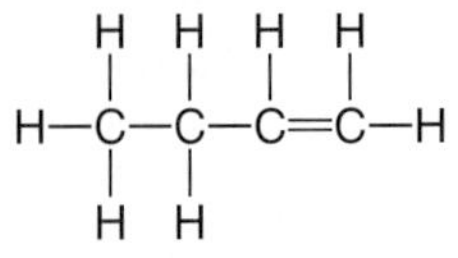

$$H-\underset{\underset{H}{|}}{\overset{\overset{H}{|}}{C}}-\underset{\underset{H}{|}}{\overset{\overset{H}{|}}{C}}-\overset{\overset{H}{|}}{C}=\overset{\overset{H}{|}}{C}-H$$

1-Butene

b. Since the alcohol is treated with a mixture of sodium dichromate and sulfuric acid, this represents a possible oxidation reaction. When a strucutral formula for 2-methyl-2-pentanol is drawn, it appears obvious that this alcohol is tertiary, and therefore, no reaction takes place.

$$CH_3-\underset{\underset{OH}{|}}{\overset{\overset{CH_3}{|}}{C}}-CH_2CH_2CH_3$$

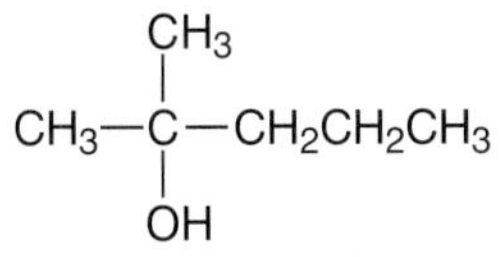

2-Methyl-2-pentanol
(a tertiary alcohol)

---

**EXERCISE 14.2**

Complete the following equations by supplying a structural formula for every organic compound formed. If no reaction takes place, write "no reaction."

a. propyl alcohol $\xrightarrow[\Delta]{H_2SO_4}$

b. cyclohexanol $\xrightarrow[H_2SO_4]{Na_2Cr_2O_7}$

c. propyl alcohol $\xrightarrow[H_2SO_4]{Na_2Cr_2O_7}$

d. 2-pentanol $\xrightarrow[\Delta]{H_2SO_4}$

e. 3-methyl-3-hexanol $\xrightarrow[H_2SO_4]{Na_2Cr_2O_7}$

**14.5
THIOLS AND
DISULFIDES**

*Thiols* (RSH) are sulfur analogues of alcohols. Thiols are called *mercaptans* in common nomenclature. Simply name the organic group bonded to the sulfur of the mercapto

---

*We will assume, in this chapter, that rearrangements do not occur. For a discussion of rearrangement that accompanies dehydration, refer to Sec. 12.5.

group (—SH), and add the word *mercaptan*. To name a thioalcohol using the IUPAC system, name the corresponding alkane. Then add the suffix *-thiol*. Since the mercapto group is a functional group, we assign the lowest possible number to the carbon bearing the mercapto group.

The term *mercaptan* comes from the Latin meaning "trapping mercury."

$$CH_3CH_2SH$$

$$\begin{array}{c} H \\ | \\ CH_3CCH_3 \\ | \\ SH \end{array}$$

Ethyl mercaptan (common)
Ethanethiol (IUPAC)

Isopropyl mercaptan (common)
2-Propanethiol (IUPAC)

Expanding our priority system of functional-group nomenclature to include the mercapto group, we get

$$OH > SH > C{=}C > C{\equiv}C > Cl(Br)(I)$$

Use of this priority system can be seen from the following examples:

$$HSCH_2CH_2CH_2OH \qquad ClCH_2CH_2CH_2SH$$

3-Mercapto-1-propanol     3-Chloro-1-propanethiol

When the —SH group is used as a substituent group, the prefix used is *mercapto-*.

## BIOCHEMICAL BOXED READING 14.3

The amino acid cysteine contains a mercapto group, whereas cystine, a disulfide-containing amino acid, is present in keratin, a protein found in human hair.

$$\begin{array}{c} O \\ \| \\ H_2N-CH-C-OH \\ | \\ CH_2SH \end{array}$$

Cysteine

$$\begin{array}{c} H \quad O \\ | \quad \| \\ NH_2-C{-\!\!-\!\!-}C-OH \\ | \\ CH_2 \\ | \\ S \\ | \\ S \\ | \\ CH_2 \quad O \\ | \quad \| \\ H_2N-C{-\!\!-\!\!-}C-OH \\ | \\ H \end{array}$$

Cystine

When hair is in the natural cystine form, it cannot be waved or shaped in any way. However, when hair is treated with a mild reducing agent, the disulfide bond in cystine is broken to give cysteine. Hair in the cysteine form can be and is waved or shaped as desired. Finally, the cysteine hair is reacted with a mild oxidizing agent, regenerating cystine hair and keeping the waved hair in place.

$$\text{Keratin—S—S—keratin} \xrightarrow{\text{Reduction}} 2 \text{ keratin—SH}$$

Cystine hair          Cysteine hair
(cannot be waved)     (can be waved)

$$2 \text{ Keratin—SH} \xrightarrow{\text{Oxidation}} \text{keratin—S—S—keratin}$$

The reversibility of the processes is represented as follows:

$$\text{Keratin—S—S—keratin} \underset{\text{Oxidation}}{\overset{\text{Reduction}}{\rightleftarrows}} 2 \text{ keratin—SH}$$

Thus the hair can be waved time after time.

Another useful mercaptan is D-2-amino-3-mercapto-3-methylbutanoic acid (penicillamine).

$$\begin{array}{c} OH \\ | \\ O{=}C \\ | \\ H-C-NH_2 \\ | \\ CH_3-C-SH \\ | \\ CH_3 \end{array}$$

Penicillamine

This compound has been used in the treatment of Wilson's disease (a genetic disease that is due to a patient's inability to synthesize ceruloplasmin, a binding transport protein for copper; this lack of ceruloplasmin results in a buildup of copper in the liver, kidneys, and brain) and mercury poisoning, where the compound binds with the metal and the resulting combination is flushed out by means of the urine.

In addition, penicillamine has potential as an anti-AIDS (acquired immune deficiency syndrome) drug based on its ability to inhibit the reproduction of the virus that causes AIDS.

The most striking physical property of the thiols is the very disagreeable odor characteristic of the lower homologues. The odor of garlic, for example, is due to allyl mercaptan, among other compounds, while one of the compounds that cause the odor responsible for halitosis, or "bad breath," is methyl mercaptan.

To prepare a thiol, treat the corresponding primary or secondary halide with sodium hydrosulfide. In general,

$$RX + {}^-SH \longrightarrow RSH + X^-$$

where X = Cl, Br, or I. For example:

$$CH_3CH_2CH_2Br + {}^-SH \longrightarrow CH_3CH_2CH_2SH + Br^-$$
$$\text{1-Propanethiol}$$

Mild oxidation of a thiol with iodine ($I_2$) or hydrogen peroxide produces the corresponding disulfide. In general,

$$2\ RS{-}H \xrightarrow{\text{[O]}} RS{-}SR$$

Examples:

$$2\ CH_3CH_2SH + I_2 \longrightarrow CH_3CH_2SSCH_2CH_3 + 2HI$$
$$2\ CH_3SH + H_2O_2 \longrightarrow CH_3SSCH_3 + 2\ H_2O$$

A disulfide, in turn, is easily reduced back to the corresponding thiol with a reducing agent such as zinc and acetic acid. In general,

$$RS{-}SR \xrightarrow{\text{[H]}} 2\ RS{-}H$$

Example:

$$CH_3CH_2SSCH_2CH_3 + Zn + 2\ CH_3COOH \longrightarrow 2\ CH_3CH_2SH + Zn(C_2H_3O_2)_2$$

The reversibility of these processes can be represented as follows:

$$2\ RS{-}H \underset{\text{[H]}}{\overset{\text{[O]}}{\rightleftarrows}} RS{-}SR$$

---

## ▶ CHAPTER ACCOMPLISHMENTS

### 14.1 Introduction, Structure, and Nomenclature of Alcohols

☐ Draw a structural formula for a monohydric alcohol that is
   a. vinylic.
   b. benzylic.
   c. primary.
   d. secondary.
   e. tertiary.

☐ Draw a structural formula for
   a. 2-butanol.
   b. 3-chloro-1-propanol.
   c. propargyl alcohol.
   d. 1,2,3-propanetriol.
   e. triphenylcarbinol.

☐ List four rules for IUPAC alcohol nomenclature.

### 14.2 Physical Properties of Alcohols

☐ Explain why ethanol boils at a temperature over 50°C higher than propane, even though both molecules have about the same molecular weight.

☐ List the three atoms to which hydrogen must be bonded in order to make hydrogen bonding a reality.

☐ Without looking at Table 14.3, qualitatively predict the solubility of *n*-hexyl alcohol in water.

### 14.3 The More Important Alcohols

☐ Explain why ethyl alcohol is also called *grain alcohol*.

☐ Explain what is meant by *absolute ethyl alcohol*.

### 14.4 Representative Reactions of the Alcohols

☐ Write a general equation to illustrate
   a. dehydration of an alcohol.
   b. oxidation of a primary alcohol.
   c. oxidation of a secondary alcohol.

☐ Furnish the class of alcohol that does not react with a mixture of sodium dichromate and sulfuric acid.

## 14.5 Thiols and Disulfides

☐ Name the following thiols using both the common and IUPAC systems of nomenclature.

 a. $CH_3CH_2SH$.

 b. $CH_3CH(SH)CH_3$.

☐ Complete the following equations by supplying a structural formula for each organic product produced and a molecular formula or formula unit for each inorganic substance formed.

 a. $2\ CH_3CH_2SH + I_2 \longrightarrow$

 b. $CH_3CH_2CH_2Br + {}^-SH \longrightarrow$

 c. $CH_3CH_2SSCH_2CH_3 + Zn + CH_3COOH \longrightarrow$

## ▶ KEY TERMS

phenol (14.1)
carbinol system of alcohol nomenclature (14.1)
dihydric alcohol (14.1)
trihydric alcohol (14.1)
monohydric alcohol (14.1)
wood alcohol (14.1)
grain alcohol (14.1)
glycerol (14.1)

azeotrope or constant-boiling mixture (14.3)
absolute ethyl alcohol (14.3)
oxidizing agent (14.4, 15.1)
reduction (14.4, 15.1)
reducing agent (14.4, 15.1)
oxidation (14.4, 15.1)
thiol (14.5)
mercaptan (14.5)

proof or proof of spirit (Biochemical Boxed Reading 14.1)
perfume (Industrial Boxed Reading 14.2)
fixative (Industrial Boxed Reading 14.2)
cologne (Industrial Boxed Reading 14.2)
splash (Industrial Boxed Reading 14.2)

## ▶ PROBLEMS

1. Name the following compounds using the IUPAC method of nomenclature. Also, if possible, supply a common name for each compound.

 a. $(CH_3)_2CHCH_2OH$

 b.

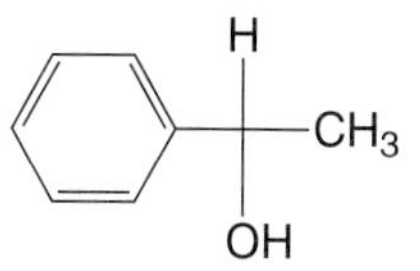

 c.

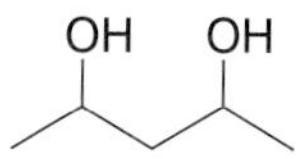

 d.

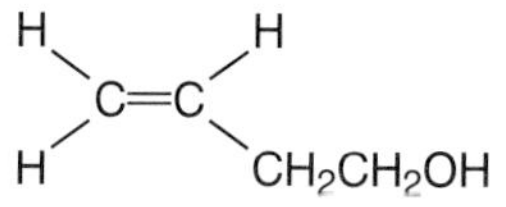

 e.

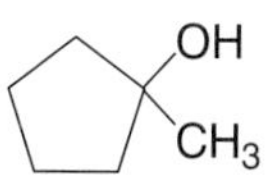

 f. 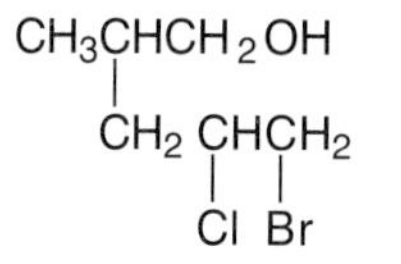 $CH_3CHCH_2OH$
  $\qquad\ \ |$
  $\qquad CH_2\ CHCH_2$
  $\qquad\qquad\ |\quad\ |$
  $\qquad\qquad Cl\ \ Br$

 g. $CH_3CH(OH)CH_2CH_2CH_3$

 h. 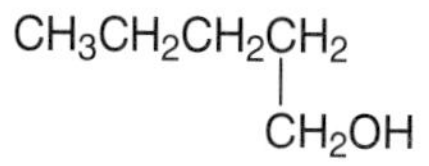 $CH_3CH_2CH_2CH_2$
  $\qquad\qquad\qquad |$
  $\qquad\qquad\quad CH_2OH$

2. Classify the alcohols in Problem 1 as primary, secondary, tertiary, allylic, or benzylic. Provide another common name for benzyl alcohol.

3. Draw a structural formula for each of the following compounds.

 a. 1,3-propanediol

 b. diphenylcarbinol

 c. (S)-2-butanol (Fischer projection)

 d. vinyl alcohol (an unstable compound)

 e. glycerol

 f. carbinol

 g. m-bromobenzyl alcohol

 h. dimethylcarbinol

4. Why is sec-pentyl alcohol an ambiguous name and therefore an incorrect name for the compound in Exercise 14.1b?

5. Draw a structural formula for each of the following compounds. Classify each alcohol.

 a. propargyl alcohol

 b. isohexyl alcohol

 c. (Z)-3-penten-2-ol

 d. (S)-3-eicosanol (wedge projection)

 e. 2-methyl-3-pentanol

 f. cyclohexyl alcohol

6. Draw a diagram to show hydrogen bonding in grain alcohol.

7. Which of the following would you expect to have the greatest water solubility? Explain.
   (i) $CH_3(CH_2)_4CH_2OH$
   (ii) $CH_3(CH_2)_3CH(OH)CH_2OH$
   (iii) $CH_3(CH_2)_2CH(OH)CH(OH)CH_2OH$

8. Supply the reagent that would produce the product(s) provided for each of the following.

   a. $? \xrightarrow[\Delta]{H_2SO_4} CH_3CH_2CH_2CH_2CH{=}CH_2 + H_2O$

   b. $? \xrightarrow[H_2SO_4]{Na_2Cr_2O_7} CH_3CH_2CH_2\overset{\overset{O}{\|}}{C}CH_2CH_3$

   c. $? + I_2 \longrightarrow (CH_3)_2CH{-}S{-}S{-}HC(CH_3)_2 + 2\,HI$

   d. $? + {}^-SH \longrightarrow CH_3CH_2CH_2CH_2SH + Br^-$

9. Would you expect ethanethiol to boil at a higher or lower temperature than 1-propanol? Explain.

10. Small amounts of 1-butanethiol are periodically added to natural gas supplies before the gas is shipped to homes for use. Why?

11. Provide an IUPAC name for each of the following compounds.
   a. $ClCH_2CH_2SH$
   b. $HC{\equiv}CCH_2CH_2CH_2CH_2OH$

   c. 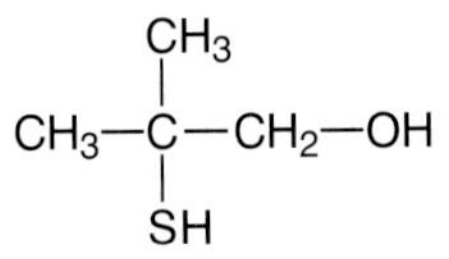 $CH_3{-}\overset{\overset{\displaystyle CH_3}{|}}{\underset{\underset{\displaystyle SH}{|}}{C}}{-}CH_2{-}OH$

# CHAPTER 15

# Alcohols

## SYNTHESIS AND CHEMICAL PROPERTIES

Certainly, organic halogen compounds are the leading industrial intermediates. However, not far behind are the alcohols. Unlike many organic halogen compounds, a number of alcohols can be obtained from natural products. Alcohols undergo a variety of reactions, and we will discuss both the synthesis and reactions of the alcohols in this chapter.

Reacting an alkene with cold concentrated sulfuric acid followed by water produces an alcohol (see Secs. 5.6 and 6.2). In general,

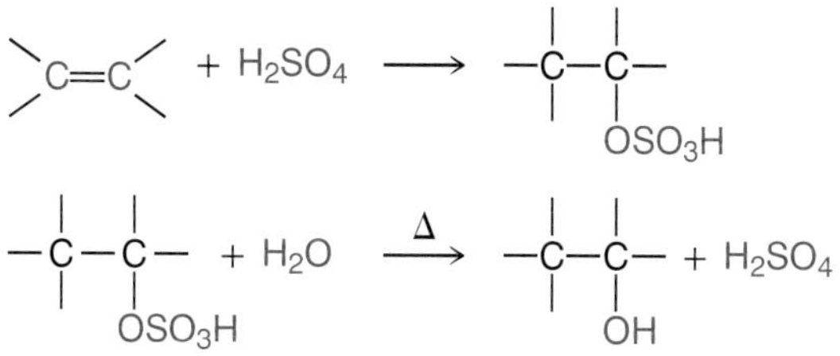

## 15.1 SYNTHESIS

### Hydration of Alkenes

$$\begin{array}{c} \diagdown \\ C=C \\ \diagup \end{array} + H_2SO_4 \longrightarrow \begin{array}{c} | \; | \\ -C-C- \\ | \; | \\ OSO_3H \end{array}$$

$$\begin{array}{c} | \; | \\ -C-C- \\ | \; | \\ OSO_3H \end{array} + H_2O \xrightarrow{\Delta} \begin{array}{c} | \; | \\ -C-C- \\ | \; | \\ OH \end{array} + H_2SO_4$$

Examples:

$$CH_3(H)C-C(H)CH_3 + H_2SO_4 \longrightarrow \begin{array}{c} H \\ | \\ CH_3CCH_2CH_3 \\ | \\ OSO_3H \end{array}$$

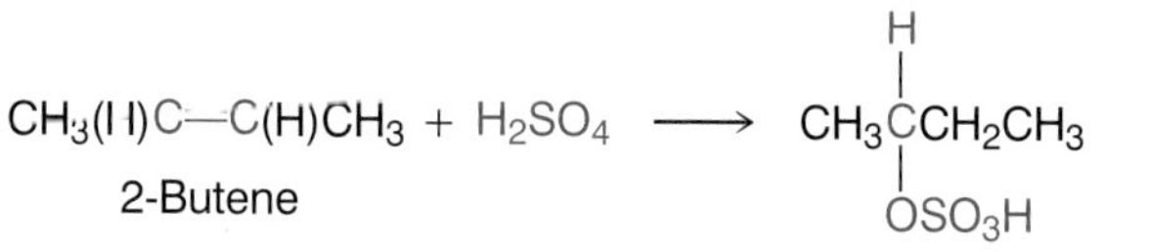

2-Butene

sec-Butyl hydrogen sulfate

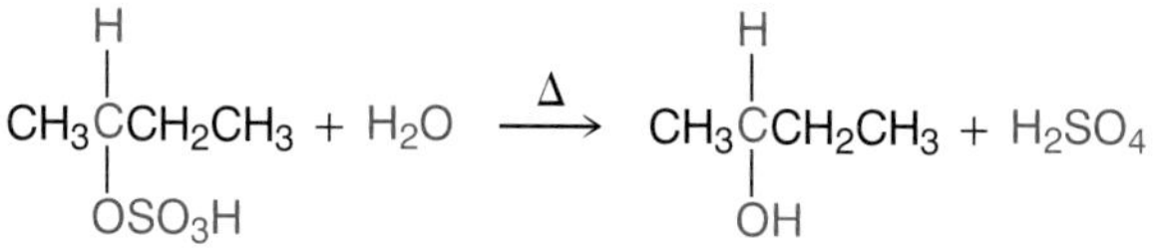

$$\begin{array}{c} H \\ | \\ CH_3CCH_2CH_3 \\ | \\ OSO_3H \end{array} + H_2O \xrightarrow{\Delta} \begin{array}{c} H \\ | \\ CH_3CCH_2CH_3 \\ | \\ OH \end{array} + H_2SO_4$$

sec-Butyl alcohol

## Reaction of Organic Halides with Hydroxide Ion

This substitution reaction is somewhat limited in scope due to the competition of elimination. Primary halides substitute best by means of an $S_N2$ mechanism (see Sec. 12.1); tertiary halides tend to eliminate (E1) (see Sec. 12.4) to form alkene, and little substitution actually takes place. Dilute rather than concentrated base is used, and the temperature is kept as low as is practical to favor substitution.

$$CH_3CH_2I + {}^{-}OH \xrightarrow[\text{Dilute aqueous}]{S_N2} CH_3CH_2OH + I^-$$

## Hydroboration-Oxidation

Hydroboration-oxidation is a two-step reaction (see Sec. 6.2). The first step is the quantitative and rapid addition of diborane, $B_2H_6$ (for convenience, represented as $BH_3$), across the double bond of the alkene to give an alkylborane and ultimately a trialkylborane. Note that the —$BH_2$ group adds to the least substituted carbon of the double bond so as to ultimately produce an alcohol with an anti-Markownikoff orientation. In general,

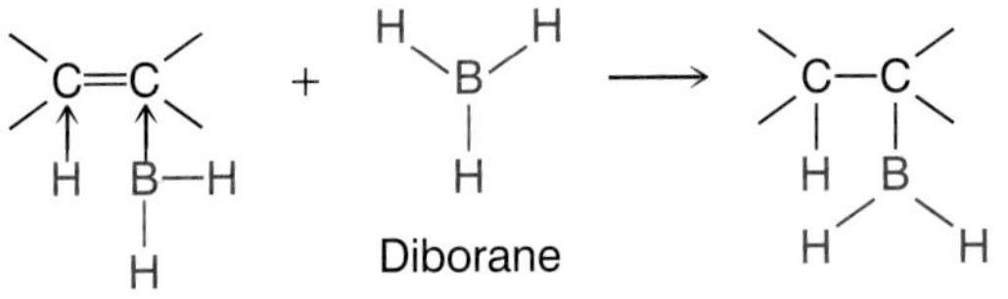

Diborane

Alkylborane

Two additional mol of alkene react with the alkylborane to give a trialkylborane.

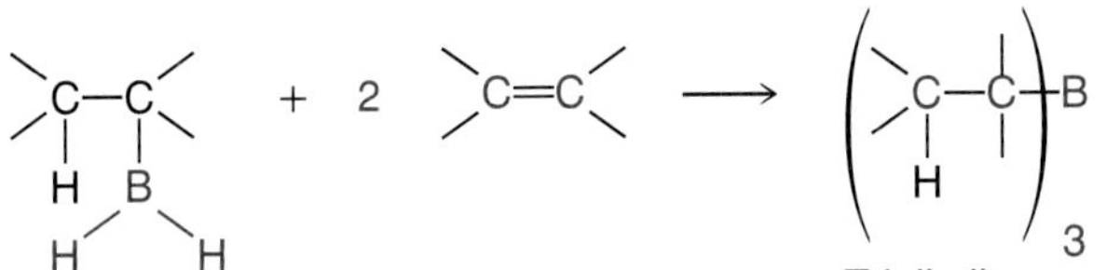

Trialkylborane

The second step of the reaction is an oxidation with hydrogen peroxide and sodium hydroxide to produce an alcohol.

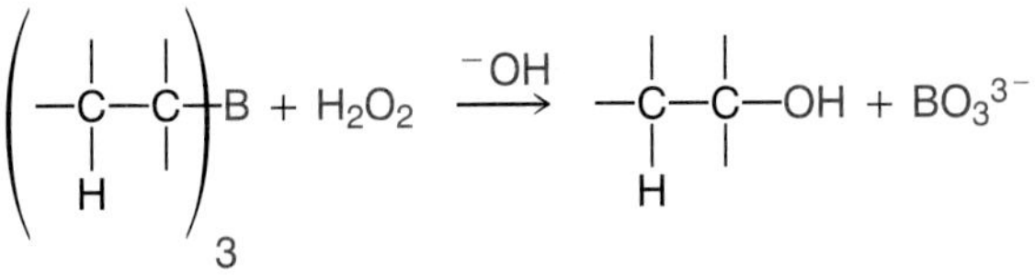

Example:

$$3\ CH_3CH{=}CH_2 + BH_3 \longrightarrow (CH_3CH_2CH_2)_3B$$

tri(*n*-propyl)borane

$$(CH_3CH_2CH_2)_3B + H_2O_2 \xrightarrow{{}^{-}OH} 3\ CH_3CH_2CH_2OH + BO_3^{3-}$$

1-Propanol

## Fermentation

This method of preparation of ethyl alcohol has been used for thousands of years and consists of reacting an aqueous solution of a sugar with yeast.

$$C_6H_{12}O_6 \xrightarrow{\text{Yeast}} 2\ C_2H_5OH + 2\ CO_2$$

A sugar

## BIOCHEMICAL BOXED READING 15.1

Ethyl alcohol is oxidized by enzymes [an *enzyme* is a biological catalyst (Sec. 11.14)] in the body to start the metabolic process. Alcohol dehydrogenase is the catalyst in the conversion of ethyl alcohol to acetaldehyde, a toxic substance. Acetaldehyde, in turn, is instantaneously oxidized to acetic acid with *cytochrome P-450*, a *coenzyme* that is the oxidizing agent. (A coenzyme is an organic molecule that

can exist on its own or as a part of an enzyme.) Thus toxic acetaldehyde cannot accumulate in the body.

One treatment for alcoholics to ensure a state of sobriety is based on the drug *Antabuse*. This drug complexes with cytochrome P-450. Thus acetaldehyde cannot be converted to acetic acid, toxic acetaldehyde accumulates in the body of the patient, and he or she becomes violently ill.

$$CH_3CH_2OH \xrightarrow{\text{Alcohol dehydrogenase}} CH_3\overset{O}{\overset{\|}{C}}-H \qquad \xrightarrow{\text{Cytochrome P-450}} CH_3-\overset{O}{\overset{\|}{C}}-OH$$

Note that without Antabuse, innocuous acetic acid is produced. In the presence of Antabuse, acetic acid cannot form because cytochrome P-450 is complexed with the drug, and

the complex does not catalyze the oxidation of acetaldehyde to acetic acid.

$$CH_3CH_2OH \xrightarrow{\text{Alcohol dehydrogenase}} CH_3\overset{O}{\overset{\|}{C}}-H \xrightarrow{\text{Cytochrome P-450–Antabuse complex}} \text{no reaction}$$

Aldehydes and ketones are reduced to produce primary and secondary alcohols, respectively.

### Reduction of Carbonyl Compounds

$$R-\overset{O}{\overset{\|}{C}}-H \xrightarrow{[H]} R-\overset{H}{\underset{H}{\overset{|}{C}}}-OH$$

$$R-\overset{O}{\overset{\|}{C}}-R' \xrightarrow{[H]} R-\overset{H}{\underset{R'}{\overset{|}{C}}}-OH$$

The symbol [H] represents any reducing agent.

Because the reaction of a carbonyl compound to prepare an alcohol is a reduction (Table 15.1), we need to employ a substance that produces a reduction—a reducing agent. Of course, a substance that causes an oxidation is an oxidizing agent.

Many different reducing agents are available to convert a carbonyl compound to the corresponding alcohol. The most often used are lithium aluminum hydride (LiAlH$_4$) and sodium borohydride (NaBH$_4$). Sodium borohydride is soluble in water, so this reagent tends to be used with water-soluble carbonyl compounds of low molecular weight. Lithium aluminum hydride, on the other hand, is soluble in ether and is used with higher-molecular-weight aldehydes and ketones that are insoluble in water. Usually, either ethyl alcohol or water is used as a solvent with sodium borohydride. Since lithium aluminum hydride reacts explosively with water, anhydrous diethyl ether is the solvent of choice. In general,

$$R-\overset{O}{\overset{\|}{C}}-H \xrightarrow[\text{2. H}_2\text{O, H}^+]{\text{1. NaBH}_4\text{ (LiAlH}_4)} R-\overset{H}{\underset{H}{\overset{|}{C}}}-OH$$

| TABLE 15.1 | Oxidation State of Selected Functional Groups |
| --- | --- |

| Name of Functional Group | Structure of Functional Group | |
| --- | --- | --- |
| Carboxylic acid | R—C(=O)—OH | Highest oxidation state |
| Aldehyde | R—C(=O)—H | ↑ |
| Ketone | R—C(=O)—R′ | ↑ ↑ |
| Primary alcohol | R—C(H)(H)—OH | ↑ ↑ ↑ ↑ |
| Secondary alcohol | R—C(H)(OH)—R′ | Lowest oxidation state |

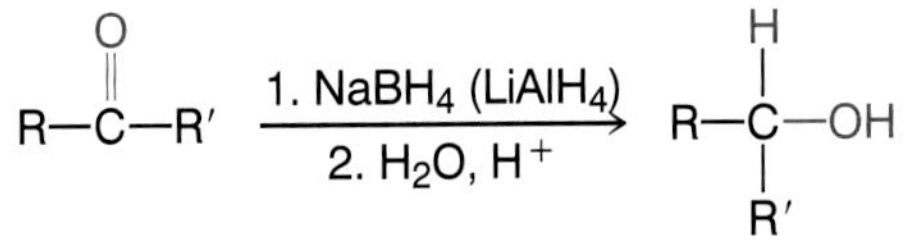

$$R-\overset{O}{\overset{\|}{C}}-R' \xrightarrow[\text{2. H}_2\text{O, H}^+]{\text{1. NaBH}_4\ (\text{LiAlH}_4)} R-\overset{H}{\underset{R'}{\overset{|}{\underset{|}{C}}}}-OH$$

Examples:

$$CH_3CH_2\overset{O}{\overset{\|}{C}}CH_3 \xrightarrow[\text{2. H}_2\text{O, H}^+]{\text{1. NaBH}_4,\ \text{ethanol}} CH_3CH_2\overset{OH}{\underset{H}{\overset{|}{\underset{|}{C}}}}CH_3$$

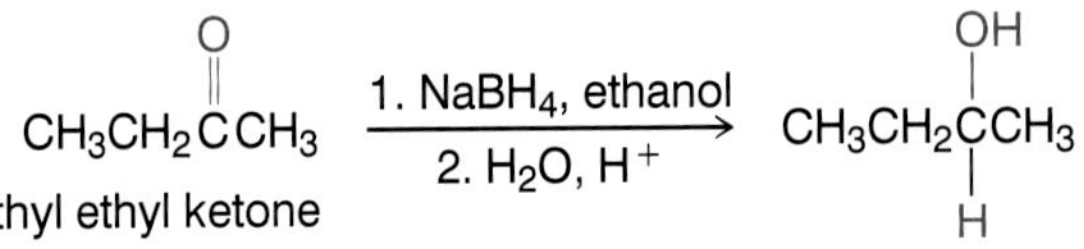

Methyl ethyl ketone

sec-Butyl alcohol

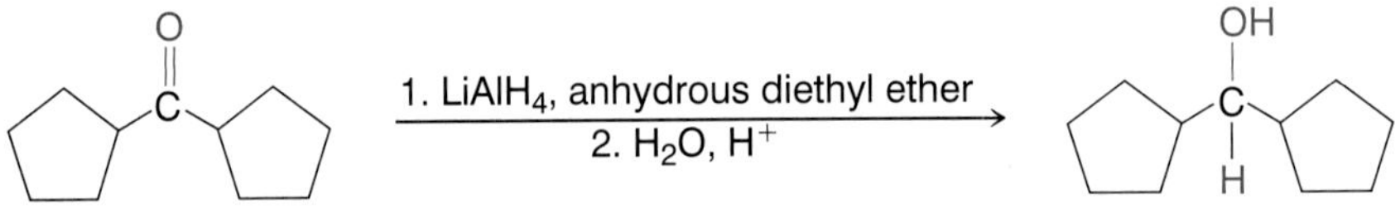

Dicyclopentyl ketone    1. LiAlH₄, anhydrous diethyl ether    2. H₂O, H⁺    Dicyclopentylcarbinol

The reaction begins when hydride ion from one of the reagents attacks the positive end of the carbonyl dipole. At the same time, a pair of electrons from the carbonyl attacks the boron (or aluminum) to give ion I. The reaction is completed when ion I is reacted with aqueous acid to yield the alcohol.

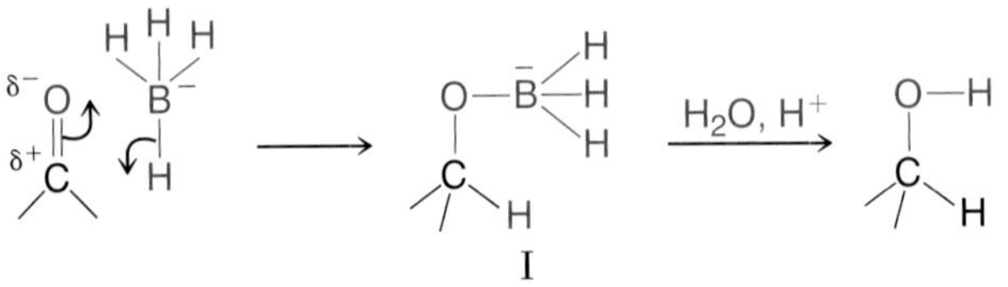

Since neither sodium borohydride nor lithium aluminum hydride reduces a carbon-carbon double bond, these reagents are used to prepare various carbon-carbon unsaturated alcohols.

$$CH_3CH=CHC\overset{\displaystyle O}{\overset{\|}{-}}H \quad \xrightarrow[\text{2. H}_2\text{O, H}^+]{\text{1. NaBH}_4\text{, ethanol}} \quad CH_3CH=CHCH_2OH$$

Note that the reduction of carbonyl compounds to produce alcohols is also discussed in Sections 18.6 and 19.3.

This reaction is studied in Section 19.2. In general,

**Addition of a Grignard Reagent to a Carbonyl Compound**

$$-\overset{\displaystyle O}{\overset{\|}{\underset{\displaystyle }{C}}}- \;+\; RMgX \;\longrightarrow\; -\overset{\displaystyle O^-\,{}^+MgX}{\underset{\displaystyle }{\overset{\|}{C}}}-R \quad \xrightarrow{\text{HCl}(aq)} \quad -\overset{\displaystyle OH}{\underset{\displaystyle }{\overset{\|}{C}}}-R$$

where X = Cl, Br, or I. For example:

$$CH_3-\overset{\displaystyle O}{\overset{\|}{C}}-H \;+\; CH_3MgBr \;\longrightarrow\; CH_3-\overset{\displaystyle O^-\,{}^+MgBr}{\underset{\displaystyle H}{\overset{\displaystyle |}{C}}}-CH_3 \quad \xrightarrow{\text{HCl}(aq)} \quad CH_3-\overset{\displaystyle OH}{\underset{\displaystyle H}{\overset{\displaystyle |}{C}}}-CH_3$$

Primary and secondary alcohols react with a number of different oxidizing reagents; tertiary alcohols do not react. Note that each step in the oxidation involves either a loss of hydrogen or an addition of oxygen. The symbol [O] represents any oxidizing agent.

## 15.2 CHEMICAL PROPERTIES

### Oxidation

$$R-CH_2OH \xrightarrow{[O]} R-\overset{\displaystyle H}{\underset{\displaystyle }{\overset{\displaystyle |}{C}}}=O \xrightarrow{[O]} R-\overset{\displaystyle O}{\overset{\|}{C}}-OH$$

Primary alcohol        Aldehyde        Carboxylic acid

$$R_2CHOH \xrightarrow{[O]} R_2C=O$$

Secondary alcohol        Ketone

$$R_3COH \xrightarrow{[O]} \text{no reaction}$$

Tertiary alcohol

The PCC reagent is prepared by dissolving chromium trioxide in hydrochloric acid. Pyridine is then added to the resulting solution.

With certain special oxidizing agents, e.g., pyridinium chlorochromate (PCC), the oxidation of a primary alcohol can be stopped at the aldehyde stage. In general,

$$RCH_2OH \xrightarrow[\text{CH}_2\text{Cl}_2]{\text{PCC}} R\overset{\displaystyle H}{\underset{\displaystyle }{\overset{\displaystyle |}{C}}}=O$$

$$CrO_3 \;+\; HCl \;+\; \text{(pyridine)}$$

Pyridine

Example:

$$CH_3(CH_2)_5CH_2OH \xrightarrow[\text{CH}_2\text{Cl}_2]{\text{PCC}} CH_3(CH_2)_5\overset{\displaystyle H}{\underset{\displaystyle }{\overset{\displaystyle |}{C}}}=O$$

1-Heptanol        Heptanal

$$\longrightarrow \quad \text{(pyridinium)} \; CrO_3Cl^-$$

Pyridinium chlorochromate

Pyridinium chlorochromate is also useful in that the reagent does not attack carbon-carbon double or triple bonds.

The solvent usually employed is methylene chloride.

$$\text{(E)-2-Octen-1-ol} \xrightarrow[\text{CH}_2\text{Cl}_2]{\text{PCC}} \text{(E)-2-Octenal}$$

(E)-2-Octen-1-ol        (E)-2-Octenal

On the other hand, with an oxidizing agent such as sodium dichromate and sulfuric acid, oxidation of the primary alcohol continues to the carboxylic acid stage. In general,

$$RCH_2OH \xrightarrow[\text{H}_2\text{SO}_4]{\text{Na}_2\text{Cr}_2\text{O}_7} R-\overset{\displaystyle O}{\overset{\|}{C}}-OH + Cr^{3+}$$

Example:

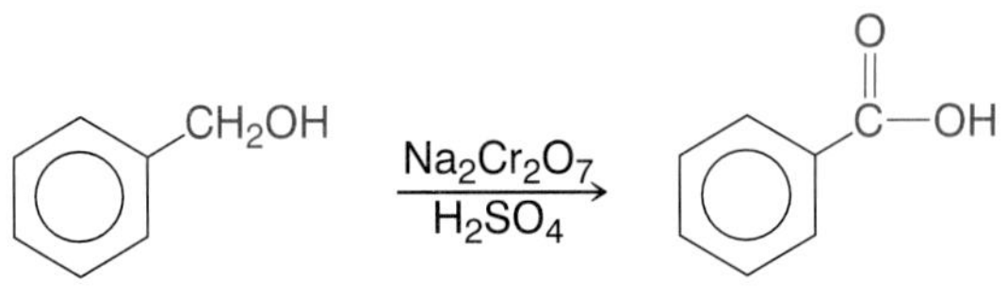

Benzyl alcohol $\xrightarrow[\text{H}_2\text{SO}_4]{\text{Na}_2\text{Cr}_2\text{O}_7}$ Benzoic acid

Reacting a secondary alcohol with a strong oxidizing agent produces a ketone. In general,

$$R_2CHOH \xrightarrow[\text{H}_2\text{SO}_4]{\text{Na}_2\text{Cr}_2\text{O}_7} R_2C{=}O + Cr^{3+}$$

Example:

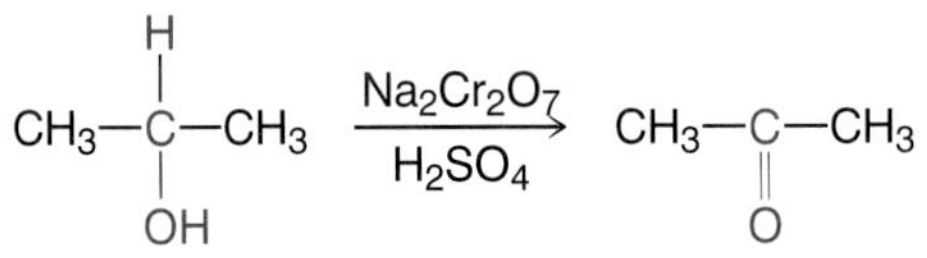

2-Propanol $\xrightarrow[\text{H}_2\text{SO}_4]{\text{Na}_2\text{Cr}_2\text{O}_7}$ Acetone

Since the chromic acid reagent ($Na_2Cr_2O_7 + H_2SO_4$) is orange in color and the chromic ion ($Cr^{3+}$) product is blue-green, this color change can be used to distinguish a primary or secondary alcohol from a tertiary alcohol in a test tube. When an unknown alcohol is reacted with chromic acid in a test tube, formation of the blue-green color indicates a primary or secondary alcohol, while no change in the orange color represents a tertiary alcohol.

A simplified process to predict the product in the oxidation of a primary or secondary alcohol can be written based on the following two rules:

1. Replace an —H on the carbinol carbon with an —OH.[*]

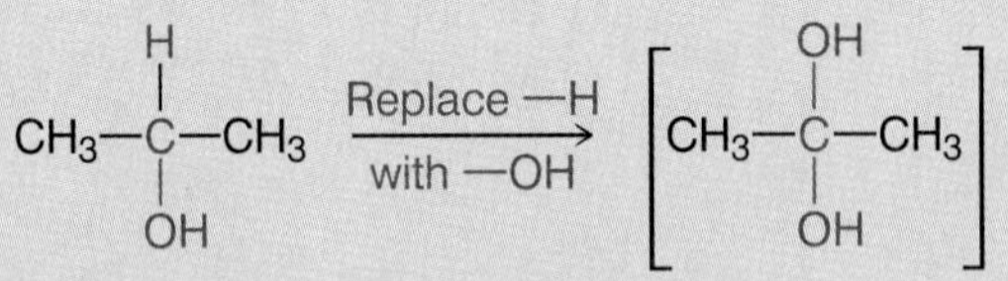

Note that brackets around a chemical species denote instability.

$$CH_3-\underset{\underset{\displaystyle OH}{|}}{\overset{\overset{\displaystyle H}{|}}{C}}-CH_3 \xrightarrow[\text{with —OH}]{\text{Replace —H}} \left[ CH_3-\underset{\underset{\displaystyle OH}{|}}{\overset{\overset{\displaystyle OH}{|}}{C}}-CH_3 \right]$$

2. A *gem*-diol (a compound with two —OH groups on the same carbon) is unstable in most cases and spontaneously loses water.

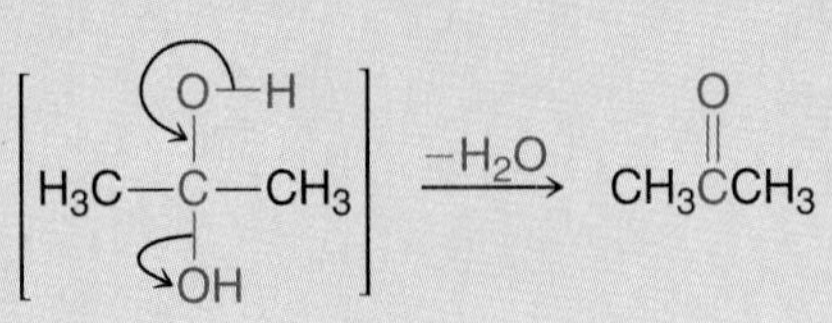

$$\left[ H_3C-\overset{\displaystyle O-H}{\underset{\displaystyle OH}{C}}-CH_3 \right] \xrightarrow{-H_2O} CH_3\overset{\displaystyle O}{\overset{\|}{C}}CH_3$$

Let us designate this as the *two-rule oxidation process*.

---

[*]This is just a formulation and is employed to simplify the oxidation process. Mechanistically, the replacement does not occur.

## SOLVED PROBLEM 15.1

Complete the following equations by supplying a structural formula for each organic compound formed. If no reaction takes place, write "no reaction."

a. $CH_3CH_2CH_2CH_2OH \xrightarrow[CH_2Cl_2]{PCC}$

b. 1-methylcyclopentanol $\xrightarrow[H_2SO_4]{Na_2Cr_2O_7}$

**SOLUTION**

a. Since pyridinium chlorochromate is the reagent used, the oxidation stops at the aldehyde stage, and the product is

$$CH_3CH_2CH_2\overset{\overset{\displaystyle O}{\|}}{C}—H$$

b. This alicyclic alcohol is a tertiary alcohol. Therefore, no reaction.

## EXERCISE 15.1

Complete the following equations by supplying a structural formula for each organic compound formed. If no reaction takes place, write "no reaction."

a. isobutyl alcohol $\xrightarrow[H_2SO_4]{Na_2Cr_2O_7}$

b. benzyl alcohol $\xrightarrow[CH_2Cl_2]{PCC}$

(Omit any by-products derived from PCC.)

## SOLVED PROBLEM 15.2

Use the two-rule oxidation process to predict the products of the oxidation of methyl alcohol.

**SOLUTION**

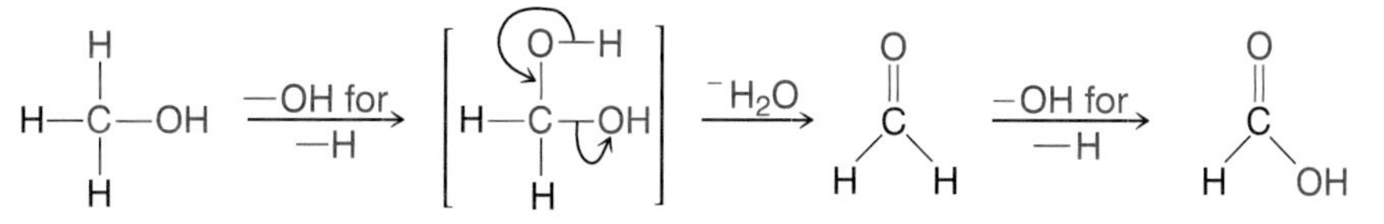


*gem*-Diol    Formaldehyde    Formic acid

$$H—\overset{\overset{\displaystyle O}{\|}}{C}—OH \xrightarrow[—H]{—OH\ for} HO—\overset{\overset{\displaystyle O}{\|}}{C}—OH \rightleftharpoons H_2O + CO_2$$

Carbonic acid

Thus carbon dioxide and water are the products in the oxidation of methyl alcohol.

## EXERCISE 15.2

Use the two-rule oxidation process to predict the products of the oxidation of

a. *sec*-butyl alcohol.

b. *n*-propyl alcohol.

## Reaction with Active Metals: Acidity of Alcohols

Alcohols are stronger acids than terminal alkynes but weaker acids than water (see Table 1.6). Thus alcohols are too weakly acidic to react with strong bases such as sodium hydroxide but are acidic enough to react with active metals such as sodium or potassium. In general,

$$ROH + M \longrightarrow RO^-M^+ + \tfrac{1}{2}H_2$$

An alkoxide

Examples:

$$CH_3OH + Na \longrightarrow CH_3O^-Na^+ + \tfrac{1}{2}H_2$$

Methyl alcohol        Sodium methoxide

$$(CH_3)_3COH + K \longrightarrow (CH_3)_3CO^-K^+ + \tfrac{1}{2}H_2$$

*tert*-Butyl alcohol        Potassium *tert*-butoxide

This reaction takes place at room temperature with progress monitored by the rate of hydrogen evolution. The rate of reaction is

$$1° > 2° > 3°$$

Because potassium is a more reactive metal than sodium, it is often used with the less reactive tertiary alcohols. Conversely, treating a reactive primary alcohol with potassium metal can result in an explosion because the reaction is so rapid. Thus sodium is usually the metal of choice.

To formulate the common name of an alkoxide, first, name the cation, and then replace the suffix -*yl* by the suffix -*oxide*. For example, $CH_3CH_2O^-Na^+$ is named sodium ethoxide. The name of the corresponding alcohol is ethyl alcohol. Dropping the -*yl* suffix, we get the fragment *eth*; addition of the suffix -*oxide* gives us sodium ethoxide.

Since alcohols are weak acids, alkoxide ions are strong bases—strong enough to pull a proton from a water molecule and produce the corresponding alcohol. In general,

$$RO^- + H_2O \longrightarrow ROH + {}^-OH$$

Example:

$$CH_3O^- + H_2O \longrightarrow CH_3OH + {}^-OH$$

## Esterification of Alcohols

Alcohols can be esterified with either carboxylic acids or inorganic acids.[*] At this point we will limit our discussion to esterification with inorganic acids. Reacting an alcohol with nitric acid in the presence of sulfuric acid as a catalyst yields an alkyl nitrate. In general,

$$RO(H) + (HO)-NO_2 \xrightarrow{H_2SO_4} RO-NO_2 + H_2O$$

Example:

$$\begin{array}{c} CH_2OH \\ | \\ HCOH \\ | \\ CH_2OH \end{array} + 3\,HNO_3 \xrightarrow{H_2SO_4} \begin{array}{c} CH_2ONO_2 \\ | \\ HCONO_2 \\ | \\ CH_2ONO_2 \end{array} + 3\,H_2O$$

Glycerol        Glyceryl trinitrate (nitroglycerin)

A powerful explosive, nitroglycerin is sensitive to the slightest jar. When introduced by Alfred Nobel in the 1850s, it caused a great many deaths in the construction and mining industries due to its instability. This haunted Nobel, who was able to increase its stability tremendously by absorbing it in diatomaceous earth. This resulting mixture is called

---

[*]The reaction of an inorganic acid with an alcohol is an equilibrium process, but for pedagogical reasons, this text will not treat it as such.

*dynamite*. Along with amyl nitrite, glyceryl trinitrate is used for relief of the pain of angina pectoris. In general,

$$R-O(H) + (HO)-N{=}O \xrightarrow{H_2SO_4} R-O-N{=}O + H_2O$$

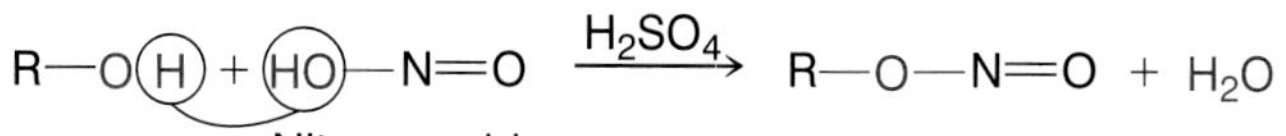

Nitrous acid

$$NaNO_2 + H_2SO_4 \longrightarrow HNO_2 + NaHSO_4$$

Since nitrous acid is unstable, it is prepared in the reaction mixture by treating sodium nitrite with excess sulfuric acid; the slight excess of sulfuric acid serves as a catalyst.

Example:

$$(CH_3)_2CHCH_2CH_2O-H + HO-N{=}O \xrightarrow{H_2SO_4} (CH_3)_2CHCH_2CH_2ON{=}O + H_2O$$

Isoamyl alcohol        Isoamyl nitrite
(isopentyl alcohol)      (isopentyl nitrite)

Alcohols react with sulfuric and phosphoric acids in a similar manner. In general,

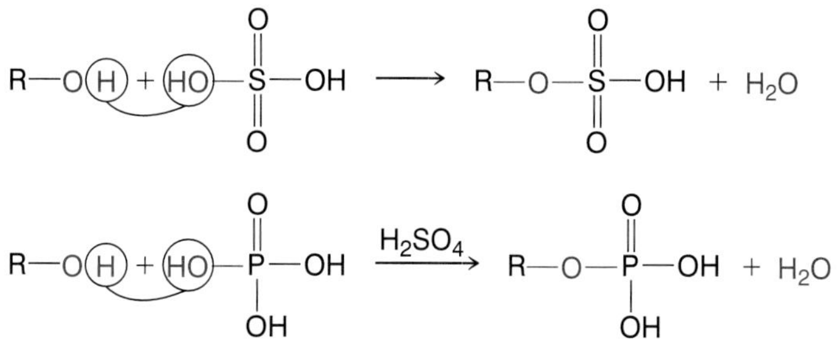

$$R-O(H) + (HO)-\overset{\displaystyle O}{\underset{\displaystyle O}{\overset{\|}{\underset{\|}{S}}}}-OH \longrightarrow R-O-\overset{\displaystyle O}{\underset{\displaystyle O}{\overset{\|}{\underset{\|}{S}}}}-OH + H_2O$$

$$R-O(H) + (HO)-\overset{\displaystyle O}{\underset{\displaystyle OH}{\overset{\|}{P}}}-OH \xrightarrow{H_2SO_4} R-O-\overset{\displaystyle O}{\underset{\displaystyle OH}{\overset{\|}{P}}}-OH + H_2O$$

Examples:

$$C_2H_5OH + HO-\overset{\displaystyle O}{\underset{\displaystyle O}{\overset{\|}{\underset{\|}{S}}}}-OH \longrightarrow C_2H_5O-\overset{\displaystyle O}{\underset{\displaystyle O}{\overset{\|}{\underset{\|}{S}}}}-OH + H_2O$$

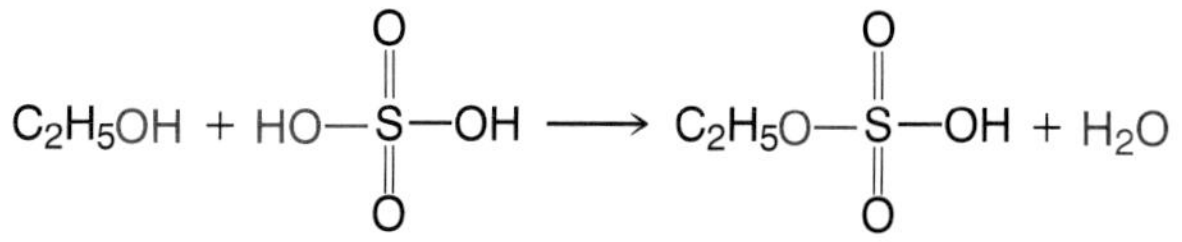

Ethyl hydrogen sulfate

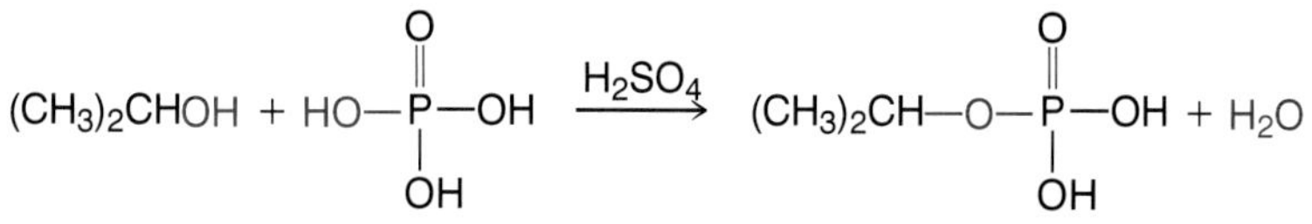

$$(CH_3)_2CHOH + HO-\overset{\displaystyle O}{\underset{\displaystyle OH}{\overset{\|}{P}}}-OH \xrightarrow{H_2SO_4} (CH_3)_2CH-O-\overset{\displaystyle O}{\underset{\displaystyle OH}{\overset{\|}{P}}}-OH + H_2O$$

Isopropyl dihydrogen phosphate

Inorganic esters are prepared at relatively low temperatures (10°C is typical). A temperature of 80°C, for example, would lead to alkene rather than ester formation.

An alkyl hydrogen sulfate ester is also formed when an alkene is treated with sulfuric acid. Usually, the alkyl hydrogen sulfate is not isolated but is hydrolyzed to produce the corresponding alcohol (see Secs. 5.6 and 6.2).

Alcohols are weak bases, just as they are weak acids. An alcohol is basic enough to dissolve in a strong acid such as cold concentrated sulfuric acid to produce an oxonium ion. In general,

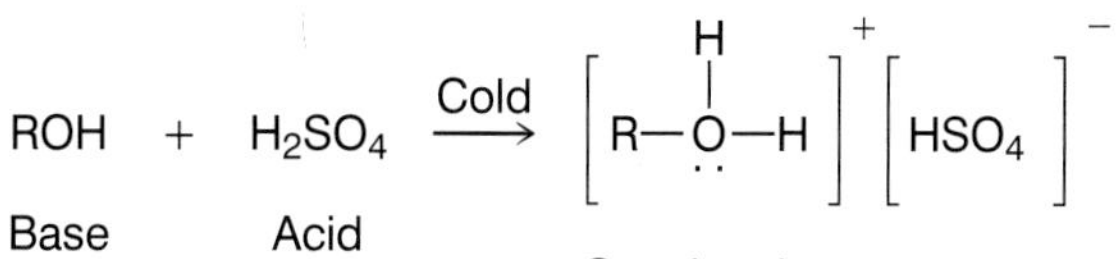

**Basicity of Alcohols**

$$ROH + H_2SO_4 \xrightarrow{Cold} \left[ R-\overset{\displaystyle H}{\underset{\displaystyle \cdot\cdot}{O}}-H \right]^{+} \left[ HSO_4 \right]^{-}$$

Base     Acid
Oxonium ion

Example:

$$CH_3CH_2OH \ + \ H_2SO_4 \ \xrightarrow{\text{Cold}} \ \left[ CH_3CH_2\!-\!\overset{\overset{\displaystyle H}{|}}{\underset{\displaystyle ..}{O}}\!-\!H \right]^{+} \ \left[ HSO_4 \right]^{-}$$

Ethyloxonium ion

Oxonium ions are unstable and cannot be isolated.[*] In time, an alkyl hydrogen sulfate would be produced. In general,

$$ROH + H_2SO_4 \ \xrightarrow{\text{Cold}} \ [ROH_2]^{+} \, [HSO_4]^{-} \ \longrightarrow \ ROSO_3H + H_2O$$

Example:

$$CH_3OH + H_2SO_4 \ \xrightarrow{\text{Cold}} \ [CH_3OH_2]^{+} \, [HSO_4]^{-} \ \longrightarrow \ CH_3OSO_3H + H_2O$$

Methyl hydrogen<br>sulfate

## SOLVED PROBLEM 15.3

How could you distinguish between *n*-butyl alcohol and pentane using a test tube reaction? Describe the test and any changes that occur due to the test.

### SOLUTION

Add each compound to a test tube containing cold concentrated sulfuric acid. The alcohol dissolves; the alkane does not dissolve, and two immiscible layers are formed, with the alkane as the top layer. An alternate procedure is to add each compound to a test tube containing a mixture of concentrated sulfuric acid and sodium dichromate (the chromic acid reagent). The contents of the test tube containing the alcohol change color from orange to blue-green, while the contents of the test tube containing the alkane remain orange.

## EXERCISE 15.3

How could you distinguish between each of the following pairs of compounds using a test tube reaction? Describe the test and any changes that occur due to the test.

a. heptane and 1-heptanol

b. 1-heptanol and 1-heptene

c. 1-heptanol and 1-heptyne

## Dehydration of Alcohols

For detailed discussions of the dehydration of an alcohol to yield an alkene, refer to Sections 6.1 and 12.5.

$$-\overset{\displaystyle |}{\underset{\displaystyle H}{C}}-\overset{\displaystyle |}{\underset{\displaystyle OH}{C}}- \ \xrightarrow[\Delta]{H_2SO_4 \text{ or } H_3PO_4} \ \diagup\!\!\diagdown C\!=\!C\diagdown\!\!\diagup \ + \ H_2O$$

## Reaction of Alcohols to Produce Halides: The Lucas Qualitative Test for Alcohols

Note that when HCl is the reagent, $ZnCl_2$ is also used.

The reaction of an alcohol with a hydrogen halide to produce an organic halide is discussed in Section 12.2.

$$ROH + HX \longrightarrow RX + H_2O$$

where X = Cl, Br or I. This same reaction is the basis of the Lucas qualitative test for alcohols. The reagent combination of hydrochloric acid and zinc chloride is often used to distinguish the classes of alcohols. When an alcohol reacts with an aqueous hydrochloric acid–zinc chloride mixture in a test tube, one of the following events can take

---

[*]The formation of an unstable oxonium ion is, in actuality, an equilibrium process.

place: If the contents of the test tube immediately turn cloudy, the alcohol is tertiary, allylic, or benzylic; in 3 to 5 minutes, secondary; if no cloudiness forms in 30 minutes, the alcohol is primary or vinylic. This is called the *Lucas test*. The cloudiness is caused by formation of a water-insoluble alkyl halide.

Organic halides also can be prepared by reacting an alcohol with a phosphorus trihalide. In general,

$$3\ ROH + PX_3 \xrightarrow{\Delta} 3\ RX + H_3PO_3$$

where X = Cl, Br, or I. For example:

$$3\ (CH_3)_2CHCH_2OH + PBr_3 \xrightarrow{\Delta} 3\ (CH_3)_2CHCH_2Br + H_3PO_3$$

Since phosphorous acid (the by-product of the preceding reaction) is stable up to 200°C, low-boiling organic halides are easily distilled off, in good yield, with no interference from high-boiling phosphorous acid ($H_3PO_3$).

A particularly convenient reagent, useful in preparing an organic chloride from a primary or secondary alcohol, is thionyl chloride:

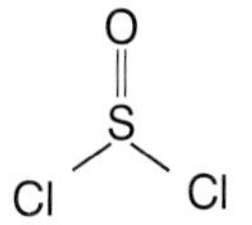

Thionyl chloride

In general,

$$ROH + SOCl_2 \xrightarrow{\Delta} RCl + SO_2 + HCl$$

Example:

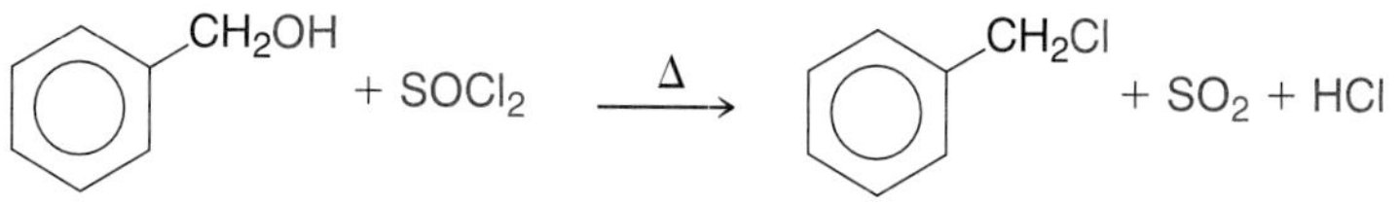

This reaction is especially useful for laboratory synthesis because the two by-products, sulfur dioxide ($SO_2$) and hydrogen chloride (HCl), are gases. Thus no time-consuming laboratory procedures need be employed to separate the product from the by-products.

---

**SOLVED PROBLEM 15.4**

Complete the following equations by writing a structural formula for each organic product and a molecular formula or formula unit for each inorganic product formed.

a. $CH_3CH_2CH_2CH_2OH + HBr \xrightarrow{\Delta}$

b. $CH_3CH(OH)CH_3 + SOCl_2 \xrightarrow{\Delta}$

**SOLUTION**

a. $CH_3CH_2CH_2CH_2Br + H_2O$

b. $CH_3CH(Cl)CH_3 + SO_2 + HCl$

---

**EXERCISE 15.4**

Complete the following equations by drawing a structural formula for each organic product and writing a molecular formula or formula unit for each inorganic product formed.

a.  $CH_3CH_2OH + HI \xrightarrow{\Delta}$

b.  $CH_3CH_2CH_2CH_2OH + PCl_3 \xrightarrow{\Delta}$

c.  $CH_3CH_2CH_2OH + HCl \xrightarrow[\Delta]{ZnCl_2}$

## ▶ CHAPTER ACCOMPLISHMENTS

### 15.1 Synthesis

☐ Write a general equation for
   a. the hydration of an alkene.
   b. the $S_N2$ substitution reaction of an organic halide with hydroxide ion.
   c. the hydroboration-oxidation of an alkene.
   d. a fermentation.
   e. the reduction of an aldehyde.
   f. the reduction of a ketone.

☐ Explain why reacting a tertiary halide with hydroxide ion does not produce an alcohol.

### 15.2 Chemical Properties

☐ Complete the following equations by supplying a structural formula for each organic product and, where relevant, a molecular formula or formula unit for every inorganic compound produced. If no reaction occurs, write "no reaction."

   a.  $CH_3CH_2OH \xrightarrow[CH_2Cl_2]{PCC}$

   b.  2-methyl-1-propanol $\xrightarrow[H_2SO_4]{Na_2Cr_2O_7}$

   c.  cyclopentanol $\xrightarrow[H_2SO_4]{Na_2Cr_2O_7}$

   d.  1-methylcyclopentanol $\xrightarrow[H_2SO_4]{Na_2Cr_2O_7}$

   e.  *tert*-butyl alcohol + K $\longrightarrow$
   f.  $CH_3O^- + H_2O \longrightarrow$
   g.  glycerol + 3 $HNO_3 \longrightarrow$
   h.  $CH_3OH + H_2SO_4 \xrightarrow{Cold}$
   i.  $CH_3CH_2OH \xrightarrow[\Delta]{H_3PO_4}$
   j.  $CH_3CH_2CH_2OH + HBr \longrightarrow$
   k.  3 $(CH_3)_2CHCH_2OH + PBr_3 \longrightarrow$
   l.  benzyl alcohol + $SOCl_2 \longrightarrow$

☐ Draw a structural formula for
   a. pyridinium chlorochromate.
   b. (*E*)-2-octen-1-ol.
   c. sodium ethoxide.
   d. glyceryl trinitrate.
   e. ethyl hydrogen sulfate.
   f. isopropyl dihydrogen phosphate.
   g. ethyloxonium ion.

☐ Summarize the two-rule oxidation process.
☐ Distinguish between *n*-butyl alcohol and pentane using a test tube reaction.
☐ List the reagents used in the Lucas test.
☐ Explain why thionyl chloride is a particularly useful reagent that is used to prepare a primary or secondary organic chloride from the corresponding alcohol.

## ▶ KEY TERMS

*two-rule oxidation process* (15.2)
*dynamite* (15.2)
*Lucas test* (15.2)

*cytochrome P-450* (Biochemical Boxed Reading 15.1)
*coenzyme* (Biochemical Boxed Reading 15.1)

*Antabuse* (Biochemical Boxed Reading 15.1)

## ▶ PROBLEMS

1. Explain why *n*-propyl alcohol cannot be prepared by treating propylene with water and sulfuric acid. Name the alcohol that can be prepared.

2. Prepare each of the following from *n*-propyl alcohol using a one-step synthesis. You may use any inorganic reagents necessary and pyridinium chlorochromate.

a. propanoic acid ($CH_3CH_2COOH$)

b. propionaldehyde ($CH_3CH_2CHO$)

c. propene

d. sodium *n*-propoxide

e. *n*-propyl nitrate

f. *n*-propyl dihydrogenphosphate

g. 1-chloropropane

3. Each of the following represents a synthesis involving at least two steps. You may use any inorganic and organic reagents necessary.

a. benzaldehyde from benzene

b. propane from 2-propanol (two ways)

c. diethyl disulfide from ethyl alcohol

d. isopentyl nitrite from 3-methyl-1-butene

e. potassium benzoate from benzene

f. sodium *tert*-butoxide from isobutyl alcohol

4. Use the two-rule oxidation process to predict the product of the oxidation of 1-hexanol.

5. Any ether (R—O—R′) dissolves in cold concentrated sulfuric acid for the same reason that any alcohol does. Write an equation that illustrates why any ether dissolves.

6. Complete the following by drawing a structural formula for each organic product formed and writing a molecular formula or formula unit for each inorganic product formed. If no reaction occurs, write "no reaction."

a. $(CH_3)_2CHCH_2OH \xrightarrow{PCC}$ (no products necessary based on PCC)

b. $CH_3CH_2CH_2CH_2OH + NaBr + H_2SO_4 \longrightarrow$

c. $CH_3CH_2CH_2OH + Na \longrightarrow$

d. $(CH_3)_2C(OH)CH_2CH_3 \xrightarrow[\Delta]{H_2SO_4}$

e. $(CH_3)_2CHO^- + H_2O \longrightarrow$

f. $(CH_3CH_2)_2S_2 + Zn + H_2SO_4 \longrightarrow$

g. $CH_3CH_2CH_2OH + PBr_3 \xrightarrow{\Delta}$

h. $CH_3CH_2CH_2OH + SOCl_2 \xrightarrow{\Delta}$

i. $CH_3(CH_2)_7CH_2CH \underset{\underset{2.\ H_2O,\ H^+}{}}{\overset{\overset{1.\ LiAlH_4,}{\text{anhydrous diethyl ether}}}{\longrightarrow}}$ (no inorganic products necessary) (with $\overset{\|}{O}$)

j. $CH_3CH_2CH_2CH_2OH \xrightarrow[H_2SO_4]{Na_2Cr_2O_7}$ (no inorganic products necessary)

k. $(CH_3)_3COH \xrightarrow[H_2SO_4]{Na_2Cr_2O_7}$ (no inorganic products neccessary)

7. a. An alcohol, $C_4H_{10}O$, gave a positive Lucas test and a negative test with sodium dichromate and sulfuric acid. Draw a structural formula for the alcohol.

b. An alcohol, $C_4H_{10}O$, gave a positive Lucas test and a positive test with sodium dichromate and sulfuric acid. Draw a structural formula for the alcohol.

c. Can an alcohol, $C_4H_{10}O$, give a negative Lucas test and a negative test with sodium dichromate and sulfuric acid? Explain.

8. Prepare the following compounds. You may use any organic or inorganic reagents necessary.

a. benzyl alcohol from benzene

b. β-phenylethyl alcohol from benzene

9. Compound *A*, $C_5H_{12}O$, produced hydrogen when reacted with metallic sodium and gave both a positive Lucas test and a positive test with sodium dichromate and sulfuric acid. Heating the compound with a catalytic amount of concentrated sulfuric acid produced compound *B*, $C_5H_{10}$. When compound *B* was reacted with ozone and then zinc and water, both acetone ($CH_3—\overset{\overset{O}{\|}}{C}—CH_3$) and acetaldehyde ($CH_3—\overset{\overset{O}{\|}}{C}—H$) were formed.

Write a structural formula for compound *A*. Show your reasoning.

10. Write an equation for the reaction of *n*-butyl alcohol, then *sec*-butyl alcohol, and finally, *tert*-butyl alcohol with each of the following substances. If no reaction occurs, write "no reaction."

a. $H_2SO_4$ (1 mol, 10°C)

b. $H_2SO_4$ (catalytic amount, heat)

c. HBr

d. Na

e. PCC (no inorganic products)

f. $Na_2Cr_2O_7$ and $H_2SO_4$ (no inorganic products)

11. Draw a structural formula for each of the following substances.

a. sodium methoxide

b. sodium *tert*-butoxide

c. isopropyloxonium ion

d. methyl hydrogen sulfate

e. ethyl dihydrogen phosphate

12. Group IIA metals (in the periodic table) such as calcium and magnesium also react with alcohol as follows:

$$2\ ROH + M \longrightarrow (RO)_2M + H_2 \quad (M = metal)$$

a. Write an equation illustrating the reaction of ethyl alcohol with magnesium. You need not balance the equation.

b. Predict the rate of the reaction in Problem 15.12a compared with the rate of the reaction of the same alcohol with sodium.

13. Name each of the following compounds.

a. $(CH_3)_2\overset{\overset{\displaystyle}{|}}{C}H$ with $\overset{|}{O}^-Na^+$

b. $CH_3—\overset{\overset{\displaystyle H}{|}}{\underset{\underset{\displaystyle CH_3CH_2}{|}}{C}}—O—\overset{\overset{\displaystyle O}{\|}}{\underset{\underset{\displaystyle O}{\|}}{S}}—OH$

c.

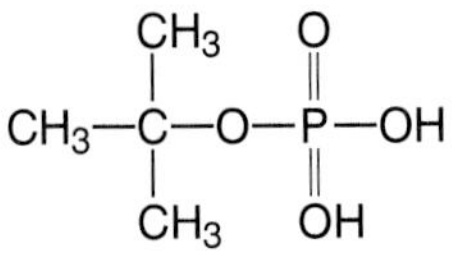

$$CH_3-\underset{\underset{CH_3}{|}}{\overset{\overset{CH_3}{|}}{C}}-O-\underset{\underset{OH}{||}}{\overset{\overset{O}{||}}{P}}-OH$$

14. Give a simple test tube reaction that would distinguish each of the following (describe the test used and any changes that occur).
    a. ethyl alcohol and cyclohexyl alcohol
    b. pentane and 1-pentanol
    c. 2-pentanol and 1-pentanol
    d. *n*-butyl alcohol and 1-butene
    e. *tert*-butyl alcohol and *tert*-butyl bromide

15. Draw a structural formula for the compound $C_4H_{10}O$ with the following proton NMR spectrum:

$\delta = 0.9$ (6H) doublet

$\delta = 1.8$ (1H) nonet (multiplet)

$\delta = 3.4$ (2H) doublet

$\delta = 4.0$ (1H) singlet

16. A compound, $C_7H_8O$, gave the following proton NMR absorption peaks:

$\delta = 3.7$ (1H) singlet

$\delta = 4.4$ (2H) singlet

$\delta = 7.3$ (5H) singlet

Draw a structural formula for this compound.

# CHAPTER 16

# Phenols, Ethers, and Epoxides (I)

STRUCTURE, NOMENCLATURE, PHYSICAL PROPERTIES,
PHENOLS AS ANTISEPTICS AND DISINFECTANTS,
ACIDITY OF PHENOLS, IDENTIFICATION OF PHENOLS,
ETHER AS AN ANESTHETIC, CROWN ETHERS

Phenols, ethers, and epoxides (along with alcohols) contain a single oxygen atom that is bonded to two other groups by means of single bonds. Some of the derivatives of carboxylic acids (see Chaps. 22 and 23) also contain a single oxygen atom, but it is bonded to carbon by means of a double bond.

Although phenols contain a hydroxyl group, as do alcohols, the physical and chemical properties of phenols differ from those of the alcohols. Both ethers and epoxides are characterized by an R—O—R grouping. Epoxides are a family of heterocyclic ethers characterized by a three-membered ring.

Like an alcohol, a phenol also contains a hydroxyl group. Unlike an alcohol, in a phenol the hydroxyl group is bonded directly to an aromatic ring (Fig. 16.1).

Today, the vast majority of what were trivial or common names of the phenols are now accepted by the IUPAC as systematic names. A significant number of the phenols are named as derivatives of phenol. Refer to Fig. 16.2 for a few examples of phenol nomenclature.

**PHENOLS**

**16.1
STRUCTURE AND
NOMENCLATURE**

**SOLVED PROBLEM 16.1**

Name the following compounds.

a.

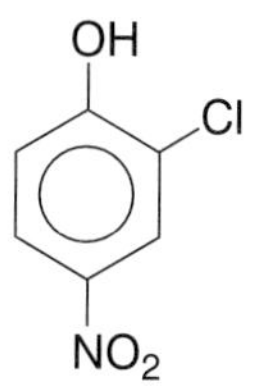

b.

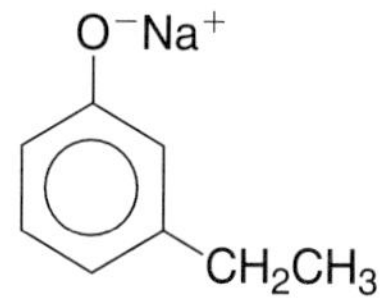

**FIGURE 16.1** Examples of the structural difference between alcohols and phenols.

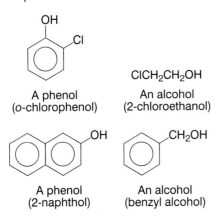

A phenol
(o-chlorophenol)

An alcohol
(2-chloroethanol)

$ClCH_2CH_2OH$

A phenol
(2-naphthol)

An alcohol
(benzyl alcohol)

$CH_2OH$

**FIGURE 16.2** Examples of phenol nomenclature.

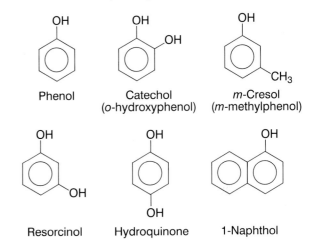

Phenol

Catechol
(o-hydroxyphenol)

m-Cresol
(m-methylphenol)

Resorcinol

Hydroquinone

1-Naphthol

**SOLUTION**

a. Since more than two groups are located on the benzene ring, numbers are used, and we get

2-chloro-4-nitrophenol

b. First, name the metal ion, in this case sodium. Then, name the phenol: *m*-ethylphenol. Drop the *-ol* suffix of the phenol, and add the *-oxide* suffix to give

sodium *m*-ethylphenoxide (see Sec. 16.4)

---

**EXERCISE 16.1**

Name the following compounds.

a.

b.

c.

---

## 16.2 PHYSICAL PROPERTIES

Phenols are low-melting solids. They boil at about the same temperature as the corresponding alcohols (phenol, 182°C; cyclohexyl alcohol, 161°C) because of intermolecular hydrogen bonding between phenol molecules.

$$Ar—O—H$$
←—Hydrogen bond
$$H—O—Ar$$

$$Ar—O—H$$

Since phenols contain, at the very least, a six-carbon nonpolar component, we would expect only slight water solubility of monohydric phenols (Table 16.1). However, if another hydroxyl group is introduced, as in resorcinol, water solubility increases appreciably (see Table 16.1).

## 16.3 PHENOLS AS ANTISEPTICS AND DISINFECTANTS

Let us begin by clarifying the distinction between an antiseptic and a disinfectant. An *antiseptic*, when applied to the skin, is used to control the growth of microorganisms on and in living cells, whereas a *disinfectant* is employed to control the microorganism population in nonliving matter.

| TABLE 16.1 | Physical Properties of Phenols | | | |
|---|---|---|---|---|
| Name | Formula | Melting Point (°C) | Boiling Point (°C) | Water Solubility (g/100 g $H_2O$) |
| Phenol | $C_6H_5OH$ | 43 | 182 | 9 |
| p-Cresol | p-$CH_3$—$C_6H_4$—OH | 35 | 202 | 2 |
| o-Cresol | o-$CH_3$—$C_6H_4$—OH | 31 | 191 | 3 |
| p-Nitrophenol | p-$NO_2$—$C_6H_4$—OH | 116 | 279 | 2 |
| o-Nitrophenol | o-$NO_2$—$C_6H_4$—OH | 46 | 216 | 0.2 |
| p-Chlorophenol | p-Cl—$C_6H_4$—OH | 44 | 220 | 3 |
| o-Chlorophenol | o-Cl—$C_6H_4$—OH | 9 | 175 | 3 |
| Resorcinol | m-OH—$C_6H_4$—OH | 110 | 276 | 147 |

**FIGURE 16.3** Other phenolic antiseptics and disinfectants.

o-Phenylphenol          Thymol          Hexachlorophene

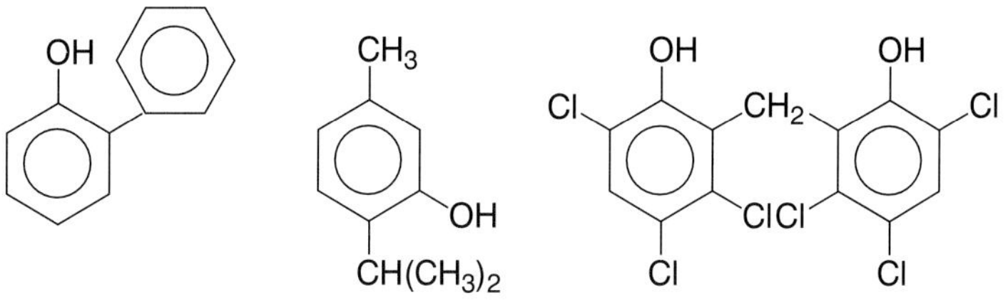

A number of phenols show some bactericidal and fungicidal activity. However, the solutions applied to the skin must be dilute to avoid tissue damage. The cresols are used as a mixture of the three isomers as an antiseptic. This cresol mixture is five times more effective an antiseptic than phenol, whereas resorcinol is three times less effective than phenol.

Other phenolic antiseptics and disinfectants are o-phenylphenol, thymol, and hexachlorophene, as shown in Fig. 16.3. o-Phenylphenol is used to control mold and mildew and is the disinfectant in Lysol. Thymol, because of its pleasant thyme odor, has been a constituent of some mouthwashes. Hexachlorophene is a powerful antiseptic. At one time, newborn infants in many hospitals were bathed daily in a hexachlorophene solution to prevent *Staphylococcus* infections. This practice was stopped when it was reported, based on animal studies, that brain damage could result from the daily baths. The Food and Drug Administration banned the use of hexachlorophene, both as an antiseptic and as a disinfectant, in 1972.

## 16.4 ACIDITY

Phenols ($pK_a \approx 10$) (see Table 1.6) are considerably more acidic than alcohols ($pK_a \approx 17$) and terminal alkynes ($pK_a \approx 25$). Both alcohols and terminal alkynes are too weakly acidic to react with aqueous sodium hydroxide; phenols, however, react to produce stable isolable phenoxide salts.[*] In general,

$$ArOH + NaOH(aq) \longrightarrow ArO^- Na^+ + H_2O \quad (Ar\!\!-\!\! \text{ represents an aryl group})$$

[*]This textbook assumes that the student realizes that inorganic compounds such as NaOH, NaHCO₃, and HCl are ionic and can be written just as well as Na⁺OH⁻, Na⁺HCO₃⁻, and H⁺Cl⁻. The text is emphasizing the ionic nature of organic salts such as the phenoxides by showing the formula as ArO⁻M⁺, where M is a metal.

Example:

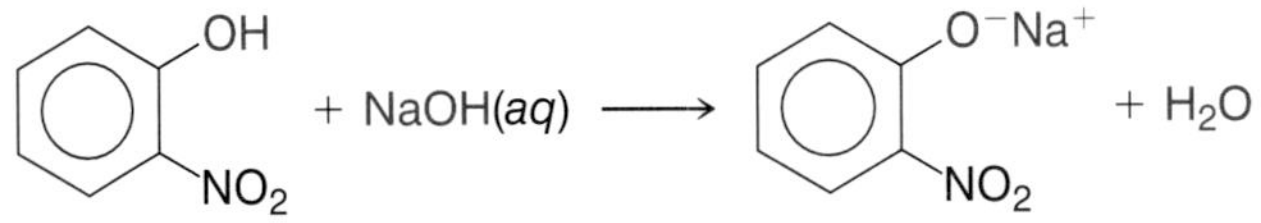

*o*-Nitrophenol                 Sodium
                           *o*-nitrophenoxide

The salts of phenols are named in the same manner as are the alkoxides (page 294). First, name the cation. Then, replace the suffix -*ol* of the corresponding phenol with the suffix -*oxide*.

The greater acidity of phenols versus alcohols can be predicted. Consider the Brønsted-Lowry model for acidity for phenol and cyclohexanol in aqueous solution.

a. 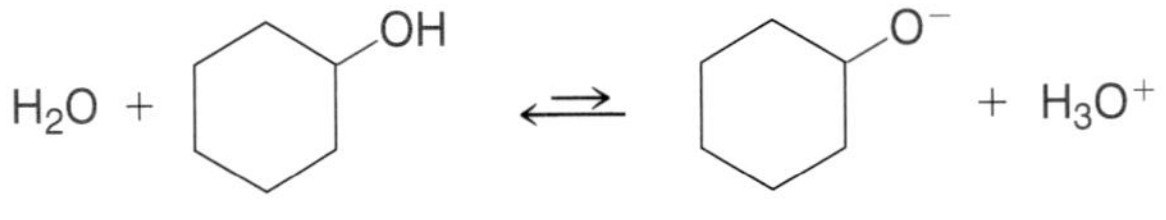   $H_2O$ +

Cyclohexanol        Cyclohexoxide ion -
                         (localized electrons on
                           oxygen, unstable)

b.    $H_2O$ +

Phenol            Phenoxide ion
       (delocalized electrons, stable)

The equilibrium in (b) lies further to the right than in (a) due to resonance stabilization of the phenoxide ion because of electron delocalization into and around the benzene ring from the oxygen.

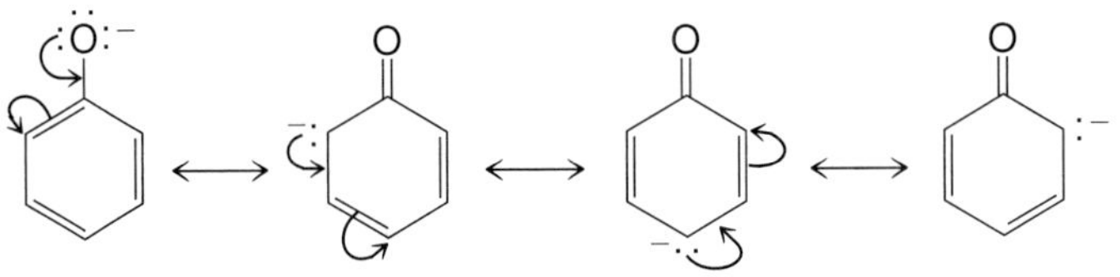

The cyclohexoxide ion shows no resonance stabilization and is less stable. Thus the equilibrium in (a) lies further to the left, resulting in less acidity in cyclohexanol.

The acidity of a phenol depends on the structure of the phenol. An electron-releasing group such as methyl (—$CH_3$) decreases the acidity of the phenol (Table 16.2) due to a +I inductive effect. Sigma electrons shift toward the more electronegative oxygen atom of a methylphenoxide anion and destabilize the anion. On the other hand, an electron-withdrawing group such as nitro (—$NO_2$) or chloro (—Cl) increases the acidity of the phenol due to a combination of the inductive and resonance effects, as you can see in Table 16.2.

## SOLVED PROBLEM 16.2

Explain why *p*-nitrophenol is a stronger acid than phenol.

### SOLUTION

The greater acid strength of *p*-nitrophenol is due to the fact that the nitro group stabilizes the *p*-nitrophenoxide ion by means of additive inductive and resonance effects. The nitrogen bears a positive formal charge, so a −I inductive movement of electrons toward the nitrogen exists that stabilizes the anion. No such effect exists in the phenoxide ion.

| TABLE 16.2 | The pKₐ Value of Some Selected Phenols | |
|---|---|---|

| Phenol | | $pK_a$ |
|---|---|---|
| Phenol | | 9.9 |
| p-Cresol | | 10.2 |
| p-Nitrophenol | | 7.2 |
| p-Chlorophenol | | 9.2 |

Remember: The higher the $pK_a$ value, the weaker an acid the phenol is.

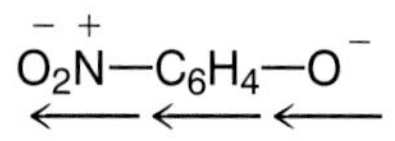

$$\overset{-}{O_2}\overset{+}{N}—C_6H_4—O^-$$

$\longleftarrow \quad \longleftarrow \quad \longleftarrow$

Inductive effect electron flow

In addition, the nitro group stabilizes the p-nitrophenoxide anion by increasing the electron delocalization throughout the structure, including the oxygen of the nitro group (structure V), by means of resonance. Again, no comparable resonance structures for the phenoxide anion can be drawn.

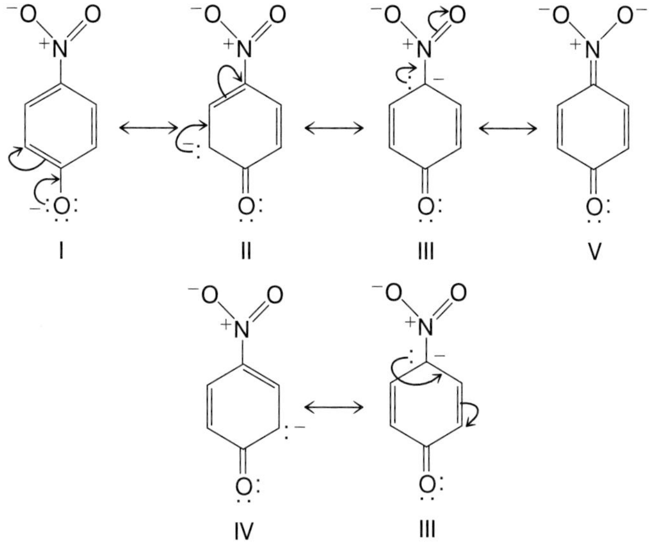

Select the stronger acid, and explain your selection.

a.  phenol or o-nitrophenol

b.  phenol or p-ethylphenol

**EXERCISE 16.2**

Phenols react with aqueous sodium hydroxide but not with aqueous sodium bicarbonate. The reason: Phenols are too weakly acidic to react with the bicarbonate ion, a weaker base than the hydroxide ion. In general,

$$ArOH + NaOH(aq) \longrightarrow ArO^-Na^+ + H_2O$$

Example:

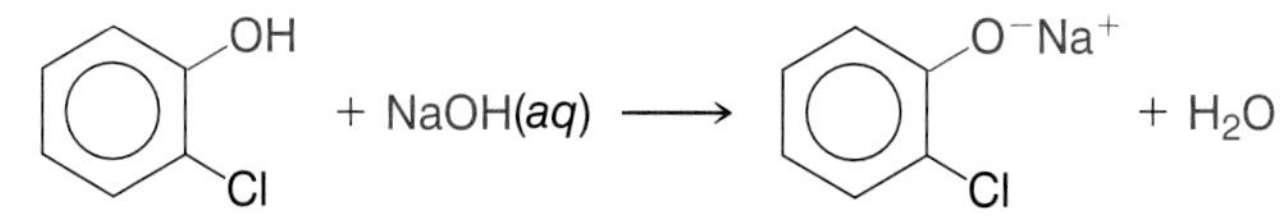

o-Chlorophenol        Sodium
o-chlorophenoxide

Since carboxylic acids are more acidic than phenols, carboxylic acids do react with aqueous sodium bicarbonate. In general,

$$RCOOH + NaHCO_3(aq) \longrightarrow RCOO^-Na^+ + H_2O + CO_2$$

Example:

$$CH_3CH_2COOH + NaHCO_3(aq) \longrightarrow CH_3CH_2COO^-Na^+ + CO_2 + H_2O$$

We can distinguish between a carboxylic acid and a phenol using a test tube reaction because the carboxylic acid dissolves in aqueous sodium bicarbonate solution to produce a water-soluble salt and bubbles of carbon dioxide, whereas the phenol does not dissolve in the aqueous bicarbonate solution.

Phenols can be regenerated from the corresponding phenolic salts by reacting the salt with a 10% solution of hydrochloric acid. In general,

$$ArO^-Na^+ + HCl \longrightarrow ArOH + NaCl$$

Example:

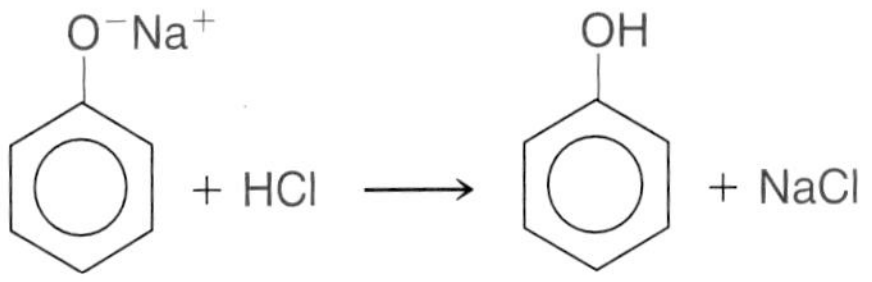

Thus the conversion of a phenol to the corresponding salt represents a reversible process. In general,

$$phenol \underset{HCl}{\overset{NaOH(aq)}{\rightleftharpoons}} phenolic\ salt$$
$$(Water\ insoluble) \qquad (Water\ soluble)$$

Example:

$$p\text{-}NO_2\text{—}C_6H_4\text{—}OH \underset{HCl}{\overset{NaOH(aq)}{\rightleftharpoons}} p\text{-}NO_2\text{—}C_6H_4\text{—}O^-Na^+$$

This reversibility is often used to separate a phenol from other classes of compounds, as illustrated in Solved Problem 16.3.

## SOLVED PROBLEM 16.3

Create a flow diagram showing the separation of each component of an ether solution of p-chlorophenol and chlorobenzene.

### SOLUTION
A mixture containing p-chlorophenol and chlorobenzene in ether (a solvent) is extracted with a 10% solution of aqueous sodium hydroxide, and the aqueous layer is separated from the ether layer. The aqueous layer, containing sodium p-chlorophenoxide, is then reacted with a 10% solution of hydrochloric acid to regenerate solid p-chlorophenol (Fig. 16.4) that is isolated by filtration. The ether in the chlorobenzene solution is removed by evaporation. Both components of the mixture are isolated pure with little loss.

## EXERCISE 16.3

Create a flow diagram showing the separation of each component of an ether solution of p-nitrophenol and bromobenzene.

## 16.5
## IDENTIFICATION

Reacting a phenol with aqueous ferric chloride instantly produces a complex that is blue, violet, or red.

**FIGURE 16.4** A flow diagram showing the separation of p-chlorophenol from chlorobenzene.

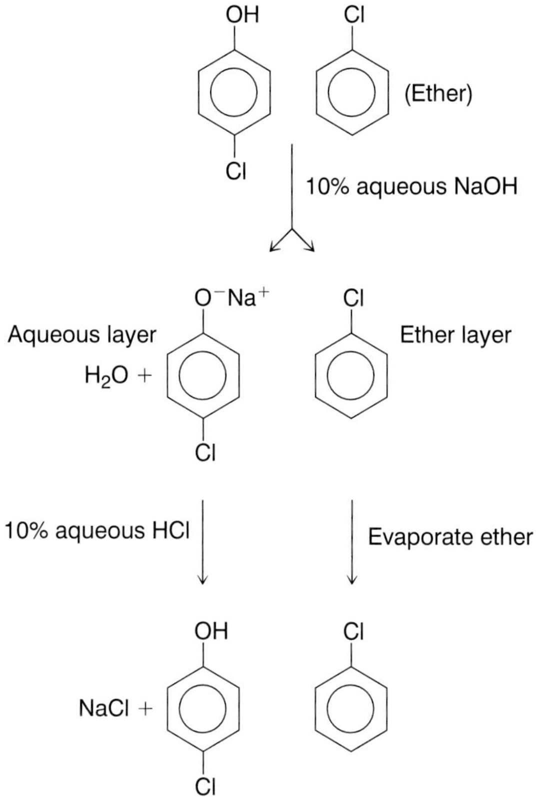

**FIGURE 16.5** Examples of common nomenclature of ethers.

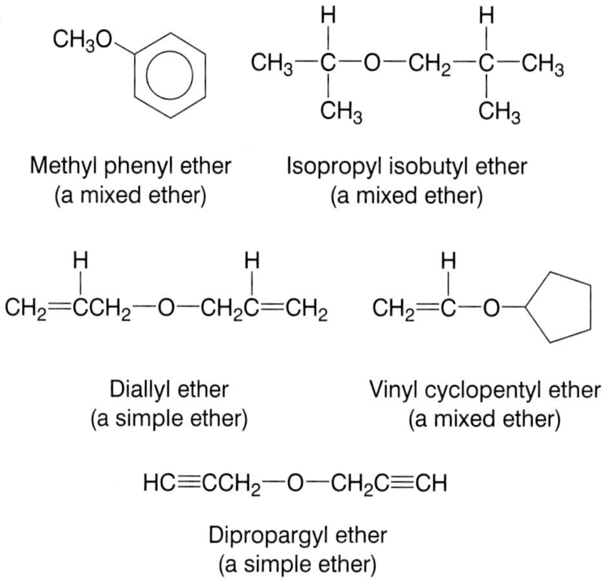

Methyl phenyl ether
(a mixed ether)

Isopropyl isobutyl ether
(a mixed ether)

Diallyl ether
(a simple ether)

Vinyl cyclopentyl ether
(a mixed ether)

Dipropargyl ether
(a simple ether)

---

Ethers can be *simple* (symmetrical) or *mixed* (unsymmetrical).

R—O—R        R—O—R′

Simple ether        Mixed ether

Ethers are named using the common system of nomenclature by naming the organic group(s) and adding the word ether (Fig. 16.5). With symmetrical ethers, sometimes the *di-* prefix is omitted.

CH₂=CH—CH₂—O—CH₂—CH₂=CH₂    CH₃CH₂—O—CH₂CH₃

Allyl ether                      Ethyl ether

When the ether increases in complexity, the IUPAC system comes into use. Ethers are named as alkoxyalkanes, alkoxyarenes, and so on in Fig. 16.6. To name an alkoxy group, name the corresponding alkyl group. Then, drop the *-yl* suffix and add the *-oxy* suffix. For example, ethyl becomes ethoxy:

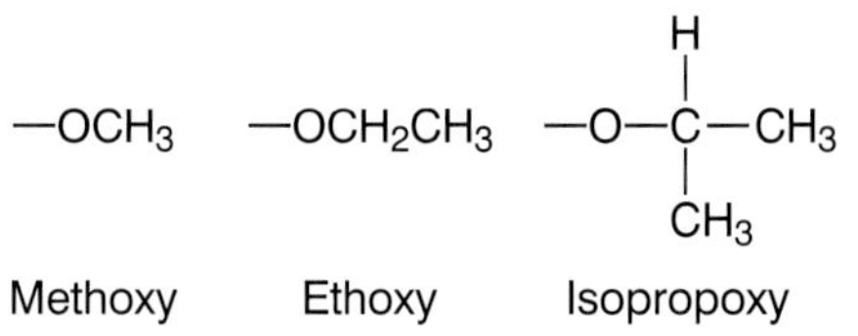

—OCH₃        —OCH₂CH₃        —O—C(H)(CH₃)—CH₃

Methoxy        Ethoxy        Isopropoxy

## ETHERS

## 16.6 STRUCTURE AND NOMENCLATURE

A wide variety of groups can be bonded to the ether oxygen: straight-chain aliphatic, branched-chain aliphatic, alicyclic, alkenyl, alkynyl, and aromatic.

Ethyl ether is the only compound designated simply as *ether*.

**FIGURE 16.6** Examples of IUPAC nomenclature of ethers.

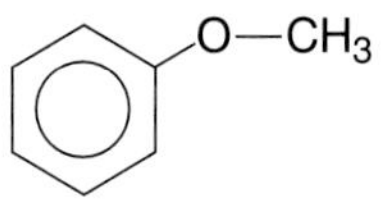

Methoxybenzene
(anisole)

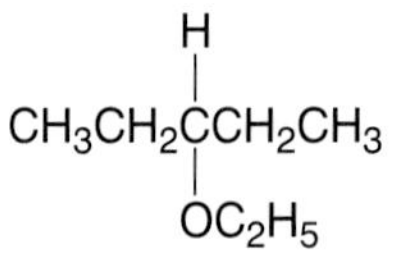

3-Ethoxypentane 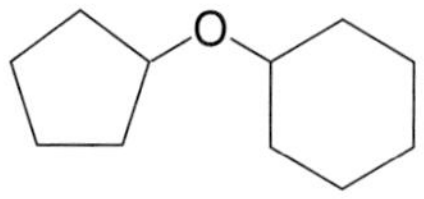          Cyclopentoxycyclohexane

---

**SOLVED PROBLEM 16.4**    Provide a common and an IUPAC name for

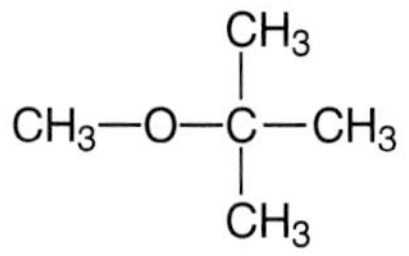

$$CH_3-O-\underset{\underset{CH_3}{|}}{\overset{\overset{CH_3}{|}}{C}}-CH_3$$

**SOLUTION**

Common name: When we name both alkyl groups and add the word *ether*, we get

methyl *tert*-butyl ether

IUPAC name:

*Step 1.* Select the parent chain—propane—a three-carbon chain.

*Step 2.* Number the carbons of the parent chain in the usual manner.

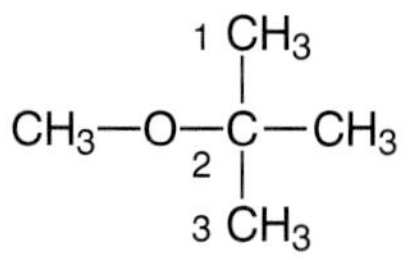

$$CH_3-O-\underset{2}{\underset{\underset{3\ CH_3}{|}}{\overset{\overset{1\ CH_3}{|}}{C}}}-CH_3$$

Note that the parent chain or ring    *Step 3.* Name the compound:
contains the greater number of
carbons.

2-methoxy-2-methylpropane

---

**EXERCISE 16.4**    Provide a common name (when possible) and an IUPAC name for the following compounds.

a.  $CH_3-O-CH_2CH_2CH_2CH_3$

b.

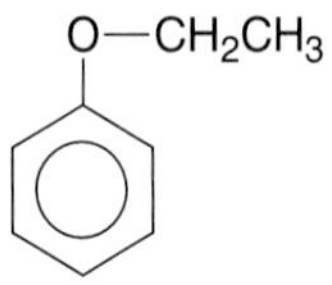

c.

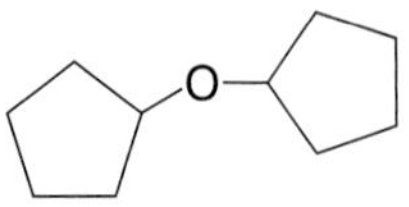

| TABLE 16.3 | A Comparison of the Boiling Points of Ethyl Ether and Pentane | |
|---|---|---|
| | Boiling Point (°C) | Molecular Weight |
| Ethyl ether | 35 | 74 |
| Pentane | 36 | 72 |

d.
$$CH_3CH_2\overset{\overset{\displaystyle H}{|}}{\underset{\underset{\displaystyle OC_2H_5}{|}}{C}}OCH_2CH_3$$

The alcohol group dominates the ether group in nomenclature and thus is assigned the lower number.

$$\overset{2}{C_2H_5O}CH_2\overset{1}{C}H_2OH$$

2-Ethoxyethanol

Let us review and summarize priority of functional-group numbering in compounds that contain two or more functional groups using the IUPAC method of nomenclature:

$$OH > SH > C{=}C > C{\equiv}C > OR > Cl(Br)(I)$$

## 16.7 PHYSICAL PROPERTIES

Ethers boil at about the same temperature as alkanes of about the same molecular weight (Table 16.3) because ethers—like alkanes—cannot show intermolecular hydrogen bonding, since hydrogen is not bonded to oxygen, nitrogen, or fluorine, only to carbon (see also Table 16.4).

Hydrogen bonding between ethers and water does occur. Thus the water solubility of ethers and alcohols with the same carbon content is similar.

| TABLE 16.4 | Physical Properties of Selected Ethers | | |
|---|---|---|---|
| Name | Formula | Boiling Point (°C) | Water Solubility (g/100 g water) |
| Methyl ether | $CH_3OCH_3$ | − 25 | Highly soluble |
| Ethyl ether | $CH_3CH_2OCH_2CH_3$ | 35 | 8 |
| n-Propyl ether | $CH_3CH_2CH_2OCH_2CH_2CH_3$ | 91 | 0.3 |
| Isopropyl ether | $CH_3{-}\overset{\overset{\displaystyle CH_3}{|}}{\underset{\underset{\displaystyle H}{|}}{C}}{-}O{-}\overset{\overset{\displaystyle CH_3}{|}}{\underset{\underset{\displaystyle H}{|}}{C}}{-}CH_3$ | 68 | 0.2 |

## 16.8 ETHER AS AN ANESTHETIC

Ether was first used as an anesthetic in 1846 at the Massachusetts General Hospital. Before that time, surgery was done only as a last resort because the pain was terrible. Before that day in 1846, some (but relatively little) relief from pain was provided by whiskey, opium, or morphine (after 1805). If a drug were not available, the patient could be knocked unconscious or strangled, but more often than not, the patient was held still by four or more strong men.

Ether is a powerful anesthetic yet not too toxic to the patient. Recently, ether has been replaced as an inhalation anesthetic because it is flammable and explosive and a significant number of patients showed symptoms of nausea and vomiting after surgery. The ether substitute is halothane, which is a superb anesthetic and yet does not demonstrate the disadvantages of ether.

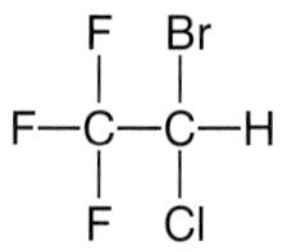

$$\begin{array}{ccc} & F & Br \\ & | & | \\ F\!-\!C\!-\!C\!-\!H \\ & | & | \\ & F & Cl \end{array}$$

Halothane
(2-bromo-2-chloro-1,1,1-trifluoroethane)

## CYCLIC ETHERS: EPOXIDES

### 16.9 STRUCTURE AND NOMENCLATURE

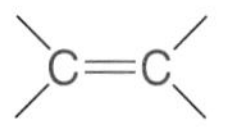

The simplest of the cyclic ethers is a three-member ring compound known as an *epoxide*.

Parent alkene 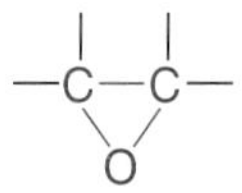      An epoxide

To name an epoxide using the common system, name the parent alkene [ethylene, propylene, etc. (App. 4)], followed by the word *oxide*. The IUPAC name of ethylene oxide is *oxirane* (see Fig. 16.7).

## SOLVED PROBLEM 16.5

Name the following compound.

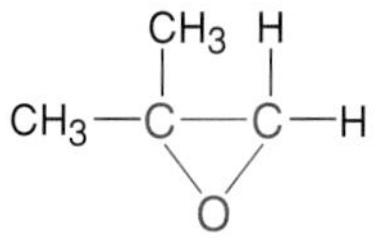

$$\begin{array}{cc} CH_3 & H \\ | & | \\ CH_3\!-\!C\!-\!-\!-\!C\!-\!H \\ \diagdown \ \diagup \\ O \end{array}$$

### SOLUTION

Common name: By removing the oxygen from the structure and adding a second bond between the two carbons that were originally bonded to the oxygen, we get the parent alkene—isobutylene (see App. 4).

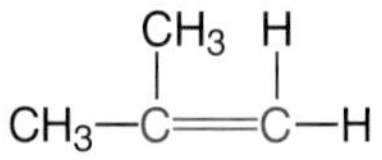

$$\begin{array}{cc} CH_3 & H \\ | & | \\ CH_3\!-\!C\!=\!\!=\!C\!-\!H \end{array}$$

Isobutylene

Finally, we add the word *oxide* to give

isobutylene oxide

## HISTORICAL BOXED READING 16.1

Ethylene oxide is a vital component in a relatively new type of bomb called a *fuel-air explosive*. First, a bomb that contains ethylene oxide under great pressure is dropped from a plane. Thirty feet or so above the ground, an explosive charge breaks the bomb casing into pieces and spreads a cloud of an ethylene oxide–air mixture over a wide area. This cloud of gas settles onto and into tanks, artillery pieces, bunkers, trenches, and buildings. Then a second bomb containing conventional explosives is dropped, exploding the ethylene oxide–air mixture. The resulting shock wave creates havoc over a large area of the battlefield; buildings cave in, vehicles are destroyed, pressure-sensitive land mines are set off, and many soldiers are killed and wounded. This bomb was used by U.S. forces in both the Vietnam and Persian Gulf wars.

**FIGURE 16.7** Examples of common and IUPAC nomenclature of epoxides.

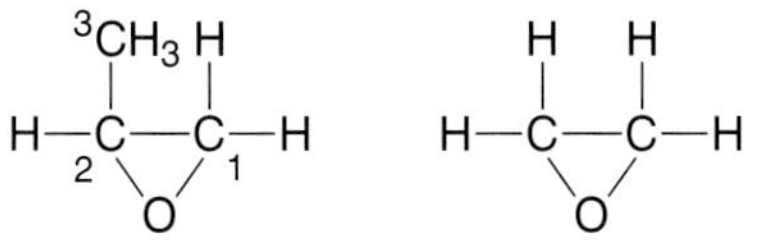

A substituted epoxide

Common: Propylene oxide
IUPAC: 2-Methyloxirane

An unsubstituted epoxide

Common: Ethylene oxide
IUPAC: Oxirane

**FIGURE 16.8** Other important cyclic ethers.

Tetrahydrofuran     Tetrahydropyran     Furan

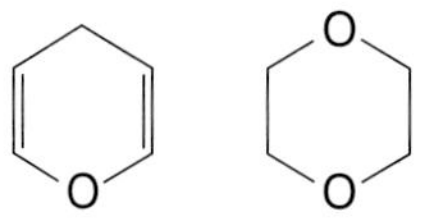

Pyran     1, 4-Dioxane

These compounds are fine solvents and, with the exception of 1,4-dioxane, play an important role in the nomenclature of carbohydrates (Chap. 27) since carbohydrates can be considered as complex cyclic ethers.

IUPAC name: Since this compound is a substituted oxirane with two methyl groups on carbon 2, we have

2,2-dimethyloxirane

Note that in the IUPAC system of nomenclature, the oxygen atom of the oxirane ring is assigned the number 1.

Name the following compound using the IUPAC system of nomenclature.

$$H-\underset{\underset{O}{|}}{C}-\underset{\underset{O}{|}}{C}-CH_2CH_2CH_3$$

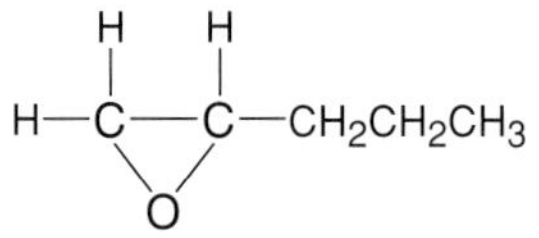

Some other important cyclic ethers are provided in Fig. 16.8.

**EXERCISE 16.5**

Recently, large-ring polyethers have come into their own. These compounds are a multiple of the —$CH_2CH_2O$— unit and are called *crown ethers*. Such an ether is named by listing the total number of carbon and oxygen atoms, followed by a dash and the word *crown*, followed by another dash and the number of oxygen atoms in the molecule. Note the example in Fig. 16.9.

Crown ethers can form stable complexes with a variety of metal ions such as $Na^+$, $K^+$, and $Hg^{2+}$. These ionic complexes dissolve readily in nonpolar solvents. This represents the power of the crown ethers. An inorganic compound such as $KMnO_4$, which under ordinary circumstances would not dissolve in a nonpolar solvent, readily dissolves (as a complex) in the presence of a crown ether in a nonpolar solvent such as benzene.

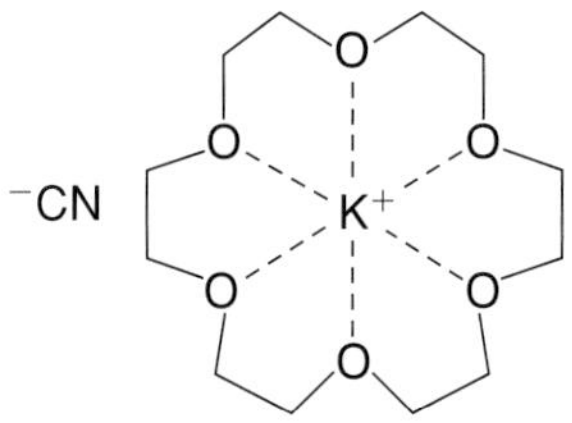

18-Crown-6 potassium cyanide complex

The larger the number of oxygens in the crown ether, the larger is the diameter of the cation that can be accommodated in the cavity of the ether. For example, 18-crown-6 can best accommodate a potassium cation.

## 16.10 CROWN ETHERS

**FIGURE 16.9** 12-crown-4.

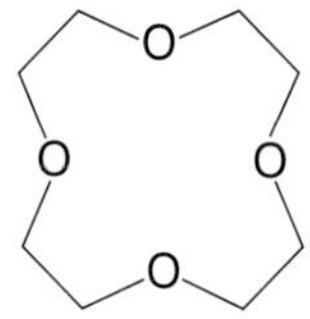

Note that in the example there are present eight carbons and four oxygens. Thus the crown ether is named 12-crown-4.

Note that the dashed lines represent ion dipole bonds between potassium ion and the oxygens of the crown ether.

**SOLVED PROBLEM 16.6**   Draw a structural formula for 15-crown-5.

**SOLUTION**

Since there are 5 oxygens present per molecule, 10 carbons also must be present (10 + 5 = 15). Using the —$CH_2CH_2O$— structural pattern, we get

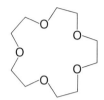

**EXERCISE 16.6**   Draw a structural formula for 21-crown-7.

## ▶ CHAPTER ACCOMPLISHMENTS

### 16.1 Structure and Nomenclature of Phenols
☐ Distinguish between an alcohol and a phenol.
☐ Draw a structural formula for
   a. phenol.
   b. *m*-cresol.
   c. catechol.
☐ Furnish an alternative name for
   a. *m*-cresol.
   b. catechol.
   c. hydroquinone.
☐ Draw a structural formula for sodium *m*-methylphenoxide.

### 16.2 Physical Properties of Phenols
☐ Predict, in general terms, the water solubility of *p*-cresol.
☐ Draw a hydrogen bond interaction between two molecules of *o*-cresol.

### 16.3 Phenols as Antiseptics and Disinfectants
☐ Draw a structural formula for *o*-phenylphenol.
☐ Explain why the Food and Drug Administration banned the use of hexachlorophene as an antiseptic in 1972.

### 16.4 Acidity of Phenols
☐ Write an equation showing that phenol can be classified as a Brønsted acid in aqueous solution.
☐ Draw the stabilizing resonance structures for phenoxide ion.
☐ Explain why *p*-nitrophenol is a stronger acid than phenol.
☐ Write an equation showing the reaction of phenol with aqueous sodium hydroxide.
☐ Describe experimentally how to distinguish between a carboxylic acid and a phenol.
☐ Write an equation showing the reaction of sodium phenoxide with hydrochloric acid.

☐ Draw a flow diagram showing the separation of *p*-chlorophenol from chlorobenzene.

### 16.5 Identification of Phenols
☐ Supply the colors produced when a phenol is reacted with an aqueous ferric chloride solution.

### 16.6 Structure and Nomenclature of Ethers
☐ Classify and name the following compounds using both the common and IUPAC systems of nomenclature.
   a. $CH_3CH_2OCH_2CH_3$
   b. $CH_3OCH_2CH_2CH_3$
☐ Draw a structural formula for
   a. diallyl ether.
   b. anisole.
   c. dipropargyl ether.
   d. allyl ether.
   e. cyclopentoxycyclohexane.

### 16.7 Physical Properties of Ethers
☐ Explain why ethyl ether boils at a far lower temperature than ethyl alcohol even though ethyl ether has a higher molecular weight than ethyl alcohol.
☐ Explain why methyl ether is completely miscible (soluble) in water.

### 16.8 Ether as an Anesthetic
☐ List two disadvantages of using ether as an anesthetic.
☐ Draw a structural formula for halothane.

### 16.9 Structure and Nomenclature of Epoxides (Cyclic Ethers)
☐ Supply both a common and an IUPAC name for each of the following compounds.

a.

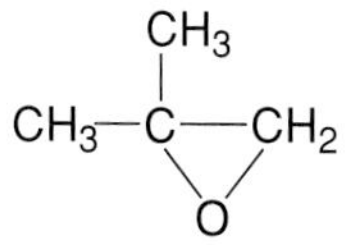

b.

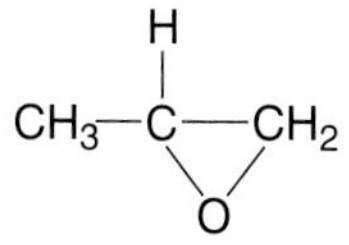

☐ Draw a structural formula for
   a. furan.
   b. pyran.
   c. 1,4-dioxane.

### 16.10 Crown Ethers

☐ Draw a structural formula for 12-crown-4.
☐ Explain the huge advantage of using crown ethers in organic chemistry.

---

## ▶ KEY TERMS

*antiseptic* (16.3)         *mixed ether* (16.6)        *crown ether* (16.10)
*disinfectant* (16.3)       *epoxide* (16.9)            *fuel-air explosive* (Historical Boxed
*simple ether* (16.6)       *oxirane* (16.9)                Reading 16.1)

---

## ▶ PROBLEMS

1. Draw a structural formula for each of the following compounds.
   a. *o*-cresol
   b. 2,2-dimethyloxirane
   c. 1,3-dioxane
   d. 2-isopropoxypentane
   e. isobutyl ether
   f. phenyl ether
   g. vinyl ether
   h. 2-ethoxypropane
   i. 2-ethyloxirane

2. Draw a structural formula for each of the following compounds.
   a. salicylic acid (*o*-hydroxybenzoic acid)
   b. sodium *p*-methylphenoxide
   c. 2-naphthol

3. Name the following compounds.

a.

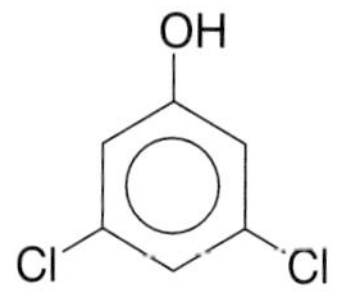

b.

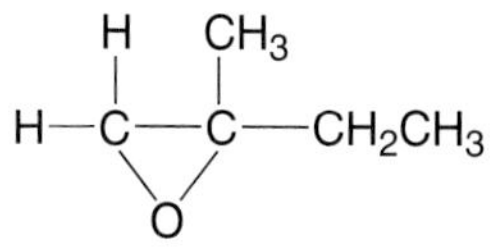

c.  CH₃CH₂CH₂CH₂OCH₂CH₃

d.  CH₃CHCH₂CH₂CHCH₂CH₃
        |            |
       OCH₃        OCH₃

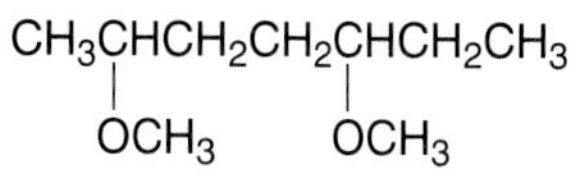

e.

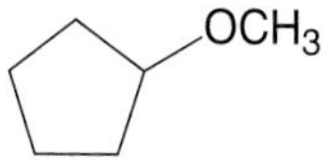

f.

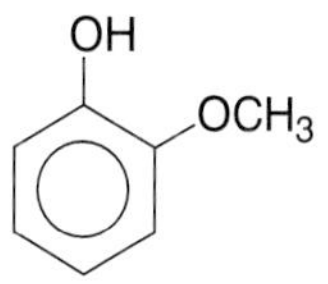

g.  CH₃CH₂OCH₂CCH₃
                |
               OH

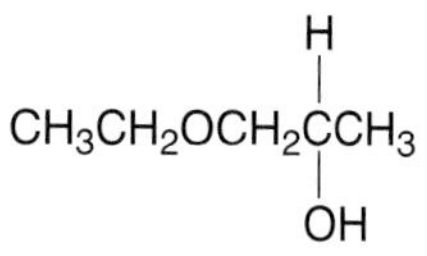

h.

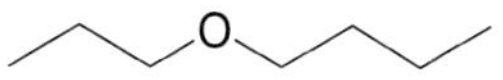

i.

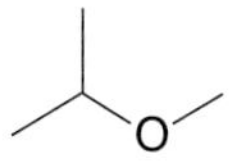

4. Predict the higher-boiling compound of each pair of compounds, and explain your prediction.
   a. ethyl ether or *n*-butyl alcohol
   b. propyl ether or ethyl ether
   c. *m*-cresol or resorcinol

5. Predict whether 2,4-dinitrophenol is a stronger or weaker acid than *p*-nitrophenol. Explain.

6. Draw a structural formula for each of the following compounds.

a. 2-isopropoxyhexane
b. phenyl benzyl ether
c. methyl ether
d. ether
e. allyl propargyl ether

7. Why does isopropyl ether boil at a lower temperature than *n*-propyl ether?

8. Write a Brønsted-Lowry acidity equation for
   a. *p*-cresol in water.
   b. 4-methylcyclohexanol in water.

   Label each reactant (Brønsted acid, base; conjugate acid, base).

9. Draw resonance structures for the *p*-cresoxide ion.

10. Explain why resonance structure V (on page 305) is particularly stable.

11. 1,4-Dioxane can be named as a crown ether even though it is not a crown ether.

a. Name it as a crown ether.
b. Why can it be named as a crown ether?

12. Draw a structural formula for
    a. 2,2-diethyloxirane.
    b. 2-ethyl-3-*n*-propyloxirane (disregard stereoisomerism). Draw the structural formula as in Exercise 16.5.

13. There are four isomeric oxiranes, $C_4H_8O$. Draw a structural formula for each isomer (include geometric stereoisomers).

14. Complete the following equations by drawing a structural formula for every organic compound formed and writing a molecular formula or formula unit for each inorganic compound formed. If no reaction occurs, write "no reaction."
    a. *m*-cresol + NaOH(*aq*) $\longrightarrow$
    b. *m*-cresol + NaHCO$_3$(*aq*) $\longrightarrow$
    c. cyclopentanol + NaOH(*aq*) $\longrightarrow$
    d. sodium *p*-nitrophenoxide + HCl $\longrightarrow$

# Phenols, Ethers, and Epoxides (II)

## SYNTHESIS AND CHEMICAL PROPERTIES

Arenesulfonic acids are fairly strong acids that react with aqueous sodium hydroxide to give sodium arenesulfonates. In general,

$$ArSO_3H + NaOH(aq) \longrightarrow ArSO_3^- Na^+ + H_2O$$

Example:

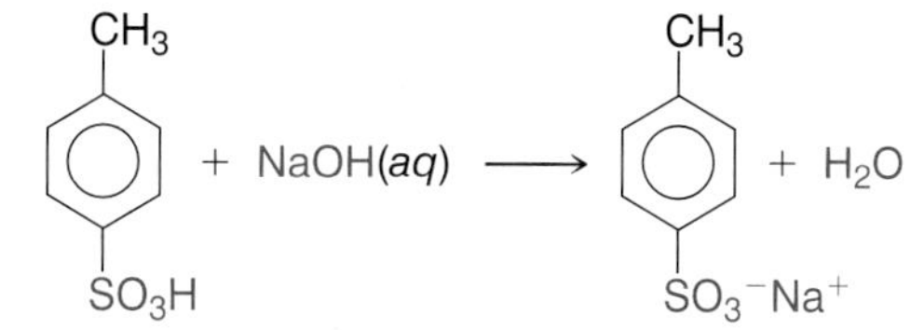

A benzene ring with $CH_3$ at top and $SO_3H$ at bottom $+ \; NaOH(aq) \longrightarrow$ a benzene ring with $CH_3$ at top and $SO_3^- Na^+$ at bottom $+ \; H_2O$

Fusing the arenesulfonate with solid sodium hydroxide produces the phenolic salt. In general,

$$ArSO_3^- Na^+ + 2\,NaOH \xrightarrow{\Delta} ArO^- Na^+ + Na_2SO_3 + H_2O$$

Example:

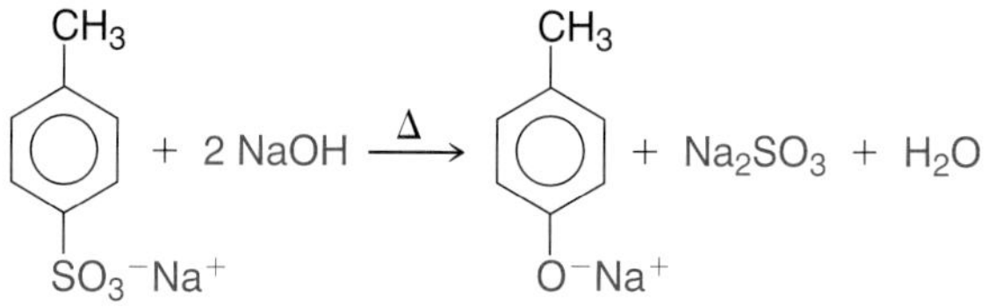

A benzene ring with $CH_3$ at top and $SO_3^- Na^+$ at bottom $+ \; 2\,NaOH \xrightarrow{\Delta}$ a benzene ring with $CH_3$ at top and $O^- Na^+$ at bottom $+ \; Na_2SO_3 \; + \; H_2O$

Finally, the phenolic salt is treated with a stronger acid to produce the phenol. In general,

$$ArO^- Na^+ + HCl \longrightarrow ArOH + NaCl$$

**PHENOLS**

## 17.1
## PREPARATION

**Hydrolysis of a Diazonium Salt (see Sec. 25.4)**

**From Sodium Arenesulfonates**

**315**

Example:

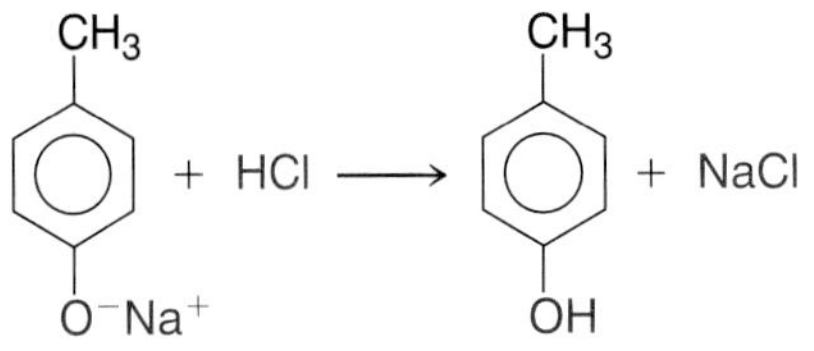

## 17.2 AROMATIC SUBSTITUTION

### Halogenation

Because the hydroxyl group in phenols is strongly ring activating (see Sec. 10.2), mild conditions are satisfactory for reaction. Refer to Chapter 10 for a detailed discussion of electrophilic aromatic substitution.

Monobromination or monochlorination of phenol or a derivative can be accomplished by using a nonpolar solvent such as carbon disulfide ($CS_2$) or chloroform ($CHCl_3$) at a low temperature.

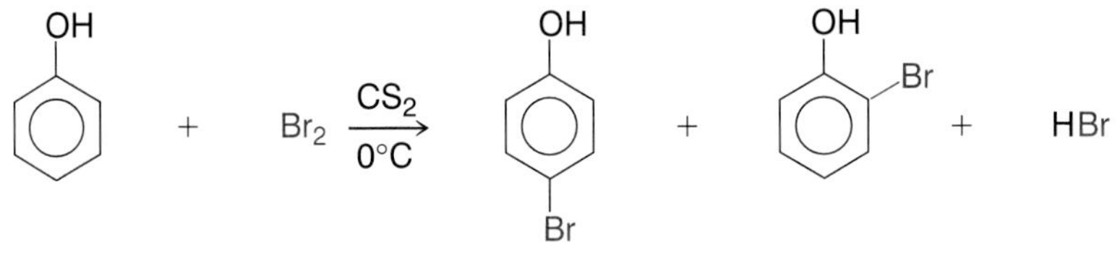

p-Bromophenol   o-Bromophenol
(major product)  (minor product)

Other phenols can be brominated or chlorinated in the same manner.

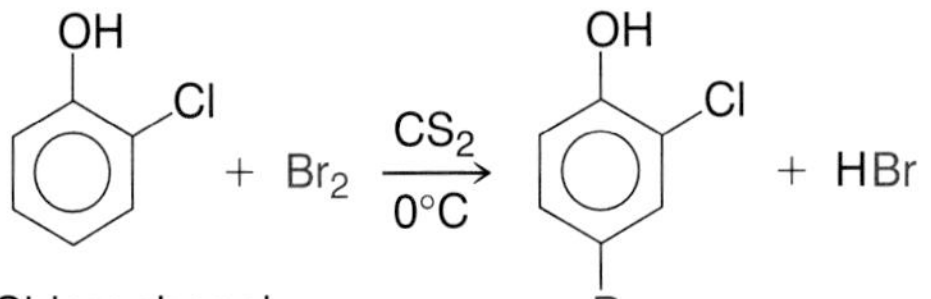

o-Chlorophenol

4-Bromo-2-chlorophenol

When a polar solvent such as water is used, polysubstitution takes place.

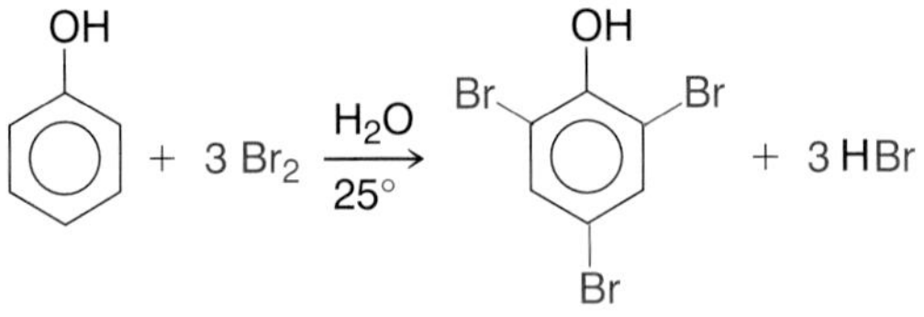

2,4,6-Tribromophenol

### Nitration

Phenol reacts with dilute nitric acid (no concentrated nitric acid and catalytic amount of concentrated sulfuric acid are needed here) to produce a mixture of o- and p-nitrophenols.

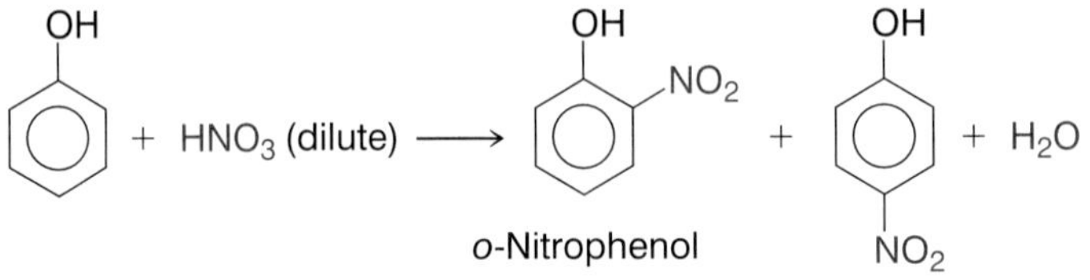

o-Nitrophenol

p-Nitrophenol

Note that o-and p-nitrophenols are prepared from phenol with dilute nitric acid, 2,4-dinitrophenol from o- and p-nitrophenols using concentrated nitric acid, and picric acid from 2,4-dinitrophenol by means of concentrated nitric acid and heat.

Since the nitro group is deactivating, only one nitro group at a time can be introduced in the benzene ring. Harsher reaction conditions are needed for each successive group introduced.

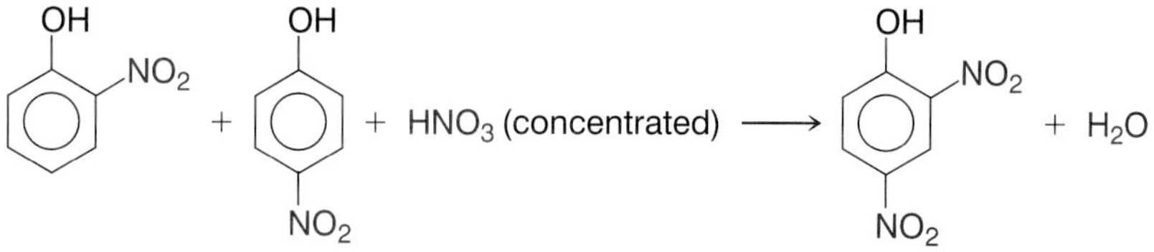

2,4-Dinitrophenol

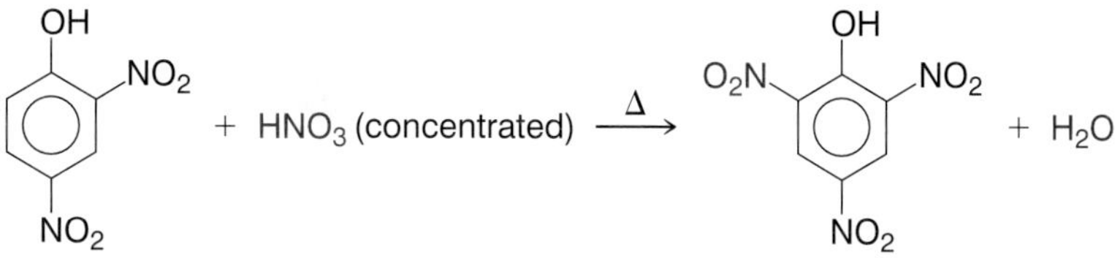

2,4,6-Trinitrophenol
(picric acid)

## 17.3
## OXIDATION: QUINONES

Phenols containing two hydroxyl groups are easily oxidized with silver oxide (many other oxidizing agents may be used) to produce *quinones* that are structurally identified as cyclohexadienediones (six-membered rings with two carbon-carbon double bonds and two carbon-oxygen double bonds).

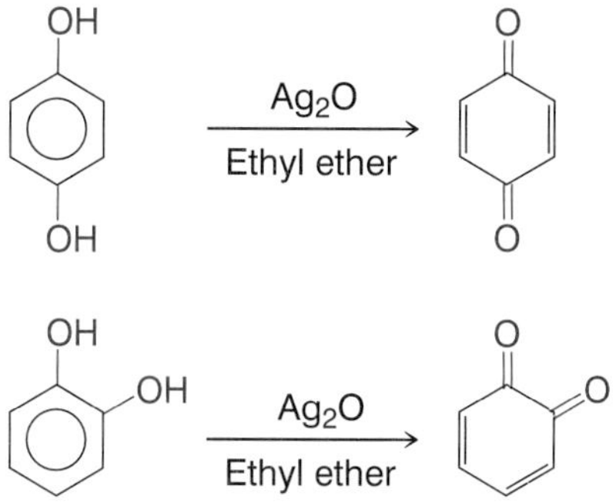

Yields, however, are not uniformly good due to the formation of a variety of by-products that we will not discuss.

To name a quinone, first, name the corresponding aromatic hydrocarbon. Then, drop the suffix *-ene* for benzene or *-alene* for naphthalene and add the suffix *-oquinone*. Finally, locate the two carbon-oxygen double bonds by use of the prefixes *ortho-* or *para-* or by the use of numbers.

Name the following compound.

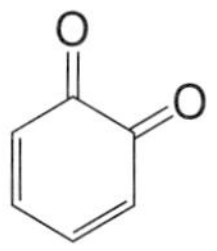

### SOLVED PROBLEM 17.1

**SOLUTION**

The corresponding aromatic hydrocarbon is benzene. Dropping the *-ene* suffix and adding the *-oquinone* suffix, we get benzoquinone. Since the carbonyl groups are located on carbons 1 and 2, we name this compound 1,2-benzoquinone, or *o*-benzoquinone.

## BIOCHEMICAL BOXED READING 17.1

The bombardier beetle has a superbly fashioned chemical defensive system. When a leg of the insect is seized, the bombardier
beetle, by means of muscle con-
traction, combines a mixture of hydroquinone, methylhydroquinone,
and hydrogen peroxide and forces
the mixture into a chamber containing an oxidizing enzyme that instantaneously converts the hydroquin-
ones to quinones, liberating a great

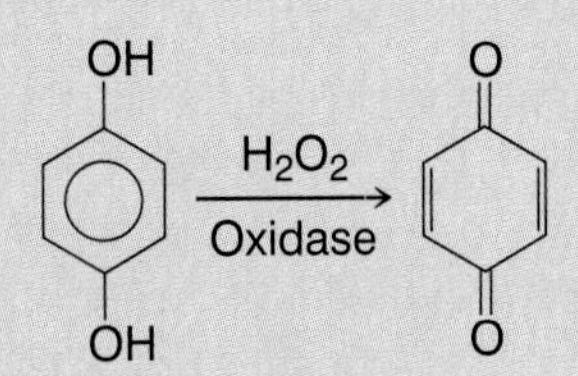

Hydroquinone    1,4-Benzoquinone

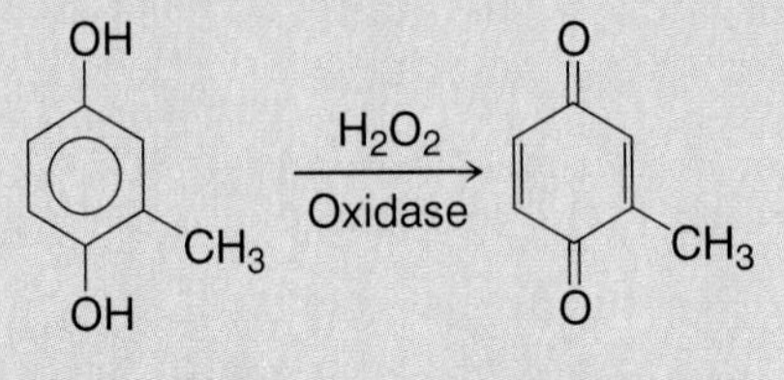

4-Hydroxy-
3-methylphenol
(methylhydroquinone)

2-Methyl-1,
4-benzoquinone

deal of heat under considerable pressure. The pressure blows out the
scalding hot mixture through a hole
on the tip of the beetle's abdomen.
This hot, irritating mixture of quinones
is sure to cause some distress to the
animal attacking the beetle.

---

### EXERCISE 17.1

If one or more substituent groups
are located on the quinone,
numbers must be used to name the
quinone.

**FIGURE 17.1** Carminic acid.

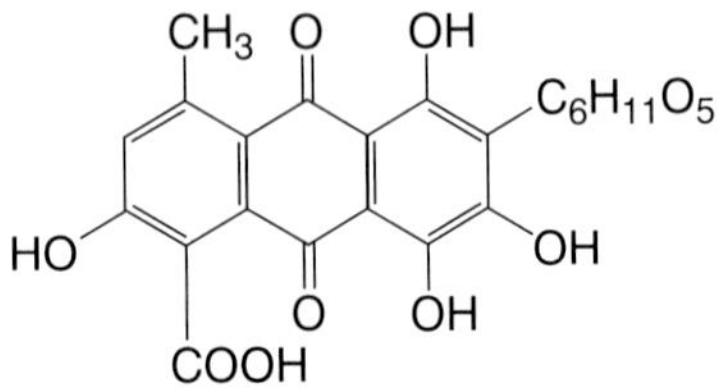

The group —$C_6H_{11}O_5$
is a sugar

Quinones are found extensively in both the plant and animal kingdoms. A number of
*para*-quinones have been found to be exuded or sprayed by cockroaches, termites,
earwigs, and beetles as a defense against predators. These quinones are toxic to the
predators and discourage further aggressive behavior. The following quinones are found
in the insects listed above. Name them.

a. 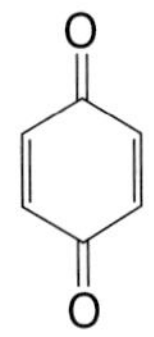            b. 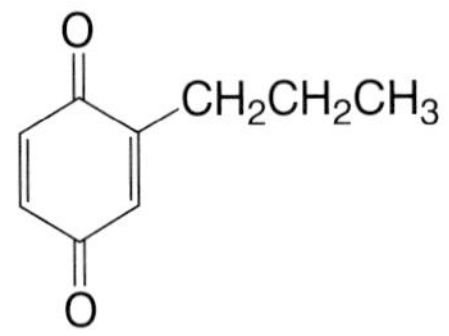    $CH_2CH_2CH_3$            c. 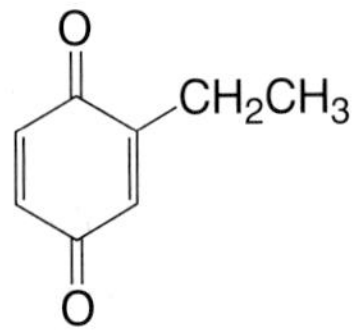    $CH_2CH_3$

Quinones are colored solids; most are yellow or red, although orange, green, and
gold quinones are known. The naturally occurring red dye carminic acid (Fig. 17.1) is
found in cochineal, a mixture of the dried bodies of the red female insect *Dactylopius
coccus*. Cochineal has been used for hundreds of years to color clothing, such as the
British "redcoats" of Revolutionary War notoriety. At the present time, cochineal is used
as a biological stain. An important quinone obtained from the plant family is vitamin K
(Fig. 17.2), which is isolated from alfalfa and several vegetable oils. Vitamin K is essential
for the normal clotting of blood. Humans cannot biosynthesize the vitamin and must
obtain it from the diet. For this reason, there has been a considerable search for a
synthetic substitute for the natural vitamin. It was found that menadione, a synthetic
compound, has clotting ability comparable with that of the natural vitamin (Fig. 17.3).

FIGURE 17.2  Vitamin K.

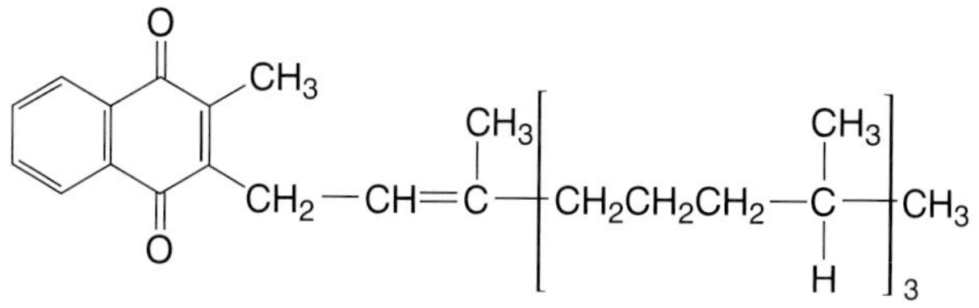

FIGURE 17.3  Menadione (2-methyl-1,4-naphthoquinone).

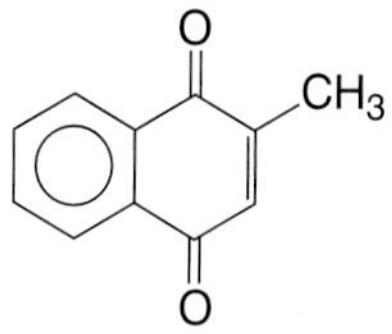

Note that both vitamin K and menadione contain a 1,4-naphthoquinone structural unit.

In general,

$$2\ ROH \xrightarrow[\Delta]{H^+} R\text{—}O\text{—}R + H_2O$$

Example:

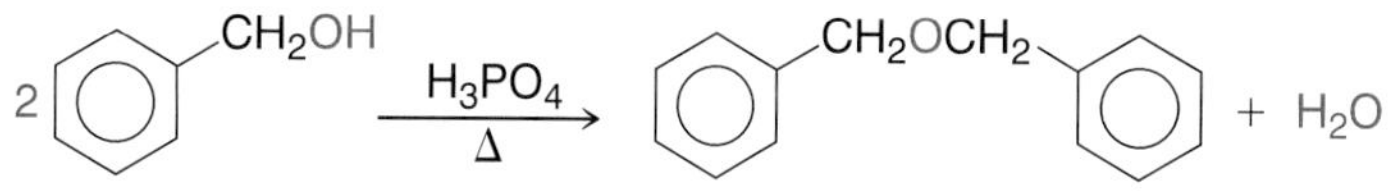

Benzyl ether

Either sulfuric acid or phosphoric acid serves as a catalyst.

There are two differences in these preparations. First, less heat is usually needed to prepare the ether than the alkene. For example, when ethyl alcohol is converted to ethyl ether, a temperature of 150°C is maintained; to produce ethylene, we need a temperature of 180°C. Second, only 1 mol of alcohol is needed to produce 1 mol of alkene, whereas 2 mol of alcohol is required to form 1 mol of ether. Mixed ethers usually cannot be prepared in good yield by this method because a total of three possible ethers can be produced.

## ETHERS

## 17.4 PREPARATION

### Dehydration of an Alcohol

Note the similarity of this preparation of ethers with that of alkenes in Section 6.1.

$$RCH_2CH_2OH \xrightarrow[\Delta]{H_2SO_4}$$
$$RCH\text{=}CH_2 + H_2O$$

## SOLVED PROBLEM 17.2

Formulate the three products that are produced when a mixture of methyl and ethyl alcohols is dehydrated under mild conditions.

SOLUTION:

$$CH_3\text{—}OH + HO\text{—}CH_3 \xrightarrow[\Delta]{H_3PO_4} CH_3\text{—}O\text{—}CH_3 + H_2O$$

$$CH_3\text{—}OH + HO\text{—}CH_2CH_3 \xrightarrow[\Delta]{H_3PO_4} CH_3\text{—}O\text{—}CH_2CH_3 + H_2O$$

$$CH_3CH_2\text{—}OH + HO\text{—}CH_2CH_3 \xrightarrow[\Delta]{H_3PO_4} CH_3CH_2\text{—}O\text{—}CH_2CH_3 + H_2O$$

## EXERCISE 17.2

Formulate the three products that are produced when a mixture of methyl and *n*-propyl alcohols is dehydrated under mild conditions.

The intermolecular (between molecules) dehydration of two molecules of alcohol proceeds to form an ether by means of an $S_N2$ mechanism. Consider the intermolecular dehydration of ethyl alcohol to produce ethyl ether.

**Envision the Reaction**

### $S_N2$ Mechanism of the Intermolecular Dehydration of Ethyl Alcohol to Ethyl Ether

First step:

$$C_2H_5OH + H^+ \longrightarrow CH_3CH_2\overset{\overset{\displaystyle H}{|}}{\underset{\cdot\cdot}{O}}\!\!-\!\!H$$

Second step:

Transition state

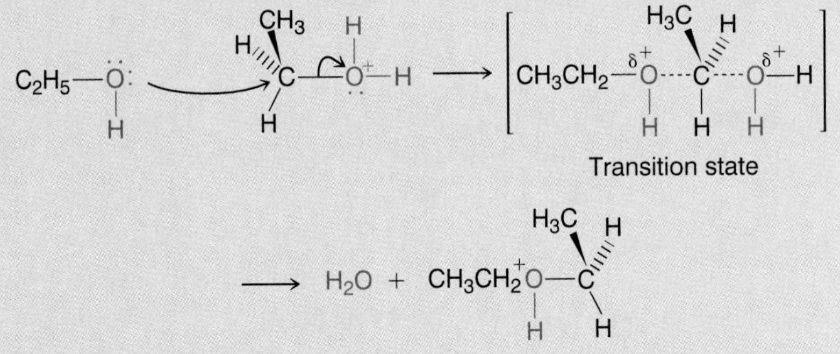

Third step:

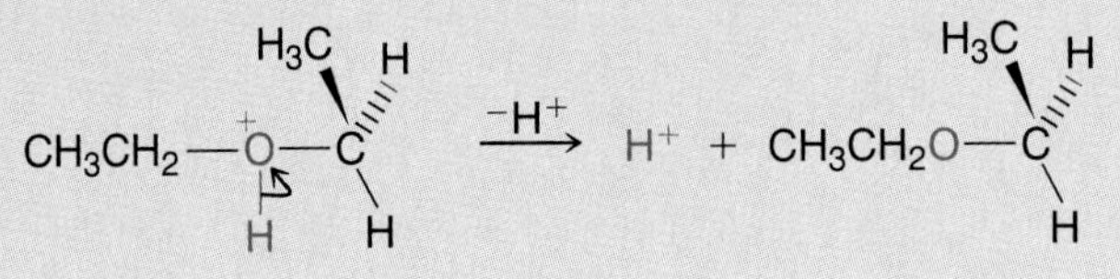

This method of preparing ethers cannot be used with secondary and tertiary alcohols because they undergo intramolecular (within a molecule) dehydration with relative ease to form the corresponding alkene. Thus any attempt to prepare *tert*-butyl ether from *tert*-butyl alcohol produces only isobutylene via an E1 mechanism.

**The Williamson Synthesis**

A more versatile method for making ethers than the dehydration of alcohols is the *Williamson synthesis*. In the Williamson synthesis, an organic halide is reacted with an alkoxide. In general,

$$R'O^-Na^+ + RX \longrightarrow R'\!\!-\!\!O\!\!-\!\!R + NaX$$

where X = Cl, Br, or I. For example:

$$CH_3CH_2O^-Na^+ + BrCH_3 \longrightarrow CH_3CH_2OCH_3 + NaBr$$

Mixed ethers containing one secondary or tertiary group can be prepared if the proper synthetic route is taken. Consider the preparation of ethyl *tert*-butyl ether. Two methods of preparation using the Williamson synthesis are possible. First, sodium ethoxide is allowed to react with *tert*-butyl bromide (method 1). Second, sodium *tert*-butoxide is permitted to react with ethyl bromide (method 2).

*This method is superior to the dehydration of alcohols in that mixed ethers can be prepared in good yield.*

$$CH_3CH_2O^-Na^+ + Br\!\!-\!\!C(CH_3)_3 \longrightarrow$$

Method 1

$$(CH_3)_3CO^-Na^+ + Br\!\!-\!\!CH_2CH_3 \longrightarrow$$

Method 2

Since the halide in method 1 is tertiary, elimination rather than substitution occurs to give isobutylene. Method 2, on the other hand, using primary ethyl bromide, produces ethyl *tert*-butyl ether in good yield.

$$CH_3CH_2-O^-Na^+ + Br-C(CH_3)_3 \xrightarrow{E2} NaBr + CH_3CH_2OH + (CH_3)_2C=CH_2$$

(Sec. 12.4)

$$(CH_3)_3C-O^-Na^+ + Br-CH_2CH_3 \xrightarrow{S_N2} NaBr + (CH_3)_3C-O-CH_2CH_3$$

(Sec. 12.1)

Aryl and vinyl halides are too unreactive to produce an ether by means of the Williamson synthesis. However, an aromatic group can be introduced into an ether by the following Williamson scheme. In general,

$$Ar-O^-Na^+ + R-X \longrightarrow Ar-O-R + NaX$$

where X = Cl, Br, or I
     R = primary or methyl
   Ar = an aromatic group

Example:

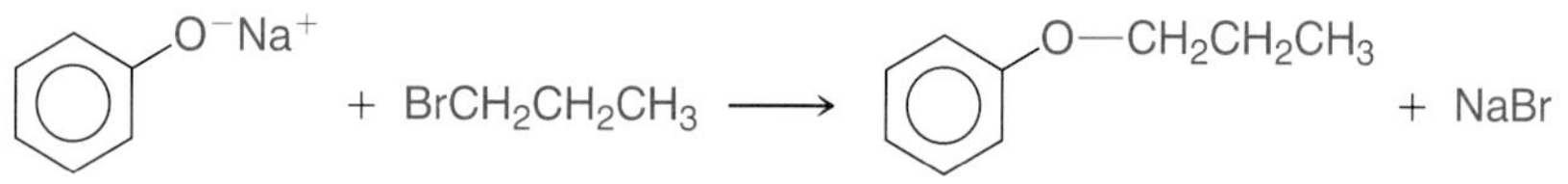

In conclusion, the halide used must be methyl or primary; the sodium salt can be primary, secondary, tertiary, or aryl.

---

Select the proper reagents to obtain a good yield of *p*-nitrophenyl ethyl ether using the Williamson procedure. Then write an equation. Finally, explain why the other reagent combination will not produce the corresponding ether.

**SOLVED PROBLEM 17.3**

**SOLUTION**

Since the aromatic component of the ether must be in the form of the sodium salt for a good yield of ether to result, we must use sodium *p*-nitrophenoxide and ethyl bromide.

$$p-O_2N-C_6H_4-O^-Na^+ + CH_3CH_2Br \longrightarrow p-O_2N-C_6H_4-O-CH_2CH_3 + NaBr$$

Selecting the other possible reagent combination, *p*-bromonitrobenzene and sodium ethoxide, to produce the ether would lead to no reaction because of the lack of reactivity of the aromatic halide.

$$p-O_2N-C_6H_4-Br + CH_3CH_2O^-Na^+ \longrightarrow \text{no reaction}$$

---

Select the proper reagents to obtain a good yield of each of the following ethers using the Williamson procedure. Then write an equation for each. Finally, explain why the other reagent combination in each case does not produce the corresponding ether.

**EXERCISE 17.3**

a.  ethyl *tert*-pentyl ether

b.  phenyl ethyl ether

---

Ethers are a relatively unreactive family of compounds. They do not react with oxidizing agents such as potassium permanganate ($KMnO_4$), sodium dichromate, and sulfuric acid ($Na_2Cr_2O_7 + H_2SO_4$); reducing agents such as iron(II) sulfate ($FeSO_4$) and tin(II) chloride ($SnCl_2$); reactive metals such as sodium (Na) and potassium (K); or strong bases such as sodium hydroxide (NaOH) or potassium hydroxide (KOH).

**17.5
CHEMICAL
PROPERTIES**

## MEDICAL BOXED READING 17.2

Most elderly people seem to appreciate "the good old days" when shrimp sold for 5 cents a pound and we traveled by horse and carriage rather than by car. Edith and Archie Bunker captured this feeling when they sang "Those Were the Days" at the start of the television program "All in the Family." Those horse-and-buggy days were valued because of the lack of frenetic activity and inflation—with good reason.

However, the skeleton in the closet of the good old days is medical care. It was awful. The practice of medicine during most of the 1800s was sheer guesswork. Physicians could only diagnose disease on visible symptoms (fever, skin breakout, swelling, coughing, and sneezing). Because they knew relatively little, they felt compelled to produce observable results by the use of bleeding, purging, and blistering.

Bleeding was standard therapy for disease because many physicians thought that all disease was due to "morbid excitement induced by capillary tension." The standard bleeding (phlebotomy) tool used was the lancet. When the lancet proved awkward, leeches were employed.

Physicians drained large amounts of blood—often bleeding patients to unconsciousness. The results, on the surface, were good. The red flush of fever gave way to paleness, and the patient began to perspire, breaking the fever. The weakness and anemia suffered by the patient were not noticed.

The most often used purgative was calomel—a chloride of mercury ($Hg_2Cl_2$). The laxative effect of the calomel was obvious. Continued doses, however, produced mercury poisoning. It would start with the gums turning white and softening and a burning pain in the mouth. In a more advanced stage of the disease, the teeth would drop out. Finally, the jaw bones would rot, and the kidneys would fail. Nineteenth-century physicians, as a rule, thought calomel was a safe and gentle drug.

Blistering consisted of the use of cantharides. These are preparations of dried beetles that contain a bitter crystalline compound called *cantharidin* which irritates the skin and forms a blister. The pus in the blister indicated that the infection was on its way out of the patient.

Cantharidin

Before the use of ether as an anesthetic, surgery consisted primarily of the amputation of limbs necessitated by gangrene. The patient was urged to swallow a full bottle of whiskey beforehand. The pain was so terrible, patients would pass out after 30 seconds of the most horrible screaming. A good surgeon would finish the amputation in no more than 2 minutes; a longer time would usually mean death of the patient from shock.

Diphtheria, typhoid, malaria, dengue fever, syphilis, gonorrhea, and yellow fever ran wild. Physicians were powerless to effect a cure. A person 50 years of age was considered old. Children died like flies.

Most Americans of that day were lucky. Physicians charged a lot for services, so a great majority of people could not afford the services of a physician and used folk remedies and patent medicines instead. In addition, with the American Medical Association in its infancy, many people claiming to be physicians were quacks.

## Cold Concentrated Sulfuric Acid

Ethers are soluble in cold concentrated sulfuric acid; alkanes are insoluble. This is a convenient method of distinguishing ethers from alkanes. The solubility of any ether in concentrated sulfuric acid is due to formation of the oxonium salt of that ether. Just as in the case of alcohols (page 295), the oxonium salts derived from ethers are unstable and cannot be isolated.* In general,

$$R-O-R(R') + H_2SO_4 \xrightarrow{\text{Cold}} \left[ R-\overset{\overset{\displaystyle H}{|}}{\overset{+}{O}}-R(R') \right] [HSO_4^-]$$

Unstable oxonium salt

Example:

$$(CH_3)_2CH-O-CH(CH_3)_2 + H_2SO_4 \xrightarrow{\text{Cold}} \left[ (CH_3)_2CH-\overset{\overset{\displaystyle H}{|}}{\overset{+}{O}}-CH(CH_3)_2 \right] [HSO_4^-]$$

## Hot Hydrobromic, Hydriodic, and Sulfuric Acids

Heating an ether with an excess of hot concentrated hydrobromic, hydriodic, or sulfuric acid causes the ether to cleave by breaking two carbon-oxygen bonds. In general,

$$R-O-R + 2\,HA \xrightarrow{\Delta} 2\,R-A + H_2O$$

$$R-O-R' + 2\,HA \xrightarrow{\Delta} R-A + R'-A + H_2O$$

*In actuality, these reactions are equilibria, but for the sake of simplicity, they are shown as going to completion.

where A = Br, I, or $OSO_3H$. For example:

$$CH_3CH_2OCH_2CH_3 + 2HBr \xrightarrow{\Delta} 2\ CH_3CH_2Br + H_2O$$

$$CH_3CH_2OCH_3 + 2\ HBr \xrightarrow{\Delta} CH_3CH_2Br + CH_3Br + H_2O$$

Note that aromatic ethers do not react, whereas mixed aliphatic-aromatic ethers produce a phenol and aliphatic halide (or alkyl hydrogen sulfate). In general.

$$Ar{-}O{-}R + HX \xrightarrow{\Delta} Ar{-}OH + R{-}X$$

where X = Br, I, or $HSO_4$. For example:

Anisole + $H_2SO_4$ $\xrightarrow{\Delta}$ Phenol + $CH_3OSO_3H$ Methyl hydrogen sulfate

**Peroxide Formation**

Ethers slowly react with oxygen in the air to form peroxides. In general,

$$2\ R{-}O{-}R + O_2\ (air) \longrightarrow 2\ R{-}O{-}O{-}R$$

Example:

$$2\ CH_3CH_2CH_2{-}O{-}CH_2CH_2CH_3 + O_2\ (air) \longrightarrow 2\ CH_3CH_2CH_2{-}O{-}O{-}CH_2CH_2CH_3$$

These organic peroxides are highly explosive and must be eliminated from the ether before the ether can be used. The presence of a white precipitate (peroxide) in the bottom of a bottle of an ether is a sign of deadly danger. In this event, do not touch the bottle, but *immediately* call the nearest bomb squad. To detect peroxide in a sample of an ether with no precipitate in the bottom of the bottle, treat a small sample of the ether with potassium iodide (KI) solution and dilute hydrochloric acid. The formation of a brown color due to iodine ($I_2$) signifies a positive test. In general,

*Most ethers form peroxides, but isopropyl ether is one of the worst offenders.*

$$2\ H^+ + 2\ I^- + ROOR \longrightarrow I_2 + ROR + H_2O$$

Example:

$$2\ H^+ + 2\ I^- + CH_3CH_2OOCH_2CH_3 \longrightarrow I_2 + CH_3CH_2OCH_2CH_3 + H_2O$$

The peroxide is removed from the ether by reacting the ether with a dilute solution of ferrous sulfate ($FeSO_4$). In general,

$$2\ Fe^{2+} + 2\ H^+ + ROOR \longrightarrow 2\ Fe^{3+} + ROR + H_2O$$

Example:

$$2\ Fe^{2+} + 2\ H^+ + (CH_3)_2CHOOCH(CH_3)_2 \longrightarrow 2\ Fe^{3+} + (CH_3)_2CHOCH(CH_3)_2 + H_2O$$

**EXERCISE 17.4**

Complete the following equations by supplying a structural formula for each organic product produced and a molecular formula or formula unit for each inorganic compound formed. If no reaction occurs, write "no reaction." Do not balance.

a. ethyl ether + $H_2SO_4$ $\xrightarrow{Cold}$

b. methyl ethyl ether + 2 HI $\xrightarrow{\Delta}$

c. phenyl ether + 2 HBr $\xrightarrow{\Delta}$

d. methyl ether + $O_2$ $\longrightarrow$

e. $CH_3CH_2{-}O{-}O{-}CH_2CH_3 + H^+ + I^- \longrightarrow$

f. butyl ether + Na $\longrightarrow$

## CYCLIC ETHERS: EPOXIDES

### 17.6 PREPARATION

Epoxides are prepared by reacting an alkene with a *peracid*. A peracid contains one more oxygen atom than the corresponding carboxylic acid (Fig. 17.4). Peracetic acid is often used in the laboratory to prepare an epoxide from the parent alkene. In general,

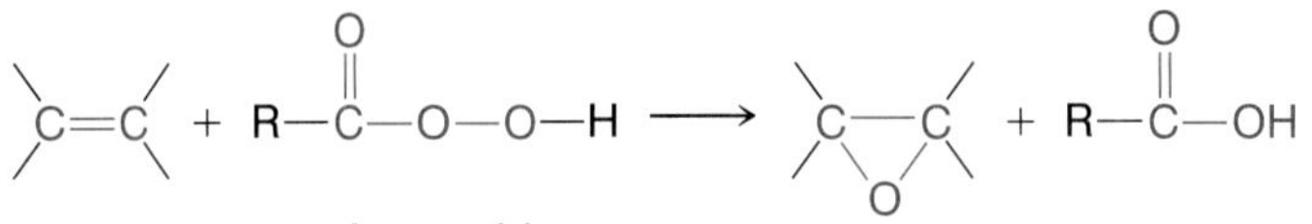

A peracid

Example:

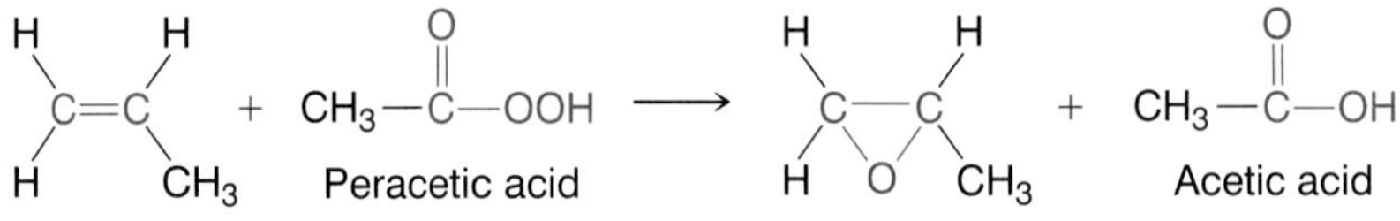

2-Methyloxirane

### 17.7 USE OF CROWN ETHERS IN CHEMICAL REACTIONS

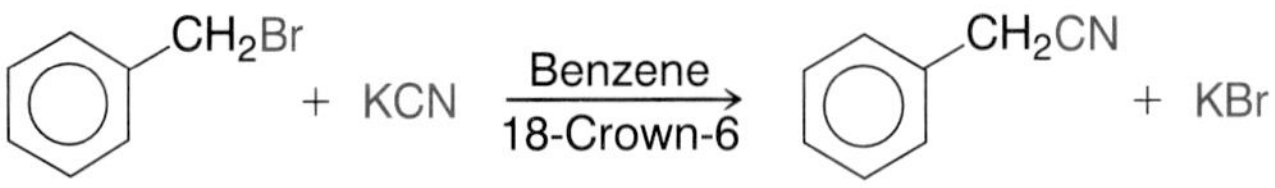

**FIGURE 17.4** The structural difference between a carboxylic acid and a peracid.

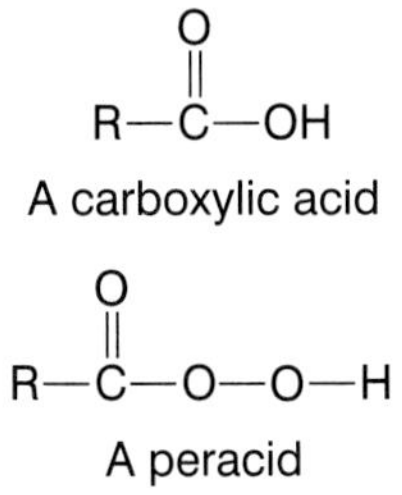

A carboxylic acid

A peracid

The fact that crown ethers make it possible to dissolve ionic salts such as KF in nonpolar solvents such as benzene enables the chemist to run reactions in excellent yield that would not occur without the presence of the crown ether. Thus:

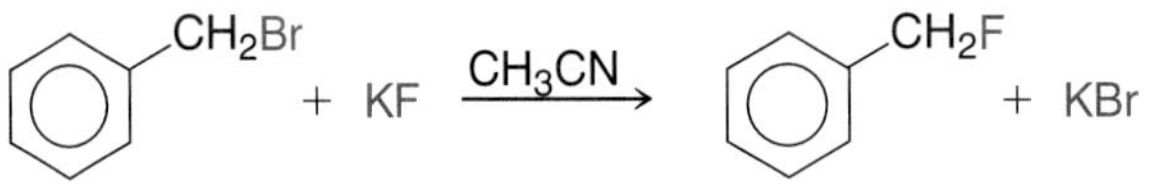

Yield < 5%

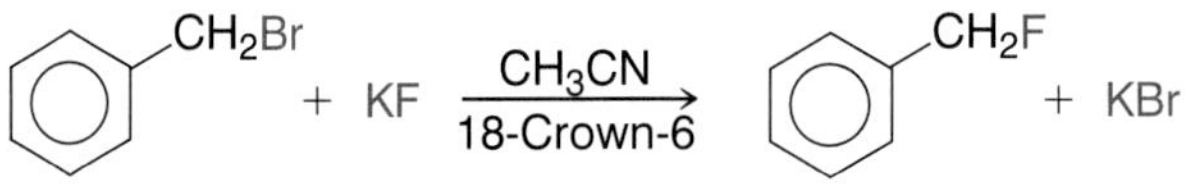

Yield = 100%

Acetonitrile ($CH_3CN$) serves as a solvent for the reaction. Note that when benzyl bromide reacts with potassium fluoride without 18-crown-6, the yield is less than 5%. However, when a small amount of 18-crown-6 is introduced into the reaction mixture, the yield improves dramatically. In the same way, cyanide ion can substitute for chloride, bromide, or iodide ion.

### 17.8 CHEMICAL PROPERTIES

The epoxides react with a variety of reagents; the strained three-member ring is easily broken, resulting in the formation of a straight-chain compound. In general,

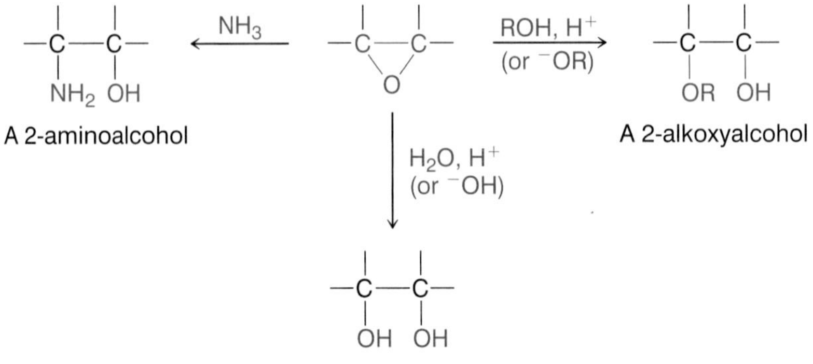

A 2-aminoalcohol                    A 2-alkoxyalcohol

A glycol

Examples: Consider the reaction of ethylene oxide with a variety of compounds in either acid or alkaline solution.

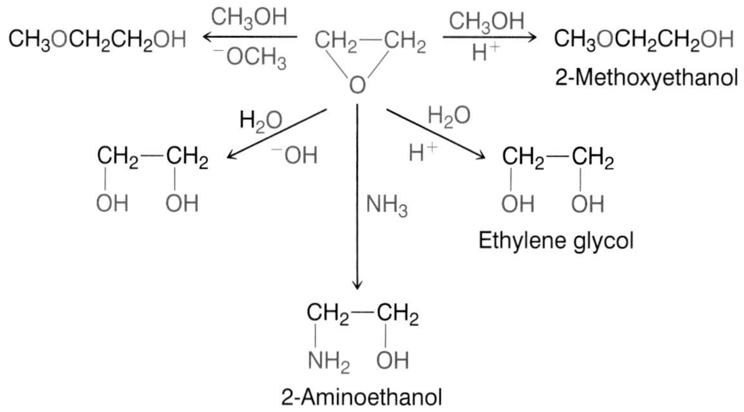

$$CH_3OCH_2CH_2OH \xleftarrow[-OCH_3]{CH_3OH} \quad CH_2\!-\!CH_2 \xrightarrow[H^+]{CH_3OH} CH_3OCH_2CH_2OH$$

2-Methoxyethanol

(epoxide center with O)

$$\xrightarrow[{}^-OH]{H_2O} \quad CH_2\!-\!CH_2 \text{ (OH  OH)}$$

$$\xrightarrow[H^+]{H_2O} \quad CH_2\!-\!CH_2 \text{ (OH  OH)}$$

Ethylene glycol

$$\xrightarrow{NH_3} \quad CH_2\!-\!CH_2 \text{ (NH}_2 \text{  OH)}$$

2-Aminoethanol

Note that in IUPAC nomenclature the amino group shows a lower priority than the hydroxyl group in an amino alcohol. We will study amines in Chapters 24 and 25.

---

**SOLVED PROBLEM 17.4**

Assuming reaction in acid solution occurs by an $S_N1$ (Sec. 12.1) mechanism, predict a product for

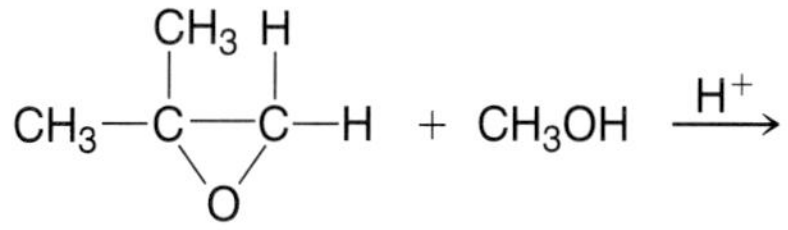

$$CH_3\!-\!\underset{O}{\overset{CH_3 \quad H}{C\!-\!C}}\!-\!H \;+\; CH_3OH \xrightarrow{H^+}$$

**SOLUTION**

The first step of the mechanism is addition of the proton to the oxygen of the ether to form an oxonium ion:

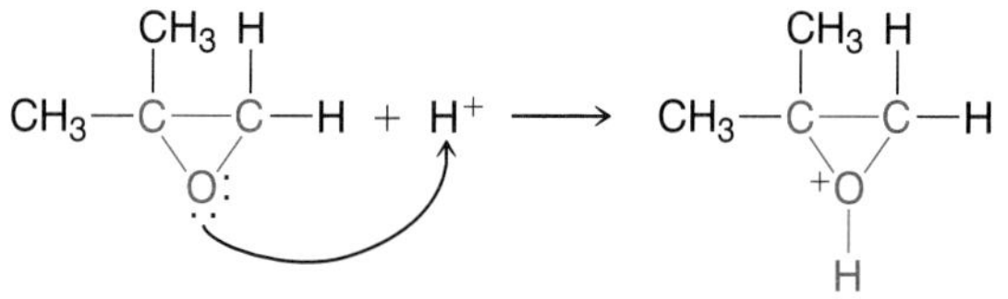

$$CH_3\!-\!\underset{\overset{\cdot\cdot}{O}:}{\overset{CH_3 \quad H}{C\!-\!C}}\!-\!H \;+\; H^+ \longrightarrow CH_3\!-\!\underset{\overset{+O}{\underset{H}{}}}{\overset{CH_3 \quad H}{C\!-\!C}}\!-\!H$$

Now, since the ether is unsymmetrical, one of two possible carbon-oxygen bonds is broken to form a carbocation:

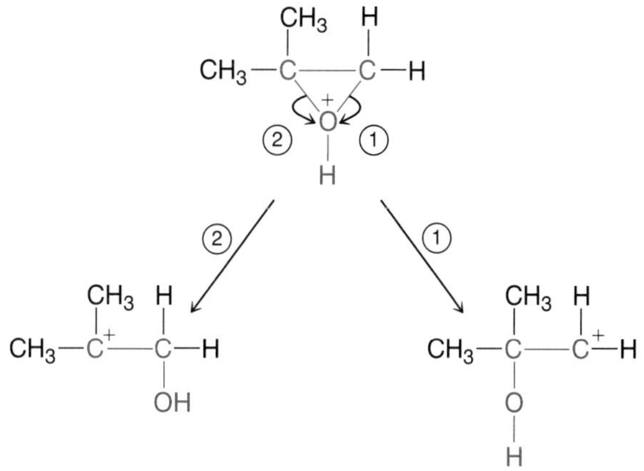

$$CH_3\!-\!\underset{\overset{+O}{\underset{H}{}}}{\overset{CH_3 \quad H}{C\!-\!C}}\!-\!H$$
② ①

② 
$$CH_3\!-\!\overset{CH_3 \quad H}{\underset{OH}{C^+\!-\!C}}\!-\!H$$

① 
$$CH_3\!-\!\overset{CH_3 \quad H}{\underset{\underset{H}{O}}{C\!-\!C^+}}\!-\!H$$

Route 2 is preferred because the tertiary carbocation formed is more stable than the primary carbocation produced from route 1. The mechanism is completed as follows:

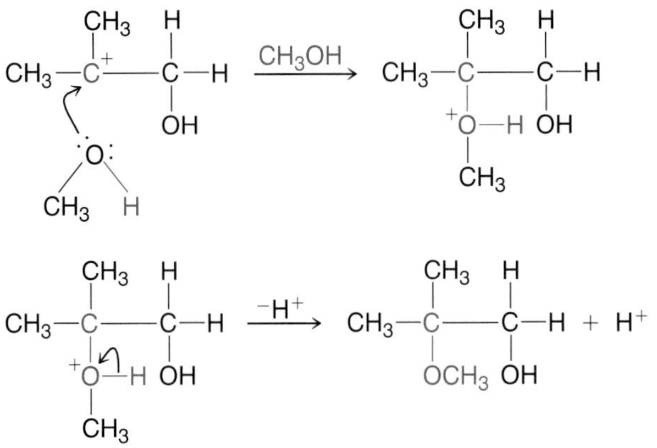

**SOLVED PROBLEM 17.5**

Assuming reaction in alkaline solution occurs by an $S_N2$ mechanism, predict a product for

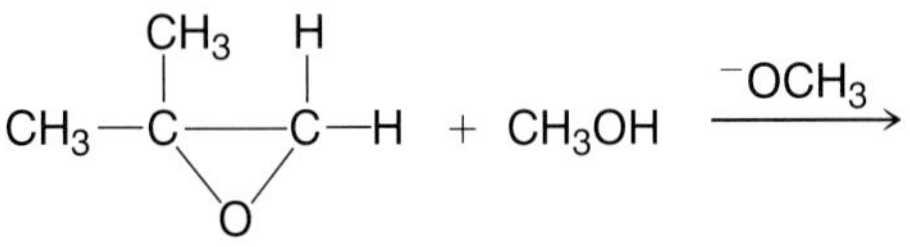

**SOLUTION**

Because the oxirane is unsymmetrical, the problem here is to decide which carbon is attacked by methoxide ion, the nucleophile. We know that $S_N2$ reactivity is governed by steric hindrance around the halogen-bonded carbon (page 239). Since carbon 3 is less crowded than carbon 2, attack takes place at carbon 3, and we get

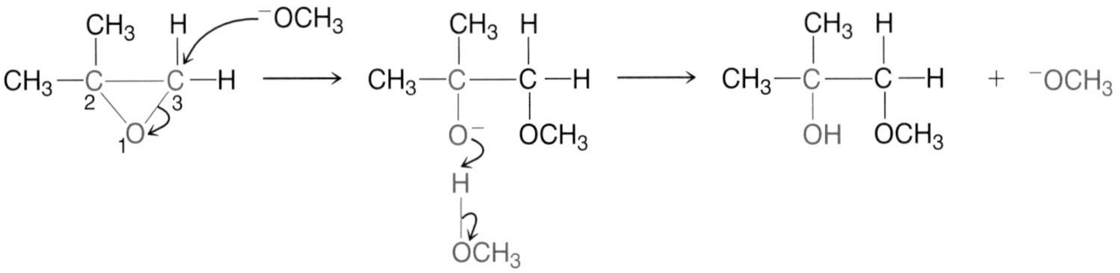

In conclusion, the point of attack on an epoxide by a nucleophile ($^-OR$, $^-OH$, $NH_3$) in basic solution takes place at the less substituted carbon on the epoxide skeleton. This makes sense because the reaction mechanism is always $S_N2$. In acid solution, when one or more carbons of the epoxide are substituted, attack of the nucleophile ($ROH$, $H_2O$) occurs on the more substituted carbon due to formation of the more stable carbocation ($S_N1$).

**EXERCISE 17.5**

Draw a detailed, step-by-step mechanism for each of the following transformations.

a.  Reaction of ethylene oxide with water in the presence of an acid catalyst to produce ethylene glycol ($S_N2$)

b.  Reaction of propylene oxide with methyl alcohol in the presence of acid to give 2-methoxyethanol ($S_N1$)

c.  Reaction of propylene oxide with methyl alcohol in the presence of a catalytic amount of methoxide ion ($S_N2$)

Complete the following equations by drawing a structural formula for each organic product. Name each product.

a.  oxirane + ethyl alcohol ($H^+$ catalyst) $\longrightarrow$

b.  2-methyloxirane + water ($^-OH$ catalyst) $\longrightarrow$

## ▶ CHAPTER ACCOMPLISHMENTS

### 17.1  Preparation of Phenols
☐ Prepare *p*-cresol from *p*-methylbenzenesulfonic acid.

### 17.2  Aromatic Substitution of Phenols
☐ Complete the following equations by drawing a structural formula for each organic product formed and writing a molecular formula or formula unit for each inorganic compound produced.

a.  phenol + $Br_2 \xrightarrow[\text{0°C}]{\text{CS}_2}$

b.  phenol + $3 Br_2 \xrightarrow[\text{25°C}]{\text{H}_2\text{O}}$

c.  phenol + $HNO_3$ (dilute) $\longrightarrow$

d.  2,4-dinitrophenol + $HNO_3$ (concentrated) $\xrightarrow{\Delta}$

### 17.3  Oxidation of Phenols: Quinones
☐ Complete the following equation by drawing a structural formula for each organic product formed and writing a molecular formula or formula unit for each inorganic compound produced.

$$\text{hydroquinone} \xrightarrow[\text{ethyl ether}]{\text{Ag}_2\text{O}}$$

(Neglect inorganic products)

Draw a structural formula for *o*-benzoquinone.

### 17.4  Preparation of Ethers
☐ Complete the following equations by drawing a structural formula for each organic product formed and writing a molecular formula or formula unit for each inorganic compound produced.

a.  $2 CH_3CH_2OH \xrightarrow[\Delta]{\text{H}_2\text{SO}_4}$

b.  $CH_3CH_2OH \xrightarrow[\Delta]{\text{H}_2\text{SO}_4}$

c.  $CH_3CH_2O^-Na^+ + BrCH_3 \longrightarrow$

d.  $CH_3O^-Na^+ + BrC(CH_3)_3 \longrightarrow$

e.  sodium phenoxide + propyl bromide $\longrightarrow$

☐ Explain why intermolecular dehydration of a mixture of methyl alcohol and ethyl alcohol does not represent a satisfactory method of preparation for the mixed ether methyl ethyl ether.

☐ Explain why an ether is not produced when an alkoxide is reacted with a tertiary halide in a Williamson synthesis.

### 17.5  Chemical Properties of Ethers
☐ Distinguish diethyl ether from pentane using a test tube reaction.

☐ Complete the following equations by drawing a structural formula for each organic product formed and writing a molecular formula or formula unit for each inorganic compound produced. If no reaction occurs, write "no reaction."

a.  $CH_3CH_2OCH_2CH_3 + 2 HI \longrightarrow$

b.  cyclohexyl ether + $2 H_2SO_4 \longrightarrow$

c.  phenyl ether + $2 HI \longrightarrow$

d.  $CH_3CH_2CH_2OCH_2CH_2CH_3 + O_2 \longrightarrow$

☐ Explain why the presence of a white precipitate in the bottom of a bottle of ether is a matter of great concern to laboratory workers.

☐ Write an equation to represent the detection of peroxide in a sample of ether.

☐ Name a reagent that is used to remove peroxide from an ether sample.

### 17.6  Preparation of Epoxides
☐ Draw a structural formula for peracetic acid.

☐ Write an equation showing the preparation of 2-methyloxirane from propene and peracetic acid.

### 17.7  Use of Crown Ethers in Chemical Reactions
☐ Write an equation showing the preparation of benzyl cyanide from benzyl bromide and KCN in the presence of 18-crown-6.

### 17.8  Chemical Properties of Epoxides
☐ Complete the following equations by drawing a structural formula for each organic product formed and writing a molecular formula or formula unit for each inorganic compound produced.

a.  ethylene oxide + $CH_3OH \xrightarrow{^-OCH_3}$

b.  ethylene oxide + $H_2O \xrightarrow{H^+}$

c.  ethylene oxide + $CH_3OH \xrightarrow{H^+}$

d.  ethylene oxide + $NH_3 \longrightarrow$

☐ Write a detailed mechanism for each of the following reactions.

a.  2,2-dimethyloxirane + $CH_3OH \xrightarrow{H^+}$ ($S_N1$)

b.  2,2-dimethyloxirane + $CH_3OH \xrightarrow{^-OCH_3}$ ($S_N2$)

## ► KEY TERMS

*quinone* (17.3)
*Williamson synthesis* (17.4)
*peracid* (17.6)

*cantharidin* (Medical Boxed Reading
17.2)

## ► PROBLEMS

1. Explain why each of the following syntheses is not practical.

   a. $CH_3-O-CH_2CH_2CH_3$   (using the alcohol dehydration method)
      Methyl *n*-propyl ether

   b. $CH_3-\langle\bigcirc\rangle-O-\langle\bigcirc\rangle-CH_3$
      *p*-Tolyl ether
      (using the Williamson method)

2. Prepare each of the following. (You may use any organic or inorganic reagents necessary.)

   a. *p*-cresol from benzene
   b. allyl ether from *n*-propyl alcohol
   c. propylene oxide from isopropyl chloride
   d. methyl *tert*-butyl ether from methyl alcohol and isobutyl alcohol

3. Phenol reacts with dilute nitric acid to produce a mixture of the *ortho*- and *para*-nitrophenols. Predict the product obtained when phenol reacts with concentrated nitric acid.

4. Draw a structural formula for each of the following defensive quinones used by the insects listed in Exercise 17.1.

   a. 2-methoxy-3-methyl-1,4-benzoquinone (the methoxy group is $-OCH_3$)
   b. 6-methyl-1,4-naphthoquinone

5. Complete the following equations by drawing a structural formula for every organic compound formed and writing a molecular formula or formula unit for each inorganic compound formed. If no reaction occurs, write "no reaction."

   a. *m*-cresol + HNO₃ (dilute) $\xrightarrow{0°C}$

   b. *m*-cresol + Br₂ $\xrightarrow{H_2O}$

   c. ether + 2HBr $\xrightarrow{\Delta}$

   d. sodium *p*-nitrophenoxide + *tert*-butyl chloride $\longrightarrow$

   e. catechol $\xrightarrow[\text{Ethyl ether}]{Ag_2O}$

   f. chlorobenzene + sodium methoxide $\longrightarrow$

   g. $ClCH_2CH_2OCH_2CH_2Cl$ + 2 KOH $\xrightarrow[\Delta]{\text{Alcohol}}$

   h. product g + Br₂ $\xrightarrow{CCl_4}$

   i. 2 mol methyl alcohol $\xrightarrow[\Delta]{H_2SO_4}$

   j. cyclohexyl ether + H₂SO₄ (cold, concentrated) $\longrightarrow$ (unstable form)

   k. isopropyl ether + KMnO₄ $\longrightarrow$

   l. 1-butene + peracetic acid $\longrightarrow$

   m. isopropyl ether + O₂ $\longrightarrow$

   n. *n*-butyl bromide + KCN $\xrightarrow[\text{18-Crown-6}]{\text{Pentane}}$

   o. 2,3-dimethyloxirane + water $\xrightarrow{^-OH}$ (Disregard stereoisomerism)

   p. methyl isopropyl ether + 2 HI $\xrightarrow{\Delta}$

   q. $Fe^{2+}$ + $CH_3CH_2CH_2OOCH_2CH_2CH_3$ + $H^+$ $\longrightarrow$

6. Write a detailed, step-by-step mechanism for

   $$4\ CH_3CH_2CH_2OH \xrightarrow{H^+} (CH_3CH_2CH_2)_2O$$

   $$+\ CH_3-\overset{\overset{\displaystyle H}{|}}{C}-O-\overset{\overset{\displaystyle H}{|}}{C}-CH_3$$
   $$\qquad\ \ \underset{\underset{\displaystyle CH_3}{|}}{}\qquad \underset{\underset{\displaystyle CH_3}{|}}{}$$

7. How would you prepare 1,4-dioxane using the alcohol dehydration method? What by-product would you expect?

8. Consider the following reaction:

   $$CH_3-\overset{\overset{\displaystyle CH_3}{|}}{C}\underset{\underset{\displaystyle O}{\diagdown\diagup}}{}\overset{\overset{\displaystyle H}{|}}{C}-CH_3 + CH_3OH \xrightarrow{H^+}$$

   Assuming reaction occurred by an $S_N1$ mechanism, write a detailed, step-by-step mechanism for the reaction, and predict the product.

9. Use a simple test tube reaction to distinguish each of the following pairs of compounds. [State what reagent(s) you add and what you see.]

   a. pentane and 1,4-dioxane
   b. *o*-cresol and *n*-heptyl alcohol
   c. phenol and benzoic acid ($C_6H_5COOH$)
   d. ethyl ether and ethyl alcohol

10. An optically active compound, $C_6H_{14}O$ (A), dissolved in cold concentrated H₂SO₄ and did not react with Na₂Cr₂O₇

and $H_2SO_4$. Treating compound A with HI yielded $C_4H_9I$ and $C_2H_5I$. Draw a structural formula for compound A.

11. Ethylene oxide and other epoxides also undergo a ring-opening reaction with HCl.
    a. Draw a structural formula for the product obtained when ethylene oxide reacts with HCl.
    b. Name the product.
    c. Write a detailed mechanism for this reaction assuming an $S_N2$ mechanism.

12. A compound, $C_3H_8O$, gave the following decoupled $^{13}C$ NMR spectrum:

$$\delta = 14.7 \ (3H)$$
$$\delta = 57.6 \ (3H)$$
$$\delta = 67.9 \ (2H)$$

Draw a structural formula for the compound consistent with the data.

13. An IR spectrum of compound A, $C_6H_6O$, gave a broad peak at $3250 \ cm^{-1}$ and peaks at approximately 1500 and $1600 \ cm^{-1}$. Classify compound A, i.e., determine what functional group or groups are present.

# CHAPTER 18

# Aldehydes and Ketones (I)

STRUCTURE, BONDING, NOMENCLATURE, PHYSICAL PROPERTIES, NATURALLY OCCURRING CARBONYL COMPOUNDS, SOME IMPORTANT CARBONYL COMPOUNDS AND REPRESENTATIVE REACTIONS OF THE CARBONYL COMPOUNDS

Both aldehydes and ketones are characterized by the presence of the carbonyl group.

An aldehyde has one (or two) hydrogen(s) bonded to the carbon of the carbonyl group; a ketone has carbons bonded to the carbonyl carbon.

$$(H)R-\overset{\overset{\displaystyle H}{|}}{C}=O \qquad R-\overset{\overset{\displaystyle O}{||}}{C}-R'$$

Aldehyde                    Ketone

The carbon atom of the carbonyl group is $sp^2$ hybridized. Thus the three atoms bonded to the carbonyl carbon and the carbonyl carbon itself lie on the same plane (Fig. 18.1). Bond angles are 120 degrees. The carbonyl carbon is bonded by means of an $sp^2$ hybridized orbital to the oxygen of the carbonyl group on the one hand and to either hydrogen or carbon using two $sp^2$ hybridized orbitals on the other hand. Thus the carbonyl carbon uses a total of three hybridized orbitals to produce three sigma bonds: $sp^2$–$sp^2$ (to oxygen), $sp^2$–$s$ (to hydrogen), and $sp^2$–$sp^3$ (to a saturated carbon atom) in an aldehyde and $sp^2$–$sp^2$ (to oxygen) and two $sp^2$–$sp^3$ bonds (each bond to a saturated carbon atom) in a ketone (Fig. 18.2). The double bond consists of an $sp^2$–$sp^2$ sigma bond and a $p$–$p$ pi bond. The pi bond is produced by overlap of a $2p$ orbital of carbon with a $2p$ orbital of oxygen.

The main difference between the carbon-carbon double bond and the carbon-oxygen double bond is that the double bond of the alkene lies between two atoms of the same element, carbon, whereas that of the carbonyl is positioned between two different elements, carbon and oxygen. Since oxygen is a more electronegative element than

## 18.1 STRUCTURE AND BONDING

**FIGURE 18.1** Geometry of the carbonyl group.

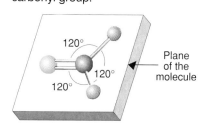

As you previously learned (pages 33 and 35), the greatest electron density in a sigma bond is between the atoms concerned, while the greatest electron density in a pi bond is above and below the carbon-oxygen atoms (see Fig. 18.2).

**FIGURE 18.2** Bonding in a carbonyl compound.

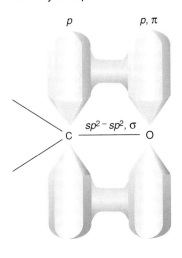

Note the close similarity of a carbon-carbon double bond (alkene) to the carbon-oxygen double bond (carbonyl compound). Both double bonds contain both a sigma bond and a pi bond.

| TABLE 18.1 | Structures and Common Names of Several Selected Aldehydes |
|---|---|
| **Structure** | **Common Name** |
| $\overset{H}{\underset{H}{\diagdown}}C{=}O$ | Formaldehyde |
| $CH_3\overset{H}{\underset{}{C}}{=}O$ | Acetaldehyde |
| $CH_3CH_2\overset{H}{\underset{}{C}}{=}O$ | Propionaldehyde |
| $CH_3CH_2CH_2\overset{H}{\underset{}{C}}{=}O$ | Butyraldehyde |
| $CH_3CH_2CH_2CH_2\overset{H}{\underset{}{C}}{=}O$ | Valeraldehyde |

carbon, the carbon-oxygen double bond is polar. That is, there exists a slight shift of electron density, both in the sigma and pi bonds, toward the oxygen. Thus we can draw the following more realistic structure for the carbonyl group:

$$\overset{\diagdown}{\underset{\diagup}{}}\overset{\delta^+}{C}{=}\overset{\delta^-}{O}$$

The carbon bears a partial positive charge, while the oxygen acquires a partial negative charge. This greatly influences the reactivity of carbonyl compounds. This will be discussed in Section 19.2.

## 18.2 NOMENCLATURE

The common names of the straight-chain aliphatic aldehydes are derived from the corresponding carboxylic acids and must be learned. (Table 18.1 lists five of the more common aliphatic straight-chain aldehydes.)

To name a substituted aliphatic aldehyde using the common system of nomenclature, groups are located by alphabetizing carbons (other than the aldehyde carbon) with Greek letters and always designating the carbon next to the carbonyl carbon as the alpha carbon (Fig. 18.3). Ketones are named by means of the common system of nomenclature by naming the alkyl, aryl, or cycloalkyl group(s) followed by the word *ketone*

**FIGURE 18.3** Examples of common nomenclature of substituted aliphatic aldehydes.

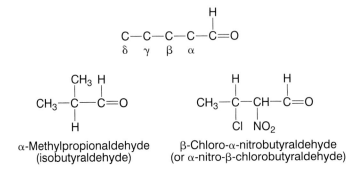

**FIGURE 18.4** Examples of common nomenclature of ketones.

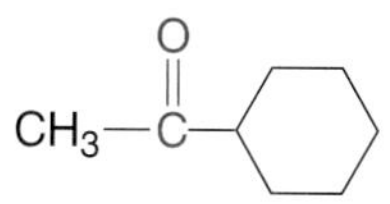

Methyl cyclohexyl ketone
(or cyclohexyl methyl ketone)

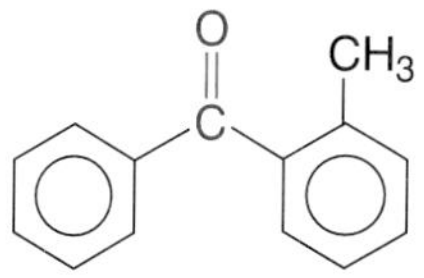

Phenyl o-tolyl ketone
(or o-tolyl phenyl ketone)

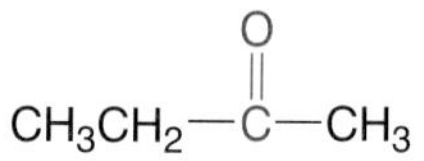

$$CH_3CH_2-\overset{\overset{\displaystyle O}{\|}}{C}-CH_3$$
Methyl ethyl ketone
(or ethyl methyl ketone)

(Fig. 18.4). If the groups bonded to the carbonyl of the ketone are the same, the prefix *di-* is used.

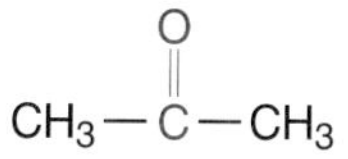

$$CH_3-\overset{\overset{\displaystyle O}{\|}}{C}-CH_3$$
Dimethyl ketone

Dimethyl ketone is also known as *acetone* (a trivial name).

In IUPAC nomenclature, drop the *-e* ending of the corresponding alkane, and replace it with the *-al* suffix for an aldehyde or the *-one* suffix for a ketone.

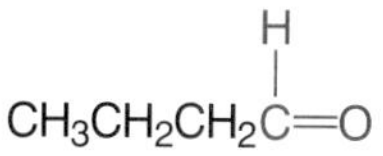

$$CH_3CH_2CH_2\overset{\overset{\displaystyle H}{|}}{C}{=}O$$
Butanal

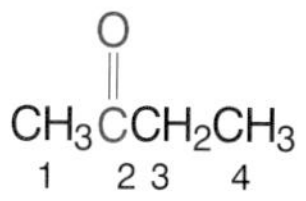

$$CH_3\overset{\overset{\displaystyle O}{\|}}{C}CH_2CH_3$$
$$\underset{1\ \ \ 2\,3\ \ \ \ 4}{}$$
2-Butanone

Figure 18.5 gives examples of IUPAC nomenclature of substituted aldehydes and ketones. The structural formulas and IUPAC names of two important aromatic aldehydes are given in Fig. 18.6. Substituted aromatic aldehydes are named using the rules given in Section 9.2 (Fig. 18.7).

Alicyclic ketones, i.e., ketones in which the carbon of the carbonyl group is a part of the ring skeleton, are named, using the IUPAC system of nomenclature, by dropping the *-e* fragment from the corresponding cycloalkane and adding the suffix *-one*. The carbon

**FIGURE 18.6** Two important aromatic aldehydes.

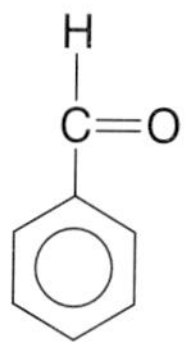

Benzaldehyde

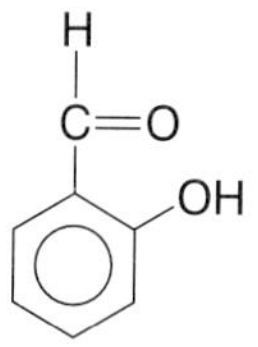

Salicylaldehyde

**FIGURE 18.5** Examples of IUPAC nomenclature of substituted aldehydes and ketones.

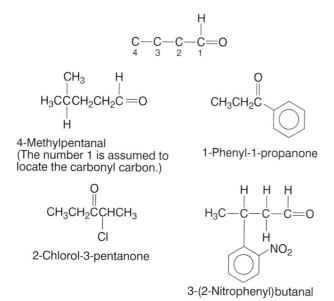

$$\underset{4\ \ \ \ 3\ \ \ \ 2\ \ \ \ 1}{C-C-C-\overset{\overset{\displaystyle H}{|}}{C}{=}O}$$

4-Methylpentanal
(The number 1 is assumed to locate the carbonyl carbon.)

1-Phenyl-1-propanone

2-Chlorol-3-pentanone

3-(2-Nitrophenyl)butanal

For substituted aldehydes, the carbonyl carbon is always assigned the number 1.

**FIGURE 18.7** The nomenclature of
selected substituted aromatic
aldehydes.

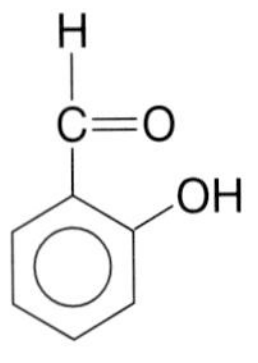

o-Hydroxybenzaldehyde

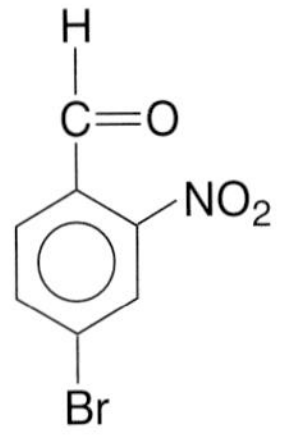

4-Bromo-2-nitrobenzaldehyde

**FIGURE 18.8** IUPAC nomenclature of selected alicyclic ketones.

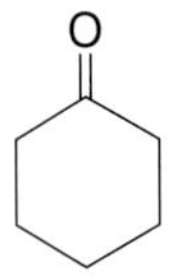

Cyclohexanone 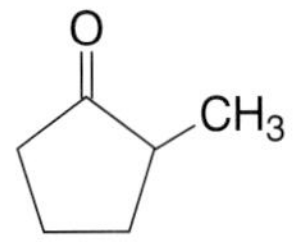        2-Methylcyclopentanone

(The number 1 is assumed to be assigned
to each carbonyl carbon.)

of the carbonyl group is always assigned the lowest possible number. When no number
is given, assume the number one. Examples are given in Fig. 18.8.

A unique common name is sometimes used when a ketone contains an aromatic ring
that is bonded to the carbon of the carbonyl group. To understand this common name, we
need to name a number of *acyl groups* that we will study with more detail in Chapter 22.
Each acyl group is imaginary, derived from the corresponding carboxylic acid, and is used
extensively in nomenclature. Let us tabulate a number of simple carboxylic acids and the
corresponding acyl group derived from each carboxylic acid (Table 18.2). To name an acyl
group, drop the *-ic* suffix of the corresponding acid, and replace it with the suffix *-yl*.

Let us name

$$CH_3-\overset{\overset{\displaystyle O}{\|}}{C}-C_6H_5$$

Methyl phenyl ketone

using this unique method. Note that the acetyl group is present.

$$CH_3-\overset{\overset{\displaystyle O}{\|}}{C}-$$

Thus we write the word *acetyl*, drop the *-yl* suffix, and then add the suffix *-ophenone* to give
*acetophenone*. The nomenclature of propiophenone is an exception to the rule in that the
suffix *-onyl* of the acyl group is dropped (rather than *-yl*) before adding the suffix *-ophenone*.

$$C_6H_5-\overset{\overset{\displaystyle O}{\|}}{C}-CH_2CH_3$$

Propiophenone

| TABLE 18.2 | A Listing of Some Simple Carboxylic Acids and the Corresponding Acyl Group |
|---|---|
| **Carboxylic Acid** | **Acyl Group** |
| $CH_3-\overset{\overset{O}{\|}}{C}-OH$ <br> Acetic acid | $CH_3-\overset{\overset{O}{\|}}{C}-$ <br> Acetyl group |
| $CH_3CH_2-\overset{\overset{O}{\|}}{C}-OH$ <br> Propionic acid | $CH_3CH_2-\overset{\overset{O}{\|}}{C}-$ <br> Propionyl group |
| $C_6H_5-\overset{\overset{O}{\|}}{C}-OH$ <br> Benzoic acid | $C_6H_5-\overset{\overset{O}{\|}}{C}-$ <br> Benzoyl group |

## SOLVED PROBLEM 18.1

Name the following compounds using the common system of nomenclature. Supply an IUPAC name for the compound in part (a).

a. 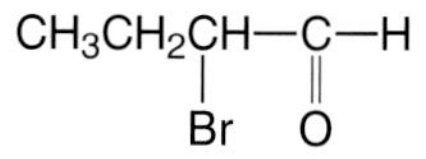
CH3CH2CH—C—H
Br O

b. 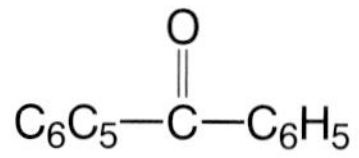
O
C6C5—C—C6H5

**SOLUTION**

a. Common name: The name of the parent compound is butyraldehyde. Since a bromine atom is bonded to the alpha carbon, we have:

α-bromobutyraldehyde

IUPAC name: The name of the parent is butanal. Since a bromine atom is bonded to carbon 2, we have

2-bromobutanal

b. Common name: The $C_6H_5\overset{\overset{O}{\|}}{C}$— group is the benzoyl group. We have one more exception (the last exception for our purposes) to the rule here in that the suffix *-oyl* is dropped before adding *-ophenone* to give

benzophenone

## EXERCISE 18.1

Provide a common and an IUPAC name for the following compounds.

a. 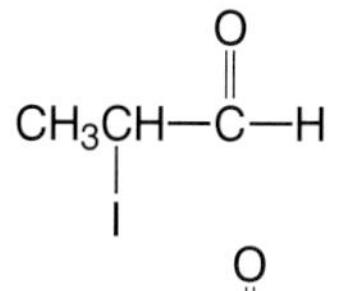
O
CH3CH—C—H
I

b. 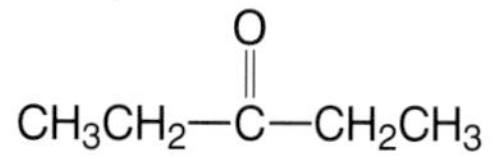
O
CH3CH2—C—CH2CH3

## 18.3 PHYSICAL PROPERTIES OF CARBONYL COMPOUNDS

Other than formaldehyde, which is a gas, the simple aldehydes and ketones are liquids. Carbonyl compounds boil at temperatures between those of alcohols and alkanes of similar molecule weight (Table 18.3). Thus propionaldehyde and dimethyl ketone (acetone) boil at temperatures higher than butane but lower than *n*-propyl alcohol. Since aldehydes and ketones are polar compounds because of the polar carbonyl group, they form intermolecular dipole-dipole interactions that cannot be formed by alkanes.

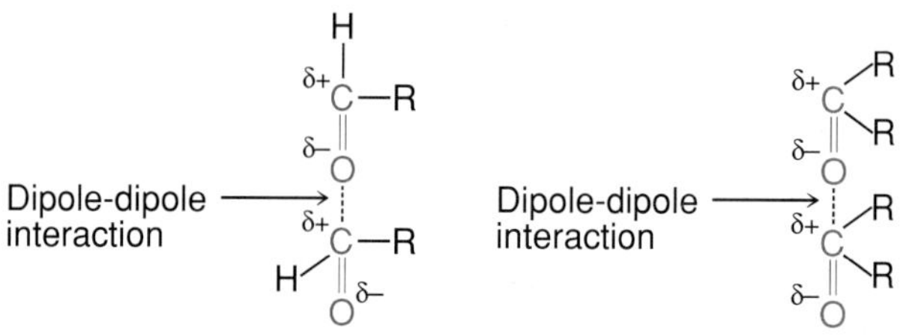
Dipole-dipole interaction
Dipole-dipole interaction
δ+
δ−

However, these dipole-dipole interactions are considerably weaker than the hydrogen bonds in an alcohol. Thus the typical aldehyde or ketone boils at a lower temperature than the typical alcohol of a similar molecular weight but at a higher temperature than the corresponding alkane.

| TABLE 18.3 | A Comparison of Boiling Points of Compounds with Similar Molecular Weights but Different Functional Groups | | |
|---|---|---|---|
| Structural Formula | Common Name | Molecular Weight | Boiling Point (°C) |
| $CH_3CH_2C=O$ (with H) | Propionaldehyde | 58 | 50 |
| $CH_3CCH_3$ ($\parallel$ O) | Dimethyl ketone | 58 | 56 |
| $CH_3CH_2CH_2CH_3$ | $n$-Butane | 58 | $-0.5$ |
| $CH_3CH_2CH_2OH$ | $n$-Propyl alcohol | 60 | 97 |

The solubility of carbonyl compounds in water decreases with increasing molecular weight (Table 18.4). Carbonyl compounds with six or more carbons are only slightly soluble in water.

| TABLE 18.4 | Physical Properties of Selected Carbonyl Compounds | | |
|---|---|---|---|
| Name of Compound | Structure | Boiling Point (°C) | Water Solubility g/100 ml $H_2O$ |
| Formaldehyde | $H_2C=O$ | $-21$ | $\infty$ |
| Acetaldehyde | $CH_3C=O$ (with H) | 21 | $\infty$ |
| Propionaldehyde | $CH_3CH_2C=O$ (with H) | 49 | 20 |
| Butyraldehyde | $CH_3CH_2CH_2C=O$ (with H) | 76 | 4 |
| Isobutyraldehyde | $(CH_3)_2CHC=O$ (with H) | 62 | 11 |
| Benzaldehyde | 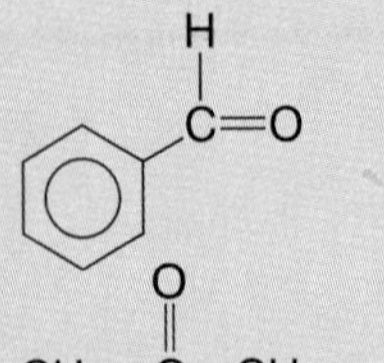 C$_6$H$_5$C=O (with H) | 178 | 0.3 |
| Acetone | $CH_3-C-CH_3$ ($\parallel$ O) | 58 | $\infty$ |
| Methyl ethyl ketone | $CH_3-C-CH_2CH_3$ ($\parallel$ O) | 80 | 37 |
| Diethyl ketone | $CH_3CH_2CCH_2CH_3$ ($\parallel$ O) | 102 | 4.7 |

Select the compound in each pair with the higher boiling point. Explain. You should not use Table 18.4.

a.  $CH_3(CH_2)_3CHO$ or $CH_3(CH_2)_2CHO$

b.  $CH_3COCH_3$ or $CH_3CH_2COCH_2CH_3$

c.  $CH_3CH_2CH_2CHO$ or $(CH_3)_2CHCHO$

**SOLUTION**

a.  $CH_3(CH_2)_3CHO$. Pentanal has the higher molecular weight with greater London forces between molecules. Dipole-dipole interactions in both molecules are the same and therefore cancel out.

b.  $CH_3CH_2COCH_2CH_3$. The same reason as in part (a).

c.  $CH_3CH_2CH_2CHO$. This straight-chain compound contains more surface area than $(CH_3)_2CHCHO$. Thus London forces are greater and the boiling point is higher. Again, dipole-dipole interactions in both molecules are the same and therefore cancel out.

Which compound of the following pairs boils at a lower temperature? Explain. You should not need to use Table 18.4.

a.  acetaldehyde or propionaldehyde

b.  2-pentanone or 3-methyl-2-butanone

## 18.4 NATURALLY OCCURRING CARBONYL COMPOUNDS

A significant number of carbonyl compounds are found in nature, many of which have pleasant aromas. Refer to Fig. 18.9 for some examples. Civetone, a component of civet, is a substance that can be isolated from the scent glands of the African civet cat. This compound is used as a fixative (page 280) in perfumes in that it prolongs the pleasant odor of the essential oils in the perfume by decreasing the rate of evaporation of the oils. However, the compound does not have a particularly pleasant odor.

The carbonyl group is found in a significant number of pheromones. A queen bee secretes 9-oxo-2-($E$)-decenoic acid (Fig. 18.10), which keeps the insects in the hive on task and therefore preserves the colony. The compound 2-heptanone is an alarm pheromone of the ant *Iridomyrmex priunosus*, whereas 4-methyl-3-heptanone sounds the alarm for the ant *Atta texana*.

**FIGURE 18.9** Some naturally occurring carbonyl compounds.

Vanillin
(vanilla)

Cinnamaldehyde
(cinnamon)

Civetone

Benzaldehyde
(almonds)

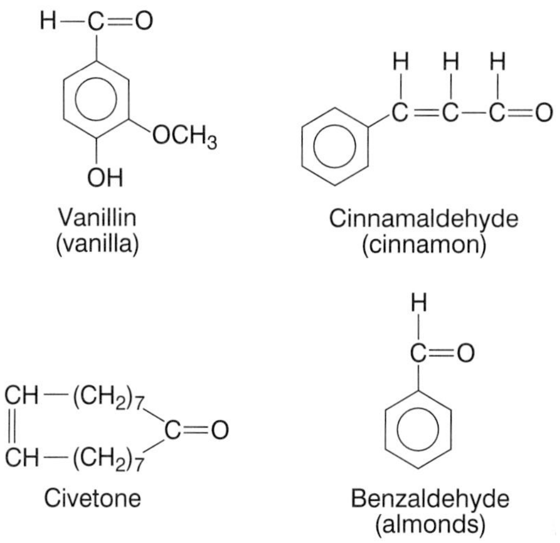

**FIGURE 18.10** A pheromone secreted by a queen bee.

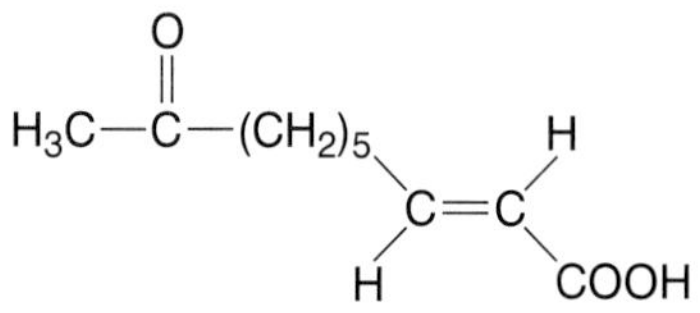

9-Oxo-($E$)-2-decenoic acid

The sense of smell is often vitally important for living things. A dog without a sense of smell would be in deep trouble. Insects rely on a sensitive sense of smell to find members of the opposite sex for mating and to find food, whereas a great white shark can smell blood for incredible distances from the source.

What about human beings? Let's start with mouth odors. An unpleasant odor of the breath can be a sign of upper respiratory infection, tooth cavities, or excessive smoking. A fishy breath odor is associated with liver failure, an odor of ammonia with kidney failure, while the sweet odor of acetone is present in an advanced case of diabetes.

The odor of sweat is due to a malodorous series of organic acids formed by the action of skin bacteria on material secreted by the sweat glands, while fecal odor is due to presence of nitrogen-containing compounds that are formed by decomposition by assorted bacteria in the large intestine.

Our sense of smell gradually fades as we get older, probably because the nerves in the olfactory system break down and die with age. When we smell something, a stimulant molecule reacts with a number of *receptors*, sensory neurons located up in the nose. This chemical reaction causes the sensory neuron to send an electrical signal to the brain. In the brain, the signal is decoded by first classifying it as an odor (as opposed to taste, which is also a chemical phenomenon); classifying the odor as pleasant, unpleasant, or neutral; and finally, identifying the odor as gasoline, eau de skunk, or violets. In the 1980s, the *single-channel theory of smell* was accepted. This theory held that one recepter reacted with a specific odorant. Today,* research data suggest that more than one receptor reacts with a given odorant. Say, for example, that eau de cologne could activate as many as ten receptors; an anosmic (a person with an impaired sense of smell) might only have two functioning receptors and thus could not smell the odor.

The sense of smell we possess is a complex chemical system and, therefore, is open to malfunction. About 9 million Americans suffer from an impaired sense of smell called *anosmia*. Although skunks and paper mills do not bother anos-

mics, they have their share of problems. Just imagine not being able to smell that magnificent aroma of bacon and eggs in the morning or baking bread or a magnolia or rose in bloom. Anosmics cannot know if they have bad breath or if they used too little or too much perfume.

On a more serious and dangerous note, anosmics cannot detect gas leaks or the odor of burning material or spoiled food. In an article written by Phyllis Lehmann,[†] Dr. Robert Henkin of the Georgetown University Medical Center indicated that, in his patients, blocking of the sense of smell can be brought on by influenza, a head injury, and allergies. Other causes of the disease are surgery and pregnancy, but these causes are less common. In addition, drugs in nose drops can cause anosmia, along with certain chemicals such as formaldehyde.

Formaldehyde, the chemical used to preserve biological specimens and as an embalming agent, interferes with our sense of smell by masking the odors of other substances. Thus formaldehyde is often used as a commercial room deodorant in the form of heated paraformaldehyde. The effect of formaldehyde on our sense of smell is temporary, but anosmia can be terribly permanent.

Dr. Henkin treats his anosmic patients with drugs that make nerve transmission easier and more effectively trigger the electrical signal from the odor molecule-receptor neuron combination.

To show that anosmia is no laughing matter, victims of this disease can qualify for disability grants through the Workmen's Compensation Law and the Veterans Administration.

The sense of smell is vital to human beings. Food becomes tasteless when the nose is blocked by a cold. Odor is essential in the perfume, tobacco, and deodorant industries and important in many other industries and facets of life.

*"What Noses Don't Know" *USA Today (Magazine)* 120 (October 1991):16.
†Phyllis Lehmann, "Common Senses: Taste and Smell" *Sciquest* 54 (March 1981): 7–11.

## 18.5
## SOME IMPORTANT CARBONYL COMPOUNDS

Formaldehyde is the main constituent of *formalin*, a solution of 37% formaldehyde in water. Formalin is used both to preserve tissue and as an embalming fluid. Formaldehyde in high concentrations is toxic to humans, but in lower concentrations it is an antiseptic. The drug hexamethylene tetramine (see Sec. 19.2) owes its activity as a urinary antiseptic to formaldehyde.

$$4\,H^+ + (CH_2)_6N_4 + 6\,H_2O \xrightarrow[\substack{\text{in the}\\ \text{urinary tract}}]{pH = 5} 4\,{}^+NH_4 + 6\,CH_2O$$

When formalin is evaporated at a reduced pressure, the solid polymer paraformaldehyde is produced. Formaldehyde can be regenerated from paraformaldehyde by heating.

$$n = 30$$
Paraformaldehyde

Gaseous formaldehyde is too reactive to be stored. Thus both liquid formalin and solid paraformaldehyde represent convenient, commercially available forms of formaldehyde.

Like formaldehyde, volatile acetaldehyde is not easily stored. The problems are a boiling point of only 21°C and high reactivity. To bypass these storage problems, acetaldehyde is converted to the cyclic trimer paraldehyde, which is a stable liquid (bp = 128°C) and a convenient form for the storage of acetaldehyde. To regenerate acetaldehyde, simply distill paraldehyde in acid.

Paraldehyde has been used in medicine as a sleep-inducing agent and in managing alcoholic patients suffering from delirium tremens.

$$3\ CH_3\overset{H}{\underset{}{C}}{=}O \quad \underset{\Delta,\ H_2SO_4}{\overset{H_2SO_4,\ 25°C}{\rightleftharpoons}}$$

Paraldehyde

Acetone is an excellent solvent for lacquers, resins, and paints. In addition, acetone is a fine sterilizing agent in medicine. It is a germicide that is used in skin testing for allergies due to its high volatility combined with its germicidal effect.

In a healthy individual, acetone is found in very low concentration in blood. However, uncontrolled diabetes or starvation creates an abnormal metabolism of fats, which, in turn, leads to a buildup of acetone in the body. Since acetone is volatile, it travels to the lungs, where it is expelled as "acetone breath." By the time acetone can be detected in the breath as opposed to the urine, the diabetic is very sick and needs immediate hospital treatment.

Just as the reactions of the alkenes are characterized by breaking of the $p—p$ pi bond and the formation of two sigma bonds, so addition reactions of carbonyl compounds proceed with the breaking of the $p—p$ pi bond of the double bond of the carbonyl group and the formation of two sigma bonds, one bonded to the carbon of the carbonyl and the other to the oxygen. Thus

## 18.6 REPRESENTATIVE REACTIONS OF THE CARBONYL GROUP

### Addition

$$\overset{\pi}{\underset{\sigma}{C}}{=}O \ + \ A—B \ \longrightarrow \ —\overset{|}{\underset{A}{C}}—\overset{|}{\underset{B}{O}}$$

**Reduction**   Aldehydes and ketones are reduced to primary and secondary alcohols, respectively, when reacted with sodium borohydride (NaBH$_4$) and the resulting intermediate compound is hydrolyzed with an aqueous solution of a mineral acid.

Carbonyl compound → Break the pi bond of the double bond → Add two hydrogens → Primary or secondary alcohol

In general,

$$R—\overset{O}{\overset{||}{C}}—H \quad \xrightarrow[\text{2. } H_2O,\ H^+]{\text{1. NaBH}_4} \quad R—\overset{H}{\underset{H}{\overset{|}{C}}}—OH$$

$$R\overset{\overset{\displaystyle O}{\|}}{-}C-R' \quad \xrightarrow[\text{2. H}_2\text{O, H}^+]{\text{1. NaBH}_4} \quad R\overset{\overset{\displaystyle H}{|}}{\underset{\underset{\displaystyle R'}{|}}{C}}-OH$$

Examples:

$$CH_3\overset{\overset{\displaystyle O}{\|}}{\underset{\underset{\displaystyle H}{|}}{C}}-H \quad \xrightarrow[\text{2. H}_2\text{O, H}^+]{\text{1. NaBH}_4} \quad CH_3\overset{\overset{\displaystyle H}{|}}{\underset{\underset{\displaystyle H}{|}}{C}}-OH$$

Acetaldehyde                    Ethyl alcohol

$$CH_3\overset{\overset{\displaystyle O}{\|}}{-}C-CH_2CH_3 \quad \xrightarrow[\text{2. H}_2\text{O, H}^+]{\text{1. NaBH}_4} \quad CH_3\overset{\overset{\displaystyle H}{|}}{\underset{\underset{\displaystyle CH_2CH_3}{|}}{C}}-OH$$

Methyl ethyl ketone

*sec*-Butyl alcohol

**Addition of Alcohols**   An aldehyde reacts with 1 mol of alcohol, in the presence of a catalytic amount of hydrochloric acid, to give an unstable *hemiacetal*; with 2 mol of alcohol, a stable *acetal* is produced. In general,

$$R\overset{\overset{\displaystyle H}{|}}{-}C{=}O + R'OH \xrightarrow{\text{H}^+} \left[ R\overset{\overset{\displaystyle OH}{|}}{\underset{\underset{\displaystyle H}{|}}{C}}-OR' \right] \xrightarrow{R'OH,\ \text{H}^+} R\overset{\overset{\displaystyle H}{|}}{\underset{\underset{\displaystyle OR'}{|}}{C}}-OR' + H_2O$$

Unstable hemiacetal              Stable acetal

Example:

$$CH_3\overset{\overset{\displaystyle H}{|}}{-}C{=}O + C_2H_5OH \xrightarrow{\text{H}^+} \left[ CH_3\overset{\overset{\displaystyle H}{|}}{\underset{\underset{\displaystyle OC_2H_5}{|}}{C}}-OH \right] \xrightarrow{CH_3CH_2OH,\ \text{H}^+} CH_3\overset{\overset{\displaystyle H}{|}}{\underset{\underset{\displaystyle OC_2H_5}{|}}{C}}-OC_2H_5 + H_2O$$

Common name:  acetaldehyde diethyl acetal
IUPAC name:  1,1-diethoxyethane

You also can think of a hemiacetal as a *gem*-hydroxy ether, whereas an acetal is a *gem*-diether in that both alkoxy groups are bonded to the same carbon.

Ketones react with alcohols in a similar way to give unstable hemiacetals[*] with 1 mol of alcohol and acetals with 2 mol of alcohol.[†] In general,

$$R\overset{\overset{\displaystyle R'}{|}}{-}C{=}O + R''OH \xrightarrow{\text{H}^+} \left[ R\overset{\overset{\displaystyle R'}{|}}{\underset{\underset{\displaystyle OH}{|}}{C}}-OR'' \right] \xrightarrow{R''OH,\ \text{H}^+} R\overset{\overset{\displaystyle R'}{|}}{\underset{\underset{\displaystyle OR''}{|}}{C}}-OR'' + H_2O$$

Unstable hemiacetal              Stable acetal

[*]Any hemiacetal prepared from a ketone was previously known as a *hemiketal*, while the corresponding acetal was called a *ketal*.
[†]This chapter will treat any acetal derived from a ketone (a ketal) as stable. In actuality, only cyclic acetals derived from ketones are stable. For more information, refer to Section 19.2.

Example:

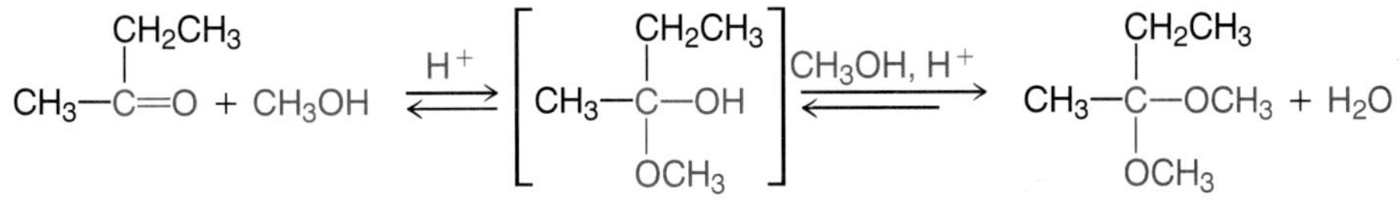

$$
\underset{\substack{|\\ CH_3-C=O}}{\overset{\substack{CH_2CH_3 \\ |}}{}} + CH_3OH \quad \underset{}{\overset{H^+}{\rightleftharpoons}} \quad \left[ \underset{\substack{|\\ OCH_3}}{\overset{\substack{CH_2CH_3 \\ |}}{CH_3-C-OH}} \right] \underset{}{\overset{CH_3OH,\ H^+}{\rightleftharpoons}} \quad \underset{\substack{|\\ OCH_3}}{\overset{\substack{CH_2CH_3 \\ |}}{CH_3-C-OCH_3}} + H_2O
$$

Common name: methyl ethyl ketone dimethyl acetal
IUPAC name: 2,2-dimethoxybutane

To name an acetal using the common system of nomenclature, name the aldehyde or ketone from which the acetal is derived. This is followed by the prefix *di-* and the name of the alkyl group of the alcohol employed to form the acetal. The separate word *acetal* completes the name. For example:

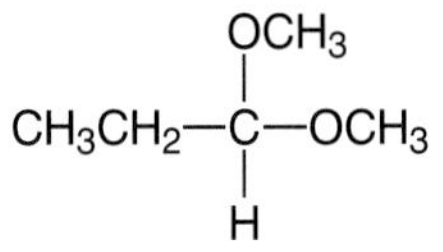

$$
\underset{\substack{|\\ H}}{\overset{\substack{OCH_3 \\ |}}{CH_3CH_2-C-OCH_3}}
$$

Propionaldehyde dimethyl acetal

To name an acetal using the IUPAC system of nomenclature, name the compound as a disubstituted alkane.

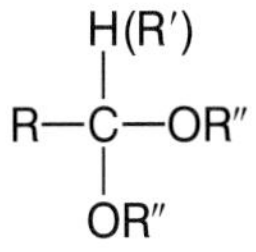

$$
\underset{\substack{|\\ OR''}}{\overset{\substack{H(R') \\ |}}{R-C-OR''}}
$$

The —OR group is called an *alkoxy group*. To convert an alkyl group into an alkoxy group, drop the *-yl* suffix, and substitute an *-oxy* suffix (see Sec. 16.6). For example:

—CH$_3$ (methyl)     becomes     —OCH$_3$ (methoxy)
—CH$_2$CH$_3$ (ethyl)     becomes     —OCH$_2$CH$_3$ (ethoxy)

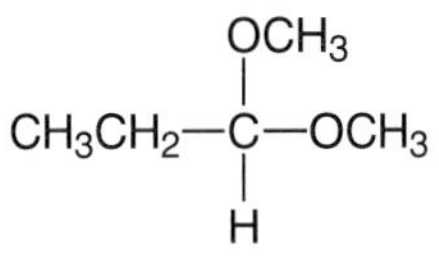

$$
\underset{\substack{|\\ H}}{\overset{\substack{OCH_3 \\ |}}{CH_3CH_2-C-OCH_3}}
$$

1,1-Dimethoxypropane

Note that the reaction is reversible. Thus *LeChâtelier's principle* applies, and good yields of acetal are obtained when an excess of alcohol is used.

Since the reaction is reversible, reacting an acetal with an excess of water in acid solution produces the corresponding carbonyl compound and alcohol.

$$
\text{carbonyl compound + alcohol} \quad \underset{}{\overset{H^+}{\rightleftharpoons}} \quad \text{acetal + water}
$$
$$
\text{(Excess)}
$$

$$
\text{acetal + H}_2\text{O} \quad \underset{}{\overset{H^+}{\rightleftharpoons}} \quad \text{carbonyl compound + alcohol}
$$
$$
\text{(Excess)}
$$

In general,

$$
\underset{\substack{|\\ OR''}}{\overset{\substack{H(R') \\ |}}{R-C-OR''}} + H_2O \quad \underset{}{\overset{H^+}{\rightleftharpoons}} \quad \underset{}{\overset{\substack{H(R') \\ |}}{R-C=O}} + 2\ R''OH
$$
$$
\text{(Excess)}
$$

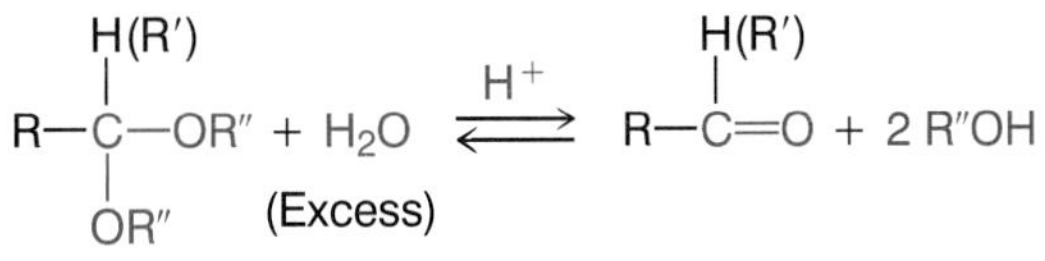

Example:

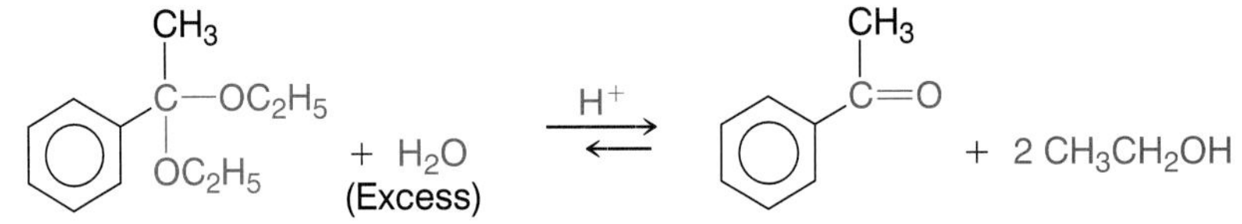

Acetophenone diethyl acetal          Acetophenone

## Oxidation

Aldehydes are readily oxidized; ketones, on the other hand, are resistant to oxidation. A simple qualitative test to distinguish an aldehyde from a ketone is the *Tollens test*. When treated with a silver ammonia complex in a clean test tube, aldehydes react to produce a silver mirror coating on the interior of the tube. Ketones do not react. In general,

$$R-\overset{O}{\overset{\|}{C}}-H + 2\ ^+Ag(NH_3)_2 + 3\ ^-OH \longrightarrow R-\overset{O}{\overset{\|}{C}}-O^- + 4\ NH_3 + 2\ Ag + 2\ H_2O$$

Example:

$$H-\overset{O}{\overset{\|}{C}}-H + 2\ ^+Ag(NH_3)_2 + 3\ ^-OH \longrightarrow H-\overset{O}{\overset{\|}{C}}-O^- + 4\ NH_3 + 2\ Ag + 2\ H_2O$$

---

**SOLVED PROBLEM 18.3**

Classify the following compounds as hemiacetal, acetal, or unrelated. Then characterize each hemiacetal or acetal as derived from an aldehyde or a ketone.

a. 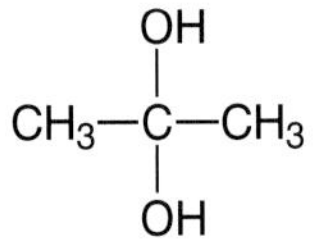   $CH_3-\overset{OH}{\underset{OH}{\overset{|}{\underset{|}{C}}}}-CH_3$

b. 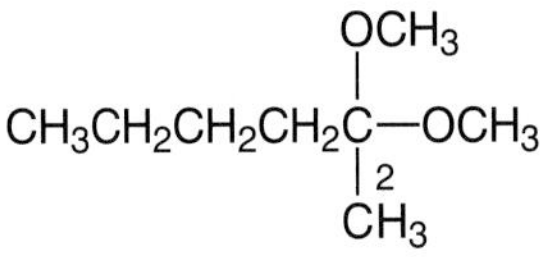   $CH_3CH_2CH_2CH_2\overset{OCH_3}{\underset{\underset{CH_3}{|\,2}}{\overset{|}{C}}}-OCH_3$

**SOLUTION**

a.   Because the central carbon is bonded to two hydroxyl groups, the compound is unrelated (neither a hemiacetal nor an acetal).

b.   Note that carbon 2 is bonded to two methoxyl (OCH₃) groups. Thus the compound is an acetal. Observe that carbon 2 is the original carbonyl carbon, and besides the two methoxyl groups, the carbon is bonded to two alkyl groups. Thus the acetal is derived from a ketone.

---

**EXERCISE 18.3**

Classify the following compounds as hemiacetal, acetal, or unrelated. Then characterize each hemiacetal or acetal as derived from an aldehyde or ketone.

a.   $CH_3CH_2CH_2CH(OCH_3)_2$

b.

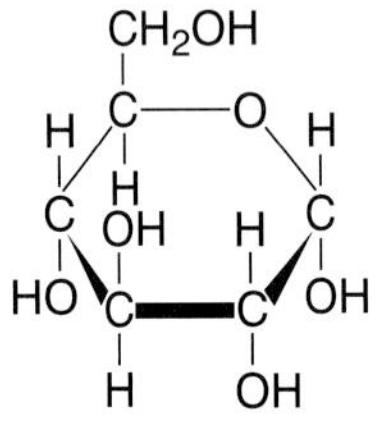

c.   $CH_3OCH_2CH_2OH$

Complete the following equations by drawing a structural formula for each organic product and writing a molecular formula for each inorganic product when appropriate. If no reaction occurs, write "no reaction."

a. 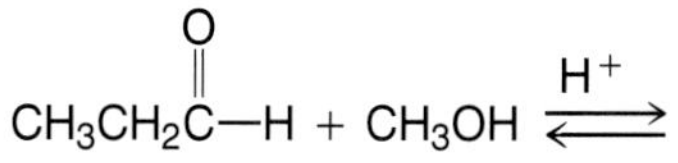 $\underset{\displaystyle \overset{O}{\|}}{CH_3CH_2C}-H + CH_3OH \underset{}{\overset{H^+}{\rightleftharpoons}}$

b. 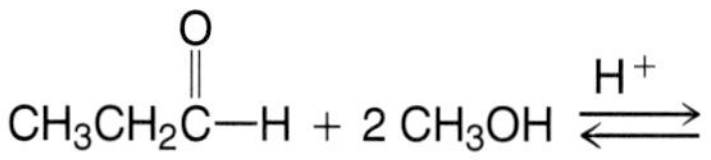 $\underset{\displaystyle \overset{O}{\|}}{CH_3CH_2C}-H + 2\ CH_3OH \underset{}{\overset{H^+}{\rightleftharpoons}}$

c. 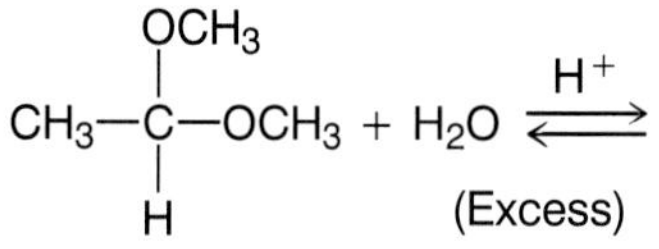 $\underset{\displaystyle \underset{H}{|}}{CH_3-\overset{\displaystyle \overset{OCH_3}{|}}{C}-OCH_3} + H_2O \underset{}{\overset{H^+}{\rightleftharpoons}}$
(Excess)

d. 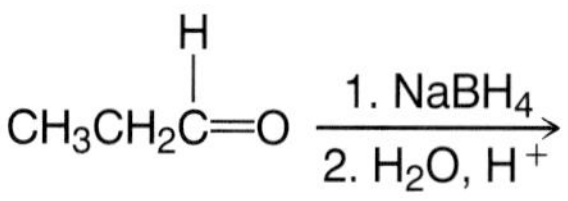 $CH_3CH_2\overset{\displaystyle \overset{H}{|}}{C}{=}O \xrightarrow[\text{2. H}_2\text{O, H}^+]{\text{1. NaBH}_4}$

(Neglect inorganic products.)

## SOLUTION

a. An important clue to deduce the product of this reaction is the fact that only 1 mol of alcohol is reacting with the 1 mol of the aldehyde. The product molecule is therefore a hemiacetal. Thus, to formulate the product, we proceed as follows:

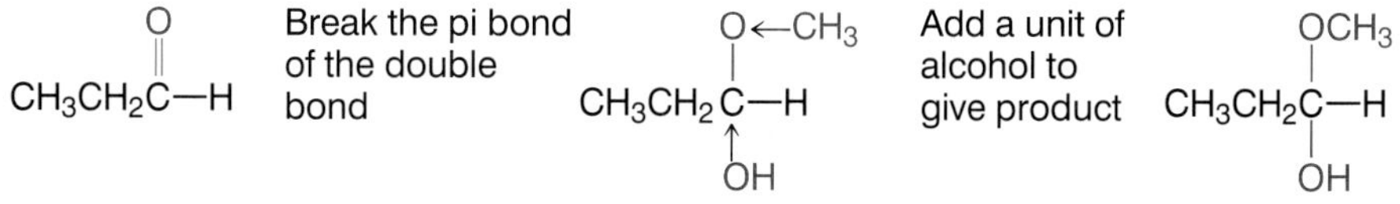

$\underset{\displaystyle \overset{O}{\|}}{CH_3CH_2C}-H$  Break the pi bond of the double bond  $CH_3CH_2\overset{\displaystyle \overset{O \leftarrow CH_3}{|}}{\underset{\displaystyle \underset{OH}{|}}{C}}-H$  Add a unit of alcohol to give product  $CH_3CH_2\overset{\displaystyle \overset{OCH_3}{|}}{\underset{\displaystyle \underset{OH}{|}}{C}}-H$

b. Since 2 mol of alcohol is reacting with 1 mol of the aldehyde, the product is an acetal.

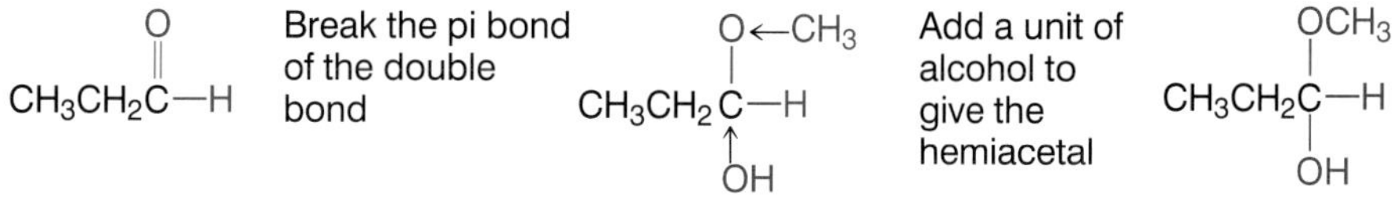

$\underset{\displaystyle \overset{O}{\|}}{CH_3CH_2C}-H$  Break the pi bond of the double bond  $CH_3CH_2\overset{\displaystyle \overset{O \leftarrow CH_3}{|}}{\underset{\displaystyle \underset{OH}{|}}{C}}-H$  Add a unit of alcohol to give the hemiacetal  $CH_3CH_2\overset{\displaystyle \overset{OCH_3}{|}}{\underset{\displaystyle \underset{OH}{|}}{C}}-H$

Remove the —OH group and combine it with H from a second molecule of alcohol to give water. Finally, add $OCH_3$ to structure I to give product.

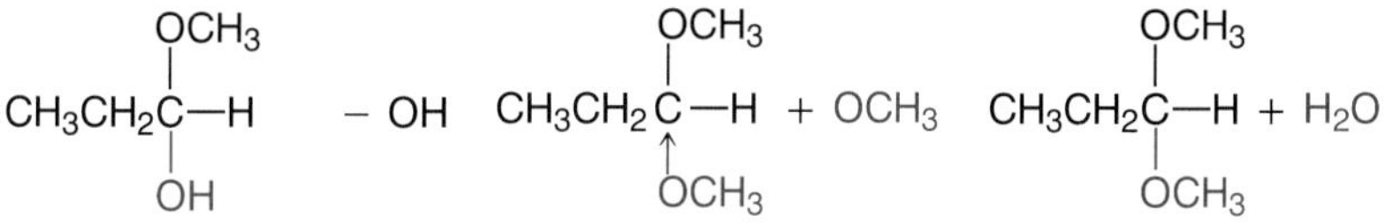

$CH_3CH_2\overset{\displaystyle \overset{OCH_3}{|}}{\underset{\displaystyle \underset{OH}{|}}{C}}-H \quad - OH \quad CH_3CH_2\overset{\displaystyle \overset{OCH_3}{|}}{\underset{\displaystyle \underset{OCH_3}{|}}{C}}-H + OCH_3 \quad CH_3CH_2\overset{\displaystyle \overset{OCH_3}{|}}{\underset{\displaystyle \underset{OCH_3}{|}}{C}}-H + H_2O$

I

c. Since an acetal is reacting with excess water, the corresponding carbonyl compound and alcohol are produced. To formulate the products:

*Step 1.* Remove the two methoxyl groups from the acetal:

$CH_3-\overset{\displaystyle \overset{OCH_3}{|}}{\underset{\displaystyle \underset{H}{|}}{C}}-OCH_3 \quad - 2\ OCH_3 \quad CH_3-\overset{\displaystyle \underset{H}{|}}{C}$

II

*Step 2.* Combine the two methoxyls with two hydrogens from water to form the alcohol ($2\ OCH_3 + 2\ H$ gives $2\ CH_3OH$).

*Step 3.* Add =O to the bond-poor carbon in structure II to formulate the carbonyl compound:

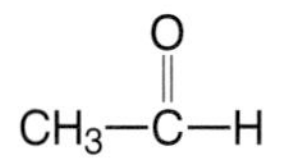

$$\overset{\displaystyle O}{\underset{\displaystyle}{CH_3-\overset{\|}{C}-H}}$$

d. We formulate the product as follows:

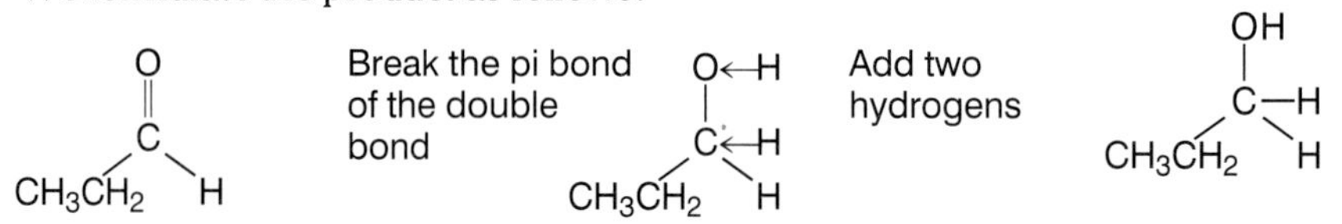

$$\underset{CH_3CH_2}{\overset{O}{\diagdown}}\overset{\|}{C}\diagup H \quad \xrightarrow[\text{bond}]{\text{Break the pi bond}\atop\text{of the double}} \quad \underset{CH_3CH_2}{\overset{O\leftarrow H}{\diagdown}}\overset{\cdot}{C}\leftarrow H \diagup H \quad \xrightarrow[\text{hydrogens}]{\text{Add two}} \quad \underset{CH_3CH_2}{\overset{OH}{\diagup}}\overset{|}{C}\diagdown H \diagup H$$

---

## EXERCISE 18.4

Complete the following equations by drawing a structural formula for each organic product and writing a molecular formula or formula unit for each inorganic product when appropriate. If no reaction occurs, write "no reaction."

a. 2-methylpropanal + $CH_3OH$ $\underset{}{\overset{H^+}{\rightleftharpoons}}$

b. 2-methylpropanal + 2 $CH_3OH$ $\underset{}{\overset{H^+}{\rightleftharpoons}}$

c. cyclohexanone $\xrightarrow[\text{2. } H_2O,\ H^+]{\text{1. } NaBH_4}$

d. 2-butanone + $^+Ag(NH_3)_2$ + $^-OH$

---

# ▶ CHAPTER ACCOMPLISHMENTS

## 18.1  Structure and Bonding
☐ List the orbitals that make up the carbon-oxygen double bond.
☐ Explain why the carbon-oxygen double bond is polar.

## 18.2  Nomenclature
☐ Name the following compounds using the common system of nomenclature.
   a. $CH_3CH_2CHO$
   b. $CH_3CH_2CH_2CH_2CHO$
   c. $CH_2O$
☐ Draw a structural formula for
   a. acetone.
   b. ethyl phenyl ketone.
   c. benzaldehyde.
   d. salicylaldehyde.
   e. cyclohexanone.
   f. acetophenone.

☐ Supply both a common and an IUPAC name for

a. 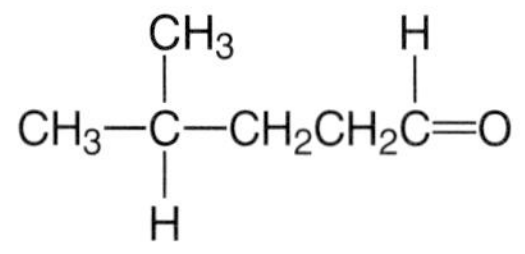 $CH_3-\overset{\displaystyle CH_3}{\underset{\displaystyle H}{\overset{|}{\underset{|}{C}}}}-CH_2CH_2\overset{\displaystyle H}{\overset{|}{C}}=O$

b. 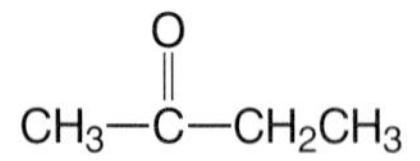 $CH_3-\overset{\displaystyle O}{\overset{\|}{C}}-CH_2CH_3$

## 18.3  Physical Properties of Carbonyl Compounds
☐ Draw a dipole-dipole interaction between two molecules of dimethyl ketone.
☐ Explain why propionaldehyde boils at a lower temperature than *n*-propyl alcohol, even though the molecular weights of the different molecules are almost the same.

## 18.4  Naturally Occurring Carbonyl Compounds
☐ Draw a structural formula of
   a. civetone.
   b. vanillin.

☐ Name the functional group represented in an organic compound when the prefix *-oxo* is used in IUPAC nomenclature.

## 18.5 Some Important Carbonyl Compounds

☐ Give two ways in which formalin is used.
☐ Draw a structural formula for paraldehyde.
☐ Explain why paraldehyde can be used as a convenient form for the storage of acetaldehyde.

## 18.6 Representative Reactions of the Carbonyl Group

☐ Complete the following equations by supplying a structural formula for each organic product and a molecular formula or formula unit for each inorganic product when applicable. If no reaction occurs, write "no reaction."

a.
$$CH_3-\overset{\overset{\displaystyle O}{\|}}{C}-H \xrightarrow[\text{2. } H_2O,\ H^+]{\text{1. } NaBH_4}$$

b.
$$CH_3-\overset{\overset{\displaystyle O}{\|}}{C}-CH_2CH_3 \xrightarrow[\text{2. } H_2O,\ H^+]{\text{1. } NaBH_4}$$

c.
$$CH_3-\overset{\overset{\displaystyle O}{\|}}{C}-H + CH_3CH_2OH \overset{H^+}{\rightleftharpoons}$$

d.
$$CH_3-\overset{\overset{\displaystyle O}{\|}}{C}-H + 2\ CH_3CH_2OH \overset{H^+}{\rightleftharpoons}$$

e.
$$CH_3-\overset{\overset{\displaystyle O}{\|}}{C}-CH_2CH_3 + CH_3OH \overset{H^+}{\rightleftharpoons}$$

f.
$$CH_3-\overset{\overset{\displaystyle O}{\|}}{C}-CH_2CH_3 + 2\ CH_3OH \overset{H^+}{\rightleftharpoons}$$

g.
$$C_6H_5-\overset{\overset{\displaystyle CH_3}{|}}{\underset{\underset{\displaystyle OCH_2CH_3}{|}}{C}}-OCH_2CH_3 + H_2O \overset{H^+}{\rightleftharpoons}$$

h.
$$CH_3-\overset{\overset{\displaystyle H}{|}}{\underset{\underset{\displaystyle OCH_2CH_3}{|}}{C}}-OCH_2CH_3 + H_2O \overset{H^+}{\rightleftharpoons}$$

i.
$$H-\overset{\overset{\displaystyle O}{\|}}{C}-H + {}^+Ag(NH_3)_2 + {}^-OH \longrightarrow$$

j.
$$CH_3-\overset{\overset{\displaystyle O}{\|}}{C}-CH_3 + {}^+Ag(NH_3)_2 + {}^-OH \longrightarrow$$

---

## ► KEY TERMS

*acetone* (18.2)
*acyl group* (18.2)
*formalin* (18.5)
*hemiacetal* (18.6, 19.2)
*acetal* (18.6, 19.2)
*alkoxy group* (18.6)

*hemiketal* (18.6, 19.2)
*ketal* (18.6, 19.2)
*LeChâtelier's principle* (18.6, 19.2, 20.5, 23.6)
*Tollens test* (18.6)

*receptor* (Biochemical Boxed Reading 18.1)
*single-channel theory of smell* (Biochemical Boxed Reading 18.1)
*anosmia* (Biochemical Boxed Reading 18.1)

---

## ► PROBLEMS

1. Name the following compounds using the common and IUPAC systems of nomenclature.

a.
$$CH_3CH_2\overset{\overset{\displaystyle H}{|}}{C}=O$$

b.
structure: phenyl ring attached to $\overset{\overset{\displaystyle O}{\|}}{C}-CH_3$

c.
$$CH_3CH_2\overset{\overset{\displaystyle CH_3}{|}}{\underset{\underset{\displaystyle H}{|}}{C}}-\overset{\overset{\displaystyle O}{\|}}{C}-H$$

d.
cyclobutanone ring with Cl substituent, =O        (IUPAC only)

e.
$$Cl-\overset{\overset{\displaystyle Cl}{|}}{\underset{\underset{\displaystyle Cl}{|}}{C}}-\overset{\overset{\displaystyle H}{|}}{C}=O$$

f.
$$CH_3-\overset{\overset{\displaystyle H}{|}}{C}=\overset{\overset{\displaystyle H}{|}}{C}-\overset{\overset{\displaystyle H}{|}}{C}=O$$
(IUPAC only)

g.
$$CH_3-\overset{\overset{\displaystyle O}{\|}}{C}-\overset{\overset{\displaystyle CH_3}{|}}{\underset{\underset{\displaystyle H}{|}}{C}}-CH_3$$

h.
structure: $CH_2=CH-CH_2-CH_2-C(=O)-CH_3$        (IUPAC only)

i.

j.

(IUPAC only)

2. Draw a structural formula for
   a. 3-hexanone.
   b. phenyl cyclopentyl ketone.
   c. salicylaldehyde.
   d. dimethyl ketone dimethyl acetal.
   e. formaldehyde diethyl acetal.
   f. diisobutyl ketone.
   g. paraldehyde.
   h. 1-phenyl-2-pentanone.
   i. valerophenone.

3. a. Given that toluene is

   draw a structural formula for o-tolualdehyde.
   b. In 1985, the U.S. State Department accused the Soviet Union of tracing the movements of personnel in the U.S. embassy in Moscow by means of a dust called 5-(4-nitrophenyl)-2,4-pentadienal. Draw a structural formula for this alleged tracing dust.

4. Draw a structural formula for each of the following compounds.
   a. propiophenone
   b. di-n-propyl ketone
   c. 3-bromo-2-methylheptanal
   d. 3-ethyl-2-hexanone
   e. cyclopentanone
   f. butyrophenone

5. List the following compounds in order of increasing boiling point.
   a. acetone, isopropyl alcohol, butane
   b. hexane, 1-pentanol, pentanal

6. Draw a dipole-dipole interaction using two molecules of acetone as an illustration.

7. Identify the following as a hemiacetal derived from an aldehyde, an acetal derived from an aldehyde, a hemiacetal derived from a ketone, or an acetal derived from a ketone.
   a.

   b.

c.

d.

8. Which aldehyde does not produce a primary alcohol on reduction?

9. Draw a structural formula for each of the following pheromones.
   a. trans-2-hexenal (myrmicine ant alarm pheromone)
   b. hendecanal (greater wax moth sex pheromone)
   c. 4-methyl-3-heptanone (component of alarm pheromone) of leaf cutting ant

10. When providing the IUPAC name of the compound 2-butenal, the aldehyde group is not assigned a specific number. Yet in the compound 2-buten-1-ol the hydroxyl group is assigned a specific number—the number 1. Explain.

11. Provide a common name for
    a. 1,1-dimethoxypropane.
    b. 2,2-diethoxypropane.

12. Which of the following compounds represents a hemiacetal?
    a. 2-methoxy-2-pentanol
    b. 1,1-dimethoxybutane
    c. 1,1-propanediol

13. Write an equation representing the acid hydrolysis of the following compounds.
    a. butyraldehyde dimethyl acetal
    b. 1,1-diethoxycyclopentane

14. Complete the following equations by supplying a structural formula for the appropriate reactant.

a. $? + 2\,CH_3OH \underset{}{\overset{H^+}{\rightleftharpoons}} CH_3-\overset{\overset{\textstyle H}{|}}{\underset{\underset{\textstyle OCH_3}{|}}{C}}-OCH_3$

b. $? \xrightarrow[\text{2. } H_2O,\, H^+]{\text{1. } NaBH_4} \text{2-pentanol}$

c. $? + CH_3OH \underset{}{\overset{H^+}{\rightleftharpoons}} CH_3-\overset{\overset{\textstyle OH}{|}}{\underset{\underset{\textstyle H}{|}}{C}}-OCH_3$

d. $? + {}^+Ag(NH_3)_2 + {}^-OH \rightarrow CH_3CH_2\overset{\overset{\textstyle O}{\|}}{C}-O^- + Ag + H_2O + NH_3$

# CHAPTER 19

# Aldehydes and Ketones (II)

## SYNTHESIS AND CHEMICAL PROPERTIES

In general,

$$R-CH_2OH \xrightarrow[CH_2Cl_2]{PCC} R-\overset{\overset{\displaystyle O}{\|}}{C}-H$$

Since the oxidizing agent pyridinium chlorochromate (PCC) is an anhydrous reagent, oxidation stops at the aldehyde stage. If water were present, the aldehyde would be converted to a carboxylic acid. In general,

$$R-\overset{\overset{\displaystyle H}{|}}{\underset{\underset{\displaystyle OH}{|}}{C}}-R' \xrightarrow[H_2SO_4]{Na_2Cr_2O_7} R-\overset{\overset{\displaystyle O}{\|}}{C}-R' + Cr^{3+}$$

Examples:

$$\text{(cyclopentyl)}-CH_2OH \xrightarrow[CH_2Cl_2]{PCC} \text{(cyclopentyl)}-\overset{\overset{\displaystyle H}{|}}{C}=O$$

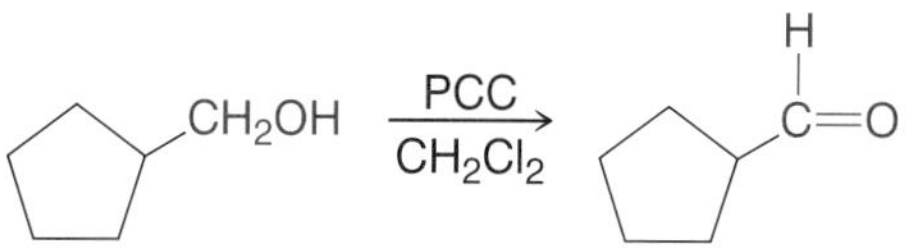

$$CH_3CH(OH)CH_2CH_3 \xrightarrow[H_2SO_4]{Na_2Cr_2O_7} CH_3-\overset{\overset{\displaystyle O}{\|}}{C}-CH_2CH_3$$

In general,

$$R-C\equiv CH + H_2O \xrightarrow[HgSO_4]{H_2SO_4} R-\overset{\overset{\displaystyle O}{\|}}{C}-CH_3$$

## 19.1 SYNTHESIS OF ALDEHYDES AND KETONES

### Oxidation of Alcohols

Primary alcohols are readily oxidized to aldehydes with pryidinium chlorochromate (PCC), whereas secondary alcohols are converted to ketones with sodium dichromate in sulfuric acid (see Secs. 14.4 and 15.2).

### Hydrolysis of Alkynes

Reacting an alkyne with water in the presence of mercuric sulfate and sulfuric acid gives the corresponding carbonyl compound (see Sec. 8.7).

347

Example:

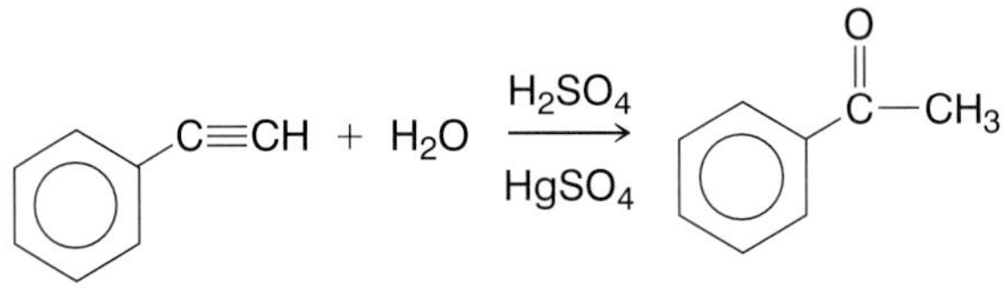

$$\text{C}_6\text{H}_5\text{—C}{\equiv}\text{CH} + \text{H}_2\text{O} \xrightarrow[\text{HgSO}_4]{\text{H}_2\text{SO}_4} \text{C}_6\text{H}_5\text{—C(=O)—CH}_3$$

## Ozonolysis of Alkenes

Reacting an alkene with ozone affords the corresponding ozonide. In general,

$$\begin{array}{c}\text{R}\\\text{H}\end{array}\text{C}{=}\text{C}\begin{array}{c}\text{R}'\\\text{R}''\end{array} + \text{O}_3 \longrightarrow \text{Ozonide}$$

Ozonide

Since ozonides are unstable compounds and could explode, they are not isolated; instead, the ozonides are decomposed with metallic zinc and water to give the corresponding carbonyl compound (see Sec. 6.3).

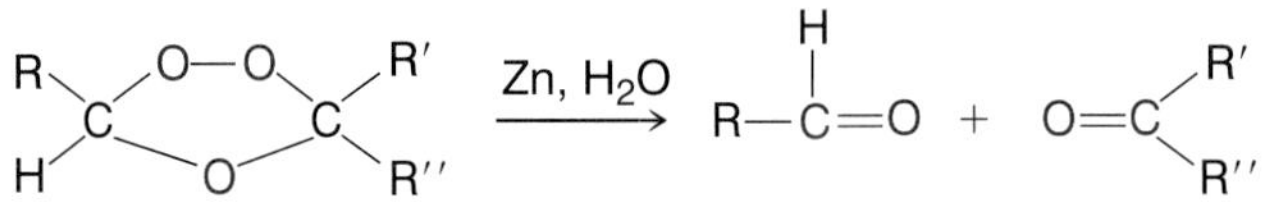

$$\text{ozonide} \xrightarrow{\text{Zn, H}_2\text{O}} \text{R—C(H)=O} + \text{O=C}\begin{array}{c}\text{R}'\\\text{R}''\end{array}$$

Example:

$$\text{CH}_2{=}\overset{\text{H}}{\text{C}}\text{CH}_2\overset{\text{H}}{\text{C}}{=}\text{CH}_2 \xrightarrow{\text{O}_3 \text{ (excess)}} \text{(diozonide)}$$

$$\text{(diozonide)} \xrightarrow{\text{Zn, H}_2\text{O}} 2\ \text{CH}_2{=}\text{O} + \text{O}{=}\overset{\text{H}}{\text{C}}\text{CH}_2\overset{\text{H}}{\text{C}}{=}\text{O}$$

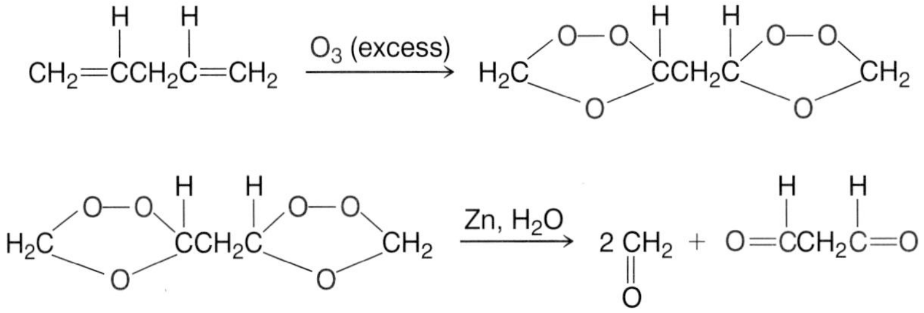

Putting both equations together, we get

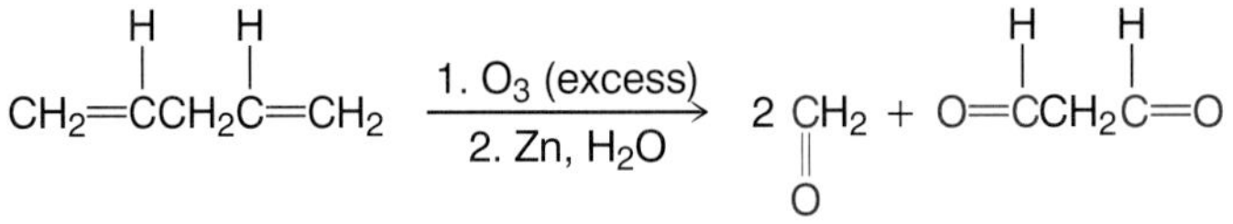

$$\text{CH}_2{=}\overset{\text{H}}{\text{C}}\text{CH}_2\overset{\text{H}}{\text{C}}{=}\text{CH}_2 \xrightarrow[\text{2. Zn, H}_2\text{O}]{\text{1. O}_3 \text{ (excess)}} 2\ \text{CH}_2{=}\text{O} + \text{O}{=}\overset{\text{H}}{\text{C}}\text{CH}_2\overset{\text{H}}{\text{C}}{=}\text{O}$$

## Friedel-Crafts Acylation

Aromatic ketones are prepared by treating an aromatic hydrocarbon with an acyl chloride

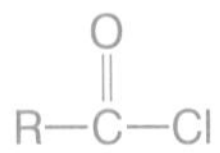

$$\text{R—C(=O)—Cl}$$

Acyl chloride

or an acid anhydride

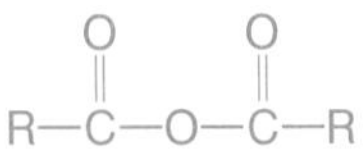

$$\text{R—C(=O)—O—C(=O)—R}$$

Acid anhydride

in the presence of aluminum chloride (see Sec. 10.1).

In general,

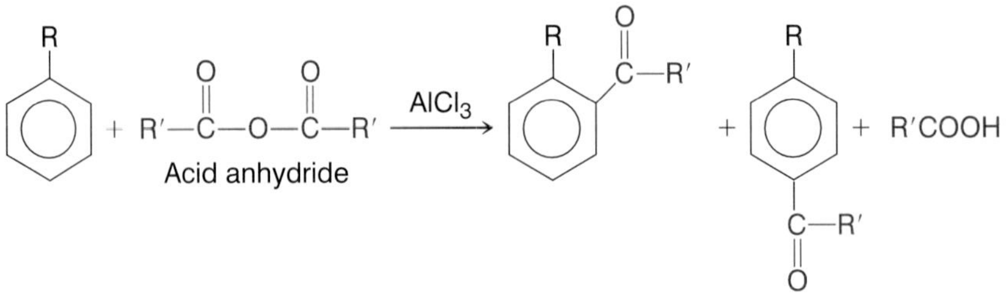

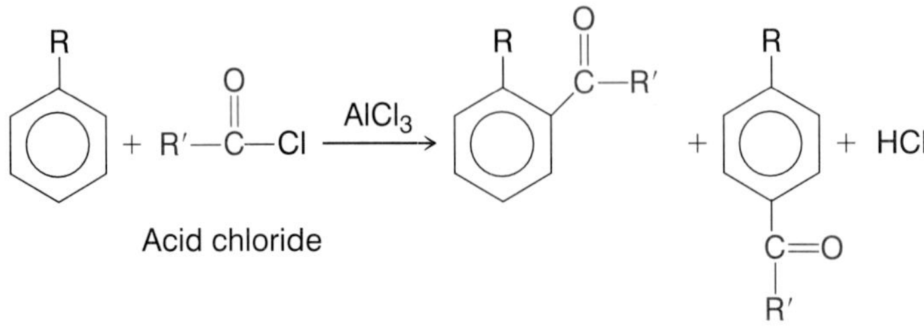

$$\text{Ar(R)} + \text{R'—C(=O)—O—C(=O)—R'} \xrightarrow{\text{AlCl}_3} \text{product} + \text{product} + \text{R'COOH}$$

Acid anhydride

$$\text{Ar(R)} + \text{R'—C(=O)—Cl} \xrightarrow{\text{AlCl}_3} \text{product} + \text{product} + \text{HCl}$$

Acid chloride

Example:

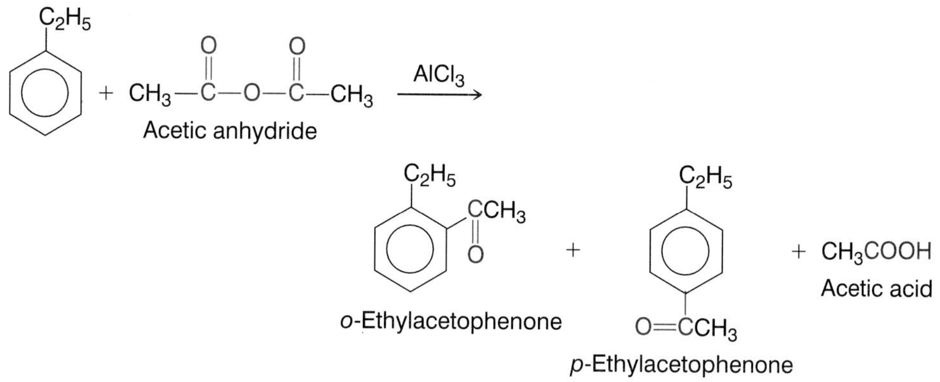
$C_2H_5$
+ $CH_3-C(=O)-O-C(=O)-CH_3$
$AlCl_3$
Acetic anhydride
$C_2H_5$
$CCH_3$
O
+
$C_2H_5$
+ $CH_3COOH$
Acetic acid
o-Ethylacetophenone
$O=CCH_3$
p-Ethylacetophenone

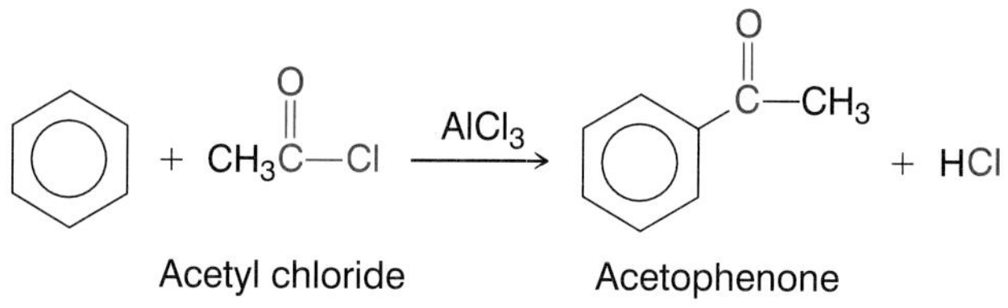
+ $CH_3C(=O)-Cl$
$AlCl_3$
$C(=O)-CH_3$
+ HCl
Acetyl chloride
Acetophenone

## EXERCISE 19.1

Prepare the following compounds. You may use any organic or inorganic reagents necessary.

a. 1-octanal from 1-octene

b. pentanedial [$H-\underset{\underset{O}{\|}}{C}-(CH_2)_3-\underset{\underset{O}{\|}}{C}-H$] from cyclopentanol

## 19.2 ADDITION REACTIONS OF THE CARBONYL GROUP

### Introduction and General Mechanism of Addition

Just as most of the addition reactions of alkenes commence with an electrophilic species attacking the double bond (electrophilic attack), so addition reactions of carbonyl compounds proceed with the attack of a nucleophile on the double bond of the carbonyl group.

To understand why nucleophilic attack on the carbonyl group takes place, consider two resonance contributing structures of the carbonyl group:

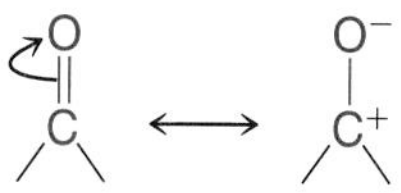
O
C
$O^-$
$C^+$

Because oxygen is more electronegative than carbon, a shift of both sigma and pi electron density toward the oxygen atom takes place. A more realistic structure of the carbonyl group is the dipole structure (see Sec. 18.1):

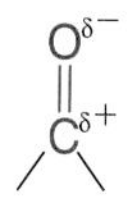
$O^{\delta-}$
$C^{\delta+}$

The partial positive charge on carbon is the driving force for nucleophilic addition.

**Envision the Reaction**

### Mechanism of the Addition Reaction of a Nucleophile ($^-$Nc) with a Carbonyl Compound

First step:

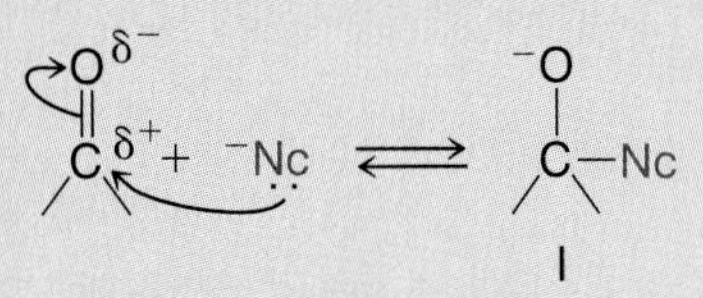

Second step:

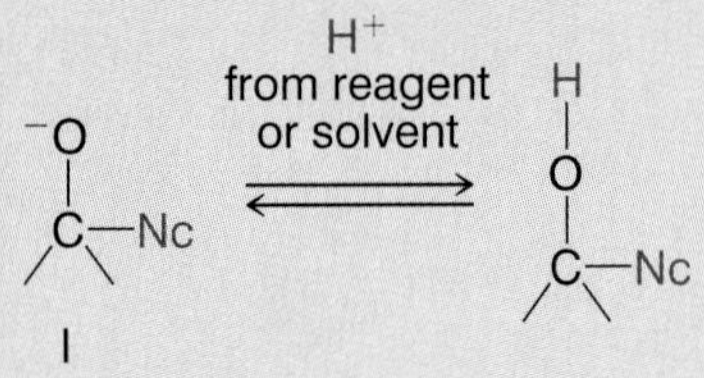

It is the partially positively charged carbon that attracts the electron-rich nucleophilic species $^-$Nc (see Sec. 12.1) to give the intermediate I. The last step of this general mechanism involves the negatively charged oxygen ion I accepting a proton from the reagent or solvent to produce product.

These addition reactions are often catalyzed by acid (represented by a proton) that adds to a pair of unshared electrons on the oxygen of the carbonyl, producing a unit positive charge on carbon and thus increased reactivity of the carbonyl to a nucleophile.

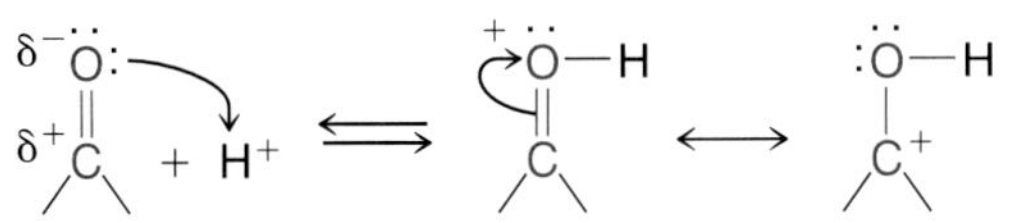

### Addition of Water: Hydrate Formation

For our purposes, we will assume that unhalogenated ketones do not form stable hydrates; very few aldehydes do. A hydrate is produced by reacting a particular carbonyl compound with water.

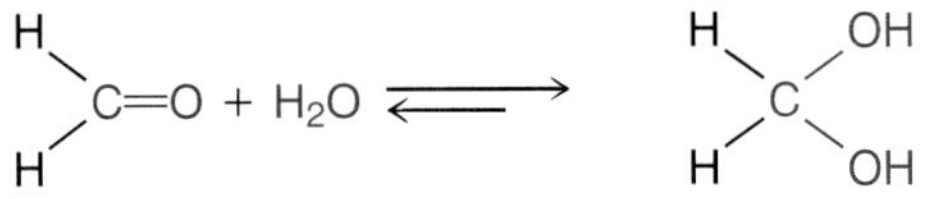

Formaldehyde hydrate

Hydrates of aldehydes and ketones are *gem*-diols and therefore are usually unstable (see Sec. 15.2) with the equilibrium favoring the carbonyl compound.

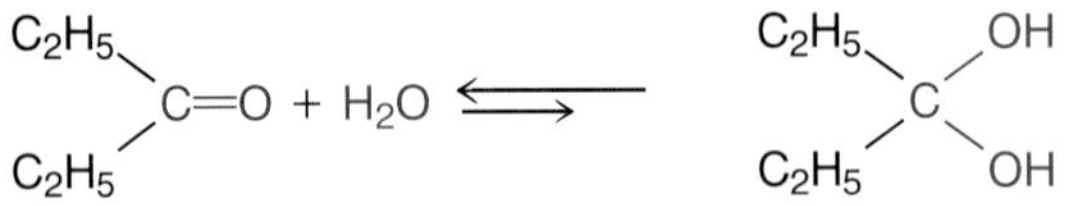

Diethyl ketone hydrate

A reasonable question to ask is, Why does formaldehyde form a stable hydrate, whereas diethyl ketone does not? It is a matter of the relative stability of the carbonyl compound versus the hydrate. Consider diethyl ketone (Fig. 19.1). Since two alkyl groups are dispersing the partial positive charge on the carbonyl carbon (by electron release a $+I$ effect, as shown by the two arrows), this ketone is more stable than a typical aldehyde with only one such stabilizing alkyl group (Fig. 19.2). Least stable of the carbonyl compounds is formaldehyde with no stabilizing alkyl groups bonded to the carbonyl carbon (see Fig. 19.2). As a result of the relative instability of formaldehyde, this compound would have the greatest tendency to form a hydrate that is more stable than the aldehyde. Conversely, more stable ketones form relatively less stable hydrates.

**FIGURE 19.1** Inductive effect analysis of the relative stability of diethyl ketones.

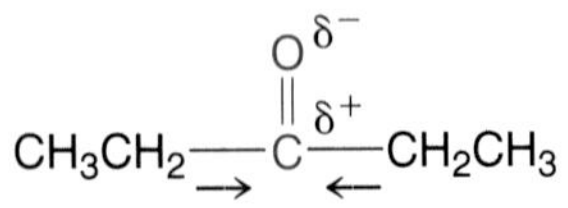

## HISTORICAL BOXED READING 19.1

Chloral was first prepared by Justus von Liebig in 1831 when he bubbled a stream of gaseous chlorine into ethyl alcohol. Besides being unstable, this colorless liquid had a disagreeable odor. There seemed to be no use for it, so chloral sat in a bottle on the shelf gathering dust.

Forty years later, Dr. Oscar Liebreich believed chloral could possibly serve as a sleep-inducing drug (*soporific*) that was both safe and effective. When he tried it on a few patients, he found that it actually worked. The drug smelled bad and tasted worse, however, and patients did not want to take it. Fortunately, chloral hydrate, a stable solid, was as good a soporific as chloral itself and had neither the disagreeable odor nor the awful taste of chloral. Thus chloral hydrate was used as a soporific for many years.

Both chloral hydrate and chloral, however, had a dark side. Both compounds were found at a later date to be addictive. In addition, one dose too many could result in death of the patient; he or she simply would not wake up. Also, chloral hydrate and chloral were used by many a sea captain to "shanghai" sailors as seamen on voyages that could last for as long as 3 years. The odor and taste of chloral were masked by the whiskey the sailors consumed. This use of chloral became so pervasive that chloral became well-known as "knockout drops." Furthermore, since an overdose resulted in death, the drug was often used in murders.

An important stable hydrate is chloral hydrate.

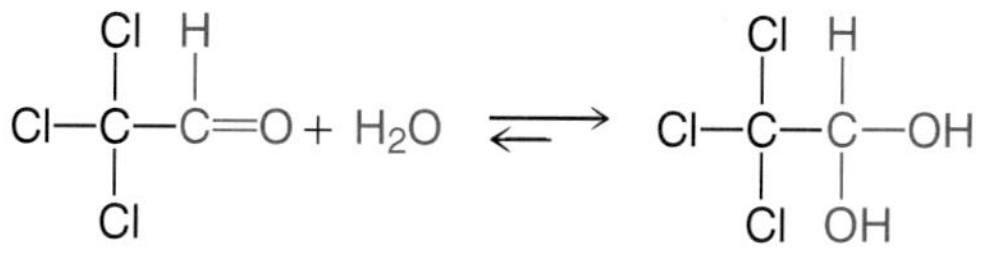

Chloral
(trivial name)

Chloral hydrate
(trivial name)

This hydrate is particularly stable because of the relative instability of chloral, which is caused by repulsive forces due to partial positive charges on adjacent carbons (Fig. 19.3). Because chlorine is more electronegative than carbon, there is a slight shifting of electron density toward each chlorine (a −I inductive effect), giving the carbon bonded to the chlorines a partial positive charge. Because this partially positively charged carbon is adjacent to the partially positively charged carbonyl carbon as shown in the dipole structure, a carbon-carbon bond exists with each carbon bearing a partial positive charge. Since like charges repel, this carbon-carbon bond is weakened, destabilizing the chloral molecule.

To name a hydrate using the common system of nomenclature, name the carbonyl compound, and then add the word *hydrate*. The compounds are named as diols using the IUPAC system of nomenclature in Fig. 19.4 (see Sec. 14.1).

**FIGURE 19.2** Inductive effect analysis of the relative stability of propionaldehyde compared to formaldehyde.

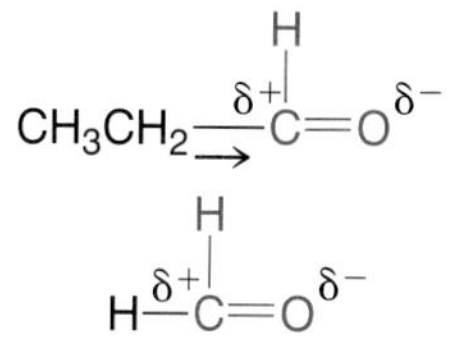

### Mechanism of the Reaction of Formaldehyde with Water

First step:

Second step:

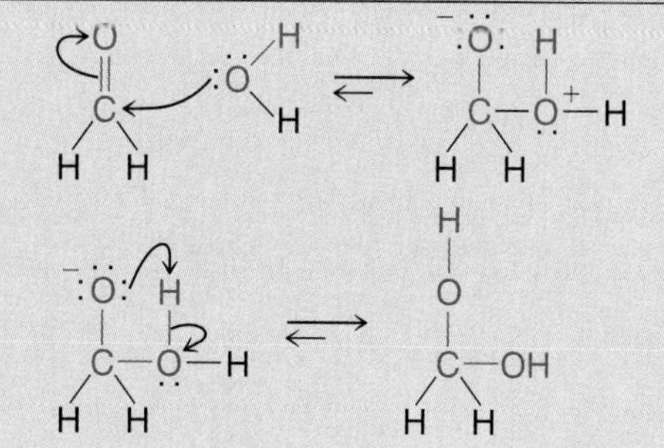

The mechanism of hydrate formation begins with addition of the nucleophile, water, to the carbonyl carbon of formaldehyde. The nucleophilic addition is followed by proton transfer.

**Envision the Reaction**

**FIGURE 19.3** The destabilization of chloral.

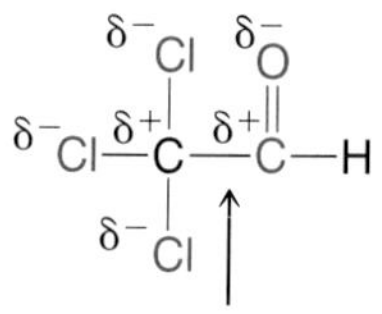

This bond is weakened by two adjacent partial positive charges

**FIGURE 19.4** Common and IUPAC nomenclature of selected hydrates.

$$CH_3-\overset{\overset{\displaystyle H}{|}}{\underset{\underset{\displaystyle OH}{|}}{C}}-OH \qquad CH_3-\overset{\overset{\displaystyle OH}{|}}{\underset{\underset{\displaystyle OH}{|}}{C}}-CH_3$$

Common: Acetaldehyde hydrate     Acetone hydrate
IUPAC:    1, 1-Ethanediol         2, 2-Propanediol

## Addition of Hydrogen Cyanide

In the presence of a basic catalyst, hydrogen cyanide (H—C≡N) reacts with aldehydes and certain ketones to give cyanohydrins. In general,

$$R-\overset{\overset{\displaystyle O}{||}}{C}-H + HC\equiv N \xrightarrow{\ ^-OH\ } R-\overset{\overset{\displaystyle OH}{|}}{\underset{\underset{\displaystyle H}{|}}{C}}-C\equiv N$$

$$R-\overset{\overset{\displaystyle O}{||}}{C}-R(R') + HC\equiv N \xrightarrow{\ ^-OH\ } R-\overset{\overset{\displaystyle OH}{|}}{\underset{\underset{\displaystyle R(R')}{|}}{C}}-C\equiv N$$

Since hydrogen cyanide is extremely toxic, it is usually generated in the reaction mixture by reacting sodium cyanide with a suitable mineral acid such as sulfuric acid.

2 NaCN + $H_2SO_4$ ⟶
2 HCN + $Na_2SO_4$

Examples:

$$CH_3-\overset{\overset{\displaystyle O}{||}}{C}-H + HC\equiv N \xrightarrow{\ ^-OH\ } CH_3-\overset{\overset{\displaystyle OH}{|}}{\underset{\underset{\displaystyle H}{|}}{C}}-C\equiv N$$

Acetaldehyde                  Acetaldehyde cyanohydrin

$$CH_3-\overset{\overset{\displaystyle O}{||}}{C}-CH_3 + HC\equiv N \xrightarrow{\ ^-OH\ } CH_3-\overset{\overset{\displaystyle OH}{|}}{\underset{\underset{\displaystyle CH_3}{|}}{C}}-C\equiv N$$

Acetone                      Acetone cyanohydrin

Once hydrogen cyanide is produced in the reaction mixture, the reaction proceeds as follows:

## Envision the Reaction

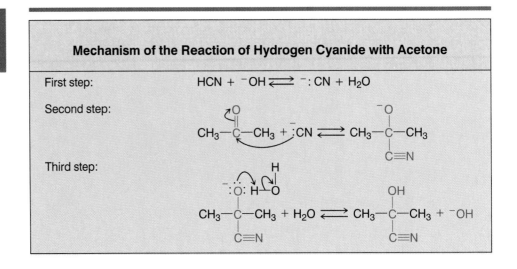

**Mechanism of the Reaction of Hydrogen Cyanide with Acetone**

First step:           HCN + ⁻OH ⇌ ⁻:CN + $H_2O$

Second step:

Third step:

Cyanohydrins are intermediates in the synthesis of many important compounds. For example, hydrolysis of acetaldehyde cyanohydrin produces lactic acid, a key compound in glucose metabolism.

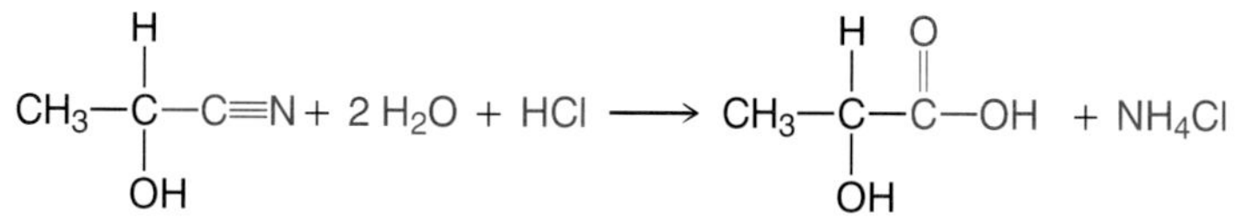

$$CH_3-\overset{\overset{\text{H}}{|}}{\underset{\underset{\text{OH}}{|}}{C}}-C\equiv N + 2\,H_2O + HCl \longrightarrow CH_3-\overset{\overset{\text{H}}{|}}{\underset{\underset{\text{OH}}{|}}{C}}-\overset{\overset{\text{O}}{\|}}{C}-OH + NH_4Cl$$

Lactic acid

Methyl and alicyclic ketones react with hydrogen cyanide. However, sterically hindered ketones such as aromatic ketones do not react.

To name a cyanohydrin using the common system of nomenclature, provide the common name of the carbonyl compound, and then add the word *cyanohydrin*, as in Fig. 19.5.

---

Would 3-pentanone be likely to form a stable hydrate? Explain.

**SOLVED PROBLEM 19.1**

**SOLUTION**

The ketone 3-pentanone contains two alkyl groups that stabilize the compound by dispersing the partial positive charge on the carbonyl carbon due to $+\,$I inductive effects. Thus we can assume that the ketone is stable with respect to the hydrate, and a stable hydrate should not form.

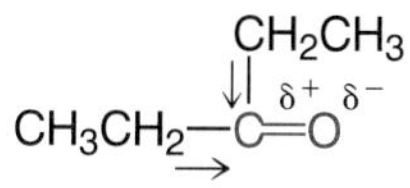

$$CH_3CH_2-\overset{\overset{\overset{\text{CH}_2\text{CH}_3}{\Downarrow}}{|}}{\underset{\longrightarrow}{C}}\!\!=\!\!\overset{\delta+}{\phantom{C}}\overset{\delta-}{O}$$

---

If each of the following compounds reacts with HCN, write the letter *R*; if not, write the letters *NR*. Explain each choice.

**SOLVED PROBLEM 19.2**

a.

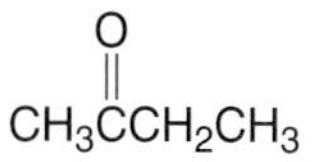

$$CH_3\overset{\overset{\text{O}}{\|}}{C}CH_2CH_3$$

b.

$$\text{(phenyl ring)}\overset{\overset{\text{O}}{\|}}{C}-C_2H_5$$

c.

$$\text{(cyclohexanone ring)}\!=\!O$$

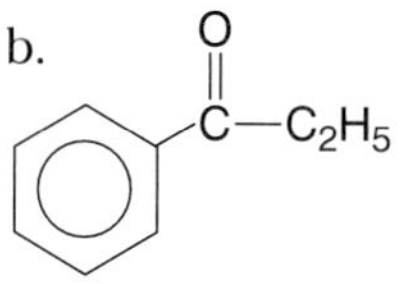

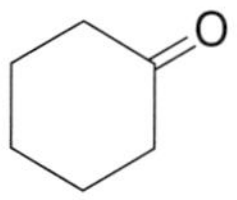

**SOLUTION**

a. *R*. This is a methyl ketone.

b. *NR*. The combination of bulky phenyl and ethyl groups adds up to no reaction.

c. *R*. This is an alicyclic ketone.

---

Would α,α,α-trifluoroacetaldehyde be likely to form a stable hydrate? Explain.

**EXERCISE 19.2**

---

**FIGURE 19.5** Common nomenclature of selected cyanohydrins.

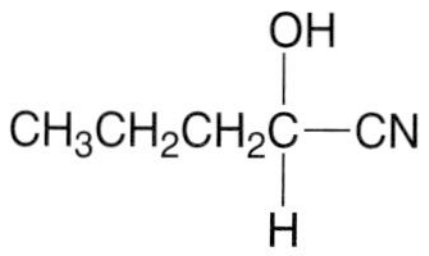

$$CH_3CH_2CH_2\overset{\overset{\text{OH}}{|}}{\underset{\underset{\text{H}}{|}}{C}}-CN$$

$$CH_3-\overset{\overset{\text{OH}}{|}}{\underset{\underset{\text{CN}}{|}}{C}}-CH_2CH_3$$

Butyraldehyde
cyanohydrin

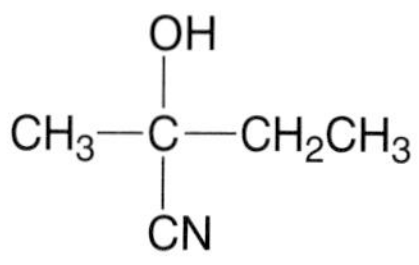

Methyl ethyl ketone
cyanohydrin

**EXERCISE 19.3**

Complete the following equations by drawing a structural formula for each organic product produced. If no reaction takes place, write "no reaction."

a.   propionaldehyde + HCN $\longrightarrow$

b.   formaldehyde + $H_2O$ $\rightleftharpoons$

c.   benzophenone + HCN $\longrightarrow$

## Addition of Alcohols

*Note that a hemiacetal is a gem-hydroxy ether, whereas an acetal is a gem-diether; i.e., the functional group(s) are bonded to the same carbon.*

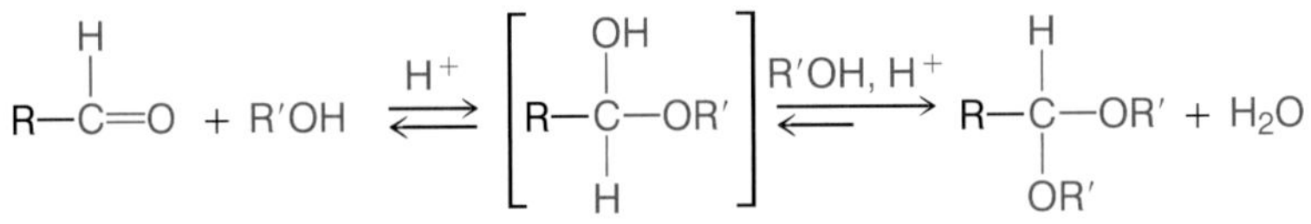

In the presence of a catalytic amount of hydrochloric acid (HCl) or any other mineral acid, an aldehyde, when treated with 1 mol of alcohol, yields an unstable hemiacetal; with 2 mol of alcohol, a stable acetal is obtained (see Sec. 18.6). In general,

$$R-\underset{\underset{H}{|}}{C}=O \ + \ R'OH \ \overset{H^+}{\rightleftharpoons} \ \left[ R-\underset{\underset{H}{|}}{\overset{\overset{OH}{|}}{C}}-OR' \right] \ \overset{R'OH,\ H^+}{\rightleftharpoons} \ R-\underset{\underset{OR'}{|}}{\overset{\overset{H}{|}}{C}}-OR' \ + \ H_2O$$

Unstable hemiacetal          Stable acetal

Example:

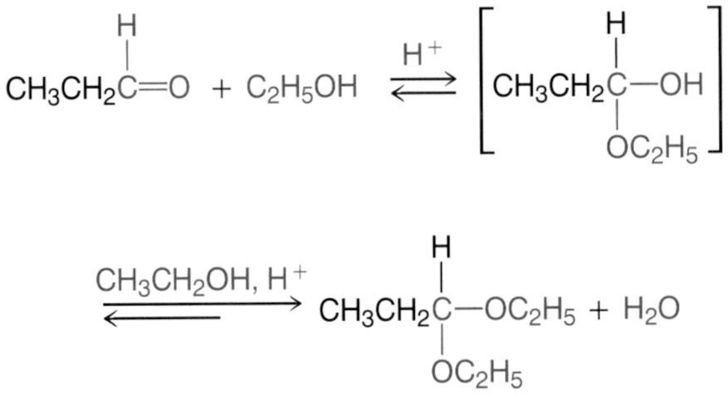

$$CH_3CH_2\underset{\underset{H}{|}}{C}=O \ + \ C_2H_5OH \ \overset{H^+}{\rightleftharpoons} \ \left[ CH_3CH_2\underset{\underset{OC_2H_5}{|}}{\overset{\overset{H}{|}}{C}}-OH \right]$$

$$\overset{CH_3CH_2OH,\ H^+}{\rightleftharpoons} \ CH_3CH_2\underset{\underset{OC_2H_5}{|}}{\overset{\overset{H}{|}}{C}}-OC_2H_5 \ + \ H_2O$$

Common name: propionaldehyde diethyl acetal
IUPAC name: 1,1-diethoxypropane

Ketones react with alcohols in a similar fashion to give unstable hemiacetals (with 1 mol of alcohol) and acetals* (with 2 mol of alcohol). Most acetals prepared from ketones are unstable. Acetals from ketones are stable when the alcohol used is a diol and the acetal produced is cyclic.

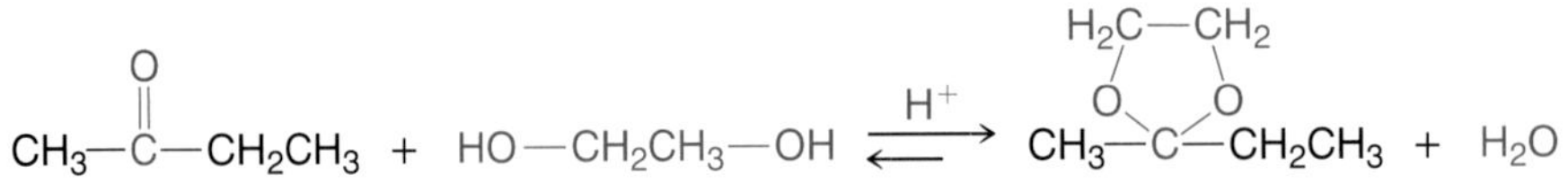

$$CH_3-\underset{\underset{}{\overset{\overset{O}{||}}{C}}}-CH_2CH_3 \ + \ HO-CH_2CH_3-OH \ \overset{H^+}{\rightleftharpoons} \ \underset{2\text{-Ethyl-2-methyl-1,3-dioxolane}}{CH_3-\overset{\overset{H_2C-CH_2}{\underset{O\quad\quad O}{\diagdown\quad\diagup}}}{C}-CH_2CH_3} \ + \ H_2O$$

2-Ethyl-2-methyl-1,3-dioxolane
(IUPAC name)
(an acetal)

The mechanism for acetal formation is well established.

---

*Any cyclic hemiacetal prepared from a ketone was previously known as a *hemiketal*, whereas an acetal prepared from a ketone was known as a *ketal*.

**Envision the Reaction**

### Mechanism of the Reaction of One Mole of Acetaldehyde with Two Moles of Methyl Alcohol to Form Acetaldehyde Dimethyl Acetal (a Stable Compound)

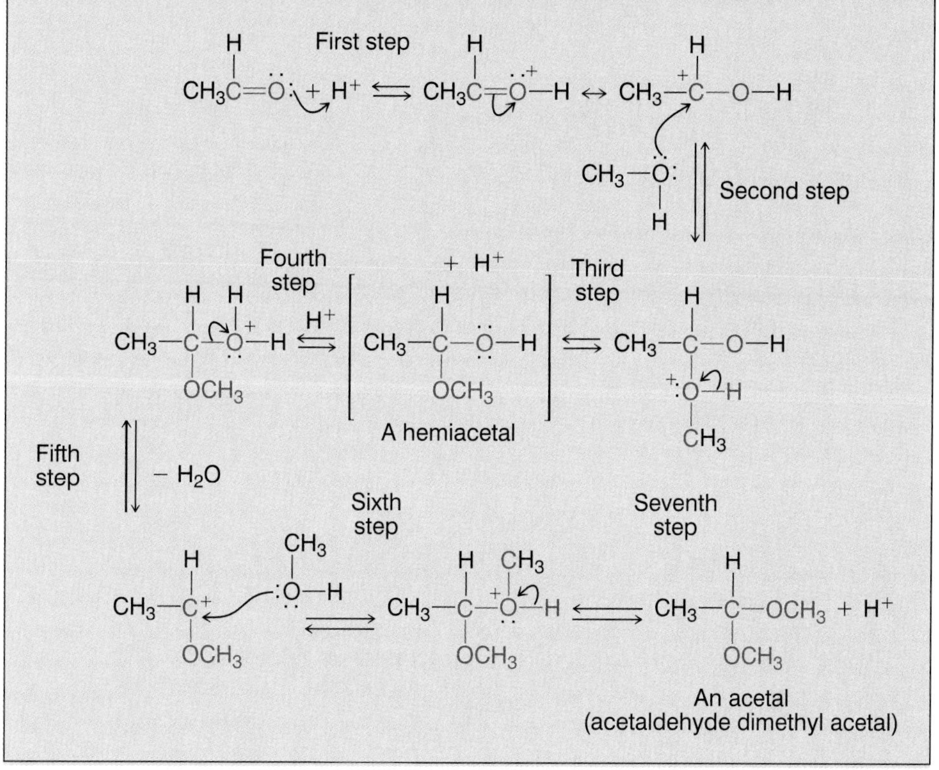


Note that each step in the mechanism is reversible. It is reasonable that this should be so because the overall reaction is itself reversible. Therefore, acetals can be hydrolyzed in acid solution to produce the corresponding carbonyl compound and alcohol. In general,

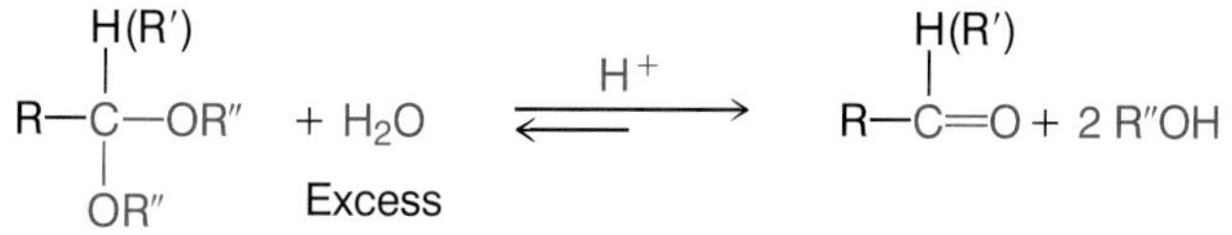


Example:

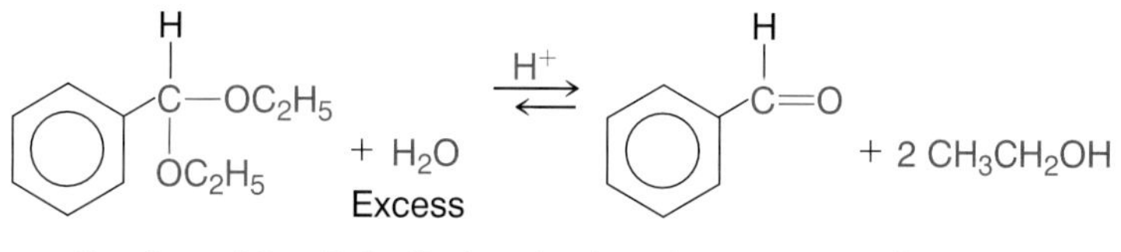


This is an application of Le Châtelier's principle (see Sec. 18.6). Hydrolysis of acetals does *not* occur in basic or neutral solution, only in acid.

Hemiacetal and acetal chemistry is of vital importance in the chemistry of carbohydrates, as we will learn in Chapter 27.

**SOLVED PROBLEM 19.3**

Characterize each of the following as a hemiacetal or an acetal. Then classify each as derived from an aldehyde or a ketone. Explain each choice.

a. 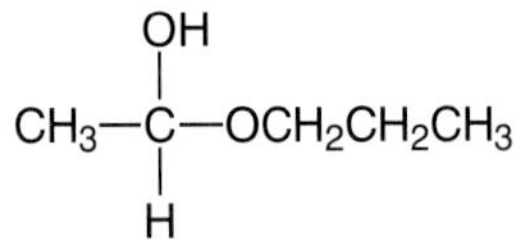
$$CH_3-\underset{\underset{H}{|}}{\overset{\overset{OH}{|}}{C}}-OCH_2CH_2CH_3$$

b.

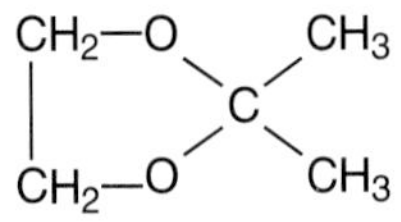

### SOLUTION

a. The compound is a *gem*-hydroxy ether; i.e., the hydroxyl (—OH) and *n*-propoxyl (—OCH₂CH₂CH₃) groups are both bonded to the same carbon. Thus the compound is a hemiacetal. Note that the carbon bonded to the —OH and —OCH₂CH₂CH₃ groups was originally the carbonyl carbon. Since that carbon is also bonded to an alkyl group and a hydrogen, the compound must be classified as an hemiacetal derived from an aldehyde.

$$CH_3-\underset{\underset{H}{|}}{\overset{\overset{O}{\|}}{C}} \qquad CH_3-\underset{\underset{H}{|}}{\overset{\overset{OH}{|}}{C}}-OCH_2CH_2CH_3$$

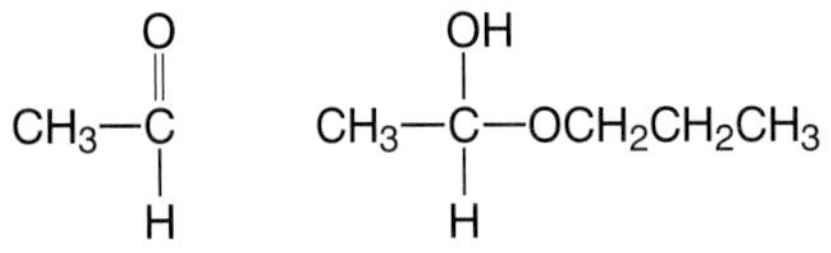

  Aldehyde              Hemiacetal

b. The key structural feature of this cyclic structure is as follows:

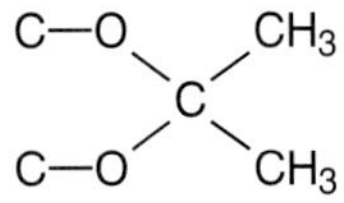

The compound is a *gem*-diether and is therefore an acetal. Since two alkyl (methyl) groups are also attached to what was the carbonyl carbon, this acetal is derived from a ketone.

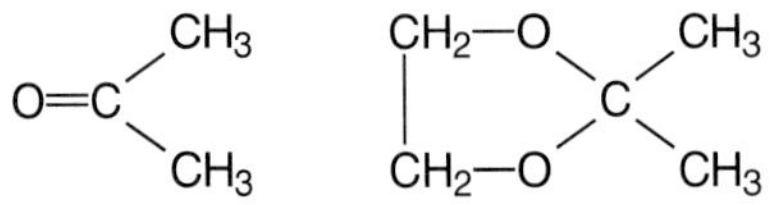

  Ketone              Acetal

**EXERCISE 19.4**

Characterize each of the following as a hemiacetal or an acetal. Then classify each as derived from an aldehyde or a ketone. Explain each choice.

a. 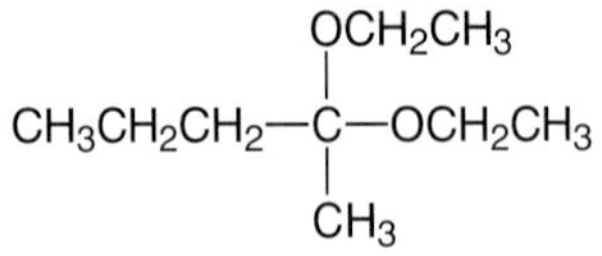
$$CH_3CH_2CH_2-\underset{\underset{CH_3}{|}}{\overset{\overset{OCH_2CH_3}{|}}{C}}-OCH_2CH_3$$

b.

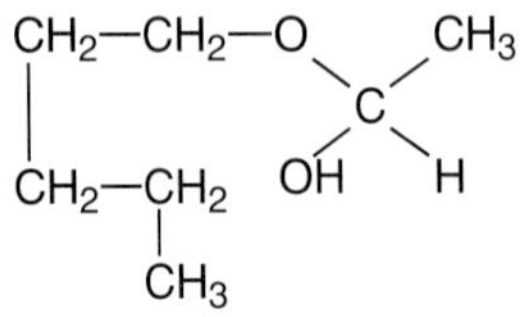

c.

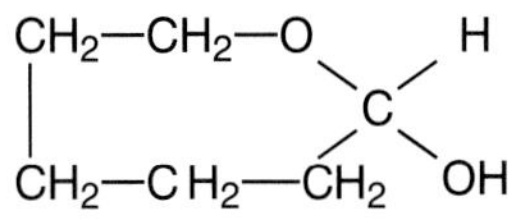

$$CH_2-CH_2-O \diagdown \diagup H$$
$$CH_2-CH_2-CH_2 \diagup C \diagdown OH$$

For a review of the Grignard reagent, refer to Section 4.1. The Grignard reagent adds to a carbonyl compound to give an alkoxide salt that is hydrolyzed to the corresponding alcohol. In general,

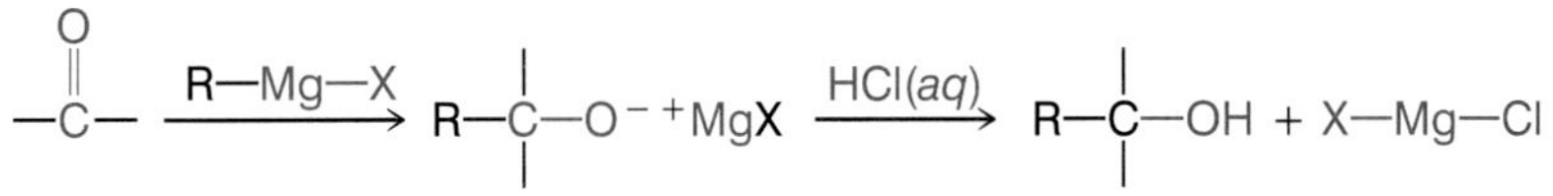

**Addition of a Grignard Reagent**

$$-\overset{O}{\overset{\|}{C}}- \xrightarrow{R-Mg-X} R-\overset{|}{\underset{|}{C}}-O^-{}^+MgX \xrightarrow{HCl(aq)} R-\overset{|}{\underset{|}{C}}-OH + X-Mg-Cl$$

The class of alcohol produced depends on the carbonyl compound used, as shown in Table 19.1.

Examples:

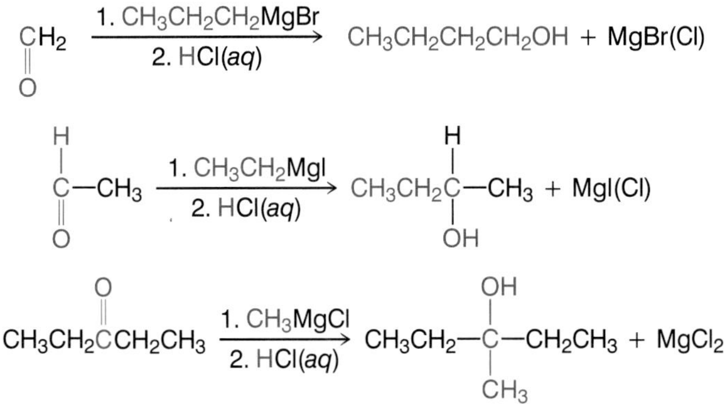

$$\overset{CH_2}{\underset{O}{\|}} \xrightarrow[\text{2. HCl}(aq)]{\text{1. } CH_3CH_2CH_2MgBr} CH_3CH_2CH_2CH_2OH + MgBr(Cl)$$

$$\overset{H}{\underset{O}{\overset{|}{\underset{\|}{C}}}}-CH_3 \xrightarrow[\text{2. HCl}(aq)]{\text{1. } CH_3CH_2MgI} CH_3CH_2\overset{H}{\underset{OH}{\overset{|}{\underset{|}{C}}}}-CH_3 + MgI(Cl)$$

$$CH_3CH_2\overset{O}{\overset{\|}{C}}CH_2CH_3 \xrightarrow[\text{2. HCl}(aq)]{\text{1. } CH_3MgCl} CH_3CH_2-\overset{OH}{\underset{CH_3}{\overset{|}{\underset{|}{C}}}}-CH_2CH_3 + MgCl_2$$

**Envision the Reaction**

### General Mechanism for the Addition of a Grignard Reagent to a Carbonyl Compound to Produce an Alcohol

First step:

$$RMgX \underset{\text{Anhydrous ether}}{\rightleftharpoons} R:^- + {}^+MgX$$

where X = Cl, Br, or I

Second step:

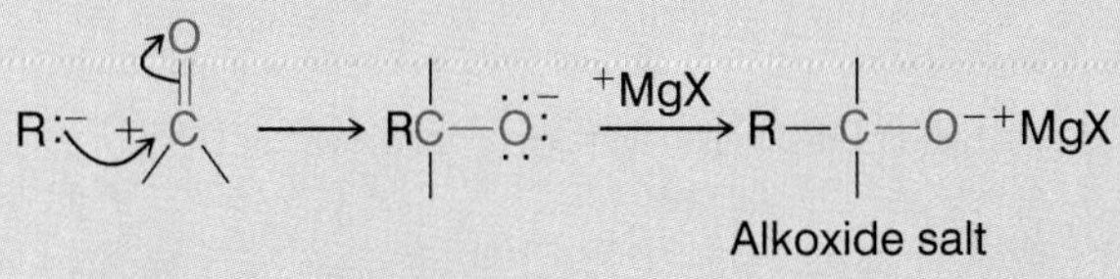

$$R:^- + \overset{O}{\overset{\curvearrowleft}{\underset{}{C}}} \longrightarrow R\overset{}{\underset{}{C}}-\ddot{\underset{\cdot\cdot}{O}}:^- \xrightarrow{{}^+MgX} R-\overset{|}{\underset{|}{C}}-O^-{}^+MgX$$

Alkoxide salt

Third step:

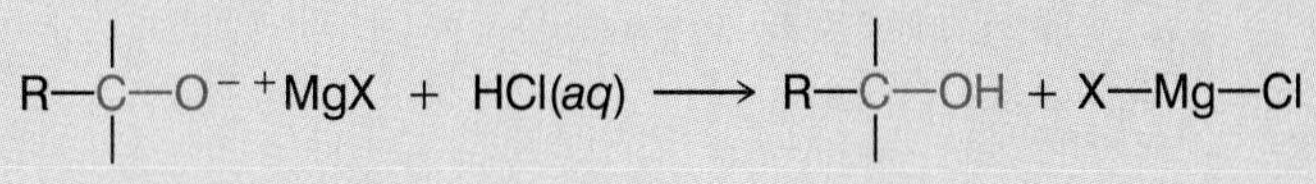

$$R-\overset{|}{\underset{|}{C}}-O^-{}^+MgX + HCl(aq) \longrightarrow R-\overset{|}{\underset{|}{C}}-OH + X-Mg-Cl$$

The first step of the mechanism involves the dissociation of the Grignard reagent to form a carbanion R:$^-$. In the second step, the nucleophile R:$^-$ attacks the electropositive carbon of the carbonyl group. The alkoxide salt is, in turn, hydrolyzed with acid to afford the alcohol in the third step.

| TABLE 19.1 | The Relationship Between the Kind of Carbonyl Compound Reacted and the Class of Alcohol Produced Using a Grignard Synthesis |
| --- | --- |
| **Carbonyl Compound Used** | **Class of Alcohol Produced** |
| Formaldehyde | Primary |
| Any aldehyde (but not formaldehyde) | Secondary |
| Any ketone | Tertiary |

## SOLVED PROBLEM 19.4

Select a Grignard reagent and a carbonyl compound that would produce 3-pentanol.

### SOLUTION

The main thing to understand in a Grignard alcohol synthesis is that the carbonyl carbon of the aldehyde or ketone is converted into the carbon bearing the hydroxyl group of the alcohol produced as a result of the Grignard synthesis. Since 3-pentanol is a secondary alcohol, we need an aldehyde for the carbonyl compound. Let us first draw a structural formula for 3-pentanol.

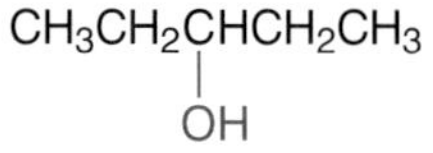

$$CH_3CH_2CHCH_2CH_3$$
$$|$$
$$OH$$

Now let us break this molecule into two parts: the carbonyl component and the Grignard component.

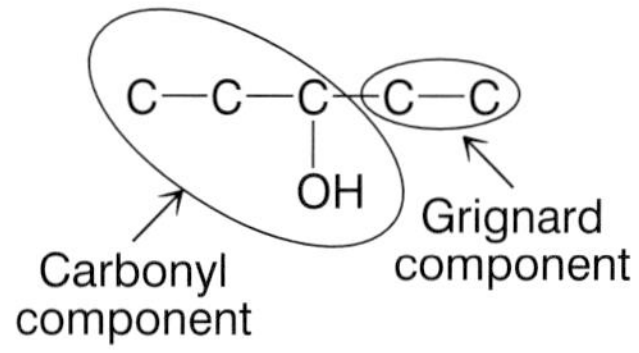

Since the carbon bearing the —OH of the alcohol corresponds to the carbon of the carbonyl group, and since the alcohol is secondary, the carbonyl compound must be a three-carbon aldehyde: propionaldehyde ($CH_3CH_2CHO$). The remaining two carbons must represent the Grignard component, say, ethylmagnesium bromide ($CH_3CH_2MgBr$). Thus we have

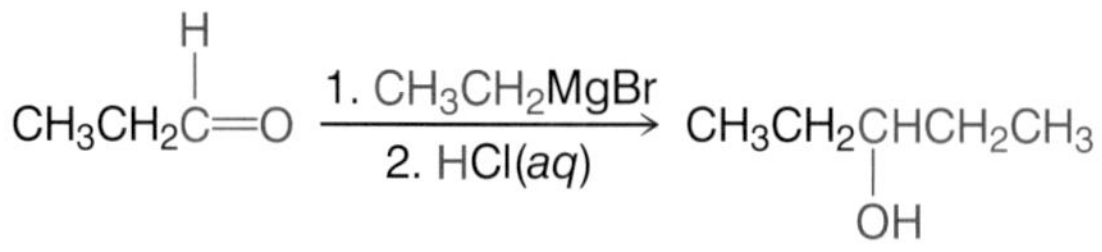

## EXERCISE 19.5

Select one or more Grignard reagent–carbonyl compound combinations that would produce

a. *tert*-pentyl alcohol (note that two combinations are possible here).

b. *n*-pentyl alcohol.

c. isopropyl alcohol.

d. isobutyl alcohol.

e. 2-hexanol (note that two combinations are possible here).

f. triphenylcarbinol.

These alcohol-producing Grignard reactions represent an excellent synthetic tool. The following solved problem will illustrate.

Synthesize 2-pentanone from alcohols containing three carbons or less. You may use any other organic and inorganic reagents necessary.

**SOLUTION**

As usual, let us work backwards until we finally satisfy the demands of the problem. First, prepare 2-pentanone by oxidation of 2-pentanol.

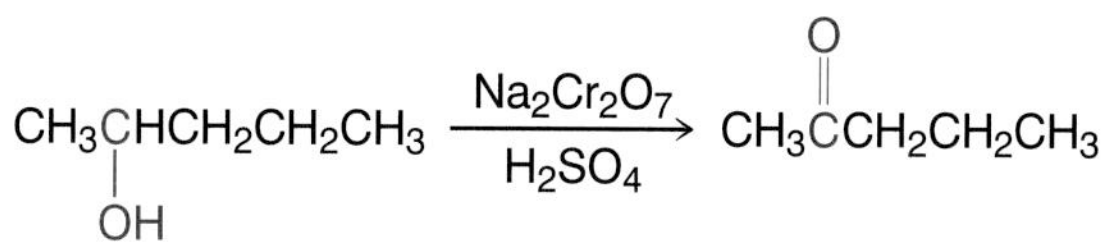

$$CH_3CHCH_2CH_2CH_3 \ \xrightarrow[H_2SO_4]{Na_2Cr_2O_7} \ CH_3\overset{O}{\overset{\|}{C}}CH_2CH_2CH_3$$
$$|$$
$$OH$$

Now, to prepare 2-pentanol, we react acetaldehyde with *n*-propylmagnesium iodide and then hydrolyze the resulting salt with aqueous hydrochloric acid.

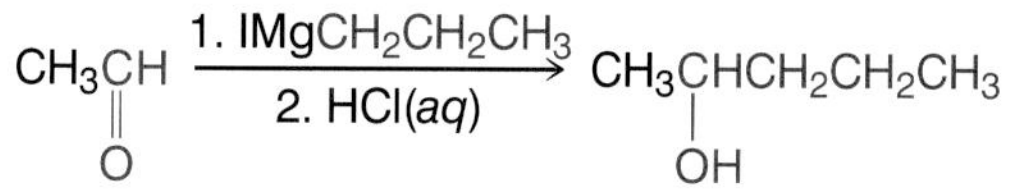

$$CH_3\overset{}{\underset{O}{\overset{\|}{C}}}H \ \xrightarrow[2.\ HCl(aq)]{1.\ IMgCH_2CH_2CH_3} \ CH_3CHCH_2CH_2CH_3$$
$$|$$
$$OH$$

To prepare acetaldehyde, we oxidize ethyl alcohol.

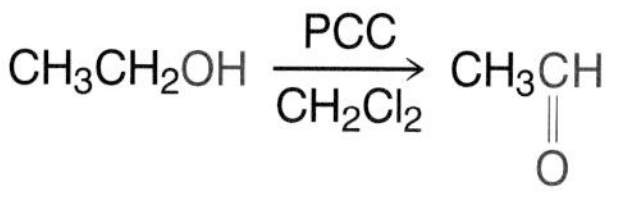

$$CH_3CH_2OH \ \xrightarrow[CH_2Cl_2]{PCC} \ CH_3\overset{}{\underset{O}{\overset{\|}{C}}}H$$

To prepare propylmagnesium iodide, we treat propyl iodide with magnesium in anhydrous ether.

$$CH_3CH_2CH_2I \ \xrightarrow[\text{Anhydrous ether}]{Mg} \ CH_3CH_2CH_2MgI$$

Finally, we prepare propyl iodide by reacting propyl alcohol with hydriodic acid.

$$CH_3CH_2CH_2OH \ \xrightarrow{HI} \ CH_3CH_2CH_2I$$

By starting the synthesis with ethyl alcohol and propyl alcohol, we have satisfied the conditions of the problem. Putting the complete synthesis together, we get

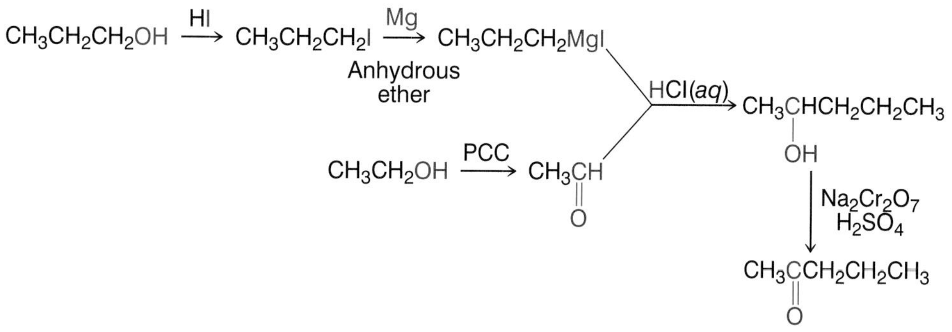

$$CH_3CH_2CH_2OH \ \xrightarrow{HI} \ CH_3CH_2CH_2I \ \xrightarrow[\substack{\text{Anhydrous} \\ \text{ether}}]{Mg} \ CH_3CH_2CH_2MgI$$

$$CH_3CH_2OH \ \xrightarrow{PCC} \ CH_3\overset{}{\underset{O}{\overset{\|}{C}}}H$$

$$\xrightarrow{HCl(aq)} \ CH_3CHCH_2CH_2CH_3$$
$$|$$
$$OH$$

$$\xrightarrow[H_2SO_4]{Na_2Cr_2O_7}$$

$$CH_3\overset{}{\underset{O}{\overset{\|}{C}}}CH_2CH_2CH_3$$

Prepare each of the following from alcohols of two carbons or less. You may use any other organic or inorganic reagents necessary.

a.  propyl alcohol
b.  2-butanone
c.  *tert*-butyl alcohol

## Addition of Ammonia and Its Derivatives

Ammonia reacts with formaldehyde to give the urinary antiseptic hexamethylene-tetramine.

$$6\ CH_2O\ +\ 4\ NH_3\ \longrightarrow\ \text{[Hexamethylenetetramine structure]}\ +\ 6\ H_2O$$

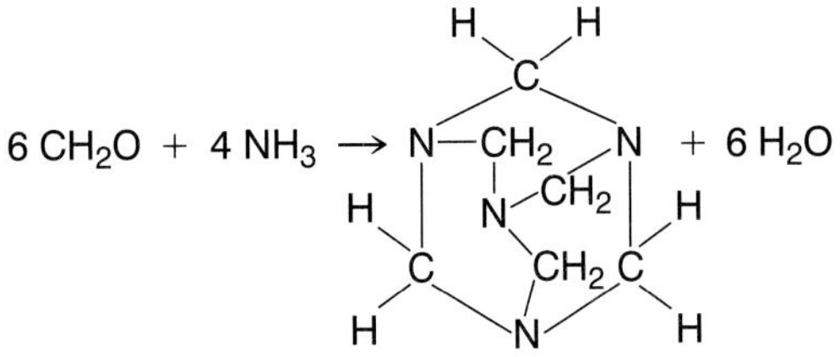

Hexamethylenetetramine

Other aldehydes and ketones, when treated with ammonia, form *imines*. An imine is an unsaturated form of an amine (amine, $RCH_2NH_2$; imine, $RCH{=}NH$). An imine is also known as a *Schiff base*.

## Envision the Reaction

### General Mechanism of the Reaction of an Aldehyde to Produce an Imine

First step:

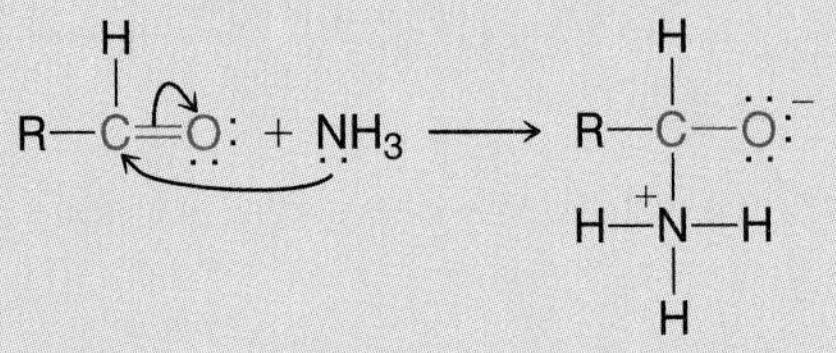

Second step:

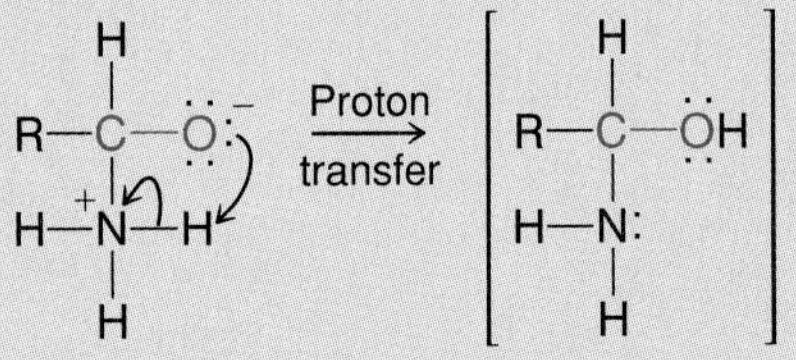

Unstable *gem-aminoalcohol*

Third step:

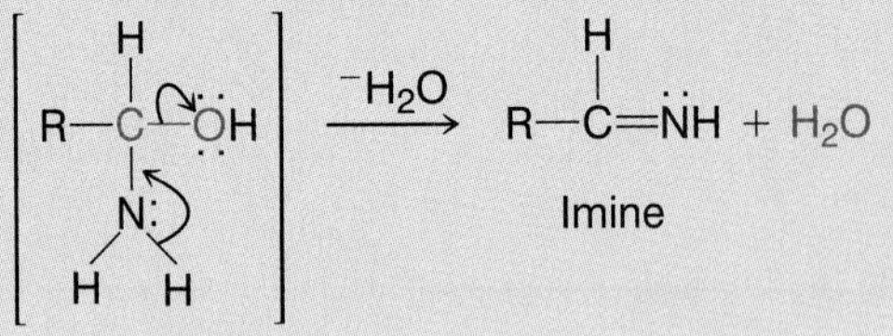

Imine

The unstable species loses water in the third step of the mechanism to form the imine. For example:

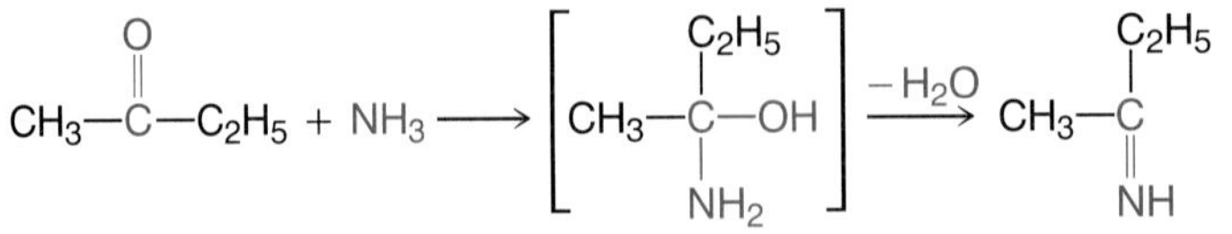

Methyl ethyl ketone imine

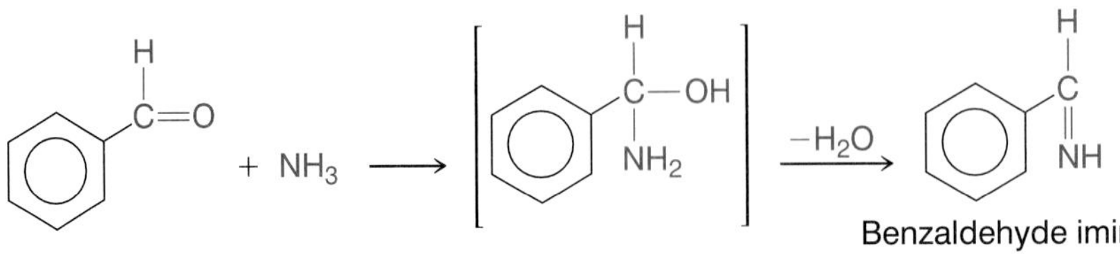

Benzaldehyde imine

| TABLE 19.2 | Reaction of Derivatives of Ammonia with Aldehydes and Ketones | |
|---|---|---|
| **Carbonyl Compound** | **Name and Formula of Derivatives of Ammonia** | **Name and Formula of Carbonyl Derivative** |

$$\underset{\substack{|\\H(R')}}{R-C}=O \quad + \quad \underset{\text{Hydroxylamine}}{NH_2OH} \quad \xrightarrow{H^+} \quad \underset{\substack{|\\H(R')}}{R-C}=N-OH + H_2O$$

Oxime

$$\underset{\substack{|\\H(R')}}{R-C}=O + NH_2-N \overset{NO_2}{\underset{}{\bigcirc}} -NO_2 \xrightarrow{H^+} \underset{\substack{|\\H(R')}}{R-C}-\underset{\substack{|\\H}}{N}-N \overset{NO_2}{\underset{}{\bigcirc}} -NO_2 + H_2O$$

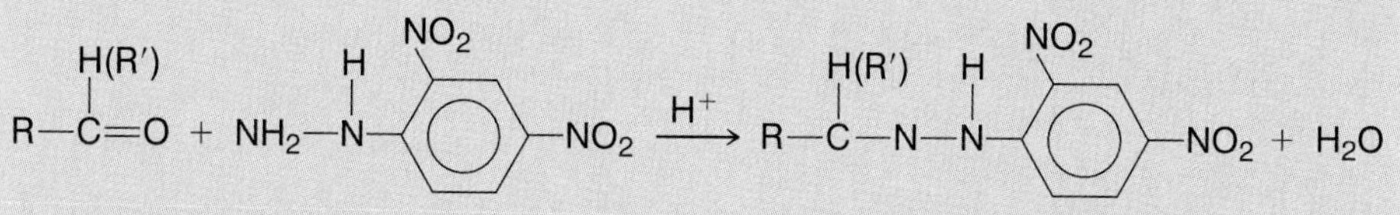

2,4-Dinitrophenylhydrazine       2,4-Dinitrophenylhydrazone

$$\underset{\substack{|\\H(R')}}{R-C}=O \quad + \quad \underset{\substack{|\\H}}{NH_2-N}-\overset{O}{\underset{}{C}}-NH_2 \xrightarrow{H^+} \underset{\substack{|\\H(R')}}{R-C}=\underset{\substack{|\\H}}{N}-\overset{O}{\underset{}{C}}-NH_2 + H_2O$$

Semicarbazide       Semicarbazone

A number of derivatives of ammonia react with carbonyl compounds in a similar way as ammonia. In these cases, the nucleophile is added to a protonated carbonyl species with subsequent loss of water. The acid catalyst usually used is acetic acid ($CH_3COOH$). These nucleophiles (derivatives of ammonia) along with the products of reaction are summarized in Table 19.2.

To name an ammonia derivative of a carbonyl compound using the common system of nomenclature, give the common name of the carbonyl compound followed by the name of the derivative as another word. For example:

$$(CH_3)_2C=N-OH$$

Acetone oxime

To name such a compound using the IUPAC system of nomenclature, provide the IUPAC name of the carbonyl compound followed by the name of the derivative as another word. Thus the IUPAC name for acetone oxime is propanone oxime.

All carbonyl compounds (for our purposes) react with these derivatives of ammonia as follows:

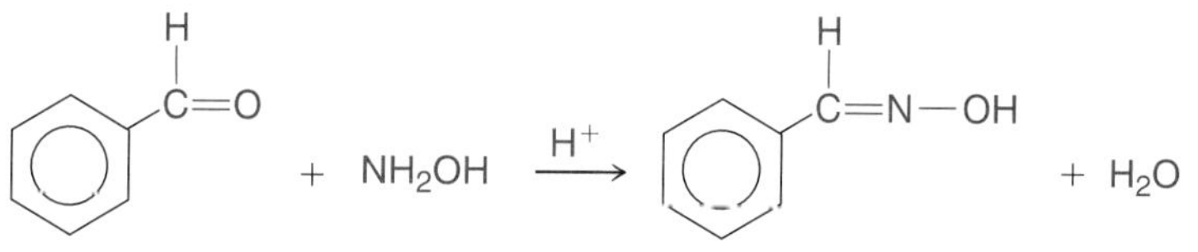

These solid crystalline carbonyl derivatives are useful in the laboratory because the vast majority have a sharp, relatively high, fixed melting point that can be used to identify unknown carbonyl compounds.

Benzaldehyde       Benzaldehyde oxime (common and IUPAC)

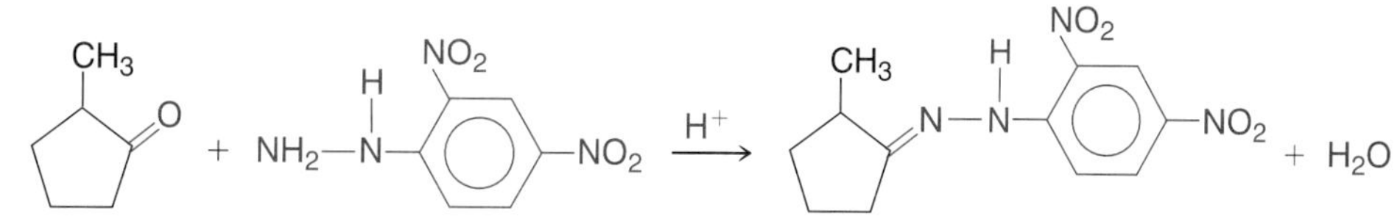

2-Methylcyclopentanone

2-Methylcyclopentanone
2,4-Dinitrophenylhydrazone (IUPAC)

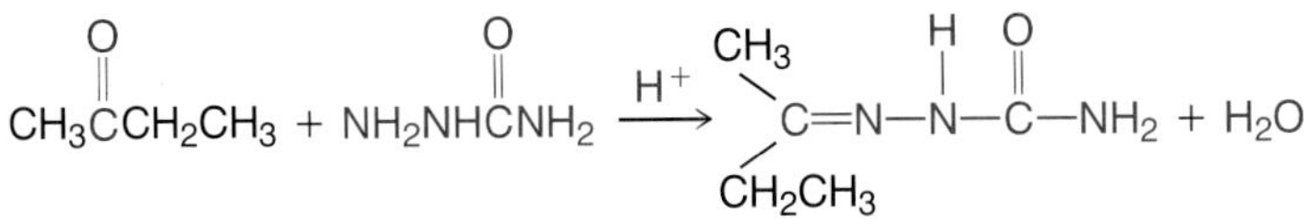

$$CH_3\overset{O}{\overset{\|}{C}}CH_2CH_3 + NH_2NH\overset{O}{\overset{\|}{C}}NH_2 \xrightarrow{H^+} \underset{CH_2CH_3}{\overset{CH_3}{C}}=N-\overset{H}{\underset{}{N}}-\overset{O}{\overset{\|}{C}}-NH_2 + H_2O$$

Methyl ethyl ketone semicarbazone (common)
2-Butanone semicarbazone (IUPAC)

---

**EXERCISE 19.7**

Complete the following equations by drawing a structural formula for each organic product and writing a molecular formula for each inorganic product.

a.  cyclohexanone + semicarbazide $\xrightarrow{H^+}$

b.  pentanal + hydroxylamine $\xrightarrow{H^+}$

c.  2-butanone + 2,4-dinitrophenylhydrazine $\xrightarrow{H^+}$

---

## 19.3
## OTHER REACTIONS OF THE CARBONYL GROUP

### Halogenation: The Haloform Reaction

Carbonyl compounds react with halogens, in the presence of an acid catalyst, to give α-halogenated carbonyl compounds. In general,

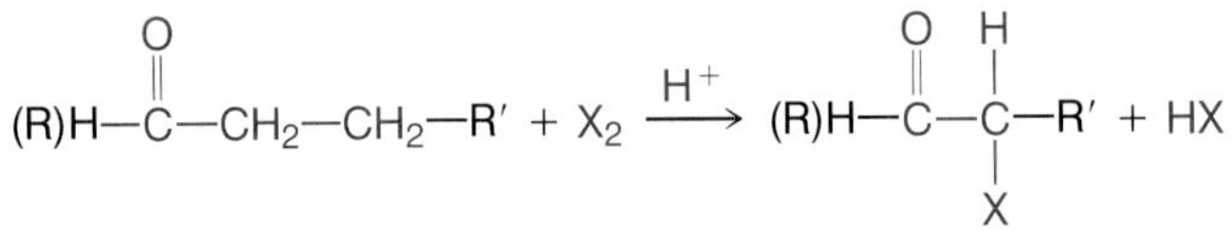

$$(R)H-\overset{O}{\overset{\|}{C}}-CH_2-CH_2-R' + X_2 \xrightarrow{H^+} (R)H-\overset{O}{\overset{\|}{C}}-\overset{H}{\underset{X}{C}}-R' + HX$$

where $X_2 = Cl_2$, $Br_2$, or $I_2$. The net effect of the reaction is the substitution of a halogen for an alpha hydrogen. For example:

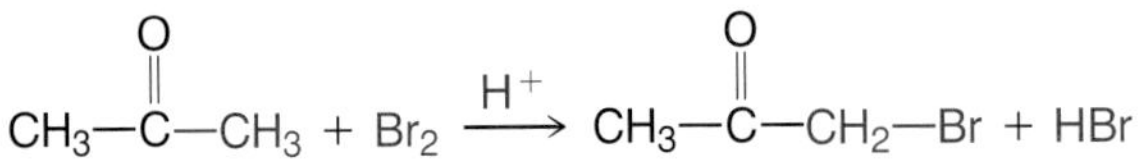

$$CH_3-\overset{O}{\overset{\|}{C}}-CH_3 + Br_2 \xrightarrow{H^+} CH_3-\overset{O}{\overset{\|}{C}}-CH_2-Br + HBr$$

α-Bromoacetone

The compound α-bromoacetone is a *lacrimal agent* (a tear-producing liquid).

When a halogenation is run in the presence of a basic *promoter*, rather than with an acid catalyst, polysubstitution takes place. In general,

A *promoter* is a substance that is used to increase the rate of a reaction, but unlike a catalyst, a promoter is used up in the course of the reaction.

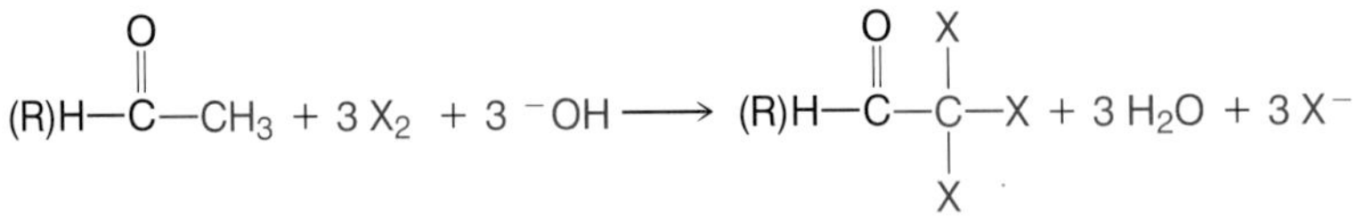

$$(R)H-\overset{O}{\overset{\|}{C}}-CH_3 + 3\,X_2 + 3\,{}^-OH \longrightarrow (R)H-\overset{O}{\overset{\|}{C}}-\overset{X}{\underset{X}{C}}-X + 3\,H_2O + 3\,X^-$$

where $X_2 = Cl_2$, $Br_2$, or $I_2$. For example:

Note that —C$_6$H$_5$ is a phenyl group.

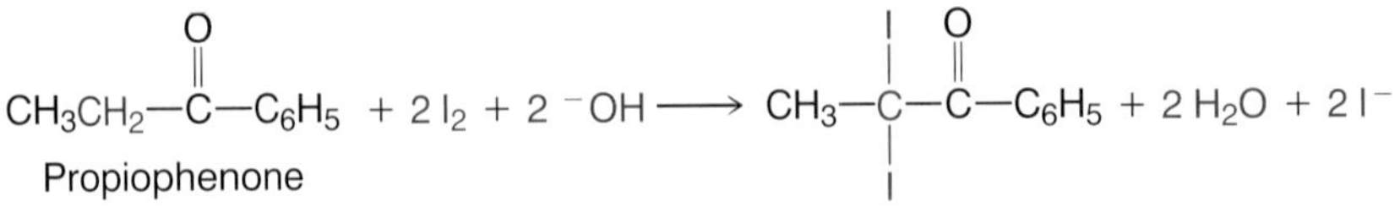

$$CH_3CH_2-\overset{O}{\overset{\|}{C}}-C_6H_5 + 2\,I_2 + 2\,{}^-OH \longrightarrow CH_3-\overset{I}{\underset{I}{C}}-\overset{O}{\overset{\|}{C}}-C_6H_5 + 2\,H_2O + 2\,I^-$$

Propiophenone

α,α-Diiodopropiophenone

This reaction has particular importance for methyl ketones.

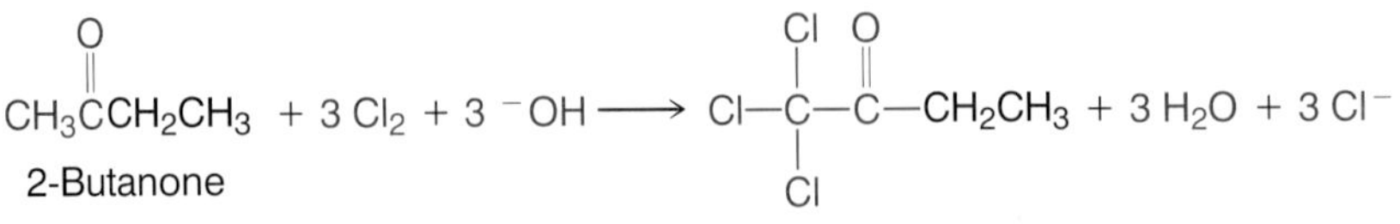

$$CH_3\overset{O}{\overset{\|}{C}}CH_2CH_3 + 3\,Cl_2 + 3\,{}^-OH \longrightarrow Cl-\overset{Cl}{\underset{Cl}{C}}-\overset{O}{\overset{\|}{C}}-CH_2CH_3 + 3\,H_2O + 3\,Cl^-$$

2-Butanone

1,1,1-Trichloro-2-butanone

Once the trihalo derivative is formed, it is decomposed by excess base to give a haloform and the salt of a carboxylic acid with one carbon less than the original carbonyl compound.

$$CH_3CH_2C(=O)CCl_3 + \ ^-OH \longrightarrow HCCl_3 + CH_3CH_2C(=O)-O^-$$

Chloroform    Propionate ion

All together, we have

$$CH_3CH_2CCH_3 + 4\ ^-OH + 3\ Cl_2 \longrightarrow CHCl_3 + CH_3CH_2C(=O)-O^- + 3\ Cl^- + 3\ H_2O$$

Butanone

In general,

$$(H)R-C(=O)-CH_3 + 4\ NaOH(aq) + 3\ X_2 \longrightarrow H(R)-C(=O)-O^-Na^+ + CHX_3 + 3\ NaX + 3\ H_2O$$

where $X_2 = Cl_2$, $Br_2$, or $I_2$. This reaction is known as the *haloform reaction* and is used in two different ways in the laboratory. First, it can be used as a synthetic tool to produce a carboxylic acid from a methyl ketone with one more carbon atom. Second, the reaction is employed as a qualitative test for a number of functional groups (see page 367).

---

Draw a structural formula for the ketone that would give each of the following carboxylic acids by the haloform reaction.

**SOLVED PROBLEM 19.6**

a.   $CH_3-C(=O)-OH$

b. 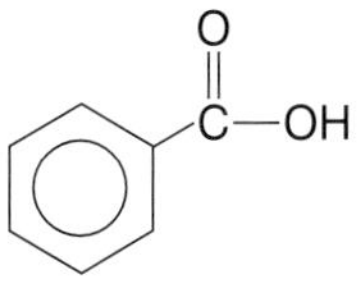   $C_6H_5-C(=O)-OH$

**SOLUTION**
The ketone must contain one carbon more than the carboxylic acid produced from the haloform reaction. Thus the ketone must contain three carbon atoms in (a) and eight carbons in (b). Since a methyl group also must be present in each ketone, we have

a.   $CH_3-C(=O)-CH_3$

b. 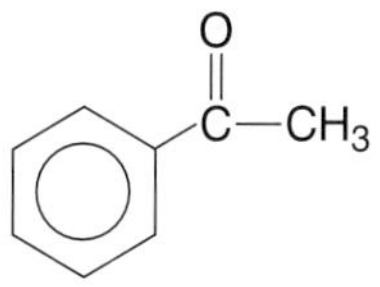   $C_6H_5-C(=O)-CH_3$

You also can obtain the structure of the ketone by replacing the —OH group of the corresponding carboxylic acid with a methyl group.

**EXERCISE 19.8**

Draw the structural formula of a ketone that yields the following acids using the iodoform reaction.

a. $CH_3CH_2COOH$

b. *p*-methylbenzoic acid (*p*-toluic acid)

In this text, if the haloform reaction is asked for in an equation, as in Exercise 19.9, base will be shown in excess; if not, the reaction is to stop at the polyhalogen-substituted ketone stage.

**EXERCISE 19.9**

Complete the following equations by drawing a structural formula for each organic product and writing a molecular or formula unit for each inorganic product formed. Do not balance these equations.

a. acetone + $I_2$ + NaOH(*aq*)(excess) $\longrightarrow$

b. acetophenone + $Br_2$ + NaOH(*aq*)(excess) $\longrightarrow$

c. acetophenone + $Cl_2$ + NaOH(aq) $\longrightarrow$

d. acetaldehyde + $Cl_2$ $\xrightarrow{H^+}$

Only one aldehyde shows the haloform reaction—acetaldehyde.

$$CH_3-\overset{\overset{\displaystyle O}{\|}}{C}-H \ + 3\,Br_2 + 4\ ^-OH \longrightarrow \ ^-O-\overset{\overset{\displaystyle O}{\|}}{C}-H + HCBr_3 + 3\,Br^- + 3\,H_2O$$

Acetaldehyde                                     Formate ion

First, the tribromo derivative of acetaldehyde is produced, followed by formate ion and bromoform.

The scope of the haloform reaction is expanded by the fact that certain alcohols (ethanol and methylcarbinols) also undergo the haloform reaction because they are oxidized by halogen and base to give acetaldehyde or a methyl ketone, respectively. In general,

$$CH_3CH(OH)R \xrightarrow[\text{NaOH}(aq)]{X_2} CH_3\overset{\overset{\displaystyle O}{\|}}{C}R \xrightarrow[\text{NaOH}(aq)(\text{excess})]{X_2} CHX_3 + \ ^+Na^-O\overset{\overset{\displaystyle O}{\|}}{C}R$$

where $X_2 = Cl_2$, $Br_2$, or $I_2$. For example:

$$CH_3CH(OH)CH_2CH_3 \xrightarrow[\text{NaOH}(aq)]{Cl_2} CH_3\overset{\overset{\displaystyle O}{\|}}{C}CH_2CH_3 \xrightarrow[\text{NaOH}(aq)(\text{excess})]{Cl_2} CHCl_3 + \ ^+Na^-O\overset{\overset{\displaystyle O}{\|}}{C}CH_2CH_3$$

The acetaldehyde (or methyl ketone) produced, in turn, reacts with halogen and base to yield the corresponding haloform and the salt of a carboxylic acid with one less carbon atom than the original carbonyl compound.

**SOLVED PROBLEM 19.7**

Which of the following alcohols undergo the haloform reaction? Explain.

a. $CH_3CH_2CH(OH)CH_2CH_3$

b. $CH_3CH_2CH_2OH$

c. 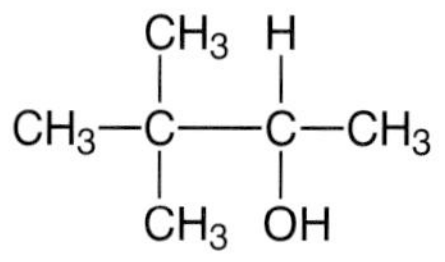
$$
\begin{array}{c}
\quad\;\; CH_3 \;\; H \\
\quad\;\; | \qquad | \\
CH_3-C-\!\!-C-CH_3 \\
\quad\;\; | \qquad | \\
\quad\;\; CH_3 \;\; OH
\end{array}
$$

## SOLUTION

a.  The alcohol is not a methylcarbinol and thus does not undergo the haloform reaction.

b.  The only primary alcohol to react is ethyl alcohol. Thus propyl alcohol does not undergo the haloform reaction.

c.  This alcohol is a methylcarbinol and does undergo the haloform reaction.

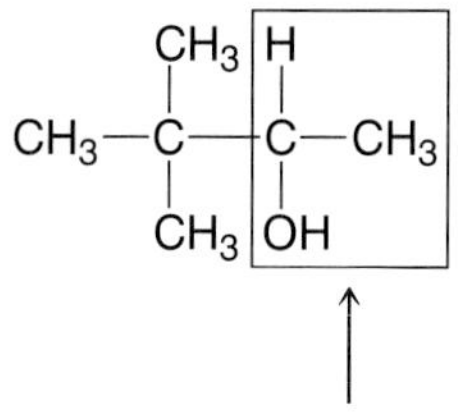

$$
\begin{array}{c}
\quad\;\; CH_3 \;\; H \\
\quad\;\; | \qquad | \\
CH_3-C-\!\!-C-CH_3 \\
\quad\;\; | \qquad | \\
\quad\;\; CH_3 \;\; OH
\end{array}
$$

A methylcarbinol

---

Which of the following alcohols undergo the haloform reaction? Explain.

EXERCISE 19.10

a.  dimethylcarbinol

b.  ethyl alcohol

c.  2-pentanol

d.  3-pentanol

---

Prepare benzoic acid ($C_6H_5COOH$) from 2-phenylethanol ($C_6H_5CH_2CH_2OH$). You may use any inorganic reagents necessary.

SOLVED PROBLEM 19.8

## SOLUTION

Since the starting compound contains one carbon more than the product acid, this is an instance where the haloform reaction can be used. But since the starting alcohol does not give the haloform reaction, we need to convert it to an alcohol that does show the reaction.

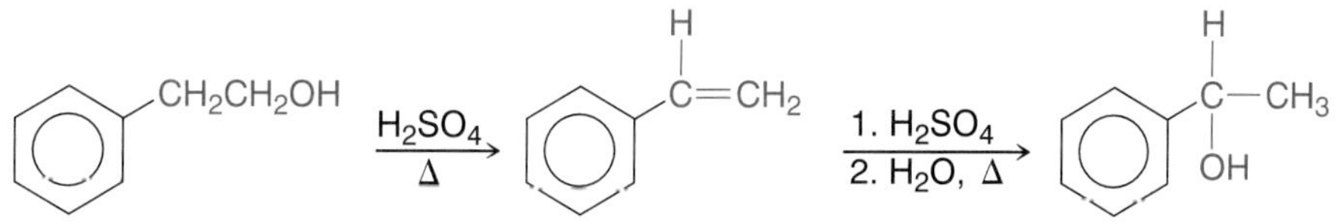

Now we have a methylcarbinol that will show the haloform reaction.

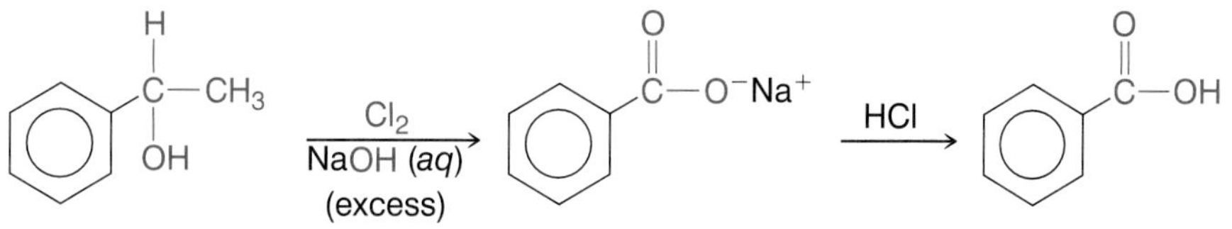

We can solve this synthesis problem by working backwards in the usual manner as follows: We need a methylcarbinol that contains one carbon more than benzoic acid in order to use the haloform reaction. Methylphenylcarbinol fills the bill nicely. Thus we have

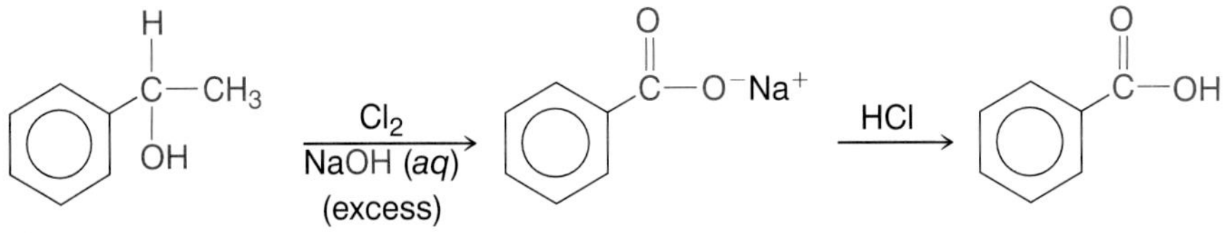

Now we simply need to produce methylphenylcarbinol from 2-phenylethanol.

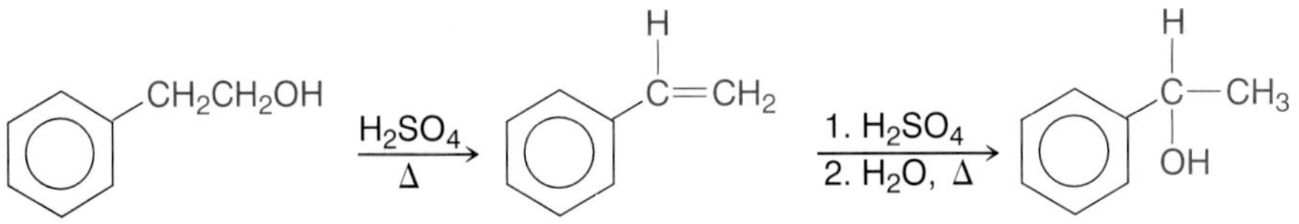

Putting this synthesis all together, we get

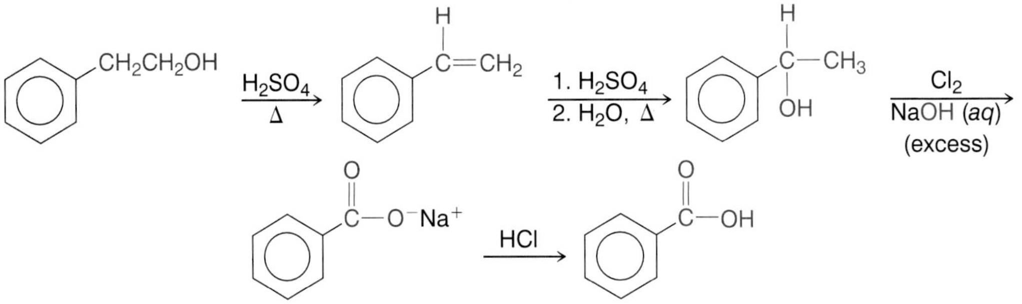

**EXERCISE 19.11**

Prepare acetic acid ($CH_3COOH$) from 2-propanol. You may use any inorganic reagents necessary.

The second way the haloform reaction can be used in the laboratory is as a qualitative test tube test for the following functional groups (Fig. 19.6) when the reagents used are iodine and sodium hydroxide [$I_2$ + NaOH($aq$)(excess)]. When one of these functional groups is present, iodoform, a yellow solid (mp = 119°C) precipitates out in an iodoform reaction. We call this qualitative test an *iodoform test*. Neither chloroform nor bromoform can serve as a qualitative indicator of the preceding groups, since both substances are colorless liquids.

**EXERCISE 19.12**

Which of the following compounds gives a positive iodoform test? Explain.

a.  3-pentanol

b.  2-pentanone

c.  formaldehyde

d.  2-butanol

**FIGURE 19.6** Structures of specific compounds and classes of compounds that undergo the haloform reaction.

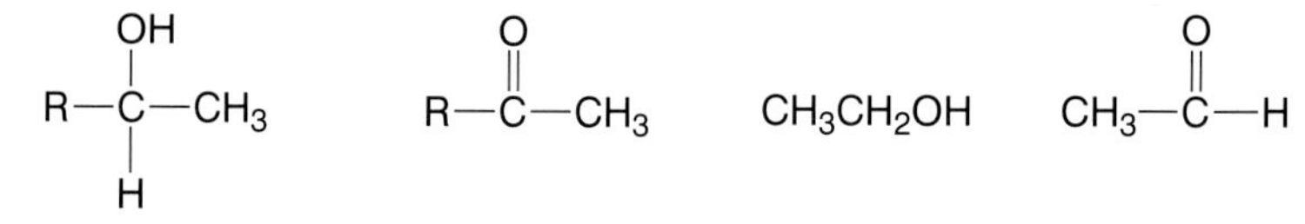

$$\underset{\text{A methylcarbinol}}{R-\overset{\overset{\displaystyle OH}{|}}{\underset{\underset{\displaystyle H}{|}}{C}}-CH_3} \qquad \underset{\text{A methyl ketone}}{R-\overset{\overset{\displaystyle O}{\|}}{C}-CH_3} \qquad \underset{\text{Ethanol}}{CH_3CH_2OH} \qquad \underset{\text{Acetaldehyde}}{CH_3-\overset{\overset{\displaystyle O}{\|}}{C}-H}$$

Aldehydes are easily oxidized. Even as weak an oxidizing agent as air converts an aldehyde to the corresponding carboxylic acid.

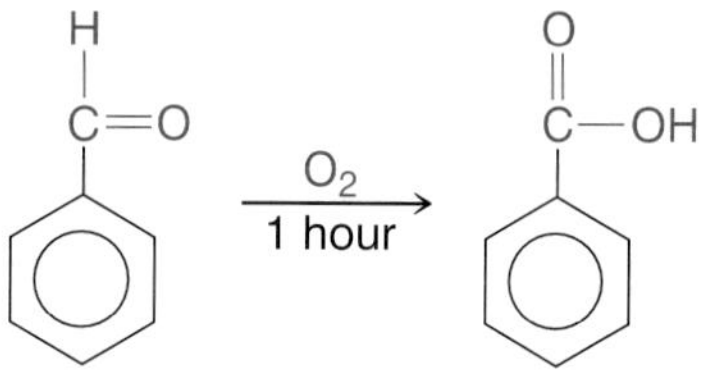

Oil
(benzaldehyde)

Solid
(benzoic acid)

**Oxidation**

Ketones, on the other hand, are resistant to oxidation. Thus an oxidation test can be used to distinguish an aldehyde from a ketone. The standard qualitative test used is the Tollens test (see Sec. 18.6). In general,

$$R-\overset{\overset{\displaystyle O}{\|}}{C}-H \;+\; 2\,{}^{+}Ag(NH_3)_2 \;+\; 3\,{}^{-}OH \longrightarrow R-\overset{\overset{\displaystyle O}{\|}}{C}-O^{-} \;+\; 4\,NH_3 \;+\; 2\,Ag \;+\; 2\,H_2O$$

Example:

$$CH_3-\overset{\overset{\displaystyle O}{\|}}{C}-H \;+\; 2\,{}^{+}Ag(NH_3)_2 \;+\; 3\,{}^{-}OH \longrightarrow CH_3-\overset{\overset{\displaystyle O}{\|}}{C}-O^{-} \;+\; 4\,NH_3 \;+\; 2\,Ag \;+\; 2\,H_2O$$

Aldehydes react with a silver ammonia complex to give a silver mirror coating on a clean test tube. Ketones do not react.

Aldehydes are reduced to primary alcohols and ketones to secondary alcohols by reacting sodium borohydride ($NaBH_4$) with water-soluble compounds. Ether-soluble carbonyl compounds are reduced by lithium aluminum hydride ($LiAlH_4$) (see Sec. 15.1). In general,

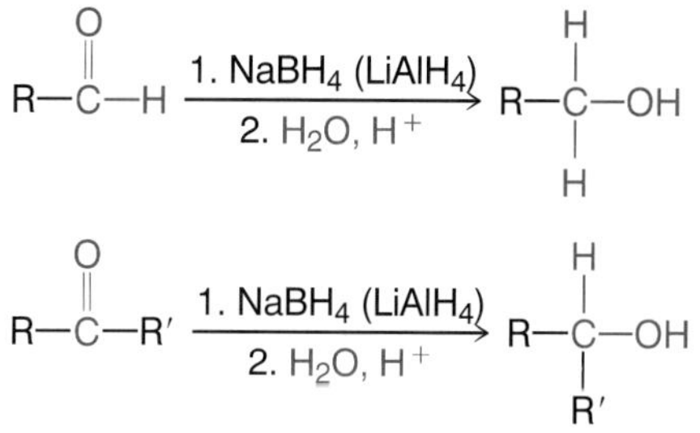

**Reduction of Aldehydes and Ketones**

$$R-\overset{\overset{\displaystyle O}{\|}}{C}-H \xrightarrow[\text{2. H}_2\text{O, H}^+]{\text{1. NaBH}_4\ (\text{LiAlH}_4)} R-\overset{\overset{\displaystyle H}{|}}{\underset{\underset{\displaystyle H}{|}}{C}}-OH$$

$$R-\overset{\overset{\displaystyle O}{\|}}{C}-R' \xrightarrow[\text{2. H}_2\text{O, H}^+]{\text{1. NaBH}_4\ (\text{LiAlH}_4)} R-\overset{\overset{\displaystyle H}{|}}{\underset{\underset{\displaystyle R'}{|}}{C}}-OH$$

Lithium aluminum hydride reacts explosively with water and alcohols and cannot be used with these solvents.

$$LiAlH_4 \;+\; 4\,H_2O \longrightarrow LiOH \;+\; Al(OH)_3 \;+\; 4\,H_2$$

$$LiAlH_4 \;+\; 4\,CH_3OH \longrightarrow LiOCH_3 \;+\; Al(OCH_3)_3 \;+\; 4\,H_2$$

Examples:

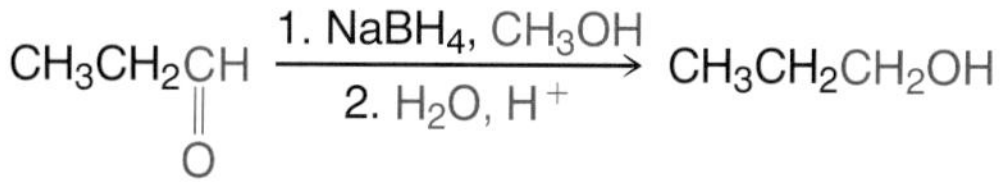

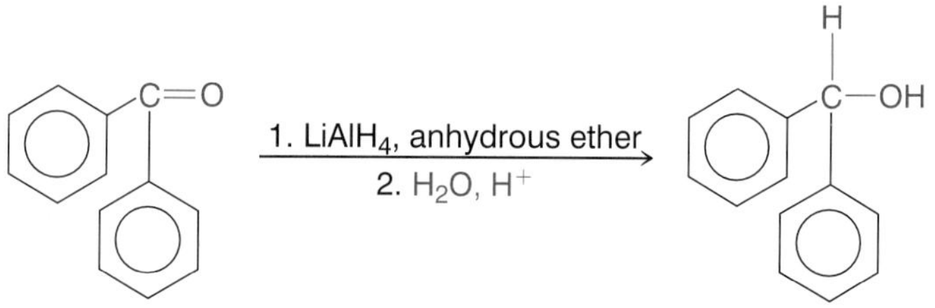

$$CH_3CH_2\underset{\underset{\displaystyle O}{\|}}{C}H \xrightarrow[\text{2. H}_2\text{O, H}^+]{\text{1. NaBH}_4,\ CH_3OH} CH_3CH_2CH_2OH$$

Benzophenone

Diphenylcarbinol

## Aldol Condensation

In the presence of a 10% solution of sodium hydroxide, 1 mol of an aldehyde or ketone with at least one alpha hydrogen condenses with another mol of the same carbonyl compound to give a β-hydroxyaldehyde or ketone containing twice the number of carbons as the original aldehyde or ketone. The simplest aldehyde to react is acetaldehyde; the product compound is called *aldol*. This general reaction, for this reason, is known as the *aldol condensation*. The word *aldol* is also used in a general way to describe any β-hydroxyaldehyde produced as a result of an aldol condensation because each contains an *ald*ehyde group and an alcoh*ol* group.

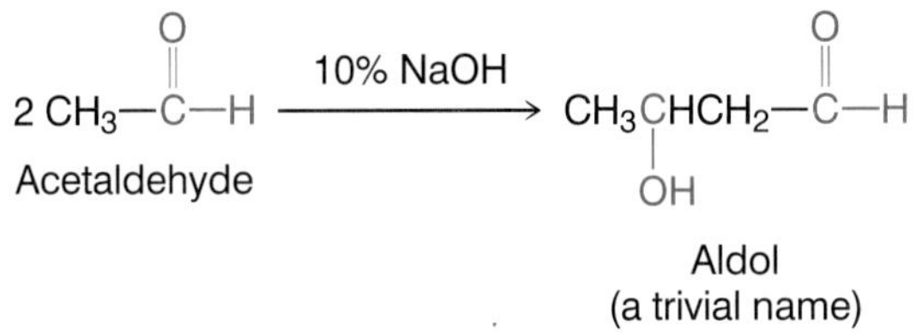

**FIGURE 19.7** The alpha carbon and the alpha hydrogens in acetaldehyde.

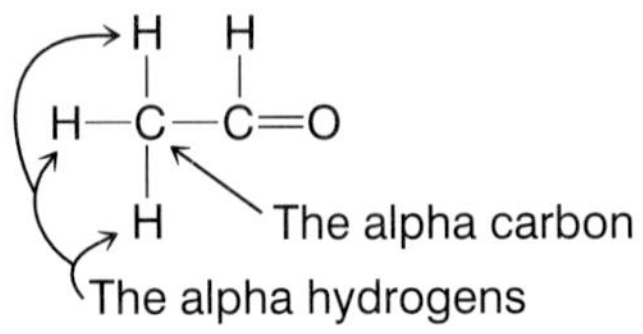

In order for this reaction to take place, at least one alpha hydrogen must be present in the carbonyl compound; acetaldehyde contains three alpha hydrogens (Fig. 19.7).

$$2\ CH_3-\overset{\underset{\|}{O}}{C}-H \xrightarrow{\text{10\% NaOH}} CH_3\underset{\underset{OH}{|}}{C}HCH_2-\overset{\underset{\|}{O}}{C}-H$$

Acetaldehyde

Aldol
(a trivial name)

## Envision the Reaction

Note that each step of this three step mechanism is reversible.

### Mechanism of the Aldol Condensation of Acetaldehyde

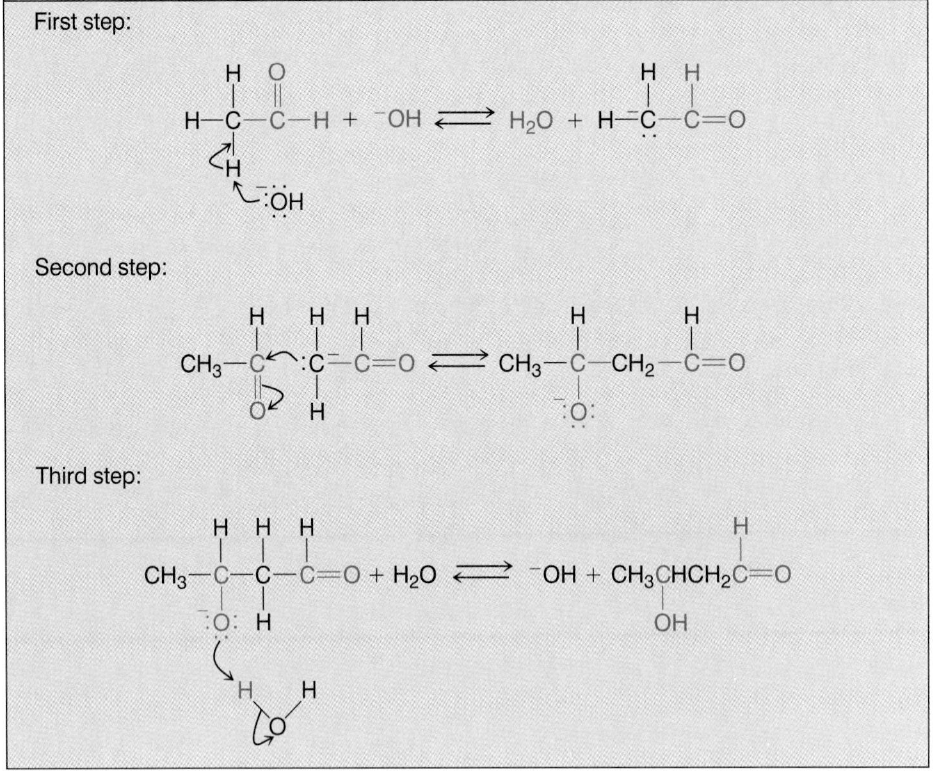

First step:

Second step:

Third step:

The first step of the mechanism shows that carbonyl compounds are slightly acidic due to the presence of one or more alpha hydrogens, in that hydroxide ion abstracts an alpha hydrogen (proton) from the carbonyl compound to give a carbanion. Aldehydes and ketones containing at least one alpha hydrogen are slightly acidic due to resonance stabilization of the corresponding carbanion. This resonance stabilization provides a driving force to promote the loss of a proton, i.e., acidity.

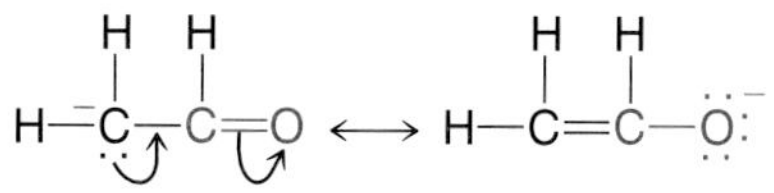

Now a reasonable question to ask is, How acidic is the typical aldehyde or ketone containing one or more alpha hydrogens? The typical $pK_a$ of such a compound is 20. Thus these compounds are less acidic than water and the alcohols but more acidic than the typical alkyne containing a terminal hydrogen (see Table 1.6). In the second step of the mechanism, the nucleophilic carbanion produced attacks the carbonyl carbon of another molecule of acetaldehyde in the usual manner to produce an alkoxide ion ($^-OR$). In the final step of this mechanism, the alkoxide ion abstracts a proton from water to generate aldol and regenerate hydroxide ion.

a. How many alpha hydrogens does a molecule of propionaldehyde contain?

b. Draw a detailed, step-by-step mechanism of an aldol condensation with two molecules of propionaldehyde.

**SOLUTION**

a. Two alpha hydrogens:

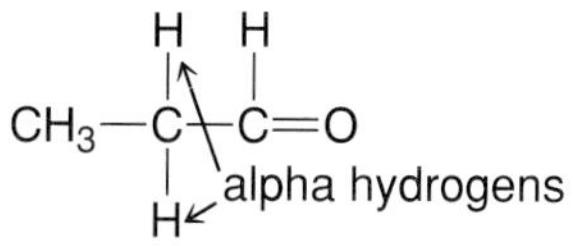

b. *Step 1:* Abstract an alpha hydrogen (proton) from the carbonyl compound with base.

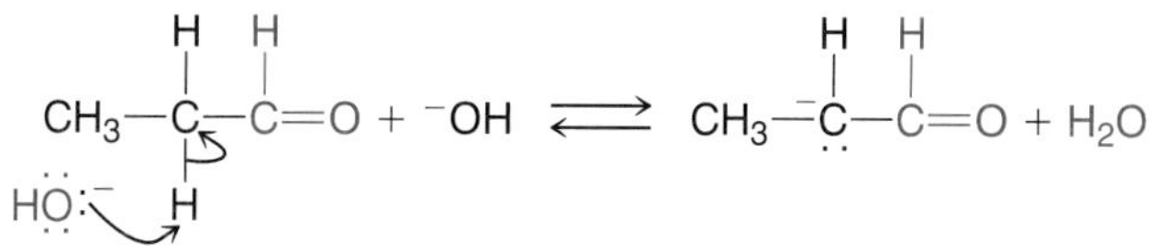

*Step 2:* Attack the carbonyl carbon of another molecule with the nucleophilic carbanion just produced.

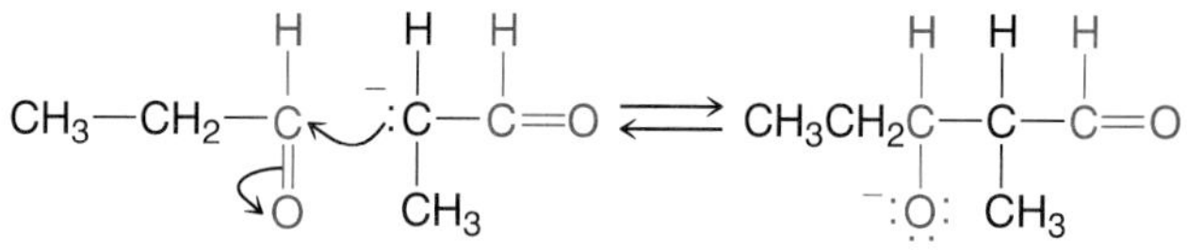

*Step 3:* The alkoxide ion just produced abstracts a proton from water to yield the aldol and regenerate hydroxide ion.

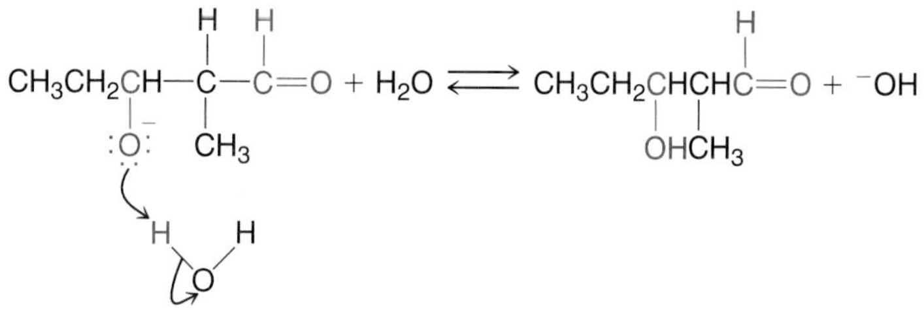

a. How many alpha hydrogens does a molecule of butanal contain?

b. Draw a detailed, step-by-step mechanism of an aldol condensation with two molecules of butanal.

The aldol condensation has great synthetic utility. Consider the following starting with aldol:

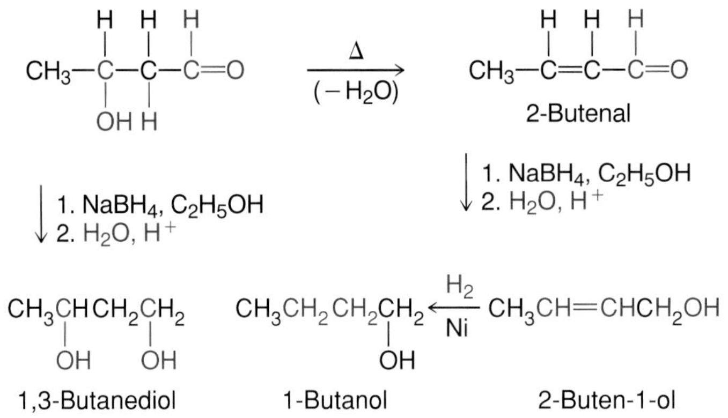

$$
\underset{\underset{\text{OH H}}{|} \quad |}{CH_3-\overset{H}{\underset{|}{C}}-\overset{H}{\underset{|}{C}}-\overset{H}{\underset{|}{C}}=O} \quad \xrightarrow[(-H_2O)]{\Delta} \quad \underset{\text{2-Butenal}}{CH_3-\overset{H}{C}=\overset{H}{C}-\overset{H}{C}=O}
$$

| 1. NaBH$_4$, C$_2$H$_5$OH
| 2. H$_2$O, H$^+$

1. NaBH$_4$, C$_2$H$_5$OH
2. H$_2$O, H$^+$

$$
\underset{\underset{\text{OH}\quad\text{OH}}{|\quad\quad|}}{CH_3CHCH_2CH_2} \qquad \underset{\underset{\text{OH}}{|}}{CH_3CH_2CH_2CH_2} \xleftarrow[\text{Ni}]{H_2} CH_3CH=CHCH_2OH
$$

1,3-Butanediol               1-Butanol               2-Buten-1-ol

Since the aldol condensation yields compounds containing two different functional groups, let us further expand our listing of functional-group priorities in IUPAC nomenclature as follows:

$$
\overset{O}{\underset{||}{C}}-H \;>\; \overset{O}{\underset{||}{C}} \;>\; OH \;>\; SH \;>\; C{=}C \;>\; C{\equiv}C \;>\; OR \;>\; Cl(Br)(I)
$$

Consider the IUPAC name of the following compound:

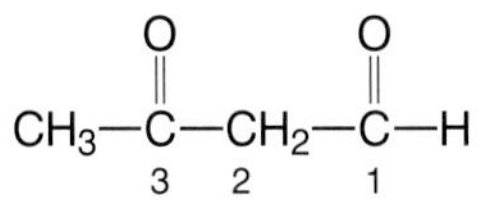

$$
CH_3-\underset{3}{\overset{O}{\underset{||}{C}}}-\underset{2}{CH_2}-\underset{1}{\overset{O}{\underset{||}{C}}}-H
$$

The prefix *oxo-* represents a keto group at the carbon number designated when the keto group does not show the higher priority in the molecule. Since the aldehyde group shows a greater priority than the keto group, we have

3-oxobutanal

---

**SOLVED PROBLEM 19.10**

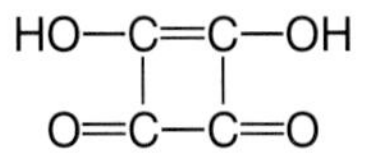

$$
\begin{array}{c}
HO-C{=}C-OH \\
|\quad\quad| \\
O{=}C-C{=}O
\end{array}
$$

The trivial name of this compound is squaric acid because the four carbon atoms contained in the molecule take the shape of a square. Name the compound using the IUPAC system of nomenclature.

**SOLUTION**

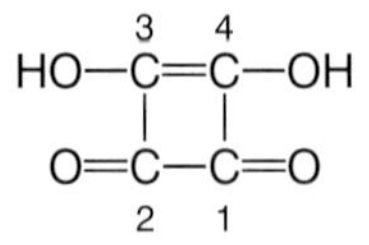

$$
\begin{array}{c}
\overset{3\quad\;4}{HO-C{=}C-OH} \\
|\quad\quad| \\
\underset{2\quad\;1}{O{=}C-C{=}O}
\end{array}
$$

The functional group order of priority is

$$
\overset{O}{\underset{||}{C}} \;>\; OH \;>\; C{=}C
$$

Thus we have

3,4-dihydroxy-3-cyclobutene-1,2-dione

Provide an IUPAC name for each of the following compounds.

a.  aldol (Supply another common name for aldol.)

b.  CH₃CH(OH)CH₂CCH₃ (with O double-bonded to the final C)

$$CH_3CH(OH)CH_2\overset{\overset{\displaystyle O}{\|}}{C}CH_3$$

c.  The aldol formed from the aldol condensation of two mol of propionaldehyde in Solved Problem 19.9. Supply a common name for this aldol.

When two alpha hydrogen–containing carbonyl compounds are treated with base in an aldol condensation, four different products are obtained, and the reaction has no synthetic utility. However, if one carbonyl compound without an alpha hydrogen and one with at least one alpha hydrogen are mixed in base, a smooth mixed aldol condensation takes place.

**Mixed Aldol Condensation**

Predict the main product of the following mixed aldol condensation. What by-product is possible?

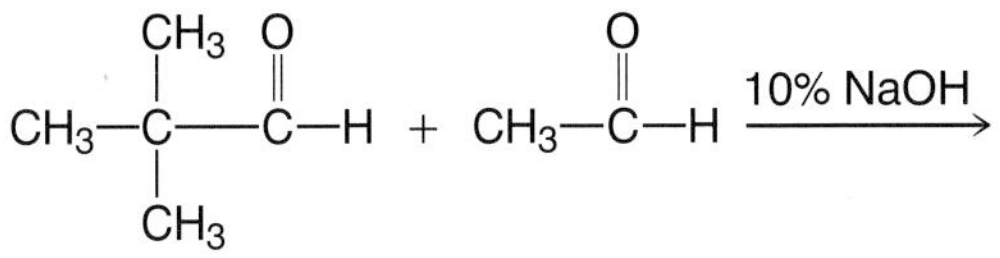

**SOLUTION**

Since acetaldehyde is the only aldehyde with alpha hydrogens, only acetaldehyde can form the carbanion.

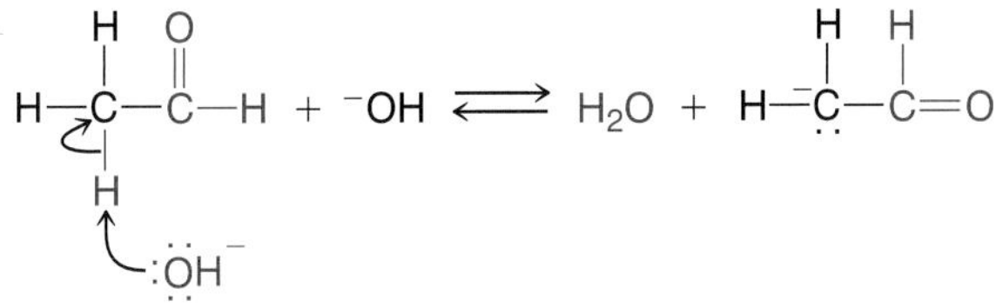

This carbanion attacks the carbonyl carbon of 2,2-dimethylpropanal.

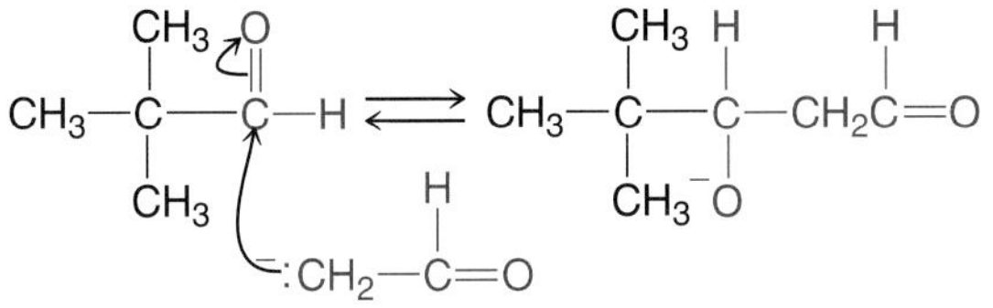

The alkoxide ion accepts a proton from water to yield the aldol and hydroxide ion.

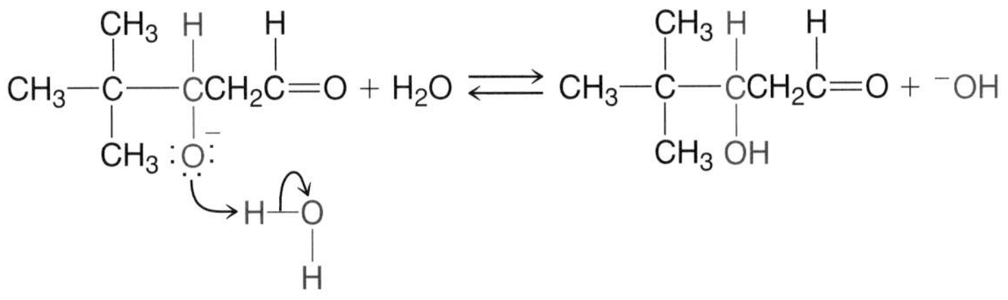

A limited amount of 3-hydroxybutanal forms and is the by-product. This occurs when the carbanion derived from acetaldehyde attacks the carbonyl carbon of another molecule of acetaldehyde.

**EXERCISE 19.15**

Predict the main product of the following mixed aldol condensations. What by-product(s) is (are) possible in each condensation?

a. 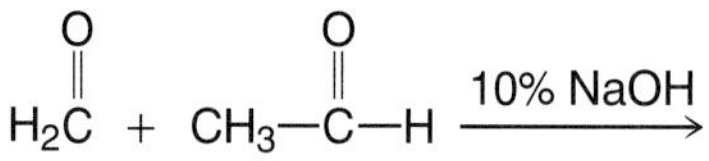   $H_2C$  +  $CH_3-\overset{O}{\underset{\|}{C}}-H$  $\xrightarrow{\text{10\% NaOH}}$

b.   acetaldehyde + benzaldehyde  $\xrightarrow{\text{10\% NaOH}}$

## 19.4
## MORE SYNTHESES

Since our supply of reactions used for synthesis is growing, let us look at some more complex synthesis problems.

**SOLVED PROBLEM 19.12**

Prepare 2-methyl-2-butanol from any alcohol containing three carbons or less. You may use any inorganic reagents necessary.

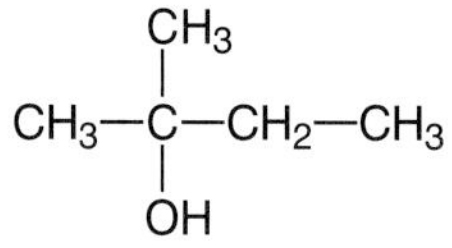

$$CH_3-\underset{\underset{OH}{|}}{\overset{\overset{CH_3}{|}}{C}}-CH_2-CH_3$$

**SOLUTION**

Since the product molecule is a tertiary alcohol, and since the product molecule contains more carbons than any of the reactant molecules, this suggests the use of a Grignard reagent synthesis. Let us start with our product molecule and work backward in the usual manner.

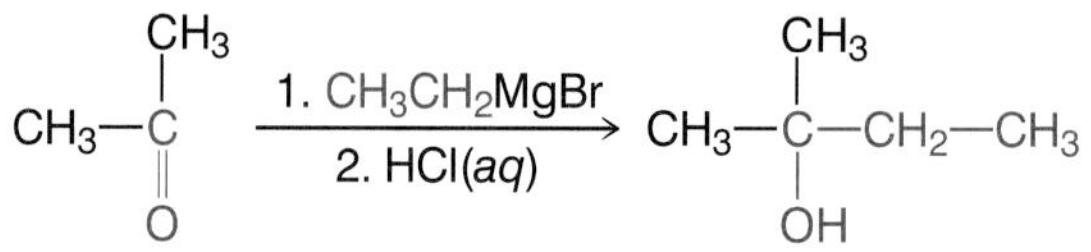

$$CH_3-\underset{\underset{O}{\|}}{\overset{\overset{CH_3}{|}}{C}} \xrightarrow[\text{2. HCl}(aq)]{\text{1. CH}_3\text{CH}_2\text{MgBr}} CH_3-\underset{\underset{OH}{|}}{\overset{\overset{CH_3}{|}}{C}}-CH_2-CH_3$$

Now we must prepare acetone from a three-carbon alcohol.

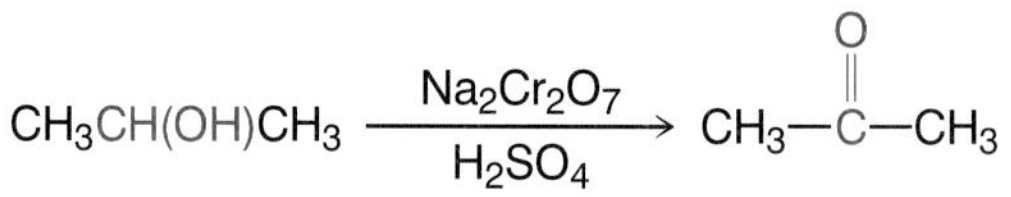

$$CH_3CH(OH)CH_3 \xrightarrow[\text{H}_2\text{SO}_4]{\text{Na}_2\text{Cr}_2\text{O}_7} CH_3-\overset{O}{\underset{\|}{C}}-CH_3$$

Finally, we have to produce ethylmagnesium bromide from an alcohol.

$$CH_3CH_2OH \xrightarrow{\text{HBr}} CH_3CH_2Br \xrightarrow[\substack{\text{Anhydrous}\\\text{ether}}]{\text{Mg}} CH_3CH_2MgBr$$

Putting all this together, we get

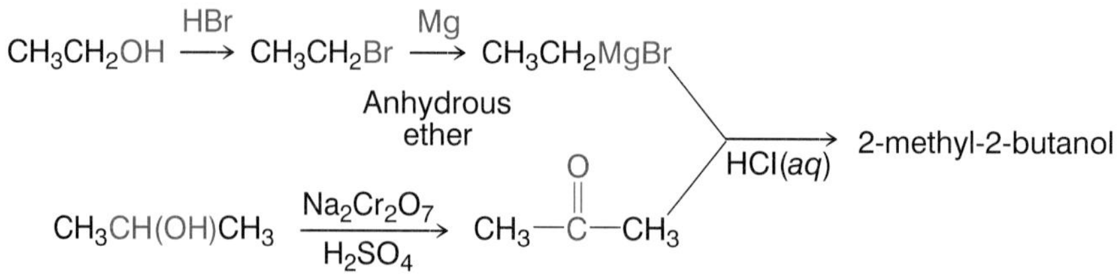

$$CH_3CH_2OH \xrightarrow{\text{HBr}} CH_3CH_2Br \xrightarrow[\substack{\text{Anhydrous}\\\text{ether}}]{\text{Mg}} CH_3CH_2MgBr$$

$$CH_3CH(OH)CH_3 \xrightarrow[\text{H}_2\text{SO}_4]{\text{Na}_2\text{Cr}_2\text{O}_7} CH_3-\overset{O}{\underset{\|}{C}}-CH_3 \xrightarrow{\text{HCl}(aq)} \text{2-methyl-2-butanol}$$

Prepare

$$CH_3CH_2\overset{\overset{\displaystyle H}{|}}{\underset{\underset{\displaystyle C\equiv N}{|}}{C}}{-}OH$$

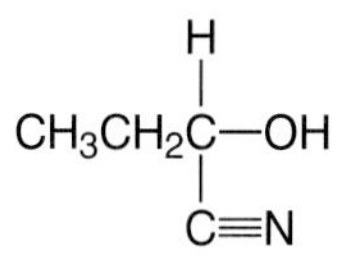

from $CH_3CH_2CH_2OH$. You may use any inorganic and organic reagents necessary.

**SOLUTION**

First, we need to recognize that $CH_3CH_2CH(OH)(CN)$ represents the cyanohydrin of propionaldehyde. Thus we have

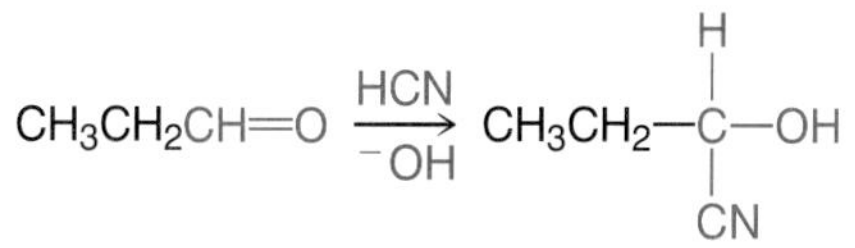

$$CH_3CH_2CH{=}O \xrightarrow[\ ^-OH]{HCN} CH_3CH_2{-}\overset{\overset{\displaystyle H}{|}}{\underset{\underset{\displaystyle CN}{|}}{C}}{-}OH$$

Finally, we can prepare propionaldehyde from *n*-propyl alcohol.

$$CH_3CH_2CH_2OH \xrightarrow[\substack{\text{Methylene} \\ \text{chloride}}]{PCC} CH_3CH_2CH{=}O$$

Putting it all together, we have

$$CH_3CH_2CH_2OH \xrightarrow[\substack{\text{Methylene} \\ \text{chloride}}]{PCC} CH_3CH_2CH{=}O \xrightarrow[\ ^-OH]{HCN} CH_3CH_2CH(OH)(CN)$$

Prepare 2-methyl-2-butanol from 2-butanol and methanol. Use any inorganic and organic reagents necessary.

Prepare propionaldehyde cyanohydrin from isopropyl alcohol. Use any inorganic and organic reagents necessary.

## ▶ CHAPTER ACCOMPLISHMENTS

### 19.1  Synthesis of Aldehydes and Ketones
☐ Write a general equation to illustrate each method of synthesis of a carbonyl compound.
☐ Explain why ozonides are not isolated.

### 19.2  Addition Reactions of the Carbonyl Group
☐ Draw the dipole structure of the carbonyl group.
☐ Explain why formaldehyde produces a stable hydrate and why diethyl ketone does not.
☐ Explain why hydrogen cyanide gas is not directly bubbled into a solution of an aldehyde to form a cyanohydrin.
☐ List the kinds of carbonyl compounds that do not react with HCN.

☐ Write a detailed mechanism for the reaction of 1 mol of acetaldehyde with 2 mol of ethyl alcohol in the presence of a mineral acid to produce the acetal.
☐ Provide a piece of experimental evidence that proves that acetal formation is a reversible process.
☐ Assign a class to the alcohol produced from a Grignard synthesis when the carbonyl compound used is

  a. formaldehyde.

  b. any other aldehyde (with the exception of formaldehyde).

  c. any ketone.

☐ Select a Grignard reagent and a carbonyl compound that would produce 3-pentanol using the Grignard synthesis.

☐ Draw a structural formula for
  a. benzaldehyde imine.
  b. acetone oxime.
  c. 2-methylcyclopentanone 2,4-dinitrophenylhydrazone.
  d. methyl ethyl ketone semicarbazone.

## 19.3 Other Reactions of the Carbonyl Group

☐ Complete the following equations.

a. $CH_3-\overset{\displaystyle O}{\overset{\|}{C}}-CH_3 \; + \; Br_2 \; \xrightarrow{\;H^+\;}$

b. $CH_3CH_2-\overset{\displaystyle O}{\overset{\|}{C}}-C_6H_5 \; + \; 2\,I_2 \; \xrightarrow{\;^-OH\;}$

c. $C_6H_5-\overset{\displaystyle O}{\overset{\|}{C}}-CH_3 \; + \; I_2 \; + \; ^-OH(\text{excess}) \longrightarrow$

☐ Prepare benzoic acid from acetophenone.
☐ Prepare sodium propionate from *sec*-butyl alcohol.
☐ Write an equation representing the Tollens test for an aldehyde.

☐ Tell why lithium aluminum hydride cannot be used as a reducing agent in any reaction in which ethyl alcohol is a solvent.
☐ Write a detailed mechanism for the following reaction.

$$2\,CH_3-\overset{\displaystyle O}{\overset{\|}{C}}-H \; \xrightarrow{\;10\%\ NaOH\;} \; CH_3CH(OH)CH_2-\overset{\displaystyle O}{\overset{\|}{C}}-H$$

☐ Explain why an alpha hydrogen in acetaldehyde is slightly acidic.
☐ Write an equation (giving both the major product and the minor product) illustrating a mixed aldol condensation in which acetaldehyde reacts with 2,2-dimethylpropanal in the presence of 10% NaOH.

## 19.4 More Syntheses

☐ Prepare 2-methyl-2-butanol from any alcohol containing three carbons or less.
☐ Prepare propionaldehyde cyanohydrin from *n*-propyl alcohol.

---

### ▶ KEY TERMS

*hydrate* (19.2)
*cyanohydrin* (19.2)
*imine* (19.2)
*Schiff base* (19.2)

*lacrimal agent* (19.3)
*haloform reaction* (19.3)
*iodoform test* (19.3)
*aldol* (19.3)

*aldol condensation* (19.3)
*soporific* (Historical Boxed Reading 19.1)

---

### ▶ PROBLEMS

1. Name the following compounds using the common and IUPAC systems of nomenclature.

a. $CH_3CH_2\overset{\displaystyle H}{\underset{\displaystyle C\equiv N}{\overset{|}{\underset{|}{C}}}}-OH$
   (Common only)

b. $CH_3-\overset{\displaystyle H}{\overset{|}{C}}=NH$

c. $CH_3CH_2CH_2CH_2-\overset{\displaystyle H}{\overset{|}{C}}=N-OH$

d. $CH_3-\underset{\underset{\displaystyle NHCNH_2}{\underset{\displaystyle |\quad\|}{N\;\;O}}}{\overset{|}{C}}-CH_2CH_2CH_3$

e. $CH_3CH_2\overset{\displaystyle H}{\underset{\displaystyle OH}{\overset{|}{\underset{|}{C}}}}-OH$

2. Draw a structural formula for each of the following compounds.
  a. an acetal derived from diethyl ketone and 1,3-propanediol
  b. a hemiacetal derived from propionaldehyde and methyl alcohol

3. Draw a structural formula for
  a. acetone 2,4-dinitrophenylhydrazone.
  b. pentanal semicarbazone.
  c. 2-ethoxy-2-propanol.
  d. 2,2 butanediol. (This compound is unstable.)
  e. 2-oxohexanal.

4. a. Given that hydrazine is $NH_2NH_2$, draw a structural formula for butyraldehyde hydrazone.
  b. Given that acetophenone is

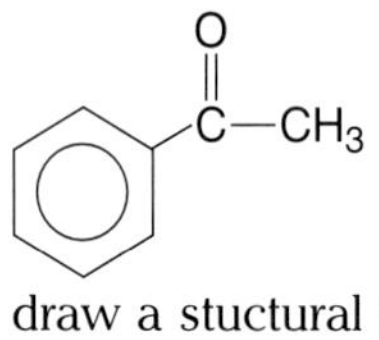

$$\text{C}_6\text{H}_5-\overset{\displaystyle O}{\overset{\|}{C}}-CH_3$$

draw a stuctural formula for acetophenone hydrate.

c. Given that

$$CH_3\overset{\underset{|}{H}}{C}{=}O \ + \ NH_3 \longrightarrow CH_3-\overset{\underset{|}{H}}{C}{=}NH \ + \ H_2O$$

then $CH_3\overset{\underset{|}{H}}{C}{=}O \ + \ HNCH_3 \overset{\underset{|}{H}}{\longrightarrow}$

5. Provide the nucleophile in each of the following reactions with acetaldehyde.
   a. hydration
   b. acetal formation with $CH_3OH$ and $H^+$
   c. semicarbazone formation
   d. $CH_3MgBr$
   e. cyanohydrin formation
   f. reduction with $NaBH_4$

6. Write an equation for the reaction of propionaldehyde with each of the following (if no reaction occurs, write "no reaction"). You need not balance each equation.
   a. 2,4-dinitrophenylhydrazine $\overset{H^+}{\longrightarrow}$
   b. semicarbazide $\overset{H^+}{\longrightarrow}$
   c. ammonia $\longrightarrow$
   d. HCN $\overset{^-OH}{\rightleftarrows}$
   e. $H_2O \rightleftarrows$
   f. $^+Ag(NH_3)_2 + {^-}OH \longrightarrow$
   g. $I_2 + NaOH(aq)(excess) \longrightarrow$
   h. $C_2H_5OH(1\ mol) \overset{H^+}{\rightleftarrows}$
   i. $I_2 \overset{H^+}{\longrightarrow}$
   j. $2\ C_2H_5OH \overset{H^+}{\longrightarrow}$
   k. $I_2 + NaOH(aq) \longrightarrow$

7. Write an equation for the reaction of methyl ethyl ketone with each of the reagents in Problem 6 (if no reaction occurs, write "no reaction"). You need not balance each equation.

8. Complete the following equations by drawing a structural formula for each organic product and writing a molecular formula or formula unit for each inorganic compound produced.

   a. $CH_3CH_2\overset{\underset{|}{H}}{C}{=}\overset{\underset{|}{H}}{C}-\overset{\underset{|}{H}}{C}{=}O \quad \overset{1.\ NaBH_4}{\underset{2.\ H_2O,\ H^+}{\longrightarrow}}$

   (Organic product only)

   b. $CH_3CH_2\overset{\underset{|}{H}}{\underset{\underset{|}{OCH_2CH_3}}{C}}{-}OCH_2CH_3 \ + \ \overset{(Excess)}{H_2O} \overset{H^+}{\rightleftarrows}$

   c. $CH_3CH_2MgBr \overset{1.\ CH_3\overset{\underset{||}{O}}{C}CH_2CH_3}{\underset{2.\ HCl(aq)}{\longrightarrow}}$

d. $6\ \overset{\underset{||}{O}}{C}H_2 + 4\ NH_3 \longrightarrow$

e. 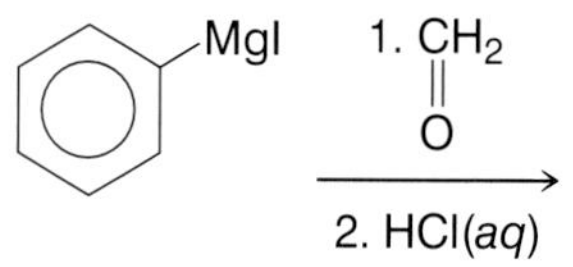 MgI $\overset{1.\ \overset{\underset{||}{O}}{C}H_2}{\underset{2.\ HCl(aq)}{\longrightarrow}}$

f. $(CH_3)_2C{=}C(CH_3)_2 \overset{1.\ O_3}{\underset{2.\ Zn,\ H_2O}{\longrightarrow}}$

   (Organic product only)

g. $CH_3CH_2C{\equiv}CH + H_2O \overset{H_2SO_4}{\underset{HgSO_4}{\longrightarrow}}$

   (Organic product only)

h. 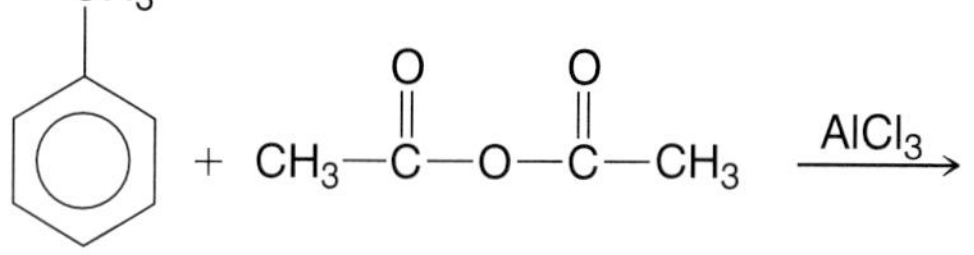 $+ CH_3-\overset{\underset{||}{O}}{C}-O-\overset{\underset{||}{O}}{C}-CH_3 \overset{AlCl_3}{\longrightarrow}$

i. $CH_3CH_2CH_2OH \overset{PCC}{\underset{CH_2CL_2}{\longrightarrow}}$

   (No products derived from PCC)

j. 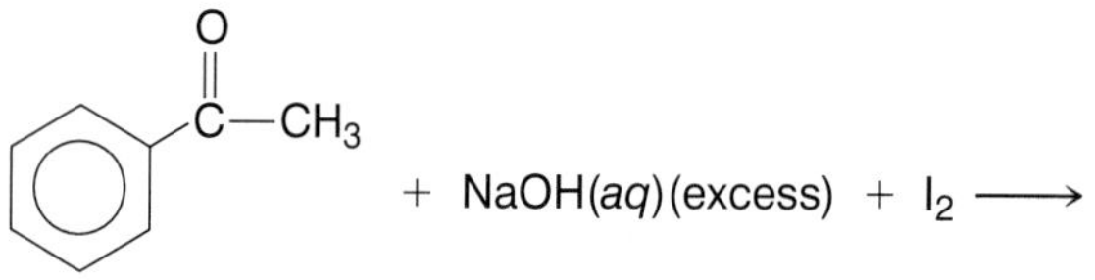 $+ NaOH(aq)(excess) + I_2 \longrightarrow$

9. a. Draw a Fischer projection for (i) (R)-aldol and (ii) (S)-propionaldehyde cyanohydrin.
   b. Prepare cinnamaldehyde ($C_6H_5-CH{=}CH-CH{=}O$) from benzyl chloride, ethyl chloride, and any inorganic reagents necessary.
   c. Why is it fairly easy to dehydrate the product of an aldol condensation? For example, in the dehydration of aldol itself, no acid is needed. Explain.

$$CH_3CH(OH)CH_2CH{=}O \overset{\Delta}{\underset{(-\ H_2O)}{\longrightarrow}} CH_3CH{=}CHCH{=}O$$

10. Write a detailed, step-by-step mechanism of
    a. the aldol condensation. Use 2 mol of acetone.
    b. the addition of $HC{\equiv}N$ to a methyl ketone. Use methyl ethyl ketone.
    c. the formation of formaldehyde dimethyl acetal.
    d. the formation of acetone semicarbazone.

11. a. Supply an IUPAC name for formaldehyde hydrate and an IUPAC name for chloral hydrate.

b. Which hydrate is most likely to form? Explain. Assign a common and an IUPAC name to each hydrate.
   (i) $CH_2(OH)_2$
   (ii) $CH_3CH(OH)_2$
   (iii) $(CH_3)_2C(OH)_2$

12. Prepare each of the following compounds. You may use any inorganic reagents necessary.

    a. 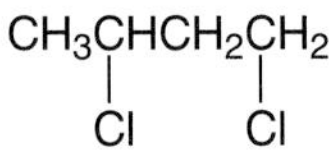 $CH_3CHCH_2CH_2$ from $CH_3CH_2OH$
       with Cl and Cl substituents

    b. 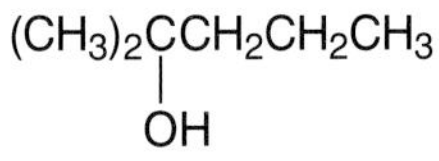 $(CH_3)_2CCH_2CH_2CH_3$ from $CH_3CH_2CH_2OH$
       with OH substituent

    c. 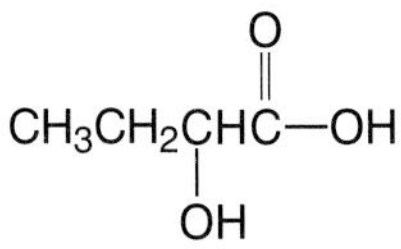 $CH_3CH_2CHC$—OH (with C=O) from $CH_3CH_2CH$ (with C=O)
       with OH substituent

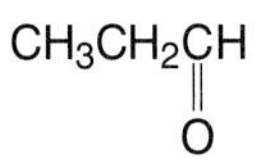

    d. $CH_3CH_2CH_2COOH$ from $CH_3CH_2CH_2CH=CH_2$
       (two ways)

    e. 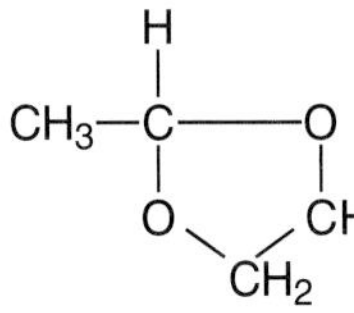 $CH_3-C$ (with H) ——O from $CH_3CH_2OH$ and $CH_2CH_2$ (with OH OH)
       ring with O, $CH_2$, $CH_2$

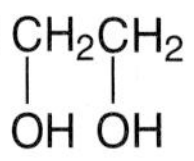

13. a. Why must a catalytic amount of base be present in cyanohydrin formation?
    b. How is hydrogen cyanide ($HC{\equiv}N$) produced in the laboratory for cyanohydrin formation?

14. Using a test tube reaction, how would you distinguish between each of the following pairs of compounds. Indicate what reagents should be used and what results you would expect to see.

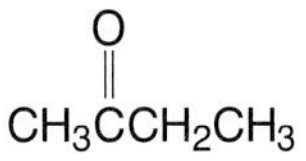

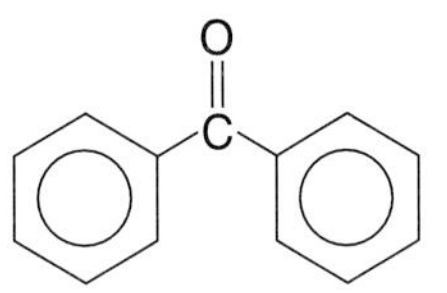

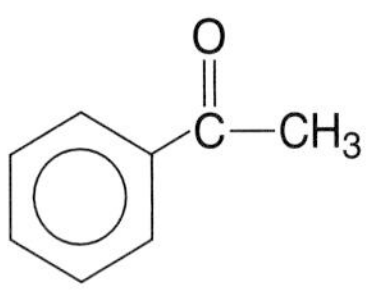

    a. 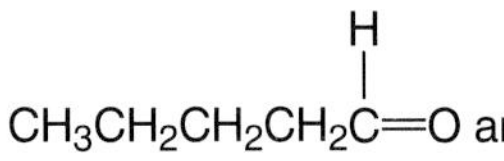 $CH_3CH_2CH_2CH_2C$ (with H) $=O$ and $CH_3CCH_2CH_3$ (with C=O)

    b. diphenyl ketone and acetophenone

    c. $CH_3CH_2CCH_2CH_3$ (with C=O) and $CH_3CH_2OCH_2CH_2CH_3$

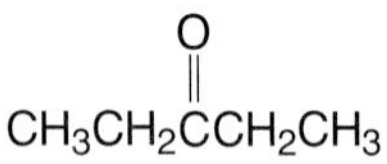

    d. 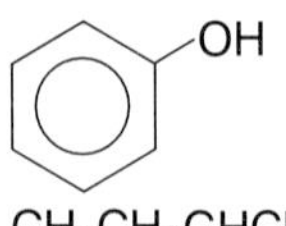 phenol (ring-OH) and benzyl alcohol (ring-$CH_2OH$)

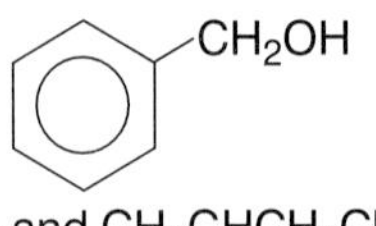

    e. 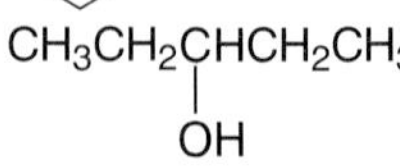 $CH_3CH_2CHCH_2CH_3$ and $CH_3CHCH_2CH_2CH_3$
       both with OH substituent

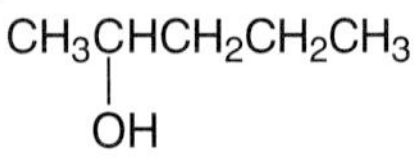

15. Prepare each of the following from propionaldehyde. You may use any inorganic reagents necessary.
    a. $CH_3CH_2CH_2OH$
    b. $CH_3CH_2COOH$
    c. 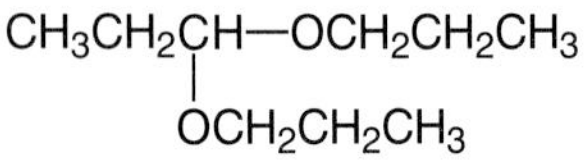 $CH_3CH_2CH-OCH_2CH_2CH_3$
       with $OCH_2CH_2CH_3$ substituent
    d. 3-hexanol

e. 2-methyl-2-pentenal
f. 2-methyl-2-penten-1-ol
g. $CH_3CH_2CH_2-O-CH_2CH_2CH_3$

16. Provide a structural formula for the Grignard reagent and carbonyl compound combination(s) needed to prepare each of the following alcohols.

    a. 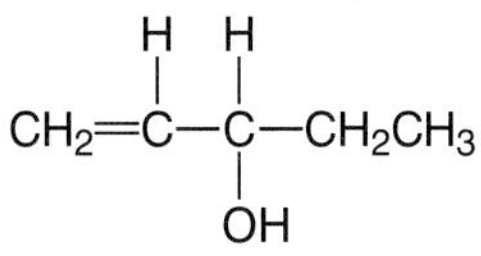 $CH_2{=}C-C-CH_2CH_3$ (with H H and OH)

    b. 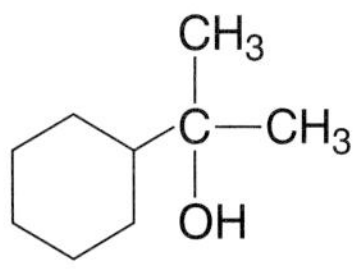 cyclohexyl-$C(CH_3)_2$ with OH

    c. $CH_3CH_2CH_2CH_2OH$

17. Prepare each of the following compounds from benzyl alcohol. You may use any necessary organic and inorganic compounds.
    a. benzaldehyde cyanohydrin
    b. mandelic acid [$C_6H_5CH(OH)COOH$]
    c. benzaldehyde semicarbazone
    d. benzoic acid
    e. benzaldehyde dimethyl acetal

18. A compound, $C_5H_{10}O$, produced a white precipitate when treated with semicarbazide, a negative Tollens test and a negative iodoform test. Draw a structural formula for $C_5H_{10}O$. Show your reasoning.

19. Compound A, $C_5H_{10}O$, (A) produced a red precipitate when treated with 2,4-dinitrophenylhydrazine. It also gave a positive Tollens test. Compound A was reacted with $LiAlH_4$ and then $H_2O$ and $H^+$ to give $C_5H_{12}O$ (compound B). Compound B also could be produced by treating isobutyl magnesium bromide with formaldehyde, followed by HCl($aq$). Draw a structural formula for compounds A and B. Show your reasoning.

20. It was observed that a bottle of an aqueous solution of formalin gradually formed a white precipitate. Draw a likely structural formula for that precipitate.

21. Write a detailed, step-by-step mechanism for the following reaction:

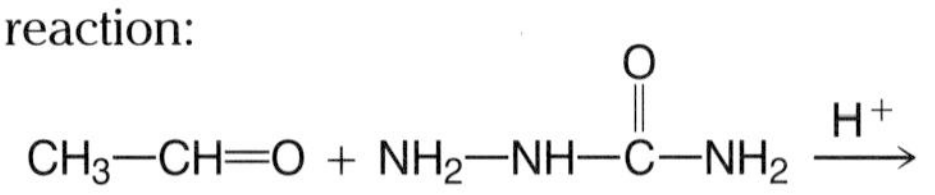

    $CH_3-CH{=}O + NH_2-NH-C(=O)-NH_2 \xrightarrow{H^+}$

    a. What is the purpose of the acid catalyst?
    b. These reactions are usually run at a pH of 5.0 with acetic acid ($C_2H_3O_2$—H) as the acid catalyst. If the pH was set at 3.0, the reaction could not occur. Explain.

22. When a mixture of 1 mol of propionaldehyde and 1 mol of acetaldehyde is treated with a 10% aqueous solution of sodium hydroxide, four products are isolated. Using the mechanism of the aldol condensation as a guide, account for each of the four products formed by drawing a structural formula for each.

23. A sample of liquid benzaldehyde was converted into a white crystalline solid on exposure to air. Explain, giving a reaction.

24. Compound A, $C_4H_8O$, reacted with excess NaOH and chlorine to give compound B, $C_3H_5O_2Na$, and a compound C that produced a proton NMR spectrum with a singlet, corresponding to one hydrogen, at $\delta = 7.3$. Identify compounds A, B, and C. Show your reasoning.

25. Predict the number of peaks, peak splitting, and the downfield absorption in the proton NMR spectrum of acetaldehyde.

# Carboxylic Acids

## STRUCTURE AND NOMENCLATURE, PHYSICAL PROPERTIES, ACIDITY, SOME OTHER IMPORTANT CARBOXYLIC ACIDS, AND REPRESENTATIVE REACTIONS

Carboxylic acids are distinguished by the presence of the carboxyl group:

The *carboxyl group* is a combination of the *carb*onyl and hydr*oxyl* groups. The carboxylic acids are found extensively in nature (more so than aldehydes) because the carbon of the carboxyl group has the highest oxidation state assigned to any carbon atom contained in any organic compound we have encountered up to now (see Table 15.1) and we live in an oxidizing atmosphere, in that the blanket of air around our planet contains 21% oxygen.

The carboxyl group is a chemically versatile functional group that can be found in aliphatic (straight-chain, branched), alicyclic, and aromatic compounds as the only functional group. It may be present with one or more other functional groups. In addition, compounds in which each molecule contains two or more carboxyl groups are not uncommon.

The straight-chain carboxylic acids were isolated by organic chemists many years ago. The common name of many of these carboxylic acids is taken from the Latin or Greek name of the source of the acid. For example, a carboxylic acid containing one carbon, i.e.,

was originally isolated from ants. The Latin for ant is *formica*. Thus the acid is called *formic acid* (Table 20.1).

A number of the acids listed in Table 20.1 (butyric acid and higher-molecular-weight homologues) are incorporated with glycerol as fats. Therefore, these acids are known as *fatty acids*

## 20.1 STRUCTURE AND NOMENCLATURE

Note that the carboxyl group is frequently condensed to —COOH. See Table 20.1.

| TABLE 20.1 | Selected Straight-Chain Aliphatic Carboxylic Acids | | |
|---|---|---|---|
| Formula | Common Name | Source | Other Comments |
| HCOOH | Formic acid | Latin: *formica* (ant) | Formic acid is found in the business end of an ant's or bee's stinger and is injected during the sting. |
| $CH_3COOH$ | Acetic acid | Latin: *acetum* (vinegar) | Acetic acid is the essence of vinegar, is produced in wine by bacteria, and is the source of sour wine. |
| $CH_3CH_2COOH$ | Propionic acid | Greek: *protos* (first); *pion* (fat) | Propionic acid has not yet been isolated from a naturally occurring fat or oil. It was named "first fat" because its chemical behavior approached that of the higher fatty acid homologues. |
| $CH_3(CH_2)_2COOH$ | Butyric acid | Latin: *butyrum* (butter) | Butyric acid is the cause of the odor of rancid butter. It is the lowest acid homologue isolated from naturally occurring fats and oils. |
| $CH_3(CH_2)_3COOH$ | Valeric acid | Latin: *valere* (strong) | Valeric acid smells bad. |
| $CH_3(CH_2)_4COOH$ | Caproic acid | Latin: *caper* (goat) | The bad odor of goats is partially due to the presence of caproic, caprylic, and capric acids. |
| $CH_3(CH_2)_6COOH$ | Caprylic acid | Latin: *caper* (goat) | |
| $CH_3(CH_2)_8COOH$ | Capric acid | Latin: *caper* (goat) | |
| $CH_3(CH_2)_{10}COOH$ | Lauric acid | | Lauric acid was first isolated from berries of the laurel tree. |
| $CH_3(CH_2)_{12}COOH$ | Myristic acid | Latin: *myristica* (fragrant) | Myristic acid was isolated from nutmeg oil. |
| $CH_3(CH_2)_{14}COOH$ | Palmitic acid | Latin: *palma* (palm) | Palmitic acid was obtained from palm oil. |
| $CH_3(CH_2)_{16}COOH$ | Stearic acid | Greek: *stear* (tallow) | Stearic acid was isolated from animal fats such as tallow. |

**FIGURE 20.1** Use of the prefix iso- in common nomenclature.

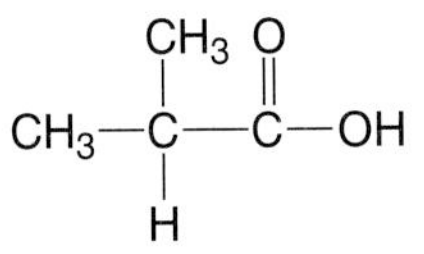

Isobutyric acid

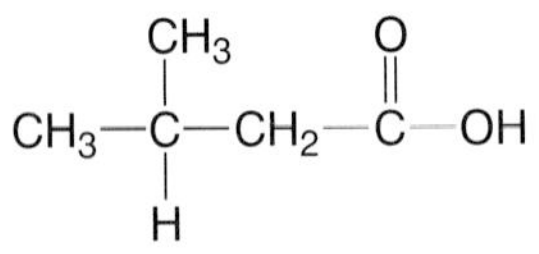

Isovaleric acid

In the common system of nomenclature, the carboxyl carbon is not assigned a Greek letter; the adjacent carbon is designated alpha, the next carbon is designated beta, and so on (see Sec. 18.2).

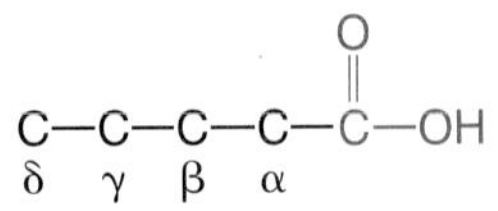

The prefix *iso-* can be used. Refer to Fig. 20.1.

**SOLVED PROBLEM 20.1**

Name the following compounds.

a.

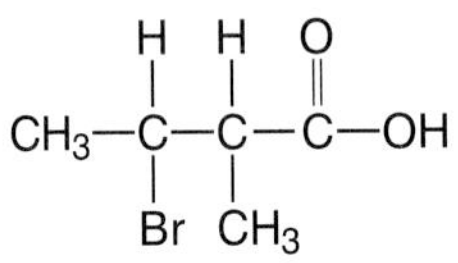

## MEDICAL BOXED READING 20.1

The apocrine sweat glands are located just under the skin in the underarm and pubic areas of the body. These glands secrete a fluid, rich in protein, on which bacteria feed to produce body odor. The odor is due to a mixture of 36 different carboxylic acids. However, the compound causing the most odor is (E)-3-methyl-2-hexenoic acid.

Since carboxylic acids containing 4 to 10 carbons have disagreeable odors, the fact that (E)-3-methyl-2-hexenoic acid contains 7 carbon atoms and smells bad makes sense.

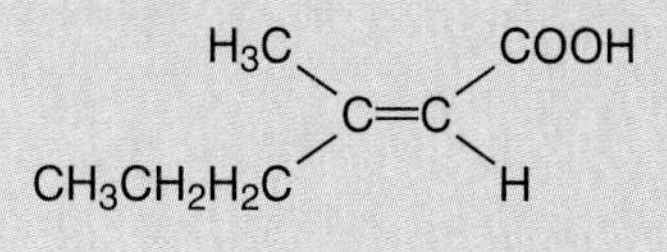

(E)-3-Methyl-2-hexenoic acid

b. 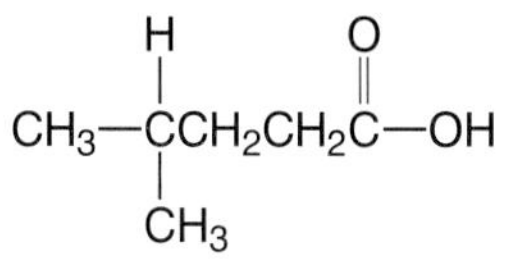
$$CH_3-\overset{\overset{\displaystyle H}{|}}{C}CH_2CH_2\overset{\overset{\displaystyle O}{\|}}{C}-OH$$
$$\underset{\displaystyle CH_3}{|}$$

### SOLUTION

a.  α-methyl-β-bromobutyric acid (or β-bromo-α-methylbutyric acid)

b.  isocaproic acid (the total number of carbons is six), or γ-methylvaleric acid.

---

Supply a common name for each of the following compounds.

a. 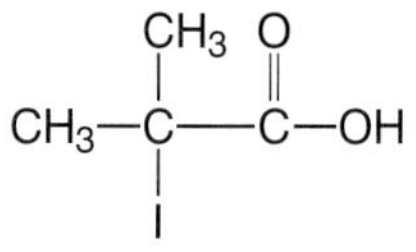
$$CH_3-\overset{\overset{\displaystyle CH_3}{|}}{\underset{\underset{\displaystyle I}{|}}{C}}-\overset{\overset{\displaystyle O}{\|}}{C}-OH$$

b. 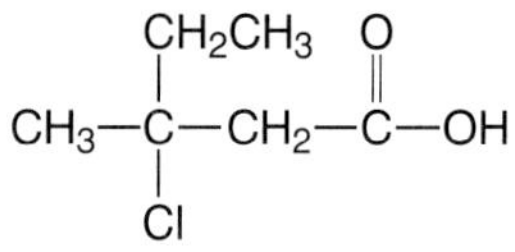
$$CH_3-\overset{\overset{\displaystyle CH_2CH_3}{|}}{\underset{\underset{\displaystyle Cl}{|}}{C}}-CH_2-\overset{\overset{\displaystyle O}{\|}}{C}-OH$$

---

To name a carboxylic acid using the IUPAC system of nomenclature, drop the -e from the corresponding alkane, and then add the suffix -oic and the word *acid*. (For some basic rules of IUPAC nomenclature, refer to Sec. 3.2.) Consider the following structural formula:

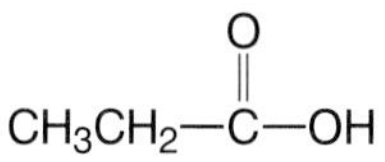

$$CH_3CH_2-\overset{\overset{\displaystyle O}{\|}}{C}-OH$$

Since the corresponding alkane is propane, drop the -e to obtain the fragment *propan*. Finally, add the suffix -oic and the word *acid* to give *propanoic acid*. A few more examples are shown in Fig. 20.2.

---

**FIGURE 20.2**
IUPAC nomenclature of selected straight-chain carboxylic acids.

$$CH_3(CH_2)_7COOH \qquad HCOOH$$
Nonanoic acid          Methanoic acid

$$CH_3(CH_2)_{12}COOH$$
Tetradecanoic acid

To name a substituted carboxylic acid using the IUPAC system of nomenclature, numbers rather than Greek letters are used to locate a particular carbon atom, with the carbon of the carboxyl group always designated as carbon 1.

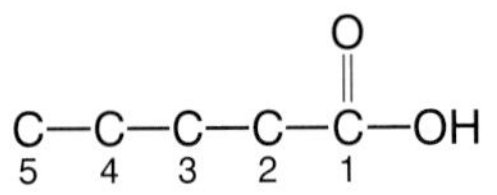

$$
\underset{5}{C}-\underset{4}{C}-\underset{3}{C}-\underset{2}{C}-\overset{\displaystyle \overset{O}{\|}}{\underset{1}{C}}-OH
$$

Substituent groups are listed alphabetically. For example:

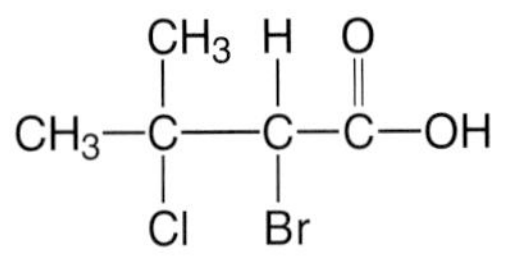

$$
CH_3-\overset{\displaystyle CH_3}{\underset{\displaystyle Cl}{C}}-\overset{\displaystyle H}{\underset{\displaystyle Br}{C}}-\overset{\displaystyle \overset{O}{\|}}{C}-OH
$$

*2-bromo-3-chloro-3-methylbutanoic acid*

Alicyclic acids are named by means of the IUPAC system of nomenclature by naming the alicyclic hydrocarbon and then adding the fragment *-carboxylic* and the word *acid*. Note that it is understood that the carbon of the ring bonded to the carboxyl carbon is always assigned the number 1.

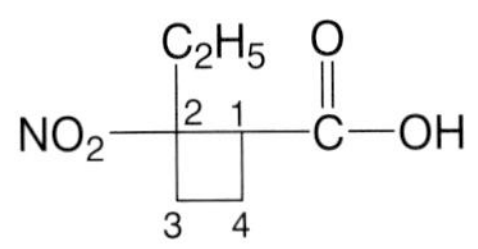

*2-Ethyl-2-nitrocyclobutanecarboxylic acid*

Aromatic acids can be named using the IUPAC system in two different ways. First, name the aromatic hydrocarbon and then add the fragment *-carboxylic* and the word *acid*. The second method involves use of the former trivial nomenclature as discussed in Section 9.2. For examples of the nomenclature of aromatic acids, refer to Fig. 20.3.

**FIGURE 20.3** Nomenclature of selected aromatic acids.

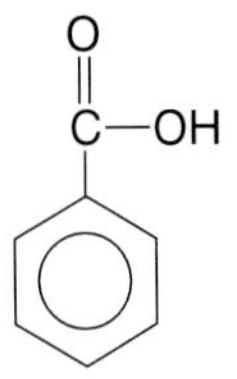

Benzenecarboxylic acid
(benzoic acid)

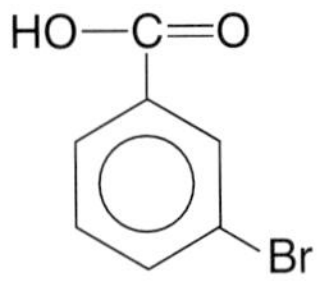

*m*-Bromobenzenecarboxylic acid
(*m*-bromobenzoic acid)

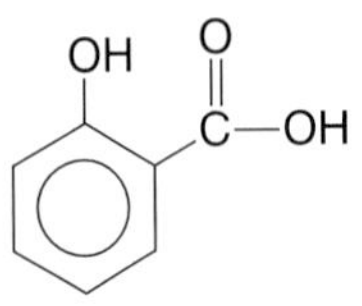

*o*-Hydroxybenzenecarboxylic acid
(salicylic acid)
(*o*-hydroxybenzoic acid)

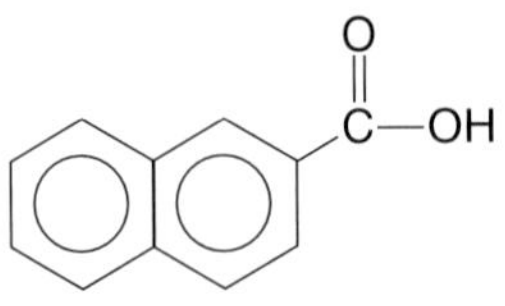

2-Naphthalenecarboxylic acid
(2-naphthoic acid)

Note: Since the carbons of the naphthalene ring are not equivalent, the carbon of the ring bonded to the carboxyl group may be assigned a number other than 1.

The carboxyl group has the highest priority in functional-group nomenclature:

$$\overset{O}{\overset{\|}{C}}-OH > \overset{O}{\overset{\|}{C}} > OH > SH > C{=}C > C{\equiv}C > OR > Cl(Br)(I)$$

Some examples of this prioritized IUPAC nomenclature are given in Fig. 20.4.

The nomenclature of a few important diacids is shown in Table 20.2. As usual, the common names must be memorized. The IUPAC name of a diacid is formulated by writing the name of the corresponding alkane (including both carboxyl groups) and then adding the suffix *-dioic* and the word *acid*. Some acids of particular biological importance are given in Fig. 20.5.

---

**SOLVED PROBLEM 20.2**

Name the following compounds using the IUPAC and common systems of nomenclature (disregard any optical isomerism).

a. 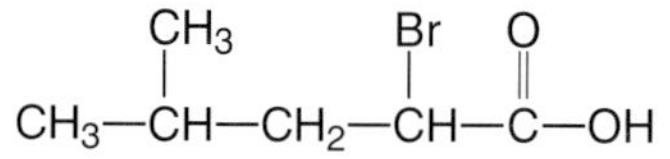 $CH_3-\overset{CH_3}{\overset{|}{CH}}-CH_2-\overset{Br}{\overset{|}{CH}}-\overset{O}{\overset{\|}{C}}-OH$

b. HOOCCH(Br)CH₂CH₂COOH

**SOLUTION**

a. IUPAC: This carboxylic acid contains a five-carbon parent chain. Assigning the carboxyl carbon the number 1, and listing the substituent groups alphabetically, we get

2-bromo-4-methylpentanoic acid

Common: Assigning the carbon next to the carboxyl carbon the Greek letter alpha, we get

α-bromo-γ-methylvaleric acid
(or γ-methyl-α-bromovaleric acid)

b. IUPAC: This dicarboxylic acid contains a five-carbon parent chain. Assigning the carboxyl carbon closer to the bromine the number 1, we get

2-bromopentanedioic acid

Common: Assigning the carbon bonded to the bromo group next to the carboxyl carbon the Greek letter alpha, we get

α-bromoglutaric acid

---

**FIGURE 20.4** Selected examples of prioritized IUPAC nomenclature of carboxylic acids.

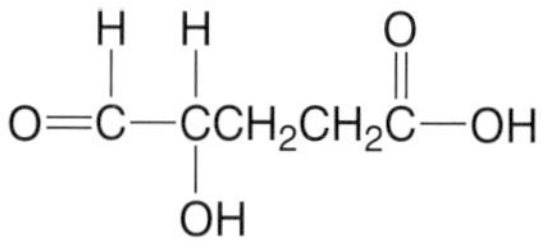

$$O{=}\overset{H}{\overset{|}{C}}-\overset{H}{\overset{|}{\underset{|}{\underset{OH}{C}}}}CH_2CH_2\overset{O}{\overset{\|}{C}}-OH$$

4-Formyl-4-hydroxybutanoic acid

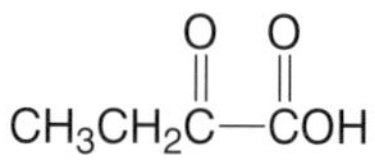

$$CH_3CH_2\overset{O}{\overset{\|}{C}}-\overset{O}{\overset{\|}{C}}OH$$

2-Oxobutanoic acid

$$CH_3CH_2OCH_2\overset{O}{\overset{\|}{C}}-OH$$

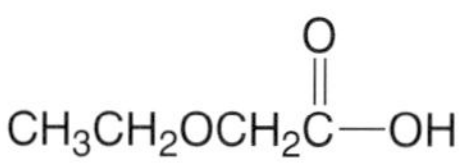

2-Ethoxyethanoic acid

Notice that when the aldehyde group is designated as a substituent group, the prefix formyl- is used to name the group.

**FIGURE 20.5** Two biologically important carboxylic acids.

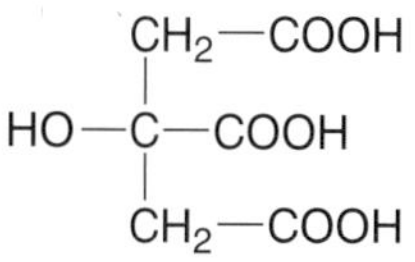

$$\overset{CH_2-COOH}{\underset{CH_2-COOH}{HO-\overset{|}{\underset{|}{C}}-COOH}}$$

Citric acid
(found in citrus fruits)
(a tricarboxylic acid)

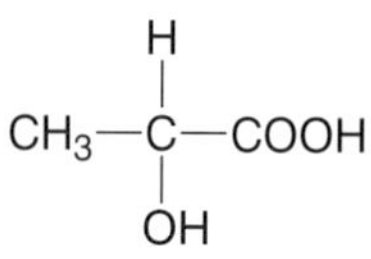

$$CH_3-\overset{H}{\underset{OH}{\overset{|}{\underset{|}{C}}}}-COOH$$

Lactic acid
(found in sour milk)

Note: these carboxylic acids often have another functional group in addition to the carboxyl group.

| TABLE 20.2 | Common and IUPAC Nomenclature of Some Selected Dicarboxylic Acids | |
|---|---|---|
| **Structure** | **Common Name** | **IUPAC Name** |
| HOOCCOOH | Oxalic acid | Ethanedioic acid |
| HOOCCH$_2$COOH | Malonic acid | Propanedioic acid |
| HOOCCH$_2$CH$_2$COOH | Succinic acid | Butanedioic acid |
| HOOCCH$_2$CH$_2$CH$_2$COOH | Glutaric acid | Pentanedioic acid |
| HOOCCH$_2$CH$_2$CH$_2$CH$_2$COOH | Adipic acid | Hexanedioic acid |
| 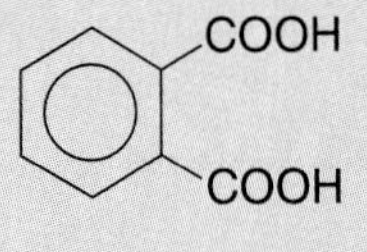 (benzene ring with two COOH groups at 1,2 positions) | | 1,2-Benzenedicarboxylic acid (phthalic acid) |
| 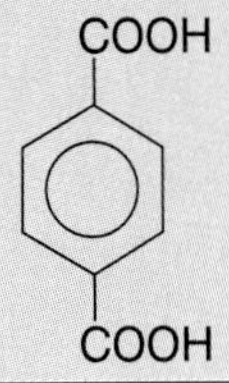 (benzene ring with two COOH groups at 1,4 positions) | | 1,4-Benzenedicarboxylic acid (terephthalic acid) |

---

**EXERCISE 20.2**

Name the following compounds using the IUPAC and common systems of nomenclature (disregard any optical isomerism).

a.  C$_6$H$_5$—CH$_2$CH$_2$COOH

b.  CH$_3$CH$_2$CH(NO$_2$)CH$_2$COOH

c.  HOOCCH(I)CH$_2$CH$_2$COOH

---

## 20.2
## PHYSICAL PROPERTIES

Alcohols hydrogen bond but do not form a similar stable molecular dimer. The alcohol dimer shown below probably does not form. Even if it did, the four-membered ring formed would mandate instability (see Sec. 4.5).

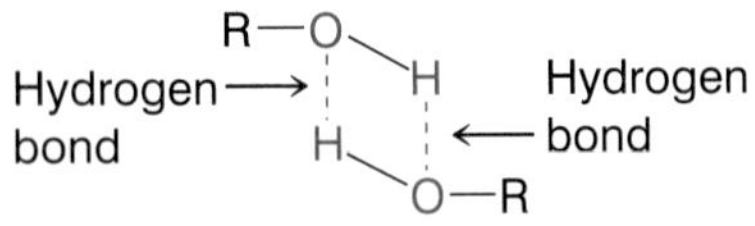

Hydrogen bond / Hydrogen bond

The first nine compounds in the straight-chain aliphatic monocarboxylic acid homologous series are liquids; higher-molecular-weight acids are waxy solids. The dicarboxylic acids are solids.

These monocarboxylic acids boil at a higher temperature than the corresponding alcohols of the same molecular weight, 20 to 30°C higher (Table 20.3). A carboxylic acid boils at a higher temperature than the corresponding alcohol of the same molecular weight because the acid forms a stable cyclic hydrogen-bonded molecular dimer (Fig. 20.6).

Butyric acid and lower homologues are completely miscible in water. Valeric through caproic acids are slightly miscible, whereas higher homologues are immiscible in water.

The first three homologues of the straight-chain carboxylic acids have sharp odors; the acid homologues containing 4 to 10 carbon atoms have disagreeable odors (see Table 20.1). Lauric acid and higher homologues are odorless.

## 20.3
## ACIDITY

**Acidity of Carboxylic Acids versus Other Functional Groups**

Carboxylic acids are weak acids. The typical acidity constant ($K_a$) is $10^{-4}$ to $10^{-5}$. Consider the carboxylic acid RCOOH in aqueous solution.

$$\text{RCOOH} + \text{H}_2\text{O} \rightleftharpoons \text{RCOO}^- + \text{H}_3\text{O}^+$$

$$K_a = \frac{[\text{RCOO}^-][\text{H}_3\text{O}^+]}{[\text{RCOOH}]} = 10^{-4} \text{ to } 10^{-5} \text{ typically}$$

where [RCOO$^-$] and [H$_3$O$^+$] are the concentrations in mol per liter of carboxylate ion and hydronium ion, respectively, while [RCOOH] represents the concentration in mol per liter of undissociated carboxylic acid. Remember, the greater the value of $K_a$, the

| TABLE 20.3 | A Comparison of the Boiling Points of Selected Carboxylic Acids with Alcohols of the Same Molecular Weight | | |
|---|---|---|---|
| **Structure and Name** | **Molecular Weight** | **Boiling Point (°C)** | |
| $CH_3C-OH$ (with O double bond) Acetic acid | 60 | 118 | |
| $CH_3CH_2CH_2OH$ n-Propyl alcohol | 60 | 97 | |
| $CH_3(CH_2)_6C-OH$ (with O double bond) Actanoic acid | 144 | 240 | |
| $CH_3(CH_2)_7CH_2OH$ 1-Nonanol | 144 | 214 | |

FIGURE 20.6 A stable cyclic hydrogen-bonded molecular dimer formed by a typical carboxylic acid.

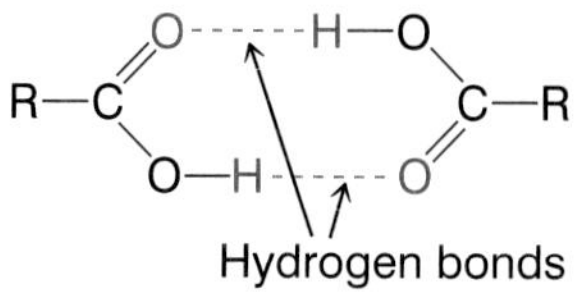

Hydrogen bonds

ionization constant of the acid, the stronger is the acid. Acid strength is usually experimentally determined as the $pK_a$, where

$$pK_a = -\log K_a$$

a. Given that the $K_a$ of propionic acid is $1.34 \times 10^{-5}$, calculate the $pK_a$.

b. Given that the $pK_a$ of benzoic acid is 4.19, calculate the $K_a$.

**SOLUTION**

a. *Step 1:* Take the log of the number. The log of $1.34 \times 10^{-5}$ is $-4.87$. You will need a calculator or a log table to obtain the logarithm of a number. To accomplish this step on a standard calculator, enter the number 1.34. Press the EE or EXP key. The screen shows 1.34 00 (the two zeros represent an exponent). Now press the $+/-$ key. The screen shows 1.34 $-00$. Press the number 5, and the screen displays 1.34 $- 05$. Finally, press the log key to give $-4.87$.

*Step 2:* Reverse the sign of the number.

$$-\log K_a = 4.87 = pK_a$$

b. *Step 1:* Change the sign of the $pK_a$.

$$-pK_a = -4.19$$

*Step 2:* Take the antilogarithm of this number ($-4.19$), which is equal to $6.46 \times 10^{-5}$ = $K_a$. You will need a calculator or a table of logarithms to obtain the antilogarithm of a number. To take the antilogarithm of a number using a standard calculator, use the $10^x$ key. First, enter the number 4.19, and then press the $+/-$ key. The display will read $-4.19$. Now press the $10^x$ key and read $6.46 \times 10^{-5}$. If your calculator is not equipped with a $10^x$ key, use the $y^x$ key. First, enter the number 10. Then press the $y^x$ key. Now enter the number 4.19, and press the $+/-$ key. The screen shows $-4.19$. Finally, press the $y^x$ key again to get $6.46 \times 10^{-5}$.

Calculate the $pK_a$ of each of the following acids.

a. acetic acid, $K_a = 1.76 \times 10^{-5}$

b.　glycolic acid, $K_a = 1.48 \times 10^{-4}$

c.　chloroacetic acid, $K_a = 1.40 \times 10^{-3}$

---

**EXERCISE 20.4**

Calculate the $K_a$ of each of the following acids.

a.　formic acid, $pK_a = 3.75$

b.　octanoic acid, $pK_a = 4.89$

---

　　　Carboxylic acids are weaker than mineral acids (HCl, $H_2SO_4$, $HNO_3$), yet carboxylic acids are far stronger acids than phenols, which, in turn, are stronger acids than alcohols. Now the key to predicting acid strength is the degree of stability of the corresponding anion obtained when the proton of the acid is removed. Consider each of the following Brønsted-Lowry acid-base expressions for the ionization of a carboxylic acid, phenol, and an alcohol, respectively:

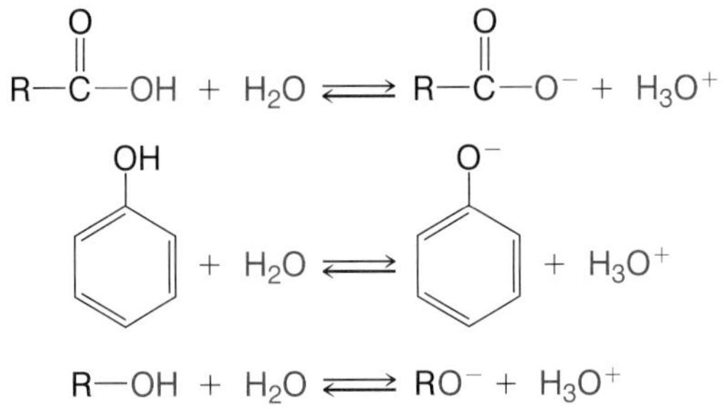

The carboxylate anion is significantly stabilized by equivalent resonance-contributing structures.

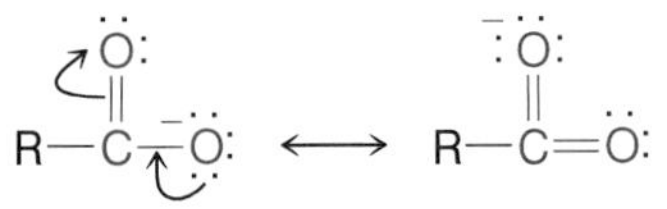

The phenoxide ion, on the other hand, is only moderately stabilized by resonance because the majority of the contributing resonance structures are not equivalent.

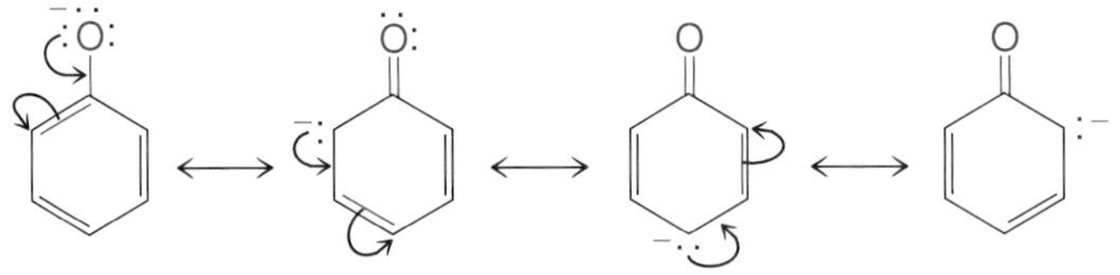

The alkoxide ion shows no resonance stabilization and therefore is a relatively unstable species compared with the carboxylate and phenoxide anions.

---

**SOLVED PROBLEM 20.4**

How can you tell that structures I, II, III, and IV above (contributing resonance structures for phenoxide ion) are not equivalent?

**SOLUTION**

There are several ways to show the structures are not equivalent:

1. Structure I has the negative charge on oxygen; structures II, III, and IV have the negative charge on carbon—a less electronegative element.

2. In the ring, structure III has the negative charge *para* to the oxygen; in structures II and IV, it is *ortho*.

3. One structure has a single bond between carbon and oxygen; the other three structures have a double bond between carbon and oxygen.

---

How can you tell the following two contributing resonance structures for ethyl acetate (an ester) are not equivalent?

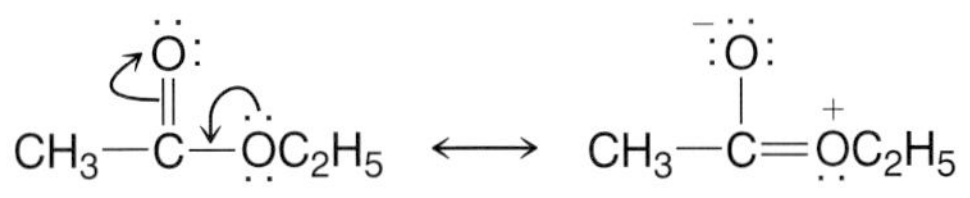

$$CH_3-C-OC_2H_5 \longleftrightarrow CH_3-C=OC_2H_5$$

EXERCISE 20.5

Consider a carboxylic acid with the following structure:

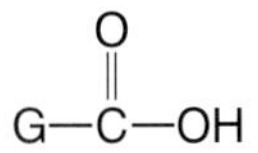

$$G-C-OH$$

**Acidity and Structure of Carboxylic Acids**

If G is electron-attracting due to the inductive effect $-I$, the anion of the acid is stabilized because the negative charge is delocalized and acidity is increased. On the other hand, if G is electron-releasing $+I$, the anion of the acid is destabilized because the negative charge is concentrated and acidity is reduced.

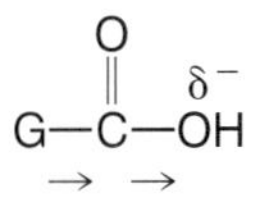

$$G-C-OH \quad \delta^- \qquad\qquad G-C-OH \quad \delta^+$$

When G— is electron-releasing, the oxygen of the —OH group acquires a partial negative charge that inhibits the loss of a proton by the acid. Thus the acidity of the acid is decreased.

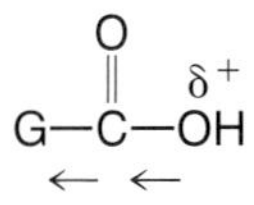

When G— is electron-attracting, the oxygen of the —OH group acquires a partial positive charge that favors the loss of a proton. Thus the acidity of the acid is increased.

Alkyl groups represent electron-releasing groups. Some examples of electron-attracting groups are as follows: $-NO_2$, $-Cl$, $-Br$, $-I$, $-NH_2$, $-OH$, $-OR$ (an atom of high electronegativity is bonded to a carbon in the structure).

The acidity of a carboxylic acid is most conveniently measured by the $pK_a$. The lower the $pK_a$, the more acidic is the carboxylic acid (Table 20.4). Note that the effect of a substituent group on acidity is a function of the distance of the group from the carboxyl carbon. Substituting —Cl for —H on the alpha carbon increases the acidity of the compound (Table 20.5). But moving the —Cl to the beta carbon decreases the acidity compared with the alpha isomer. Thus the magnitude of the effect of the electronegative —Cl on acidity decreases as the —Cl moves a greater distance from the carboxyl carbon.

The acidity is also affected by the number of substituent groups present. For example, dichloroacetic acid is a stronger acid than chloroacetic acid.

Structure:  $ClCH_2COOH$     $Cl_2CHCOOH$     $Cl_3CCOOH$
$pK_a$:         2.85                  1.48                 0.7

Because dichloroacetic acid contains two electron-withdrawing chloro groups, the partial positive charge on the hydroxyl oxygen is increased, and the compound loses a proton more easily—making it a stronger acid. Because trichloroacetic acid contains three electron-withdrawing chloro groups, it is a stronger acid than dichloroacetic acid. Finally, increasing the size of the alkyl group has virtually no effect on acidity (see Table 20.4).

| TABLE 20.4 | Structural Formulas and p$K_a$ Values for Some Selected Carboxylic Acids |
|---|---|
| **Structure of Acid** | **p$K_a$** |
| H—CH$_2$—C(=O)—OH | 4.75 |
| Cl—CH$_2$—C(=O)—OH | 2.85 |
| Br—CH$_2$—C(=O)—OH | 2.69 |
| I—CH$_2$—C(=O)—OH | 3.12 |
| CH$_3$—CH$_2$—C(=O)—OH | 4.87 |
| CH$_3$—CH(Cl)—C(=O)—OH | 2.83 |
| Cl—CH$_2$—CH$_2$—C(=O)—OH | 3.98 |
| CH$_3$—CH$_2$—CH$_2$—C(=O)—OH | 4.81 |

---

**EXERCISE 20.6**

Predict the stronger acid in each of the following pairs.

a. CH$_3$CH$_2$CH$_2$CH(Br)—COOH  and  CH$_3$CH(Br)CH$_2$CH$_2$—COOH

b. CH$_3$CH$_2$CH$_2$C(Br)(Br)—COOH  and  CH$_3$CH(Br)CH$_2$CH$_2$—COOH

---

## 20.4 SOME OTHER IMPORTANT CARBOXYLIC ACIDS

A number of carboxylic acids serve as plant hormones (Fig. 20.7). 3-Indolacetic acid is a naturally occurring hormone found in plants that catalyzes fruit development and stem elongation, stimulates ethylene synthesis in plants, and begins root formation in cuttings. 1-Naphthalenacetic acid is a synthetic hormone used to initiate root formation in cuttings and decrease fruit loss and spoilage by the fruit dropping from the tree to the ground.

| TABLE 20.5 | The Effect on Acidity of the Distance of a Substituent Group from the Carboxyl Carbon of a Carboxylic Acid | | |
|---|---|---|---|
| Structure of acid | CH$_3$CHCOOCH<br>\|<br>H | CH$_3$CHCOOH<br>\|<br>Cl | CH$_2$CH$_2$COOH<br>\|<br>Cl |
| p$K_a$ | 4.87 | 2.83 | 3.98 |

**FIGURE 20.7** Some carboxylic acids that function as plant hormones.

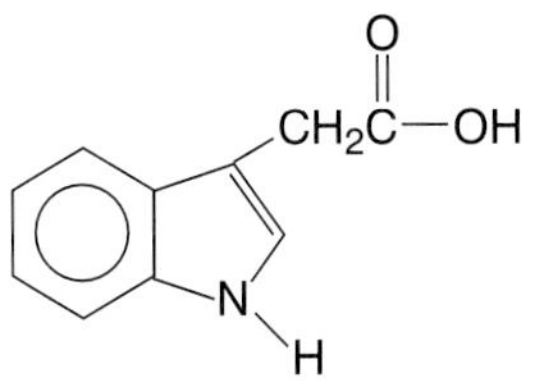

3-Indolacetic acid 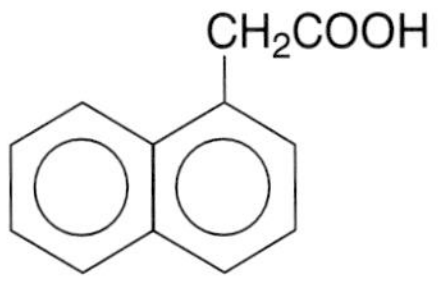    1-Naphthalenacetic acid

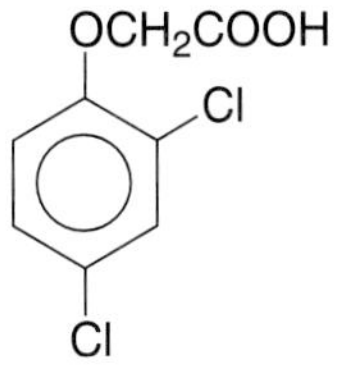

2, 4-Dichlorophexoxyacetic acid
(2, 4-D)

**FIGURE 20.8** Ibuprofen and aspirin.

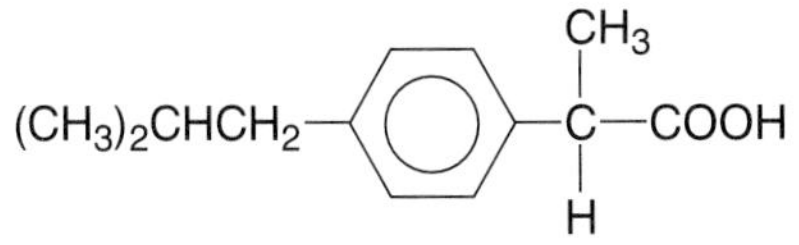

Ibuprofen (Motrin) 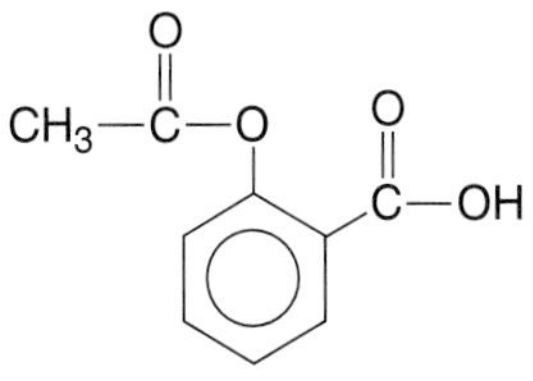    Aspirin
(acetylsalicylic acid)

**FIGURE 20.9** Pyruvic acid.

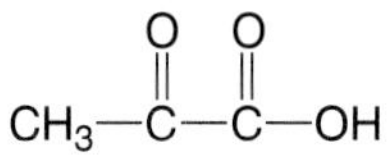

2,4-Dichlorophenoxyacetic acid (2,4-D) is a synthetic hormone that produces the same effects in plants as indoleacetic acid. However, in greater concentrations, this compound turns toxic to certain plants. Thus it is an excellent *herbicide* for broadleaf weeds, not affecting the cereal plants or grasses but destroying many vegetable and fruit crops. It is used to control weeds on highways, railroad tracks, and drainage ditches. The compound destroys plants by creating uncontrolled growth (something like cancer in humans). Obviously, this substance must be used with great care.

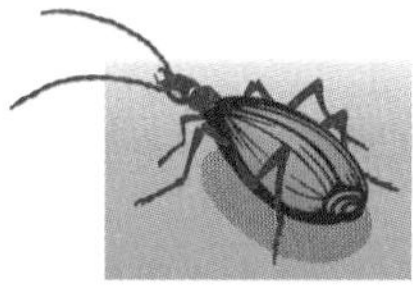

Ibuprofen (Fig. 20.8) is a carboxylic acid that shows great promise as a substitute for *aspirin,* as an anti-inflammatory agent, for patients who cannot tolerate aspirin yet need the anti-inflammatory activity of a drug for arthritis.

Pyruvic acid (Fig. 20.9) is an important intermediate in the metabolism of glucose in human beings to ultimately produce $CO_2$, $H_2O$, and energy.

Carboxylic acids completely react with aqueous solutions of strong bases such as sodium hydroxide and potassium hydroxide. A usually water-insoluble carboxylic acid is converted to a water-soluble sodium or potassium salt. In general,

## 20.5 REPRESENTATIVE REACTIONS

**Salt Formation**

$$R-\overset{O}{\overset{\|}{C}}-OH \ + \ NaOH(aq) \longrightarrow \ R-\overset{O}{\overset{\|}{C}}-O^- Na^+ \ + \ H_2O$$

Aspirin is a white powder that is odorless but has a bitter taste. It is not very soluble in water but soluble enough to permit a 5-grain tablet (1 grain = 64.8 mg) to dissolve in a glass of water. Like the winner of the NCAA basketball tournament, aspirin is number 1—the most used drug in the United States today. The yearly production of aspirin is so large that it corresponds to over 200 five-grain tablets for every man, woman, and child in the United States.

The story of aspirin started in the 1860s. A problem at that time was a shortage of quinine extracted from ground-up Cinchona bark that could only be found in certain areas of South America. Quinine was very valuable to the 19th-century physician as an *analgesic* (pain reliever) and an *antipyretic* (fever reducer). This lack of quinine-containing Cinchona bark created the impetus to find synthetic substitutes.

By 1865, physicians became very interested in a compound called *salicylic acid*, as a substitute for quinine, for two reasons. First, a chemist named Kolbe, in 1860, synthesized the compound, ensuring a substantial supply. Second, salicylic acid was a fine analgesic and antipyretic. The zinger was that the salicylic acid had an awful taste and an irritating effect on the stomach. People could not tolerate the salicylic acid. To solve this problem, chemists began to prepare derivatives (close relatives) of salicylic acid by changing the structure of the salicylic acid molecule just a bit.

In 1893, Felix Hofmann, a chemist employed by the Bayer Company, treated salicylic acid with sulfuric acid and acetic anhydride to give acetylsalicylic acid. This compound, which we know as aspirin, tasted far better and was much less acidic than salicylic acid so that a patient could tolerate taking the drug by mouth. In addition, Dr. Heinrich Dreser of the Bayer Company found that it was a fine analgesic and antipyretic. Thus he introduced aspirin into medicine in 1899.

It is ironic that Dreser was responsible for this wonderfully helpful drug being available to the public. About 1 year earlier (1898), he also introduced heroin to medicine.

The aspirin tablet is designed to dissolve quickly and rapidly reach the small intestine, where it is chemically broken down to harmless acetic acid (a constituent of vinegar) and salicylic acid. The salicylic acid then causes the analgesic and antipyretic effects of the drug.

Most aspirin is sold as a 5-grain tablet. Five grains of aspirin is too small a quantity to be compressed into a tablet by machine, so an inert substance such as starch or milk sugar (lactose) is added to increase the size and weight of the granulated aspirin so that the material can be compressed by machine and is a convenient size for patients to use. In addition to providing bulk, the starch also functions as a binder to ensure the tablet will not break up after compression and acts as a disintegrator to guarantee that the tablet physically breaks down after ingestion.

Aspirin is used for relief of mild headache, neuralgia, and rheumatism pain and is used extensively to lower body temperature of patients with fever. Recently, it has gained attention as a heart attack preventive.

Aspirin was named by Felix Hofmann. Hofmann used the little known fact that in 1840 a chemist isolated salicylic acid from plants of the *Spiraea* family. Only he didn't realize it was salicylic acid, so he called it spiric acid—naming it for the plant. Hofmann proposed the name acetyspiric acid or, in a shortened form, aspirin.

Aspirin
(acetylsalicylic acid)

$CH_3-C-O-C-CH_3$
Acetic anhydride

Example:

Phenylacetic acid + NaOH(aq) ⟶ Sodium phenylacetate + $H_2O$

To name a salt by either the common or IUPAC system of nomenclature, first name the cation, and then drop the *-ic* suffix of the corresponding acid and add the suffix *-ate*.

**SOLVED PROBLEM 20.5**

Supply both a common and an IUPAC name for

$CH_3CH_2CH_2C-O^-Na^+$

**SOLUTION**

Common: First, name the cation—*sodium*. The common name of the corresponding acid is butyric acid. Thus we have *sodium butyric*. Now, drop the *-ic* suffix and substitute *-ate* to give *sodium butyrate*.

IUPAC: First, name the cation—*sodium*. The IUPAC name of the corresponding acid is butanoic acid. Thus we have *sodium butanoic*. Now, drop the *-ic* suffix and substitute *-ate* to give *sodium butanoate*.

Provide both a common and an IUPAC name for the following.

a.   $CH_3CH_2COO^-K^+$

b.   $(CH_3CH_2COO)_2^-Ca^{2+}$

**EXERCISE 20.7**

   Carboxylic acids are strong enough acids to react with an aqueous solution of the weaker bases sodium bicarbonate and potassium bicarbonate (see Sec. 16.4). In general,

$$R-\overset{\overset{\displaystyle O}{\|}}{C}-OH + KHCO_3(aq) \longrightarrow R-\overset{\overset{\displaystyle O}{\|}}{C}-O^-K^+ + H_2O + CO_2$$

Example:

$$CH_3-\overset{\overset{\displaystyle O}{\|}}{C}-OH + KHCO_3(aq) \longrightarrow CH_3-\overset{\overset{\displaystyle O}{\|}}{C}-O^-K^+ + H_2O + CO_2$$

Potassium acetate

The evolution of bubbles of carbon dioxide when an unknown acidic compound is treated with a bicarbonate identifies the unknown as a carboxylic acid (or a sulfonic acid) as opposed to a phenol.

   Carboxylic acids can be regenerated from the salts by means of a mineral acid such as HCl. In general,

$$R-\overset{\overset{\displaystyle O}{\|}}{C}-O^-Na^+ + HCl \longrightarrow R-\overset{\overset{\displaystyle O}{\|}}{C}-OH + NaCl$$

Example:

$$CH_3-\overset{\overset{\displaystyle CH_3}{|}}{\underset{\underset{\displaystyle H}{|}}{C}}-\overset{\overset{\displaystyle O}{\|}}{C}-O^-Na^+ + HCl \longrightarrow CH_3-\overset{\overset{\displaystyle CH_3}{|}}{\underset{\underset{\displaystyle H}{|}}{C}}-\overset{\overset{\displaystyle O}{\|}}{C}-OH + NaCl$$

Thus the conversion of a carboxylic acid to the corresponding salt represents a reversible process shown as follows:

$$\underset{\text{(Usually water-insoluble)}}{\text{carboxylic acid}} \underset{\text{HCl}}{\overset{\text{NaOH}(aq) \text{ or NaHCO}_3(aq)}{\rightleftharpoons}} \underset{\text{(Water-soluble)}}{\text{carboxylate salt}}$$

Since phenols are too weakly acidic to react with bicarbonate (see Sec. 16.4), we can use this difference in chemical behavior as a means of separating a carboxylic acid from a phenol and other functional groups.

Calcium propionate and sodium benzoate are often used as food preservatives, effectively preventing the formation of bacteria and mold in food. Both the salts and their parent acids cause no damage when present within the human body.

**SOLVED PROBLEM 20.6**

Create a flow diagram to show how you would separate and isolate the components of a mixture of phenol and decanoic acid.

**SOLUTION**

Dissolve the mixture in ether, extract with a 10% solution of aqueous sodium bicarbonate, and separate the aqueous layer from the ether layer. The aqueous layer containing sodium decanoate is then reacted with a 10% solution of hydrochloric acid to precipitate decanoic acid. The ether in the phenol solution is removed by evaporation. Both components of the mixture are isolated in good yield (Fig. 20.10).

**EXERCISE 20.8**

Create a flow diagram to show how you would separate and isolate the components of a mixture of benzoic acid (a solid) and benzophenone (a solid).

## Ester Formation

An ester is produced when a carboxylic acid reacts with an alcohol in the presence of a catalytic amount of sulfuric acid. This reaction is called *Fischer esterification* and is reversible. In general,

$$R-\overset{\overset{\displaystyle O}{\|}}{C}-OH + HO-R' \underset{\longleftarrow}{\overset{H_2SO_4}{\rightleftharpoons}} R-\overset{\overset{\displaystyle O}{\|}}{C}-O-R' + H_2O$$

To formulate the product, first remove a molecule of water (OH from the acid; H from the alcohol) to give

$$R-\overset{\overset{\displaystyle O}{\|}}{C} \quad O-R' + H_2O$$

Finally, link the two organic fragments together by means of a single bond to give the ester:

$$R-\overset{\overset{\displaystyle O}{\|}}{C}-O-R'$$

Example:

$$CH_3-\overset{\overset{\displaystyle O}{\|}}{C}-OH + H-O-CH_3 \overset{H_2SO_4}{\rightleftharpoons} CH_3-\overset{\overset{\displaystyle O}{\|}}{C}-O-CH_3 + H_2O$$

Acetic acid    Methyl alcohol       Methyl acetate

**FIGURE 20.10** A flow diagram showing the separation of decanoic acid from phenol.

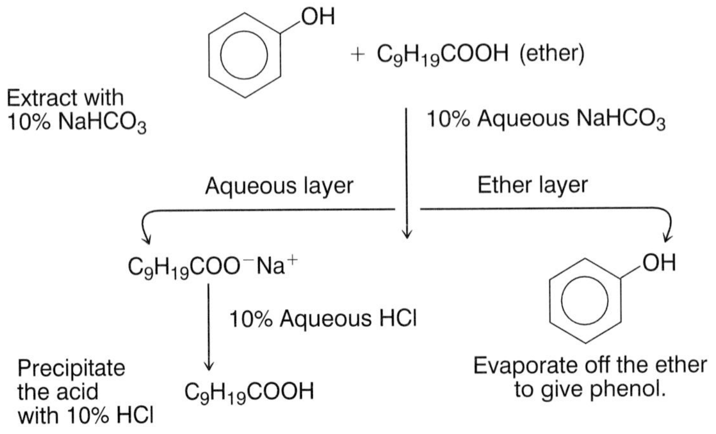

Since this reaction is reversible, yields of ester can be increased by the use of LeChâtelier's principle by adding an excess of acid or alcohol used or by decreasing the concentration of ester or water formed (see Sec. 23.6).

---

Complete the following equations by drawing a structural formula for each organic product and a molecular formula for each inorganic product.

a. 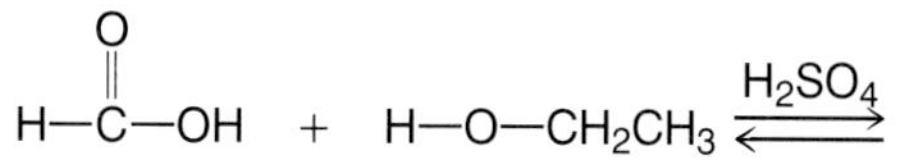  $H-\overset{\overset{\displaystyle O}{\|}}{C}-OH \ + \ H-O-CH_2CH_3 \ \overset{H_2SO_4}{\rightleftharpoons}$

b. 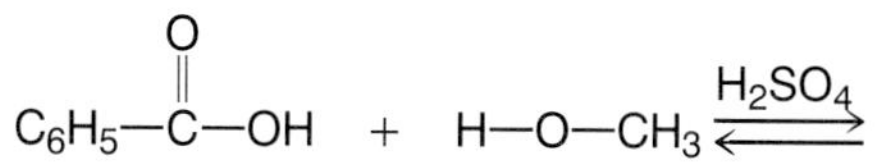  $C_6H_5-\overset{\overset{\displaystyle O}{\|}}{C}-OH \ + \ H-O-CH_3 \ \overset{H_2SO_4}{\rightleftharpoons}$

---

**Amide Formation**

Carboxylic acids react with amines (see Chaps. 24 and 25) to yield amides. A necessary substance that must be present in order to increase the rate of the reaction to an acceptable level is *di*cyclohexylcarbodiimide (DCC).

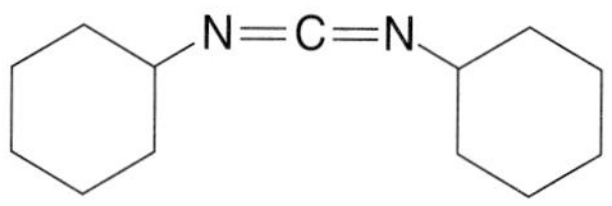

Dicyclohexylcarbodiimide
(DCC)

In general,

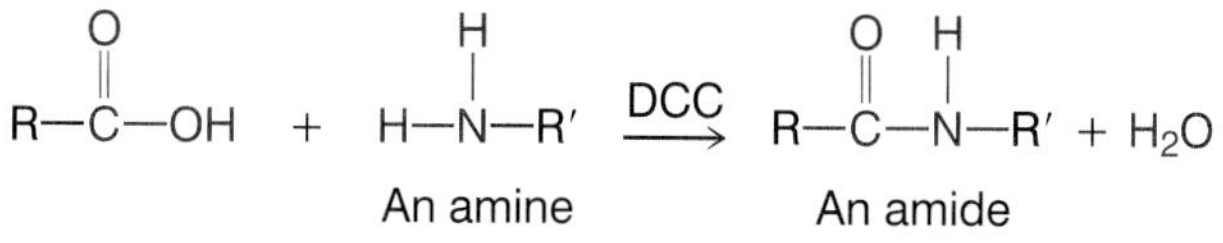

$$R-\overset{\overset{\displaystyle O}{\|}}{C}-OH \ + \ H-\overset{\overset{\displaystyle H}{|}}{N}-R' \ \overset{DCC}{\longrightarrow} \ R-\overset{\overset{\displaystyle O}{\|}}{C}-\overset{\overset{\displaystyle H}{|}}{N}-R' + H_2O$$

An amine        An amide

To formulate the product, first remove a molecule of water to give

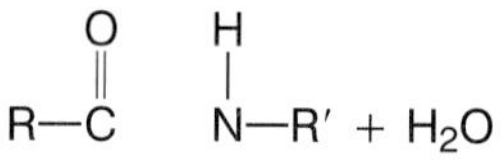

$$R-\overset{\overset{\displaystyle O}{\|}}{C} \qquad \overset{\overset{\displaystyle H}{|}}{N}-R' + H_2O$$

Finally, link the two organic fragments together by means of a single bond to give the amide:

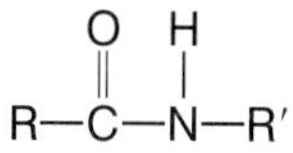

$$R-\overset{\overset{\displaystyle O}{\|}}{C}-\overset{\overset{\displaystyle H}{|}}{N}-R'$$

Example:

$$CH_3-\overset{\overset{\displaystyle O}{\|}}{C}-OH \ + \ H-\overset{\overset{\displaystyle H}{|}}{N}-CH_2CH_3 \ \overset{DCC}{\longrightarrow} \ CH_3-\overset{\overset{\displaystyle O}{\|}}{C}-\overset{\overset{\displaystyle H}{|}}{N}-CH_2CH_3 + H_2O$$

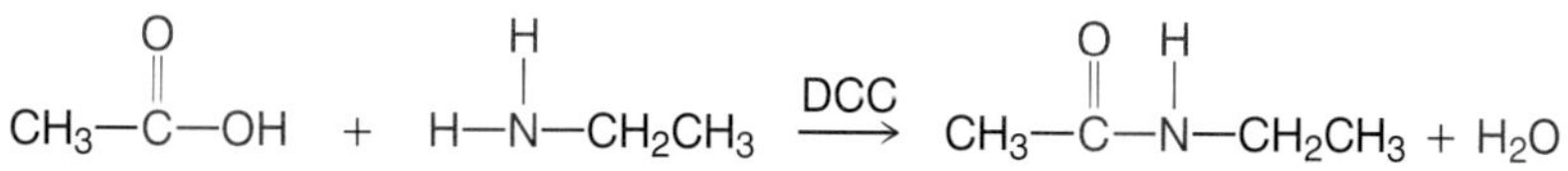

This reaction is frequently employed in the preparation of *peptides* (see Sec. 28.11).

---

Complete the following equations by drawing a structural formula for each organic product and a molecular formula for each inorganic product.

a. 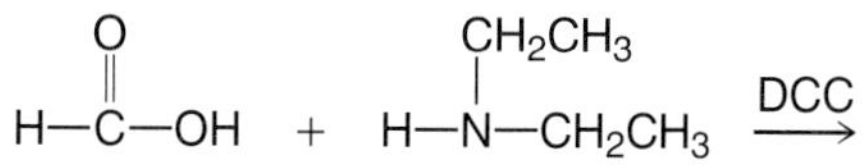  $H-\overset{\overset{\displaystyle O}{\|}}{C}-OH \ + \ H-\overset{\overset{\displaystyle CH_2CH_3}{|}}{N}-CH_2CH_3 \ \overset{DCC}{\longrightarrow}$

b. 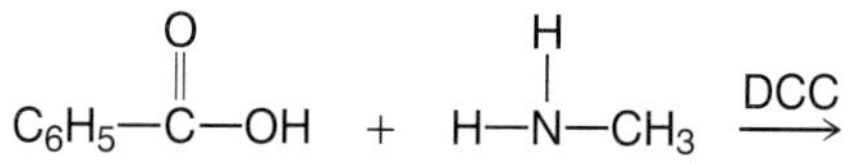  $C_6H_5-\overset{\overset{\displaystyle O}{\|}}{C}-OH \ + \ H-\overset{\overset{\displaystyle H}{|}}{N}-CH_3 \ \overset{DCC}{\longrightarrow}$

## ▶ CHAPTER ACCOMPLISHMENTS

**Introduction.**

☐ Explain why the carboxylic acids are extensively found in nature.

### 20.1  Structure and Nomenclature

☐ Draw a structural formula for

  a. formic acid.
  b. propanoic acid.
  c. caprylic acid.
  d. palmitic acid.
  e. tetradecanoic acid.
  f. γ-methylvaleric acid.
  g. 2-ethyl-2-nitrocyclobutanecarboxylic acid.
  h. benzoic acid.
  i. 2-oxobutanoic acid.
  j. succinic acid.
  k. lactic acid.
  l. hexanedioic acid.

### 20.2  Physical Properties

☐ Draw a hydrogen-bonded molecular dimer for acetic acid.
☐ Predict the odor (pleasant or unpleasant) of a six-carbon aliphatic straight-chain carboxylic acid.

### 20.3  Acidity

☐ Calculate the p$K_a$ of propionic acid given that the $K_a$ is 1.34 × $10^{-5}$.
☐ Calculate the $K_a$ of benzoic acid given that the p$K_a$ is 4.19.
☐ Explain why the carboxylic acids are stronger acids than the phenols.

☐ Explain why

  a. dichloroacetic acid is a stronger acid than acetic acid.
  b. α-chloropropionic acid is a stronger acid than β-chloropropionic acid.

### 20.4  Some Other Important Carboxylic Acids

☐ Explain how 2,4-dichlorophenoxyacetic acid destroys plants.

### 20.5  Representative Reactions

☐ Complete the following equations by drawing a structural formula for each organic compound produced and a molecular formula or formula unit for each inorganic compound formed.

  a. $CH_3COOH + NaHCO_3(aq) \longrightarrow$
  b. sodium 2-methylpropanoate + HCl $\longrightarrow$

☐ Provide an advantage to reacting a carboxylic acid with aqueous sodium bicarbonate as opposed to aqueous sodium hydroxide.
☐ Provide a use for calcium propionate in various kinds of foods.
☐ Draw a structural formula for cyclohexylcarbodiimide.
☐ Complete the following equations.

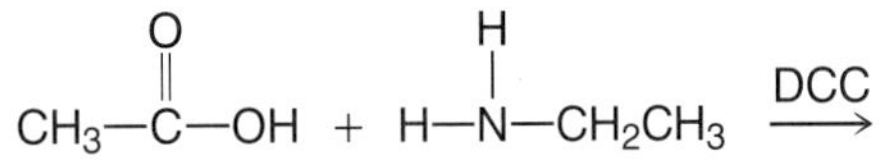

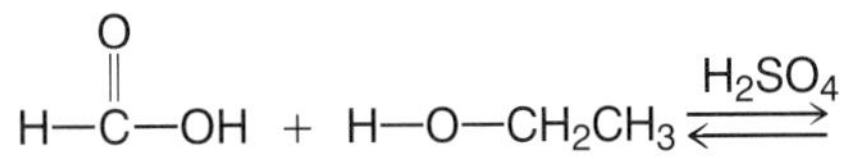

## ▶ KEY TERMS

*carboxyl group* (Introduction)
*formic acid* (20.1)
*fatty acid* (20.1)
*herbicide* (20.4)
*aspirin* (20.4, Historical Boxed Reading 20.2)

*Fischer esterification* (20.5, 23.6)
*peptide* (20.5, 28.4)
*analgesic* (Historical Boxed Reading 20.2)

*antipyretic* (Historical Boxed Reading 20.2)
*salicylic acid* (Historical Boxed Reading 20.2)

## ▶ PROBLEMS

1. Why are so many carboxylic acids found in nature, while so few aldehydes are naturally occurring?

2. Name each of the following using the common method (when possible) and the IUPAC method.

  a. $CH_3(CH_2)_{13}COOH$

  b.

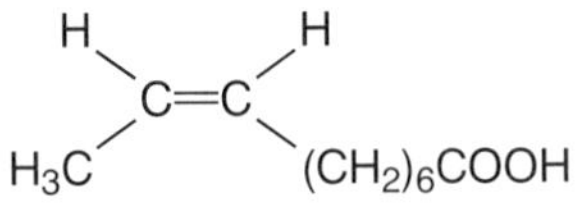

  c. $(CH_3)_2C(H)CH_2COOH$

  d.

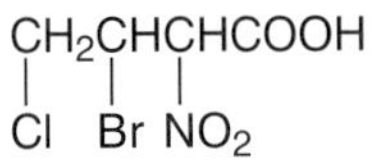

e.

f. $CH_3 \overset{\overset{\displaystyle H}{|}}{\underset{\underset{\displaystyle OH}{|}}{C}} COOH$

g.

h. $CH_3C(H)CH_2COO^-K^+$    disregard any optical isomerism
  $\quad\quad\quad |$
  $\quad\quad\;\; OH$

i.

j. $CH_3CH_2CH_2\overset{\overset{\displaystyle O}{||}}{C}$

k. $CH_3\overset{}{\underset{\underset{\displaystyle O}{||}}{C}}CH_2COOH$

l.    disregard any optical isomerism

m. $Br \overset{\overset{\displaystyle H}{|}}{\underset{\underset{\displaystyle CH_3}{|}}{C}} COOH$

3. Draw a structural formula for each of the following compounds.
   a. salicylic acid
   b. 2,2-dimethylpentanoic acid
   c. 2,4-D
   d. 3-indolacetic acid
   e. ibuprofen
   f. 2-chloro-4-methylhexanoic acid
   g. *m*-methoxyphenylacetic acid
   h. 2,4,5-trichlorophenoxyacetic acid
   i. crotonic acid (*trans*-2-butenoic acid)
   j. maleic acid (*cis*-2-butenedioic acid)
   k. palmitic acid
   l. octadecanoic acid
   m. aspirin

4. Draw a structural formula for each of the following compounds.
   a. caprylic acid
   b. β-bromopropionic acid

   c. α,γ-dinitrovaleric acid
   d. isocapric acid
   e. stearic acid
   f. formic acid

5. Draw a structural formula for each of the following compounds.
   a. *o*-benzylbenzoic acid
   b. cyclobutanecarboxylic acid
   c. *cis*-2-tetradecenoic acid
   d. 2,2-dimethylhexanoic acid
   e. 3-chloropentanedioic acid

6. Draw a flow diagram showing how would you isolate the carboxylic acid in a mixture of stearic acid and cyclohexanone.

7. Draw stabilizing resonance structures for the palmitate ion.

8. Which of the following pairs of compounds is the stronger acid? Explain.

a.

b.

c. $CH_3CH_2\overset{\overset{\displaystyle Cl}{|}}{\underset{\underset{\displaystyle H}{|}}{C}}\overset{\overset{\displaystyle O}{||}}{C}-OH$ or $CH_3CH_2\overset{\overset{\displaystyle Cl}{|}}{\underset{\underset{\displaystyle Cl}{|}}{C}}\overset{\overset{\displaystyle O}{||}}{C}-OH$

d. $H-\overset{\overset{\displaystyle O}{||}}{C}-OH$ or $CH_3-\overset{\overset{\displaystyle O}{||}}{C}-OH$

9. An acid A has a p$K_a$ of 4.5. An acid B has a $K_a$ of $6 \times 10^{-4}$. Which acid is stronger?

10. a. Calculate the p$K_a$ for each of the following acids.
       (i) *o*-cresol, $K_a = 6.3 \times 10^{-11}$
       (ii) cyanoacetic acid, $K_a = 3.7 \times 10^{-3}$.
    b. Calculate the $K_a$ for each of the following acids.
       (i) caproic acid, p$K_a = 4.8$
       (ii) 1-naphthol, p$K_a = 9.3$

11. Draw a structure for the hydrogen bond–stabilized dimer of caproic acid.

12. Draw a flow diagram to illustrate the separation and isolation of octanoic acid from a mixture of
    a. octanoic acid and *p*-cresol.
    b. octanoic acid and 1-octanol.
    (Show each step as in Solved Problem 20.6.)

13. Criticize each of the following statements.
    a. Aqueous sodium hydroxide is used to convert water-insoluble acetic acid to water-soluble sodium acetate.
    b. Propanedioic acid is a liquid.
    c. Potassium stearate is a water-insoluble solid.

d. Salicylic acid is *p*-hydroxybenzoic acid.

e. Bromoacetic acid is a stronger acid than dibromoacetic acid.

14. Draw a structural formula for the following salts.

a. lithium cyclopentanecarboxylate

b. potassium *o*-chlorobenzoate

c. sodium eicosanoate

15. Complete the following equations by supplying a structural formula for each organic reactant and a molecular formula or formula unit for each inorganic reagent or catalyst designated by a question mark.

a. $? + ? \longrightarrow CH_3CH_2CH_2CH_2COO^-K^+ + H_2O$

b. $? + K_2CO_3(aq) \longrightarrow (CH_3)_2CHCH_2COO^-K^+ + CO_2 + H_2O$

c. $? + ? \overset{?}{\rightleftharpoons} CH_3(CH_2)_6COOCH_2CH_2CH_3 + H_2O$

d. $? + ? \xrightarrow{DCC} CH_3(CH_2)_6CONHCH_2CH_3 + H_2O$

# CHAPTER 21

# Carboxylic Acids and Nitriles

## SYNTHESIS AND CHEMICAL PROPERTIES

In general,

$$RCH_2OH \xrightarrow[H_2SO_4]{Na_2Cr_2O_7} R-\overset{\displaystyle O}{\underset{\displaystyle \|}{C}}-OH + (Cr^{3+})$$

$$R-\overset{\displaystyle O}{\underset{\displaystyle \|}{C}}-H + 2\,{}^+Ag(NH_3)_2 + 3\,{}^-OH \longrightarrow R-\overset{\displaystyle O}{\underset{\displaystyle \|}{C}}-O^- + 4\,NH_3 + 2\,Ag + 2\,H_2O$$

To generate the carboxylic acid, the acid salt is reacted with a mineral acid such as hydrochloric acid (see Sec. 20.5).

$$R-\overset{\displaystyle O}{\underset{\displaystyle \|}{C}}-O^- + H^+ \longrightarrow R-\overset{\displaystyle O}{\underset{\displaystyle \|}{C}}-OH$$

Examples:

$$\text{C}_6\text{H}_5\text{CH}_2OH \xrightarrow[H_2SO_4]{Na_2Cr_2O_7} \text{C}_6\text{H}_5\overset{\displaystyle O}{\underset{\displaystyle \|}{C}}-OH$$

$$CH_3\overset{\displaystyle H}{\underset{\displaystyle \|}{C}}=O + 2\,{}^+Ag(NH_3)_2 + 3\,{}^-OH \longrightarrow CH_3\overset{\displaystyle O}{\underset{\displaystyle \|}{C}}-O^- + 4\,NH_3 + 2\,Ag + 2\,H_2O$$

$$CH_3-\overset{\displaystyle O}{\underset{\displaystyle \|}{C}}-O^- + H^+ \longrightarrow CH_3-\overset{\displaystyle O}{\underset{\displaystyle \|}{C}}-OH$$

## 21.1
## SYNTHESIS OF CARBOXYLIC ACIDS

Oxidation of Primary Alcohols or Aldehydes (see Secs. 15.2, 18.6, and 19.3)

## Oxidation of Alkylbenzenes

An arene with at least one benzylic hydrogen per alkyl group reacts smoothly with a basic aqueous solution of potassium permanganate. When next treated with hydrochloric acid, the salt of the acid gives the corresponding carboxylic acid (see Sec. 10.7).

Since the *tert*-butyl group does not contain a benzylic hydrogen, it does not react.

Example:

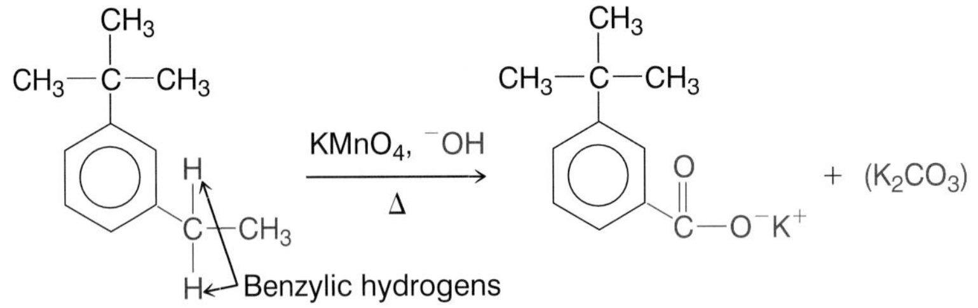


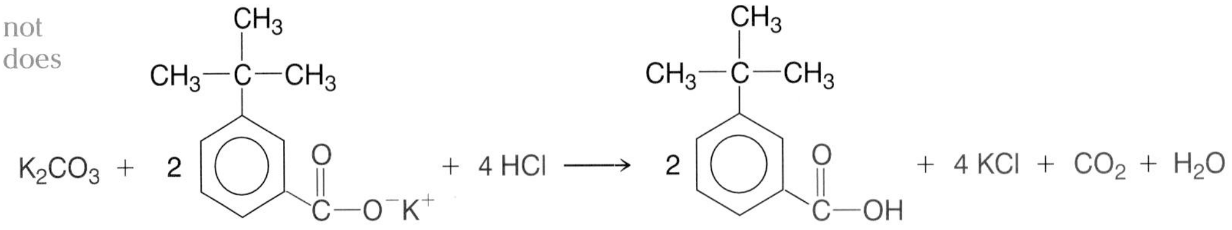


## The Haloform Reaction

Refer to Sec. 19.3. In general,

$$\text{(H) R}-\overset{\text{O}}{\overset{\|}{\text{C}}}-\text{CH}_3 + 4\ \text{NaOH}(aq) + 3\ \text{X}_2 \longrightarrow \text{H(R)}-\overset{\text{O}}{\overset{\|}{\text{C}}}-\text{O}^-\ \text{Na}^+ + \text{CHX}_3 + 3\ \text{NaX} + 3\ \text{H}_2\text{O}$$

where $X_2 = Cl_2$, $Br_2$, or $I_2$.

$$\text{H(R)}-\overset{\text{O}}{\overset{\|}{\text{C}}}-\text{O}^-\ \text{Na}^+ + \text{HCl} \longrightarrow \text{H(R)}-\overset{\text{O}}{\overset{\|}{\text{C}}}-\text{OH} + \text{NaCl}$$

Example:

$$\underset{\text{3-methyl-2-butanone}}{\text{CH}_3-\overset{\text{H}}{\underset{\text{CH}_3}{\text{C}}}-\overset{\text{O}}{\overset{\|}{\text{C}}}-\text{CH}_3} + 4\ \text{NaOH}\ (aq) + 3\ \text{I}_2 \longrightarrow \underset{\text{sodium 2-methylpropanoate}}{\text{CH}_3-\overset{\text{H}}{\underset{\text{CH}_3}{\text{C}}}-\overset{\text{O}}{\overset{\|}{\text{C}}}-\text{O}^-\ \text{Na}^+}$$
$$+ \text{CHI}_3 + 3\ \text{NaI} + 3\ \text{H}_2\text{O}$$

$$\text{CH}_3-\overset{\text{H}}{\underset{\text{CH}_3}{\text{C}}}-\overset{\text{O}}{\overset{\|}{\text{C}}}-\text{O}^-\ \text{Na}^+ + \text{HCl} \longrightarrow \underset{\text{2-methylpropanoic acid}}{\text{CH}_3-\overset{\text{H}}{\underset{\text{CH}_3}{\text{C}}}-\overset{\text{O}}{\overset{\|}{\text{C}}}-\text{OH}} + \text{NaCl}$$

In addition, ethanol and methylcarbinols share this reaction (see Sec. 19.3).

## Carbonation of the Grignard Reagent

Treating a Grignard reagent with carbon dioxide followed by hydrochloric acid yields a carboxylic acid with one carbon more than present in the original Grignard reagent. In general,

$$\text{RMgX} + \text{CO}_2 \longrightarrow \text{R}-\overset{\text{O}}{\overset{\|}{\text{C}}}-\text{O}^-\ {}^+\text{MgX}$$

where X = Cl, Br, or I.

$$\text{R}-\overset{\text{O}}{\overset{\|}{\text{C}}}-\text{O}^-\ {}^+\text{MgX} + \text{HCl}(aq) \longrightarrow \text{R}-\overset{\text{O}}{\overset{\|}{\text{C}}}-\text{OH} + \text{MgCl(X)}$$

Example:

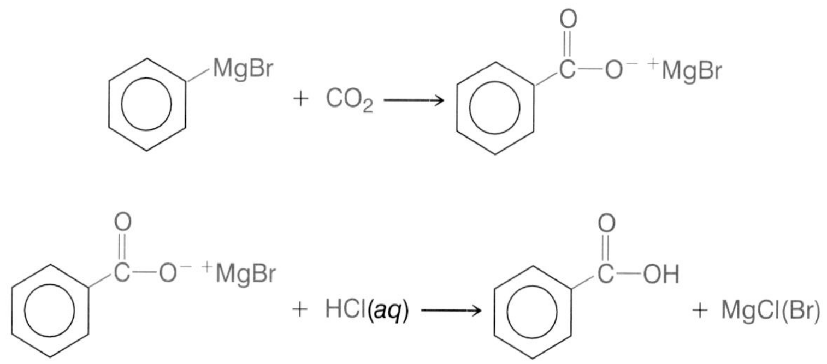

The mechanism of this reaction is similar to the addition of a Grignard reagent to a carbonyl compound to produce an alcohol (see Sec. 19.2).

**Envision the Reaction**

**Mechanism of the Reaction of Ethylmagnesium Iodide with Carbon Dioxide Followed by Hydrochloric Acid to Produce Propionic Acid**

First step:

$$CH_3CH_2MgI \rightleftharpoons CH_3\ddot{C}^{-}H_2 + {}^{+}MgI$$

Second step:

$$CH_3\ddot{C}^{-}H_2 + \overset{\delta^-}{O}=\overset{\delta^+}{C}=\overset{\delta^-}{O} \longrightarrow CH_3CH_2C(=O)-O^-$$

$$CH_3CH_2C(=O)-O^- + {}^{+}MgI \longrightarrow CH_3CH_2C(=O)O^-\ {}^{+}MgI$$

Third step:

$$CH_3CH_2C(=O)-O^-\ {}^{+}MgI + HCl(aq) \longrightarrow CH_3CH_2C(=O)-OH + MgCl(I)$$

First, a Grignard reagent dissociates to form the corresponding carbanion and a positively charged magnesium halide species. In the second step of the mechanism, the carbanion attacks the partially positively charged carbon in carbon dioxide to form a halomagnesium salt of the acid. The final step involves hydrolysis of the salt with hydrochloric acid to produce a carboxylic acid that contains one carbon atom more than the original Grignard reagent.

## Hydrolysis of Nitriles (Nomenclature and Methods of Preparation of Nitriles)

Before we consider the hydrolysis of *nitriles*, let us examine the nomenclature and some methods of preparation of this important class of compounds.

The common name of a nitrile is based on the carboxylic acid obtained from hydrolysis of the nitrile. Drop the *-ic* suffix of the acid formed, and then add the suffix *-onitrile*. For example, consider $CH_3CN$. Since the compound contains two carbon atoms, the acid produced by hydrolysis also contains two carbon atoms—acetic acid. We drop the *-ic* suffix to get the fragment *acet*. Adding the suffix *-onitrile* gives us *acetonitrile*. Two important exceptions to the *-onitrile* method of nomenclature are illustrated in Fig. 21.1.

**FIGURE 21.1** Some important exceptions to the -onitrile method of common nomenclature.

Benzonitrile
(−*oic* dropped)

Propionitrile
(−*onic* dropped)

To assign an IUPAC name to a nitrile, name the corresponding alkane (including the carbon in the nitrile group) and add the suffix -*nitrile*). For example, consider the compound $CH_3CH_2CH_2CN$. Since the compound contains a total of four carbon atoms, the corresponding alkane is butane. Adding the word *nitrile*, we get *butanenitrile*.

To name substituted nitriles using the IUPAC system, the carbon of the cyano group (—CN) is assigned the number 1. For example, the compound

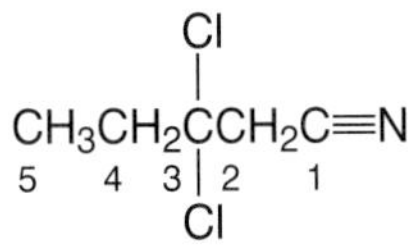

is named 3,3-dichloropentanenitrile.

---

**EXERCISE 21.1**

Name the following compounds.

a. $CH_3CH_2CH_2CN$ (common)

b. $CH_3CN$ (IUPAC)

---

Nitriles are prepared in a variety of ways:

**The Reaction of an Alkyl Halide with Cyanide Ion** Yields are good with primary halides (see Sec. 12.1), poor with secondary halides. The reaction does not occur with tertiary halides because cyanide ion is a strong base—resulting in elimination rather than substitution (see Sec. 12.4). Refer to Solved Problem 12.12 for an example.

---

**SOLVED PROBLEM 21.1**

When *tert*-butyl iodide is treated with cyanide ion, 2,2-dimethylpropanenitrile is not produced. Why? What compound is formed?

**SOLUTION**
Reaction cannot take place using an $S_N2$ route because the bulky *tert*-butyl group makes it impossible for cyanide ion to attack the carbon bearing the halogen from the back side (Fig. 21.2). Substitution using the $S_N1$ route appears to be more reasonable, but cyanide ion is too strong a base. Before dissociation to produce a *tert*-butyl carbocation and iodide ion can take place, cyanide ion, a strong base, attacks and abstracts (removes) a more accessible hydrogen, and E2 elimination occurs to produce the most stable alkene (in this case, the only possible alkene).

**FIGURE 21.2** The bulky methyl groups of *tert*-butyl iodide prohibiting $S_N2$ attack of cyanide ion.

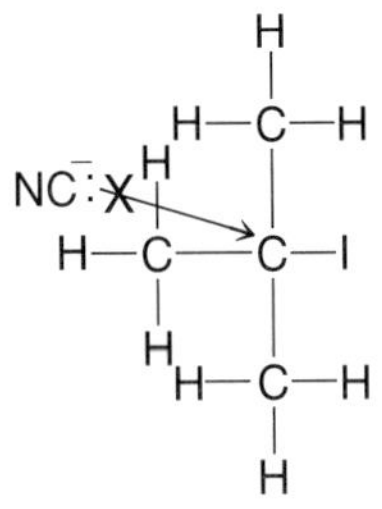

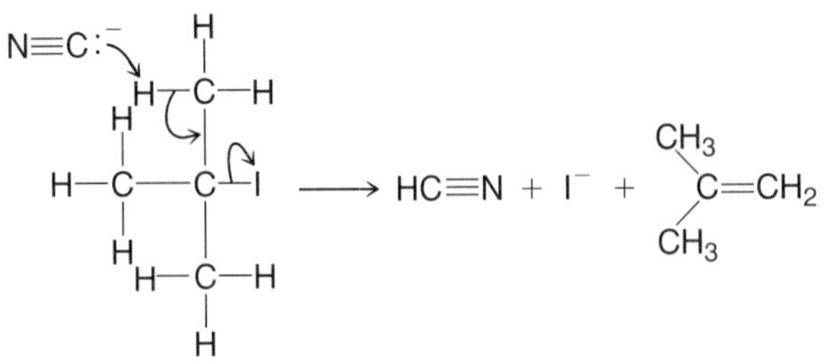

---

**EXERCISE 21.2**

When *tert*-pentyl iodide is treated with cyanide ion, 2,2-dimethylbutanenitrile is not produced. Why? What organic products are formed? What compound is the main product? Explain.

**Reacting Hydrogen Cyanide with an Aldehyde or Certain Ketones (see Sec. 19.2)**    In general,

$$R-\overset{\overset{\displaystyle O}{\|}}{C}-H(R') + HCN \xrightarrow{^-OH} R-\overset{\overset{\displaystyle OH}{|}}{\underset{\underset{\displaystyle CN}{|}}{C}}-H(R')$$

A cyanohydrin

Example:

$$CH_3-\overset{\overset{\displaystyle O}{\|}}{C}-CH_2CH_3 + HCN \xrightarrow{^-OH} CH_3-\overset{\overset{\displaystyle OH}{|}}{\underset{\underset{\displaystyle CN}{|}}{C}}-CH_2CH_3$$

Methyl ethyl ketone

Methyl ethyl ketone
cyanohydrin

**Dehydration of an Aldoxime**    In the presence of acetic anhydride, aldoximes can be dehydrated to form nitriles. In general,

$$R-\overset{\overset{\displaystyle H}{|}}{C}=N-OH + CH_3-\overset{\overset{\displaystyle O}{\|}}{C}-O-\overset{\overset{\displaystyle O}{\|}}{C}-CH_3 \longrightarrow R-C\equiv N + 2\ CH_3COOH$$

Acetic anhydride                Acetic acid

Example:

$$CH_3CH_2\overset{\overset{\displaystyle H}{|}}{C}=N-OH + CH_3-\overset{\overset{\displaystyle O}{\|}}{C}-O-\overset{\overset{\displaystyle O}{\|}}{C}-CH_3 \longrightarrow CH_3CH_2C\equiv N + 2\ CH_3COOH$$

Propanal oxime                Propanenitrile

**Reactions of Nitriles: Hydrolysis**    The most important reaction of nitriles is hydrolysis. Nitriles can be hydrolyzed in both acidic and basic solutions. In acidic solution, the corresponding carboxylic acid and an ammonium salt (ammonium chloride when hydrochloric acid is used) are produced. In basic solution, the salt of the corresponding carboxylic acid and ammonia are produced. In general,

$$R-C\equiv N + 2\ H_2O + H^+ \longrightarrow R-\overset{\overset{\displaystyle O}{\|}}{C}-OH + {}^+NH_4$$

$$R-C\equiv N + H_2O + {}^-OH \longrightarrow R-\overset{\overset{\displaystyle O}{\|}}{C}-O^- + NH_3$$

Examples:

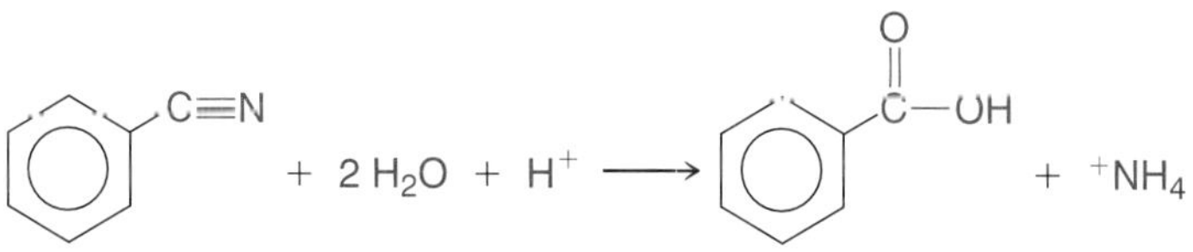

Benzoic acid

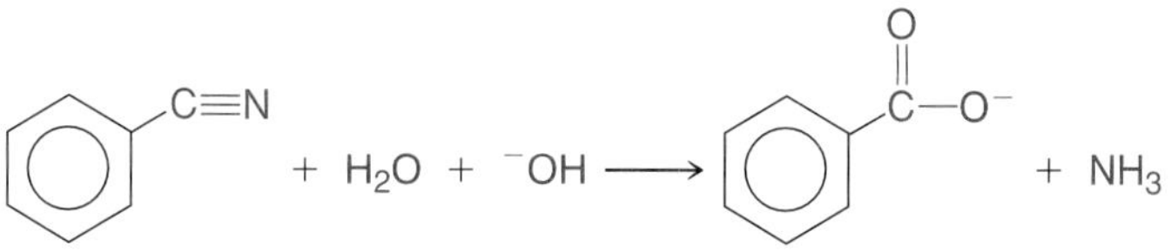

Benzoate ion

The carboxylate salt can be reacted with a mineral acid (e.g., hydrochloric acid) in the usual manner to produce the carboxylic acid (see Sec. 20.5). For example,

$$\text{C}_6\text{H}_5\text{C(=O)O}^- + \text{H}^+ \longrightarrow \text{C}_6\text{H}_5\text{C(=O)OH}$$

---

**SOLVED PROBLEM 21.2**

Prepare butyric acid from *n*-propyl bromide (two ways). You may use any inorganic reagents necessary.

**SOLUTION**

Because *n*-propyl bromide contains three carbons and butyric acid contains four carbons, we must increase the carbon content in the chain by one. As usual, let us work backward:

*Method 1:* Butyric acid can be produced by hydrolyzing butyronitrile in acid solution.

$$\text{CH}_3\text{CH}_2\text{CH}_2\text{C}\equiv\text{N} \xrightarrow[\text{HCl}]{\text{H}_2\text{O}} \text{CH}_3\text{CH}_2\text{CH}_2\text{COOH}$$

Let us complete this synthesis by reacting *n*-propyl bromide with potassium cyanide.

$$\text{CH}_3\text{CH}_2\text{CH}_2\text{Br} \xrightarrow[\text{Ethanol-water}]{\text{KCN}} \text{CH}_3\text{CH}_2\text{CH}_2\text{C}\equiv\text{N}$$

*Method 2:* React *n*-propyl magnesium bromide with carbon dioxide, and then hydrolyze the magnesium bromide salt of the carboxylic acid with hydrochloric acid to give butyric acid.

$$\text{CH}_3\text{CH}_2\text{CH}_2\text{MgBr} \xrightarrow[\text{2. HCl}(aq)]{\text{1. CO}_2} \text{CH}_3\text{CH}_2\text{CH}_2\text{COOH}$$

Reacting *n*-propyl bromide with magnesium in anhydrous ether to obtain the Grignard reagent completes the synthesis.

$$\text{CH}_3\text{CH}_2\text{CH}_2\text{Br} \xrightarrow[\substack{\text{Anhydrous} \\ \text{ether}}]{\text{Mg}} \underset{\substack{n\text{-propylmagnesium} \\ \text{bromide}}}{\text{CH}_3\text{CH}_2\text{CH}_2\text{MgBr}}$$

All together, we have

$$\text{CH}_3\text{CH}_2\text{CH}_2\text{C}\equiv\text{N} \xleftarrow[\text{C}_2\text{H}_5\text{OH}-\text{H}_2\text{O}]{\text{KCN}} \text{CH}_3\text{CH}_2\text{CH}_2\text{Br} \xrightarrow[\substack{\text{Anhydrous} \\ \text{ether}}]{\text{Mg}} \text{CH}_3\text{CH}_2\text{CH}_2\text{MgBr}$$

$$\downarrow \substack{\text{H}_2\text{O} \\ \text{HCl}} \qquad\qquad\qquad\qquad\qquad\qquad \downarrow \substack{\text{1. CO}_2 \\ \text{2. HCl}(aq)}$$

$$\text{CH}_3\text{CH}_2\text{CH}_2\overset{\text{O}}{\overset{\|}{\text{C}}}\text{-OH} \qquad\qquad\qquad\qquad\qquad \text{CH}_3\text{CH}_2\text{CH}_2\overset{\text{O}}{\overset{\|}{\text{C}}}\text{-OH}$$

---

**EXERCISE 21.3**

Prepare benzoic acid from benzene. You may use any inorganic reagents necessary.

**MEDICAL BOXED READING 21.1**

Today, a majority of Americans are well aware of the damage the sun can do to the skin of the face. Leathery-looking skin, fine lines, and roughness are characteristic of a person spending a great deal of time exposed to the rays of the sun. Today, a medical technique employing carboxylic acids exists that can slough off several layers of epidermal skin. In time, the old skin is replaced by fresh skin that is more pleasing to the eye and also less likely to develop cancer. To accomplish this loss of skin, an aqueous solution of one or more organic acids is applied to the skin, causing the top layers to peel off. Acids used are glycolic, lactic, salicylic, and trichloroacetic. Phenol, a weaker organic acid than the carboxylic acids, is also used.

Skin stripping can be shallow (only a few layers are removed) or deep (many layers are removed). Phenol, used for the deep peels, is toxic. Patients obtaining a deep peel must have good kidney function (phenol is excreted via the kidneys) and good heart function (phenol puts stress on the heart) before the peeling is attempted.

$$HO{-}CH_2{-}\overset{\overset{\large O}{\|}}{C}{-}OH$$

Glycolic acid

## 21.2 CHEMICAL PROPERTIES OF CARBOXYLIC ACIDS

### Salt Formation

In general,

$$R{-}\overset{\overset{\large O}{\|}}{C}{-}OH + MOH(aq) \longrightarrow R{-}\overset{\overset{\large O}{\|}}{C}{-}O^-M^+ + H_2O$$

$$2\,R{-}\overset{\overset{\large O}{\|}}{C}{-}OH + M_2CO_3(aq) \longrightarrow 2\,R{-}\overset{\overset{\large O}{\|}}{C}{-}O^-M^+ + H_2O + CO_2$$

$$R{-}\overset{\overset{\large O}{\|}}{C}{-}OH + MHCO_3(aq) \longrightarrow R{-}\overset{\overset{\large O}{\|}}{C}{-}O^-M^+ + H_2O + CO_2$$

where M = Na or K.

Refer to Section 20.5 for a detailed discussion. A carboxylic acid reacts with a variety of bases to produce water-soluble salts.

This reaction can be reversed by treating the salt with a mineral acid such as hydrochloric acid. In general,

$$R{-}\overset{\overset{\large O}{\|}}{C}{-}O^-Na^+ + HCl \longrightarrow R{-}\overset{\overset{\large O}{\|}}{C}{-}OH + NaCl$$

### Reduction

Lithium aluminum hydride reduces a carboxylic acid to the corresponding alcohol by adding $H_2$ across the carbon-oxygen double bond in the carboxylic acid. In general,

$$R{-}\overset{\overset{\large O}{\|}}{C}{-}OH \xrightarrow[\text{2. } H_2O,\ H^+]{\text{1. } LiAlH_4} R{-}CH_2OH$$

Example:

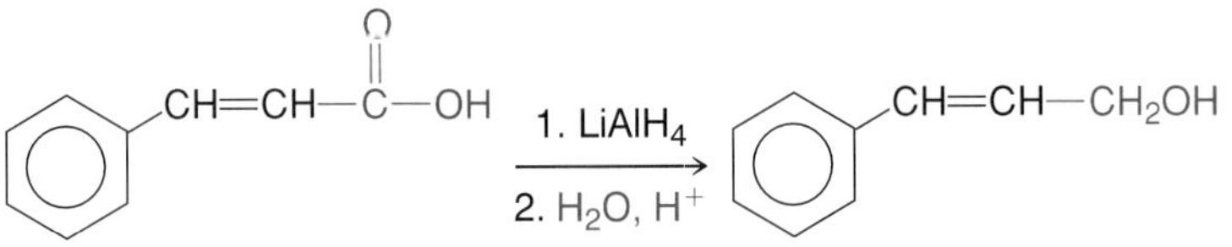


Note, that lithium aluminum hydride does not react with (hydrogenate) unsaturated carbon-carbon bonds.

### The Hell-Volhard-Zelinsky Reaction

In the presence of phosphorus trichloride as a catalyst, chlorine or bromine reacts with a carboxylic acid containing an alpha hydrogen to produce an α-haloacid. In general,

$$RCH_2COOH + X_2 \xrightarrow{PCl_3} R\underset{\underset{\large X}{|}}{C}HCOOH + HX$$

where $X_2 = Cl_2$ or $Br_2$. For example,

$$CH_3CH_2CH_2COOH + Br_2 \xrightarrow{PCl_3} CH_3CH_2\underset{\underset{Br}{|}}{C}HCOOH + HBr$$

## Ester Formation

For a brief discussion of esterification, refer to Section 20.5. For a more detailed discussion, see Section 23.6.

## Amide Formation

In the presence of dicyclohexylcarbodiimide (DCC) (see Sec. 20.5) as a promoter, a carboxylic acid can react with an amine (see Chaps. 24 and 25) to produce an amide. This reaction is often used in the preparation of peptides (see Sec. 28.11). In general,

*A promoter* is a substance that increases the rate of a reaction. Unlike a catalyst, however, a promoter is destroyed during the course of a reaction.

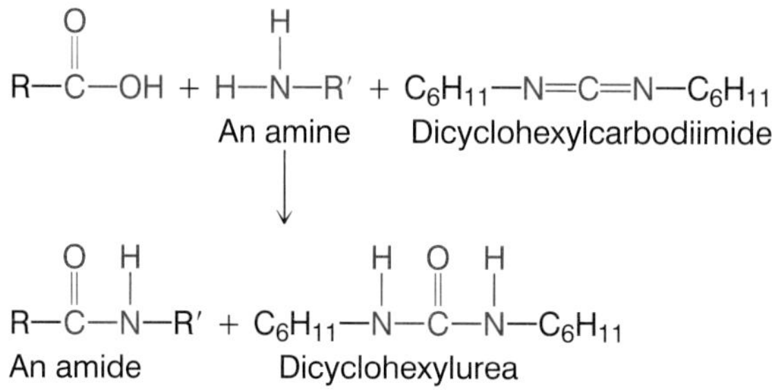

$$\underset{}{R-\overset{\overset{O}{\|}}{C}-OH} + \underset{\text{An amine}}{H-\overset{\overset{H}{|}}{N}-R'} + \underset{\text{Dicyclohexylcarbodiimide}}{C_6H_{11}-N=C=N-C_6H_{11}}$$

$$\downarrow$$

$$\underset{\text{An amide}}{R-\overset{\overset{O}{\|}}{C}-\overset{\overset{H}{|}}{N}-R'} + \underset{\text{Dicyclohexylurea}}{C_6H_{11}-\overset{\overset{H}{|}}{N}-\overset{\overset{O}{\|}}{C}-\overset{\overset{H}{|}}{N}-C_6H_{11}}$$

---

**EXERCISE 21.4**

Provide a structural formula for each organic reagent and a molecular formula or formula unit for each inorganic reagent designated by a question mark in the following equations.

a.  $? + ? \longrightarrow CH_3CH_2CH_2C\equiv N + 2 CH_3COOH$

b.  $? + {}^-CN \longrightarrow CH_3CH_2CH_2CN + Br^-$

c.  $? + HCN \xrightarrow{{}^-OH} CH_3-\underset{\underset{CN}{|}}{\overset{\overset{H}{|}}{C}}-OH$

d.  $? + H_2O + ? \longrightarrow CH_3CH_2CH_2COOH + NH_4Cl$

e.  $? + H_2O + ? \longrightarrow CH_3CH_2COO^-Na^+ + NH_3$

f.  $? + K_2CO_3 \longrightarrow CH_3COO^-K^+ + H_2O + CO_2$

g.  $CH_3CH_2COO^-Na^+ + ? \longrightarrow CH_3CH_2COOH + NaBr$

h.  $CH_3CH_2COOH + ? \xrightarrow{DCC} CH_3CH_2CONHCH_2CH_3 + H_2O$

---

# ► CHAPTER ACCOMPLISHMENTS

### 21.1 Synthesis of Carboxylic Acids
☐ Write a complete equation showing the oxidation of acetaldehyde with $^+Ag(NH_3)_2 + {}^-OH$.
☐ Explain why *tert*-butylbenzene does not react with a hot basic solution of potassium permanganate.
☐ Write a detailed, step-by-step mechanism for the following reaction:

$$CH_3CH_2MgI \xrightarrow[\text{2. HCl}(aq)]{\text{1. } CO_2}$$

☐ Name the following nitriles.
  a. $CH_3CH_2C\equiv N$
  b. $CH_3CH_2\underset{\underset{Cl}{|}}{\overset{\overset{Cl}{|}}{C}}CH_2C\equiv N$

☐ Draw a structural formula for
  a. benzonitrile.
  b. butanenitrile.

☐ Explain why a nitrile is not produced when *tert*-butyl iodide is reacted with cyanide ion. Name and draw a structural formula for the compound that is produced.

☐ Draw a structural formula for a cyanohydrin.

☐ Prepare butyric acid from *n*-propyl bromide using two different methods.

## 21.2 Chemical Properties of Carboxylic Acids

☐ Complete the following equations.

a. $CH_3(CH_2)_3COOH \xrightarrow[\text{2. } H_2O, H^+]{\text{1. LiAlH}_4}$

b. $RCOOH + KOH(aq) \longrightarrow$

☐ Identify or name the structural feature that a carboxylic acid must show in order to undergo the Hell-Volhard-Zelinsky reaction.

---

## ▶ KEY TERM

*nitrile* (21.1)

---

## ▶ PROBLEMS

1. Draw a structural formula for the following compounds.
   a. 2-methylbutanenitrile
   b. acrylonitrile (2-propenenitrile)
   c. pentanenitrile
   d. propanenitrile
   e. *m*-chlorobenzonitrile
   f. 3-methylbutanenitrile

2. Complete each of the following equations by supplying a structural formula for each organic product and a molecular formula or formula unit for each inorganic product.

a. $+ H_2O + H^+ \longrightarrow$

b. $+ NaHCO_3(aq)\text{(excess)} \longrightarrow$

c. $+ NaOH(aq)\text{(excess)} \longrightarrow$

d. acrylic acid ($CH_2{=}CHCOOH$) $+ H_2 \xrightarrow{\text{Pd}}$

e. acrylic acid $\xrightarrow[\text{2. } H_2O, H^+]{\text{1. LiAlH}_4}$ (organic product only)

f. $CH_3CN + H_2O + {}^-OH \longrightarrow$

g. $(CH_3)_2CHCN + H_2O + H^+ \longrightarrow$

h. $CH_3CH_2CH_2OH \xrightarrow[H_2SO_4]{Na_2Cr_2O_7}$ (organic product only)

i. $CH_3CH_2CH_2OH \xrightarrow[CH_2Cl_2]{PCC}$ (organic product only; omit product derived from PCC)

j. $CH_3(CH_2)_{10}COO^-Na^+ + HCl \longrightarrow$

k. malononitrile ($NCCH_2CN$) $+ H_2O\text{(excess)} + H^+ \longrightarrow$

l. acetaldehyde oxime + acetic anhydride $\longrightarrow$

m. phenylacetic acid $+ Br_2 \xrightarrow{PCl_3}$

n. *n*-butyl bromide $+ KCN \xrightarrow{C_2H_5OH - H_2O}$

o. $CH_3CH_2MgI \xrightarrow[\text{2. HCl}(aq)]{\text{1. } CO_2}$

p. 2-iodo-2-methylpentane $+ KCN \longrightarrow$

3. a. A carboxylic acid, $C_5H_{10}O_2$, is prepared by treating isobutyl magnesium chloride with $CO_2$ and then hydrolyzing the salt produced with hydrochloric acid. Supply a structural formula for this acid, and write the relevant equations.

   b. A carboxylic acid, $C_5H_{10}O_2$, can be resolved into two enantiomers. Draw a structural formula for this acid, and explain.

   c. The carboxylic acid, $C_5H_{10}O_2$, cannot be made by treating an alkyl halide, $C_4H_9Br$, with potassium cyanide. Draw a structural formula for this acid, and explain.

   d. A carboxylic acid, $C_5H_{10}O_2$, can be prepared by treating 2-hexanone with $NaOCl(aq)$ [$NaOH(aq) + Cl_2$] and acidifying the resulting sodium salt. Draw a structure for this acid, and explain using chemical reaction equations.

4. Write a detailed, step-by-step mechanism for

   $$CH_3CH_2MgBr \xrightarrow[\text{2. HCl}(aq)]{\text{1. } CO_2}$$

5. Prepare each of the following compounds. Use any organic or inorganic reagents necessary.

   a. $CH_3CH_2\overset{\displaystyle O}{\overset{\displaystyle \|}{C}}{-}OH$ from $CH_3CH{=}CH_2$

   b. succinic acid from ethylene

c. $CH_3CH_2C{\equiv}N$ from $CH_3CH_2CH_2OH$

d.

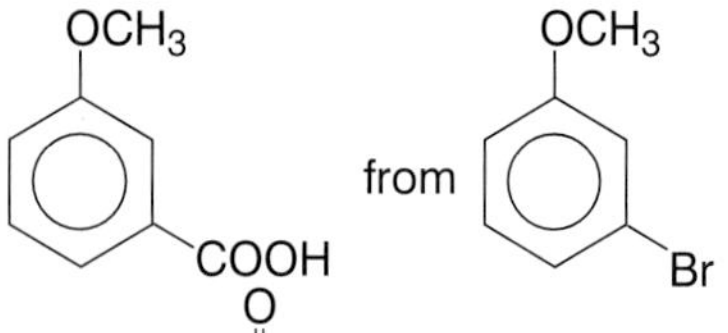

$\underset{\text{COOH}}{\overset{\text{OCH}_3}{\bigodot}}$  from  $\underset{\text{Br}}{\overset{\text{OCH}_3}{\bigodot}}$

e. $HOCH_2CH_2\overset{O}{\overset{\|}{C}}{-}OH$ from ethylene

f. pyruvic acid from acetaldehyde

6. Prepare hexanoic acid from each of the following. You may use any necessary reagents.
   a. hexanal
   b. 1-hexanol
   c. 2-heptanol
   d. 1-pentanol (two different ways)

7. 2,4,5-Trichlorophenoxyacetic acid (2,4,5-T) is an excellent herbicide. Unfortunately, it defoliates both cultivated plants and weeds together.
   a. Draw a structural formula for the compound.
   b. Synthesize 2,4,5-T from 2,4,5-trichlorophenol and acetic acid. Use any inorganic reagents necessary.

8. Compound A, $C_8H_{10}$, when heated with an alkaline solution of $KMnO_4$ and then treated with HCl, gave a compound B. Reacting compound B with aqueous $NaHCO_3$ produced compound C, which was identified as sodium benzoate. What is the structure of compounds A, B, and C. Explain your reasoning.

9. One carboxylic acid, when reacted with $LiAlH_4$, then $H_2O$, and $H^+$, did not produce a primary alcohol. Name the carboxylic acid and the corresponding alcohol.

10. Draw a structural formula for a compound, $C_4H_8O_2$, the proton NMR spectrum of which is as follows:
    $\delta = 1.0$ (3H) triplet
    $\delta = 1.8$ (2H) hextet
    $\delta = 2.3$ (2H) triplet
    $\delta = 11.2$ (1H) singlet

11. Spectroscopically, how could you determine when the following reaction is complete?

$$CH_3(CH_2)_3\overset{O}{\overset{\|}{C}}{-}OH \xrightarrow[\text{2. H}_2\text{O, H}^+]{\text{1. LiAlH}_4} CH_3(CH_2)_3CH_2OH$$

# CHAPTER 22

# Derivatives of Carboxylic Acids (I)

## CLASSIFICATION AND NOMENCLATURE, PHYSICAL PROPERTIES, NATURAL OCCURRENCE AND USES OF SELECTED ACID DERIVATIVES, AND REPRESENTATIVE REACTIONS

Several families of compounds, the acid chlorides, acid anhydrides, esters, and acid amides (Fig. 22.1), are closely related to their corresponding carboxylic acids and are collectively known as *carboxylic acid derivatives.*

To formulate an acid chloride, for example, replace the —OH group of the corresponding carboxylic acid with —Cl.

$$CH_3-\overset{\overset{\displaystyle O}{\|}}{C}-OH \qquad CH_3-\overset{\overset{\displaystyle O}{\|}}{C}-Cl$$

Acetic acid     Acetyl chloride

These formulations are summarized in Table 22.1.

Acid chlorides (also called *acyl chlorides*) that are derived from aliphatic carboxylic acids are named using both the common and IUPAC systems of nomenclature by replacing the suffix *-ic* of the corresponding acid with *-yl* and adding the word *chloride*. With aromatic and alicyclic acid chlorides, the term *carboxylic acid* is replaced by *carbonyl chloride* in IUPAC nomenclature. An IUPAC name more often used for aromatic

**22.1
CLASSIFICATION AND
NOMENCLATURE**

| TABLE 22.1 | Formulation of Acid Derivatives from the Corresponding Carboxylic Acid | | $\left[R-\overset{\overset{\displaystyle O}{\|}}{C}-\right]$ |
|---|---|---|---|
| | Group Removed from the Carboxylic Acid | Group Added to the Acyl Group | |
| Acid chloride | —OH | —Cl | |
| Acid anhydride | —OH | $-O-\overset{\overset{\displaystyle O}{\|}}{C}-R$ | |
| Ester | —OH | —O—R′ | |
| Amide | —OH | —NH$_2$ | |

**FIGURE 22.1** General structural formulas of the carboxylic acid derivatives.

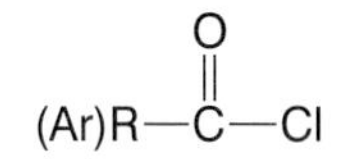

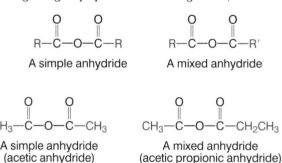

To name an acid anhydride using both the common and IUPAC systems of nomenclature, name the corresponding acid(s), and then replace the word *acid* with the word *anhydride*. An IUPAC name more often used for aromatic acid anhydrides involves the use of the former trivial name, as discussed in Section 9.2. Refer to Fig. 22.3 for examples of the nomenclature of the acid anhydrides.

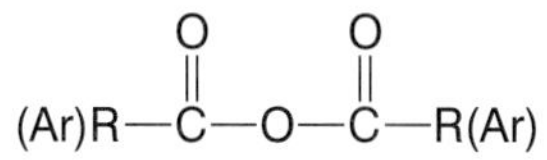

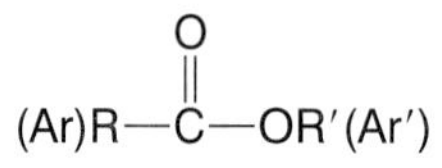

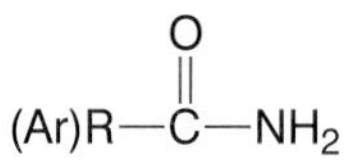

Esters are named by both methods of nomenclature by first naming the group contributed by the alcohol (R′) as one word.

**FIGURE 22.2** Nomenclature of selected acid chlorides.

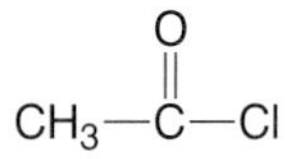

$$CH_3-\overset{\overset{\displaystyle O}{\|}}{C}-Cl$$

Common: Acetyl chloride
IUPAC: Ethanoyl chloride

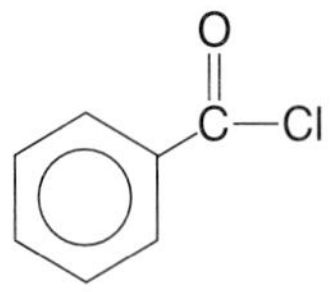

$$\text{(benzene ring)}-\overset{\overset{\displaystyle O}{\|}}{C}-Cl$$

Benzoyl chloride
(benzenecarbonyl
chloride)

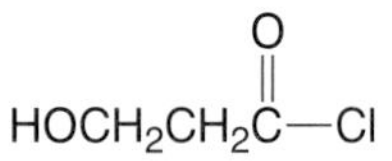

$$HOCH_2CH_2\overset{\overset{\displaystyle O}{\|}}{C}-Cl$$

β-Hydroxypropionyl chloride
3-Hydroxypropanoyl chloride

acid chlorides (aroyl chlorides) involves the use of the former trivial name, as discussed in Section 9.2. Refer to Fig. 22.2 for examples of the nomenclature of the acid chlorides.

Anhydrides are classified as simple or mixed. A *simple anhydride* contains two identical organic groups in each molecule; a *mixed anhydride*, on the other hand, contains two different organic groups per molecule. In general,

$$R-\overset{\overset{\displaystyle O}{\|}}{C}-O-\overset{\overset{\displaystyle O}{\|}}{C}-R \qquad R-\overset{\overset{\displaystyle O}{\|}}{C}-O-\overset{\overset{\displaystyle O}{\|}}{C}-R'$$

A simple anhydride          A mixed anhydride

Examples:

$$CH_3-\overset{\overset{\displaystyle O}{\|}}{C}-O-\overset{\overset{\displaystyle O}{\|}}{C}-CH_3 \qquad CH_3-\overset{\overset{\displaystyle O}{\|}}{C}-O-\overset{\overset{\displaystyle O}{\|}}{C}-CH_2CH_3$$

A simple anhydride          A mixed anhydride
(acetic anhydride)          (acetic propionic anhydride)

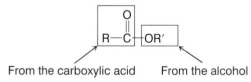

From the carboxylic acid          From the alcohol

Then, as a second word, the *-ic* suffix of the acid is replaced by the suffix *-ate*. With aromatic and alicyclic esters, replace *carboxylic acid* with *carboxylate* in IUPAC nomenclature. An IUPAC name more often used for aromatic esters (in which the acid contributing group is aromatic) involves the use of the former trivial name, as discussed in Section 9.2. Examples of ester nomenclature are found in Fig. 22.4.

To name a *simple amide* (an unsubstituted amide) using the common method of nomenclature, replace the *-ic* suffix of the corresponding carboxylic acid with the *-amide* suffix. For the IUPAC system of nomenclature, replace the *-oic* suffix of the corresponding carboxylic acid with the *-amide* suffix. A simple amide contains two hydrogens bonded to nitrogen.

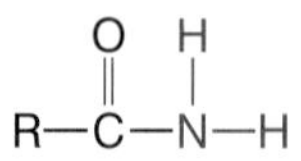

**FIGURE 22.3** Nomenclature of selected acid anhydrides.

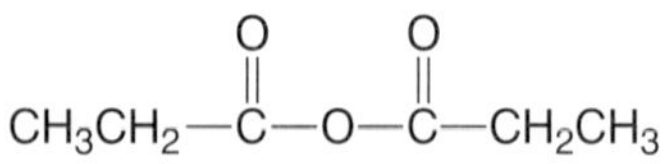

$$CH_3CH_2-\overset{\overset{\displaystyle O}{\|}}{C}-O-\overset{\overset{\displaystyle O}{\|}}{C}-CH_2CH_3$$

Common: Propionic anhydride
IUPAC: Propanoic anhydride

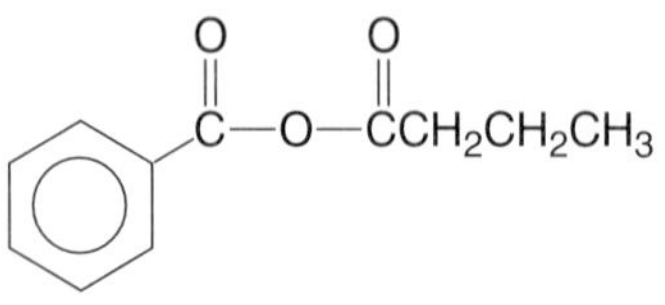

$$\text{(benzene ring)}-\overset{\overset{\displaystyle O}{\|}}{C}-O-\overset{\overset{\displaystyle O}{\|}}{C}CH_2CH_2CH_3$$

Benzoic butanoic anhydride

$$R-\overset{\overset{\displaystyle O}{\|}}{C}-\overset{\overset{\displaystyle H}{|}}{N}-H$$

Simple amide

With aromatic and alicyclic amides, replace *carboxylic acid* with *carboxamide* in IUPAC nomenclature. An IUPAC name more often used for aromatic amides involves the use of the former trivial name, as discussed in Section 9.2. Refer to Fig. 22.5 for examples of simple acid amide nomenclature.

*Substituted amides* are compounds in which one or more organic groups replace one or more hydrogens bonded to the nitrogen of the amide.

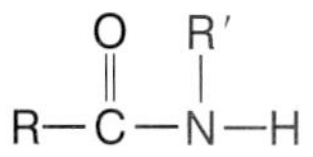

Monosubstituted amide 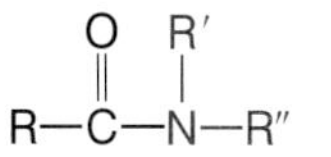　　　Disubstituted amide

In nomenclature, an organic group bonded to nitrogen is designated by an uppercase italic *N* (meaning the group is bonded to nitrogen) followed by a hyphen and the name of the group, as shown in Fig. 22.6.

Using the common system of nomenclature, *N*-phenylamides also can be named as a derivative of the corresponding amine—aniline. Consider

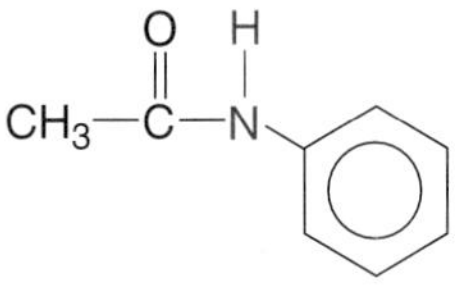

Acetanilide

To name an *N*-phenylamide as an anilide, first name the acyl (or aroyl) group (in the preceding example, acetyl).

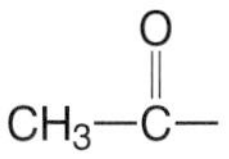

The acetyl group

Finally, drop the *-yl* suffix and add the fragment *-anilide* to get *acetanilide.* Other examples of anilide nomenclature are given in Fig. 22.7.

Name the following compounds.

a.　 (common; disregard optical isomerism)

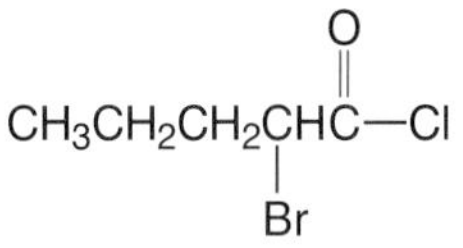

b.　$CH_3CH_2\overset{O}{\overset{\|}{C}}-OCH_2CH_3$ (IUPAC)

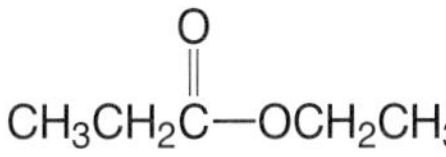

c.　 (IUPAC and common)

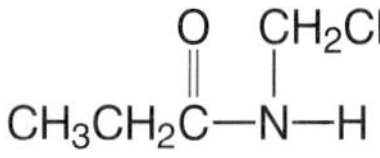

**FIGURE 22.6** Nomenclature of selected substituted amides.

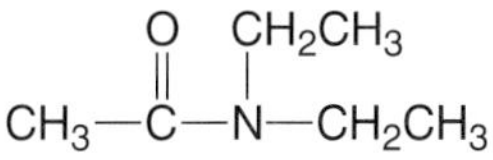

Common: *N, N*-Diethylacetamide
IUPAC: *N, N*-Diethylethanamide

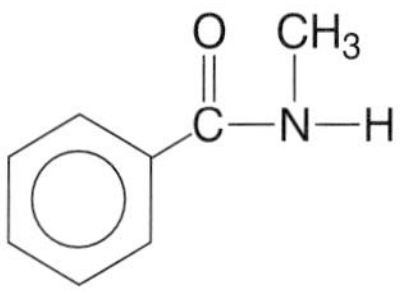

*N*-Methylbenzamide
(*N*-methylbenzenecarboxamide)

**FIGURE 22.4** Nomenclature of selected esters.

**FIGURE 22.4** Nomenclature of selected esters.

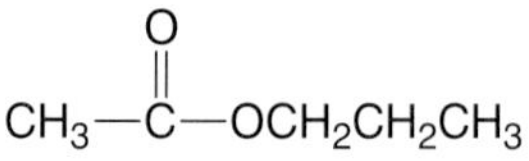

Common: *n*-Propyl acetate
IUPAC: *n*-Propyl ethanoate

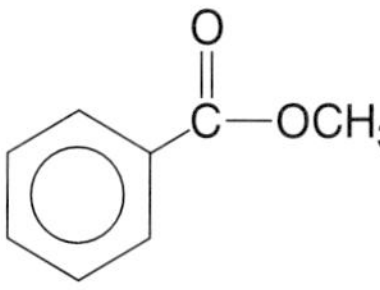

IUPAC:　Methyl benzoate

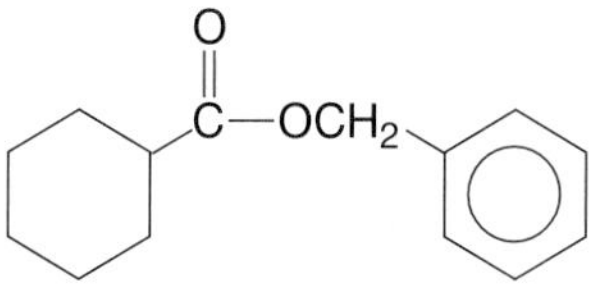

Benzyl cyclohexanecarboxylate

**SOLVED PROBLEM 22.1**

**FIGURE 22.5** Nomenclature of selected simple amides.

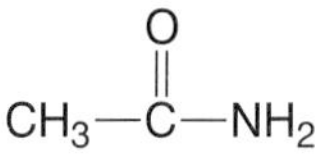

Common: Acetamide
IUPAC: Ethanamide

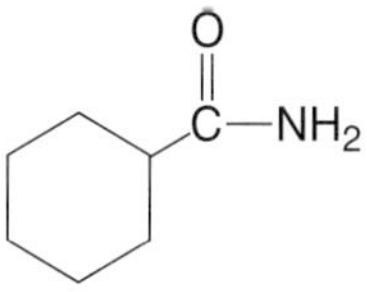

Cyclohexanecarboxamide

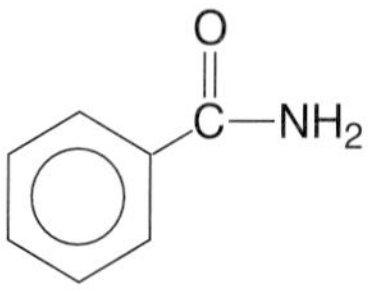

IUPAC:　　Benzamide
(benzenecarboxamide)

**FIGURE 22.7** Selected examples of anilide nomenclature.

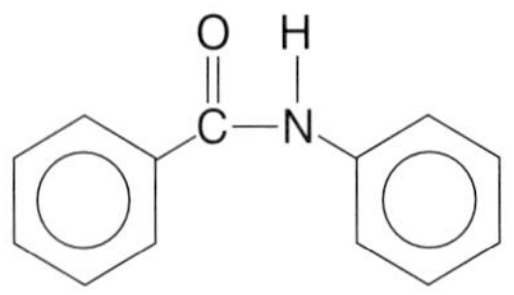

Benzanilide
(drop -oyl here)

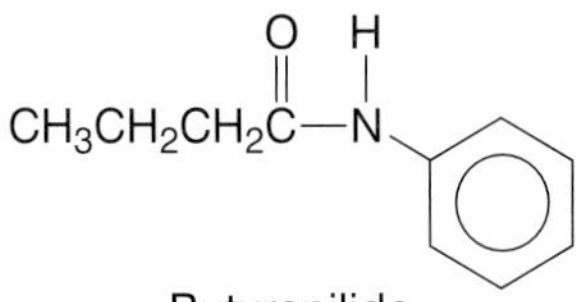

Butyranilide

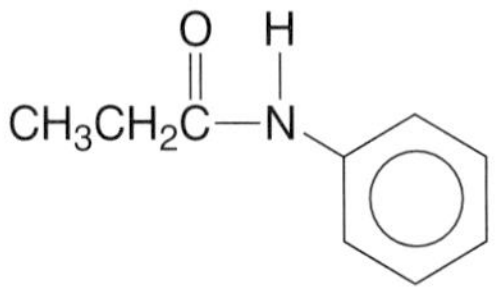

Propionanilide

**SOLUTION**

a.  Since the corresponding carboxylic acid consists of a five-carbon chain, we have valeric acid. Dropping the -ic suffix gives us the fragment valer. We then add the suffix -yl and the word chloride to give valeryl chloride. Finally, since a bromo group (—Br) is bonded to the alpha carbon, we have

α-bromovaleryl chloride

b.  First, we name the group contributed by the alcohol, i.e., ethyl, as one word. The other carbons pertain to the parent carboxylic acid, i.e., propanoic acid (a parent three-carbon chain). Dropping the -ic suffix and adding -ate, we get

ethyl propanoate

c.  IUPAC: The group bonded to the nitrogen is ethyl. To show the location of the ethyl group bonded to nitrogen, we write N-ethyl. The parent acid is propanoic acid. Replacing the -oic suffix with -amide gives us

N-ethylpropanamide

Common: The group bonded to the nitrogen is ethyl. To show the location of the ethyl group bonded to nitrogen, we write N-ethyl. The parent acid is propionic acid. Replacing the -ic suffix with -amide gives us

N-ethylpropionamide

---

**EXERCISE 22.1**

Name the following compounds.

a.  $CH_3CH_2CH_2CH_2CH_2C-Cl$ (IUPAC)
             $\parallel$
              $O$

b.  $CH_3-C-OCH_2CH_3$ (common and IUPAC)
          $\parallel$
           $O$

c.  $CH_3CH_2C-O-CCH_2CH_3$ (common and IUPAC)
             $\parallel$      $\parallel$
              $O$       $O$

d.  $CH_3CH_2CH_2C-N-CH_3$ (common and IUPAC)
               $\parallel$  $\mid$
                $O$   $CH_3$

---

**FIGURE 22.8** Hydrogen bonding in a simple amide.

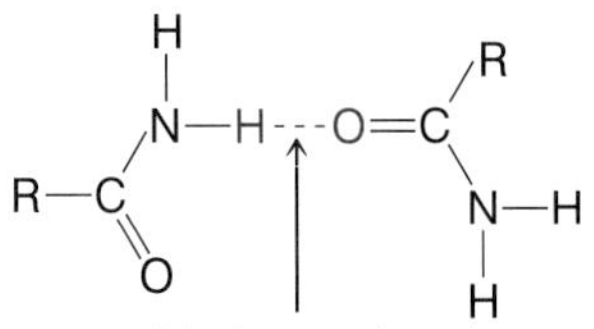

Hydrogen bond

## 22.2
## PHYSICAL PROPERTIES

For example, acetyl chloride boils at 51°C, while acetamide melts at 82°C and boils at 221°C.

The acid chlorides and acid anhydrides irritate the mucous membranes of the nose and throat. For this reason, these substances should only be handled in a laboratory hood.

The first few members of the acid chloride, acid anhydride, and ester homologous series of compounds are low-boiling liquids due to a lack of hydrogen bonding. Acid amides, on the other hand, are high-boiling solids because intermolecular hydrogen bonding is present (Fig. 22.8).

Trends of water solubility of esters and amides are similar to those of other functional groups. Water insolubility occurs in those homologues which contain five carbons or more.

The acid chlorides and anhydrides are irritating and hazardous to human beings. Naturally occurring volatile esters often have fragrant odors, which are the basis of the aromas of the pleasant-smelling fruits, vegtables, flowers, and perfumes in Fig. 22.9.

---

**EXERCISE 22.2**

Predict the higher-boiling compound: N-methylacetamide (molecular weight = 73 g/mol) or N,N-dimethylacetamide (molecular weight = 87 g/mol). Explain.

**FIGURE 22.9** Selected examples of ester nomenclature (and flavors).

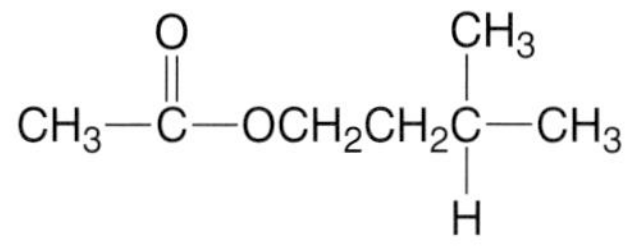

Isopentyl acetate
(banana)

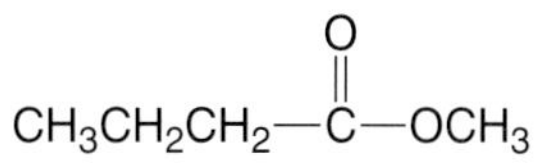

Methyl butyrate
(pineapple)

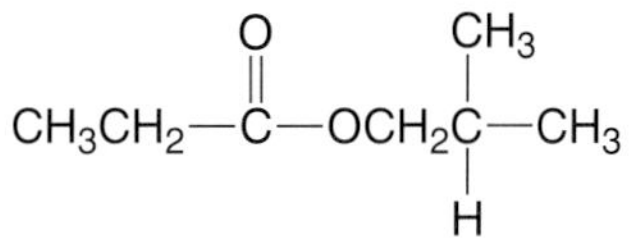

Isobutyl propionate
(rum)

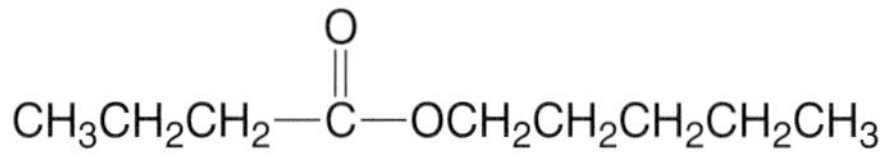

N-Pentyl butyrate
(apricot)

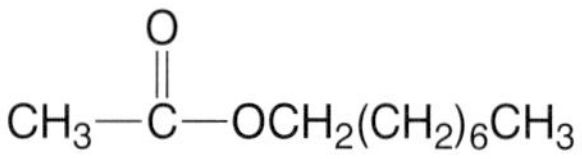

N-Octyl acetate
(orange)

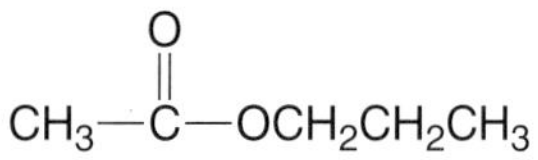

N-Propyl acetate
(pear)

Acid chlorides and acid anhydrides do not occur in nature because each class of compound reacts with water at room temperature to form the corresponding carboxylic acid (see Secs. 23.3 and 23.5).

Esters are found to a great extent in nature. Animal fats and other compounds of biological significance are esters of the alcohol glycerol (see Chap. 26).

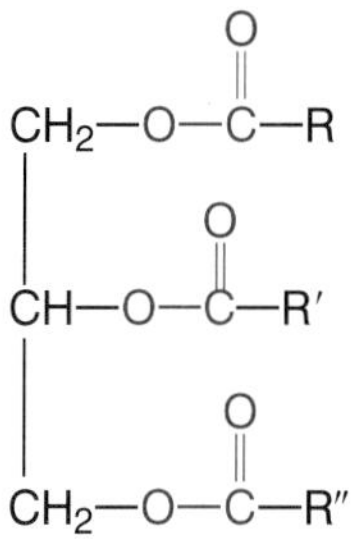

## 22.3 NATURAL OCCURRENCE AND USES OF SELECTED ACID DERIVATIVES

A triester of glycerol

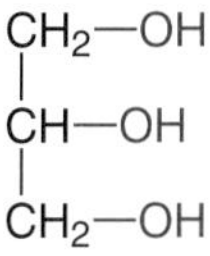

Glycerol

A significant number of cyclic esters (*lactones*) (see Sec. 23.6) are found in nature.

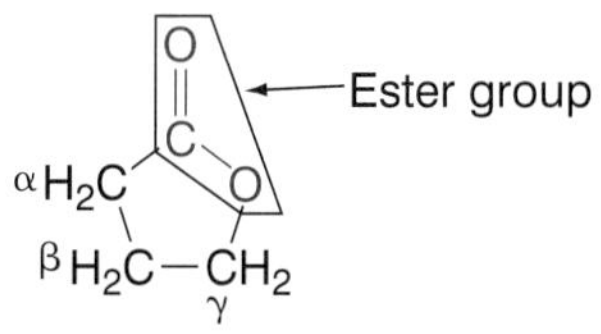

Ester group

γ-Butyrolactone

**FIGURE 22.10** Coumarin.

Coumarin (Fig. 22.10) is found in tonka beans and sweet clover. Ascorbic acid (vitamin C) (Fig. 22.11) is obtained from oranges, berries, cabbage, potatoes, and other fruits and vegetables.

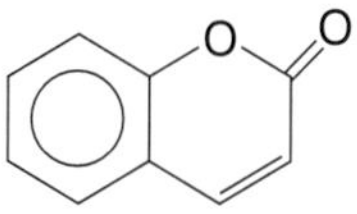

Coumarin has a fragrant odor (similar to vanilla) and is included in pharmaceuticals as a flavoring agent.

Although a number of naturally occurring lactones are innocuous when ingested, there exist a small number of toxic "black sheep" lactones. For example, aflatoxin B-1 (Fig. 22.12), reported in peanuts and corn, is a toxic and carcinogenic compound produced by the fungus *Aspergillus flavus*.

**EXERCISE 22.3**

Locate the lactone group in vitamin C and aflatoxin B-1.

**FIGURE 22.11** Vitamin C (ascorbic acid).

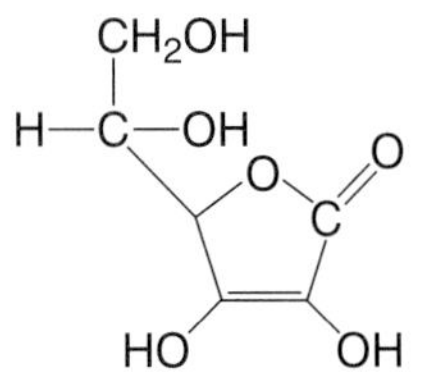


Prevents and cures scurvy.

**FIGURE 22.12** Aflatoxin B-1.

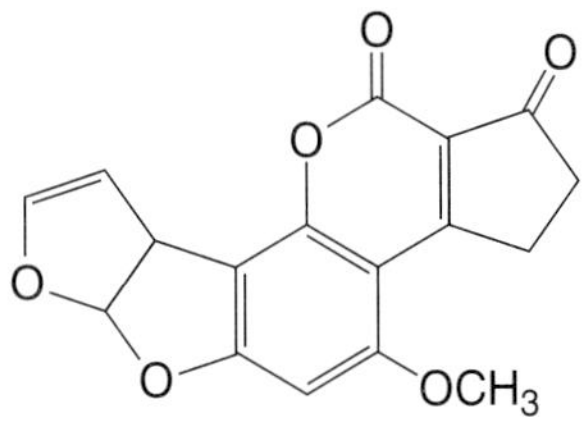


A number of amides are very useful compounds. Nicotinamide, a B-complex vitamin (Fig. 22.13), is used to relieve the symptoms of *pellagra*. Acetanilide (Fig. 22.13) was used as an analgesic and antipyretic before the introduction of aspirin. The compound was soon abandoned because it was shown to be too toxic for humans, causing a type of anemia called *methemoglobinemia*. Essentially what happens in this blood disease is that the Fe(II) ion in *hemoglobin* is oxidized to Fe(III), producing *methemoglobin*. Unfortunately, the methemoglobin cannot function as an oxygen carrier to the cells, and anemia results. A number of derivatives of acetanilide were tested for the combination of significant analgesia accompanied by low toxicity. The compound *p*-hydroxyacetanilide (Fig. 22.13) fit the bill and is the main ingredient in Tylenol. Nylon-6,6 is one of a series of polyamides that are used in clothing, carpets, and a variety of plastic products.

$$\left[ \overset{\displaystyle O}{\overset{\|}{C}}-CH_2CH_2CH_2CH_2-\overset{\displaystyle O}{\overset{\|}{C}}-\overset{\displaystyle H}{\overset{|}{N}}-CH_2CH_2CH_2CH_2CH_2CH_2-\overset{\displaystyle H}{\overset{|}{N}} \right]_n$$

Nylon-6,6

where $n$ is a large number.

**EXERCISE 22.4**

Locate the amide group in nicotinamide, acetanilide, and *p*-hydroxyacetanilide.

## 22.4 REPRESENTATIVE REACTIONS

### Hydrolysis and Saponification of Esters

Esters undergo a reaction in acid solution to produce the corresponding acid and alcohol. This reaction is the opposite of the Fischer esterification that we discussed in Section 20.5 and thus is an equilibrium process. Yields of acid and alcohol are increased by using an excess of water. To formulate the product, replace OR′ with OH to give the acid. To formulate the alcohol by-product, combine the remaining H (from $H_2O$) with OR′. In general,

$$R-\overset{\displaystyle O}{\overset{\|}{C}}-OR' + H_2O \overset{H^+}{\rightleftharpoons} R-\overset{\displaystyle O}{\overset{\|}{C}}-OH + R'OH$$

**FIGURE 22.13** Some medically related amides.

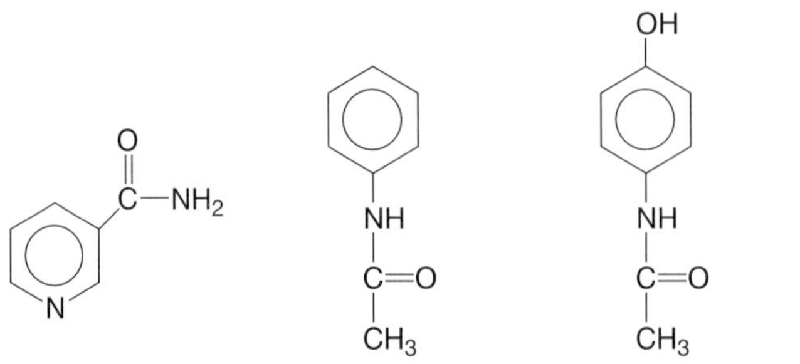


Nicotinamide    Acetanilide    *p*-Hydroxyacetanilide (acetaminophen)

## MEDICAL BOXED READING 22.1

A few years ago, some nut decided to put cyanide in Tylenol capsules. As a result of this act, a number of people were killed, Johnson & Johnson stock was temporarily hurt on the New York Stock Exchange, and tons of copy have been written about the incidents—creating a great deal of fear about over-the-counter (OTC) drugs.

In laboratories all over the country, analysts had to test for the presence of cyanide ion in Tylenol to ensure that our supplies of this important drug were safe. To accomplish this, the suspect capsule was dissolved in distilled water, and ferrous sulfate was added to give ferrocyanide ion (if $^{-}CN$ was present).

$$FeSO_4 + 6\ ^{-}CN \longrightarrow \underset{\text{Ferrocyanide ion}}{^{4-}Fe(CN)_6} + SO_4^{2-}$$

The resulting solution was boiled to produce some ferric ion.

$$\underset{\text{Ferrous ion}}{Fe^{2+}} \xrightarrow{\text{Boiling}} \underset{\text{Ferric ion}}{Fe^{3+}}$$

The ferrocyanide ion spontaneously reacts with ferric ion to produce ferric ferrocyanide, a deep blue precipitate that represents a positive test for cyanide ion.

$$3\ ^{4-}Fe(CN)_6 + 4\ Fe^{3+} \longrightarrow \underset{\text{Ferric ferrocyanide}}{Fe_4[Fe(CN)_6]_3}$$

Tylenol has long been popular as a substitute for aspirin. Because aspirin is an acidic compound, it can cause some stomach distress. Acetaminophen (*para*-hydroxyacetanilide), the active ingredient in Tylenol, Bromo Seltzer, and other products, is not acidic and does not cause stomach irritation, so Tylenol and its cousins are useful drugs.

Eventually, the police apprehended some of the perpetrators. However, because the copycat syndrome is alive and well in America, we will probably have to face more of this sort of thing in the future.

This does not mean you should stop taking all OTC drugs. They are far too useful for us to abandon. What you can do is take some reasonable precautions before you consume these drugs.

First, check the expiration date printed on the label. If that date has passed, do not take the drug. Some medicines do decompose with time. Second, make sure the drug bottle has some "tamper-free" feature, perhaps a piece of plastic or aluminum foil across the top of the bottle or a blister pack, where each tablet or capsule is contained in its own plastic bed. If this plastic or aluminum foil is disturbed, do not take the drug. Third, capsules are much easier to tamper with than tablets. Avoid them; choose tablets or caplets. Fourth, use your eyes and nose when you open the bottle. Does the drug look discolored? Are there traces of powder on the capsules? Does the drug smell different? If so, do not use it.

Example:

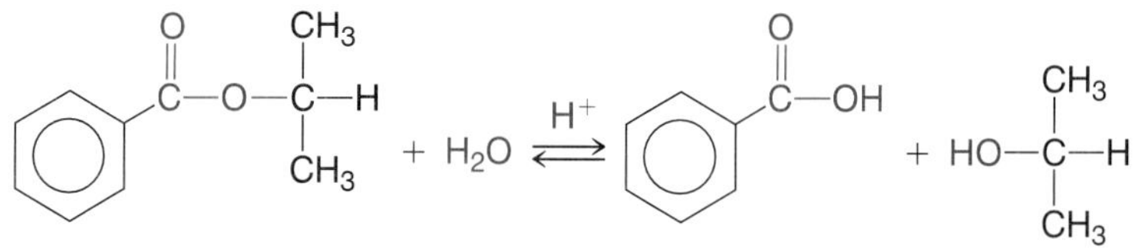

To avoid the complications of an equilibrium reaction, chemists often hydrolyze an ester in basic solution (NaOH or KOH), a process known as *saponification*. The reaction goes to completion, and yields are good. To formulate the product, replace the OR′ group with $O^{-}M^{+}$, where M is Na or K. Then combine the OR′ group with H (from MOH) to create the alcohol by-product. In general,

$$R{-}\overset{O}{\overset{\|}{C}}{-}OR' + NaOH(aq) \rightleftharpoons R{-}\overset{O}{\overset{\|}{C}}{-}O^{-}Na^{+} + R'OH$$

Example:

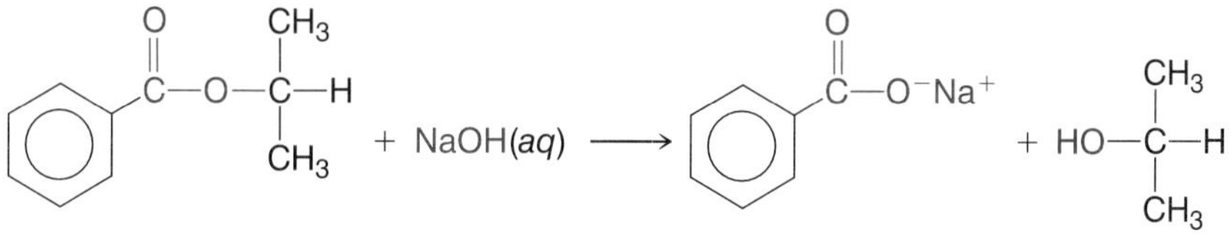

As shown earlier, the corresponding carboxylic acid can be obtained from the sodium or potassium salt by using a mineral acid such as HCl (see Sec. 20.5). In general,

$$R{-}\overset{O}{\overset{\|}{C}}{-}O^{-}K^{+} + HCl \longrightarrow R{-}\overset{O}{\overset{\|}{C}}{-}OH + KCl$$

Example:

$$C_6H_5{-}\overset{\overset{\displaystyle O}{\|}}{C}{-}O^-Na^+ + HCl \longrightarrow C_6H_5{-}\overset{\overset{\displaystyle O}{\|}}{C}{-}OH + NaCl$$

**Hydrolysis of Amides**

Both simple and substituted amides can be hydrolyzed in either acidic or basic solution to give the corresponding acid (in acid solution) or acid salt (in basic solution). In general (for a simple amide in acid solution),

$$R{-}\overset{\overset{\displaystyle O}{\|}}{C}{-}\overset{\overset{\displaystyle H}{|}}{N}{-}H + H_2O + HCl \longrightarrow R{-}\overset{\overset{\displaystyle O}{\|}}{C}{-}OH + H{-}\overset{\overset{\displaystyle H}{|}}{\underset{\underset{\displaystyle H}{|}}{N}}{\overset{+}{-}}HCl^-$$

To understand this reaction, let us formulate it as a two-step process. First, form the acid by replacing the $NH_2$ group with OH. Ammonia is obtained by combining the remaining H (from $H_2O$) with $NH_2$ as follows:

$$R{-}\overset{\overset{\displaystyle O}{\|}}{C}{-}\overset{\overset{\displaystyle H}{|}}{N}{-}H + H_2O \longrightarrow R{-}\overset{\overset{\displaystyle O}{\|}}{C}{-}OH + H{-}\overset{\overset{\displaystyle H}{|}}{N}{-}H$$

Finally, combine ammonia with HCl by means of a proton transfer to give ammonium chloride.

$$H{-}\overset{\overset{\displaystyle H}{|}}{\underset{\underset{\displaystyle H}{|}}{N}} + HCl \longrightarrow H{-}\overset{\overset{\displaystyle H}{|}}{\underset{\underset{\displaystyle H}{|}}{N}}{\overset{+}{-}}HCl^-$$

In general (for a substituted amide in acid solution),

$$R{-}\overset{\overset{\displaystyle O}{\|}}{C}{-}\overset{\overset{\displaystyle R'}{|}}{N}{-}R'' + H_2O + HCl \longrightarrow R{-}\overset{\overset{\displaystyle O}{\|}}{C}{-}OH + H{-}\overset{\overset{\displaystyle R'}{|}}{\underset{\underset{\displaystyle H}{|}}{N}}{\overset{+}{-}}R''Cl^-$$

Just as with the simple amide, let us subdivide this reaction into two reactions. First, produce the acid and amine.

$$R{-}\overset{\overset{\displaystyle O}{\|}}{C}{-}\overset{\overset{\displaystyle R'}{|}}{N}{-}R'' + H_2O \longrightarrow R{-}\overset{\overset{\displaystyle O}{\|}}{C}{-}OH + H{-}\overset{\overset{\displaystyle R'}{|}}{N}{-}R''$$

Finally, combine the amine with HCl to give the substituted ammonium salt.

$$HCl + H{-}\overset{\overset{\displaystyle R'}{|}}{N}{-}R'' \longrightarrow Cl^- \; H{-}\overset{\overset{\displaystyle R'}{|}}{\underset{\underset{\displaystyle H}{|}}{N}}{\overset{+}{-}}R''$$

Examples:

$$CH_3{-}\overset{\overset{\displaystyle O}{\|}}{C}{-}\overset{\overset{\displaystyle H}{|}}{N}{-}H + H_2O + HCl \longrightarrow CH_3{-}\overset{\overset{\displaystyle O}{\|}}{C}{-}OH + H{-}\overset{\overset{\displaystyle H}{|}}{\underset{\underset{\displaystyle H}{|}}{N}}{\overset{+}{-}}HCl^-$$

$$CH_3{-}\overset{\overset{\displaystyle O}{\|}}{C}{-}\overset{\overset{\displaystyle CH_2CH_2CH_3}{|}}{N}{-}CH_2CH_3 + H_2O + HCl \longrightarrow CH_3{-}\overset{\overset{\displaystyle O}{\|}}{C}{-}OH + H{-}\overset{\overset{\displaystyle CH_2CH_2CH_3}{|}}{\underset{\underset{\displaystyle H}{|}}{N}}{\overset{+}{-}}CH_2CH_3Cl^-$$

In general (for a simple amide in basic solution),

$$CH_3-\overset{\overset{\displaystyle O}{\|}}{C}-\overset{\overset{\displaystyle H}{|}}{N}-H + {}^-OH \longrightarrow R-\overset{\overset{\displaystyle O}{\|}}{C}-O^- + \overset{\overset{\displaystyle H}{|}}{\underset{\underset{\displaystyle H}{|}}{N}}-H$$

To formulate the products, first replace the $NH_2$ of the amide with an O. Then add the remaining hydrogen (from OH) to $NH_2$ to give ammonia.

In general (for a substituted amide in basic solution),

$$R-\overset{\overset{\displaystyle O}{\|}}{C}-\overset{\overset{\displaystyle R'}{|}}{N}-R'' + {}^-OH \longrightarrow R-\overset{\overset{\displaystyle O}{\|}}{C}-O^- + \overset{\overset{\displaystyle R'}{|}}{\underset{\underset{\displaystyle H}{|}}{N}}-R''$$

Examples:

$$CH_3-\overset{\overset{\displaystyle O}{\|}}{C}-\overset{\overset{\displaystyle H}{|}}{N}-H + {}^-OH \longrightarrow CH_3-\overset{\overset{\displaystyle O}{\|}}{C}-O^- + H-\overset{\overset{\displaystyle H}{|}}{N}-H$$

$$CH_3-\overset{\overset{\displaystyle O}{\|}}{C}-\overset{\overset{\displaystyle CH_2CH_2CH_3}{|}}{N}-CH_2CH_3 + {}^-OH \longrightarrow CH_3-\overset{\overset{\displaystyle O}{\|}}{C}-O^- + H-\overset{\overset{\displaystyle CH_2CH_2CH_3}{|}}{N}-CH_2CH_3$$

---

Complete the following equations by supplying a structural formula for each organic reagent and a molecular formula or formula unit for each inorganic reagent or catalyst designated by a question mark.

a.   $? + ? \overset{H^+}{\rightleftharpoons} CH_3CH_2CH_2COOH + (CH_3)_2CHOH$

b.   $? + ? \longrightarrow HCOO^-K^+ + CH_3CH_2CH_2CH_2OH$

c.   $? + ? \longrightarrow C_6H_5COO^- + CH_3CH_2OH$

d.   $? + ? \longrightarrow$ sodium propionate + ammonia

e.   $? + ? \longrightarrow CH_3CH_2COOH + (CH_3)_2\overset{+}{N}H_2Cl^-$

f.   $? + ? \longrightarrow CH_3COO^-Na^+ + NH_3$

---

## ▶ CHAPTER ACCOMPLISHMENTS

### Introduction
☐ Formulate a structural formula for each acid derivative of acetic acid.

### 22.1 Classification and Nomenclature
☐ Draw a structural formula for
   a. benzoyl chloride.
   b. a mixed anhydride.
   c. acetic anhydride.
   d. *n*-propyl acetate.
   e. a monosubstituted amide.
   f. acetamide.

☐ Name the following compounds.

a.   $CH_3CH_2CH_2\overset{\overset{\displaystyle Br}{|}}{\underset{}{CH}}\overset{\overset{\displaystyle O}{\|}}{C}-Cl$ (common; disregard optical isomerism)

b.   $CH_3CH_2\overset{\overset{\displaystyle O}{\|}}{C}-OCH_2CH_3$ (IUPAC)

c.   $CH_3CH_2\overset{\overset{\displaystyle O}{\|}}{C}-\overset{\overset{\displaystyle CH_2CH_3}{|}}{N}-H$ (IUPAC and common)

### 22.2 Physical Properties

☐ Explain why the acid chlorides, acid anhydrides, and esters are low-boiling liquids, whereas the amides are high-boiling solids.

☐ Predict the water solubility (in a qualitative way) of

a. acetamide.

b. *n*-hexyl hexanoate.

☐ Point out the families of acid derivatives that have irritating odors and are hazardous to human beings.

## 22.3 Natural Occurrence and Uses of Selected Acid Derivatives

☐ Explain why acid chlorides and acid anhydrides are not found in nature.

☐ Draw a structural formula for

a. a lactone.

b. acetanilide.

c. acetaminophen.

## 22.4 Representative Reactions

☐ Complete the following equations.

a. $\underset{\displaystyle CH_3-\overset{\textstyle O}{\overset{\|}{C}}-OCH_3}{} + H_2O \underset{}{\overset{H^+}{\rightleftharpoons}}$

b. $CH_3-\overset{\textstyle O}{\overset{\|}{C}}-OCH_3 + {}^-OH \longrightarrow$

c. $CH_3-\overset{\textstyle O}{\overset{\|}{C}}-\overset{\textstyle H}{\overset{|}{N}}-H + H_2O + HCl \longrightarrow$

d. $CH_3-\overset{\textstyle O}{\overset{\|}{C}}-\overset{\textstyle CH_2CH_2CH_3}{\overset{|}{N}}-CH_2CH_3 + {}^-OH \longrightarrow$

---

## ▶ KEY TERMS

carboxylic acid derivative (Introduction)
acyl chloride (22.1)
simple anhydride (22.1)
mixed anhydride (22.1)

simple amide (22.1)
substituted amide (22.1)
lactone (22.3, 23.6)
pellagra (22.3)

methemoglobinemia (22.3)
hemoglobin (22.3)
methemoglobin (22.3)
saponification (22.4, 23.7)

---

## ▶ PROBLEMS

1. Draw a structural formula for each of the following compounds.

a. benzoic anhydride

b. isopropyl formate (W-style geometric condensed)

c. butanamide

d. pentanoyl chloride

e. *N*-ethylformamide (W-style geometric condensed)

f. cyclohexanecarbonyl chloride

g. *N,N*-dimethylpropionamide

h. *N*-phenylcyclohexanecarboxamide

2. Name the following compounds.

a. $CH_3-\overset{\textstyle CH_3}{\overset{|}{\underset{|}{\underset{\textstyle H}{C}}}}-\overset{\textstyle O}{\overset{\|}{C}}-Cl$ (common and IUPAC)

b. $CH_3CH_2\overset{\textstyle O}{\overset{\|}{C}}-O-\overset{\textstyle H}{\underset{\textstyle CH_3}{\overset{|}{\underset{|}{C}}}}-CH_3$ (common and IUPAC)

c. $H-\overset{\textstyle O}{\overset{\|}{C}}-O-\overset{\textstyle O}{\overset{\|}{C}}-H$ (common and IUPAC)

d. (a benzene ring attached to) $\overset{\textstyle O}{\overset{\|}{C}}-\overset{\textstyle H}{\overset{|}{N}}-CH_3$ (common and IUPAC)

e. (a benzene ring with Cl attached to) $\overset{\textstyle O}{\overset{\|}{C}}-Cl$ (IUPAC)

f. $CH_3(CH_2)_{10}\overset{\textstyle O}{\overset{\|}{C}}-NH_2$ (IUPAC)

3. Draw a structural formula for each isomeric ester with the molecular formula $C_5H_{10}O_2$. Then provide each isomer with an IUPAC name.

4. Draw a structural formula for and classify the following compounds.

a. *N,N*-dimethylformamide

b. propionic butyric anhydride

c. *N*-phenylpropanamide

d. benzoic anhydride

5. Draw a structural formula for each of the following.

a. methyl cyclohexanecarboxylate

b. acetic anhydride
c. ethanoyl chloride
d. *N*-methyl-p-chlorobenzamide
e. phenyl benzoate
f. *p*-nitroacetanilide

6. a. Consider the following data:

| | Boiling Point (°C) | Molecular Weight (g/mol) |
|---|---|---|
| $CH_3\overset{O}{\underset{\|\|}{C}}{-}Cl$ | 51 | 78.5 |
| $CH_3CH_2\overset{O}{\underset{\|\|}{C}}{-}OH$ | 141 | 74.1 |

Explain why acetyl chloride boils at a lower temperature than propionic acid even though their molecular weights are similar.

b. Would you predict that propionyl chloride would be found in nature? Explain.

c. Can disubstituted acid amides show intermolecular hydrogen bonding? Explain.

7. Each of the following names is incorrect. Draw a structural formula, explain why the compound is named incorrectly, and write a correct name for each compound.
a. cyclobutanoic benzoic anhydride
b. β-bromopropanoyl chloride
c. 2-ethylpropanamide
d. ethyl 2-bromopropionate
e. diethylethanamide

8. Why aren't water solubilities of acyl halides and acid anhydrides discussed in Section 22.2?

9. In the southern states of the United States, Skin So Soft has been somewhat successful in repelling gnats and mosquitoes. The main ingredient of this cosmetic product is isopropyl palmitate. Draw a structural formula for isopropyl palmitate.

10. Complete the following equations by drawing a structural formula for each organic product and writing a molecular formula or formula unit for each inorganic product formed.

a. 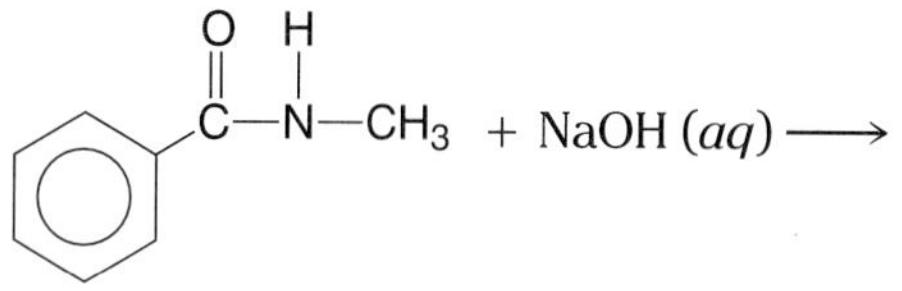 $\phantom{xx}\overset{O\ \ H}{\underset{\phantom{x}}{\text{Ph}{-}\underset{\|\|\phantom{xx}\|}{C}{-}N{-}CH_3}}$ + NaOH $(aq) \longrightarrow$

b. $(CH_3)_2CH\overset{O}{\underset{\|\|}{C}}{-}OH + CH_3OH \overset{H^+}{\rightleftharpoons}$

c. $CH_3CH_2\overset{O\ \ H}{\underset{\|\|\phantom{xx}\|}{C}}{-}\underset{}{N}{-}C_2H_5 + H_2O + HCl \longrightarrow$

d. 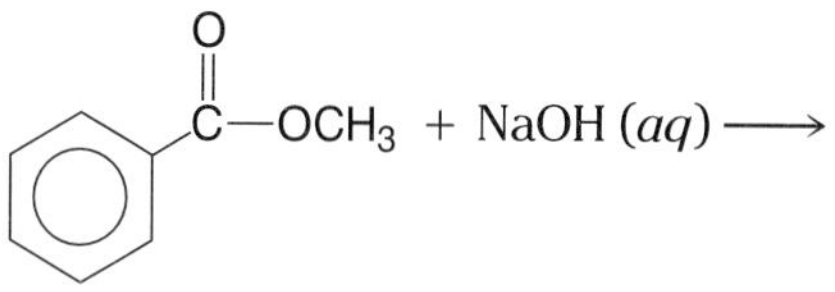 $\phantom{xx}\text{Ph}{-}\overset{O}{\underset{\|\|}{C}}{-}OCH_3$ + NaOH $(aq) \longrightarrow$

e. $CH_2{=}\overset{H\ \ O}{\underset{\|\phantom{xx}\|\|}{C}{-}C}{-}OCH(CH_3)_2 + H_2O \overset{H^+}{\rightleftharpoons}$

f. Product (d.) + HCl $\longrightarrow$

# CHAPTER 23

# Derivatives of Carboxylic Acids (II)

## SYNTHESIS AND CHEMICAL PROPERTIES

Acid chlorides are readily prepared from carboxylic acids by heating the acid with phosphorus trichloride ($PCl_3$) or thionyl chloride ($SOCl_2$). In general,

$$3\ R-\overset{\overset{\displaystyle O}{\|}}{C}-OH + PCl_3 \longrightarrow 3\ R-\overset{\overset{\displaystyle O}{\|}}{C}-Cl + H_3PO_3$$

$$R-\overset{\overset{\displaystyle O}{\|}}{C}-OH + SOCl_2 \longrightarrow R-\overset{\overset{\displaystyle O}{\|}}{C}-Cl + SO_2 + HCl$$

**23.1**
**PREPARATION OF ACID CHLORIDES**

Examples:

$$3\ CH_3(CH_2)_3-\overset{\overset{\displaystyle O}{\|}}{C}-OH + PCl_3 \longrightarrow 3\ CH_3(CH_2)_3-\overset{\overset{\displaystyle O}{\|}}{C}-Cl + H_3PO_3$$

Valeric acid  

Valeryl chloride
(decomposes at 200°C)
(bp–128°C)

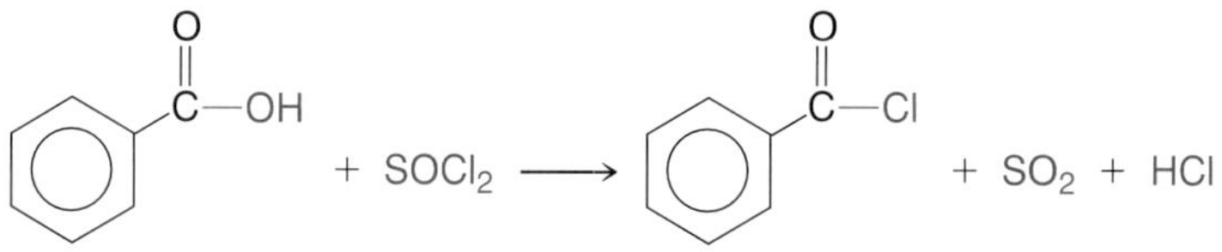

Benzoic acid  Benzoyl chloride

The method with thionyl chloride is superior because sulfur dioxide and hydrogen chloride are gases and bubble out of the reaction mixture and therefore do not contaminate the acid chloride product. Acid chlorides are low-boiling liquids that can be removed by distillation. If the difference between the decomposition point of phosphorus acid and the boiling point of the acid halide is less than 70°C, isolation of the product is troublesome because the acid chloride cannot be separated from phosphorous acid by simple distillation. Thus the reaction loses some utility.

## 23.2
## REACTION RATES OF THE CARBOXYLIC ACID DERIVATIVES: A COMPARISON

Carboxylic acid derivatives react primarily by means of *acyl nucleophilic substitution* reactions. In acyl nucleophilic substitution, we have an acid derivative,

$$R-\overset{\overset{\displaystyle O}{\|}}{C}-X$$

that reacts with a nucleophile $Y^-$ to produce

$$R-\overset{\overset{\displaystyle O}{\|}}{C}-Y \quad \text{and} \quad X\!:^-$$

In general,

$$R-\overset{\overset{\displaystyle O}{\|}}{C}-X + Y\!:^- \longrightarrow R-\overset{\overset{\displaystyle O}{\|}}{C}-Y + X\!:^-$$

The methods of preparation and the reactions of the acid derivatives are all intertwined and are based on the relative order of reactivity of the acid derivatives in acyl nucleophilic substitution reactions, which is

$$R-\overset{\overset{\displaystyle O}{\|}}{C}-Cl > R-\overset{\overset{\displaystyle O}{\|}}{C}-O-\overset{\overset{\displaystyle O}{\|}}{C}-R > R-\overset{\overset{\displaystyle O}{\|}}{C}-OR' > R-\overset{\overset{\displaystyle O}{\|}}{C}-NH_2$$

Acid chloride        Acid anhydride         Ester          Acid amide

Why is the order of reactivity the way it is? That is, why is the acid chloride most reactive, while the amide is least reactive? To understand this order of reactivity, we need to study the mechanism of this type of reaction. Consider the general mechanism of an acid derivative,

$$R-\overset{\overset{\displaystyle O}{\|}}{C}-X$$

reacting with a negatively charged nucleophile $Y\!:^-$.

---

**General Acyl Nucleophilic Substitution Mechanism of RCOX Reacting with a Nucleophile $Y\!:^-$ to Give RCOY and $X\!:^-$**

Unstable intermediate

---

Remember that the weaker the base ($:X^-$) is, the better it is as a leaving group, and the faster is the corresponding acyl nucleophilic substitution.

The rate of reaction of an acid derivative with a nucleophile depends on how well $:X^-$ functions as a leaving group. Consider Table 23.1.

The acid chloride with its excellent leaving group, $Cl^-$, is the most reactive of the carboxylic acid derivatives, whereas the amide, with its poor leaving group, is the least reactive. One more generalization comes in handy here; i.e., a more reactive acid derivative reacts via acyl nucleophilic displacement to form a less reactive acid derivative but not the reverse. Note that in actual practice, ammonia is used in excess. In general,

$$R-\overset{\overset{\displaystyle O}{\|}}{C}-Cl + 2\,NH_3 \longrightarrow R-\overset{\overset{\displaystyle O}{\|}}{C}-NH_2 + NH_4Cl$$

| TABLE 23.1 | Relative Basicity of Some Leaving Groups in Certain Acyl Nucleophilic Substitution Reactions of Carboxylic Acid Derivatives | | |
|---|---|---|---|
| **Leaving Group (:X⁻)** | **Basicity** | **Leaving-Group Character** | |
| $^-:\ddot{C}l:$ | Weak | Excellent | |
| $^-:\ddot{O}-\overset{\overset{\displaystyle O}{\|}}{C}-R$ | Moderate | Good | |
| $^-:\ddot{O}-R^1$ | Strong | Satisfactory | |
| $^-:NH_2$ | Very strong | Poor | |

Example:

$$CH_3CH_2-\overset{\overset{\displaystyle O}{\|}}{C}-Cl + 2\,NH_3 \longrightarrow CH_3CH_2-\overset{\overset{\displaystyle O}{\|}}{C}-NH_2 + NH_4Cl$$

Propanoyl chloride                Propanamide

But not, in general,

$$R-\overset{\overset{\displaystyle O}{\|}}{C}-NH_2 + Cl^- \overset{\times}{\longrightarrow} R-\overset{\overset{\displaystyle O}{\|}}{C}-Cl + {}^-NH_2$$

Less reactive                More reactive

or

$$R-\overset{\overset{\displaystyle O}{\|}}{C}-OR' + {}^-O\overset{\overset{\displaystyle O}{\|}}{C}R'' \overset{\times}{\longrightarrow} R-\overset{\overset{\displaystyle O}{\|}}{C}-O-\overset{\overset{\displaystyle O}{\|}}{C}R'' + {}^-OR'$$

Less reactive                More reactive

Since the acid chlorides are the most reactive of the four acid derivatives we have, in general,

## 23.3 REACTIONS OF ACID CHLORIDES

$$R-\overset{\overset{\displaystyle O}{\|}}{C}-Cl + Na^+\,{}^-O-\overset{\overset{\displaystyle O}{\|}}{C}-R' \longrightarrow R-\overset{\overset{\displaystyle O}{\|}}{C}-O-\overset{\overset{\displaystyle O}{\|}}{C}-R' + NaCl$$

$$R-\overset{\overset{\displaystyle O}{\|}}{C}-Cl + H-O-R' \longrightarrow R-\overset{\overset{\displaystyle O}{\|}}{C}-O-R' + HCl$$

$$R-\overset{\overset{\displaystyle O}{\|}}{C}-Cl + 2\,NH_3 \longrightarrow R-\overset{\overset{\displaystyle O}{\|}}{C}-NH_2 + NH_4Cl$$

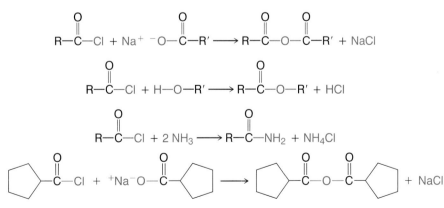

Cyclopentanecarbonyl        Sodium                Cyclopentanecarboxyic anhydride
chloride        cyclopentanecarboxylate

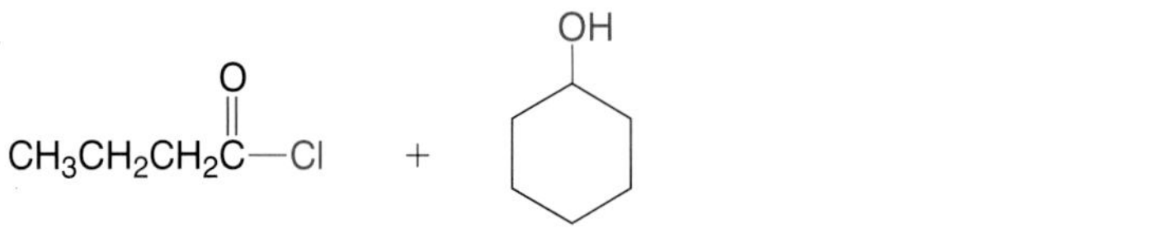

$$CH_3CH_2CH_2\overset{\overset{\displaystyle O}{\|}}{C}-Cl \;\; + \;\;$$

Butyryl chloride          Cyclohexanol

$$\longrightarrow CH_3CH_2CH_2\overset{\overset{\displaystyle O}{\|}}{C}-O-$$ $$+ \;HCl$$

Cyclohexyl butyrate

$$CH_3CH_2CH_2\overset{\overset{\displaystyle O}{\|}}{C}-Cl \; + \; 2\,NH_3 \longrightarrow CH_3CH_2CH_2\overset{\overset{\displaystyle O}{\|}}{C}NH_2 \; + \; NH_4Cl$$

Butyryl chloride                          Butyramide

**SOLVED PROBLEM 23.1**

Write a detailed, step-by-step mechanism for the following reaction.

$$CH_3CH_2-\overset{\overset{\displaystyle O}{\|}}{C}-Cl \; + \; {}^-O-\overset{\overset{\displaystyle O}{\|}}{C}-CH_3 \longrightarrow CH_3CH_2-\overset{\overset{\displaystyle O}{\|}}{C}-O-\overset{\overset{\displaystyle O}{\|}}{C}-CH_3 \; + \; Cl^-$$

**SOLUTION**

When the nucleophile is acetate ion, a negatively charged species, this acyl nucleophilic substitution is similar to the general mechanism shown on page 000.

This mechanism is also followed for nucleophilic attack of Y⁻, a negatively charged species, on acid anhydrides, esters, or acid amides.

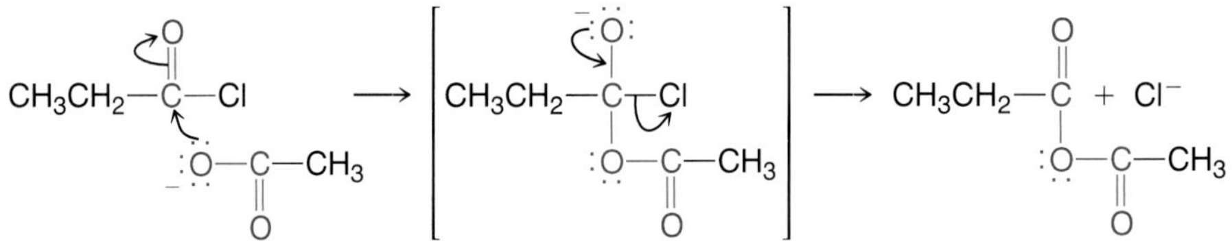

Unstable intermediate

**EXERCISE 23.1**

Write a detailed mechanism for the reaction of acetyl chloride with amide ion ($^-NH_2$).

Notice that in some cases the nucleophile is a neutral molecule rather than a negatively charged species. In this particular case, the mechanism is a bit more complex. Consider the following general scheme.

$$R-\overset{\overset{\displaystyle O}{\|}}{C}-X \; + \; :Y-H \longrightarrow R-\overset{\overset{\displaystyle O}{\|}}{C}-Y \; + \; H-X$$

**General Acyl Nucleophilic Substitution Mechanism of RCOX Reacting with a Nucleophile :Y—H to Give RCOY and H—X**

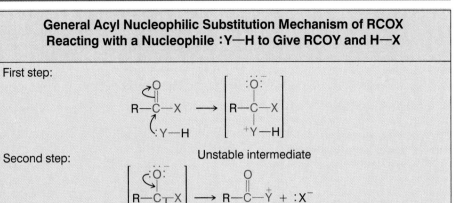

First step:

Second step:                    Unstable intermediate

Third step:

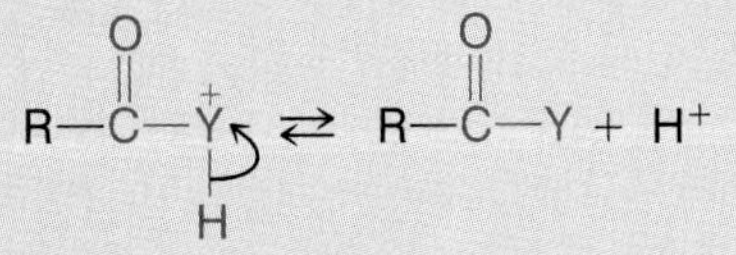

This mechanism is also followed for nucleophilic attack of Y—H, a neutral molecule, on acid anhydrides, esters, or acid amides.

As is usual, :Y—H attacks the carbonyl carbon of the acid derivative in the first step to give an unstable intermediate, which, in turn, expels X⁻ ion in the second step to form the positively charged species I. The third step of this process is the removal of a proton from the positively charged product I. Finally, we have the proton removed from I combining with X⁻ to give H—X

**SOLVED PROBLEM 23.2**

Write a detailed, step-by-step mechanism for the following reaction.

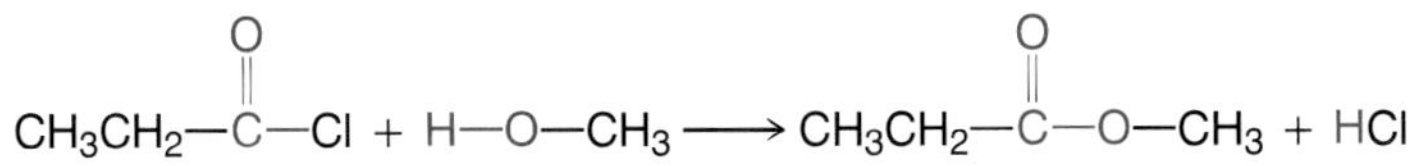

$$CH_3CH_2-\overset{O}{\underset{||}{C}}-Cl + H-O-CH_3 \longrightarrow CH_3CH_2-\overset{O}{\underset{||}{C}}-O-CH_3 + HCl$$

**SOLUTION**

We employ an acyl nucleophilic substitution reaction in which the nucleophile—methanol—attacks the carbonyl carbon of propionyl chloride to form an unstable intermediate. This intermediate, in turn, undergoes an electron shift to produce a protonated ester and the leaving group—chloride ion.

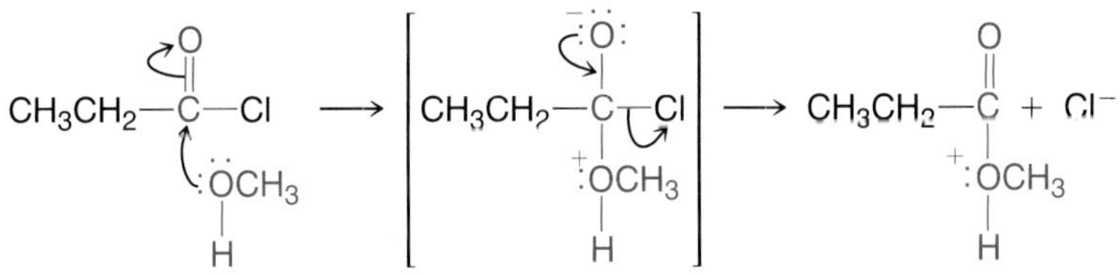

Unstable intermediate

The protonated ester, in turn, loses a proton to give the ester.

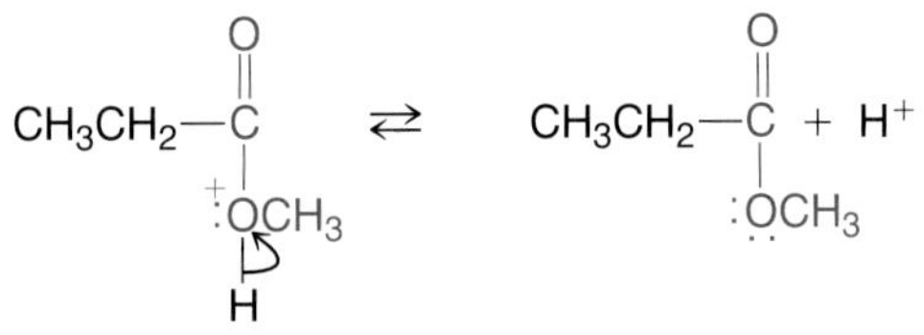

Finally, we can write

$$H^+ + Cl^- \longrightarrow HCl$$

**EXERCISE 23.2**

Write a detailed, step-by-step mechanism for the reaction of acetyl chloride with ammonia.

A reaction of acid chlorides that we have studied is the Friedel-Crafts acylation (see Secs. 10.1 and 19.1). In general,

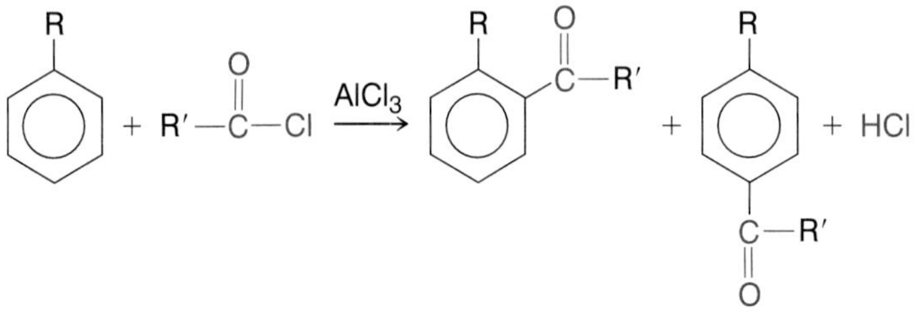

Example:

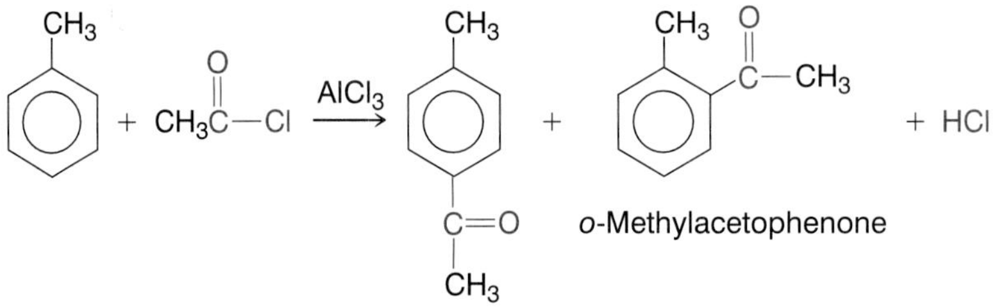

*o*-Methylacetophenone

*p*-Methylacetophenone

Acid chlorides also can be hydrolyzed to give the corresponding carboxylic acids (see Sec. 22.3). In general,

$$R{-}\overset{O}{\overset{\|}{C}}{-}Cl + H_2O \longrightarrow R{-}\overset{O}{\overset{\|}{C}}{-}OH + HCl$$

Example:

The mechanism of this reaction is similar to that of the reaction of propionyl chloride with methyl alcohol (see Solved Problem 23.2).

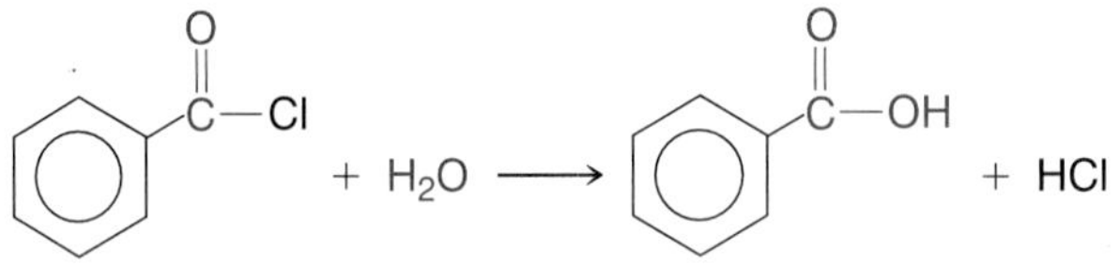

**EXERCISE 23.3**

Write a detailed, step-by-step mechanism for the following reaction.

$$CH_3{-}\overset{O}{\overset{\|}{C}}{-}Cl + H_2O \longrightarrow$$

The method used to prepare simple or mixed anhydrides in the laboratory is a general method that involves reacting an acid chloride with the sodium or potassium salt of a carboxylic acid (see Sec. 20.5). Yields are excellent. In general,

**23.4
PREPARATION OF
ACID ANHYDRIDES**

$$R-\overset{O}{\overset{\|}{C}}-Cl + {}^+Na^-O-\overset{O}{\overset{\|}{C}}-R' \longrightarrow R-\overset{O}{\overset{\|}{C}}-O-\overset{O}{\overset{\|}{C}}-R' + NaCl$$

Example:

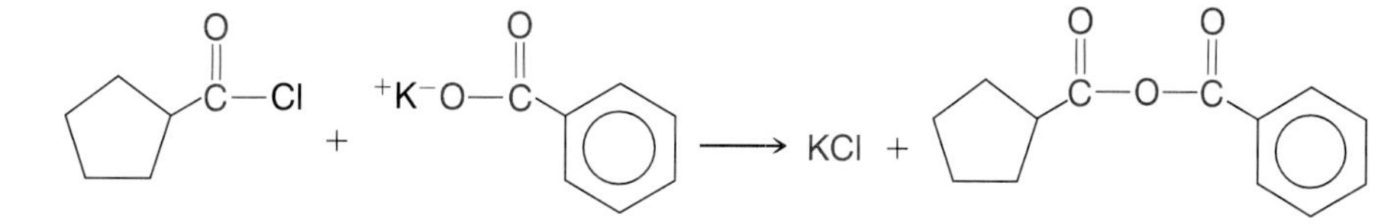

Cyclopentanecarbonyl          Potassium                    Cyclopentanecarboxylic
chloride                     benzoate                     benzoic anhydride

Acid anhydrides react with the vast majority of reagents with which acid chlorides react—only at a slower rate. In general,

**23.5
REACTIONS OF
ACID ANHYDRIDES**

$$R-\overset{O}{\overset{\|}{C}}-O-\overset{O}{\overset{\|}{C}}-R + HOR' \longrightarrow R-\overset{O}{\overset{\|}{C}}-O-R' + R-\overset{O}{\overset{\|}{C}}-OH$$

Example:

$$CH_3-\overset{O}{\overset{\|}{C}}-O-\overset{O}{\overset{\|}{C}}-CH_3 + HOCH_2CH_3 \longrightarrow CH_3-\overset{O}{\overset{\|}{C}}-O-CH_2CH_3 + CH_3-\overset{O}{\overset{\|}{C}}-OH$$

Acetic anhydride                          Ethyl acetate              Acetic acid

In general,

$$R-\overset{O}{\overset{\|}{C}}-O-\overset{O}{\overset{\|}{C}}-R + 2\,NH_3 \longrightarrow R-\overset{O}{\overset{\|}{C}}-NH_2 + R-\overset{O}{\overset{\|}{C}}-O^{-\,+}NH_4$$

Example:

$$CH_3\overset{O}{\overset{\|}{C}}-O-\overset{O}{\overset{\|}{C}}CH_3 + 2\,NH_3 \longrightarrow CH_3\overset{O}{\overset{\|}{C}}-NH_3 + CH_3\overset{O}{\overset{\|}{C}}-O^{-\,+}NH_4$$

Acetic anhydride                          Acetamide          Ammonium acetate

Anhydrides, like acid chlorides, react with water, but again, at a slower, more manageable rate. In general,

$$R-\overset{O}{\overset{\|}{C}}-O-\overset{O}{\overset{\|}{C}}-R + H_2O \longrightarrow 2\,R-\overset{O}{\overset{\|}{C}}-OH$$

Example:

$$CH_3\overset{O}{\overset{\|}{C}}-O-\overset{O}{\overset{\|}{C}}CH_3 + H_2O \longrightarrow 2\,CH_3\overset{O}{\overset{\|}{C}}-OH$$

An acid anhydride can be used to replace the corresponding acid chloride in Friedel-Crafts acylations in order to moderate the reaction rate (see Sec. 19.1).

Provide one other benefit of using acetic anhydride in a Friedel-Crafts acylation as a replacement for acetyl chloride besides that of slowing down or moderating the reaction.

## 23.6 PREPARATION OF ESTERS

Esters are prepared by reacting an acid chloride or the corresponding acid anhydride with an alcohol (see Secs. 23.3 and 23.5). Esters are also prepared by reacting a carboxylic acid with an alcohol in the presence of a catalytic amount of sulfuric acid (see Sec. 21.2). In general,

*This reaction is known as the Fischer esterification (see Sec. 20.5).*

$$R-\overset{\overset{\textstyle O}{\|}}{C}-OH + H-OR' \underset{}{\overset{H_2SO_4}{\rightleftharpoons}} R-\overset{\overset{\textstyle O}{\|}}{C}-OR' + H_2O$$

Example:

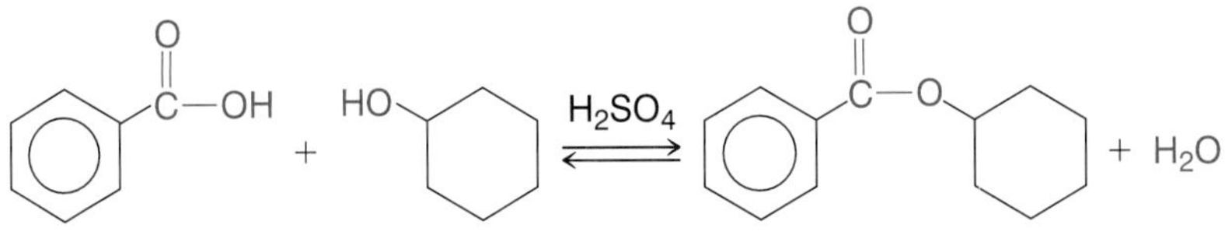

Benzoic acid          Cyclohexyl                Cyclohexyl benzoate
                       alcohol

Note that this esterification reaction is reversible. Thus yields can vary depending on the value of the equilibrium constant. Consider a word equation representing such an equilibrium.

$$\text{acid} + \text{alcohol} \overset{H_2SO_4}{\rightleftharpoons} \text{ester} + \text{water}$$

Writing the equilibrium constant expression, we get

$$K_{eq} = \frac{[\text{ester}][\text{water}]}{[\text{carboxylic acid}][\text{alcohol}]}$$

where $K_{eq}$ represents the equilibrium constant of the reaction, and each of the brackets represents the concentration of a reactant or product in moles per liter. Values of the equilibrium constant vary, but 4.0 represents a typical value. Thus the position of equilibrium slightly favors formation of product, and yields of about 66% can be expected if equimolar quantities of acid and alcohol are used. Yields of ester can be increased by the use of LeChâtelier's principle (see Secs. 18.6, 19.2, and 20.5). This principle states that when a system in equilibrium is subject to stress (e.g., an increase in the concentration of alcohol), the equilibrium shifts so as to counteract the stress (the equilibrium shifts to the right so as to decrease the concentration of alcohol). Thus, by increasing the concentration of acid or alcohol or by decreasing the concentration of ester or water produced, the esterification equilibrium is shifted to the right, and the yield of ester is increased.

*For example, a threefold excess of acid or alcohol will increase the yield to 90%.*

The mechanism of the Fischer esterification is well established.

---

### Mechanism of the Fischer Esterification of Acetic Acid with Ethyl Alcohol in the Presence of Sulfuric Acid to Give Ethyl Acetate and Water

**Envision the Reaction**

First step:

$$CH_3\overset{\overset{\displaystyle :O:}{\|}}{C}-OH \;+\; H^+ \;\rightleftharpoons\; CH_3-\overset{\overset{\displaystyle \overset{+}{O}-H}{\|}}{C}-O-H \;\longleftrightarrow\; CH_3-\overset{\overset{\displaystyle O-H}{|}}{\overset{+}{C}}-O-H$$

(From $H_2SO_4$)

Second step:

$$CH_3-\overset{\overset{\displaystyle O-H}{|}}{\overset{+}{C}}-O-H \;\;\overset{C_2H_5\overset{..}{O}H}{\rightleftharpoons}\;\; CH_3-\overset{\overset{\displaystyle OH}{|}}{\underset{\underset{\displaystyle C_2H_5}{\overset{+}{O}-H}}{C}}-OH$$

Third step:

$$CH_3-\overset{\overset{\displaystyle OH}{|}}{\underset{\underset{\displaystyle C_2H_5}{\overset{+}{O}-H}}{C}}-\overset{..}{O}H \;\;\rightleftharpoons\;\; CH_3-\overset{\overset{\displaystyle :\overset{..}{O}H}{|}}{\underset{\underset{\displaystyle C_2H_5}{O\quad H}}{C}}-\overset{+}{O}-H$$

Fourth step:

$$CH_3-\overset{\overset{\displaystyle :\overset{..}{O}H}{|}}{\underset{\underset{\displaystyle C_2H_5}{O\quad H}}{C}}-\overset{+}{O}-H \;\;\rightleftharpoons\;\; H_2O \;+\; CH_3-\overset{\overset{\displaystyle \overset{+}{O}-H}{\|}}{C}-OC_2H_5$$

Fifth step:

$$CH_3-\overset{\overset{\displaystyle \overset{+}{O}-H}{\|}}{C}-OC_2H_5 \;\;\rightleftharpoons\;\; H^+ \;+\; CH_3-\overset{\overset{\displaystyle \overset{..}{O}:}{\|}}{C}-OC_2H_5$$

---

In the first step, the carbonyl oxygen of the acid attacks a proton donated by sulfuric acid, producing a species with a positive charge on oxygen. The positive charge is then delocalized to the carbonyl carbon by resonance, creating a positive charge on the carbonyl carbon and thereby increasing its reactivity toward nucleophilic attack by ethyl alcohol, which takes place in the second step. Following a proton shift in the third step, a molecule of water is eliminated in the fourth step. Finally, in the fifth step, the protonated ester loses a proton to yield the ester.

When an acyclic ester is produced, the loss of water is *intermolecular* (between two molecules) (Fig. 23.1). However, when a cyclic ester (also called a *lactone*) is formed, the loss of water is said to be *intramolecular* (within one molecule) (see Fig. 23.1), and a lactone or cyclic ester is produced. A popular method of producing a lactone uses the Fischer esterification. Either a γ- or δ-hydroxyacid is reacted with an acid catalyst. You can therefore consider a cyclic ester or a lactone to be a compound formed by means of an intramolecular Fischer esterification. The equilibrium constants for the formation of γ- and δ-lactones are large, and these reactions take place with excellent yields.

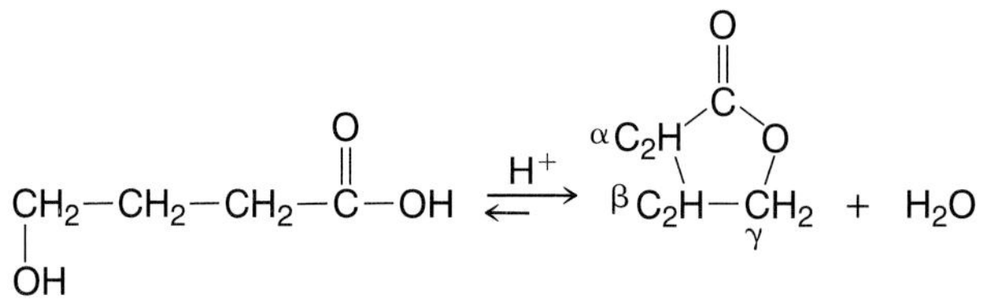

γ-Hydroxybutyric acid              γ-Butyrolactone

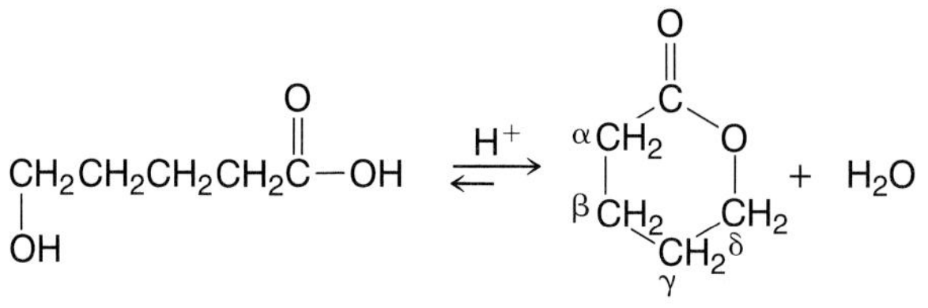

δ-Hydroxyvaleric acid              δ-Valerolactone

When a β-hydroxyacid is reacted with a catalytic amount of sulfuric acid, a lactone is not produced. Explain.

**FIGURE 23.1**  Illustration of intermolecular and intramolecular loss of water in ester formation.

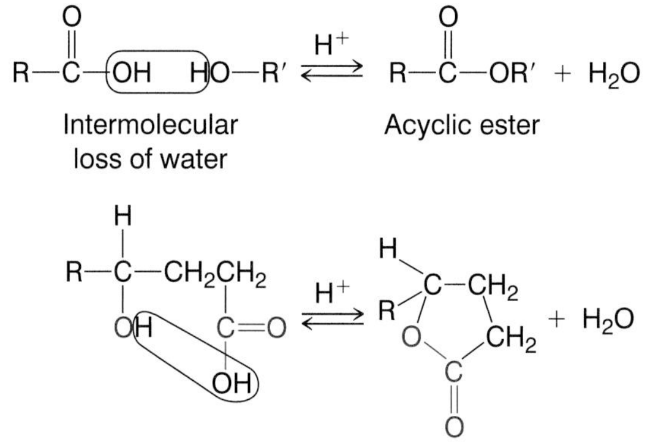

Intermolecular              Acyclic ester
loss of water

Intramolecular              Cyclic ester
loss of water

## 23.7
## REACTIONS OF ESTERS

Since esters are less reactive than anhydrides and acid chlorides, esters cannot be used to prepare these classes of compounds.

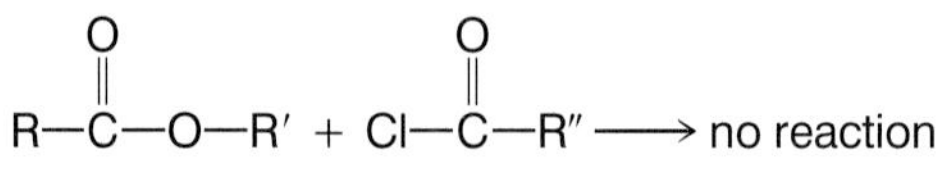

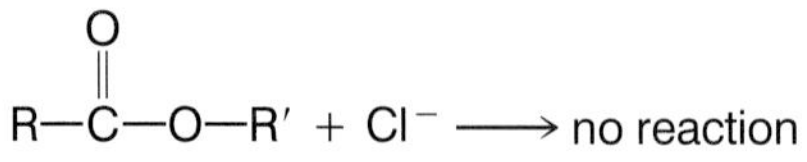

However, because esters are more reactive than amides, esters can be employed to prepare amides. In general,

$$R-\overset{O}{\underset{\|}{C}}-OR' + NH_3 \longrightarrow R-\overset{O}{\underset{\|}{C}}-NH_2 + R'OH$$

Example:

$$CH_3-\overset{O}{\underset{\|}{C}}-OCH_2CH_3 + NH_3 \longrightarrow CH_3-\overset{O}{\underset{\|}{C}}-NH_2 + CH_3CH_2OH$$

When an ester is reacted with 2 mol of Grignard reagent and hydrolyzed, a tertiary alcohol is produced. The tertiary alcohol contains two identical groups that are derived from the Grignard reagent; the third group is derived from the acid moiety (part) of the ester. In general,

$$R-\overset{O}{\underset{\|}{C}}-OR' \xrightarrow[\text{2. HCl}(aq)]{\text{1. 2 R''MgX}} R-\overset{R''}{\underset{R''}{\overset{|}{\underset{|}{C}}}}-OH + R'OMgX + Mg(X)Cl$$

where X = Cl, Br, or I. For example:

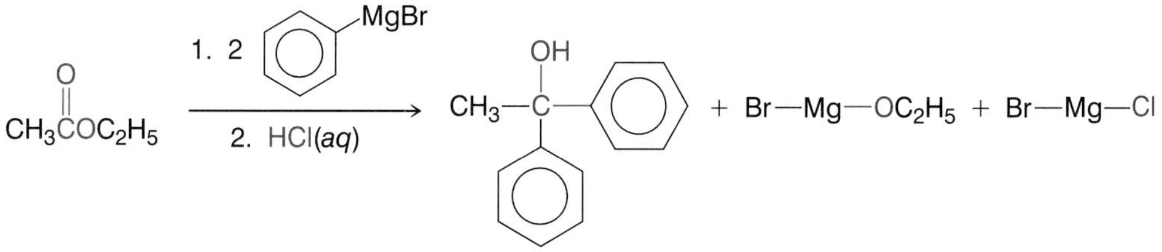

$$CH_3\overset{O}{\underset{\|}{C}}OC_2H_5 \xrightarrow[\text{2. HCl}(aq)]{\text{1. 2 } \phantom{x}MgBr} \text{[tertiary alcohol product]} + Br-Mg-OC_2H_5 + Br-Mg-Cl$$

---

**SOLVED PROBLEM 23.3**

Prepare 2-methyl-2-butanol from methyl alcohol and *n*-propyl alcohol using an ester-Grignard reaction. You may use any inorganic reagents necessary.

**SOLUTION**

Let us first draw a structural formula for 2-methyl-2-butanol.

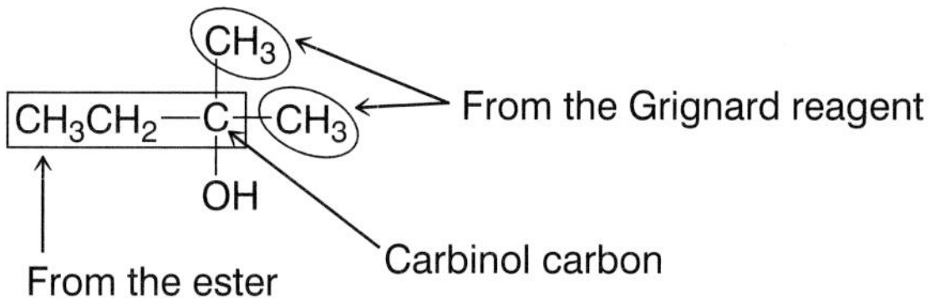

On close inspection, it is reasonable to assume that the two methyl groups bonded to the carbinol carbon must be supplied by the Grignard reagent, whereas the remaining three carbons are derived from the ester. We can satisfy these conditions by reacting methyl propionate with 2 mol of methylmagnesium bromide, followed by acid hydrolysis to give the product alcohol.

$$CH_3CH_2-\overset{O}{\underset{\|}{C}}-OCH_3 \xrightarrow[\text{2. HCl}(aq)]{\text{1. 2 CH}_3\text{MgBr}} CH_3CH_2-\overset{CH_3}{\underset{OH}{\overset{|}{\underset{|}{C}}}}-CH_3$$

2-Methyl-2-butanol

The Grignard reagent is formed from the alcohol as follows:

$$CH_3OH \xrightarrow{HBr} CH_3Br \xrightarrow[\substack{\text{Anhydrous} \\ \text{ethyl ether}}]{Mg} CH_3MgBr$$

To complete the problem, we need to prepare the ester from *n*-propyl alcohol.

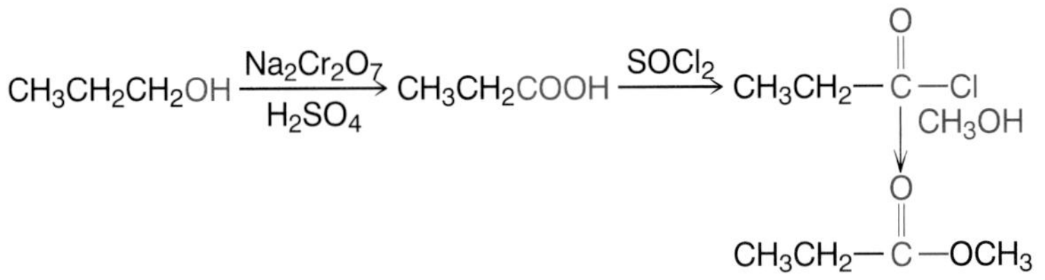

$$CH_3CH_2CH_2OH \xrightarrow[\text{H}_2\text{SO}_4]{\text{Na}_2\text{Cr}_2\text{O}_7} CH_3CH_2COOH \xrightarrow{\text{SOCl}_2} CH_3CH_2\overset{\overset{\displaystyle O}{\|}}{C}-Cl$$

$$\downarrow CH_3OH$$

$$CH_3CH_2-\overset{\overset{\displaystyle O}{\|}}{C}-OCH_3$$

Putting it all together, we have

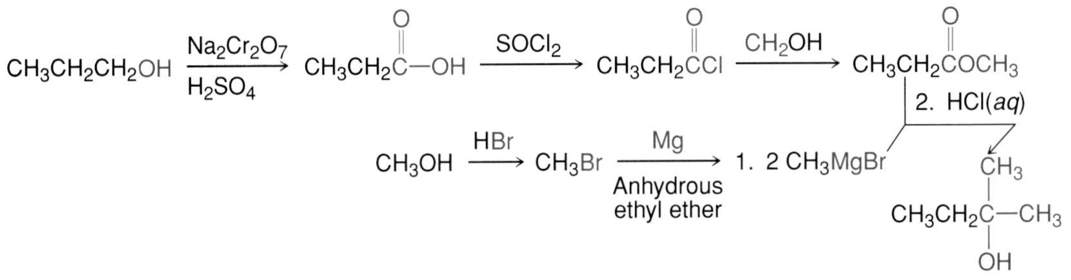

$$CH_3CH_2CH_2OH \xrightarrow[\text{H}_2\text{SO}_4]{\text{Na}_2\text{Cr}_2\text{O}_7} CH_3CH_2\overset{\overset{\displaystyle O}{\|}}{C}-OH \xrightarrow{\text{SOCl}_2} CH_3CH_2\overset{\overset{\displaystyle O}{\|}}{C}Cl \xrightarrow{\text{CH}_2\text{OH}} CH_3CH_2\overset{\overset{\displaystyle O}{\|}}{C}OCH_3$$

$$CH_3OH \xrightarrow{\text{HBr}} CH_3Br \xrightarrow[\substack{\text{Anhydrous}\\ \text{ethyl ether}}]{\text{Mg}} \text{1. } 2\,CH_3MgBr \quad \overset{\text{2. HCl}(aq)}{\searrow}$$

$$CH_3CH_2\overset{\overset{\displaystyle CH_3}{|}}{\underset{\underset{\displaystyle OH}{|}}{C}}-CH_3$$

---

**EXERCISE 23.6**

Prepare the following compounds. You may use any organic and inorganic reagents necessary.

a. 3-ethyl-3-pentanol from ethyl alcohol
b. propionic anhydride from propyl alcohol
c. propyl alcohol from propionic anhydride

---

It is sometimes not practical to hydrolyze an ester to the corresponding carboxylic acid in an acid-water solution. Under such conditions, the equilibrium favors formation of ester and water rather than carboxylic acid and alcohol.[*]

$$R-\overset{\overset{\displaystyle O}{\|}}{C}-OR' + H_2O \rightleftharpoons R-\overset{\overset{\displaystyle O}{\|}}{C}-OH + R'OH$$

This problem is solved by hydrolyzing the ester in basic (NaOH or KOH) solution, a process known as *saponification* (see Sec. 22.4). The beauty of saponification is that it is not an equilibrium process[†]; the reaction goes to completion, and the yields are good. In general,

$$R-\overset{\overset{\displaystyle O}{\|}}{C}-OR' + NaOH(aq) \longrightarrow R-\overset{\overset{\displaystyle O}{\|}}{C}-O^-Na^+ + R'OH$$

Example:

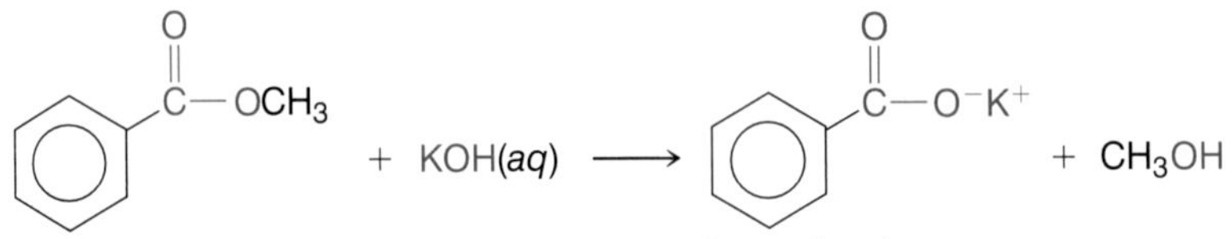

Methyl benzoate    + KOH(aq) ⟶    Potassium benzoate    + CH_3OH

---

[*]In reality, there are a number of published procedures in which an ester is hydrolyzed using an acid catalyst. A large excess of water shifts the equilibrium toward the acid and alcohol.
[†]Strictly speaking, saponification is an equilibrium process with a very large $K_{eq}$, in contrast to, for example, a Fischer esterification in which the ester is formed from a reacting mixture of acid and alcohol.

### Mechanism of the Saponification of an Ester (RCOOR′) with Hydroxide Ion to Give an Acid Anion (RCOO⁻) and an Alcohol (ROH)

First step:

R—C(=O)—OR′  with :OH attacking  ⟶  [ R—C(O⁻)(OR′)—OH ]

Second step:

[ R—C(O⁻)—OR′ with OH ]  ⟶  R—C(=O)—OH  +  ⁻OR′

Third step:

R—C(=O)—O—H  +  :ÖR′  ⟶  R—C(=O)—O⁻  +  R′OH

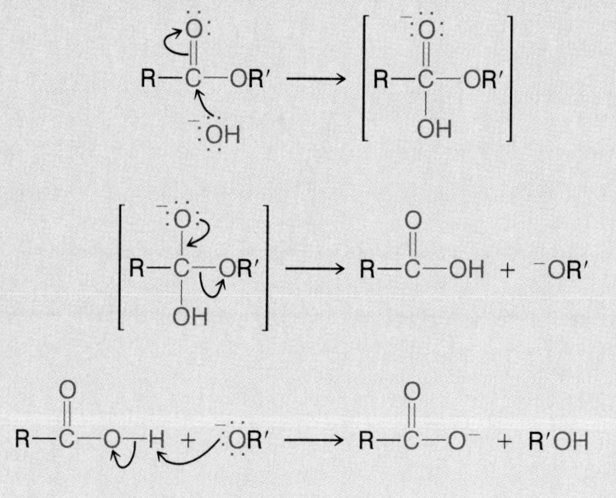

The reaction proceeds when the negatively charged hydroxide ion attacks the partially positive carbon of the carbonyl (first step), with the subsequent loss of alkoxide ion in the second step. Since the saponification is run in basic solution, an acid species cannot exist. Thus we get a proton transfer to complete the mechanism in the third step.

As mentioned many times before, a carboxylic acid can be produced easily by treating the salt with a mineral acid (see Sec. 20.5). In general,

$$R-C(=O)-O^- M^+ + HCl \longrightarrow R-C(=O)-OH + MCl$$

where M = Na or K. For example:

$$C_7H_{15}-C(=O)-O^- K^+ + HCl \longrightarrow C_7H_{15}-C(=O)-OH + KCl$$

Potassium caprylate                     Caprylic acid

In general,

$$R-C(=O)-Cl + 2 NH_3 \longrightarrow R-C(=O)-NH_2 + NH_4Cl$$

$$R-C(=O)-O-C(=O)-R + 2 NH_3 \longrightarrow R-C(=O)-NH_2 + R-C(=O)-O^- {}^+NH_4$$

Esters react with ammonia to produce amides, but in actual practice, the reaction is very slow and thus is seldom used. In general,

$$R-C(=O)-OR' + NH_3 \longrightarrow R-C(=O)-NH_2 + R'OH$$

## 23.8 PREPARATION OF AMIDES

Simple amides can be prepared from acid chlorides (see Sec. 23.3) or acid anhydrides (see Sec. 23.5) by reacting the appropriate acid derivative with ammonia.

**FIGURE 23.2** General formulas of a monosubstituted and disubstituted amine compared with ammonia.

$NH_3$

Ammonia

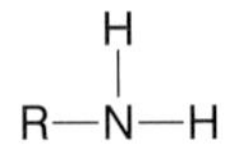

$$R-\overset{\overset{\displaystyle H}{|}}{N}-H$$

A monosubstituted amine

$$R-\overset{\overset{\displaystyle H}{|}}{N}-R(R')$$

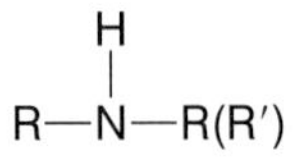

A disubstituted amine

Monosubstituted amides can be prepared in a similar way by reacting the acid derivative with a monosubstituted amine (Fig. 23.2), whereas disubstituted amides are produced by treating the acid derivative with a disubstituted amine (see Sec. 24.1). In general,

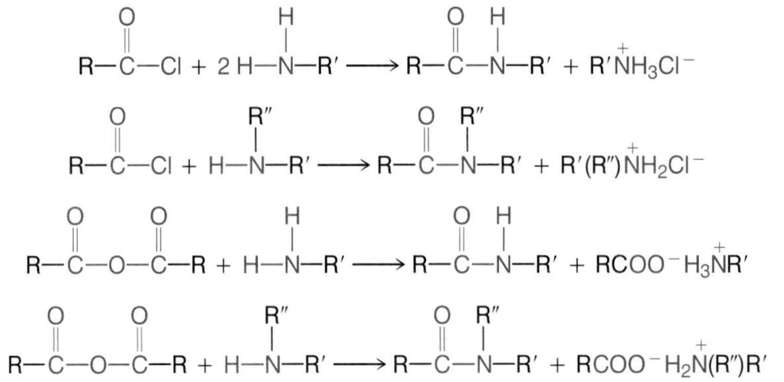

$$R-\overset{\overset{\displaystyle O}{\|}}{C}-Cl + 2\ H-\overset{\overset{\displaystyle H}{|}}{N}-R' \longrightarrow R-\overset{\overset{\displaystyle O}{\|}}{C}-\overset{\overset{\displaystyle H}{|}}{N}-R' + R'\overset{+}{N}H_3Cl^-$$

$$R-\overset{\overset{\displaystyle O}{\|}}{C}-Cl + H-\overset{\overset{\displaystyle R''}{|}}{N}-R' \longrightarrow R-\overset{\overset{\displaystyle O}{\|}}{C}-\overset{\overset{\displaystyle R''}{|}}{N}-R' + R'(R'')\overset{+}{N}H_2Cl^-$$

$$R-\overset{\overset{\displaystyle O}{\|}}{C}-O-\overset{\overset{\displaystyle O}{\|}}{C}-R + H-\overset{\overset{\displaystyle H}{|}}{N}-R' \longrightarrow R-\overset{\overset{\displaystyle O}{\|}}{C}-\overset{\overset{\displaystyle H}{|}}{N}-R' + RCOO^-H_3\overset{+}{N}R'$$

$$R-\overset{\overset{\displaystyle O}{\|}}{C}-O-\overset{\overset{\displaystyle O}{\|}}{C}-R + H-\overset{\overset{\displaystyle R''}{|}}{N}-R' \longrightarrow R-\overset{\overset{\displaystyle O}{\|}}{C}-\overset{\overset{\displaystyle R''}{|}}{N}-R' + RCOO^-H_2\overset{+}{N}(R'')R'$$

Examples:

$$CH_3\overset{\overset{\displaystyle O}{\|}}{C}-Cl + H-\overset{\overset{\displaystyle CH_3}{|}}{N}-C_2H_5 \longrightarrow CH_3\overset{\overset{\displaystyle O}{\|}}{C}-\overset{\overset{\displaystyle CH_3}{|}}{N}-C_2H_5 + H-\overset{\overset{\displaystyle \overset{+}{N}}{|}}{\underset{|}{H}}-C_2H_5Cl^-$$

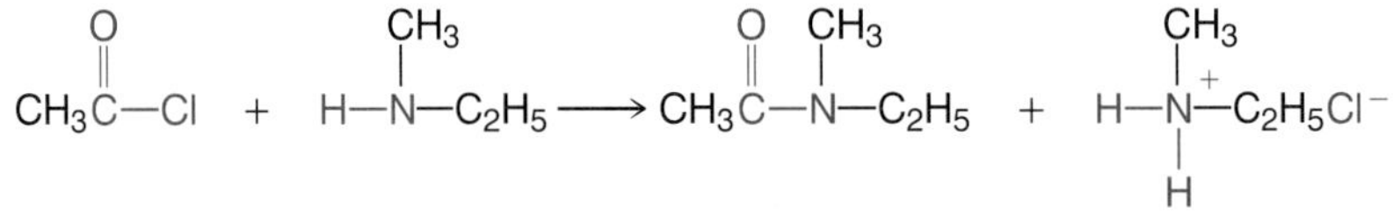

Acetyl chloride    Methylethylamine    *N*-Methyl-*N*-ethylacetamide    Ethylmethylammonium chloride

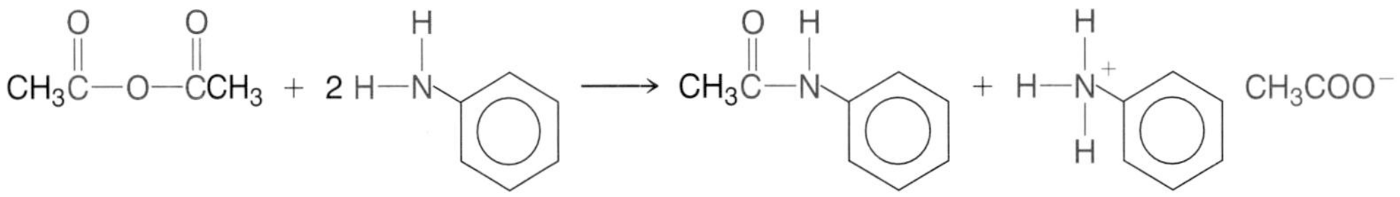

$$CH_3\overset{\overset{\displaystyle O}{\|}}{C}-O-\overset{\overset{\displaystyle O}{\|}}{C}CH_3 + 2\ H-\overset{\overset{\displaystyle H}{|}}{N}-\text{(phenyl)} \longrightarrow CH_3\overset{\overset{\displaystyle O}{\|}}{C}-\overset{\overset{\displaystyle H}{|}}{N}-\text{(phenyl)} + H-\overset{\overset{\displaystyle \overset{+}{N}}{|}}{\underset{|}{H}}-\text{(phenyl)}\ CH_3COO^-$$

Acetic anhydride    Aniline    *N*-Phenylacetamide    Anilinium acetate

Cyclic amides are called *lactams*, and γ- and δ-lactams are prepared from the corresponding amino acid by heating to cause an intramoleculer dehydration (a δ-amino acid produces a δ-lactam). An amino acid contains both an amino group and a carboxyl group.

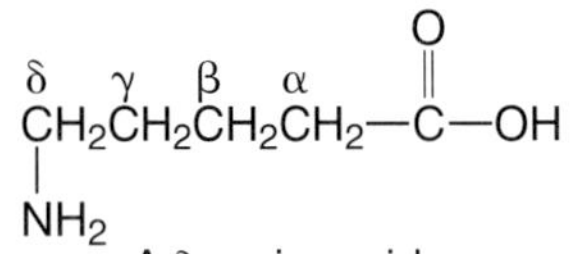

$$\underset{\underset{\displaystyle NH_2}{|}}{\overset{\displaystyle \delta\quad\ \gamma\quad\ \beta\quad\ \alpha}{CH_2CH_2CH_2CH_2}}-\overset{\overset{\displaystyle O}{\|}}{C}-OH$$

A δ-aminoacid

$$\underset{\underset{\displaystyle NH_2}{|}}{CH_2CH_2CH_2}\overset{\overset{\displaystyle O}{\|}}{C}-OH \overset{\Delta}{\longrightarrow} \ \text{γ-lactam ring} \ + H_2O$$

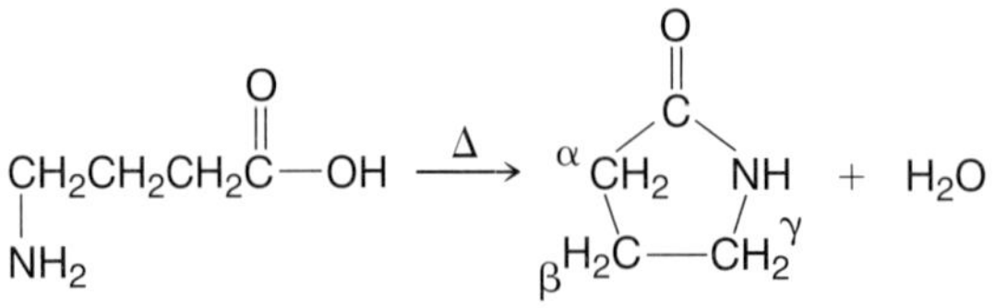

A γ-amino acid      A γ-lactam

Although γ- and δ-lactams are most stable, β-lactams are relatively stable (unlike β-lactones, which are highly reactive). Both the penicillin antibiotics and the more recently discovered cephalosporin antibiotics contain the β-lactam ring (Fig. 23.3).

**EXERCISE 23.7**

Write an equation showing the preparation of a cyclic δ-lactam from 5-aminopentanoic acid. The amino group is —$NH_2$.

## INDUSTRIAL BOXED READING 23.1

A number of important compounds that have an impact on the lives of a great many people are part of a series of synthetic polymers called *polyesters*. One frequently used polyester is made by condensing a mixture of terephthalic acid and ethylene glycol as follows:

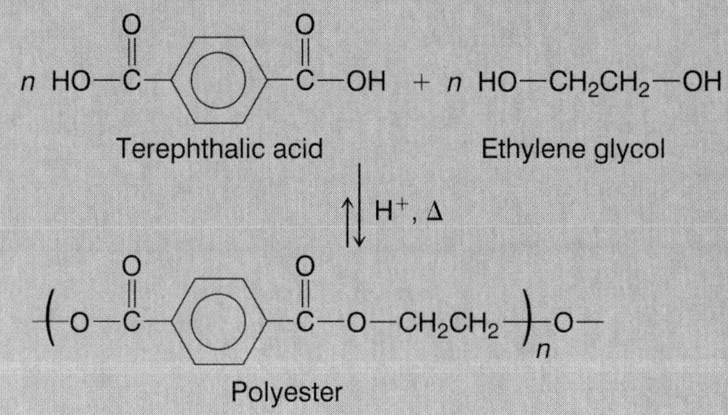

n HO—C(=O)—[benzene ring]—C(=O)—OH  +  n HO—CH$_2$CH$_2$—OH

Terephthalic acid          Ethylene glycol

↑↓ H$^+$, Δ

[(O—C(=O)—[benzene ring]—C(=O)—O—CH$_2$CH$_2$)]$_n$ O—

Polyester

This particular polyester (which is designated *polyester*) is used heavily in clothing. However, there are some disadvantages associated with its use. First, polyester is composed of long chains of polymer. Because there are no branching polymer chains present (side chains), the surface area of the polymer is relatively low, and thus the polymer is difficult to dye. Second, the polymeric fabric displays a lack of breathability; i.e., it does not absorb or permit easy passage of perspiration.

On the other hand, the use of polyester fabric has some powerful advantages, namely, durability, crease resistance, and no shrinking.

Clothing manufacturers tend to use a mixture of cotton and polyester fabrics in the garments made. These permanent-press fabrics highlight the ease of dying, comfort, and breathability of cotton with the durability, crease resistance, and lack of shrinking of the polyester.

**FIGURE 23.3** Two antibiotics containing a β-lactam ring.

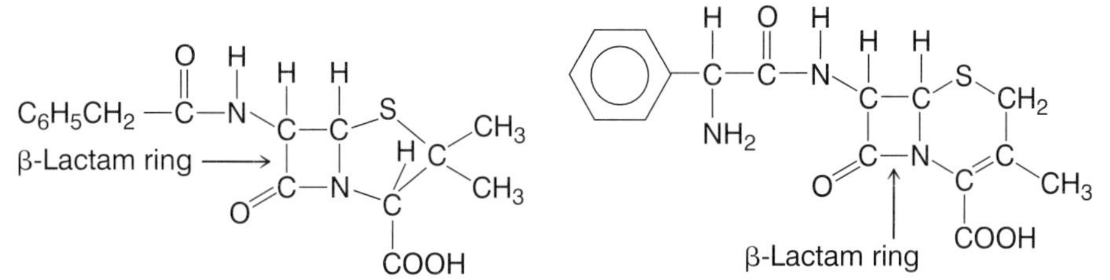

Penicillin G
(a penicillin)

Cephalexin,
Keflex (Lilly)
(a cephalosporin)

In general,

## 23.9 REACTIONS OF AMIDES

Amides are hydrolyzed to give the corresponding carboxylic acids either in acidic or basic solution (see Sec. 22.4).

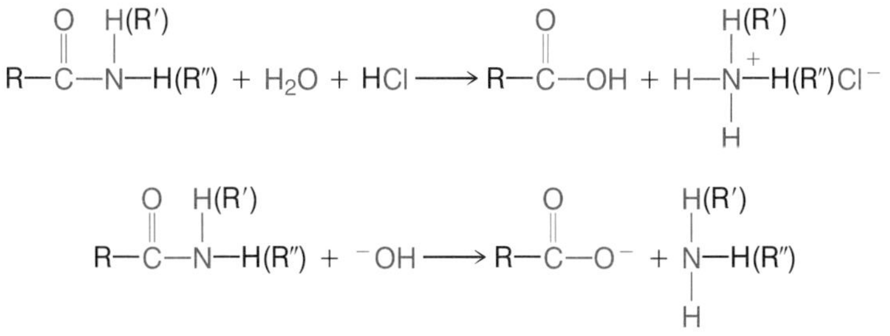

Salts of carboxylic acids are readily converted to the corresponding carboxylic acid by reacting the salt with hydrochloric acid (see Sec. 20.5).

Examples:

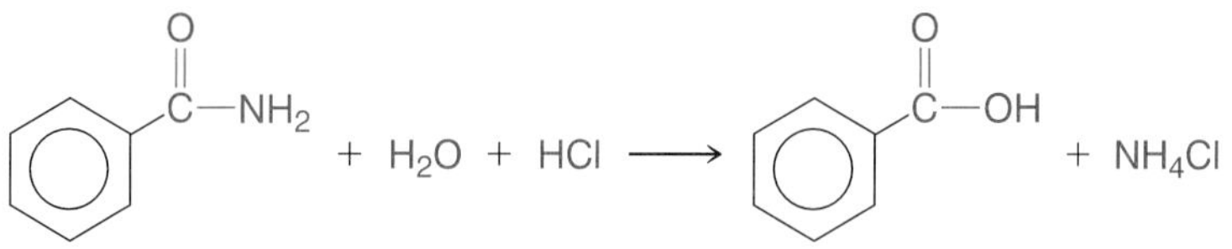

Benzamide          +  H$_2$O  +  HCl  ⟶          Benzoic acid          +  NH$_4$Cl

$$CH_3CH_2\overset{\overset{\displaystyle O}{\|}}{C}-NH_2 + NaOH(aq) \longrightarrow CH_3CH_2\overset{\overset{\displaystyle O}{\|}}{C}-O^-Na^+ + NH_3$$

Propionamide                                    Sodium propionate

$$H-\overset{\overset{\displaystyle O}{\|}}{\underset{\underset{\displaystyle CH_3}{|}}{C}}-\overset{\overset{\displaystyle CH_3}{|}}{N} + H_2O + HCl \longrightarrow H-\overset{\overset{\displaystyle O}{\|}}{C}-OH \quad + \quad H-\overset{\overset{\displaystyle CH_3}{|}}{\underset{\underset{\displaystyle CH_3}{|}}{N}}\overset{+}{-}HCl^-$$

N,N-Dimethylformamide            Formic acid      Dimethylammonium
                                                    chloride

Amides are reduced to amines in excellent yield by the use of lithium aluminum hydride. In general,

$$R-\overset{\overset{\displaystyle O}{\|}}{C}-\overset{\overset{\displaystyle H(R')}{|}}{N}-H(R'') \xrightarrow[\text{2. } H_2O,\, ^-OH]{\text{1. } LiAlH_4} R-\overset{\overset{\displaystyle H}{|}}{\underset{\underset{\displaystyle H}{|}}{C}}-\overset{\overset{\displaystyle H(R')}{|}}{N}-H(R'')$$

Examples:

$$CH_3(CH_2)_8\overset{\overset{\displaystyle O}{\|}}{C}-NH_2 \xrightarrow[\text{2. } H_2O,\, ^-OH]{\text{1. } LiAlH_4} CH_3(CH_2)_8CH_2NH_2$$

Decanamide                            1-Decanamine

$$CH_3-\overset{\overset{\displaystyle O}{\|}}{C}-\overset{|}{N}-C_2H_5 \xrightarrow[\text{2. } H_2O,\, ^-OH]{\text{1. } LiAlH_4} CH_3-CH_2-\overset{|}{N}-C_2H_5$$

N-Ethyl-N-phenylacetamide            Phenyldiethylamine

Nylon is a polyamide that can be produced by the reaction of a diamine with a diacyl chloride.

$$n\ Cl-\overset{\overset{\displaystyle O}{\|}}{C}CH_2CH_2CH_2CH_2\overset{\overset{\displaystyle O}{\|}}{C}-Cl + n\ H-\overset{\overset{\displaystyle H}{|}}{N}CH_2CH_2CH_2CH_2CH_2CH_2\overset{\overset{\displaystyle H}{|}}{N}-H$$

Adipyl chloride                            Hexamethylenediamine

$$Cl-\overset{\overset{\displaystyle O}{\|}}{C}CH_2CH_2CH_2CH_2\overset{\overset{\displaystyle O}{\|}}{C}-(-\overset{\overset{\displaystyle H}{|}}{N}(CH_2)_6\overset{\overset{\displaystyle H}{|}}{N}-\overset{\overset{\displaystyle O}{\|}}{C}(CH_2)_4\overset{\overset{\displaystyle O}{\|}}{C}-)_{n-1}-\overset{\overset{\displaystyle H}{|}}{N}CH_2CH_2CH_2CH_2CH_2CH_2\overset{\overset{\displaystyle H}{|}}{N}-H + n\ HCl$$

Nylon-6,6

The name nylon-6,6 means that a six-carbon diacyl chloride and a six-carbon diamine are to be used to prepare the compound. It follows, then, that a family of nylon polymers can be produced by varying the carbon content of the diamine and the diacyl chloride.

**EXERCISE 23.8**

Write an equation for the preparation of nylon-4,4. Use the preceding equation for the preparation of nylon-6,6 as a guide.

# ► CHAPTER ACCOMPLISHMENTS

### 23.1 Preparation of Acid Chlorides
☐ Explain why the use of thionyl chloride is better than the use of phosphorus trichloride in the preparation of an acyl chloride from a carboxylic acid.

### 23.2 Reaction Rates of the Carboxylic Acid Derivatives: A Comparison
☐ List the acid derivatives in order of decreasing reactivity.

☐ Explain why the acyl chlorides are the most reactive and the acid amides the least reactive of the acid derivatives.

### 23.3 Reactions of Acid Chlorides
☐ Write a detailed, step-by-step mechanism for the following reaction.

$$CH_3CH_2-\overset{\overset{\displaystyle O}{\|}}{C}-Cl \ + \ ^-O-\overset{\overset{\displaystyle O}{\|}}{C}-CH_3 \longrightarrow$$

☐ Write a detailed, step-by-step mechanism for the following reaction.

$$CH_3CH_2-\overset{\overset{\displaystyle O}{\|}}{C}-Cl \ + \ H-O-CH_3 \longrightarrow$$

### 23.4 Preparation of Acid Anhydrides

☐ Complete the following equation by drawing a structural formula for each organic product and writing a formula unit for each inorganic compound formed.

$$CH_3-\overset{\overset{\displaystyle O}{\|}}{C}-Cl \ + \ ^+Na^-O-\overset{\overset{\displaystyle O}{\|}}{C}-CH_2CH_3 \longrightarrow$$

☐ Name the organic compound formed in the preceding equation using both the common and IUPAC systems of nomenclature.

### 23.5 Reactions of Acid Anhydrides
☐ Complete the following equations by drawing a structural formula for each organic product produced and writing a formula unit for each inorganic compound formed.

a.   $CH_3-\overset{\overset{\displaystyle O}{\|}}{C}-O-\overset{\overset{\displaystyle O}{\|}}{C}-CH_3 \ + \ 2\ NH_3 \longrightarrow$

b.   $CH_3-\overset{\overset{\displaystyle O}{\|}}{C}-O-\overset{\overset{\displaystyle O}{\|}}{C}-CH_3 \ + \ CH_3OH \longrightarrow$

☐ Explain why acetic anhydride is often used as a reagent in Friedel-Crafts acylations in place of acetyl chloride.

### 23.6 Preparation of Esters
☐ Draw a detailed, step-by step mechanism for a Fischer esterification in which acetic acid reacts with ethyl alcohol in the presence of concentrated sulfuric acid as a catalyst.

☐ Draw a structural formula for a lactone.

☐ Illustrate an intramolecular loss of water in a γ-hydroxyacid to produce a lactone.

### 23.7 Reactions of Esters
☐ Explain why an ester cannot be used to prepare an acyl halide by reacting an ester such as ethyl acetate with chloride ion.

☐ Prepare 2-methyl-2-butanol from methyl alcohol and *n*-propyl alcohol using an ester-Grignard reaction (you may use any inorganic reagents necessary).

☐ Suggest a reason why saponification represents a more convenient method of hydrolyzing an ester than acid hydrolysis.

### 23.8 Preparation of Amides
☐ Draw a structural formula for a γ-lactam.

☐ Complete the following equation by drawing a structural formula for each organic product formed.

$$CH_3-\overset{\overset{\displaystyle O}{\|}}{C}-O-\overset{\overset{\displaystyle O}{\|}}{C}-CH_3 \ + \ (CH_3CH_2)_2NH \longrightarrow$$

### 23.9 Reactions of Amides

☐ Complete the following equations by drawing a structural formula for each organic product and writing a molecular formula or formula unit for each inorganic compound formed.

a.   $CH_3CH_2\overset{\overset{\displaystyle O}{\|}}{C}-NH_2 \ + \ NaOH(aq) \longrightarrow$

b.   $H-\overset{\overset{\displaystyle O}{\|}}{C}-N(CH_3)_2 \ + \ HCl \ + \ H_2O \longrightarrow$

c.   $CH_3(CH_2)_6\overset{\overset{\displaystyle O}{\|}}{C}-NH_2 \ \xrightarrow[\text{2. }H_2O,\ ^-OH]{\text{1. LiAlH}_4} \ \text{(no inorganic products)}$

---

# ► KEY TERMS

*acyl nucleophilic substitution* (23.2)
*intermolecular dehydration in ester formation* (23.6)

*intramolecular dehydration in ester formation* (23.6)
*lactam* (23.8)

*polyester* (Industrial Boxed Reading 23.1)

## ▶ PROBLEMS

1. Which of the following reactions do not occur? Explain why.

   a. $CH_3-\overset{O}{\overset{\|}{C}}-Cl + {}^-OCH_2CH_3 \longrightarrow$

   b. $CH_3-\overset{O}{\overset{\|}{C}}-NH_2 + {}^-Cl \longrightarrow$

   c. $CH_3-\overset{O}{\overset{\|}{C}}-OC_2H_5 + {}^-O-\overset{O}{\overset{\|}{C}}-CH_3 \longrightarrow$

2. Which reaction of the given pair proceeds faster? Explain why.

   $CH_3-\overset{O}{\overset{\|}{C}}-Cl + H_2O \longrightarrow$

   $CH_3-\overset{O}{\overset{\|}{C}}-NH_2 + H_2O \longrightarrow$

3. Complete the following equations by drawing a structural formula for each organic product and writing a molecular formula or formula unit for each inorganic product formed.

   a. 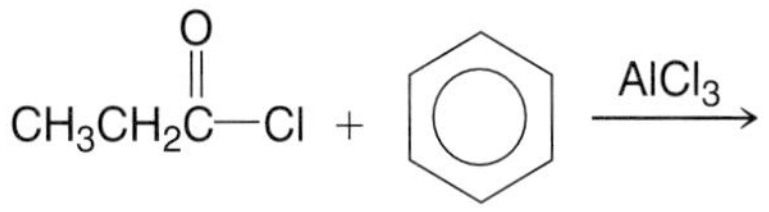 $CH_3CH_2\overset{O}{\overset{\|}{C}}-Cl +$ $\xrightarrow{AlCl_3}$

   b. $CH_3\overset{CH_3}{\underset{H}{\overset{|}{C}}}-\overset{O}{\overset{\|}{C}}-Cl + H_2O \longrightarrow$

   c. 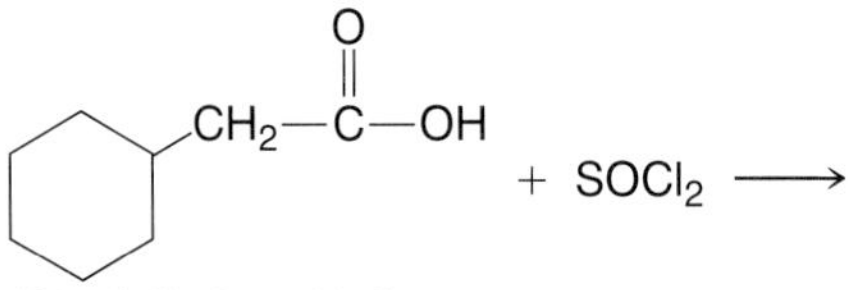 $+ SOCl_2 \longrightarrow$

   d. $(CH_3CO)_2O + H_2O \longrightarrow$

   e. $CH_3CH_2\overset{O}{\overset{\|}{C}}-Cl + {}^+Na^-O-\overset{O}{\overset{\|}{C}}CH_3 \longrightarrow$

   f. $CH_3\overset{O}{\overset{\|}{C}}-O-\overset{O}{\overset{\|}{C}}-CH_3 + CH_3OH \longrightarrow$

   g. $CH_3\overset{O}{\overset{\|}{C}}-Cl + \overset{H}{\underset{CH_3}{\overset{|}{N}}}-CH_3 \longrightarrow$

   h. 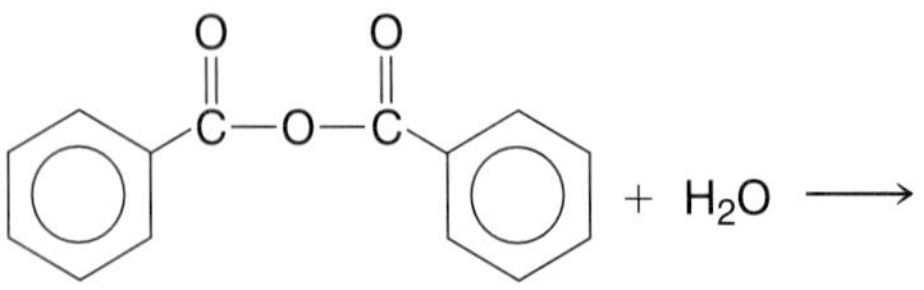 $+ H_2O \longrightarrow$

   i. $CH_3\overset{O}{\overset{\|}{C}}-OH + HOCH_3 \underset{\phantom{H_2SO_4}}{\overset{H_2SO_4}{\rightleftharpoons}}$

   j. $CH_3-\overset{CH_3}{\underset{OH}{\overset{|}{C}}}-CH_2CH_2-\overset{O}{\overset{\|}{C}}-OH \underset{}{\overset{H+}{\rightleftharpoons}}$

   k. 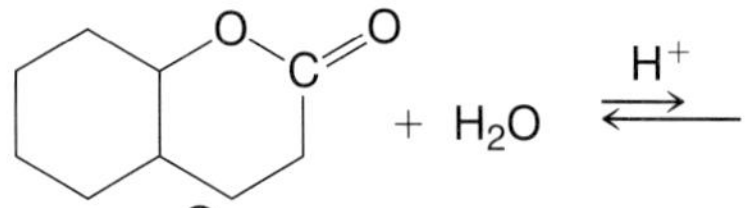 $+ H_2O \underset{}{\overset{H^+}{\rightleftharpoons}}$

   l. $CH_3CH_2\overset{O}{\overset{\|}{C}}-OC_2H_5 + NH_3 \longrightarrow$

   m. $CH_3-\overset{O}{\overset{\|}{C}}-OCH_3 \xrightarrow[\text{2. HCl}(aq)]{\text{1. 2 CH}_3\text{MgBr}}$

   n. $CH_3\overset{O}{\overset{\|}{C}}-OCH_2CH_3 + H_2O \underset{}{\overset{H^+}{\rightleftharpoons}}$

   o. $CH_3(CH_2)_{16}\overset{O}{\overset{\|}{C}}-OCH_3 + NaOH(aq) \longrightarrow$

   p. product o + HCl $\longrightarrow$

   q. $CH_3CH_2\overset{O}{\overset{\|}{C}}-OH + NaOH(aq) \longrightarrow$

   r. $CH_3\overset{O}{\overset{\|}{C}}-NH_2 + H_2O + HCl \longrightarrow$

   s. $CH_3\overset{O}{\overset{\|}{C}}-NH_2 + NaOH(aq) \longrightarrow$

   t. 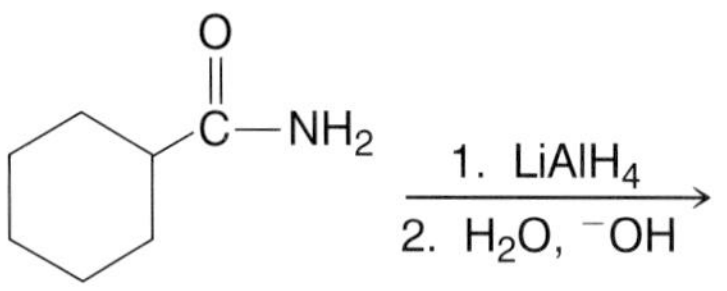 $\xrightarrow[\text{2. H}_2\text{O, }^-\text{OH}]{\text{1. LiAlH}_4}$ no inorganic products

   u. $CH_3\overset{O}{\overset{\|}{C}}-O-\overset{O}{\overset{\|}{C}}CH_3 + C_2H_5NH_2 \longrightarrow$

4. Prepare the following compounds. You may use any inorganic reagents necessary.

   a. Nicotinamide from

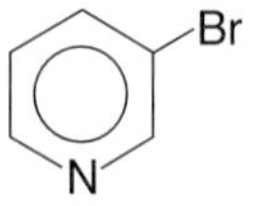

   b. Acetanilide from benzene and ethyl alcohol
   c. $CH_3CH_2NH_2$ from ethanoic acid
   d. From benzene and propionic acid

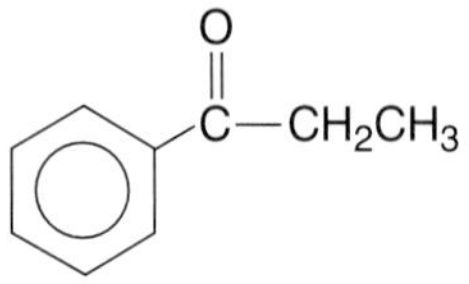

   e. tert-butyl alcohol from acetic acid and methyl alcohol

f. ethyl acetate from ethyl alcohol

5. Write a detailed, step-by-step mechanism for the following.

a.

$$\text{(benzoyl chloride, } C_6H_5\text{-C(=O)-Cl)} + H\text{-NCH}_3\text{(H)} \longrightarrow$$

b. $$H\text{-C(=O)-OH} + CH_3OH \overset{H^+}{\rightleftharpoons}$$

c. $$CH_3\text{-C(=O)-O-C(=O)-CH}_3 + {}^-OCH_2CH_3 \longrightarrow$$

6. The compound β-propiolactone is one of several β-lactones that have been synthesized (and isolated) by methods other than, in the case of β-propiolactone, the intramolecular dehydration of β-hydroxypropionic acid. Draw a structural formula for β-propiolactone.

7. Even though acetyl chloride is a better acylating agent than acetic anhydride when treated with benzene and aluminum chloride, many chemists prefer acetic anhydride. Explain.

8. Acid chlorides do react with phenols in a manner similar to alcohols. Write an equation showing acetyl chloride reacting with o-cresol.

9. In the reaction of an acid chloride with ammonia, 2 mol of ammonia is usually added to the acid chloride. Why is the ammonia added in excess?

10. Why in Solved Problem 23.3 was propionic acid converted to methyl propionate in two steps rather than by means of a one-step Fischer esterification?

11. Compound A ($C_6H_{12}O_6$) was saponified to give $C_3H_7OH$ (compound B) and $C_2H_5CO_2K$ (compound C). Compound C was treated with aqueous HCl to give $C_3H_6O_2$

(compound D), which formed bubbles of gas when mixed with solid $NaHCO_3$. Compound B gave a positive iodoform test. Draw a structure for each compound, and show your reasoning.

12. What combination of acid chloride and amine would produce N,N-diethyl-m-toluamide, an excellent insect repellent?

13. Why is the nylon molecule shown in Section 23.9 named nylon-6,6?

14. Write a detailed, step-by-step mechanism for

$$CH_3\text{-C(=O)-OCH}_2CH_2CH_3 \xrightarrow[\text{2. HCl}(aq)]{\text{1. 2 CH}_3CH_2MgBr}$$

Hint: Add the first mole of ethyl carbanion to the carbonyl carbon of the ester to produce 2-butanone and $CH_3CH_2CH_2O^-Mg^+Br$. Then add a second mole of ethyl carbanion to the ketone in the usual way.

15. Produce a reasonable synthesis for p-hydroxyacetanilide from aniline and acetic acid. Use any inorganic reagents necessary.

16. Write a detailed, step-by-step mechanism for

$$CH_3CH_2C\text{-OH} + CH_3OH \overset{H^+}{\rightleftharpoons} CH_3CH_2C\text{-OCH}_3 + H_2O$$

(with carbonyl O on each acid/ester)

17. The infrared spectrum of a compound showed an absorption at 1745 cm$^{-1}$ but no absorption at 3300 cm$^{-1}$. What class of acid derivative cannot be represented by these data? Explain.

18. How could you distinguish an acid from an ester using an infrared spectrum?

# CHAPTER 24

# Amines (I)

## CLASSIFICATION, STRUCTURE, NOMENCLATURE, PHYSICAL PROPERTIES, BASICITY OF AMINES, ALKALOIDS, AND REPRESENTATIVE REACTIONS OF AMINES

Amines represent an unusual family of organic compounds for two reasons. First, they are the only family of organic bases we will study. Second, classifying amines is done by a somewhat different method.

Amines can be thought of as derivatives of ammonia where one or more hydrogens is replaced by one or more organic groups:

$$RNH_2 \qquad R_2NH \qquad R_3N$$

Primary (1°) amine    Secondary (2°) amine   Tertiary (3°) amine

**24.1 CLASSIFICATION**

Replacing one ammonia hydrogen with an organic group produces a primary amine, replacing two hydrogens produces a secondary amine, and replacing three hydrogens produces a tertiary amine. This different method of classification is clarified by considering *tert*-butylamine and *tert*-butyl chloride (Fig. 24.1). Note that the nitrogen of the primary amine is bonded to a tertiary alkyl group. This can be confusing, so be on guard for this sort of thing.

Amines, like ammonia, are pyramidal in shape (Fig. 24.2a). Each of the three groups bonded to the nitrogen occupies a corner of a regular tetrahedron (the base of a pyramid), with the two unshared electrons located on the fourth corner of the tetrahedron

**24.2 STRUCTURE**

---

**FIGURE 24.1** Examples of a primary amine and a tertiary chloride using the same alkyl group.

Primary amine
Tertiary alkyl group

Tertiary chloride
Tertiary alkyl group

**FIGURE 24.2** Geometry and orbital descriptions of amines.

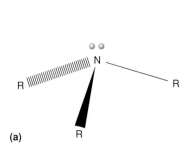

(a)

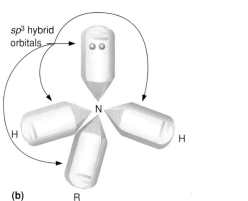

(b)

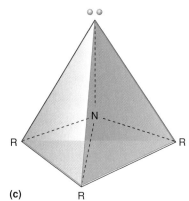

(c)

For example, 107 degrees for a CNC bond in trimethylamine, 113 degrees for a CNH bond of methylamine (Fig. 24.2b).

(the top of the pyramid) (Fig. 24.2c). The nitrogen atom is $sp^3$ hybridized (Fig. 24.2b). Bond angles vary slightly but are consistent with $sp^3$ hybridization.

Consider an amine with three different groups on the nitrogen atom. If you consider the unshared electron pair to be a fourth different group with the pair of electrons a part of the tetrahedral geometry, you have satisfied the conditions for chirality (see Sec. 11.2). Thus you would predict that such amines would be chiral molecules and that two enantiomeric forms for each amine can be isolated. In actuality, no enantiomeric forms are isolated because, at room temperature, these amines are rapidly undergoing inversion of the pyramid.

Transition state
(planar)

The conversion of compound I to compound II and back to compound I is extremely rapid.

# 24.3
# NOMENCLATURE

**Common Nomenclature**

Organic groups bonded to the nitrogen of the amine can be aliphatic, alicyclic, or aromatic.

First, name the organic group(s), and then add the suffix -*amine*. For some examples, refer to Fig. 24.3. If two or three units of the same group are bonded to the nitrogen of an amine, use the prefix *di-* or *tri-*.

**FIGURE 24.3** Common nomenclature of selected amines.

$$CH_3-\overset{\overset{\displaystyle H}{|}}{N}-H$$

Methylamine,
a 1° amine

Phenylcyclopentylmethylamine,
a 3° amine

$$CH_3CH_2CH_2-\overset{\overset{\displaystyle H}{|}}{N}-H \qquad H-\overset{\overset{\displaystyle CH_3}{|}}{\underset{\displaystyle CH_3}{C}}-CH_3$$

Isopropylpropylamine,
a 2° amine

Note: It makes no difference which organic group is named first.

**FIGURE 24.4** IUPAC nomenclature of two primary amines.

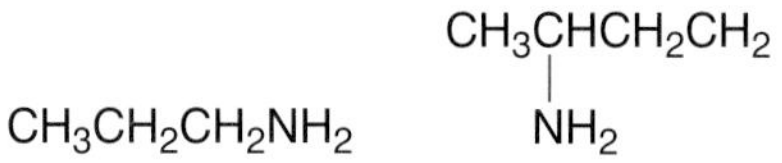

$$CH_3CH_2CH_2NH_2$$

1-Propanamine
(1°)

$$CH_3CHCH_2CH_2$$
$$|$$
$$NH_2$$

2-Butanamine
(1°)

**FIGURE 24.5** IUPAC nomenclature of secondary and tertiary amines.

$$CH_3\overset{\overset{\displaystyle H}{|}}{N}CH_2CH_3$$

*N*-Methylethanamine
(2°)

$$\overset{1}{CH_3}-\overset{\overset{\displaystyle H}{|}}{\underset{2}{C}}-\overset{3}{CH_3}$$
$$CH_3-N-CH_3$$

2-(*N*, *N*-Dimethyl)propanamine
(3°)

The capital Ns are used to show the organic groups are bonded to the nitrogen. The parent chain selected is the longest chain of carbons.

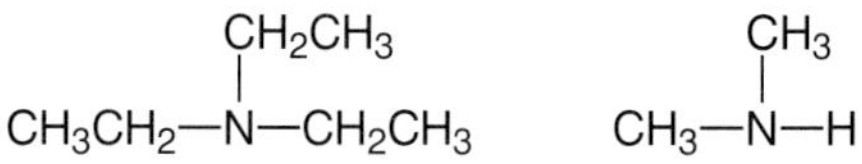

$$CH_3CH_2-\overset{\overset{\displaystyle CH_2CH_3}{|}}{N}-CH_2CH_3$$

Triethylamine
(a 3° amine)

$$CH_3-\overset{\overset{\displaystyle CH_3}{|}}{N}-H$$

Dimethylamine
(a 2° amine)

Simple diamines are named by using the prefix for the number of methylene groups in the molecule and then adding the fragment *-methylenediamine*.

|  | $$H_2NCH_2CH_2CH_2CH_2NH_2$$ | $$H_2NCH_2CH_2CH_2CH_2CH_2NH_2$$ |
|---|---|---|
| Common: | Tetramethylenediamine | Pentamethylenediamine |
| Trivial: | Putrescine | Cadaverine |

The trivial names *putrescine* and *cadaverine* stem from the fact that these two compounds can be isolated from decomposing human corpses. The odors associated with these compounds are terrible.

Amines are named by dropping the *-e* of the corresponding alkane and adding *-amine* (Fig. 24.4). Secondary and tertiary amines are named as alkyl (aryl) substituted amines (Fig. 24.5). Diamines are named (Fig. 24.6) by numbering the two amino groups on the parent chain, separating the numbers with a comma, and then adding a dash. Finally, add the name of the corresponding hydrocarbon followed by the fragment *-diamine*. Primary aromatic amines are named as derivatives of aniline in the manner explained in

**IUPAC nomenclature**

**FIGURE 24.6** IUPAC nomenclature of selected diamines.

$$\overset{\overset{\displaystyle CH_2CH_2CH_2CH_2}{|\qquad\qquad|}}{\underset{NH_2\qquad\quad NH_2}{}}$$

1, 4-Butanediamine

$$\overset{\overset{\displaystyle CH_2CH_2CH_2CH_2CH_2}{|\qquad\qquad\qquad|}}{\underset{NH_2\qquad\qquad NH_2}{}}$$

1, 5-Pentanediamine

**FIGURE 24.7** Nomenclature of selected primary aromatic amines.

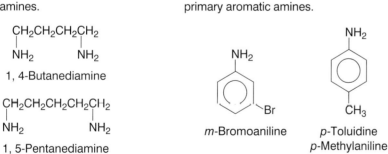

*m*-Bromoaniline

*p*-Toluidine
*p*-Methylaniline

**FIGURE 24.8** Nomenclature of selected secondary and tertiary aromatic amines.

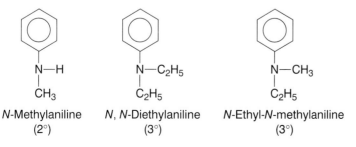

*N*-Methylaniline
(2°)

*N*, *N*-Diethylaniline
(3°)

*N*-Ethyl-*N*-methylaniline
(3°)

$$CH_2{=}\overset{\overset{\displaystyle H}{|}}{C}{-}CH_2NH_2$$

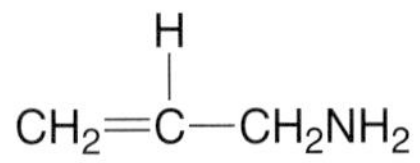

2-Propene-1-amine

$$H_2N{-}\overset{\overset{\displaystyle H}{|}}{\underset{\underset{\displaystyle H}{|}}{C}}{-}\overset{\overset{\displaystyle H}{|}}{\underset{\underset{\displaystyle H}{|}}{C}}{-}\overset{\overset{\displaystyle O}{\|}}{C}{-}OH$$

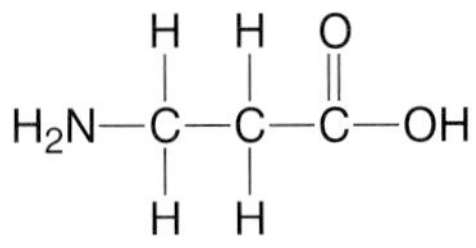

3-Aminopropanoic acid
(*not* 3-carboxy-1-propanamine)

Section 9.2. Some examples are found in Fig. 24.7. Secondary and tertiary aromatic amines also can be named as derivatives of aniline. The italic capital *N* is used to show that the organic group(s) are bonded to the nitrogen atom. If two different alkyl groups are involved, list the alkyl groups alphabetically. Note the examples given in Fig. 24.8 illustrating this type of nomenclature.

Now let us incorporate the amino group into the list of functional-group priorities that we have previously developed.

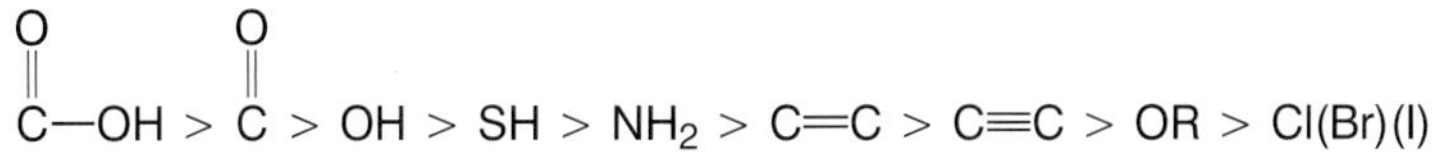

$$\overset{\overset{\displaystyle O}{\|}}{C}{-}OH > \overset{\overset{\displaystyle O}{\|}}{C} > OH > SH > NH_2 > C{=}C > C{\equiv}C > OR > Cl(Br)(I)$$

As usual, what this listing means is that if we have a compound in which both an —NH$_2$ and —OH group are incorporated, the —OH group is always assigned the lower number and the amino group is listed as a substituent. For example,

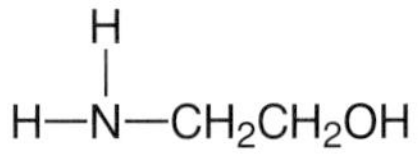

$$H{-}\overset{\overset{\displaystyle H}{|}}{N}{-}CH_2CH_2OH$$

2-Aminoethanol
(*not* 2-hydroxy-1-ethanamine)

Some other examples are in Figure 24.9.

---

**SOLVED PROBLEM 24.1**

Name the following compounds using the method of nomenclature asked for.

a. 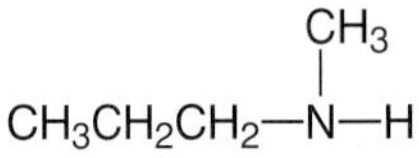   $CH_3CH_2CH_2{-}\overset{\overset{\displaystyle CH_3}{|}}{N}{-}H$ (common)

b.   $Cl{-}CH_2CH_2CH_2{-}NH_2$ (IUPAC)

**SOLUTION**

a.   This compound is a secondary amine. Since a common name is called for, we simply name each group (in no particular order) and then add the fragment -*amine* to get

methyl *n*-propylamine or (*n*-propylmethylamine)

b.   Since two functional groups are present in this particular compound, we need to use the listing of priorities to establish that the amino group has a greater priority than the chloro group. The chain of carbons contains three carbons, and we get, assigning the lower number to the carbon bonded to the amino group,

$$Cl{-}\underset{3}{C}H_2\underset{2}{C}H_2\underset{1}{C}H_2{-}NH_2$$

3-chloro-1-propanamine

---

**EXERCISE 24.1**

Name the following compounds using the method of nomenclature asked for.

a.   $CH_3CH_2CH_2CH_2CH_2NH_2$ (common and IUPAC)

b.   $C_6H_5CH_2NH_2$ (common)

c.   $H_2NCH_2CH_2CH_2CH_2CH_2NH_2$ (common and IUPAC)

d. 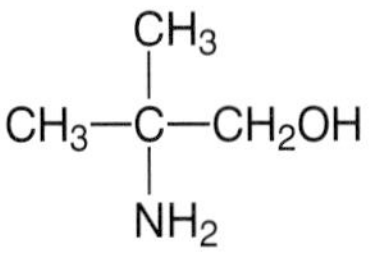   $CH_3{-}\overset{\overset{\displaystyle CH_3}{|}}{\underset{\underset{\displaystyle NH_2}{|}}{C}}{-}CH_2OH$ (IUPAC)

Classify the following compounds as primary, secondary, or tertiary.

a.  triethylamine

b.  isopropylamine

c.  *N*-ethylpentanamine

d.  *p*-chloroaniline

## 24.4
## PHYSICAL PROPERTIES

Most amines boil at a temperature between that of an alkane and an alcohol of approximately the same molecular weight. Since primary and secondary amines can form intermolecular hydrogen bonds, whereas alkanes cannot, the fact that primary and secondary amines boil at a higher temperature than an alkane of about the same molecular weight is not unexpected (Table 24.1).

$$
\begin{array}{ccc}
\text{R} & \text{R} & \text{H} \\
| & | & | \\
\text{R--N--H----N--R} & & \text{R--N} \quad \text{R} \\
\text{Hydrogen bond} \uparrow | \text{H} & & \text{H----N--H} \\
& \text{Hydrogen bond} \uparrow | \\
& & \text{H}
\end{array}
$$

Hydrogen bonding in     Hydrogen bonding in
a secondary amine       a primary amine

Tertiary amines cannot form intermolecular hydrogen bonds. Thus a tertiary amine and an alkane of about the same molecular weight have similar boiling points (Table 24.2).

The nitrogen-associated hydrogen bond (in an amine) is weaker than the oxygen-associated hydrogen bond (in an alcohol). This makes sense because nitrogen is less electronegative than oxygen, and therefore, the N—H bond is less polar than the O—H bond. Thus the less positively charged hydrogen of the N—H bond is less attracted to the electrons of nitrogen in another molecule. Because the amine has a weaker intermolecular interaction than an alcohol of about the same molecular weight, the amine boils at a lower temperature than the alcohol (see Table 24.1).

Primary, secondary, and tertiary amines form hydrogen bonds with water. Thus all classes of amines, which contain up to five carbon atoms, are water-soluble.

The higher-molecular-weight amines have a "fishy" odor, whereas low-molecular-weight amines such as methylamine and ethylamine smell something like ammonia. The obnoxious odors of putrescine and cadaverine can be deduced from their names. Many aromatic amines are toxic and need to be handled with great care because they are easily absorbed through the skin. Some aromatic and alicyclic amines are carcinogenic, e.g., 2-naphthylamine and cyclohexylamine (Fig. 24.10).

| TABLE 24.1 | A Comparison of the Boiling Point of a Selected Amine with an Alcohol and Alkane of About the Same Molecular Weight | |
|---|---|---|
| Compound | Molecular Weight | Boiling Point (°C) |
| $C_2H_5NH_2$ Ethylamine | 45 | 17 |
| $CH_3CH_2CH_3$ Propane | 44 | −42 |
| $C_2H_5OH$ Ethyl alcohol | 46 | 78 |

**FIGURE 24.10** Two carcinogenic amines.

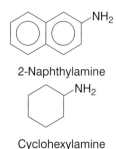

2-Naphthylamine

Cyclohexylamine

| TABLE 24.2 | A Comparison of the Boiling Points of Different Classes of Amines | | |
|---|---|---|---|
| Compound and Class | Molecular Weight | Boiling Point (°C) |
| $C_2H_5NH_2$ Ethylamine (1°) | 45 | 17 |
| $(C_2H_5)_2NH$ Diethylamine (2°) | 73 | 56 |
| $CH_3(CH_2)_5CH_3$ Heptane | 100 | 98 |
| $(C_2H_5)_3N$ Triethylamine (3°) | 101 | 89 |
| $(C_3H_7)_2NH$ Di-$n$-propylamine (2°) | 101 | 111 |
| $CH_3(CH_2)_4CH_2NH_2$ $n$-Hexylamine (1°) | 101 | 129 |

## 24.5 BASICITY OF AMINES

### Basicity of Amines Compared with Other Functional Groups

Refer to Section 1.14 for a general discussion of acidity and basicity. All classes of amines react with mineral acids to give stable isolable salts. In general,

$$RNH_2 + HX \longrightarrow R\overset{+}{N}H_3X^-$$

Examples:

$$CH_3CH_2CH_2NH_2 + HCl \longrightarrow CH_3CH_2CH_2\overset{+}{N}H_3Cl^-$$
$$\text{\textit{n}-Propylamine} \qquad \text{\textit{n}-Propylammonium chloride}$$

$$(CH_3CH_2)_2NH + HCl \longrightarrow (CH_3CH_2)_2\overset{+}{N}H_2Cl^-$$
$$\text{Diethylamine} \qquad \text{Diethylammonium chloride}$$

Consider a Lewis base such as an ether or alcohol reacting with a mineral acid (HX):

$$R\ddot{O}H + HX \rightleftarrows R\overset{\overset{\displaystyle H}{|}}{\underset{}{\overset{+}{O}}}H\ X^-$$

The presence of one or more donatable electron pairs makes these compounds Lewis bases.

$$R\ddot{O}R' + HX \rightleftarrows R\overset{\overset{\displaystyle H}{|}}{\underset{}{\overset{+}{O}}}R'\ X^-$$

These salts are not isolable due to the presence of a positive charge on an electronegative oxygen. Thus the alcohols and ethers are not considered to be basic compounds. Nitrogen is less electronegative, and the salt ($R\overset{+}{N}H_3X^-$) is therefore more stable in the presence of a positive charge.

## SOLVED PROBLEM 24.2

Consider an alkyl fluoride (RF) as a possible Lewis base. Explain.

### SOLUTION

*Step 1.* Write the equation showing the reaction of RF with a mineral acid (HX).

$$R-\ddot{F}: + HX \rightleftarrows R-\overset{\overset{\displaystyle H}{|}}{\underset{}{\overset{+}{F}}}:X^-$$

*Step 2.* Since very electronegative fluorine (see Figure 1.5) bears a positive charge, the ionic species is unstable, not isolable, and an alkyl fluoride is not a base.

Consider an alkyl chloride (RCl) as a possible Lewis base. Explain. Assume any organic species containing an atom with an electronegativity greater than that of nitrogen (3.0) that bears a positive charge to be unstable.

The key to base strength, it seems reasonable to state, is the degree of stability of the cation obtained when a proton is captured by the base from an acid.

Consider an amine G—$NH_2$, where G— represents any organic group. Let us now react the amine, a Brønsted base, with water, a Brønsted acid, as follows:

## Basicity and Structure of Organic Amines

$$GNH_2 + H_2O \rightleftharpoons G\overset{+}{N}H_3 + {}^-OH$$

Just as an acidity-constant expression can be written for a carboxylic acid, a basicity constant $K_b$ can be written for the preceding equation.

$$K_b = \frac{[G\overset{+}{N}H_3][{}^-OH]}{[GNH_2]}$$

where $[G\overset{+}{N}H_3]$ and $[{}^-OH]$ are the concentrations in mol per liter of substituted ammonium ion and hydroxide ion, respectively, while $[GNH_2]$ represents the concentration in mol per liter of undissociated amine. Comparable expressions can be written for secondary and tertiary amines. The greater the value of $K_b$, the ionization constant of the base, the stronger is the base. Base strength is usually determined expermentally as the $pK_a$. The $pK_b$ value is then calculated from

$$pK_b = 14 - pK_a$$

$pK_b$ values vary from about 3 to 5 for a typical aliphatic amine to 8 to 13 for a typical aromatic amine (Table 24.3).

As a matter of convenience, base strength is usually given in the form of a $pK_b$ where $pK_b = -\log K_b$ (see Table 24.3). Thus the larger the value of the $pK_b$, the weaker is the base.

The $pK_a$ of a dilute aqueous solution of m-bromoaniline was determined to be 3.58. Calculate the $pK_b$.

**SOLUTION**

Using the expression $pK_b = 14 - pK_a$, we get

$$14 - 3.58 = 10.42$$

| TABLE 24.3 | Structure and $pK_b$ Values for Some Selected Organic Amines | |
|---|---|---|
| **Amine** | **$pK_b$** | |
| $NH_3$ | 4.74 | |
| $CH_3NH_2$ | 3.36 | |
| $(CH_3)_2NH$ | 3.28 | |
| $(CH_3)_3N$ | 4.19 | |
| $CH_3CH_2NH_2$ | 3.19 | |
| $CH_3CH_2CH_2NH_2$ | 3.32 | |
| $CH_3CH_2CH_2CH_2NH_2$ | 3.32 | |
| $C_6H_5NH_2$ | 9.40 | |
| $p\text{-}O_2N\text{—}C_6H_4\text{—}NH_2$ | 13.00 | |
| $p\text{-}CH_3\text{—}C_6H_4\text{—}NH_2$ | 8.92 | |

**EXERCISE 24.4**

Given the $pK_a$ of each of the following dilute aqueous solutions, calculate the $pK_b$.

a.  o-iodoaniline, $pK_a = 2.60$

b.  ethylenediamine, $pK_a = 10.71$

**EXERCISE 24.5**

Calculate the $K_b$ for each compound in Exercise 24.4 (see Solved Problem 20.3).

Now let us focus our attention on $G\overset{+}{N}H_3$, the conjugate acid of the amine $GNH_2$. If G is an alkyl group, it is an electron donor due to the inductive effect. The electrons therefore slightly decrease the positive charge of the cation, stabilizing the cation and increasing the basicity of the amine. Thus dimethylamine, with two alkyl groups releasing electrons, is a stronger base than methylamine, which, in turn, is a stronger base than ammonia.

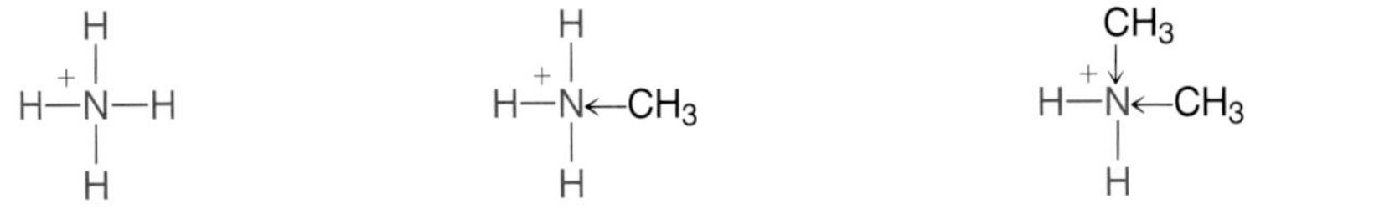

Ammonium ion
($pK_b$ ammonia $= 4.74$)

Methylammonium ion
($pK_b$ methylamine $= 3.36$)

Dimethylammonium ion
($pK_b$ dimethylamine $= 3.28$)

Increasing the size of the alkyl group has virtually no effect on the basicity of amines (see Table 24.3).

If G is an electron-attracting group such as $-OH$, $-NO_2$, $-Cl$, $-Br$, $-I$, or $-NH_2$, it withdraws electrons from the positively charged nitrogen of $G-\overset{+}{N}H_3$ via the inductive effect. Thus the positive charge on nitrogen is increased, destabilizing the cation and decreasing the basicity of the amine. Therefore, hydroxylamine is a weaker base than ammonia.

$$NH_2OH \qquad\qquad NH_3$$

Hydroxylamine      ($pK_b = 4.74$)
($pK_b = 7.97$)

A reasonable question to ask is, Why are aromatic amines less basic than aliphatic amines (Table 24.3)? The reason is that resonance decreases the electron density on nitrogen (remember, a base is an electron-pair donor) on the amine. Thus the nitrogen atom in an aromatic amine contains less electron density available to donate to a proton. Figure 24.11 shows aniline as an example.

**EXERCISE 24.6**

Draw two more resonance structures for aniline.

## 24.6 ALKALOIDS

Alkaloids are basic organic amines that occur naturally in plants. Alkaloids have been isolated for the most part from the seeds, leaves, and bark of plants but are also found in the roots, flowers, and fruit of plants. Many of the alkaloids (but not all) show some kind of biological activity. Some representative alkaloids are shown in Fig. 24.12.

**FIGURE 24.11** Three resonance structures of aniline.

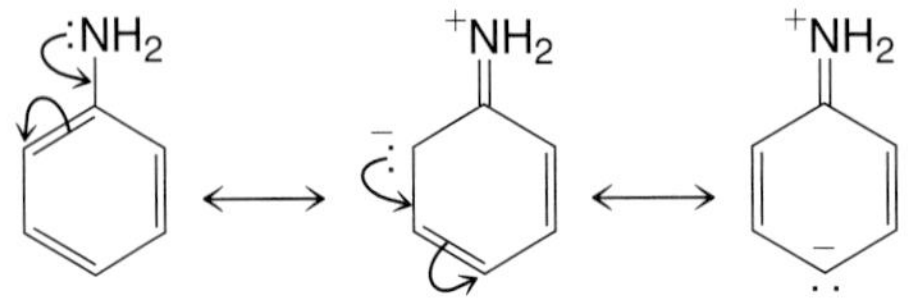

**FIGURE 24.12** Some representative alkaloids.

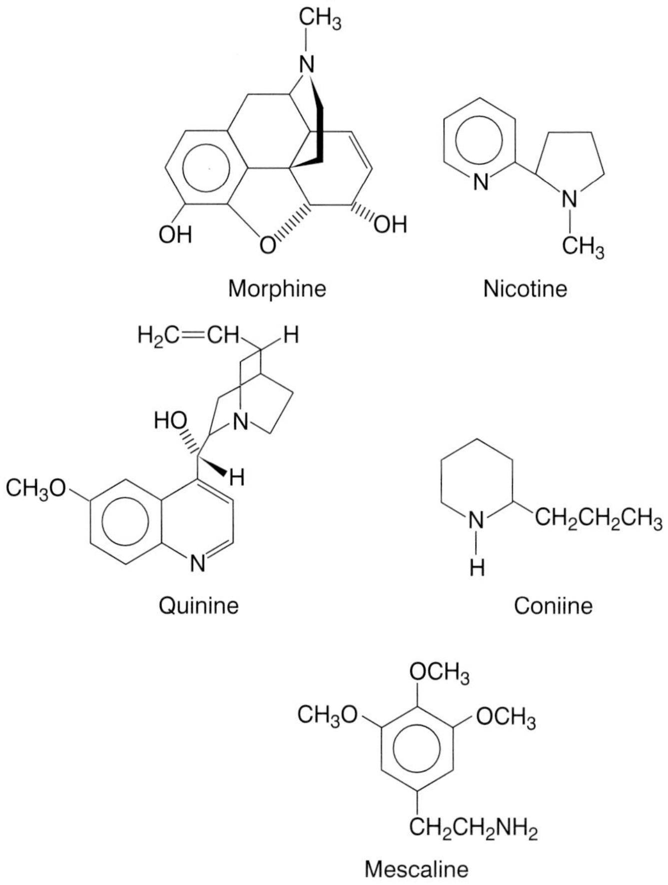

Morphine

Nicotine

Quinine

Coniine

Mescaline

*Morphine* was first isolated from the seed pod of the opium poppy in 1803. Used to relieve pain, the drug was terribly abused. By the early 1900s, about 1 in 10 Americans was using opiates. Elixirs containing morphine and heroin (the diacetate of morphine) were sold over the counter in both drug and grocery stores. This rampant abuse of opium, morphine, and heroin led to passage of the Harrison Narcotics Act in 1914.

The Harrison Narcotics Act regulated the use of these dangerous drugs.

Using the structural formula of morphine in Fig. 24.12 as a guide, draw a structural formula of heroin.

**EXERCISE 24.7**

In relatively large amounts, the drug *nicotine* is toxic to insects. The substance is present in milligram quantities in cigars and cigarettes and causes a psychological addiction and physiological effects in people who smoke. *Quinine* was isolated from the bark of the cinchona tree. This drug effectively suppresses the symptoms of malaria. *Coniine* is a toxic substance found in the fruit of the hemlock plant. The philosopher Socrates died when he drank a mixture containing the fruit of the hemlock plant. Isolated from peyote cactus, *mescaline* is a potent psychedelic (hallucination-causing) drug.

## BIOCHEMICAL BOXED READING 24.1

One example of a free base much preferred, in a negative way, to the hydrochloride salt is the free base cocaine. The cocaine free base that is so widely abused today is known as *"crack."* Crack is preferred to *cocaine hydrochloride* (known as cocaine) because crack can be snorted directly into the nostrils, whereas cocaine (cocaine hydrochloride) must be injected for the addict to obtain a high in which the central nervous system is stimulated.

Vaporization of the free base is due to the relatively weak dipole-dipole interactions and London forces holding the molecules together. Cocaine (cocaine hydrochloride), on the other hand, represents an ionic compound with strong interionic forces. Thus cocaine (cocaine hydrochloride) decomposes before vaporization takes place because of the strong interionic forces. Cocaine (cocaine hydrochloride) can readily be injected into the bloodstream because it is soluble in the aqueous solution that is our bloodstream. Injecting crack would not produce a high because crack is insoluble in the bloodstream.

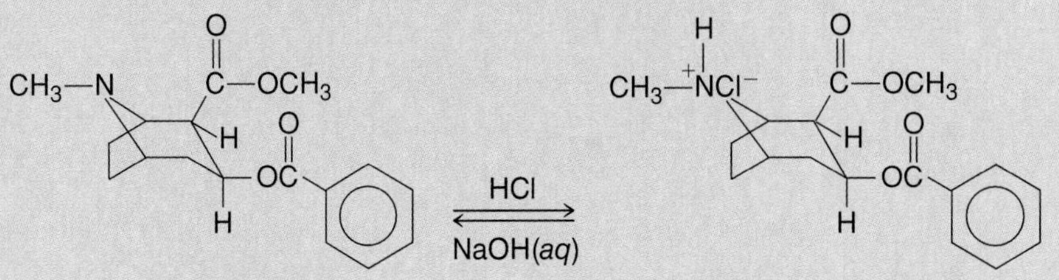


Crack (cocaine) — Cocaine (cocaine hydrochloride)

## 24.7 REPRESENTATIVE REACTIONS OF AMINES

### Salt Formation (see Sec. 24.5)

Amines are the only class of organic compound that when treated with a mineral acid yields an isolable salt. In general,

$$R-\underset{H(R'')}{\overset{H(R')}{N}} + HX \longrightarrow R-\underset{H(R'')}{\overset{H(R')}{\overset{+}{N}}}-HX^-$$

Examples:

$$(CH_3)_3N + HCl \longrightarrow (CH_3)_3\overset{+}{N}H\overset{-}{Cl}$$

Dimethylammonium chloride

$$2\,H_3C-\underset{NH_2}{\overset{H}{C}}-CH_3 + H_2SO_4 \longrightarrow \left( H_3C-\underset{\overset{+}{N}H_3}{\overset{H}{C}}-CH_3 \right)_2 SO_4^{2-}$$

Isopropylammonium sulfate

Amine salts are usually named as ammonium salt derivatives. Name the group or groups bonded to the nitrogen, and add the fragment *-ammonium*. Finally, name the anion as a separate word. Aromatic amine salts can be named as anilinium salts.

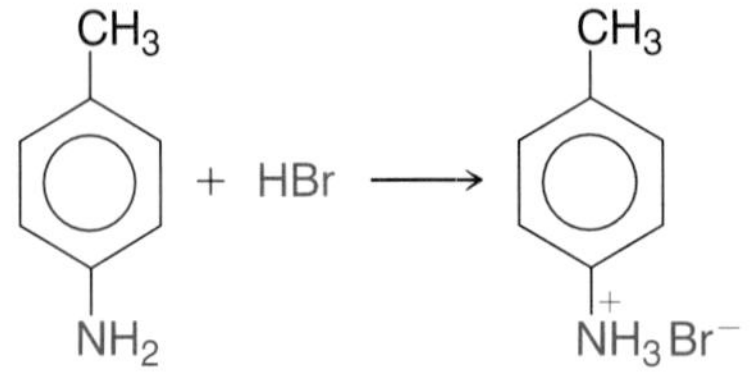


*p*-Toluidine — *p*-Methylanilinium bromide or *p*-Methylphenylammonium bromide

Note that the ordinary ammonium salt contains three or fewer organic groups bonded to nitrogen; the quaternary ammonium salt contains four.

$$R\overset{+}{N}H_3Cl^- \qquad R_4N^+Cl^-$$

Ordinary ammonium salt      Quaternary ammonium salt

These ordinary ammonium salts can be treated with a strong base such as sodium hydroxide to regenerate the amine. In general,

$$R\overset{\underset{|}{H(R'')}}{\underset{\overset{|}{H(R')}}{\overset{+}{N}}}HX^- + NaOH(aq) \longrightarrow R\overset{\underset{|}{H(R'')}}{\underset{\overset{|}{\;}}{N}}H(R') + H_2O + NaX$$

where X = Cl, Br, or I. For example:

$$(CH_3)_2\overset{+}{N}H_2Cl^- + NaOH(aq) \longrightarrow (CH_3)_2NH + NaCl + H_2O$$

(benzene ring)—$\overset{+}{N}H_3Br^-$ + NaOH(aq) ⟶ (benzene ring)—$NH_2$ + NaBr + H₂O

Anilinium bromide          Aniline

Thus the conversion of an amine to the corresponding salt represents a reversible process shown as follows.

$$\text{amine} \underset{NaOH(aq)}{\overset{HCl}{\rightleftharpoons}} \text{amine salt}$$

(Usually water-insoluble)      (water-soluble)

In many cases, the differences in properties between amine and amine salt prove very useful (Table 24.4).

Since most amines (particularly primary and secondary aromatic amines) are sensitive to oxidation, a particular amine can be kept on the shelf in the form of the stable salt. When the amine is needed, it can be regenerated easily from the amine salt. The amine–amine salt relationship also can be used in the separation and isolation of the components of a complex mixture (see Solved Problem 24.4).

| TABLE 24.4 | Properties of an Amine versus an Amine Salt | |
|---|---|---|
| | **Amine** | **Amine Salts** |
| State of matter | Usually liquid | Solid |
| Water solubility | With six carbons or more, water-insoluble | Water-soluble |
| Stability | Decompose by oxidation | Stable |
| Odor | Usually "fishy" or unpleasant | Odorless |

**SOLVED PROBLEM 24.4**

Create a flow diagram to show how you would separate and isolate the components of a mixture of benzoic acid, *p*-bromoaniline, and anthracene (each compound is a solid).

**SOLUTION**

Dissolve the mixture in ether, and extract with a 10% solution of hydrochloric acid. The aqueous layer containing the amine salt (*p*-bromoanilinium chloride) is then reacted with a 10% solution of sodium hydroxide to precipitate *p*-bromoaniline. The remaining ether layer containing benzoic acid and anthracene is extracted with a 10% solution of sodium hydroxide. The aqueous layer containing sodium benzoate is then treated with a 10% solution of hydrochloric acid to precipitate benzoic acid. The ether in the anthracene solution is removed by evaporation. All three components of the mixture are isolated in good yield.

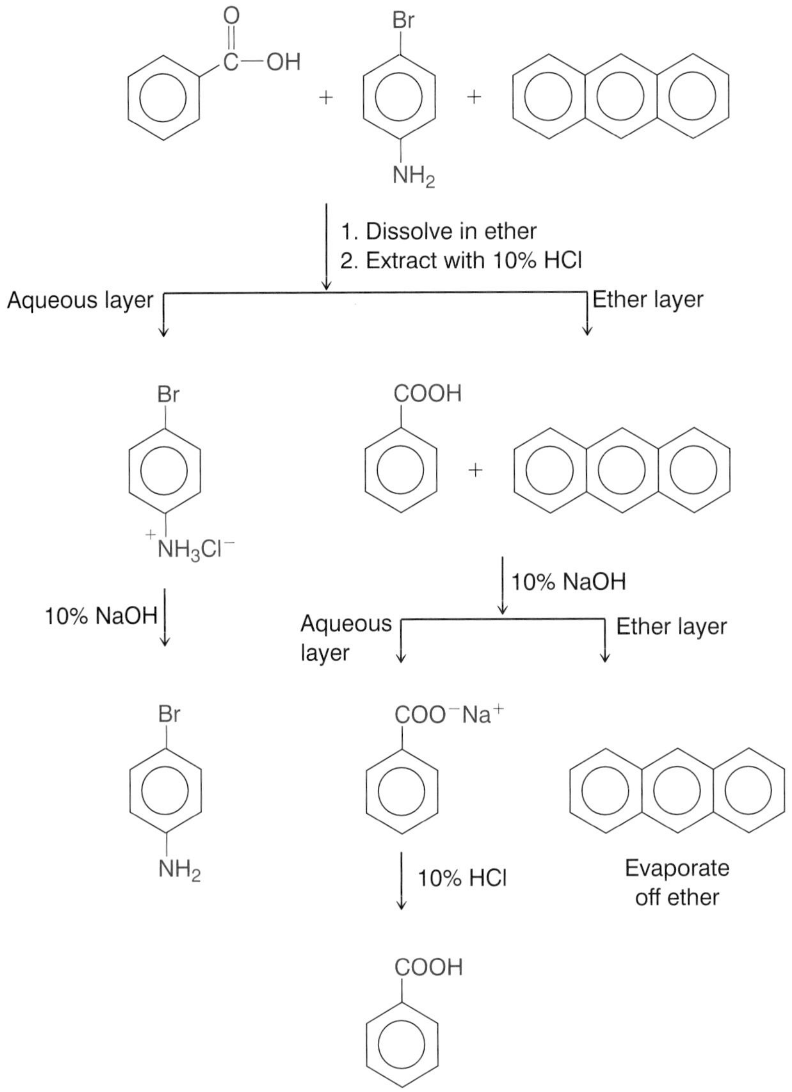

<table>
<tr><td>EXERCISE 24.8</td><td>Create a flow diagram to show how you would separate and isolate the components of a mixture of o-methylbenzoic acid, m-iodoaniline, and naphthalene (each compound is a solid).</td></tr>
</table>

**FIGURE 24.13** Dopamine.

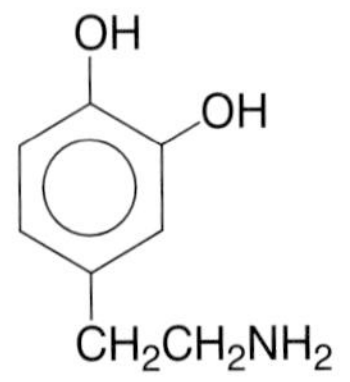

Because the free amine is not sufficiently soluble in blood, a number of drugs containing the amino group are administered to patients as the more soluble salt. For example, dopamine [4-(2-aminoethyl) catechol], a drug used to increase the heartbeat of a patient after a heart attack, is administered as the hydrochloride rather than as the free base (Fig. 24.13).

<table>
<tr><td>EXERCISE 24.9</td><td>Draw a structural formula for dopamine hydrochloride.</td></tr>
</table>

**Amide Formation from Carboxylic Acids**

In the presence of dicyclo hexyl-carbodiimide (DCC), primary and secondary amines react with carboxylic acids to produce amides. Refer to Section 20.5 for a detailed discussion.

## MEDICAL BOXED READING 24.2

Read this box if you want to get acquainted with a schizophrenic chemical compound. It has a "split personality" in that in one way it hurts us humans, while in another way it is very helpful. The name of this strange beast is *para-aminobenzoic acid* (PABA).

In order to understand the way this compound hurts us, we need to begin with a very sick baby. In December of 1932, a baby boy was burning with fever in a hospital in Germany, suffering from a nasty *Staphylococcus* infection. The child was dying, and the physicians knew it. One physician, Dr. Schreus, had an idea. A few days before, a representative of the research branch of the I. G. Farben dye conglomerate gave him some bright red pills that the company research group developed that could be useful against a *Staphylococcus* infection. The research section of the company had conducted experiments that indicated that this red dye—called *Prontosil*—stopped the growth of the deadly *Streptococcus* bacteria in mice. In desperation, since streptococci and staphylococci are similar, Dr. Schreus decided to dose the baby with Prontosil.

Fortunately the child quickly recovered once given the Prontosil. It was soon discovered that Prontosil decomposed in the human body to give *sulfanilamide* (Fig. 24.14), the compound that actually destroyed the bacteria (and the first of the well-known sulfa drugs).

Now where does PABA fit in? Well, it seems that sulfanilamide and PABA are closely related, chemical brother and sister so to speak. *Staphylococcus* must have PABA because it is necessary in the synthesis of folic acid, which, in turn, is essential for growth of the microorganism. The bacillus converts PABA to folic acid by means of several enzymes. The PABA molecule fits onto one particular enzyme the same way that a key fits into a lock. What happens is that the enzyme is fooled by the sulfanilamide. The sulfanilamide molecule looks so much like the PABA molecule (Fig. 24.15) that the bacillus enzyme cleaves to the sulfanilamide rather than to PABA. Thus folic acid cannot be produced, and the bacillus dies. Thus PABA is bad news in that it is used by certain nasty microorganisms to promote growth of these organisms and consequently make humans sick.

The good news begins in the springtime when the sun worshippers flow out of dormitories and apartments, take off as much clothing as society will allow, and soak up sun in order to obtain a suntan. Apparently, the suntan is used by humans (mostly young humans) the same way that the peacock uses its plumage. Anyway, sun worship is a mixed blessing. On the one hand, sunlight, particularly the ultraviolet portion of it, is good for our health. Sunlight converts a steroid in our skin, 7-dehydrocholesterol, to vitamin D, which does all sorts of good things for us, such as helping the bones to soak up important calcium and phosphorus, stimulating enzymes, and preventing rickets in children (osteomalacia in adults).

Too much vitamin D, however, leads to an excess of calcium and phosphorus in the bones. This excess eventually will result in kidney failure.

Now when the ultraviolet rays from the sun strike the skin, besides vitamin D formation, a complicated molecule called *melanin* is converted to a dark brown color, resulting in a suntan.

However, too much tanning hurts our skin. Not so much the pain, blistering, and peeling of a sunburn, but too much tan itself. The skin thickens, dries, and wrinkles like a prune and remains that way permanently. In addition, with continued exposure to the sun, dark spots called *liver spots* and precancerous gray lesions form.

This is where the good news about PABA fits in. It is the best FDA-approved sunscreen known to chemical science. Now a sunscreen blocks out most of the ultraviolet rays that cause skin thickening, pruning, and drying and lets in the remainder of the sun rays that are beneficial. Thus the use of a sunscreen that contains PABA usually provides our suntan lover with a good golden tan yet blocks the dangerous ultraviolet light that causes the undesirable skin changes.

**FIGURE 24.14** The enzymatic conversion of Prontosil to sulfanilamide.

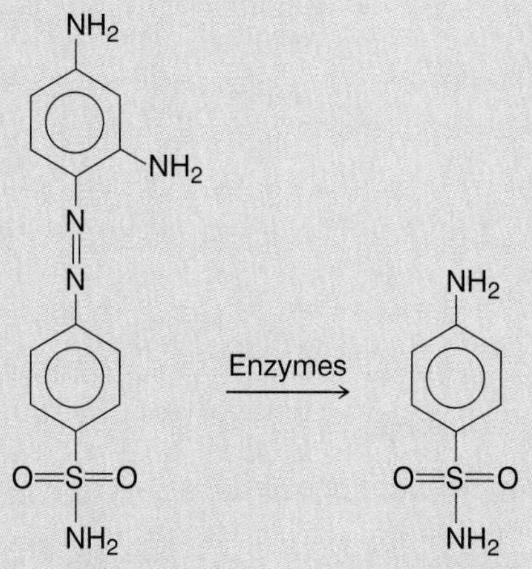

**FIGURE 24.15** The striking similarity between *p*-aminobenzoic acid and sulfanilamide to a bacillus.

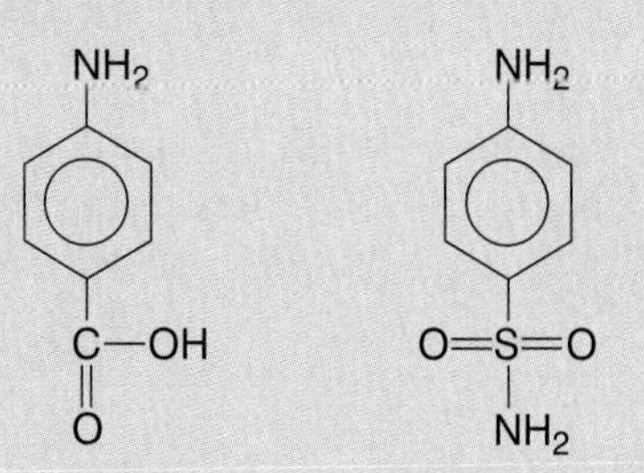

Amounts of vitamin D are expressed in units. One unit of the vitamin is equivalent to 0.25 μg of the D vitamin—a very small amount indeed. A child needs about 400 units (100 μg) each day. A long day in the sun could result in the formation of over 750,000 units of the vitamin—far too much. Therefore, just a small amount of exposure to sunlight (particularly the ultraviolet part) per day is in order.

| **EXERCISE 24.10** | Complete the following equations by drawing a structural formula for each organic product and writing a molecular formula for each inorganic product. If no reaction takes place, write "no reaction." |

a.  butanoic acid + dimethylamine $\xrightarrow{\text{DCC}}$

b.  formic acid + triethylamine $\xrightarrow{\text{DCC}}$

c.  benzylamine + HBr $\longrightarrow$

d.  product Exercise 24.10c + NaOH($aq$) $\longrightarrow$

---

# ▶ CHAPTER ACCOMPLISHMENTS

## 24.1 Classification
☐ Classify the following amines.

a.
$$CH_3-\underset{\underset{CH_3}{\overset{\overset{CH_3}{|}}{|}}{\overset{|}{C}}-NH_2$$
with H below

b.
$$H-\underset{\underset{CH_3}{|}}{\overset{\overset{CH_3}{|}}{N}}$$

☐ Explain why *tert*-butyl chloride is a tertiary chloride, while *tert*-butylamine is a primary amine.

## 24.2 Strcture
☐ Explain why an amine containing three different organic groups with a pair of electrons making up the fourth group satisfies the condition for chirality, yet two enantiomeric forms of such a compound cannot be isolated.
☐ Supply the geometry of a typical amine.

## 24.3 Nomenclature
☐ Draw a structural formula for
  a. methylamine.
  b. isopropyl *n*-propylamine.
  c. dimethylamine.
  d. cadaverine.
  e. propanamine.
  f. N-methylethanamine.
  g. 1,4-butanediamine.
  h. 2-aminobutanol.

☐ Provide names for the following compounds.
  a. $CH_3CH_2CH_2NH_2$ (common and IUPAC)
  b. $NH_2CH_2CH_2COOH$ (common and IUPAC)
  c. $(CH_3)_3N$ (common and IUPAC)

## 24.4 Physical Properties
☐ Show a hydrogen bond linking two molecules of ethylamine.

☐ Provide the class of amine that cannot display intermolecular hydrogen bonding.
☐ Explain why although one molecule of ethylamine and one molecule of ethyl alcohol have about the same molecular weight, ethylamine boils at a lower temperature (17°C) than ethyl alcohol (78°C).

## 24.5 Basicity of Amines
☐ Explain why methyl fluoride is not a Lewis base.
☐ Calculate the $pK_b$ of a dilute aqueous solution of *m*-bromoaniline given that the $pK_a$ of the compound is 3.58.
☐ Calculate the $K_b$ of a dilute aqueous solution of methylammonium ion given that the $pK_b$ is 3.36.
☐ Explain why aniline is less basic than *n*-hexylamine.

## 24.6 Alkaloids
☐ Define an alkaloid.
☐ Classify the following alkaloids as amines.
  a. coniine
  b. mescaline
  c. morphine

## 24.7 Representative Reactions of Amines
☐ Complete the following equations.
  a. $(CH_3)_2NH + HCl \longrightarrow$
  b. α-phenethylamine + $HNO_3 \longrightarrow$
  c. $(CH_3)_2\overset{+}{N}H_2Cl^- + NaOH(aq) \longrightarrow$

☐ Create a flow diagram to show how you would separate and isolate the components of a mixture of benzoic acid, *p*-bromoaniline, and anthracene (each compound is a solid).
☐ Draw a structural formula for the following compounds.
  a. isopropylammonium sulfate
  b. dimethylammonium chloride

## ▶ KEY TERMS

*putrescine* (24.3)
*cadaverine* (24.3)
*morphine* (24.6)
*nicotine* (24.6)
*quinine* (24.6)
*coniine* (24.6)
*mescaline* (24.6)

*"crack" cocaine* (Biochemical Boxed
    Reading 24.1)
*cocaine hydrochloride* (Biochemical
    Boxed Reading 24.1)
para-*aminobenzoic acid* (Medical
    Boxed Reading 24.2)

*Prontosil* (Medical Boxed Reading
    24.2)
*sulfanilamide* (Medical Boxed
    Reading 24.2, 25.3)
*melanin* (Medical Boxed Reading
    24.2)

## ▶ PROBLEMS

1. Draw a structural formula for each of the eight isomeric
   amines with the formula $C_4H_{11}N$.
   a. Name each isomer.
   b. Classify each isomer as a primary, secondary, or ter-
      tiary amine.

2. Draw a structural formula for each of the following compounds.
   a. 3-amino-1-propanol
   b. putrescine
   c. anilinium sulfate
   d. dibenzylamine
   e. *n*-butylisobutyl-*tert*-butylamine
   f. *sec*-pentylamine (two structures)
   g. phenylbenzylamine
   h. dimethylamine
   i. ethylammonium bisulfate
   j. (*S*)-2-butanamine (Fischer projection)
   k. allylamine
   l. triethylamine
   m. *p*-aminobenzoic acid
   n. *p*-aminoacetanilide
   o. 1,6-hexanediamine
   p. *N,N*-diethylaniline
   q. l-hexadecanamine

3. Name the following compounds.
   a. $(CH_3CH_2)_2\overset{+}{N}H_2Cl^-$

   b. 
   $$CH_2CH_2NH_2$$
   Mescaline, a
   hallucinogen
   $CH_3O$        $OCH_3$
           $OCH_3$

   c. $(CH_3)_4N^+I^-$

   d. 
   $NH_2$
   $NO_2$
   $NO_2$

   e. $CH_3CH_2CH_2CHCH_2CH_3$
             |
          $H-N-CH_3$

   f. $CH_3CH_2CH_2CHCH_2CH_3$
             |
            $N-C_2H_5$
             |
            $C_2H_5$

   g. $CH_2COOH$
       |
      $NH_2$

   h. (structure: propyl-N-ethyl amine skeletal)

   i. (structure: cyclopentyl-$NH_2$)

4. Which compound of each of the following pairs of compounds boils at a higher temperature? Explain.
   a. $CH_3CH_2NH_2$ or $(CH_3)_2NH$
   b. $CH_3CH_2NH_2$ or $CH_3CH_2CH_2CH_2NH_2$
   c. $CH_3CH_2OH$ or $CH_3CH_2NH_2$
   d. $CH_3CH_2NH_2$ or $CH_3CH_2CH_3$

5. Which compound of the following pairs would you expect to be more water soluble? Explain.
   a. $CH_3CH_2NH_2$ or $CH_3(CH_2)_6CH_2NH_2$
   b. $(C_2H_5)_3N$ or $(C_2H_5)_4N^+Cl^-$

6. Trimethylamine and *n*-propylamine are isomers and thus
   have the same molecular weight. Which compound boils
   at the higher temperature? Why?

7. An amine such as $CH_3(CH_2)_{10}CH_2NH_2$ is odorless compared with $CH_3CH_2NH_2$. Why?

8. Why is compound I a stronger base than compound II?

   $CH_2NH_2$        $NH_2$
   (benzene ring)     (benzene ring)

        I                II

9. Draw the structural formula for the conjugate acid of
   a. $(CH_3CH_2)_2NH$
   b. $CH_3CH_2NH_2$
   c. $(CH_3CH_2)_3N$

10. Which amine is the stronger base, compound I or compound II? Explain.

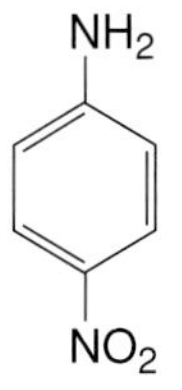

11. Devise a procedure by which you would be able to separate and isolate each component of the following mixtures.
    a. a mixture of p-chloroaniline and p-toluic acid (each compound is a solid)

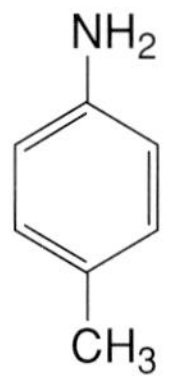

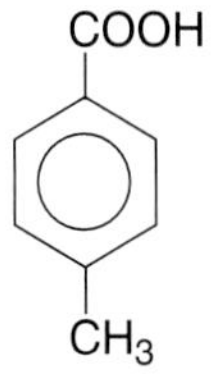

p-Toluic acid

b. a mixture of naphthalene and diphenylamine (each substance is a solid)

12. Amphetamine (β-phenylisopropylamine) is a powerful stimulant to the central nervous system.
    a. Draw a structural formula for amphetamine.
    b. Furnish an IUPAC name for the compound.

13. Draw a structural formula of a(an)
    a. secondary amine containing primary alkyl groups.
    b. alkaloid.
    c. quaternary ammonium salt.
    d. primary amine containing a secondary group.

14. Locate the chiral center in the structural formula of nicotine and coniine and the chiral centers in the structural formula of quinine. Refer to Fig. 24.12 for the structural formulas.

CHAPTER 25

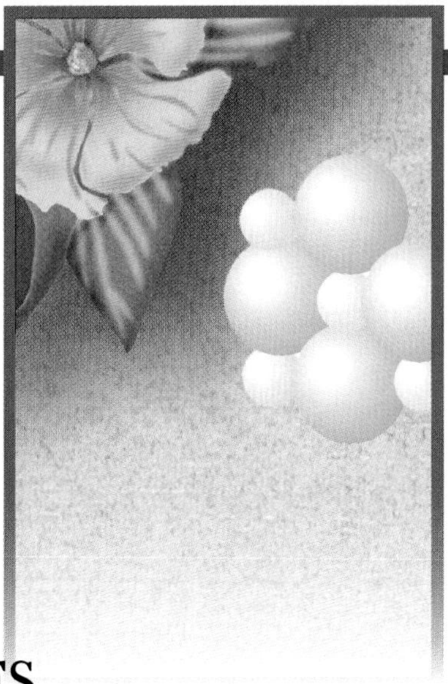

# Amines (II)

## SYNTHESIS, CHEMICAL PROPERTIES OF AMINES, AND REACTIONS OF DIAZONIUM SALTS

### 25.1 SYNTHESIS OF AMINES

#### Alkylation of Ammonia

An alkyl halide reacts with ammonia (in huge excess, i.e., as much as 17 mol ammonia to 1 mol alkyl halide) to give a primary alkylammonium salt that is converted into the corresponding amine when the salt reacts with some of the excess ammonia present (see Sec. 24.7). In general,

$$X{-}R + :NH_3 \xrightarrow{} R\overset{+}{N}H_3X^- \quad \text{(Huge excess)}$$

$$R\overset{+}{N}H_2X^- \,(H) + :NH_3 \longrightarrow RNH_2 + {}^{+}NH_4 + X^-$$

where R = primary or secondary alkyl group, and X = Cl, Br, or I. For examples:

$$\underset{\text{Ethyl bromide}}{CH_3CH_2Br} + NH_3\text{(huge excess)} \longrightarrow CH_3CH_2\overset{+}{N}H_3Br^-$$

$$\underset{\text{Ethylammonium bromide}}{CH_3CH_2\overset{+}{N}H_3Br^-} + NH_3 \longrightarrow \underset{\text{Ethylamine}}{CH_3CH_2NH_2} + NH_4Br$$

In a similar manner, we can prepare a secondary amine from a primary amine and a tertiary amine from a secondary amine. In general,

$$X{-}R' + :NH_2R \longrightarrow R(R')\overset{+}{N}H_2X^- \quad \text{(Huge excess)}$$

$$R(R')\overset{+}{N}HX^- \,(H) + :NH_2R \longrightarrow R(R')NH + R\overset{+}{N}H_3 + X^-$$

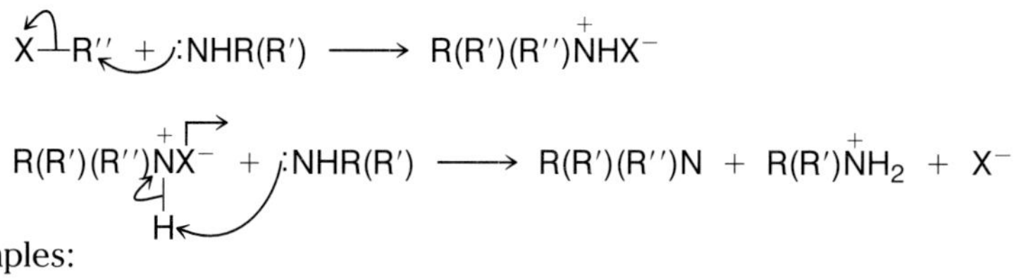

(Huge excess)

$$X\!-\!R'' + :NHR(R') \longrightarrow R(R')(R'')\overset{+}{N}HX^-$$

$$R(R')(R'')\overset{+}{N}X^- + :NHR(R') \longrightarrow R(R')(R'')N + R(R')\overset{+}{N}H_2 + X^-$$
$$\overset{|}{H}$$

Examples:

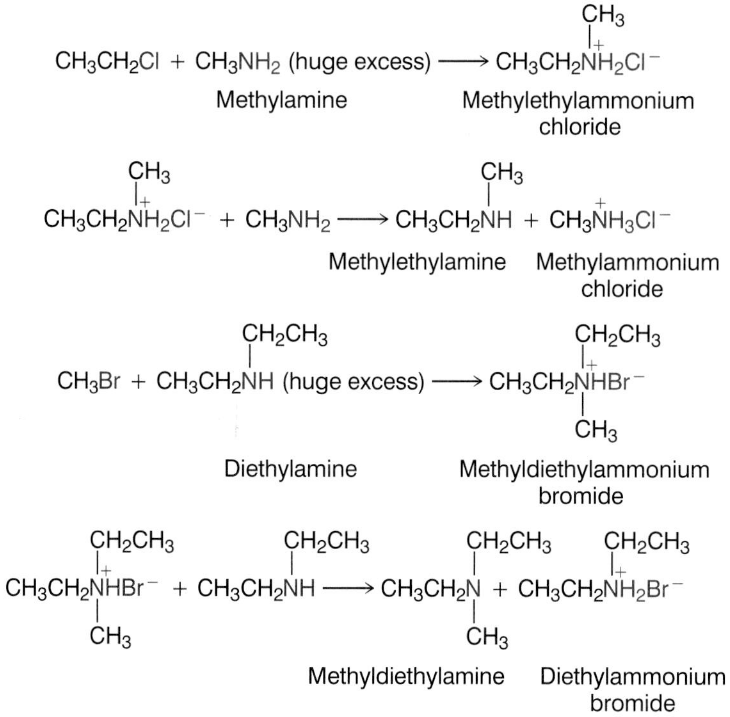

$$CH_3CH_2Cl + CH_3NH_2 \text{ (huge excess)} \longrightarrow CH_3CH_2\overset{+}{\underset{|}{\overset{CH_3}{N}}}H_2Cl^-$$

Methylamine          Methylethylammonium chloride

$$CH_3CH_2\overset{+}{\underset{|}{\overset{CH_3}{N}}}H_2Cl^- + CH_3NH_2 \longrightarrow CH_3CH_2\overset{CH_3}{\underset{|}{N}}H + CH_3\overset{+}{N}H_3Cl^-$$

Methylethylamine    Methylammonium chloride

$$CH_3Br + CH_3CH_2\overset{CH_2CH_3}{\underset{|}{N}}H \text{ (huge excess)} \longrightarrow CH_3CH_2\overset{CH_2CH_3}{\underset{|}{\overset{+}{N}}}HBr^-$$
$$\underset{CH_3}{}$$

Diethylamine         Methyldiethylammonium bromide

$$CH_3CH_2\overset{CH_2CH_3}{\underset{CH_3}{\overset{+}{N}}}HBr^- + CH_3CH_2\overset{CH_2CH_3}{\underset{|}{N}}H \longrightarrow CH_3CH_2\overset{CH_2CH_3}{\underset{CH_3}{N}} + CH_3CH_2\overset{CH_2CH_3}{\underset{|}{\overset{+}{N}}}H_2Br^-$$

Methyldiethylamine    Diethylammonium bromide

Limitations of this reaction and stem from several sources. First, although yields are good when a huge excess of amine is employed, amines are expensive. Therefore, it is not practical to use this method of preparation to prepare secondary or tertiary amines. Since ammonia is an inexpensive reagent, the reaction is used to prepare primary amines (in particular, a number of amino acids; see Chap. 28). Second, since the reaction is an example of an $S_N2$ nucleophilic substitution, it fails with tertiary halides.

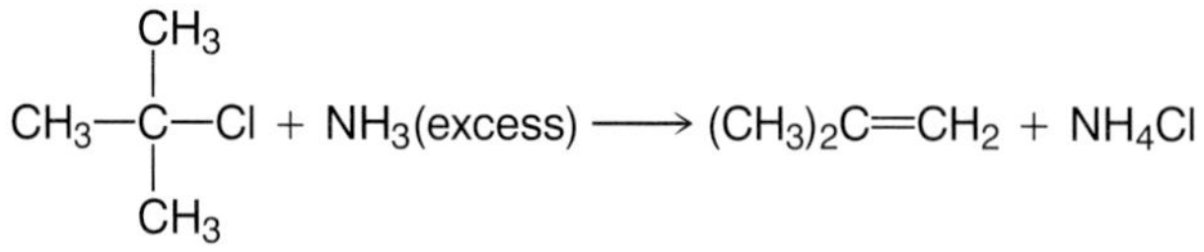

$$CH_3\!-\!\overset{CH_3}{\underset{CH_3}{\overset{|}{C}}}\!-\!Cl + NH_3(\text{excess}) \longrightarrow (CH_3)_2C\!=\!CH_2 + NH_4Cl$$

**EXERCISE 25.1**

Write a detailed mechanism for the preceding reaction. Assume E2 conditions.

Third, if ammonia is not in great excess, a mixture of the primary amine with the corresponding secondary and tertiary amines is produced along with the *quaternary ammonium salt*. A quaternary ammonium salt contains four organic groups bonded to nitrogen. These salts are ionic compounds. Refer to Section 24.7 for a discussion of the nomenclature of the amine salts.

Example:

$$CH_3CH_2Br + NH_3 \longrightarrow CH_3CH_2\overset{+}{N}H_3Br^-$$

$$CH_3CH_2\overset{+}{N}H_3Br^- + NH_3 \longrightarrow CH_3CH_2NH_2 + NH_4Br$$
Ethylamine

Ethyl bromide reacts with ethylamine rather than ammonia to produce diethylamine.

$$CH_3CH_2Br + CH_3CH_2NH_2 \longrightarrow (CH_3CH_2)_2\overset{+}{N}H_2Br^-$$

$$(CH_3CH_2)_2\overset{+}{N}H_2Br^- + CH_3CH_2NH_2 \longrightarrow (CH_3CH_2)_2NH + CH_3CH_2\overset{+}{N}H_3Br^-$$
Diethylamine    Ethylammonium bromide

Ethyl bromide reacts with diethylamine to obtain triethylamine.

$$CH_3CH_2Br + (CH_3CH_2)_2NH \longrightarrow (CH_3CH_2)_3\overset{+}{N}HBr^-$$

$$(CH_3CH_2)_3\overset{+}{N}HBr^- + (CH_3CH_2)_2NH \longrightarrow (CH_3CH_2)_3N + (CH_3CH_2)_2\overset{+}{N}H_2Br^-$$
Diethylamine                          Diethylammonium bromide

Finally, the quaternary salt is produced when ethyl bromide reacts with triethylamine.

$$CH_3CH_2Br + (CH_3CH_2)_3N \longrightarrow (CH_3CH_2)_4N^+Br^-$$
Tetraethylammonium<br>bromide

A quaternary ammonium salt also can be prepared in better yield by treating a tertiary amine with an alkyl halide. In general,

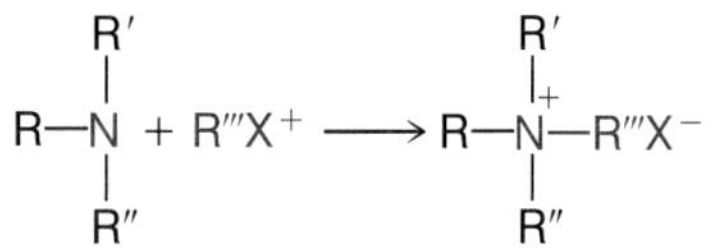

$$\begin{array}{c} R' \\ | \\ R-N \\ | \\ R'' \end{array} + R'''X^+ \longrightarrow \begin{array}{c} R' \\ |+ \\ R-N-R'''X^- \\ | \\ R'' \end{array}$$

where X = Cl, Br, or I. For example:

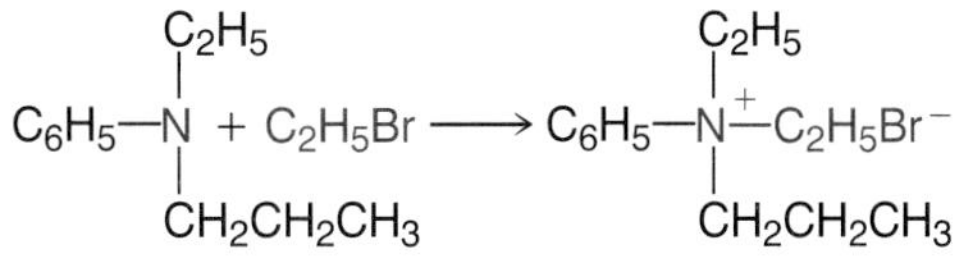

$$\begin{array}{c} C_2H_5 \\ | \\ C_6H_5-N \\ | \\ CH_2CH_2CH_3 \end{array} + C_2H_5Br \longrightarrow \begin{array}{c} C_2H_5 \\ |+ \\ C_6H_5-N-C_2H_5Br^- \\ | \\ CH_2CH_2CH_3 \end{array}$$
Diethylpropylphenylammonium bromide<br>(a quaternary ammonium salt)

Because *phthalimide* is intimately involved in the Gabriel synthesis, let us first examine the structure and relevant chemical properties of the imides. Each imide contains the grouping

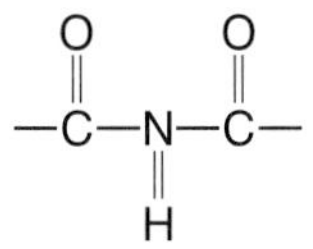

$$\begin{array}{ccc} O & & O \\ \| & & \| \\ -C-N-C- \\ & \| \\ & H \end{array}$$

The imides most often used in the laboratory are cyclic and are similar in structure to the corresponding anhydrides. To convert an anhydride to an imide in terms of structure, replace the —O— atom with an —NH— group. An example is given in Fig. 25.1. The importance of phthalimide to the Gabriel synthesis is that the compound (along with other cyclic imides) is acidic enough to react with aqueous sodium hydroxide.

*Quaternary ammonium salts are stable, isolable, and function like ionic compounds.*

## Gabriel Phthalimide Synthesis

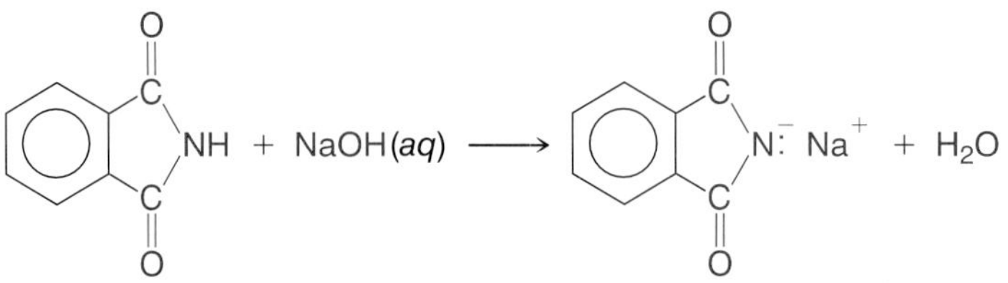

Sodium phthalimide

The phthalimide anion reacts with a primary alkyl halide by means of an $S_N2$ substitution to produce an *N*-substituted imide. The halide used can be a chloride, bromide, or iodide.

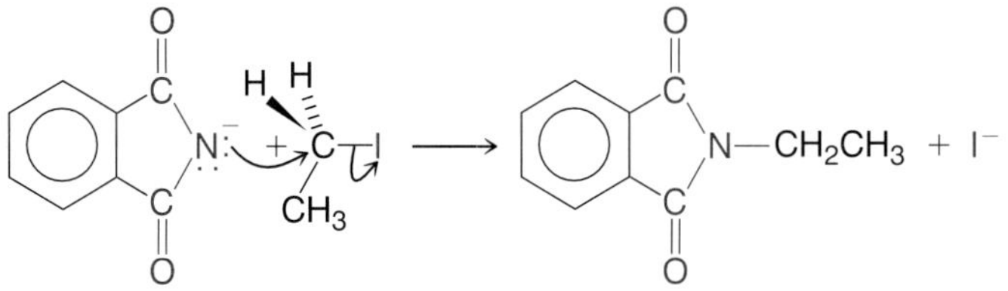

*N*-Ethylphthalimide

Note that both the amine produced and the primary alkyl halide used contain the same number of carbon atoms.

Finally, the substituted imide is hydrolyzed in aqueous base to produce the amine (see Sec. 24.7).

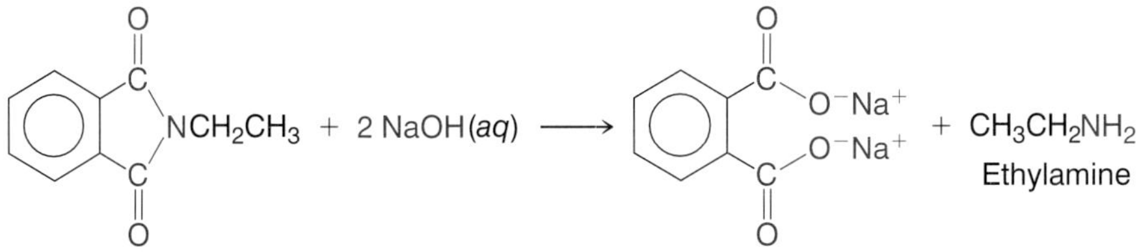

Sodium phthalate

Ethylamine

---

**FIGURE 25.1** Two examples of the structural difference between a cyclic acid anhydride and a cyclic imide.

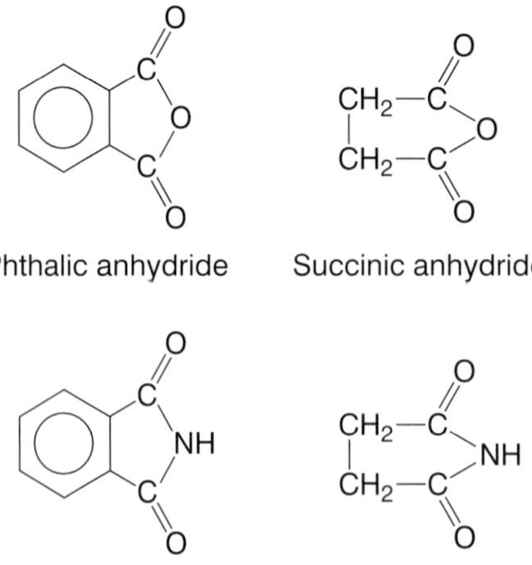

Phthalic anhydride         Succinic anhydride

Phthalimide                Succinimide

## EXERCISE 25.2

The $pK_a$ of phthalimide is 8.4.

a. Calculate the $K_a$ (see Solved Problem 20.3b)

b. Phthalimide is as strong an acid as it is due to resonance stabilization of the phthalimide anion. Draw the stabilizing resonance structures of the phthalimide anion.

## Reduction of Nitrogen-Containing Compounds

**Aromatic Nitro Compounds (see Sec. 10.5)**

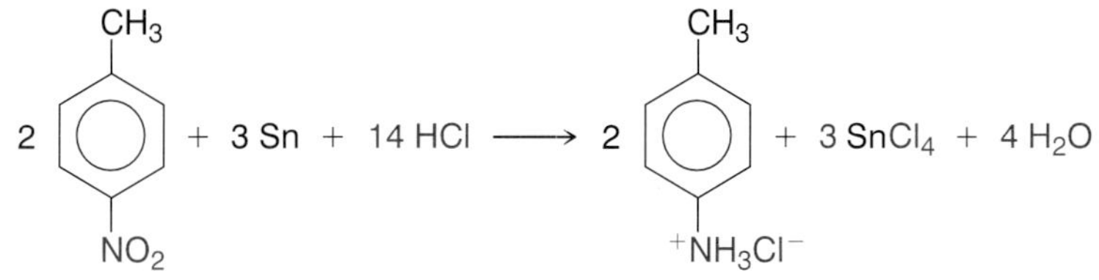

*p*-Nitrotoluene

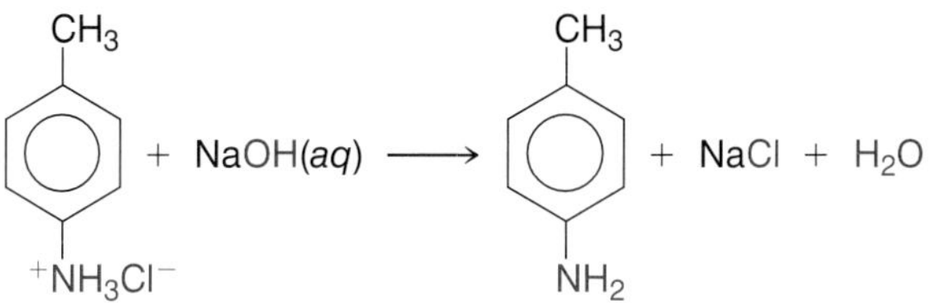

*p*-Toluidine

Catalytic hydrogenation (see Secs. 4.1 and 5.6) also conveniently accomplishes the reduction of aromatic nitro compounds to arylamines.

**Reduction of Amides** Amides are converted in good yield to amines by the use of lithium aluminum hydride (see Sec. 23.9), followed by hydrolysis in aqueous basic solution (see Sec. 24.7). Ether is used as a solvent. Note that simple amides, monosubstituted amides, and disubstituted amides produce primary, secondary, and tertiary amines, respectively. In general,

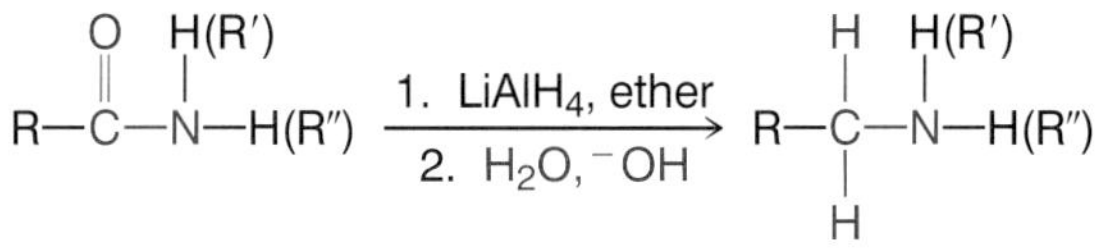

Examples:

$$CH_3(CH_2)_6\overset{\overset{O}{\|}}{C}{-}NH_2 \xrightarrow[2.\ H_2O,\ ^-OH]{1.\ LiAlH_4,\ ether} CH_3(CH_2)_6CH_2NH_2$$

Octanamide $\qquad$ 1-Octanamine

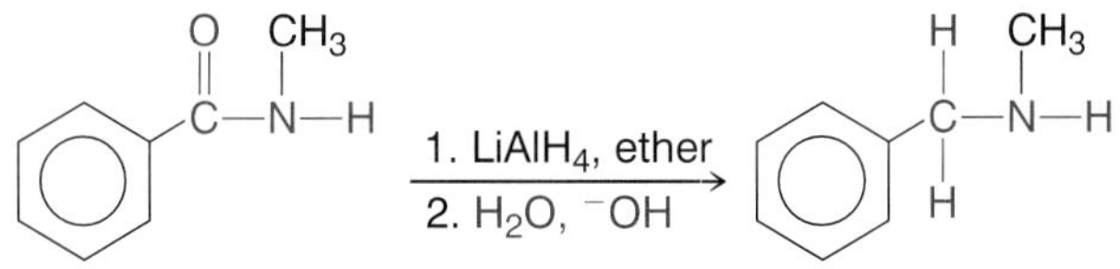

*N*-Methylbenzamide $\qquad$ Methylbenzylamine

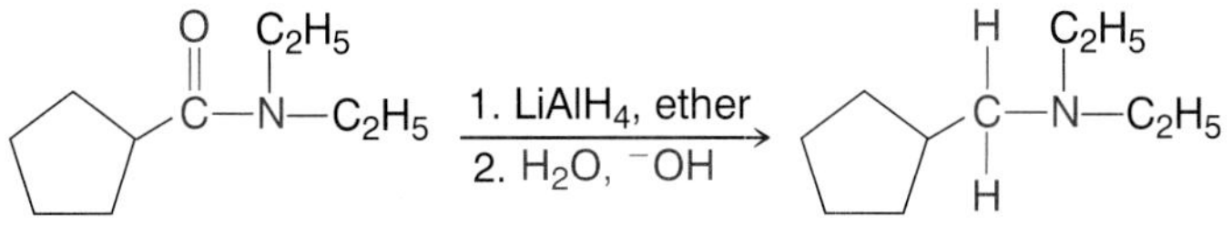

*N-N*-Diethylcyclopentanecarboxamide          *N-N*-Diethylcyclopentylmethanamine

**Reduction of Nitriles**   Amines are converted in good yield from nitriles by the use of lithium aluminum hydride, followed by hydrolysis in aqueous basic solution (see Sec. 24.7). In general,

$$R-C\equiv N \xrightarrow[\text{2. H}_2\text{O, }^-\text{OH}]{\text{1. LiAlH}_4\text{, ether}} RCH_2NH_2$$

Example:

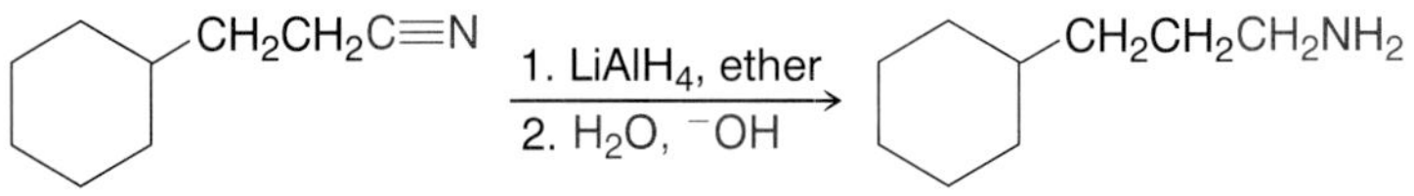

3-Cyclohexylpropanenitrile                    3-Cyclohexyl-1-propanamine

**Reductive Amination**

Treating an aldehyde or ketone with ammonia in the presence of hydrogen and a catalytic amount of platinum gives a primary amine. In general,

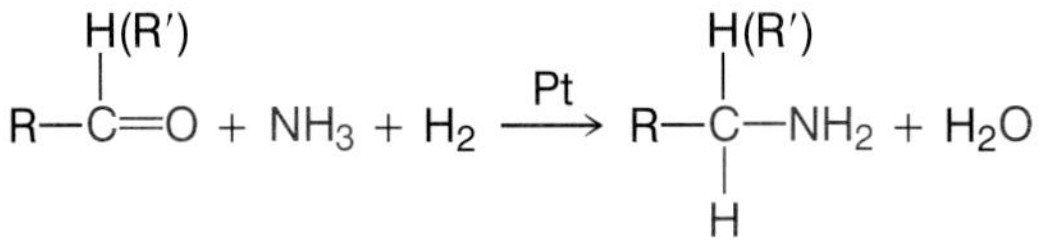

A primary amine

Example:

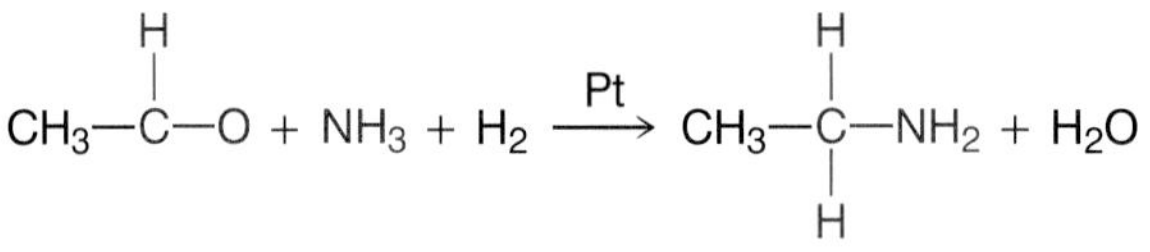

Acetaldehyde                              Ethylamine

The reaction proceeds as follows (see Sec. 19.2).

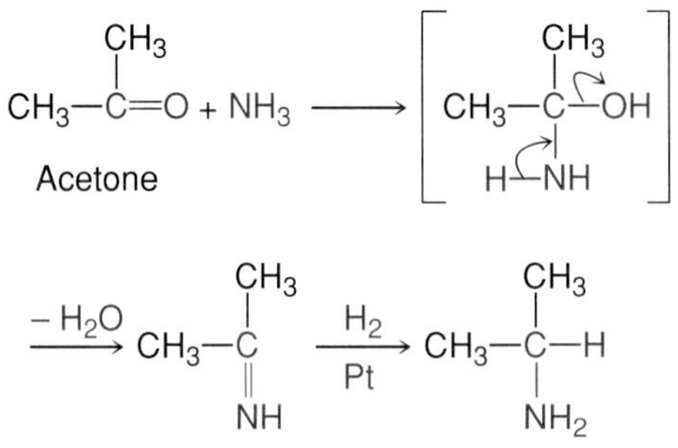

Acetone

Acetone imine     Isopropylamins

The imine prepared from a carbonyl compound and ammonia is unstable and is not isolated.

A similar reaction takes place with a primary amine or a secondary amine instead of ammonia to produce a secondary or tertiary amine, respectively.* In general,

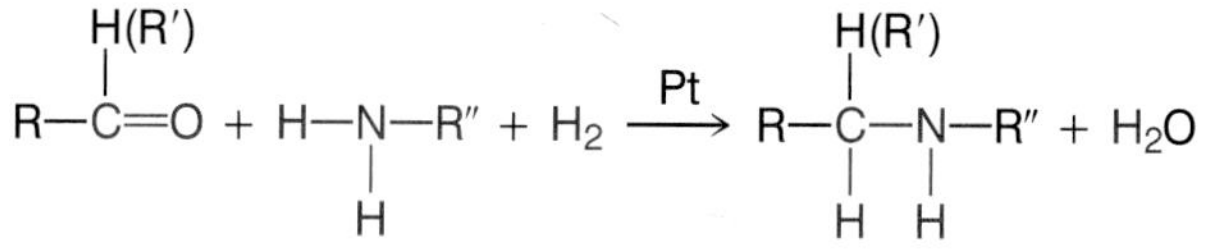

$$R-\overset{\overset{\displaystyle H(R')}{|}}{C}=O + H-\overset{\overset{}{\underset{\underset{\displaystyle H}{|}}{N}}}{N}-R'' + H_2 \xrightarrow{Pt} R-\overset{\overset{\displaystyle H(R')}{|}}{\underset{\underset{\displaystyle H}{|}}{C}}-\overset{}{\underset{\underset{\displaystyle H}{|}}{N}}-R'' + H_2O$$

A primary amine          A secondary amine

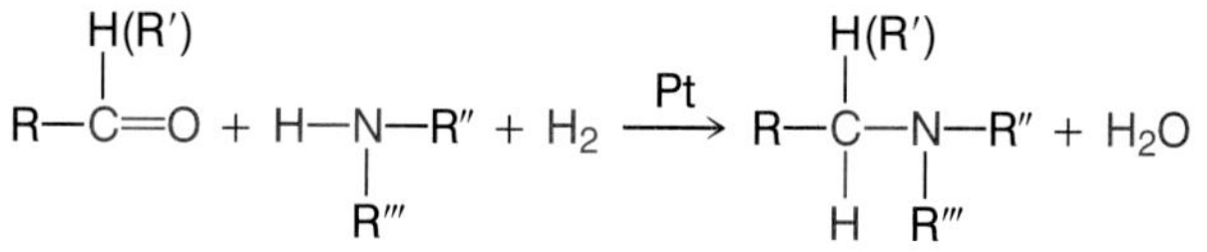

$$R-\overset{\overset{\displaystyle H(R')}{|}}{C}=O + H-\overset{}{\underset{\underset{\displaystyle R'''}{|}}{N}}-R'' + H_2 \xrightarrow{Pt} R-\overset{\overset{\displaystyle H(R')}{|}}{\underset{\underset{\displaystyle H}{|}}{C}}-\overset{}{\underset{\underset{\displaystyle R'''}{|}}{N}}-R'' + H_2O$$

A secondary amine          A tertiary amine

Examples:

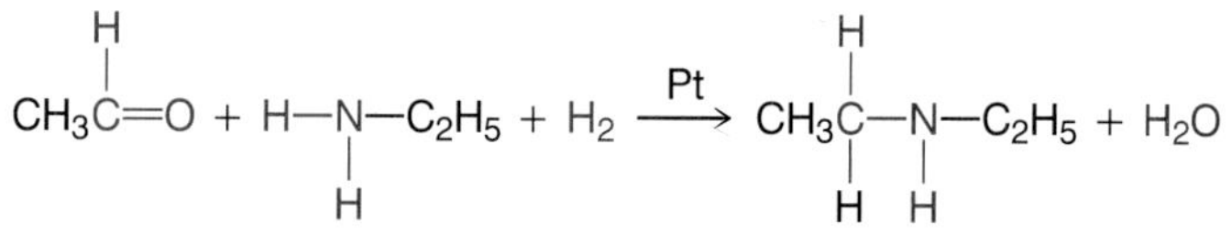

$$CH_3\overset{\overset{\displaystyle H}{|}}{C}=O + H-\overset{}{\underset{\underset{\displaystyle H}{|}}{N}}-C_2H_5 + H_2 \xrightarrow{Pt} CH_3\overset{\overset{\displaystyle H}{|}}{\underset{\underset{\displaystyle H}{|}}{C}}-\overset{}{\underset{\underset{\displaystyle H}{|}}{N}}-C_2H_5 + H_2O$$

Acetaldehyde   Ethylamine          Diethylamine

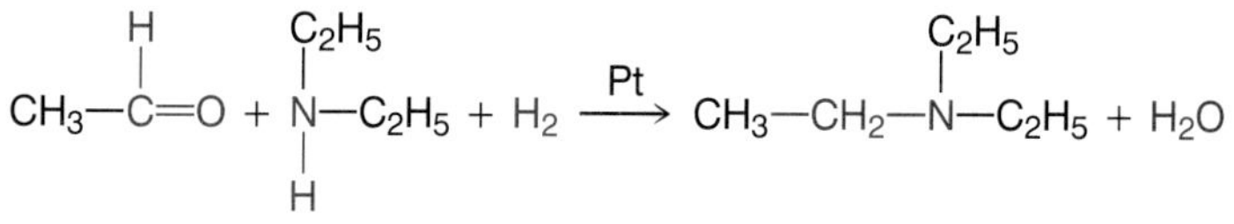

$$CH_3-\overset{\overset{\displaystyle H}{|}}{C}=O + \overset{\overset{\displaystyle C_2H_5}{|}}{\underset{\underset{\displaystyle H}{|}}{N}}-C_2H_5 + H_2 \xrightarrow{Pt} CH_3-CH_2-\overset{\overset{\displaystyle C_2H_5}{|}}{N}-C_2H_5 + H_2O$$

Diethylamine          Triethylamine

When an amide is reacted with sodium hydroxide and a halogen (NaOH + $X_2$, where $X_2$ is $Br_2$ or $Cl_2$), a rearrangement takes place to yield an amine with one less carbon atom than the original amide. In general,

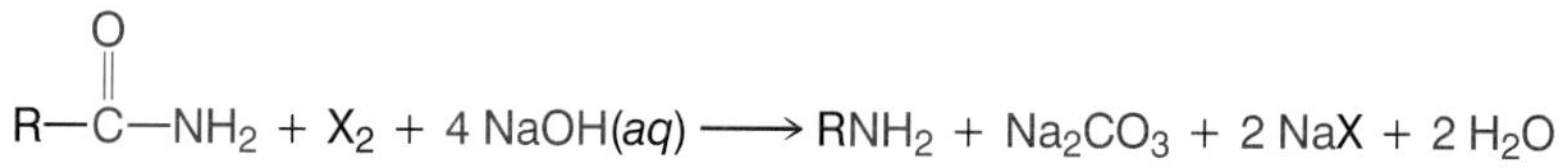

$$R-\overset{\overset{\displaystyle O}{\|}}{C}-NH_2 + X_2 + 4\,NaOH(aq) \longrightarrow RNH_2 + Na_2CO_3 + 2\,NaX + 2\,H_2O$$

where $X_2 = Br_2$ or $Cl_2$. For example:

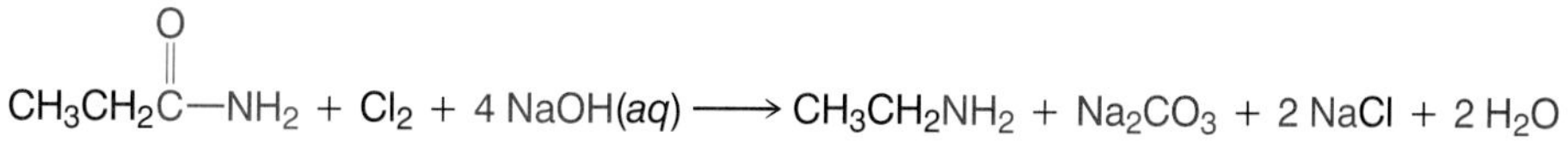

$$CH_3CH_2\overset{\overset{\displaystyle O}{\|}}{C}-NH_2 + Cl_2 + 4\,NaOH(aq) \longrightarrow CH_3CH_2NH_2 + Na_2CO_3 + 2\,NaCl + 2\,H_2O$$

The mechanism of the *Hoffmann rearrangement* should give you a better understanding of the reaction.

## Hoffmann Rearrangement

Note the loss of one carbon from the amide to the amine. Because yields are excellent, the reaction has a significant amount of synthetic utility.

---

*When a carbonyl compound reacts with a secondary amine, an imine cannot form; the intermediate produced instead is an iminium ion that leads to the formation of a tertiary amine.

**Envision the Reaction**

---

## Mechanism of the Hoffmann Rearrangement with Propanamide

**First step:**

$$2\ Br_2 + 4\ NaOH \longrightarrow 2\ NaOBr + 2\ NaBr + 2\ H_2O$$

**Second step:**

$$CH_3CH_2\overset{\displaystyle O}{\overset{\|}{C}}-NH_2 + NaOBr \longrightarrow CH_3CH_2\overset{\displaystyle O}{\overset{\|}{C}}-\overset{\displaystyle H}{\underset{}{N}}-Br + NaOH$$

**Third step:**

$$CH_3CH_2\overset{\displaystyle O}{\overset{\|}{C}}-\overset{\displaystyle H\leftarrow\ :OH}{\underset{}{N}}-Br \xrightarrow{-H^+} H_2O + CH_3CH_2\overset{\displaystyle O}{\overset{\|}{C}}-\overset{..\ -}{\underset{..}{N}}-Br$$

**Fourth step:**

$$CH_3CH_2\overset{\displaystyle O}{\overset{\|}{C}}-\overset{..}{\underset{..}{N}}-Br \xrightarrow{-Br^-} CH_3CH_2\overset{\displaystyle O}{\overset{\|}{C}}-\overset{..}{N} + Br^-$$

**Fifth step:**

$$CH_3CH_2-\overset{\displaystyle O}{\overset{\|}{C}}-\overset{..}{N} \longrightarrow O{=}C{=}N-CH_2CH_3$$
Ethyl isocyanate

**Sixth step:**

$$CH_3CH_2N{=}C{=}O + H_2O \longrightarrow \left[ CH_3CH_2\overset{}{\underset{\displaystyle H}{N}}-\overset{\displaystyle O}{\overset{\|}{C}}-OH \right]$$
Ethylcarbamic acid

**Seventh step:**

$$\left[ CH_3CH_2\overset{}{\underset{\displaystyle H}{N}}-\overset{\displaystyle O}{\overset{\|}{C}}-OH \right] \xrightarrow{-CO_2} CH_3\overset{\displaystyle H}{\underset{}{N}}H + CO_2$$
Ethylcarbamic acid           Methylamine

---

In the first step of the mechanism, bromine or chlorine reacts with sodium hydroxide to give sodium hypohalite and sodium halide (in this case, sodium hypobromite and sodium bromide, respectively). In the second step, sodium hypobromite then reacts with the amide (in this case, propanamide) to yield the corresponding N-bromoamide. These bromoamides are isolable species. However, they are not isolated during the course of the reaction for convenience. Hydroxide ion then attacks the N-bromoamide in the third step, abstracting a proton from the N-bromoamide. Bromide ion is then spontaneously lost in the fourth step. The ethyl carbanion rearranges to the electron-deficient nitrogen in a 1,2 shift in the fifth step of the mechanism. Simultaneously, two unshared electrons on nitrogen move to form a double bond between carbon and nitrogen to produce ethyl isocyanate. Isocyanates, like N-bromoamides, are isolable but usually are not isolated during the course of the reaction, again for convenience. The isocyanate is then attacked by water in the sixth step to produce a substituted carbamic acid. Substituted carbamic acids and carbamic acid itself are unstable and decompose in the seventh step to give the corresponding amine (or ammonia) and carbon dioxide.

$$[H_2N{-}COOH] \longrightarrow NH_3 + CO_2$$
Carbamic acid

In general,

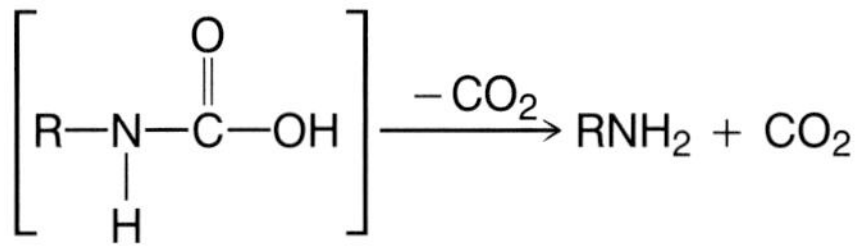

$$\left[ \begin{array}{c} \overset{\displaystyle O}{\underset{\displaystyle H}{R-N-C-OH}} \end{array} \right] \xrightarrow{-CO_2} RNH_2 + CO_2$$

A substituted carbamic acid

An example of this decomposition can be found in the preceding mechanism.

Let us consider in more detail the decomposition of a substituted carbamic acid (in this case, ethylcarbamic acid) to form the amine.

### Mechanism of the Decomposition of Ethylcarbamic Acid to Form Ethylamine

First step:

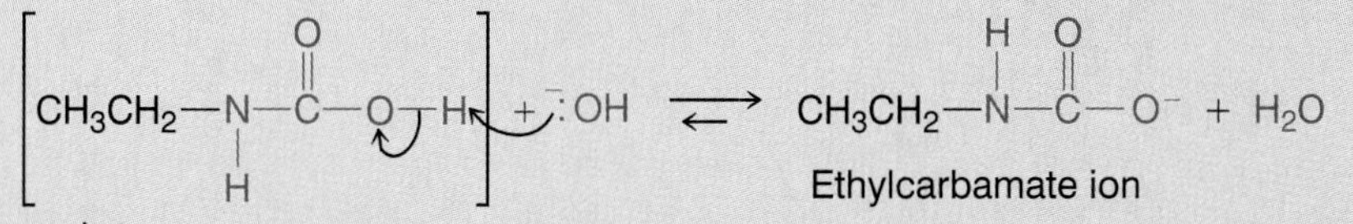

$$\left[ CH_3CH_2-\overset{O}{\underset{H}{N-C}}-O-H \right] + \colon OH \;\rightleftharpoons\; CH_3CH_2-\overset{H\;\;O}{N-C}-O^- + H_2O$$

Ethylcarbamate ion

Second step:

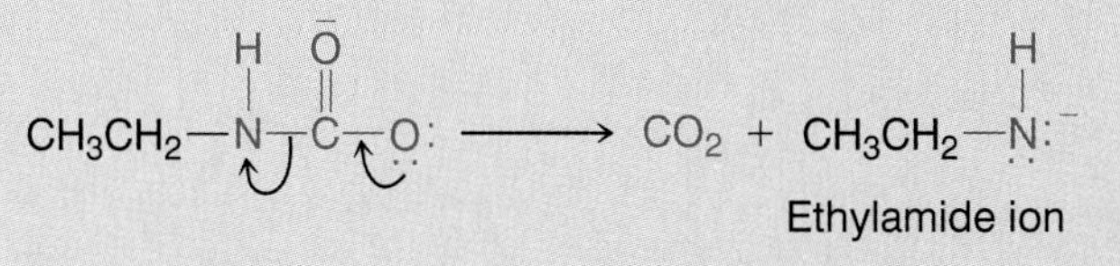

$$CH_3CH_2-\overset{H\;\;\bar{O}}{N-C-O}\colon \longrightarrow CO_2 + CH_3CH_2-\overset{H}{N}\colon^-$$

Ethylamide ion

Third step:

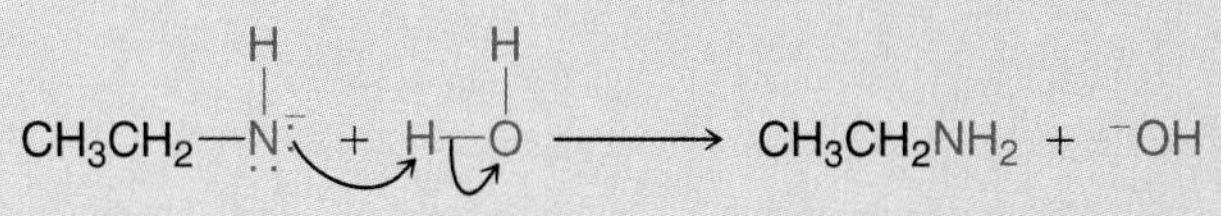

$$CH_3CH_2-\overset{H}{N}\colon + H-O \longrightarrow CH_3CH_2NH_2 + {}^-OH$$

Fourth step:

$$H_2CO_3 + NaOH \longrightarrow NaHCO_3 + H_2O$$
$$NaHCO_3 + NaOH \longrightarrow Na_2CO_3 + H_2O$$

The first step of the mechanism shows hydroxide ion abstracting a proton from the carbamic acid to produce ethylcarbamate ion. The second step shows the loss of carbon dioxide from the ethylcarbamate ion to produce ethylamide ion. Finally, ethylamide ion abstracts a proton from water to produce ethylamine and regenerate hydroxide ion. Now, since the Hoffmann rearrangement is carried out in base, acidic carbon dioxide cannot exist and therefore instantaneously reacts with base in the fourth step to eventually produce sodium carbonate to complete the mechanism.

Now that we have developed a number of different methods to prepare amines, let us incorporate these reactions into syntheses.

| | |
|---|---|
| **SOLVED PROBLEM 25.1** | Synthesize the following compounds. You may use any inorganic reagents necessary.<br><br>a.  ethylamine from propionic acid<br><br>b.  1-propanamine from ethyl bromide |

**SOLUTION**

a.  Because ethylamine contains one less carbon atom than propionic acid, this suggests the use of the Hoffmann rearrangement. Thus we have

$$CH_3CH_2\overset{\overset{\displaystyle O}{\|}}{C}-NH_2 \xrightarrow[-OH]{Br_2} CH_3CH_2NH_2$$

Now, we must produce propionamide from propionic acid.

$$CH_3CH_2\overset{\overset{\displaystyle O}{\|}}{C}-OH \xrightarrow{SOCl_2} CH_3CH_2\overset{\overset{\displaystyle O}{\|}}{C}-Cl \xrightarrow{NH_3} CH_3CH_2\overset{\overset{\displaystyle O}{\|}}{C}-NH_2$$

Putting this all together, we have

$$CH_3CH_2\overset{\overset{\displaystyle O}{\|}}{C}-OH \xrightarrow{SOCl_2} CH_3CH_2\overset{\overset{\displaystyle O}{\|}}{C}-Cl \xrightarrow{NH_3} CH_3CH_2\overset{\overset{\displaystyle O}{\|}}{C}-NH_2 \xrightarrow[NaOH(aq)]{Br_2} CH_3CH_2NH_2$$

b.  Since 1-propanamine contains one more carbon atom than ethyl bromide, this suggests the use of a reduction of a nitrile to prepare the amine. Thus we have

$$CH_3CH_2C{\equiv}N \xrightarrow[\text{2. } H_2O, \ ^-OH]{\text{1. } LiAlH_4, \text{ ether}} CH_3CH_2CH_2NH_2$$

Now, we prepare the propionitrile from ethyl bromide.

$$CH_3CH_2Br \xrightarrow{KCN} CH_3CH_2C{\equiv}N$$

Putting this all together, we have

$$CH_3CH_2Br \xrightarrow{KCN} CH_3CH_2C{\equiv}N \xrightarrow[\text{2. } H_2O, \ ^-OH]{\text{1. } LiAlH_4, \text{ ether}} CH_3CH_2CH_2NH_2$$

| | |
|---|---|
| **EXERCISE 25.3** | Synthesize the following compounds. You may use any inorganic reagents necessary.<br><br>a.  1-butanamine from 1-iodopropane<br><br>b.  1-pentanamine from hexanoic acid<br><br>c.  2-propanamine from 2-propanol (use reductive amination)<br><br>d.  ethanamine from ethanol (use either phthalimide or reductive amination) |

| | |
|---|---|
| **EXERCISE 25.4** | Draw structures for a carbonyl compound and an amine that would give the following amines using a reductive amination procedure.<br><br>a.  dipropylamine<br><br>b.  propyldiethylamine (two possible combinations) |

A huge excess of a primary or secondary amine reacts smoothly with an alkyl halide to yield an amine of one higher class, i.e., secondary from primary. This method is used in the laboratory, where cost of the amine is not a factor. However, it is very rarely used on an industrial scale because amines are relatively expensive reagents. In general,

$$R'X + RNH_2 \text{ (huge excess)} \longrightarrow \underset{\underset{R'}{|}}{R\overset{+}{N}H_2X^-}$$

$$\underset{\underset{R'}{|}}{R\overset{+}{N}H_2X^-} + RNH_2 \longrightarrow \underset{\underset{R'}{|}}{RNH} + R\overset{+}{N}H_3X^-$$

where X = Cl, Br, or I. For examples:

$$CH_3Br + CH_3CH_2CH_2NH_2 \text{ (huge excess)} \longrightarrow \underset{\underset{CH_3}{|}}{CH_3CH_2CH_2\overset{+}{N}H_2Br^-}$$

$$\underset{\underset{CH_3}{|}}{CH_3CH_2CH_2\overset{+}{N}H_2Br^-} + CH_3CH_2CH_2NH_2 \longrightarrow \underset{\underset{CH_3}{|}}{CH_3CH_2CH_2NH} + CH_3CH_2CH_2\overset{+}{N}H_3Br^-$$

Comparable equations can be written for a tertiary amine prepared from a secondary amine.

In general,

Amines react with mineral acids to produce isolable salts.

$$\underset{\underset{H(R'')}{|}}{\overset{\overset{H(R')}{|}}{R-N}} + HX \longrightarrow \underset{\underset{H(R'')}{|}}{\overset{\overset{H(R')}{|}}{R-\overset{+}{N}-HX^-}}$$

Example:

$$CH_3CH_2CH_2NH_2 + HBr \longrightarrow CH_3CH_2CH_2\overset{+}{N}H_3Br^-$$
Propylammonium bromide

These ordinary ammonium salts, in turn, can be reacted with a strong base such as potassium hydroxide or sodium hydroxide to regenerate the amine. In general,

$$\underset{\underset{H(R'')}{|}}{\overset{\overset{H(R')}{|}}{R-\overset{+}{N}-HX^-}} + KOH\,(aq) \longrightarrow \underset{\underset{H(R'')}{|}}{\overset{\overset{H(R')}{|}}{R-N}} + H_2O + KX$$

where X = Cl, Br, or I. For example:

$$CH_3CH_2CH_2\overset{+}{N}H_3Br^- + NaOH\,(aq) \longrightarrow CH_3CH_2CH_2NH_2 + NaBr + H_2O$$

Amines, when treated with an appropriate acid chloride or anhydride, produce amides (see Secs. 23.3 and 23.5), i.e., replace ammonia with a primary or secondary amine.

Note that tertiary amines do not react with either an acid chloride or an acid anhydride.

*m*-Methylaniline → *m*-Methylacetanilide + CH₃COOH

$$CH_3NH_2 + \underset{}{Cl-\overset{\overset{O}{\|}}{C}-CH_3} \longrightarrow \underset{}{CH_3\overset{\overset{H}{|}}{N}-\overset{\overset{O}{\|}}{C}-CH_3} + HCl$$
*N*-Methylacetamide

$$\underset{\underset{}{}}{\overset{\overset{CH_3}{|}}{CH_3-N}-H} + \underset{}{Cl-\overset{\overset{O}{\|}}{C}-CH_2CH_3} \longrightarrow \underset{}{\overset{\overset{CH_2}{\|}}{CH_3-N}-\overset{\overset{O}{\|}}{C}-CH_2CH_3} + HCl$$
*N,N*-Dimethylpropionamide

**EXERCISE 25.5**

What structural feature of tertiary amines is the cause of their lack of reactivity with acid chlorides or acid anhydrides?

**FIGURE 25.2** Benzenesulfonyl chloride.

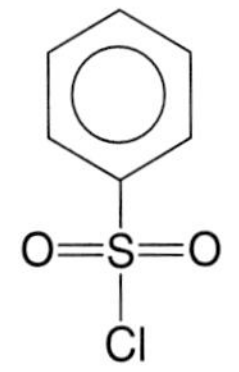

Amines, in addition to reacting with acid chlorides derived from carboxylic acids, also react with acid chlorides derived from aromatic sulfonic acids. One such often used acid chloride is benzenesulfonyl chloride (Fig. 25.2). Amines react with benzenesulfonyl chloride to produce sulfonamides. In the presence of base, this reaction, the basis of the *Hinsberg test*, is used to distinguish between primary, secondary, and tertiary amines. In general,

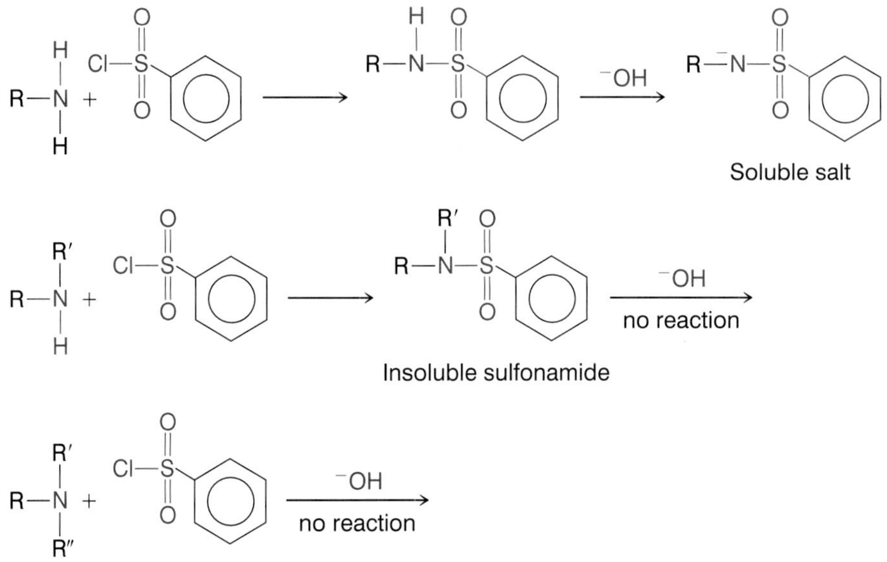

Soluble salt

Insoluble sulfonamide

Examples:

*N*-Methylbenzenesulfonamide

*N*-Methyl-*N*-phenylbenzenesulfonamide

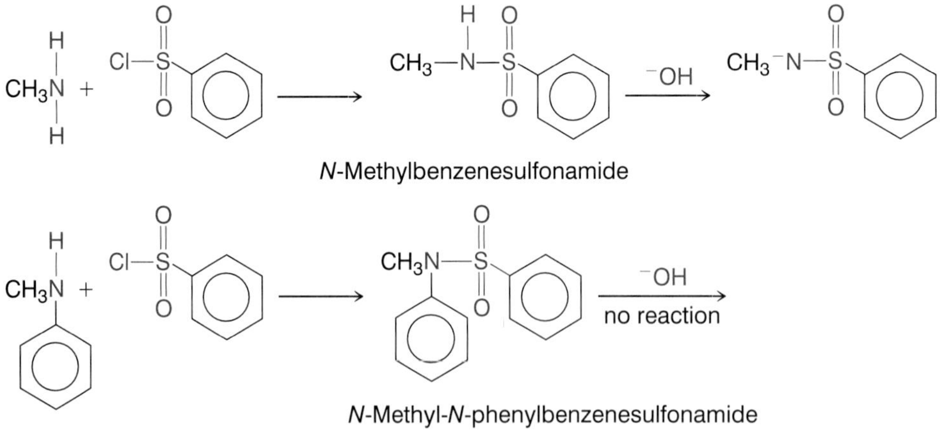

A primary amine reacts with benzenesulfonyl chloride and base to produce a sulfonamide that is soluble in base. Thus a primary amine is distinguished by the heat evolved in the reaction and the formation of a solution. A sulfonamide derived from a primary amine is far more acidic ($pK_a \approx 10$) than the corresponding carboxylic acid amide because the anion of the sulfonamide is resonance stabilized to a greater extent than the carboxylic acid amide.

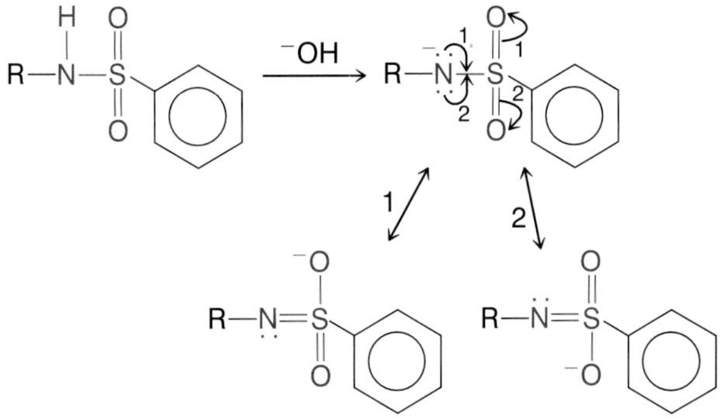

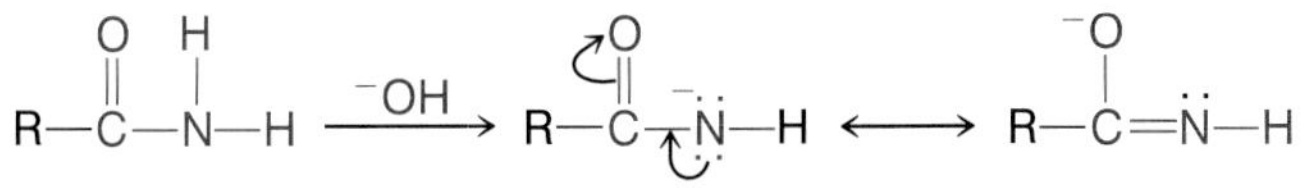

Note that the sulfonamide anion is stabilized by three resonance structures, whereas the carboxylic acid amide anion is stabilized by only two resonance structures.

A secondary amine reacts with benzenesulfonyl chloride and base to produce an *N,N*-substituted benzenesulfonamide that is an insoluble solid. Thus a secondary amine is distinguished by the heat evolved in the reaction and the formation of a precipitate. A tertiary amine does not react with benzenesulfonyl chloride and base. Thus a tertiary amine is distinguished by the lack of heat evolved and the fact that the state of matter of the amine is not changed.

Sulfanilamide (Medical Boxed Reading 24.2), the first antibiotic, is a sulfonamide (Fig. 25.3).

Nitrous acid is a weak, unstable acid. The acid, therefore, must be produced in the reaction mixture. The reagents often used to accomplish this are sodium nitrite and hydrochloric acid.

## Reaction with Nitrous Acid

$$NaNO_2 + HCl \longrightarrow HNO_2 + NaCl$$

Nitrous<br>acid

**FIGURE 25.3** The first antibiotic.

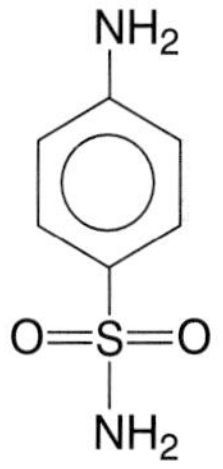

*p*-Aminobenzenesulfonamide
(sulfanilamide)

Primary aliphatic amines react with nitrous acid to give nitrogen gas, water, and a mixture of the corresponding alcohol, alkene, and alkyl chloride. In general,

$$RCH_2CH_2NH_2 + NaNO_2 + HCl \longrightarrow$$
$$RCH_2CH_2OH + RCH_2CH_2Cl + RCH{=}CH_2 + N_2 + NaCl + H_2O$$

(This equation is not balanced.) For example:

$$CH_3CH_2NH_2 + NaNO_2 + HCl \longrightarrow$$
$$CH_3CH_2OH + CH_3CH_2Cl + CH_2{=}CH_2 + N_2 + NaCl + H_2O$$

This reaction has no synthetic utility because a wide variety of organic products is obtained, each one in poor yield. However, a quantitative amount of nitrogen is evolved. This is the basis of an analysis of nitrogen in amino acids and proteins (see Chap. 28).

Synthetically important diazonium salts are produced when a primary aromatic amine is treated with nitrous acid. The temperature of the reaction mixture is kept low, about 5°C, because aryl diazonium salts are reasonably stable at that temperature; at higher temperatures, the diazonium salts decompose to produce the corresponding phenol (see Sec. 25.4). In general,

$$ArNH_2 + NaNO_2 + 2\,HCl \xrightarrow{5°C} Ar{-}\overset{+}{N}{\equiv}NCl^- + NaCl + 2\,H_2O$$

Example:

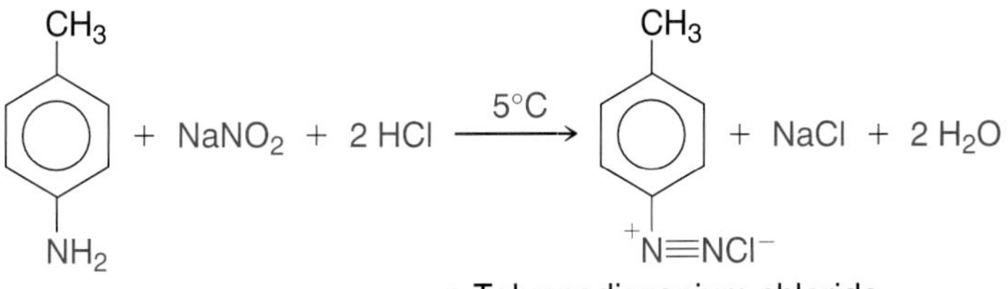

*p*-Toluenediazonium chloride

The reactions of diazonium salts are discussed in Section 25.4. Diazonium salts are prepared in the laboratory by treating the amine with hydrochloric acid, chilling the reaction mixture to 5°C, and adding a solution of sodium nitrite.

$$ArNH_2 + HCl \longrightarrow Ar\overset{+}{-}NH_3Cl^- \xrightarrow[5°C]{NaNO_2} Ar\overset{+}{-}N\equiv NCl^-$$

An aryldiazonium
chloride

The formation of a telltale yellow oil of a *N*-nitrosamine results when a secondary amine reacts with nitrous acid. In general,

$$\underset{\underset{\overset{|}{R}}{}}{R}\overset{\overset{R'}{|}}{-}N-H + NaNO_2 + HCl \longrightarrow R-\overset{\overset{R'}{|}}{N}-N=O + NaCl + H_2O$$

Example:

$$CH_3-\overset{\overset{CH_3}{|}}{N}-H + NaNO_2 + HCl \longrightarrow CH_3-\overset{\overset{CH_3}{|}}{N}-N=O + NaCl + H_2O$$

Dimethylamine                Dimethylnitrosamine
(*N*-nitrosodimethylamine)

The reaction of secondary amines with nitrous acid has serious negative health implications. Sodium nitrite is added to meat to prevent the growth of *Clostridium botulinum*, the bacterium responsible for botulism. Also, it maintains the red color of the meat to make it more attractive to the consumer. In the stomach, nitrite ion is converted to nitrous acid by means of acidic gastric juice.

$$NO_2^- + H^+ \text{ (gastric juice)} \longrightarrow HNO_2$$

The nitrous acid, in turn, combines with secondary amines in the meat to give carcinogenic nitrosamines.

$$\underset{\underset{R}{|}}{\overset{\overset{R'}{|}}{N}}H + HNO_2 \longrightarrow \underset{\underset{R}{|}}{\overset{\overset{R'}{|}}{N}}N=O + H_2O$$

Because it has been well established that nitrosamines are carcinogenic in animals, and because it is likely that nitrosamines are carcinogenic in humans, it would be prudent for you to consider a limit on the amount of processed meat (bacon, hot dogs, ham, bologna, salami, etc.) you consume.

We will not study the reaction of tertiary amines with nitrous acid because this topic is of lesser importance.

**SOLVED PROBLEM 25.2**

Complete the following equations by drawing a structural formula for each organic product and writing a molecular formula or formula unit for each inorganic product. If no reaction takes place, write "no reaction."

a.  $(C_2H_5)_2NH + NaNO_2 + HCl \longrightarrow$

b.  $CH_3CH_2CH_2-NH_2 + CH_3-\overset{\overset{}{\underset{\underset{O}{\|}}{}}}{C}-Cl \longrightarrow$

c.  $(CH_3)_3N + CH_3-\overset{\overset{}{\underset{\underset{O}{\|}}{}}}{C}-O-\overset{\overset{}{\underset{\underset{O}{\|}}{}}}{C}-CH_3 \longrightarrow$

**SOLUTION**

a. Diethylamine is a secondary amine. Since this secondary amine is reacted with nitrous acid, we get

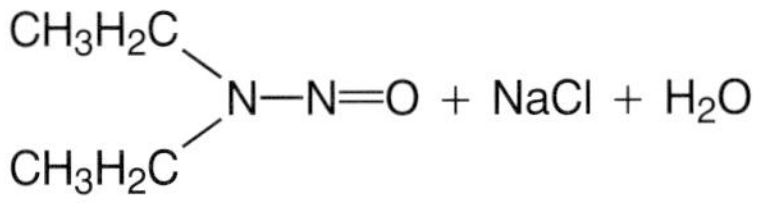

Diethylnitrosamine
(*N*-nitrosodiethylamine)

b. Primary amines react with acyl halides to give the corresponding amide.

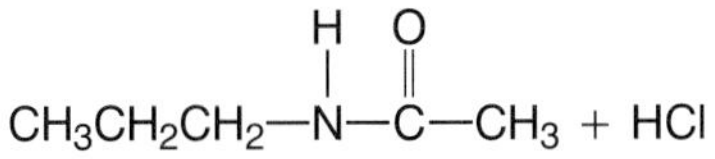

$$CH_3CH_2CH_2-N-C-CH_3 + HCl$$

c. Since there is no hydrogen bonded to the nitrogen of the amine, a tertiary amine does not react with an acid anhydride, and we have no reaction.

---

**EXERCISE 25.6**

Complete the following equations by drawing a structural formula for each organic product and writing a molecular formula or formula unit for each inorganic product. If no reaction takes place, write "no reaction."

a. methylamine + acetic anhydride $\longrightarrow$

b. diphenylamine (huge excess) + methyl iodide $\longrightarrow$

c. triethylamine + methyl bromide $\longrightarrow$

d. tribenzylamine + butyryl chloride $\longrightarrow$

---

**25.3
REACTIONS OF
AROMATIC AMINES**

Aromatic amines undergo all the characteristic reactions of aliphatic and alicyclic amines and in addition participate in reactions involving the aromatic ring as follows.

**Bromination**

Because the amino group is a powerful activating group (see Table 10.1), reacting an aromatic amine with bromine water leads to a polysubstituted brominated product, substituting in every position ortho or para to the amino group. For example, reacting *p*-ethylaniline with bromine water gives a precipitate of 2,6-dibromo-4-ethylaniline.

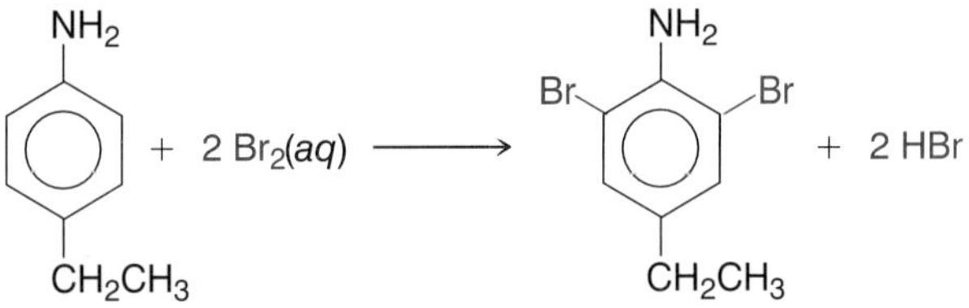

In order to limit the number of bromines entering the ring, the structure of the amino group must be modified. This is accomplished by reacting the aromatic amine with acetyl chloride or acetic anhydride to produce an amide.

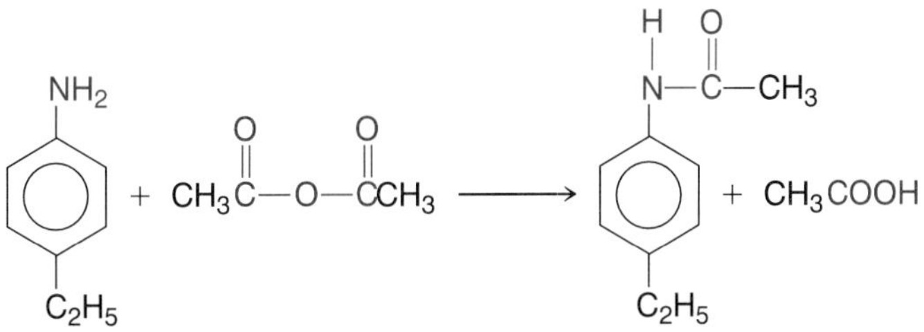

The acetamido group (—NHCOCH₃) (see Table 10.1) is a less activating ortho-para orienting group than the amino group. Thus a single bromine can be introduced into the aromatic ring. This process of replacing the amino group with the acetamido group is known as *"protecting" the amino group.*

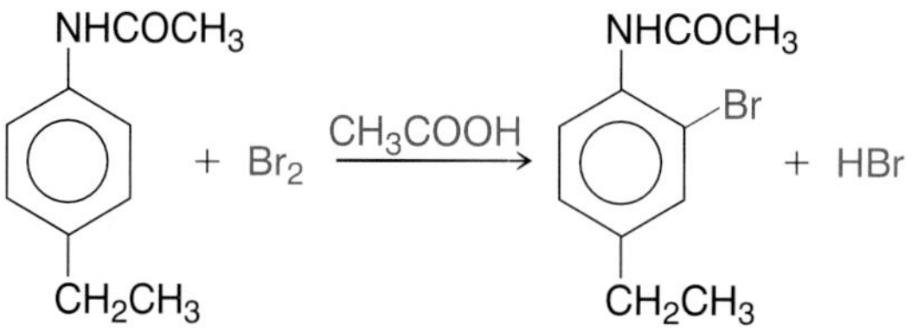

Once the bromine group has been introduced into the ring, the protecting group is removed to produce the desired amine.

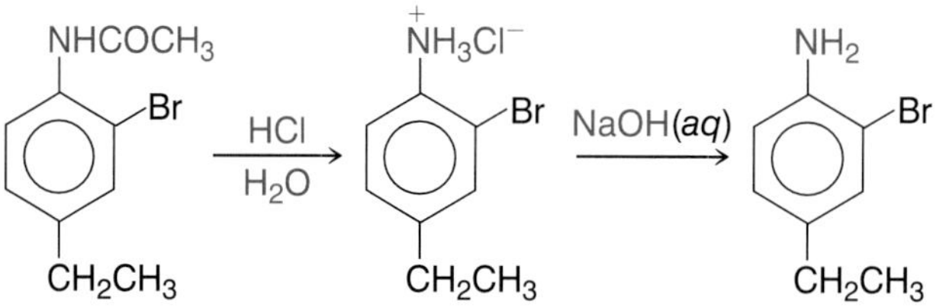

| **SOLVED PROBLEM 25.3** |
|---|

Reacting an aromatic primary or secondary amine with nitric acid produces significant oxidation of the amino group to produce black tarlike material. Thus, to introduce a nitro group into the benzene ring, we must "protect" the amino group. With this in mind, prepare *p*-nitroaniline from aniline. You may use any organic and inorganic compounds necessary.

**SOLUTION**

First, protect the amino group by reacting aniline with acetic anhydride.

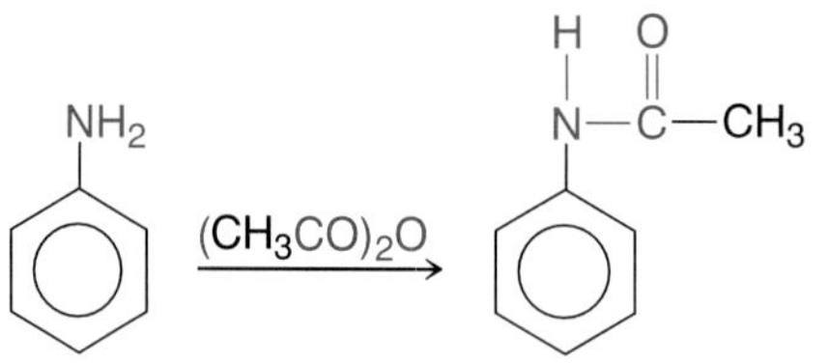

Now the nitro group is introduced by reacting acetanilide with a mixture of nitric and sulfuric acids.

As usual, assume that we can separate an ortho-substituted compound from its para isomer. Thus we assume we can separate *p*-nitroacetanilide from its isomer.

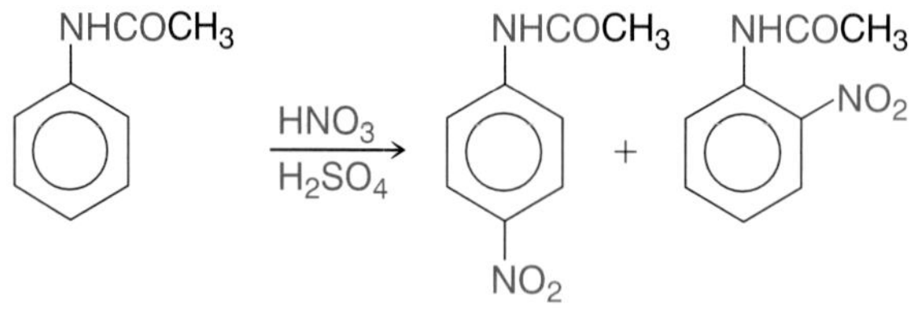

The para isomer is then hydrolyzed to produce *p*-nitroaniline.

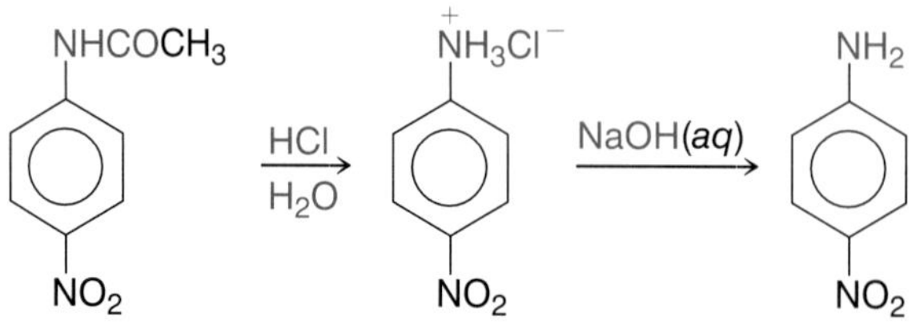

Putting all these steps together, we have

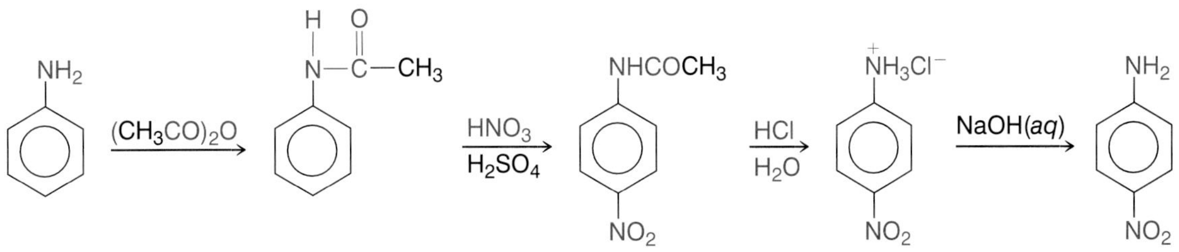

Since all our previously solved synthesis problems have been solved by working backwards, let us do the same for this problem. We begin by preparing p-nitroaniline from p-nitroacetanilide, since we need to eliminate the "protecting" acetamido group.

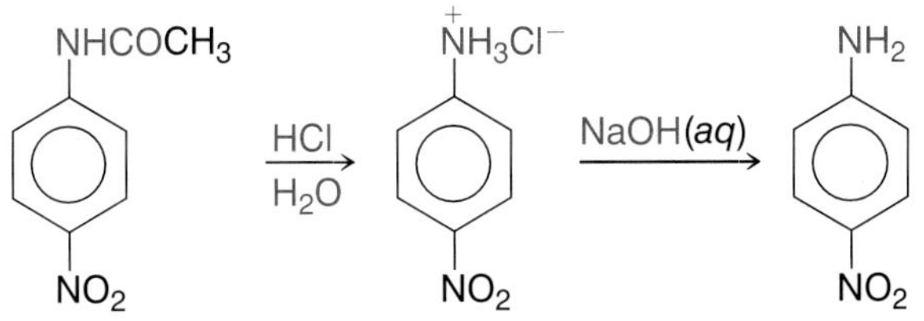

We produce p-nitroacetanilide by reacting acetanilide with nitric acid in the presence of sulfuric acid. We can assume that the ortho isomer can be removed.

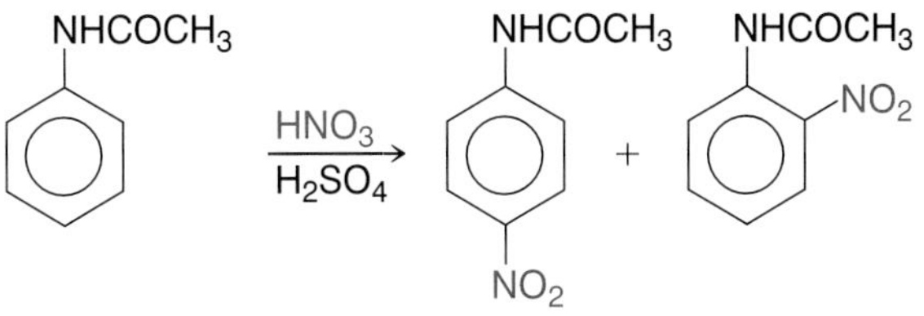

Finally, acetanilide is prepared by reacting aniline with acetic anhydride in order to protect the amino group.

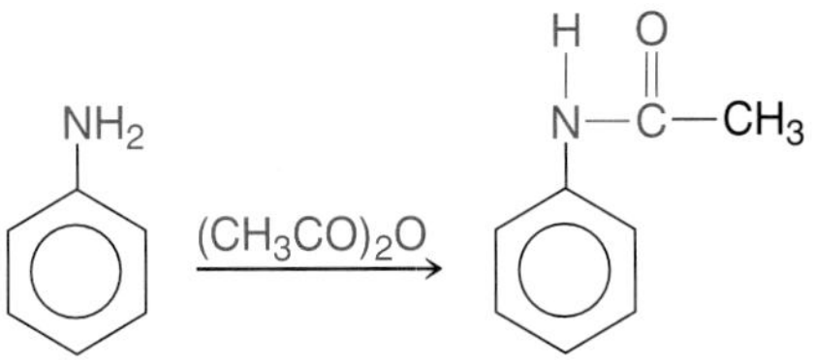

**EXERCISE 25.7**

Synthesize 2-bromo-4-methylaniline from toluene. You may use any organic and inorganic reagents necessary.

**Sulfonation**

When aniline is heated with sulfuric acid, sulfanilic acid is produced.

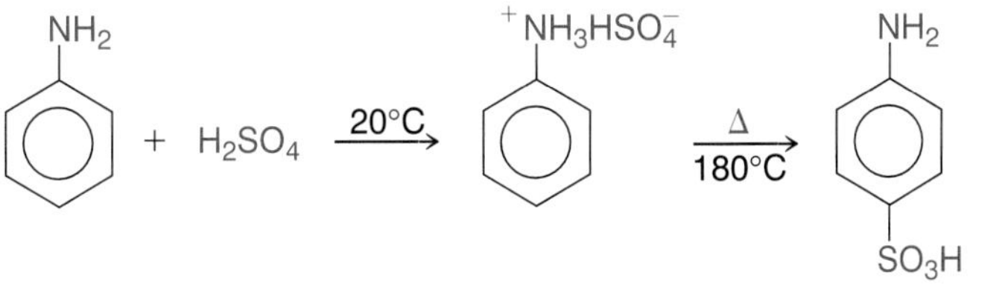

Anilinium hydrogen sulfate                 Sulfanilic acid

Sulfanilic acid is interesting in that both an acid group ($-SO_3H$) and a basic group ($-NH_2$) are present in the molecule. The consequences of this type of group presence will be discussed in depth in Chapter 28. Note that the more powerful fuming sulfuric acid reagent (see Sec. 10.1) is not needed here because the amino group is a strongly activating group. In a similar way, we can see that by reacting acetanilide, containing the moderately activating acetamido group —$NHCOCH_3$, with sulfuric acid, then heating gives *p*-acetamidobenzenesulfonic acid.

Acetanilide *p*-Acetamidobenzenesulfonic acid

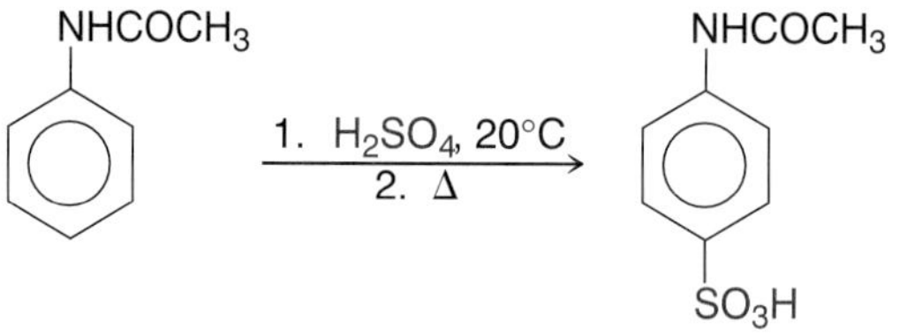

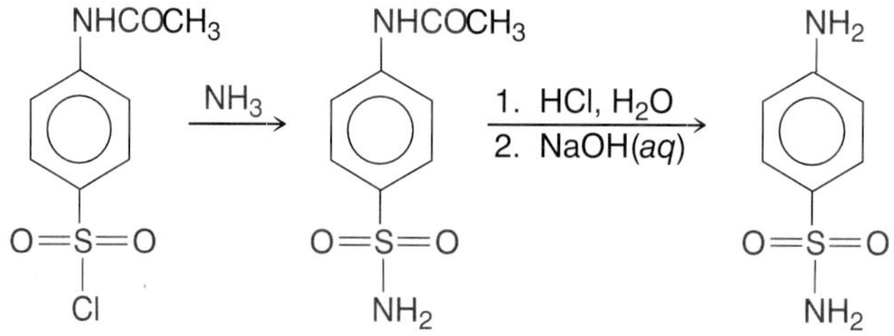

Using the reactions we have discussed, the drug sulfanilamide can now be prepared.

Acetanilide

*p*-Acetamidobenzenesulfonamide

*p*-Acetamidobenzenesulfonyl chloride

Sulfanilamide

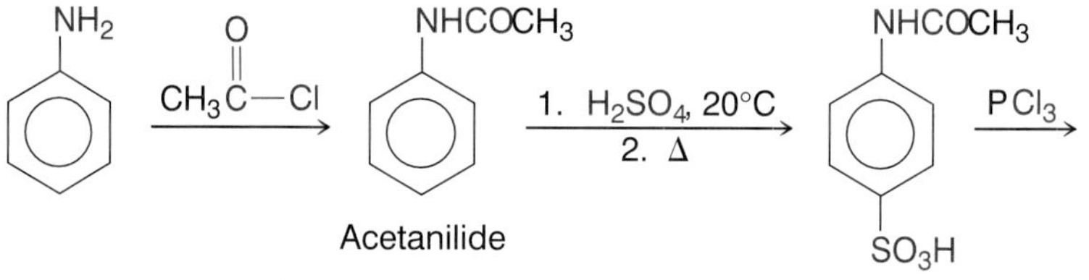

Note that an arylsulfonyl chloride is prepared from an arenesulfonic acid using the same reagents ($SOCl_2$ or $PCl_3$) that are used to prepare a carboxylic acid chloride from a carboxylic acid.* In general,

$$Ar{-}SO_3H + SOCl_2 \longrightarrow Ar{-}SO_2Cl + HCl + SO_2$$
$$3\ Ar{-}SO_3H + PCl_3 \longrightarrow 3\ Ar{-}SO_2Cl + H_3PO_3$$

Example:

$$C_6H_5{-}SO_3H + SOCl_2 \longrightarrow C_6H_5{-}SO_2Cl + HCl + SO_2$$

*For the sake of convenience and clarity, this text assumes that an arenesulfonyl chloride is prepared from the aromatic sulfonic acid using thionyl chloride or phosphorus trichloride. In actuality, the preferred method of preparation involves reaction of an aromatic compound with chlorosulfonic acid ($ClSO_3H$). For example, consider the reaction of acetanilide with chlorosulfonic acid.

Acetanilide *p*-Acetamidobenzenesulfonyl chloride

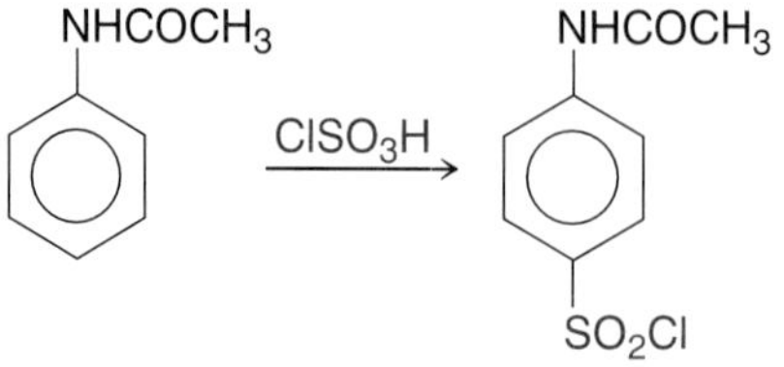

Aromatic primary and secondary amines are very sensitive to oxidation. Thus oxidation takes place with as mild an oxidizing agent as oxygen in air. When reacted with an oxidizing agent, these amines give a variety of different colored products. For example, aniline darkens gradually in air due to oxidation. Alphatic amines are more stable to oxidation.

Diazonium salts undergo a wide variety of substitution reactions that have great synthetic importance. Let us first examine the reactions and then consider the importance of the reactions in synthesis. The reactions are, in general, as follows.

$$Ar-\overset{+}{N}\equiv NCl^- \xrightarrow[95°C]{KI} Ar-I + N_2$$

Heating a diazonium salt with potassium iodide gives the corresponding aryl iodide.

$$Ar-\overset{+}{N}\equiv NCl^- \xrightarrow{CuCl(Br)(CN)} Ar-Cl(Br)(CN) + N_2$$

$$Ar-\overset{+}{N}\equiv NCl^- \xrightarrow[40°C]{H_2O} Ar-OH + N_2$$

**Reaction with Water**

Note: The reaction, when cuprous chloride, bromide, or cyanide is used, is called the *Sandmeyer reaction*.

Warming the aqueous diazonium salt gives the corresponding phenol.

$$Ar-\overset{+}{N}\equiv NCl^- \xrightarrow{R-OH} Ar-O-R + N_2$$

Reacting the diazonium salt with an alcohol produces the corresponding ether. Methyl and ethyl alcohols are most often used.

$$Ar-\overset{+}{N}\equiv NCl^- \xrightarrow{H_3PO_2} Ar-H + N_2$$

This reaction represents an effective method of removing an amino group from the benzene ring.

Examples:

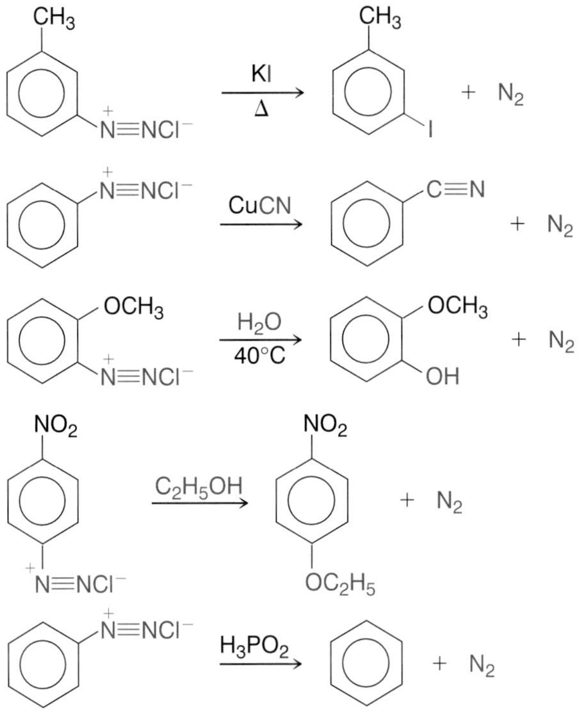

## Coupling with an Arylamine or Phenol

*Coupling also takes place in the ortho position.*

A substitution reaction that diazonium salts undergo during which nitrogen gas is not evolved is called *coupling* or *azo dye formation*. A coupling reaction takes place when a diazonium salt is treated with an arylamine or phenol. In general,

$$Ar-\overset{+}{N}\equiv NCl^- + Ar'NH_2 \longrightarrow Ar-N=N-Ar'-\textit{p}-NH_2 + HCl$$

An azo dye

$$Ar-\overset{+}{N}\equiv NCl^- + Ar'OH \longrightarrow Ar-N=N-Ar'-\textit{p}-OH + HCl$$

An azo dye

Examples:

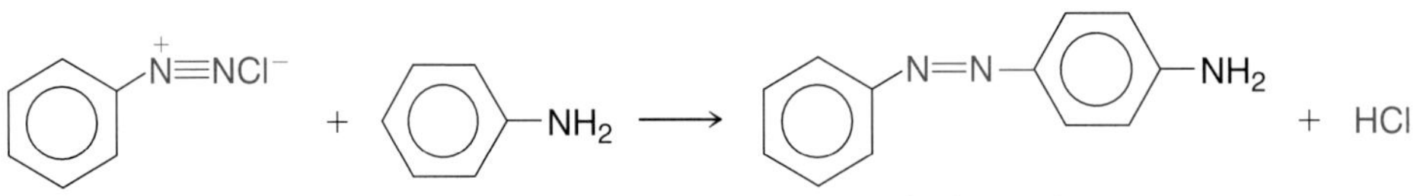

*p*-Aminoazobenzene

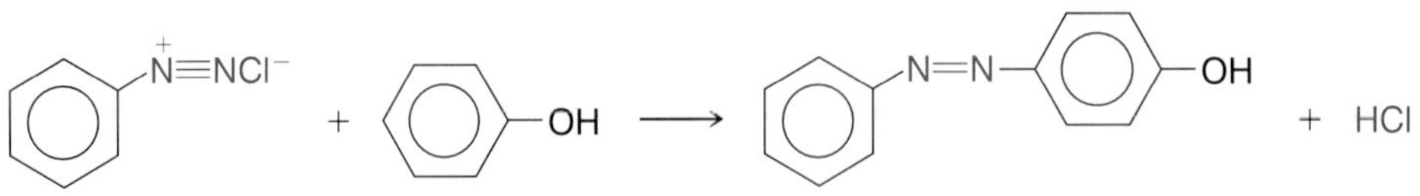

**FIGURE 25.4** Azobenzene.

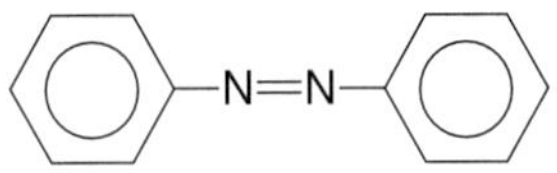

*p*-Hydroxyazobenzene

Note that each of the preceding uncomplicated azo dyes is named as a derivative of azobenzene (Fig. 25.4).

**EXERCISE 25.8**

Given the structural formula of azobenzene, draw a structural formula for azomethane. Also draw the formula for the azo group.

Azo dyes are intensely colored (red, orange, yellow, and blue). These deep colors exist due to extensive conjugation within the different dye molecules.

A few points of interest are in order about the coupling reaction:

1. Coupling occurs para to the —NH$_2$ or —OH group. If the para position is occupied by another group, coupling ortho to the —NH$_2$ or —OH group takes place.

2. The mechanism of the coupling reaction is as follows.

**Envision the Reaction**

### Condensed Mechanism of a Coupling Reaction in Which Benzenediazonium Chloride Reacts with Phenol

First step:

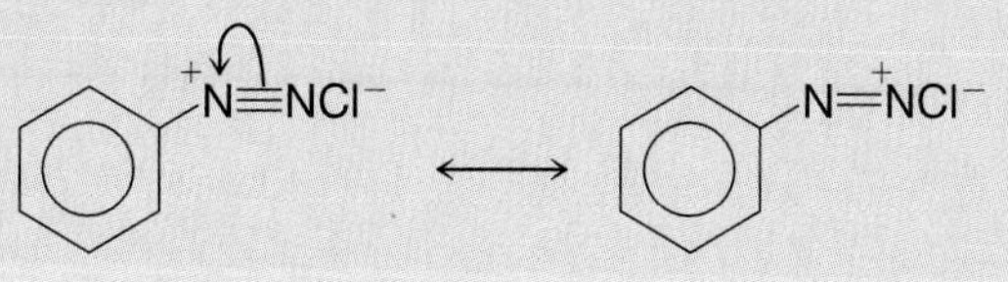

Second step:

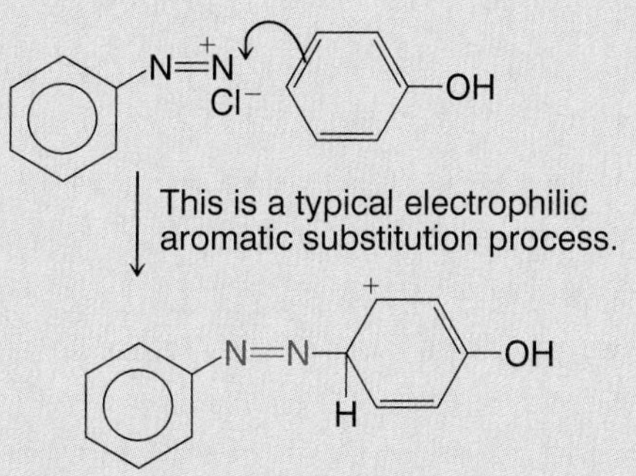

This is a typical electrophilic aromatic substitution process.

Third step:

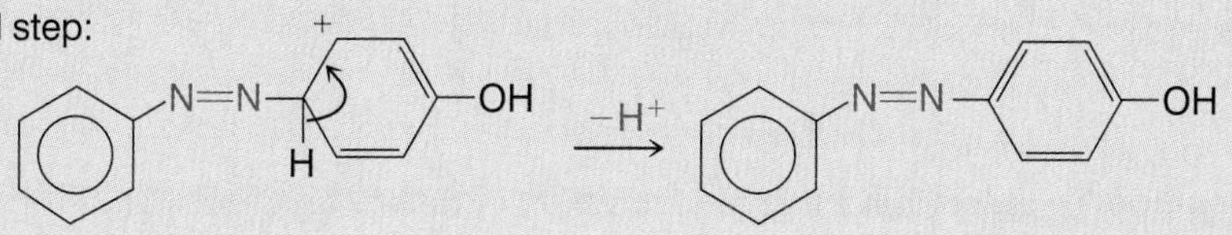

Note that coupling only takes place when a diazonium salt is treated with an aromatic amine of the type Ar—NH$_2$, Ar—NH(R), and Ar—NR$_2$ or a phenol. This is due to the fact that the Ar—N≡N$^+$ species is a weak electrophile that requires an increased electron density in the benzene ring for attack to occur. Only strong activating groups provide the increased electron density necessary for reaction.

The use of diazonium salts in organic synthesis is extensive because synthetic products are obtained with group placement that cannot be provided in any other way. Consider Solved Problem 25.4.

**SOLVED PROBLEM 25.4**

Prepare *m*-chloroiodobenzene from benzene. You may use any inorganic reagents necessary.

#### SOLUTION

This is an example of a synthesis where the diazonium salt technique is found to be so valuable. Both —Cl and —I groups orient ortho-para, yet the groups are meta to each other in the compound to be prepared. This is the power of a diazonium salt synthesis, i.e., to enable a given compound to be prepared with a group placement that cannot be obtained by ordinary electrophilic substitution. The plan of the synthesis is as follows: By preparing nitrobenzene, we have a meta-orienting group on the benzene ring. Reacting nitrobenzene with chlorine in the presence of iron affords *m*-chloronitrobenzene, which, in turn, is reduced to the corresponding amine. Diazotizing *m*-chloroaniline, reacting the diazonium salt with KI, and heating give *m*-chloroiodobenzene.

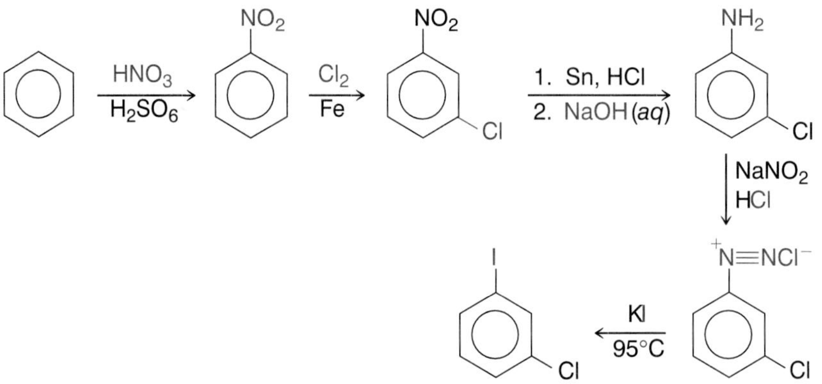

$$\text{benzene} \xrightarrow[\text{H}_2\text{SO}_6]{\text{HNO}_3} \text{nitrobenzene} \xrightarrow[\text{Fe}]{\text{Cl}_2} \text{m-chloronitrobenzene} \xrightarrow[\text{2. NaOH}(aq)]{\text{1. Sn, HCl}} \text{m-chloroaniline}$$

$$\xrightarrow[\text{HCl}]{\text{NaNO}_2}$$

$$\text{m-chlorobenzenediazonium} \xrightarrow[95°\text{C}]{\text{KI}} \text{m-chloroiodobenzene}$$

Now, let us reverse this procedure and work this problem backwards. Since iodine can only be introduced into the benzene ring by diazotization, let us diazotize *m*-chloroaniline and heat the diazonium salt with KI to give product.

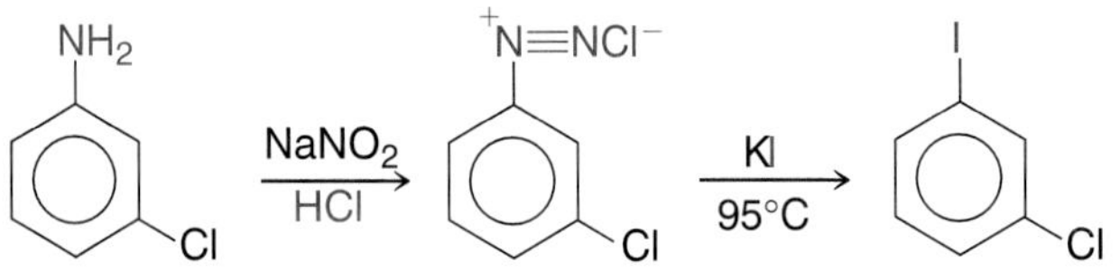

$$\text{m-chloroaniline} \xrightarrow[\text{HCl}]{\text{NaNO}_2} \text{diazonium salt} \xrightarrow[95°\text{C}]{\text{KI}} \text{product}$$

To produce *m*-chloroaniline, reduce *m*-chloronitrobenzene with tin and hydrochloric acid, followed by hydrolysis with aqueous sodium hydroxide.

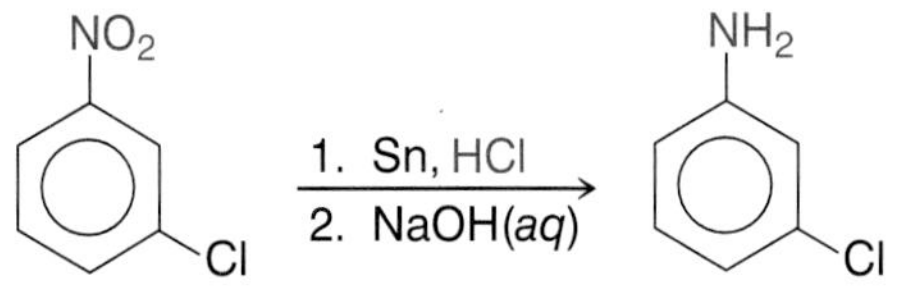

$$\text{m-chloronitrobenzene} \xrightarrow[\text{2. NaOH}(aq)]{\text{1. Sn, HCl}} \text{m-chloroaniline}$$

The compound, *m*-chloronitrobenzene, in turn, is produced by reacting nitrobenzene with chlorine in the presence of metallic iron.

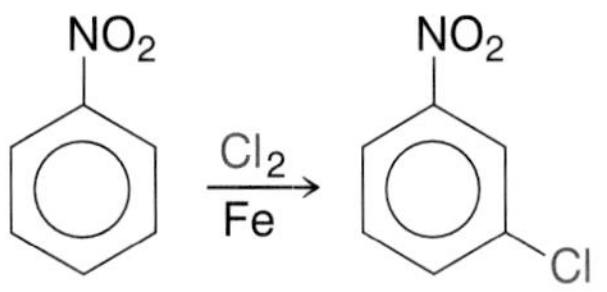

$$\text{nitrobenzene} \xrightarrow[\text{Fe}]{\text{Cl}_2} \text{m-chloronitrobenzene}$$

Finally, benzene is treated with nitric acid in the presence of sulfuric acid to yield nitrobenzene.

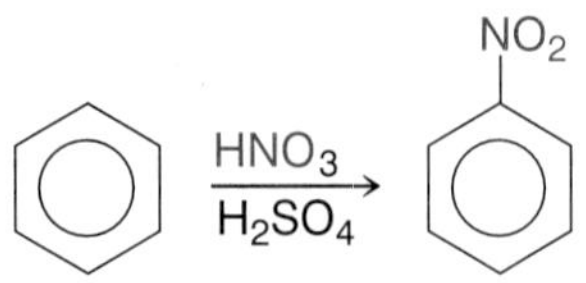

$$\text{benzene} \xrightarrow[\text{H}_2\text{SO}_4]{\text{HNO}_3} \text{nitrobenzene}$$

---

**EXERCISE 25.9**

Prepare the following compounds. You may use any organic and inorganic reagents necessary.

a. benzene from aniline

b. benzoic acid from nitrobenzene

c. *o*-bromophenol from 2-bromo-4-nitroaniline

# CHAPTER ACCOMPLISHMENTS

## 25.1 Synthesis of Amines
☐ Prepare ethylamine from ethyl bromide and ammonia.
☐ List the limitations of the alkylation method of preparing amines.
☐ Prepare tetraethylammonium bromide from triethylamine and ethyl bromide.
☐ Draw a structural formula for phthalimide.
☐ Prepare ethylamine from ethyl iodide using the Gabriel phthalimide method of synthesis.
☐ Write an equation illustrating the reduction of each type of nitrogen-containing compound to obtain an amine.
☐ Draw a structural formula for acetone imine.
☐ Write a detailed, step-by-step mechanism of the following reaction.

$$CH_3CH_2-\overset{\overset{\displaystyle O}{\|}}{C}-NH_2 + Br_2 + {}^-OH$$
$$\longrightarrow CH_3CH_2NH_2 + Na_2CO_3 + Br^- + H_2O$$

## 25.2 Chemical Properties of Amines
☐ Complete the following equations by drawing a structural formula for each organic product and writing a molecular formula or formula unit for each inorganic product.

a. $CH_3Br + CH_3CH_2CH_2NH_2$ (huge excess) $\longrightarrow$

b. $CH_3CH_2CH_2NH_2 + HBr \longrightarrow$

c. $CH_3CH_2CH_2\overset{+}{N}H_3Br^- + NaOH(aq) \longrightarrow$

d. $CH_3NH_2 + CH_3-\overset{\overset{\displaystyle O}{\|}}{C}-Cl \longrightarrow$

e. $(CH_3)_2NH + CH_3CH_2-\overset{\overset{\displaystyle O}{\|}}{C}-O-\overset{\overset{\displaystyle O}{\|}}{C}-CH_2CH_3 \longrightarrow$

f. $C_6H_5SO_2Cl + CH_3CH_2NH_2 \xrightarrow{NaOH(aq)}$

g. $CH_3CH_2NH_2 + NaNO_2 + HCl \longrightarrow$

h. $C_6H_5NH_2 \xrightarrow[\text{2. NaNO}_2]{\text{1. HCl}}$

i. $(CH_3CH_2)_2NH + NaNO_2 + HCl \longrightarrow$

## 25.3 Reactions of Aromatic Amines
☐ Prepare p-nitroaniline from aniline using any necessary inorganic and organic reagents.
☐ Draw a structural formula for
  a. sulfanilic acid.
  b. sulfanilamide.

## 25.4 Reactions of Diazonium Salts
☐ Write an equation representing each substitution reaction that $Ar-\overset{+}{N}\equiv NCl^-$ undergoes.
☐ Draw a structural formula for p-aminoazobenzene, an azo dye.
☐ Draw a detailed, step-by-step mechanism for the reaction of benzenediazonium chloride and phenol.
☐ Prepare m-chloroiodobenzene from benzene using any inorganic reagents necessary.

# KEY TERMS

quaternary ammonium salt (25.1)      "protecting" the amino group (25.3)      coupling (25.4)
Hoffmann rearrangement (25.1)      Sandmeyer reaction (25.4)      azo dye formation (25.4)
Hinsberg test (25.2)

# PROBLEMS

1. Write an equation to show the reaction of n-propylamine with each of the following. Draw a structural formula for each organic product produced.
  a. HNO₂

  b. 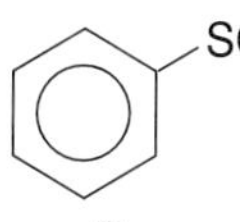  SO₂Cl, ⁻OH

  c. $CH_3\overset{\overset{\displaystyle O}{\|}}{C}-Cl$

  d. $CH_3\overset{\overset{\displaystyle O}{\|}}{C}-O-\overset{\overset{\displaystyle O}{\|}}{C}-CH_3$

  e. HCl

2. Write an equation to show the reaction of aniline with each of the following. Draw a structural formula for each organic product produced.
  a. 1. H₂SO₄, 20°C; 2. Δ

  b. $CH_3\overset{\overset{\displaystyle O}{\|}}{C}-Cl$

c. $CH_3\overset{\overset{O}{\|}}{C}-O-\overset{\overset{O}{\|}}{C}CH_3$

d. $H_2SO_4$, 20°C

e. $HNO_2$, 5°C

f.
benzene ring with $SO_2Cl$, $^-OH$

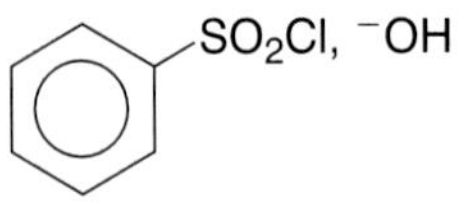

3. How would you distinguish between each compound of the following pairs using a test tube reaction. Write what reagents you would add and the experimental results you would observe.

   a. $(CH_3CH_2)_3N$ and $(CH_3CH_2)_2NH$

   b.
   aniline ($NH_2$) and benzylamine ($CH_2NH_2$)

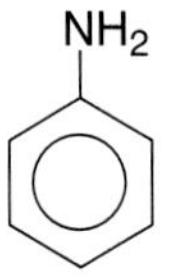

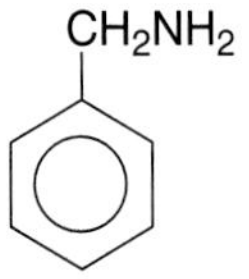

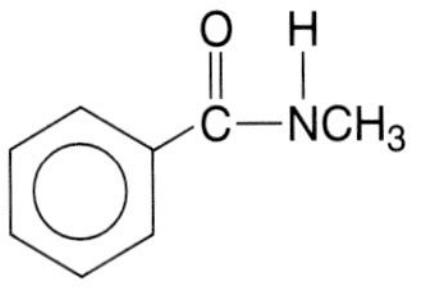

   c.
   N-methylbenzamide and aniline ($NH_2$)

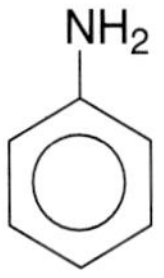

4. Prepare each of the following, starting with o-nitrotoluene, using a diazonium salt reaction. You may use any organic or inorganic reagents necessary.

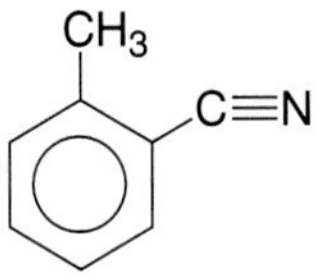

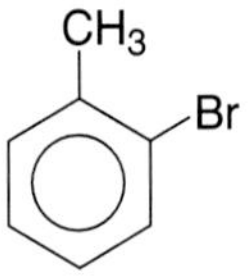

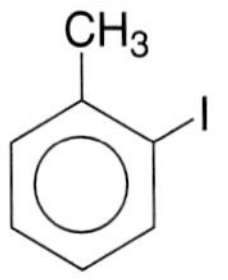

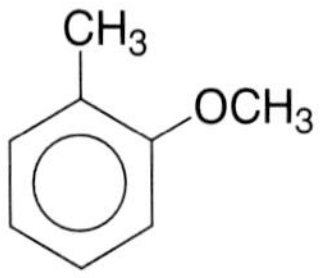

   a.
   toluene with $C{\equiv}N$

   b.
   toluene with I

   c.
   toluene

   d.
   toluene with Br

   e.
   toluene with $OCH_3$

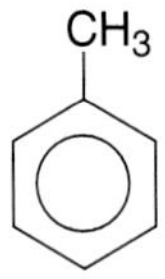

   f.
   toluene with OH

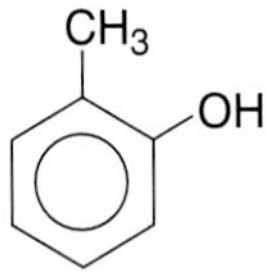

5. Explain why the nitration of aniline represents a poor method of preparing either o-nitroaniline or p-nitroaniline.

6. Synthesize each of the following. You may use any reagents necessary.

   a. 2,4-dinitroaniline from aniline 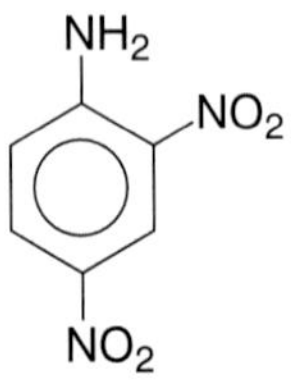
   2,4-Dinitroaniline

   b. acetylcholine chloride (salt of acetylcholine, a neurotransmitter) from 2-aminoethanol
   $(CH_3)_3\overset{+}{N}C\,H_2CH_2O\,\overset{\overset{O}{\|}}{C}\,CH_3Cl^-$
   Acetylcholine chloride

   c. putrescine from ethylene.

   d.
   from

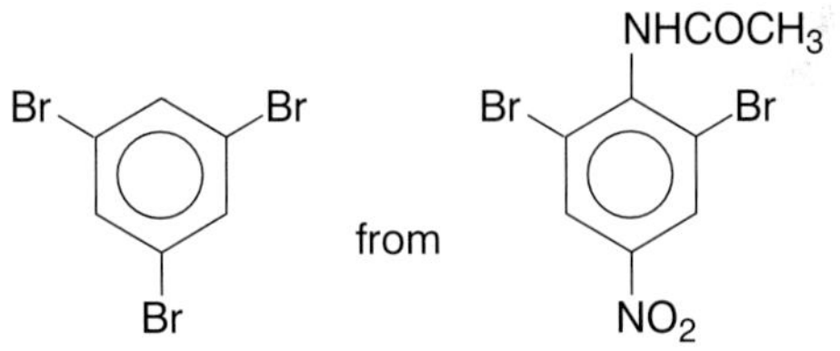

   e. acetaminophen from benzene. Use a method to produce —OH that is different from the one used in Problem 23.15.
   Acetaminophen 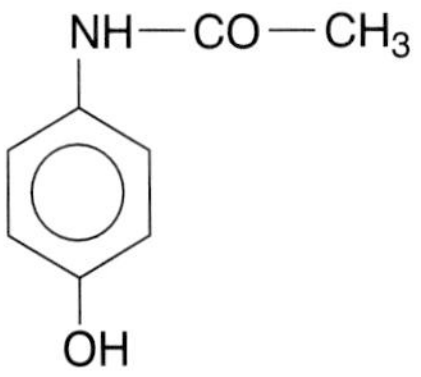

   f.
   from

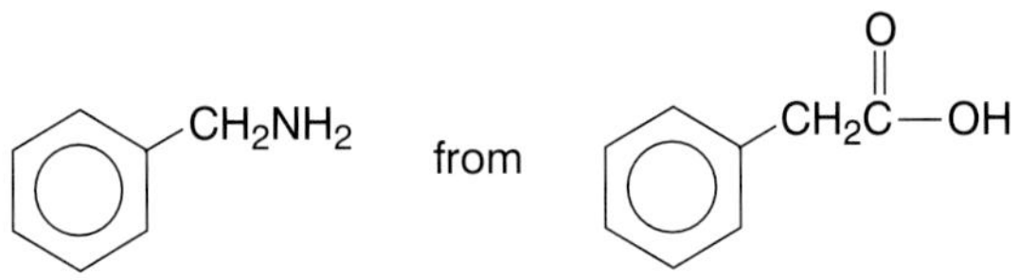

   g. ethylamine from ethyl bromide using the Gabriel synthesis.

   h. tri-n-propylamine from propylene

7. An amine ($C_3H_9N$) gave a yellow oil when treated with nitrous acid. Draw the structural formula of this amine.

8. Prepare propanamine from each of the following compounds. You may use any inorganic reagents necessary.

   a. n-propyl alcohol (Use three different methods; one of which requires PCC or phthalimide.)
   b. propionitrile
   c. n-butyl alcohol
   d. ethyl alcohol
   e. propanamide
   f. propylene and phthalimide

9. Synthesize each of the following compounds. You may use any inorganic and organic compounds necessary.

   a. 2,6-dibromo-4-methylaniline from toluene
   b. tetraethylammonium bromide from ethyl bromide
   c. sulfanilamide from benzene

10. In the laboratory, 1 mol of aniline reacts with 3 mol of bromine in water to produce 3 mol of hydrogen bromide and an organic product. Draw a structural formula for the organic product and name the compound.

11. An amine ($C_3H_9N$) did not react with benzenesulfonyl chloride and then base. Draw a structural formula for the amine, and show your reasoning.

12. When aniline was reacted with a mixture of nitric and sulfuric acids, a substantial amount of black tarry material was produced. However, to the surprise of the chemist, all three mononitro compounds were isolated, each in very low yield. Explain why all three mononitro isomers were isolated.

13. Synthesize each of the following. You may use any organic and inorganic reagents necessary.
    a. *m*-nitrotoluene from benzene and methyl chloride
    b. phenacetin from acetaminophen (Problem 25.6e) (Phenacetin, *p*-ethoxyacetanilide, is a mild analgesic and antipyretic.)

14. Draw a structural formula of a(an)
    a. cyclic imide.
    b. imine.
    c. *N*-nitrosamine.

15. Butter yellow [*p*-(dimethylamino)azobenzene] is an azo dye that at one time was used to color margarine. Because it was proven to be carcinogenic, its use was prohibited by the Food and Drug Administration.
    a. Draw a Kekulé structural formula for butter yellow.
    b. Verify for yourself that there exists a significant amount of conjugation in this molecule to classify it as a dye.

16. Consider again the nitration of acetanilide as a part of Solved Problem 25.3.

a. Can you suggest a reason that the nitric acid does not hydrolyze the amide?
   b. Write an equation to show this hydrolysis.
   c. Name each organic product of the hydrolysis.

17. Consider the general reaction of an organic aliphatic primary amine with nitrous acid (page 467).

$$RCH_2CH_2NH_2 + NaNO_2 + HCl \longrightarrow RCH_2CH_2OH$$
$$+ RCH_2CH_2Cl + RCH{=}CH_2 + N_2 + NaCl + H_2O$$

Given that during the course of the reaction a carbocation, $RCH_2\overset{+}{C}H_2$, is produced, provide a structural formula for two more products of the general reaction. Explain.

18. A compound ($C_7H_9N$) gave two peaks in the 3000 to 3700 cm$^{-1}$ range in the infrared spectrum. In addition, the infrared spectrum showed the two groups present to be para to each other. A proton NMR spectrum of the compound yielded the following data:

$$\delta = 2.2 \ (3H) \ singlet$$
$$\delta = 3.4 \ (2H) \ singlet$$
$$\delta = 6.6 \ (4H) \ multiplet$$

Deduce a structural formula for the compound. Show your reasoning.

19. A compound ($C_6H_{15}N$) gave the following proton NMR spectrum:

$$\delta = 0.9 \ (6H) \ triplet$$
$$\delta = 1.1 \ (1H) \ singlet$$
$$\delta = 1.5 \ (4H) \ hextet$$
$$\delta = 2.6 \ (4H) \ triplet$$

Deduce a structural formula for the compound. Show your reasoning.

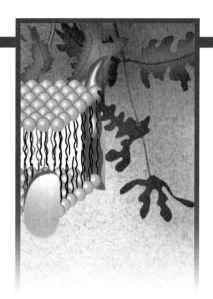

# CHAPTER 26

# Lipids

Lipids are naturally occurring compounds, isolated from plants or animals, that are soluble in nonpolar solvents such as carbon tetrachloride and low-polarity solvents such as ether and chloroform. Because this is a definition based on physical properties rather than on chemistry, we will observe that lipids are characterized by a variety of functional groups rather than on a particular functional group.

The five main classes of lipids that we will discuss are

1. Waxes
2. Triglycerides (fats and oils)
3. Phosphoglycerides (phospholipids)
4. Steroids
5. Prostaglandins

*Waxes* and *triglycerides* are *simple lipids*. These simple lipids contain at least one ester group derived from a carboxylic acid and an alcohol. In addition, simple lipids, like esters in general, can be saponified. *Compound lipids*, such as the *phospholipids*, are tetraesters and contain two ester groups derived from phosphoric acid and two alcohols in addition to two carboxylic acid esters. Compound lipids, like simple lipids, undergo saponification. The *steroids* and *prostaglandins* are examples of *derived lipids*. Both types of derived lipids contain a wide variety of functional groups and, if an ester group is not present, cannot be saponified.

A wax is a mixture of high-molecular-weight esters, each ester formed from a monocarboxylic fatty acid (RCOOH) and a monohydric alcohol (R′OH). Both the acyl group of the carboxylic acid and the alkyl group of the alcohol are straight-chain compounds; no branching is present in wax molecules.

## 26.1 WAXES

$$
\begin{array}{c} O \\ \parallel \\ R-C-OR' \end{array}
$$

Saturated acids varying in composition from $C_{15}H_{31}COOH$ to $C_{37}H_{75}COOH$ and saturated alcohols varying in composition from $C_{15}H_{31}CH_2OH$ to $C_{35}H_{71}CH_2OH$ have been isolated from waxes. For example, three of the main constituents of beeswax are triacontanyl hexacosanoate, dotriacontanyl hexacosanoate, and hexadecanyl hexadecanoate (see Table 3.5).

Draw a structural formula for each of the three main constituents of beeswax.

Waxes are solid and slick and are present in both plants and animals. Typical melting ranges are 42 to 47°C (*spermaceti*) and 83 to 91°C (*carnauba*). Since each wax consists of a mixture of compounds, we would expect to observe a wide melting range for each wax. Spermaceti is isolated from sperm oil, which is found in the head and blubber of sperm whales, whereas carnauba wax is present on the leaves of a number of different species of palm trees.

Leaves of plants, feathers of birds, and hair or skin of animals are coated with wax both to repel excess moisture and to prevent the loss of moisture. In addition, the wax coating serves to protect the leaf, bird, or animal from injury. Wool wax, isolated from sheep, is known as *lanolin* when purified. A number of pharmaceuticals and cosmetics use lanolin as the chief ingredient because it is easily absorbed by the skin. Lanolin is a mixture of cholesterol esters (see Sec. 26.6) of higher-molecular-weight fatty acids.

Waxes can be saponified, but with difficulty (see Secs. 22.4 and 23.7). In general,

$$R-\overset{\overset{\displaystyle O}{\|}}{C}-OR' + MOH(aq) \xrightarrow{\Delta} R-\overset{\overset{\displaystyle O}{\|}}{C}-O^-M^+ + R'OH$$

where M = Na or K. For example:

$$CH_3(CH_2)_{16}\overset{\overset{\displaystyle O}{\|}}{C}-OCH_2(CH_2)_{20}CH_3 + KOH(aq) \xrightarrow{\Delta} CH_3(CH_2)_{16}\overset{\overset{\displaystyle O}{\|}}{C}-O^+K^+ + CH_3(CH_2)_{20}CH_2OH$$

Docosanyl octadecanoate                    Potassium octadecanoate    1-Docosanol

## 26.2 STRUCTURE, COMPOSITION, AND NOMENCLATURE OF TRIGLYCERIDES (FATS AND OILS)

Triglycerides, or *triacylglycerols*, are esters derived from the trihydric alcohol glycerol. A trihydric alcohol contains three hydroxyl groups per molecule, a dihydric alcohol contains two groups per molecule, and a monohydric alcohol contains one group per molecule (see Sec. 14.1).

$$\begin{array}{c} \text{H} \quad \text{H} \quad \text{H} \\ | \quad\; | \quad\; | \\ \text{H}-\text{C}-\text{C}-\text{C}-\text{H} \\ | \quad\; | \quad\; | \\ \text{OH} \; \text{OH} \; \text{OH} \end{array}$$

Glycerol
(a trihydric alcohol)

$$\begin{array}{c} \text{H} \quad \text{H} \\ | \quad\; | \\ \text{H}-\text{C}-\text{C}-\text{H} \\ | \quad\; | \\ \text{OH} \; \text{OH} \end{array}$$

Ethylene glycol
(a dihydric alcohol)

In naturally occurring triglycerides, the acyl groups associated with the triglyceride can be the same or different, always straight-chain, and almost always containing an even number of carbons. If the acyl groups are the same, the triglyceride is said to be *simple*; if different, the triglyceride is *mixed*.

$$\begin{array}{c} \overset{\displaystyle O}{\|} \\ CH_2-O-C-R \\ | \quad\;\; \overset{\displaystyle O}{\|} \\ HC-O-C-R \\ | \quad\;\; \overset{\displaystyle O}{\|} \\ CH_2-O-C-R \end{array}$$

A simple triglyceride

$$\begin{array}{c} \overset{\displaystyle O}{\|} \\ CH_2-O-C-R \\ | \quad\;\; \overset{\displaystyle O}{\|} \\ HC-O-C-R' \\ | \quad\;\; \overset{\displaystyle O}{\|} \\ CH_2-O-C-R'' \end{array}$$

A mixed triglyceride

Refer to Table 20.1 for an additional listing of the saturated carboxylic acids. Note that all the carbon-carbon double bonds listed in Table 26.1 are in a cis confuguration in natural triglycerides.

Table 26.1 lists the carboxylic acids most often incorporated into a given triglyceride.

| TABLE 26.1 | Structures of the More Important Carboxylic Acids Found in Fats and Oils | | |
|---|---|---|---|
| Common Name | Number of Carbons | Structural Formula | Number of Carbon-Carbon Double Bonds |
| Lauric | 12 | $CH_3(CH_2)_{10}COOH$ | 0 |
| Myristic | 14 | $CH_3(CH_2)_{12}COOH$ | 0 |
| Palmitic | 16 | $CH_3(CH_2)_{14}COOH$ | 0 |
| Stearic | 18 | $CH_3(CH_2)_{16}COOH$ | 0 |
| Oleic | 18 | $CH_3(CH_2)_7CH{=}CH(CH_2)_7COOH$ | 1 |
| Linoleic | 18 | $CH_3(CH_2)_4CH{=}CHCH_2CH{=}CH(CH_2)_7COOH$ | 2 |
| Linolenic | 18 | $CH_3CH_2CH{=}CHCH_2CH{=}CHCH_2CH{=}CH(CH_2)_7COOH$ | 3 |
| Arachidonic | 20 | $CH_3(CH_2)_3(CH_2CH{=}CH)_4(CH_2)_3COOH$ | 4 |

**SOLVED PROBLEM 26.1**

Draw a structural formula for oleic acid showing the cis geometry around the carbon-carbon double bond. Then draw a structural formula for the product ester when 1 mol of glycerol is reacted with 1 mol of oleic acid (show the cis configuration of the oleyl group), 1 mol of palmitic acid, and 1 mol of lauric acid. Note that more than one structural formula can be drawn for the product ester.

**SOLUTION**

In the cis configuration, both vinyl hydrogens must be on the same side of the molecule. Remember that a vinyl hydrogen is bonded to the carbon bearing the double bond. Thus we have

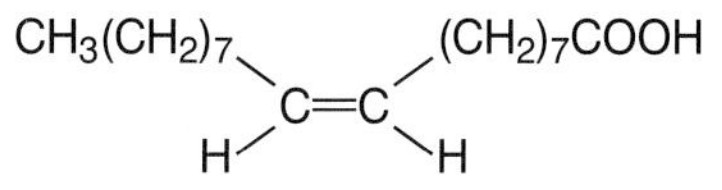

Now, we formulate a structural formula for one possible product triglyceride.

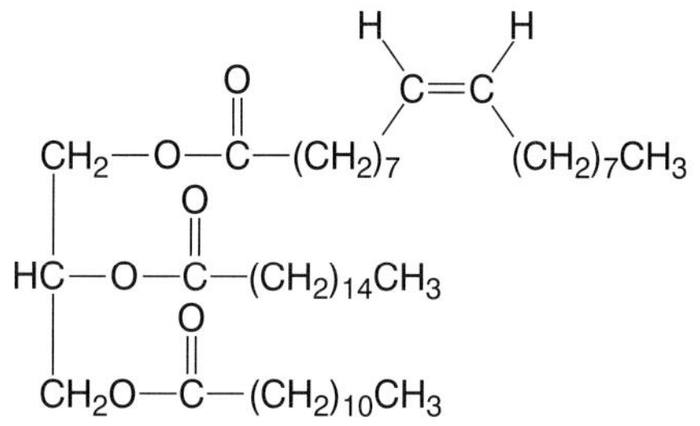

**EXERCISE 26.2**

Draw a structural formula for linoleic acid showing the cis geometry around the carbon-carbon double bonds.

**EXERCISE 26.3**

Locate the chiral carbon in the lipid described in Solved Problem 26.1.

**EXERCISE 26.4**

Draw another structural formula to represent the lipid in Solved Problem 26.1.

Examples of naturally occurring simple and mixed triglycerides are found in Fig. 26.1.

**FIGURE 26.1** Glyceryl trimyristate (a typical naturally occurring simple triglyceride) and glyceryl palmitooleostearate (a typical naturally occurring mixed triglyceride).

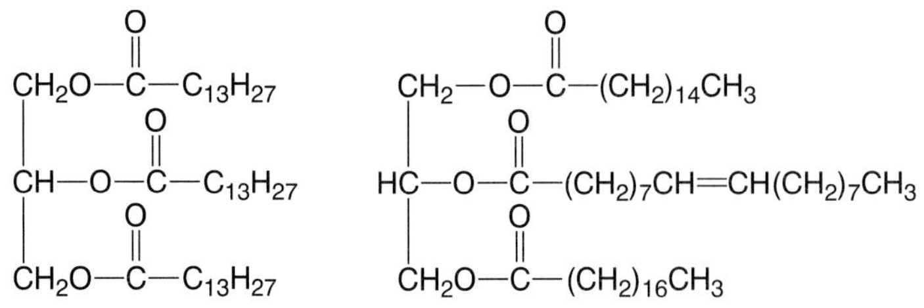

Glyceryl trimyristate                  Glyceryl palmitooleostearate

Note in the example of the mixed glyceride that the *cis* stereochemistry around the carbon-carbon double bond in the oleyl group is not shown but is understood to exist.

Fats are solids and are isolated from various animals, whereas oils are liquids and are obtained from plants. Both fats and oils are triglycerides, but usually fats have mostly saturated carboxylic acids associated with glycerol, whereas oils have mostly unsaturated carboxylic acids combined with glycerol.

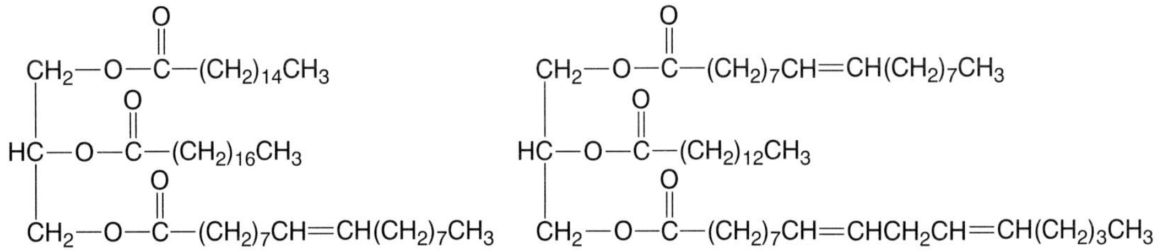

Glyceryl palmitostearooleate          Glyceryl oleomyristolinoleate
(a fat)                               (an oil)

Fats and oils are named the same way as esters (see Sec. 22.1) with two exceptions: (1) the first word of the name is always *glyceryl*, and (2) in the second word of the name, the *-ic* suffixes of the first and second *acid residues* are dropped and replaced by the letter *o*. An acid residue is an acyl group incorporated in a triglyceride.

**SOLVED PROBLEM 26.2**

Name this triglyceride:

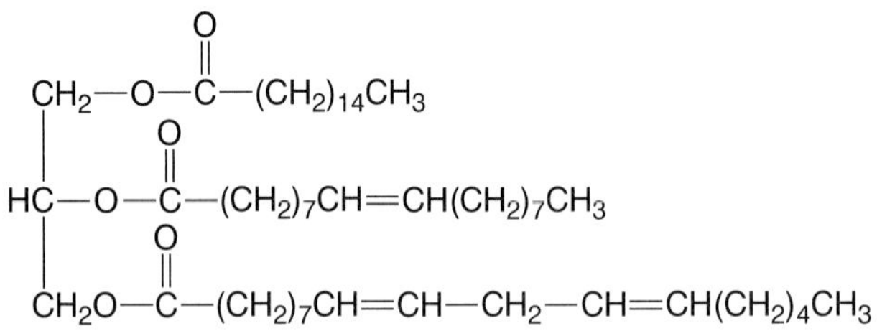

**SOLUTION**

The first word of the name is *glyceryl*, since the compound is a triglyceride. Now the uppermost acid incorporated in the triester structure is palmitic, the second is oleic, and the bottom is linoleic. Thus we have

Glyceryl palmitooleolinoleate

Name this triglyceride:

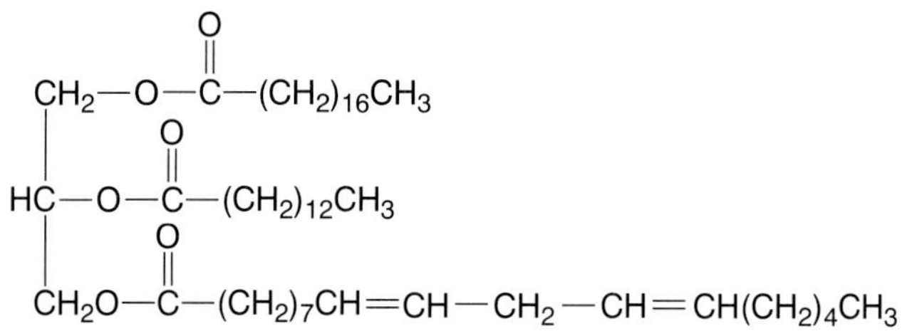

$$CH_2-O-\overset{\overset{\displaystyle O}{\|}}{C}-(CH_2)_{16}CH_3$$
$$HC-O-\overset{\overset{\displaystyle O}{\|}}{C}-(CH_2)_{12}CH_3$$
$$CH_2O-\overset{\overset{\displaystyle O}{\|}}{C}-(CH_2)_7CH=CH-CH_2-CH=CH(CH_2)_4CH_3$$

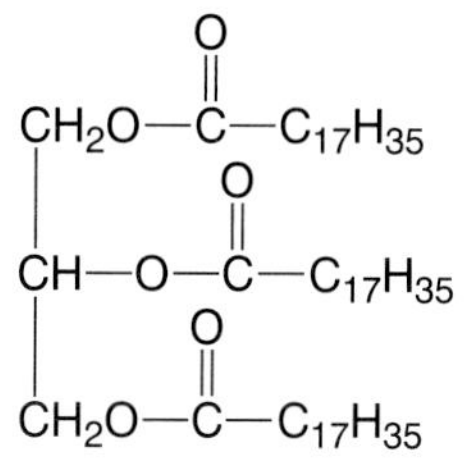

An alternate method of naming only simple triglycerides is to use the prefix *tri-* followed by the common name of the incorporated acid with the *-ic* suffix replaced by the suffix *-in*, as is shown in the triglyceride in Fig. 26.2.

**FIGURE 26.2**  Tristearin (or glyceryl tristearate).

$$CH_2O-\overset{\overset{\displaystyle O}{\|}}{C}-C_{17}H_{35}$$
$$CH-O-\overset{\overset{\displaystyle O}{\|}}{C}-C_{17}H_{35}$$
$$CH_2O-\overset{\overset{\displaystyle O}{\|}}{C}-C_{17}H_{35}$$

a. Draw a structural formula for tripalmitin.

b. Give an alternate name for tripalmitin.

Actually, the fats and oils isolated from animals and plants are mixtures of simple and mixed triglycerides. The composition of the acids in the particular fat or oil, in turn, depends on the diet of the animal and the composition of the surrounding soil for the plant. Thus the organic acid composition for these fats and oils is variable even though typical values are given in Table 26.2.

Note the greater percentage of unsaturated acids incorporated in the oils as opposed to the fats.

Draw a structural formula for a molecule that would best represent depot fat given the data in Table 26.2.

**SOLUTION**

Select the acid residue present to the greatest extent in the triglyceride, in this case, oleic acid (47%). Then choose the two acid residues representing the next two highest percentage compositions, in this case, palmitic acid (24%) and linoleic acid (10%). Finally, draw a triglyceride structural formula containing these three acid residues.

| TABLE 26.2 | Typical Composition of Fatty Acids in Selected Fats and Oils | | | | | | | |
|---|---|---|---|---|---|---|---|---|
|  | **Lauric** | **Myristic** | **Palmitic** | **Stearic** | **Oleic** | **Linoleic** | **Linolenic** | **Other** |
| *Fats* | | | | | | | | |
| Depot (human) | — | 3 | 24 | 8 | 47 | 10 | — | 8 |
| Butter | 3 | 11 | 29 | 9 | 27 | 4 | — | 17 |
| Lard | — | 1 | 28 | 12 | 48 | 6 | — | 5 |
| Tallow | — | 6 | 27 | 14 | 50 | 3 | — | — |
| *Oils* | | | | | | | | |
| Olive | — | — | 7 | 2 | 84 | 4 | — | 3 |
| Corn | — | 1 | 10 | 3 | 50 | 34 | — | 2 |
| Soybean | — | — | 10 | 2 | 29 | 51 | 7 | 1 |
| Linseed | — | — | 6 | 3 | 19 | 24 | 47 | 1 |

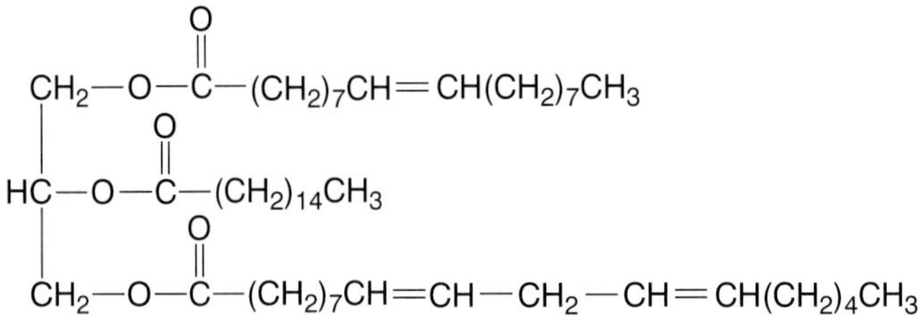

$$CH_2-O-\overset{\displaystyle O}{\overset{\|}{C}}-(CH_2)_7CH=CH(CH_2)_7CH_3$$

$$HC-O-\overset{\displaystyle O}{\overset{\|}{C}}-(CH_2)_{14}CH_3$$

$$CH_2-O-\overset{\displaystyle O}{\overset{\|}{C}}-(CH_2)_7CH=CH-CH_2-CH=CH(CH_2)_4CH_3$$

A structural formula of depot fat

---

**EXERCISE 26.7**

Draw a structural formula for a molecule that would best represent lard given the data in Table 26.2.

---

## 26.3 PHYSICAL PROPERTIES OF TRIGLYCERIDES (FATS AND OILS)

Fats and oils are soluble in the nonpolar solvent carbon tetrachloride and low-polarity solvents such as ether and chloroform. Both fats and oils leave grease spots on paper and are greasy to the touch.

The density of fats and oils varies from 0.90 to 0.96 g/ml—less dense than water. Note the globules of fat floating on the surface in a typical soup.

Melting points of fats are low, 30 to 45°C. Even so, fats melt at a higher temperature than oils because oils are liquids at room temperature.

## 26.4 CHEMICAL PROPERTIES OF TRIGLYCERIDES (FATS AND OILS)

### Saponification: Soap Formation, Detergents

Note that $RCOO^-K^+$, $R'COO^-K^+$, and $R''COO^-K^+$ are soaps.

Fats and oils are easily saponified by hot aqueous potassium hydroxide or sodium hydroxide solution to produce glycerol and potassium or sodium carboxylates ($C_{12}$ to $C_{20}$) known as *soaps* (see Secs. 22.4 and 23.7). In general,

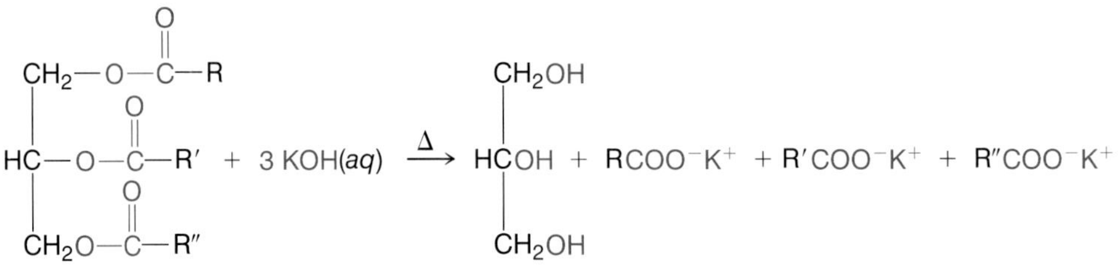

$$\begin{array}{l} CH_2-O-\overset{O}{\overset{\|}{C}}-R \\ HC-O-\overset{O}{\overset{\|}{C}}-R' \\ CH_2O-\overset{O}{\overset{\|}{C}}-R'' \end{array} + 3\,KOH(aq) \overset{\Delta}{\longrightarrow} \begin{array}{l} CH_2OH \\ HCOH \\ CH_2OH \end{array} + RCOO^-K^+ + R'COO^-K^+ + R''COO^-K^+$$

Example:

$$\begin{array}{l} CH_2-O-\overset{O}{\overset{\|}{C}}-C_{17}H_{35} \\ HC-O-\overset{O}{\overset{\|}{C}}-C_{17}H_{35} \\ CH_2O-\overset{O}{\overset{\|}{C}}-C_{17}H_{35} \end{array} + 3\,KOH(aq) \overset{\Delta}{\longrightarrow} \begin{array}{l} CH_2OH \\ HCOH \\ CH_2OH \end{array} + 3\,C_{17}H_{35}COO^-K^+$$

Tristearin                                      Glycerol

Potassium stearate (a soap)

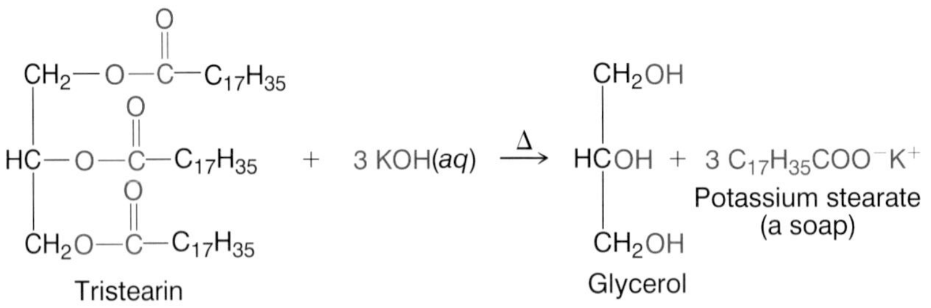

**FIGURE 26.3** Sodium palmitate (a typical soap).

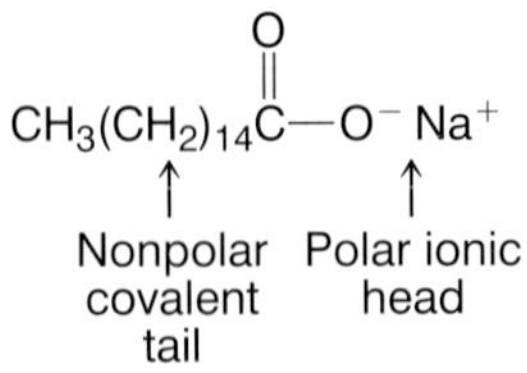

$$CH_3(CH_2)_{14}\overset{\displaystyle O}{\overset{\|}{C}}-O^-\ Na^+$$

Nonpolar covalent tail / Polar ionic head

**FIGURE 26.4** The formation of an ion-dipole bond between sodium ion (in sodium palmitate) and water.

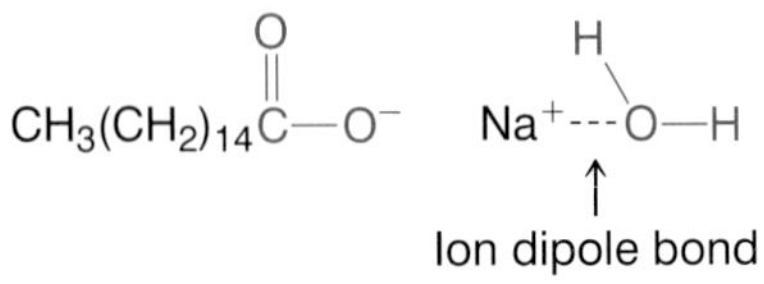

$$CH_3(CH_2)_{14}\overset{\displaystyle O}{\overset{\|}{C}}-O^-\qquad Na^+\text{---}\overset{\displaystyle H}{\overset{\mid}{O}}-H$$

Ion dipole bond

Soap cuts grease because the soap molecule has both nonpolar covalent and polar ionic character (Fig. 26.3). Although we write sodium palmitate as $CH_3(CH_2)_{14}COO^-Na^+$, it is important to understand that the sodium ion is free to roam in aqueous solution, forming ion-dipole bonds with water, and is not necessarily paired with the carboxylate ion, as shown in Fig. 26.4. It is the nonpolar hydrocarbon-like tail of the soap molecules that penetrates into the nonpolar grease (Fig. 26.5). The polar heads of the soap molecule remain dissolved in water at the same time. The result of this is the formation of a *micelle*.

Putting it another way, a micelle is created when a group of soap molecules enters a grease particle. The micelle is spherical in shape. Stirring in a washing machine or

## HISTORICAL BOXED READING 26.1

Mention the word *fat* in a conversation, and someone is sure to giggle. Of course, to be fat in our society, where to be thin is to be in, is highly undesirable. Heavy people as a rule are laughed at as children and often ignored as adults.

Fats are one of the "big three" of proteins, carbohydrates, and fats. This big three, found in foods, are the sources of energy (carbohydrates and fats) and muscle, skin, etc. (proteins) for our bodies.

But fats have other uses. The fats we do not burn are stored by our bodies to be used for energy when no food is available. Thus fats are a storehouse of energy. Without fat in our bodies, we would not be able to use the fat-soluble vitamins (A, D, E, and K) to the benefit of our bodies.

Under our skins, a layer of fat exists to insulate our bodies from the cold. By the way, this layer of fat is thicker in women than in men. Also, a number of vital organs in our bodies—the heart, liver, and kidneys—are surrounded by fat to protect them from bumps and shocks.

In days gone by, a fat person represented a person whose belly was filled—a successful person. Genesis 45:18 pulled no punches as it stated, "And ye shall eat the fat of the land." This almost reverence for fat continued through the years. Consider this gem from the poem "The Irish Mimic" (1795), "Fat, fair and forty were all the toasts of the young men."

The reason fat was desirable was that food was scarce. In addition, people needed a great deal of fat in their food because in that day they did much more physical work than is done today. Fat, when oxidized in the body, supplies about 9000 calories of energy per gram, whereas the oxidation of carbohydrates yields only 4000 calories of energy per gram. Thus our ancestors knew exactly what they were doing when they stuffed themselves full of fat.

Sometime between 1910 and 1920, obesity was kicked off its pedestal, and thin came in. At about that time lives became more sedentary, mechanical devices became available to help people do their work, and the need for a high-calorie diet decreased.

This "thin is in" philosophy permeates our society today, particularly after the relationship between fats in the diet and heart disease was established. Men and women who perform for the television or movie camera or for nightclub audiences are almost uniformly thin. So are athletes. When Jack Nicklaus broke into professional golf a while back, he was heavy. Some sports announcers taunted him about his weight, as good as he was.

Models must be tall and very thin. More and more young women are suffering from a condition called *anorexia nervosa*. Affected women simply do not eat because they perceive themselves as being very fat. This can lead to death by starvation.

Cyril Connolly caught the pulse beat of our society when he wrote, "Imprisoned in every fat man a thin one is wildly signaling to be let out."

It seems fair to state that a substantial number of adult Americans are dieters. Unfortunately, most dieters (95%) inevitably gain back the weight they worked so hard to lose.

There is new hope for dieters. Fat-free fat is almost a reality for the American consumer. Chemically, one fat-free fat product is sucrose polyester (SPE), i.e., one molecule of sucrose (table sugar) (see Chap. 27) acting as an alcohol with anywhere from six to eight molecules of one or more fatty acid molecules. It looks and tastes just like fat, only this artificial fat cannot be broken down by the body. Thus zero calories are taken in. SPE can be substituted for real fats in just about anything from ice cream to margarine. In addition, this remarkable substance can be substituted for fats and oils in cooking, baking, and frying.

The FDA has recently approved Olestra, an SPE, as a fat substitute. However, the decision is controversial because Olestra may strip the fat-soluble vitamins A, D, E, and K from the body and additionally may cause diarrhea and gastrointestinal distress. A consumer group, Center for Science in the Public Interest, is convinced the product is dangerous and probably will appeal the FDA approval.

scrubbing by hand breaks the grease globule into smaller globules that are emulsified (colloidally dispersed) in water. These small globules cannot coalesce back to the original large globule because the layers of negative charge in the micelles repel one another (Fig. 26.6) (remember some cations are forming ion-dipole bonds with water and are not necessarily paired with the negative ions). Once the grease globules are dispersed in water, the grease is then drained out with the water.

Experts are not sure when soap was first made and used. One story has it that soap was first made by chance about 3000 years ago in Rome. Often the people sacrificed animals to the gods. Animal fat dripped from the altars through the wood ashes into the clay soil. Gradually, the women became aware that this clay (containing soap) worked very well in washing clothes.

When the Roman Empire fell (about A.D. 500), knowledge of the art of soapmaking gradually disappeared in Europe. For the next thousand years, until the art of soapmaking resurfaced, Europe was almost stripped of its population by a number of terrible diseases, including typhoid fever and others. Certainly a number of these diseases, but not all, were due to the complete lack of body and home cleanliness, which, in turn, stemmed from the lack of soap.

When soapmaking was rediscovered, the soap produced was appreciated and used by the people. But soap was a rare commodity because of the scarcity of lye—one of the two main ingredients in the recipe for soap. Lye was originally defined as

**FIGURE 26.5** Soap interactions with grease-micelle formation.

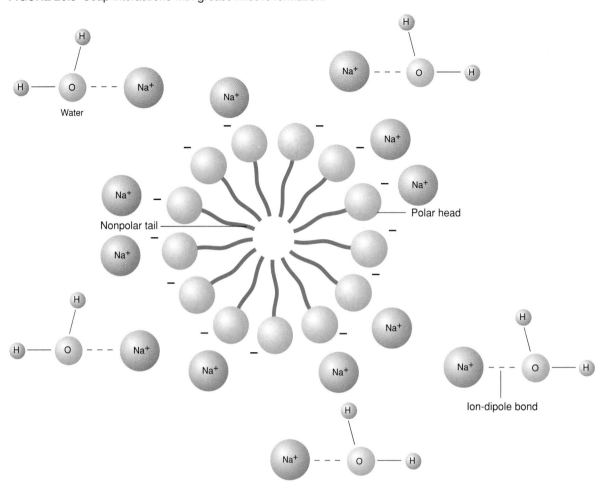

**FIGURE 26.6** Repulsion between micelles.

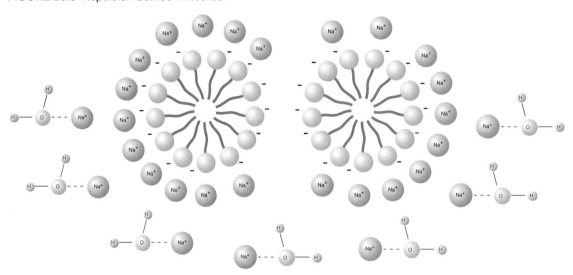

an alkaline solution rich in potassium carbonate. Today, it is also considered to be an alkaline solution of sodium hydroxide or potassium hydroxide. Lye could only be made by the laborious process of filtering water through wood ashes. Soap was so scarce that the household staff of Queen Elizabeth I could find only enough to permit Her Majesty to take three baths per month. It seems reasonable to assume that the lady used plenty of perfume to compensate for the lack of soap.

The breakthrough in soap manufacture came in 1791 when Nicholas Leblanc, a French chemist, invented a process to prepare sodium carbonate (a substitute for potassium carbonate) cheaply from readily available sodium chloride (household salt). Very quickly, technology was developed to permit the manufacture of as much as several thousand pounds of soap at a time in huge kettles.

The process works this way: Animal fats (usually tallow) or vegetable oils are mixed with sodium hydroxide, and the mixture is heated in a kettle for a few hours in an alcohol-water solvent. The reaction is complete when the fat or oil has dissolved in the alcohol-water mixture, resulting in a clear solution. To this "soap" is added a concentrated solution of table salt in water. This brine effectively precipitates out the soap and causes it to separate from the glycerol, a by-product of this chemical reaction. The soap is finally filtered and washed with cold water to remove traces of alkali that would burn the skin.

The soap can be modified to reflect a variety of different uses. For example, pumice (an abrasive) can be added to make a scouring soap, perfumes can be added to produce a toilet soap, and antiseptic (phenol) can be added to make a medicated soap.

In addition, soap can be filled with air to make a floating soap, dyed to form a colored soap for those who want a splash of color in the bathroom, and mixed with alcohol to give an almost transparent soap.

Soap and civilization go hand in hand. If the soap-manufacturing plants in the world would shut down, a rash of disease and death would follow.

Sodium soaps are less water-soluble than potassium soaps and are less expensive. Ordinary soaps are sodium soaps. Potassium soaps are usually found in liquid soaps and shaving cream.

For the last 20 years or so the use of soap has been on the decline due to the emergence of *synthetic detergents* with better cleaning properties and the ineffectiveness of soap in hard water. Hard water contains magnesium, calcium, iron(II), and iron(III) salts that replace the sodium or potassium in the soap and form an insoluble soap. (In an insoluble soap, cations and anions are completely associated.) This is the famous (or infamous) bathtub ring. In general,

$$2\ RCOO^-Na^+ + Ca^{2+} \longrightarrow (RCOO^-)_2Ca^{2+} + 2\ Na^+$$

Water-soluble             Water-insoluble
(bathtub ring)

## HISTORICAL BOXED READING 26.2

There is an old saying to the effect that "politics makes strange bedfellows." So do detergents make strange bedfellows—with sharks. To see the relationship between the two, read on. Nothing is as frightening to a number of people as a shark attack. That fear was masterfully manipulated in the motion picture *Jaws*. During World War II, a shark repellent was developed by the U.S. Navy to protect downed pilots and the crews of ships that were sunk. Unfortunately, this repellent proved to be ineffective. No promising shark repellent turned up for many years. However, in the 1970s it was found that the moses sole, a fish found in the Red Sea, secretes a substance that stops members of some shark species "in their tracks." The substance, called *pardoxin*, is a protein (see Chap. 28) composed of 162 amino acid residues. Since the structure of pardoxin is so complex, it is difficult and costly to synthesize. Thus producing enough pardoxin to distribute this product for general use seems unlikely.

In the early 1980s, it was proposed that pardoxin worked on sharks the way it did because of its surfactant properties. Sure enough, it was found that the inexpensive, easily prepared detergent sodium lauryl sulfate repelled sharks at one-quarter the concentration required by pardoxin. Although detergents seem full of promise as an inexpensive and effective shark repellent, further research is necessary in order to "settle the matter."

**FIGURE 26.7** A branched alkyl benzenesulfonate detergent (nonbiodegradable).

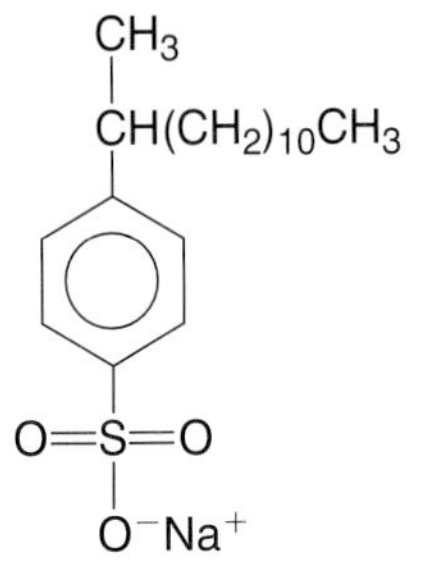

Detergents clean using the same micelle mechanism as do soaps.

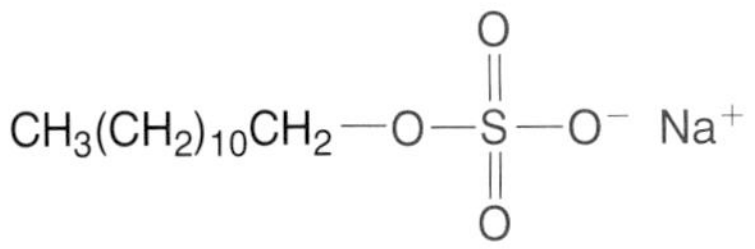

Example:

$$2\ CH_3(CH_2)_{14}COO^-Na^+ + Mg^{2+} \longrightarrow (CH_3(CH_2)_{14}COO^-)_2Mg^{2+} + 2\ Na^+$$

Soaps were not effective in commercial laundering operations for two reasons. First, the soap scum that formed due to the use of hard water wasted both time and money in cleanup. Second, when dissolved in water, soaps produced an alkaline solution that destroyed delicate fabrics. In general, .

$$RCOO^-Na^+ + H_2O \rightleftharpoons RCOOH + {}^-OH$$

Example:

$$CH_3(CH2)_{10}COO^-Na^+ + H_2O \rightleftharpoons CH_3(CH_2)_{10}COOH + {}^-OH$$

To remedy these situations, chemists developed a number of synthetic detergents, also known as *syndets*. These synthetic detergents do not react with the ions responsible for hard water and, in addition, produce a neutral solution when dissolved in water.

There exist three different structural types of detergents. An *anionic detergent* contains a negative charge on the long organic chain. Here are two such detergents:

Sodium lauryl sulfate 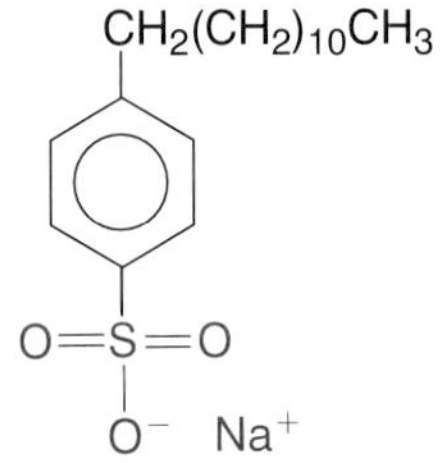            Sodium p-dodecylbenzenesulfonate

**FIGURE 26.8** Octadecyltrimethyl-ammonium chloride (a cationic detergent).

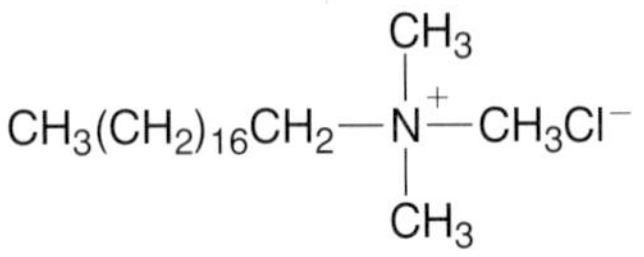

These detergents are effective when the long-chain alkyl group contains 16 to 18 carbon atoms.

Both long-chain sodium alkyl sulfates ($C_{12}$ to $C_{18}$) and sodium *p*-alkylbenzenesulfonates ($R = C_{10}$ to $C_{14}$) used today are *biodegradable*. This means that microorganisms in the water can decompose the pollution-causing detergent to benign substances that cause no harm to the environment. Unfortunately, detergents containing a branched alkyl chain (Fig. 26.7) are not biodegradable and, when used in the 1950s, caused problems such as sudsing and fish kills when they accumulated in our lakes.

A *cationic detergent* contains a positive charge on the long organic chain. An example is given in Fig. 26.8. Both anionic and cationic detergents are similar to soaps in that each contains a long-chain nonpolar group and a polar ionic species.

*Nonionic detergents* do not contain an ionic species. An example is given in Fig. 26.9.

**EXERCISE 26.8**

Would you expect the nonionic detergent in Figure 26.9 to be biodegradable? Explain.

Nonionic detergents have gained recently in market share for two reasons. First, sudsing or frothing is minimal, so these detergents are well suited to automatic dishwashers. Second, synthetic fiber garments (such as polyester) become cleaner when washed with nonionic detergents.

**EXERCISE 26.9**

The sodium and potassium salts of oleic, linoleic, and linolenic acids are soaps. Draw a structural formula of

a.  potassium oleate.

b.  sodium linolenate.

**FIGURE 26.9** A nonionic detergent.

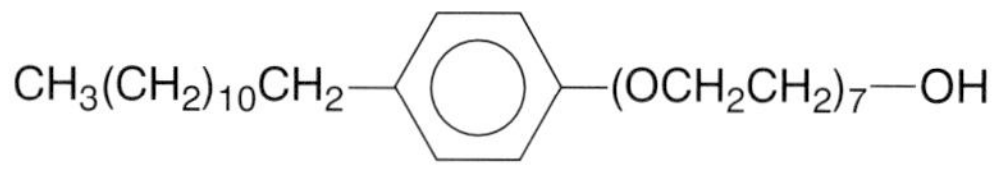

$$CH_3(CH_2)_{10}CH_2 \!-\!\!\bigcirc\!\!-(OCH_2CH_2)_7\!-\!OH$$

Complete the following equations.

a. $CH_3(CH_2)_{12}COO^-Na^+ + HCl \longrightarrow$

b. $2\ CH_3(CH_2)_{14}COO^-K^+ + MgCl_2 \longrightarrow$

c. $CH_3(CH_2)_{16}COO^-Na^+ + H_2O \rightleftarrows$

Synthesize an alkylbenzenesulfonate detergent from benzene and decanoyl chloride. You may use any inorganic reagents necessary.

Both soaps and detergents are *surfactants*; i.e., they lower the surface tension of water. This property enables the soap or detergent solution to soak into textile pores and loosen and remove grease and grime more effectively. To sum up, we can say that soaps and detergents clean by a combination of micelle formation and the ability to act as a surfactant.

## Hydrogenation (Hardening) of Oils

There is an oversupply of oils available commercially as compared with fats. In order to prevent vegetable oils from going to waste, excess oils are partially hydrogenated to give smoothly textured solid fat products such as Crisco, Spry, and margarine. These oils are not completely hydrogenated because the resulting fat would be too brittle for packaging and cooking. Note the following equation representing partial hydrogenation of a liquid oil to a solid fat.

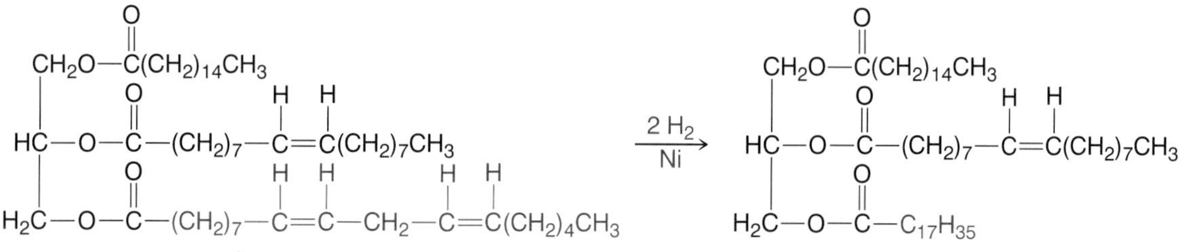

This process causes some cis bonds to change to trans bonds, and these acids are a nutritional detriment leading to clogging of the arteries.

A liquid oil  →(2 H₂ / Ni)→  A solid fat

## Oxidation—Rancidity and Drying Oils

Oils tend to be more susceptible to oxidation by air or water than fats due to the greater numbers of carbon-carbon double bonds in oils. This oxidation, a complex process, produces foul-smelling carboxylic acids that contaminate the oil or fat when one or more carbon-carbon double bonds in the acid residue is broken. The fat or oil is said to turn *rancid*. For example:

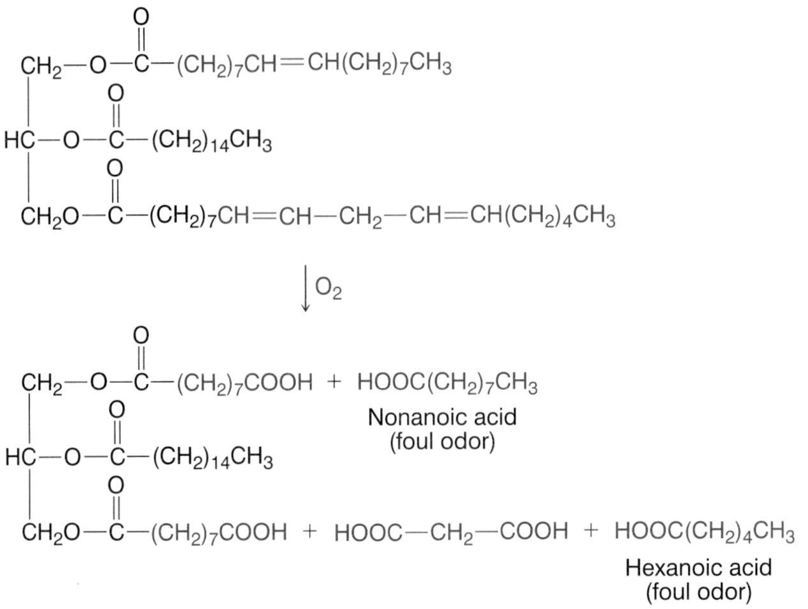

$$CH_2-O-\overset{\displaystyle O}{\overset{\|}{C}}-(CH_2)_7CH=CH(CH_2)_7CH_3$$

$$HC-O-\overset{\displaystyle O}{\overset{\|}{C}}-(CH_2)_{14}CH_3$$

$$CH_2O-\overset{\displaystyle O}{\overset{\|}{C}}-(CH_2)_7CH=CH-CH_2-CH=CH(CH_2)_4CH_3$$

$$\downarrow O_2$$

$$CH_2-O-\overset{\displaystyle O}{\overset{\|}{C}}-(CH_2)_7COOH \; + \; HOOC(CH_2)_7CH_3$$

Nonanoic acid
(foul odor)

$$HC-O-\overset{\displaystyle O}{\overset{\|}{C}}-(CH_2)_{14}CH_3$$

$$CH_2O-\overset{\displaystyle O}{\overset{\|}{C}}-(CH_2)_7COOH \; + \; HOOC-CH_2-COOH \; + \; HOOC(CH_2)_4CH_3$$

Hexanoic acid
(foul odor)

The $-CH_2CH_2\overset{+}{N}(CH_3)_3$ group is a derivative of choline, $HOCH_2CH_2\overset{+}{N}(CH_3)_3$, while $-CH_2CH_2\overset{+}{N}H_3$ stems from 2-aminoethanol (ethanolamine), $HOCH_2CH_2NH_2$, and

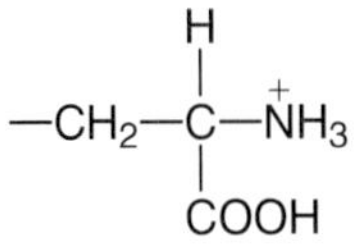

$$-CH_2-\overset{\displaystyle H}{\underset{\displaystyle COOH}{\overset{|}{\underset{|}{C}}}}-\overset{+}{N}H_3$$

is formulated from the amino acid serine.

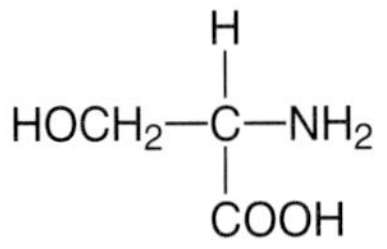

$$HOCH_2-\overset{\displaystyle H}{\underset{\displaystyle COOH}{\overset{|}{\underset{|}{C}}}}-NH_2$$

Serine

The rancidity process can be slowed by introducing into the oil a compound called an *antioxidant*. It is the antioxidant that is oxidized in the presence of air or water rather than the oil. Two antioxidants often used in cooking oils are 2,6-di-*tert*-butyl-4-methylphenol, known as *BHT* (Fig. 26.10a)(BHT is an acronym for *butylated hydroxy toluene*), and 3-*tert*-butyl-4-hydroxylanisole, known as *BHA* (Fig. 26.10b)(BHA is an acronym for *butylated hydroxy anisole*).

An oil (such as linseed oil) that contains a significant percentage of linoleic and linolenic acid residues can be used as a *drying oil*. The drying process begins when the oil undergoes oxidation at the allylic positions on the linoleic and linolenic acid residue chains. Ultimately, a polymeric hard film is formed on the surface of the oil. Drying oils are used in paints and varnishes to protect the painted or varnished surface.

**FIGURE 26.10** Two often-used antioxidants.

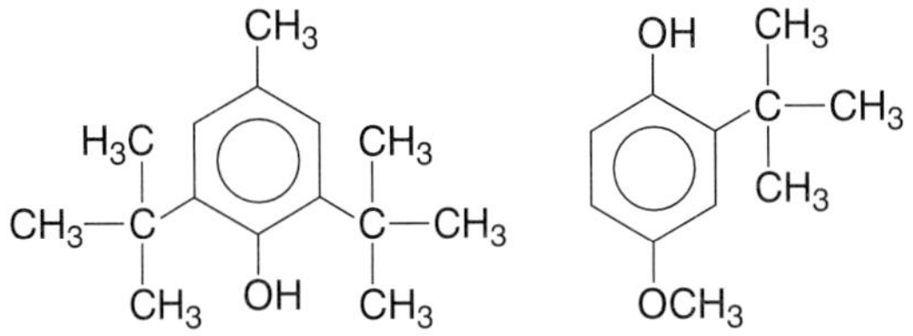

BHT
(a)

BHA
(b)

## 26.5 PHOSPHOGLYCERIDES (PHOSPHOLIPIDS)

Structurally, phospholipids are closely related to the fats and oils, the difference being one

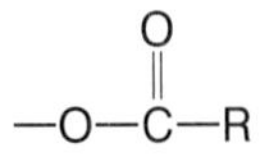

$$-O-\overset{\displaystyle O}{\overset{\|}{C}}-R$$

group in the triglyceride is replaced by a

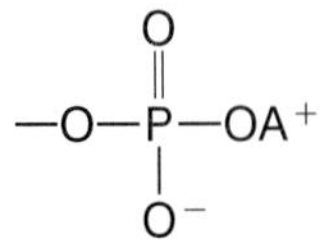

$$-O-\overset{\displaystyle O}{\underset{\displaystyle O^-}{\overset{\|}{\underset{|}{P}}}}-OA^+$$

group, as in Fig. 26.11. When $A^+$ is $-CH_2CH_2\overset{+}{N}(CH_3)_3$, the phospholipid is a lecithin; when it is $-CH_2CH_2\overset{+}{NH}_3$ or

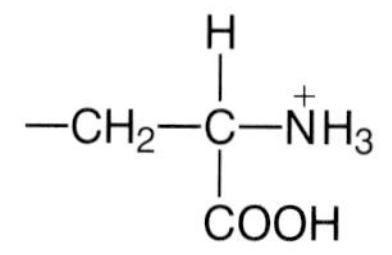

**FIGURE 26.11** General formula for a phospholipid.

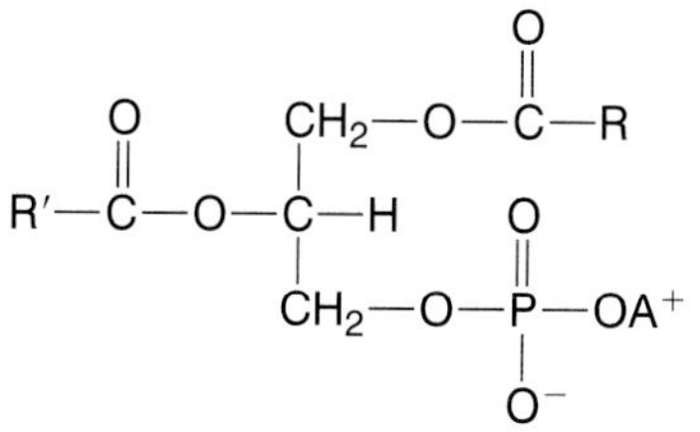

$$-CH_2-\underset{\underset{COOH}{|}}{\overset{\overset{H}{|}}{C}}-\overset{+}{NH}_3$$

the phospholipid is a cephalin. Two examples are given in Fig. 26.12.

Phospholipids can be formulated in another way. Think of the phospholipids as resulting from the combination of a phosphatidic acid with either choline, ethanolmine, or serine. The $-R$ and $-R'$ groups are usually $-C_{17}H_{35}$, $-C_{17}H_{33}$, or $-C_{15}H_{31}$.

$-R$ and $R'$ are usually $-C_{17}H_{35}$, $-C_{17}H_{33}$ or $-C_{15}H_{31}$.

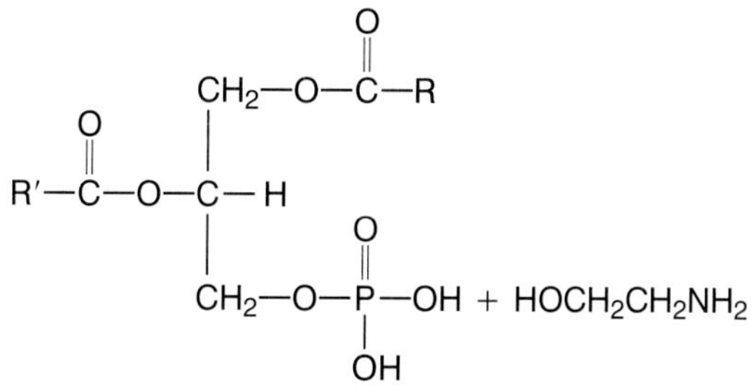

Phosphatidic acids

Loss of water and transfer of a proton give a cephalin we designate as *phosphatidyl ethanolamine* (Fig. 26.13).

**EXERCISE 26.12**

Locate the chiral carbon atom in phosphatidic acid. Then draw a Fischer projection of an (*S*)-phosphatidic acid (R = $-C_{17}H_{35}$, R' = $-C_{17}H_{33}$).

Note that a phospholipid is like a soap in that each has a nonpolar "tail" and a polar "head." Thus phospholipids are good emulsifying agents. They are, in addition, good wetting agents (surfactants).

**FIGURE 26.12** Two types of phospholipids.

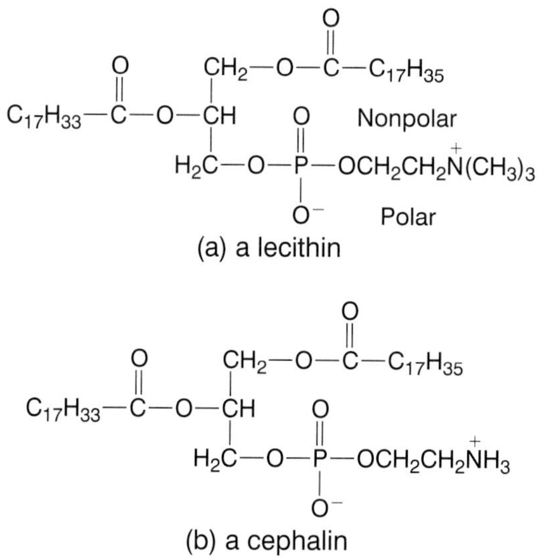

(a) a lecithin

(b) a cephalin

**FIGURE 26.13** Phosphatidyl ethanolamine—a cephalin.

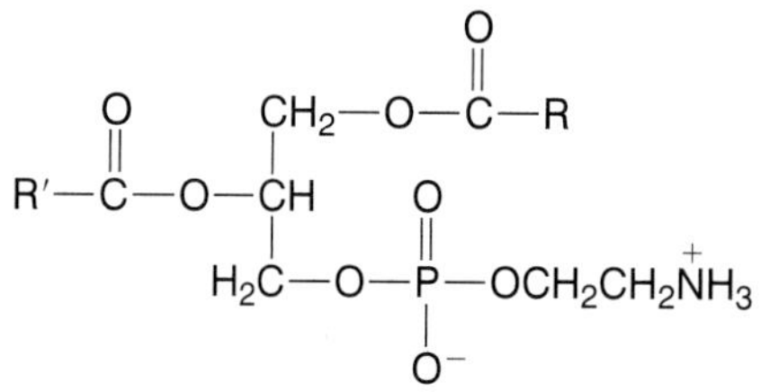

Cell membranes consist mainly of phospholipids, cholesterol (40%) (see Sec. 26.6), and proteins (60%) (see Chap. 28). Since the cell membrane controls the passage of compounds into and out of the cell, the biological importance of phospholipids is significant.

In the cell membrane, the phospholipids are aligned as a *bilayer* with the nonpolar tails facing each other. This puts the polar ionic heads facing into the cell on the one hand and out of the cell on the other hand (Fig. 26.14). The nonpolar tails within the bilayer prevent water, ions, and low-molecular-weight water-soluble compounds from diffusing through the cell membrane. It is the proteins placed in certain locations throughout the bilayer that selectively pass water, ions, and low-molecular-weight water-soluble compounds into and out of the cell (see Fig. 26.14).

Phospholipids, like triglycerides, can be saponified. In general,

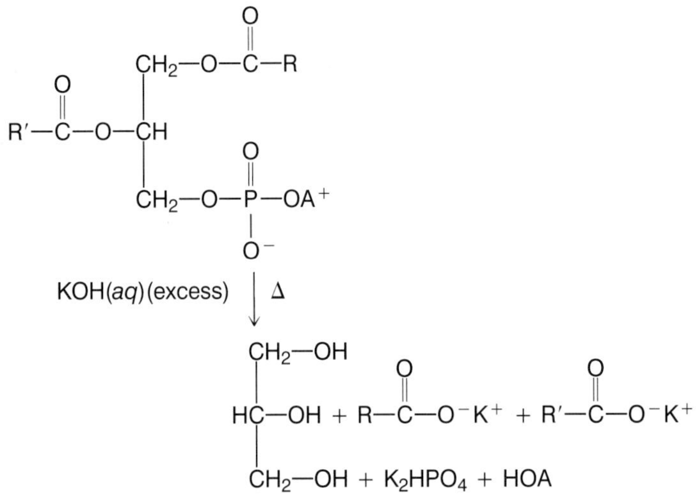

Example:

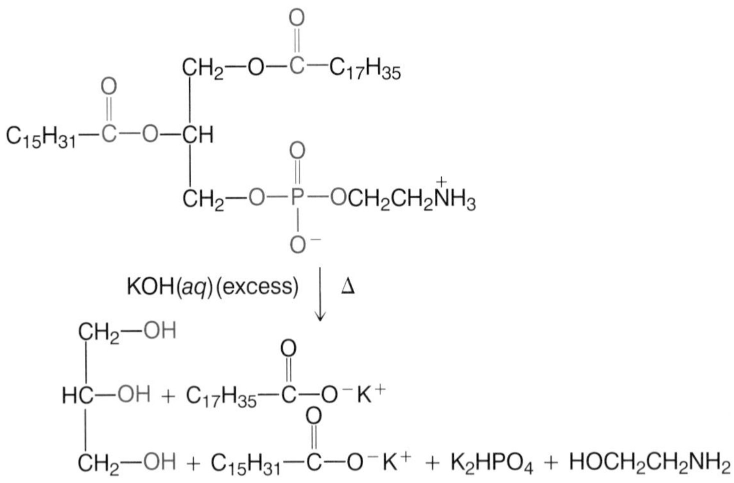

---

**EXERCISE 26.13**

Complete the following equations by drawing a structural formula for each organic product and writing a molecular formula or formula unit for every inorganic product.

a.   tripalmitin + 3 KOH($aq$) $\xrightarrow{\Delta}$

b.   triolein + 3 H$_2$ $\xrightarrow{\text{Ni}}$

c.   phosphatidyl choline (use —R and —R′) $\xrightarrow[\Delta]{\text{KOH }(aq)(\text{excess})}$

d.   trilinolein + Br$_2$ (excess) $\longrightarrow$

e.   glyceryl palmitolinolenostearate + 3 KOH($aq$) $\xrightarrow{\Delta}$

**FIGURE 26.14** The phospholipid bilayer (including protein and cholesterol placement) in a cell membrane.

## 26.6 STEROIDS

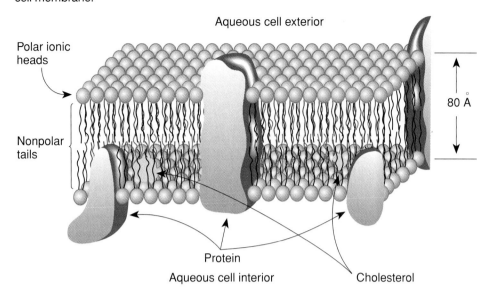

A major class of lipid is the steroids, a series of compounds commonly, but not always, found in plants and animals, many of which perform a variety of different biological functions—for good or ill. The "parent" structure of the steroid class of lipid is a saturated four-ring system somewhat related to phenanthrene (Fig. 26.15a). Note that the four rings of a given steroid are designated *A, B, C,* and *D* (Fig. 26.15b).

Some of the more important steriods are *cholesterol, cortisone, progesterone, testosterone,* and *vitamin D₂*.

Cholesterol

Cortisone

Progesterone

Vitamin D₂ (calciferol)

Testosterone

▬ Bond projects toward the reader
ıııı Bond projects away from the reader

**FIGURE 26.15** The relationship of phenanthrene to the "parent" steroid.

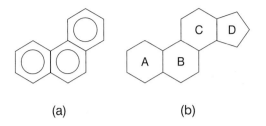

(a)                              (b)

Cholesterol is widely present within the human body and is of particular importance for a number of reasons. First, the deposition of water-insoluble cholesterol on the inner wall of arteries results in a decrease in the diameter in the major arteries of the body. This results in a condition called *arteriosclerosis*. Regrettably, more problems can result due to the arterial deposition of cholesterol. Since the major arteries are narrower, the heart must work harder to pass the blood through the artery. Thus blood pressure increases and so does the possibility of a heart attack or stroke. Second, cholesterol plays an important role in controlling the degree of solidification of the cell membrane (see Fig. 26.14). The greater the amount of cholesterol incorporated in the cell membrane between nonpolar tails of phospholipids, the more solid is the state of the cell membrane.

Cortisone is a superb anti-inflammatory drug. It is often used to control itching and rashes due to various allergies.

Progesterone is a female hormone that is vital in regulating the menstrual cycle and in preparing the uterus for a fertilized ovum.

Testosterone is an important male sex hormone. It is testosterone that is responsible for the secondary sex characteristics of the male along with tissue and muscle growth. Note that the very slight structural difference between testosterone, a male hormone, and progesterone, a female hormone, leads to huge physiological differences.

Vitamin $D_2$ provides protection from a vitamin-deficiency disease called *rickets*. Rickets is a disease of the young characterized by soft and deformed bones.

Each of the other classes of lipids thus far discussed always has one or more ester functional groups present. Not the steroids. In the steroids, the ester functional group may or may not be present in any particular molecule. When an ester group is not present, steroids cannot undergo saponification.

We ingest about 800 mg of cholesterol per day, while our bodies produce another 1 g or so. To decrease the amount of cholesterol ingested, avoid red meat, eggs, and cheese.

## 26.7 PROSTAGLANDINS

Prostaglandins are a family of recently discovered compounds biochemically derived from arachidonic acid and other C-20 unsaturated fatty acids.

Arachidonic acid    →(Several steps)→    Prostaglandin $E_2$

➤ Bond projects toward the reader
∥∥∥ Bond projects away from the reader

Prostaglandins are classified by a letter, a subscript number, and sometimes a subscript Greek letter. The letter designates the functional groups substituted on C-9 and C-11 of the cyclopentane ring. For example, a prostaglandin of the E type contains a keto group on C-9 and a hydroxyl group on C-11, and a prostaglandin of the F type contains two hydroxyl groups. The number refers to the number of carbon-carbon double bonds on the carbon side chains in the molecule. The subscript Greek letter designates the configuration of the hydroxyl group on C-9 with respect to that of the carboxyl side chain. If the groups are cis, use α; if trans, β. Some examples follow:

Prostaglandin $E_1$        Prostaglandin $F_{2\alpha}$

These compounds are excellent chemical regulators within the human body. Prostaglandin $E_2$ is a bronchodilator, prostaglandin $F_{2\alpha}$ initiates uterine contractions, whereas prostaglandin $E_1$ induces fever. It is thought that aspirin is effective as an antipyretic (fever reducer) because it inhibits the formation of prostaglandin $E_1$ from arachidonic acid. In addition, prostaglandins can regulate heart rate, blood clotting and pressure, immune responses, and kidney function—and that is not all.

*As expected, prostaglandin research is high in priority with drug companies today because no other family of compound administered as produced in very small amounts creates such a wide variety of substantial physiological responses.*

## ► CHAPTER ACCOMPLISHMENTS

### Introduction
☐ Give a definition for a lipid.
☐ List the five main classes of lipids.
☐ Name the two classes of lipids that do not undergo saponification if an ester group is not present.

### 26.1 Waxes
☐ Predict if the following compound is a wax.

$$C_{15}H_{31}-\overset{\displaystyle O}{\overset{\|}{C}}-OC_{32}H_{65}$$

☐ Explain the reason that lanolin is often used in pharmaceuticals and cosmetics.

### 26.2 Structure, Composition, and Nomenclature of Triglycerides (Fats and Oils)
☐ Name the following mixed triglyceride:

$$
\begin{array}{l}
CH_2O-\overset{\displaystyle O}{\overset{\|}{C}}-C_{13}H_{27}\\
\quad\quad\overset{\displaystyle O}{\phantom{X}}\\
CHO-\overset{\displaystyle O}{\overset{\|}{C}}-C_{13}H_{27}\\
\quad\quad\overset{\displaystyle O}{\phantom{X}}\\
CH_2O-\overset{\displaystyle O}{\overset{\|}{C}}-C_{13}H_{27}
\end{array}
$$

☐ Draw a structural formula for tristearin.
☐ Draw a structural formula for a molecule that would best represent depot fat given the data in Table 26.2.

### 26.3 Physical Properties of Triglycerides (Fats and Oils)
☐ Predict the solubility (soluble or insoluble) of tristearin in water and in carbon tetrachloride.
☐ State which class of triglyceride is a solid at room temperature.

### 26.4 Chemical Properties of Tryglycerides (Fats and Oils)
☐ Explain how a soap cuts grease.
☐ List one advantage and one disadvantage of using soap.
☐ List two reasons that synthetic detergents are increasing market share at the expense of soap.
☐ Draw a diagram showing soap molecules interacting with grease and the formation of a micelle.
☐ Write an equation that illustrates the formation of bathtub ring.
☐ Draw a structural formula for a(an)

    a. anionic detergent.
    b. cationic detergent.
    c. nonionic detergent.

☐ Define the term *surfactant*.
☐ Complete the following equations by drawing a structural formula for each organic product and writing a molecular formula or formula unit for each inorganic product.

    a. tristearin + 3 KOH(*aq*) $\longrightarrow$
    b. glyceryl palmitooleolinoleate + $H_2$ $\overset{Ni}{\longrightarrow}$
    c. glyceryl oleopalmitolinoleate + $O_2$ $\longrightarrow$

### 26.5 Phosphoglycerides (Phospholipids)
☐ Draw a structural formula for a lecithin.
☐ Draw a structural formula for a cephalin.
☐ Write an equation illustrating the saponification of a phospholipid.

### 26.6 Steroids
☐ Explain how cholesterol plays a role in the generation of a heart attack or stroke.
☐ Draw a structural formula for cholesterol.

### 26.7 Prostaglandins
☐ Explain why prostaglandin research is a high priority with drug companies today.
☐ Explain why aspirin is an effective antipyretic.

## ▶ KEY TERMS

*wax* (Introduction, 26.1)
*triglyceride* (Introduction, 26.2)
*simple lipid* (Introduction)
*compound lipid* (Introduction)
*phospholipid* (Introduction, 26.5)
*steroid* (Introduction, 26.6)
*prostaglandin* (Introduction, 26.7)
*derived lipid* (Introduction)
*spermaceti wax* (26.1)
*carnauba wax* (26.1)
*lanolin* (26.1)
*triacylglycerol* (26.2)
*simple triglyceride* (26.2)

*mixed triglyceride* (26.2)
*acid residue* (26.2)
*soap* (26.4)
*micelle* (26.4)
*synthetic detergent* (26.4)
*syndet* (26.4)
*anionic detergent* (26.4)
*biodegradable* (26.4)
*cationic detergent* (26.4)
*nonionic detergent* (26.4)
*surfactant* (26.4)
*rancid* (26.4)
*antioxidant* (26.4)

*drying oil* (26.4)
*bilayer* (26.5)
*cholesterol* (26.6)
*cortisone* (26.6)
*progesterone* (26.6)
*testosterone* (26.6)
*vitamin $D_2$* (26.6)
*arteriosclerosis* (26.6)
*rickets* (26.6)
*pardoxin* (Historical Boxed Reading 26.2)
*anorexia nervosa* (Historical Boxed Reading 26.1)

## ▶ PROBLEMS

1. Draw a structural formula that would best represent each of the following given the data in Table 26.2.
   a. olive oil
   b. butter
   c. linseed oil
   d. tallow

2. Draw a structural formula for each of the following mixed triglycerides.
   a. glyceryl linolenomyristooleate
   b. glyceryl oleolauropalmitate

3. Convert triolein to each of the following compounds. You may use any inorganic reagents necessary.
   a. oleic alcohol
   b. stearyl alcohol
   c. stearic acid
   d. stearamide
   e. stearyl stearate
   f. $CH_3(CH_2)_7\overset{\displaystyle H}{\underset{|}{C}}=O$

4. Tung oil is an excellent drying oil. Like any other fat or oil, its composition varies and depends on a number of factors. Typically, tung oil contains 10% esterified oleic acid, 5% esterified linoleic acid, and 81% esterified eleosteric acid. Given that eleostearic acid is (9*Z*, 11*E*, 13*E*)-octadecatrienoic acid, draw a structural formula for eleostearic acid. Then draw a structural formula that would best represent tung oil.

5. Define or explain
   a. rancidity.
   b. hard water.
   c. an oil.
   d. simple triglyceride.
   e. mixed triglyceride.
   f. hardening of an oil.
   g. a fat.
   h. a wax.
   i. a drying oil.

6. Which compound melts at a higher temperature, tristearin or trilinolein? Explain.

7. Why does the addition of a concentrated brine to a soap solution precipitate out the soap?

8. Classify the following as a soap, detergent, or neither.
   a. sodium butyrate
   b. sodium myristyl sulfate
   c. sodium *p*-tetradecylbenzenesulfonate
   d. potassium palmitate

9. Could a phospholipid replace a soap for cleaning? Explain.

10. Explain why a phospholipid is more soluble in water than a triglyceride.

11. Complete the following saponification equations by drawing a structural formula for each organic compound formed. If no reaction occurs, write "no reaction."

    a. $C_{15}H_{31}\overset{\displaystyle O}{\overset{\|}{C}}-OC_{26}H_{53} + NaOH(aq) \xrightarrow{\Delta}$

b.

$$CH_2-O-\overset{\overset{\displaystyle O}{\|}}{C}-(CH_2)_7CH=CH(CH_2)_7CH_3$$

$$HC-O-\overset{\overset{\displaystyle O}{\|}}{C}-(CH_2)_7CH=CH-CH_2-CH=CH(CH_2)_4CH_3 \ + \ 3\ NaOH(aq) \xrightarrow{\Delta}$$

$$CH_2-O-\overset{\overset{\displaystyle O}{\|}}{C}-(CH_2)_{10}CH_3$$

c. cholesterol + NaOH(aq) $\xrightarrow{\Delta}$

12. Name the triglyceride in Problem 11b.

13. Locate each chiral carbon in the cholesterol molecule, and designate every one with a star.

14. Comment on the physical properties of a typical steroid such as cholesterol. Explain each conclusion you reach with respect to
    a. physical state.
    b. water solubility.
    c. color.
    d. odor.

15. Draw a structural formula for cholesteryl stearate, a component of lanolin.

16. Prednisone is a derivative of cortisone used as a substitute for cortisone to treat rheumatoid arthritis because of fewer disagreeable side effects. Given that prednisone differs from cortisone by the presence of a carbon-carbon double bond on C-1, draw a structural formula for prednisone.

17. Draw a structural formula for
    a. tripalmitin.
    b. arachidonic acid.
    c. a typical oil.
    d. a soap.
    e. a wax.
    f. glyceryl myristooleopalmitate.
    g. prostaglandin $E_1$.
    h. phosphatidyl choline.
    i. a cephalin.
    j. BHT.

18. a. Given the structure of cortisone in Section 26.6 and that hydrocortisone represents a reduction of the non-conjugated alicyclic ketone in cortisone, draw two possible structural formulas for hydrocortisone.
    b. Given the prostaglandin structures in Section 26.7 and that a C prostaglandin contains the following cyclic structure

    draw a structural formula for prostaglandin $C_1$.

19. Criticize each of the following.
    a. A sample of beeswax melted at 70°C sharp.
    b. A sample of cortisone was easily saponified.
    c. Glycerol is a lipid.
    d. Fats are isolated from plants.
    e. It takes a large amount of a given prostaglandin to obtain a physiological response from a subject.

20. Spermaceti is a wax obtained from the head cavity and blubber of sperm whales and was used, years ago, in candles. In modern times, it is used in cosmetics. The main constituent of spermaceti is hexadecanyl hexadecanoate. Synthesize spermaceti from tripalmitin. You may use any inorganic reagents necessary.

21. Write an equation representing the reaction of oleic acid with each of the following (neglect stereoisomerism).
    a. $KMnO_4 + H_2O$ (cold)
    b. $H_2$ (Pd)
    c. $Br_2$ ($CCl_4$, $-20°C$)
    d. $Cl_2$, $PCl_3$
    e. NaOH(aq)
    f. $NaHCO_3(aq)$

# CHAPTER 27

# Carbohydrates

A substance was originally designated a *carbohydrate* if it was a hydrate of carbon, i.e., if it had a formula corresponding to $C_m(H_2O)_n$. One hundred years ago, glucose ($C_6H_{12}O_6$) was considered a carbohydrate, whereas starch ($C_6H_{10}O_5)_n$, where $n$ is a large number, was not. Today, a substance is considered to be a carbohydrate if it is (1) a *sugar*, (2) *starch* or *glycogen* and their derivatives, or (3) *cellulose* and its derivatives.

Carbohydrates can be classified in several ways. We will consider three of these classifications at this time.

The first classification is based on the hydrolysis of the carbohydrate. A carbohydrate that does not undergo hydrolysis is known as a *monosaccharide*. A *disaccharide* hydrolyzes to yield two monosaccharide units; a *trisaccharide* hydrolyzes to yield three monosaccharide units, etc. An *oligosaccharide* gives anywhere from two to ten monosaccharide units, whereas a *polysaccharide* furnishes eleven or more monosaccharide units on hydrolysis. In general,

**27.1
CLASSIFICATION OF
CARBOHYDRATES
AND SOME
STRUCTURES OF THE
MONOSACCHARIDES**

$$\text{monosaccharide} + H_2O \xrightarrow{H^+} \text{no reaction}$$

$$\text{disaccharide} + H_2O \xrightarrow{H^+} \text{2 monosaccharide units}$$

$$\text{polysaccharide} + (n-1)H_2O \xrightarrow{H^+} n \text{ monosaccharide units}$$

where n = several thousands. For example:

$$\text{maltose} + H_2O \xrightarrow{H^+} \text{2 glucose units}$$
A disaccharide

$$\text{lactose} + H_2O \xrightarrow{H^+} \text{1 glucose unit} + \text{1 galactose unit}$$
A disaccharide

$$\text{amylose} + 1999\ H_2O \xrightarrow{H^+} \text{2000 glucose units}$$
A component of
starch and a
polysaccharide

**501**

Before we discuss a second classification of monosaccharide carbohydrates, we need to examine the open-chain structure of these compounds. Monosaccharides are polyhydroxyketones (*ketoses*) or polyhydroxyaldehydes (*aldoses*), as shown in Fig. 27.1.

Monosaccharides are classified according to (1) the nature of the carbonyl group and (2) the number of carbon atoms in the molecule. Thus compound I in Fig. 27.1 is a *ketopentose* because it contains *five* carbons and a keto group, whereas compound II is an *aldotetrose* because it has *four* carbons and an aldehyde group. The -*ose* suffix is used in the nomenclature of carbohydrates. Let us sum this up:

| aldo-<br>(aldehyde) | tri-<br>(3 carbons) | -ose |
|---|---|---|
| keto-<br>(ketone) | tetr-<br>(4 carbons) | -ose |
| | pent-<br>(5 carbons) | |
| | hex-<br>(6 carbons) | |
| | hept-<br>(7 carbons) | |

The third method of classification is stereochemical. Let us consider an aldotriose, glyceraldehyde.

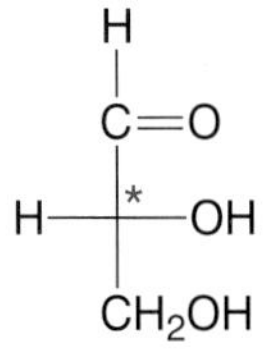

H                                        H

C=O        is equivalent to        C=O

H—*—OH                     H—C*—OH

$CH_2OH$                       $CH_2OH$

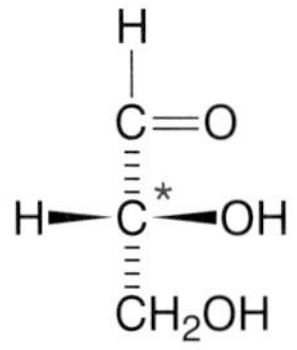

D-Glyceraldehyde          (R)-(+)-Glyceraldehyde

The chiral carbon in the Fischer projection (see Sec. 11.8) of glyceraldehyde is designated with an asterisk. When the chiral carbon farthest away from the aldehyde (or keto group) has a hydroxyl group to the right in glyceraldehyde or any other sugar, it is designated a *D-sugar*; when the farthest hydroxyl group is oriented to the left, the compound is called an *L-sugar*. For an example, see Fig. 27.2.

A mixture of 50% D and 50% L sugars represents a racemic mixture. The chiral carbons in erythrose are designated with asterisks. An important point to understand is that there is no relationship between the D- and L-stereochemical families and direction of rotation [(+), dextrorotary; (−), levorotary] of the compound. Thus you can find D-(+) and D-(−) sugars along with L-(+) and L-(−) sugars.

*Note that the D- stereoisomer in Fig. 27.2 is a mirror image of the L- stereoisomer.*

---

**FIGURE 27.1** General structures of a ketose and aldose.

$CH_2OH$            H

C=O             C=O

CHOH           CHOH

CHOH           CHOH

$CH_2OH$        $CH_2OH$

A polyhydroxyketone (I)  A polyhydroxyaldehyde (II)

(a)                (b)

**FIGURE 27.2** Fischer projections of two enantiomeric aldotetroses.

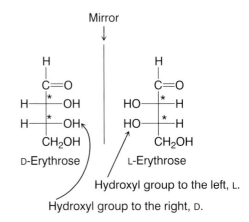

The D- and L- prefixes denote stereochemical families traditionally derived from D-(+)-glyceraldehyde and L-(−)-glyceraldehyde. These D- and L- stereochemical designations are still used with sugars and amino acids (see Chap. 28).

---

**SOLVED PROBLEM 27.1**

Classify the following monosaccharides.

a. (As an aldose or ketose)

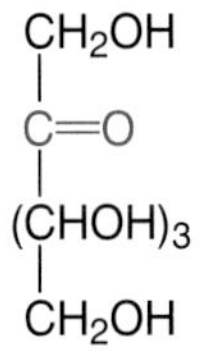

$$CH_2OH$$
$$|$$
$$C{=}O$$
$$|$$
$$(CHOH)_3$$
$$|$$
$$CH_2OH$$

b. (Stereochemically)

```
        H
        |
        C=O
        |
   H————OH
        |
   H————OH
        |
  HO————H
        |
  HO————H
        |
       CH2OH
```

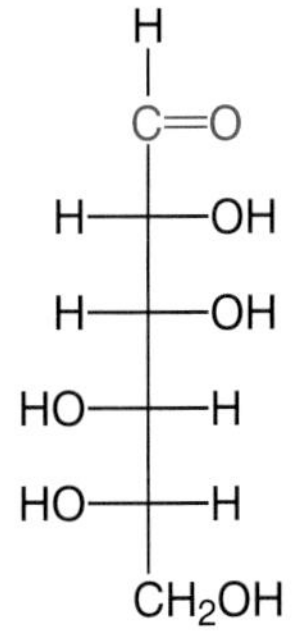

**SOLUTION**

a. Since the functional group of this sugar is a ketone and the compound contains six carbons, this compound is classified as a ketohexose.

b. Locate the chiral carbon farthest from the aldehyde group. Since the hydroxyl group bonded to that carbon is located on the left-hand side of the molecule, this is an L-sugar.

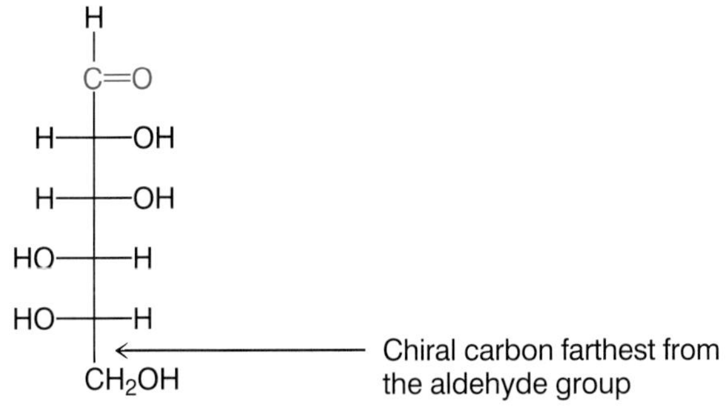

```
        H
        |
        C=O
        |
   H————OH
        |
   H————OH
        |
  HO————H
        |
  HO————H          ←———————— Chiral carbon farthest from
       CH2OH                    the aldehyde group
```

---

**EXERCISE 27.1**

Classify the following sugars.

a. (As an aldose or ketose)

$$H$$
$$|$$
$$C{=}O$$
$$|$$
$$(CHOH)_4$$
$$|$$
$$CH_2OH$$

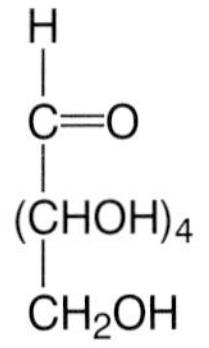

b.  (Stereochemically)

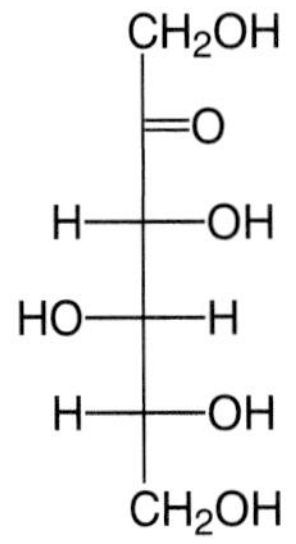

$$
\begin{array}{c}
CH_2OH \\
\quad = O \\
H\!-\!\!-\!OH \\
HO\!-\!\!-\!H \\
H\!-\!\!-\!OH \\
CH_2OH
\end{array}
$$

## 27.2
## MORE ABOUT MONOSACCHARIDE STRUCTURE

We now know that the stereochemical family of a carbohydrate is based on the orientation of the farthest hydroxyl group from the carbonyl: To the right, the sugar is designated D; to the left, L. It is the pattern of placement of each hydroxyl group on its particular chiral carbon in a Fischer projection that identifies a given sugar. Here are a few examples:

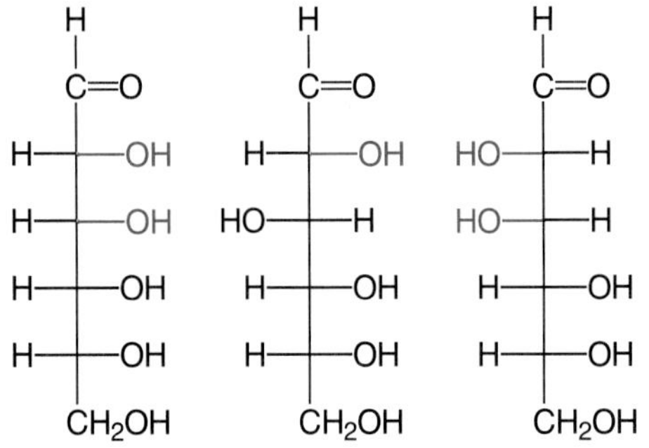

$$
\begin{array}{ccc}
\text{H} & \text{H} & \text{H} \\
| & | & | \\
\text{C}=\text{O} & \text{C}=\text{O} & \text{C}=\text{O} \\
\text{H}-\text{OH} & \text{H}-\text{OH} & \text{HO}-\text{H} \\
\text{H}-\text{OH} & \text{HO}-\text{H} & \text{HO}-\text{H} \\
\text{H}-\text{OH} & \text{H}-\text{OH} & \text{H}-\text{OH} \\
\text{H}-\text{OH} & \text{H}-\text{OH} & \text{H}-\text{OH} \\
\text{CH}_2\text{OH} & \text{CH}_2\text{OH} & \text{CH}_2\text{OH} \\
\text{D-Allose} & \text{D-Glucose} & \text{D-Mannose}
\end{array}
$$

Note the mirror image relationship of the groups bonded to the chiral carbons.

To draw an enantiomeric Fischer projection, only swap hydrogen and hydroxyls bonded to chiral carbons. Thus we have for L-allose

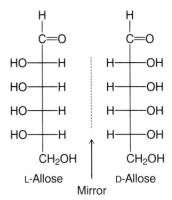

L-Allose            D-Allose

Mirror

Table 27.1 displays Fischer projections of the D-stereochemical series of sugars.

## SOLVED PROBLEM 27.2

Draw a Fischer projection formula for L-glucose.

**SOLUTION**

Use Table 27.1 to obtain the Fischer projection formula of D-glucose. In L-glucose, each of the four hydroxyl groups is swapped with the hydrogen bonded to each chiral carbon to give the mirror image of the D-isomer.

**TABLE 27.1** A Series of Fischer Projection Formulas of the D-Stereochemical Family of Sugars Derived from D-Glyceraldehyde

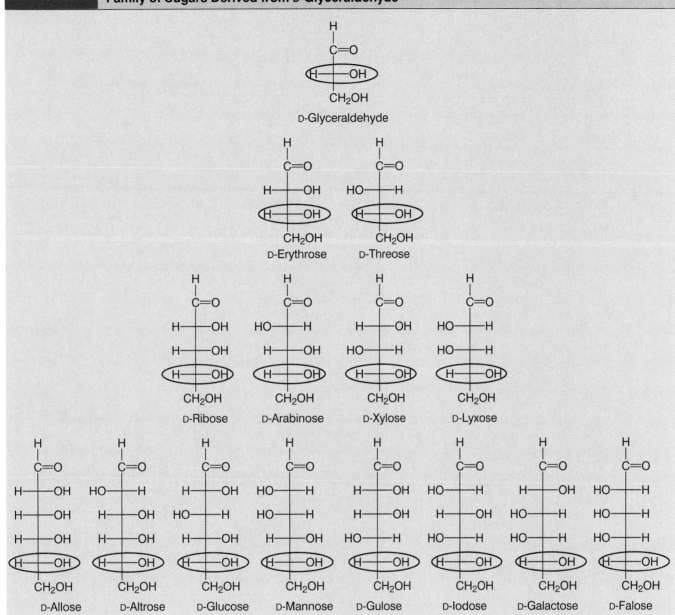

Note that an oval locates the hydroxyl group on the chiral center (the chiral carbon atom) farthest from the carbonyl group.

Draw a Fischer projection of L-altrose. Use Table 27.1.

Just as glyceraldehyde is the parent compound of the two stereochemical aldose series of sugars, so dihydroxyacetone, a ketotriose, is the parent compound of the two stereochemical families of ketoses (both D- and L-series).

$$
\begin{array}{c}
CH_2OH \\
| \\
C=O \\
| \\
CH_2OH
\end{array}
$$

Dihydroxyacetone (a ketotriose)

A significant stereochemical difference exists between glyceraldehyde and dihydroxy-acetone. What is this difference?

Some examples of ketoses are provided in Fig. 27.3.

## 27.3 PHYSICAL PROPERTIES OF MONOSACCHARIDES

Sugars are polar compounds. Thus you would expect solubility in polar solvents (e.g., water) and insolubility in nonpolar solvents (e.g., benzene, carbon tetrachloride). Sugars are solids and are sweet to the taste, the degree of sweetness depending on the sugar.

## 27.4 UNEXPECTED REACTIONS OF THE MONOSACCHARIDES: MUTAROTATION— CYCLIC FORMS

Some of the reactions of the monosaccharides that we will discuss in Section 27.11 are based on the functional groups present in the compounds. However, the monosaccharides displayed some chemical properties that simply did not make sense in terms of the open-chain structure of the sugars.

1. Two separate forms of D-glucose (and other monosaccharides) were isolated. One form was named α-D-glucose because it had the higher specific rotation ($+112$ degrees). The other form, with a lower specific rotation of $+19$ degrees, was called β-D-glucose. This was puzzling because there could not be two open-chain forms of D-glucose.

2. D-Glucose (or other monosaccharides) in aqueous solution showed a change of specific rotation with time, a phenomenon known as *mutarotation*. For example, the specific rotation of α-D-glucose in water changed from 112 to 53 degrees in about 13 hours. In the same way, the specific rotation of β-D-glucose changed from 19 to 53 degrees in about 13 hours.

3. Simple aldehydes reacted with 2 mol of an alcohol to produce an acetal (see Secs. 18.6 and 19.2). Yet D-glucose (or other monosaccharides) only reacted with 1 mol of an alcohol to produce a stable product. For example, D-glucose reacted with only 1 mol of methyl alcohol.

**FIGURE 27.3** Fischer projections of two ketoses.

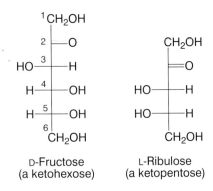

D-Fructose
(a ketohexose)

L-Ribulose
(a ketopentose)

These three unexpected properties of the monosaccharides (using D-glucose as an example) were eventually explained, but it took years to completely solve the structural puzzle. The reasoning to contribute to the solution of the problem of structures went something like this: When both alpha and beta forms of a given monosaccharide were dissolved in water, the fact that the resulting solutions ultimately gave the same specific rotation is indicative that an equilibrium is involved. When α-D-glucose is added to water, the specific rotation of the sugar begins to decrease as more and more β-D-glucose is produced. Eventually, the rate of β-D-glucose formation slows down, and β-D-glucose starts to re-form α-D-glucose. Eventually, the rates become equal (in about 13 hours), and the system reaches a state of equilibrium with 36% α-D-glucose and 64% β-D-glucose present. The amount of open-chain aldehyde present at equilibrium is very small (less than 0.2%). Thus chemists hypothesized an equilibrium for the monosaccharide involving the α-sugar, β-sugar, and the open-chain form of the carbohydrate. For D-glucose, we have

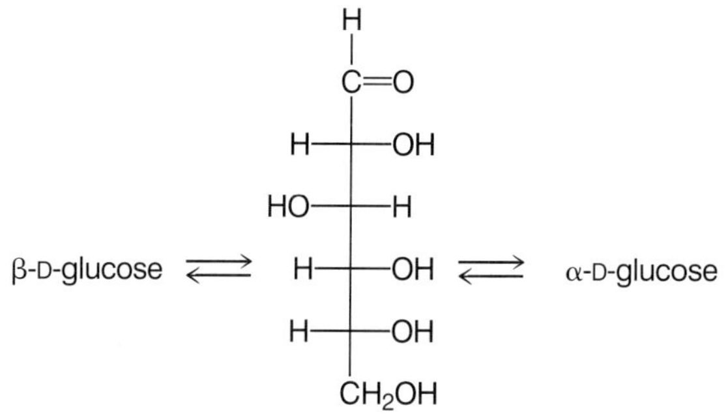

Open-chain form of D-glucose

Structurally, both α-D-glucose and β-D-glucose had to be stable hemiacetals. This way, only 1 mol of alcohol would add to each isomer to form the acetal. Ultimately, a cyclic hemiacetal was shown to be the correct structure for the alpha (Fig. 27.4a) and beta (Fig. 27.4b) forms of glucose and other monosaccharides. Each cyclic structure is known as a *Haworth projection* and often is abbreviated as in Fig. 27.5.

Note that the alpha and beta forms of glucose are similar; the only difference is that the hydroxyl bonded to C-1 in the alpha form is projected down, whereas it is projected up in the beta form. Since α-D-glucose and β-D-glucose only differ in configuration about C-1, these stereoisomers are known as *anomers*.

**FIGURE 27.4** Haworth projections of α-D-glucose and β-D-glucose.

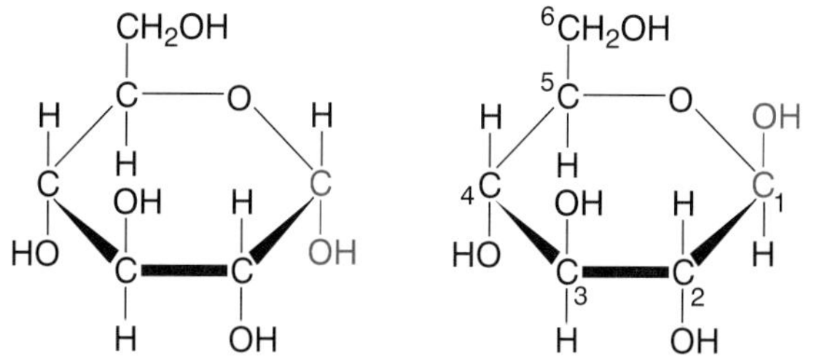

α-D-Glucose
(a)

β-D-Glucose
(b)

**FIGURE 27.5** Geometric condensed Haworth projections of α-D-glucose and β-D-glucose.

In the α form, the hydroxyl group on C-1 is projected down.

In the β form, the hydroxyl group on C-1 is projected up.

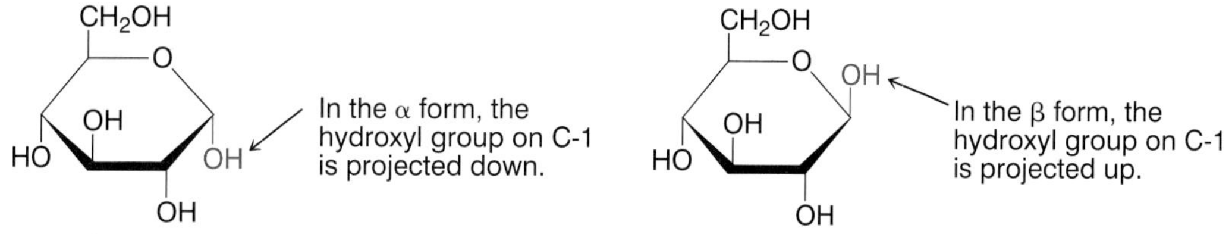

α-D-Glucose
(a)

β-D-Glucose
(b)

**Envision the Reaction**

## Mechanism of the Conversion of the Open-Chain Form (Fischer Projection) of D-Glucose to the Haworth Projection of the Alpha or Beta Form

**First step:**

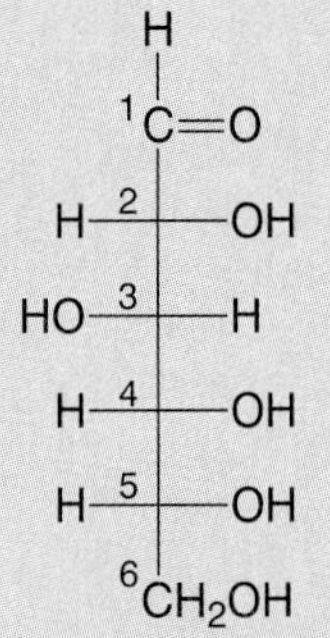

Turn structure I
on its side to get structure II

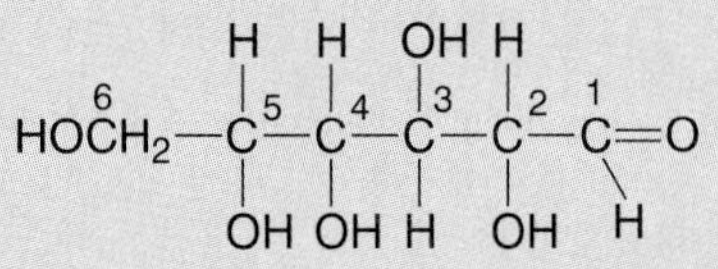

Open-chain form of
D-Glucose
(Fischer projection)
I

Now rotate bonds 1—2, 3—4, 4—5,
and 5—6 to obtain structure IIa

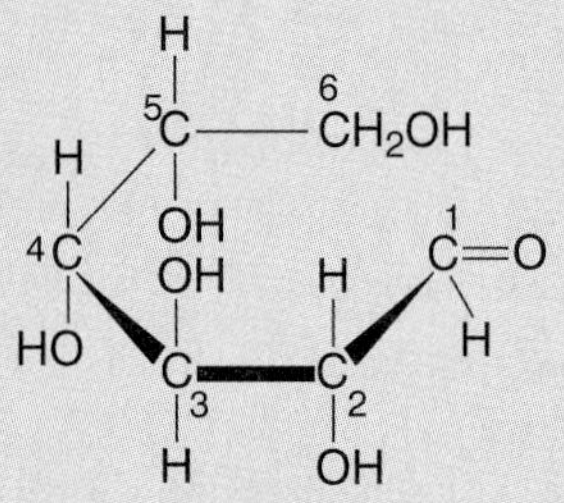

IIa

**Second step:**

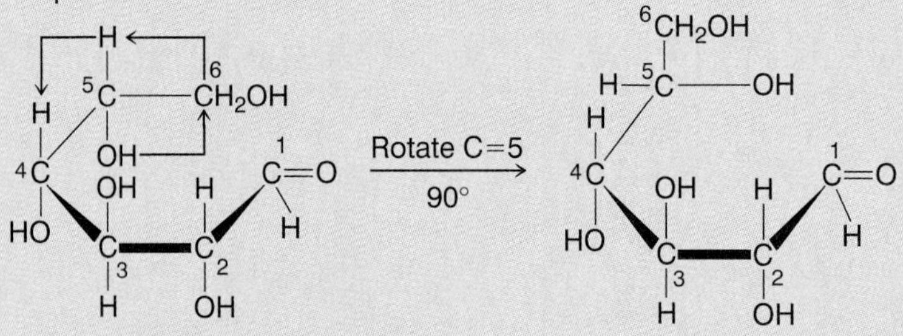

Rotate C=5
90°

IIa                                                IIb

**Third step:**

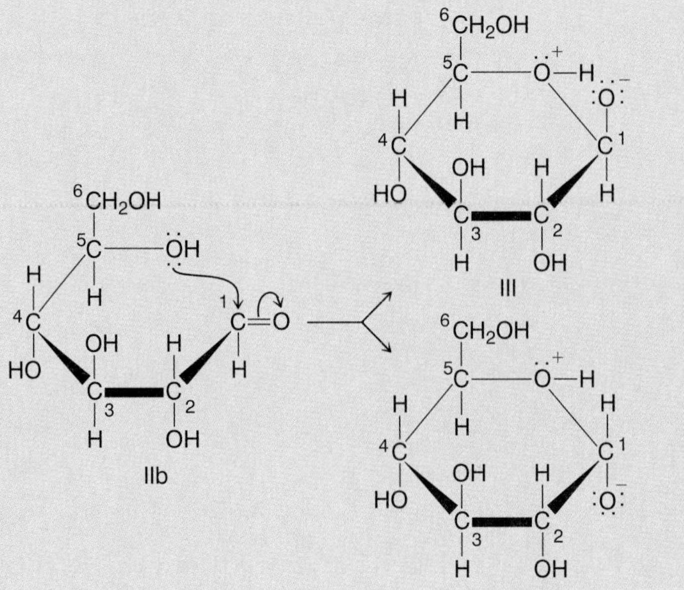

IIb

III

IV

**Mechanism of the Conversion of the Open-Chain Form
(Fischer Projection) of D-Glucose to the Haworth Projection
of the Alpha or Beta Form (*continued*)**

Fourth step:

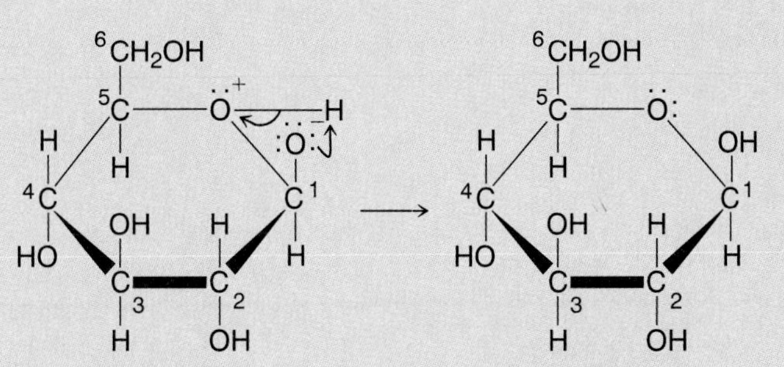

III

β-D-Glucose

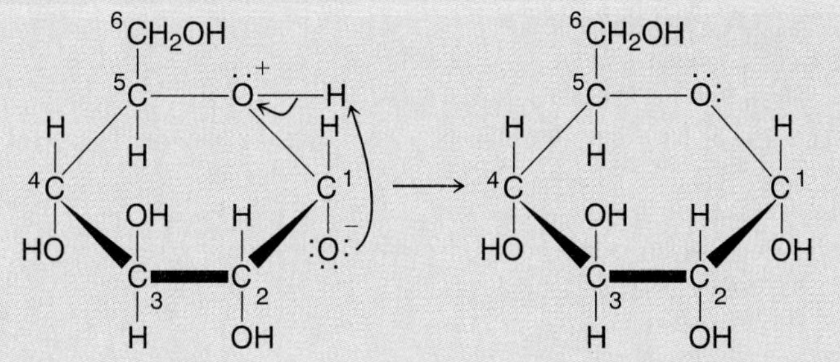

IV

α-D-Glucose

The mechanism of the conversion of the open-chain form (Fischer projection) of D-glucose to the Haworth projection of the alpha or beta form begins by orienting the glucose molecule in such a manner that C-1 and the hydroxyl bonded to C-5 are near each other (first step). Constructing a molecular model of IIa will demonstrate that the hydroxyl bonded to C-5 is relatively far from C-1. To remedy this, rotate C-5 90 degrees to the right (second step) to obtain a conformer (see Sec. 2.3) of IIa, IIb. This will locate the hydroxyl group on C-5 to a position nearer to the carbonyl carbon. Now the hydroxyl oxygen on C-5 is in a position to attack the carbonyl carbon by nucleophilic addition in the third step of the mechanism. When the substituted alkoxide ion (III) is oriented up, β-D-glucose is ultimately produced when the hydroxyl proton on C-5 migrates to the oxygen bonded to C-1 (fourth step); when the substituted alkoxide ion (IV) is projecting down, α-D-glucose is the resulting product.

Two simple rules help in converting a Fischer projection to a Haworth projection:

1. A hydroxyl bonded to the right in an open-chain Fischer projection is oriented *down* in the Haworth projection. A hydroxyl bonded to the left in an open-chain Fischer projection is oriented *up* in a Haworth projection.

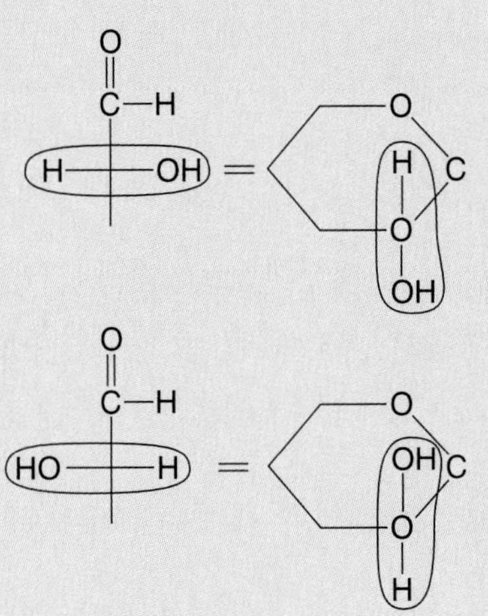

2. In the Haworth projection of a D-sugar, the —CH$_2$OH group is oriented *up*.

---

**SOLVED PROBLEM 27.3**

Draw a Haworth projection for α-D-galactose.

**SOLUTION**

The only difference between a Fischer projection of D-galactose as opposed to D-glucose is that in D-galactose the C-4 hydroxyl projects to the left. Thus the hydroxyl on C-4 in the Haworth projection juts up. Because this is a D-sugar the —CH$_2$OH group projects up and the hydroxyl on C-1 is oriented down because this is an alpha anomer. Putting all this together, we have

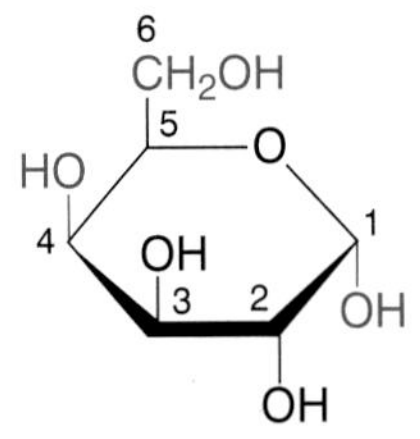

α-D-Galactose

---

**EXERCISE 27.4**

Draw a Haworth projection formula for each of the following.

a.  β-D-mannose

b.  α-D-allose

c.  β-D-galactose

---

Now a reasonable question to ask at this point is, Why is a greater amount of β-D-glucose (64%, page 507) present in the equilibrium mixture? Putting it another way, Why doesn't the equilibrium mixture consist of 50% α-D-glucose and 50% β-D-glucose? The reason, simply put, is that β-D-glucose is more stable than α-D-glucose. To demonstrate this, we need to draw the anomers in a chair conformation that a number of monosaccharides actually adopt (see Sec. 3.10) rather than the convenient, but less sterically accurate, planar Haworth projection (page 507).

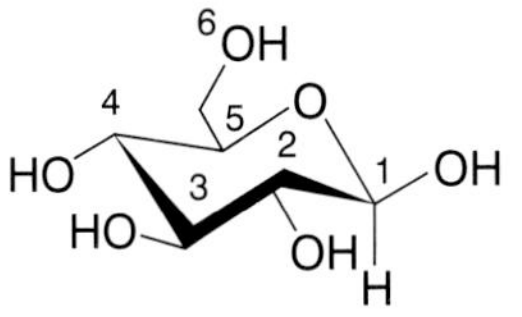

β-D-Glucose
(All the large groups are
equatorial. Thus, this
anomer is more stable.)

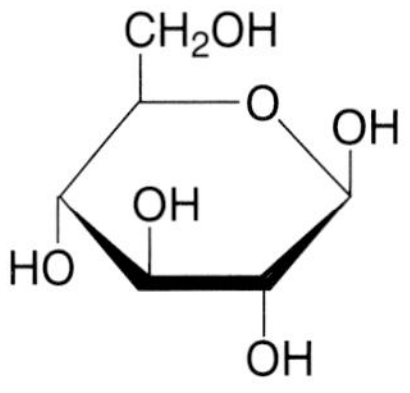

β-D-Glucose

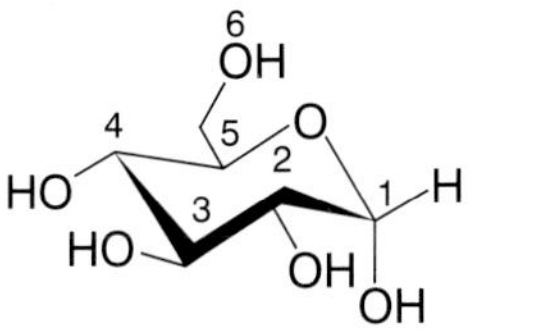

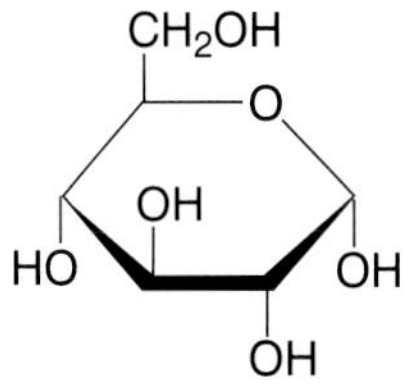

α-D-Glucose
(All the large groups are
equatorial but one. Since
the hydroxyl on C-1 is in an
axial position, this anomer
is less stable.)

α-D-Glucose

Note that β-D-glucose is the more
stable isomer because every bulky
—OH group and —CH₂OH group is
located in an equatorial position in
the molecule. Since the alpha
anomer contains one —OH group
on C-1 that is located in an axial
position, destabilizing 1,3-diaxial
interactions come into play (see
Sec. 3.10).

---

Draw a structural formula of the chair conformation adopted by α-D-galactose.

**SOLVED PROBLEM 27.4**

**SOLUTION**

Since each bulky substituent in the β-D-glucose ring is in an equatorial position in the ring, and since the only differences between α-D-galactose and β-D-glucose are in C-1 and C-4 hydroxyl placement, then in α-D-galactose the C-1 and C-4 hydroxyls are in axial positions on the ring. Putting it another way, each β-monosaccharide contains an equtorial hydroxyl on C-1, whereas each α-monosaccharide includes an axial hydroxyl on C-1. Furthermore, since the hydroxyl on C-4 in glucose is in an equatorial position, the hydroxyl on C-4 in galactose must be in an axial position.

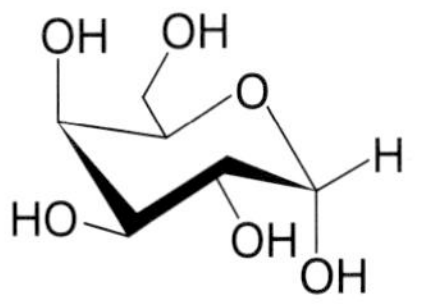

α-D-Galactose

---

Draw a structural formula of the chair conformation adopted by each of the following.

a.   α-D-mannose

b.   β-D-galactose

**EXERCISE 27.5**

---

Finally, let us use the structures of α-D-glucose and β-D-glucose and the general equilibrium hypothesized to explain the unexpected reactions of D-glucose (and other monosaccharides).

α-D-Glucose and β-D-glucose are cyclic hemiacetals and only react with 1 mol of alcohol to produce the acetal. Cyclic hemiacetals of this type are more stable than the corresponding hemiacetals produced from the simple aldehydes. Thus β-D-glucose and α-D-glucose react to produce the corresponding glucosides.

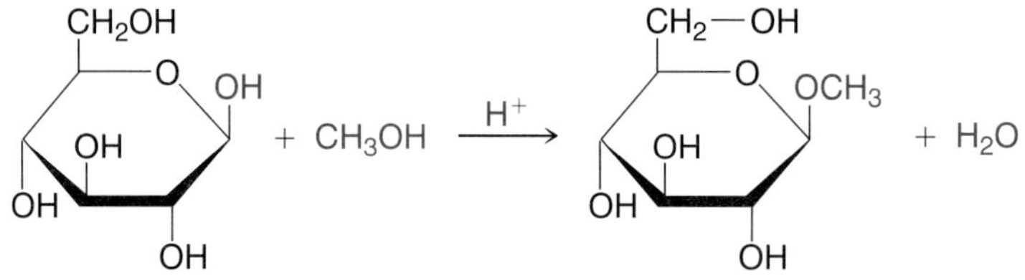

β-D-Glucose
(a hemiacetal)

Methyl-β-D-glucoside
(an acetal)

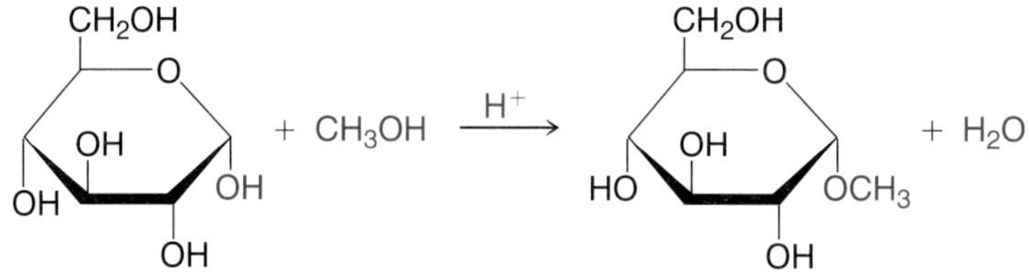

α-D-Glucose
(a hemiacetal)

Methyl-α-D-glucoside
(an acetal)

The Haworth projections are planar and are convenient to use in the reactions of the cyclic hemicetalic forms of the monosaccharides. However, experimental work has proved that these anomeric forms of D-glucose (or any other hexose) actually exist in the chair form. Thus we can more accurately write

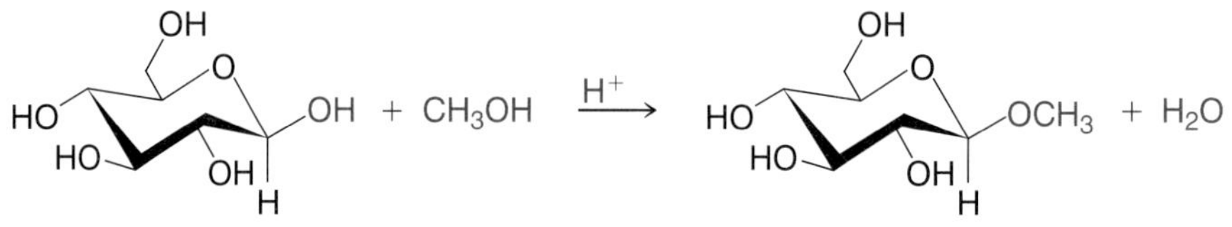

β-D-Glucose

Methyl-β-D-glucoside

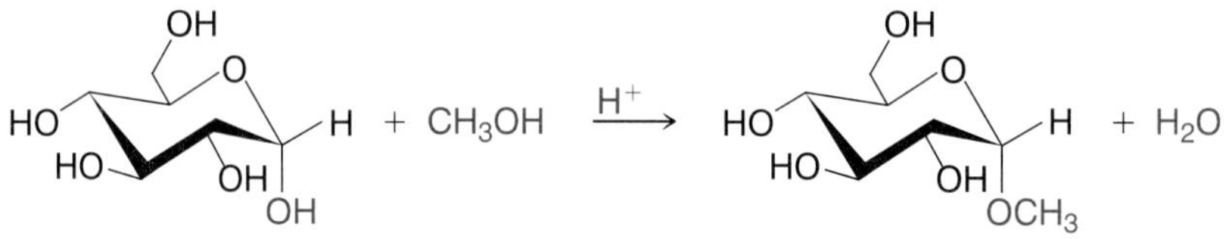

α-D-Glucose

Methyl-α-D-glucoside

One other unexpected reaction of the monosaccharides (osazone formation) is discussed in Section 27.11.

## 27.5
## MORE ABOUT CYCLIC SUGARS: STRUCTURE AND NOMENCLATURE

For example, D-ribose, an aldopentose, exists as an equilibrium mixture of the open-chain form and two cyclic five-membered ring forms.

Usually, aldohexoses form six-membered rings, and aldopentoses form five-membered rings.

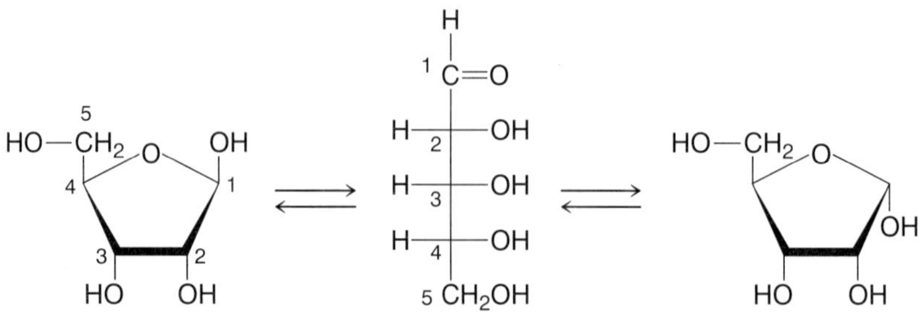

β-D-Ribose

D-Ribose

α-D-Ribose

To distinguish the monosaccharides containing a five-membered ring from the sugars including a six-membered ring, an addendum to monosaccharide nomenclature is necessary. Since a six-membered ring resembles pyran, a heterocyclic compound, such a sugar is called a *pyranose* (Fig. 27.6).

In the same manner, since a five-membered ring resembles furan, a heterocyclic compound, such a sugar is called a *furanose* (Fig. 27.7). To name a sugar as a pyranose or furanose, drop the *-se* ending of the monosaccharide, and add *pyranose* or *furanose*. An example is given in Fig. 27.8.

Fructose, a ketohexose, is the most abundant ketose (Fig. 27.9). It is more complex structurally. By itself, fructose exists in a pyranose form. However, when combined with another monosaccharide to form a disaccharide (see Sec. 27.8), fructose exists as a furanose.

β-D-Fructopyranose

α-D-Fructopyranose

β-D-Fructofuranose

α-D-Fructofuranose

**FIGURE 27.6** Pyran and a pyranose.

Pyran

β-D-Glucose
(a pyranose)

**FIGURE 27.7** Furan and a furanose.

Furan

β-D-Ribose
(a furanose)

---

The monosaccharides D-2-deoxyribose and D-galactose are of some biological importance. Draw a Haworth projection for each cyclic form, and name each cyclic form as a pyranose or furanose.

**SOLUTION**

Let us first draw a Fischer projection for each monosaccharide.

D-2-Deoxyribose
(Fischer projection)

D-Galactose
(Fischer projection)

Now, since D-2-deoxyribose is an aldopentose, the sugar exists in the furanose form. Furthermore, we can use the the rules of Haworth projection formation from the corresponding Fischer projection to get

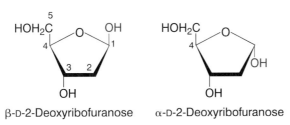

β-D-2-Deoxyribofuranose                α-D-2-Deoxyribofuranose

**SOLVED PROBLEM 27.5**

**FIGURE 27.8** Examples of pyranose and furanose nomenclature.

β-D-Glucopyranose

β-D-Ribofuranose

**FIGURE 27.9** Fischer projection of D-fructose.

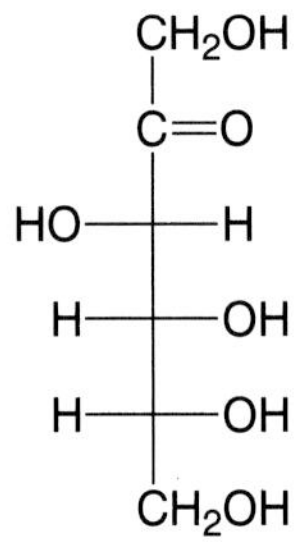

$$CH_2OH$$
$$C=O$$
$$HO—H$$
$$H—OH$$
$$H—OH$$
$$CH_2OH$$

Again, since D-galactose is an aldohexose, the sugar exists in the pyranose form. Furthermore, we can use the rules of Haworth projection formation from the corresponding Fischer projection to get

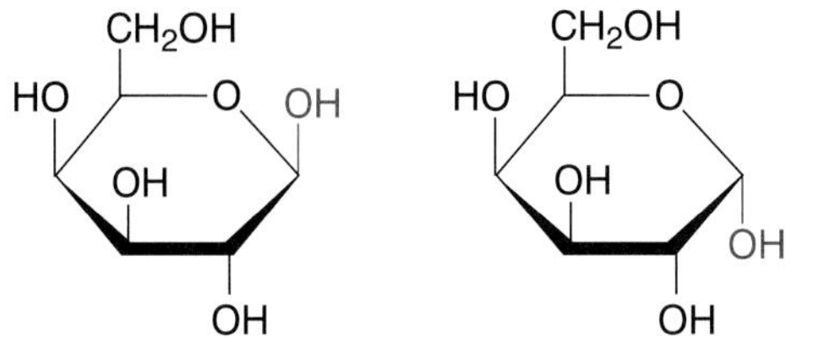

β-D-Galactopyranose              α-D-Galactopyranose

---

**EXERCISE 27.6**

Draw a Haworth projection for and name each cyclic form of the following as a pyranose or furanose.

a.   D-arabinose

b.   D-allose

---

Monosaccharides and all carbohydrates also can be classified as *reducing* or *nonreducing* based on the reaction of the carbohydrate with Benedict's or Fehling's solution (see Sec. 27.6). A positive test is indicated by the precipitation of red copper(I) oxide in the test tube. All aldoses and C-2 ketoses* are reducing sugars because the pyranose or furanose form contains a hemiacetal group.

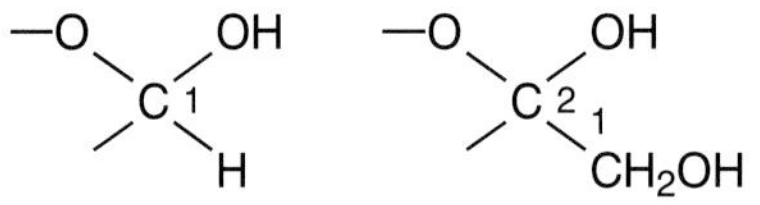

Hemiacetal group          Hemiacetal group
derived from              derived from
an aldehyde               a ketone

Should C-1 in the case of an aldose or C-2 in the case of a ketose contain a full acetal, the carbohydrate is a nonreducing sugar.

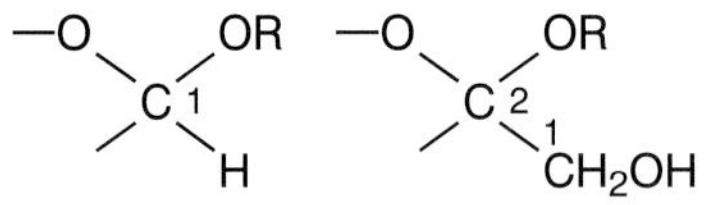

Acetal group             Acetal group
derived from             derived from
an aldehyde              a ketone

Consider Solved Problem 27.6 to get practice in identifying reducing and nonreducing carbohydrates.

---

**SOLVED PROBLEM 27.6**

Explain why D-glucose and D-fructose are classified as reducing sugars. Draw a structure for a nonreducing sugar and explain.

**SOLUTION**

A reducing sugar must contain a hemiacetal group. This structure is in equilibrium with the open-chain aldehyde or ketone compound, and oxidation proceeds. For D-glucose, we have

---

*In actuality, C-2 ketoses do not react with Benedict's or Fehling's solution. What takes place is that the ketose is converted to an aldose in basic solution, and it is the aldose that actually reacts with Benedict's or Fehling's solution.

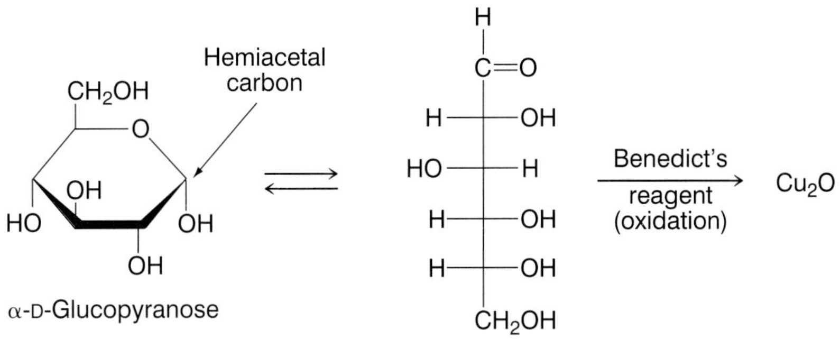

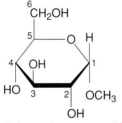

α-D-Glucopyranose

D-Glucose

and for D-fructose, we have

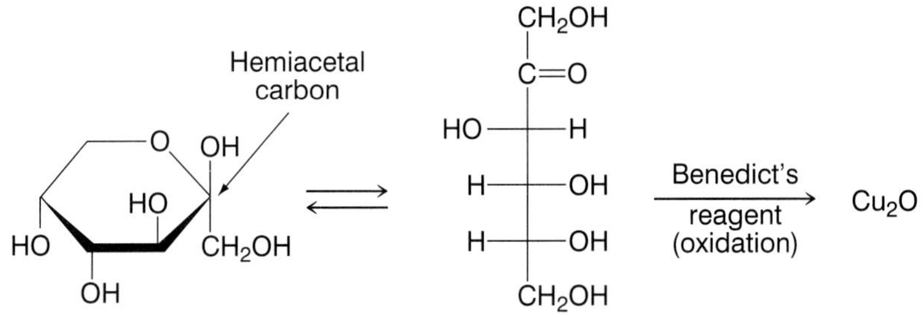

β-D-Fructopyranose

D-Fructose

The glucoside, methyl α-D-glucopyranoside, is a nonreducing sugar because the pyranose form contains a full acetal (at C-1) that is not in equilibrium with the open-chain aldose, and oxidation cannot occur. To name a glycoside (a general term including a glucoside, mannoside, galactoside, etc.), name the alkyl group bonded to the C-1 oxygen as one word. Finally, drop the *-e* fragment and substitute the suffix *-ide*.

Methyl α-D-glucopyranoside

Note the two ether groups bonded to C-1. This *gem*-diether represents a full acetal. Thus this compound is a nonreducing carbohydrate.

**SOLVED PROBLEM 27.7**

Name the following compound.

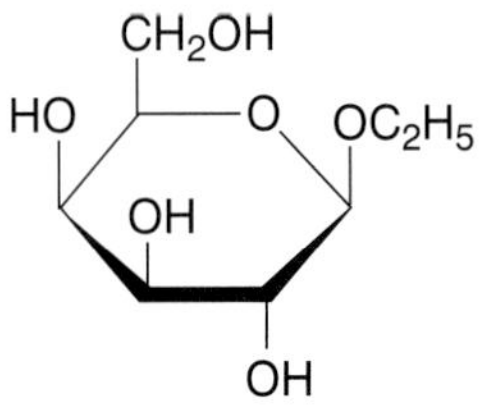

**SOLUTION**

The alkyl group on the C-1 oxygen is ethyl. Thus we have

ethyl

Now, name the sugar as an aldose to give

ethyl β-D-galactopyranose

Finally, replace the *-e* fragment with *-ide* to give

ethyl β-D-galactopyranoside

---

**EXERCISE 27.7**

Classify the following compounds as a reducing or nonreducing sugar. Justify your answer in each case by drawing the appropriate projection (Haworth or Fischer).

a.   D-arabinose

b.   β-D-mannopyranose

c.   ethyl α-D-galactopyranoside

d.   methyl β-D-ribofuranoside

---

## 27.6
## DIAGNOSTIC TESTS
## FOR CARBOHYDRATES

The oxidation of aldoses in basic solution gives a variety of organic products. Thus these reactions have no synthetic utility. However, they are excellent diagnostic tests to differentiate a reducing carbohydrate from a nonreducing carbohydrate.

Sugars can be categorized by reaction with $Ag^+$ (Tollens solution; see Sec. 18.6), $Cu^{2+}$ complex, and $^-OH$ (Fehling's solution and Benedict's solution). A reducing carbohydrate reacts with Tollens, Benedict's, or Fehling's solutions to give a telltale silver mirror (Tollens test) or red precipitate of copper(I) oxide (Benedict's or Fehling's test). Both simple aldoses and ketoses are reducing carbohydrates (see Sec. 27.5 footnote).

aldose or ketose + $^+Ag(NH_3)_2$ + $^-OH$ $\longrightarrow$ Ag (a silver mirror forms on the test tube)

Tollens solution

aldose or ketose + $Cu^{2+}$ (citrate complex) $\longrightarrow$ $Cu_2O$ (red precipitate)

Benedict's solution

aldose or ketose + $Cu^{2+}$ (tartrate complex) $\longrightarrow$ $Cu_2O$ (red precipitate)

Fehling's solution

---

## 27.7
## SOME IMPORTANT
## D-MONOSACCHARIDES

D-Glucose is also known as *blood sugar* because it is found in the blood (typically, 90 mg/100 ml) and it is intimately involved in human carbohydrate metabolism.

D-Glucose (also known as *dextrose*) is certainly the most abundant of the monosaccharides. D-Fructose (also called *levulose*) is found in honey and in assorted fruits, whereas D-2-deoxyribose is a constituent of DNA (deoxyribonucleic acid), the storehouse of genetic information about an individual. D-Ribose is a constituent of RNA (ribonucleic acid) that is present in almost all cells, where it, RNA, directs the synthesis of proteins.

---

**EXERCISE 27.8**

Speculate as to why D-glucose is known as *dextrose* and D-fructose is also called *levulose*?

---

## 27.8
## DISACCHARIDES

**Maltose**

A disaccharide consists of two monosaccharide units linked together and can be thought of as an acetal where one monosaccharide unit represents the alcohol and the other represents the hemiacetal.

Hydrolyzing 1 mol of maltose with water and acid produces 2 mol of D-glucose.

$$\text{maltose} + H_2O \xrightarrow{H^+} \text{D-glucose} + \text{D-glucose}$$

Maltose is found in germinating barley and other grains. Maltose is usually prepared by the enzymatic (partial) hydrolysis of starch (see Sec. 27.9).

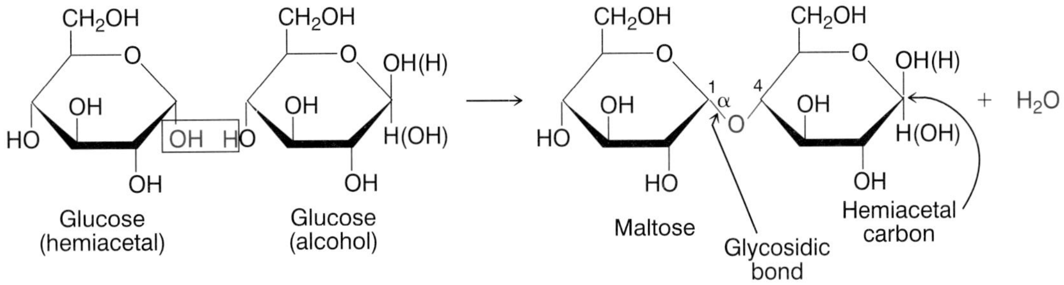

Glucose (hemiacetal)   Glucose (alcohol)   →   Maltose   Glycosidic bond   Hemiacetal carbon   $+ H_2O$

α-D-Glucopyranosyl-(1, 4)-β-D-glucopyranose or
α-D-glucopyranosyl-(1, 4)-α-D-glucopyranose

Two points need to be made here to explain the systematic nomenclature of maltose (a trivial name) in the preceding formulation:

1. The glycosidic bond linking the monosaccharide units is always alpha in maltose, whereas the "alcohol" glucose C-1 hydroxyl can be projected alpha or beta. We say that maltose is an *α-glucoside*.
2. The nonreducing unit of any disaccharide by convention is displayed on the left and is named by converting the suffix *-ose* to *-osyl*. Let us explore the systemic name of maltose in some depth. The ring on the left (the nonreducing unit) has the oxygen at C-1 projecting down (an α-glucosidic bond), so we have

α-D-glucopyranosyl

Since the link between the two rings is at C-1 in the nonreducing unit and C-4 in the reducing unit, we have

α-D-glucopyranosyl-(1,4)

Finally, we name the reducing unit of the carbohydrate:

α-D-glucopyranosyl-(1,4)-α-D-glucopyranose   or
α-D-glucopyranosyl-(1,4)-β-D-glucopyranose

Since maltose contains a hemiacetal carbon, this carbohydrate is reducing and undergoes mutarotation in the same manner as the monosaccharides.

α-maltose ⇌ open-chain form of maltose ⇌ β-maltose

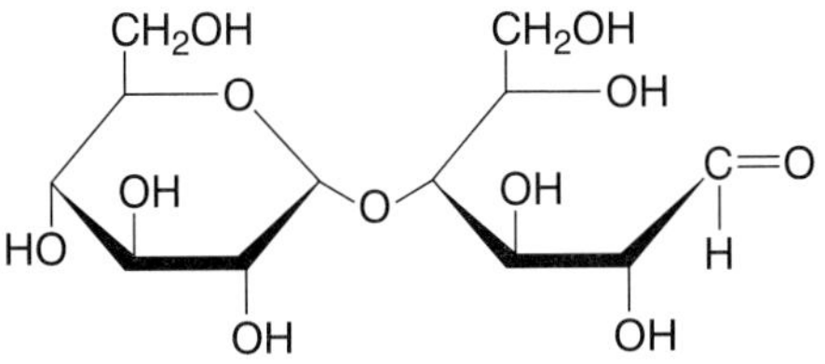

Maltose and all other disaccharides also can be pictured by a more accurate chair projection (Fig. 27.10b) along with the more convenient Haworth projection (Fig. 27.10a).

**FIGURE 27.10** Various projections of maltose.

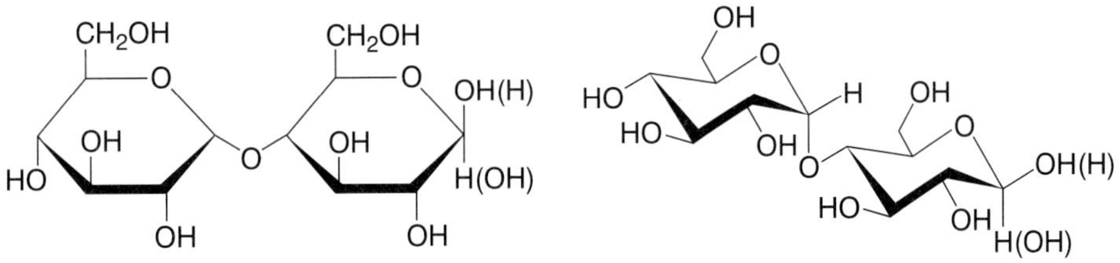

(a) Haworth projection                    (b) Chair projection

## Cellobiose

Hydrolyzing 1 mol of cellobiose with water and acid produces 2 mol of D-glucose.

$$\text{cellobiose} + H_2O \xrightarrow{H^+} \text{D-glucose} + \text{D-glucose}$$

*Note that cellobiose, like maltose, is a reducing sugar.*

Cellobiose is a slightly sweet sugar prepared by the enzymatic (partial) hydrolysis of the polysaccharide cellulose (see Sec. 27.9). The only difference between maltose and cellobiose is the glucosidic bond in cellobiose; it is beta. We say that cellobiose is a β-*glucoside*.

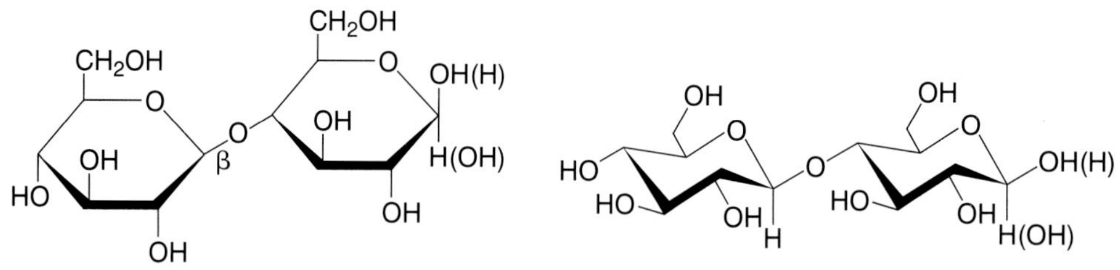

Cellobiose
β-D-Glucopyranosyl-(1, 4)-β-D-glucopyranose or
β-D-glucopyranosyl-(1, 4)-α-D-glucopyranose

## Lactose (Milk Sugar)

Hydrolyzing 1 mol of lactose with water and acid produces 1 mol of D-glucose and 1 mol of D-galactose.

$$\text{lactose} + H_2O \xrightarrow{H^+} \text{D-glucose} + \text{D-galactose}$$

Lactose is found in cow's milk and human milk in a concentration of about 6%.

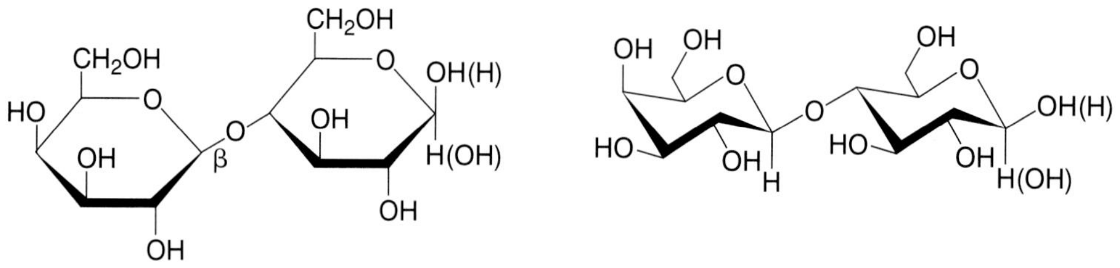

Lactose
β-D-Galactopyranosyl-(1, 4)-β-D-glucopyranose or
β-D-galactopyranosyl-(1, 4)-α-D-glucopyranose

**EXERCISE 27.9**

Would you predict lactose to be a reducing sugar? Explain.

## Sucrose

Hydrolyzing 1 mol of sucrose with water and acid yields 1 mol of D-fructose and 1 mol of D-glucose.

$$\text{sucrose} + H_2O \xrightarrow{H^+} \text{D-glucose} + \text{D-fructose}$$

*This mixture is used in the manufacture of syrups and candies because it is about as sweet as sugar but does not crystallize as readily.*

Hydrolysis of sucrose produces a 50-50 mixture of fructose and glucose known as *invert sugar*. The mixture is called *invert sugar* due to the inversion of specific rotation from dextrorotary (+) (sucrose) to levorotary (−) (the mixture of fructose and glucose).

$$\text{sucrose } [\alpha] = +66.5 \text{ degrees}$$
$$\text{glucose } [\alpha] = +52.7 \text{ degrees}$$
$$\text{fructose} [\alpha] = -92.4 \text{ degrees}$$

Calculate the specific rotation of invert sugar.

Sucrose is easily isolated from sugar cane and sugar beets and is the "table sugar" used to sweeten coffee or tea. This disaccharide is one of the least expensive foods available. For this reason, a wide variety of commercially sold foods are made to include sucrose—from potato salad and cole slaw to breakfast cereals and puddings.

The structure of sucrose is formulated as follows using Haworth projections:

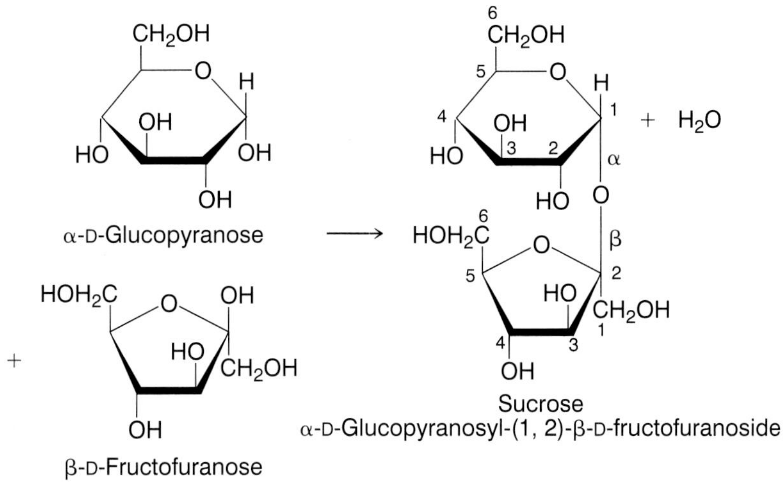

α-D-Glucopyranose

β-D-Fructofuranose

Sucrose
α-D-Glucopyranosyl-(1, 2)-β-D-fructofuranoside

Of course, sucrose is also represented more accurately as a nonplanar species using a chair projection for the pryanose ring (Fig. 27.11).

Sucrose is a nonreducing disaccharide and thus does not show mutarotation because both C-1 of the glucose unit and C-2 of the fructose unit represent sites of a full acetal. In other words, neither carbon is the site of a hemiacetal group that would be in equilibrium with an open-chain form of sucrose.

**FIGURE 27.11** Sucrose.

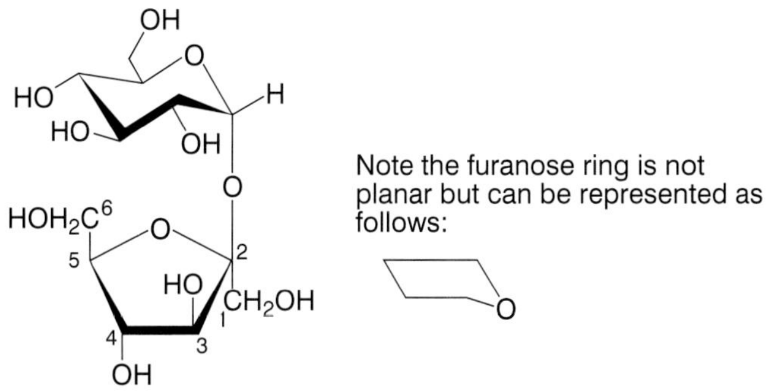

Note the furanose ring is not planar but can be represented as follows:

The polysaccharides differ from monosaccharides and disaccharides in many significant ways. The polysaccharides are not sweet, are usually insoluble in water, and are nonreducing carbohydrates and thus do not undergo mutarotation.

Starch is made up of two polymeric components: a water-insoluble *amylose* fraction (20%) and a water-soluble *amylopectin* fraction (80%). Amylopectin does not form a true solution with water but forms a colloidal dispersion instead due to the large size of the amylopectin polymer.

Hydrolysis of both components gives only D-glucose. Both amylose and amylopectin contain long chains of glucose units each with an α1–4 linkage to a neighboring unit. Amylose polymers contain no branching. In amylopectin, on the other hand, the chains are branched with one α1–6 linkage present in every 20 to 25 glucose units.

## 27.9 POLYSACCHARIDES

Amylose, Amylopectin, and Glycogen

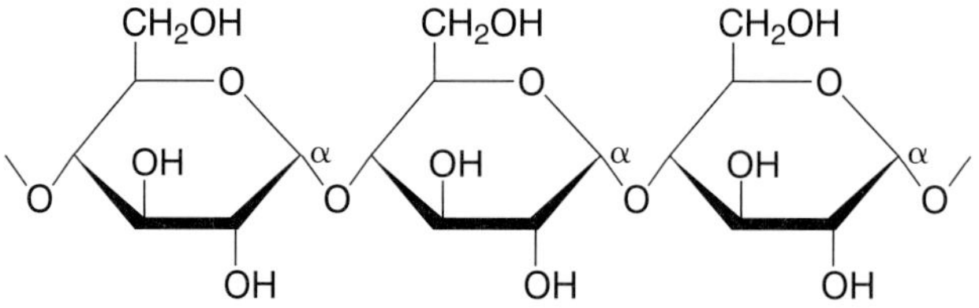

Amylose (three glucose units are shown)
(Haworth projection)

The sensitive starch-iodine test is positive when a blue-black color is observed in the mixture. This color is due to absorption of the iodine reagent by the amylose fraction of starch.

Amylose (three glucose units are shown)
(chair projection)

Starch is most conveniently isolated from potatoes, corn, rice, and wheat and is the reserve carbohydrate in plants and a source of food for animals.

Both amylose and amylopectin are nonreducing carbohydrates. Certainly, a hemiacetal group exists at the end of each polymeric chain, but this represents such a small portion of the molecule as a whole that it has no effect on its chemical properties.

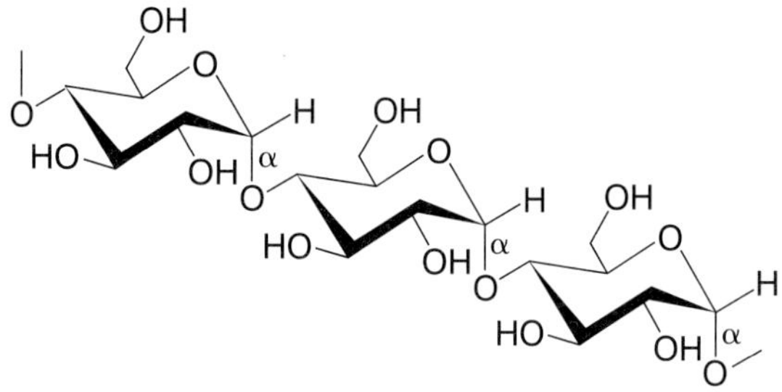

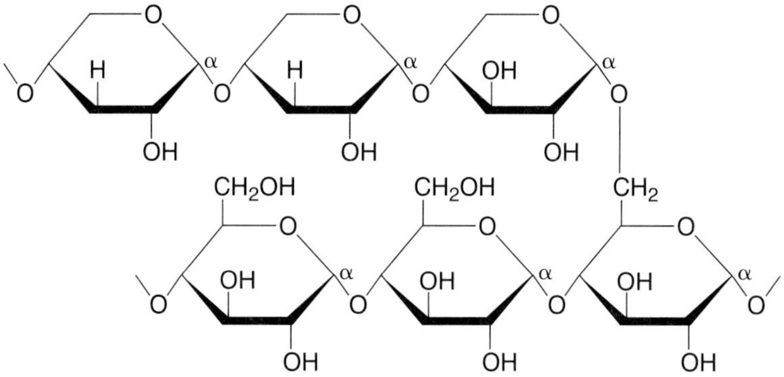

Amylopectin and glycogen (one branch is shown)
(Haworth projection)

Amylopectin and glycogen (one branch is shown)
(chair projection)

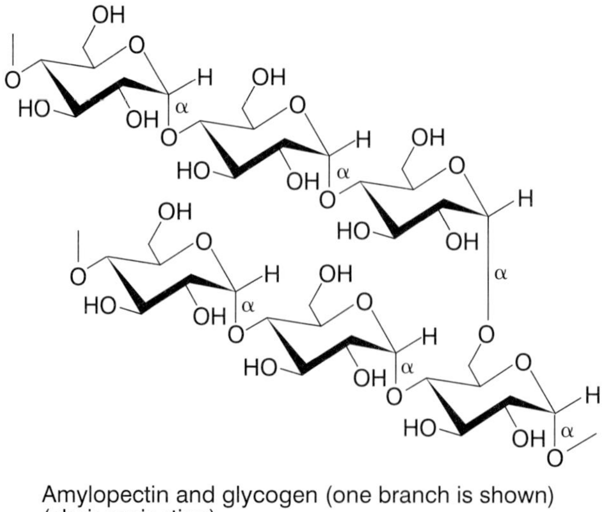

Glycogen is very similar structurally to amylopectin. The sole difference is the 1–6 branching that occurs in glycogen every 8 to 10 glucose units. The polymer serves as a storehouse of excess glucose ingested by an animal. This glucose is converted to glycogen by enzymes, and the glycogen is stored in the liver and muscles until it is needed by the animal.

---

Would you expect glycogen to be a reducing carbohydrate? Explain.

**EXERCISE 27.11**

**Cellulose**

---

Just as amylose, amylopectin, and glycogen are polymeric α-glucosides, cellulose is a polymeric β-glucoside. This simple difference has a profound effect. Starch, the α-glucoside, can be digested by humans; cellulose, the β-glucoside, cannot be digested. Structurally, cellulose, like amylose, is a straight-chain polysaccharide.

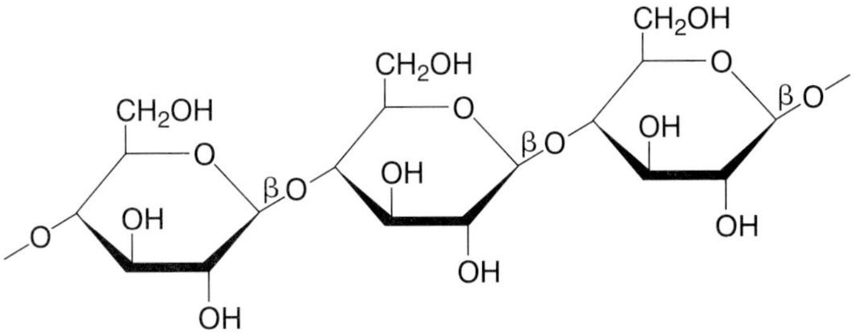

Cellulose (three glucose units are shown)
(Haworth projection)

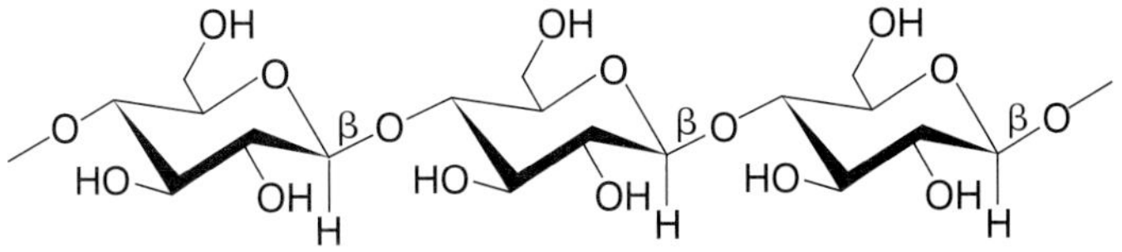

Cellulose (three glucose units are shown)
(Chair projection)

Cellulose is the substance that makes up the "skeleton" of plants. Also, cotton and wood are rich in cellulose, and as you might predict, cellulose is more abundant than sucrose.

The Kiliani-Fischer synthesis is used to increase the carbon content of an aldose by one. In other words, the method is employed to prepare an aldotetrose from an aldotriose, an aldohexose from an aldopentose, etc. First, the aldose is treated with HCN to give two diastereomeric cyanohydrins (see Sec. 19.2).

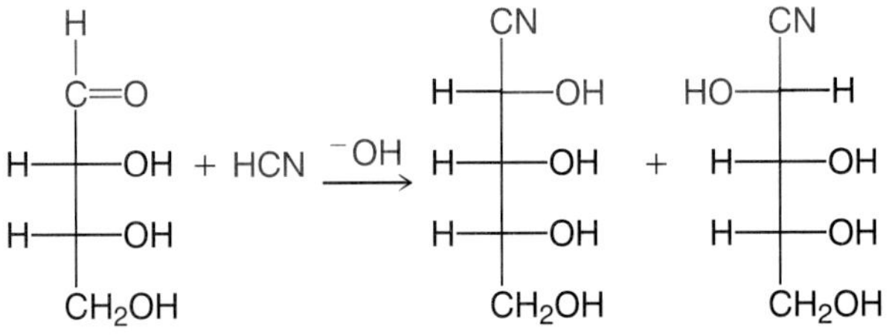

**27.10
SYNTHESIS OF
MONOSACCHARIDES**

Kiliani-Fischer Synthesis

D-Erythrose

These diastereomeric cyanohydrins can be separated and isolated, and one or both of the cyanohydrins are hydrolyzed in aqueous acid to yield the corresponding carboxylic acid (or acids). Isolating one of the cyanohydrins, we have

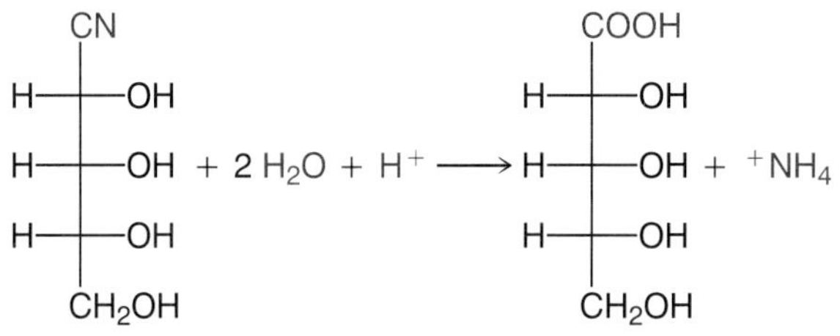

D-Ribonic acid

Finally, the carboxylic acid is reduced with a sodium amalgam under mildly acid conditions to give the chain-lengthened aldose.

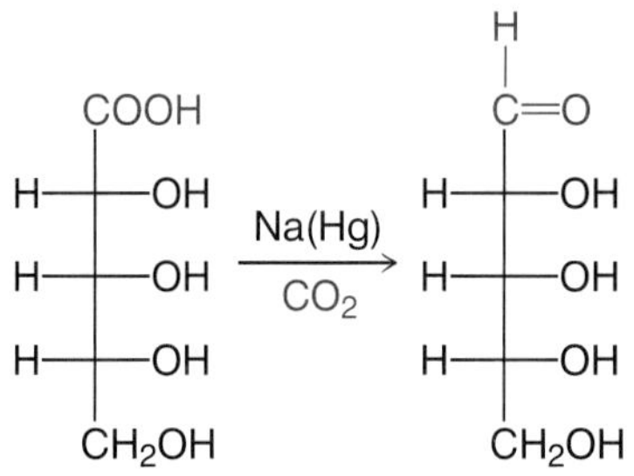

D-Ribose

**Ruff Degradation**

The Ruff degradation has the opposite effect of the Kiliani-Fischer synthesis in that the length of the carbon chain of the aldose is reduced by one. The first step of the process involves oxidation of the aldose with bromine water to give *aldonic acid* (see Sec. 27.11).

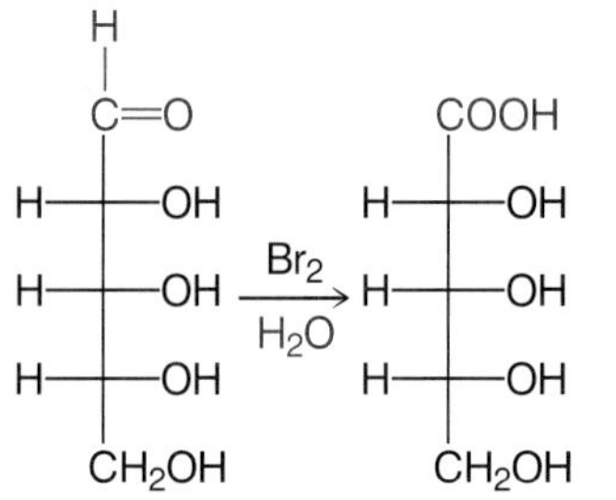

D-Ribonic acid

The aldonic acid, in turn, is both oxidized and decarboxylated with hydrogen peroxide and ferric sulfate to yield the aldose with one less carbon.

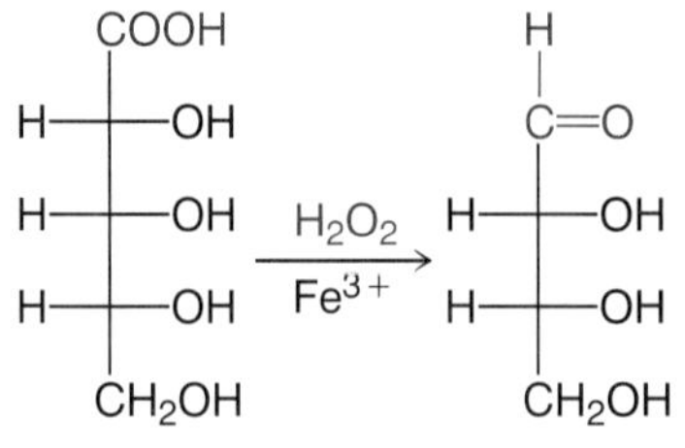

D-Erythrose

**27.11
REACTIONS OF
MONOSACCHARIDES**

Cyanohydrin formation, oxidation, and reduction represent characteristic reactions of the carbonyl group, whereas osazone formation is an unexpected reaction.

### Cyanohydrin Formation (see Secs. 19.2 and 27.10)

Both aldoses and ketoses react.

### Oxidation

**Reaction with Bromine Water (see Sec. 27.10)** Bromine water oxidizes the aldehyde group of an aldose of both the D- and L- stereochemical families to the corresponding carboxylic acid without affecting any of the hydroxyl groups present. Yields are good. Ketoses do not react. In general,

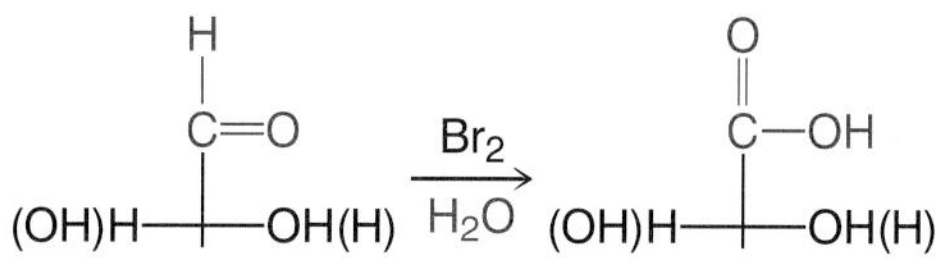

An aldose → An aldonic acid

Example:

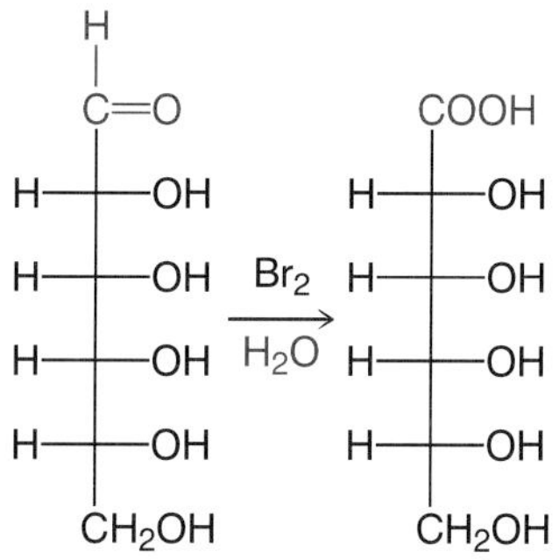

D-Allose → D-Allonic acid

To name an aldonic acid, replace the *-ose* suffix with *-onic*, and add the word *acid*. For example:

| Aldose | Aldonic acid |
|---|---|
| D-glucose | D-gluconic acid |
| L-ribose | L-ribonic acid |

### EXERCISE 27.12

Draw a Fischer projection formula for D-gluconic acid and L-ribonic acid.

**Reaction with Nitric Acid** Aldoses of both the D- and L- stereochemical families are oxidized with nitric acid to give dicarboxylic acids, where the aldehyde and primary alcohol groups of the aldose are attacked by the reagent. In general,

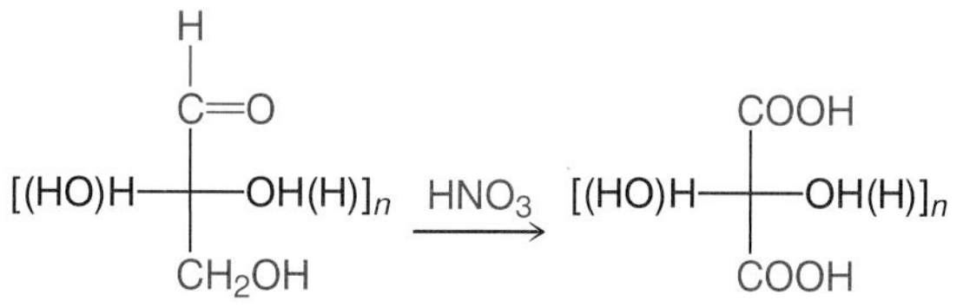

An aldose → An aldaric acid

Example:

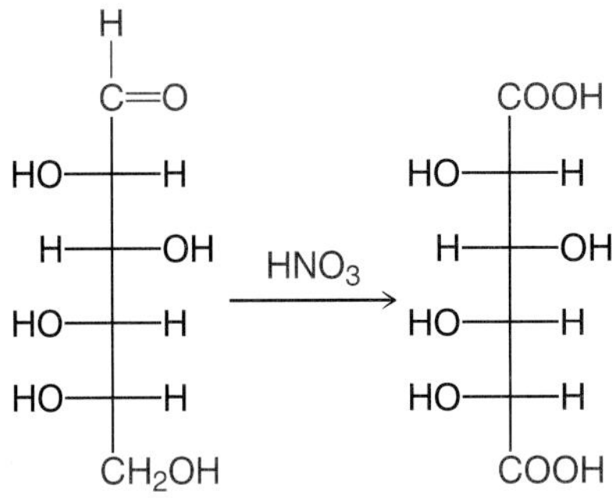

$$
\begin{array}{ccc}
\text{L-Glucose} & \xrightarrow{\text{HNO}_3} & \text{L-Glucaric acid}
\end{array}
$$

To name an *aldaric acid*, replace the *-ose* suffix with the suffix *-aric*, and add the word *acid*. For example:

| Aldose | Aldaric acid |
|--------|--------------|
| D-mannose | D-mannaric acid |
| L-ribose | L-ribaric acid |

Yields of aldonic and aldaric acids are good. Thus the oxidation of aldoses with bromine water and nitric acid represents a good preparative method.

---

**EXERCISE 27.13**

Draw a Fischer projection formula for D-mannaric acid and L-ribaric acid.

---

**Reduction**

Treating a monosaccharide with sodium borohydride gives the corresponding alcohol. In general,

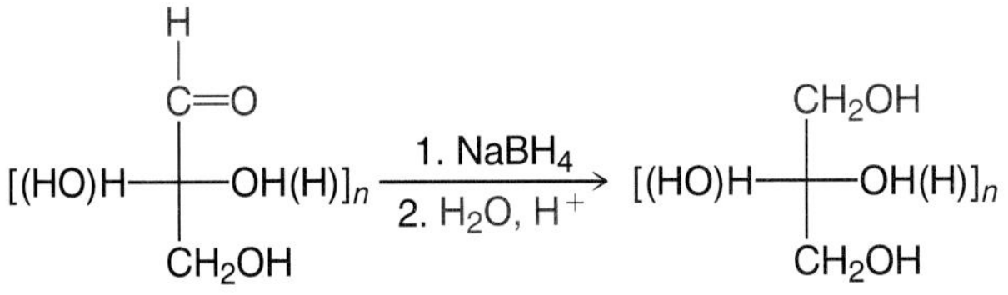

$$
\text{An aldose} \xrightarrow[\text{2. H}_2\text{O, H}^+]{\text{1. NaBH}_4} \text{An alditol}
$$

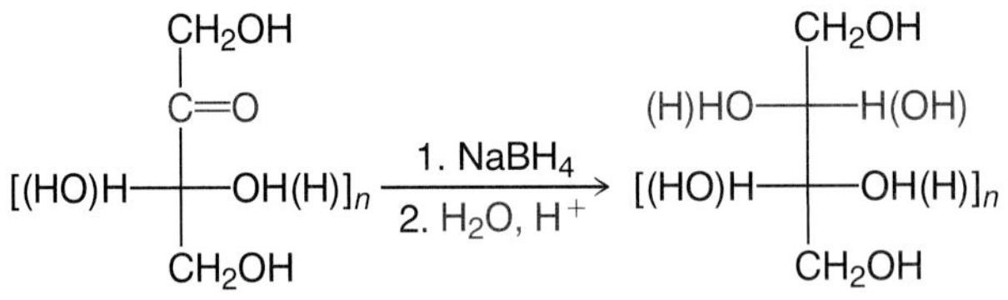

$$
\text{A ketose} \xrightarrow[\text{2. H}_2\text{O, H}^+]{\text{1. NaBH}_4} \text{Two ketitols}
$$

Example:

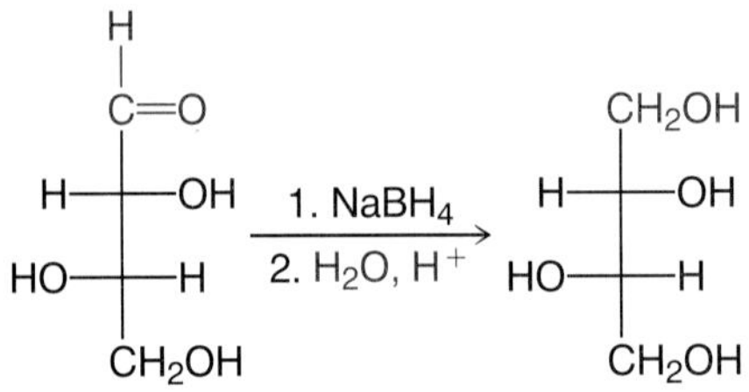

$$
\text{L-Threose} \xrightarrow[\text{2. H}_2\text{O, H}^+]{\text{1. NaBH}_4} \text{L-Threitol}
$$

To name an *alditol* or *ketitol*, replace the suffix *-ose* with the suffix *-itol*. For example:

| Aldose | Alditol |
|--------|---------|
| D-mannose | D-mannitol |
| L-ribose | L-ribitol |

Draw a Fischer projection for D-galactitol.

a. The compound D-glucitol (also known as D-sorbitol) is used as an artificial sweetener. Draw a Fischer projection for D-sorbitol.

b. Reducing D-fructose with sodium borohydride yields two diastereomeric ketitols. Draw a Fischer projection for each ketitol.

## HISTORICAL BOXED READING 27.2

**FIGURE 27.12** o-Toluenesulfonamide.

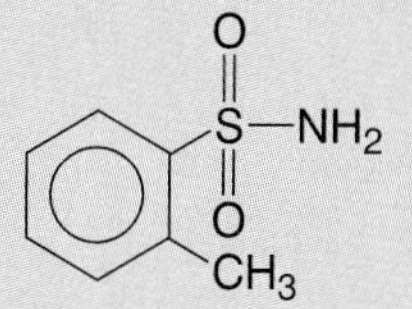

**FIGURE 27.13** Saccharin.

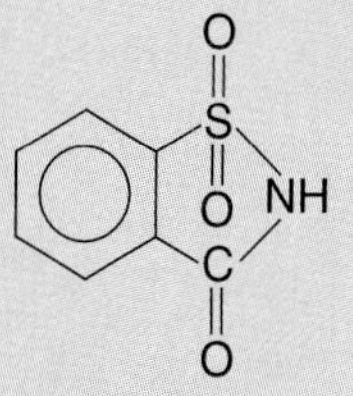

*Saccharin* has been used heavily by dieting and diabetic Americans for almost 100 years. Today it shares the market with aspartame (Industrial Boxed Reading 28.2), with saccharin's market share decreasing year by year. Most of the saccharin is in food and drink; the remainder is in mouthwashes, toothpastes, and medicines where sugar would cause rapid mold growth.

Saccharin is about 300 times sweeter than table sugar (sucrose), and the ¼-grain tablet produced by the pharmaceutical industry is equivalent to a level teaspoon of table sugar.

So important is saccharin to Americans that when the compound was banned by the Food and Drug Administration (FDA) for being a possible carcinogen on March 9, 1977, intense public displeasure forced the FDA to withdraw the ban for at least 18 months. As of now, saccharin is still "alive and kicking." However, there are real concerns about the fact that it causes cancer in laboratory animals.

Saccharin was the brainchild of Ira Remsen, a well-respected professor of chemistry at the Johns Hopkins University in Baltimore. Professionally, Remsen was top-notch, with many chemistry awards to his credit. However, he was a typical university professor in that he supervised the laboratory work of several graduate and postdoctoral students; each student would have a particular problem to solve.

In 1879, Remsen became interested in the behavior of a compound known as o-toluenesulfonamide (Fig. 27.12). He assigned this problem to a postdoctoral student named Constantine Fahlberg and told him to treat the compound with potassium permanganate to see what it would do.

Fahlberg went to work. Much to his surprise, he found that his food and drink started to taste sweet. He gradually began to suspect that the sweetness came from the laboratory when he discovered small spots of sweetness on his hands and arms. Using his sense of taste, he found the source of the sweetness to be the product of the reaction he was investigating.

Fahlberg had a moral obligation to report these observations to Dr. Remsen. After all, he was introduced to the problem by Remsen. But he chose not to do so. Instead, recognizing the vast amount of money to be made by preparing this sweet-tasting compound on a commercial scale, he quickly applied for and was granted a patent for the manufacture of saccharin (Fig. 27.13).

He completely blocked Remsen from participating in the financial gold mine. To make matters worse, he also tried to prove (without success) that Remsen had nothing at all to do with the discovery. This hurt Remsen deeply, for he later said to some of his students, "I did not want his money, but I did feel that I ought to have received a little credit for the discovery."

Fahlberg prepared saccharin using the following synthesis:

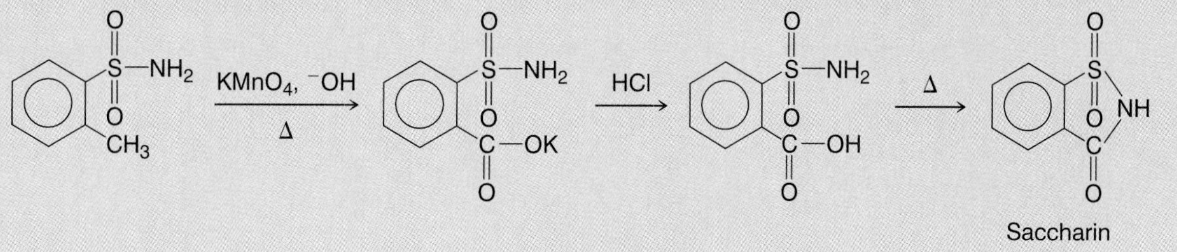

Saccharin

## Reaction with Phenylhydrazine

Monosaccharides along with aldehydes and ketones react with phenylhydrazine to give phenylhydrazones (see Sec. 19.2). Unfortunately, these phenylhydrazones are difficult to isolate. Instead, a sugar is treated with an excess (3 mol) of phenylhydrazine to give a yellow crystalline isolable *osazone*. In general,

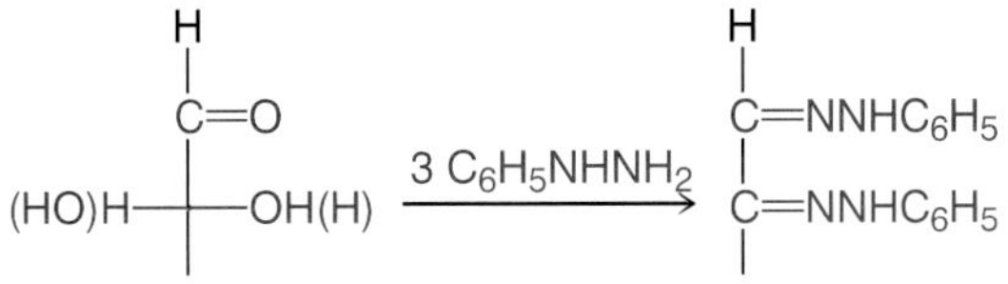

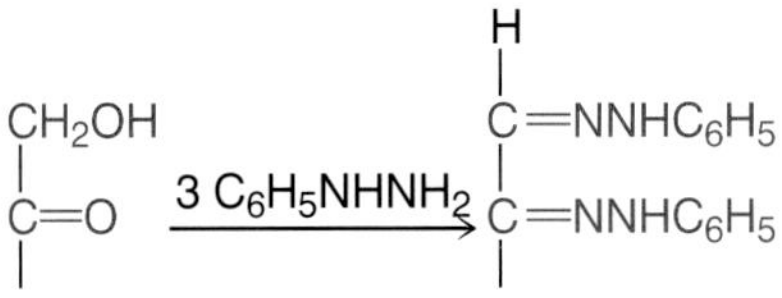

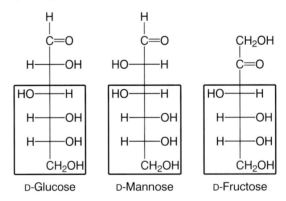

Example:

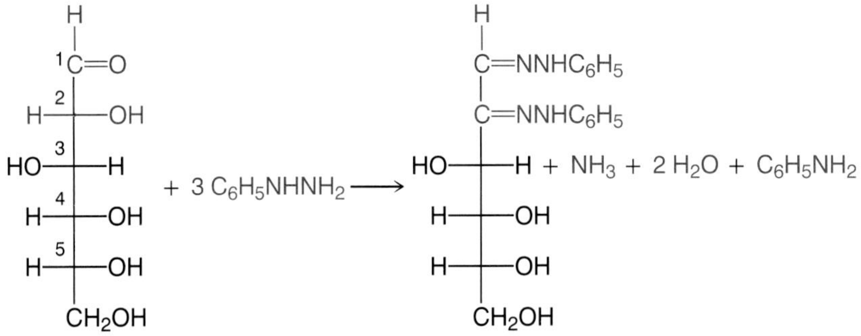

Think of the osazone as a diphenylhydrazone. Note that carbons 3 to 6 in an aldohexose or ketohexose are not affected by this reaction; only carbons 1 and 2 are chemically changed. Thus sugars that show the same configuration in carbons 3 to 6 form the same osazone. For example, D-glucose, D-fructose, and D-mannose form the same osazone.

Putting it another way, monosaccharides with the same configuration in carbons other than C-1 or C-2 must form the same osazone.

**SOLVED PROBLEM 27.8**

Draw the structure of each monosaccharide that forms the following osazone:

H
1C=NNHC6H5
2C=NNHC6H5
3
H
OH
4CH2OH

**SOLUTION**

Select monosaccharides with the same configuration at C-3. Two aldoses are possible. Finally, select the one possible ketose with the same configuration at C-3.

H
C=O
H
OH
H
OH
CH2OH
H
C=O
HO
H
H
OH
CH2OH
CH2OH
C=O
H
OH
CH2OH

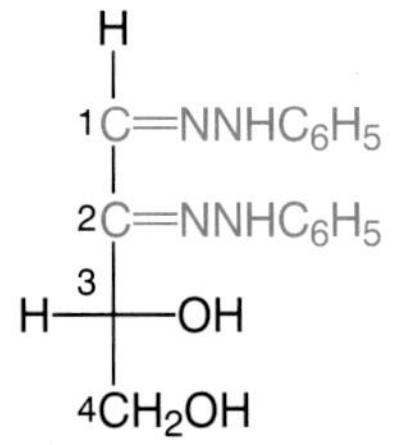

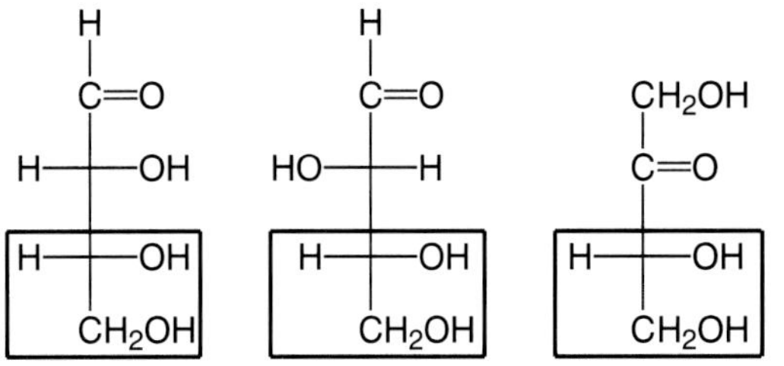

D-Erythrose D-Threose D-Erythrulose

**EXERCISE 27.16**

Draw a Fischer projection for each of the three pentoses that react with excess phenylhydrazine to yield the following osazone:

H
1C=NNHC6H5
2C=NNHC6H5
3
H
OH
4
HO
H
CH2OH

Aldoses that differ in configuration only at C-2 are called *epimers*. Thus D-erythrose and D-threose are epimers.*

**EXERCISE 27.17**

Prove to yourself that D-glucose and D-mannose are epimers by drawing a Fischer projection for each aldohexose.

*Recently, the term *epimer* has been expanded to include any diastereomeric pair of compounds that differ in configuration at only one carbon. Thus, using this definition, D-glucose and D-galactose are also epimers.

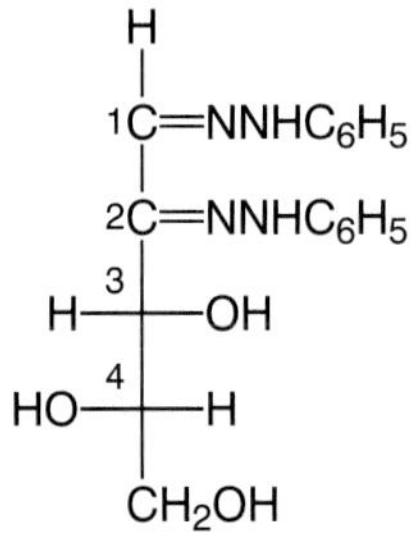

**EXERCISE 27.18**

Complete each of the following equations by drawing a structural formula for each organic product and writing a molecular formula for each inorganic product formed.

a.   α-D-mannose (Haworth) + $C_2H_5OH$ $\xrightarrow{H^+}$

b.   L-mannose (Fischer) $\xrightarrow{HNO_3}$ (no inorganic products)

c.   L-iodose (Fischer) $\xrightarrow[H_2O]{Br_2}$ (no inorganic products)

---

## ▶ CHAPTER ACCOMPLISHMENTS

**Introduction**
- [ ] List the three types of substances that are considered to be carbohydrates today.

**27.1  Classification of Carbohydrates and Some Structures of the Monosaccharides**
- [ ] Define a *disaccharide*.
- [ ] Draw a structural formula for a(an)
  - a. aldopentose.
  - b. ketotetrose.
- [ ] Draw a structural formula for
  - a. D-erythrose.
  - b. L-erythrose.
- [ ] Classify amylose as a saccharide.

**27.2  More About Monosaccharide Structure**
- [ ] Draw a Fisher projection for
  - a. D-lyxose.
  - b. L-ribose.
  - c. D-fructose.
- [ ] Draw a structural formula for dihydroxyacetone.

**27.3  Physical Properties of Monosaccharides**
- [ ] Predict the solubility (soluble or insoluble) of a monosaccharide in water.

**27.4  Unexpected Reactions of the Monosaccharides: Mutarotation—Cyclic Forms**
- [ ] Define *mutarotation*.
- [ ] Draw a Haworth projection for α-D-glucose.
- [ ] Draw a structural formula of the chair conformation adopted by α-D-glucose.
- [ ] Explain why more β-D-glucose than α-D-glucose is present in the α-D-glucose ⇌ D-glucose (open chain) ⇌ β-D-glucose equilibrium.
- [ ] Explain why β-D-glucose only reacts with 1 mol of alcohol.

**27.5  More About Cyclic Sugars: Structure and Nomenclature**
- [ ] Distinguish between a pyranose and a furanose.
- [ ] Draw a Haworth projection for α-D-fructopyranose and α-D-fructofuranose.

- [ ] Classify the following sugars as reducing or nonreducing.
  - a. α-D-glucopyranose
  - b. β-D-glucopyranoside
- [ ] Draw a Haworth projection for α-D-glucose and locate the structural feature that indicates that this substance is a reducing carbohydrate.

**27.6  Diagnostic Tests for Carbohydrates**
- [ ] Name the diagnostic test(s) for a reducing carbohydrate that give(s) as a product
  - a. $Cu_2O$.
  - b. Ag.

**27.7  Some Important D-Monosaccharides**
- [ ] Supply an alternate name for
  - a. blood sugar.
  - b. levulose.

**27.8  Disaccharides**
- [ ] Draw a Haworth projection for
  - a. α-D-glucopyranosyl-(1,4)-β-D-glucopyranose.
  - b. maltose (open chain form).
  - c. sucrose.
- [ ] Draw a chair projection for
  - a. β-D-glucopyranosyl-(1,4)-β-D-glucopyranose.
  - b. β-D-galactopyranosyl-(1,4)-α-D-glucopyranose.
- [ ] Give the composition of invert sugar.

**27.9  Polysaccharides**
- [ ] Supply a chair projection for three glucose units of
  - a. amylose.
  - b. cellulose.
- [ ] Explain how glycogen and amylopectin differ structurally.

**27.10  Synthesis of Monosaccharides**
- [ ] Synthesize D-ribose from D-erythrose.
- [ ] Prepare D-erythrose from D-ribose.

## 27.11 Reactions of Monosaccharides

☐ Complete the following equations by drawing a structural formula for each organic product.

a. D-glucose (open chain, Fischer projection) $+ Br_2 \longrightarrow$

b. L-glucose (open chain, Fischer projection) $+ HNO_3 \longrightarrow$

c. D-mannose (open chain, Fischer projection) $\xrightarrow[\text{2. } H_2O]{\text{1. } NaBH_4}$

d. L-mannose (open chain, Fischer projection) $\xrightarrow{3\ C_6H_5NHNH_2}$

---

## ▶ KEY TERMS

carbohydrate (Introduction)
sugar (Introduction, 27.1)
starch (Introduction, 27.9)
glycogen (Introduction, 27.9)
cellulose (Introduction, 27.9)
monosaccharide (27.1)
disaccharide (27.1)
trisaccharide (27.1)
oligosaccharide (27.1)
polysaccharide (27.1)
ketose (27.1)
aldose (27.1)
ketopentose (27.1)
aldotetrose (27.1)

D-sugar (27.1)
L-sugar (27.1)
mutarotation (27.4)
Haworth projection (27.4)
anomer (27.4)
pyranose (27.5)
furanose (27.5)
reducing carbohydrate (27.5)
nonreducing carbohydrate (27.5)
dextrose (27.7)
levulose (27.7)
blood sugar (27.7)
α-glucoside (27.8)
β-glucoside (27.8)

invert sugar (27.8)
amylose (27.9)
amylopectin (27.9)
aldonic acid (27.10)
aldaric acid (27.11)
alditol (27.11)
ketitol (27.11)
osazone (27.11)
epimer (27.11)
diabetes (Medical Boxed Reading 27.1)
saccharin (Historical Boxed Reading 27.2)
guncotton (Problem 27.18)

---

## ▶ PROBLEMS

1. Draw a structural formula for
   a. an aldotetrose.
   b. a D-ketotetrose.
   c. L-glyceraldehyde.

2. Draw a Fischer projection for
   a. an aldooctose.
   b. a ketopentose.
   c. L-altrose.
   d. D-mannonic acid.
   e. L-galactaric acid.
   f. D-mannitol.
   g. a threose racemic mixture (name the racemate of D-threose).
   h. a ketohexose (other than D-fructose).

3. Why would you expect an aldoheptose to be more soluble in water than
   a. 1-heptanol?
   b. heptanal?

4. Draw a Haworth projection for
   a. α-D-mannopyranose.
   b. β-D-galactopyranose.
   c. α-D-arabinofuranose.
   d. α-D-fructopyranose.
   e. α-D-fructofuranose.
   f. β-D-2-deoxyribofuranose.
   g. ethyl α-D-mannopyranoside.

5. Write a detailed, step-by-step mechanism for the following transformation:

   D-mannose $\longrightarrow$ β-D-mannopyranose
   (Fischer projection)    (Haworth projection)

6. Draw a chair projection for
   a. α-D-galactopyranose.
   b. β-D-allopyranose.
   c. methyl α-D-mannopyranoside.

7. Which of the compounds listed in Problem 27.4 is a nonreducing carbohydrate? Explain.

8. Draw a Haworth projection and a chair projection for
   a. sucrose.
   b. maltose.
   c. cellobiose.
   d. lactose.
   e. methyl-β-D-galactopyranoside.

9. Draw an open-chain structural formula for cellobiose.

10. Why is maltose a reducing sugar (structurally), while sucrose is not?

11. What is(are) the structural difference(s) between
    a. cellulose and amylose?
    b. amylose and amylopectin?
    c. amylopectin and glycogen?

12. A systematic name for the disaccharide isomaltose is α-D-glucopyranosyl-(1,6)-β-D-glucopyranose.
    a. Draw a Haworth projection for this carbohydrate.
    b. Supply another systematic name for this disaccharide.

13. Draw the structural formula of a D-tetrose that forms an achiral aldaric acid?

14. Sucrose is a nonreducing sugar, yet if sucrose is warmed with Benedict's solution, eventually a telltale precipitate of $Cu_2O$ forms. Explain.

15. Describe a simple test tube test that will distinguish between each of the following pairs of compounds. Describe the reagents you would use and the color changes you would observe.
    a. D-glucose and methyl α-D-glucopyranoside
    b. D-glucose and D-fructose
    c. D-glucitol and D-glucose
    d. starch and cellulose

16. Define or explain each of the following.
    a. oligosaccharide
    b. anomers
    c. mutarotation
    d. invert sugar
    e. reducing sugar
    f. epimers
    g. D-sugar

17. Draw a Fischer projection for each D-2-ketopentose.

18. When cellulose is completely nitrated with a mixture of nitric and sulfuric acids at 30°C for 30 minutes, cellulose trinitrate is produced in good yield.
    a. Draw a Haworth projection for cellulose trinitrate showing three glucose units.
    b. Speculate why cellulose trinitrate is also called *guncotton*.

19. Synthesize
    a. D-arabinitol from D-erythrose.
    b. D-arabinaric acid from D-mannose.

20. A monosaccharide was treated with HCN and then with HCl and $H_2O$ to give

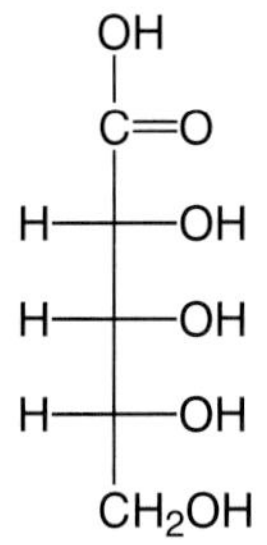

$$
\begin{array}{c}
OH \\
| \\
C=O \\
H-\!\!-OH \\
H-\!\!-OH \\
H-\!\!-OH \\
| \\
CH_2OH
\end{array}
$$

Draw a Fischer projection of the monosaccharide.

21. Complete the following reactions of D-ribose with each of the following reagents by supplying a Fischer projection for each organic product formed.
    a. $Br_2$, $H_2O$
    b. $HNO_3$
    c. excess $C_6H_5\overset{\displaystyle H}{N}NH_2$
    d. $NaBH_4$ then $H_2O$, $H^+$

22. Would you expect aldoses to react with $NH_2OH$? Explain. If reaction does take place, what product would you expect?

23. Draw a Fischer projection for each of the three monosaccharides that forms the following osazone:

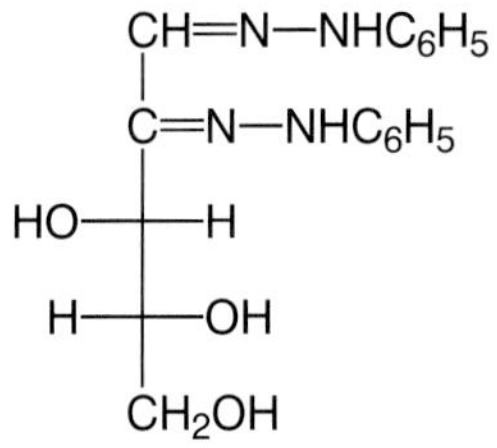

$$
\begin{array}{c}
CH=N-NHC_6H_5 \\
| \\
C=N-NHC_6H_5 \\
HO-\!\!-H \\
H-\!\!-OH \\
| \\
CH_2OH
\end{array}
$$

24. Write a complete equation to represent the reaction of α-D-ribofuranose with $CH_3OH$ and $H^+$.

# CHAPTER 28

# Amino Acids and Proteins

The term *protein* is derived from the Greek *proteios*, which means "of first importance." The meaning of the term is well taken because protein is absolutely vital to life. Protein is an integral part of skin, muscle, hair, connective tissue, nerves, and blood. In addition, enzymes, antibodies, many hormones, and some toxins are proteins.

Proteins are polymers that are derived from a number of monomers called *amino acids*. There are 20 amino acids that are incorporated into proteins. We will first study these alpha-substituted amino acids (Fig. 28.1), and then the proteins. In Fig. 28.1, R— represents a variety of organic groups, as shown in Table 28.1 (hydrogen in the case of glycine), and is referred to as a *side chain*. For glycine, R— = H—; for alanine, R— = CH₃—, etc.

Structures of the 20 amino acids that are incorporated into proteins are given in Table 28.1.

Referring to Table 28.1, note that the *neutral amino acids* have one carboxyl and one amino group, the *acid amino acids* contain two carboxyl groups and only one amino group, whereas the *basic amino acids* contain two basic groups and only one carboxyl group.

Glycine is the only achiral amino acid. Each of the remaining naturally occurring amino acids has a chiral alpha carbon, and all occur in an L- configuation. Both the compounds in Fig. 28.2 are classified as L-amino acids because each compound can be demonstrated to be related to L-glyceraldehyde by synthesis.

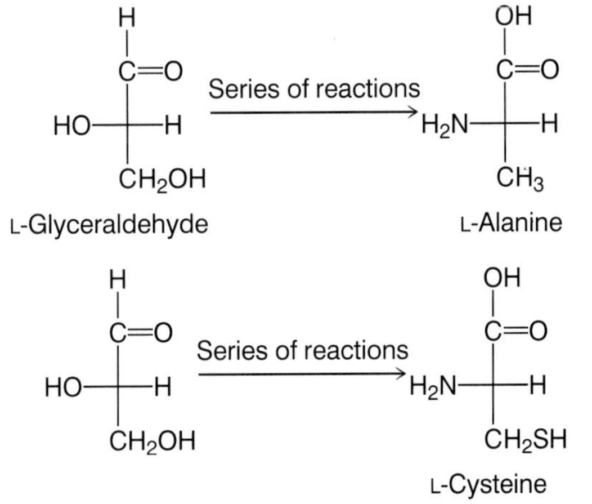

**FIGURE 28.1** An α-amino acid.

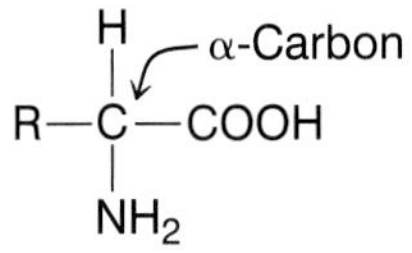

## 28.1
## STRUCTURE AND CLASSIFICATION OF AMINO ACIDS

**FIGURE 28.2** Fischer projections of two L-amino acids.

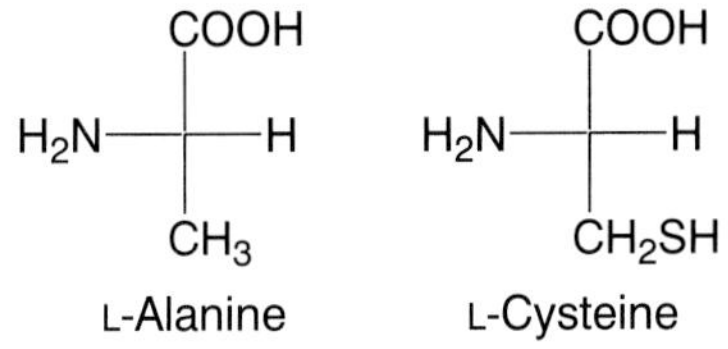

L-Glyceraldehyde → Series of reactions → L-Alanine

L-Glyceraldehyde → Series of reactions → L-Cysteine

L-Alanine    L-Cysteine

| TABLE 28.1 | Amino Acids Commonly Incorporated into Proteins | | |
|---|---|---|---|
| **Formula** | **Name** | **Abbreviation** | **Isoelectric Point (pI)** |
| **Neutral Amino Acids** | | | |
| $NH_2CH_2COOH$ | Glycine | Gly (G) | 6.0 |
| $NH_2\overset{H}{\underset{CH_3}{C}}COOH$ | Alanine | Ala (A) | 6.0 |
| $NH_2CHCOOH$, $\overset{CH}{\underset{CH_3\ \ CH_3}{}}$ | Valine | Val (V) | 6.0 |
| $NH_2CHCOOH$, $CH_2CH(CH_3)_2$ | Leucine | Leu (L) | 6.0 |
| $NH_2CHCOOH$, $CHCH_2CH_3$, $CH_3$ | Isoleucine | Ile (I) | 6.0 |
| 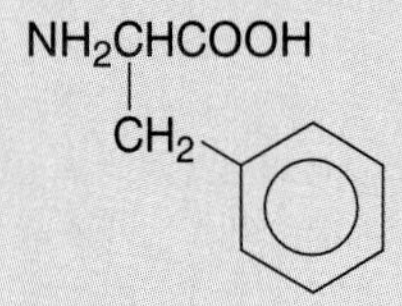 $NH_2CHCOOH$, $CH_2$— (phenyl) | Phenylalanine | Phe (F) | 5.5 |
| 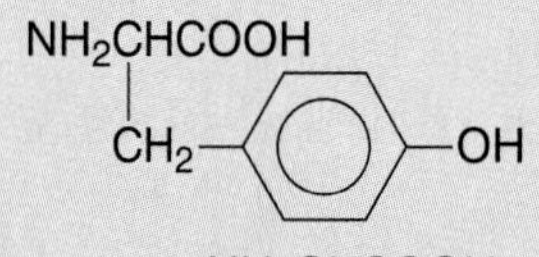 $NH_2CHCOOH$, $CH_2$—(phenyl)—OH | Tyrosine | Tyr (Y) | 5.7 |
| 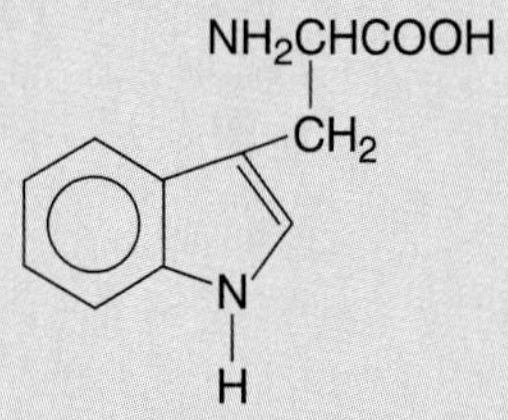 $NH_2CHCOOH$, $CH_2$—(indole, N–H) | Tryptophan | Trp (W) | 5.9 |
| 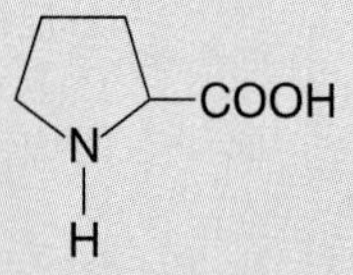 (pyrrolidine ring) —COOH, N–H | Proline | Pro (P) | 6.3 |
| $NH_2\overset{H}{\underset{CH_2SH}{C}}COOH$ | Cysteine | Cys (C) | 35.1 |
| $NH_2CHCOOH$, $CH_2CH_2SCH_3$ | Methionine | Met (M) | 5.7 |

| | | | | | |
|---|---|---|---|---|---|

**TABLE 28.1**    **Amino Acids Commonly Incorporated into Proteins** *(continued)*

| Formula | Name | Abbreviation | Isoelectric Point (pI) |
|---|---|---|---|
| NH$_2$CHCOOH<br>  \|<br>  CH$_2$OH | Serine | Ser (S) | 5.7 |
| NH$_2$CHCOOH<br>  \|<br>  CH(OH)<br>  \|<br>  CH$_3$ | Threonine | Thr (T) | 5.6 |
| NH$_2$CHCOOH<br>  \|   O<br>  \|   \|\|<br>  CH$_2$CNH$_2$ | Asparagine | Asn (N) | 5.4 |
| NH$_2$CHCOOH<br>  \|     O<br>  \|     \|\|<br>  CH$_2$CH$_2$CNH$_2$ | Glutamine | Gln (Q) | 5.7 |

**Acidic Amino Acids**

| Formula | Name | Abbreviation | Isoelectric Point (pI) |
|---|---|---|---|
| NH$_2$CHCOOH<br>  \|<br>  CH$_2$COOH | Aspartic acid | Asp (D) | 3.0 |
| NH$_2$CHCOOH<br>  \|<br>  CH$_2$CH$_2$COOH | Glutamic acid | Glu (E) | 3.2 |

**Basic Amino Acids**

| Formula | Name | Abbreviation | Isoelectric Point (pI) |
|---|---|---|---|
| NH$_2$CHCOOH<br>  \|<br>  CH$_2$CH$_2$CH$_2$CH$_2$NH$_2$ | Lysine | Lys (K) | 9.7 |
| NH$_2$CHCOOH<br>  \|<br>  CH$_2$CH$_2$CH$_2$—N—C—NH$_2$<br>          \|  \|\|<br>          H  NH | Arginine | Arg (R) | 10.8 |
| 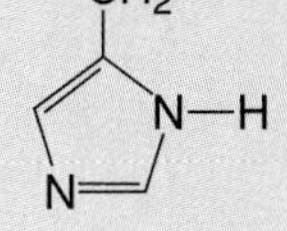 NH$_2$CHCOOH<br>  \|<br>  CH$_2$<br>  (imidazole ring)<br>  N—H<br>  N | Histidine | His (H) | 7.6 |

---

Demonstrate that a molecule of L-alanine adopts the (*S*)-configuration, whereas a molecule of L-cysteine assumes the (*R*)-configuration.

**EXERCISE 28.1**

---

The amino acid nomenclature shown in Table 28.1 is trivial. Supply a common and IUPAC name for

a.  glycine.

b.  alanine (neglect stereochemistry).

**EXERCISE 28.2**

Ten of these amino acids cannot be produced by the body yet are necessary for the maintenance of a state of good health and must be obtained from food ingested. These are called *essential amino acids* and are listed in Table 28.2.

Phenylalanine is an essential amino acid that, in the presence of the enzyme phenylalanine hydroxylase, is converted to tyrosine in a healthy person.

$$C_6H_5CH_2—\overset{\displaystyle H}{\underset{\displaystyle NH_2}{C}}—COOH \quad \xrightarrow[\text{hydroxylase}]{\text{Phenylalanine}} \quad p\text{-HO}—C_6H_4—CH_2—\overset{\displaystyle H}{\underset{\displaystyle NH_2}{C}}—COOH$$

Phenylalanine                                              Tyrosine

Unfortunately, this enzyme is not present in some people. In this event, phenylalanine accumulates in the blood, is converted to phenylpyruvic acid instead, and the person is said to suffer from a rare inherited disease called *phenylketonuria*, or *PKU*. Ultimately, when the concentration of phenylpyruvic acid rises to the point of saturation in the blood, it spills over into the urine (*-uria*: a suffix meaning the presence of a substance, in this case phenylpyruvic acid, in the urine).

---

**EXERCISE 28.3**

Given that the structure of pyruvic acid is

$$CH_3—\overset{\displaystyle O}{\overset{\displaystyle \|}{C}}—\overset{\displaystyle O}{\overset{\displaystyle \|}{C}}—OH$$

draw a structural formula for phenylpyruvic acid.

---

*Approximately 1 infant in 16,000 gets the disease.*

The disease should be and usually is detected early (at birth) so that the amount of phenylalanine ingested can be controlled; if not regulated, severe mental retardation is the grim result. At birth, each infant is given a PKU test in which a sample of blood is taken from the heel, and the concentration of phenylalanine is determined. If the concentration of phenylalanine is abnormally high, the infant must be put on a low protein diet.

Abbreviations are almost always used to designate amino acids when the amino acids are incorporated into a polymeric protein. Almost always the abbreviation is the first three letters in the name of the amino acid. The abbreviation for each amino acid is listed in Table 28.1. More recently, a one-letter abbreviation (in parentheses in Table 28.1) is appearing more and more frequently in research papers to represent amino acids included in various proteins.

| **TABLE 28.2** | **An Alphabetical Listing of the Essential Amino Acids** |
|---|---|
| | Arginine |
| | Histidine |
| | Isoleucine |
| | Leucine |
| | Lysine |
| | Methionine |
| | Phenylalanine |
| | Threonine |
| | Tryptophan |
| | Valine |

| TABLE 28.3 | An Alphabetical Listing of Certain Naturally Occurring Amino Acids |
| --- | --- |
| **Nonpolar Side Chain** | **Polar Side Chain** |
| Alanine | Arginine |
| Isoleucine | Asparagine |
| Leucine | Aspartic acid |
| Phenylalanine | Cysteine |
| Proline | Glutamic acid |
| Tryptophan | Glutamine |
| Valine | Lysine |
| | Methionine |
| | Serine |
| | Threonine |
| | Tyrosine |

Nineteen of the 20 amino acids contain a primary amine; only proline is a secondary amine.

Why isn't asparagine abbreviated Asp rather than Asn? Why isn't asparagine abbreviated A rather than N?

**EXERCISE 28.4**

An amino acid also can be classified based on the polarity of the side chain. Examine Table 28.3 for an alphabetical listing of the amino acids with a nonpolar side chain and the amino acids that contain a polar side chain.

Amino acids show some unexpected physical properties. Consider a comparison of physical properties between glycine and α-hydroxyacetic acid (glycolic acid) in Table 28.4. An examination of this table shows that although molecular weights are about the same, there is, nevertheless, a considerable difference in the physical properties of the two compounds. Note that glycine shows some unexpected physical properties such as a high temperature of decomposition, alcohol and ether insolubility, and electrolytic conductivity in solution. Since these properties are representative of an ionic compound rather than a covalent compound, we can say that glycine and other amino acids display physical properties that resemble ionic salts. Why?

Every α-amino acid structurally exists as a *dipolar ion* or *zwitterion* where a proton is transferred from a carboxyl group to an amino group.* Consider glycine.

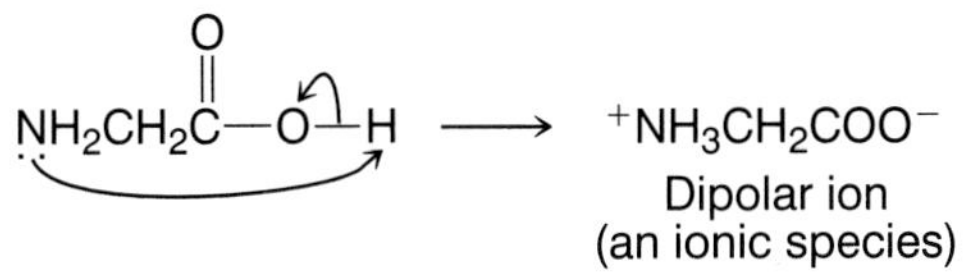

## 28.2
## PHYSICAL PROPERTIES OF THE AMINO ACIDS: ISOELECTRIC POINT— THE DIPOLAR ION

$$NH_2CH_2C\!-\!O\!-\!H \longrightarrow \ ^+NH_3CH_2COO^-$$

Dipolar ion
(an ionic species)

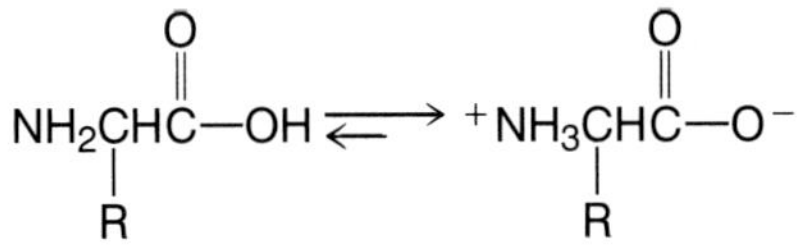

$$NH_2CHC\!-\!OH \rightleftharpoons \ ^+NH_3CHC\!-\!O^-$$
$$\quad\quad\ \ R \quad\quad\quad\quad\quad\quad R$$

Dipolar ion

Now, let us examine lysine with the view of dipolar ion formation. A reasonable question to ask here is, Which of the two amino groups in lysine gets the proton lost by the carboxyl group? The answer is the amino group that is more basic—the amino group in lysine farther away from the electron-withdrawing —COOH group.

*In actuality, an equilibrium exists between the uncharged form of an amino acid and the dipolar form that almost completely favors the dipolar form.

| TABLE 28.4 | A Comparison of the Physical Properties of Glycine and α-Hydroxyacetic Acid | | | | |
| --- | --- | --- | --- | --- | --- |
| Name | Structure | Molecular Weight | Melting Point | Solubility | Conductivity in Aqueous Solution |
| Glycine | $NH_2CH_2COOH$ | 75 | 232–236°C*d | Water-soluble; insoluble in alcohol and ether | A strong electrolyte |
| Glycolic acid | $HOCH_2COOH$ | 76 | 80°C | Soluble in water, alcohol, and ether | A weak electrolyte |

*d = decomposition; no melting takes place.

$$NH_2CHCOO-H \longrightarrow NH_2CHCOO^-$$
$$\hspace{1.2cm} | \hspace{4cm} | \hspace{1cm} +$$
$$(CH_2)_4\ddot{N}H_2 \hspace{2.5cm} (CH_2)_4\overset{+}{N}H_3$$

Dipolar form

Finally, let us consider glutamic acid with respect to dipolar ion formation. The question here is, Which carboxyl group donates the proton to the amino group to form the dipolar ion? The answer is the carboxylic group that is more acidic—the carboxyl group closer to the electron-withdrawing —$NH_2$ group.

$$\ddot{N}H_2CHCOO-H \longrightarrow {}^+NH_3CHCOO^-$$
$$\hspace{1.5cm} | \hspace{4cm} |$$
$$CH_2COOH \hspace{2.5cm} CH_2COOH$$

Dipolar ion

Every α-amino acid exists as the dipolar ion in the solid state. In aqueous solution, however, the structure of the amino acid is a function of the pH of the solution because amino acids are *amphoteric*; i.e., they react with both acids and bases. Consider glycine.

$$^+NH_3CH_2COOH \underset{^-OH}{\overset{H^+}{\rightleftharpoons}} {}^+NH_3CH_2COO^- \underset{H^+}{\overset{^-OH}{\rightleftharpoons}} NH_2CH_2COO^-$$

Glycine cation          Glycine dipolar ion          Glycine anion

---

**EXERCISE 28.5**

Draw an equilibrium system similar to the preceding one for alanine. Label each species.

---

Glycine exists primarily as the dipolar ion form when the pH is 6.0. If two drops of an aqueous solution of glycine are spotted in the center of a piece of filter paper or cellulose acetate gel that is saturated with an aqueous buffer solution at pH 6.0 and an electric field is applied to the paper or gel, it is found that the spot of glycine does not move (Fig. 28.3). If, on the other hand, the buffer were at pH 9.0, the spot would move toward the positive electrode because the glycine would be predominantly in the glycine anion form (see Fig. 28.5). In a pH 3.0 buffer, the spot would move toward the negative electrode because the glycine exists mainly as the cation (Fig. 28.4).

The pH at which no migration of an amino acid occurs in an electric field is called the *isoelectric point* (pI).* At a pH below 6.0, glycine exists mainly as the cation in aqueous solution, and migration of the cationic form to the negatively charged cathode takes place (see Fig. 28.4).

In a similar fashion, at a pH above 6.0, glycine is present predominantly as the anion, and migration to the positively charged anode takes place (see Fig. 28.5).

---

*In actuality, some migration does take place both to the anode and to the cathode. However, the predominant dipolar form does not migrate. This same state of affairs occurs in Figs. 28.4 and 28.5, and these figures represent the migration of the predominant species.

**FIGURE 28.3**  The activity of glycine in an electric field at pH 6.0.

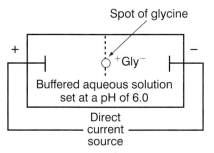

$^+$Gly$^-$ = dipolar form of glycine

**FIGURE 28.4**  The activity of glycine in an electric field at pH 3.0.

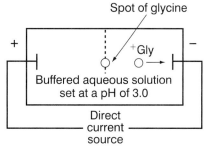

$^+$Gly = cationic form of glycine

**FIGURE 28.5**  The activity of glycine in an electric field at pH 9.0.

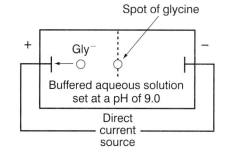

Gly$^-$ = anionic form of glycine

The isoelectric point of neutral amino acids varies from 5.1 to 6.3, acid amino acids from 3.0 to 3.2, and basic amino acids from 7.6 to 10.8 (see Table 28.1).

**SOLVED PROBLEM 28.1**

Draw a structural formula for the predominant form of valine in aqueous solution at a pH of

a.  6.0

b.  9.2

c.  2.8

**SOLUTION**

a.  Since the isoelectric point of valine is 6.0 (see Table 28.1), at that pH valine must exist mainly as the dipolar ion.

$$^+NH_3CHCOO^-$$
$$|$$
$$CH(CH_3)_2$$

b.  The pH is above the isoelectric point. Thus the amino acid is present predominantly as the anion.

$$NH_2CHCOO^-$$
$$|$$
$$CH(CH_3)_2$$

c.  Since the pH is below the isoelectric point of valine, the amino acid must exist mainly as the cation.

$$^+NH_3CHCOOH$$
$$|$$
$$CH(CH_3)_2$$

**EXERCISE 28.6**

Draw a structural formula for the predominant form of leucine in aqueous solution at a pH of

a.  6.0

b.  2.6

c.  12.3

## 28.3 SEPARATION OF AMINO ACIDS (AND PROTEINS)

**FIGURE 28.6** The separation of a mixture of glycine, aspartic acid and lysine by electrophoresis at pH 6.0.

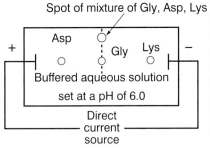

The differences in net charge exhibited by amino acids (and water-soluble proteins) along with the differences in pH of the buffer solutions in which the amino acids (or proteins) are dissolved can be used as a tool to separate mixtures of amino acids (or proteins). The experimental technique used to accomplish the separation is called *electrophoresis*. Experimentally, five drops of a mixture of two or more amino acids in water are spotted in the center of a piece of filter paper or cellulose acetate gel that is saturated with an aqueous buffer solvent at pH 6 (page 538). Two electrodes are connected, one to each end of the paper or gel, and a direct current is applied. The amino acids are separated on the basis of the net charge of the particular species (Fig. 28.6).

At a pH of 6.0, glycine exists mainly in the form of the dipolar ion. The net charge on the species is zero, so when the current is turned on, the predominant species, $^+NH_3CH_2COO^-$, remains in place (see Fig. 28.6). Using the same reasoning, since the isoelectric point of aspartic acid is 3.0, at a pH of 6.0, aspartic acid must exist in an anionic form. Since this form bears a negative charge, the species that migrates to the positive anode (Fig. 28.7) depends on the pH of the buffered solvent (see Fig. 28.6). (For the answer to the question of how we can deduce that mainly species I, rather than species II, actually is present to migrate, read the next paragraph.) Again, since the isoelectric point of lysine is 9.7, at a pH of 6.0, lysine exists for the most part in a cationic form. Since this form bears a net positive charge, the species (III) migrates to the negative cathode (see Fig. 28.6).

$$^+NH_3CHCOO^- \qquad NH_2CHCOO^- \qquad NH_2CHCOO^-$$
$$\underset{(CH_2)_4\overset{+}{N}H_3}{|} \qquad \underset{(CH_2)_4\overset{+}{N}H_3}{|} \qquad \underset{(CH_2)_4NH_2}{|}$$
$$(pH = 6.0) \qquad (pH = 9.7) \qquad (pH = 11)$$
$$\text{III} \qquad\qquad \text{IV} \qquad\qquad \text{V}$$

Note that if the solvent buffer is set at a pH of 9.7, lysine does not migrate because species IV is the predominant species present, whereas at a pH of 11, lysine migrates to the anode rather than to the cathode due to the presence of species V as the predominant species.

Now a reasonable question to ask is, How do you know what chemical species is dominant at a given pH for a nonneutral amino acid? We need the information in Table 28.5 to help us decide on the proper species.

**FIGURE 28.7** Two anionic forms of aspartic acid.

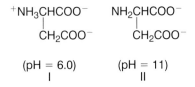

| TABLE 28.5 | Acidity Constants (as the p$K_a$) for Selected Nonneutral Amino Acids | | |
|---|---|---|---|
| **Amino Acid** | **p$K_{a1}$** | **p$K_{a2}$** | **p$K_{a3}$** |
| Aspartic acid | 2.1 | 3.9 | 9.8 |
| Glutamic acid | 2.2 | 4.2 | 9.5 |
| Lysine | 2.2 | 9.0 | 10.4 |

Consider aspartic acid in solution at a pH of 1.0. The species exists in the following cationic form:

$$\text{proton 3} \longrightarrow \; ^+NH_3CHCOOH \longleftarrow \text{proton 1}$$
$$\text{CH}_2\text{COOH} \longleftarrow \text{proton 2}$$
$$(\text{pH} = \text{lower than 2.1})$$

This species contains three ionizable protons that we will designate as proton 1, proton 2, and proton 3. Between the pH values of 2.1 and 3.9, the most acidic carboxyl (proton 1) ionizes to give

$$^+NH_3CHCOO^-$$
$$\text{CH}_2\text{COOH}$$
$$(\text{pH} = 2.1\text{–}3.9)$$

*Note:* pI = p$K_a$1 + p$K_a$2/2 = 3.0
At the pI (isoelectric point), the dipolar form is the dominant species (pages 535–537).

Between the pH values of 3.9 and 9.8, the second most acidic proton ionizes (proton 2) to give

$$^+NH_3CHCOO^- \;\; \text{Predominant species}$$
$$\text{CH}_2\text{COO}^-$$
$$(\text{pH} = 3.9\text{–}9.8)$$

Finally, when the pH reaches 9.8, the ammonium proton (proton 3) is ionized

$$NH_2CHCOO^-$$
$$\text{CH}_2\text{COO}^- \;\; \text{Major species}$$
$$(\text{pH} = 9.8 \text{ and higher})$$

A similar type of procedure can be used to obtain the dominant species in a solution of glutamic acid or lysine at a given pH. Consider Fig. 28.8, which shows the fraction of the four forms of aspartic acid as a function of pH to graphically summarize the material just discussed.

As you can see from Fig. 28.8, certain species *predominate* in certain pH regions but are not the *exclusive* species. Indeed, at pH 3.0 (pI of aspartic acid), about 80% is in the form of the dipolar ion, while 10% each exist as the anionic and cationic forms.

**FIGURE 28.8** A plot of the fraction of species versus pH for the four forms of aspartic acid.

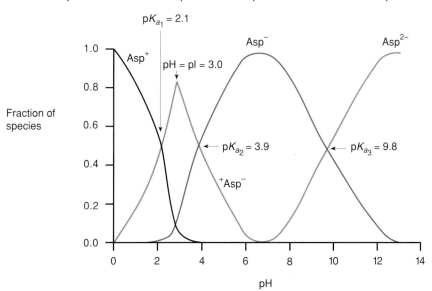

**EXERCISE 28.7**

Draw a structural formula for the dominant species of glutamic acid in solution at a pH of 9.7. Use the data in Table 28.5 as an aid.

## 28.4 PEPTIDES: CLASSIFICATIONS— THE PEPTIDE BOND

The linking together of two or more amino acids produces a *peptide*. The structural contribution of each amino acid to the peptide is called an *amino acid residue*. Consider the formation of a peptide from one molecule of glycine and one molecule of alanine.

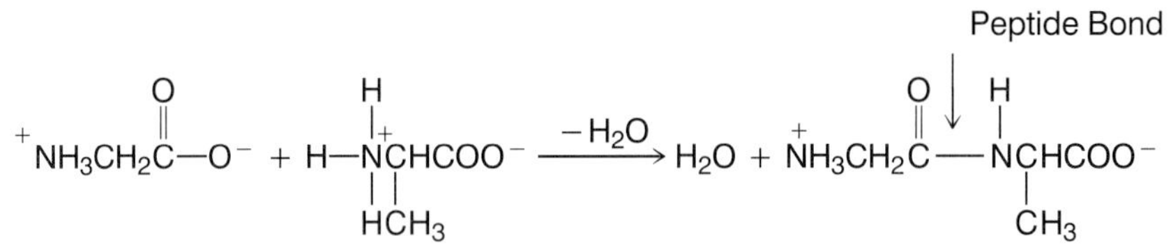

All peptides contain one or more amide bonds. Each carbon-nitrogen bond of an amide group linking up two amino acid residues is known as a *peptide bond*.

Glycine            Alanine                    Glycylalanine

This equation is an oversimplification, since it takes several steps to actually prepare glycylalanine from glycine and alanine in the laboratory, as we will see later.

---

## INDUSTRIAL BOXED READING 28.2

Artificial sweeteners are big business. Millions of dollars are spent on these sugar substitutes year after year by consumers for essentially two reasons. First, the substitute sweeteners contain very few calories, a real boon for people who are watching their weight. Second, these sweeteners are essential for people with diabetes.

From the 1950s to 1970, the market was dominated by two low-calorie sweeteners: saccharin (Fig. 28.9B) (see also Historical Boxed Reading 27.2) and cyclamate (Fig. 28.9A). In 1970, however, the Food and Drug Administration (FDA) banned cyclamate because it acted as a carcinogen in animals. Thus saccharin remained as the only FDA-approved sugar substitute.

However, even saccharin has had its troubles with the FDA (page 525). With all its problems with the FDA, saccharin is still in use today.

All of a sudden there is a relatively new kid on the block. In 1981, the FDA approved use of *aspartame* (Fig. 28.9C), an artificial sweetener that is about 200 times sweeter than sugar. Aspartame has obtained a good share of the market because it does not leave the bitter aftertaste often created with saccharin.

Both cyclamate and saccharin are substances that are foreign to the body, but aspartame is a dipeptide that represents the union of one naturally occurring amino acid (aspartic acid) with the methyl ester of another (phenylalanine) (Fig. 28.10).

**FIGURE 28.9** Three artificial sweeteners.

Cyclamate
(calcium cyclohexylsulfamate)
(a)

Saccharin
(b)

Aspartame
(c)

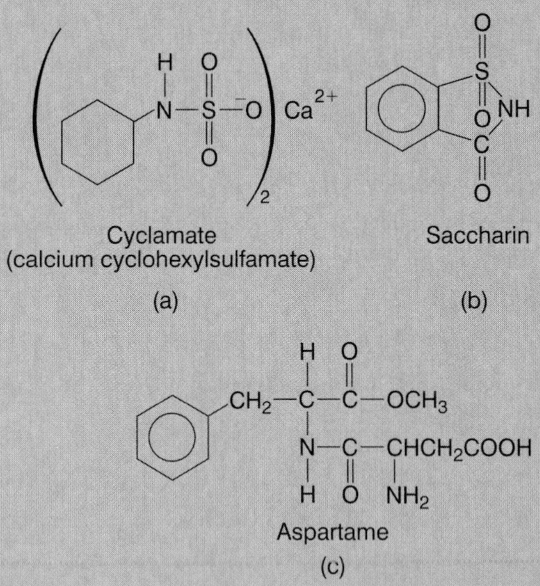

**FIGURE 28.10** A formulation of aspartame.

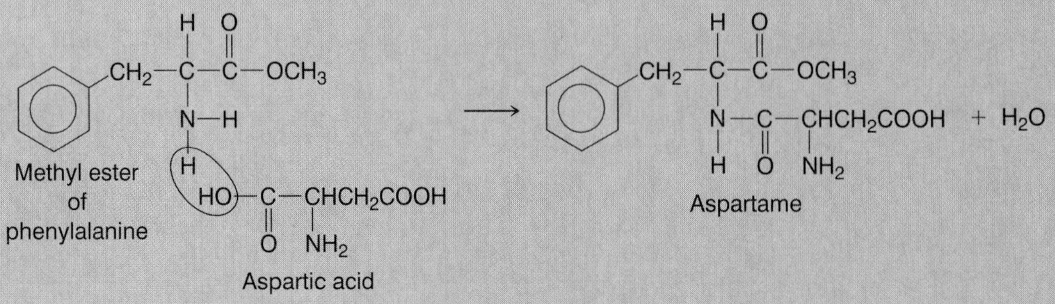

Methyl ester of phenylalanine

Aspartic acid

Aspartame

We say the *dipeptide* glycylalanine consists of 2 amino acid residues, glycine and alanine. A *tripeptide*, then, consists of 3 amino acid residues, a *tetrapeptide* consists of four residues, a *polypeptide* consists of 10 to 49 residues, and a *protein* consists of 50 or more amino acid residues (or a molecular weight over 5000).

The tripeptide *glutathione* (Fig. 28.11) is abundant in mammalian tissue. It acts as a reducing agent in a key biochemical reaction that is important in keeping red blood cells functional.

These peptides can be cumbersome to draw, particularly if a significant number of amino acids are involved. Thus either three-letter or one-letter abbreviations are used (see Table 28.1). Note that glycylalanine is written as Gly-Ala or GA, whereas glutathione is written as γ-Glu-Cys-Gly (the single-letter abbreviations are not used for unusually formulated peptides such as glutathione). It is customary to write or draw the amino acid residue containing the free amino group on the extreme left of the molecule while we write or draw the amino acid residue with the free carboxyl group on the extreme right of the molecule. We name from left to right with the *-ine* changed to *-yl* in all but the amino acid residue on the extreme right, which is given its usual name. The amino acid residue with the free amino group is called the *N-terminus*, whereas the amino acid residue with the free carboxyl group is known as the *C-terminus*.

**FIGURE 28.11** Glutathione.

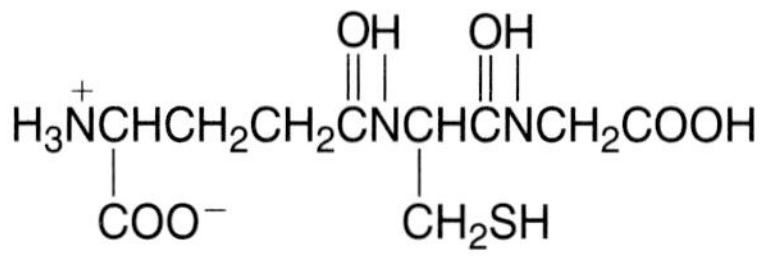

$$\underset{COO^-}{\overset{+}{H_3N}CHCH_2CH_2}\overset{O}{\overset{\|}{C}}NHCH\overset{O}{\overset{\|}{C}}NHCH_2COOH$$

γ-Glutamylcysteinylglycine

Note that the γ-carboxyl of the glutamic acid residue participates in the formation of the peptide bond rather than the α-carboxyl in glutathione.

**SOLVED PROBLEM 28.2**

Consider the following *enkephalin* (a naturally produced pain killer in human beings):

Tyr-Gly-Gly-Phe-Met

An enkephalin

a. Locate the N-terminus and C-terminus using the abbreviated structural formula of the enkephalin.

b. Draw a structural formula for the enkephalin.

c. Name this enkephalin.

d. Write an abbreviated structural formula of the pentapeptide using one-letter abbreviations for the amino acid residues.

**SOLUTION**

a. The N-terminus is the amino acid residue on the extreme left of the structural formula, while the C-terminus is the amino acid residue on the extreme right.

Tyr-Gly-Gly-Phe-Met

N-terminus    C-terminus

b. To draw the structural formula of a peptide, simply create a peptide bond between every two neighboring amino acid residues.

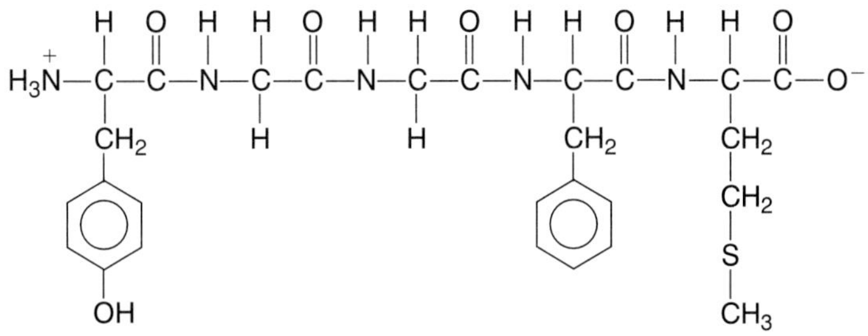

Remember to draw the pentapeptide as a dipolar ion.

c.  To name the pentapeptide, replace each *-ine* suffix with a *-yl* fragment, except for the C-terminus. Naming from left to right, we have

tyrosylglycylglycylphenylalanylmethionine

d.  Using Table 28.1, we get YGGFM.

---

**EXERCISE 28.8**

Consider the following tetrapeptide:

Gly-Ala-Val-Leu

a.  Locate the N-terminus and C-terminus using the abbreviated structural formula of the tetrapeptide.
b.  Draw a structural formula for the tetrapeptide.
c.  Name this tetrapeptide.
d.  Write an abbreviated structural formula of the tetrapeptide using one-letter abbreviations for the amino acid residues.

---

Now we previously illustrated peptide formation by coupling a glycine residue with an alanine residue to give

Gly-Ala

However, we can formulate an isomeric dipeptide by using the glycine residue as the C-terminus to give

Ala-Gly

With this in mind, the number of isomeric pentapeptides, for example, can be considerable. Consider Solved Problem 28.3.

---

**SOLVED PROBLEM 28.3**

Consider a tripeptide composed of Phe, Ala, and Gly. How many isomers are possible for this tripeptide? Write an abbreviated structural formula (three-letter) for each tripeptide.

**SOLUTION**
To calculate the number of possible isomers for a peptide composed of $n$ different neutral amino acid residues, perform an $n$ factorial ($n!$) calculation; i.e., multiply $n$ by $n - 1$, multiply that product by $n - 2$, etc. until the number 1 is reached.

$$n! = n(n - 1)(n - 2)(n - 3) \ldots (1)$$

For a tripeptide, we have

$$3! = 3 \times 2 \times 1 = 6 \text{ possible isomers}$$

Here is an abbreviated structural formula for each isomeric tripeptide.

| | |
|---|---|
| Phe-Ala-Gly | Ala-Phe-Gly |
| Phe-Gly-Ala | Gly-Phe-Ala |
| Ala-Gly-Phe | Gly-Ala-Phe |

---

**EXERCISE 28.9**

How many isomers are possible for a pentapeptide composed of five different neutral amino acid residues?

## 28.5 PEPTIDE AND PROTEIN STRUCTURE DETERMINATION

How can the structure of a peptide or protein be determined? It is a time-consuming, tiresome process, but it has been done in a number of cases.

First, the number of and nature of each amino acid residue in the peptide are determined by complete hydrolysis. In general,

$$\text{peptide} \xrightarrow[H_2O]{HCl} \text{a mixture of amino acids}$$

Example:

$$\text{Phe-Ala-Val-Leu-Trp} \xrightarrow[H_2O]{HCl} \text{Phe} + \text{Ala} + \text{Val} + \text{Leu} + \text{Trp}$$

Second, the N-terminus is determined by treating the peptide with *Sanger's reagent*, 2,4-dinitrofluorobenzene.

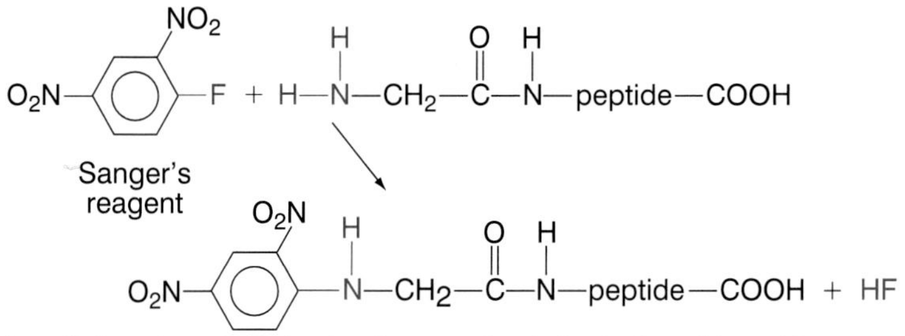

The resulting dinitrophenyl derivative is completely hydrolyzed in acid.

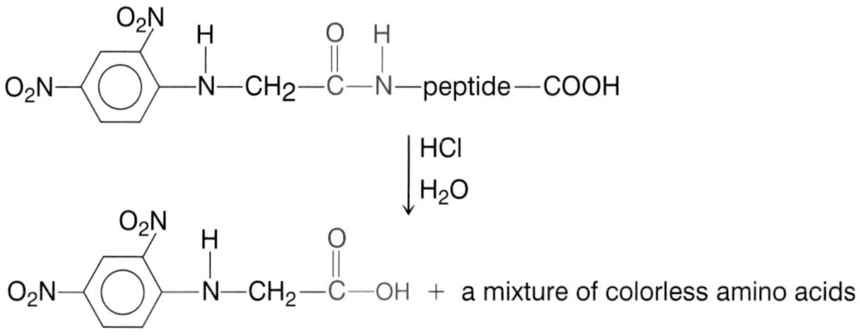

A colored compound

The colored compound produced is easily isolated, the amino acid residue identified, and thus the N-terminus determined.

In the third step, the C-terminus is determined by treating the peptide with *carboxypeptidase* (an enzyme that selectively attacks the peptide bond of the C-terminus) and identifying the first free amino acid cleaved off from the peptide. This free amino acid is the C-terminus.

The final step of the process is partial hydrolysis (usually using an enzyme catalyst) of the peptide to produce peptide fragments. For example:

$$\text{Phe-Val-Trp-Leu-Met-Gly} \xrightarrow{\text{Enzyme}} \text{Phe-Val-Trp} + \text{Trp-Leu} + \text{Val-Trp-Leu-Met} + \text{Leu-Met-Gly}$$

**SOLVED PROBLEM 28.4**

A hexapeptide is known to contain the following amino acid residues: Tyr, Ser, Met, Val, Pro, and Leu. Write an abbreviated structural formula for the hexapeptide based on the following data:

a. 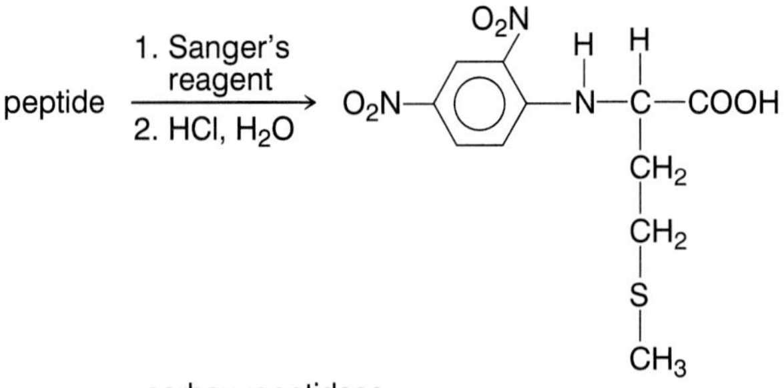  peptide $\xrightarrow[\text{2. HCl, H}_2\text{O}]{\text{1. Sanger's reagent}}$

$$O_2N\text{—}\underset{NO_2}{\overset{O_2N}{\bigcirc}}\text{—}\underset{H}{\overset{H}{N}}\text{—}\underset{\substack{CH_2 \\ | \\ CH_2 \\ | \\ S \\ | \\ CH_3}}{\overset{H}{C}}\text{—COOH}$$

b.  peptide $\xrightarrow{\text{carboxypeptidase}}$ Tyr

c.  peptide $\xrightarrow[\text{hydrolysis}]{\text{partial}}$ Met-Leu + Leu-Pro-Ser + Pro-Ser + Pro-Ser-Val

**SOLUTION**

Reaction with Sanger's reagent establishes methionine as the N-terminus, while the reaction catalyzed by carboxypeptidase proves tyrosine to be the C-terminus. Thus we have

Met __ __ __ __ Tyr

The fragments produced by partial hydrolysis can be arranged in such a way as to enable us to reason out the complete structure

Met-Leu
Leu-Pro-Ser
Pro-Ser
Pro-Ser-Val

Thus the complete structure is

Met-Leu-Pro-Ser-Val-Tyr

---

**EXERCISE 28.10**

A pentapeptide is known to contain the following amino acid residues: Gly, Ser, Ile, Val, and Asn. Write an abbreviated structural formula for the pentapeptide based on the following data:

a. 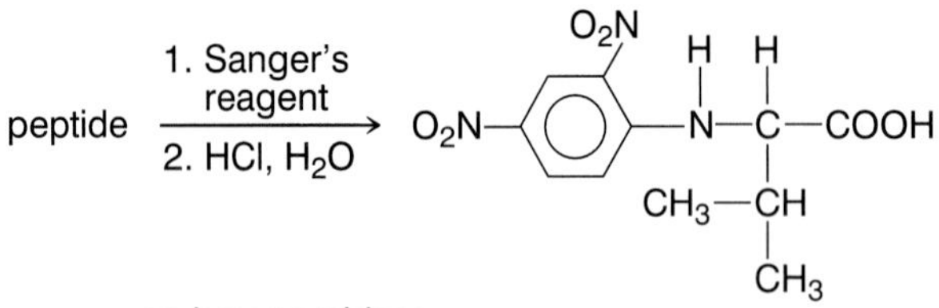  peptide $\xrightarrow[\text{2. HCl, H}_2\text{O}]{\text{1. Sanger's reagent}}$

$$O_2N\text{—}\underset{NO_2}{\overset{O_2N}{\bigcirc}}\text{—}\underset{H}{\overset{H}{N}}\text{—}\underset{\substack{CH_3\text{—CH} \\ | \\ CH_3}}{\overset{H}{C}}\text{—COOH}$$

b.  peptide $\xrightarrow{\text{carboxypeptidase}}$ Ser

c.  peptide $\xrightarrow[\text{hydrolysis}]{\text{Partial}}$ Val-Asn-Gly + Asn-Gly-Ile

---

## 28.6 CLASSIFICATION OF PROTEINS

A prosthetic group can be organic or inorganic.

Proteins can be classified in a number of different ways. We will consider two such ways. For example, on the basis of composition, proteins are classified as *simple* or *conjugated*. A simple protein yields on hydrolysis only amino acids. A conjugated protein, on the other hand, yields on hydrolysis at least one substance called a *prosthetic group*, along with the amino acids.

Examples of simple proteins (sources are in parentheses) are keratin (hair, feathers, wool, nails), albumin (eggs), zein (corn), gluten (wheat), and myosin (muscle). Some conjugated proteins are given as follows with the prosthetic group and source in brackets: casein [phosphoserine complex (Fig. 28.12), milk] and hemoglobin [heme (Fig. 28.13), blood].

Proteins also can be classified on a structural basis. *Fibrous proteins* consist of long polymeric chains, one placed alongside another. These proteins are water-insoluble and are found in muscle, nails, and hooves. On the other hand, *globular proteins* are small, spherically shaped proteins that are water-soluble. As would be expected, these proteins are found in blood (an aqueous solvent).

The *primary structure* of a protein, as we have discussed in Section 28.5, consists of the identity, number, and sequence of the amino acid residues making up the protein. The *secondary structure* of a protein refers to the stereochemistry of the protein, i.e., its arrangment in space. The two main secondary structure arrangements are the $\alpha$-helix (Fig. 28.14a) and the pleated sheet (Fig. 28.14b).

Note that the $\alpha$-helix is stabilized by intramolecular hydrogen bonds; i.e., the bonding occurs between $-C{=}O$ and $-NH$ groups in the same molecule. Figure 28.14A presents an illustration of the $\alpha$-helix. There are about 3.6 amino acids present per turn of the helix; in addition, note that the distance between turns, known as the *pitch*, is 5.4 Å. The pleated-sheet structure, on the other hand, is stabilized by intermolecular hydrogen bonding; i.e., the bonding takes place between a $-C{=}O$ from one molecule and a $-NH$ from another (see Fig. 28.14b).

Whether a protein forms the $\alpha$-helical coil (like keratin) or the planar pleated sheet (like fibroin) depends on the amino acid residues present in the primary structure. Certain amino acid residues, such as leucine, lysine, and phenylalanine, each with a large $R-$ group, prefer the roomy $\alpha$-helix arrangement, whereas serine, glycine, and cysteine, each with a smaller $R-$ group, prefer the more crowded pleated-sheet alignment.

The *tertiary structure* of proteins is a modification of the secondary structure due to forces such as hydrogen bonding, *disulfide bonds*, *salt bridges*, and London forces (Fig. 28.15). The disulfide bond is found only when two cysteine amino acid residues are in reasonable proximity to each other (see Sec. 14.5) and oxidation to the disulfide compound takes place.

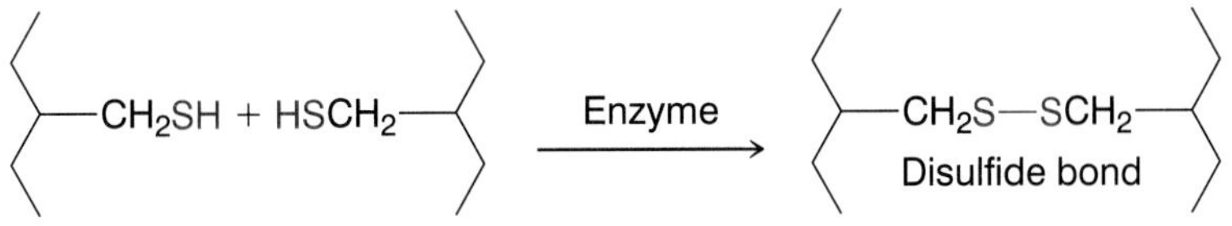

$$\text{Protein}-CH_2SH + HSCH_2-\text{Protein} \xrightarrow{\text{Enzyme}} -CH_2S-SCH_2-$$

Disulfide bond

A salt bridge is an ionic attraction produced when an acid amino acid residue is in close contact with a basic amino acid residue. The loss of a proton by the acid to the base produces the salt bridge.

Many more proteins are conjugated than are simple.

## 28.7
## THE LEVELS OF PROTEIN STRUCTURE

Insulin (regulates carbohydrate metabolism), hemoglobin (present in red blood cells and used in respiration), and antibodies (destroy dangerous bacteria) are globular proteins.

**FIGURE 28.12** Phosphoserine complex.

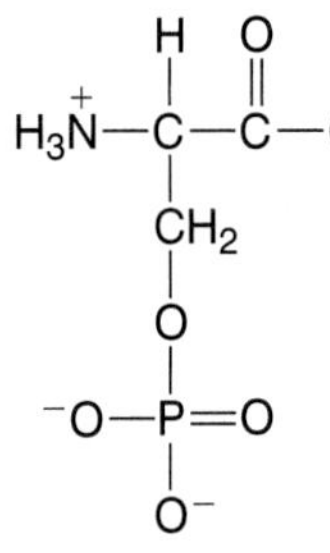

**FIGURE 28.13** Heme.

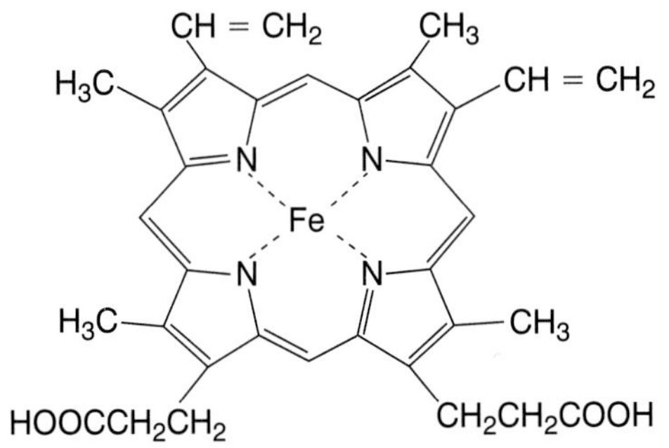

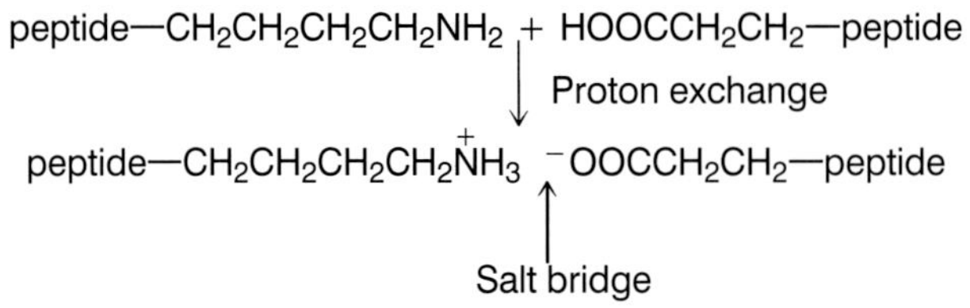

$$peptide\text{—}CH_2CH_2CH_2CH_2NH_2 + HOOCCH_2CH_2\text{—}peptide$$

$$\downarrow \text{Proton exchange}$$

$$peptide\text{—}CH_2CH_2CH_2CH_2\overset{+}{N}H_3 \quad {}^{-}OOCCH_2CH_2\text{—}peptide$$

$$\uparrow \text{Salt bridge}$$

London forces are present when two nonpolar groups are very close to each other. These are weak, short-range attractive interactions and are designated a *hydrophobic bond* or, literally, a "water-hating bond" (see Sec. 1.6).

**FIGURE 28.14** The types of secondary structure showing hydrogen bond stabilization.

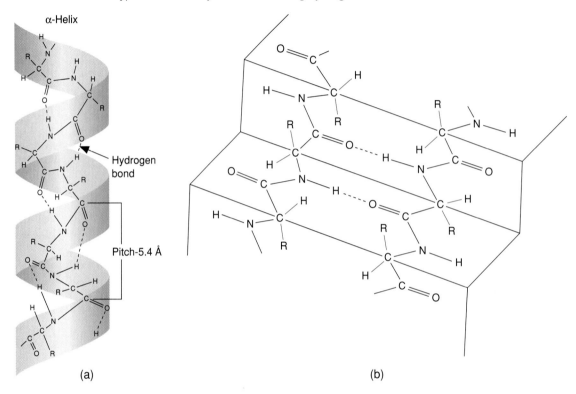

(a)                                                          (b)

**FIGURE 28.15** An illustration of the types of interactions resulting in the tertiary structure of a protein.

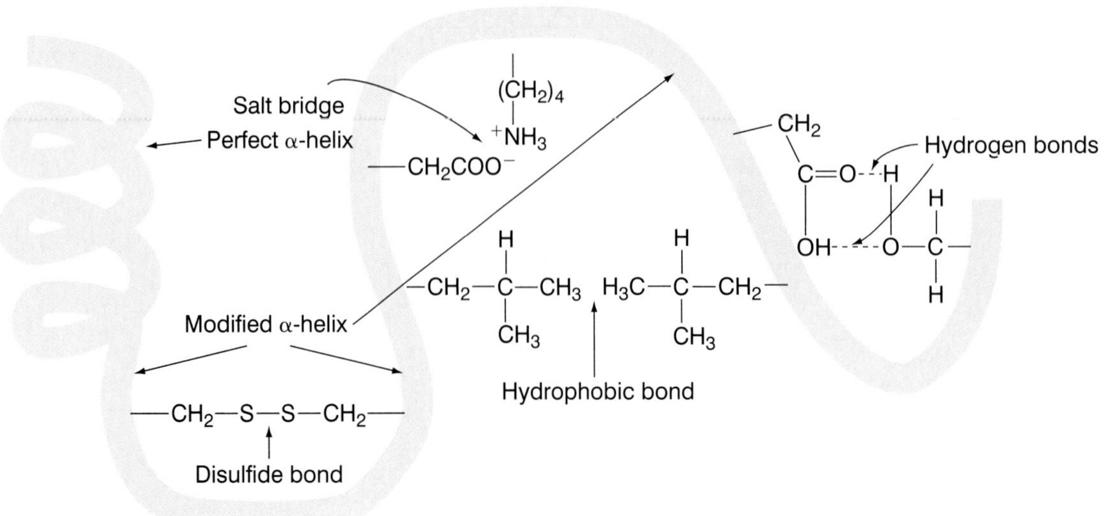

**FIGURE 28.16** The structure of hemoglobin.

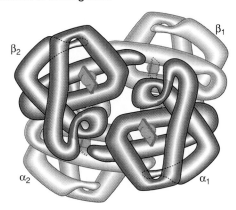

Some proteins contain two or more polypeptide chains. The *quaternary structure* of such a protein represents the manner in which these polypeptide units are arranged in space. Most (but not all) proteins displaying a quaternary structure contain one or more prosthetic groups. For example, hemoglobin, a protein in blood that carries oxygen to the cells, contains one protein chain per heme unit for a total of four of each. Two of the chains are called *alpha chains* (each with 141 amino acid residues), and the other two are known as *beta chains* (each with 146 amino acid residues). Note that each protein chain has a primary, secondary, and tertiary structure. It is the overall structure of the four protein chains that we call the quaternary structure of the protein, which in the case of hemoglobin represents a crude sphere. In Fig. 28.16, note that the four protein chains are labeled $\alpha_1$, $\alpha_2$, $\beta_1$, and $\beta_2$, while each heme unit is desigated by a rectangle.

Like amino acids, soluble proteins exist in anionic, dipolar, and cationic forms depending on the pH of the solution. Thus an isoelectric point can be determined experimentally for each protein.

The precipitation of a protein by a variety of experimental conditions or reagents such as acid or base, heat, ultraviolet light, high-formula-weight metal ions, detergents, high pressure, organic solvents, etc. is a process called *denaturation*. Denaturation causes the protein to lose biological activity along with the capability to crystallize. This loss of biological activity usually is permanent. Structurally, denaturation occurs when the secondary and tertiary structures of the protein are destroyed. The primary structure of the protein remains intact.

In general,

$$RCH_2COOH \xrightarrow[Cl_2]{P} \underset{\underset{Cl}{|}}{RCHCOOH} \xrightarrow[NH_3]{\text{Huge excess}} \underset{\underset{NH_2}{|}}{RCHCOO^-}$$

Example:

$$\underset{\underset{CH_3}{|}}{HCHCOOH} \xrightarrow[Cl_2]{PCl_3} \underset{\underset{CH_3}{|}}{ClCHCOOH} \xrightarrow[NH_3]{\text{Huge excess}} \underset{\underset{CH_3}{|}}{H_2NCHCOO^-}$$

Propionic acid    α-Chloropropionic acid    Alanine (anionic form)

For the sake of accuracy, the appropriate form of the amino acid is used depending on the pH of the solution.

Ethyl alcohol and isopropyl alcohol are good germicides because they denature bacterial protein. The use of heat and pressure of an autoclave to sterilize surgical instruments is accomplished by denaturation of bacterial protein. Tannic acid, a complex organic compound, is incorporated into burn ointments. The tannic acid, when applied to the skin, causes a protective layer of denatured protein to form that prevents water loss from the burn area.

## 28.8 PROPERTIES OF PROTEINS: DENATURATION

## 28.9 PREPARATION OF AMINO ACIDS

### Amination of an α-Haloacid

Refer to Section 25.1 for a discussion of the amination reaction.

Note that the α-haloacid is prepared from the corresponding carboxylic acid using the Hell-Volhard-Zelinsky reaction (see Sec. 21.2).

**Strecker Synthesis**

In the Strecker synthesis, an aldehyde is treated with a mixture of $NH_3$ and HCN to give the corresponding cyanoamine, which, in turn, is hydrolyzed to the amino acid. In general,

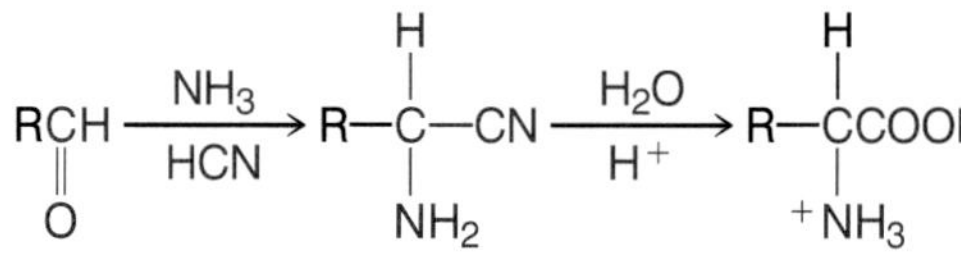

$$RCH{=}O \xrightarrow[\text{HCN}]{\text{NH}_3} R-\underset{\underset{NH_2}{|}}{\overset{\overset{H}{|}}{C}}-CN \xrightarrow[H^+]{H_2O} R-\underset{\underset{^+NH_3}{|}}{\overset{\overset{H}{|}}{C}}COOH$$

Example:

$$CH_3C{=}O \xrightarrow[\text{HCN}]{\text{NH}_3} CH_3-\underset{\underset{NH_2}{|}}{\overset{\overset{H}{|}}{C}}-CN \xrightarrow[H^+]{H_2O} CH_3\underset{\underset{^+NH_3}{|}}{\overset{\overset{H}{|}}{C}}-COOH$$

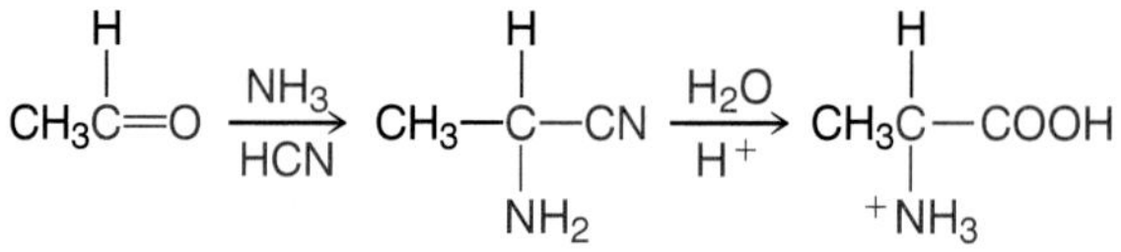

Alanine
(cationic form)

## 28.10 REACTIONS OF AMINO ACIDS

### Reactions Characteristic of the Amino Group

From this point on in this textbook, all amino acids will be drawn in the more structurally correct dipolar form.

**Reaction with Nitrous Acid (see Sec. 25.2)**   Amino groups, when treated with nitrous acid, give a quantitative yield of nitrogen, along with the hydroxy acid, other organic products, and water. In general,

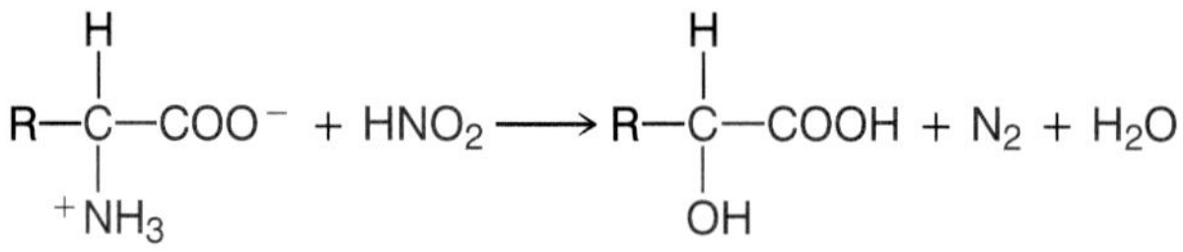

$$R-\underset{\underset{^+NH_3}{|}}{\overset{\overset{H}{|}}{C}}-COO^- + HNO_2 \longrightarrow R-\underset{\underset{OH}{|}}{\overset{\overset{H}{|}}{C}}-COOH + N_2 + H_2O$$

Example:

$$CH_3-\underset{\underset{^+NH_3}{|}}{\overset{\overset{H}{|}}{C}}-\overset{\overset{O}{\|}}{C}-O^- + HNO_2 \longrightarrow CH_3-\underset{\underset{OH}{|}}{\overset{\overset{H}{|}}{C}}-COOH + N_2 + H_2O$$

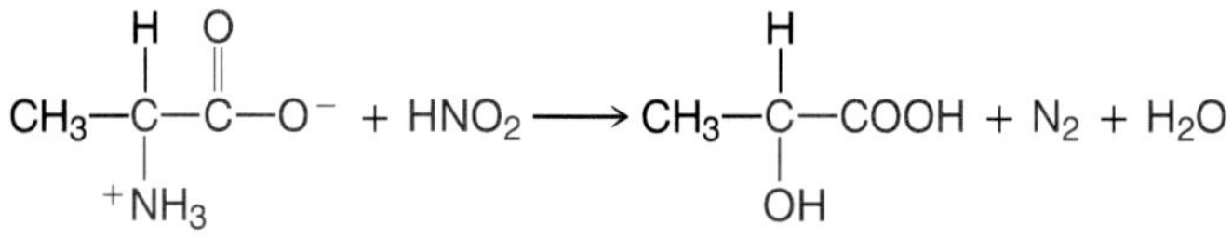

The reaction is the basis of the *Van Slyke procedure* to determine the number of primary amine functions in an amino acid, peptide, or protein. The amount of nitrogen produced is a measure of the number of primary amino groups in the compound.

**Reaction with Acetic Anhydride and Acyl Chlorides**   Acetic anhydride and acyl chlorides react with an amino acid to produce an *N*-acetylamino acid or *N*-acylamino acid, respectively (see Sec. 25.2). In general,

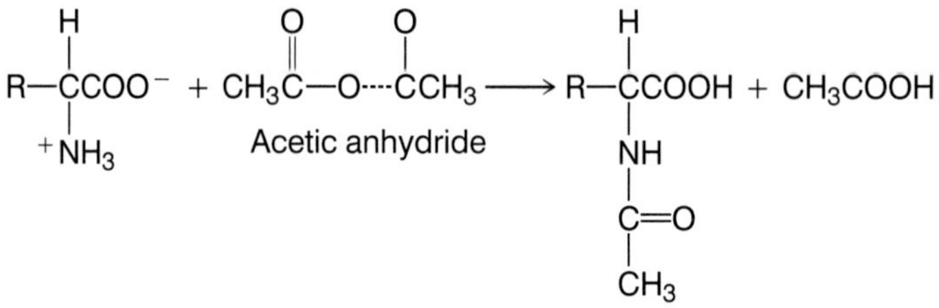

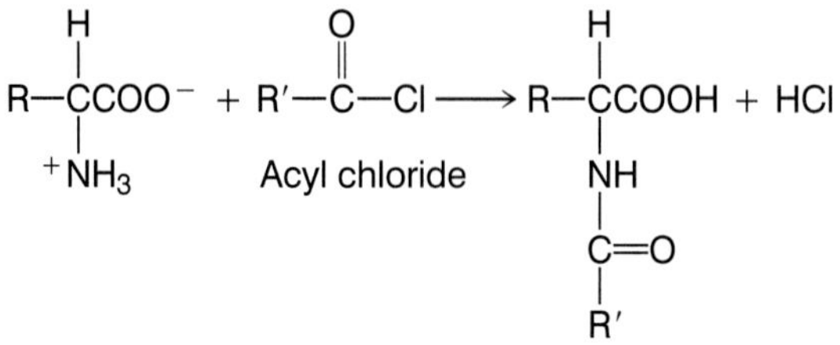

$$R-\underset{\underset{^+NH_3}{|}}{\overset{\overset{H}{|}}{C}}COO^- + CH_3\overset{\overset{O}{\|}}{C}-O\cdots\overset{\overset{O}{\|}}{C}CH_3 \longrightarrow R-\underset{\substack{| \\ NH \\ | \\ C=O \\ | \\ CH_3}}{\overset{\overset{H}{|}}{C}}COOH + CH_3COOH$$

Acetic anhydride

$$R-\underset{\underset{^+NH_3}{|}}{\overset{\overset{H}{|}}{C}}COO^- + R'-\overset{\overset{O}{\|}}{C}-Cl \longrightarrow R-\underset{\substack{| \\ NH \\ | \\ C=O \\ | \\ R'}}{\overset{\overset{H}{|}}{C}}COOH + HCl$$

Acyl chloride

Examples:

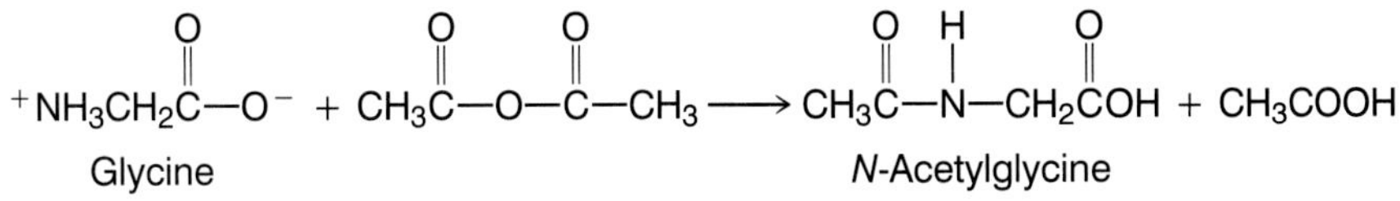

$$^+NH_3CH_2\overset{\displaystyle O}{\overset{\|}{C}}{-}O^- + CH_3\overset{\displaystyle O}{\overset{\|}{C}}{-}O{-}\overset{\displaystyle O}{\overset{\|}{C}}{-}CH_3 \longrightarrow CH_3\overset{\displaystyle O}{\overset{\|}{C}}{-}\overset{\displaystyle H}{\overset{|}{N}}{-}CH_2\overset{\displaystyle O}{\overset{\|}{C}}OH + CH_3COOH$$

Glycine                                   N-Acetylglycine

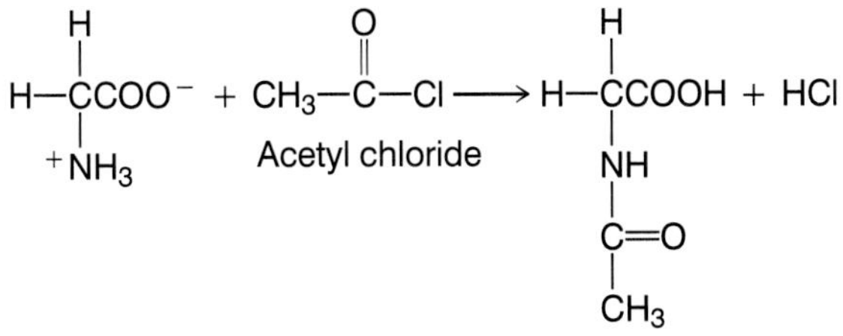

$$H{-}\overset{\displaystyle H}{\overset{|}{\underset{\underset{^+NH_3}{|}}{C}}}COO^- + CH_3\overset{\displaystyle O}{\overset{\|}{C}}{-}Cl \longrightarrow H{-}\overset{\displaystyle H}{\overset{|}{\underset{\underset{\underset{\underset{CH_3}{|}}{\overset{\|}{C}=O}}{\overset{|}{NH}}}{C}}}COOH + HCl$$

Acetyl chloride

If practical, an excess of amino acid is used to produce the ammonium chloride salt rather than hydrogen chloride.

**Reactions Characteristic of the Carboxyl Group**

**Fischer Esterification (see Secs. 20.5 and 21.2)**   To a mixture of the amino acid and excess gaseous hydrogen chloride an alcohol is added. In the laboratory, water is removed continuously during the course of the reaction so that the equilibrium is driven to the right and the ester salt is obtained in good yield. In general,

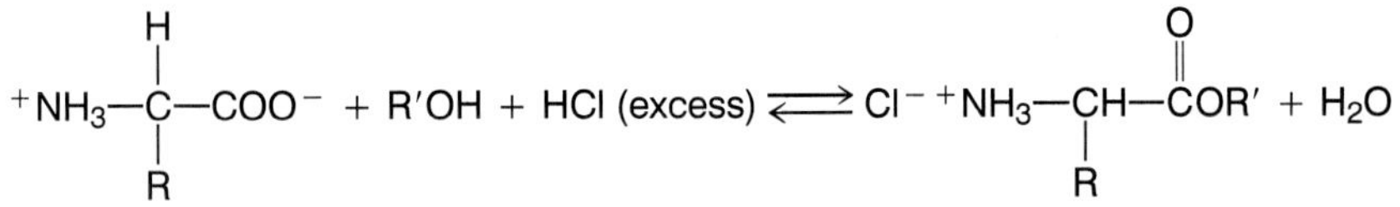

$$^+NH_3{-}\overset{\displaystyle H}{\overset{|}{\underset{\underset{R}{|}}{C}}}{-}COO^- + R'OH + HCl \text{ (excess)} \rightleftharpoons Cl^-\ ^+NH_3{-}\overset{\displaystyle H}{\overset{|}{\underset{\underset{R}{|}}{C}}}{-}\overset{\displaystyle O}{\overset{\|}{C}}OR' + H_2O$$

Example:

$$^+NH_3CH_2COO^- + C_2H_5OH + HCl \text{ (excess)} \rightleftharpoons Cl^-\ ^+NH_3CH_2\overset{\displaystyle O}{\overset{\|}{C}}OC_2H_5 + H_2O$$

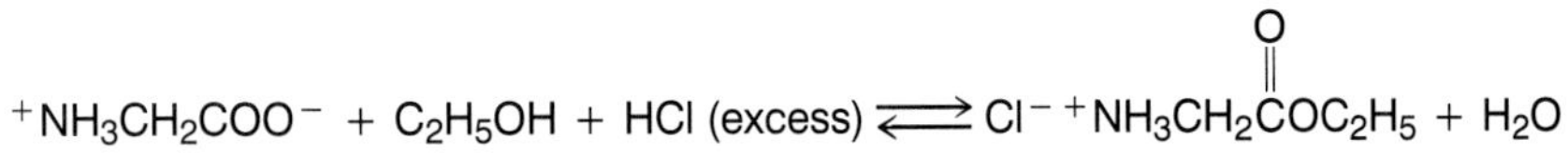

In general, a given functional group in an amino acid has the same chemical properties that it would have if it were the only functional group in a molecule.

**Reaction with Ninhydrin**

Treating *ninhydrin* with any amino acid that contains a primary amino group produces a purple dye. This reaction serves as a test for the presence of almost all amino acids. This test is sensitive and is also valid for peptides and proteins.

---

Name one amino acid that does not react with ninhydrin to form a purple dye. Explain.

**EXERCISE 28.11**

---

   The first step of the reaction involves the oxidation of the amino acid with 1 mol of ninhydrin. In general,

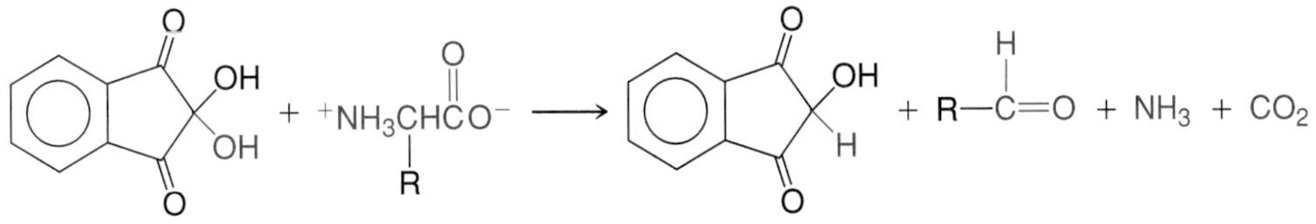

Ninhydrin                          Reduced form of
                                      ninhydrin

Example:

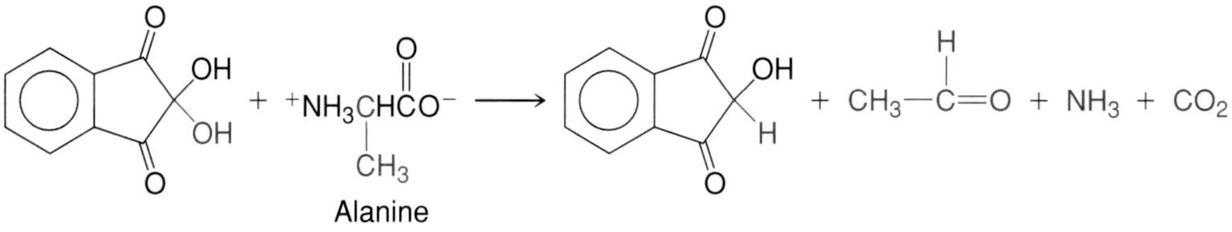

Alanine

The ammonia produced then combines with another molecule of ninhydrin and a molecule of the reduced form of ninhydrin just produced to create the keto species (VI) that undergoes a keto-enol tautomeric shift (see Sec. 8.7) to form species VII, an enol, which, in turn, loses a proton to produce the purple pigment (VIII).

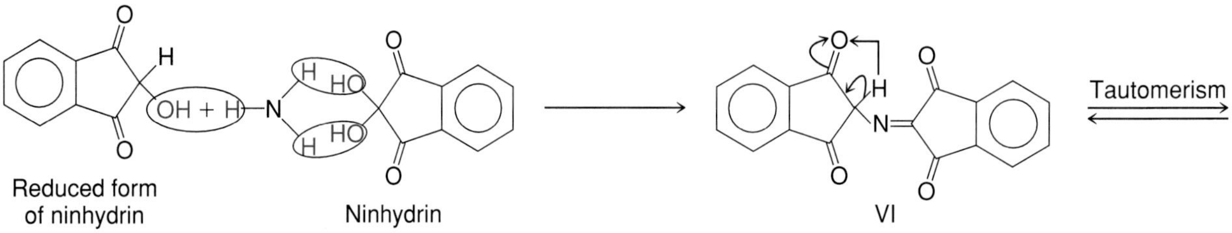

Reduced form of ninhydrin          Ninhydrin          VI          Tautomerism

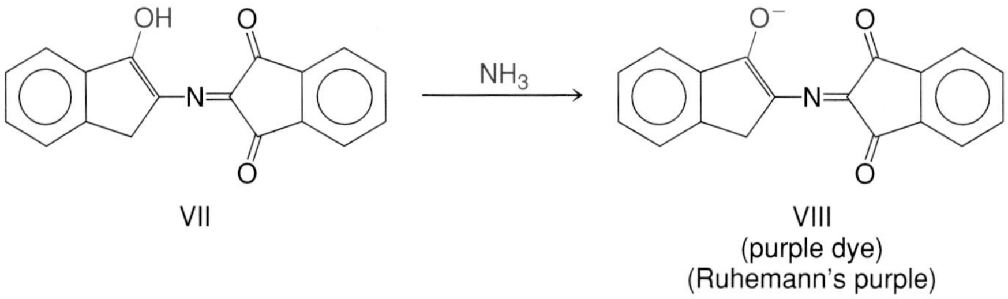

The purple color is due to the extensive amount of double-bond conjugation in the anionic species (VIII).

VII          VIII
(purple dye)
(Ruhemann's purple)

## 28.11 LABORATORY PREPARATION OF PEPTIDES

Once the structure of a naturally occurring peptide has been established, it must be confirmed by synthesizing the peptide and comparing the properties of the naturally occurring peptide with those of the synthetically prepared peptide. Now you might be inclined to prepare, for example, Phe-Ser by mixing the two amino acids and distilling off the water formed. Unfortunately, yields are poor because four possible dipeptides can form:

$$Phe + Ser \longrightarrow Phe\text{-}Ser + Ser\text{-}Phe + Ser\text{-}Ser + Phe\text{-}Phe$$

What we must do, then, to obtain the dipeptide we want (Phe-Ser) is to protect the amino group of phenylalanine and the carboxyl group of serine so that neither group can participate in a reaction. This will leave the carboxyl group of phenylalanine and the amino group of serine open to reaction to fashion the dipeptide Phe-Ser.

protect—Phe—COOH + NH$_2$—Ser—protect

↓

ultimately
Phe—Ser

Once the desired dipeptide is formed, the protecting groups are removed. A typical chemical synthesis of a dipeptide, using the preparation of phenylalanylserine as an example, proceeds as follows:

*Step 1.* Protect the amino group of phenylalanine with benzyl chloroformate. Note that the amine group is converted to an amide group.

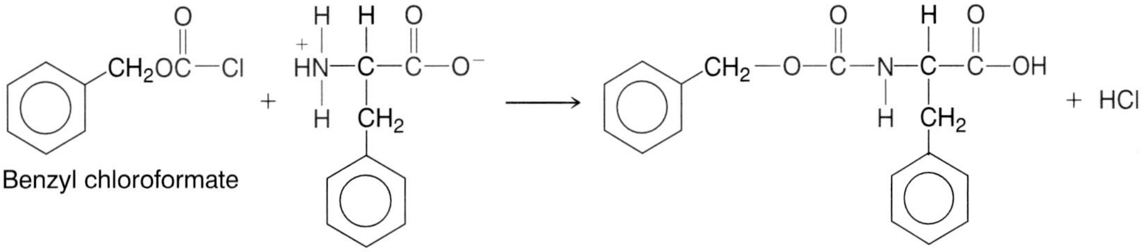

Benzyl chloroformate

*Step 2.* Protect the carboxyl group of serine by producing the benzyl ester.

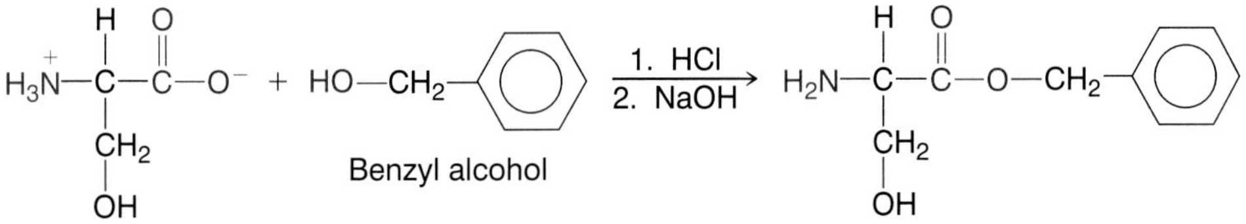

Benzyl alcohol

*Step 3.* Combine the carboxyl group of phenylalanine (the N-terminus) with the amino group of serine (the C-terminus) to prepare an amide.

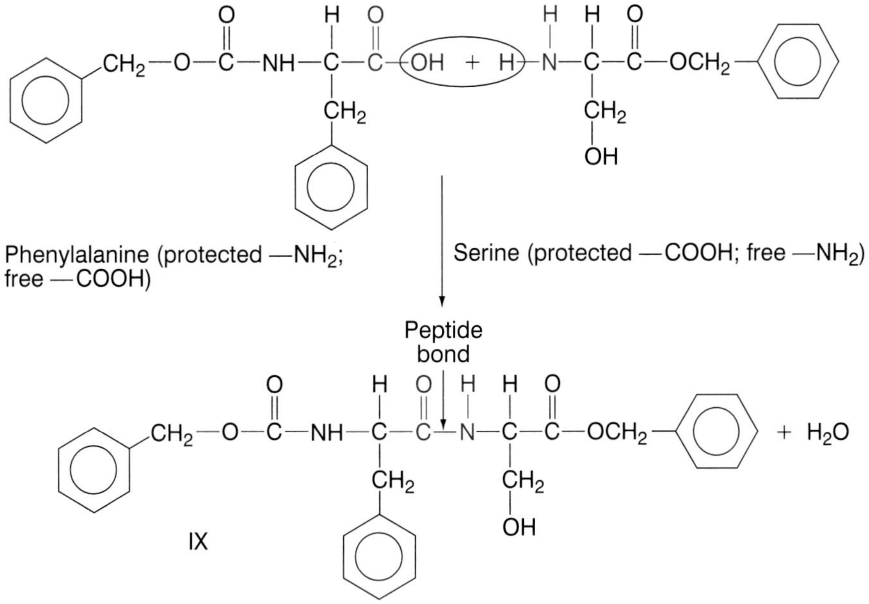

Phenylalanine (protected —NH₂; free —COOH)

Serine (protected —COOH; free —NH₂)

Peptide bond

IX

The reagent *di*cyclohexylcarbodiimide (DCC) is often used to increase the rate of the reaction.

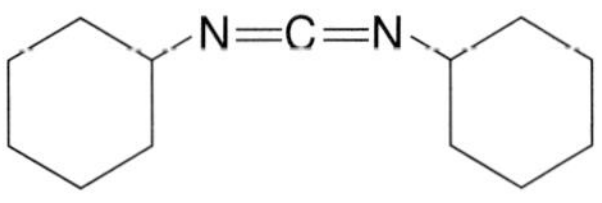

Dicyclohexylcarbodiimide

Since DCC is destroyed during the course of the reaction, we say that DCC is a *promoter* as opposed to a catalyst. Now the driving force of the preceding reaction is the reaction of water with DCC to produce dicyclohexylurea.*

---

*The mechanism used to produce dicyclohexylurea from dicyclohexylcarbodiimide is oversimplified. Both the amine and carboxylic acid participate in the accepted mechanism.

**Envision the Reaction**

## Mechanism of the Reaction of Water with Dicyclohexylcarbodiimide to Produce Dicyclohexylurea

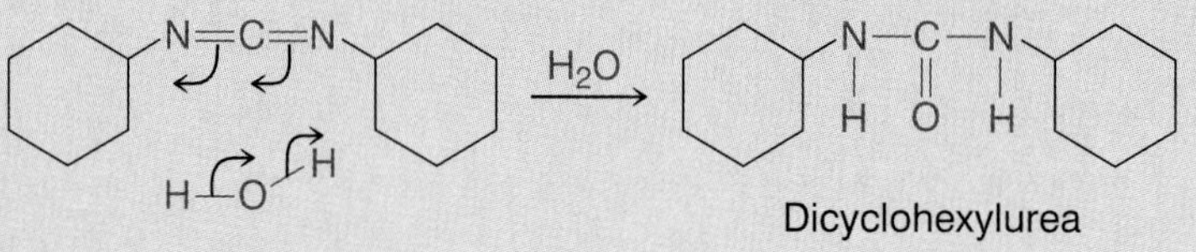

Dicyclohexylurea

Putting all this together, we have

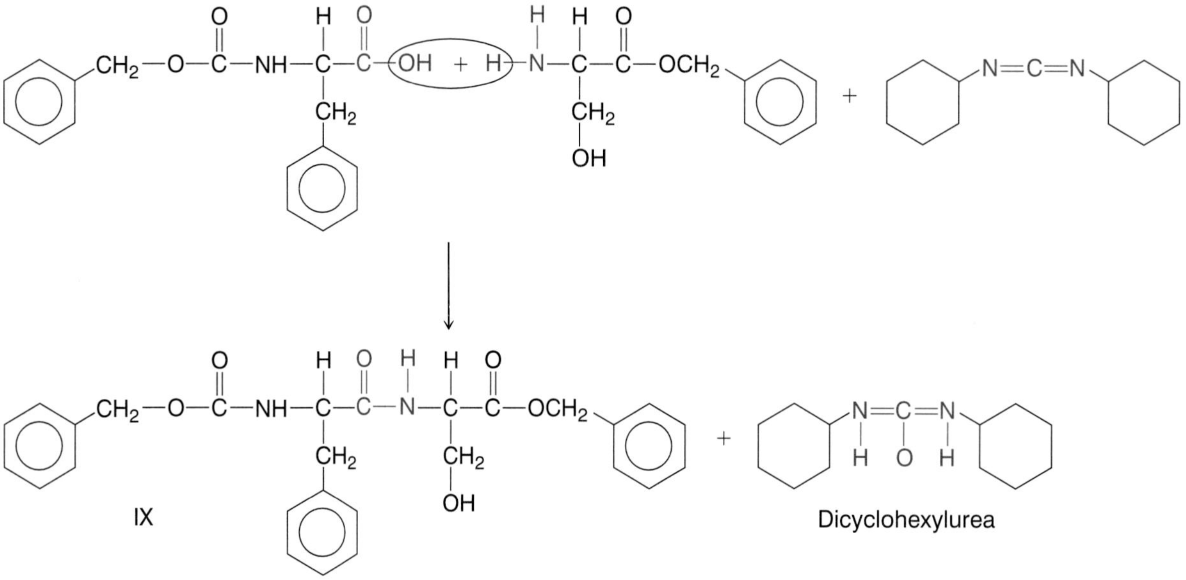

IX

Dicyclohexylurea

*Step 4.* Removal of the protecting groups from the dipeptide (IX) is accomplished by the use of catalytic hydrogenolysis (catalytic hydrogenation in which the reacting molecule is cleaved) to give the dipeptide, carbon dioxide, and toluene. Hydrolysis with aqueous hydrochloric acid cannot be used to deprotect because the newly formed peptide bond would be destroyed.

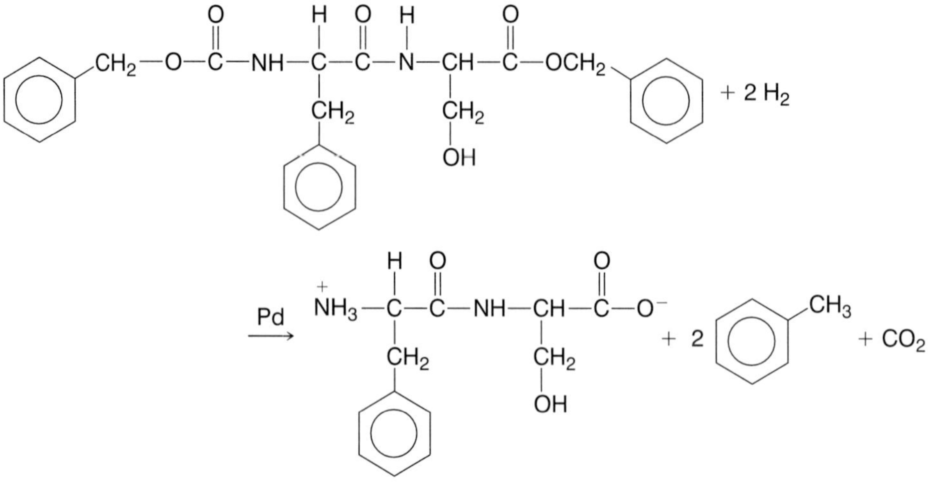

Phenylalanylserine (Phe-Ser)

To synthesize a tripeptide, first remove the acid protecting group from species IX by basic hydrolysis and then treat with a mineral acid to give species X. Note, in the formation of X, that basic hydrolysis (saponification) hydrolyzes the carboxylic acid ester protecting group but leaves the carbamic acid ester protecting group and the peptide bond intact.

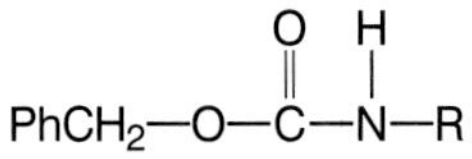

$$PhCH_2-O-\overset{\overset{\text{O}}{\|}}{C}-\overset{\overset{\text{H}}{|}}{N}-R$$

R-substituted
carbamic acid ester

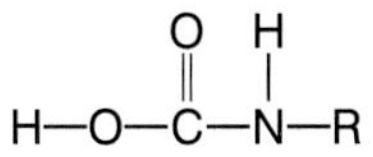

$$H-O-\overset{\overset{\text{O}}{\|}}{C}-\overset{\overset{\text{H}}{|}}{N}-R$$

R-substituted
carbamic acid

$$\left[H-O-\overset{\overset{\text{O}}{\|}}{C}-\overset{\overset{\text{H}}{|}}{N}-H\right] \longrightarrow CO_2 + NH_3$$

Carbamic acid,
unstable

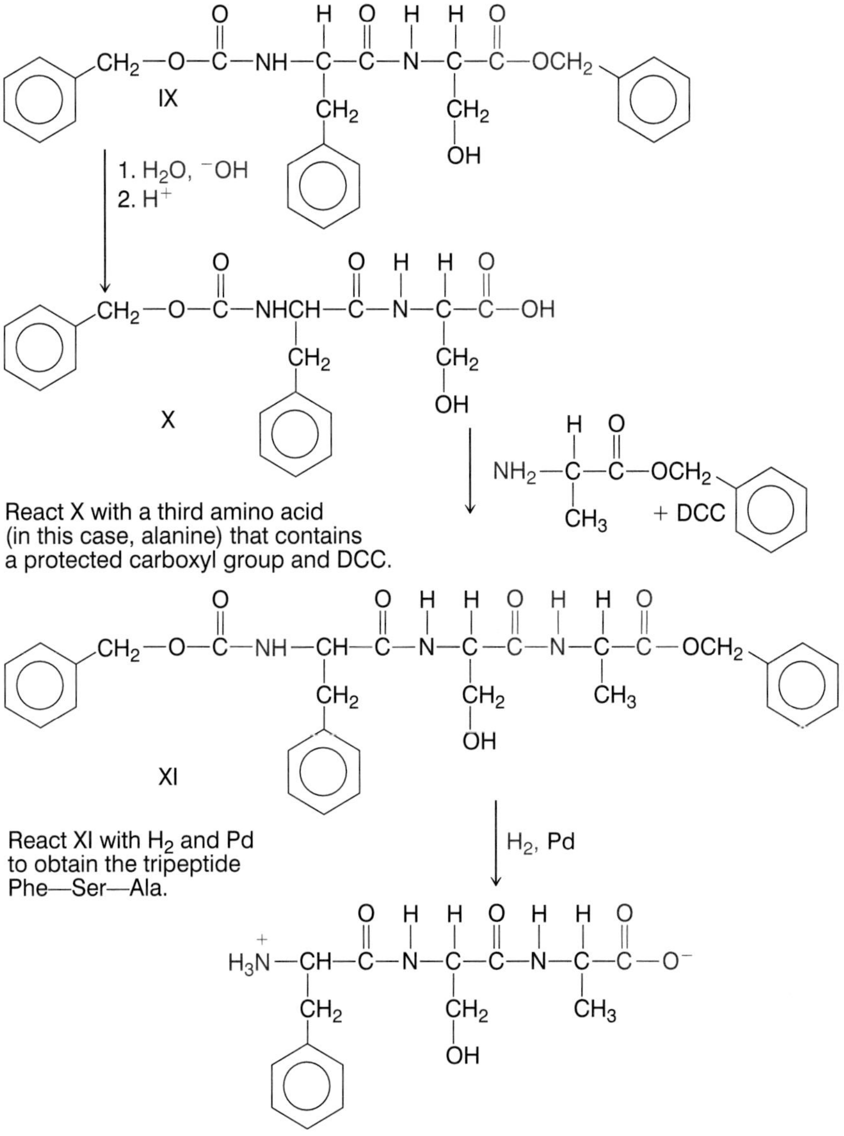

IX

1. $H_2O$, $^-OH$
2. $H^+$

X

React X with a third amino acid (in this case, alanine) that contains a protected carboxyl group and DCC.

$NH_2-\overset{\overset{\text{H}}{|}}{C}-\overset{\overset{\text{O}}{\|}}{C}-OCH_2$   + DCC
      $|$
      $CH_3$

XI

React XI with $H_2$ and Pd to obtain the tripeptide Phe—Ser—Ala.

$H_2$, Pd

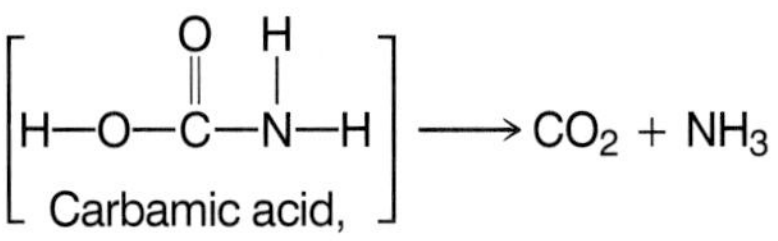

The R-substituted carbamic acid ester is less prone to basic hydrolysis than a carboxylic acid ester because an amide is far more stable to basic hydrolysis than an ester (see Sec. 23.2).

| **EXERCISE 28.12** | Prepare serylphenylalanine fron serine and phenylalanine. You may use any organic and inorganic reagents necessary. |
|---|---|

## ▶ CHAPTER ACCOMPLISHMENTS

### Introduction
☐ Explain why protein is so very important to humans.
☐ Define or explain the term *side chain* .

### 28.1 Structure and Classification of Amino Acids
☐ Draw a structural formula for:
    a. alanine.
    b. glycine.
    c. valine.
    d. tryptophan.
    e. asparagine.
    f. lysine.
    g. glutamic acid.
☐ Write a three-letter abbreviation for each of the preceding amino acids.
☐ Write a one-letter abbreviation for each of the preceding amino acids.
☐ Draw a structural formula for a(an)
    a. neutral amino acid.
    b. acid amino acid.
    c. basic amino acid.
☐ Draw a Fischer projection for L-alanine.
☐ Define an *essential amino acid*.
☐ Name the amino acid that contains a secondary amine in the alpha position.

### 28.2 Physical Properties of Amino Acids: Isoelectric Point—The Dipolar Ion
☐ List some unusual physical properties that the $\alpha$-amino acids show.
☐ Draw a structural formula for each of the following as a dipolar ion.
    a. glycine
    b. lysine
    c. glutamic acid

### 28.3 Separation of Amino Acids (and Proteins)
☐ Complete each of the following equations.
    a. $^+NH_3CH_2COO^- + {}^-OH \longrightarrow$
    b. $^+NH_3CH_2COO^- + H^+ \longrightarrow$
☐ Draw a structural formula for the predominant form of glycine present at the pI.

☐ Draw a structural formula for the predominant form of glycine present at pH 3.0.
☐ Draw a structural formula for the predominant form of glycine present at pH 9.0.
☐ Calculate the pI of aspartic acid from the data in Table 28.5.

### 28.4 Peptides: Classification—The Peptide Bond
☐ Draw a structural formula for glycylalanine.
☐ Define an N-terminus, a C-terminus.
☐ Draw an expanded structural formula for YGGFM.

### 28.5 Peptide and Protein Structure Determination
☐ Complete the following equations.
    a. Phe-Ala-Val-Leu-Trp $\xrightarrow[\text{H}_2\text{O}]{\text{HCl}}$

    b. Phe-Ala-Val-Leu-Trp $\xrightarrow{\text{Carboxypeptidase}}$

    c. Phe-Ala-Val-Leu-Trp $\xrightarrow[\text{2. HCl, H}_2\text{O}]{\text{1. Sanger's reagent}}$

### 28.6 Classification of Proteins
☐ Explain the difference between a simple protein and a conjugated protein.
☐ Explain why globular proteins are found in blood as opposed to fibrous proteins.

### 28.7 The Levels of Protein Structure
☐ List what the primary structure of a protein consists of.
☐ Name the structure level associated with the stereochemistry of a protein.
☐ List the forces involved in the tertiary structure of a protein.
☐ State which of the preceding forces represents an ionic attraction.

### 28.8 Properties of Proteins: Denaturation
☐ Define or explain denaturation.

### 28.9 Preparation of Amino Acids
☐ Prepare alanine from propionic acid.
☐ Prepare alanine from acetaldehyde.

### 28.10 Reactions of Amino Acids

☐ Complete the following equations.

a. $CH_3-\overset{\overset{\displaystyle H}{|}}{\underset{\underset{\displaystyle {}^+NH_3}{|}}{C}}-COO^- + HNO_2 \longrightarrow$

b. $CH_3-\overset{\overset{\displaystyle H}{|}}{\underset{\underset{\displaystyle {}^+NH_3}{|}}{C}}-COO^- + CH_3-\overset{\overset{\displaystyle O}{\|}}{C}-Cl \longrightarrow$

c. $CH_3-\overset{\overset{\displaystyle H}{|}}{\underset{\underset{\displaystyle {}^+NH_3}{|}}{C}}-COO^- + CH_3CH_2OH + HCl \text{ (excess)} \longrightarrow$

### 28.11 Laboratory Preparation of Peptides

☐ Prepare Phe-Ser from Phe, Ser, and any other necessary reagents.

---

## ▶ KEY TERMS

protein (Introduction)
amino acid (Introduction)
side chain (Introduction)
neutral amino acid (28.1)
acid amino acid (28.1)
basic amino acid (28.1)
essential amino acid (28.1)
phenylketonurea (PKU) (28.1)
dipolar ion (28.2)
zwitterion (28.2)
amphoteric (28.2)
isoelectric point (28.2)
electrophoresis (28.3)
peptide (28.4)
amino acid residue (28.4)
dipeptide (28.4)

tripeptide (28.4)
tetrapeptide (28.4)
polypeptide (28.4)
peptide bond (28.4)
glutathione (28.4)
N-terminus (28.4)
C-terminus (28.4)
enkephalin (28.4)
Sanger's reagent (28.5)
carboxypeptidase (28.5)
simple protein (28.6)
conjugated protein (28.6)
prosthetic group (28.6)
fibrous protein (28.6)
globular protein (28.6)
primary structure (28.7)

secondary structure (28.7)
pitch (28.7)
tertiary structure (28.7)
disulfide bond (28.7)
salt bridge (28.7)
hydrophobic bond (28.7)
quaternary structure (28.7)
alpha chain (28.7)
beta chain (28.7)
denaturation (28.8)
Van Slyke procedure (28.10)
ninhydrin (28.10)
sickle cell anemia (Medical Boxed Reading 28.1)
aspartame (Industrial Boxed Reading 28.2)

---

## ▶ PROBLEMS

1. Draw a structural formula (as in Table 28.1) and write a three-letter and a one-letter abbreviation for
   a. leucine.
   b. tryptophan.
   c. serine
   d. glutamine.
   e. lysine.

2. Given that the isoelectric point (pI) of an amino acid is 3.2, what type of amino acid must it be?

3. The compound L-DOPA [L(−)-2-amino-3-(3,4-dihydroxyphenyl)propanoic acid] is used as a treatment to reduce some of the symptoms of Parkinson's disease. At the present time, there is no known cure for Parkinson's disease, but L-DOPA serves as a "palliative." Draw a Fischer projection for L-DOPA. L-DOPA can be thought of as a derivative of what amino acid?

4. Give a common and IUPAC name for
   a. valine.
   b. leucine.
   c. phenylalanine.
   d. lysine.
   e. aspartic acid.

5. Draw a structural formula for the
   a. dipolar form of tyrosine, leucine.
   b. anionic form of serine, methionine.
   c. cationic form of valine, glutamine.

6. Which form of isoleucine is dominant at
   a. pH 2?
   b. pH 12?
   c. pH 6?

7. a. Which amino acid is achiral?
   b. Which amino acids are
      (1) aromatic?
      (2) heterocyclic?
      (3) both?
   c. Which amino acids contain sulfur?
   d. Which amino acids contain an amide group?
   e. Which amino acid does not contain a primary amine?

8. What are the nutritional consequences of an amino acid that is essential?

9. Calculate the pI of glutamic acid using the data in Table 28.5.

10. With the help of Table 28.5, draw a structural formula for the dominant species in an aqueous solution of
    a. lysine at pH 1.
    b. glutamic acid at pH 1.5.
    c. lysine at pH 9.5.
    d. glutamic acid at pH 5.6.
    e. glutamic acid at pH 3.2.
    f. lysine at pH 5.0.

11. Draw a structural formula for
    a. Val-Trp-Ser-Gln (dipolar).
    b. glycine hydrochloride.

12. Complete the following equations by supplying a structural formula for each organic compound produced and writing a molecular formula or formula unit for each inorganic compound formed.
    a. leucine (dipolar) + $^-$OH $\longrightarrow$
    b. leucine (dipolar) + H$^+$ $\longrightarrow$
    c. HO—CCH$_2$—C—C—OH + $^-$OH (1 mol) $\longrightarrow$
    d. H$_3$N—C—C—O$^-$ + H$^+$ (1 mol) $\longrightarrow$
    e. aspartic acid (dipolar) + Sanger's reagent $\longrightarrow$

13. Calculate the number of possible isomeric peptides for glutathione.

14. Calculate the number of possible isomers of a tetrapeptide. Assume that four different neutral amino acid residues make up the tetrapeptide.

15. Define or explain
    a. prosthetic groups.
    b. secondary structure of a protein.
    c. denaturation.
    d. carboxypeptidase.
    e. amphoteric.
    f. electrophoresis.

16. A tetrapeptide contains Gly, Ala, Phe, and Ser yet does not react with Sanger's reagent or carboxypeptidase. Give a possible reason for this lack of reactivity.

17. A hexapeptide contains Gly$_2$, Val, Cys, Met, and Ser. Sanger analysis gave the following colored compound:

Carboxypeptidase first produced free glycine. Partial hydrolysis gave

Met-Cys-Gly
Val-Gly-Ser
Ser-Met-Cys

Write an abbreviated (three-letter) structure for this hexapeptide.

18. Define or explain
    a. primary structure of a protein.
    b. tertiary structure of a protein.
    c. quaternary structure of a protein.

19. List the forces (interactions) that make up the tertiary structure of a protein.

20. Myoglobin is a protein found in certain mammals that contains a protein chain with 45 amino acid residues along with heme (a prosthetic group). Does myoglobin show a quaternary structure? Explain.

21. Prepare the following compounds (you may use any inorganic reagents necessary).
    a. glycine from ethyl alcohol
    b. valine from CH$_3$—C—C=O

22. Complete the following equations by supplying a structural formula for each organic compound produced and a molecular formula or formula unit for each inorganic compound formed.
    a. valine (dipolar) + HNO$_2$ $\longrightarrow$
    b. alanine (dipolar) + CH$_3$COCl $\longrightarrow$
    c. valine (dipolar) + C$_2$H$_5$OH + HCl (gas) (excess) $\rightleftharpoons$
    d.
    e. $^+$NH$_3$—C—C—O$^-$ + CH$_3$CH$_2$—C—Cl $\longrightarrow$
    f. CH$_3$C—Cl + HN—C$_2$H$_5$ $\longrightarrow$

23. Draw a structural formula for
    a. ninhydrin.
    b. the methyl ester of valine.
    c. alanine hydrochloride.
    d. dicyclohexylcarbodiimide.
    e. benzyl chloroformate.

24. Synthesize Ala-Gly-Ser from the individual amino acids and any other necessary compounds.

25. In Section 28.10, ninhydrin is represented as the hydrate of triketohydrindane, yet the compound is actually triketo-hydrindane. Why is this particular hydrate stable?

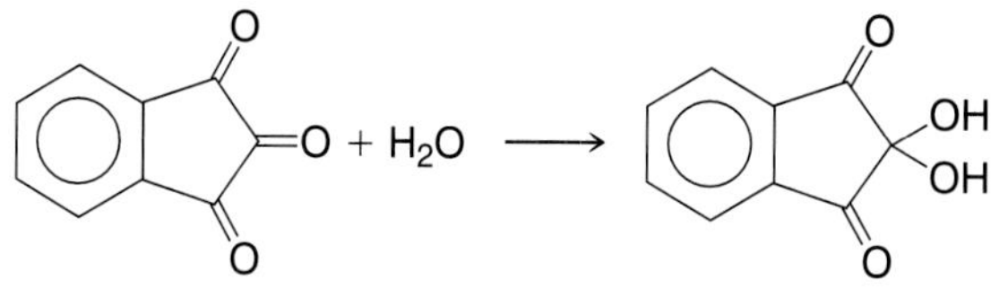

Ninhydrin
(triketohydrindane)

Ninhydrin hydrate

# CHAPTER 29

# Spectroscopy: Instrumental Methods of Structure Determination

When a new compound is obtained either by means of a chemical synthesis or by isolation from a plant or animal, the chemist must first determine the structural formula of the newly discovered compound. An even greater driving force for a rapid structure determination occurs when the compound shows promise as a potent new antibiotic, a fresh anticancer drug, or an anti-AIDS medicine.

During the first half of the 20th century, structure identification involved a series of solubility tests, chemical classification tests, and reactions that ultimately enabled the chemist to deduce the structure of the compound. The problems with the classical method of structure identification were twofold. First, the process was very time-consuming. Second, a relatively large amount of compound was required.

Fortunately, the last 40 years or so have resulted in the introduction of a variety of chemical instrumentation that has enabled chemists to determine a structure quickly and using a very small amount of material to boot. We will, in this chapter, cover the two most useful instrumental techniques employed by chemists in structure determination: *infrared spectroscopy (IR)* and *nuclear magnetic resonance (NMR) spectroscopy*. Let us start our study of instrumentation by examining electromagnetic radiation.

*Spectroscopy* is that branch of chemistry that studies the changes that occur in matter when bombarded with various types of *elecromagnetic radiation*

## 29.1 ELECTROMAGNETIC RADIATION

Electromagnetic radiation represents a continuum of energies from the highly energetic *gamma* and *x-rays* to the low-energy *microwaves* and *radio waves* (Table 29.1). An important point to realize is that the division between various types of electromagnetic radiation is not exactly defined; rather, there is some overlap in regions.

Our descriptions of electromagnetic radiation show a duality of behavior: Under certain experimental conditions, it appears to exist as a wave; under other conditions, it appears as a particle (known as a *photon*, a packet of energy). As a result of the work of Max Planck in 1900, the energy of a photon was determined to be proportional to its frequency, expressed by the equation

$$E = h\nu \qquad (29.1)$$

| TABLE 29.1 | The Electromagnetic Spectrum | |
|---|---|---|
| **Frequency ($\nu$), hertz (Hz)\*** | **Type of Radiation** | **Wavelength ($\lambda$), cm\*** |
| $3 \times 10^{21}$ | | $10^{-11}$ |
| $3 \times 10^{20}$ | Gamma rays | $10^{-10}$ |
| $3 \times 10^{19}$ | | $10^{-9}$ |
| $3 \times 10^{18}$ | X-rays | $10^{-8}$ |
| $3 \times 10^{17}$ | | $10^{-7}$ |
| $3 \times 10^{16}$ | Ultraviolet light | $10^{-6}$ |
| $3 \times 10^{15}$ | Visible light† | $10^{-5}$ |
| $3 \times 10^{14}$ | Infrared radiation | $10^{-4}$ |
| $3 \times 10^{13}$ | | $10^{-3}$ |
| $3 \times 10^{12}$ | | $10^{-2}$ |
| $3 \times 10^{11}$ | Microwaves | $10^{-1}$ |
| $3 \times 10^{10}$ | | $10^{0}$ |
| $3 \times 10^{9}$ | | $10^{1}$ |
| $3 \times 10^{8}$ | Radio waves | $10^{2}$ |
| $3 \times 10^{7}$ | | $10^{3}$ |

↑ Increasing frequency and energy          ↑ Decreasing wavelength

\* Refer to page 560 for definitions of frequency and wavelength.

† Visible light encompasses from the more energetic violet light with a wavelength of approximately 400 nm ($10^9$ nm = 1 m) to the less energetic red light with a wavelength of about 780 nm.

Note the very small portion of the electromagnetic spectrum represented by *visible light*

where $E$ represents the energy of the photon, $h$ is Planck's constant, and $\nu$ is the *frequency* of the radiation. The mathematical expression

$$\lambda = \frac{c}{\nu} \tag{29.2}$$

where $c$ is the velocity of light ($3.0 \times 10^{10}$ cm/s), $\lambda$ represents the *wavelength* of the radiation (cm), and $\nu$ is the frequency ($s^{-1}$ or Hz), directly represents electromagnetic radiation as a wave, since the wavelength term is included in Equation (29.2). Under the experimental conditions of the types of spectroscopy we will study, electromagnetic radiation appears to exist as a wave. The wavelength represents the distance from the peak (or valley) of one wave to the peak (or valley) of the following wave (Fig. 29.1), whereas the frequency is a measure of the number of waves that pass a given spot per second. From Equation 29.1, we can see that the energy of electromagnetic radiation is directly proportional to the frequency. Thus gamma radiation, with high frequencies, represents high-energy radiation, whereas radio waves, with low frequencies, illustrate low-energy radiation.

**FIGURE 29.1** One wavelength of electromagnetic radiation.

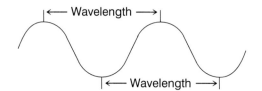

By substituting a value for frequency in Equation 29.2, we can calculate the wavelength of radiant energy, and vice versa. Consider Solved Problem 29.1.

---

Calculate the wavelength of infrared radiation with a frequency of $7 \times 10^{13}$ Hz.

**SOLVED PROBLEM 29.1**

**SOLUTION**
We substitute in Equation (29.2) as follows:

$$\lambda = \frac{c}{\nu} = \frac{3.0 \times 10^{10} \text{ cm/s}}{7.0 \times 10^{13} \text{ s}^{-1}} = 4.3 \times 10^{-4} \text{ cm}$$

---

Calculate the frequency of x-radiation with a wavelength of $6.0 \times 10^{-8}$ cm.

**EXERCISE 29.1**

---

In general terms, spectroscopy involves determining the frequencies of radiation absorbed by an organic molecule. We measure the frequencies and the amount of energy absorbed at each frequency to determine the *spectrum* of the molecule. A *spectrophotometer* measures the amount of radiation absorbed at each frequency as the frequency of the radiation is varied. Each absorption of energy results in the formation of a *peak* in the spectrum. Anywhere from 1 peak to as many as 30 or more peaks are found in the typical spectrum.

When radiant energy is passed through a sample, some of the radiation is absorbed by the molecules if the energy of the radiation corresponds to the energy-level differences in the molecule. The radiation that is not absorbed by the sample passes through unchanged and is detected by the instrument. The radiation that is absorbed may be reemitted in all directions or may cause chemical changes in the molecules but is not detected by the instrument (Fig. 29.2a and c). An absorption peak corresponds to *more radiation absorbed and hence less detected* (Fig. 29.2b).

---

The importance of infrared spectroscopy in the identification of organic compounds is that it is effective in determining the presence of a wide variety of organic functional groups. Seldom can an *infrared (IR) spectrum* by itself serve to identify an organic compound. Usually, one or more other different types of spectra are necessary (particularly nuclear magnetic resonance spectroscopy) as a complement to the infrared spectrum along with other experimental data.

When a molecule absorbs infrared radiation energy, there is an increase in the *energy of vibration* of one or more bonds in the molecule. These vibrations are of two types: bond stretching and bond bending (Fig. 29.3).

An infrared spectrum consists of a plot of percentage transmittance of radiation versus decreasing frequency of radiation, expressed as *wave number*. Typically, the wavelengths scanned by an instrument vary from $2.2 \times 10^{-4}$ to $25 \times 10^{-4}$ cm$^{-1}$ corresponding to wave numbers (frequencies) of 4500 to 400 cm$^{-1}$.

**INFRARED SPECTROSCOPY**

**29.2 INTRODUCTION**

The wave number (cm$^{-1}$) is the reciprocal of the wavelength in centimeters.

---

Convert the wavelength 2.2 μm (1 μm = $10^{-6}$ m) to the corresponding wave number in cm$^{-1}$ units.

**SOLVED PROBLEM 29.2**

**SOLUTION**
First, convert 2.2 μm to centimeters.

$$\text{cm} = 2.2 \text{ μm} \times \frac{1.0 \times 10^{-4} \text{ cm}}{1.0 \text{ μm}} = 2.2 \times 10^{-4} \text{ cm}$$

**FIGURE 29.2** The simplified operation of a typical spectrophotometer.

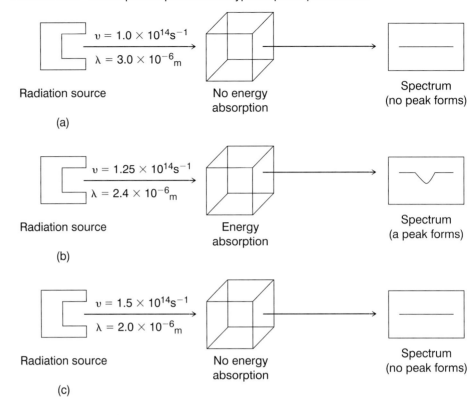

Radiation source    No energy absorption    Spectrum (no peak forms)

(a)

Radiation source    Energy absorption    Spectrum (a peak forms)

(b)

Radiation source    No energy absorption    Spectrum (no peak forms)

(c)

**FIGURE 29.3** Types of bond vibrations excited by infrared radiation.

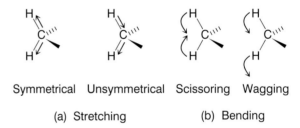

Symmetrical    Unsymmetrical    Scissoring    Wagging

(a) Stretching    (b) Bending

Finally, take the reciprocal of the preceding quotient.

$$\text{cm}^{-1} = \frac{1.0}{2.2 \times 10^{-4}\text{ cm}} = 4546\text{ cm}^{-1}\ (4500\text{ cm}^{-1}\text{ to two significant figures})$$

---

**EXERCISE 29.2**

Convert an absorption of 400 cm$^{-1}$ to μm units.

---

Note that solid KBr is transparent in the infrared spectrum.

This is particularly useful for insoluble samples.

For a liquid sample, one or two drops are placed between two KBr plates for infrared analysis. Solid samples may be dissolved in chloroform or carbon tetrachloride and the solution treated just like a liquid sample, or they may be ground up, "mulled" with mineral oil (Nujol), and smeared between two KBr plates. Through trial and error, spectroscopists have discovered a correlation between the frequency of absorption and the presence of a bond or bonds comprising a functional group. Thus a certain absorption or absorptions automatically mean the presence of one or more functional groups, and the absence of certain absorptions means the absence of one or more functional groups. This is the power of infrared spectroscopy. For example, both a simple molecule such as acetone and a more complex molecule such as 5-hexene-2-one display an absorption from 1680 to 1750 cm$^{-1}$ that is characteristic of the presence of a keto group.

$$CH_3—\overset{\overset{\displaystyle O}{\|}}{C}—CH_3 \qquad CH_3—\overset{\overset{\displaystyle O}{\|}}{C}—CH_2CH_2CH=CH_2$$

$$\nu = 1715\ cm^{-1} \qquad\qquad \nu = 1710\ cm^{-1}$$

Consider the infrared spectrum of acetone (Fig. 29.4). The series of peaks represents a pattern that is uniquely that of acetone; i.e., no other ketone or other organic compound shows this exact pattern of peaks.

The frequencies that identify a functional group are found between 4500 and 1400 cm$^{-1}$. Table 29.2 is a summary of certain infrared absorption frequencies associated with the more common bonds. The use of Table 29.2 is important in the interpretation of infrared spectra (see Sec. 29.3).

Not all the peaks in a spectrum can be utilized. For example, that segment of the spectrum between 1400 and 400 cm$^{-1}$ is called the *fingerprint region*. Usually, this region has many more peaks than the 4500 to 1400 cm$^{-1}$ region, many of which cannot be ascribed to a particular bond or group. However, this region of the spectrum in particular and the spectrum as a whole are often used to absolutely identify a particular compound (say, acetone) by comparing this spectrum with a known spectrum.

Besides the carbonyl absorption between 1680 and 1750 cm$^{-1}$ (at 1715 cm$^{-1}$), note the carbon-hydrogen absorption between 2800 and 3000 cm$^{-1}$ (see Table 29.2). This is a commmonly observed absorption in infrared spectra because many organic compounds contain $sp^3$ hybridized carbon-hydrogen bonds.

Now, let us examine the spectrum of *tert*-pentyl(amyl) alcohol (Fig. 29.5). The absorption between 3000 and 3700 cm$^{-1}$ is characteristic of an oxygen-hydrogen bond. Typically, such peaks are relatively broad, which is characteristic of extensive hydrogen bonding in the liquid state. In addition, note the presence of a carbon-oxygen adsorption peak from 900 to 1300 cm$^{-1}$. In many cases (and this case is no exception), however, this peak is difficult to pick out because it is located in the fingerprint region of the spectrum and therefore is surrounded by many other peaks. Usually, we do not attempt to locate and assign such a peak.

**FIGURE 29.4** The infrared spectrum of acetone.

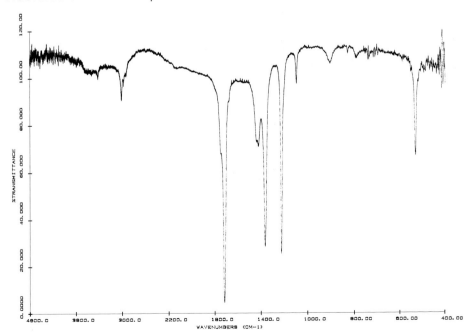

| TABLE 29.2 | A Listing of Often Used Infrared Absorption Frequencies |
|---|---|
| **Bond (Functional Group)** | **Frequency (cm⁻¹)** |
| C—H (alkane) | 2800–3000 |
| C—H (alkene) | 3000–3300 |
| C—H (alkyne) | About 3300 |
| C—H (aryl) | 3000–3300 |
| C=C (aryl) | 1450–1520, 1580–1620 |
| C—C (alkane) | Absorption too weak for use |
| C=C (alkene) | 1600–1700 |
| C≡C (alkyne) | 2100–2250 |
| O—H (alcohol and phenol) | 3000–3700 |
| O—H (carboxylic acid) | 2500–3600 |
| C—O (alcohol) | 900–1300 |
| N—H (amine) | 3000–3700 |
| C—O (ether) | 1050–1260 |
| C=O (aldehyde) | 1690–1740 |
| C=O (ketone) | 1680–1750 |
| C=O (carboxylic acid) | 1700–1780 |
| C=O (acid chloride) | 1745–1810 |
| C=O (acid anhydride) | 1730–1750, 1800–1830 |
| C=O (ester) | 1735–1750 |
| C=O (amide) | 1630–1700 |
| C—O (ester) | 1050–1250 |
| C≡N (nitrile) | 2200–2300 |

Finally, in this section let us consider the spectrum of an amine, (±)-1-phenyl-ethanamine (Fig. 29.6). The nitrogen-hydrogen absorption from 3000 to 3700 cm⁻¹ is weaker and narrower than the corresponding oxygen-hydrogen absorption of an alcohol. The class of an amine can be determined easily from the infrared spectrum: If the

**FIGURE 29.5** The infrared spectrum of tert-pentyl alcohol.

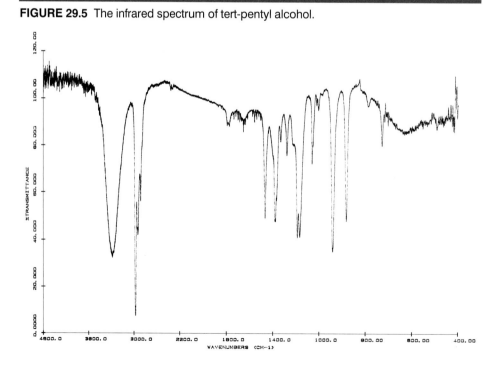

FIGURE 29.6 The infrared spectrum of (±)-1-phenylethanamine.

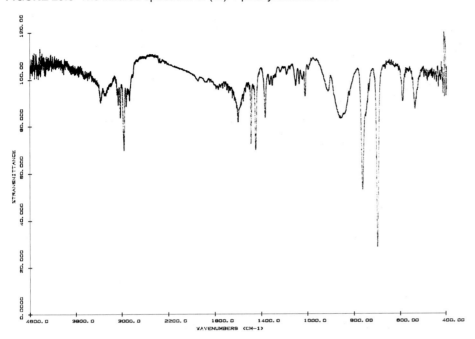

compound is a primary amine with two nitrogen-hydrogen bonds, two peaks from 3000 to 3700 cm$^{-1}$ can be found; if it is secondary, only one peak is present; whereas a tertiary amine displays no peaks in the 3000 to 3700 cm$^{-1}$ range of the spectrum.

Let us learn how to interpret a simple infrared spectrum by actually considering the spectrum of an unknown compound in Solved Problem 29.3.

## 29.3
## INTERPRETATION OF
## INFRARED SPECTRA

### SOLVED PROBLEM 29.3

Determine the functional group(s) present in the infrared spectrum shown in Fig. 29.7.

### SOLUTION
The broad peak at about 3000 cm$^{-1}$ suggests the presence of a hydroxyl group that masks the $sp^3$ carbon-hydrogen absorption between 2800 and 3000 cm$^{-1}$. The next main peak to the right in the spectrum is between 1700 and 1780 cm$^{-1}$, which is indicative of a carbonyl group of a carboxylic acid. Thus the compound is most likely a carboxylic acid. An aromatic acid such as benzoic acid can be eliminated because there are no absorptions at about 1500 and 1600 cm$^{-1}$. As a matter of fact, this is a spectrum of propionic acid.

### EXERCISE 29.3

A compound, $C_6H_{15}N$, is basic to red litmus, and an infrared spectrum of the compound is shown in Fig. 29.8. Supply a structural formula for and name the compound. Hint: Only one alkyl group is present in the molecule. Show your reasoning.

**FIGURE 29.7** Infrared spectrum of an unknown compound.

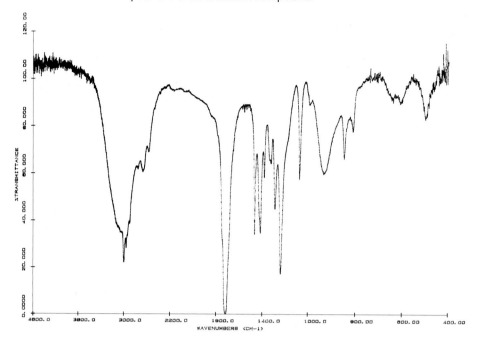

**FIGURE 29.8** Infrared spectrum of an unknown compound.

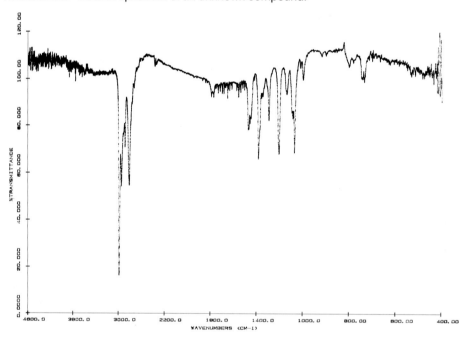

**EXERCISE 29.4**

List the main peak(s) you would expect in terms of frequency (and width when applicable) for

a. acetic acid.

b. acetophenone.

c. butanal.

d. diethylamine.

e. methyl butyrate.

While the infrared spectrum of an organic compound is useful in identifying a number of functional groups, its weakness is that it provides very little information about the hydrocarbon group(s) located in the compound. The power of nuclear magnetic resonance (NMR) spectroscopy is that this instrumental technique can identify a wide variety of alkyl, alkenyl, alkynyl, cycloalkyl, and aromatic groups. Thus nuclear magnetic resonance spectroscopy serves as a complement to infrared spectroscopy in the determination of molecular structures. Let us begin our study of NMR by exploring *nuclear spin*.

**29.5 NUCLEAR SPIN**

Some nuclei of atoms of the elements display a property called *spin*; others do not. A nucleus with spin spins like a top. As a result of the spin the nucleus undergoes, a small magnetic field is created; in other words, each spinning nucleus behaves like a tiny bar magnet. To the organic chemist, $^1H$ and $^{13}C$ are the most useful nuclei. The $^{12}C$ isotope plays no role in nuclear magnetic resonance spectroscopy because the nucleus does not spin.

Now, when we place a sample containing an $^1H$ nucleus (a proton) or a $^{13}C$ nucleus within a strong external magnetic field (say 15,000 gauss), slightly more than half of the hydrogen nuclei align themselves in a *parallel orientation* to the strong external magnetic field because the parallel orientation represents greater stability (Fig. 29.9), while slightly less than half the hydrogen nuclei take an *antiparallel orientation*. When the spinning nuclei are bombarded with radio waves of increasing frequency, a *spin-flip* takes place at the proper frequency, and some of the nuclei with a parallel orientation align themselves in an antiparallel orientation to the external magnetic field. The antiparallel orientation is a high-energy unstable state, and the resulting loss of energy back to the parallel state shows up as a peak (often called a *signal*) on a spectrum. When a proton or other spinning nucleus undergoes a spin-flip as a result of the unique blend of radio frequency and external magnetic field applied to the sample, the proton (a *nucleus*) is said to be in *resonance* with the particular radio frequency and *magnetic* field strength utilized. Thus we have the term *nuclear magnetic resonance*.

**FIGURE 29.9** Parallel and antiparallel alignments of spinning nuclei to a strong external magnetic field.

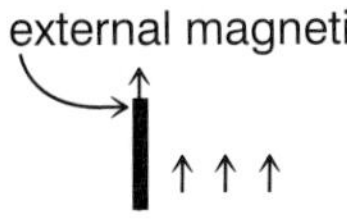

external magnetic field

↑ ↑ ↑

Parallel alignment of the magnetic field produced by each proton to the external magnetic field.

external magnetic field

↓ ↓ ↓

Antiparallel alignment of the magnetic field produced by each proton to the external magnetic field.

A sample is prepared for analysis by dissolving it in a solvent that does not contain hydrogen. Most often used are carbon tetrachloride and deuterochloroform ($CDCl_3$). A small amount of a standard, tetramethylsilane (TMS)[$(CH_3)_4Si$], when added, completes the process of sample preparation. Consider the nuclear magnetic resonance spectrum of acetone (Fig. 29.10) dissolved in deuterochloroform with tetramethylsilane as the standard. Note the peak on the extreme right of the spectrum. This peak, set at $\delta = 0$, is due to the absorption by the 12 hydrogens (or protons)[*] of TMS, the standard used in nuclear magnetic resonance spectroscopy. The $\delta$ unit is expressed in parts per million (ppm) and represents a unit of frequency. A $\delta$ value of 0 represents a relatively high frequency, and we say that it is in the *upfield* portion of the spectrum. The single absorption at $\delta = 2.1$ is further *downfield* and represents an absorption at a lower frequency. This single absorption is due to the 6 equivalent hydrogens (or protons)[*] in acetone and is called the *chemical shift*, which is defined as follows:

**29.6 THE NUCLEAR MAGNETIC RESONANCE SPECTRUM**

---

[*]In NMR parlance, the proton and hydrogen atom are synonymous, and both terms are used in this text.

**FIGURE 29.10** Proton nuclear magnetic resonance spectrum of acetone (in deuterochloroform).*

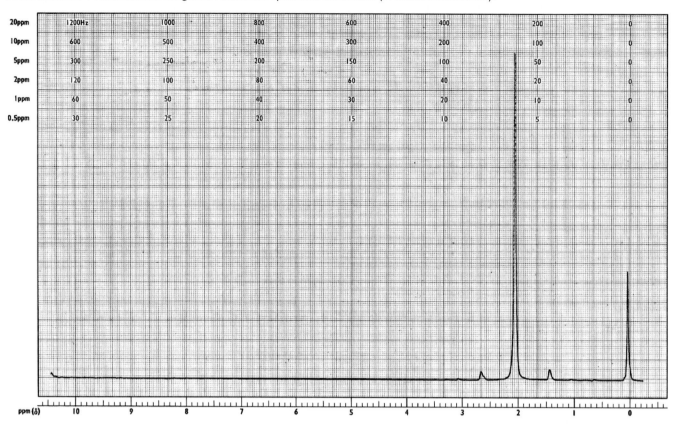

$$\delta \text{ (in ppm)} = \frac{\text{peak distance from TMS (in Hz)}}{\text{spectrometer frequency (in MHz)}} \times 10^6$$

Chemical shifts are expressed in parts per million rather than hertz, although both units are included on the chart paper, because the values in parts per million are constant regardless of the radio frequency employed. Some approximate proton chemical shifts are given in Table 29.3. TMS is used as a standard because its chemical shift is upfield of most other protons in organic molecules and hence does not interfere with other peaks in the spectrum. In addition, it has 12 equivalent protons and only this one type of proton, so only one peak is observed to represent the compound.

Sometimes, it is useful to convert an absorption frequency from parts per million to hertz, and vice versa. For a 60-MHz radio frequency instrument, we have

$$\delta \text{ value of } 1 = 1 \text{ ppm} = 60 \text{ Hz}$$

In a similar fashion, for a more powerful 100-MHz radio frequency instrument,

$$\delta \text{ value of } 1 = 1 \text{ ppm} = 100 \text{ Hz}$$

Consider Solved Problem 29.4.

## SOLVED PROBLEM 29.4

Note that we can only convert from parts per million to hertz when we know the operating frequency of the instrument.

The only absorption peak in the NMR spectrum of acetone appears at 2.1 ppm ($\delta = 2.1$) using an instrument with 60 MHz radio frequency.

a. Convert this to hertz.

b. If this same absorption, 2.1 ppm ($\delta = 2.1$), was observed using a 100-MHz radio frequency instrument, calculate this absorption in hertz.*

---

*The very small peaks at $\delta = 1.4$ and $\delta = 2$ in this spectrum and in other spectra are spinning side bonds that can be ignored.

| TABLE 29.3 | Approximate Proton Chemical Shifts |
|---|---|
| **Kind of Proton** | **Chemical Shift ($\delta$), ppm** |
| Primary alkyl: $-CH_2-\underline{H}$ | 0.8–1.1 |
| Secondary alkyl: $-CH-\underline{H}$ | 1.2–1.5 |
| Tertiary alkyl: $-\overset{\mid}{\underset{\mid}{C}}-\underline{H}$ | 1.5–1.7 |
| Benzylic: $Ar-CH_2-\underline{H}$ | 2.2–2.5 |
| Alkyl chloride: $-C\underline{H}_2Cl$ | 3.6–3.8 |
| Alkyl bromide: $-C\underline{H}_2Br$ | 3.4–3.6 |
| Alkyl iodide: $-C\underline{H}_2I$ | 3.1–3.3 |
| Ether: $-C\underline{H}_2O-$ | 3.3–3.9 |
| Alcohol: $-C\underline{H}_2OH$ | 3.3–4.0 |
| Alcohol: $-CH_2O\underline{H}$ | 0.5–6.0 |
| Aldehyde: $-C\underline{H}_2-\overset{O}{\underset{\parallel}{C}}-H$ | 1.5–2.5 |
| Aldehyde: $-\overset{O}{\underset{\parallel}{C}}-\underline{H}$ | 9.5–10.0 |
| Ketone: $-C\underline{H}_2-\overset{O}{\underset{\parallel}{C}}-R$ | 1.5–2.5 |
| Aromatic: $Ar-\underline{H}$ | 6.5–8.0 |
| Carboxylic acid: $-\overset{O}{\underset{\parallel}{C}}-O\underline{H}$ | 11.0–12.0 |
| amine: $-N\underline{H}_2$ | 1.0–5.0 |

## SOLUTION

a.  Hertz units $= 2.1 \times 60 = 126$ Hz

b.  Hertz units $= 2.1 \times 100 = 210$ Hz

---

**EXERCISE 29.5**

An NMR spectrum of chloroform, operating at a radio frequency of 60 MHz, contained a peak at 436 Hz.

a.  Convert 436 Hz into ppm ($\delta$) units.

b.  Use your answer to part (a) to calculate the absorption of the peak in hertz when an instrument operating at a radio frequency of 100 MHz is employed.

---

Now let us consider the proton NMR spectrum of methyl alcohol (Fig. 29.11). Note that the peak at $\delta = 3.3$ can be assigned to the protons on a carbon, while the peak at $\delta = 4.6$ can be assigned to a proton of a hydroxyl group. The area under the peaks is generally proportional to the number of equivalent protons. A measurement of the peak areas shows a 3:1 ratio between the $\delta = 3.3$ and $\delta = 4.6$ peaks, suggesting that there are three protons on a carbon for every proton on a hydroxyl group. Since no other peaks appear, these are the only protons in the molecule. If we know from other evidence that we are dealing with an alcohol, we may conclude that possible formulas are $CH_3OH$ or some multiple of this formula containing the same ratio of methyl to hydroxyl protons. In general, then, we can state that the area under a peak is directly proportional to the number of hydrogens producing the peak—a powerful tool. Note that the very small peaks are spinning side bonds that can be ignored.

Now a reasonable question to ask is, Why do these two nonequivalent groups of hydrogens absorb at different frequencies and thus can be differentiated? The answer is

**FIGURE 29.11** Proton nuclear magnetic resonance spectrum of methyl alcohol (in deuterochloroform).

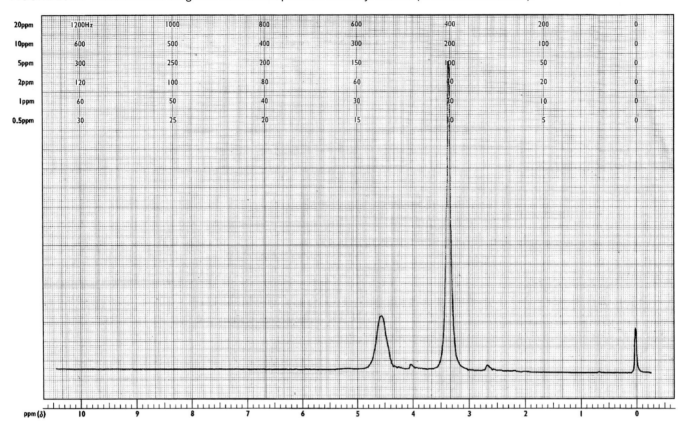

that the frequency of absorption is dependent on the electron environment of the hydrogen, i.e., sigma clouds surrounding each hydrogen. It works this way: A large external magnetic field, say 15,000 gauss, induces a small opposing magnetic field from each hydrogen. The size of this small opposing magnetic field, in turn, is dependent on the electron environment surrounding the hydrogen. Because electron environments vary somewhat, absorptions at varying magnetic field strengths are observed. The typical NMR instrument, however, uses a constant large external magnetic field and variable radio frequency.* Thus absorptions take place at slightly different radio frequencies.

We say the hydrogen nuclei are *shielded* from the large external magnetic field by the electron clouds surrounding each hydrogen. The greater the shielding, the further upfield the peak appears and the higher is the frequency of the peak. Conversely, the less the shielding (we say the hydrogen nuclei are *deshielded*), the more downfield the peak appears and the lower is the frequency of the peak (Fig. 29.12).

Deshielding of a hydrogen occurs when

1. The hydrogen is near an electronegative atom such as nitrogen, oxygen, or a halogen. This makes sense because the electronegative atom strips the electron cloud away from the hydrogen. Note that the hydroxyl hydrogen (or proton) in methyl alcohol is deshielded, absorbing at $\delta = 4.6$, while the absorption due to the methyl hydrogens appears upfield at $\delta = 3.3$ and represents greater shielding.

2. The hydrogen (or proton) is near one or more unsaturated bonds. Consider the NMR spectrum of toluene (Fig. 29.13). The five aromatic hydrogens are not equivalent chemically, but the NMR instrument in this case sees them as equivalent because of low resolution efficiency, and they are represented by the downfield peak at $\delta = 6.7$. This peak represents deshielding by the pi electron cloud of the benzene ring. The absorption of the three methyl hydrogens is upfield at $\delta = 2.1$.

*The radio frequency is held constant and the field strength is varied in some NMR instruments.

**FIGURE 29.12**  NMR terminology as applied to a spectrum.

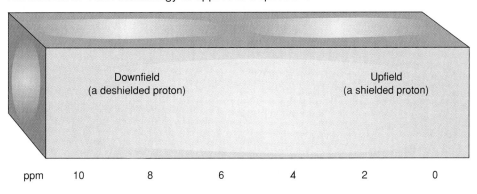

**FIGURE 29.13**  Proton nuclear magnetic resonance spectrum of toluene (in deuterochloroform).

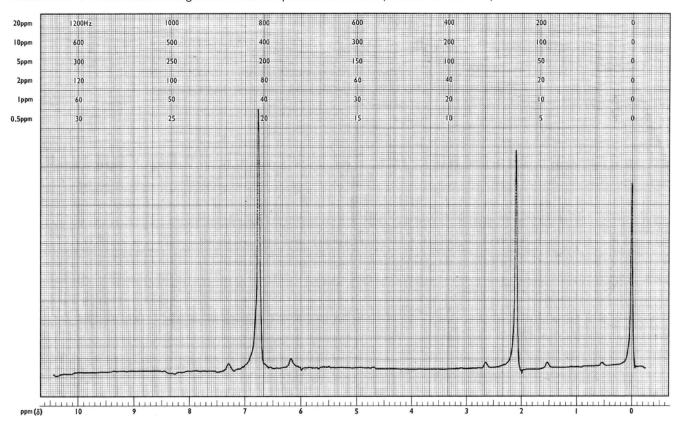

Identify the given compound from the following data:

**SOLVED PROBLEM 29.5**

Molecular formula:   $C_3H_6O_2$
NMR spectrum:      $\delta$ = 2.1 (3H)
                   $\delta$ = 3.6 (3H)

**SOLUTION**
Each of the two peaks contains three hydrogens. Thus we are dealing with two methyl groups comprising all the hydrogen in the compound. The downfield absorption implies deshielding by either an electronegative oxygen or a double bond. A carbon and two oxygens are unaccounted for and could represent an ester:

$$\overset{\displaystyle O}{\overset{\displaystyle \|}{-C-O-}}$$

Filling in the two methyl groups, we have

$$CH_3-\overset{\overset{\displaystyle O}{\|}}{C}-O-CH_3$$

δ = 3.6 downfield    δ = 2.1 upfield
due to deshielding
by the pi electrons
of the carbonyl bond

<table>
<tr><td>EXERCISE 29.6</td><td>Identify the given compound from the following data:</td></tr>
</table>

Molecular formula:  $C_2H_6O$
NMR spectrum:    δ = 3.5 (6H)

## 29.7
## SPIN-SPIN SPLITTING

Each of the spectra we have thus far examined contains one or two isolated peaks called *singlets*. However, when a compound has two (or more) neighboring carbons, all of which contain one or more hydrogens, a more complex peak pattern emerges due to *spin-spin splitting*. Consider the NMR spectrum of ethyl bromide (Fig. 29.14). What we see in the spectrum is a grouping of three related peaks at δ = 1.5 called a *triplet* and a grouping of four related peaks called a *quartet* at δ = 3.3. (A grouping of two related peaks is called a *doublet*).

We will not discuss the theory of *peak splitting*, but we will learn to predict a splitting pattern from the structure (and vice versa) using the *multiplicity rule*. The rule states that the splitting pattern of the hydrogens on a given carbon can be predicted by adding the

**FIGURE 29.14** Proton nuclear magnetic resonance spectrum of ethyl bromide (in deuterochloroform).

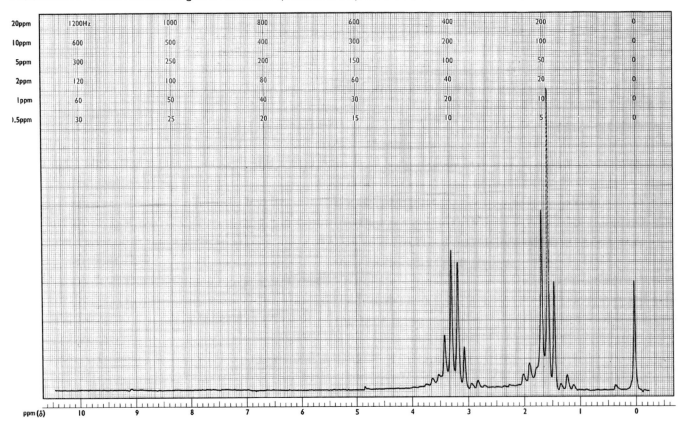

number of neighboring hydrogens (i.e., hydrogens on adjacent carbons) and then adding one. Let us consider the structure of ethyl bromide and use the multiplicity rule to predict the splitting patterns in the spectrum.

$$CH_3{-}CH_2{-}Br$$

The peak of the methyl hydrogens is split into a triplet $(2 + 1)$ by the two hydrogens on the methylene group, while the peak of the methylene hydrogens is split into a quartet $(3 + 1)$ by the three hydrogens of the methyl group.

$$CH_3{-}CH_2{-}Br$$

A 2 + 1A 3 + 1
triplet   quartet

The quartet should show up downfield because of the deshielding effect of the nearby bromine atom. Sure enough, when you check the spectrum of ethyl bromide (Fig. 29.14), this is exactly what you observe: an upfield shielded triplet at $\delta = 1.5$ representing the methyl group and a downfield deshielded quartet at $\delta = 3.3$ representing the methylene group.

SOLVED PROBLEM 29.6

Predict the NMR spectrum of $CH_3CHCl_2$. Indicate the splitting, if any, and the number of hydrogens involved, and give an approximate chemical shift for each peak or pattern of peaks. Use the chemical shifts in Table 29.3.

**SOLUTION**
Since one or more hydrogens are on adjacent carbons in 1,1-dichloroethane, we must use the multiplicity rule. The hydrogen in the $-CHCl_2$ group splits the peak of the hydrogens of the methyl group to a doublet, while the hydrogens of the methyl group split the peak of the hydrogen on $-CHCl_2$ to a quartet.

$$CH_3{-}{-}{-}{-}{-}{-}CHCl_2$$

1 + 1 = 2
A doublet:
The adjacent
carbon is bonded
to one hydrogen.

3 + 1 = 4
A quartet:
The adjacent
carbon is bonded
to three hydrogens.

Note that the quartet is downfield due to the presence of the two deshielding chlorines. Thus we have using the chemical shifts in Table 29.3:

NMR spectrum:   $\delta \approx 1.0$ (3H) doublet
$\delta \approx 6.0$ (1H) quartet
(Deshielding is increased by the presence of two chlorines rather than one.)

EXERCISE 29.7

Predict the proton NMR spectrum of each of the following compounds. Indicate the splitting, if any, and the number of hydrogens involved, and give an approximate chemical shift for each peak or pattern of peaks. Use the chemical shifts in Table 29.3 and employ Solved Problem 29.6 as a guide.

a.   $CH_3CH_2\overset{\displaystyle O}{\overset{\displaystyle \|}{C}}{-}H$

b.   $CH_3CH_2OH$

c.   $(CH_3)_3CNH_2$

## 29.8
## CARBON-13 NUCLEAR MAGNETIC RESONANCE SPECTROSCOPY

Proton NMR spectroscopy has been used as a tool by chemists for over a quarter of a century. However, carbon-13 NMR spectroscopy has only been used for the past decade or so. The reason is that the carbon-13 peak is considerably smaller than the comparable proton peak because carbon-13 is present as only 1.1% of the natural isotopic mix. This translates into the need for very sensitive equipment that can pick up the faint signal.

There are two reasons that carbon-13 NMR spectroscopy is useful. First, the chemical-shift range is broadened, from $\delta = 0.0$ to 200 ppm, as opposed to proton NMR spectroscopy with $\delta = 0.0$ to 12 ppm. This means that peak overlap is far less probable than in proton NMR spectroscopy with a range of less than 15 $\delta$ units. Second, $^{13}C$—$^{13}C$ spin-spin splitting is far less likely to occur in a carbon-13 spectrum due to the low abundance of the $^{13}C$ isotope. This means that it is not very likely that two carbon-13 nuclei are adjacent to each other. Thus it is not likely that $^{13}C$—$^{13}C$ spin-spin coupling can occur. A type of splitting that can take place is $^{13}C$—$^1H$ spin-spin splitting. When the instrument is operated in the normal mode, splitting does not take place; in the *off-resonance* mode, however, $^{13}C$—$^1H$ spin-spin splitting does occur, and doublets, triplets, and quartets are observed in the spectrum. To illustrate the importance of carbon-13 NMR spectroscopy, consider the proton nuclear magnetic resonance spectrum of *n*-butyl chloride (Fig. 29.15). Note that the triplet downfield at $\delta = 3.4$ represents the methylene hydrogens adjacent to the chlorine. However, the doublet at $\delta = 0.8$ should be a triplet representing the methyl hydrogens, and the peak at $\delta = 1.5$ representing the two internal methylene groups shows significant peak overlap and considerably lessens the value of the spectrum.

**FIGURE 29.15** Proton nuclear magnetic resonance spectrum of a *n*-butyl chloride (in deuterochloroform).

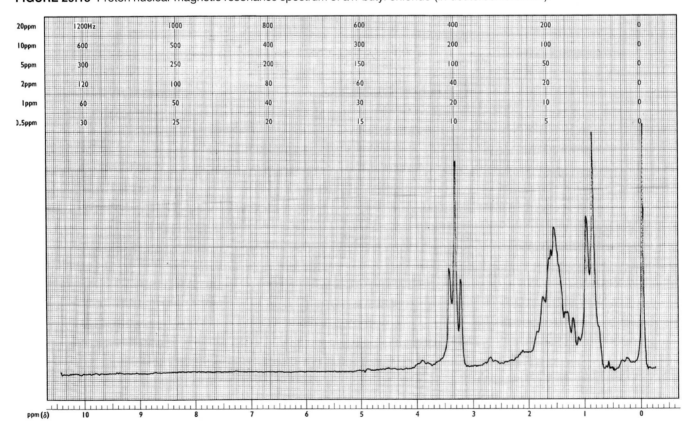

**FIGURE 29.16** (a) Ordinary decoupled spectrum. (b) Coupled spectrum.

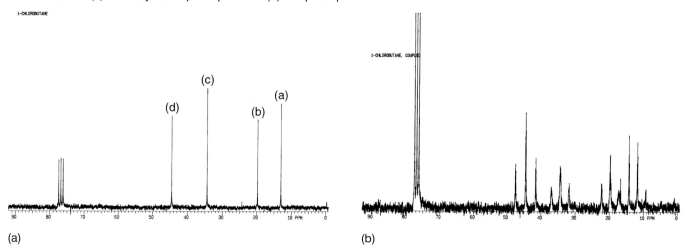

(a)                                             (b)

EXERCISE 29.8

Assuming that peak overlap does not occur, predict the proton NMR spectrum for *n*-butyl chloride. Indicate the splitting, if any, and the number of hydrogens involved, and give an approximate chemical shift for each peak or pattern of peaks. Use the chemical shifts in Fig. 29.16 and employ Solved Problem 29.6 as a guide.

Now, let us examine the spectra of *n*-butyl chloride using $^{13}$C NMR spectroscopy (Fig. 29.16). Let us first examine the decoupled $^{13}$C spectrum (Fig. 29.16a). Each peak clearly defines a particular nonequivalent carbon as follows:

$$\underset{\text{(a)}}{CH_3}-\underset{\text{(b)}}{CH_2}-\underset{\text{(c)}}{CH_2}-\underset{\text{(d)}}{CH_2}-Cl$$

As expected, the carbon of the methylene group nearest to the chlorine shows a peak that is furthest downfield. The triplet at $\delta = 78$ is due to deuterochloroform. In the coupled spectrum (Fig. 29.16b), the multiplicity rule is applied in a similar manner to proton nuclear magnetic resonance. The $^{13}$C—$^{1}$H spin-spin splitting in the methylene groups—(d), (c), and (b)—yields three triplets $(2 + 1)$ at $\delta = 45$, $\delta = 35$, and $\delta = 20$, respectively (see Table 29.4), while similar splitting in the methyl group—(a)—produces a quartet $(3 + 1)$ at $\delta = 12$. Note the overlap between the quartet at $\delta = 12$ and the triplet at $\delta = 20$.

| TABLE 29.4 | Approximate Carbon-13 Chemical Shifts |
| --- | --- |
| **Kind of Carbon** | **Chemical Shift ($\delta$), ppm** |
| Primary carbon: —$\underline{C}H_3$ | 0–35 |
| Secondary carbon: —$\underline{C}H_2$— | 15–45 |
| Tertiary carbon: —$\underset{\mid}{\overset{\mid}{\underline{C}}}$—H | 25–50 |
| Alkyl chloride: —$\underline{C}H_2Cl$ | 25–50 |
| Carbonyl: $\diagdown\underline{C}=O$ | 190–220 |

**SOLVED PROBLEM 29.7**

Predict the number of peaks in a $^{13}$C decoupled spectrum of acetone.

**SOLUTION**

Since the two methyl carbons are equivalent, we would predict only one peak ($\delta = 31$) upfield. A second downfield peak due to the carbonyl carbon ($\delta = 204$) also should be present.

**EXERCISE 29.9**

A compound, $C_3H_7Cl$, shows a $^{13}$C decoupled spectrum that contains only two peaks. Draw a structural formula for the compound, and explain.

---

## MEDICAL BOXED READING 29.1

The chest x-ray has been a staple in medical diagnosis for well over 70 years. The x-ray picture resembles a photographic negative, with opaque bone showing up clearly, while the gas-filled lungs appear transparent. Muscle, blood vessels, and various organs, which are translucent, can be observed with great difficulty.

A quantum leap in x-ray technology was accomplished with the development of *x-ray computed tomography*. In this technique, which yields data in the form of a *CT scan* or *CAT scan*, the resulting picture of a section (slice) of an organ is not complicated by the shadows of any structure in front of or in back of that particular section (slice). Thus the CT scan yields much clearer pictures of blood vessels, muscle, and organs than the standard x-ray. In addition, pictures of many sections (slices) of the organ can be taken to obtain a complete picture of the organ.

The introduction of *magnetic resonance imaging (MRI)*, during the early 1980s made the CT scan obsolete for two good reasons. First, the nonionizing radio frequency radiation used in MRI is less energetic than x-rays and thus is less likely to do damage to the patient. Second, the patient does not need to

drink an MRI preprocedure "cocktail." Such cocktails, necessary in x-ray computed tomography, are a disadvantage because the patient can become nauseous from the cocktail or suffer an allergic reaction to the cocktail. The clarity of various organs using the MRI technique is fine without the use of a cocktail to enhance the picture.

It should be emphasized that MRI is actually NMR as applied to medicine. The medical establishment eliminated the term *nuclear* because they wanted to distinguish MRI, which does not use radioisotopes, from nuclear medicine, which does employ radioisotopes.

How does MRI work? It measures protons (hydrogen atoms) in water and hydrogen-abundant lipids. In essence, there is a time difference between healthy and diseased tissue in which hydrogen, in the antiparallel alignment with the external magnetic field, relaxes to the more stable parallel alignment. This time difference is displayed as lighter and darker regions of the structure as observed in the image (Fig. 29.17). Both the external magnetic fields used and the radiofrequencies employed are comparable with those used in a laboratory instrument.

**FIGURE 29.17** A magnetic resonance image of a normal brain.

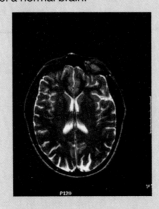

## ▶ CHAPTER ACCOMPLISHMENTS

### Introduction
☐ List two advantages of spectroscopy in structure identification as opposed to the standard chemical methods.

### 29.1 Electromagnetic Radiation
☐ Name the most energetic electromagnetic radiation.
☐ Calculate the frequency of radiation in hertz from a wavelength in centimeters.

### 29.2 Infrared Spectroscopy—Introduction
☐ Define the term *wave number*.
☐ Convert the wavelength 2.2 μm to the wave number in $cm^{-1}$ units.
☐ Locate the fingerprint region of an infrared spectrum.
☐ Explain how to determine the class of an amine using an infrared spectrum.

### 29.3 Interpretation of Infrared Spectra
☐ State the two functional-group absorptions you would expect to find in the infrared spectrum of a carboxylic acid.
☐ Give two key infrared absorptions that indicate that a compound is aromatic.

### 29.4 Nuclear Magnetic Resonance Spectroscopy—Introduction
☐ Supply an advantage in structure determination of nuclear magnetic resonance spectroscopy over infrared spectroscopy.

### 29.5 Nuclear Spin
☐ Explain why the $^{12}C$ isotope is of no use to the organic chemist in NMR spectroscopy, while the $^{13}C$ isotope is very useful.
☐ Explain the term *spin-flip*.

### 29.6 The Nuclear Magnetic Resonance Spectrum
☐ Explain the presence of a peak in a typical NMR spectrum at $\delta = 0.0$.
☐ Select the higher frequency: $\delta = 2.5$ or $\delta = 9.1$.
☐ List the functional groups that contain at least one hydrogen and then use Table 29.3 to predict the absorption characteristic of the functional group.
☐ Convert an absorption of 2.1 ppm, using an instrument with 60-MHz radio frequency, into hertz.
☐ Supply two structural conditions that lead to deshielding of a hydrogen in an organic molecule.

### 29.7 Spin-Spin Splitting
☐ State the multiplicity rule.
☐ Predict the peak splitting in the NMR spectrum of $CH_3CHCl_2$.

### 29.8 Carbon-13 Nuclear Magnetic Resonance Spectroscopy
☐ Explain why $^{13}C$—$^{13}C$ spin-spin splitting is less likely to occur in $^{13}C$ NMR spectroscopy than $^{1}H$—$^{1}H$ splitting in proton NMR spectroscopy.
☐ Interpret both the "ordinary" decoupled $^{13}C$ NMR spectrum of *n*-butyl chloride and the coupled spectrum.

## ▶ KEY TERMS

*infrared (IR) spectroscopy* (Introduction, 29.2)
*nuclear magnetic resonance (NMR) spectroscopy* (Introduction, 29.4)
*spectroscopy* (Introduction)
*electromagnetic radiation* (Introduction, 29.1)
*gamma-ray radiation* (29.1)
*x-ray radiation* (29.1)
*microwave radiation* (29.1)
*radiowave radiation* (29.1)
*visible light* (29.1)
*photon* (29.1)
*frequency* (29.1)
*wavelength* (29.1)
*spectrum* (29.1)

*spectrophotometer* (29.1)
*peak* (29.1)
*infrared (IR) spectrum* (29.2)
*energy of vibration* (29.2)
*wave number* (29.2)
*fingerprint region of the infrared* (29.2)
*nuclear spin* (29.4, 29.5)
*parallel orientation* (29.5)
*antiparallel orientation* (29.5)
*spin-flip* (29.5)
*signal* (29.5)
*upfield* (29.6)
*downfield* (29.6)
*chemical shift* (29.6)
*shielded* (29.6)
*deshielded* (29.6)

*singlet* (29.7)
*spin-spin splitting* (29.7)
*triplet* (29.7)
*quartet* (29.7)
*doublet* (29.7)
*peak splitting* (29.7)
*multiplicity rule* (29.7)
*off-resonance mode of $^{13}C$ NMR spectroscopy* (29.8)
*x-ray computed tomography* (Medical Boxed Reading 29.1)
*CT scan or CAT scan* (Medical Boxed Reading 29.1)
*magnetic resonance imaging (MRI)* (Medical Boxed Reading 29.1)

## ▶ PROBLEMS

1. Which of the following represents the more energetic radiation? Use Table 29.1 to catagorize each type of radiation?
   a. radiation with a wavelength of $5 \times 10^{-12}$ cm or radiation with a frequency of $5 \times 10^9$ Hz.
   b. radiation with a frequency of $6.5 \times 10^{12}$ Hz or radiation with a wavelength of $9.1 \times 10^{-6}$ cm.

2. Calculate the wavelength of ultraviolet radiation with a frequency of $1.5 \times 10^{17}$ Hz.

3. Calculate the wave number $(cm^{-1})$ of each of the following absorptions.
   a. 5.85 μm (a carboxylic acid)
   b. 2.80 μm (an alcohol)

4. Calculate the wavelength in microns (μm) of each of the following absorptions.
   a. 1080 $cm^{-1}$ (an ether)
   b. 1730 $cm^{-1}$ (a carboxylic acid)

5. Suggest a structure or structures for each of the following compounds.
   a. $C_2H_3N$, a neutral compound that absorbs in the infrared at 2245 $cm^{-1}$
   b. $C_7H_6O_2$, an acidic compound that absorbs in the infrared at 3000 $cm^{-1}$ (broad peak), 1680 $cm^{-1}$ (due to double-bond conjugation; without the conjugation, we would expect an absorption at 1700 $cm^{-1}$), 1460 $cm^{-1}$, and 1600 $cm^{-1}$
   c. $C_3H_6O$, a neutral compound that absorbs in the infrared at 1720 $cm^{-1}$
   d. $C_3H_8O$, a neutral compound that absorbs in the infrared at 3300 $cm^{-1}$ (This compound can be oxidized.)

6. Convert each of the following proton NMR absorptions to hertz (assume a 60-MHz instrument is used).
   a. $\delta = 3.6$
   b. $\delta = 8.2$

7. Which of the following compounds show only one peak in a proton NMR spectrum?

   a. $CH_3-CH_2-\overset{\displaystyle O}{\overset{\displaystyle \|}{C}}-H$

   b. $H-\overset{\displaystyle O}{\overset{\displaystyle \|}{C}}-O-CH_2CH_3$

   c. $CH_3CH_2CH_2NH_2$

   d. $CH_3-\overset{\displaystyle O}{\overset{\displaystyle \|}{C}}-Cl$

   e. $Cl-CH_2-\overset{\displaystyle O}{\overset{\displaystyle \|}{C}}-CH_3$

   f. $CH_3CH_2I$

   g. $(CH_3)_4C$

8. Which of the compounds in Problem 7 show more than one singlet in a proton NMR spectrum? Indicate how many singlets for each compound.

9. Which of the compounds in Problem 7 show spin-spin splitting in a proton NMR spectrum?

10. Predict the proton NMR spectrum of each of the compounds listed in Problem 7. Indicate the splitting, if any, and the number of hydrogens involved. Employ Solved Problem 29.6 as a guide. You do not need to supply approximate chemical shifts.

11. Define or explain each of the following.
    a. upfield
    b. shielded proton
    c. the multiplicity rule
    d. frequency
    e. wavelength

12. Draw a structural formula for each of the following compounds.
    a. Compound A

       Molecular formula:   $C_2H_4Cl_2$
       Proton NMR spectrum:   $\delta = 2.3$ (3H) doublet
       $\delta = 6.1$ (1H) quartet

    b. Compound B

       Molecular formula:   $C_3H_7Br$
       Proton NMR spectrum:   $\delta = 1.8$ (6H) doublet
       $\delta = 4.3$ (1H) septet

    c. Compound C

       Molecular formula:   $C_8H_{10}$
       Proton NMR spectrum:   $\delta = 1.3$ (3H) triplet
       $\delta = 2.7$ (2H) quartet
       $\delta = 7.0$ (5H) singlet

    d. Compound D

       Molecular formula:   $C_5H_{12}O$
       Proton NMR spectrum:   $\delta = 1.1$ (9H) singlet
       $\delta = 3.4$ (3H) singlet

    Compound D does not react with sodium.

13. A sample of tert-butyl alcohol, when gently heated with $Na_2Cr_2O_2$ and $H_2SO_4$, gave a positive test; i.e., the orange solution turned blue. Careful infrared analysis of the original tert-butyl alcohol sample revealed a peak at 1710 $cm^{-1}$. Explain.

14. A compound yielded the following data:
    Proton NMR spectrum:   $\delta = 1.0$ (3H) triplet
    $\delta = 1.7$ (2H) sextet
    $\delta = 2.8$ (2H) triplet
    $\delta = 7.4$ (5H) doublet (Sometimes, the vinyl hydrogens on the phenyl group do show splitting.)
    Is this compound propiophenone or n-butyrophenone?

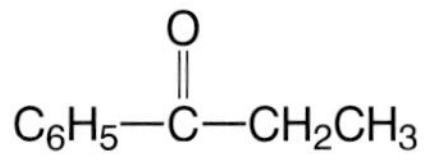

$$C_6H_5-\overset{\overset{\displaystyle O}{\|}}{C}-CH_2CH_3 \qquad C_6H_5-\overset{\overset{\displaystyle O}{\|}}{C}-CH_2CH_2CH_3$$

Propiophenone          *n*-Butyrophenone

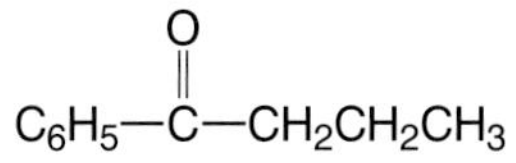

15. How many peaks (include splitting) would you expect to observe in the proton NMR spectrum of each of the following compounds? Assume that no peak overlap occurs.

a. 1,1,1,-tribromoethane

b. 3-pentanone

c. isobutyl chloride

16. A compound, $C_2H_6O$, gave the following decoupled $^{13}C$ NMR spectrum:

$$\delta = 18 \ (3H)$$
$$\delta = 57 \ (2H)$$

a. Draw a structural formula for the compound.

b. Predict the splitting in the coupled $^{13}C$ NMR spectrum of this compound.

17. Which alcohol, with the formula $C_4H_9OH$, shows only two peaks in the decoupled $^{13}C$ NMR spectrum? Explain.

Note: Additional problems dealing with spectroscopy and structure appear at the end of most problem sets beginning with Chapter 13.

# Appendixes

*APPENDIX* 1    **General Chemistry-Related Problems**

If you need help with the following problems, you should consult your general chemistry textbook. Answers to some of these problems are in the Student Study Guide.

1. Define or explain the following.
   a. empirical formula
   b. mole

2. Write a chemical formula for each of the following inorganic compounds used in organic chemistry.
   a. potassium permanganate
   b. sodium hydroxide
   c. hydrogen peroxide
   d. mercuric sulfate
   e. zinc chloride
   f. potassium dichromate
   g. ozone
   h. sodium bisulfite
   i. potassium cyanide
   j. aluminum chloride
   k. sodium bicarbonate
   l. copper(I) iodide

3. Calculate the percentage composition of
   a. morphine, $C_{17}H_{19}NO_3$, a narcotic.
   b. niacin, $C_6H_5NO_2$, a vitamin.
   c. ergotamine, $C_{33}H_{35}N_5O_5$, a powerful vasoconstrictor.
   d. sulfanilamide, $C_6H_8N_2O_2S$, an antibiotic.
      Atomic weights: C = 12.0, H = 1.01, N = 14.0, O = 16.0, S = 32.0

4. Analysis of a colorless crystalline solid isolated from horse urine gave C = 60.4%, H = 5.0%, N = 7.8%, and O = 26.8%. Calculate the empirical formula of this compound. Atomic weights: C = 12.0, H = 1.0, N = 14.0, O = 16.0.

5. A student weighed 0.025 g of an organic unknown and dissolved it in 0.500 g of camphor (freezing point = 178°C; $K_f$ = 40°C · kg/mol). He found the freezing point of the solution to be 170°C. Calculate the molecular weight of the organic solute.

6. A sample of 0.45 g of ether was vaporized and, at STP, found to have a volume of 156 ml. Calculate the molecular weight of ether.

7. An analysis of an 11.00-mg sample of an organic liquid by the Liebig method gave 37.16 mg $CO_2$ and 7.67 mg $H_2O$. Calculate (a) the percentage composition, (b) the empirical formula, and knowing that the molecular weight of the compound is 78.12, calculate (c) the molecular formula of the compound. Atomic weights: C = 12.01, H = 1.01, O = 16.00.

*APPENDIX* 2     **The Common System of Nomenclature**

This system of nomenclature is useful for low-molecular-weight alkanes but leads to problems when employed with more complex compounds. Thus its use is limited. Two simple rules can be followed:

1. The prefix *n-* (for normal) is used in naming straight-chain alkanes. This prefix is not used with methane, ethane, and propane because they have no isolable isomers.

$CH_3CH_2CH_2CH_3$     $CH_3(CH_2)_6CH_3$

*n*-Butane         *n*-Octane

2. The prefix *iso-* shows a methyl group bonded to a carbon on the chain next to the end carbon.

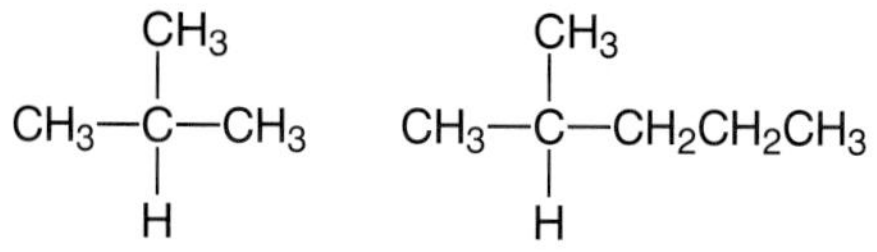

$$CH_3-\underset{\underset{H}{|}}{\overset{\overset{CH_3}{|}}{C}}-CH_3 \qquad CH_3-\underset{\underset{H}{|}}{\overset{\overset{CH_3}{|}}{C}}-CH_2CH_2CH_3$$

Isobutane         Isohexane

A common name can be assigned to a given compound for a wide variety of reasons. For example, the name *neopentane* was assigned to the most recently discovered isomer of $C_5H_{12}$.

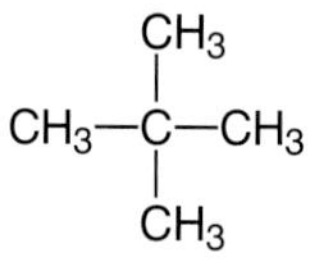

$$CH_3-\underset{\underset{CH_3}{|}}{\overset{\overset{CH_3}{|}}{C}}-CH_3$$

Neopentane

The prefix *neo-* means "new." Because there is often no systematic procedure in assigning a common name to a particular compound, these types of names are also designated as trivial names.

One limitation of common or trivial nomenclature can be seen with a compound with a different or more complex carbon branching pattern like the example below. The common system of nomenclature simply cannot work with this pattern of carbon branching.

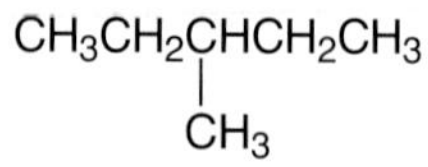

$$CH_3CH_2\underset{\underset{CH_3}{|}}{C}HCH_2CH_3$$

Another limitation of common or trivial nomenclature is that since many of these trivial names cannot be deduced from the structure of the molecules, each name derived from whatever source must be memorized. For example, $CH_3(CH_2)_4COOH$, is assigned the name *caproic acid* because this particular acid is a component of the odor of goats and the Latin name for goat is *caper*. Obviously, a systematic method of nomenclature was needed and was provided by the International Union of Pure and Applied Chemistry (IUPAC) system of nomenclature.

APPENDIX 3    # A Comparison of Common and IUPAC Nomenclature

Note in Table A3.1 that common and IUPAC names for the first three homologues of the alkane series are the same. Differences in nomenclature surface with the C-4 homologue because it is the first alkane of the series to show isomerism. Thus the compound $CH_3CH_2CH_2CH_3$ is named, with the common system, $n$-butane because it is an alkane with no branches and $(CH_3)_3CH$ is named isobutane, with the common system, due to a methyl branch bonded to a carbon next to the end carbon. On the other hand, $CH_3CH_2CH_2CH_3$ is named butane with the IUPAC system. Since $(CH_3)_3CH$ has a longest continuous chain of three carbons, using the IUPAC system of nomenclature, the parent compound is propane and the name of the compound is *2-methylpropane*.

| TABLE A3.1 | A Comparison of Common and IUPAC Nomenclature for Alkanes | |
|---|---|---|
| **Structural Formula** | **Common Name** | **IUPAC Name** |
| $CH_4$ | Methane | Methane |
| $CH_3CH_3$ | Ethane | Ethane |
| $CH_3CH_2CH_3$ | Propane | Propane |
| $CH_3CH_2CH_2CH_3$ | $n$-Butane | Butane |
| $CH_3-\overset{\overset{\displaystyle CH_3}{\mid}}{\underset{\underset{\displaystyle H}{\mid}}{C}}-CH_3$ | Isobutane | 2-Methylpropane |

*APPENDIX* 4     **Common Nomenclature of Alkenes**

Today, the common system of nomenclature is very seldom used for alkenes of five carbons or more. The IUPAC system is much preferred. However, the common system is particularly used for the two simplest alkenes. Thus let us study a few rules appropriate to this less often used system of nomenclature:

1.  The common name of an alkene is formulated by taking the corresponding alkane stem, dropping the *-ane*, and adding *-ylene*.

2.  The double bond is located with a Greek letter, and this is the letter closest to alpha of the two representing the double bond.

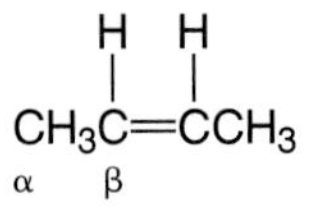

$$CH_3C\!\!=\!\!CCH_3 \qquad CH_2\!\!=\!\!CCH_2CH_2CH_3$$
α    β          α    β γ   δ   ε

β-Butylene 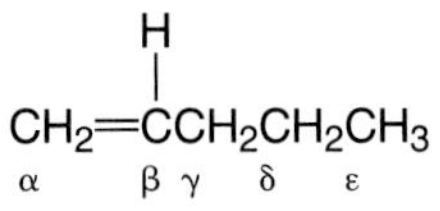          α-Pentylene
                    (α-amylene)

3.  The prefix *iso-* is used in the same way as with alkanes. It represents a methyl group bonded to a carbon next to the end carbon.

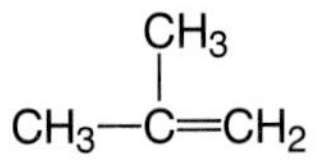

$$CH_3\!\!-\!\!C\!\!=\!\!CH_2$$

Isobutylene

Just as in alkane nomenclature, common nomenclature of alkenes fails with alkenes of higher molecular weight that have more complex branching patterns.

APPENDIX 5    # A Comparison of Common and IUPAC Nomenclature for Alkenes and Certain Alkenyl Derivatives

Table A5.1 compares common and IUPAC nomenclature for five simple alkenes. Obvious points of comparison are the *-ylene* suffix (common) versus the *-ene* suffix (IUPAC), locating the carbon-carbon double bond with a Greek letter (common) versus a number (IUPAC), and the use of all carbons (including branched carbons) as the parent compound (common) versus the longest continuous chain of carbons (IUPAC).

## ALKENES

| TABLE A5.1 | A Comparison of Common and IUPAC Nomenclature for Alkenes | |
|---|---|---|
| **Compound** | **Common** | **IUPAC** |
| $CH_2{=}CH_2$ | Ethylene | Ethene |
| $CH_3\overset{\overset{H}{\mid}}{C}{=}CH_2$ | Propylene | Propene |
| $CH_2{=}CHCH_2CH_3$ | α-Butylene | 1-Butene |
| $CH_3CH{=}CHCH_3$ | β-Butylene | 2-Butene |
| $(CH_3)_2C{=}CH_2$ | Isobutylene | 2-Methylpropene |

Two important alkenyl groups often used in nomenclature are derived from ethylene or propylene. To name a compound using an alkenyl group, name the group and then add the word *chloride* (halide) or *alcohol* (Table A5.2).

## ALKENYL GROUPS

| TABLE A5.2 | Selected Alkenes and Alkenyl Groups and Examples of Common Nomenclature Using Three Alkenyl Groups | |
|---|---|---|
| **Alkene** | **Alkenyl Group** | **Example (Common Name)** |
| $CH_2{=}CH_2$ <br> Ethylene | $CH_2{=}\overset{\overset{H}{\mid}}{C}{-}$ <br> Vinyl | $CH_2{=}\overset{\overset{H}{\mid}}{C}{-}Cl$ <br> Vinyl chloride |
| $CH_2{=}\overset{\overset{H}{\mid}}{C}{-}CH_3$ <br> Propylene | $CH_2{=}\overset{\overset{H}{\mid}}{C}{-}CH_2{-}$ <br> Allyl | $CH_2{=}\overset{\overset{H}{\mid}}{C}{-}CH_2OH$ <br> Allyl alcohol |

*Note:* Only the removal of one of the methyl hydrogens of propylene gives the allyl group.

# Glossary

**α-glucoside (27.8)** A molecule of α-glucose in which the anomeric hydroxyl group is replaced by an O—T group, where T can be an alkyl group or a far more complex group.

**absolute configuration (11.7)** The actual arrangement of groups in space around a chiral center of a molecule.

**absolute ethyl alcohol (14.3)** Ethyl alcohol that is water-free.

**acetal (18.6, 19.2)** A stable *gem*-diether derived from an aldehyde or ketone.

$$R''-\underset{\underset{H(R'')}{|}}{\overset{\overset{OR'}{|}}{C}}-OR''$$

**acetone (18.2)** The trivial name of dimethyl ketone, an important ketone.

**acetyl coenzyme A (7.9)** A thioester that plays a vital role in carbohydrate metabolism.

$$CH_3-\underset{\underset{O}{\|}}{C}-S-CoA$$

Acetyl coenzyme A

**achiral (11.2)** A molecule is achiral when it is identical to its mirror image (or superimposable on its mirror image).

**acid amino acid (28.1)** An amino acid that contains one amino group and two carboxyl groups.

**acid residue (26.2)** An acyl group incorporated in a triglyceride.

**activating group (10.2)** A group bonded to a benzene ring that increases the rate of reaction as compared with benzene.

**acyclic hydrocarbon (2.1, 3.2)** A hydrocarbon in which the carbons are arranged in a straight chain or a branched chain.

**acyl chloride (22.1)** An acid chloride.

**acyl group (18.2)** An imaginary group derived from the corresponding carboxylic acid by the loss of an —OH group. In general, we have

$$\underset{\text{Carboxylic acid}}{R-\overset{\overset{O}{\|}}{C}-OH} \qquad \underset{\text{Acyl group}}{R-\overset{\overset{O}{\|}}{C}-}$$

**acyl nucleophilic substitution (23.2)** A reaction in which an acid derivative combines with a nucleophile, with the following as an example:

$$R-\overset{\overset{O}{\|}}{C}-Cl + {}^-NH_2 \longrightarrow R-\overset{\overset{O}{\|}}{C}-NH_2 + Cl^-$$

**addition reaction (5.6, 6.2)** A reaction, shown by alkenes, alkynes, and carbonyl compounds, in which a reagent adds across a double or triple bond.

**1,2-addition (7.4)** Synonymous with an *addition reaction* (see Sec. 6.2).

**1,4-addition (7.4)** An addition reaction in which a reactant (A—B) adds across a conjugated diene in the 1 and 4 positions as follows:

$$\underset{1\ \ 2\ \ 3\ \ 4}{\overset{}{C=C-C=C}} + A-B \longrightarrow \underset{A \qquad B}{-C-C=C-C-}$$

**adduct (7.7)** The addition of any two compounds to form a new compound. An example is the product obtained in a Diels-Alder reaction.

**aldaric acid (27.11)** A dicarboxylic acid produced when an aldose is oxidized with nitric acid.

**alditol (27.11)** A polyol obtained when an aldose is reduced with sodium borohydride.

**aldol (19.3)** A trivial name for 3-hydroxybutanal, the product of the aldol condensation of 2 mol of acetaldehyde.

**aldol condensation (19.3)** A general reaction of 2 mol of an α-hydrogen–containing aldehyde or ketone in the presence of a 10% solution of sodium hydroxide that results in the formation of a β-hydroxyaldehyde or ketone with double the carbon content of the aldehyde or ketone reagent.

**aldonic acid (27.10)** A monocarboxylic acid produced when an aldose is oxidized with bromine water.

**aldose (27.1)** A polyhydroxyaldehyde, a monosaccharide.

**aldotetrose (27.1)** A monosaccharide that contains an aldehyde (formyl) group ($-\overset{\overset{H}{|}}{C}=O$) and a total of four carbons (—OH groups).

**aliphatic compound (3.6)** Any alkane that is not cyclic. Isobutane is an aliphatic compound. Nonaromatic cyclic compounds such as cyclohexane, cyclopentane, and cyclohexene are aliphatic compounds.

**alkene (5.1)** An unsaturated hydrocarbon that contains one or more carbon-carbon double bonds.

**alkoxy group (18.6)** Substituted —OH group (—OR) where R is an alkyl group. Like the alkyl group, the alkoxy group does not exist but is useful in nomenclature. Examples are —$OCH_3$, methoxy, —$OCH(CH_3)_2$, isopropoxy.

**alkyl group (3.2)** An alkyl group is formed by the loss of one hydrogen from the corresponding alkane. For example, methane (the hydrocarbon) = $CH_4$, methyl (the alkyl group) = —$CH_3$.

**allylic carbocation (7.4)** A carbocation with the positive charge located on a carbon adjacent to a carbon with a double bond.

**allylic hydrogen (6.3)** A hydrogen bonded to a carbon adjacent to a carbon with a double bond.

*alpha chain (28.7)* A protein chain, sometimes projecting from a prosthetic group, that helps to define the quaternary structure of the protein.

*amino acid (Introduction, Chap. 28)* A species that contains both the amino and carboxyl groups. If the amino group is located on the alpha carbon, and the amino acid is called an α-amino acid.

*amino acid residue (28.4)* The structural contribution of an amino acid to a peptide.

*amphoteric (28.2)* A chemical species that reacts with both acids and bases.

*amylopectin (27.9)* The water-soluble polymeric component of starch and a nonreducing carbohydrate.

*amylose (27.9)* The water-insoluble polymeric component of starch and a nonreducing carbohydrate.

*analgesic (Historical Boxed Reading 20.2)* A drug that provides relief from aches and pain.

*analyzer (11.3)* A component of a polarimeter, composed of Polaroid, that is rotated to obtain the observed optical rotation produced by the optically active sample.

*anion (1.4)* A negatively charged ion.

*anionic detergent (26.4)* A synthetic detergent in which the organic component bears a negative charge.

*anomer (27.4)* One of a pair of diastereomers that differ in configuration only at the hemiacetalic (or acetalic) chiral carbon of the cyclic form of a carbohydrate.

*anorexia nervosa (Historical Boxed Reading 26.1)* A disease in which young women in particular have a terrible fear of gaining weight and as a consequence often starve themselves.

*anosmia (Biochemical Boxed Reading 18.1)* An impaired sense of smell.

*Antabuse (Biochemical Boxed Reading 15.1)* A drug used as a treatment for alcoholics to ensure a state of sobriety.

*antioxidant (26.4)* A compound that inhibits oxidation in a product by itself undergoing oxidation.

*antiparallel orientation (29.5)* A condition in which the small magnetic field of the nucleus is oriented in the opposite direction from that of the external magnetic field.

*antiperiplanar geometry (12.4)* A condition in which a molecule shows periplanar geometry and the hydrogen and leaving group (usually halogen) leave from opposite sides of the molecule.

*antipyretic (Historical Boxed Reading 20.2)* A drug that reduces fever.

*antiseptic (16.3)* A substance that prevents the growth of microorganisms when applied to the skin.

*arene (9.2)* An aromatic hydrocarbon that contains an alkyl group.

*arenium ion (10.1)* A carbocation that is produced as an intermediate when an electrophile attacks an aromatic compound in an electrophilic aromatic substitution.

*aromatic (Introduction, Chap. 9)* A cyclic conjugated species that is stabilized by pi electron delocalization.

*aromatic character or aromaticity (9.7)* The chemical behavior displayed by aromatic compounds, i.e., a tendency to undergo substitution reactions rather than addition reactions and the presence of high resonance energies that show great stability.

*aromatic hydrocarbon (2.1, 9.1)* In a narrow sense, a compound that contains one or more benzene rings (six-membered rings that contain alternating single and double carbon-carbon bonds).

*Aromatic Orientation Rule 1 (10.3)* Every atom (or group) with one or more unshared electron pairs (lone pairs) on an atom bonded to a benzene or aromatic ring carbon is an ortho-para director.

*Aromatic Orientation Rule 2 (10.3)* Every atom (or group) with no unshared electron pair (lone pair) on an atom bonded to a benzene or aromatic ring carbon is a meta director.

*arteriosclerosis (26.6)* A disease characterized by significant deposition of cholesterol on the inner wall of arteries resulting in a decrease in the diameter in the major arteries of the body. This state of affairs can lead to a heart attack or stroke.

*aryl group (9.2)* An imaginary group created when a hydrogen is lost from an aromatic compound (like benzene) or an arene (like toluene).

*aspartame (Industrial Boxed Reading 28.2)* A dipeptide artificial sweetener, *N*-aspartylphenylalanine methyl ester.

*aspirin (20.4, Historical Boxed Reading 20.2)* A drug, acetylsalicylic acid, used to ease pain and control fever.

*atomic number (1.3)* The number of protons in the nucleus of an atom. The atomic number of the carbon atom is 6.

*axial bond (3.10)* A bond to the chair conformation of cyclohexane that is perpendicular to the plane of the ring.

*azeotrope or constant-boiling mixture (14.3)* A mixture of two or more compounds that boils at a fixed temperature. A constant-boiling mixture.

*azo dye formation (25.4)* The process of forming a deeply colored compound containing the azo group (—N=N—) by means of a coupling reaction.

*ball-and-stick model (2.2, 2.3)* A method of representing organic compounds in three dimensions where each carbon is a large sphere, each hydrogen a small sphere, and single bonds are represented by sticks.

*basic amino acid (28.1)* An amino acid that contains two basic groups and one carboxyl group $\left(-N\big\langle\right)$.

*β-glucoside (27.8)* A molecule of β-glucose in which the anomeric hydroxyl group is replaced by an O—R group, where R can be an alkyl group or a far more complex group.

*beta chain (29.7)* A protein chain, sometimes projecting from a prosthetic group, that helps to define the quaternary structure of the protein.

*bilayer (26.5)* The arrangement of phospholipids within a cell membrane with the nonpolar tails facing each other within the bilayer. This puts the polar ionic heads facing into the cell on the one hand and out of the cell on the other hand.

*biodegradable (26.4)* A substance which can be broken down into harmless products by microorganisms.

*blood sugar (27.7)* Another name for D-glucose.

*boat conformation (3.10)* The less stable conformation of cyclohexane, crudely shaped like a boat.

*bridged bicyclic ring (7.7)* A molecule, produced in a Diels-Alder reaction, that contains three carbon-to-carbon bridges.

**bromonium ion (6.2)** An unstable ion in which bromine is bonded to two carbons and bears a positive charge.

**Bronsted concept of acidity (1.14)** An acid donates one or more protons; a base accepts one or more protons from an acid.

**cadaverine (24.3)** 1,5-Pentanediamine or pentamethyene diamine.

**cantharidin (Medical Boxed Reading 17.2)** A crystalline substance that causes blistering when applied to the skin.

**carbanion (4.3)** A highly reactive, negatively charged ion that contains one trivalent carbon surrounded by eight electrons. For example, $^-:CH_3$ is a carbanion.

**carbinol system of alcohol nomenclature (14.1)** A common system of nomenclature in which a compound is named as a derivative of carbinol which is methyl alcohol. For example:

$$CH_3CH_2CH_2OH$$
Ethylcarbinol

**carbocation (4.3, 6.2)** A highly reactive, positively charged ion that contains one trivalent carbon surrounded by six electrons. For example, $^+CH_3$ is a carbocation.

**carbohydrate (Introduction, Chap. 27)** A compound is considered to be a carbohydrate if it is a sugar, or starch, glycogen, and their derivatives, or cellulose and its derivatives.

**carboxyl group (Introduction, Chap. 20)** A functional group that is a combination of the *carb*onyl and hyd*roxyl* groups.

$$
\begin{matrix}
& O \\
& \| \\
-&C-OH
\end{matrix}
$$

**carboxylic acid derivative (Introduction, Chap. 22)** One of four families of organic compounds closely related to the carboxylic acid as follows:

$$
\begin{matrix}
& O & & & O & & O \\
& \| & & & \| & & \| \\
R-&C-Cl & & R-&C-O-&C-R
\end{matrix}
$$
Acid chloride              Acid anhydride

$$
\begin{matrix}
& O & & & O \\
& \| & & & \| \\
R-&C-OR' & & R-&C-NH_2
\end{matrix}
$$
Ester                      Acid amide

**carboxypeptidase (28.5)** An enzyme that selectively attacks a peptide at the peptide bond of the C-terminus. Thus the enzyme is used to identify the C-terminus of a peptide.

**carnauba wax (26.1)** A wax isolated from the leaves of a number of different species of palm trees.

**cation (1.4)** A positively charged ion.

**cationic detergent (26.4)** A synthetic detergent in which the organic component bears a positive charge.

**cellulose (Introduction, Chap. 27, 27.9)** A β-glucosidic polymer.

**chain reaction (4.3)** A reaction that, once initiation takes place, maintains itself through a huge number of propagation steps. The halogenation of an alkane is a chain reaction.

**chair conformation (3.10)** The more stable conformation of cyclohexane, crudely shaped like a chair.

**chemical shift (29.6)** A site of absorption in an NMR spectrum relative to a standard (like tetramethylsilane) that absorbs higher upfield than the vast majority of protons. Each kind of proton shows a different characteristic chemical shift.

**chiral (11.2)** A molecule is chiral when it is not identical to its mirror image (or not superimposable on its mirror image).

**chiral center (11.2)** Usually a carbon atom (although it can be a nitrogen atom) that is bonded to four different groups in a molecule.

**cholesterol (26.6)** A steroidal alkenol responsible for arteriosclerosis.

**Clemmensen reduction (10.1)** A method by which a carbonyl compound is reduced to the corresponding hydrocarbon by the use of a mixture of a zinc amalgam, Zn(Hg), and concentrated hydrochloric acid.

**cocaine hydrochloride (Biochemical Boxed Reading 24.1)** A water-soluble salt of the alkaloid, cocaine.

**coenzyme (Biochemical Boxed Reading 15.1)** An organic molecule that can exist on its own or as a part of an enzyme.

**cologne (Industrial Boxed Reading 14.2)** A perfume-like mixture that only contains about 4% essential oils.

**compound lipid (Introduction, Chap. 26)** Tetraesters that contain two ester groups derived from phosphoric acid and two alcohols in addition to two carboxylic acid esters.

**concerted reaction mechanism (12.1)** A mechanism in which one or more bonds are breaking and one or more bonds are forming at the same time.

**condensed formula (1.8)** An abbreviated structural formula in which all the bonds are not shown.

**conformational isomer** or **conformation (2.3, 3.1)** An arrangement in space of a molecule that is due to free rotation about a carbon-carbon single bond.

**coniine (24.6)** A toxic alkaloid that is found in the fruit of the hemlock plant.

**conjugated diene (5.1)** A diene that contains alternating carbon-carbon double bonds and carbon-carbon single bonds. For example:

$$CH_2=CH-CH=CH-CH=CH_2$$
A conjugated diene

**conjugated protein (28.6)** A protein that yields on hydrolysis at least one substance called a *prosthetic group* along with the amino acids.

**conjugated system of bonds (2.6)** A series of alternating single and double (or triple) bonds.

**contributing structure** or **resonance structure (1.11, 7.2)** A structural formula obtained by delocalizing pi and unshared electrons.

**coordinate covalent bond** or **dative bond (1.5)** A covalent bond in which both electrons of the bond are donated by one atom.

**Corey-House synthesis (4.1)** A synthetic method used to prepare an alkane containing a greater number of carbons than either reactant. A methyl or primary organic halide is reacted with a lithium dialkyl cuprate ($R_2CuLi + R'X \longrightarrow R-R' + LiX + RCu$).

**cortisone (26.6)** A superb steroidal anti-inflammatory drug along with hydrocortisone. These drugs are often used to control itching and rashes due to various allergies.

**coupling (25.4)** A substitution reaction in which a diazonium salt reacts with an arylamine or a phenol to form an azo compound ($-N=N-$), often a dye.

*covalent bond (1.4)*   A chemical bond formed by the sharing of one or more electron pairs between two atoms.

*"crack" cocaine (Biochemical Boxed Reading 24.1)*   An alkaloid, so widely abused today, that can be snorted.

*crown ether (16.10)*   A cyclic polyether of the type —$CH_2CH_2O$— that can form stable complexes with a variety of metal ions.

*C-terminus (28.4)*   The amino acid residue of a peptide or protein with the free carboxyl group. By convention, the C-terminus is written or drawn on the extreme right of the molecule.

*CT scan or CAT scan (Medical Boxed Reading 29.1)*   The data (picture) obtained from an x-ray computed tomographic scan.

*cumulated diene (5.1)*   A diene (also called an allene) that contains two double bonds bonded to the same carbon. For example:

$$CH_3—CH=C=CH—CH_3$$

A cumulated diene

*curved arrow symbol (4.3)*   A symbol used to show the movement of one or more electrons.

*cyanohydrin (19.2)*   A *gem*-cyanoalcohol that is prepared by reacting a carbonyl compound with hydrogen cyanide in the presence of a basic catalyst.

A carbonyl compound   + HCN   $\xrightarrow{^-OH}$   A cyanohydrin

*cyclic hydrocarbon (2.1, 3.7)*   A hydrocarbon in which the skeletal carbon atoms constitute a ring.

*cytochrome P-450 (Biochemical Boxed Reading 15.1)*   A coenzyme that oxidizes acetaldehyde to acetic acid in the body.

*deactivating group (10.2)*   A group bonded to a benzene ring that decreases the rate of reaction as compared with benzene.

*dehydration (6.1, 7.3, 12.5, 15.2)*   An elimination reaction in which a unit of water is lost from adjacent carbons in an alcohol to produce an alkene.

*dehydrohalogenation (6.1, 7.3, 8.7)*   An elimination reaction in which a unit of HX (where X = Cl, Br, or I) is lost from adjacent carbons in an organic halide to produce an alkene.

*delocalized (1.11)*   Unshared and pi electrons are delocalized when they are dispersed over a number of atoms by resonance.

*denaturation (28.8)*   The precipitation of a protein by a variety of reagents and experimental conditions that produces a loss of biological activity.

*derived lipid (Introduction, Chap. 26)*   Steroids and prostaglandins are classified as derived lipids because they usually cannot be saponified.

*deshielded (29.6)*   A proton (or hydrogen nucleus) is deshielded from the external magnetic field by a lack of electron clouds surrounding the proton. The more the deshielding, the further downfield the peak appears.

*dextrorotary (11.3)*   A substance that rotates the plane of polarized light to the right.

*dextrose (27.7)*   Another name for D-glucose. The name is derived from dextro- (to the right) because D-glucose is a dextrorotary compound.

*diabetes (Medical Boxed Reading 27.1)*   A disease characterized by excess glucose in the blood that spills into the urine due to a lack of insulin.

*diastereomers (11.10)*   Stereoisomers that are not enantiomers, i.e., stereoisomers that are not mirror images of each other.

*1,3-diaxial interaction (3.10, 5.3)*   A repulsive destabilizing force due to crowding between, for example, any axial group other than hydrogen on C-1 of the cyclohexane ring with axial hydrogens on C-3 and C-5 of the ring.

*diene (5.1)*   An alkene that contains two carbon-carbon double bonds.

*dienophile (7.7)*   An alkene usually containing another functional group that reacts with a diene in a Diels-Alder reaction.

*dihydric alcohol (14.1)*   An alcohol that contains two hydroxyl groups per molecule. A diol.

*dipeptide (28.4)*   A compound produced when two amino acids are joined together.

*dipolar ion (28.2)*   The actual form of existence of an amino acid. This form is created as a result of the transfer of a proton from a carboxyl group to an amino group.

$$R—\overset{\overset{\displaystyle H}{|}}{\underset{\underset{\displaystyle NH_3}{+|}}{C}}—COO^-$$

Dipolar ion

*dipole (1.6)*   A polar covalent molecule with oppositely charged ends.

*dipole-dipole attraction (1.6)*   The attraction of the negative end of one dipole to the positive end of another dipole.

*disaccharide (27.1)*   A carbohydrate that hydrolyzes to yield two monosaccharide units.

*disinfectant (16.3)*   A substance that prevents the growth of microorganisms when applied to nonliving matter.

*disubstituted alkene (6.1)*   An alkene in which two organic groups are bonded to one or more carbons of the double bond.

*disulfide bond (28.7)*   Produced when two cysteine molecules, in reasonable proximity to each other, are oxidized to the disulfide compound.

*double-headed arrow (1.11, 7.2)*   A symbol ($\leftrightarrow$) used to indicate the presence of two contributing structures.

*doublet (29.7)*   A grouping of two related peaks.

*downfield (29.6)*   A relatively low frequency of absorption, on the left side of the spectrum.

*drying oil (26.4)*   An oil, containing a large percentage of linoleic and linolenic acid residues, that forms a hard film on the surface when exposed to air.

*D-sugar (27.1)*   A molecule in which the highest numbered chiral carbon has the same configuration as (+)-glyceraldehyde is designated a D-sugar.

*dynamite (15.2)*   A stable explosive with a variety of industrial uses that consists of a mixture of nitroglycerine and diatomaceous earth (a porous substance).

*E1 mechanism (12.4)*   A two-step (usually) elimination reaction in which the reaction rate is dependent on the concentration of only the substrate.

*E2 mechanism (12.4)* A one-step (usually) elimination reaction in which the reaction rate is dependent on the concentration of both nucleophile and substrate.

*eclipsed conformation (2.3)* The conformation of a compound where adjacent carbon-substituent bonds (e.g., carbon-hydrogen bonds) are parallel to one another (see Fig. 2.6).

*electromagnetic radiation (Introduction, Chap. 29, 29.1)* A continuum of energies from the highly energetic gamma and x-rays to the low-energy microwaves and radiowaves.

*electron-dot structure (1.8)* A structural formula showing every bonded electron pair and unshared electron pair as two dots.

*electronegativity (1.5)* The tendency of a bonded atom to attract a pair of electrons to itself.

*electrophilic aromatic substitution (10.1)* A reaction characteristic of aromatic compounds in which an electrophile, $E^+$, attacks an aromatic ring and substitutes for a hydrogen bonded to the ring as follows:

$$Ar\!-\!H + E^+ \longrightarrow Ar\!-\!E + H^+$$

where Ar—H is an aromatic compound.

*electrophilic reagent or electrophile (6.2)* A Lewis acid; a species (a compound or positively charged ion) that seeks a pair of electrons.

*electrophoresis (28.3)* An experimental technique by which two or more amino acids or proteins can be separated by passing an electric current through a solution of the substances.

*elimination reaction (6.1)* A reaction in which a pi bond is produced by the loss of two atoms or groups on adjacent carbon atoms.

*enantiomers (11.2)* Two substances that are nonsuperimposable mirror images of each other. Optical isomers.

*endo (7.7)* A stereochemical isomer produced in a Diels-Alder reaction in which the functional group of the adduct that originally was found in the dienophile is located on the opposite side of the adduct molecule as the shortest bridge.

*endothermic (Problem 4.3)* A chemical reaction or a physical change that absorbs heat.

*energy of vibration (29.2)* The energy absorbed by a molecule during an infrared scan.

*enkephalin (28.4)* A peptide, found in humans, that is a pain killer.

*enol (8.7)* A usually unstable compound [RCH=CHOH or RC(OH)=CH$_2$] that contains an alcohol group that is bonded to a carbon bearing a carbon-carbon double bond.

*enzyme (11.14)* A protein that catalyzes one or more chemical reactions in living systems.

*epimer (27.11)* One of two aldoses that differs in configuration only at C-2.

*epoxide (16.9)* A cyclic ether of the type

$$-\overset{|}{C}-\overset{|}{C}-$$
$$\diagdown \!\! \underset{O}{} \!\! \diagup$$

*equatorial bond (3.10)* A bond to the chair conformation of cyclohexane that lies along the plane of the ring.

*equivalent (3.2)* Hydrogens that are bonded to carbons that, in turn, are bonded to the same atoms or groups. The four methane and six ethane hydrogens are equivalent.

*essential amino acid (28.1)* An amino acid that cannot be produced by the body yet is necessary for the maintenance of a state of good health and must be obtained from food ingested.

*essential oils (7.9)* Fragrant-smelling oils used in the perfume industry.

*exo (7.7)* A stereochemical isomer produced in a Diels-Alder reaction in which the functional group of the adduct that originally was found in the dienophile is located on the same side of the molecule as the shortest bridge.

*exothermic (4.2)* A chemical reaction or a physical change that liberates heat.

*fatty acid (20.1)* One of several carboxylic acids (RCOOH) containing four carbons or more that is incorporated with glycerol as a fat.

*fibrous protein (28.6)* Water-insoluble protein consisting of long polymeric chain(s) placed along side each other.

*fingerprint region of the infrared (29.2)* The segment in an infrared spectrum between 1400 and 400 cm$^{-1}$.

*Fischer esterification (20.5, 23.6)* An equilibrium reaction in which a carboxylic acid reacts with an alcohol in the presence of sulfuric acid to produce an ester and water.

*Fischer projection (11.8)* A two-dimensional configurational representation of a three-dimensional chiral molecule. The central, chiral carbon atom is represented by the intersection of two lines with the horizontally bonded groups projecting toward the reader and the vertically bonded groups projecting away from the reader.

*fixative (Industrial Boxed Reading 14.2)* A component of perfume that functions to slow the rate of evaporation of the essential oils and thereby extend the life of the perfume.

*formal charge (1.10)* An imaginary charge that can be calculated for each atom in a structural formula by assuming that each atom shares half its bonded electrons and keeps all its unshared electrons.

*formalin (18.5)* An aqueous solution of 37% formaldehyde. Since formaldehyde is a gas, an aqueous solution represents a more conveniently employed form of the compound.

*formic acid (20.1)* The simplest of the carboxylic acids containing only one carbon atom.

$$\overset{\displaystyle O}{\underset{\displaystyle H-C-OH}{\|}}$$

Formic acid

*free radical (4.3)* A highly reactive, uncharged chemical species that contains one trivalent carbon surrounded by seven electrons. For example, ·CH$_3$ is a free radical.

*frequency (29.1)* The number of waves that pass a set point per second.

*fuel-air explosive (Historical Boxed Reading 16.1)* A mixture of ethylene oxide and air that causes great military damage over a wide area when set off by conventional explosives.

*functional group (1.12)* The reactive portion of a molecule, consisting of an atom or a group of atoms, that differentiates one class of compound from another.

*functional-group isomers (1.13)* Two or more isomers that have the same molecular formula but each contains a different functional group.

*furanose (27.5)* A carbohydrate that exists as a five-membered ring in its anomeric form.

*gamma-ray radiation (29.1)* The most energetic type of electromagnetic radiation.

*geminal (gem-) dihalide (8.6)* A dihalide ($RCX_2R'$) in which both halogen groups (X) are positioned on the same carbon.

*generic formula (2.1)* A general formula used to calculate the number of hydrogens in a hydrocarbon containing *n* carbon atoms. For example, since the generic formula of an alkene is $C_nH_{2n}$, an alkene containing four carbons has the molecular formula $C_4H_{2(4)} = C_4H_8$.

*geometric or cis-trans isomerism (5.3)* A phenomenon shown by alkenes and cycloalkanes wherein groups display different orientations in space around adjacent atoms of a double bond or ring.

*geometric condensed structure (2.6, 3.7, 3.11)* A condensed structure that is represented by a series of lines (one line, single bond; two parallel lines, double bond; three parallel lines, triple bond) in which no atoms are shown.

*globular protein (28.6)* A small water-soluble protein that is spherically shaped.

*glutathione (28.4)* A tripeptide, present in mammalian tissue, that acts as a reducing agent in a key biochemical reaction that is important in keeping red blood cells functional.

*glycerol (14.1)* A trivial name for 1,2,3-propanetriol.

*glycogen (Introduction, Chap. 27, 27.9)* A polymeric α-glucoside that is more branched than amylopectin.

*grain alcohol (14.1)* A trivial name for ethanol.

*Grignard reagent (4.1)* A very valuable synthetic reagent (RMgX) made by reacting an organic halide with magnesium in anhydrous diethyl ether.

*guncotton (Problem 27.18)* A synonym for cellulose trinitrate. An explosive substance used in smokeless powder.

*haloform reaction (19.3)* A reaction that produces a haloform when certain carbonyl compounds and alcohols are reacted with aqueous sodium hypohalite (NaOX or NaOH + $X_2$).

*Haworth projection (27.4)* A planar representation of cyclic carbohydrate structure that is not sterically accurate but is convenient to use.

*hemiacetal (18.6, 19.2)* A *gem*-hydroxyether derived from an aldehyde or ketone, usually unstable.

$$
\begin{array}{c}
\text{OH} \\
| \\
\text{R}-\text{C}-\text{OR}' \\
| \\
\text{H(R}')
\end{array}
$$

*hemiketal (18.6, 19.2)* A *gem*-hydroxyether derived from a ketone. Usually known as a *hemiacetal* today.

$$
\begin{array}{c}
\text{OH} \\
| \\
\text{R}-\text{C}-\text{OR}' \\
| \\
\text{R}'
\end{array}
$$

*hemoglobin (22.3)* A conjugated protein that contains $Fe^{2+}$ and is found in red blood cells of vertebrates.

*herbicide (20.4)* A substance that suppresses plant growth or destroys the plant.

*heterocyclic compound (9.7)* A cyclic compound in which the ring skeleton contains one or more atoms other than carbon.

*heterocyclic pi electron rule (9.7)* If a heteroatom in a heterocyclic species has a double bond, two unshared electrons on the heteroatom are not counted as pi electrons; if not, count two unshared electrons on the heteroatom as pi electrons.

*heterolytic bond breaking (4.3)* A process of bond breakage where one participating atom of the bond donates both electrons to the other atom. For example: $A{:}B \longrightarrow A^+ + {:}B^-$.

*heterolytic bond formation (4.3)* The formation of a bond where the participating anion donates both electrons to the bond, while the participating cation donates none. For example, $B{:}^- + A^+ \longrightarrow A{:}B$.

*Hinsberg test (25.2)* A test used to distinguish between the various classes of amines. When reacted with benzenesulfonyl chloride and base, primary amines form a soluble salt, secondary amines give an insoluble sufonamide, while tertiary amines do not react.

*Hoffmann rearrangement (25.1)* A reaction in which a simple acid amide is converted to a primary amine with one less carbon.

*homocyclic or carbocyclic species (9.7)* A cyclic species in which the ring skeleton contains only one carbon.

*homologous series (2.1)* Compounds containing the same functional group that differ from member to member by a —$CH_2$— (methylene) group.

*homolytic bond formation (4.3)* The formation of a bond where each participating atom donates one electron to the bond. For example: $R{\cdot} + {\cdot}Cl \longrightarrow R{:}Cl$.

*homolytic dissociation (bond breaking) (2.8, 4.3)* The breaking of a bond in which each atom of the bond keeps one electron.

*hormone (5.5)* A substance produced, in very small amounts, by a plant or animal. A hormone may be produced in one part of an organism yet performs its function in another part of the organism.

*Huckel's rule (9.7)* A species must contain $4n + 2$ delocalized pi electrons in order to be classified as aromatic.

*Hund's rule (1.3)* Before any orbital in a subshell is completely filled with two electrons, each orbital in the subshell must contain one electron.

*hybrid orbital (2.2)* An orbital formed by merging two or more different orbitals in an atom.

*hybridization (2.2, 2.4, 2.5, 2.6, 2.7)* The process of producing an orbital by merging two or more different orbitals in an atom.

*hydrate (19.2)* A *gem*-diol, usually unstable, that is produced by reacting a carbonyl compound with water.

$$
\begin{array}{ccc}
\begin{array}{c}\text{O} \\ \| \\ -\text{C}- \end{array} & + \, H_2O \rightleftharpoons & \begin{array}{c}\text{OH} \\ | \\ -\text{C}- \\ | \\ \text{OH}\end{array} \\
\text{A carbonyl} & & \\
\text{compound} & & \text{A hydrate}
\end{array}
$$

*hydration (5.6, 6.2)* An addition reaction in which a unit of water is added to adjacent carbon atoms in an alkene to produce alcohol.

*hydrocarbon (2.1)* A compound that contains only the elements carbon and hydrogen.

*hydrogen bond (1.6, 14.2)* An intermolecular attractive force that is created when a hydrogen bonded to an electronegative element (N, O, or F) interacts with a lone pair of electrons (unshared electron pair) on an electronegative atom in another molecule.

*hydrogen deficit (5.1)* The number of hydrogens that are subtracted from the corresponding alkane to obtain the formula of a particular hydrocarbon. For example, consider $C_3H_6$. The hydrogen deficit is calculated to be $C_3H_8 - C_3H_6 = 2$.

*hydrophobic bond (28.7)* A water-hating bond, a nonpolar bond.

*imine (19.2)* An unsaturated amine with the general structure $RCH\!=\!NH$ or $RCH\!=\!NR'$ that is produced by the reaction of a carbonyl compound with ammonia or a primary amine $R'NH_2$, respectively. Imines are also known as Schiff bases.

*inductive effect (6.2, 10.2)* A slight electron attraction or release across a sigma bond.

*infrared (IR) spectroscopy (Introduction, Chap. 29, 29.2)* A kind of spectroscopy that measures the absorptions of infrared radiation by various organic functional groups.

*infrared (IR) spectrum (29.2)* A plot of percentage transmittance of radiation versus decreasing frequency of radiation, expressed as wave number. Typically, the wavelengths scanned by an instrument vary from $2.2 \times 10^{-4}$ to $25 \times 10^{-4}$ cm corresponding to wavenumbers (frequencies) of 4500 to 400 cm$^{-1}$.

*initiation step (4.3)* A proposed step in a free-radical mechanism that produces a reactive free-radical species. For example, the initiation step in the chlorination of methane is $Cl_2 \longrightarrow 2Cl\cdot$.

*instantaneous dipole (1.6)* This dipole arises when for a split second a greater electron density exists on one side of a nonpolar molecule than on the other.

*intermolecular attractive forces (1.6)* The forces of attraction between molecules.

*intermolecular dehydration in ester formation (23.6)* The process by which an acyclic ester is produced; i.e., the loss of water to produce the ester is between two molecules, a molecule of carboxylic acid and a molecule of alcohol.

*internal alkyne (8.1)* An alkyne ($RC\!\equiv\!CR$) that contains the triple bond between two interior carbons of the main chain.

*International Union of Pure and Applied Chemistry (3.2)* A group of chemists that met in Geneva in 1892 to create a system where each organic compound is assigned a specific name that can be deduced from the structure of the compound.

*intramolecular dehydration in ester formation (23.6)* The process by which a cyclic ester, or lactone, is produced; i.e., the loss of water to produce the ester is within one molecule, a molecule of certain substituted hydroxyacids.

*intramolecular hydrogen bond (8.7)* An attractive force that is created when a hydrogen bonded to an electronegative element (N, O, or F) interacts with a lone pair of electrons (unshared electron pair) on an electronegative atom within the same molecule.

*inversion of configuration (Walden inversion) (12.1)* A characteristic of $S_N2$ reaction in which the nucleophile attacks, for example, an ($R$) substrate from the backside resulting in an inverted molecular configuration and formation of an ($S$) product.

*invert sugar (27.8)* A 50:50 mixture of D-fructose and D-glucose.

*iodoform test (19.3)* Iodoform is a yellow solid, the formation of which represents a positive test for a methylcarbinol, methyl ketone, ethanol, or acetaldehyde when the test compound is treated with aqueous sodium hypoiodite (NaOI or NaOH + $I_2$).

*ionic bond (1.4)* A chemical bond formed between ions of opposite charge.

*isoelectric point (28.2)* The pH at which no migration of an amino acid occurs in an electric field.

*isolated diene (5.1)* A diene in which at least one methylene group ($-CH_2-$) is located between the two carbon-carbon double bonds. For example:

$$CH_2\!=\!CH(CH_2)_3CH\!=\!CH_2$$
An isolated diene

*isomer (1.13)* One of two or more compounds that have the same molecular formula but differ in the way the atoms are linked together.

*isomerism (1.13)* The phenomenon where two or more compounds have the same molecular formula but differ in the way the atoms are linked together.

*Kekulé structural formula (2.6)* A geometric condensed structure used for benzene and its derivatives.

*ketal (18.6, 19.2)* A *gem*-diether derived from a ketone. Usually known as an *acetal* today.

$$R-\overset{\overset{\displaystyle OR'}{|}}{\underset{\underset{\displaystyle H(R')}{|}}{C}}-OR'$$

*ketitol (27.11)* A polyol obtained when a ketose is reduced with sodium borohydride.

*keto-enol tautomerism (8.7)* An equilibrium mixture of an aldehyde or ketone and the corresponding enol.

*ketopentose (27.1)* A monosaccharide that contains a keto group and a total of five carbons.

*ketose (27.1)* A polyhydroxyketone, a monosaccharide.

*lacrimal agent (19.3)* A tear-producing substance.

*lactam (23.8)* A cyclic amide.

*lactone (22.3)* A cyclic ester.

*lanolin (26.1)* Purified wool wax, isolated from sheep.

*leaving group (12.1)* The group (usually a halogen) that is emitted in the form of an anion from the substrate when the substrate is reacted with a nucleophile ($N:^-$).

$$\underset{\text{Substrate}}{R-Cl} \quad + \quad \underset{\text{Nucleophile}}{N:^-} \quad \longrightarrow R-N + \underset{\text{Leaving group}}{Cl^-}$$

*LeChâtelier's principle (6.1, 18.6, 19.2, 20.5, 23.6)* When a system in equilibrium is stressed by a change of concentration, pressure, or temperature, the equilibrium shifts so as to minimize the stress.

*levorotary (11.3)* A substance that rotates the plane of polarized light to the left.

*levulose (27.7)* Another name for D-fructose. The name is derived from levo- (to the left) because D-fructose is a levorotary compound.

*Lewis definition of acidity (1.14)* An acid accepts a pair of electrons; a base donates a pair of electrons.

*Lindlar's catalyst (8.7)* A catalyst used to prepare a *cis*-alkene from an alkyne.

*line-bond structure (1.8)* A structural formula in which each bonded electron pair is represented by a dash. Unshared electron pairs (lone pairs) are not shown.

*London forces (1.6, 3.3)* Very weak intermolecular attractive forces that arise from instantaneous dipoles flickering on and off like fireflies.

*London forces of repulsion (2.2)* Repulsive forces that exist when electron pairs on adjacent carbon-hydrogen bonds are too close together.

*L-sugar (27.1)* A molecule in which the highest numbered chiral carbon has the same configuration as (−)-glyceraldehyde is designated an L-sugar.

*Lucas test (15.2)* A qualitative test tube test to establish the class of an alcohol. When a mixture of zinc chloride and hydrochloric acid is reacted with an alcohol, the presence of instantaneous cloudiness means the alcohol is tertiary, allylic, or benzylic; cloudiness in 3 to 5 minutes means the alcohol is secondary; and no cloudiness in 30 minutes means the alcohol is primary or vinylic.

*magnetic resonance imaging (MRI) (Medical Boxed Reading 29.1)* An instrumental technique used in medicine, using NMR, in which the protons in water and hydrogen-abundant lipids are measured. There is a time difference between healthy and diseased tissue in which hydrogen in the antiparallel alignment with the external magnetic field relaxes to the more stable parallel alignment. This time difference is displayed as lighter and darker regions of the structure as observed in the image (see Fig. 29.17).

*main chain (3.2)* A synonym for the term *parent compound*.

*major contributor (7.2)* The most stable contributing structure (resonance structure) that can be drawn for a chemical species.

*Markownikoff-type addition (6.2)* A reaction that conforms to Markownikoff's rule but takes place by means of a different mechanism.

*Markownikoff's rule (6.2, 8.7)* If an unsymmetrical reagent, H—A, adds to an alkene (or alkyne), the hydrogen adds to the carbon of the double bond (or triple bond) with the greater number of hydrogens.

*melanin (Medical Boxed Reading 24.1)* A polymeric quinone-like molecule present in human skin that is the cause of different skin colors in different races.

*mercaptan (14.5)* The common name for a thiol.

*mescaline (24.6)* A potent psychedelic alkaloid.

*meso form (11.11)* An optically inactive compound even though each of its molecules contains two or more similar chiral carbons (centers). Each molecule contains a plane of symmetry, so the molecule as a whole is achiral.

*meta- (9.2)* Substituents are meta when they are located in a 1,3 position on a benzene ring.

*methemoglobin (22.3)* An oxidation product of hemoglobin in which iron is present as $Fe^{3+}$.

*methemoglobinemia (22.3)* A type of anemia caused by the ingestion of certain compounds such as acetanilide.

*micelle (26.4)* A spherical group of soap ions in aqueous solution in which the nonpolar tail of each soap species is within the sphere (embedded in a grease globule) and the polar ionic head of each soap species remains dissolved in water at the same time.

*microwave radiation (29.1)* A type of low-energy electromagnetic radiation, slightly more energetic than radio waves.

*minor contributor (7.2)* A contributing structure (resonance structure) less stable than the major contributor that can be drawn for a chemical species.

*mixed anhydride (22.1)* An anhydride and R' that contains two different organic groups per molecule. R can be either alkyl or aryl group.

$$R-\overset{\overset{\displaystyle O}{\|}}{C}-O-\overset{\overset{\displaystyle O}{\|}}{C}-R'$$

A mixed anhydride

*mixed ether (16.6)* A compound of the type R—O—R', Ar—O—Ar', or Ar—O—R (Ar— is an aryl group) where both organic groups are different.

*mixed triglyceride (26.2)* A triglyceride in which at least two of the acyl groups are different.

*molecular formula (1.8)* A formula that provides the actual number of atoms of each element in the molecule.

*monochromatic light (11.3)* Light that is composed of waves with the same wavelength.

*monocyclic aromatic compound (9.6)* An aromatic compound (like benzene) that contains one aromatic ring.

*monoene (5.1)* An alkene that contains one carbon-carbon double bond.

*monohydric alcohol (14.1)* An alcohol that contains one hydroxyl group per molecule.

*monomer (5.6, 6.3)* A simple, starting reactant used in the preparation of a polymer. For example, ethylene and propylene are monomers used in the preparation of polyethylene and polypropylene, respectively.

*monosaccharide (27.1)* A carbohydrate that does not undergo hydrolysis.

*monosubstituted alkene (6.1)* An alkene in which one organic group is bonded to a carbon of the double bond.

*morphine (24.6)* The first pain-killing alkaloid that was isolated.

*multiline arrow (4.3, 8.7)* Two or more straight lines culminating with an arrowhead to designate the movement of an atom by itself or an atom and a pair of electrons.

*multiplicity rule (29.7)* The splitting pattern of the hydrogens on a given carbon can be predicted by adding the number of neighboring hydrogens (i.e., hydrogens on adjacent carbons) and then adding one.

*mutarotation (27.4)* The change of specific rotation of an aqueous solution of a reducing carbohydrate with time.

*natural gas (3.4)* A mixture of mostly methane with some ethane and very little propane mainly used as fuel.

*natural product (7.9)* An organic compound isolated from a plant or animal.

*neutral amino acid (28.1)* An amino acid that contains one amino group and one carboxyl group.

*Newman projection (2.3)* A method of representing organic compounds in three dimensions on a two-dimensional page in which the carbon-carbon bond is seen from the front to the rear of the molecule (see Fig. 2.6).

*nicotine (24.6)* An alkaloid present in cigars and cigarettes.

*ninhydrin (28.10)* When ninhydrin is reacted with almost all the amino acids, a deep purple color is observed. Thus this reaction serves as a positive test for almost all the amino acids.

*nitrile (21.1)* A functional group $-C\equiv N$ as in RCN (an alkyl nitrile) or ArCN (an aryl nitrile).

*noble gas electron configuration (1.4)* An electron configuration that exists in a typical noble gas: $ns^2np^6$. Helium has an atypical noble gas electron configuration of $1s^2$.

*nonionic detergent (26.4)* A synthetic detergent that does not contain an ionic species.

*nonpolar covalent bond (1.5)* A covalent bond in which the electron density is equally placed between the atoms participating in the bond.

*nonreducing carbohydrate (27.5)* A carbohydrate that does not react when treated with Benedict's or Fehling's solution.

*N-terminus (28.4)* The amino acid residue of a peptide or protein with the free amino group. By convention, the N-terminus is written or drawn on the extreme left of the molecule.

*nuclear magnetic resonance (NMR) spectroscopy (Introduction, Chap. 29, 29.4)* A kind of spectroscopy that measures the absorption of radiofrequency radiation by certain nuclei with nuclear spin in the presence of a strong magnetic field. This instrumentation can identify a wide variety of alkyl, alkenyl, alkynyl, cycloalkyl, and aromatic groups and is thus a complement to infrared spectroscopy.

*nuclear spin (29.4, 29.5)* The capacity of certain atomic nuclei to spin like a top, each nucleus around its own axis. As a result of the spin, a small magnetic field is created.

*nucleophile (12.1)* A nucleophilic reagent.

*nucleophilic reagent (10.1, 12.1)* A Lewis base. A compound or an anion that contains one or more unshared electron pairs that attack a partially positively charged carbon in a molecule.

*nucleophilic substitution reaction (12.1)* A reaction in which a nucleophile, N:$^-$, replaces a group (usually a halogen) in a reagent as follows:

$$R-Cl \ + \ N:^- \longrightarrow R-N + Cl^-$$

Reagent    Nucleophile

*observed optical rotation ($\alpha$) (11.3)* The number of degrees to the left or right the analyzer in a polarimeter rotates.

*off-resonance mode of $^{13}C$ NMR spectroscopy (29.8)* A form of $^{13}C$ NMR spectroscopy in which $^{13}C-^1H$ spin-spin splitting does take place and doublets, triplets, and quartets are observed.

*olefin (5.1)* An alkene. Alkenes are known as *olefins*, or "oil formers," because they react with the halogens, along with other reagents, to produce oils.

*oligomer (7.9)* A molecule produced from the "replication" of anywhere from 2 to 10 monomer molecules.

*oligosaccharide (27.1)* A carbohydrate that hydrolyzes to yield anywhere from 2 to 10 monosaccharide units.

*optical isomer (11.1)* A stereoisomer that is composed of chiral molecules.

*optical isomerism (11.1)* The study of stereoisomers that are mirror images of each other yet are not identical.

*optically active (11.3)* A substance that rotates the plane of polarized light in a polarimeter.

*orbital (1.3)* A segment of space that can hold a maximum of two electrons.

*organ (1.1)* A body part that performs a specific function.

*organic (1.1)* An organic compound originally was thought to be derived from an organ of a living or once-living thing. Today, organic compounds represent the vast majority of the compounds of carbon (see Sec. 1.8).

*ortho- (9.2)* Substituents are ortho when they are located in a 1,2 position on a benzene ring.

*osazone (27.11)* A diphenylhydrazone produced when a reducing carbohydrate is reacted with an excess of phenylhydrazine. Osazones are usually yellow solids.

*oxidation (14.4, 15.1)* A substance undergoes oxidation when it shows a gain of oxygen (or loss of hydrogen) during the course of a chemical reaction.

*oxidizing agent (14.4, 15.1)* A substance that undergoes reduction and thereby produces an oxidation.

*oxirane (16.9)* The IUPAC name of ethylene oxide.

*oxonium ion (10.1, 12.1, 12.2)* A species bearing a positive charge on oxygen formed by addition of water or an alcohol to a carbocation (see Secs. 6.2 and 12.1) or by the addition of a proton to an alcohol (see Secs. 10.1 and 12.2).

*para- (9.2)* Substituents are para when they are located in a 1,4 position on a benzene ring.

*para-aminobenzoic acid (Medical Boxed Reading 24.2)* A compound necesarry for the growth of a number of microorganisms.

*paraffin (3.5, 4.2)* An alkane.

*parallel orientation (29.5)* A condition in which the small magnetic field of the nucleus is oriented in the same direction as that of the external magnetic field.

*pardoxin (Historical Boxed Reading 26.2)* A protein, composed of 162 amino acid residues, that was found to repel certain shark species.

*parent compound (3.2)* The longest continuous chain of carbon atoms in the compound to be named. Sometimes called the *main chain*.

*peak (29.1)* A V-shaped portion of a spectrum that represents radiation absorbed by a molecule.

*peak splitting (29.7)* The observed result of spin-spin splitting.

*pellagra (22.3)* A disease caused by a deficiency of niacin (or failure to convert tryptophan to niacin). Nicotinamide, a B-complex vitamin, is used to relieve the symptoms.

*peptide (20.5, 28.4)* A compound produced when two or more amino acids are joined together by means of one or more peptide bonds.

*peptide bond (28.4)* A carbon-nitrogen bond formed by the amine of one amino acid to the carbonyl carbon of a second amino acid to form an amide bond.

*peracid (17.6)* A class of oxidizing agent used to convert an alkene to the corresponding oxirane. Note that a peracid contains one more oxygen atom than a carboxylic acid.

$$R-\overset{\overset{\displaystyle O}{\|}}{C}-O-O-H$$
A peracid

*perfume (Industrial Boxed Reading 14.2)* A mixture of compounds (essential oils, solvent, and fixative) that emits a pleasant odor. Perfumes contain 20–40% essential oils.

*periplanar geometry (12.4)* A conformation necessary for elimination in which two adjacent carbons and the groups (hydrogen and halogen or other leaving group) bonded to these carbons must lie in the same plane.

*phenol (14.1)* A compound containing one or more hydroxyl group that is bonded to an aromatic (benzene) ring.

*phenylketonurea (PKU) (28.1)* An inherited disease in humans that stems from an inability to metabolize phenylalanine to tyrosine.

*pheromone (3.5)* An organic compound secreted by a variety of insects in as little as nanogram quantities that serves as a sex or defense message.

*phospholipid (Introduction, Chap. 26, 26.5)* A tetraester that yields glycerol, an aminoalcohol, carboxylate salts of fatty acids, and monohydrogen phosphate ion ($HPO_4^{2-}$) when saponified.

*photon (29.1)* A particle of electromagnetic energy.

*pi (π) bond (2.4)* A bond in which the greatest electron density is above and below an imaginary straight line linking the two atoms participating in the bond.

*pitch (28.7)* The distance between turns in a protein with an α-helical secondary structure.

*plane of symmetry (11.2, 11.11)* A plane of symmetry exists in a molecule when an imaginary plane drawn through a molecule creates two halves of a molecule that are mirror images.

*plane-polarized light (11.3)* Ordinary or monochromatic light that vibrates on only one plane perpendicular to the direction the light comes from.

*polar covalent bond (1.5)* A covalent bond in which the electron density is closer to one atom participating in the bond than to the other.

*polarimeter (11.3)* An instrument used to measure optical rotation.

*polarizability (13.2)* The tendency of an electron cloud around an atom to be distorted due to the presence of an electric charge.

*polarizer (11.3)* A component of a polarimeter, composed of Polaroid, in which monochromatic light is converted to plane-polarized light.

*Polaroid (11.3)* A substance that produces plane-polarized light when ordinary or monochromatic light is passed through.

*polycyclic aromatic compound (9.6)* An aromatic compound (like naphthalene) that contains more than one aromatic ring.

*polyester (Industrial Boxed Reading 23.1)* One of a group of polymers that is composed of many ester groups present in a systematic repeating pattern. An important polyester, poly(ethylene terephthalate), is an ingredient in plastic soft drink bottles and permanent-press fabrics.

$$\left(\!\!-O-\overset{\overset{\displaystyle O}{\|}}{C}-\!\!\bigcirc\!\!-\overset{\overset{\displaystyle O}{\|}}{C}-O-CH_2CH_2-\!\!\right)_{\!\!n}\!\!-O-$$
Poly (ethylene terephthalate)

*polymer (5.6, 6.3)* A large molecule produced from the "replication" of several monomer molecules. For example, the polymers polypropylene and polyethylene are prepared from the respective monomers, propylene and ethylene.

*polypeptide (28.4)* A compound produced when from 10 to 49 amino acids are joined together.

*polysaccharide (27.1)* A carbohydrate that hydrolyzes to yield 11 or more monosaccharide units.

*positional isomerism (1.13)* The phenomenon in which two or more isomers have the same molecular formula and each contains the same functional group but in different locations within the molecule.

*primary (3.2)* A hydrogen or other atom or group that is bonded to a carbon that, in turn, is bonded to one carbon. Each of the six hydrogens in ethane is primary.

*primary structure (28.7)* The identity, number, and sequence of the amino acid residues making up the protein.

*progesterone (26.6)* A steroidal female hormone that is vital in regulating the menstrual cycle and in preparing the uterus for a fertilized ovum.

*promoter (19.3)* A substance that increases the rate of a chemical reaction but, unlike a catalyst, is used up in the reaction.

*Prontosil (Medical Boxed Reading 24.2)* A red dye that, when acted on by enzymes of the body, produces sulfanilamide.

*proof* or *proof of spirit (Biochemical Boxed Reading 14.1)* A term that represents double the percentage of ethanol in a drinkable alcoholic solution. Thus a solution of 20% by volume is 40 proof.

*propagation step (4.3)* One or more steps in a free-radical mechanism that continues the reaction by producing one free radical for every free radical destroyed. For example, in the chlorination of methane,

$$Cl\cdot + H-CH_3 \longrightarrow H-Cl + \cdot CH_3 \quad \text{and}$$
$$\cdot CH_3 + Cl_2 \longrightarrow Cl\cdot + Cl-CH_3$$

*prostaglandin (Introduction, Chap. 26, 26.7)* A lipid that usually cannot be saponified that is biochemically derived from arachidonic acid and other C-20 unsaturated fatty acids.

*prosthetic group (28.6)* That part of a conjugated protein that does not consist of amino acid residues.

*"protecting" the amino group (25.3)* The process of replacing an amino group on a benzene ring with an acetamido group. The resulting compound can then be brominated or nitrated without any damage to the easily brominated and oxidized amino group. Once the bromo or nitro group has been introduced into the ring, the acetamido group (the protecting group) is removed to liberate the desired amine.

*protein (Introduction, Chap. 28)* A polymeric natural product typically composed of hundreds of amino acid residues.

*putrescine (24.3)* 1,4-Butanediamine or tetramethylene diamine.

*pyranose (27.5)* A carbohydrate that exists as a six-membered ring in its anomeric form.

*quaternary ammonium salt (25.1)* A salt in which four organic groups are bonded to nitrogen.

$$R-\overset{\displaystyle R'}{\underset{\displaystyle R''}{N^+}}-R'''\ Br^-$$

A quarternary ammonium bromide salt

*quarternary carbon (Problem 3.11)* A carbon that is bonded to four carbons.

*quaternary structure (28.7)* The manner in which two or more protein chains in the same protein molecule are arranged in space with respect to each other.

*quartet (29.7)* A grouping of four related peaks.

*quinine (24.6)* An alkaloid that was useful during World War II in suppressing the symptoms of malaria.

*quinone (17.3)* Quinones, resulting from the oxidation of phenols, are cyclohexadiendiones such as

P-benzoquinone
(1,4-benzoquinone)

*racemic mixture or racemate (11.5)* A 50:50 mixture of two enantiomers.

*radiowave radiation (29.1)* The least energetic type of electromagnetic radiation.

*rancid (26.4)* A fat or oil turns rancid when it acquires an offensive odor due to oxidation that produces fatty acids containing from 4 to 10 carbons.

*rate law (12.1)* A mathematical expression that relates the rate of the reaction to the concentration of each reactant raised to an appropriate power:

$$\text{Rate} = k[A]^m[B]^n$$

*rearrangement (6.1, 10.1, 12.5)* A reaction that produces an unexpected product; either a double bond or group (Cl, OH) is located in an unanticipated place or there is a change in the carbon skeleton of the reactant.

*receptor (Biochemical Boxed Reading 18.1)* A sensory neuron, located up in the nose, that reacts with an odor- or taste-causing molecule to produce an electrical signal to the brain.

*rectus (R) (11.6)* The (R) configuration is assigned to a chiral substance as follows: When the group of lowest priority is away from the reader, arranging the remaining three groups in order of decreasing priority takes place in a clockwise direction.

*reducing agent (14.4, 15.1)* A substance that undergoes oxidation and thereby produces a reduction.

*reducing carbohydrate (27.5)* A carbohydrate that gives a red precipitate of copper(I) oxide when treated with Benedict's or Fehling's solution. A reducing carbohydrate must contain a hemiacetal.

*reduction (4.1, 14.4, 15.1)* A substance undergoes reduction when it shows a loss of oxygen (or gain of hydrogen) during the course of a chemical reaction.

*refined (3.4)* Liquid petroleum is refined when it is separated into fractions by distillation.

*reforming (9.4)* A variety of processes which convert straight chain alkanes to the branched chain compounds or to the corresponding cycloalkane or the corresponding aromatic hydrocarbon. The purpose of the reforming processes is to develop a fuel that prevents engine knock (i.e., a jerky stroke of the piston in an internal combusion engine rather than a smooth stroke).

*resolution (11.13)* A procedure by which a racemic mixture is separated into its enantiomeric components.

*resonance (1.11, 7.4)* A means of more precisely describing the electron structure of a species by delocalizing pi and unshared electron pairs (lone pairs) to produce a number of different structures.

*resonance effect (10.2)* The movement of pi electrons, either into the benzene ring or out of the benzene ring, produced by a substituent group on the ring.

*resonance energy (7.2, 9.7)* The difference in internal energy between a substance that is resonance stabilized (like 1,3-pentadiene) and an isomer that is not resonance stabilized (like 1,4-pentadiene).

*resonance hybrid (1.11, 7.2)* A blended structure that is a blend of all the contributing structures and thus is the most accurate depiction of a chemical species that we know of.

*retention of configuration (12.1)* A stereochemical option in which both the reactant and product show the same configuration.

*rickets (26.6)* A vitamin-deficiency disease in the young characterized by soft and deformed bone.

*ring flip (3.10)* A process whereby a chair conformation of cyclohexane is converted into another chair conformation by way of a boat conformation. An axial bond is converted to an equatorial bond by a ring flip, and vice versa.

*saccharin (Historical Boxed Reading 27.2)* A synthetically produced sweetener, about 300 times sweeter than table sugar.

*salicylic acid (20.1, Historical Boxed Reading 20.2)* o-Hydroxybenzoic acid. The laboratory precursor of aspirin. When ingested, aspirin is metabolized to produce salicylic acid that is responsible for the analgesic and antipyretic effects of the drug.

*salt bridge (28.7)* An ionic attraction produced when an acid amino acid residue loses a proton to a basic amino acid residue.

*Sandmeyer reaction (25.4)* A powerful synthetic tool that enables a chemist to replace an aromatic diazonum group ($-\overset{+}{N}\equiv N$) with a chloro ($-Cl$), bromo ($-Br$) or cyano ($-CN$) group.

*Sanger's reagent (28.5)* 2,4-Dinitrofluorobenzene, used to identify the N-terminus of a peptide.

*saponification (22.4, 23.7)* The basic hydrolysis of an ester that produces the salt of an acid and an alcohol.

*saturated hydrocarbon (2.1)* A hydrocarbon that contains only carbon-carbon single bonds.

*sawhorse diagram (2.3)* A method of representing organic compounds in three dimensions on a two-dimensional page in which the intersection of three lines represents a carbon (see Fig. 2.6).

*Saytzeff's rule (6.1)* A rule that is used to predict the major product in an elimination reaction in which two or more alkenes can be formed. The elimination process that produces the most highly substituted (and most stable) alkene is favored.

*Schiff base (19.2)* An imine that results from the reaction of an aldehyde or ketone with a primary amine or ammonia.

$$R-\overset{\overset{\displaystyle H(R')}{|}}{C}=O \; + \; H_2N-R'' \longrightarrow R-\overset{\overset{\displaystyle H(R')}{|}}{C}=N-R'' \; + \; H_2O$$

An imine

*second-order kinetics (12.1)* The rate of a reaction is dependent on the concentration of two reactants taken to the first power as given in the rate law:

$$\text{Rate} = k[A]^1[B]^1$$

The rate follows first-order kinetics with respect to [A] and first-order kinetics with respect to [B]. Thus the reaction rate follows second-order kinetics overall.

*second-order reaction (12.1)* A second-order reaction shows second-order kinetics.

*secondary (3.2)* A hydrogen or other atom or group that is bonded to a carbon that, in turn, is bonded to two carbons. The middle carbon, not the end carbons, in propane is secondary.

*secondary structure (28.7)* The arrangement of the protein in space or its stereochemistry.

*semicondensed formula (1.8)* An abbreviated structural formula in which most of the bonds are not shown.

*shielded (29.6)* A proton (or hydrogen nucleus) is shielded from the external magnetic field by electron clouds surrounding the proton. The more the shielding, the further upfield the peak appears.

*sickle cell anemia (Medical Boxed Reading 28.1)* A genetic disease in which red blood cells change shape (are said to sickle) and thereby block the flow of blood through the capillaries causing anemia, shortness of breath, and muscle and bone pain.

*side chain (Introduction, Chap. 28)* An organic group bonded to the carbon bearing the amino and carboxyl groups.

$$R-\overset{\overset{\displaystyle H}{|}}{\underset{\underset{\displaystyle NH_2}{|}}{C}}-COOH$$

R— is a side chain in the preceding α-amino acid

*sigma (σ) bond (2.2)* A sigma bond is a bond where the greatest electron density is directly between the atoms forming the bond.

*signal (29.5)* Synonymous with a peak. A peak is produced in NMR spectroscopy when the antiparallel oriented nuclei lose energy back to the more stable parallel orientation. This lost energy shows up as a signal or peak.

*simple amide (22.1)* An amide in which nitrogen is bonded to two hydrogens.

$$R-\overset{\overset{\displaystyle O}{\|}}{C}-\overset{\overset{\displaystyle H}{|}}{N}-H$$

A simple amide

*simple anhydride (22.1)* An anhydride that contains two identical organic groups per molecule.

$$R-\overset{\overset{\displaystyle O}{\|}}{C}-O-\overset{\overset{\displaystyle O}{\|}}{C}-R \quad \text{or} \quad Ar-\overset{\overset{\displaystyle O}{\|}}{C}-O-\overset{\overset{\displaystyle O}{\|}}{C}-Ar$$

A simple anhydride        A simple anhydride
                                    (Ar— is an aryl group)

*simple ether (16.6)* A compound of the type R—O—R or Ar—O—Ar (Ar— is an aryl group) where both organic groups are the same.

*simple lipid (Introduction, Chap. 26)* A triester that yields only glycerol and carboxylate salts of fatty acids when saponified.

*simple protein (28.6)* A protein that yields on hydrolysis only amino acids.

*simple triglyceride (26.2)* A triglyceride in which the acyl groups are the same.

*single-channel theory of smell (Biochemical Boxed Reading 18.1)* The theory that one receptor was associated with a specific odor or taste.

*singlet (29.7)* An isolated peak or signal.

*sinister (S) (11.6)* The (S) configuration is assigned to a chiral substance as follows: When the group of lowest priority is away from the reader, arranging the remaining three groups in order of decreasing priority takes place in a counterclockwise direction.

*soap (26.4)* A mixture of sodium or potassium salts of long-chain ($C_{12}$–$C_{20}$) fatty acids produced by the saponification of fats or oils.

*solvolysis (12.1)* An $S_N1$ reaction in which the nucleophile also functions as a solvent.

*soporific (Historical Boxed Reading 19.1)* A drug employed to induce sleep.

*specific rate constant (12.1)* A constant ($k$) present in the rate-law expression that represents the rate of the reaction when the concentration of each reactant is 1 mol/L.

*specific rotation (11.4)* An intensive property characteristic of the substance that can be defined as the observed optical rotation when the tube length is 1 dm and the concentration 1 g/ml:

$$[\alpha] = \frac{\alpha}{l \cdot C}$$

where $[\alpha]$ is the specific rotation, $\alpha$ is the observed rotation, $l$ is the length of the polarimeter tube in dm, and $C$ is the concentration of the solution in g/ml.

*spectrophotometer (29.1)* An instrument that measures the amount of radiation absorbed at each frequency as the frequency of radiation is varied.

*spectroscopy (Introduction, Chap. 29)* That branch of chemistry which studies the changes that occur in matter when bombarded by various types of electromagnetic radiation.

*spectrum (29.1)* The end product of a spectrophotometer, usually in the form of one or more peaks on a graphed paper, that is interpreted to yield data about molecular structure.

*spermaceti wax (26.1)* A wax obtained from the head and blubber of the sperm whale.

*spin-flip (29.5)* A process in which at a certain frequency a few nuclei with a parallel orientation to the external magnetic field are excited to an antiparallel orientation.

*spin-spin splitting (29.7)* The splitting of a singlet into a multiplet due to the coupling of nuclear spins of neighboring atoms.

*splash (Industrial Boxed Reading 14.2)* A perfume-like mixture that only contains about 2% essential oils.

*stable octet (1.4)* An electron configuration, characteristic of a noble gas, that shows little tendency to react.

*staggered conformation (2.3)* The conformation of a compound where adjacent carbon-substituent bonds (e.g., carbon-hydrogen bonds) are the farthest away from each other (see Fig. 2.6).

*starch (Introduction, Chap. 27, 27.9)* A mixture composed of two α-glucosidic polymeric components, amylose and amylopectin.

*stereoisomerism (11.1)* The study of the spacial arrangements of the atoms (or groups) in a molecule (stereoisomer).

*stereoisomers (5.3)* Two or more substances that have the same molecular formula, the same structural formula, but different orientations in space.

*steric hindrance (10.4, 12.1)* The presence of atoms and groups around a reaction site that causes crowding and prevents one reactant from colliding with another. This, in turn, decreases the rate of reaction.

*steroid (Introduction, Chap. 26, 26.6)* A lipid that usually cannot be saponified. A steroid, like cholesterol, contains three six-membered rings and a five-membered ring.

*structural formula (structure) (1.8)* A formula that describes the way in which the atoms are bonded to each other.

*substituent group (3.2)* An alkyl or other group such as I— (iodo) that is bonded to an organic compound.

*substituted amide (22.1)* An amide in which nitrogen is bonded to one or no hydrogens.

$$\underset{\text{A monosubstituted amide}}{\overset{\displaystyle\text{O}\quad\text{R}'}{\text{R—C—N—H}}} \qquad \underset{\text{A disubstituted amide}}{\overset{\displaystyle\text{O}\quad\text{R}'}{\text{R—C—N—R}''}}$$

*substitution nucleophilic first-order reaction ($S_N1$ reaction) (12.1)* A two-step (usually) nucleophilic substitution reaction in which the reaction rate is dependent on the concentration of only the substrate.

*substitution nucleophilic second-order reaction ($S_N2$ reaction) (12.1)* A one-step (usually) nucleophilic substitution reaction in which the reaction rate is dependent on the concentration of both nucleophile and substrate.

*substitution reaction (4.2)* A chemical reaction in which one atom is substituted for another. For example, in halogenation, we have

$$\text{R—H} + \text{Cl}_2 \longrightarrow \text{R—Cl} + \text{HCl (Cl substituted for H)}$$

*substrate (12.1)* The reagent that is attacked by a nucleophile.

*sugar (Introduction, Chap. 27, 27.1)* A carbohydrate.

*sulfanilamide (Medical Boxed Reading 24.2, 25.3)* The first of the antibiotics.

*surfactant (26.4)* A substance that lowers the surface tension of water. This enables the substance (usually a soap or detergent) to soak into cloth and loosen and remove grease and grime more effectively.

*syndet (26.4)* An abbreviated form of synthetic detergent.

*synthetic detergent (26.4)* A compound prepared by synthetic means. A synthetic detergent is an alternative to soap as a cleaning agent that is effective in hard water.

*tautomers (8.7)* Functional-group isomers that are in equilibrium.

*terminal alkyne (8.1)* An alkyne (RC≡CH) that contains the triple bond between carbons 1 and 2 of the main chain.

*termination step (4.3)* One or more steps in a free-radical mechanism that stops the reaction by eliminating reactive free radicals. For example, in the chlorination of methane,

$$\text{H}_3\text{C}\cdot + \cdot\text{CH}_3 \longrightarrow \text{CH}_3\text{—CH}_3,$$
$$\text{Cl}\cdot + \cdot\text{CH}_3 \longrightarrow \text{CH}_3\text{—Cl} \quad \text{and} \quad \text{Cl}\cdot + \cdot\text{Cl} \longrightarrow \text{Cl—Cl}$$

*terpene (7.9)* A compound that can be formulated by combining units of isoprene.

*tertiary (3.2)* A hydrogen or other atom or group that is bonded to a carbon that, in turn, is bonded to three carbons. The lone hydrogen in isobutane that is not part of a methyl group is tertiary.

*tertiary structure (28.7)* A modification of the secondary structure due to forces such as hydrogen bonding, disulfide bonds, salt bridges, and London forces.

*testosterone (26.6)* A male sex hormone responsible for secondary sex characteristics and tissue and muscle growth.

*tetraene (5.1)* An alkene that contains four carbon-carbon double bonds.

*tetrahedral carbon (2.4)* A carbon atom bonded to four other atoms arranged in the shape of a tetrahedron.

*tetrahedral geometry (2.2)* An alkane shows tetrahedral geometry; i.e., it takes the shape of a regular tetrahedron (see Fig. 2.2).

*tetrapeptide (28.4)* A compound produced when four amino acids are joined together.

*tetrasubstituted alkene (6.1)* An alkene in which four organic groups are bonded to the carbons of the double bond.

*thiol (14.5)* A compound that contains the —SH functional group.

*Tollens test (18.6)* A qualitative test tube reaction shown by aldehydes; also known as the "silver mirror" test. The reagent used is a silver ammonia complex, $^+\text{Ag(NH}_3)_2\ ^-\text{OH}$.

*transition state (12.1)* An unstable species formed during the course of a chemical reaction in which bonds are in the process of being broken and formed.

*triacylglycerol (26.2)* Synonymous with a triglyceride.

*triene (5.1)* An alkene that contains three carbon-carbon double bonds.

*triglyceride (Introduction, Chap. 26, 26.2)* A derivative of glycerol in which three acyl groups replace the three hydroxyl hydrogens of glycerol.

*trigonal carbon (2.4, 5.1)* A carbon atom bonded to three other atoms arranged in the shape of a triangle on a plane.

*trihydric alcohol (14.1)* An alcohol that contains three hydroxyl groups per molecule. A triol.

*tripeptide (28.4)* A compound produced when three amino acids are joined together.

*triplet (29.7)* A grouping of three related peaks.

*trisaccharide (27.1)* A carbohydrate that hydrolyzes to yield three monosaccharide units.

*trisubstituted alkene (6.1)* An alkene in which three organic groups are bonded to the carbons of the double bond.

*two-rule oxidation process (15.2)* Two rules to predict the product in the oxidation of a primary or secondary alcohol.

*unsaturated hydrocarbon (2.1, 5.1, 7.1, 8.1)* A hydrocarbon that contains one or more carbon-carbon double or triple bonds.

*upfield (29.6)* A relatively high frequency of absorption, on the right side of the spectrum.

*valence electron (1.3)* An electron in an atom that participates in bonding.

*van der Waal's forces (1.6)* The combination of London forces and dipole-dipole attractions in a polar covalent molecule.

*Van Slyke procedure (28.10)* An amino acid, peptide, or protein is reacted with nitrous acid to produce a quantitative yield of nitrogen that is used to determine the number of primary amino groups in the compound.

*van't Hoff's rule (11.10)* A mathematical expression by means of which the number of optically active stereoisomers can be calculated for a molecule with two or more dissimilar chiral centers: number of stereoisomers = $2^n$, where $n$ is the number of dissimilar chiral centers in the molecule.

*vicinal dihalide (vic-) (6.2)* A vicinal dihalide (or other disubstituted compound) contains one halogen atom bonded to each of two adjacent carbons.

*vinyl polymer (6.3)* A polymer in which the monomer contains a vinyl group ($H_2C=CH-$) in the form $CH_2=CHG$ or $CH_2=CG_2$.

*visible light (29.1)* Electromagnetic radiation that can be seen by the human eye.

*vital force theory (1.2)* The theory that predicted that organic compounds could only be isolated from living or once-living things.

*vitamin $D_2$ (26.6)* A steroidal fat-soluble vitamin that provides protection from rickets.

*wave number (29.2)* The reciprocal of wave length in $cm^{-1}$. A unit of frequency.

*wavelength (29.1)* The distance between the peak of one wave and the peak of the following wave.

*wax (Introduction, 26.1)* A mixture of esters in which each ester in the mixture contains an acyl fragment ($C_{16}$ to $C_{38}$) and an alkyl fragment ($C_{16}$ to $C_{36}$).

*wedge projection (2.2, 2.3)* A method of representing organic compounds in three dimensions on a two-dimensional page in which a wedged bond projects toward the reader, while a series of dashes represents a bond projecting away from the reader.

*Williamson synthesis (17.4)* A method of preparing ethers in which a primary alkyl halide reacts, via an $S_N2$ mechanism, with alkoxide ion as follows:

$$RX + {}^-OR' \longrightarrow ROR' + X^- \quad \text{(where X = Cl, Br, or I)}$$

*wood alcohol (14.1)* A trivial name for methanol.

*x-ray computed tomography (Medical Boxed Reading 29.1)* A more advanced form of x-ray technology in which much clearer pictures of blood vessels, muscle, and organs can be obtained than in the standard x-ray.

*x-ray radiation (29.1)* An energetic type of electromagnetic radiation but a little less energetic than gamma radiation.

*(Z)-(E) zusammen-entgegen method of specifying configuration (5.3)* A means of specifying configuration, more general than the *cis-trans* method, based on a system of priority rankings.

*zwitterion (28.2)* A synonym for a dipolar ion.

# Answers to Selected Chapter Exercises and Problems

## Chapter 1 Exercises

1.1. $H = 1$; $C = 4$; $Ne = 8$

1.3. a. 

$$H-N\overset{\displaystyle O}{\underset{\displaystyle O^-}{}} \qquad H-\overset{+}{N}\overset{\displaystyle O^-}{\underset{\displaystyle O}{}}$$

b.

$$H-\overset{\overset{\displaystyle H}{|}}{\underset{\underset{\displaystyle H}{|}}{C}}-\overset{\overset{\displaystyle H}{|}}{\underset{\underset{\displaystyle H}{|}}{C}}-\overset{\overset{\displaystyle H}{|}}{\underset{\underset{\displaystyle H}{|}}{C}}-Cl \qquad H-\overset{\overset{\displaystyle H}{|}}{\underset{\underset{\displaystyle H}{|}}{C}}-\overset{\overset{\displaystyle H}{|}}{\underset{\underset{\displaystyle Cl}{|}}{C}}-\overset{\overset{\displaystyle H}{|}}{\underset{\underset{\displaystyle H}{|}}{C}}-H$$

c.

$$\overset{\displaystyle H}{\underset{\displaystyle H}{}}N-N\overset{\displaystyle H}{\underset{\displaystyle H}{}}$$

1.5. Ionic charge of molecule: $0 \cdot 3(H) + 0 \cdot 1(C) + 0 \cdot 1(N) + 0 \cdot (O_1) + 0 \cdot 1 (O_2) = 0$

1.7. a. Functional group isomers
b. Positional isomers

1.9. a. $RC{\equiv}CH + {}^-NH_2 \longrightarrow R-C{\equiv}C^- + NH_3$
b. $HC_2H_3O_2 + CH_3O^- \longrightarrow C_2H_3O_2{}^- + CH_3OH$
c. $C_6H_5OH + Cl^- \longrightarrow$ No reaction

## Chapter 1 Problems

1. Sugar cane and sugar beets, from which sucrose is obtained, are both from living sources, the cane sugar and beet sugar plants. Chemists in the 1700s believed that organic compounds could be obtained only from living or once-living sources, because a vital force was present in living things that enabled them to synthesize organic compounds.

3. a. Carbon: $1s^2 2s^2 2p^2$ $\cdot\overset{..}{C}\cdot$
   Nitrogen: $1s^2 2s^2 2p^3$ $\cdot\overset{..}{N}:$
   Oxygen: $1s^2 2s^2 7p^4$ $:\overset{..}{O}\cdot$

5. a. 5
   b. 2
   c. 7
   d. 1

7. a. C is greater than Si.
   b. F is greater than C.
   c. Br is greater than I.

9. a. $C^{\delta+}-O^{\delta-}$
   b. $H^{\delta+}-S^{\delta-}$
   c. $B^{\delta+}-I^{\delta-}$
   d. $C^{\delta+}-N^{\delta-}$
   e. $C^{\delta+}-S^{\delta-}$

11. a. Organic; b. inorganic; c. inorganic; d. organic; e. organic

13. a. Dipole-dipole; b. London forces; c. London forces

15. 
   | | Semicondensed | Condensed |
   |---|---|---|
   | a. | $CH_3-\underset{\underset{\displaystyle Cl}{|}}{CH}-CH_3$ | $(CH_3)_2CHCl$ |
   | b. | $CH_3-CH_2CH{=}CH_2$ | $CH_3CH_2CH{=}CH_2$ |

17. a.

$$H-\overset{\overset{\displaystyle H}{|}}{C}-\overset{\overset{\displaystyle H}{|}}{\underset{\underset{\displaystyle H-\overset{\displaystyle C}{|}-H}{}}{C}}-\overset{\overset{\displaystyle H}{|}}{C}-\overset{\overset{\displaystyle H}{|}}{C}-\overset{\overset{\displaystyle H}{|}}{C}-\overset{\overset{\displaystyle H}{|}}{C}-H$$

b.

$$H-\overset{\overset{\displaystyle H}{|}}{\underset{\underset{\displaystyle H}{|}}{C}}-\overset{\overset{\displaystyle H}{|}}{\underset{\underset{\displaystyle H}{|}}{C}}-\overset{\overset{\displaystyle H}{|}}{\underset{\underset{\displaystyle H}{|}}{C}}-\overset{\overset{\displaystyle H}{|}}{\underset{\underset{\displaystyle H}{|}}{C}}-\overset{\overset{\displaystyle H}{|}}{\underset{\underset{\displaystyle H}{|}}{C}}-\overset{\overset{\displaystyle H}{|}}{\underset{\underset{\displaystyle H}{|}}{C}}-\overset{\overset{\displaystyle H}{|}}{\underset{\underset{\displaystyle H}{|}}{C}}-\overset{\overset{\displaystyle H}{|}}{\underset{\underset{\displaystyle H}{|}}{C}}-H$$

19. a. $O{=}N-O-O-N{=}O$

b.

$$H-\overset{\overset{\displaystyle H}{|}}{\underset{\underset{\displaystyle H}{|}}{C}}-C\overset{\displaystyle O}{\underset{\displaystyle OH}{}} \qquad \overset{\displaystyle H}{\underset{\displaystyle H}{}}C{=}C\overset{\displaystyle OH}{\underset{\displaystyle OH}{}} \qquad H-\overset{\overset{\displaystyle H}{|}}{\underset{\underset{\displaystyle OH}{|}}{C}}-C\overset{\displaystyle O}{\underset{\displaystyle H}{}}$$

c.

$$\overset{\displaystyle H}{\underset{\displaystyle H}{}}C{=}\overset{\overset{\displaystyle H}{|}}{C}-\overset{\overset{\displaystyle H}{|}}{C}{=}C\overset{\displaystyle H}{\underset{\displaystyle H}{}} \qquad \overset{\displaystyle H}{\underset{\displaystyle H}{}}\underset{\underset{\underset{\displaystyle H}{|}}{H-C-H}}{\overset{\overset{\displaystyle H}{|}}{H-C{-}H}}{}$$

(cyclobutene structure)

$$H-C{\equiv}C-\overset{\overset{\displaystyle H}{|}}{\underset{\underset{\displaystyle H}{|}}{C}}-\overset{\overset{\displaystyle H}{|}}{\underset{\underset{\displaystyle H}{|}}{C}}-H \qquad \overset{\displaystyle H}{\underset{\displaystyle H}{}}C{=}C{=}\overset{\overset{\displaystyle H}{|}}{C}-\overset{\overset{\displaystyle H}{|}}{\underset{\underset{\displaystyle H}{|}}{C}}-H$$

$$H-\overset{\overset{\displaystyle H}{|}}{\underset{\underset{\displaystyle H}{|}}{C}}-C{\equiv}C-\overset{\overset{\displaystyle H}{|}}{\underset{\underset{\displaystyle H}{|}}{C}}-H \qquad \underset{\underset{\displaystyle H}{|}\underset{\displaystyle H}{|}}{\overset{\overset{\displaystyle H-C-H}{}}{\underset{\underset{\displaystyle H-C-C-H}{}}{C}}}$$

(cyclopropene structure)

d. $H-\overset{\overset{\displaystyle H}{|}}{\underset{\underset{\displaystyle H}{|}}{C}}-C{\equiv}N \qquad H-C{\equiv}C-\overset{\overset{\displaystyle H}{|}}{\underset{\underset{\displaystyle H}{|}}{N}} \qquad \overset{\displaystyle H}{\underset{\displaystyle H}{}}C{=}C{=}N\overset{\displaystyle H}{\underset{\displaystyle}{}}$$

e. None are possible.

f.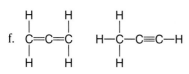

g.

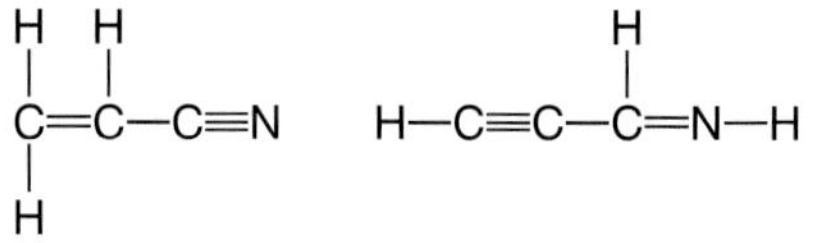

h.

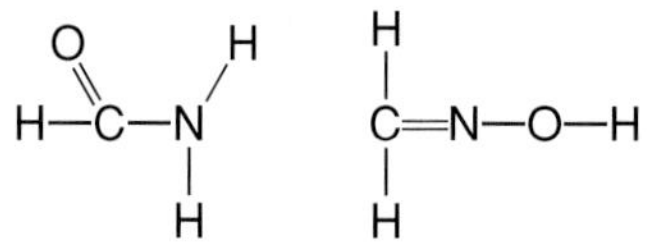

i.

21.

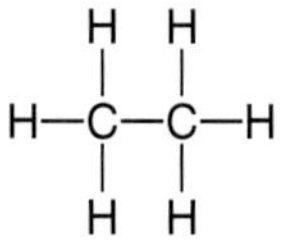

Nonpolar

23.  a. ion; b. ion; c. ion; d. ion

25.  False.

27.  $0 \cdot 1(C) + 0 \cdot 1(O) = 0 \therefore$ neutral compound

29.  a. Alkene or olefin

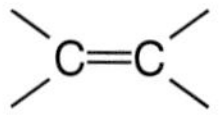

   Alcohol          —OH

   b. Aromatic hydrocarbon

   Amine            —NH₂

   c. Aldehyde

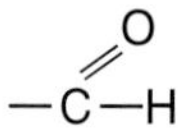

   Alcohol          —OH

   d. Aromatic hydrocarbon

   Ether            C—O—C

   Amine

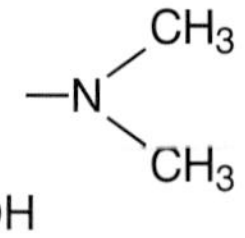

   e. Phenol

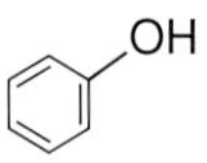

   Amine            N—CH₃

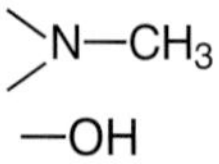

   Alcohol          —OH

   Ether            C—O—C

   Alkene or olefin

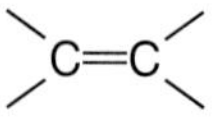

f.  Amide

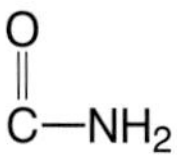

   Amine            —NH₂

g.  Carboxylic acid    —C—OH

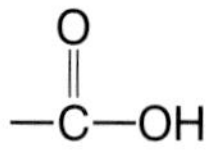

h.  Ester    C—C—O—C

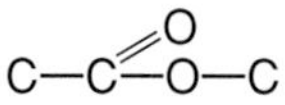

   Note: Every compound shown (except urea) also can be listed as having alkane function.

31.  a. 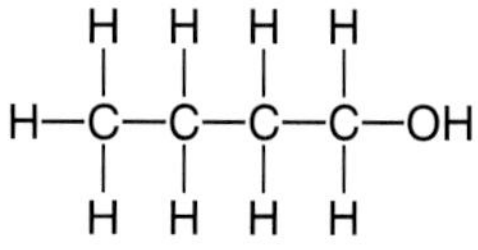 H—C—C—C—C—OH   CH₃(CH₂)₂CH₂OH

   H—C—C—C—C—H   CH₃CH(OH)CH₂CH₃

   H—C—C—C—H   (CH₃)₃COH

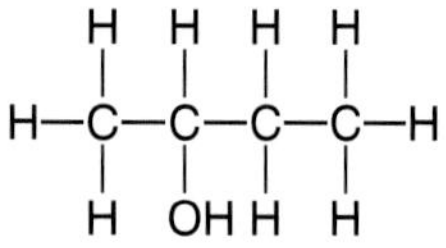

   H—C—C—C—OH   (CH₃)₂CHCH₂OH

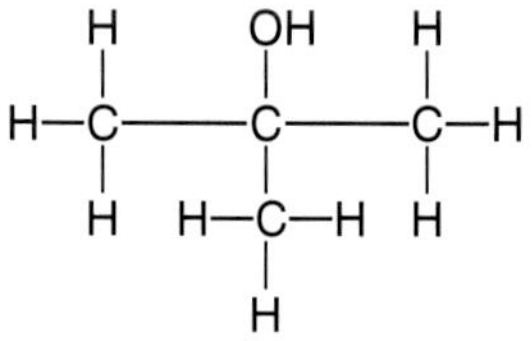

   b. 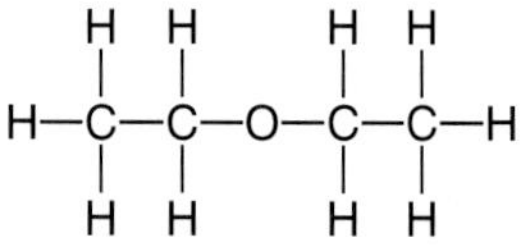 H—C—C—O—C—C—H   (CH₃CH)₂O

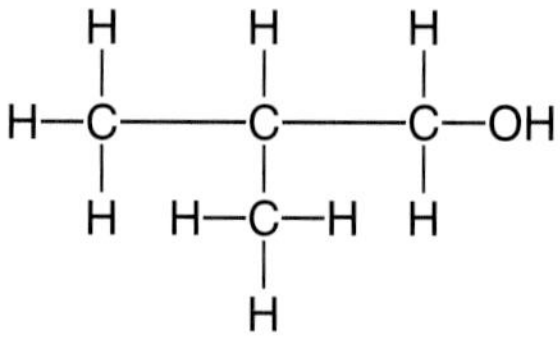

   H—C—O—C—C—C—H   CH₃—O—CH₂CH₂CH₃

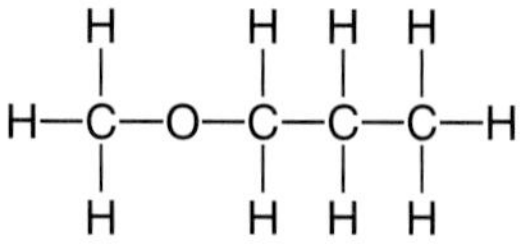

   H—C—O—C—C—H   CH₃—O—CH(CH₃)₂

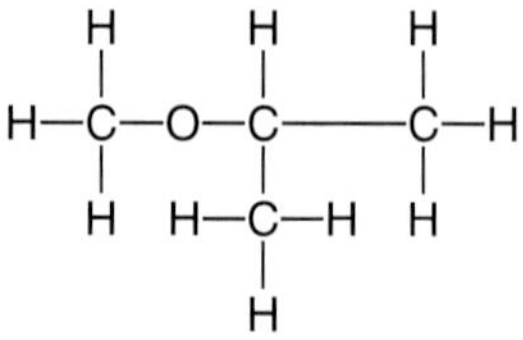

33.

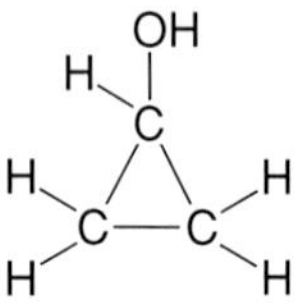

H—C—O
|
H—C—C—H
|     |
H    H

(square with O)

H   H   H H
 \ /   \  |
  C—C    C—H
 /|  \   |
H H   H  H

(epoxide structure) —CH₃

H       H H
|       |  |
H—C—O—C≡C        CH₃—O—CH=CH₂
|       |
H       H

35. a. Unrelated; b. Identical; c. Unrelated

37. a. $HC_2H_3O_2$ is a stronger acid than $CH_3CH_2OH$.
    b. $NH_3$ is a stronger acid than $CH_3CH_2CH_3$.

39. a. $CH_3OH$
    b. HBr

41. a. $-1$; $+1$; b. $-1$; $+3$

## Chapter 2 Exercises

2.1.
H   H   H  H   H
|   |   |  |   |
H—C=C—C—C—C—H
  1   2 | 3 |4 |5
        H   H  H

a. 2; b. 1; c. 1; d. 7; e. 1; f. 2

## Chapter 2 Problems

1. a. Saturated; b. Unsaturated—alkene or olefin; c. Unsaturated—alkyne

3. a. No pi bonds; b. One pi bond; c. Two pi bonds

5. The $2s^2$ electrons are mixed with the $2p^2$ electrons to form four bonds with equal energy. These are known as $sp^3$ bonds. For the hydrogen atoms of methane to be equidistant and as far away from each other as possible, they must be at the corners of a tetrahedron. It is not possible for all four hydrogen to be equidistant from each other in a planar structure.

7.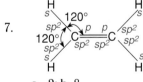

a. 2; b. 8

9.

| a. Bond Type | 1-Butene | 2-Butene |
|---|---|---|
| $p-p$ | 1 ($C_1-C_2$) | 1 ($C_2-C_3$) |
| $sp^2-s$ | 3 (C—H$_{0n}$ C$_1$ + C$_2$) | 2 (C—H$_{0n}$ C$_2$—C$_3$) |
| $sp^3-s$ | 5 (C—H$_{0n}$ C$_3$ + C$_4$) | 6 (C—H$_{0n}$) C$_1$ + C$_4$ |
| $sp^2-sp^2$ | 1 ($C_1-C_2$) | 1 ($C_2-C_3$) |
| $sp^2-sp^3$ | 1 ($C_2$) | 2 ($C_2$ + C) |
| $sp^3-sp^3$ | 1 ($C_3$ to $C_4$) | 0 |

11. The double bond is made up of an $sp^2=sp^2$ (sigma) bond and a π bond. The $sp^2=sp^2$ has end-to-end overlap and is very strong; the π bond ($2p-2p$) is divided into two lobes, one lobe below the sigma bond and the other above the sigma bond.

13. a. Sigma bonds
    b. π bonds
    c. Sigma bonds

15.
① C≡② C—③ C (with H substituents)

H——C≡C—C(H)(H)(H)          $C_1=C_2-C_3≡C_4$

a. 1; b. 1; c. 1; d. 3; e. 1; f. 3

## Chaper 3 Exercises

3.1.

CH₃  60°
H        CH₃        Newman
                    projections
H        H
   H

Less stable
(a)

CH₃  180°
H        H
H        H
   CH₃

More stable
(b)

Staggered conformations of butane: Conformation (a) is less stable than conformation (b). In conformation (a), the methyl groups are at a 60-degree angle, which allows for more overlap (London repulsion) than conformation (b), in which the methyl groups are 180 degrees apart.

3.3 a. 3-Methylhexane; b. 1-Bromoheptane

3.5 a. 2,3,3-Tribromo-2-chloropentane
    b. 5-Bromo-2,6-dichloro-4,4-diiodo-2-nitrooctane

3.7 a. Ethane; b. 2,2-Dimethylpentane;
    c. 3,3-Dimethylhexane

3.9 a. 1-Cyclopropylpentane
    b. 1-Bromo-4-methyl-2-isopropylcyclohexane

3.11 a. 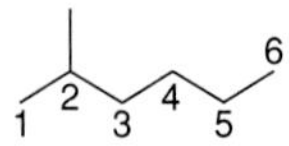 　2-methylhexane

　　b. 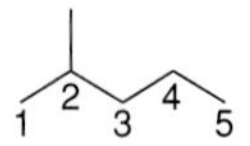 　isohexane

　　c. 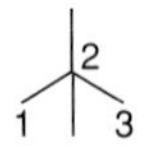 　2,2-dimethyl propane

## Chapter 3 Problems

1.
Newman projections

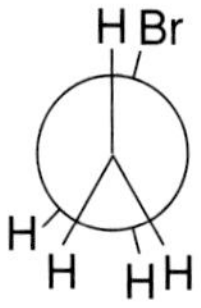

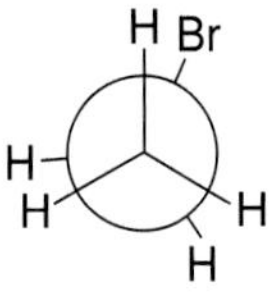

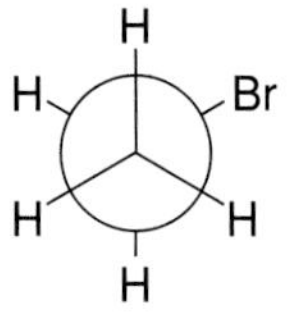

Wedge projections

 　　　 　c. awkward to draw

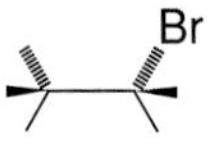

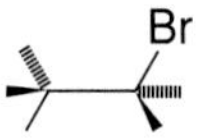

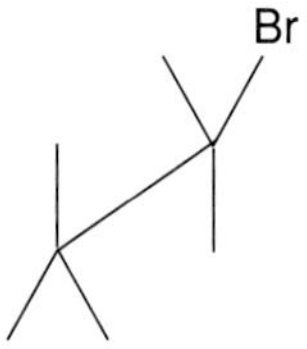

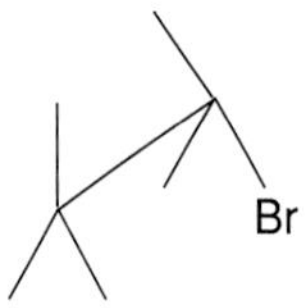

3.　Three.

5.　a, b.

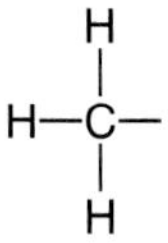

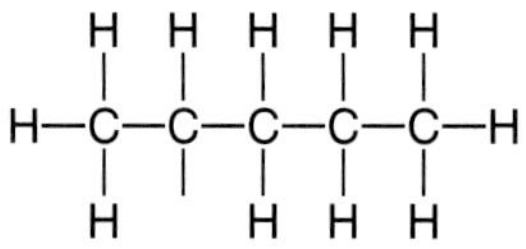

_sec_-pentyl (secondary)

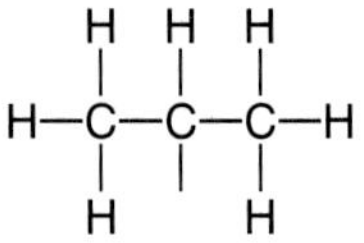

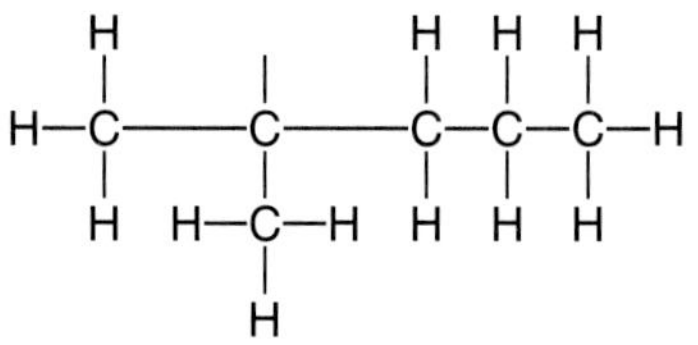

_tert_-isohexyl
(tertiary)

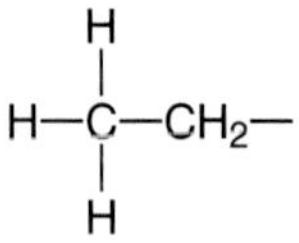

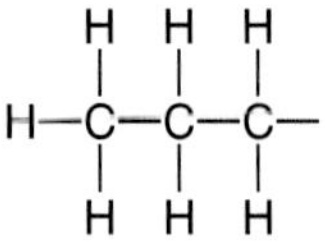

7.　Yes.

9.　a. 9 primary, 4 secondary, and 1 tertiary; b. 9 primary, 2
　　secondary, and 1 tertiary

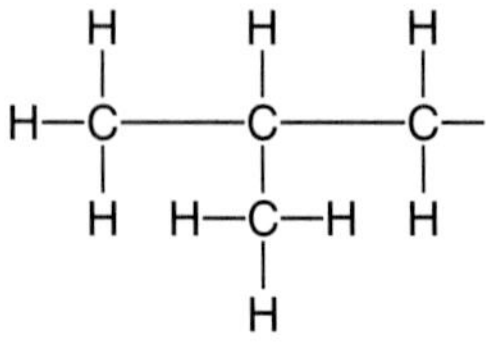

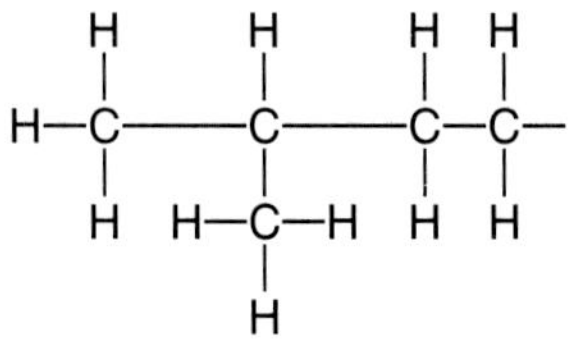

11.　a. 9 primary, 2 secondary, and 1 tertiary; b. 12 primary;
　　c. 6 primary and 6 secondary

13.　a.

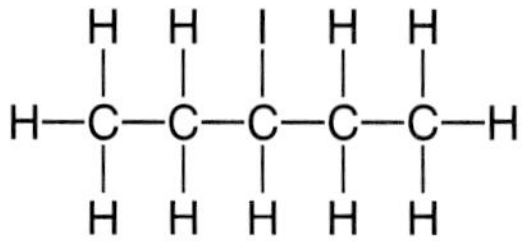

　　b.

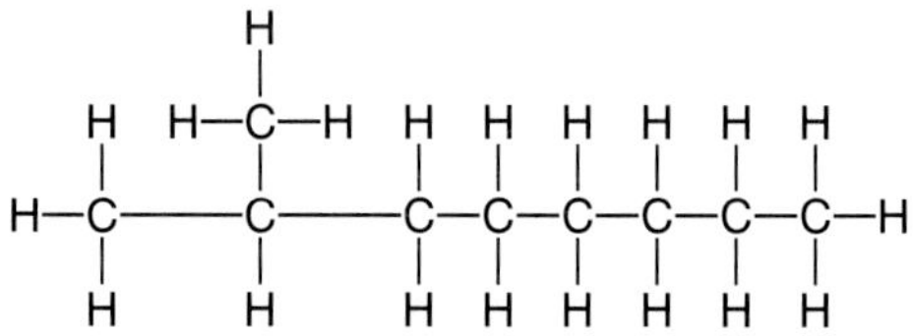

15.　a. 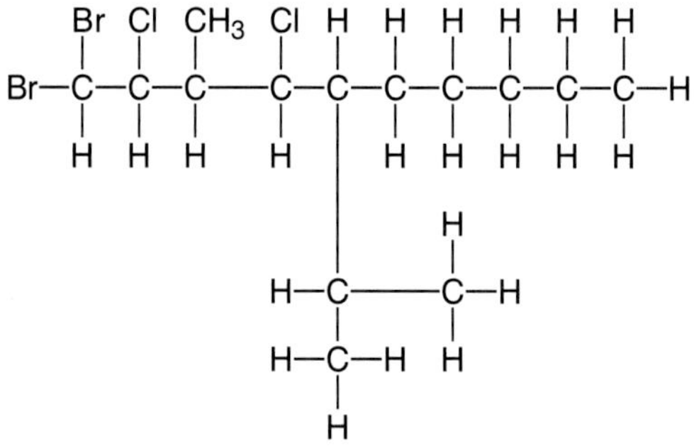 $Br-$

　　b.

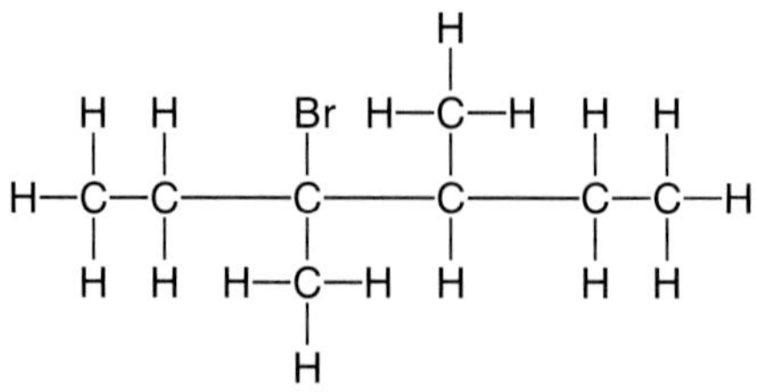

17.　a. (lowest) $(CH_3)_3CCH_2CH_3$ < $(CH_3)_2CHCH_2CH_2CH_3$
　　　< $CH_3(CH_2)_4CH_3$ (highest)
　　b. (lowest) Ethane < _n_-butane < _n_-heptane (highest)
　　c. (lowest) Ethane < isobutane < neopentane < iso-
　　　hexane < _n_-hexane (highest)

19.　a. Eclipsed conformation: The formation in which the
　　　hydrogen atom or a substituent one-one carbon is
　　　directly behind the hydrogen atom or a substituent
　　　of the front (adjacent) carbon atom.
　　b. Alkyl group: An alkyl group results when one hydro-
　　　gen is removed from an alkane.

21.  a.  $CH_3(CH_2)_{21}CH_3$ or

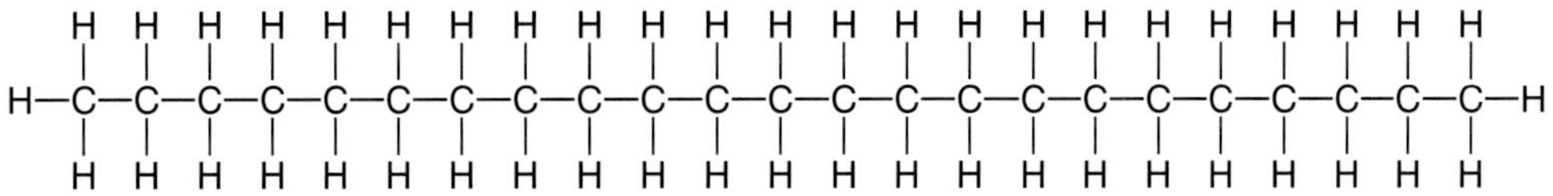

b.  $CH_3(CH_2)_{34}CH_3$ or

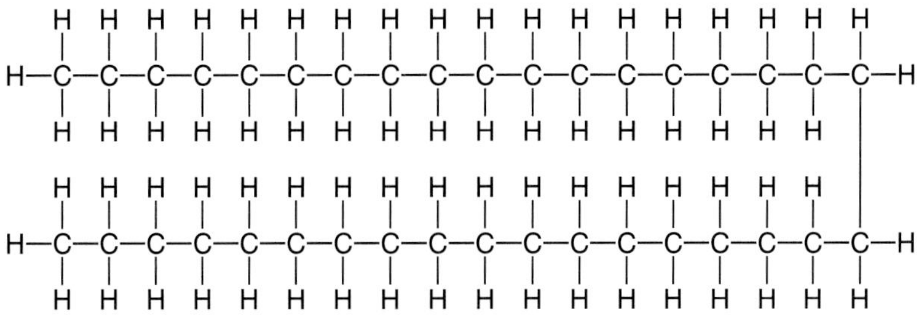

c.  $CH_3(CH_2)_{39}CH_3$ or

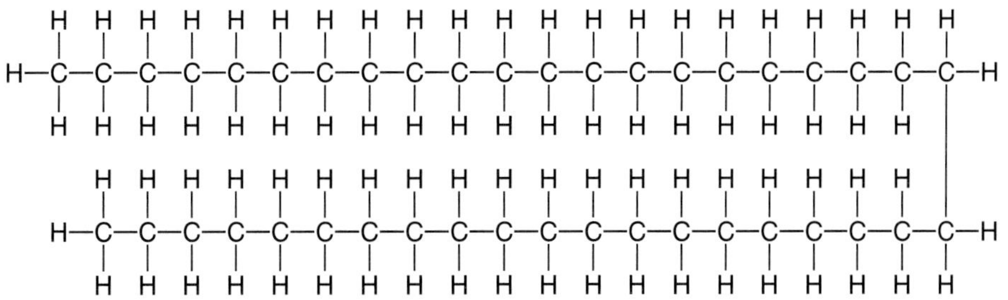

23.  Alkanes are generally thought to be ingested as part of our diet. Food may be coated with paraffin wax as a preservative. Mineral oil is used as a laxative. Alkanes also can be absorbed through the skin. Vasoline, for example, has been used for many years as an ingredient in cleansing creams, as a moisturizer, to prevent chapping, and to treat diaper rash.

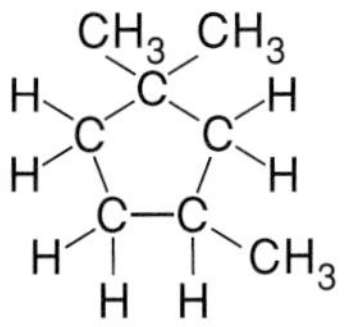

b.

25.  a.

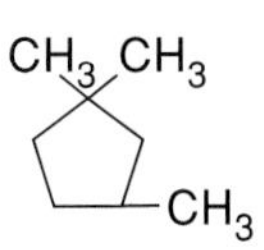

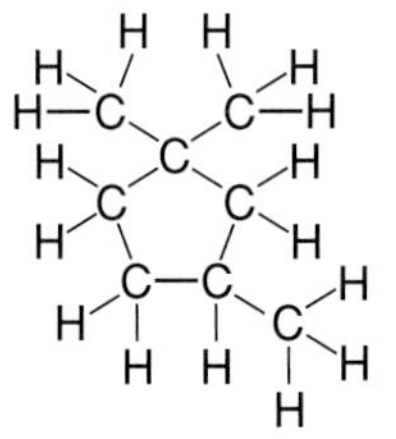

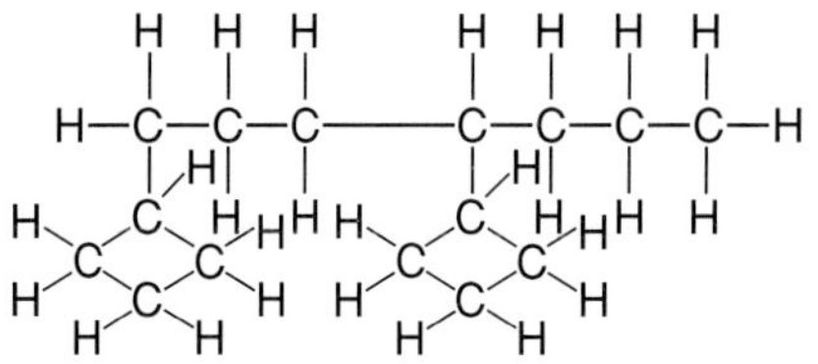

c.

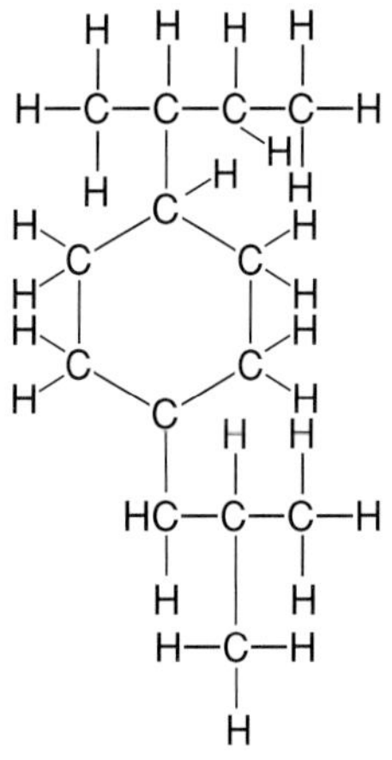

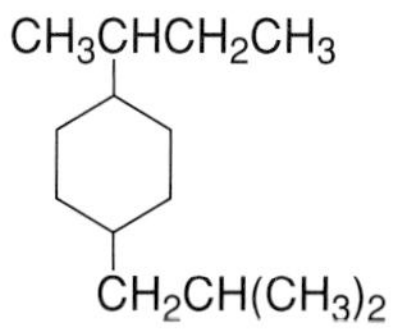

**27.**　**a.**

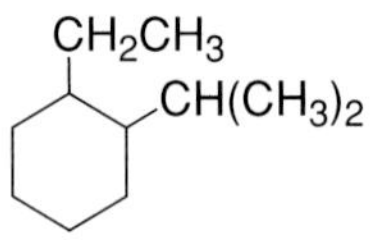

or

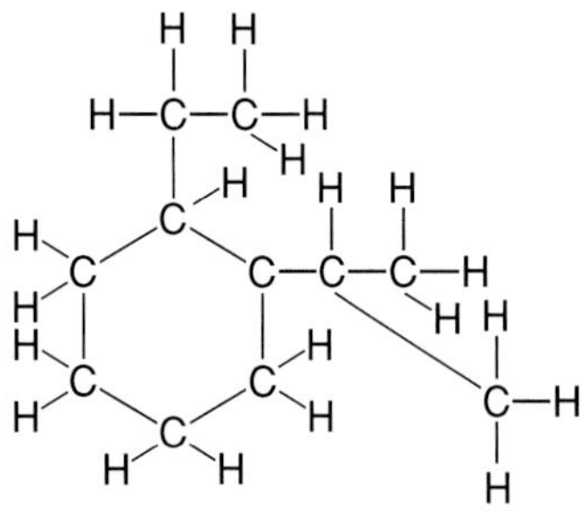

**b.**

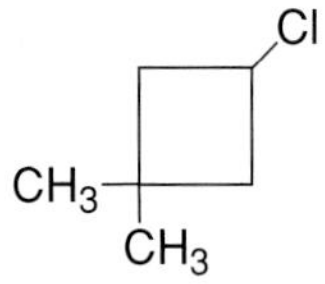

or

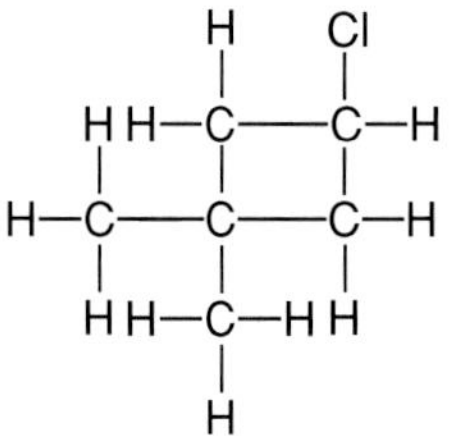

**c.**

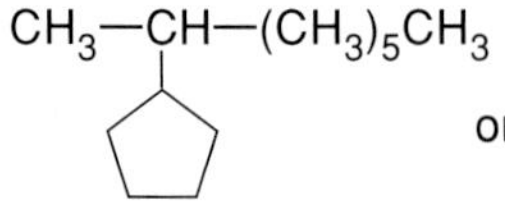

$CH_3-CH-(CH_3)_5CH_3$

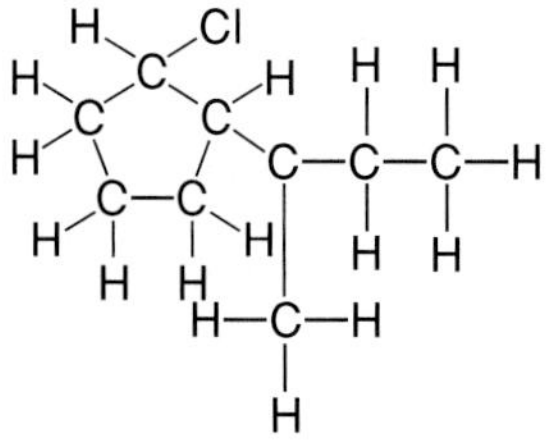

or

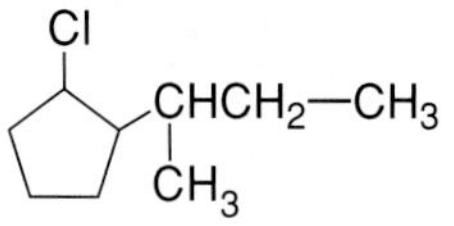

*2-sec* Butyl-1-chlorocyclopentane

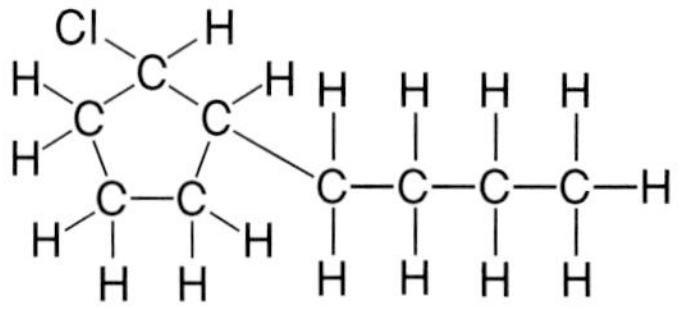

or

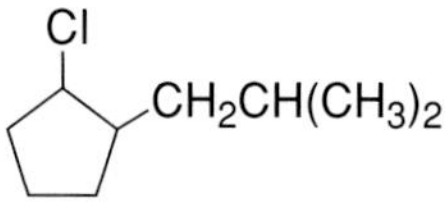

*2-iso*-Butyl-1-chlorocyclopentane

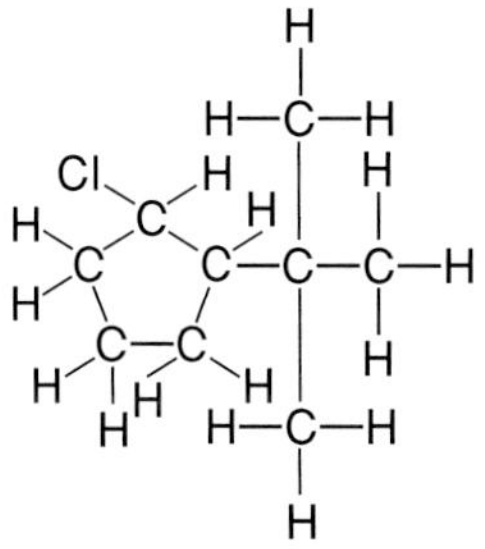

or

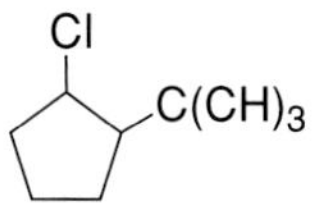

*2-tert*-Butyl-1-chlorocyclopentane

Without a prefix for the butyl group, it is not possible to know which of the preceding compounds is the correct one.

**c.** *n*-Pentylcyclopropane should be 1-cyclopropyl-pentane.

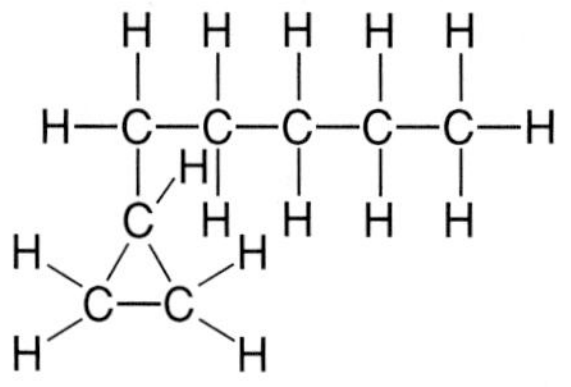

or

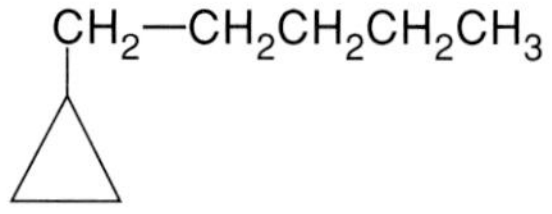

If the ring compound has fewer carbons than the alkane, the ring compound is the substitutent, and the name is based on the alkane.

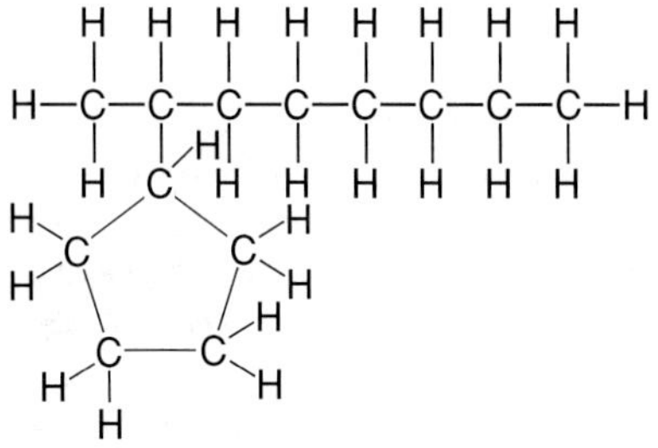

**29.**　**a.** 2-Methyl-1-ethylcycloheptane should be 2-ethyl-1-methylcycloheptane.

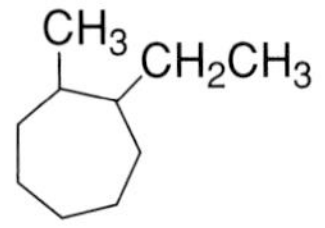

or

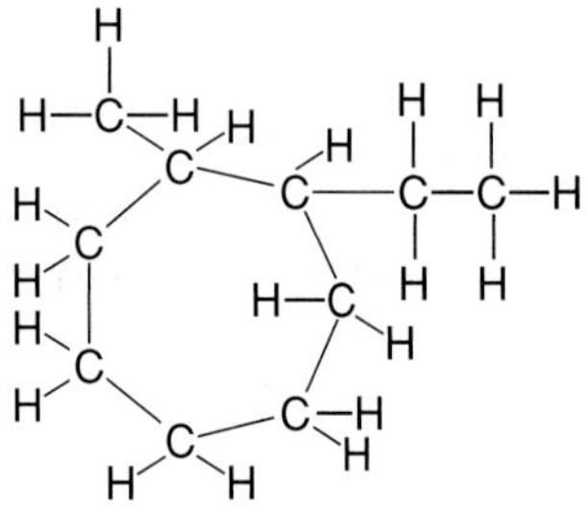

The substituents must be listed alphabetically for the name to be correct.

**b.** 2-butyl-1-chlorocyclopentane could be one of the following compounds:

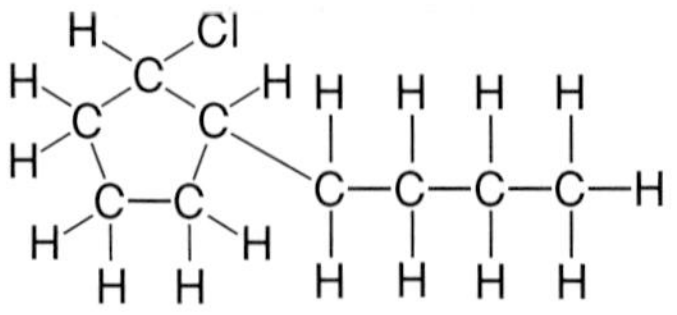

or

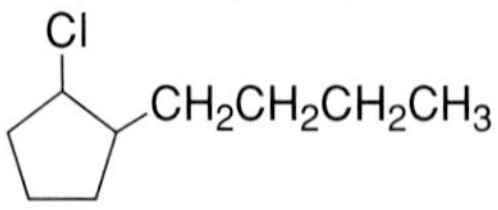

2*n*-Butyl-1-chlorocyclopentane

**31.**　**a.**

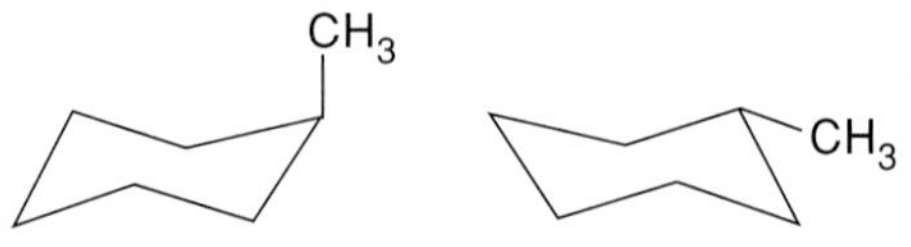

Axial

Equatorial

**33.**　**a.**

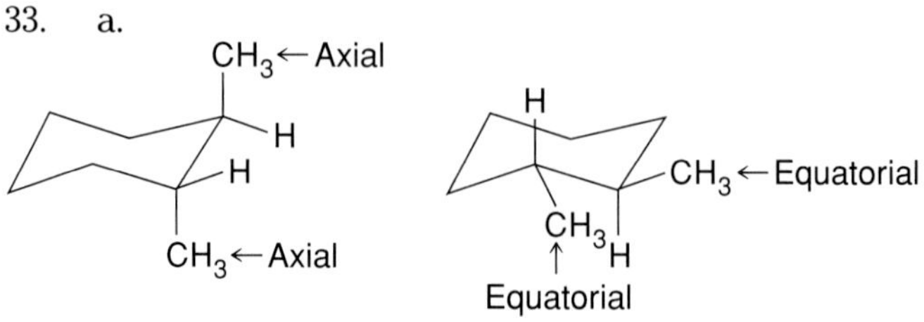

Diaxial

Diequatorial

35. a. [structure] Pentane

    b. [structure] Isopentane

    c. [structure] Cyclopropylcyclopentane

37. a. [structure] Octane

    b. [structure] 3-Ethylpentane

## Chapter 4 Exercises

4.1. a. $2\,CH_3CH_2Br \xrightarrow{\text{Li}} 2\,CH_3CH_2Li \xrightarrow{\text{Cu}} (CH_3CH_2)_2CuLi$

     $CH_3CH_2CH_2Br \longrightarrow CH_3CH_2Cu + CH_3CH_2CH_2CH_2CH_3$

    b. $2\,CH_3CH(I)CH_3 \xrightarrow{\text{Li}} 2[(CH_3)_2CH]Li \xrightarrow{\text{Cu}}$
     $[CH_3)_2CH]_2CuLi \xrightarrow{CH_3Cl}$

     $(CH_3)_2CH\,Cu + (CH_3)_2CHCH_3$

4.3. a. $(CH_3)_2CHCH_2^+$

    b. $CH_3{-}CH$
          $\;\;\;\;\;\;|$
          $\;\;\;CH_2CH_3$

    c. $CH_3CH_2CH_2CH_2CH_2{:}^{-}$

## Chapter 4 Problems

1. a. $CH_3CH_2{:}^{-}$
    b. $CH_3\overset{.}{C}HCH_3$

    c. $CH_3{-}\overset{\overset{\textstyle CH_3}{|}}{\underset{\underset{\textstyle CH_3}{|}}{C}}{-}CH_2^+$

3. For a fuel to be useful, energy must be produced when the compound being used as a fuel reacts with the oxygen in air. If a reaction is endothermic, energy (heat) is required from the surroundings.

5. a. Carbocation: A carbon atom that shares only six electrons and therefore has a positive charge.
    b. Exothermic reaction: A reaction during which heat (energy) is produced.
    c. Free radical: A neutral species in which a carbon has one unshared electron.

7. [reaction mechanism scheme]

$Cl{:}Cl \xrightarrow[h\nu]{\Delta} 2\,Cl\cdot$

$Cl\cdot + H{:}CH_3 \longrightarrow HCl + \cdot CH_3$

$Cl{\cdot}\,CH_3 \longrightarrow Cl{-}CH_3$

$\cdot CH_3 + \cdot CH_3 \longrightarrow CH_3{-}CH_3$

9. a. [cyclopentene] $+\ H_2 \xrightarrow[\text{or}\,(Pt,Pd)]{\text{Ni}}$ [cyclopentane]

    b. [cyclopentyl]–Br $+\ Mg \xrightarrow{\text{Anhydrous diethyl ether}}$ [cyclopentyl]–MgBr $\xrightarrow[\text{HCl}]{H_2O}$ [cyclopentane]

    c. [cyclohexyl]–Br $+\ Li \xrightarrow{\text{Anhydrous diethyl ether}}$ [cyclohexyl]–Li $+\ CuI \longrightarrow$

     [cyclohexyl]–$CH_3 \xleftarrow{CH_3Br}$ ([cyclohexyl])$_2$CuLi

11. a. [cyclopentane] $+\ O_2 \longrightarrow CO_2 + H_2O$

    b. [cyclopentene] $+\ H_2 \xrightarrow{\text{Ni}}$ [cyclopentane]

    c. [cyclobutane] $+\ H_2 \xrightarrow[\Delta]{\text{Pt}} CH_3CH_2CH_2CH_3$

    d. [cyclohexyl]–Br $+\ Li \xrightarrow{\text{Anhydrous diethyl ether}}$ [cyclohexyl]–Li $+\ LiBr$

    e. $BrCH_2CH_2CH_2Br + Zn \xrightarrow{\Delta}$ [cyclopropane] $+\ ZnBr_2$

13. [cyclopropane] $+\ Br_2 \xrightarrow{h\nu}$ [cyclopropyl]–Br $+\ HBr$ major

    [cyclopropane] $+\ Br_2 \longrightarrow CH_2{=}CH{-}CH_2Br + HBr$ minor

15. a. *sec*-Butyl iodide
    b. Methylene bromide
    c. *tert*-Butyl chloride

## Chapter 5 Exercises

5.1. a.

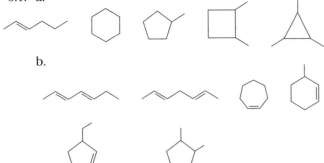

b.

5.3   a. Isolated; b. Conjugated; c. Conjugated

5.5   a. no; b. no; c. yes; d. no

5.7   a. *trans*-a,e; b. *trans*-a,e; c. *trans*-e,e; d. *trans*-a,a

5.9. a.   ⬠ + H$_2$  →(Ni)→  ⬠

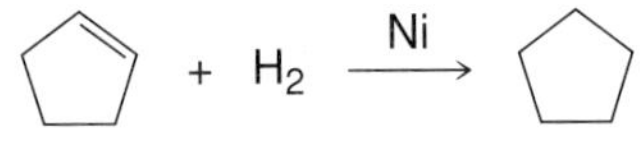

b.   CH$_2$=CH$_2$  →(Polymerization)→  $\left(\text{C-C}\right)_n$

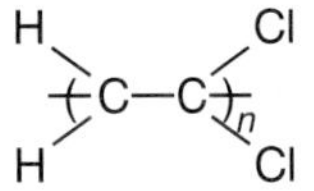

c.   CH$_3$(H)C=C(H)CH$_3$  →(1. H$_2$SO$_4$)→  [CH$_3$—CH$_2$—CH—CH$_3$ ; OSO$_3$H]

→(2. H$_2$O / Δ)→  CH$_3$—CH$_2$CH(OH)CH$_3$

d.   CH$_3$(H)C=C(H)CH$_3$  →(H$_2$SO$_4$)→  CH$_3$—CH$_2$—CH—CH$_3$ ; OSO$_3$H

e.   CH$_3$CH$_2$CH$_2$OSO$_3$H + H$_2$O →(Δ)→ CH$_3$CH$_2$CH$_2$OH + H$_2$SO

## Chapter 5 Problems

1.   a. Isolated; b. conjugated; c. isolated; d. conjugated;
     e. conjugated; f. isolated; g. isolated and conjugated

3.   a. C$_7$H$_{10}$; b. C$_7$H$_{10}$; c. C$_7$H$_{10}$

5.   C$_7$H$_{12}$

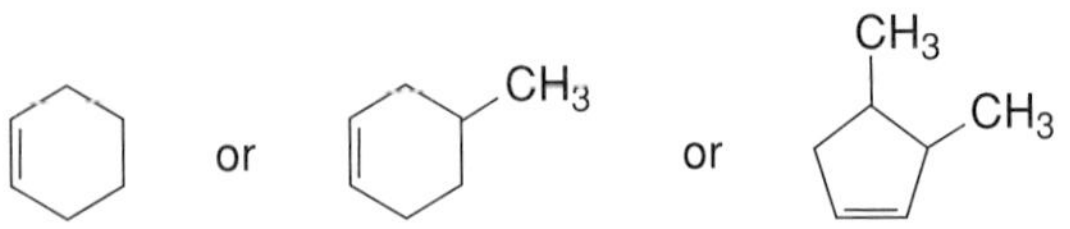

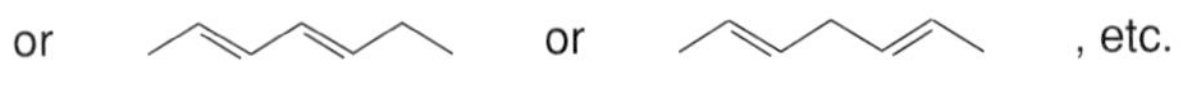

   or ... or ..., etc.

7.   a.

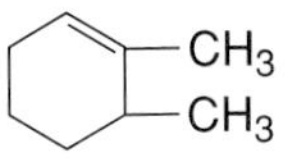

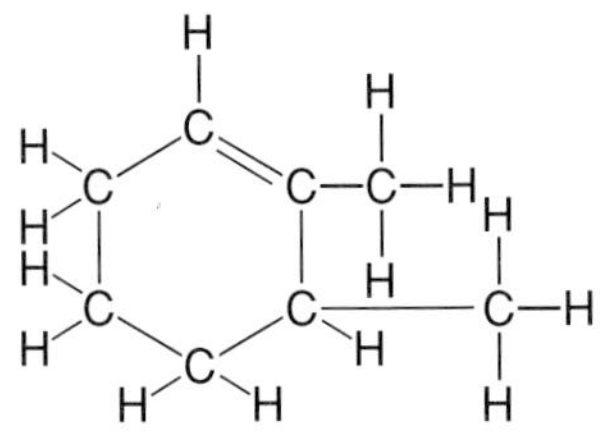

b.   CH$_2$=CHBr   or

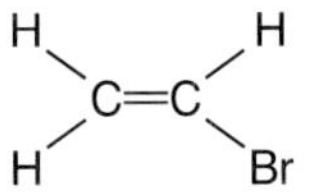

c.   CH$_3$—CH$_2$CH=CH(CH$_2$)(CH$_2$)$_{11}$CH$_3$   or

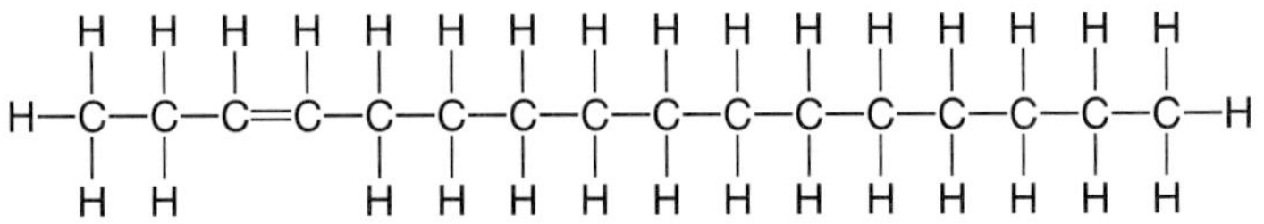

d.   CH$_2$=C—CH—CH$_2$CH$_2$CH$_3$   or
         |      |
        CH$_3$  CH$_3$

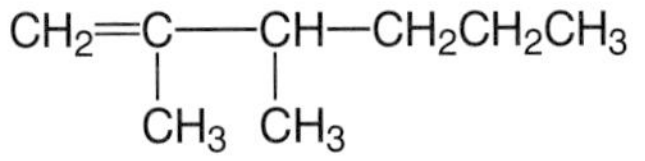

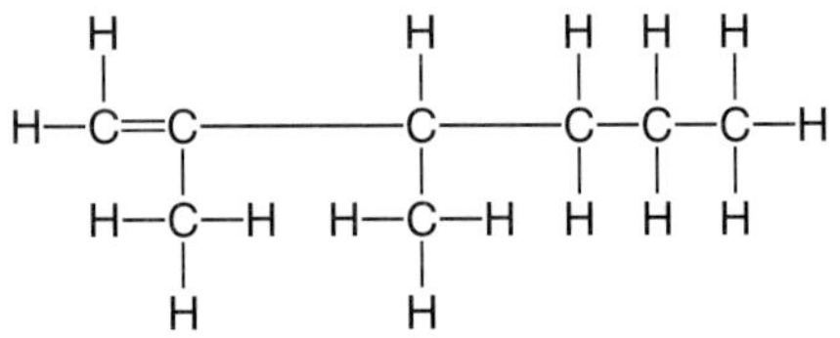

e. 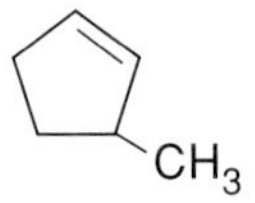   or

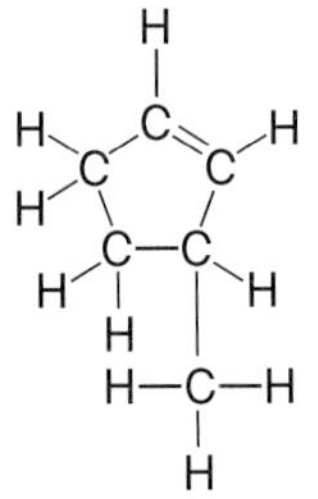

f.   CH$_2$=CH—CH$_2$I   or

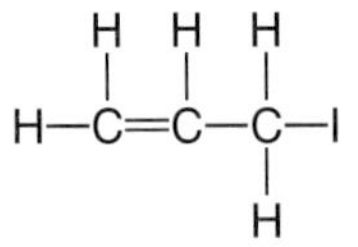

g.   CH$_2$=CH(CH$_2$)$_{19}$CH$_3$   or

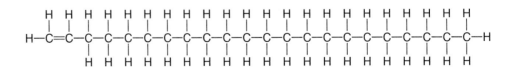

9.   a. All vinylic    b. 1,2,3,6,7,8 Allylic
                          4,5 vinylic
     c. 1,2,3,5,6,9,10 allylic
        4 vinylic
        7,8 secondary

11.  b. *cis*-1-Bromo-2-nitrocyclohexane; d. (*E*)-1-Bromo-1-
        chlorobutene; e. (*z*)-3-Nitro-2-pentene

13.  (3). (*E*)-1-Chloro-1-fluoro-2- iodopropene
     (4). (*cis*)-2-pentene
     (9). *trans*-7-Methyl-3- tridecene

15.  a.

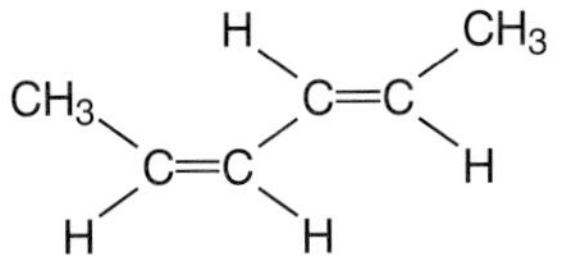

     b.

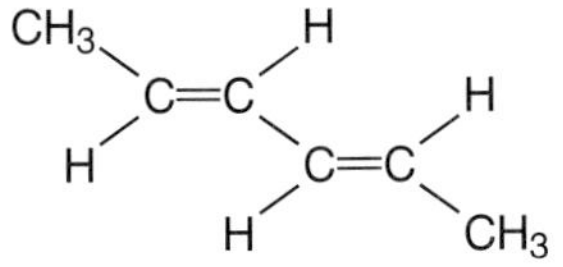

     c.

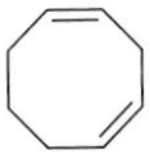

17.  a.

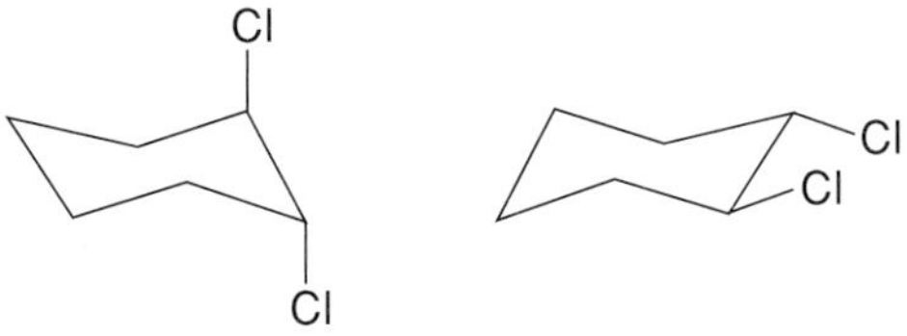

     Diaxial (*trans*)          Diequatorial (*trans*)

     Diequatorial is the preferred conformation.

     b.
        Equatorial        Cl ← Axial

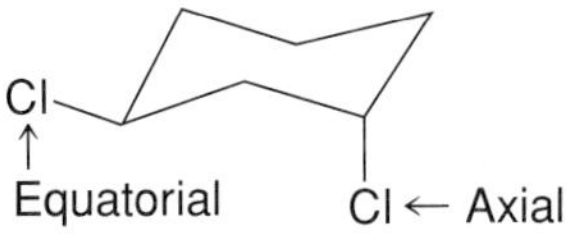

     c.

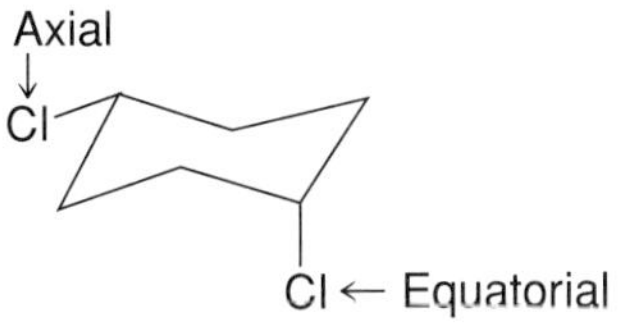

19.  a. *cis*-3-Hexene; b. 1-Octene; c. 2-Methyl-2-butene.

21.  a. CH$_3$—CH=C(CH$_3$)—CH$_2$—CH$_3$  or

        CH$_2$=CH—CH(CH$_3$)—CH$_2$—CH$_3$

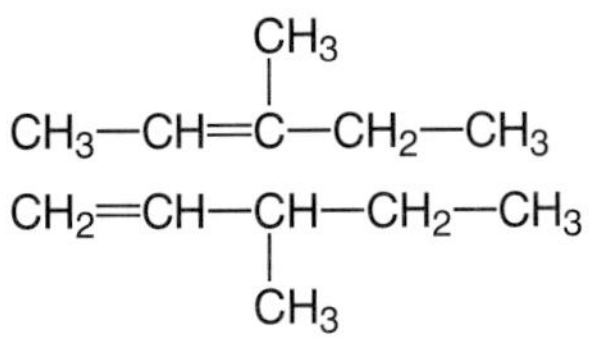

     b. CH$_3$—CH(OSO$_3$H)—CH$_3$

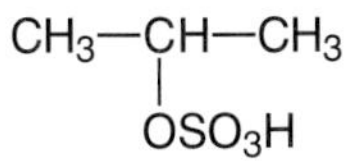

c.

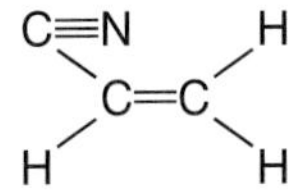

## Chapter 6 Exercises

6.1. a.
CH$_3$CH(OH)CH$_2$CH$_2$CH$_2$CH$_3$ $\xrightarrow[\Delta]{H_2SO_4}$ CH$_2$=CH—CH$_2$CH$_2$CH$_2$CH$_3$
                                                            Minor
+ H$_2$O + CH$_3$—CH=CH—CH$_2$CH$_2$CH$_3$
                          Major

b.

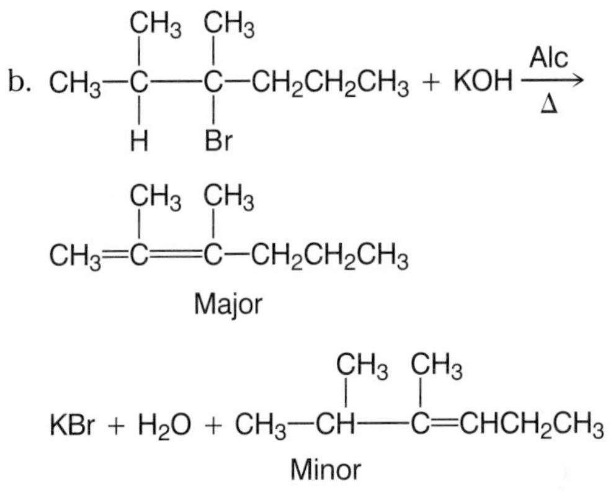

c.

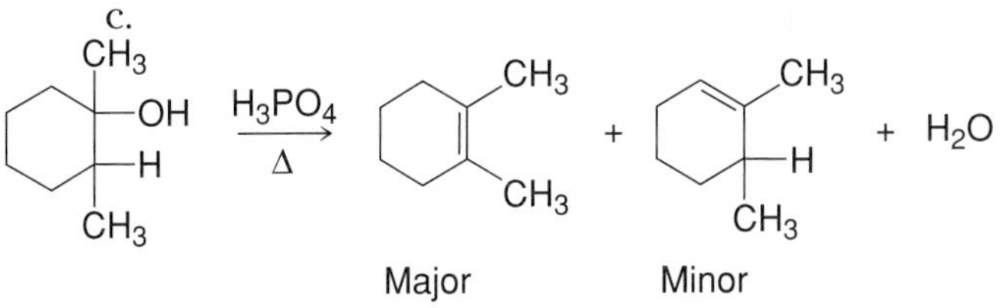

     Major          Minor

d.

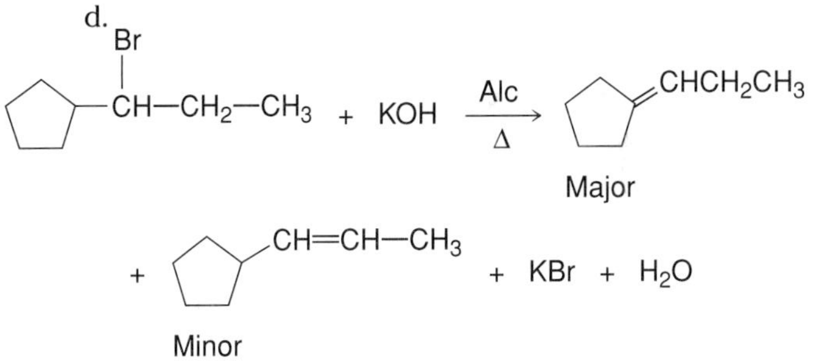

6.3. a.

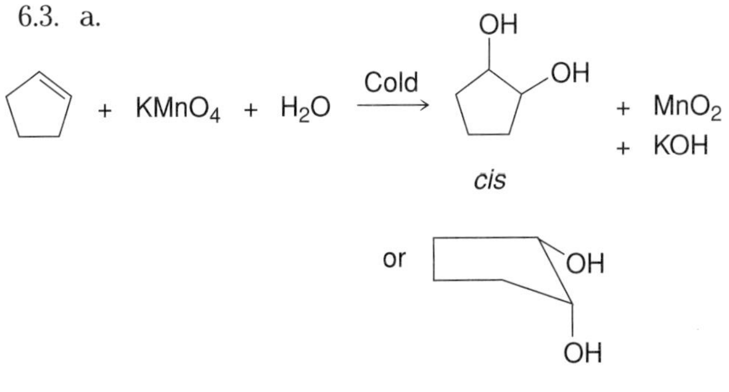

b. CH$_2$=CH—CH$_2$CH$_2$CH$_2$CH$_3$ + Br$_2$ $\xrightarrow[20°C]{CCl_4}$

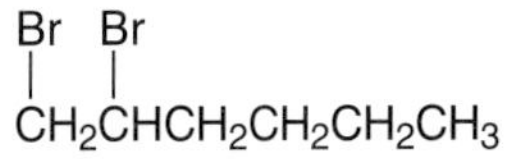

CH$_2$(Br)CH(Br)CH$_2$CH$_2$CH$_2$CH$_3$

c. CH$_2$=CH—CH$_2$CH$_3$ + H$_2$ $\xrightarrow{Pd}$ CH$_3$CH$_2$CH$_2$CH$_3$

d.

+ H$_2$ $\xrightarrow{\text{Ni}}$

or

c.

$$CH_2{=}\overset{\overset{\displaystyle CH_3}{|}}{C}{-}CH_2CH_2CH_2CH_3 \xrightarrow[\text{2. Zn/H}_2\text{O}]{\text{1. O}_3} \overset{\overset{\displaystyle O}{\|}}{HC}{-}H$$

$$+ \ CH_3CH_2CH_2CH_2\overset{\overset{\displaystyle O}{\|}}{C}{-}CH_3$$

6.5

+ HBr

6.11 a. $CH_3{-}\overset{\overset{\displaystyle OH}{|}}{CH}{-}CH_3 \xrightarrow[\Delta]{\text{H}_2\text{SO}_4} CH_3CH{=}CH_2 \xrightarrow[\text{H}_2]{\text{Ni}} CH_3CH_2CH_3$

b. $CH_3{-}\overset{\overset{\displaystyle OH}{|}}{CH}{-}CH_3 \xrightarrow[\Delta]{\text{H}_2\text{SO}_4} CH_3CH{=}CH_2 \xrightarrow{\text{HBr}} \underset{\overset{\displaystyle |}{\underset{\displaystyle Br}{}}}{CH_2}CH{-}CH_3$

6.7. a. $CH_3{-}\overset{\overset{\displaystyle CH_3}{}}{C}{=}CHCH_3 + Cl_2 \xrightarrow[\text{20°C}]{\text{CCl}_4} CH_3{-}\overset{\overset{\displaystyle CH_3}{|}}{\underset{\underset{\displaystyle Cl}{|}}{C}}{-}\overset{}{\underset{\underset{\displaystyle Cl}{|}}{CH}}{-}CH_3$

c.

$$CH_3{-}\overset{\overset{\displaystyle OH}{|}}{CH}{-}CH_3 \xrightarrow[\Delta]{\text{H}_2\text{SO}_4} CH_3{-}CH{=}CH_2 \xrightarrow[\text{H}_2\text{O}]{\text{Br}_2} CH_3{-}\overset{}{\underset{\underset{\displaystyle OH}{|}}{CH}}{-}CH_2Br$$

b. $CH_3{-}\overset{\overset{\displaystyle CH_3}{}}{C}{=}CHCH_3 + HCl \longrightarrow CH_3{-}\overset{\overset{\displaystyle CH_3}{|}}{\underset{\underset{\displaystyle Cl}{|}}{C}}{-}CH_2CH_3$

**Major**

d.

$$CH_3CH{-}CH_3 \xrightarrow[\Delta]{\text{H}_2\text{SO}_4} CH_3{-}CH{=}CH_2 \xrightarrow{\text{HBr}} CH_3{-}\overset{\overset{\displaystyle Br}{|}}{CH}{-}CH_3$$
(with OH on first carbon)

$$\overset{\overset{\displaystyle Li}{|}}{} \xrightarrow{\text{2 Li}} \text{LiBr} + CH_3{-}CH{-}CH_3$$

+ $CH_3{-}\overset{\overset{\displaystyle CH_3}{|}}{CH}{-}\overset{}{\underset{\underset{\displaystyle Cl}{|}}{CH}}{-}CH_3$

**Minor**

c. $CH_3{-}CH{=}CH{-}CH_2CH_3 \xrightarrow[\text{2. H}_2\text{O, }\Delta]{\text{1. H}_2\text{SO}_4} CH_3{-}\overset{}{\underset{\underset{\displaystyle OH}{|}}{CH}}{-}CH_2CH_2CH_3$

**Slightly more than other product**

e.

$$CH_3{-}\overset{\overset{\displaystyle OH}{|}}{CH}{-}CH_3 \xrightarrow[\Delta]{\text{H}_2\text{SO}_4} CH_3CH{=}CH_2 \xrightarrow[\text{2. H}_2\text{O}_2]{\text{1. BH}_3} CH_3CH_2CH_2O$$

+ $CH_3{-}CH_2{-}\overset{}{\underset{\underset{\displaystyle OH}{|}}{CH}}{-}CH_2{-}CH_3$

**Slightly less than preceding product**

f. $CH_3{-}\overset{\overset{\displaystyle OH}{|}}{CH}{-}CH_3 \xrightarrow[\Delta]{\text{H}_3\text{PO}_4}$

d. $CH_2{=}CHCH_2CH_3 + H_2 \xrightarrow{\text{Ni}} CH_3CH_2CH_2CH_3$

e. $CH_2{=}CHCH_3 \xrightarrow[\text{2. H}_2\text{O}_2, \text{OH}^-]{\text{1. BH}_3} HOCH_2CH_2CH_3 + BO_3{}^{-3}$

$$CH_3{-}CH{=}CH_2 \xrightarrow{\text{H}_2\text{SO}_4} CH_3{-}\overset{}{\underset{\underset{\displaystyle HO_3SO}{|}}{CH}}{-}CH_3$$

## Chapter 6 Problems

1. a. 2-Pentene; b. 2-Methyl-2-pentene; c. 2,3-Dimethyl-2-butene

6.9. a.

$CH_3CH_2CH_2CH{=}CHCH_2CH_2CH_3 \xrightarrow[\text{2. Zn,H}_2\text{O}]{\text{1. O}_3} 2CH_3CH_2CH_2\overset{\overset{\displaystyle O}{\|}}{C}{-}H$

3. a. $CH_3{-}\overset{\overset{\displaystyle CH_3}{|}}{\underset{\underset{\displaystyle I}{|}}{C}}{-}CH_2CH_2CH_3 + KOH \xrightarrow[\Delta]{\text{Alc}} CH_3{-}\overset{\overset{\displaystyle CH_3}{}}{C}{=}CHCH_2CH_3$

(1)   **Major**

b.

$\xrightarrow[\text{2. Zn/H}_2\text{O}]{\text{1. O}_3} H{-}\overset{\overset{\displaystyle O}{\|}}{C}CH_2CH_2CH_2CH_2\overset{\overset{\displaystyle O}{\|}}{C}{-}H$

+ $H_2O + KI + CH_3{-}\overset{\overset{\displaystyle CH_2}{\|}}{C}{-}CH_2CH_2CH_3$

(2)   **Minor**

b.

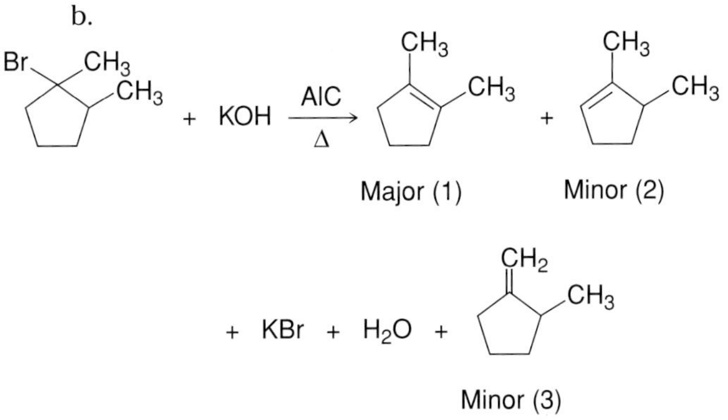

+ KOH $\xrightarrow[\Delta]{AlC}$

Major (1)     Minor (2)

+ KBr + H$_2$O +

Minor (3)

c. $CH_3CH_2CH_2OH \xrightarrow[\Delta]{H_2SO_4} CH_3-CH=CH_2$

5. 

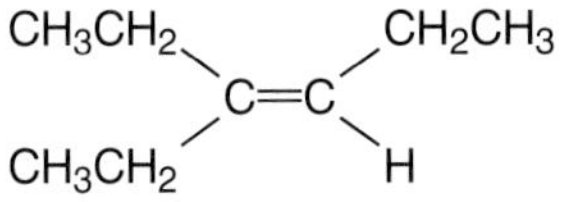

*trans*-3-Ethyl-3-hexene

7. $BrCH_2CH_2Br$

Ethylene Dibromide (EDB)

9. Methyl carbocation: $H-\overset{\overset{\displaystyle H}{|}}{\underset{\underset{\displaystyle H}{|}}{C}}{}^+$

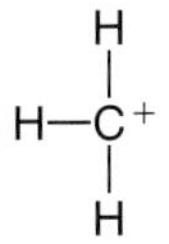

11. a. $Br^+$
    b. $H^+$
    c. $H^+$

13. a. $(CH_3CH_2)_3B$

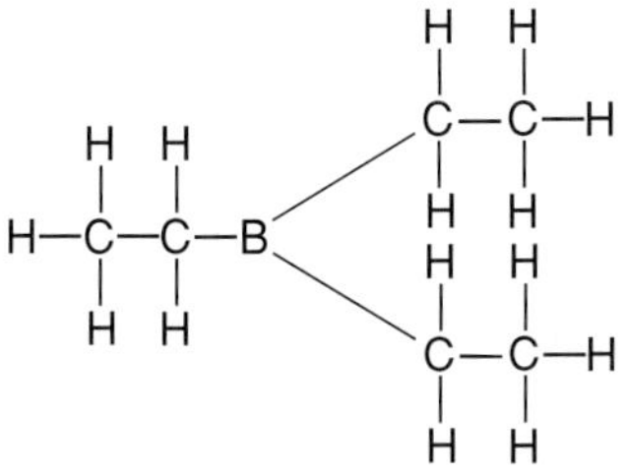

b. $O_3$ $[:\ddot{O}-\overset{+}{\ddot{O}}=\ddot{O}: \longleftrightarrow :\ddot{O}=\overset{+}{O}-\ddot{O}:^-]$

c.

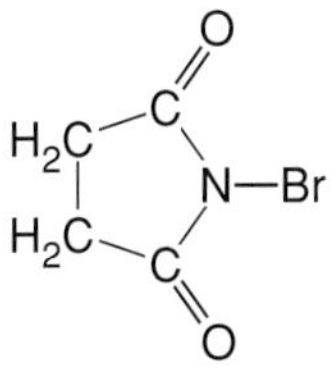

d. 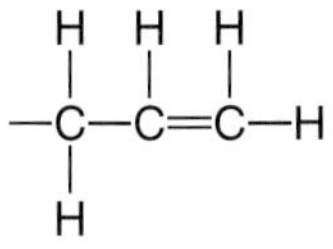 or $-CH_2CH=CH_2$

e. or $-CH=CH_2$

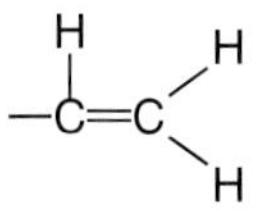

15. a. Homolytic; b. homolytic; c. heterolytic

17. a. Teflon

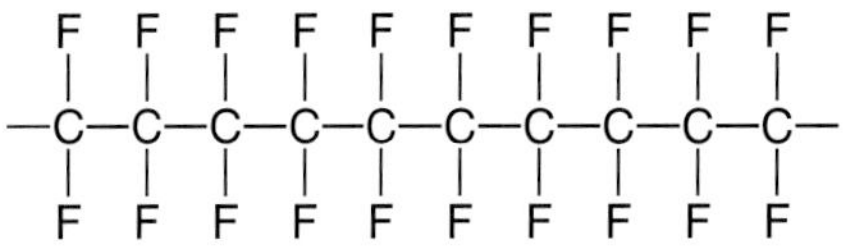

b.

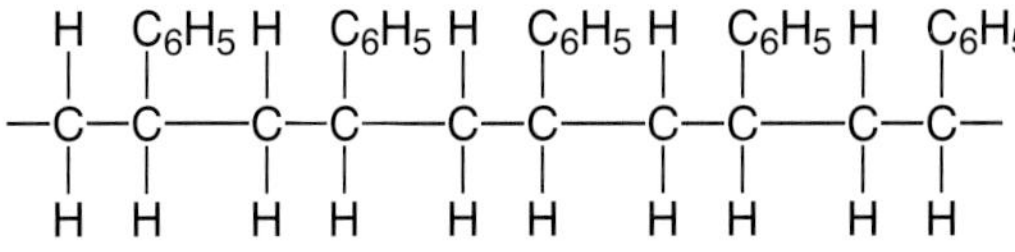

c.

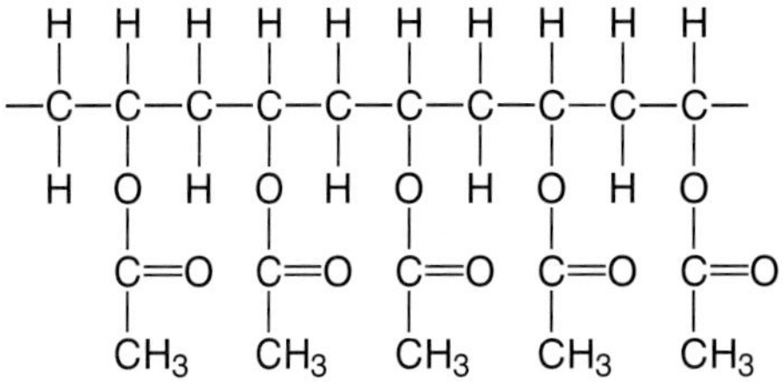

19. a. 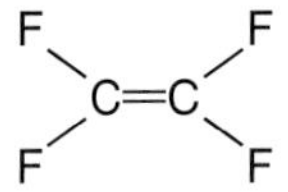     c.

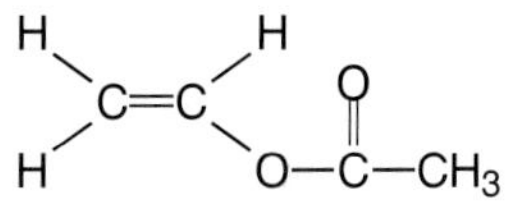

b.

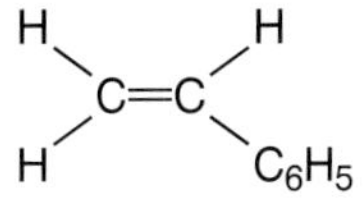

21. a.

$CH_3\overset{\overset{\displaystyle OH}{|}}{C}H-CH_3 \xrightarrow[\Delta]{H_2SO_4} CH_3CH=CH_2 \xrightarrow[\text{2. }H_2O_2/OH^-]{\text{1. }BH_3} CH_3CH_2CH_2O$

b. $CH_3CH_2CH_2CH_2Cl \xrightarrow[\text{Alc, }\Delta]{KOH} CH_3CH_2CH=CH_2$

$\downarrow$ HBr

$CH_3CH_2\overset{\overset{\displaystyle}{}}{C}H-CH_3$
$\qquad\quad |$
$\qquad\quad Br$

c. $CH_3CH_2Br \xrightarrow{Li} CH_3CH_2Li \xrightarrow{CuI} [CH_3CH_2]_2CuLi$

$\downarrow$ $CH_3CH_2Br$

$CH_3CH_2CH_2CH_3$

d. $CH_3CH_2CH_2CH_2Br \xrightarrow[\text{Alc, }\Delta]{KOH} CH_3CH_2CH=CH_2$

$\downarrow$ $Br_2/H_2O$

$CH_3CH_2\overset{\overset{\displaystyle}{}}{C}H-CH_2Br$
$\qquad\quad |$
$\qquad\quad OH$

23. a.

$CH_3CH_2CH_2CH_2CH_2OH \xrightarrow[\Delta]{H_3PO_4} CH_3CH_2CH_2CH=CH_2 + H_2O$

b. $+ H_2 \xrightarrow{Ni} CH_3-CH_2\overset{\overset{\displaystyle}{}}{C}H-CH_3$
$\qquad\qquad\qquad\qquad\qquad\qquad |$
$\qquad\qquad\qquad\qquad\qquad\quad CH_3$

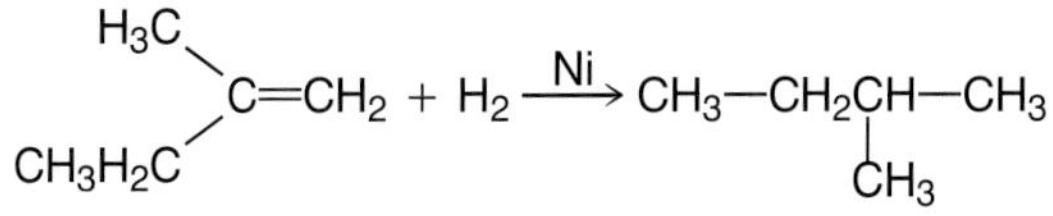

c. $CH_3CH_2CH_2I + Mg \xrightarrow[\text{Ethyl ether}]{\text{Anhydrous}} CH_3CH_2CH_2MgI$

d.

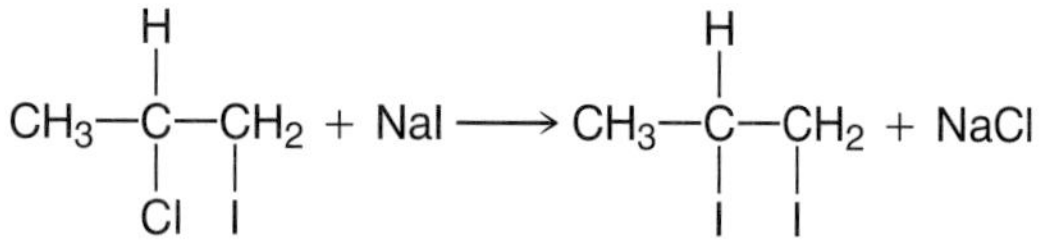

e.

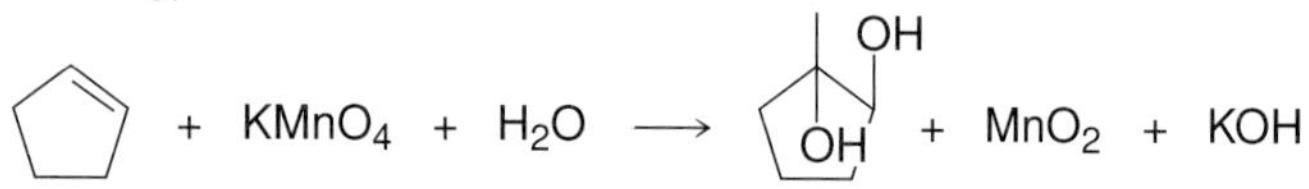

f.

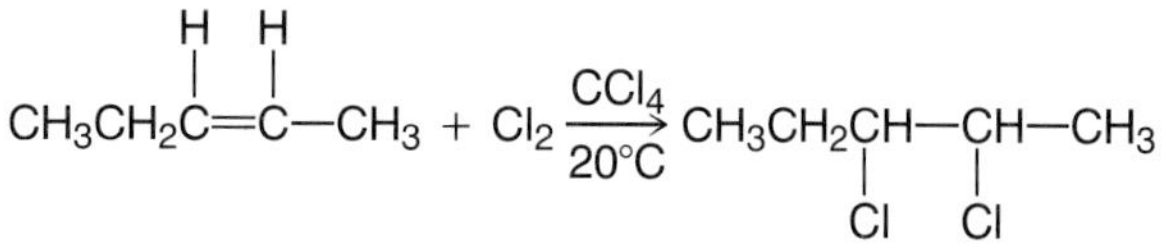

g.

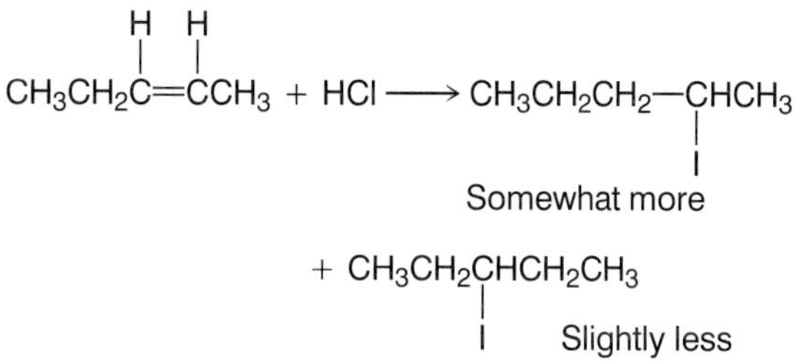

Somewhat more

$+ CH_3CH_2CHCH_2CH_3$    Slightly less

h.

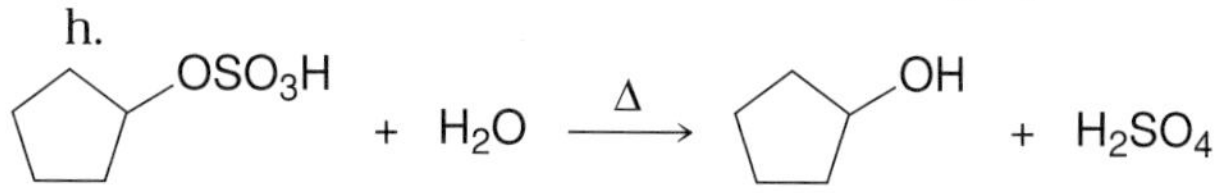

i.

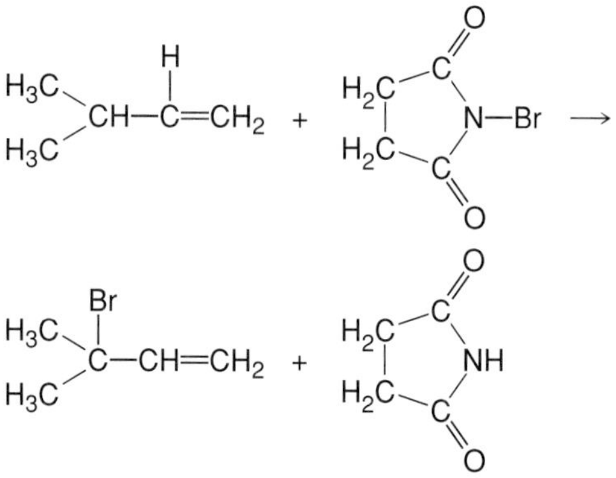

j.

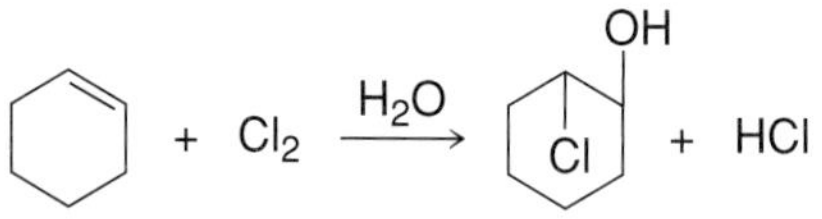

k. 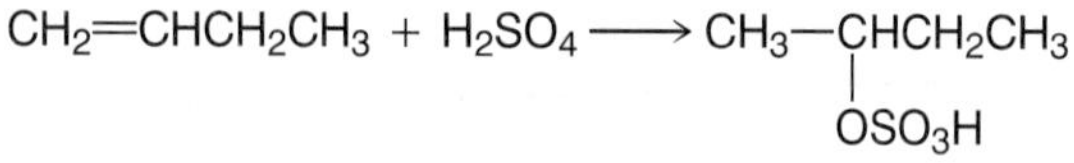 $CH_2=CHCH_2CH_3 + H_2SO_4 \longrightarrow CH_3-CHCH_2CH_3$ with $OSO_3H$

l.

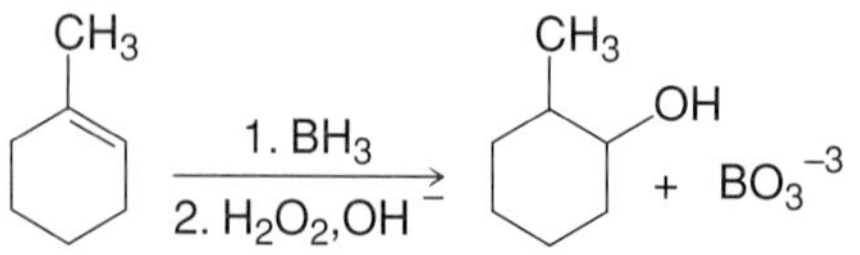

m.

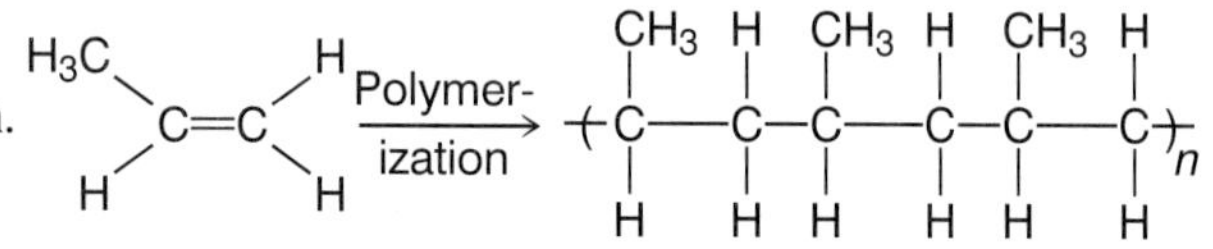

## Chapter 7 Exercises

7.1.  a. no; b. yes; c. yes

7.3.

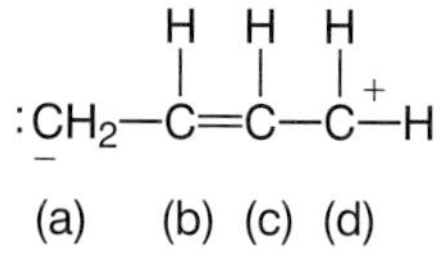

$$FC_{C_a} = 4 - \frac{6}{2} - 2 = -1$$
$$FC_{C_{b\ \&\ c}} = 4 - \frac{8}{2} - 0 = 0$$
$$FC_{C_d} = 4 - \frac{6}{2} - 0 = +1$$

7.5.

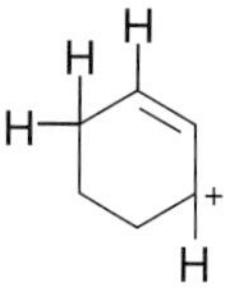

7.7.  a.

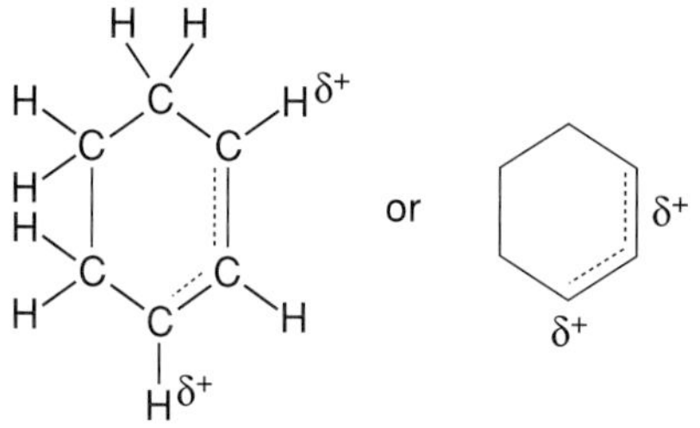

Structures 11 and 12 from Sec. 7.2 (p. 130).

b.

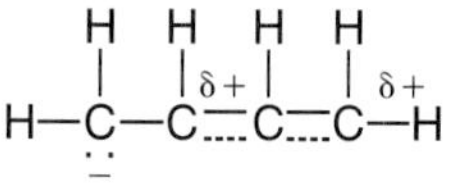

Structures 5 and 6 from Solved Problem 7.2 (p. 125).

7.9.  a.

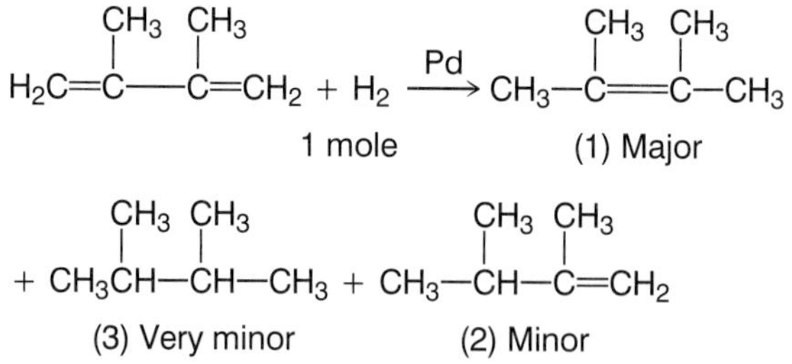

b.

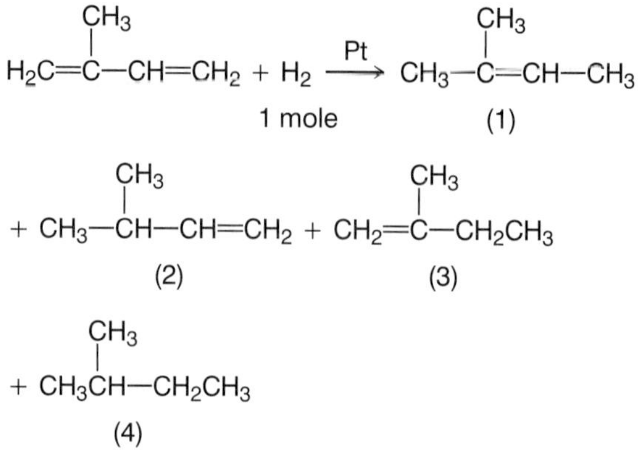

7.11. a.

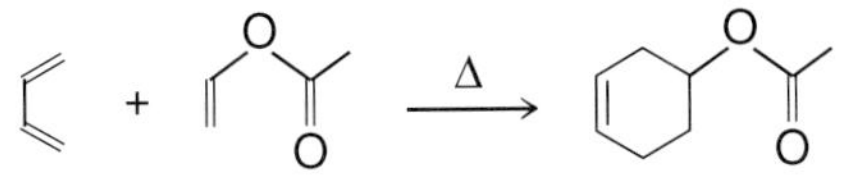

b.

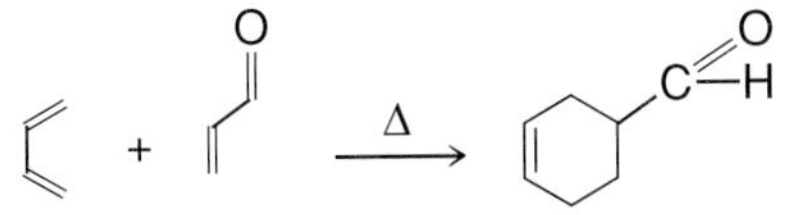

7.13. a.

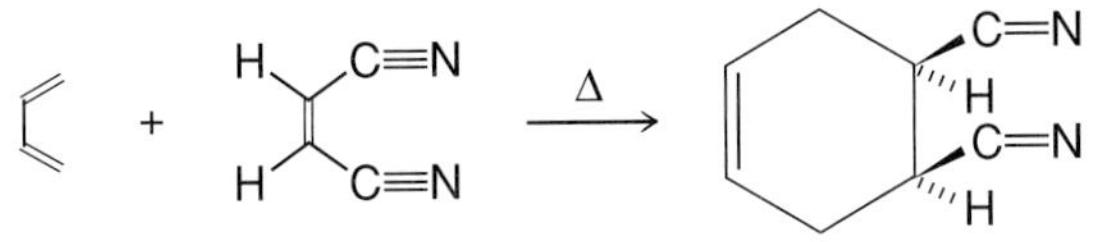

b.

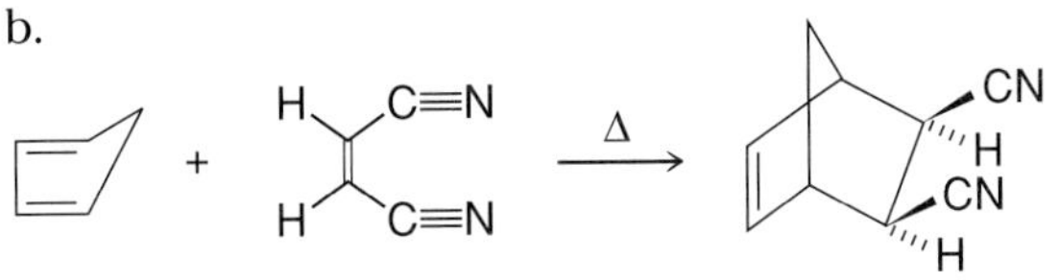

## Chapter 7 Problems

1.2,4,-Heptadiene; b. 1,3-Cyclohexadiene

3. a.

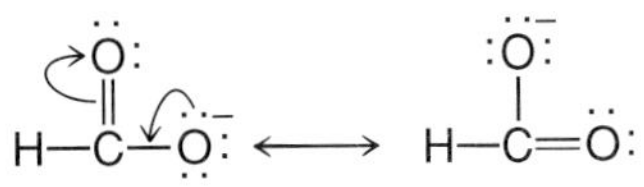

b.

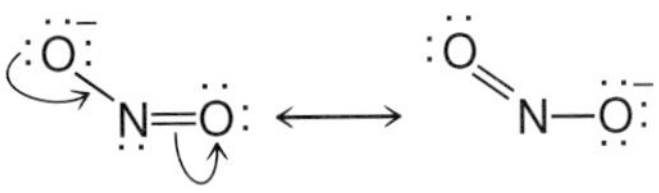

c.

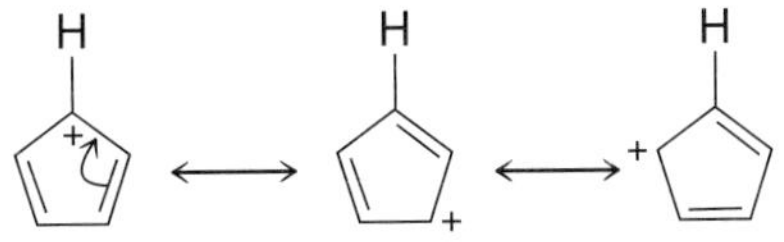

d.

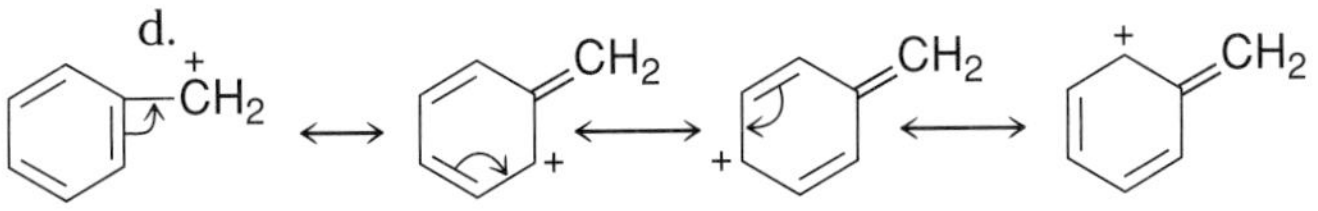

e.

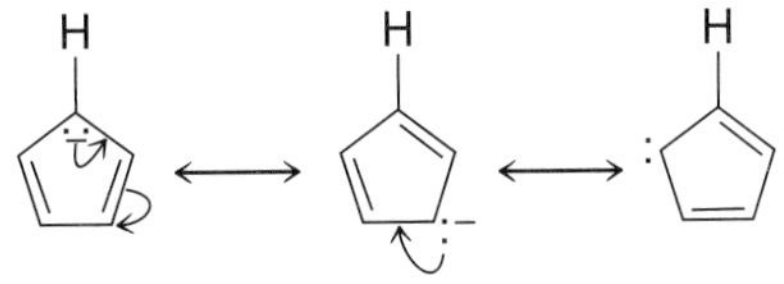

5. a.

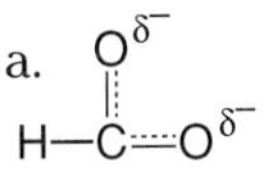

b.

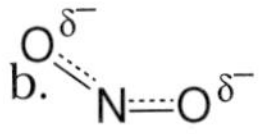

c.

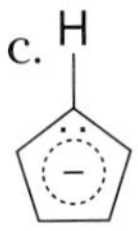

d.

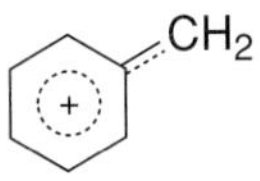

e.

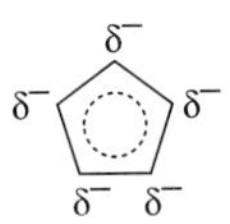

7. A resonance form cannot be written for $CH_3CH_2OH$ because any structure would violate the octet rule; for example,

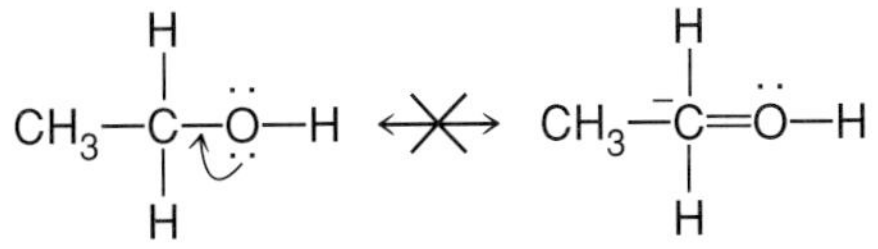

9. $FC_H = 1 - \frac{2}{2} - 0 = 0$

$FC_C = 4 - \frac{10}{2} - 0 = -1$

$FC_N = 5 - \frac{8}{2} - 0 = +1$

11. a.

$$CH{=}CH{-}CH{=}CH_2 + 2H_2 \xrightarrow[\text{Pd or Ni}]{\text{Pt or}} CH_3CH_2CH_2CH_3$$

b. $$CH_3CH_2CH_2CH_2Cl \xrightarrow[\Delta,\ \text{alcohol}]{\text{KOH}} CH_3CH_2CH{=}CH_2$$

$$\xrightarrow{H_2O_2,\ \Delta}$$

$$\underset{\displaystyle |\atop Br}{CH_3CH}{-}CH{=}CH_2 \xrightarrow[\Delta,\ \text{alcohol}]{\text{KOH}} CH_2{=}CH{-}CH{=}CH_2$$

c. $$CH_2{=}CH{-}\underset{\displaystyle |\atop Cl}{CH}{-}CH_3 \xrightarrow[\Delta,\ \text{alcohol}]{\text{KOH}} CH_2{=}CH{-}CH{=}CH_2$$

$$\xrightarrow[CCl_4,\ 20^\circ]{\text{1 mol } Br_2} BrCH_2{-}CH{=}CH{-}CH_2Br$$

$[+\ CH_2{=}CH{-}CHBr{-}CH_2Br]$

d.

e.

13.

1,2 Addition

1,4 Addition

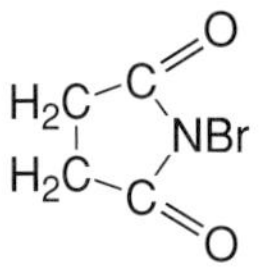

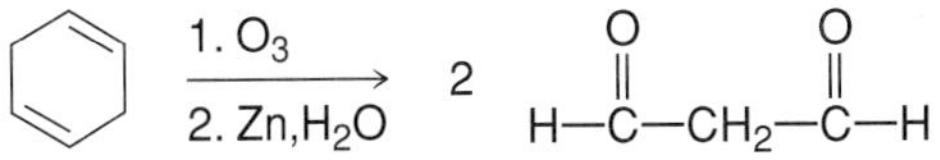

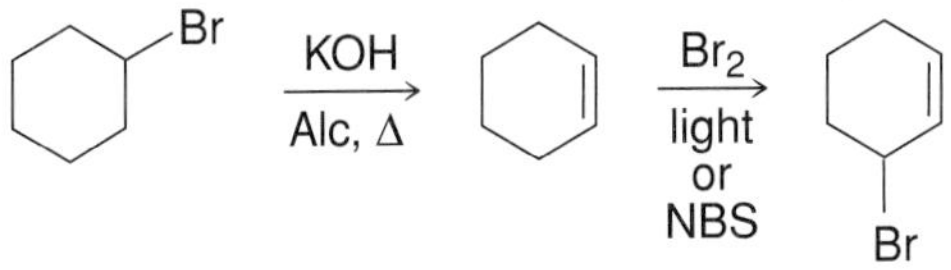

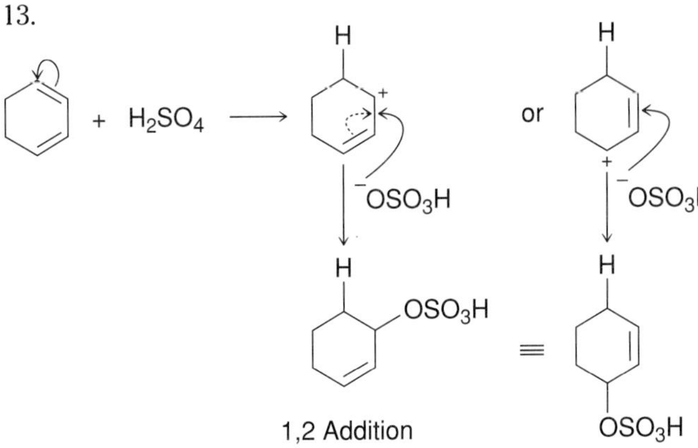

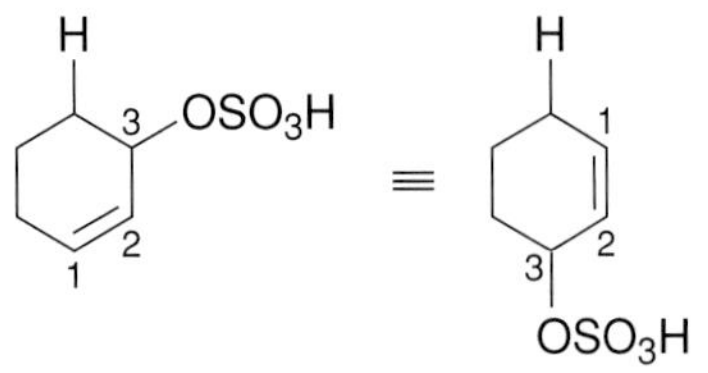

**15.** a.

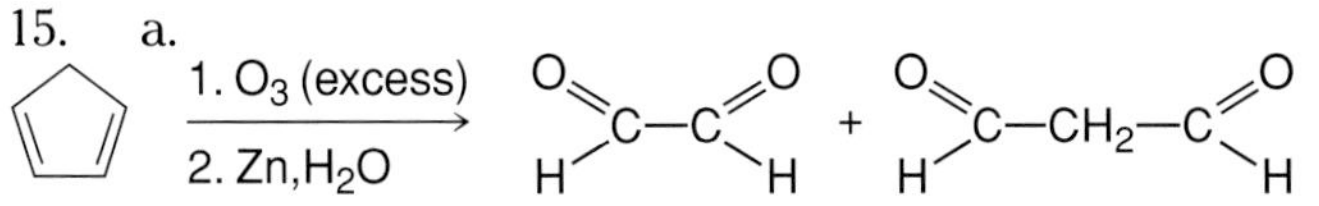

1. O$_3$ (excess)
2. Zn, H$_2$O

b.

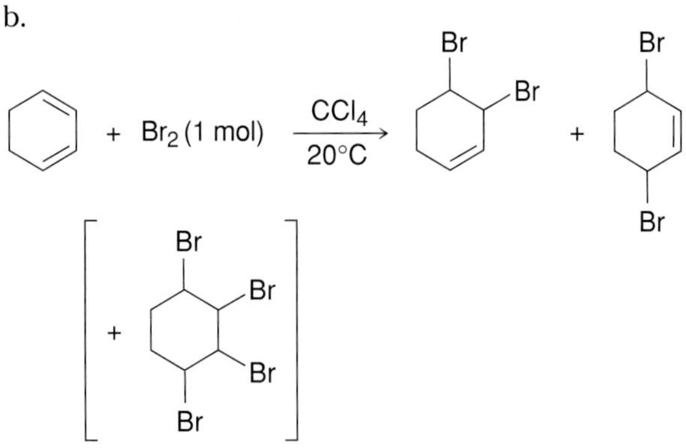

+ Br$_2$ (1 mol)   $\xrightarrow[20°C]{CCl_4}$

Trace

c.

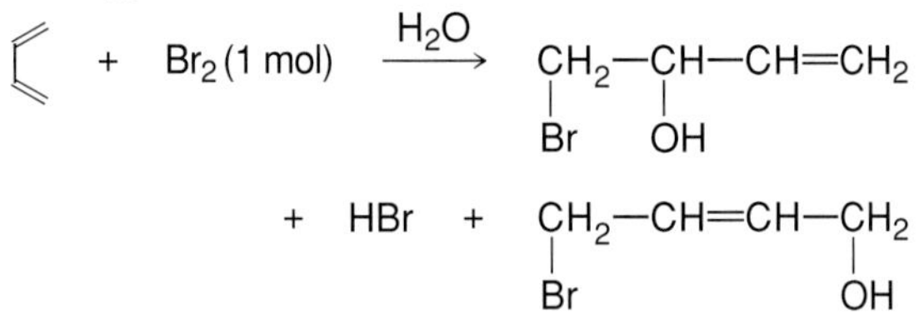

+ Br$_2$ (1 mol)  $\xrightarrow{H_2O}$  CH$_2$—CH—CH=CH$_2$ (Br, OH)

+ HBr + CH$_2$—CH=CH—CH$_2$ (Br, OH)

d.

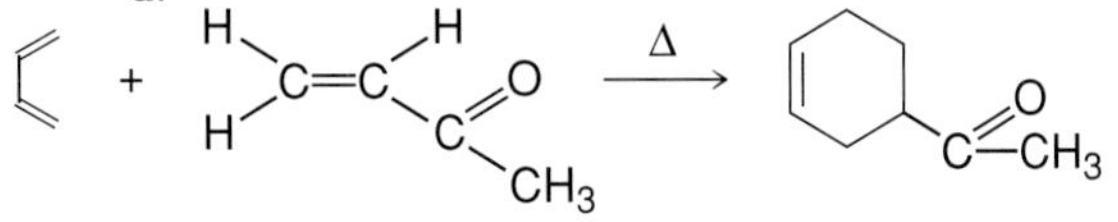

+ $\xrightarrow{\Delta}$

e.

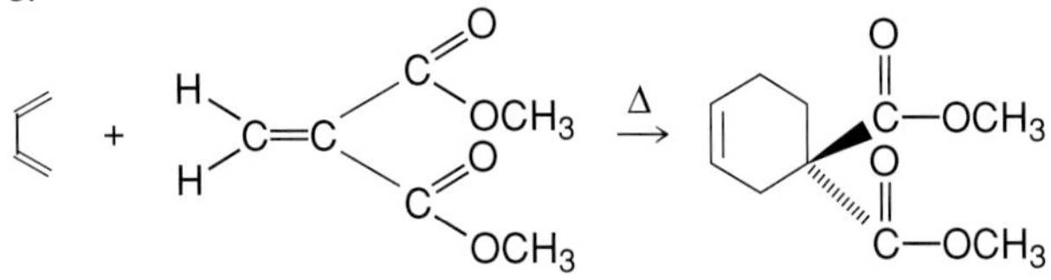

+ $\xrightarrow{\Delta}$

f.

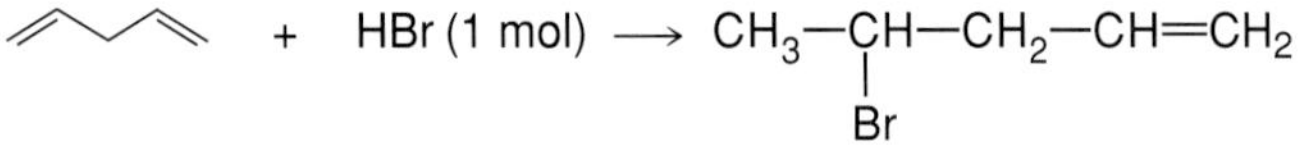

+ HBr (1 mol) ⟶ CH$_3$—CH—CH$_2$—CH=CH$_2$ (Br)

g.

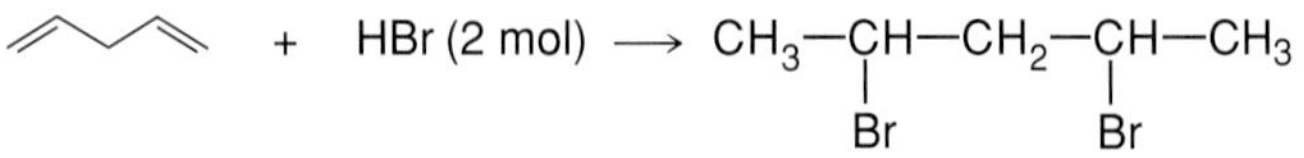

+ HBr (2 mol) ⟶ CH$_3$—CH—CH$_2$—CH—CH$_3$ (Br, Br)

h.

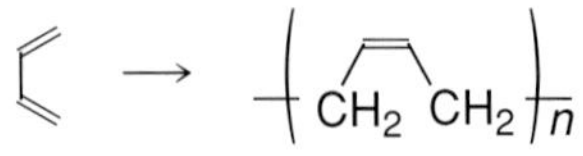

⟶ $\left(\text{CH}_2\ \text{CH}_2\right)_n$

i.

+ HI (1 mol) ⟶ CH$_3$CH=CH—CH$_2$I

+ CH$_3$—CH$_2$—CH=CH$_2$ (I)

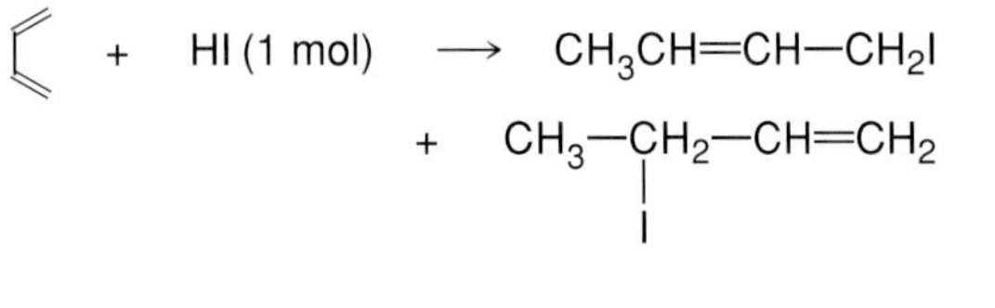

**17.** a.

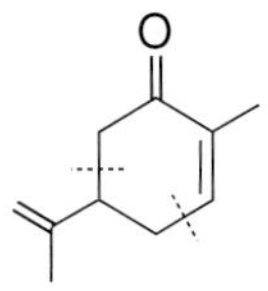

Carvone monoterpene

b.

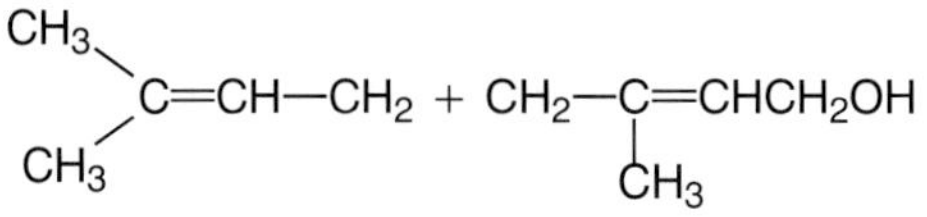

Geraniol monoterpene

c.

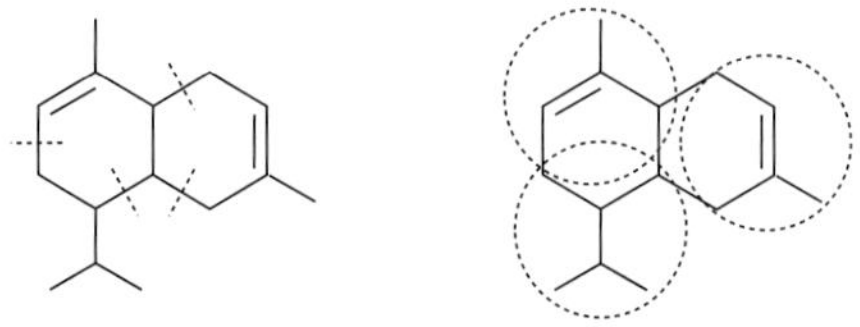

Cadinene          Sesquiterpene

**19.** a. Add bromine in carbon tetrachloride dropwise to a test tube containing 1,3-pentadiene and a test tube containing pentane. The orange-red bromine color will remain in the tube containing pentane. The 1,3-pentadiene will decolorize the bromine solution.

b. Bromine in carbon tetrachloride also can be used to determine which test tube contains 1,3-pentadiene and which contains the cyclopentane. The 1,3-pentadiene will decolorize the bromine solution immediately. The cyclopentane will decolorize the bromine solution slowly with the evolution of HBr gas. The HBr can be detected by odor or with a piece of wet litmus or pH paper.

**21.**

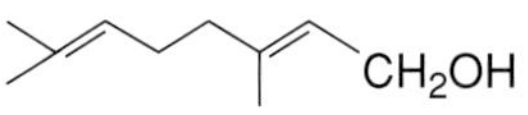

Geraniol          or

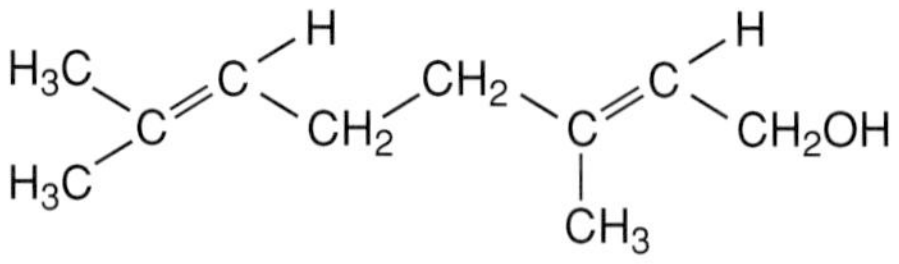

**Chapter 8 Exercises**

8.1. a. 2-Hexyne; b. 4,4-Dimethyl-2-pentyne; c. 1-Pentyne; d. 1,4,8-Decatriyne

8.3. a. Treat both 1-hexyne and 2-hexyne with a silver ammonia or copper ammonia complex. 1-Hexyne will react and form a precipitate. 1-Hexyne will not react, and a clear solution will remain.

b. Treat both 1-hexyne and 1-hexene with a silver ammonia or copper ammonia complex. 1-Hexyne will react and form a precipitate. 1-Hexene will not react because 1-hexene is not as strong an acid as 1-hexyne.

c. Treat 1-hexyne with a silver ammonia or copper ammonia complex. A precipitate of the metal salt will form. Cyclohexane will not react under the same conditions.

d. Treat both 2-hexyne and hexane with bromine in carbon tetrachloride. 2-Hexyne will decolorize the bromine solution. The red-orange bromine color will remain in the hexane.

8.5. a.

$$CH_3CH_2CH_2OH \xrightarrow[\Delta]{H_3PO_4} CH_3CH{=}CH_2 \xrightarrow[CCl_4,\ 20°C]{Cl_2} \underset{\underset{Cl}{|}\ \underset{Cl}{|}}{CH_3CH{-}CH_2}$$

$$\Big\downarrow \begin{array}{c} 2\ NaNH_2 \\ \text{Mineral oil, } \Delta \end{array}$$

$$\underset{\underset{Cl\ Cl}{|\ |}}{\overset{Cl\ Cl}{\underset{|\ |}{CH_3C{-}CH}}} \xleftarrow[CCl_4\ 20°C]{2\ Cl_2} CH_3C{\equiv}CH$$

1,1,2,2-Tetrachloropropane

b.

$$CH_3CH_2CH_2OH \xrightarrow[\Delta]{H_3PO_4} CH_3CH{=}CH_2 \xrightarrow[CCl_4,\ 20°C]{Br_2} \underset{\underset{Br}{|}}{CH_3CH{-}CH_2Br}$$

$$\Big\downarrow \begin{array}{c} 2NaNH_2 \\ \text{mineral oil} \end{array}$$

$$\underset{Br}{\overset{CH_3}{\diagdown}}C{=}C\underset{H}{\overset{Br}{\diagup}} \xleftarrow[CCl_4,\ -20°C]{Br_2} CH_3C{\equiv}CH$$

*trans*-1,2-Dibromopropene

c.

$$CH_3CH_2OH \xrightarrow[\Delta]{H_3PO_4} CH_3CH{=}CH_2 \xrightarrow[CCl_4,\ 20°C]{Br_2} \underset{\underset{Br}{|}}{CH_3CH{-}CH_2Br}$$

$$\Big\downarrow \begin{array}{c} 2NaNH_2 \\ \text{mineral oil} \end{array}$$

$$CH_3C{\equiv}C^-Na^+ \xleftarrow{NaNH_2} CH_3C{\equiv}CH$$

Sodium Propynide

8.7. a.

$$CH_3C{\equiv}CCH_3 + H_2O \xrightarrow[HgSO_4]{H_2SO_4} \left[\underset{\underset{OH\ H}{|\ |}}{CH_3C{=}C{-}CH_3}\right] \rightleftharpoons$$

$$\overset{O}{\underset{\|}{CH_3C{-}CH_2CH_3}}$$

b.

$$CH_3CH_2CH_2C{\equiv}CH + H_2O \xrightarrow[HgSO_4]{H_2SO_4} \left[\underset{\underset{OH}{|}}{CH_3CH_2CH_2C{=}CH_2}\right]$$

$$\Big\Updownarrow \overset{O}{\underset{\|}{CH_3CH_2CH_2C{-}CH_3}}$$

c. $CH_3CH_2CH_2CH_2C{\equiv}CCH_2CH_3 + H_2O \xrightarrow[HgSO_4]{H_2SO_4}$

$$\left[CH_3CH_2CH_2CH_2{-}\underset{\underset{OH}{|}}{C}{=}CCH_2CH_3 + CH_3CH_2CH_2CH_2{-}\underset{\underset{H}{|}}{C}{=}C\underset{\underset{OH}{|}}{H}CH_2CH_3\right]$$

$$\Big\Updownarrow$$

$$\overset{O}{\underset{\|}{CH_3CH_2CH_2CH_2{-}C{-}CH_2CH_2CH_3}} + \overset{O}{\underset{\|}{CH_3CH_2CH_2CH_2CH_2C{-}CH_3}}$$

8.9. a.

$$CH_3CH_2CH_2CH_2C{\equiv}CCH_2CH_3 \xrightarrow[2.\ H_2O]{1.\ O_3} \overset{O}{\underset{\|}{CH_3CH_2CH_2CH_2C{-}OH}}$$

$$+ \overset{O}{\underset{\|}{CH_3CH_2C{-}OH}}$$

b.

$$CH_3{-}CH_2\underset{\underset{CH_3}{|}}{CH}{-}C{\equiv}CH \xrightarrow[2.\ H_2O]{1.\ O_3} CH_3{-}CH_2{-}\underset{\underset{CH_3}{|}}{CH}{-}\overset{O}{\underset{\|}{C}}{-}OH + CO_2$$

c. $HC{\equiv}CH \xrightarrow[2.\ H_2O]{1.\ O_3} 2CO_2$

d. $CH_3CH_2CH_2CH_2C{\equiv}CCH_2CH_2CH_2CH_3 \xrightarrow[2.\ H_2O]{1.\ O_3}$

$$2\overset{O}{\underset{\|}{CH_3CH_2CH_2CH_2C{-}OH}}$$

## Chapter 8 Problems

1. The two carbons of the triple bond and the two atoms bonded to them are in a straight line. An alkyne contains one $sp{-}sp$ bond and two $p{-}p$ ($\pi$) bonds that are perpendicular to each other and to the $sp{-}sp$ bond.

3. Two triple bonds, e.g., $HC{\equiv}C{-}CH_2CH_2C{\equiv}C{-}CH_3$ or $HC{\equiv}C{-}CH_2CH_2CH_2C{\equiv}H$.

5. $HC{\equiv}C{-}CH_2{-}CH_2CH_3$ 1-pentyne, terminal
   $CH_3{-}C{\equiv}C{-}CH_2CH_3$ 2-pentyne, internal

7. a. $HC{\equiv}C{-}CH_2{-}C{\equiv}CH$ isolated
   b. $HC{\equiv}C{-}C{\equiv}C{-}CH_3$ conjugated

9. $CH_3{-}C{\equiv}CH$ Propyne $H{-}\underset{\underset{H}{|}}{\overset{\overset{H}{|}}{C}}{-}C{\equiv}C{-}H$

$$CH_2{=}CH{-}CH{=}CH{-}C{\equiv}N \qquad \underset{\underset{\underset{H}{|}}{\overset{H}{|}}{C}{=}C}{\overset{\overset{H}{|}}{\overset{|}{C}}{=}C}{-}C{\equiv}N$$

1-Cyano-1,3-butadiene

11. a. $CH_3C{\equiv}CCH_2CH_3$
    b. $HC{\equiv}C{-}C{\equiv}CCH_2CH_3$

13. In the ethylene carbanion, the carbon with the unshared electron pair is $sp^2$ hybridized, which has 67% $p$ character and 33% $s$ character. In acetylide carbanion,

the carbon with the unshared electron pair is *sp* hybridized, which has 50% *p* character and 50% *s* character. Since the *p* orbitals are further from the nucleus than the *s* orbitals, ethylene carbanion has a higher ratio of *p* to *s* orbitals than acetylene carbanion. Because the electron pairs of the ethylene carbanion are further from the carbon nucleus, they are more likely to be donated than the electron pairs of the acetylide carbanion, which are under more influence of the carbon of the anion. According to the Brønsted theory, a base is a electron pair donor. Ethylene is a better electron donor and therefore a stronger base and weaker acid than the acetylide anion.

The *sp* hybridized carbon of the acetylide carbanion is more electronegative than the $sp^2$ hybridized carbon of the ethylene carbanion. Therefore, the carbon-hydrogen bond of ethylene is less polar than that of acetylene. According to the electronegativity theory, there will be less partial positive charge on the hydrogen of ethylene than on the hydrogen of acetylene, and ethylene will be a weaker acid than acetylene.

15. a. $CH_3-C{\equiv}CH + Br_2 \xrightarrow[-10°C]{CCl_4}$ 

b. $CH_3C{\equiv}CH + H_2 \xrightarrow[\text{catalyst}]{\text{Lindlar's}} CH_3CH{=}CH_2$

c. $CH_3C{\equiv}CH + O_3 \xrightarrow{2.\ H_2O} CH_3\overset{O}{\overset{\|}{C}}{-}OH + CO_2$

d. $CH_3C{\equiv}CH + 2\ HBr \xrightarrow{20°C} CH_3{-}\overset{Br}{\underset{Br}{\overset{|}{\underset{|}{C}}}}{-}CH_3 {\equiv} CH_3CBr_2CH_3$

e. $CH_3C{\equiv}CH + H_2O \xrightarrow[HgSO_4]{H_2SO_4} \left[ CH_3{-}\overset{}{\underset{OH}{\overset{}{\underset{|}{C}}}}{=}CH_2 \right]$ →

$CH_3{-}\overset{O}{\overset{\|}{C}}{-}CH_3$

f. $CH_3C{\equiv}H + 2H_2 \xrightarrow[\text{Pd or Ni}]{\text{Pt or}} CH_3CH_2CH_3$

g. $CH_3C{\equiv}CH + H_2 \xrightarrow[\text{catalyst}]{\text{Lindlar}} CH_3CH{=}CH_2$

$\xrightarrow[2.\ H_2O,/OH^-]{1.\ BH} CH_3CH_2CH_2OH$

h. $CH_3C{\equiv}CH + NaNH_2 \longrightarrow CH_3C{\equiv}C{:}^- Na^+ + NH_3$

$\downarrow ICH_2CH_2CH_3$

$CH_3C{\equiv}CCH_2CH_2CH_3$

i. $CH_3C{\equiv}CH + NaNH_2 \longrightarrow CH_3C{\equiv}C{:}^- Na^+ + NH_3$

$\downarrow CH_3CH_2 I$

$\xrightarrow[\text{Lindlar}\ \text{catalyst}]{H_2} CH_3C{\equiv}CCH_2CH_3$

17. a.

$CH_3CHClCH_2CH_2CH_3 \xrightarrow[\text{ETOH, }\Delta]{KOH} CH_3CH{=}CHCH_2CH_3 \xrightarrow[CCl_4,\ 20°C]{Br_3}$

$CH_3\overset{Br}{\overset{|}{CH}}{-}\overset{Br}{\overset{|}{CH}}{-}CH_2CH_3 \xrightarrow[\text{Minerall oil}]{NaNH_2} CH_3C{\equiv}C{-}CH_2CH_3$

$\xrightarrow[\text{Lindlar}\ \text{catalyst}]{H_2}$ 

b. $CH_3CHClCH_2CH_2CH_3 \xrightarrow[\text{EtOH, }\Delta]{KOH}$

$\downarrow \overset{KOH}{\underset{\text{ETOH, }\Delta}{}}$

$CH_3CH{=}CHCH_2CH_3$ *trans*-2-Pentene (Major product)

$\downarrow Br_2/CCl_4$

$CH_3CHBrCHBrCH_2CH_3$

$\downarrow \overset{NaNH_2}{\underset{\text{mineral oil, }\Delta}{}}$

$CH_3C{\equiv}C{-}CH_2CH_3 \xrightarrow[\text{liquid }NH_3]{Na}$

c. $CH_3CHClCH_2CH_2CH_3 \xrightarrow[\text{ETOH, }\Delta]{KOH} CH_3CH{=}CHCH_2CH_3$

$\downarrow Br_2,\ CCl_4,\ 20°C$

$CH_3{-}C{\equiv}C{-}CH_2CH_3 \xleftarrow[\text{Mineral oil, }\Delta]{NaNH_2} CH_3\overset{}{\underset{Br}{\overset{}{\underset{|}{CH}}}}{-}\overset{}{\underset{Br}{\overset{}{\underset{|}{CH}}}}{-}CH_2CH_3$

$\downarrow \overset{\text{Excess}}{\underset{CCl_4,\ 20°C}{Br_2,}}$

$CH_3{-}CBr_2{-}CBr_2CH_2CH_3$

19. a. $CH_3{-}C{\equiv}CH + HCl \xrightarrow{\text{Cold}} CH_3\overset{}{\underset{Cl}{\overset{}{\underset{|}{C}}}}{=}CH_2$

b. $CH_3{-}C{\equiv}CH + 2HCl \xrightarrow{20°C} CH_3{-}\overset{Cl}{\underset{Cl}{\overset{|}{\underset{|}{C}}}}{-}CH_3$

c. $CH_3{-}C{\equiv}C{-}CH_3 + NaNH_2 \longrightarrow$ No reaction

d. $(CH_3)_3C{-}C{\equiv}CH + Ag(NH_3)_2^+ + OH^- \longrightarrow$
$(CH_3)_3C{-}C{\equiv}C\ Ag + H_2O + 2NH_3$

e. $CH_3CH{=}CHBr + KOH \xrightarrow[\Delta]{\text{Alcohol}}$ No reaction

f. $CH_3CH_2CH_2Br + CH_3{-}\overset{}{\underset{CH_3}{\overset{}{\underset{|}{C}}}}{-}C{\equiv}C^-Na^+ \longrightarrow NaBr +$

$CH_3{-}\overset{}{\underset{CH_3}{\overset{}{\underset{|}{C}}}}{-}C{\equiv}CCH_2CH_2CH_3$

g. $CH_3-C\equiv C-C_2H_5 + 2H_2 \xrightarrow{Pd} CH_3-CH_2CH_2CH_2CH_3$

h. $C_2H_5-C\equiv C-C_2H_5 \xrightarrow[\substack{Liquid \\ NH_3}]{Na}$

$$\underset{H_5C_2}{\overset{H}{}}C=C\underset{H}{\overset{C_2H_5}{}}$$
(*trans*)

i. $C_2H_5-C\equiv C-C_2H_5 + H_2 \xrightarrow[catalyst]{Lindlar's}$

$$\underset{H_5C_2}{\overset{H}{}}C=C\underset{C_2H_5}{\overset{H}{}}$$

j. $HC\equiv CH + 2Br_2 \xrightarrow[20°C]{CCl_4} Br_2CH-CHBr_2$

k. $HC\equiv CH + Br_2 \xrightarrow[20°C]{CCl_4} BrCH=CHBr$

$$\underset{Br}{\overset{H}{}}C=C\underset{H}{\overset{Br}{}}$$
(*trans*)

l. $C_2H_5-C\equiv C-C_2H_5 + H_2O \xrightarrow[HgSO_4]{H_2SO_4}$

$$\left[\underset{}{\overset{OH\ H}{C_2H_5=C=C-C_2H_5}}\right] \longrightarrow C_2H_5-\overset{O}{\overset{\|}{C}}-CH_2C_2H_5$$

m. $CH_3(CH_2)_{15}C\equiv C-CH_2CH_3 \xrightarrow[2.\ H_2O]{1.\ O_3}$

$$CH_3(CH_2)_{15}\overset{O}{\overset{\|}{C}}-OH + CH_3CH_2\overset{O}{\overset{\|}{C}}-OH$$

21. a. $CH_3CH_2\underset{OH}{\overset{}{C}}=CH_2 \rightleftharpoons CH_3CH_2-\overset{O}{\overset{\|}{C}}-CH_3$

b. $CH_2=\underset{OH}{\overset{}{CH}} \rightleftharpoons CH_3-\overset{O}{\overset{\|}{C}}-H$

c. $CH_3(CH_2)_5\underset{OH}{\overset{}{C}}=CH_2 \rightleftharpoons CH_3(CH_2)_5\overset{O}{\overset{\|}{C}}-CH_3$

23.

$$CH_3-\underset{OH}{\overset{H}{C=C}}-CH_3 + H^+ \rightleftharpoons CH_3-\overset{+}{\underset{\overset{\frown}{:OH}}{C}}-CH_2CH_3$$

$$CH_3\overset{}{\underset{O}{C}}-CH_2CH_3 + H_3\overset{+}{O} \longleftarrow CH_3-\underset{\underset{\overset{..}{:O}-H}{\overset{+}{OH}}}{\overset{}{C}}-CH_2CH_3$$

25. No.

27. a. Compound A: $CH_3CH_2C\equiv CH$ 1-butyne
Compound B: $CH_3-C\equiv C-CH_3$ 2-butyne
Compound C: $CH_2=CH-CH=CH_2$ 1,3-butadiene

Compound D: ☐ Cyclobutene
Reactions:
Compound A
(2) $CH_3CH_2C\equiv CH + Ag(NH_3)_2^+ + OH^- \longrightarrow$
$$CH_3CH_2C\equiv CAg + NH_3 + H_2O$$

(1) $CH_3CH_2C\equiv CH + Br_2 \xrightarrow{CCl_4} CH_3CH_2CBr_2CHBr_2$

Compound B

(1) $CH_3C\equiv CCH_3 + Br_2 \xrightarrow{CCl_4} CH_3CBr_2CBr_2CH_3$

(2) $CH_3C\equiv CCH_3 + Ag(NH_3)_2^+ + OH^- \longrightarrow$ No
reaction

(3) $CH_3C\equiv CCH_3 + 1.\ O_3 \xrightarrow{2.\ H_2O} 2CH_3COOH$

Compound C

(1) $CH_2=CH-CH=CH_2 + Br_2 \xrightarrow{CCl_4}$
$CH_2Br-CHBrCHBr-CH_2$

(2) $CH_2=CH-CH=CH_2 + HCl \longrightarrow$
$CH_3-CH=CH-CH_2Cl + CH_3-CHCl-CH=CH_2$
Compound D

$$☐ + Br_2 \xrightarrow{CCl_4} \text{(cyclobutene with Br, Br)}$$

$$☐ + O_3 \xrightarrow[H_2O]{Zn} \underset{CH_2-\overset{O}{\overset{\|}{C}}-H}{\overset{CH_2-\overset{O}{\overset{\|}{C}}-H}{}}$$

## Chapter 9 Exercises

9.1. a. *p*-Chloroiodobenzene; b. *m*-Ethylene nitrobenzene or 3-nitrostyrene; c. *p*-Dichlorobenzene or 1,4-dichlorobenzene

9.3. a. 1-Nitronaphthalene; b. 3-Bromo-1-nitronaphthalene; c. 1-Bromo-5-chloro-10-nitroanthracene

9.5. a. Aromatic: 1. Cyclic; 2. Resonance forms can be written; 3. Obeys Huckel's rule; b. Not aromatic: 1. Cyclic; 2. No resonance forms can be written; c. Not aromatic: 1. Cyclic; 2. Resonance forms can be written; 3. Does not conform to Huckel's rule.

## Chapter 9 Problems

1.

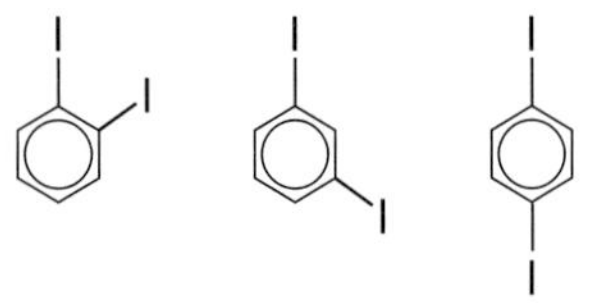

**Benzo-pyrene**

3.  a.

9.  a.
$$CH_2CH_3 \longleftrightarrow CH_2CH_3$$
(1)

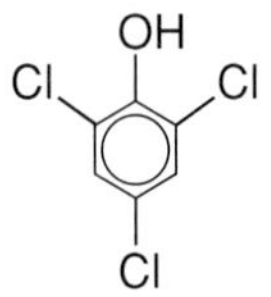

b.
$$CH_3$$
(1)

(2)
$$\overset{O}{\underset{C-H}{\parallel}} \longleftrightarrow \overset{O}{\underset{C-H}{\parallel}}$$

(2)
$$Br$$

(3)
$$CH=CH_2 \longleftrightarrow CH=CH_2$$

(3)
$$NH_2$$

b.

c. CH₂Cl

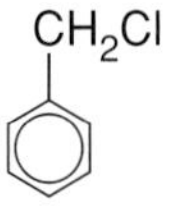

d. OH, I

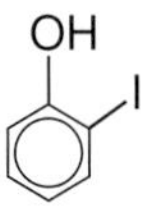

e. Br, Br

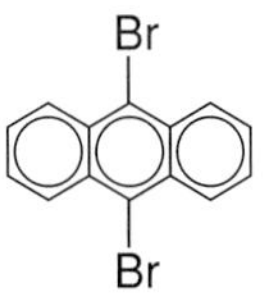

f. CH=CH₂

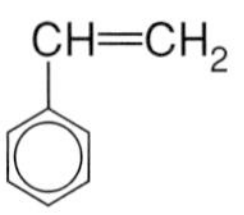

g.

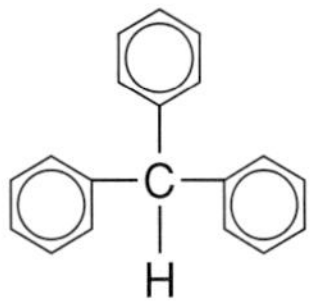

h. NO₂, NO₂

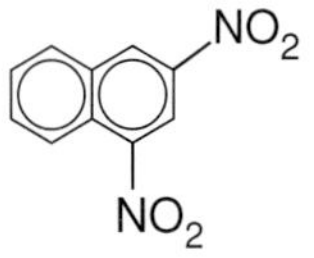

i. CH₃, CH₃

j. CH₃, SO₃H

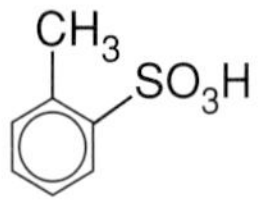

5.  a. OH, Cl, Cl, Cl

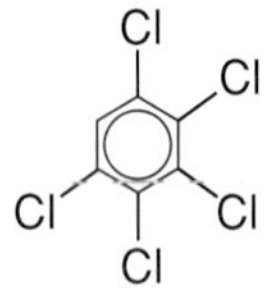

d. F, NO₂, NO₂

b. CHO, O₂N, NO₂

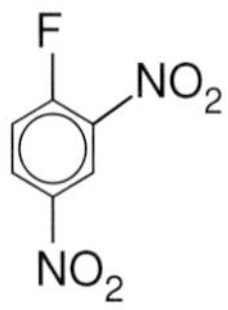

e. CH=CH₂, Br, Br

c. Cl, Cl, Cl, Cl, Cl

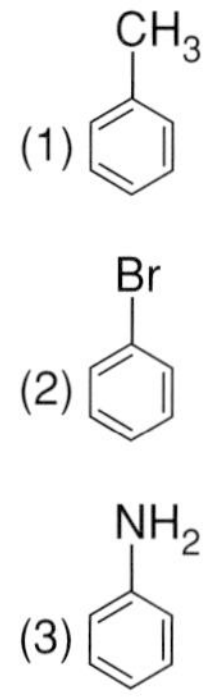

11.  a. Not aromatic; No resonance structures; Does not follow Huckel's rule b. Not aromatic; not cyclic; c. Aromatic; d. Aromatic

13.  Benzofuran

15.  a.

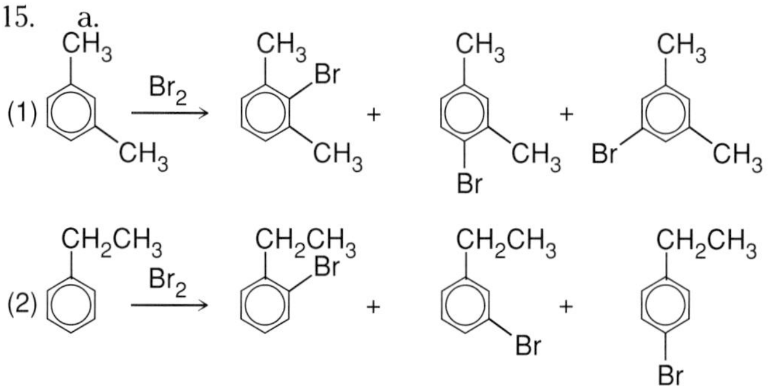

(1) CH₃ ... Br₂ → ... + ... + ...

(2) CH₂CH₃ ... Br₂ → ... + ... + ...

b.

(1) CH₃, CH₃ Br₂ → ... + ...

and

c.
CH₃ ... Br₂ → CH₃, Br
CH₃ ... CH₃

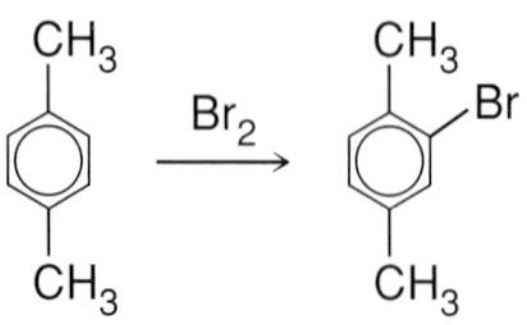

All positions on ring are equal with respect to bromine.

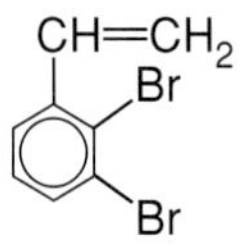

7.  a. Aromatic: For a compound to be aromatic, it must be (1) cyclic, (2) resonance stabilized, and (3) follow Huckel's rule.

  b. *Ortho-*: Substitution on adjacent carbons of an aromatic compound.

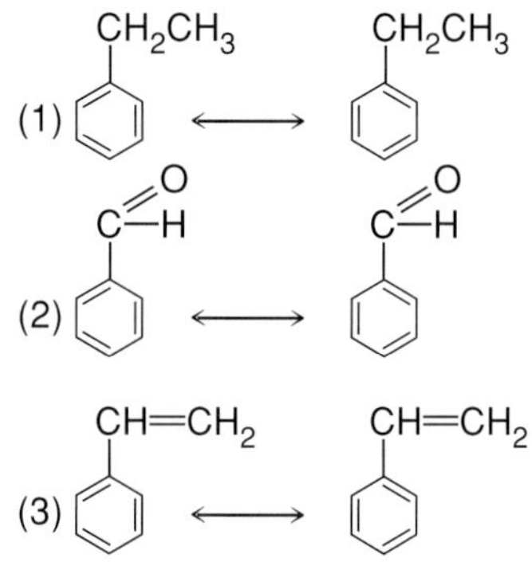

  c. Reforming: Process by which cyclohexane or other alkanes found in petroleum are converted to aromatic compounds.

  d. *benzo-*: A C₄H₄ aromatic unit bonded to two adjacent aromatic carbons to form an additional aromatic ring, e.g., *benzo*-pyrene:

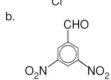

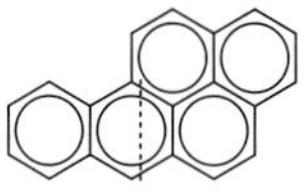

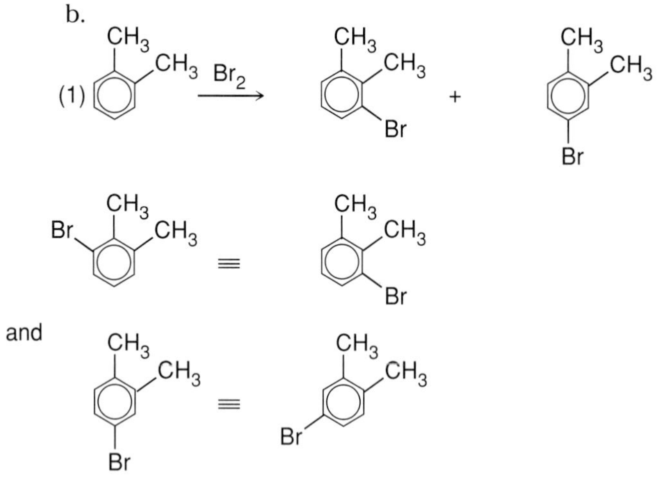

## Chapter 10 Exercises

10.1.  a.

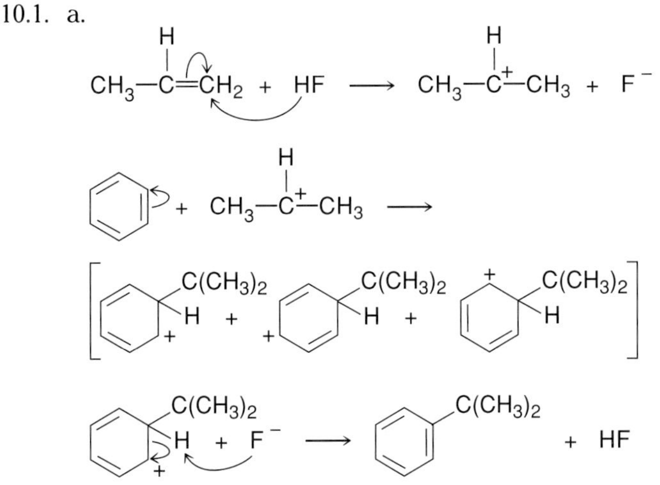

b.

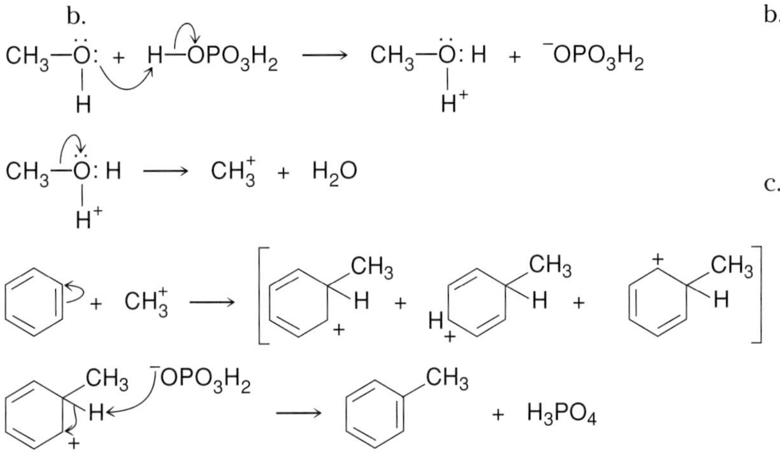

10.3.  a.

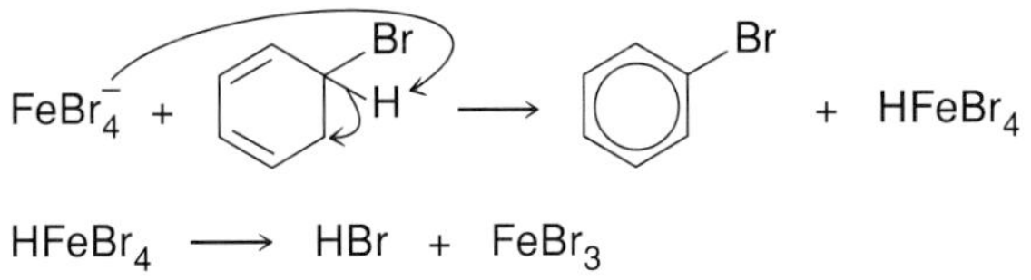

b.

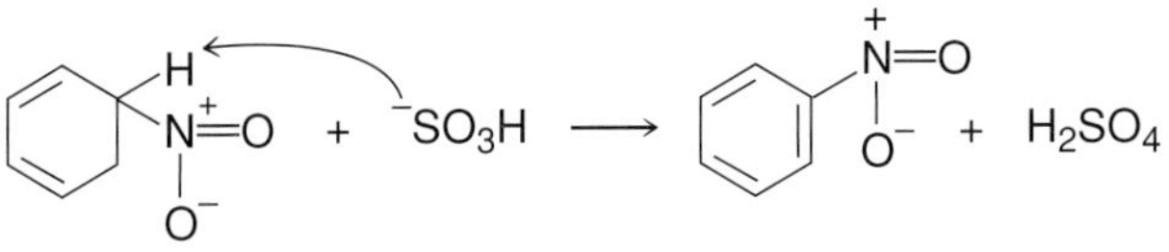

10.5.  Yes.

10.7.  a.

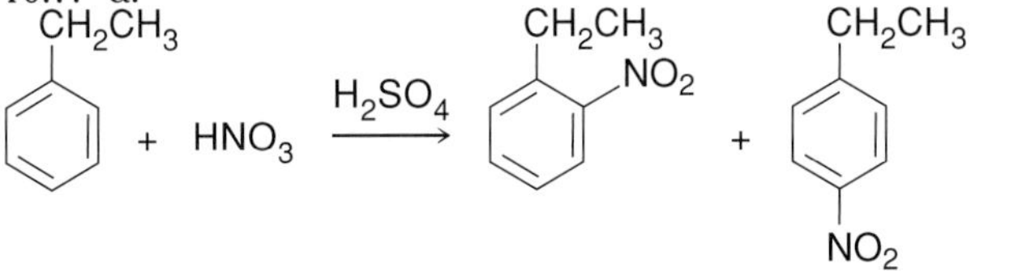

b.

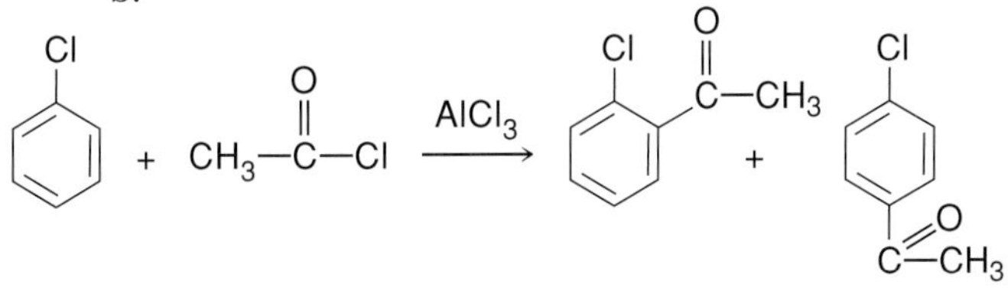

c.

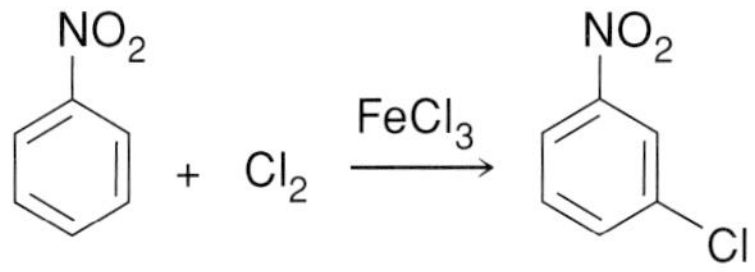

d.

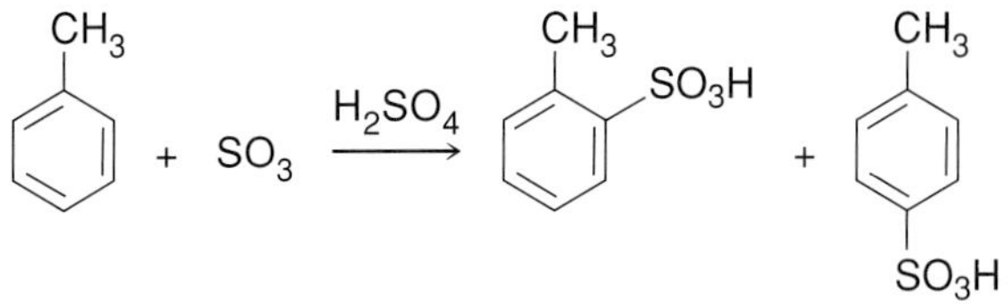

10.9.  a.

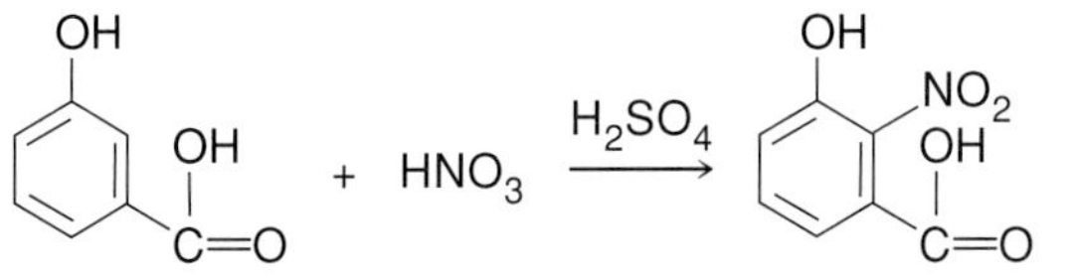

b.

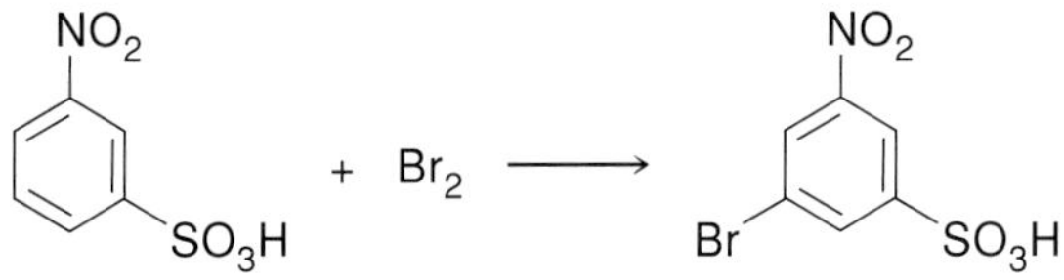

c.

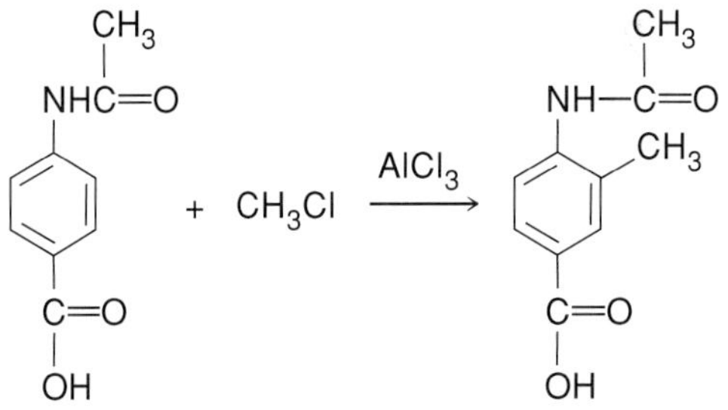

d.

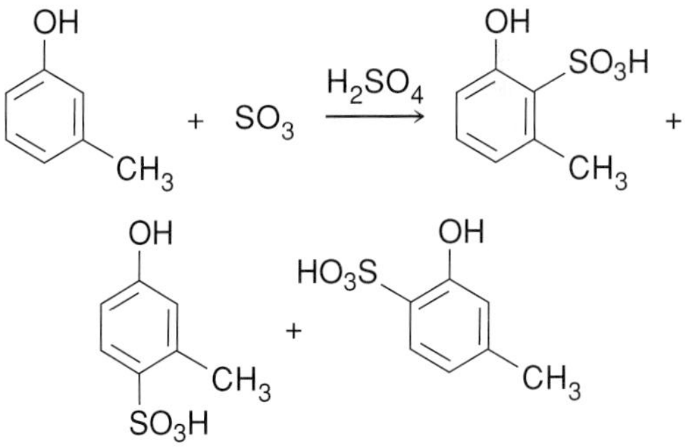

10.11. a.

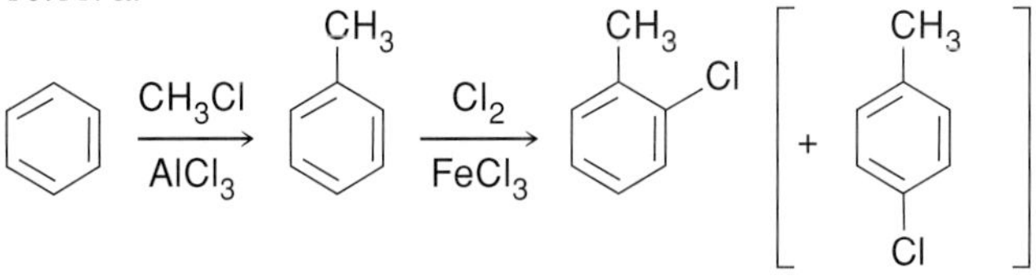

b.

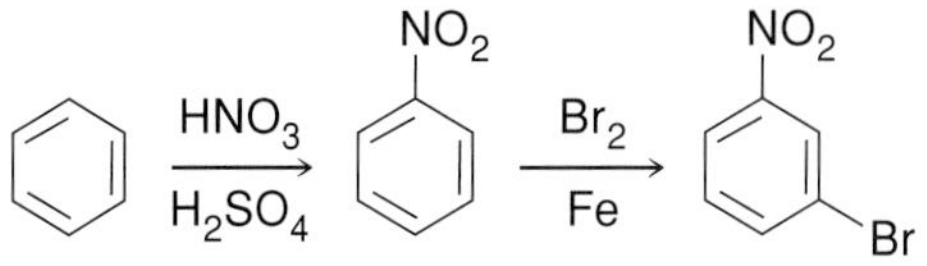

c.

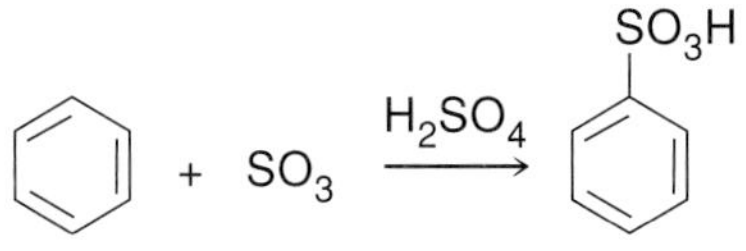

d.

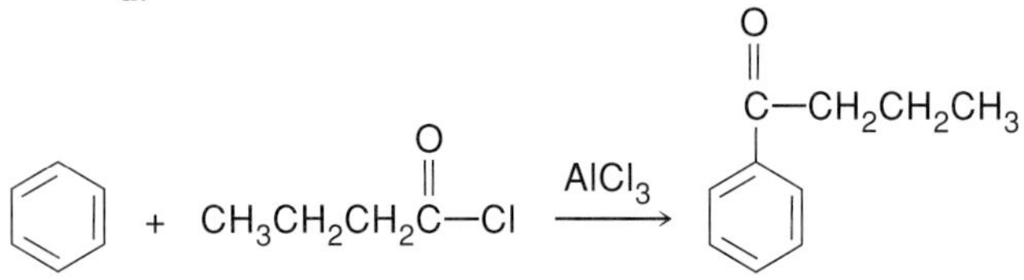

e.

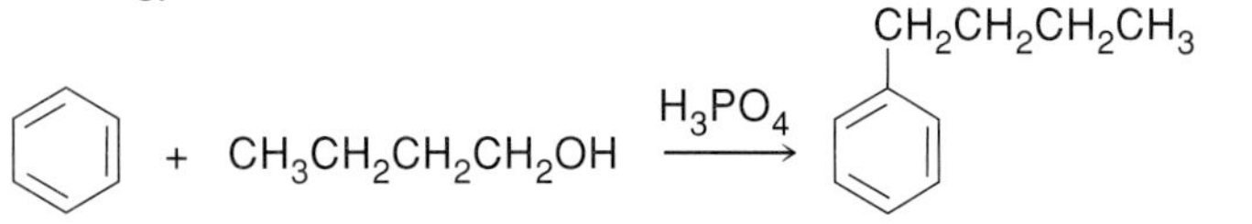

## Chapter 10 Problems

1.

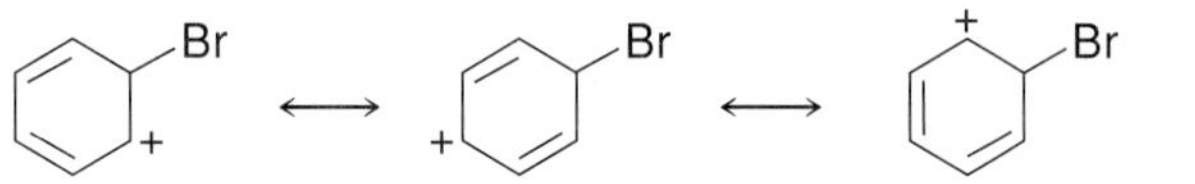

3. Both the arenium and allylic cations are stabilized by resonance structures; that is, the double bond electrons can move through more than one carbon-carbon double bond:

$$[\overset{+}{C}H_2{-}CH{=}CH_2 \longleftrightarrow CH_2{=}CH{-}CH_2^+]$$

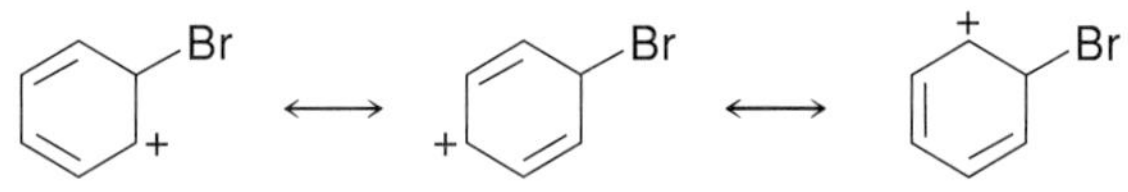

5. Yes.

7. Chlobenzene is too unreactive to form a phenyl carbocation.

9. Sulfuric acid.

11. a.

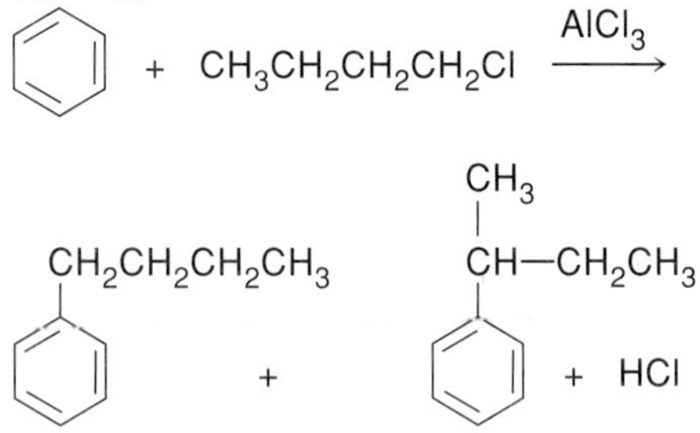

b.

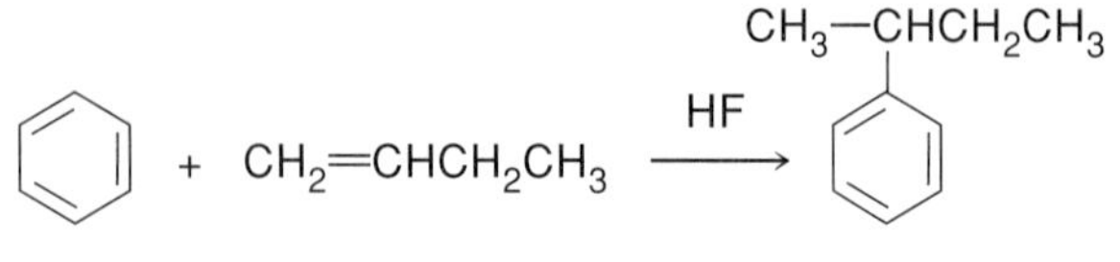

c.

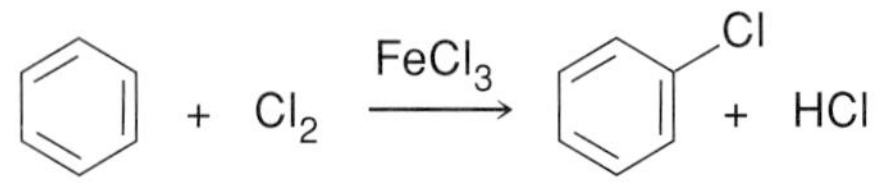

13. The formation of a resonance-stabilized aromatic ring is favored over the addition of $Cl^-$ to

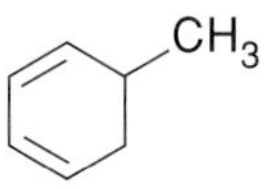

15.

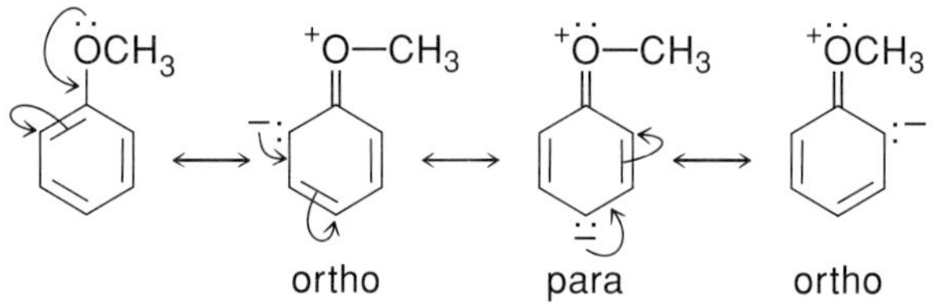

17.

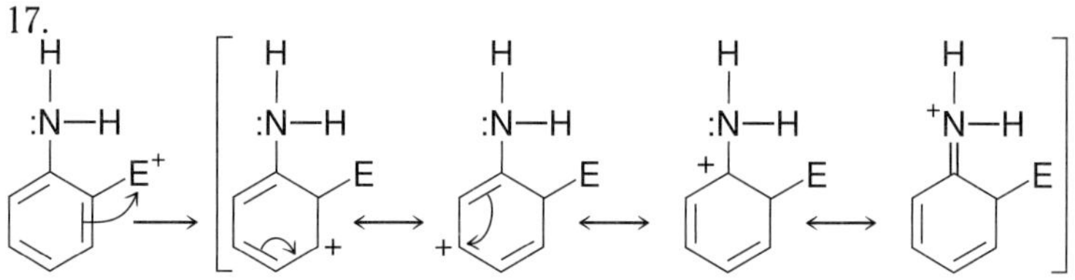

Very stable

19.

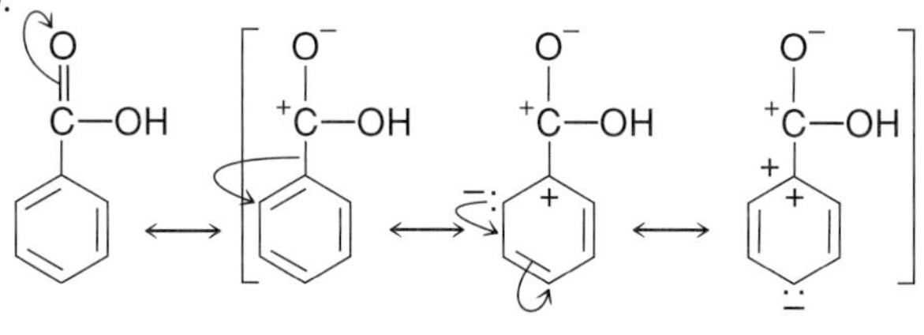

Ortho para position — Adjacement positive charges (due to inductive effect of the carbonyl group) makes the carbocation unstable.

21. a. More rapidly; b. more rapidly; c. less rapidly; d. less rapidly

23. a. *ortho-para* (unshared electron pair)
b. *ortho-para* (unshared electron pair)
c. *meta* (no unshared electron pair)
d. *ortho-para* (unshared electron pairs)

25. a.

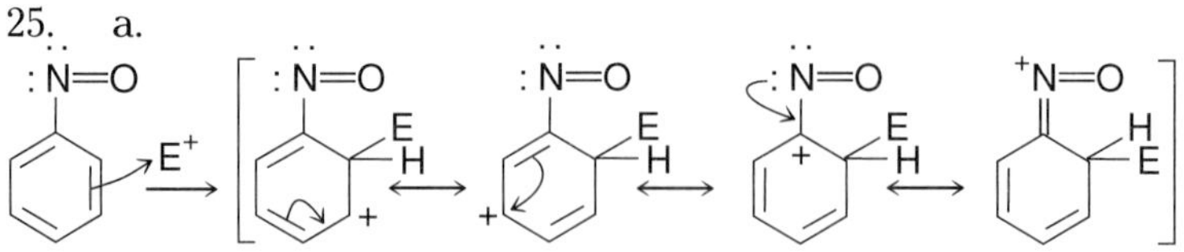

ortho

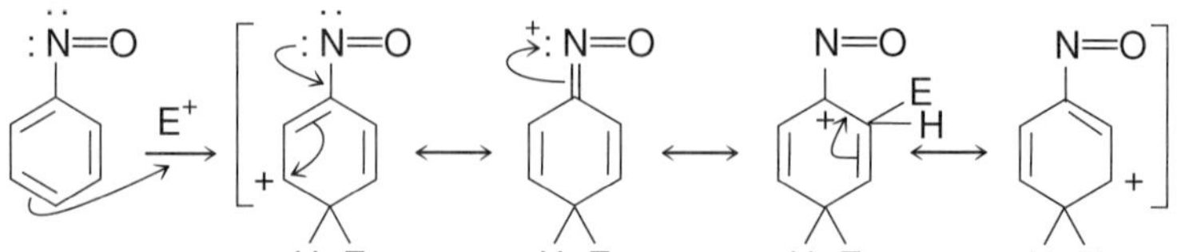

para

b.

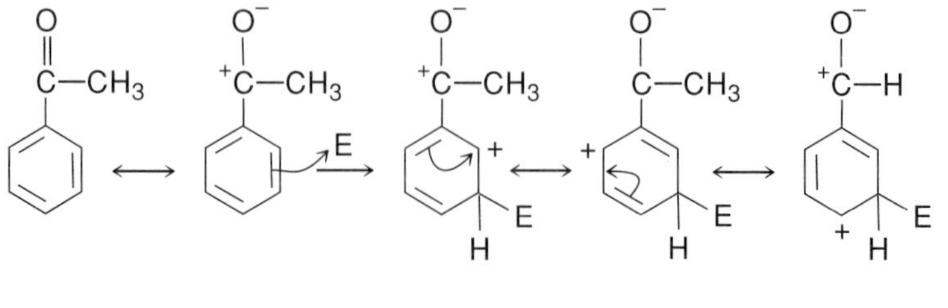

If in *ortho* or *para* position, a resonance structure would have positive charges on adjacent carbons, e.g.,

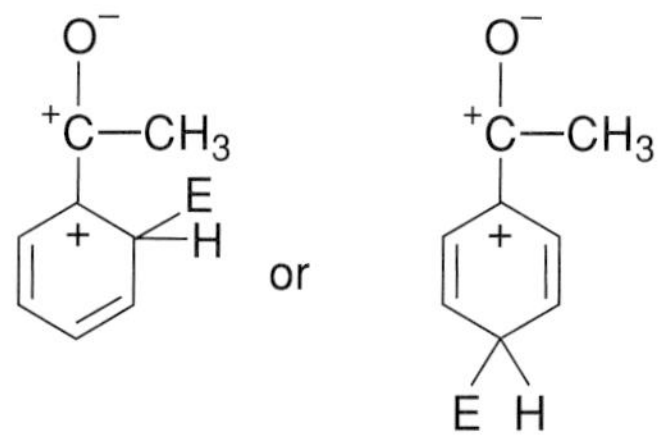

These are very unstable resonance structures.

27.

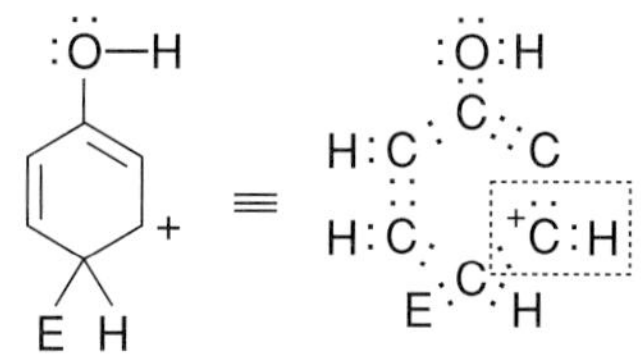

Boxed carbon only has a sextet of electrons.

29. Slowest.

31. Isolated double bonds are not stabilized by resonance energy, and therefore, milder conditions can be used than would be required to reduce an aromatic compound, which is stabilized by resonance.

33.

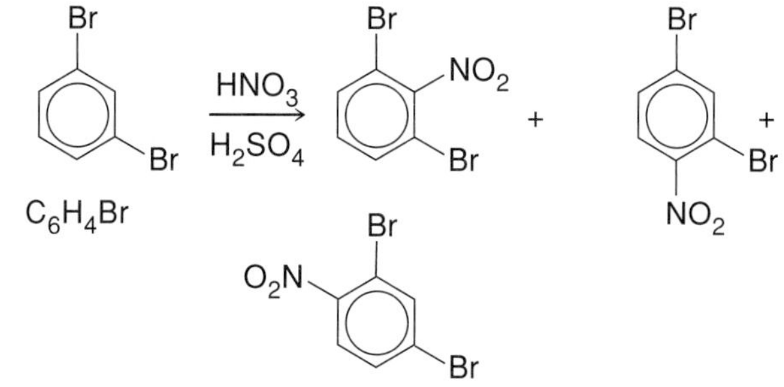

35. a.

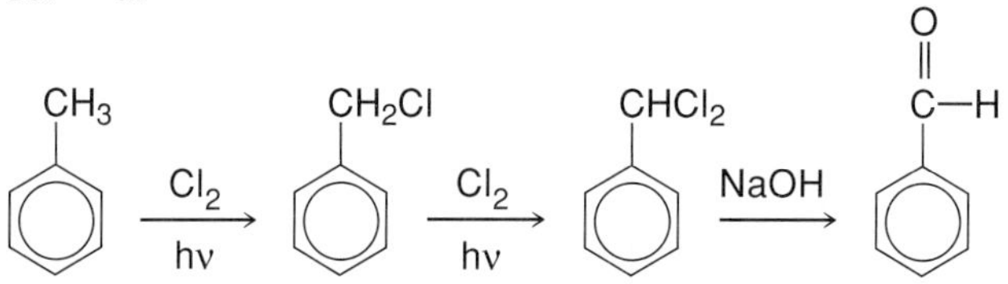

b.

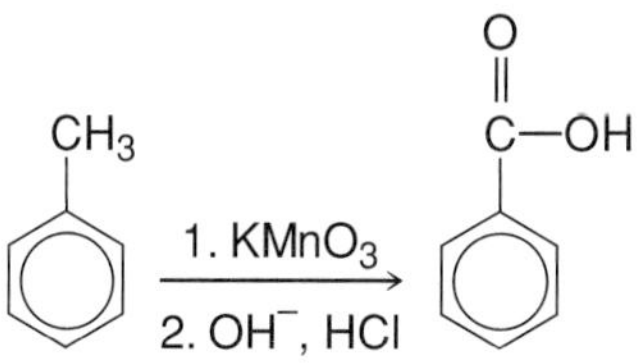

c.

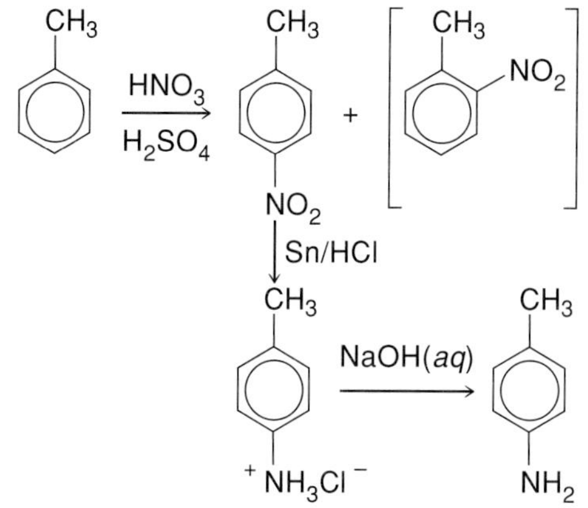

d.

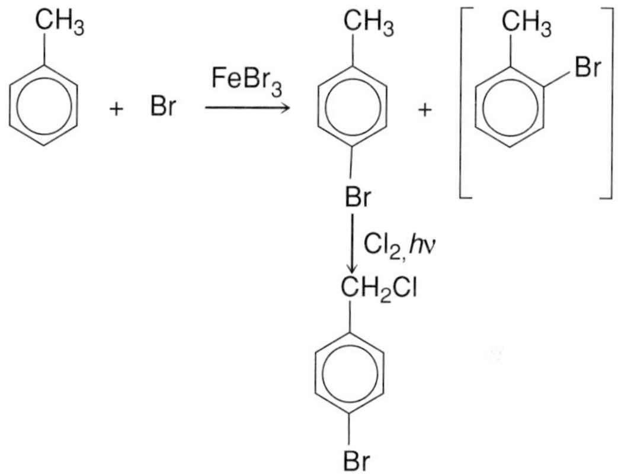

37. The Friedel Craft reactions will not take place on molecules with substituents more deactivating than halogens. Nitro group is much more deactivating than halogens.

## Chapter 11 Exercises

11.1. a.

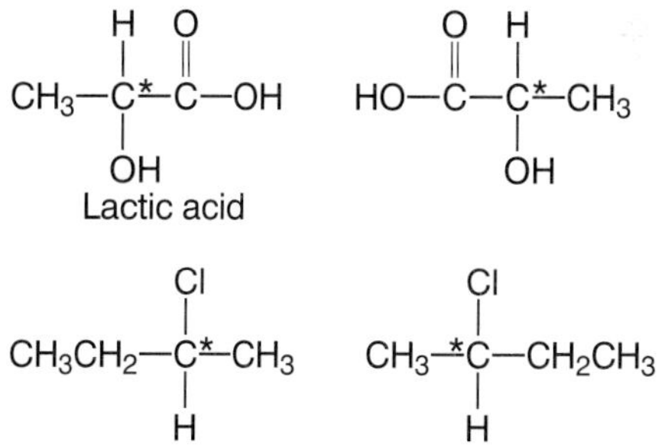

Lactic acid

b. Ethyl benzene contains no carbon with four different groups, i.e., it has no chiral center.

11.3. −31.5

11.5. a. Identical
b. Identical
c. Identical

7.   a.

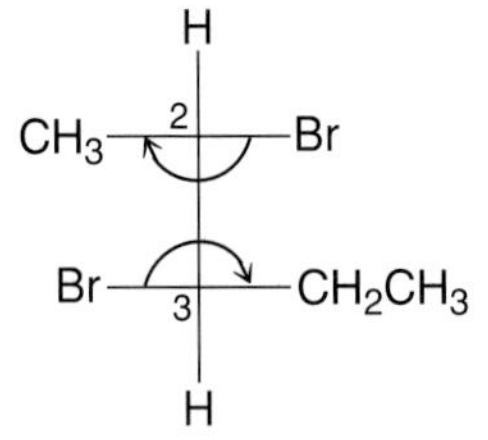

CH₃ —2— Br

Br —3— CH₂CH₃

c.

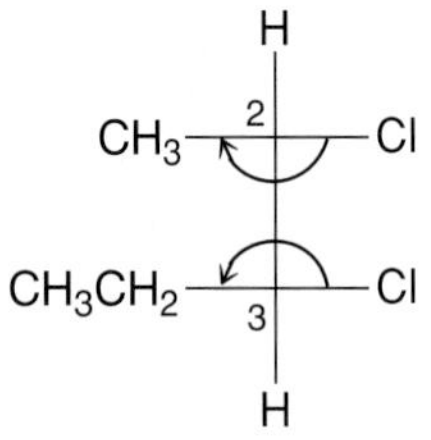

CH₃ —2— Cl

CH₃CH₂ —3— Cl

11.   a.

(R)     (S)

b.

Br —2— CH₃

CH₃ —3— Br

b.

Cl
①

CH₃

(R)

Cl —— CH₃

(S)

## Chapter 11 Problems

1.   H₃C—CH₂—CHCl₂

CH₃—C(Cl)₂—CH₃

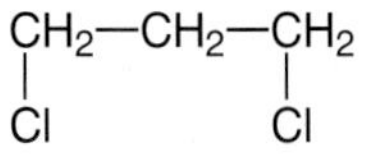

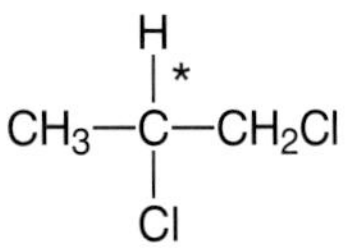

CH₂—CH₂—CH₂
|            |
Cl         Cl

H
| *
CH₃—C—CH₂Cl
|
Cl

1,2-Dichloropropane has a chiral center.

3.   a.

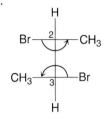

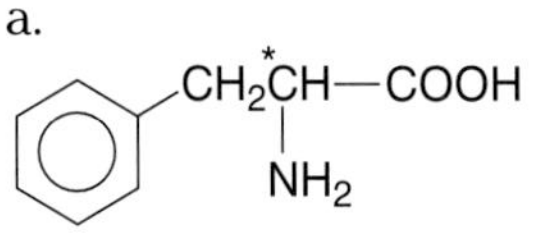

CH₂*CH—COOH
      |
      NH₂

b.

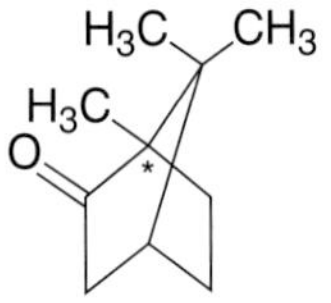

H₃C  CH₃

H₃C

O

c.

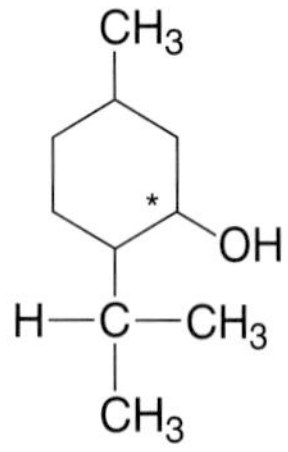

CH₃

OH

H—C—CH₃
|
CH₃

d.

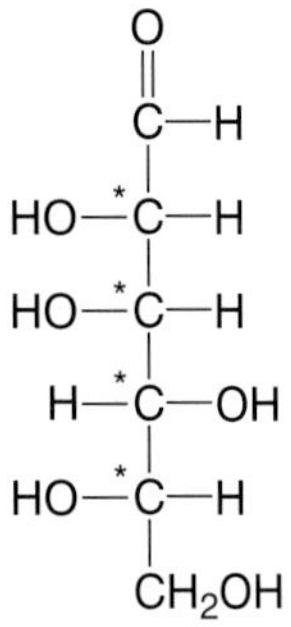

O
‖
C—H
|
HO—*C—H
|
HO—*C—H
|
H—*C—OH
|
HO—*C—H
|
CH₂OH

e.   CH₃—*C—C—*C—CH₃
        |  |  |
        Cl Cl Cl
(with H H H on top)

13.   a.
O
‖
—C—OH

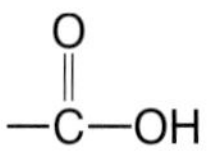

b. —C≡CH

c. —Cl

d.
H.

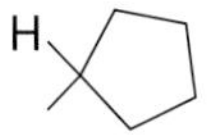

15.   a. Enantiomers
     b. Enantiomers

17.   a. (S); b. (S); c. (R); d. (R)

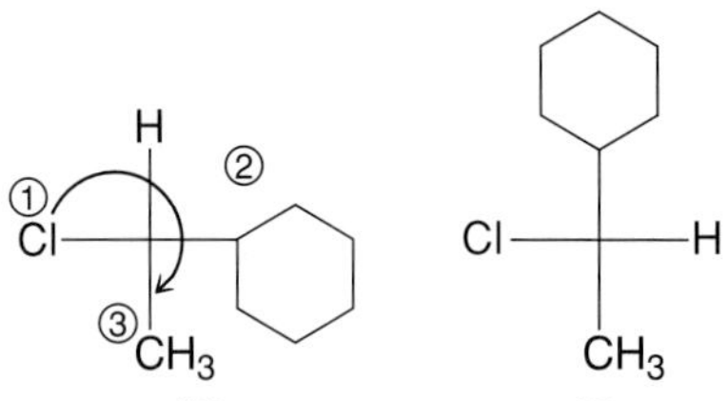

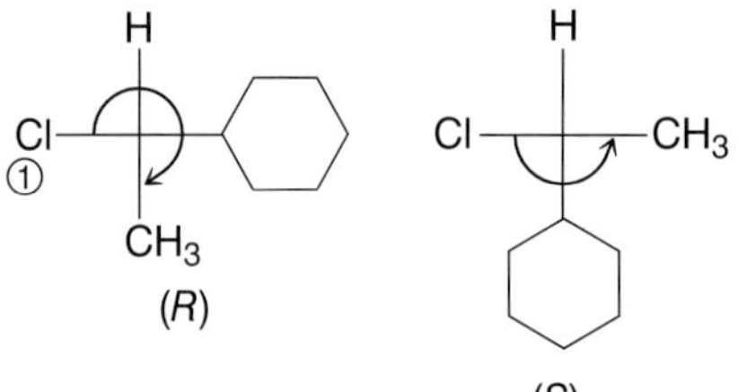

19.   a.

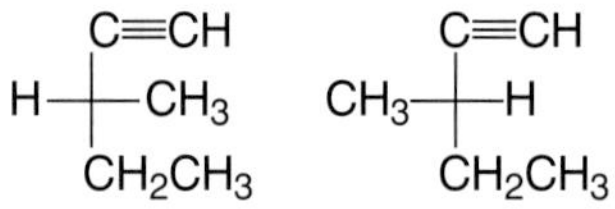

C≡CH        C≡CH
H—CH₃    CH₃—H
CH₂CH₃     CH₂CH₃

b. No stereoisomers can exist because no chiral center is present. Groups attached to central carbon are identical.

c. Two stereoisomers can exist.

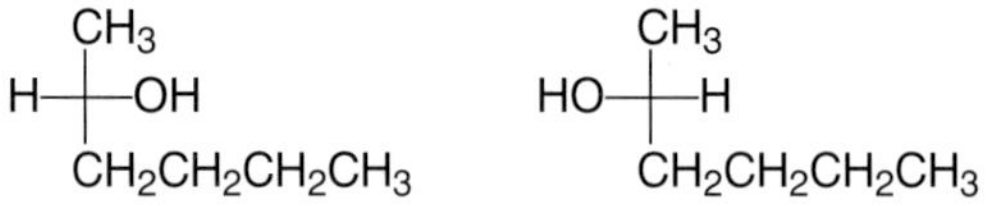

CH₃            CH₃
H—OH       HO—H
CH₂CH₂CH₂CH₃   CH₂CH₂CH₂CH₃

d.
O  OH H  OH
‖  |  |   |
H—C—C—C—C—CH₂OH
    |  |  |
    H  OH H

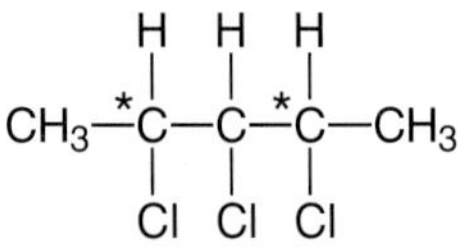

5.   3.8g cholesterol

7.   − 105.6°

9.   *sec*-Butyl chloride has a chiral center,

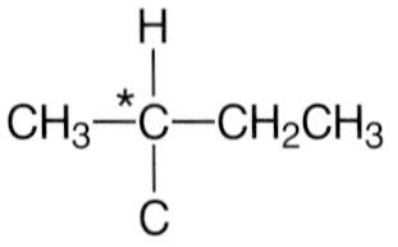

H
|
CH₃—*C—CH₂CH₃
|
C

and is optically active when resolved.

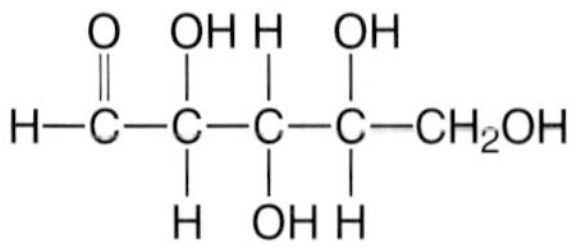

Eight stereoisomers can exist.

```
 CHO           CHO           CHO           CHO
 ──OH      HO──          ──OH      HO──
 ──OH          ──OH      HO──          HO──
 ──OH          ──OH          ──OH              ──OH
 CH₂OH         CH₂OH         CH₂OH         CH₂OH
```

Mirror images

- - - - - - - - - - - - - - - - - - - - - - - - - - - - - -

```
     CHO           CHO           CHO               CHO
HO──          ──OH      HO──              ──OH
HO──      HO──              ──OH              ──OH
HO──      HO──      HO──          HO──
 CH₂OH         CH₂OH         CH₂OH             CH₂OH
```

21.  $CH_3CH-CH-CH_2-CH_3$
      with $Cl$ and $Cl$ on the middle carbons

Mirror image

```
 CH₃          CH₃           CH₃          CH₃
H─C─Cl       H──Cl    Cl──H        Cl─C─H
H─C─Cl       H──Cl    Cl──H        Cl─C─H
 CH₂CH₃       CH₂CH₃         CH₂CH₃       CH₂CH₃
(2S, 3R)      2S, 3R    (2R, 3S)      2R, 3S
Dichloropentane           Dichloropentane
```

```
 CH₃          CH₃           CH₃          CH₃
H─C─Cl   =   H──Cl    Cl──H    =   Cl─C─H
Cl─C─H       Cl──H    H──Cl        H─C─Cl
 CH₂CH₃       CH₂CH₃         CH₂CH₃       CH₂CH₃
(2S, 3S)      2S, 3S    (2R, 3R)      2R, 3R
Dichloropentane           Dichloropentane
```

2S, 3S, and 2R, 3R  enantiomers
2S, 3R, and 2R, 3S  enantiomers

$$\left.\begin{array}{l} 2S,3S \text{ and } 2S,3R \quad 2R,3R \text{ and } 2R,3S \\ 2S,3R \text{ and } 2R,3R \quad 2S,3S \text{ and } 2S,3R \end{array}\right\} \begin{array}{l}\text{Diastereoiso-}\\ \text{meric pairs}\end{array}$$

23.

```
        CH₃①
       R②
Br────CH
- - - - - - - - - - -  Plane of symmetry
       S③
Br────H
        ④
        CH₃
```

If compound is numbered from $C_1$ to $C_4$, it is $(2R, 3S)$-dibromobutane. If the compound is numbered

from $C_4$ to $C_1$, it is $(2S, 3R)$-dibromobutane. These are meso and therefore the same.

```
     CH₃           CH₃
      R             S
Br────H       H────Br
      S             R
Br────H       H────Br
     CH₃           CH₃
```

25.  a. A 50:50 mixture of two enantiomers.

b. The rotation of an optically active compound with a concentration expressed as grams per milliliter measured in a tube whose length is measured in decimeters, i.e., $[\alpha] = \alpha/l\ (dm) \times g/ml$, where $[\alpha] =$ specific rotation, $\alpha =$ observed rotation, $l =$ length of tube, and $g/ml =$ solution concentration. $\alpha$ and $[\alpha]$ are expressed in degrees.

c. Used to specify the configuration of a molecule. The four groups attached to a chiral center are prioritized according to molecular weight. The group with the lowest priority (lowest molecular weight) is oriented away from the reader. If the remaining three goups are descending in a clockwise direction, the configuration is $R$. $R$ is from the latin word *rectus*, meaning "to the right."

d. The direction of rotation of the sample. If an optically active sample rotates plane-polarized light to the left, it is levorotary or $(-)$.

e. If an optically active compound is superimposable on its mirror image when a Fischer projection is rotated 180 degrees, the two structures, which represent the same molecule, are meso forms and are not optically active, even though each representation has chiral centers.

f. Two molecules that can be placed on top of one another and each atom of one molecule coincides with the same atom of the other molecule in its three-dimensional position.

g. The separation of a racemic mixture of $(R)$ and $(S)$ enantiomers into the pure $(R)$ and pure $(S)$ compounds.

h. The term chiral is derived from a greek word meaning "handed." To determine whether an object is "handed," look at the mirror image. If the mirror image is different from the original object/molecule, it is chiral.

i. Optically active:  If a compound rotates a plane of polarized light, it is optically active.

27.  a.

```
              H
              |
      ⬡───── C─CH₂CH₃
              |
              CH₃
```

b.

$$\text{C}_6\text{H}_5-\underset{\underset{\text{CH}_3}{|}}{\overset{\overset{\text{CH}_3}{|}}{\text{C}}}-\text{CH}_3$$

c.

$$\text{C}_6\text{H}_5-\text{CH}_2\text{CH}_2\text{CH}_2\text{CH}_3$$

d.

$$\text{C}_6\text{H}_5-\text{CH}_2\underset{\underset{\text{CH}_3}{|}}{\text{CH}}-\text{CH}_3$$

29.   a.

$$\underset{\underset{\overset{|}{\text{OH}}}{\overset{|}{\text{Br}}}}{\overset{\text{H}_3\text{C}}{\text{C}}}=\overset{\overset{\text{H}}{|}}{\underset{\underset{\text{CH}_3}{}}{\text{C}}}\text{CH}_3 \qquad\qquad \text{H}_3\text{C}\overset{\text{H}}{\text{C}}=\underset{\underset{\overset{|}{\text{OH}}}{\overset{|}{\text{Br}}}}{\text{C}}\text{CH}_3$$

*cis*              *cis*

b.

*trans*            *trans*

c.

$$(\text{CH}_3)_2\text{HC}-\underset{\underset{\text{HO}-\text{C}=\text{O}}{|}}{\overset{\overset{\text{H}}{|}}{\text{C}}}-\text{NH}_2 \qquad \text{H}_2\text{N}-\underset{\underset{\text{O}=\text{C}-\text{OH}}{|}}{\overset{\overset{\text{H}}{|}}{\text{C}}}-\text{CH(CH}_3)_2$$

Valine

31.   a.   $\text{H}_2\text{C}=\overset{\overset{\text{H}}{|}}{\text{C}}-\text{CH}_3 + \text{H}_2 \xrightarrow{\text{Pd}} \text{CH}_3\text{CH}_2\text{CH}_3$

                    No isomers

   b.   $\text{CH}_3\text{CH}_2\text{CH}=\text{CH}_2 + \text{HBr} \longrightarrow \text{CH}_3\text{CH}_2-\underset{\underset{\text{H}}{|}}{\overset{\overset{\text{Br}}{|}}{\text{C}}}-\text{CH}_3$

$$\text{CH}_3-\underset{\underset{\text{CH}_2\text{CH}_3}{|}}{\overset{\overset{|}{}}{}}\text{Br} \qquad \text{Br}-\underset{\underset{\text{CH}_2\text{CH}_3}{|}}{\overset{\overset{|}{}}{}}\text{CH}_3$$

1 Chiral center
2 Stereoisomers

   c.

$$\text{CH}_3-\underset{\underset{}{\overset{\overset{\text{CH}_3\;\text{H}}{|\;\;|}}{\text{C}}}}{}=\overset{}{\text{C}}-\text{CH}_2\text{CH}_3 + \text{HCl} \longrightarrow \text{CH}_3-\underset{\underset{\text{Cl}}{|}}{\overset{\overset{\text{CH}_3}{|}}{\text{C}}}-(\text{CH}_2)_2\text{CH}_3$$

Major product
No stereoisomers

$$+ \;\text{CH}_3-\underset{\underset{\text{H}\;\;\text{Cl}}{|\;\;\;|}}{\overset{\overset{\text{CH}_3\;\text{H}}{|\;\;\;|}}{\text{C}}}-\overset{*}{\text{C}}-\text{CH}_2\text{CH}_3$$

Minor Product
1 Chiral center
2 Stereoisomers

$$\text{CH}_3\text{CH}_2-\underset{\underset{\text{CH(CH}_3)_2}{|}}{\overset{\overset{\text{H}}{|}}{}}\text{Cl} \qquad \text{Cl}-\underset{\underset{\text{CH(CH}_3)_2}{|}}{\overset{\overset{\text{H}}{|}}{}}\text{CH}_2\text{CH}_3$$

d.

$$\text{C}_6\text{H}_5-\underset{\underset{\text{CH}_3}{|}}{\overset{\overset{\text{H}}{|}}{\text{C}}}-\text{C}_2\text{H}_5 + \text{Br}_2 \xrightarrow{h\nu} \text{C}_6\text{H}_5-\overset{*}{\underset{\underset{\text{CH}_3}{|}}{\overset{\overset{\text{Br}}{|}}{\text{C}}}}-\text{C}_2\text{H}_5$$

1 Chiral center
2 Stereoisomers

$$\text{H}_5\text{C}_2-\underset{\underset{\text{C}_6\text{H}_5}{|}}{\overset{\overset{\text{CH}_3}{|}}{}}\text{Br} \qquad \text{Br}-\underset{\underset{\text{C}_6\text{H}_5}{|}}{\overset{\overset{\text{CH}_3}{|}}{}}\text{C}_2\text{H}_5$$

33.    The structure of $\text{C}_5\text{H}_{11}\text{OH}$ is

$$\text{CH}_3-\underset{\underset{\text{CH}_3}{|}}{\text{CH}}-\underset{\underset{\text{OH}}{|}}{\text{CH}}-\text{CH}_3$$

## Chapter 12 Exercises

12.1.   a. yes; b. no; c. yes

12.3.   Polarimetric measurements can only prove the change in configuration in 12.2b, the formation of (*R*) 2-methoxybutane from (*S*) 2-chlorobutane.

12.5.   Partial double bond character of the carbon halogen bond due to resonance strengthens the carbon halogen bond and makes $S_N2$ substitution more difficult.

$$\text{CH}_2=\overset{\overset{\text{H}}{|}}{\text{C}}-\text{Cl}^- \longleftrightarrow \;\;{}^-{:}\text{CH}_2-\overset{\overset{\text{H}}{|}}{\text{C}}=\text{Cl}^+$$

12.7.   a.   $S_N2$

         $\text{CH}_3\text{CH}_2\text{CH}_2\text{I} + {}^-\text{OCH}_3 \longrightarrow \text{CH}_3\text{CH}_2\text{CH}_2\text{OCH}_3 \; {}^+\text{I}^-$

      b.   $S_N1$

         $(\text{CH}_3\text{CH}_2)_3\text{Cl} + \text{CH}_3\text{OH} \longrightarrow (\text{CH}_3\text{CH}_2)_3\,\text{C}-\text{OCH}_3 \; {}^+\text{I}^-$

12.9.   a.   $\text{CH}_3\text{CH}_2\underset{\underset{\text{OH}}{|}}{\text{CH}}-\text{CH}_3 + \text{HI} \xrightarrow{\Delta} \text{CH}_3\text{CH}_2\underset{\underset{\text{I}}{|}}{\text{CH}}\text{CH}_3$

                   (*S*)-Isomer

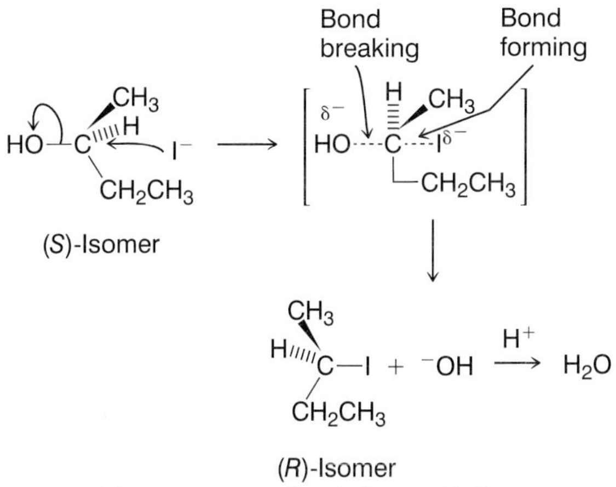

Bond breaking    Bond forming

(S)-Isomer

(R)-Isomer

b. $(CH_3)_3COH + HI \longrightarrow (CH_3)_3C-I + H_2O$

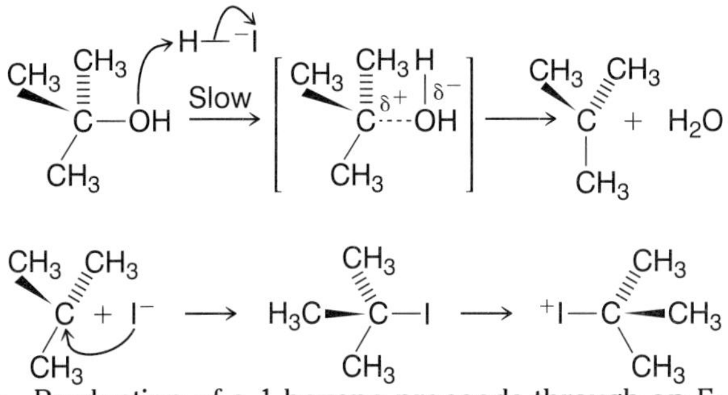

12.11. a. Production of a 1-hexene proceeds through an $E_2$ reaction.    b. Since 2-methyl-2-butanol is a tertiary alcohol, dehydration proceeds through an E1 mechanism.

12.13. a.

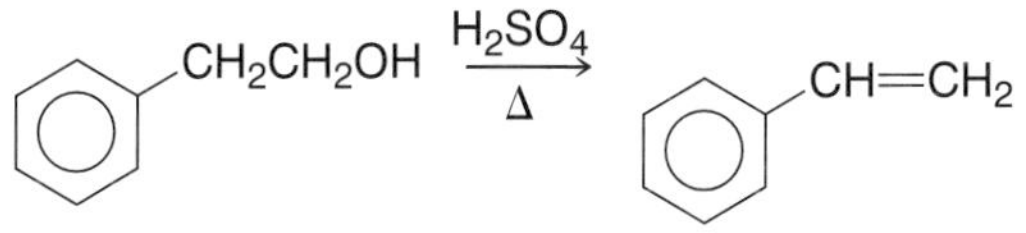

Using heat results in an elimination reaction. Since a primary alcohol is dehydrated, the reaction is E2.

b. $CH_3CH_2CH_2CH_2OH + HI \longrightarrow CH_3CH_2CH_2I$

Primary alcohols are converted to halides by an $S_N2$ process.

## Chapter 12 Problems

1. a.

$CH_3CH_2I + {}^-O-\overset{\displaystyle O}{\overset{\displaystyle \|}{C}}-CH_3 \longrightarrow CH_3CH_2-O-\overset{\displaystyle O}{\overset{\displaystyle \|}{C}}-CH_3 + I^-$

b. Rate $= k\,[CH_3CH_2I]\,[{}^-O-\overset{O}{\overset{\|}{C}}-CH_3]$

c. Substrate: $CH_3CH_2I$    Nucleophile: ${}^-O-\overset{O}{\overset{\|}{C}}-CH_3$

d.

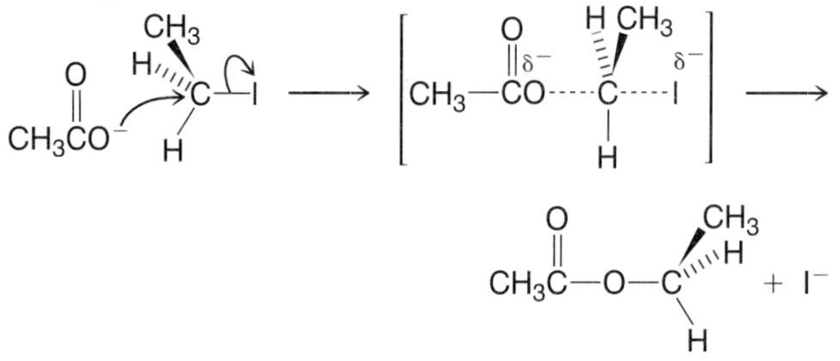

3.

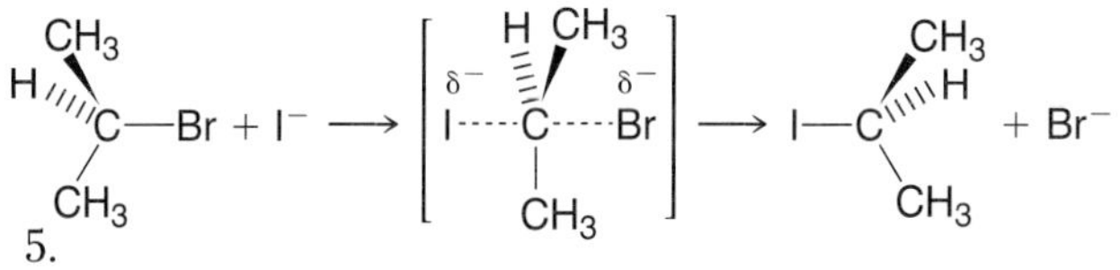

5.

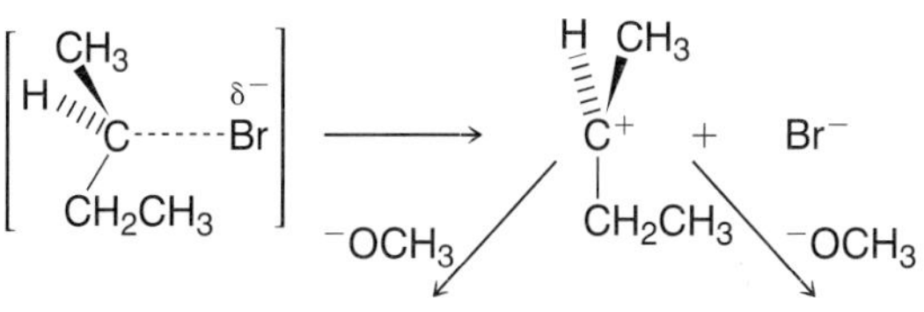

(R)-2-bromobutane            53%(S)

7. Because of steric hindrance, it is more difficult to form the $S_N2$ transition state for a tertiary halide than for a primary halide.

9. a. The benzyl carbocation is stabilized by resonance throughout the aromatic system, making it very stable.

b.

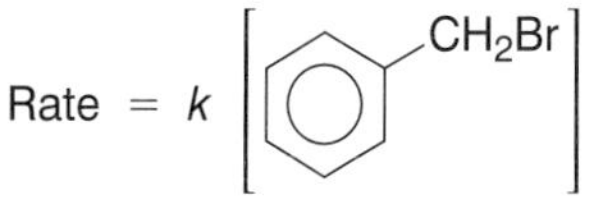

Rate $= k\left[\text{}\right]$

c.

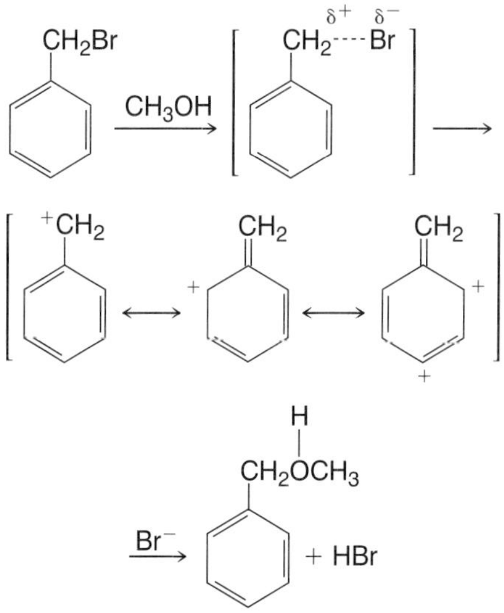

d. Because the aromatic system would be destroyed and the product would be less stable than the substitution product.

e. (i) $S_N2$

(ii) The strong nucleophile attacks the carbon before the carbocation has time to form. In $S_N1$ the formation of the carbocation is the rate-determining (slow) step.

(iii)

$$CH_2Br\text{-}C_6H_5 + {}^-OCH_3 \longrightarrow \left[ H_3CO \overset{\delta-}{\cdots} \underset{H}{\overset{H}{C}} \overset{\delta-}{\cdots} Br \right]$$

$$\longrightarrow CH_2OCH_3\text{-}C_6H_5 + Br^-$$

11. a. $H{-}C{\equiv}C^- Na^+ + Br{-}CH(CH_3)_2 \longrightarrow CH_2{=}CH{-}CH_3$ (Major)

   In the presence of a strong base, the secondary halide

   $$\left( Br{-}CH(CH_3)_2 \right)$$

   can undergo elimination to form an alkene, $CH_2{=}CH{-}CH_3$.

   b. $H{-}C{\equiv}C^- Na^+ + Br{-}CH_2CH_2CH_3 \longrightarrow$
   $HC{\equiv}C{-}CH_2CH_2CH_3$

   This reaction is the conventional $S_N2$ reaction to form $H{-}C{\equiv}C{-}CH_2CH_2CH_3$. In the case of the secondary alkyl halide, an $E1/S_N1$ reaction occurs more readily than with a primary alkyl halide.

13. a. Due to the steric hindrance of the *tert*-butyl group adjacent to the alcohol carbon, a concerted mechanism is not probable.

   b. 3-Methyl-1-butene is less highly substituted. More highly substituted alkenes are favored in rearrangements.

   c.

   $$CH_3{-}\underset{CH_3}{\overset{CH_3}{C}}{-}CH_2\ddot{O}H \quad H^+ \longrightarrow CH_3{-}\underset{CH_3}{\overset{CH_3}{C}}{-}CH_2{-}\overset{+}{O}{-}H$$

   $$CH_3{-}\overset{+}{C}{=}CHCH_3 \longleftarrow CH_3{-}\underset{CH_3}{\overset{CH_3}{C}}{-}CH_2^+$$

   $$CH_3{-}\underset{CH_3}{\overset{CH_3}{C}}{=}C\overset{H}{\underset{CH_3}{}} + H^+$$

## Chapter 13 Exercises

13.1. a. 1-Bromo-2-methyl propane (isobutyl bromide)
   b. Trichloromethane (chloroform)
   c. Tetrabromomethane (carbon tetrabromide)
   d. 1-Chloro-2-pentene
   e. 1-Iodo-2-butyne
   f. 2-Bromo-6-chloro-2,5,5-trimethyl-3-nitrononane

13.3. a. *n*-Hexyl bromide; b. *n*-Pentyl chloride; c. *n*-Pentyl iodide

## Chapter 13 Problems

1. a. Tribromomethane (IUPAC)  Primary
   Bromoform (common)
   b. 1-Chloroethene (IUPAC)  Vinylic
   Vinyl chloride (common)
   c. 1-Chloro-3-nitrobenzene (IUPAC)
   *m*-Chloronitrobenzene (common)
   d. 1-Methyl-1-iodocyclopentane (IUPAC)  Tertiary
   Cyclopentyliodomethane (common)
   e. 1-Bromo-2,2-dichloro-1-iodo-3-methylbutane (IUPAC)
   α-Bromo-β-dichloro-α-iodoisopentane (common)
   Bromo: primary    Chloro: secondary
   Iodo: primary
   f. 3-Chlorocycloheptene (IUPAC)  Allylic
   α-Chlorocycloheptene (common)
   g. 1-Bromo-2-methylpropane (IUPAC)  Primary
   Isobutylbromide (common)
   h. 1-Chloro-1,1,1-triphenylmethane (IUPAC)  Primary
   Triphenylchloromethane (common)
   i. 2,3-Diiodo-5-methylheptane (IUPAC)  Both primary
   j. Bromoethyne (IUPAC)
   Bromoacetylene (common)
   k. 1-Iodobutane (IUPAC)  Primary
   *n*-Butyliodide (common)

3. a. $CH_3{-}\underset{CH_3}{\overset{CH_3}{C}}{-}CH_2Cl$

   b. $C_6H_3(Br)(NO_2)(NO_2)$ (1-bromo-2,4-dinitrobenzene)

   c. cyclohexyl–Br

   d. $CH_3CH_2{-}\underset{CH_3}{\overset{CH_3}{C}}{-}I$

   e. $(H_3C)_2CHCH_2CH_2Cl$

   f. $CH_3CH_2{-}\underset{CH_3}{\overset{}{CH}}{-}I$

   g. $CH_3{-}\underset{Br}{\overset{}{CH}}{-}\underset{Cl}{\overset{Cl}{C}}{-}CH_2{-}\underset{CH_3}{\overset{}{CH}}{-}\underset{NO_2}{\overset{}{CH}}{-}CH_2CH_2{-}CH_3$

h. 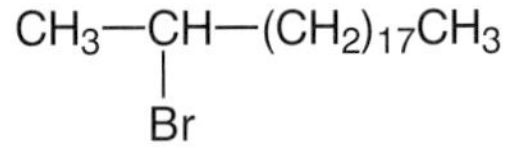 $CH_3-\underset{\underset{Br}{|}}{CH}-(CH_2)_{17}CH_3$

5.  a. Vinylic; b. Tertiary; c. Benzylic
    d. $(CH_3)_2C=C(CH_3)CH_2Br$ is allylic.

7.  a. $CH_3(CH_2)_2CH_2Br > CH_3CH(Br)CH_2CH_3 > (CH_3)_3CBr$
    b. $CH_3CH_2CH_2I > CH_3CH_2CH_2Br > CH_3CH_2CH_2Cl$

9.  a.

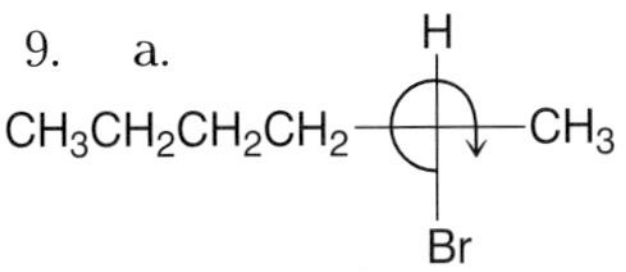

(R)-2-bromohexane 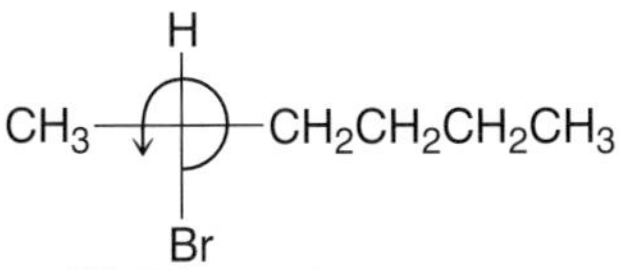    (S)-2-bromohexane

    b.

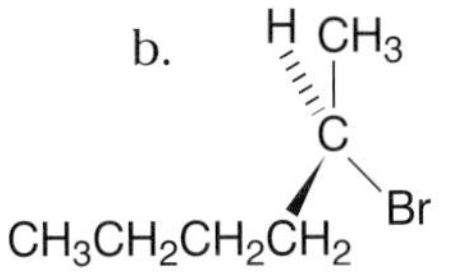

(R)-2-bromohexane 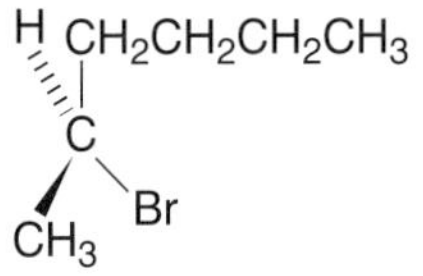    (S)-2-bromohexane

11. Isopropyl chloride. The presence of a white precipitate in 4 minutes indicates a secondary chloride.

13. The compound is *sec*-butyl bromide:

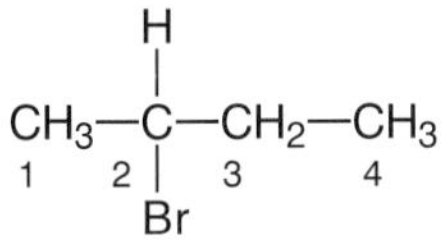

## Chapter 14 Exercises

14.1.  a. 4-Pentyn-1-ol
       b. 3-Pentanol or pentan-3-ol
       c. 3-Methylbutan-1-ol

## Chapter 14 Problems

1.  a. 2-Methylpropan-1-ol (IUPAC) or isobutyl alcohol (common)
    b. (R)-1-Phenyl-1-ethanol (IUPAC)
    c. 2,4-Pentanediol
    d. 3-Buten-1-ol
    e. 1-Methylcyclopentanol (IUPAC)
       α-Methylcyclopentyl alcohol (common)
    f. 5-Bromo-4-chloro-2-methyl-1-pentanol
    g. 2-pentanol
    h. 1-pentanol (IUPAC) or *n*-pentyl alcohol (common)

3.  a. 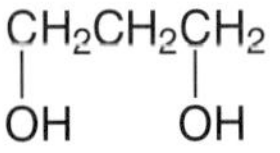 $\underset{\underset{OH}{|}}{CH_2}\underset{\underset{OH}{|}}{CH_2}CH_2$   $HOCH_2CH_2CH_2OH$

    b. 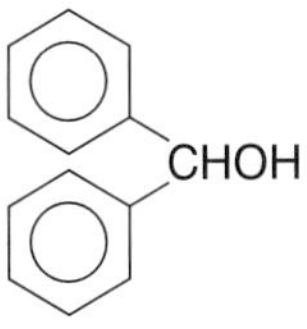                          c.

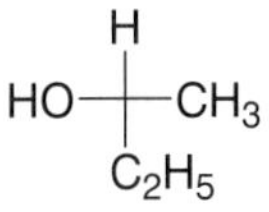

d. 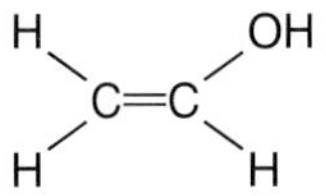   e. 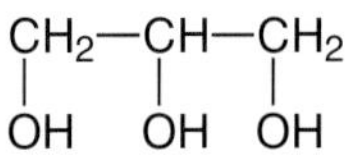 $\underset{\underset{OH}{|}}{CH_2}-\underset{\underset{OH}{|}}{CH}-\underset{\underset{OH}{|}}{CH_2}$

f. $CH_3OH$                  g.

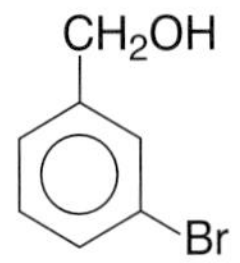

h. 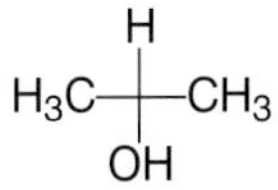 $H_3C-\underset{\underset{OH}{|}}{\overset{\overset{H}{|}}{C}}-CH_3$

5.  a. $HC\equiv C-CH_2OH$  Primary (1°)
    b. 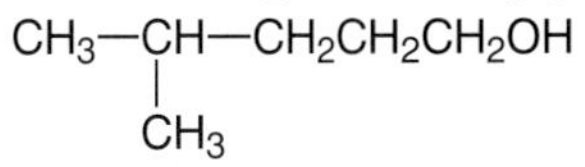 $CH_3-\underset{\underset{CH_3}{|}}{CH}-CH_2CH_2CH_2OH$  Primary (1°)

    c. 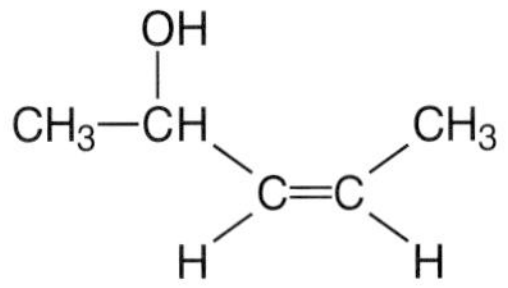 $CH_3-\underset{\underset{OH}{|}}{CH}\underset{\underset{H}{|}}{\overset{}{C}}=\underset{\underset{H}{|}}{C}CH_3$  Secondary (2°)

    d. 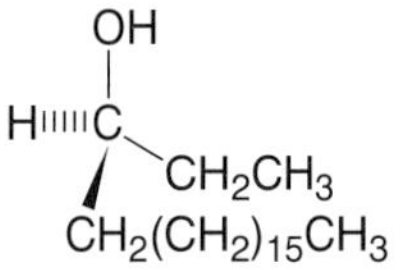   Secondary (2°)

    e. 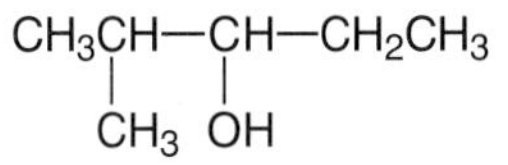 $CH_3CH-\underset{\underset{OH}{|}}{CH}-CH_2CH_3$  Secondary (2°)
       $\quad\;\underset{CH_3}{|}$

    f. 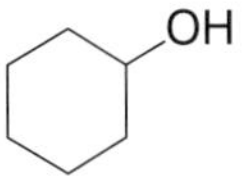   Secondary (2°)

7. 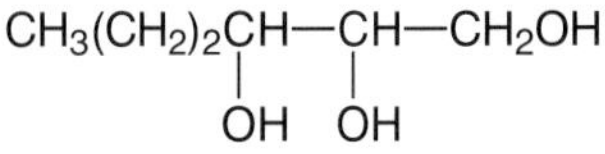  $CH_3(CH_2)_2\underset{\underset{OH}{|}}{CH}-\underset{\underset{OH}{|}}{CH}-CH_2OH$  (iii)

9.  Ethanethiol would boil lower than 1-propanol.

11. a. 2-Chloroethanethiol
    b. 5-Hexyn-1-ol
    c. 2-Mercapto-2-methyl propanol

## Chapter 15 Exercises

15.1  a.

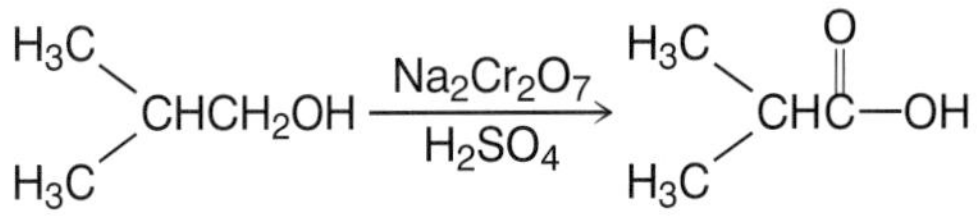

b.

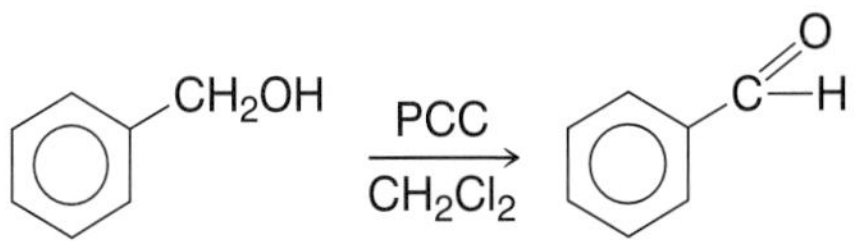

15.3.  a. Add concentrated sulfuric acid to a test tube containing heptane and to one containing heptanol. The heptanol will dissolve, but the heptane will not, and two layers will be seen, with the lighter heptane as the top layer. Also add a solution of $Na_2Cr_2O_7$ and $H_2SO_4$ to each substance. The solution containing

1-heptanol will turn from orange to green. No color change is observed for the test tube containing heptane.

b. Add a drop of bromine in carbon tetrachloride to a test tube containing 1-heptanol and to one containing 1-heptene. The bromine will be decolorized in the tube containing the 1-heptene but not in the one containing 1-heptanol.

c. The same test can be used to distinguish between 1-heptanol and 1-heptyne as that used to distinguish between 1-heptanol and 1-heptene. A second method of distinguishing between an alcohol and an unsaturated hydrocarbon is to test each with pyridinium chlorochromate (PCC). 1-Heptanol will be oxidized to 1-heptanal. The color will change. 1-Heptyne will not react.

## Chapter 15 Problems

1. Treating propylene with water and sulfuric acid will yield *iso*-propyl alcohol (2-propanol) not *n*-propyl alcohol (1-propanol).

3. a.

b.

$$CH_3-CH-CH_3 \xrightarrow[\Delta]{H_2SO_4} CH_3CH=CH \xrightarrow[Pd]{H_2} CH_3CH_2CH_3$$
with OH on middle carbon

or

$$CH_3-CH-CH_3 \xrightarrow{PCC} CH_3-\overset{O}{\overset{\|}{C}}-CH_3$$
(with OH)

$$\xrightarrow[CH_3CH_2OH, HCl]{Zn/Hg} CH_3CH_2CH_3$$

c. $CH_3CH_2OH \xrightarrow[\Delta]{H_2SO_4} CH_2=CH_2 \xrightarrow{HCl} CH_3CH_2CL$

$$\downarrow SH^-$$

$$CH_3CH_2-S-SCH_2 \xrightarrow{I_2} CH_3CH_2SH$$
with CH_3 branch

d. $CH_2=CH-CH-CH_3 \xrightarrow[2. H_2O_2,\, ^-OH]{1. B_2H_6} CH_2-CH_2CH-CH_3$
with CH_3, and OH, CH_3 substituents

Note: $B_2H_6 = 2BH_3$

$$\downarrow \begin{array}{c}HNO_2 \\ H_2SO_4\end{array}$$

$$O=N-OCH_2CH_2CH-CH_3$$
with CH_3

e.

f.

$$CH_3-CH-CH_2OH \xrightarrow[\Delta]{H_2SO_4} CH_3-C=CH_2$$
with CH_3 substituents

$$\xrightarrow[2.\ H_2O,\ H_2SO_4,\ \Delta]{1.\ H_2SO_4,\ cold} CH_3-\overset{CH_3}{\underset{CH_3}{C}}-OH \xrightarrow{Na^*} CH_3-\overset{CH_3}{\underset{CH_3}{C}}-O^-Na^+$$

5. $$R-OR' + H_3\overset{+}{O} \rightleftharpoons R-\overset{\overset{H}{|}}{\overset{+}{O}}-R' + H_2O$$

7. a. $CH_3-\overset{CH_3}{\underset{CH_3}{C}}-OH$

b. $CH_3-\overset{CH_3}{\underset{CH_3}{C}}-OH$

c. No.

9. Compound A is $CH_3-\overset{CH_3}{\underset{OH}{CH}}-CH-CH_3$ .

11. a. $Na^+\ ^-OCH_3$

b. $Na^+{}^-O-\overset{CH_3}{\underset{CH_3}{C}}-CH_3$

c. $CH_3-\overset{\overset{+}{\overset{|}{O}}-H}{\underset{H}{C}}-CH_3$

d. $CH_3-O-\overset{O}{\underset{O}{\overset{\|}{S}}}-SH$

e. $CH_3CH_2O-\overset{O}{\underset{OH}{\overset{\|}{P}}}-OH$

13. a. Sodium isopropoxide; b. *sec*-Butyl hydrogen sulfate; c. *tert*-Butyl dihydrogen phosphate

15. $CH_3-\overset{CH_3}{\underset{H}{C}}-CH_2-OH$

## Chapter 16 Exercises

16.1. a. 3-Bromo-4-nitrophenol; b. *o*-Methylphenol (O-cresol); c. Potassium-α-naphthoxide or potassium-1-naphthoxide

16.3.

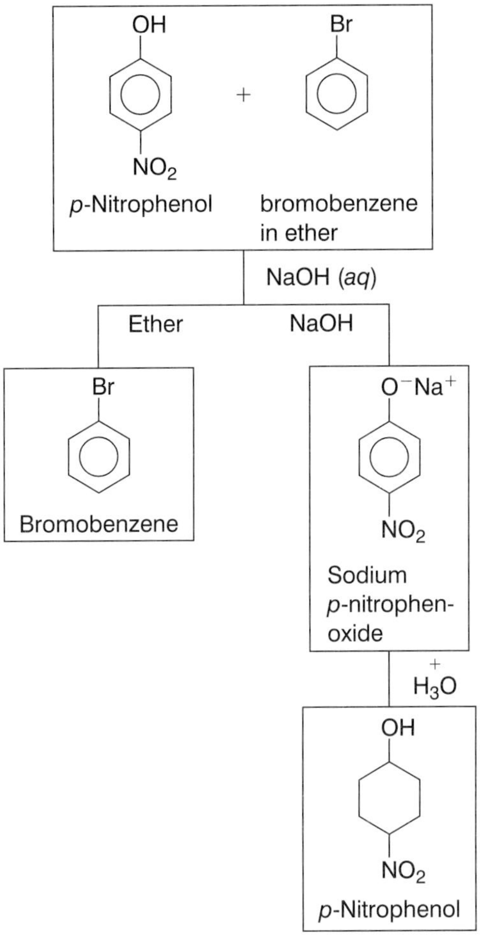

16.5. 2-*n*-Propyloxirane

## Chapter 16 Problems

1. a.

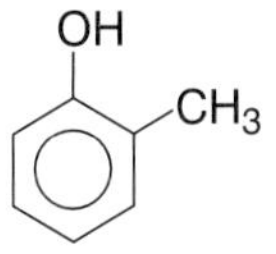

b.

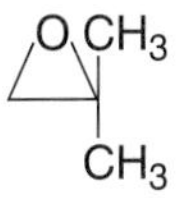

c.

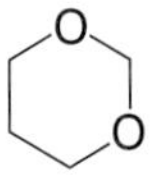

d. $CH_3-CHCH_2CH_2CH_3$

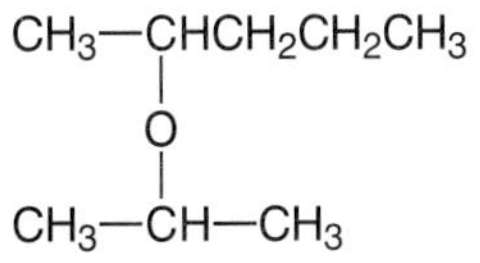

e.

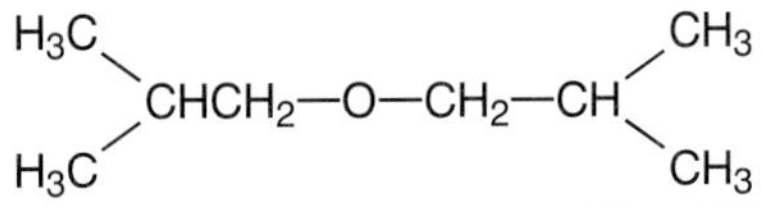

f.

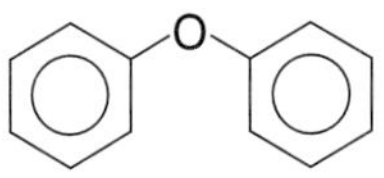

g. $CH_2=CH-O-CH=CH_2$

h. $CH_3CH_2-O-CH$

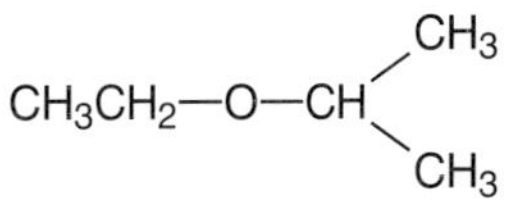

i.

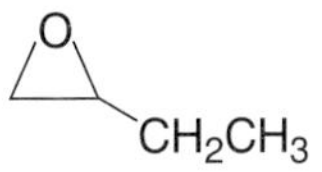

3. a. 3,5-Dichlorophenol
   b. 2-Ethyl-2-methyloxirane
   c. Ethyl *n*-butyl ether (common) 1-ethoxybutane (IUPAC)
   d. 2,5-Dimethoxyheptane
   e. Methyl cyclopentyl ether (common) Methoxycyclopentane (IUPAC)
   f. *o*-Methoxyphenol; 2-Methoxypitenol
   g. 1-Ethoxy-2-propanol
   h. *n*-propyl *n*-butylether (common) 1-propoxybutane (IUPAC)
   i. Methyl *iso*-propyl ether (common) 2-Methoxypropane (IUPAC)

5. 2,4-Dinitrophenol is a stronger acid than *p*-nitrophenol.

7. Branching lowers boiling points.

9.

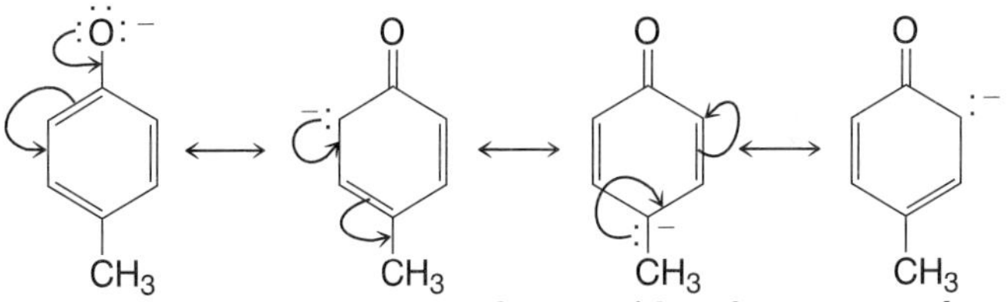

11. a. 6-crown-2; b. It can be considered a crown ether because it is a cyclic ether compound containing regularly spaced oxygen atoms ($-O-CH_2-CH_2-O-$).

13.

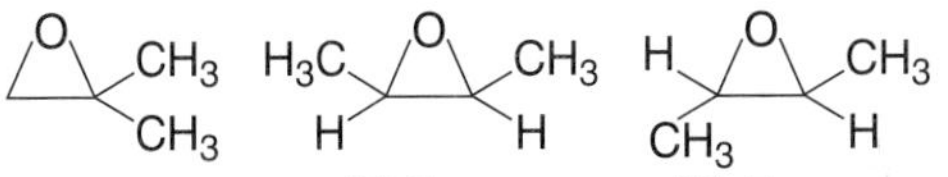

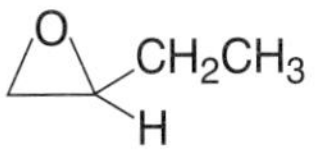

## Chapter 17 Exercises

1. a. *p*-benzoquinone or 1,4-benzoquinone
   b. 2-propyl-1,4-benzoquinone
   c. 2-ethyl-1,4-benzoquinone

3. a.

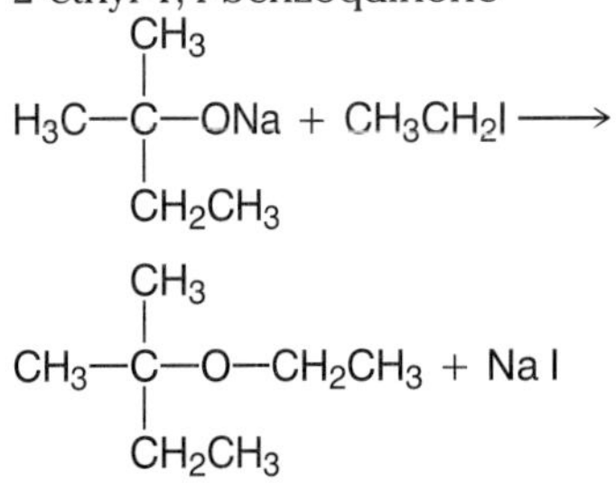

b.

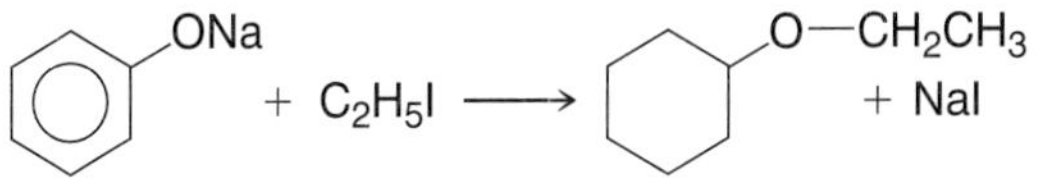

17.5 a.

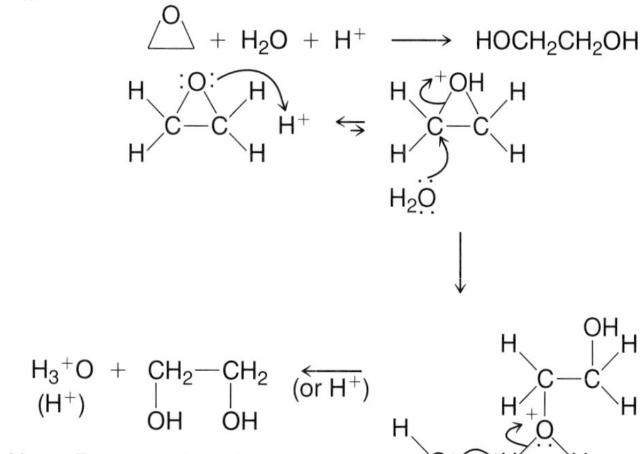

$$H_3{}^+O + CH_2\text{—}CH_2 \overset{\longleftarrow}{\text{(or H}^+)}$$
(H⁺)     |      |
         OH   OH

Note: Base catalyzed ring openings are S_N2 generally.

b.

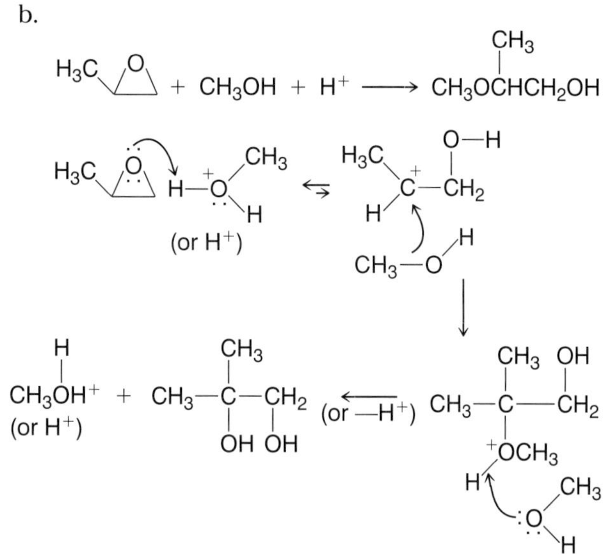

$$H_3C \overset{O}{\triangle} + CH_3OH + H^+ \longrightarrow CH_3OCHCH_2OH$$

c.

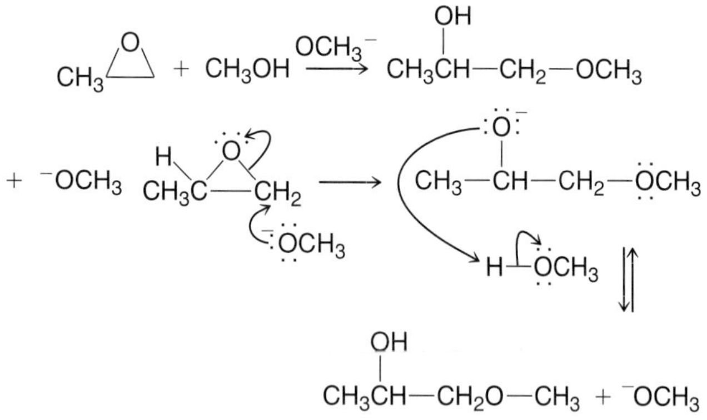

## Chapter 17 Problems

**1.** a. Using $CH_3OH$ and $CH_3CH_2CH_2OH$ to form an ether under alcohol dehydration conditions would yield a mixture of the following three products: $CH_3\text{—}O\text{—}CH_3$ $CH_3CH_2CH_2\text{—}O\text{—}CH_2CH_2CH_3$, and $CH_3\text{—}O\text{—}CH_2CH_2CH_3$. Only the last product is desired.

b.

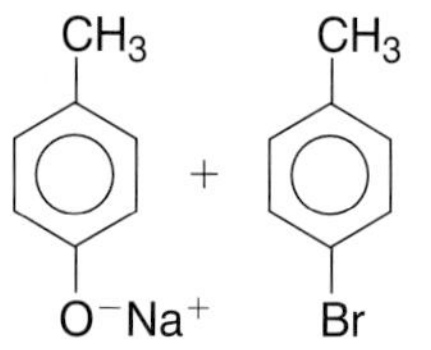

Must be used.
*p*- Bromotoluene is too unreactive ($S_N2$) and won't react.

3.

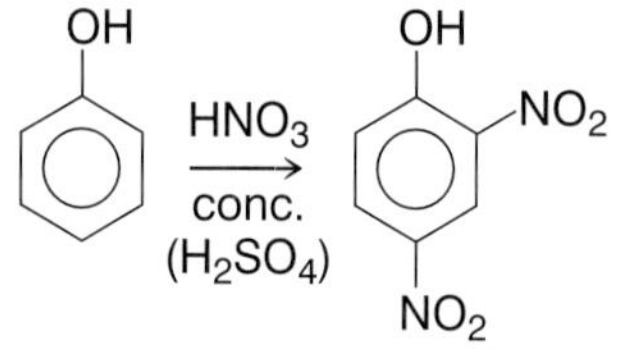

5.   a.

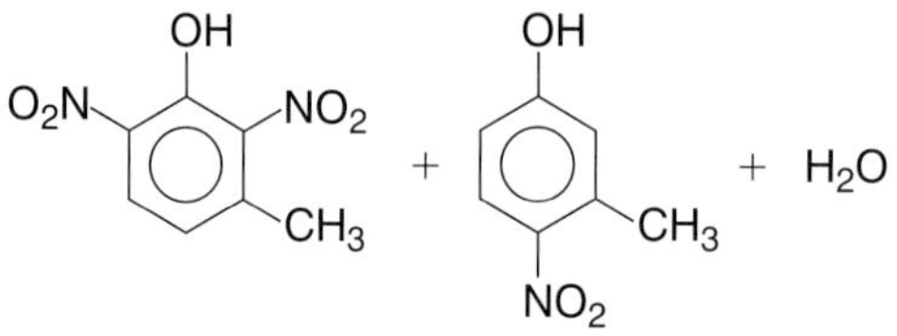

b.

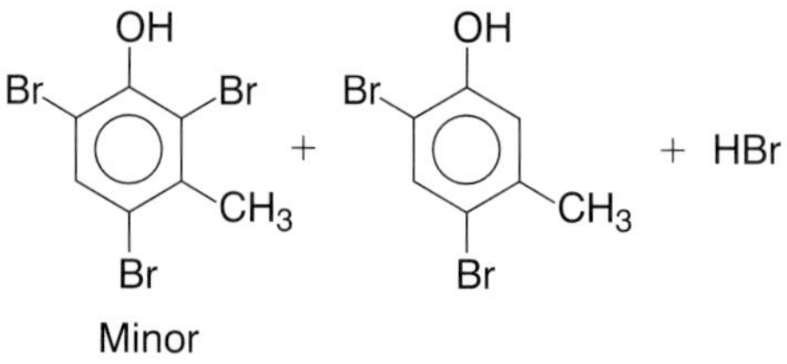

Minor

c. $CH_3CH_2\text{—}O\text{—}CH_2CH_3 + 2\ HBr \overset{\Delta}{\longrightarrow} 2\ CH_3CH_2Br + H_2O$

d.

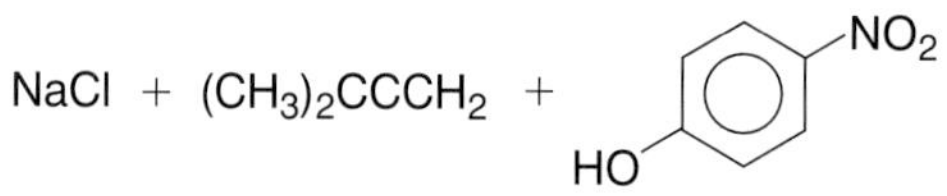

$$NaCl + (CH_3)_2CCCH_2 +$$

e.

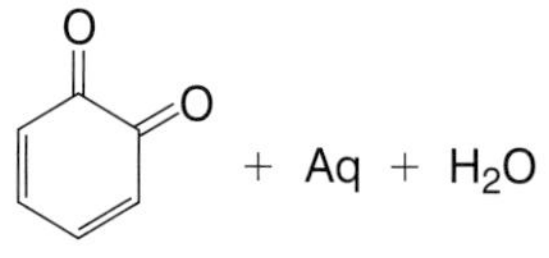

$$+ \ Aq + H_2O$$

f.

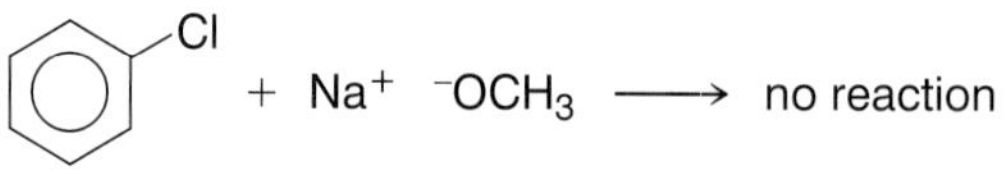

$+ Na^+ \ {}^-OCH_3 \longrightarrow$ no reaction

g. $2\ KCl + CH_2\text{=}CH\text{—}O\text{—}CH\text{=}CH_2$

h. $CH_2\text{—}CH\text{—}O\text{—}CH\text{—}CH_2$
        |      |           |      |
        Br    Br         Br    Br

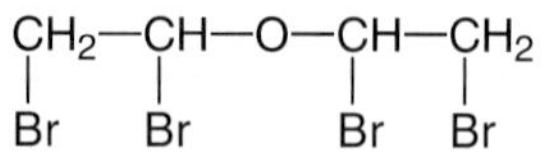

i. $CH_3\text{—}O\text{—}CH_3 + H_2O$

j.

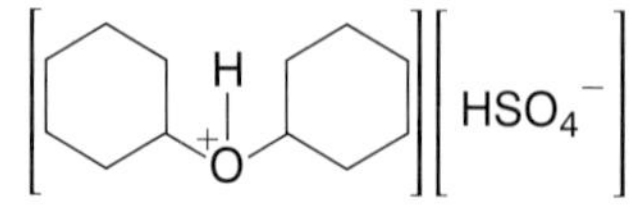

k. No reaction

l.

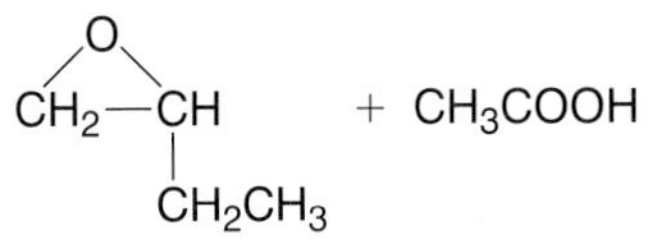

$$+ \ CH_3COOH$$

m.

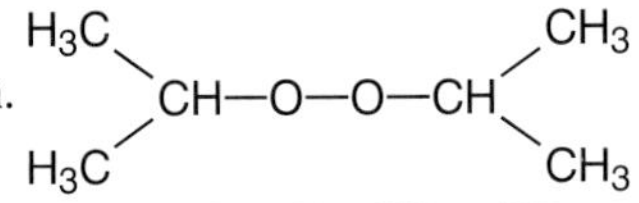

n. $CH_3CH_2CH_2CH_2CN + KBr$

o.

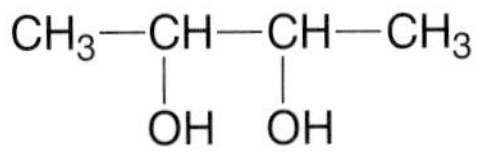

p. $CH_3I + CH_3{-}CH{-}CH_3 + H_2O$

q. $2 \ Fe^{3+} + CH_3CH_2CH_2{-}O{-}CH_2CH_2CH_3 + H_2O$

7.

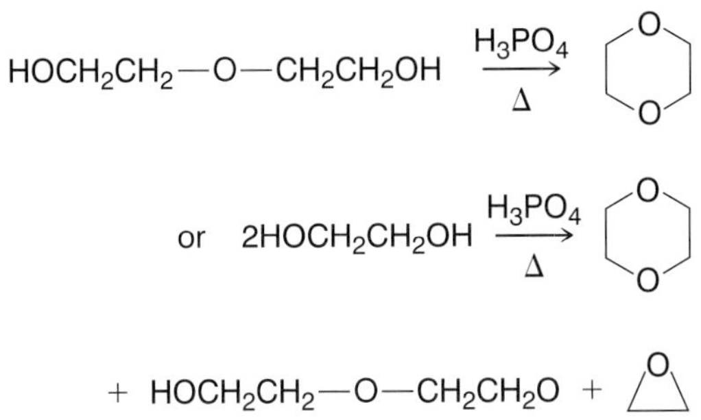

$$HOCH_2CH_2{-}O{-}CH_2CH_2OH \xrightarrow[\Delta]{H_3PO_4}$$

or $2HOCH_2CH_2OH \xrightarrow[\Delta]{H_3PO_4}$

$$+ \ HOCH_2CH_2{-}O{-}CH_2CH_2O +$$

9. a. Add water (cold con $H_2SO_4$) to samples of pentane and 1,4-dioxane. 1,4-Dioxane is soluble and pentane is not, and two layers will be observed, with pentane the top layer and water (cold con $H_2SO_4$) the bottom layer.
   b. Test both o-cresol and n-heptyl alcohol with ferric chloride. The phenol will give a deep color, normally purple or red. The heptyl alcohol will remain unchanged.
   c. Test phenol and benzoic acid with sodium bicarbonate solution. Benzoic acid is soluble in sodium bicarbonate. Phenol is not. Phenol requires a stronger base. It will dissolve in a sodium hydroxide solution.
   d. Ethyl alcohol will mix with water (con $H_2SO_4$). If water is added to ethyl ether, two layers form. The top layer is ether, and the bottom is water (con $H_2SO_4$).

11. a.  $+ \ HCl \longrightarrow HOCH_2CH_2Cl$
    b. 2-Chloroethanol or ethyl chlorohydrin

c.

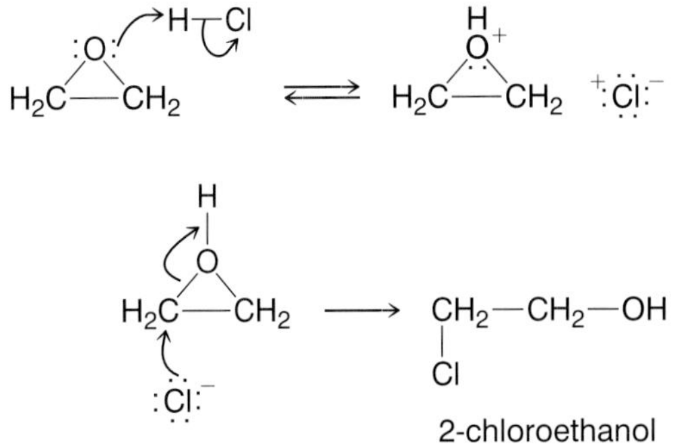

2-chloroethanol

13. $C_6H_6O$ is 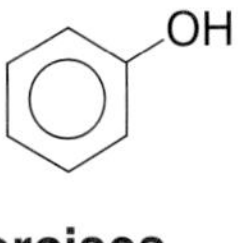 The compound is phenol.

## Chapter 18 Exercises

18.1. a. α-Iodo propionaldehyde, 2-iodopropanal
      b. Diethyl ketone, 3-pentanone

18.3. a. Acetal. There exist 2 gem —OR groups derived from an aldehyde because of one hydrogen on C bonded to 2 OR groups.
      b. Hemiacetal. Cyclic derived from one aldehyde (see a.).
      c. Unrelated. The —OH and —OR groups bonded to different carbons.

## Chapter 18 Problems

1. a. Propionaldehyde (common); Propanal (IUPAC)
   b. Acetophenone (common)/methyl phenylketone (common); 1-Phenyl-1-ethanone (IUPAC)
   c. α-Methyl butyraldehyde (common); 2-Methyl butanal (IUPAC)
   d. 2-Methylcyclobutanone (IUPAC)
   e. (α, α, α)-Trichloroacetaldehyde (common) 2,2,2-Trichloroethanal (IUPAC)
   f. 2-Butenal (IUPAC)
   g. Methyl isopropyl ketone (common) 3-Methyl-2-butanone (IUPAC)
   h. 5-Hexen-2-one (IUPAC)
   i. Methyl ethyl ketone (common); 2-Butanone (IUPAC)
   j. Heptanal (IUPAC)

3. a.

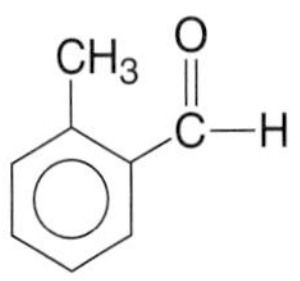

   b.

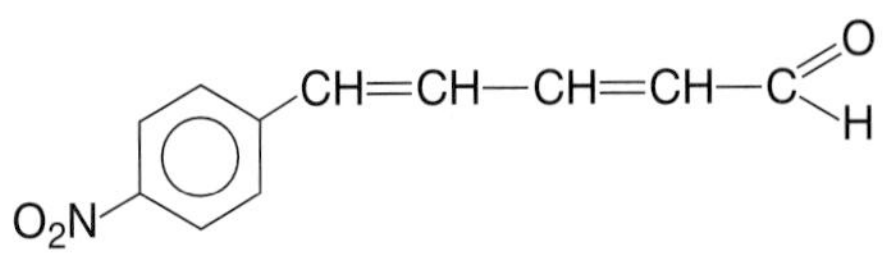

5. a. Butane; acetone; isopropyl alcohol
   b. Hexane; pentanal; 1-pentanol

7.　a. Acetal derived from an aldehyde; b. Acetal derived from a ketone; c. Hemiacetal derived from a ketone; d. Hemiacetal derived from an aldehyde.

9.　a.

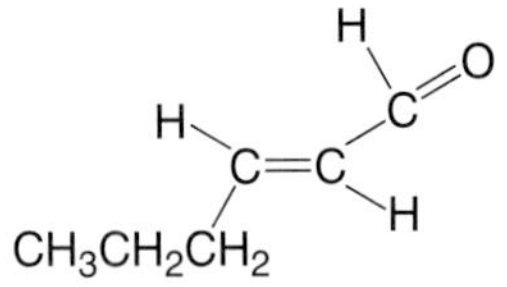

　b. $CH_3(CH_2)_9\overset{\overset{\displaystyle O}{\|}}{C}-H$

　c. $CH_3CH_2\overset{\overset{\displaystyle O}{\|}}{C}-\underset{\underset{\displaystyle CH_3}{|}}{CH}-CH_2CH_2CH_3$

11.　a. Propionaldehyde dimethyl acetal
　　b. Dimethyl ketone diethyl acetal

13.　a.

$CH_3CH_2CH_2-\underset{\underset{\displaystyle OCH_3}{|}}{\overset{\overset{\displaystyle OCH_3}{|}}{CH}} + H_2O \overset{H_3\overset{+}{O}}{\rightleftharpoons} CH_3CH_2CH_2\overset{\overset{\displaystyle O}{\|}}{C}-H$
　　　　　　　　　　　　Excess　　　　　　$+ 2CH_3OH$

　b.

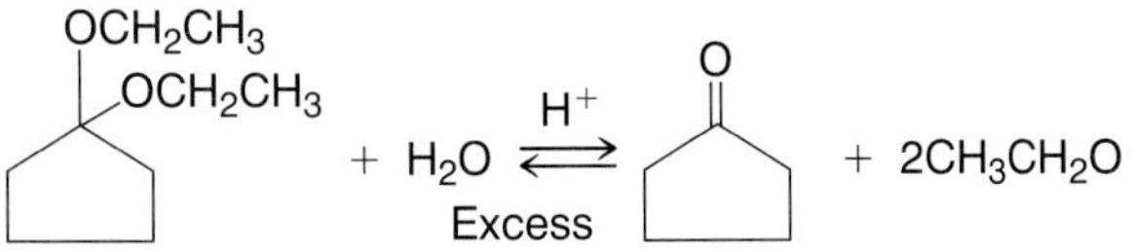

## Chapter 19 Exercises

19.1.　a. $CH_3(CH_2)_5CH=CH_2 \xrightarrow[\substack{2.\ H_2O_2,\\ OH^-/H_2O}]{1.\ BH_3} CH_3(CH_2)_5CH_2-CH_2OH$

$\xrightarrow[CH_2Cl_2]{PCC}$

$CH_3(CH_2)_6\overset{\overset{\displaystyle}{}}{C}-H$ （$\overset{O}{\|}$ below C）

　b.

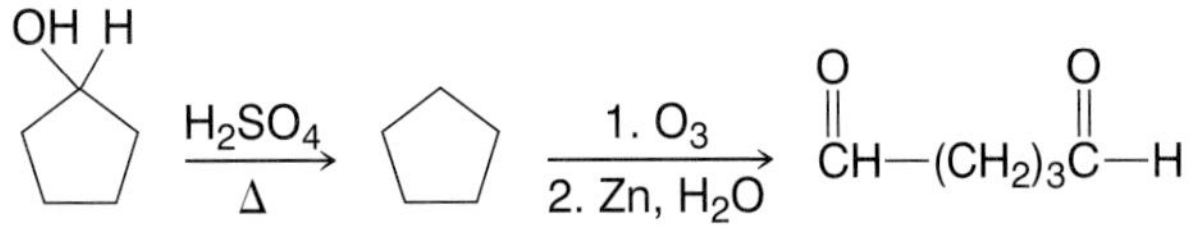

19.3.　a. $CH_3CH_2-\underset{\underset{\displaystyle OH}{|}}{CH}-CN$　　b. $H-\underset{\underset{\displaystyle OH}{|}}{\overset{\overset{\displaystyle H}{|}}{C}}-OH$

　　c. No reaction

19.5.　a. $CH_3CH_2\overset{\overset{\displaystyle O}{\|}}{C}-CH_3 + CH_3MgBr \xrightarrow[2.\ H_2O/H^+]{1.\ Ether}$

　　or $CH_3\overset{\overset{\displaystyle O}{\|}}{C}-CH_3 + CH_3CH_2MgBr \xrightarrow[2.\ H_2O/H^+]{1.\ Ether}$

　b. $H-\overset{\overset{\displaystyle O}{\|}}{C}-H + CH_3CH_2CH_2CH_2MgBr \xrightarrow[2.\ H_2O/H^+]{1.\ Ether}$

　c. $CH_3\overset{\overset{\displaystyle O}{\|}}{C}-H + CH_3MgBr \xrightarrow[2.\ H_2O/H^+]{1.\ Ether}$

　d. $H-\overset{\overset{\displaystyle O}{\|}}{C}-H + \underset{\underset{\displaystyle CH_3}{}}{\overset{\overset{\displaystyle CH_3}{}}{}}CHMgBr \xrightarrow[2.\ H_2O/H^+]{1.\ Ether}$

　e. $CH_3\overset{\overset{\displaystyle O}{\|}}{C}-H + CH_3CH_2CH_2CH_2MgBr \xrightarrow[2.\ H_2O/H^+]{1.\ Ether}$

　　or $CH_3CH_2CH_2CH_2\overset{\overset{\displaystyle O}{\|}}{C}-H + CH_3MgBr \xrightarrow[2.\ H_2O/H^+]{1.\ Ether}$

　f.

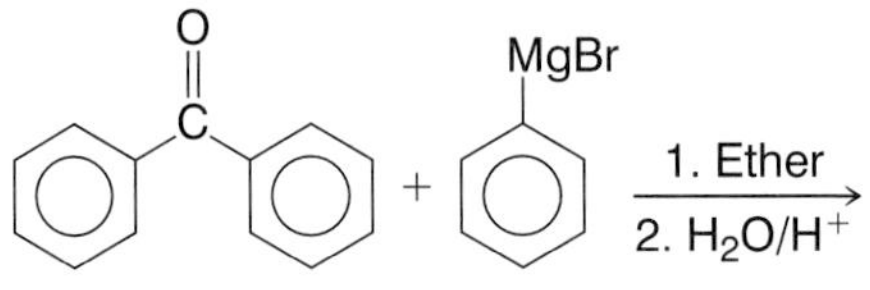

19.7.　a.

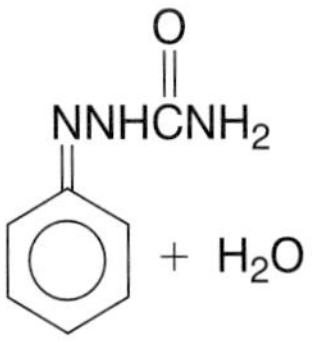

$+ H_2O$

　b. $H_2O + CH_3CH_2CH_2CH_2CH=N-OH$

　c.

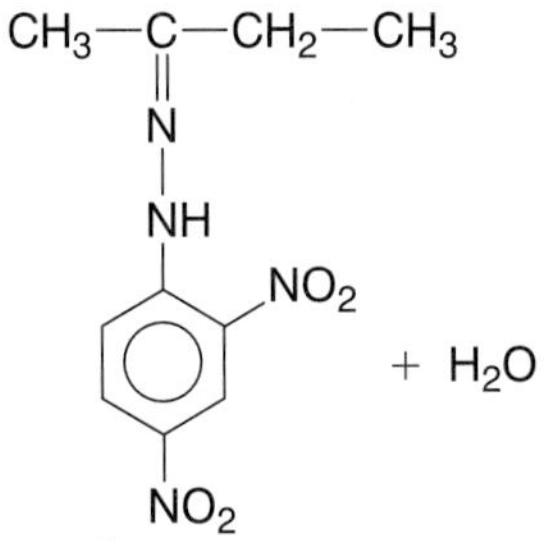

$+ H_2O$

19.9.　a. $CH_3\overset{\overset{\displaystyle O}{\|}}{C}-O^-Na^+ + CHI_3 + 3NaI + 3H_2OCH_3\overset{\overset{\displaystyle O}{\|}}{C}-O^-$

　b.

$+ CHBr_3 + 3NaBr + 3H_2O$

　c.

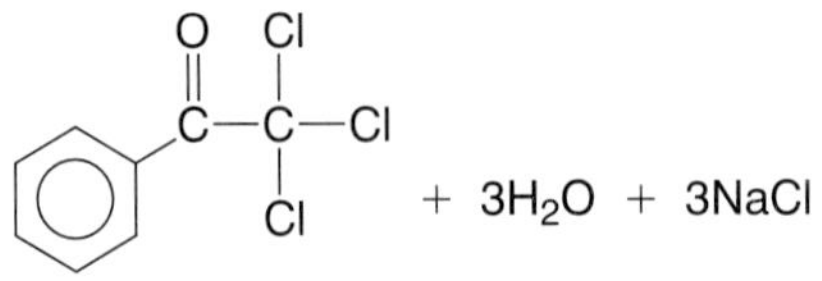

$+ 3H_2O + 3NaCl$

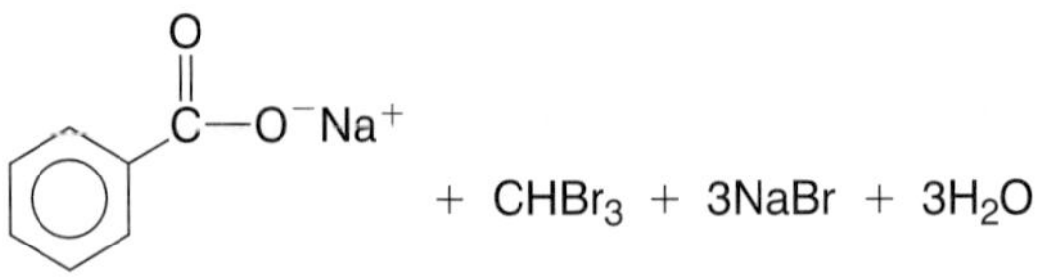

　d. $Cl-\underset{\underset{\displaystyle H}{|}}{\overset{\overset{\displaystyle H}{|}}{C}}-\overset{\overset{\displaystyle O}{\|}}{C}-H + HCl$

19.11. a.

$$CH_3-\overset{\overset{\displaystyle OH}{|}}{CH}-CH_3 + I_2 + \underset{\text{(excess)}}{NaOH} \text{ (aq)} \longrightarrow$$

$$CH_3\overset{\overset{\displaystyle O}{\|}}{C}-O^- \ Na^+ + CHI_3 + 3NaI + H_2O$$

b.

$$CH_3-\overset{\overset{\displaystyle O}{\|}}{C}-O^- \ Na^+ + H_3^+O \longrightarrow CH_3\overset{\overset{\displaystyle O}{\|}}{C}-OH + Na^+$$
$$\underset{\text{(or } H^{+)}}{\phantom{xx}} \qquad\qquad + H_2O$$

19.13. a. Butanal contains two alpha hydrogens.

b.

$$CH_3-CH_2-\overset{\overset{\displaystyle H}{|}}{\underset{\underset{:OH^-}{\overset{\displaystyle H}{\uparrow}}}{C}}-\overset{\overset{\displaystyle O}{\|}}{C}-H \rightleftharpoons CH_3CH_2\overset{\overset{\displaystyle H}{|}}{\underset{\displaystyle \cdot\cdot}{C}}-\overset{\overset{\displaystyle O}{\|}}{C}-H + H_2O$$

$$H-\overset{\overset{\displaystyle CH_2CH_3}{|}}{\underset{\displaystyle \cdot\cdot}{C}}-C\overset{\displaystyle O}{\diagdown}H + H-\overset{\overset{\displaystyle O}{\|}}{C} \ CH_2CH_2CH_3 \rightleftharpoons$$

$$\overset{H-\overset{O}{\diagup}\overset{\displaystyle H}{\phantom{x}} \ :\overset{\cdot\cdot}{O}:^-}{CH_3CH_2CH_2CH-\underset{\overset{|}{CH_2CH_3}}{CH}-\overset{\overset{\displaystyle O}{\|}}{C}-H}$$

$$\downarrow$$

$$CH_3CH_2CH_2\overset{\overset{\displaystyle OH}{|}}{CH}-\underset{\overset{|}{CH_2CH_3}}{CH}-\overset{\overset{\displaystyle O}{\|}}{C}-H$$

19.15. a.

$$HO-CH_2-CH_2-\overset{\overset{\displaystyle O}{\|}}{C}-H$$

Aldol would be the by product produced when the carbanion attacks the carbonyl carbon of another molecule of acetaldehyde.

$$CH_3\overset{\overset{\displaystyle O}{\diagup}}{C}-H + :\overset{\overset{\displaystyle H}{|}}{\underset{\overset{|}{H}}{C}}-\overset{\overset{\displaystyle O}{\|}}{C}-H \overset{\text{to give}}{\underset{\text{ultimately}}{\longrightarrow}} CH_3\overset{\overset{\displaystyle H}{|}}{\underset{\overset{|}{OH}}{C}}=H_2\overset{\displaystyle C}{}=O$$

b.

$$\underset{\overset{|}{OH}}{\overset{\overset{\displaystyle H \quad O}{\diagup \quad \|}}{C}}-CH_2\overset{\displaystyle C}{}=O$$ (benzene ring)

the by products will be aldol and

(benzene ring)$$-\overset{\overset{\displaystyle H}{|}}{C}=\overset{\overset{\displaystyle H}{|}}{C}-\overset{\overset{\displaystyle H}{|}}{\underset{\overset{\|}{O}}{C}}$$

19.17.

$$CH_3\overset{\overset{\displaystyle}{}}{\underset{\overset{|}{OH}}{CH}}-CH_3 \overset{H_2SO_4}{\underset{\Delta}{\longrightarrow}} CH_3-CH=CH_2 \overset{1.\ BH_3}{\underset{2.\ H_2O_2,\ OH}{\longrightarrow}} CH_3CH_2CH_2OH$$
$$\qquad\qquad\qquad\qquad\qquad\qquad\qquad\qquad\qquad \downarrow \underset{CH_2Cl_2}{PCC}$$

$$CH_3CH_2\overset{\overset{\displaystyle H}{|}}{\underset{\overset{|}{HO}}{C}}-C\equiv N \overset{HCN}{\underset{OH^-}{\longleftarrow}} CH_3-CH_2\overset{\overset{\displaystyle O}{\|}}{C}-H$$

## Chapter 19 Problems

1. a. *n*-Propyl aldehyde cyanohydrin (common)
   b. Ethanal imine (IUPAC)
      Acetaldehyde imine (common)
   c. Valeraldehyde oxime (common)
      Pentanal oxime (IUPAC)
   d. 2-Pentanone semicarbazone (IUPAC)
      Methyl propyl ketone semicarbazone (common)
   e. Propanal hydrate or 1,1-propanediol (IUPAC)
      Propionaldehyde hydrate

3. a.

$$CH_3-\overset{\overset{\displaystyle N-NH}{\|}}{C}-CH_3 \quad \text{(dinitrophenyl ring with } NO_2, NO_2\text{)}$$

   b. $$CH_3CH_2CH_2CH_2CH=N-NH-\overset{\overset{\displaystyle O}{\|}}{C}-NH_2$$

   c. $$CH_3-\overset{\overset{\displaystyle OCH_2CH_3}{|}}{\underset{\overset{|}{OH}}{C}}-CH_3$$

   d. $$CH_3\overset{\overset{\displaystyle OH}{|}}{\underset{\overset{|}{OH}}{C}}-CH_2CH_3$$

   e. $$CH_3CH_2CH_2CH_2\overset{\overset{\displaystyle O}{\|}}{C}CHO$$

5. a. $$H_2O \left( \delta + H \overset{\overset{\displaystyle \cdot\cdot}{\underset{\cdot\cdot}{O}}}{\diagdown} H^{\delta -} \right)$$  b. $CH_3OH$

   c. $$H_2N-NH-\overset{\overset{\displaystyle O}{\|}}{C}-NH_2$$  d. $\cdot\overset{\cdot\cdot}{C}H_3^-$

   e. $^-:C\equiv N$  f. $H:^-$

7. a.

   (nitro-substituted phenylhydrazine ring with $NO_2$, $NO_2$, $NNH$)
   $$CH_3\overset{\overset{\displaystyle NNH}{\|}}{C}CH_2CH_3 + H_2O$$

   b. $$\overset{\overset{\displaystyle O}{\|}}{NNH-C-NH_2}$$
   $$CH_3\overset{\overset{\displaystyle}{}}{C}-CH_2CH_3 + H_2O$$

   c. $$CH_3-\overset{\overset{\displaystyle NH}{\|}}{C}CH_2CH_3 + H_2O$$  d. $$CH_3-\overset{\overset{\displaystyle ON}{\|}}{\underset{\overset{|}{OH}}{C}}-CH_2CH_3$$

$$\text{e. } CH_3\overset{\overset{\displaystyle OH}{|}}{\underset{\underset{\displaystyle OH}{|}}{C}}-CH_2CH_3 \qquad \text{f. No reaction}$$

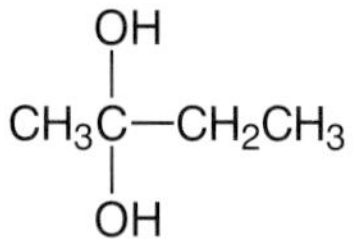

$$\text{g. } Na^+ \ {}^-O-\overset{\overset{\displaystyle O}{\|}}{C}CH_2CH_3 + CHI_3 + NaI + H_2O$$

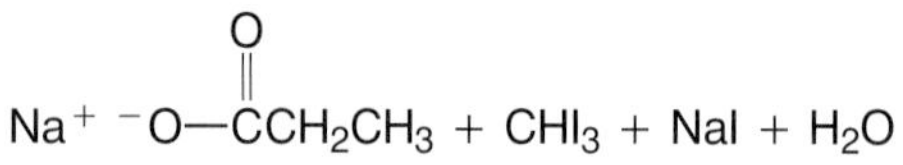

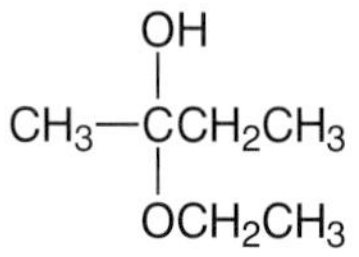

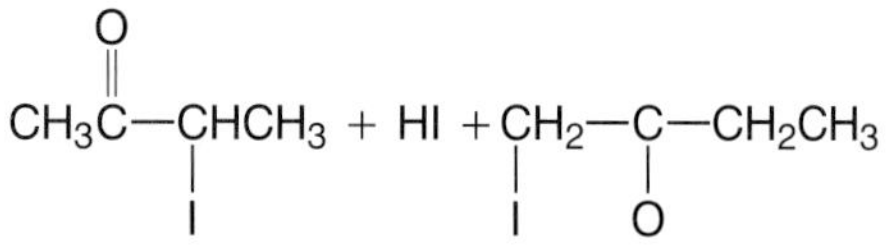

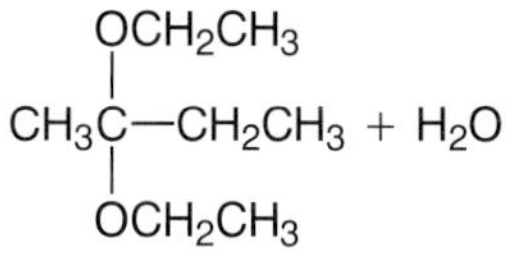

$$\text{h. } CH_3-\overset{\overset{\displaystyle OH}{|}}{\underset{\underset{\displaystyle OCH_2CH_3}{|}}{C}}CH_2CH_3$$

$$\text{i. } CH_3\overset{\overset{\displaystyle O}{\|}}{C}-\underset{\underset{\displaystyle OCH_2CH_3}{|}}{C}HCH_3 + HI + CH_2{=}\overset{}{C}-CH_2CH_3$$

(with O below)

$$\text{j. } \qquad \underset{\underset{\displaystyle OCH_2CH_3}{|}}{\overset{\overset{\displaystyle OCH_2CH_3}{|}}{CH_3C}}-CH_2CH_3 + H_2O$$

$$\text{k. } I-\underset{\underset{\displaystyle I}{|}}{\overset{\overset{\displaystyle I}{|}}{C}}-\overset{\overset{\displaystyle O}{\|}}{C}-CH_2CH_3 + NaI + H_2O$$

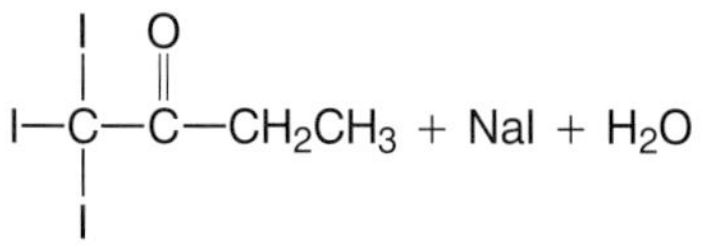

9. a. 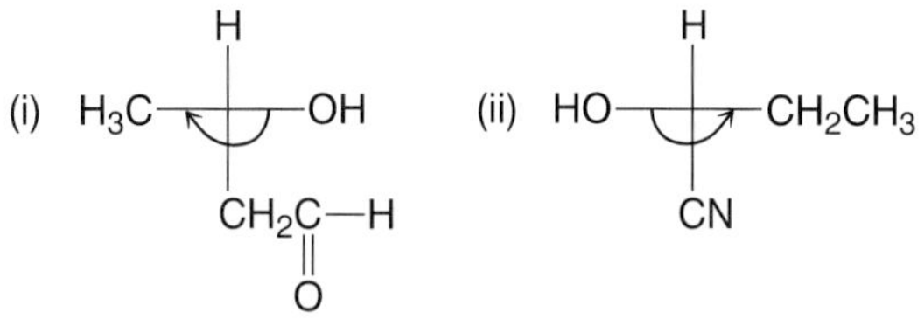 (i) $H_3C$ —OH, $CH_2C$—H, O

(ii) HO—$CH_2CH_3$, CN

b. 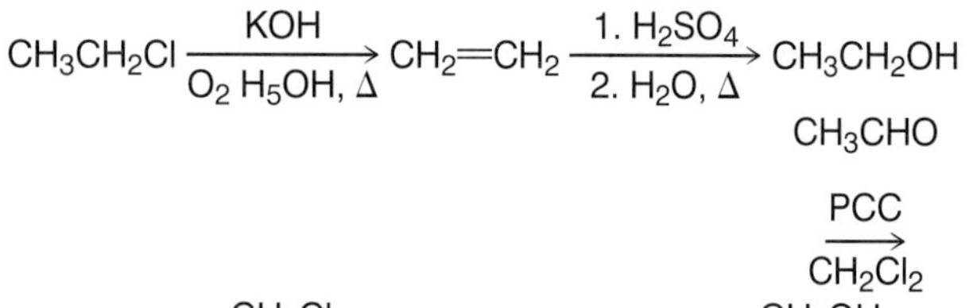 $CH_3CH_2Cl \xrightarrow[O_2H_5OH, \Delta]{KOH} CH_2{=}CH_2 \xrightarrow[2.\ H_2O,\ \Delta]{1.\ H_2SO_4} CH_3CH_2OH$

$CH_3CHO$

$\xrightarrow[CH_2Cl_2]{PCC}$ $CH_2OH$

[benzene ring with $CH_2Cl$] + NaOH $\longrightarrow$ [benzene ring with $CH_2OH$]

$\Big\downarrow \begin{array}{l} PCC \\ CH_2Cl \end{array}$

[benzene ring with C(=O)—H]

[benzene ring with C(=O)—H] + $CH_3\overset{\overset{\displaystyle O}{\|}}{C}$—H $\xrightarrow[H_2O]{NaOH}$ [benzene ring with CH=CH—C(=O)—H]

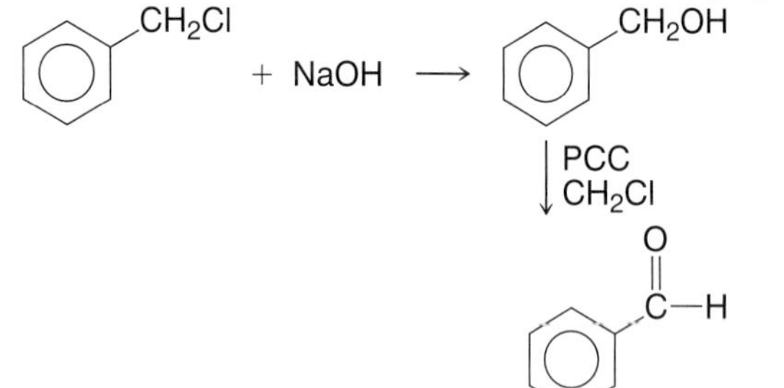

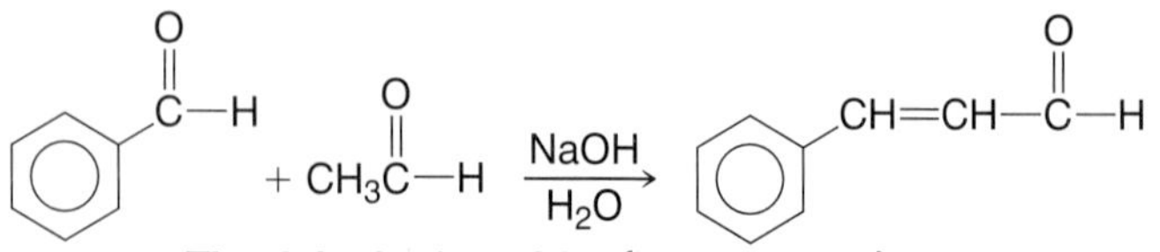

c. The dehydration of the (intermediate) product of an aldol condensation is driven by the stability of the product. The α,β-unsaturated aldehyde or ketone formed has extended conjugation, which is not present in the intermediate aldol.

11. a. Methanediol, methylenediol; Chloral hydrate (common); 2,2,2-Trichloro-1,1-ethanediol (IUPAC); b. Formaldehyde hydrate (common); Methane diol (IUPAC); Acetaldehyde hydrate (common); 1,1-Ethanediol

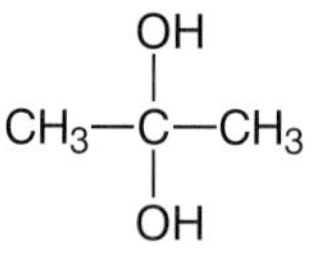

$$CH_3-\overset{\overset{\displaystyle OH}{|}}{\underset{\underset{\displaystyle OH}{|}}{C}}-CH_3$$

Acetone hydrate (common)
2,2-Propane diol (IUPAC)

13. a. A catalytic amount of base will produce $^-{:}C{\equiv}N$, the reacting species, from $H{-}C{\equiv}N$.

b. NaCN in aqueous $H_2SO_4$ is used to carry out the cyanohydrin reaction.

$$2\ NaCN + H_2SO_4 \longrightarrow 2\ HCN + Na_2SO_4$$

15. a. 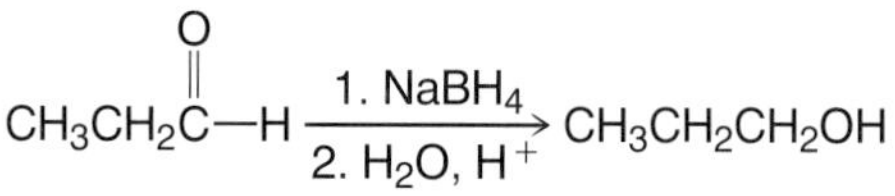 $CH_3CH_2\overset{\overset{\displaystyle O}{\|}}{C}{-}H \xrightarrow[2.\ H_2O,\ H^+]{1.\ NaBH_4} CH_3CH_2CH_2OH$

b. $CH_3CH_2\overset{\overset{\displaystyle O}{\|}}{C}{-}H \xrightarrow[H_2SO_4]{Na_2Cr_2O_7} CH_3CH_2COOH$

c. $CH_3CH_2\overset{\overset{\displaystyle O}{\|}}{C}{-}H \xrightarrow[H_2O/H^+]{NaBH_4} CH_3CH_2CH_2OH$

$2CH_3CH_2CH_2OH + CH_3CH_2\overset{\overset{\displaystyle O}{\|}}{C}{-}H \xrightarrow{H^+}$

$$CH_3CH_2\underset{\underset{\displaystyle OCH_2CH_2CH_3}{|}}{\overset{\overset{\displaystyle OCH_2CH_2CH_3}{|}}{CH}}$$

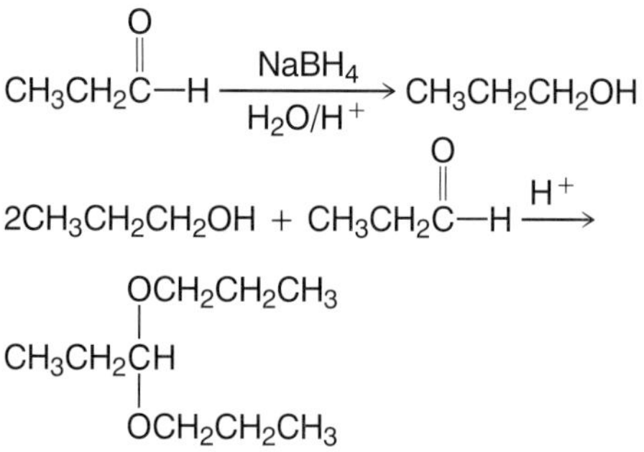

d. 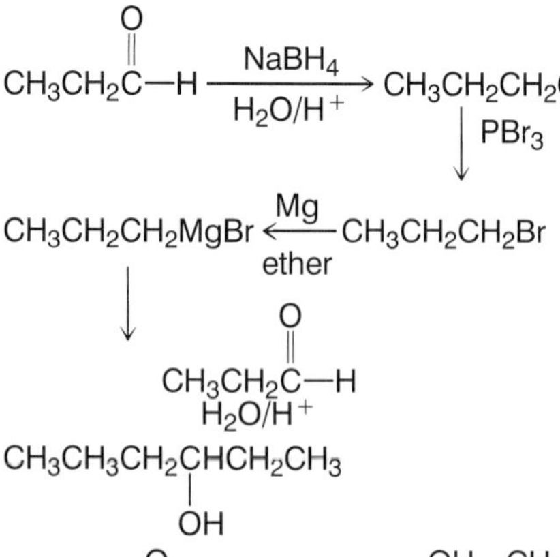 $CH_3CH_2\overset{\overset{\displaystyle O}{\|}}{C}{-}H \xrightarrow[H_2O/H^+]{NaBH_4} CH_3CH_2CH_2OH$

$\Big\downarrow PBr_3$

$CH_3CH_2CH_2MgBr \xleftarrow[ether]{Mg} CH_3CH_2CH_2Br$

$\Big\downarrow$

$CH_3CH_2\overset{\overset{\displaystyle O}{\|}}{C}{-}H$
$H_2O/H^+$

$CH_3CH_3CH_2\underset{\underset{\displaystyle OH}{|}}{CH}CH_2CH_3$

e. 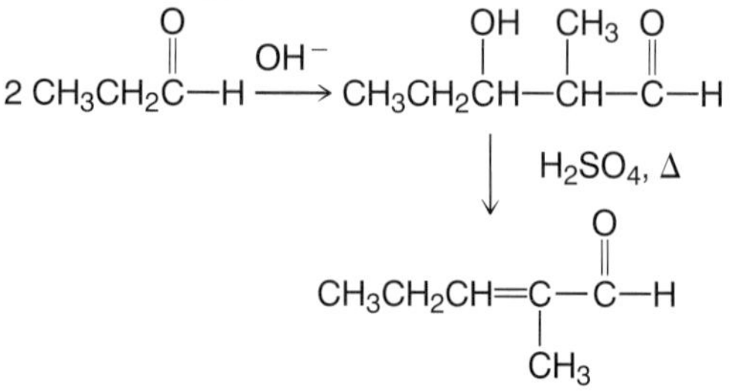 $2\ CH_3CH_2\overset{\overset{\displaystyle O}{\|}}{C}{-}H \xrightarrow{OH^-} CH_3CH_2\underset{\underset{\displaystyle OH}{|}}{CH}-\underset{\underset{\displaystyle CH_3}{|}}{CH}-\overset{\overset{\displaystyle O}{\|}}{C}{-}H$

$\Big\downarrow H_2SO_4, \Delta$

$CH_3CH_2CH{=}\underset{\underset{\displaystyle CH_3}{|}}{C}-\overset{\overset{\displaystyle O}{\|}}{C}{-}H$

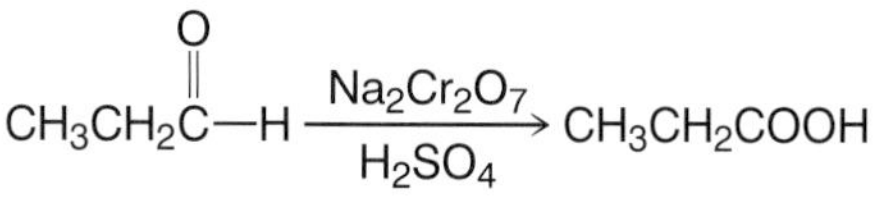

f. 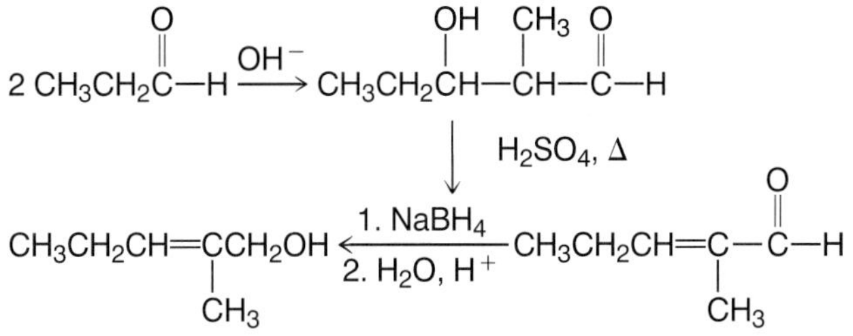  $2\ CH_3CH_2\overset{O}{\underset{}{C}}-H \xrightarrow{OH^-} CH_3CH_2\overset{OH}{\underset{}{CH}}-\overset{CH_3}{\underset{}{CH}}-\overset{O}{\underset{}{C}}-H$

$\downarrow H_2SO_4,\ \Delta$

$CH_3CH_2CH=\overset{}{\underset{CH_3}{C}}CH_2OH \xleftarrow[2.\ H_2O,\ H^+]{1.\ NaBH_4} CH_3CH_2CH=\overset{}{\underset{CH_3}{C}}-\overset{O}{\underset{}{C}}-H$

g. 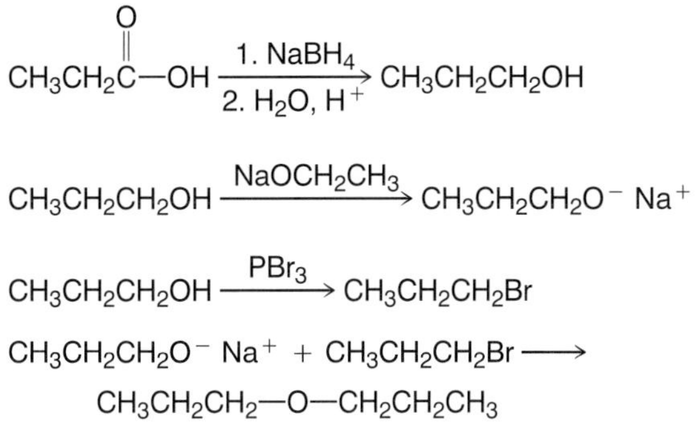  $CH_3CH_2\overset{O}{\underset{}{C}}-OH \xrightarrow[2.\ H_2O,\ H^+]{1.\ NaBH_4} CH_3CH_2CH_2OH$

$CH_3CH_2CH_2OH \xrightarrow{NaOCH_2CH_3} CH_3CH_2CH_2O^-\ Na^+$

$CH_3CH_2CH_2OH \xrightarrow{PBr_3} CH_3CH_2CH_2Br$

$CH_3CH_2CH_2O^-\ Na^+ + CH_3CH_2CH_2Br \longrightarrow$

$CH_3CH_2CH_2-O-CH_2CH_2CH_3$

17.  a.

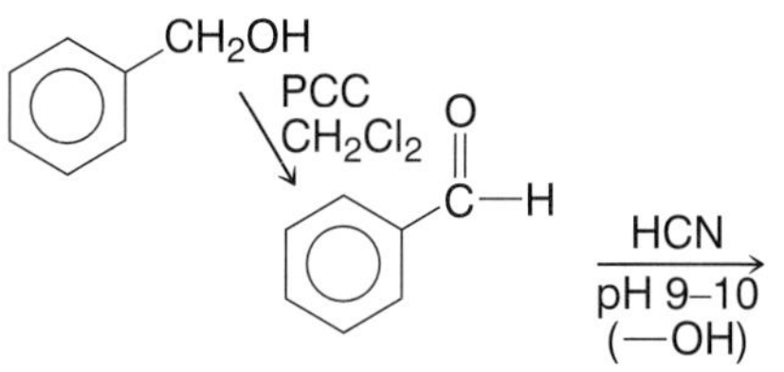

b.

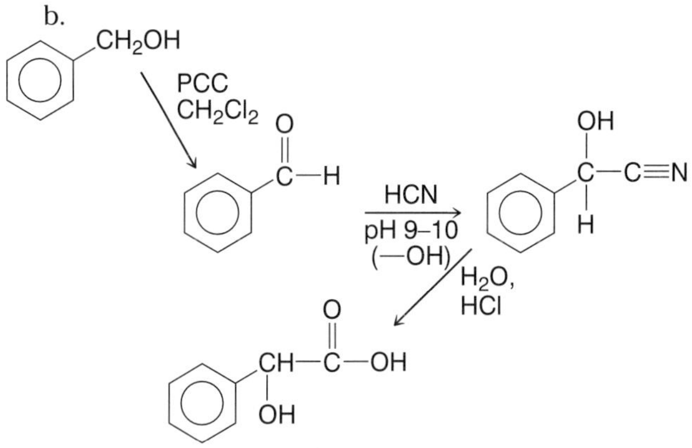

c.

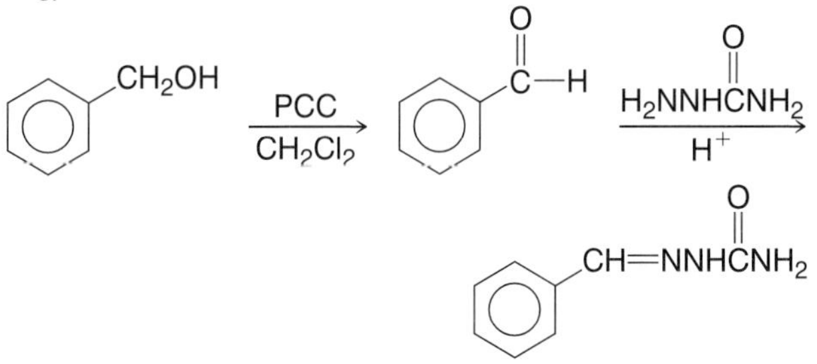

d.

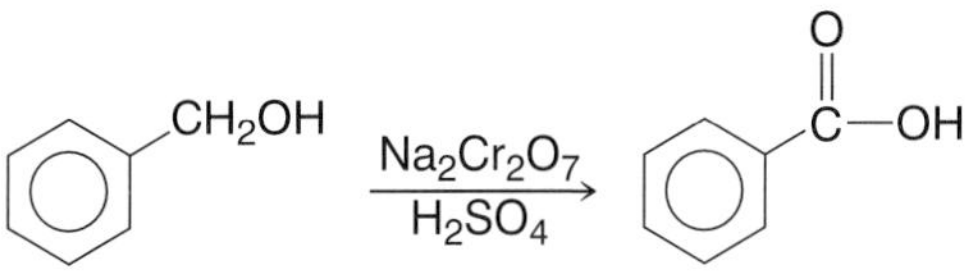

e.

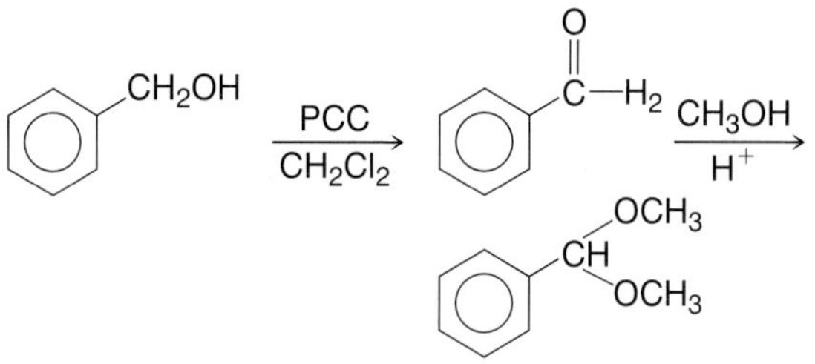

19.  Compound B = 

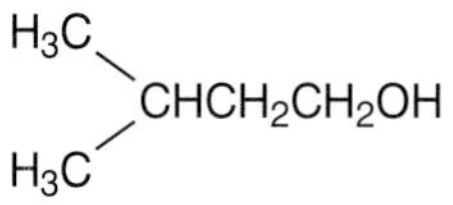

Compound A = 

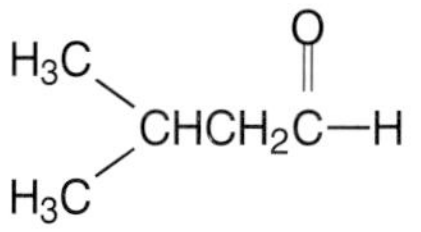

21.

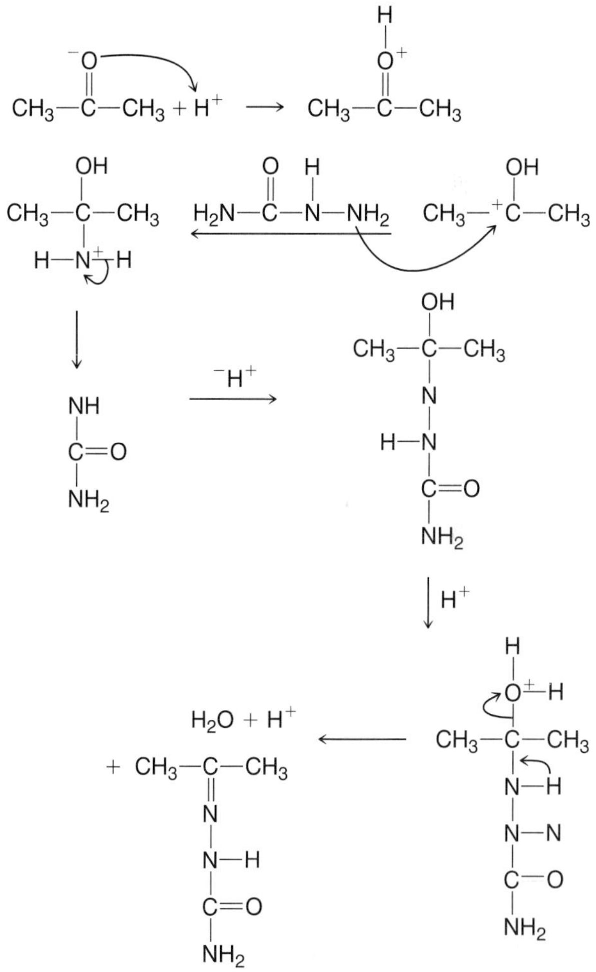

a.  Thus make it susceptible to nucleophilic attack and acid conditions are necessary so that dehydration can occur.

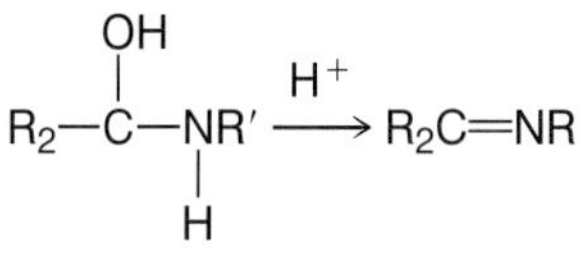

b. If the pH is too low (pH 3), the carbinol amine cannot form, because the semicarbazide will exist as

$$\overset{+}{H_3N}-NH-\overset{\overset{\displaystyle O}{\|}}{C}-NH_2$$

**23.**

$$\text{benzaldehyde} \xrightarrow{O_2} \text{benzoic acid}$$

**25.** The aldehyde proton will be observed between $\delta = 9$ and 10 and will be split into a quartet by the $3\alpha$ protons of the methyl group of acetaldehye. The protons of the methyl group ($\alpha$-protons) should be seen between $\delta = 1.5$ and 2.5. The area under the $\alpha$-proton peak should be three times that of the aldehyde hydrogen, since they are in the ratio of 3:1. All of the $\alpha$-protons are equivalent, and the peak will appear as a doublet because it is split by the hydrogen of the carbonyl group.

## Chapter 20 Exercises

**20.1.** a. α-Iodo-α-methylpropionic acid
     b. β-Chloro-β-methylvaleric acid

**20.3.** a. 4.75;   b. 3.83;   c. 2.85

**20.5.**

$$CH_3-\overset{\overset{\displaystyle O}{\|}}{C}-OC_2H_5 \rightleftharpoons CH_3-\overset{\overset{\displaystyle \cdot\cdot}{\phantom{.}}}{C}=\overset{+}{O}C_2H_5$$
          (I)                     (II)

| **Structure I** | **Structure II** |
|---|---|
| No atom in the structure bears a charge. | Oxygen bears both a positive and negative charge. |
| The $C-OC_2H_5$ bond is a single bond. | The $C=OC_2H_5$ bond is a double bond. |

The $\overset{\overset{\displaystyle O}{\|}}{C}$ bond is a double bond.    The $\overset{\overset{\displaystyle O}{\|}}{C}$ bond is a single bond.

**20.7.** a. Potassium propionate (common)
        Potassium propanoate (IUPAC)
     b. Calcium propionate (common)
        Calcium propanoate (IUPAC)

**20.9.** a. $H\overset{\overset{\displaystyle O}{\|}}{C}-OCH_2CH_3 + H_2O$    b. $C_6H_5-\overset{\overset{\displaystyle O}{\|}}{C}-O-CH_3 + H_2O$

## Chapter 20 Problems

**1.** More carboxylic acids are found in nature than aldehydes because acids have a higher oxidation state assigned to the

$$-\overset{\overset{\displaystyle O}{\|}}{C}-OH$$

carbon and our atmosphere is oxidizing (air contains 21% oxygen).

**3.** a. salicylic acid (2-hydroxybenzoic acid)

b. $CH_3-CH_2-CH_2-\overset{\overset{\displaystyle CH_3}{|}}{\underset{\underset{\displaystyle CH_3}{|}}{C}}-\overset{\overset{\displaystyle O}{\|}}{C}-OH$

c. 2,4-dichlorophenoxyacetic acid

d. indole-3-acetic acid

e. ibuprofen structure

f. $CH_3-CH_2-\overset{\overset{\phantom{|}}{|}}{\underset{\underset{\displaystyle CH_3}{|}}{CH}}-CH_2-\overset{\overset{\phantom{|}}{|}}{\underset{\underset{\displaystyle Cl}{|}}{CH}}-\overset{\overset{\displaystyle O}{\|}}{C}-OH$

g. (methoxyphenyl)acetic acid

h. trichlorophenoxyacetic acid

i. $\overset{\displaystyle H}{\phantom{x}}\,\overset{\displaystyle C}{=}\,\overset{\displaystyle\overset{O}{\|}}{C}-OH$ with $CH_3$ and $H$

j. maleic/fumaric acid (diacid, HOOC-CH=CH-COOH)

k. $CH_3(CH_2)_{14}\overset{\overset{\displaystyle O}{\|}}{C}-OH$

l. $CH_3(CH_2)_{16}\overset{\overset{\displaystyle O}{\|}}{C}-OH$

m. aspirin (acetylsalicylic acid)

**5.** a. (carboxyphenyl) structure

b. cyclobutanecarboxylic acid

c. $CH_3(CH_2)_{10}$ ... $CH=CH$ ... $\overset{\overset{\displaystyle O}{\|}}{C}-OH$

d.  $CH_3CH_2CH_2-CH_2-\overset{\overset{\displaystyle CH_3}{|}}{\underset{\underset{\displaystyle CH_3}{|}}{C}}-\overset{\overset{\displaystyle O}{||}}{C}-OH$

e.  $HO-\overset{\overset{\displaystyle O}{||}}{C}CH_2\underset{\underset{\displaystyle Cl}{|}}{C}HCH_2\overset{\overset{\displaystyle O}{||}}{C}-OH$

7.

$CH_3(CH_2)_{14}\overset{\nearrow O}{\underset{\searrow}{C}}\ddot{\overset{..}{O}}{:}^- \longleftrightarrow CH_3(CH_2)_{14}\overset{:\ddot{O}{:}^-}{\underset{}{C}}=\ddot{O}{:}$

9.  Acid B

11.  $CH_3(CH_2)_4\overset{\overset{\displaystyle \ddot{O}:-----H-\ddot{O}:}{||}}{\underset{\underset{\displaystyle :\ddot{O}-H-----:\ddot{O}}{||}}{C}}\overset{}{}C(CH_2)_4CH_3$

Hydrogen bond

13.  a.  Acetic acid is water-soluble.
b.  All dicarboxylic acids are solids.
c.  Potassium salts of carboxylic acids are water-soluble.
d.  Salicylic acid is *o*-hydroxybenzoic acid.
e.  Dibromoacetic acid has two electron-withdrawing groups ($-Rr$), which makes it a stronger acid than bromoacetic acid, which has only one electron-withdrawing group.

15.  a.  $CH_3CH_2CH_2CH_2\overset{\overset{\displaystyle O}{||}}{C}-OH + KOH$

b.  $CH_3-\underset{\underset{\displaystyle CH_3}{|}}{C}H-CH_2\overset{\overset{\displaystyle O}{||}}{C}-OH$

c.  $CH_3(CH_2)_6\overset{\overset{\displaystyle O}{||}}{C}-OH + CH_3CH_2CH_2OH \overset{H_2SO_4}{\underset{\longleftarrow}{\rightleftharpoons}}$

d.  $CH_3(CH_2)_6\overset{\overset{\displaystyle O}{||}}{C}-OH + CH_3CH_2NH_2$

## Chapter 21 Exercises

21.1.  a.  *n*-Butyronitrile (common)
b.  Ethanenitrile (IUPAC)

21.3.

## Chapter 21 Problems

1.  a.  $CH_3-CH_2-\underset{\underset{\displaystyle CH_3}{|}}{C}H-C\equiv N$     b.  $CH_2=CH-C\equiv N$

c.  $CH_3CH_2CH_2CH_2C\equiv N$     d.  $CH_2CH_2\equiv N$

e.      f.  $CH_3-\underset{\underset{\displaystyle CH_3}{|}}{C}H-CH_2-C\equiv N$

3.  a.  $\underset{\underset{\displaystyle CH_3}{}}{\overset{\overset{\displaystyle CH_3}{}}{}}CHCH_2MgCl + CO_2 \overset{Ether}{\longrightarrow} \underset{\underset{\displaystyle H_3C}{}}{\overset{\overset{\displaystyle H_3C}{}}{}}CHCH_2\overset{\overset{\displaystyle O}{||}}{C}-O^-$

$\downarrow HCl/H_2O$

$\underset{\underset{\displaystyle H_3C}{}}{\overset{\overset{\displaystyle H_3C}{}}{}}CHCH_2\overset{\overset{\displaystyle O}{||}}{C}-OH$

$C_5H_{10}O_2$

b.  $\overset{4}{C}H_3\overset{3}{C}H_2\overset{2}{\underset{\underset{\displaystyle CH_3}{|}}{C}H}-\overset{1}{\overset{\overset{\displaystyle O}{||}}{C}}-OH$

$C_5H_{10}O_2$

C-2 is a chiral carbon. Since it is the only chiral carbon in the compound, both an (*R*) and an (*S*) form exist.

c.  The alklyl halide is tertiary. Potassium cyanide is too strong a base to react with tertiary halides to form a nitrile. Elimination takes place instead.

$CH_3-\underset{\underset{\displaystyle CH_3}{|}}{\overset{\overset{\displaystyle CH_3}{|}}{C}}-\overset{\overset{\displaystyle O}{||}}{C}-OH$

The halide $C_4H_9Br$ is     $CH_3-\underset{\underset{\displaystyle CH_3}{|}}{\overset{\overset{\displaystyle CH_3}{|}}{C}}-Br$

d.  $CH_3CH_2CH_2CH_2\overset{\overset{\displaystyle O}{||}}{C}-CH_3 \overset{1.\ NaOH}{\underset{2.\ Cl_2}{\longrightarrow}}$

$CH_3CH_2CH_2CH_2\overset{\overset{\displaystyle O}{||}}{C}-O^-\ Na^+$

$\downarrow HCl/H_2O$

$CH_3CH_2CH_2CH_2\overset{\overset{\displaystyle O}{||}}{C}-OH$

$C_5H_{10}O_2$

5.  a.  $CH_3CH=CH_2 \overset{1.\ B_2H_6}{\underset{2.\ H_2O_2,\ ^-OH}{\longrightarrow}} CH_3CH_2CH_2OH$

$\overset{Na_2Cr_2O_7}{\underset{H_2SO_4}{\longrightarrow}} CH_3CH_2\overset{\overset{\displaystyle O}{||}}{C}-OH$

**b.**

$$CH_2=CH_2 + Br_2 \longrightarrow BrCH_2-CH_2Br$$

Mg/ether (or KCN)
then H₂O, H⁺

$$\underset{\substack{1.\ CO_2 \\ 2.\ HCl/H_2O}}{\longleftarrow} BrMgCH_2-CH_2MgBr$$

CH₂—CH₂ with C=O, C—OH groups

**c.**  $CH_3CH_2CH_2OH \xrightarrow{PCC} CH_3CH_2\overset{O}{\underset{}{C}}-H \xrightarrow{NH_2OH}$

$$CH_3CH_2CH=N-OH \xrightarrow{Ac_2O} CH_3CH_2C\equiv N$$

**d.**

OCH₃ benzene ring with Br  + Mg $\xrightarrow{\text{Ether}}$  OCH₃ benzene ring with MgBr

$$\downarrow \begin{array}{l} 1.\ CO_2 \\ 2.\ HCl/H_2O \end{array}$$

OCH₃ benzene ring with $\overset{O}{\underset{}{C}}-OH$

**e.**  $H_2C=CH_2 + HOCl \longrightarrow HOCH_2-CH_2Cl$

$$\downarrow \begin{array}{l} KCN \\ CH_3CH_2OH/H_2O \end{array}$$

$$HOCH_2CH_2\overset{O}{\underset{}{C}}-OH \xleftarrow{HCL/H_2O} HOCH_2-CH_2CN$$

**f.**  $CH_3\overset{O}{\underset{}{C}}-H + HCN \xrightarrow{OH^-} CH_3-\overset{OH}{\underset{}{CH}}-CN$

$$\downarrow HCl/H_2O$$

$$CH_3\overset{O}{\underset{}{C}}-\overset{O}{\underset{}{C}}-OH \xleftarrow{Na_2Cr_2O_7/H^+} CH_3-\overset{OH}{\underset{}{CH}}-\overset{O}{\underset{}{C}}-OH$$
Pyruvic acid

**7.  a.**  Cl-substituted benzene with OCH₂$\overset{O}{\underset{}{C}}$—OH

**b.**

Cl-substituted phenol with OH $\xrightarrow[(aq)]{NaOH}$ Cl-substituted phenol with O⁻Na⁺ $\xrightarrow[Cl_2,\ PCl_3]{CH_3COOH}$

Cl-substituted ring with OCH₂$\overset{O}{\underset{}{C}}$—O⁻Na⁺ $\xrightarrow[H_2O]{HCl}$ Cl-substituted ring with OCH₂$\overset{O}{\underset{}{C}}$—OH

**9.**  Formic acid,

$$H-\overset{O}{\underset{}{C}}-OH,$$

is reduced with LiAlH₄ and acid to methanol, CH₃OH. The carbon in methanol is attached to only hydrogens, not another carbon atom, as primary alcohols are.

**11.**  This reaction can be followed using infrared spectroscopy. By taking aliquots of the reaction mixture and running and IR on them, the disappearance of the carbonyl band ($\sim1710$ cm⁻) can be observed. The reaction is complete when no absorbance can be seen in the carbonyl region.

## Chapter 22 Exercises

**22.1.  a.**  Hexanoyl chloride (IUPAC)
**b.**  Ethyl acetate (common); Ethyl ethanoate (IUPAC)
**c.**  Propionic anhydride (common)
Propanoic anhydride (IUPAC)
**d.**  *N,N*-Dimethylbutyramide (common)
*N,N*-Dimethylbutanamide (IUPAC)

**22.3.  a.**

Vitamin C structure with lactone group circled.     Lactone group is circled.

Vitamin C

**b.**

Aflatoxin B-1 structure with lactone group circled.    Lactone group is circled.

Aflatoxin B-1

**22.5.  a.**  $CH_3CH_2CH_2\overset{O}{\underset{}{C}}-O-CH(CH_3)_2 + H_2O$

**b.**  $H\overset{O}{\underset{}{C}}-O-CH_2CH_2CH_2CH_3 + KOH\ (aq)$

**c.**

benzene ring with $\overset{O}{\underset{}{C}}-O-CH_2CH_3$  + OH⁻

**d.**  $CH_3CH_2\overset{O}{\underset{}{C}}-NH_2 + NaOH(aq)$

**e.**  $CH_3CH_2\overset{O}{\underset{}{C}}-N\overset{CH_3}{\underset{CH_3}{}}$  + HCl + H₂O

f. $CH_3\overset{O}{\overset{\|}{C}}-NH_2 + NaOH$

## Chapter 22 Problems

1. a. (structure: benzoic anhydride)

b. (structure with $CH_3$, $H_3C$, $\cdots H$, $O$, $O-CH$)

c. $CH_3CH_2CH_2\overset{O}{\overset{\|}{C}}-NH_2$

d. $CH_3CH_2CH_2CH_2\overset{O}{\overset{\|}{C}}-Cl$

e. $HC\overset{O}{\overset{\|}{}}-\overset{\ddot{N}}{\underset{CH_2CH_3}{}}{}^{\cdots H}$

f. (structure: cyclohexane $\overset{O}{\overset{\|}{C}}-Cl$)

g. $CH_3CH_2\overset{O}{\overset{\|}{C}}-N\overset{CH_3}{\underset{CH_3}{}}$

h. (structure: cyclohexane $\overset{O}{\overset{\|}{C}}-NH$ phenyl)

e. (structure: benzoate phenyl ester)

f. (structure with $O_2N$ phenyl $-N\overset{H}{} \overset{O}{\overset{\|}{}}-O-CH_3$)

3. $CH_3CH_2\overset{O}{\overset{\|}{C}}-O-CH_2CH_3$
   Ethyl propanoate

   $CH_3\overset{O}{\overset{\|}{C}}-O-CH_2CH_2CH_3$
   *n*-propyl ethanoate

   $CH_3-\overset{O}{\overset{\|}{C}}-O-CH\overset{CH_3}{\underset{CH_3}{}}$
   Isopropyl ethanoate

   $H-\overset{O}{\overset{\|}{C}}-O-CH_2CH_2CH_2CH_3$
   *n*-Butyl methanoate

   $H-\overset{O}{\overset{\|}{C}}-O-CH_2CH\overset{CH_3}{\underset{CH_3}{}}$
   *iso*-Butyl methanoate

   $H-\overset{O}{\overset{\|}{C}}-O-C(CH_3)_3$
   *tert*-butyl methanoate

   $CH_3CH_2CH_2\overset{O}{\overset{\|}{C}}-O-CH_3$
   Methyl butanoate

   $H_3C\overset{}{\underset{H_3C}{}}CH-\overset{O}{\overset{\|}{C}}-O-CH_3$
   Methyl 2-methyl propanoate

   $H-\overset{O}{\overset{\|}{C}}-O-\overset{H}{\underset{CH_3}{C}}-CH_2CH_3$
   *sec*-butyl methanoate

5. a. (structure: cyclohexane $\overset{O}{\overset{\|}{C}}-O-CH_3$)

   b. $CH_3\overset{O}{\overset{\|}{C}}-O-\overset{O}{\overset{\|}{C}}-CH_3$

   c. $CH_3\overset{O}{\overset{\|}{C}}-Cl$

   d. (structure: Cl phenyl $-NHCH_3$)

7. a. (structure: cyclobutane $\overset{O}{\overset{\|}{C}}-O-\overset{O}{\overset{\|}{C}}$ benzene)

   Benzoic cyclobutane carboxic anhydride

   b. $BrCH_2CH_2\overset{O}{\overset{\|}{C}}-Cl$
   3-Bromopropionyl chloride

   c. $CH_3CH\overset{O}{\overset{\|}{C}}-C-NH$, $\underset{CH_2CH_3}{}$
   2-Methyl butanamide

   d. $CH_3CH\overset{O}{\overset{\|}{C}}-C-O-CH_2CH_3$, $\underset{Br}{}$
   Ethyl α-bromopropionate

   e. $CH_3\overset{O}{\overset{\|}{C}}-N\overset{CH_2CH_3}{\underset{CH_2CH_3}{}}$
   *N,N*-Diethylethanamide

9. $CH_3(CH_2)_{14}\overset{O}{\overset{\|}{C}}-O-CH\overset{CH_3}{\underset{CH_3}{}}$

## Chapter 23 Exercises

23.3.

$CH_3-\overset{O}{\overset{\|}{C}}-Cl \longrightarrow \left[ CH_3-\overset{\ddot{O}:^-}{\overset{|}{C}}-Cl \right] \longrightarrow CH_3-\overset{O}{\overset{\|}{C}}-{}^+\overset{H}{\overset{}{O}}\overset{}{\underset{H}{}} + Cl^-$

(with $\overset{}{\underset{H}{O}}\overset{}{\underset{H}{}}$ attacking)

$CH_3-\overset{O}{\overset{\|}{C}}-{}^+O\overset{H}{\underset{H}{}} \rightleftharpoons CH_3\overset{O}{\overset{\|}{C}}-OH + H^+$

$H^+ + Cl^- \longrightarrow H-Cl$

23.5. Suppose a lactone *was* produced by reacting a β-hydroxy acid—for example, β-hydroxypropionic acid. The four-membered ring product is too unstable and the equilibrium far to the left.

23.7.

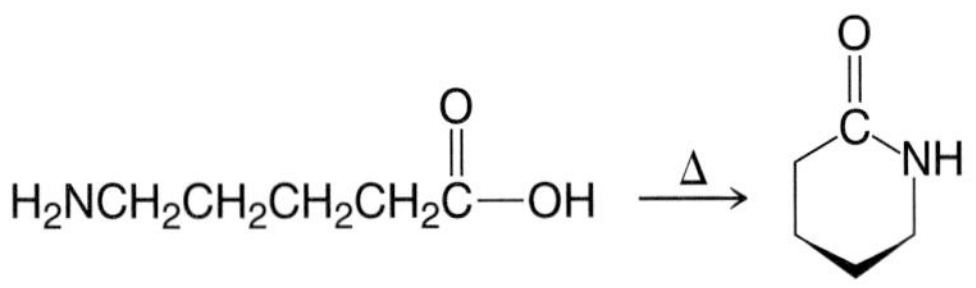

## Chapter 23 Problems

1. b. No reaction; c. no reaction

3. a.

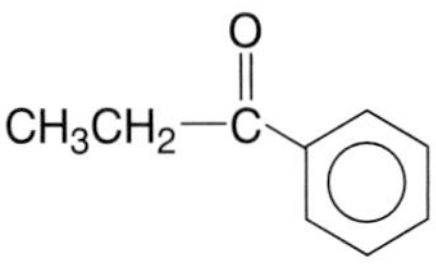

$+ HCl$

b. $CH_3{-}CH(CH_3){-}C(=O){-}OH + HCl$

c.

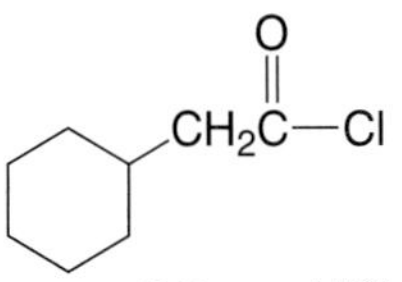

$+ SO_2 + HCl$

d. $2CH_3C(=O){-}OH$

e. $CH_3CH_2C(=O){-}O{-}C(=O){-}CH_3 + NaCl$

f. $CH_3C(=O){-}O{-}CH_3 + CH_3C(=O){-}OH$

g. $CH_3{-}C(=O){-}N(CH_3){-}CH_3 + HCl$ or $CH_3{-}C(=O){-}\overset{+}{N}H(CH_3)_2\ Cl^-$

h.

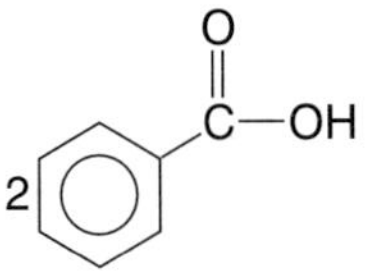

i. $CH_3C(=O){-}OCH_3 + H_2O$

j.

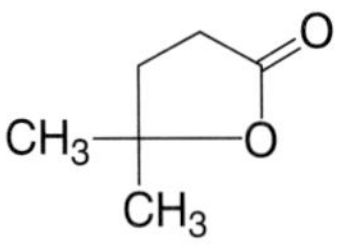

$+ H_2O$

k.

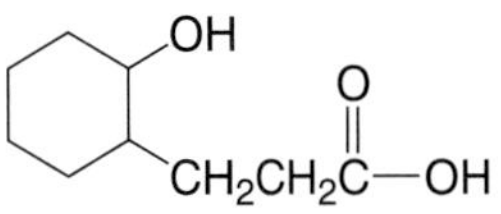

l. $CH_3CH_2C(=O){-}NH_2 + CH_3CH_2OH$

m. $CH_3{-}C(CH_3)(CH_3){-}OH + Br{-}Mg{-}OCH_3 + Br{-}Mg{-}Cl$

n. $CH_3C(=O){-}OH + CH_3CH_2OH$

o. $CH_3(CH_2)_{16}C(=O){-}O^-\ Na^+ + CH_3OH$

p. $CH_3(CH_2)_{16}C(=O){-}OH + NaCl$

q. $CH_3CH_2C(=O){-}O^-\ Na^+ + H_2O$

r. $CH_3C(=O){-}OH + NH_4^+\ Cl^-$

s. $CH_3C(=O){-}O^-\ Na^+ + NH_3$

t.

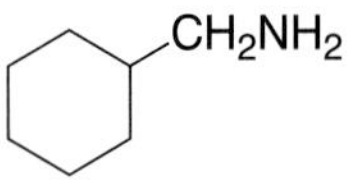

u. $CH_3C(=O){-}NH{-}CH_2CH_3$ or $+ CH_3C(=O){-}O^-\ \overset{+}{N}H_3{-}CH_2CH_3$

5. a.

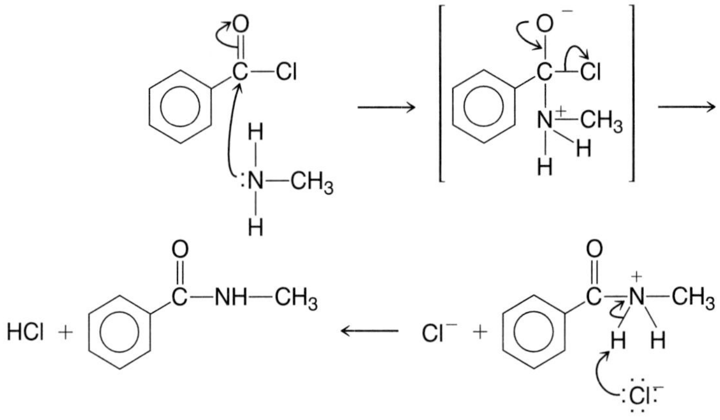

b.

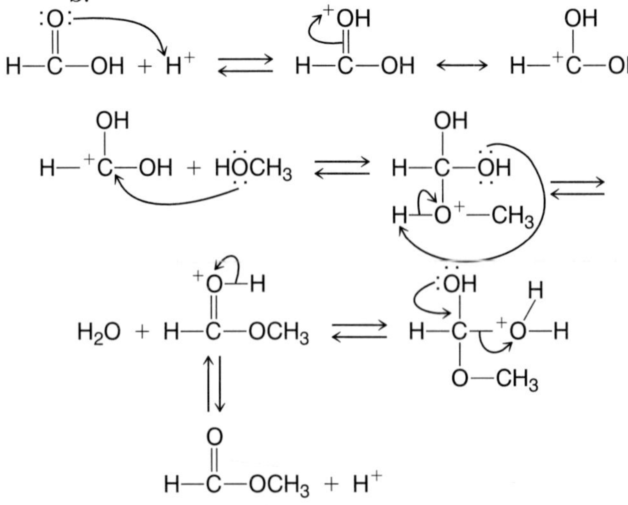

c.

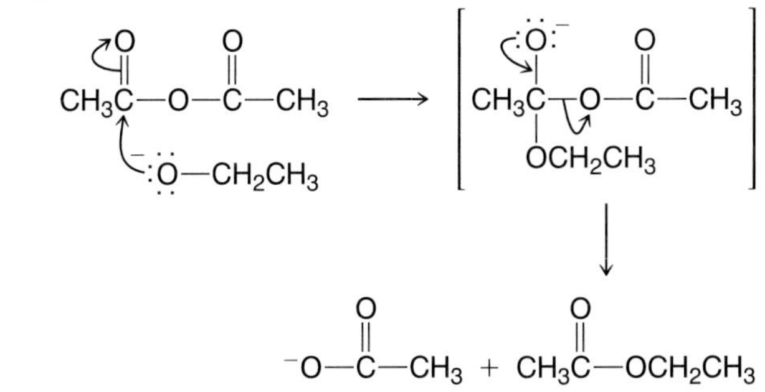

7. Using acetic anhydride rather than acetyl chloride for the acylation of benzene using $AlCl_3$ (Frieder-Crafts reaction) makes the reaction less vigorous and more manageable. Acetic anhydride is more stable (less reactive); therefore, the reactant will be purer. Also, the byproduct using acetic anhydride is innocuous acetic acid.

9. Two moles of ammonia are required per mole of acid chloride, even though only one mole of amide forms. The second mole of ammonia reacts with the hydrogen chloride released.

11.

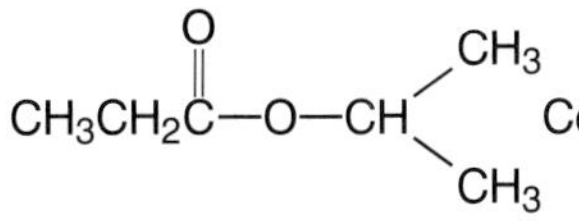

Compound A

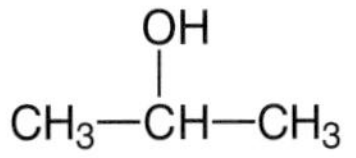

Compound B 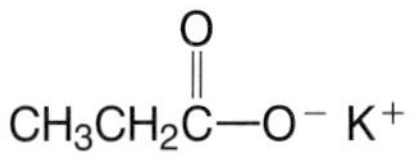     Compound C

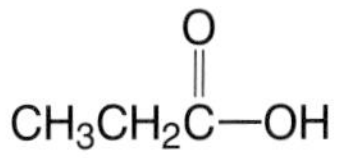

Compound D

13. Nylon-6,6 is so named because the diacyl chloride and the diamine each have six carbons. If each had four carbons, the nylon formed would be nylon-4,4.

15.

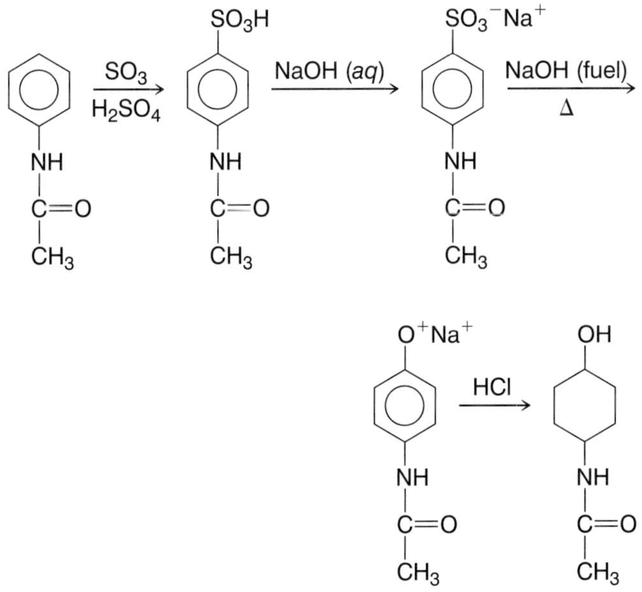

17. Amides with the

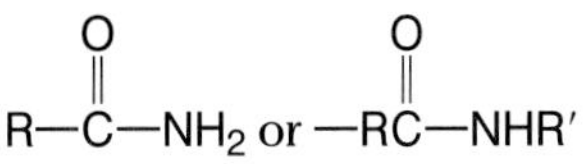

could not be represented because the $-NH_2$ or $-NHR$ absorbs in the 3300 $cm^{-1}$ region of an infrared spectrum.

## Chapter 24 Exercises

24.1.  a.  *n*-Pentylamine (common)
            1-Pentanamine (IUPAC)
        b.  Benzylamine (common)
        c.  Pentamethylenediamine (common)
            1,5-Pentanediamine (IUPAC)
        d.  2-Amino-2-methylpropanol (IUPAC)

24.3.

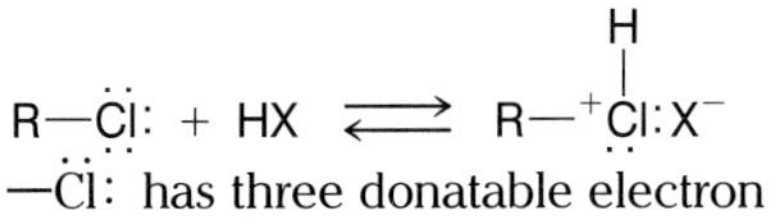

$-\ddot{C}l\!:$ has three donatable electron pairs; therefore, it has the potential to be a Lewis base. The special $R-^+Cl\!:X^-$ is unstable because the electronegativity of Cl is greater than 3.0 so an alkyl chloride is not a base.

24.5.  a.  $3.981 \times 10^{-12}$;   b.   $5.13 \times 10^{-4}$

24.7.

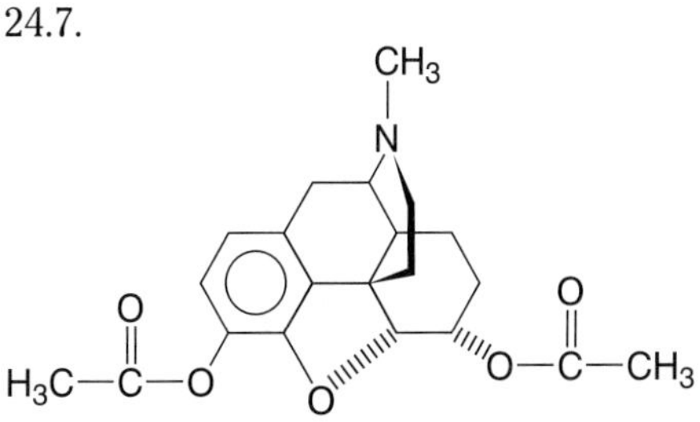

Heroin

24.9.

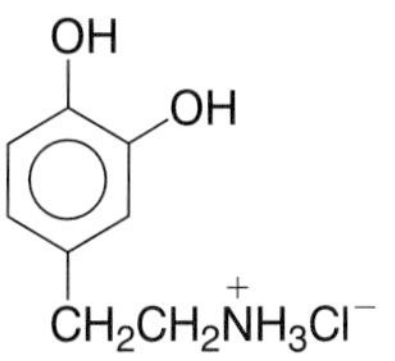

## Chapter 24 Problems

1. $CH_3CH_2CH_2CH_2NH_2$   (*n*-butylamine) (common)
                                1-Butanamine (1°)

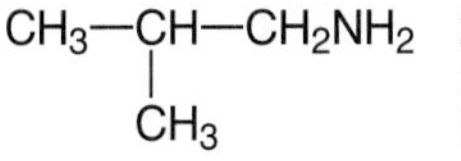

2-Methyl-1-propanamine (1°) (IUPAC)
(isobutylamine) (common)

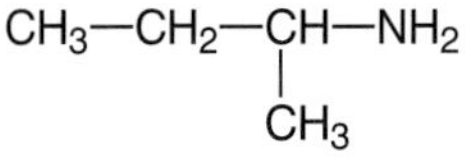

2-butanamine (1°) (IUPAC)
(sec-butylamine) (common)

CH₃—CH₂—CH₂—NH—CH₃
*N*-Methyl-1-propanamine
(IUPAC) (2°)
(methyl propylamine) (common)

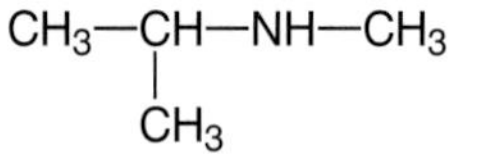

CH₃—CH—NH—CH₃  *N*-Methyl-2-propanamine (IUPAC)
       |                        (methylisopropylamine) (common)
      CH₃

CH₃—CH₂—NH—CH₂CH₃  *N*-Ethyl ethanamine
(IUPAC) (2°) (diethylamine) (common)

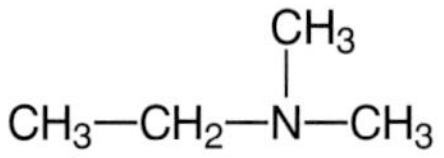

              CH₃
              |
CH₃—CH₂—N—CH₃

*N,N*-Dimethylethanamine
(3°) (IUPAC)
(ethyldimethylamine) (common)

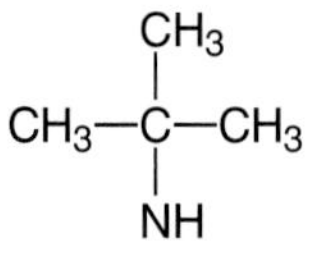

         CH₃
         |         (ethyldimethyl amine) (common)
CH₃—C—CH₃   2-methyl-2-propanamine (IUPAC)
         |         (tert-butylamine) (common) (1°)
        NH

3. a. $(CH_3CH_2)_2\ NH_2^+\ Cl^-$
      Diethyl ammonium chloride
   b. 2-(3,4,5-Trimethoxyphenyl)-1-ethanamine
   c. Tetramethylammonium iodide
   d. 2,4-Dinitroaniline
   e. N-Methyl-3-hexanamine
   f. *N,N*,-Diethyl-3-hexanamine
   g. 2-Aminoethanoic acid
   h. *N*-ethyl-1-propanamine
   i. cyclopentanamine (cyclopentylamine)

5. a. $CH_3CH_2NH_2$
   b. $(CH_3CH_2)_4\overset{+}{N}Cl^-$

7. The higher the molecular weight of the amine, the lower the vapor pressure.

9. a. $(CH_3—CH_2)_2\overset{+}{N}H_2$
   b. $CH_3CH_2\overset{+}{N}H_3$
   c. $(CH_3CH_2)_3\overset{+}{N}H$

11. a.

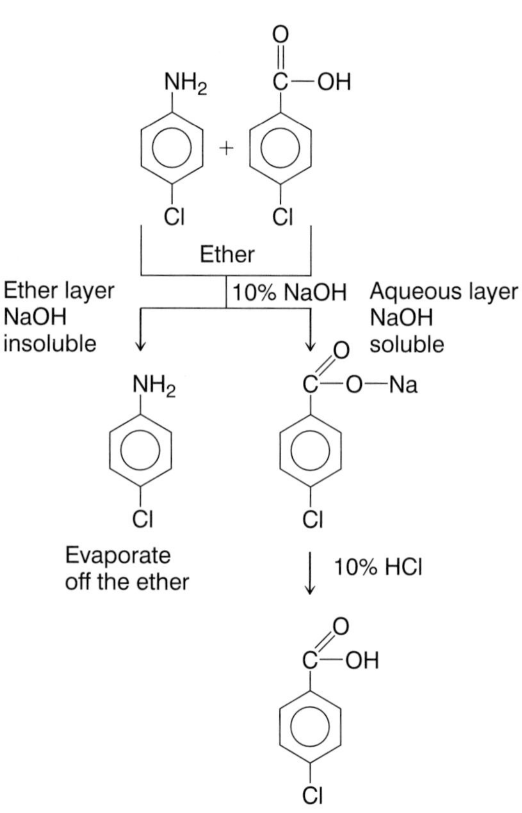

b.

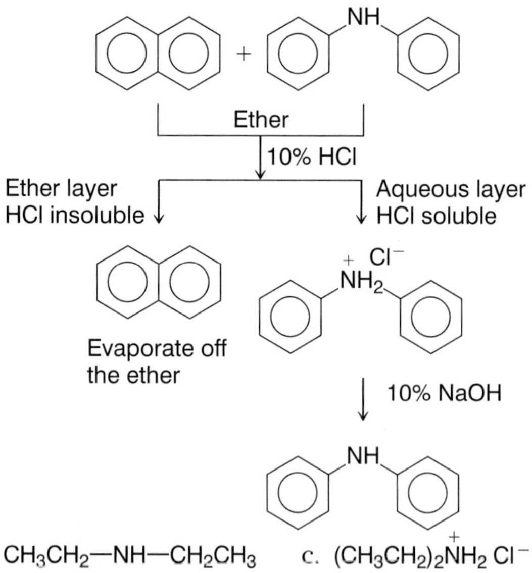

13. a. CH₃CH₂—NH—CH₂CH₃    c. $(CH_3CH_2)_2\overset{+}{N}H_2\ Cl^-$

    Diethylamine

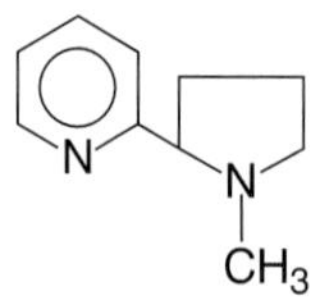

   b.

    Nicotine

   d. 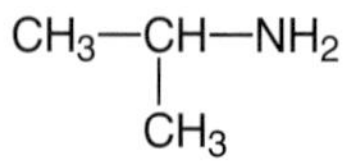 CH₃—CH—NH₂
            |
           CH₃

## Chapter 25 Exercises

**25.1.**

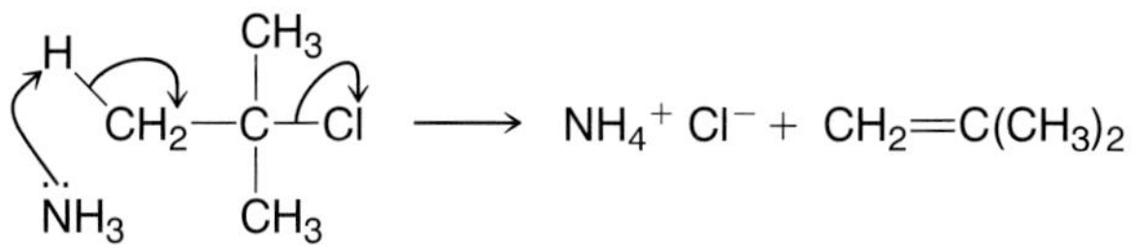

$$\text{NH}_4{}^+ \text{ Cl}^- + \text{CH}_2{=}\text{C(CH}_3)_2$$

**25.3.** **a.**

$$\text{CH}_3\text{CH}_2\text{CH}_2\text{I} \xrightarrow{\text{KCN}} \text{CH}_3\text{CH}_2\text{CH}_2{-}\text{C}{\equiv}\text{N}$$

$$\xrightarrow[\text{2. H}_2\text{O/OH}^-]{\text{1. LiAl H}_4} \text{CH}_3\text{CH}_2\text{CH}_2\text{CH}_2\text{NH}_2$$

**b.** 
$$\text{CH}_3(\text{CH}_2)_4\overset{\text{O}}{\overset{\|}{\text{C}}}{-}\text{OH} \xrightarrow{\text{SOCl}_2} \text{CH}_3(\text{CH}_2)_4\overset{\text{O}}{\overset{\|}{\text{C}}}{-}\text{Cl}$$

$$\downarrow \text{NH}_3$$

$$\text{CH}_3(\text{CH}_2)_3\text{CH}_2\text{NH}_2 \xleftarrow[\text{NaOH}(aq)]{\text{Br}_2} \text{CH}_3(\text{CH}_2)_4\overset{\text{O}}{\overset{\|}{\text{C}}}{-}\text{NH}_2$$

**c.** 
$$\underset{\underset{\text{OH}}{|}}{\text{CH}_3{-}\text{CH}{-}\text{CH}_3} \xrightarrow[\text{H}_2\text{SO}_4]{\text{Na}_2\text{Cr}_2\text{O}_7} \underset{\underset{\text{O}}{\|}}{\text{CH}_3\text{C}{-}\text{CH}_3}$$

$$\xrightarrow[\text{H}_2/\text{PT}]{\text{NH}_3} \underset{\underset{\text{NH}_2}{|}}{\text{CH}_3{-}\text{CH}{-}\text{CH}_3}$$

**d.**

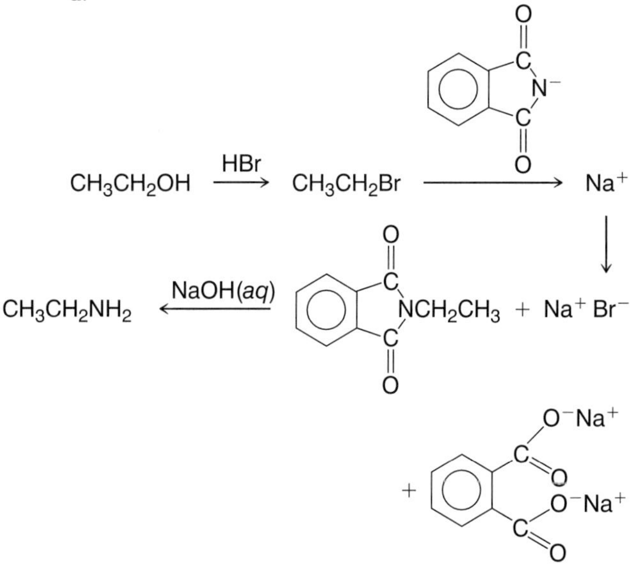

**25.5.** The reaction requires the extraction of an amine hydrogen. Since a tertiary amine has no

$$\overset{}{\underset{}{\diagup}}\text{N}{-}\text{H},$$

the reaction cannot take place.

**25.7.**

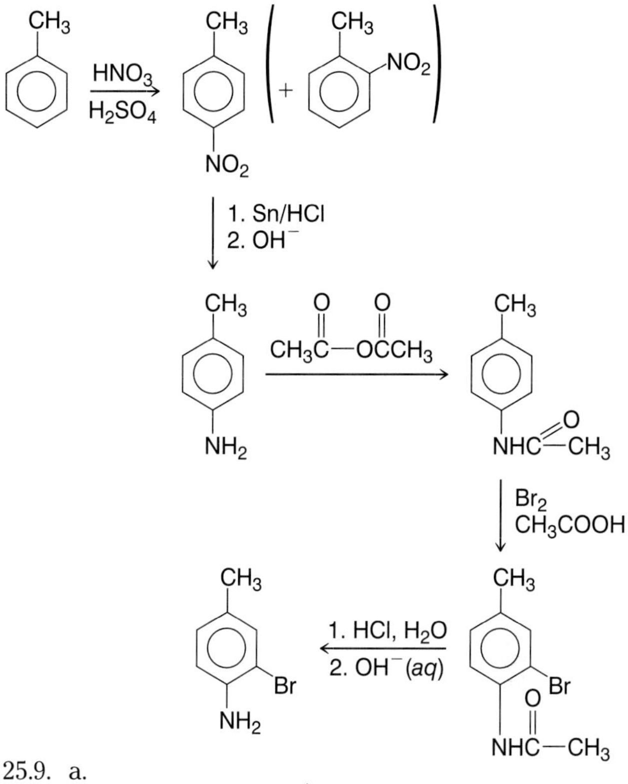

**25.9.** **a.**

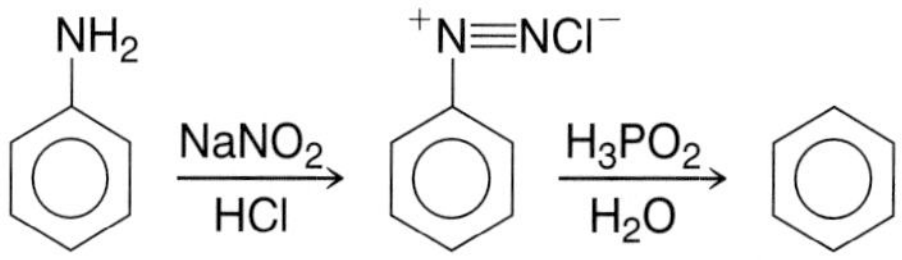

**b.**

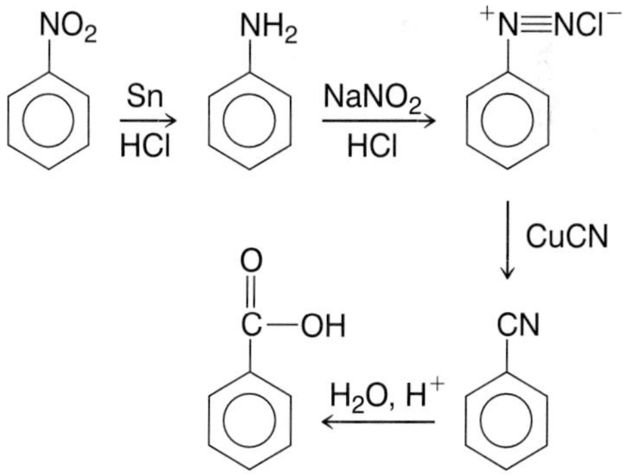

c.

[Reaction scheme: 2-bromo-4-nitroaniline → (NaNO₂/HCl) diazonium salt → (Δ, H₂O) 2-bromo-4-nitrophenol]

$$\text{NH}_2 \text{ (2-Br, 4-NO}_2\text{)} \xrightarrow[\text{HCl}]{\text{NaNO}_2} \ ^+\text{N}\equiv\text{NCl}^- \xrightarrow[\text{H}_2\text{O}]{\Delta} \text{OH}$$

1. Sn, HCl
2. NaOH(aq)

[Further scheme continuing: 2-bromo-4-aminophenol ← (NaNO₂/HCl) → diazonium → (H₃PO₂) → 2-bromophenol]

## Chapter 25 Problems

1.  a.  $CH_3CH_2CH_2OH + CH_3CH=CH_2$
    b.

$$CH_3CH_2CH_2\overset{-}{N}-\underset{\underset{O}{\parallel}}{\overset{\overset{O}{\parallel}}{S}}-C_6H_5$$

   c.  $CH_3CH_2CH_2-NH-\overset{\overset{O}{\parallel}}{C}-CH_3$

   d.  $CH_3\overset{\overset{O}{\parallel}}{C}-OH + CH_3CH_2CH_2NH-\overset{\overset{O}{\parallel}}{C}-CH_3$

   e.  $CH_3CH_2CH_3NH_2 + HCl \longrightarrow CH_3CH_2CH_2\overset{+}{N}H_3\ Cl^-$

3.  a.  Treat both substances with benzenesulfonyl chloride. The secondary amine $(CH_3CH_2)_2$ NH will form an insoluble sulfonamide, which is insoluble in base. The tertiary amine $(CH_3CH_2)_3$ N will not react at all.
    b.  Formation of the diazonium salts of aniline and benzyl amine will yield.

$$\text{NH}_2 \xrightarrow[\text{HCl}]{\text{NaNO}_2} \ ^+\text{N}\equiv\text{NCl}^-$$

$$\text{CH}_2\text{NH}_2 \xrightarrow[\text{HCl}]{\text{NaNO}_2} \text{CH}_2^+ + \text{N}_2$$

The diazonium salt of aniline is stabilized by resonance with the ring and produces nitrogen very slowly. On the other hand, the diazonium salt of benzyl amine decomposes immediately because the nitrogen is a good leaving group. If both products are treated with phenol, the aromatic diazonium salt will form a deeply colored azo dye. The benzyl carbocation will remain colorless.

   c.  To distinguish between aniline and N-methyl- benzamide, test for the basicity with indicator paper (pH paper). The aniline is basic (pH/~/10), and the

amide is almost neutral. *Or* dissolve the amine and amide in dilute HCl. The aniline will dissolve. The ammonium salt is formed.

$$\text{NH}_2 + \text{HCl} \longrightarrow \ ^+\text{NH}_3\ Cl^-$$

The protonated amide does not dissolve because it is not a strong enough base.

5.  If aniline is treated with nitric acid, the amino group is oxidized, and a black tar results. Therefore, the amino group must be converted to an amide prior to reacting with a strong oxidizing agent such as nitric acid.

7.  $C_3H_9N = CH_3CH_2-NH-CH_3$

9.  a.

$$\text{CH}_3\text{-C}_6\text{H}_5 \xrightarrow[\text{H}_2\text{SO}_4]{\text{HNO}_3} \text{(p-nitrotoluene)} + \text{[o-nitrotoluene]} \xrightarrow[\text{NaOH(aq)}]{\text{Sn, HCl}}$$

[continuing scheme]

$$\xrightarrow[\text{2. OH}^-]{\text{1. Sn, HCl}} \text{(2,6-dibromo-4-methylaniline)} \xleftarrow[\text{2. H}_2\text{O}]{\text{1. 2 Br}_2} \text{(p-toluidine)}$$

   b.  $CH_3CH_2Br \xrightarrow{\text{NH}_3} (CH_3CH_3)_3N \xrightarrow{\text{C}_2\text{H}_5\text{Br}} (CH_3CH_2)_4\overset{+}{N}Br^-$

   c.

$$\text{C}_6\text{H}_6 \xrightarrow[\text{H}_2\text{SO}_4]{\text{HNO}_3} \text{NO}_2 \xrightarrow[\text{2. OH}^-]{\text{1. Sn/HCl}} \text{NH}_2$$

[acetic anhydride: $CH_3C(=O)-O-, CH_3C(=O)$]

$$\text{NHC}(=O)\text{CH}_3 \xrightarrow[\text{SOCl}_2]{\text{1. H}_2\text{SO}_4,\ \text{2. PCl}_3\text{ or}} \text{(p-SO}_2\text{Cl acetanilide)} \xrightarrow{\text{NH}_3} \text{(p-SO}_2\text{NH}_2\text{ acetanilide)}$$

1. HCl, H₂O
2. OH⁻

$$\text{NH}_2 \text{ (SO}_2\text{NH}_2\text{)}$$

11.  $C_3H_9N = CH_3-\overset{\overset{\textstyle CH_3}{|}}{\underset{\underset{\textstyle CH_3}{|}}{N}}$

13.  a.

benzene $\xrightarrow[\text{AlCl}_3]{\text{CH}_3\text{Cl}}$ toluene $\xrightarrow[\text{H}_2\text{SO}_4]{\text{HNO}_3}$ (CH$_3$, NO$_2$ para)

$\left(\text{+ } o\text{-nitrotoluene (CH}_3,\text{ NO}_2)\right)$ $\xrightarrow[\text{2. NaOH }(aq)]{\text{1. Sn, HCl}}$ (CH$_3$, NH$_2$ para)

$\xrightarrow{\underset{\overset{\|}{\text{CH}_3\overset{\textstyle O}{C}}-\text{Cl}}{}}$ (CH$_3$, NHCOCH$_3$) $\xrightarrow[\text{H}_2\text{SO}_4]{\text{HNO}_3}$ (CH$_3$, O$_2$N, NHCOCH$_3$) $\xrightarrow[\text{2. NaOH}(aq)]{\text{1. HCl, H}_2\text{O}}$ (CH$_3$, O$_2$N, NH$_2$)

$\xrightarrow[\text{2. H}_3\text{PO}_2]{\text{1. HNO}_2}$ (CH$_3$, O$_2$N)

b.

(NHCOCH$_3$, OH on benzene) $\xrightarrow{-\text{OH}}$ (NHCOCH$_3$, O$^-$) $\xrightarrow{\text{CH}_3\text{CH}_2\text{Br}}$ (NHCOCH$_3$, OCH$_2$CH$_3$ on cyclohexane)

15.  a.

Butter yellow: phenyl–N=N–(benzene)–$\overset{\overset{\textstyle ..}{\textstyle |}}{N}$(CH$_3$)$_2$

The free electron pair on the methyl group as well as the free electron pairs on the azo nitrogen set up the possibility for very many resonance structures and a very stable resonance hybrid.

17.  During the course of the reaction, a carbocation is produced. $CH_3CH_2\overset{+}{C}H_2$. This carbocation rearranges to a more stable carbocation.

$CH_3CH-\overset{\overset{\textstyle H}{|}}{\underset{\underset{\textstyle H}{|}}{\overset{+}{C}}}-H \longrightarrow CH_3-\overset{\overset{\textstyle H}{|}}{\underset{}{\overset{+}{C}}}-CH_3$

$\underset{Cl^-}{\swarrow} \qquad \underset{H_2O^-}{\searrow}$

$CH_3-\overset{\overset{\textstyle H}{|}}{\underset{\underset{\textstyle Cl}{|}}{C}}-CH_3 \qquad CH_3-\overset{\overset{\textstyle H}{|}}{\underset{\underset{\textstyle \overset{+}{O}-H}{|}}{C}}-CH_3$

$\underset{\overset{\textstyle |}{H}}{}$

$\searrow$

$CH_3-\overset{\overset{\textstyle H}{|}}{\underset{\underset{\textstyle OH}{|}}{C}}-CH_3$

19.  $CH_3-CH_2-CH_2-\overset{\overset{\textstyle H}{|}}{N}-CH_2-CH_2-CH_3$

## Chapter 26 Exercises

26.1.  $CH_3(CH_2)_{14}\overset{\overset{\textstyle O}{\|}}{C}-O-CH_2(CH_2)_{14}CH_3$

Hexadecanyl hexadeconate

$CH_3(CH_2)_{24}\overset{\overset{\textstyle O}{\|}}{C}-O-CH_2(CH_2)_{28}CH_3$

Tricontanyl hexacosanate

$CH_3(CH_2)_{24}\overset{\overset{\textstyle O}{\|}}{C}-O-CH_2(CH_2)_{30}CH_3$

Dotriacontanyl hexacosanoate

26.3.

$CH_2-O-\overset{\overset{\textstyle O}{\|}}{C}-(CH_2)_7\overset{\overset{\textstyle H}{\diagup}}{C}=\overset{\overset{\textstyle H}{\diagdown}}{C}(CH_2)_7CH_3$

$H-\overset{*}{C}-O-\overset{\overset{\textstyle O}{\|}}{C}-(CH_2)_{14}CH_3$

$CH_2-O-\overset{\overset{\textstyle O}{\|}}{C}-(CH_2)_{10}CH_3$

*C is chiral.

26.5.  Glyceryl stearomyristolinoleate

26.7.

$CH_2-O-\overset{\overset{\textstyle O}{\|}}{C}-(CH_2)_7CH=CH(CH_2)_7CH_3$

$HC-O-\overset{\overset{\textstyle O}{\|}}{C}-(CH_2)_{14}CH_3$ Lard

$CH_2-O-\overset{\overset{\textstyle O}{\|}}{C}-(CH_2)_{16}CH_3$

Oleic 48%, palmitic 28%, stearic 12%

**26.9.**   a.   $CH_3(CH_2)_7CH{=}CH(CH_2)_7\overset{\text{O}}{\underset{}{C}}{-}O^-\ K^+$

          Potassium Oleate

   b.

$CH_3CH_2CH{=}CH{-}CH_2{-}CH{=}CHCH_2CH{=}(CH_2)_7\overset{\text{O}}{\underset{}{C}}{-}O^-\ N^+$

          Sodium linolenate

**26.11.**

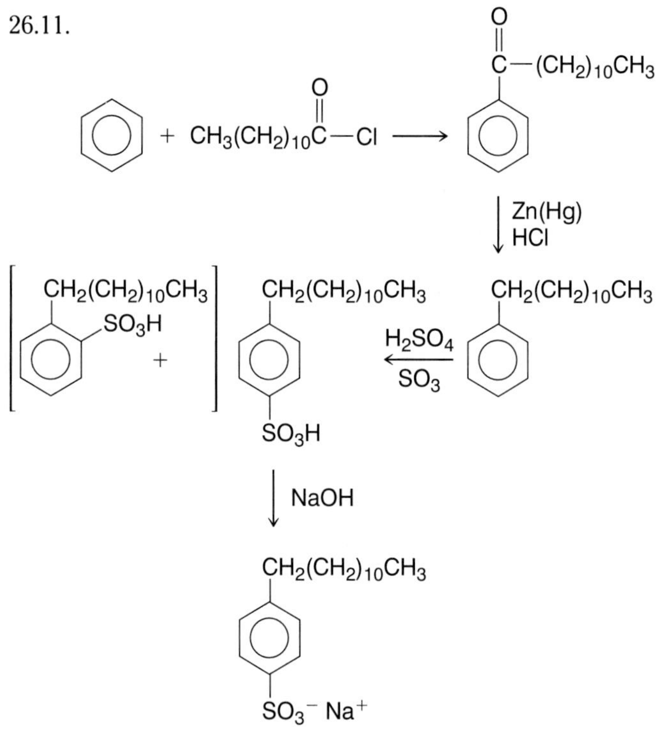

**26.13.** a.

$HOCH_2{-}\underset{\underset{OH}{|}}{CH}{-}CH_2OH\ +\ 3CH_3(CH_2)_{14}\overset{\text{O}}{\underset{}{C}}{-}O^-/K^+$

b.

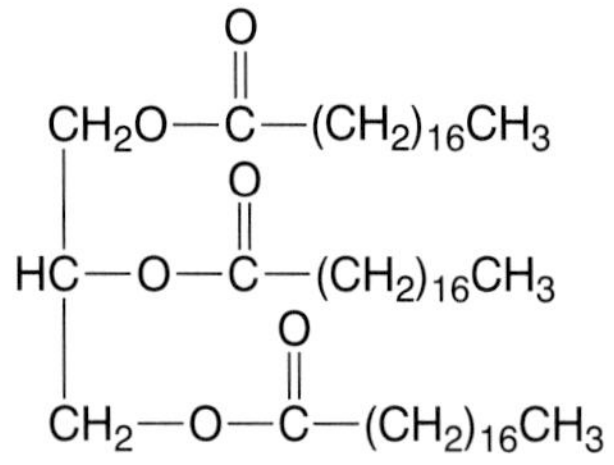

c.   $HOCH_2{-}\underset{\underset{OH}{|}}{CH}{-}CH_2OH\ +\ RCOO^-K^+\ +\ R'COO^-K^+\ +$

         $HOCH_2CH_2N(CH_3)_3\ +\ CH_3OH\ +\ K_2HPO_4$

d.

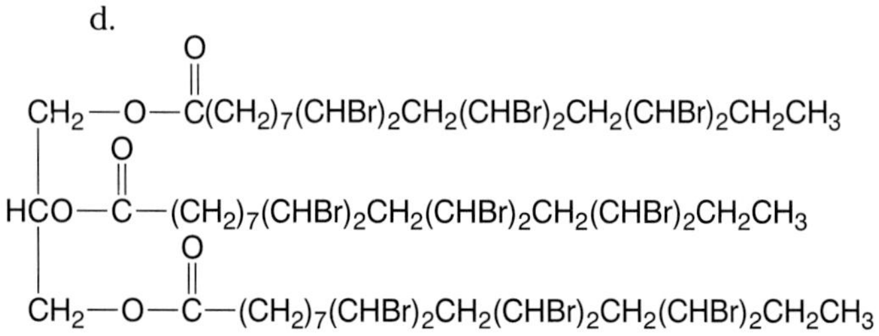

e.

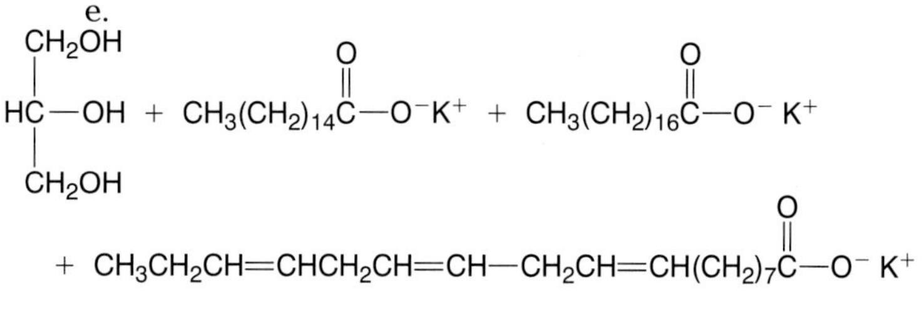

$HC{-}OH\ +\ CH_3(CH_2)_{14}\overset{\text{O}}{\underset{}{C}}{-}O^-K^+\ +\ CH_3(CH_2)_{16}\overset{\text{O}}{\underset{}{C}}{-}O^-\ K^+$

(with $CH_2OH$ above and $CH_2OH$ below)

   $+\ CH_3CH_2CH{=}CHCH_2CH{=}CH{-}CH_2CH{=}CH(CH_2)_7\overset{\text{O}}{\underset{}{C}}{-}O^-\ K^+$

## Chapter 26 Problems

1.   a.

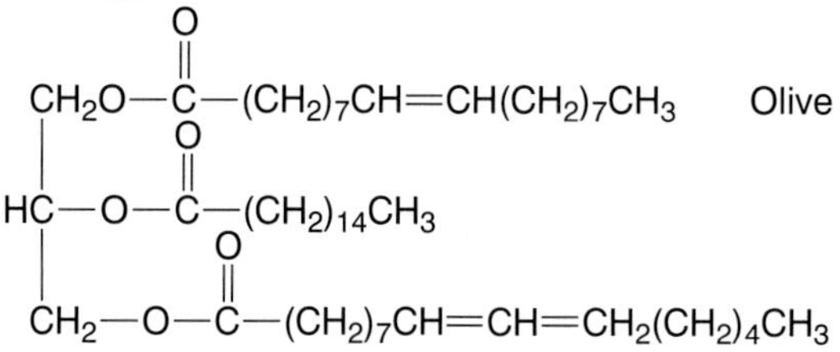

   Olive oil

b.

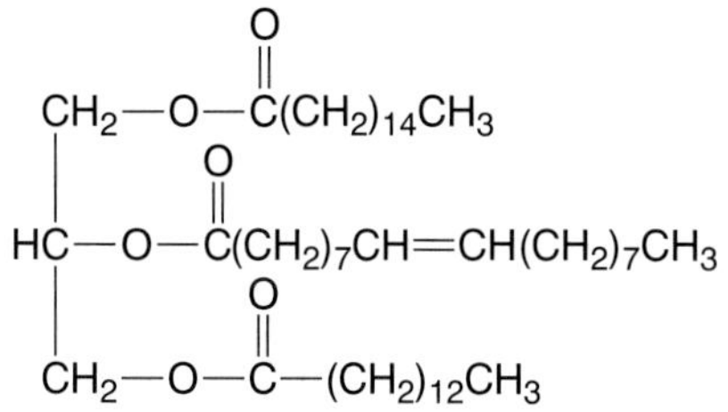

   Butter

c.

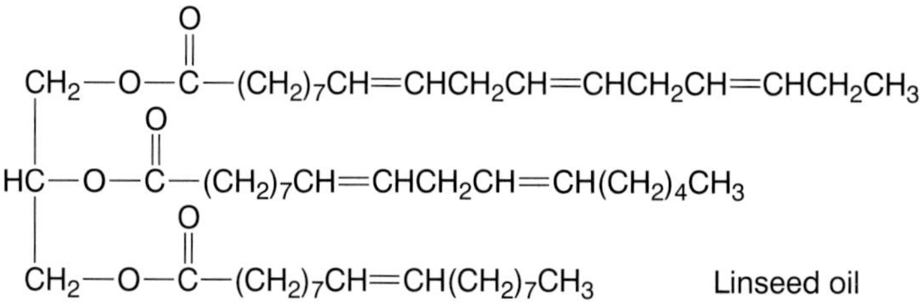

   Linseed oil

d.

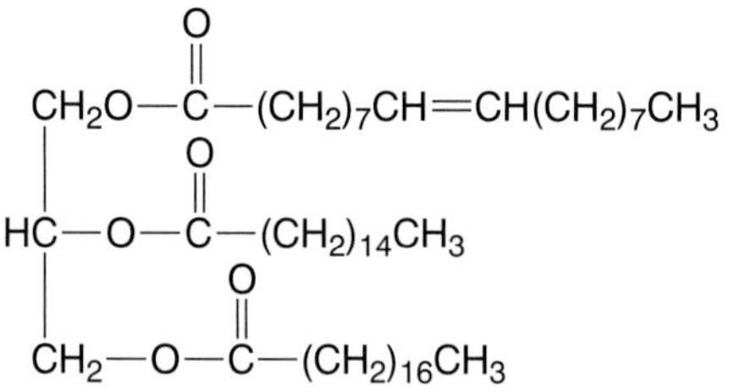

   Tallow

3.   a.

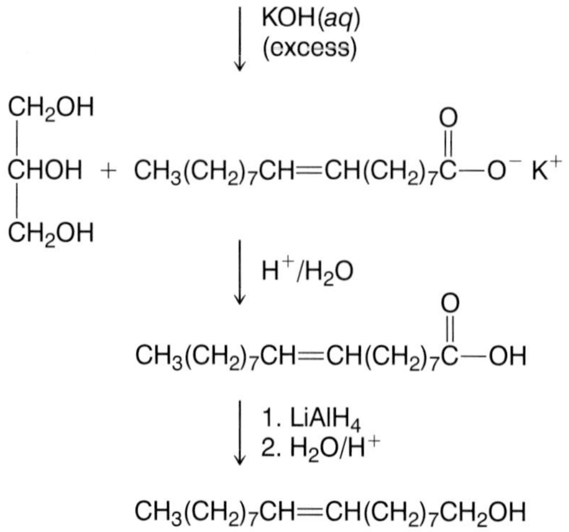

$CH_3(CH_2)_7CH{=}CH(CH_2)_7CH_2OH$

b.

$$\downarrow \begin{array}{l} \text{1. KOH}(aq), \Delta \\ \text{2. H}_2\text{O/H}^+ \end{array}$$

CH₂OH
|
CH—OH + CH₃(CH₂)₇CH=CH(CH₂)₇COOH
|
CH₂OH

$$\downarrow \begin{array}{l} \text{1. LiAlH}_4 \\ \text{2. H}_2\text{O/H}^+ \end{array}$$

CH₃(CH₂)₇CH=CH(CH₂)₇CH₂OH

$$\downarrow \text{H}_2/\text{Ni}$$

CH₃(CH₂)₁₆CH₂OH

c.

$$\downarrow \begin{array}{l} \text{1. KOH}(aq), \Delta \\ \text{2. H}_2\text{O/H}^+ \end{array}$$

CH₂OH   O
|       ‖
CHOH + HOC—(CH₂)₇CH=CH(CH₂)₇CH₃
|
CH₂OH

$$\downarrow \text{H}_2/\text{Ni or PT}$$

    O
    ‖
HO—C—(CH₂)₁₆CH₃

d.

$$\downarrow \begin{array}{l} \Delta \\ \text{1. KOH}(aq) \\ \text{2. H}_2\text{O/H}^+ \end{array}$$

CH₂OH   O
|       ‖
CHOH + HOC—(CH₂)₇CH=CH(CH₂)₇CH₃
|
CH₂OH

$$\downarrow \text{H}_2/\text{Ni or Pt}$$

    O
    ‖
HO—C—(CH₂)₁₆CH₃

$$\downarrow \text{SOCl}_2$$

    O
    ‖
Cl—C—(CH₂)₁₆CH₃

$$\downarrow \text{NH}_3$$

     O
     ‖
H₂N—C—(CH₂)₁₆CH₃

e.

$$\downarrow \begin{array}{l} \text{1. KOH}, \Delta \\ \text{2. H}_2\text{O/H}^+ \end{array}$$

CH₂OH
|
CHOH + CH₃(CH₂)₇CH=CH(CH₂)₇COOH
|
CH₂OH

$$\downarrow \text{H}_2/\text{Ni or Pt}$$

           O
           ‖
CH₃(CH₂)₁₆C—OH

$$\downarrow \begin{array}{l} \text{1. LiAlH}_4 \\ \text{2. H}_2\text{O/H}^+ \end{array}$$

CH₃(CH₂)₁₆CH₂OH

           O                                          O
           ‖                                          ‖
CH₃(CH₂)₁₆C—OH + SOCl₂ ⟶ CH₃(CH₂)₁₆C—Cl

              O
              ‖
⟶ CH₃(CH₂)₁₆C—Cl + CH₃(CH₂)₁₆CH₂OH

              O
              ‖
⟶ CH₃(CH₂)₁₆C—O—CH₂(CH₂)₁₆CH₃

f.

$$\downarrow \begin{array}{l} \text{1. O}_3 \\ \text{2. Zn, H}_2\text{O} \end{array}$$

        O          O
        ‖          ‖
CH₂O—C—(CH₂)₇C—H
        O          O
        ‖          ‖                    O
CH—O—C—(CH₂)₇C—H +    CH₃(CH₂)₇C—H
        O          O
        ‖          ‖
CH₃—O—C(CH₂)₇C—H

5. a. Oxidation of oils and fats by air or water to produce foul-smelling, carboxylic acids that contaminate the oil/fat when carbon-carbon double bonds in the acid part of the molecule are broken. Fats become rancid much more slowly than oils because they have fewer oxidizable double bonds.

   b. Water containing one or more of the following ions: magnesium, calcium, iron(II), and iron(III) that replaces the sodium and potassium in the soap. The result is salts that are insoluble.

   c. Liquid triglycerides made up mainly of esters of unsaturated acids with glycerol. Oils are liquids at room temperature.

   d. Triglyceride made up of a single carboxylic acid residue combined with each alcohol group of glycerol.

e. Triglyceride made up of two or three different carboxylic acid residues combined with the alcohol groups of glycerol.
f. Hydrogenation of the carbon-carbon bonds (all or partial) of the carboxylic acid residues of the oil. This raises the melting point of the oil; thus it is called *hardening of an oil.*
g. A mixture of triglycerides in which the acid residues are mainly saturated. Fats are solid.
h. A mixture of high-molecular-weight esters. Each ester is formed from a monocarboxylic fatty acid and a monohydric alcohol. Neither the acid nor the alcohol is branched. Waxes are dark in color and slick.
i. An oil made up of highly carbon-carbon unsaturated carboxylic acid residues combined with glycerol. The acids have three or more double bonds.

7. This is an example of the common ion effect.

$$\underset{\text{Precipitate}}{R{-}\overset{O}{\overset{\|}{C}}{-}O^-Na^+} \rightleftharpoons R{-}\overset{O}{\overset{\|}{C}}{-}O^- + Na^+$$

9. A phospholid can replace a soap for cleaning.

11. a. $\xrightarrow{\Delta} C_{15}H_{31}\overset{O}{\overset{\|}{C}}{-}O^-Na^+ + C_{25}H_{51}CH_2OH$

b.

$CH_2{-}OH$
$CH{-}OH$
$CH_2OH$

$+ CH_3(CH_2)_7CH{=}CH(CH_2)_7\overset{O}{\overset{\|}{C}}{-}O^-\ Na^+ +$

$CH_3(CH_2)_4CH{=}CH{-}CH_2{-}CH{=}CH{-}(CH_2)_7\overset{O}{\overset{\|}{C}}{-}O^-\ Na^+$

$+ CH_3(CH_2)_{10}\overset{O}{\overset{\|}{C}}{-}O^-\ Na^+$

c. No reaction

13.

15.

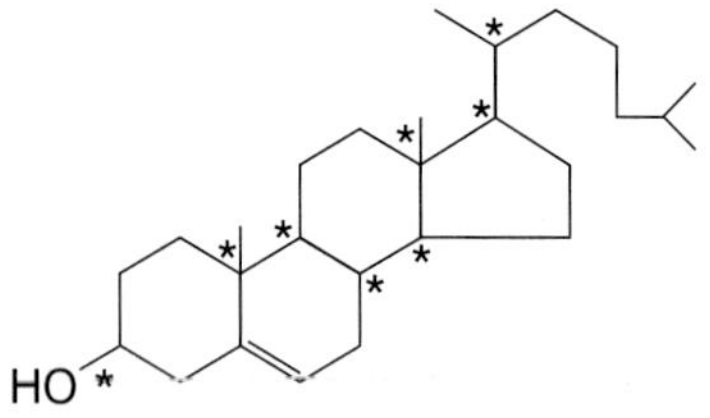

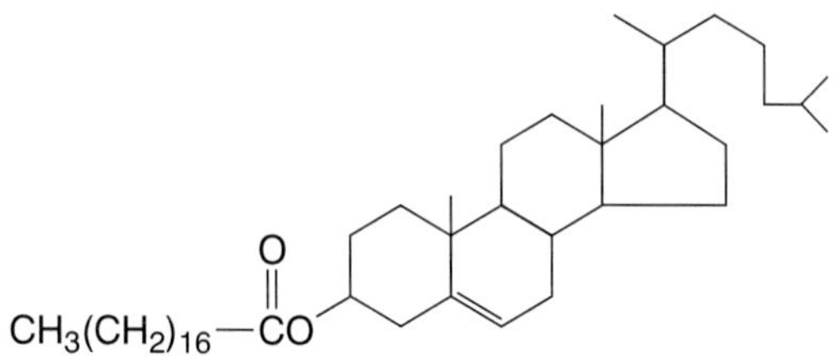

17. a.

$CH_2O{-}\overset{O}{\overset{\|}{C}}{-}(CH_2)_{14}CH_3$
$CH{-}O{-}\overset{O}{\overset{\|}{C}}{-}(CH_2)_{14}CH_3$
$CH_2{-}O{-}\overset{O}{\overset{\|}{C}}{-}(CH_2)_{14}CH_3$

b. $CH_3(CH_2)_4CH{=}CH{-}CH_2CH{=}CHCH_2CH{=}$
$CHCH_2CH{=}CH$
$HO{-}\overset{O}{\overset{\|}{C}}(CH_2)_3$

c.

$CH_2{-}O{-}\overset{O}{\overset{\|}{C}}{-}(CH_2)_7CH{=}CH(CH_2)_7CH_3$
$CH{-}O{-}\overset{O}{\overset{\|}{C}}{-}(CH_2)_{12}CH_3$
$CH_2{-}O{-}\overset{O}{\overset{\|}{C}}{-}(CH_2)_7CH{=}CHCH_2CH{=}CH(CH_2)_4CH_3$

d. $CH_3(CH_2)_{16}\overset{O}{\overset{\|}{C}}{-}O^-\ Na^+$

e. $CH_3(CH_2)_{16}\overset{O}{\overset{\|}{C}}{-}O{-}CH_2(CH_2)_{20}CH_3$

f.

$CH_2{-}O{-}\overset{O}{\overset{\|}{C}}(CH_2)_{12}CH_3$
$CH{-}O{-}\overset{O}{\overset{\|}{C}}{-}(CH_2)_7CH{=}CH(CH_2)_7CH_3$
$CH_2{-}O{-}\overset{O}{\overset{\|}{C}}{-}(CH_2)_{14}CH_3$

g.

h.

$CH_2{-}O{-}\overset{O}{\overset{\|}{C}}{-}C_{17}H_{35}$
$CH{-}O{-}\overset{O}{\overset{\|}{C}}{-}C_{17}H_{33}$
$CH_2{-}O{-}\overset{O}{\overset{\|}{P}}{-}OCH_2CH_2\overset{+}{N}(CH_3)_3$
$\quad\quad\quad\ \ |$
$\quad\quad\quad O^-$

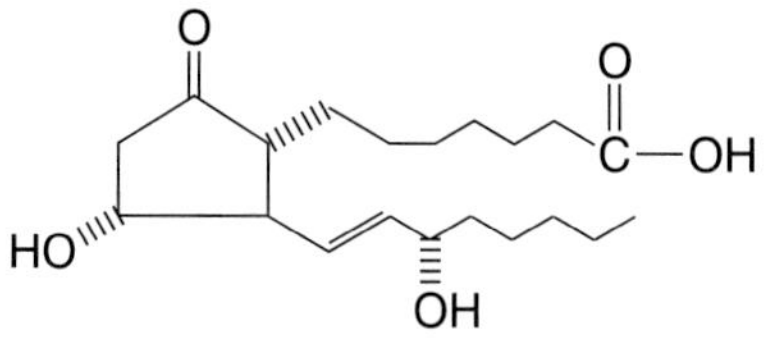

i.

$$CH_2-O-\overset{\overset{\displaystyle O}{\|}}{C}-C_{17}H_{35}$$

$$CH-O-\overset{\overset{\displaystyle O}{\|}}{C}-C_{17}H_{33}$$

$$CH_2-O-\overset{\overset{\displaystyle O}{\|}}{\underset{\underset{\displaystyle O-}{|}}{P}}-OCH_2CH_2\overset{+}{N}H_3$$

j.

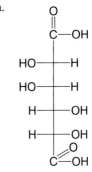

19.  a. Beeswax is a mixture, not a pure compound.
     b. Cortisone does not react when treated with sodium hydroxide or potassium hydroxide.
     c. Glycerol is a trihydroxy alcohol.
     d. Fats are isolated from animals.
     e. Prostaglandins are produced in very small quantities

21.  a.
$$CH_3(CH_2)_7\underset{\underset{\displaystyle OH}{|}}{CH}-\underset{\underset{\displaystyle OH}{|}}{CH}-(CH_2)_7\overset{\overset{\displaystyle O}{\|}}{C}-OH + MmO_2 + KOH$$

b.
$$CH_3(CH_2)_{16}\overset{\overset{\displaystyle O}{\|}}{C}-OH$$

c. $CH_3(CH_2)_7CHBrCHBr(CH_2)\overset{\overset{\displaystyle O}{\|}}{C}-OH$

d. $CH_3(CH_2)_7CH_2-\underset{\underset{\displaystyle Cl}{|}}{CH}(CH_2)_7\overset{\overset{\displaystyle O}{\|}}{C}-OH + HCl$

e. $CH_3(CH_2)_7CH{=}CH(CH_2)_7\overset{\overset{\displaystyle O}{\|}}{C}-O^-\,Na^+ + H_2O$

f. $CH_3(CH_2)_7CH{=}CH(CH_2)_7\overset{\overset{\displaystyle O}{\|}}{C}-O^-\,Na^+ + H_2O + CO_2$

## Chapter 27 Exercises

27.1.  a. Aldose;  b. D-stereochemistry

27.3.  Glyceraldehyde is an optically acitve compound because it contains a chiral center. Dihydroxyacetone does not contain a chiral center and therefore has no optical activity and exists in a single form.

27.5.  a.                    b.

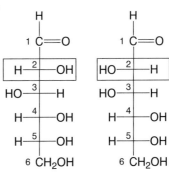

α-D-Mannose            β-D-Galactose

27.7.  a. reducing sugar;  b. reducing sugar;  c. nonreducing sugar;  d. nonreducing sugar

27.9.  Yes, lactose should be a reducing sugar because both α- and β-forms are in equilibrium with the open chain form that contains a free aldehyde group.

27.11. We would not expect glycogen to be a reducing carbohydrate. Although there exists an open-chain glucose unit at the beginning of each polymeric chain, the overwhelming majority of the glucose units exist in the nonreducing full acetal form.

27.13. a.                           b.

| | |
|---|---|
| $\overset{\overset{\displaystyle O}{\|}}{C}-OH$ | $\overset{\overset{\displaystyle O}{\|}}{C}-OH$ |
| HO——H | HO——H |
| HO——H | HO——H |
| H——OH | HO——H |
| H——OH | $\overset{\overset{\displaystyle O}{\|}}{C}-OH$ |
| $\overset{\overset{\displaystyle O}{\|}}{C}-OH$ | |

D-Mannaric acid            L-Ribaric acid

27.15. a.

CH$_2$OH
H——OH
HO——H
H——OH
H——OH
CH$_2$OH

b.

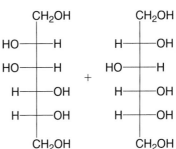

27.17.

| | |
|---|---|
| H | H |
| 1 C=O | 1 C=O |
| H—2—OH | HO—2—H |
| HO—3—H | HO—3—H |
| H—4—OH | H—4—OH |
| H—5—OH | H—5—OH |
| 6 CH$_2$OH | 6 CH$_2$OH |

D-Glucose          D-Mannose

D-Glucose and D-mannose differ only at C-2 and therefore are epimers.

## Chapter 27 Problems

1. a.    b. 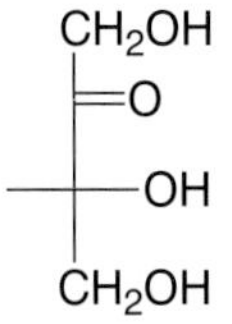    c.

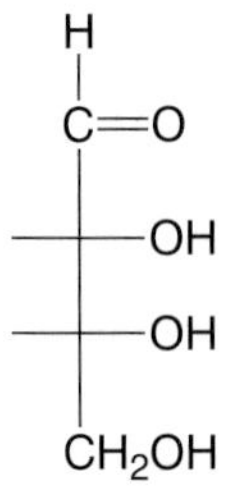

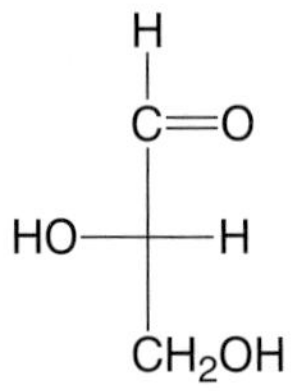

3. a. An aldoheptose is more water soluble than 1-heptanol because an aldoheptose is a polar compound and 1-heptanol is nonpolar.

   b. An aldoheptose is more water soluble because it is a polar compound and heptanal is a nonpolar compound.

5.

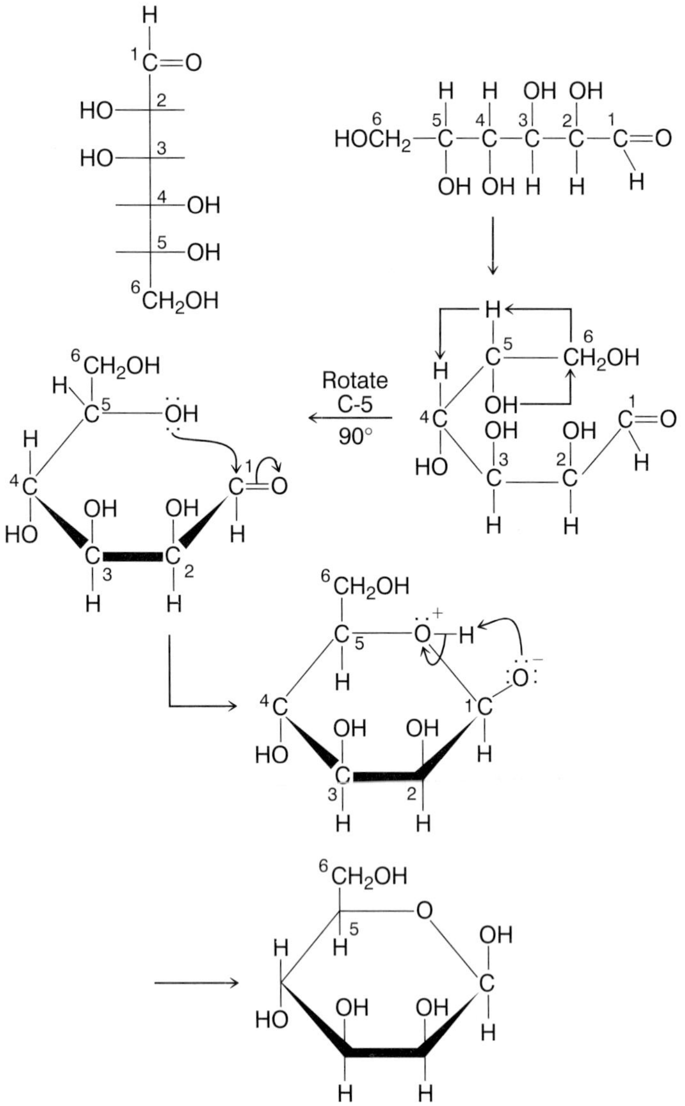

7. Ethyl α-D-mannopyranoside is the only nonreducing sugar listed in problem 27.4. because it is a full acetal.

9.

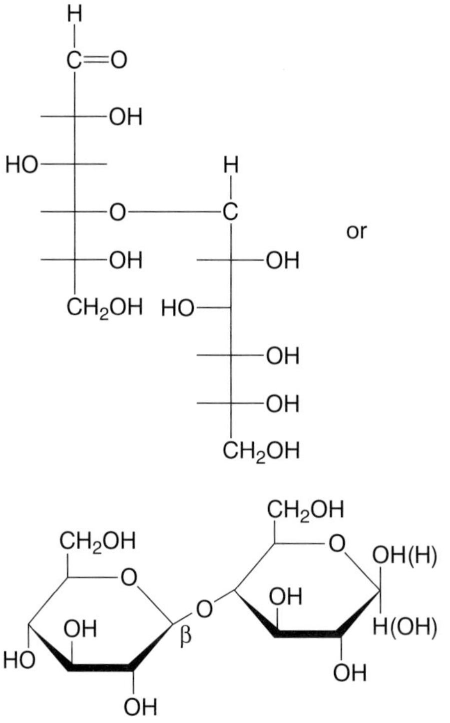

11. a. Both cellulose and amylose have glucose residues, but cellulose has 1,4-β-glycosidic linkages, whereas anylose has 1,4-α-glycosidic linkages.

    b. Amylose has glucose residues with 1,4-α-glycosidic linkages, which means that the glucose residues are linear. Amylopectin has both 1,4-α- and 1,6-α-glycosidic linkages, which means that the glucose residues branch off instead of being linear.

    c. Both amylopectin and glycogen have 1,4-α-glycosidic and 1,6-α-glycosidic linkages, but glycogen is more highly branched than amylopectin.

13.

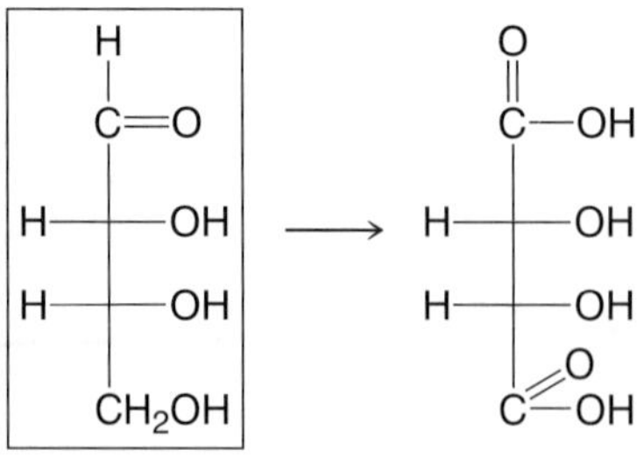

15. a. Benedict's test. With glucose, a red precipitate of $Cu_2O$ forms. No precipitate forms in the tube containing α-D-glucopyranoside.

    b. Treat glucose and fructose with bromine water. Glucose will decolorize bromine water; fructose will not react.

    c. Treat each with 3 moles of phenylhydrazine. The glucose will form a yellow crystalline ozazone. The reduced glucitol will not react.

    d. Starch gives a characteristic blue test when treated with iodine. Cellulose does not become blue.

17.

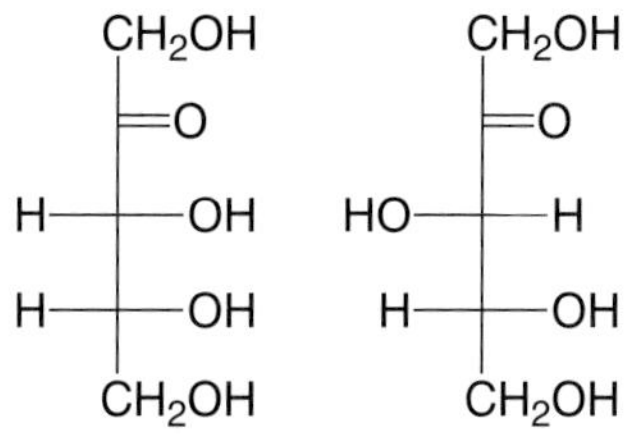

$CH_2OH$ — $=O$ — H—OH — H—OH — $CH_2OH$    $CH_2OH$ — $=O$ — HO—H — H—OH — $CH_2OH$

19. a.

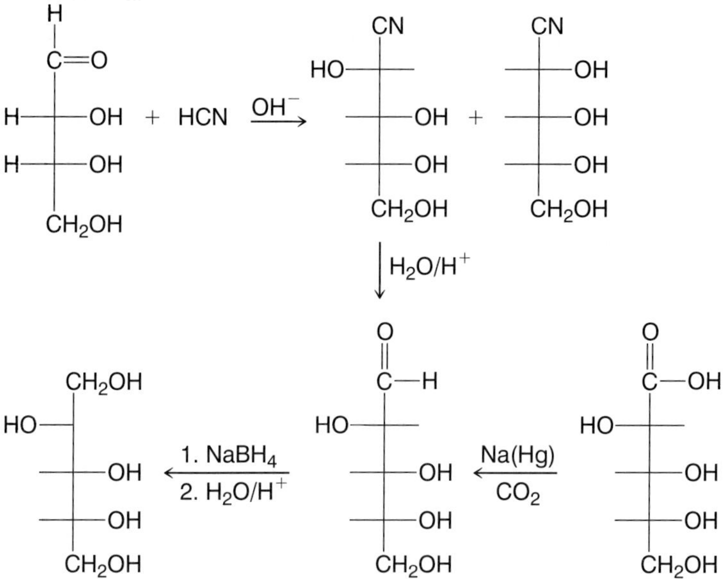

$C=O$ (H) — H—OH — H—OH — $CH_2OH$  + HCN $\xrightarrow{OH^-}$  HO—CN / —OH / —OH / $CH_2OH$  + —CN / —OH / —OH / $CH_2OH$

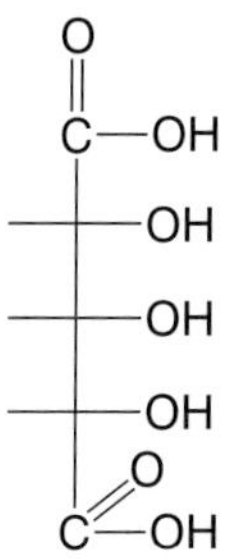

$\downarrow H_2O/H^+$

HO—$CH_2OH$ / —OH / —OH / $CH_2OH$  $\xleftarrow[2.\ H_2O/H^+]{1.\ NaBH_4}$  HO— $\overset{O}{C}$—H / —OH / —OH / $CH_2OH$  $\xleftarrow[CO_2]{Na(Hg)}$  HO— $\overset{O}{C}$—OH / —OH / —OH / $CH_2OH$

b.

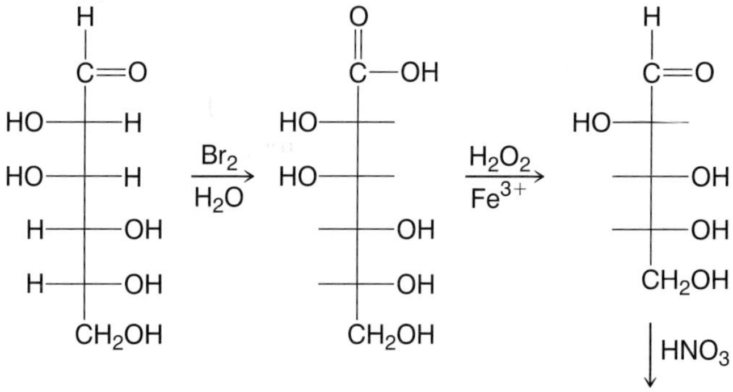

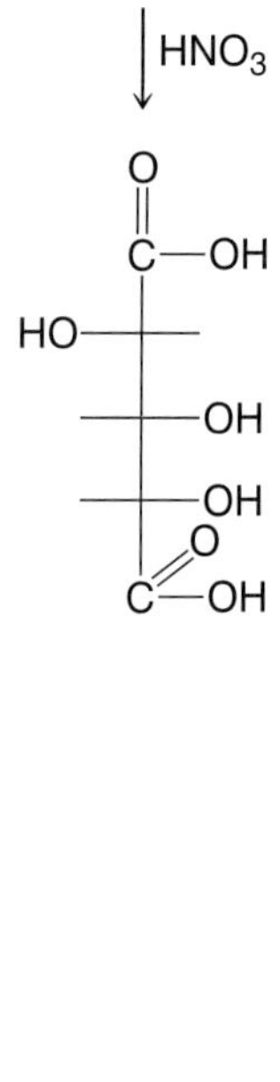

HO—H / HO—H / H—OH / H—OH / $CH_2OH$ ($C=O$, H) $\xrightarrow[H_2O]{Br_2}$  HO— $\overset{O}{C}$—OH / HO— / HO— / —OH / —OH / $CH_2OH$ $\xrightarrow[Fe^{3+}]{H_2O_2}$  HO— ($\overset{H}{C}=O$) / —OH / —OH / $CH_2OH$

$\downarrow HNO_3$

$\overset{O}{C}$—OH / HO— / —OH / —OH / $\overset{O}{C}$—OH

21. a.

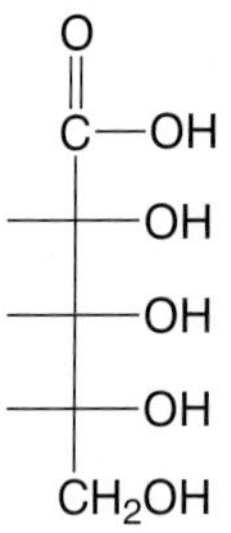

$\overset{O}{C}$—OH / —OH / —OH / —OH / $CH_2OH$

b.

$\overset{O}{C}$—OH / —OH / —OH / —OH / $\overset{O}{C}$—OH

c.

$\overset{H}{C}=N$—$NHC_6H_5$ / $C=N$—$NHC_6H_5$ / —OH / —OH / $CH_2OH$

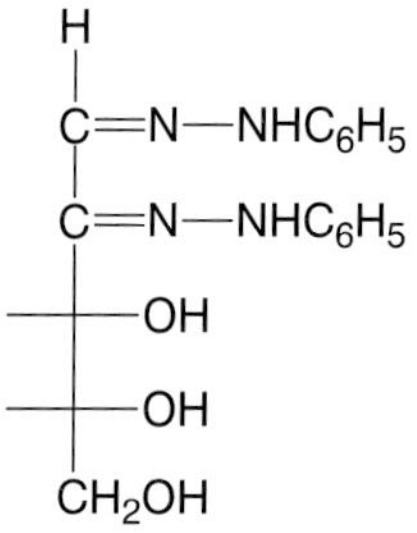

d.

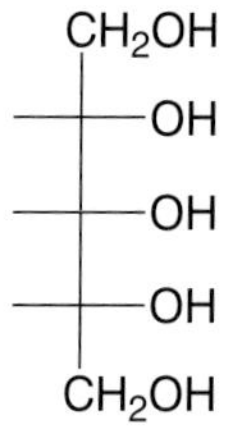

$CH_2OH$ / —OH / —OH / —OH / $CH_2OH$

23.

$\overset{H}{C}=O$ / HO— / —OH / $CH_2OH$    $\overset{H}{C}=O$ / HO— / HO— / —OH / $CH_2OH$    $CH_2OH$ / $=O$ / HO— / —OH / $CH_2OH$

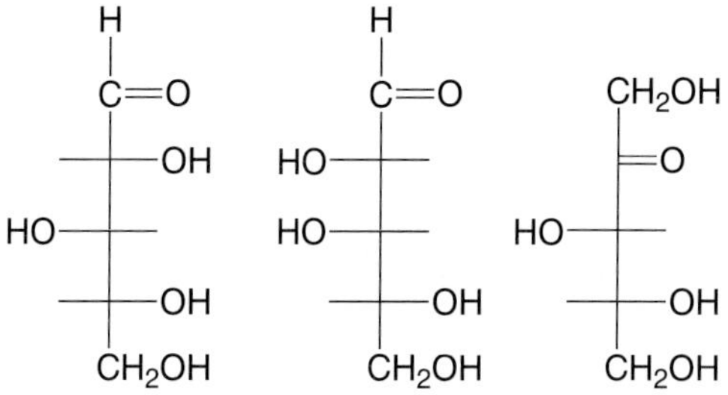

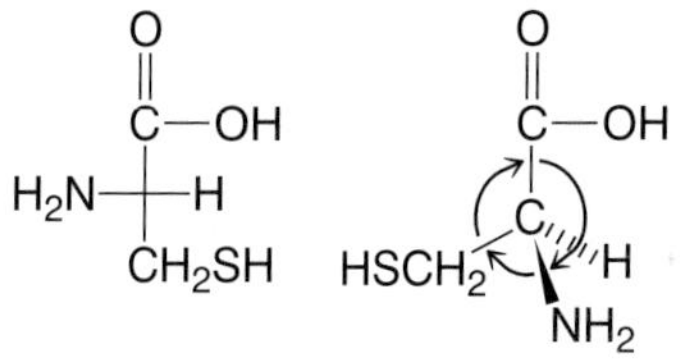

## Chapter 28 Exercises

28.1.

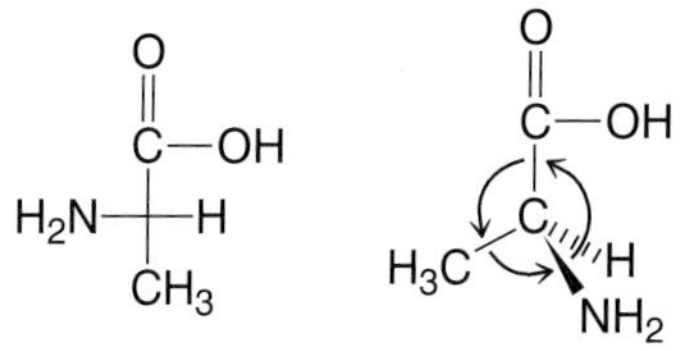

$\overset{O}{C}$—OH / $H_2N$—H / $CH_3$   L-Alanine     $\overset{O}{C}$—OH, $H_3C$—$C$—H, $NH_2$   S-Configuration

$\overset{O}{C}$—OH / $H_2N$—H / $CH_2SH$   L-Cysteine     $\overset{O}{C}$—OH, $HSCH_2$—$C$—H, $NH_2$   R-Configuration

28.3.  $C_6H_5CH_2-\overset{O}{C}-\overset{O}{C}-OH$

28.5.  $H_3\overset{+}{N}-\overset{H}{C}-\overset{O}{C}-OH$ ($CH_3$) $\underset{H^+}{\overset{OH^-}{\rightleftharpoons}}$ $H_3\overset{+}{N}-\overset{H}{C}-\overset{O}{C}-O^-$ ($CH_3$)

Alanine cation     Alanine dipolar ion

$\underset{H^-}{\overset{OH^-}{\rightleftharpoons}}$ $H_2N-\overset{H}{C}-\overset{O}{C}-O^-$ ($CH_3$)

Alanine anion

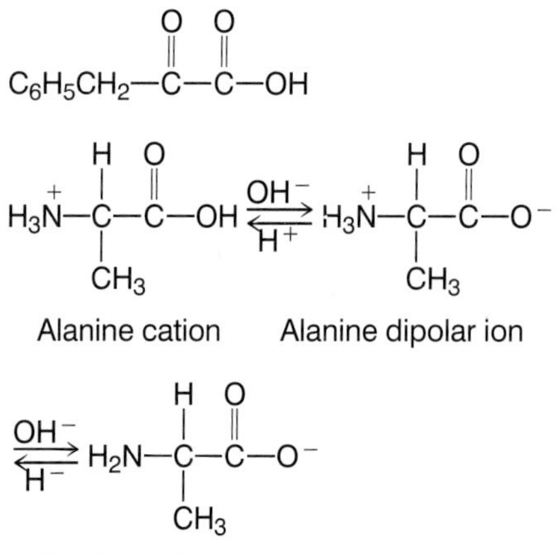

**28.7.**

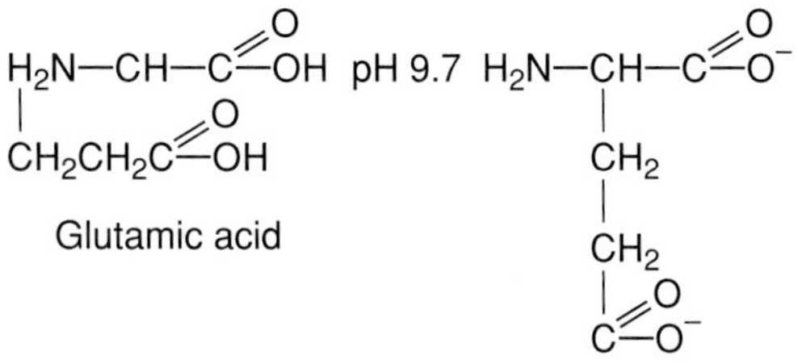

Glutamic acid

The pH of 9.7 is greater than the $pK_{a3}$. Thus, the dominant species of glutamic acid is the dianionic form.

**28.9.** $n! = 5$

$n! = n(n-1)(n-2)(n-3)(n-4)(n-5)\cdots(1)$

$n! = 5 \times 4 \times 3 \times 2 \times 1 = 120$ possible isomers

**28.11.** Proline.

## Chapter 28 Problems

1.  a.

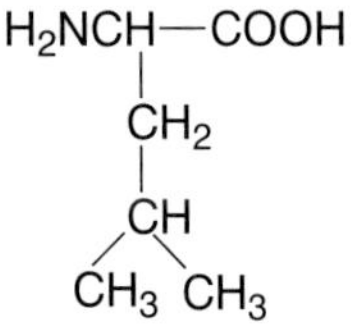

      Leu    L

   b.

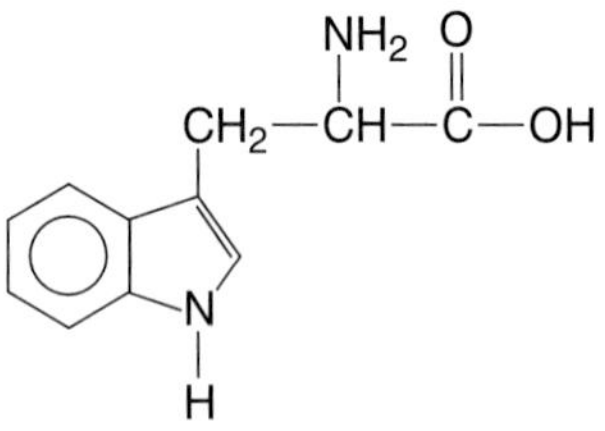

      Trp    W

   c.

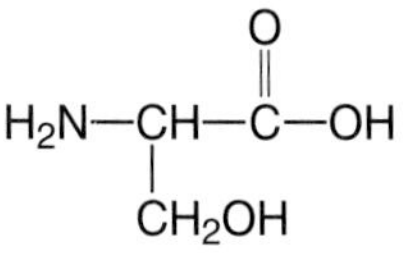

      Ser    S

   d.

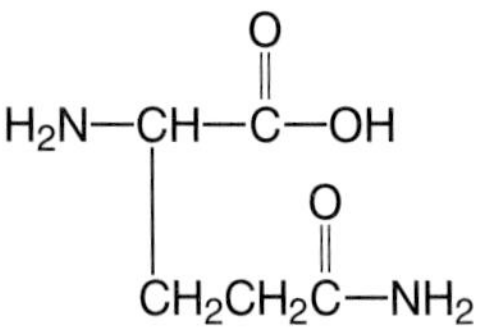

      Gln    Q

   e.

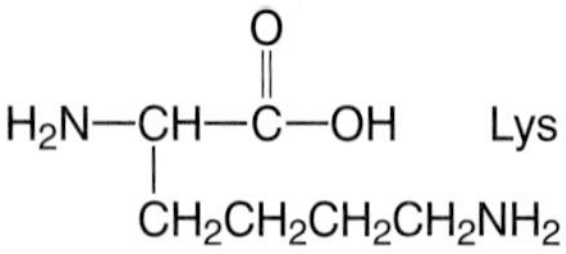

      Lys    K

3.

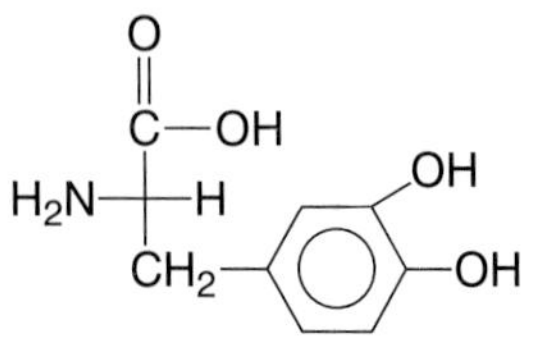

L-Dopa

L-Dopa can be considered to be a derivative of tyrosine.

5.  a. Dipolar

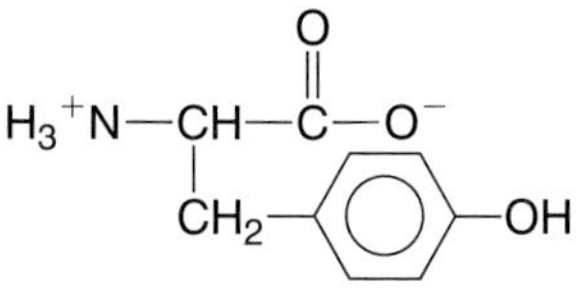

Tyrosine 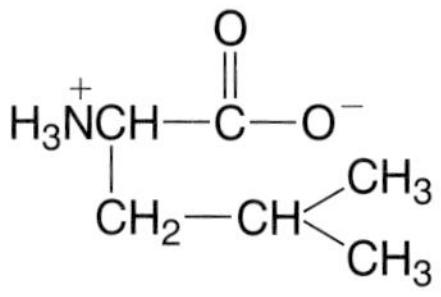            Leucine

   b. Anionic

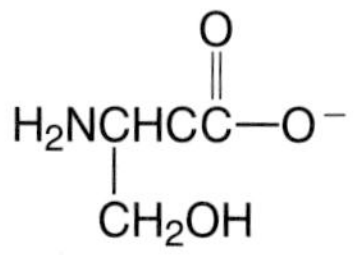

Serine 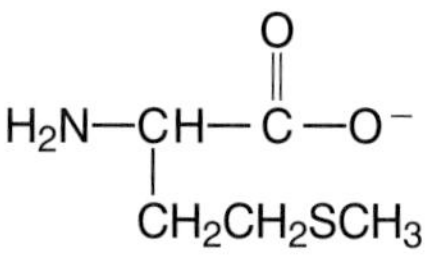            Methionone

   c. Cationic

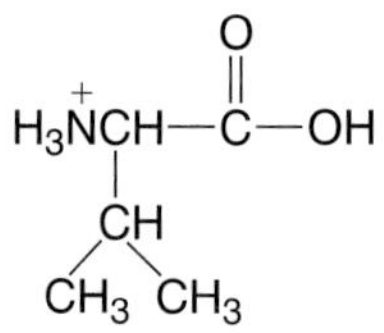

Valine 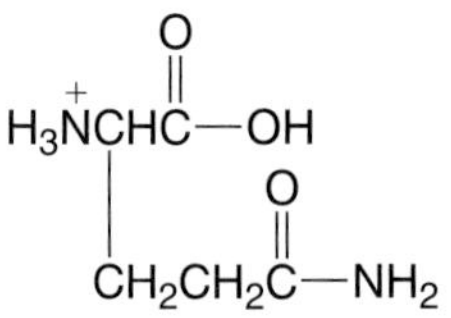            Glutamine

7.  a. Glycine does not have a chiral carbon:
   b. (1) Aromatic: Phenylalanine; Tyrosine; Tryptophan; (2) Heterocyclic: Tryptophan; Proline; Histidine; (3) Aromatic + heterocyclic: Tryptophan; Histidine; c. Sulfur: Cysteine and Methionone
   c. Amide: Asparagine; and Glutamine
   d. No primary amine: Proline

9.  3.2

11.  a.

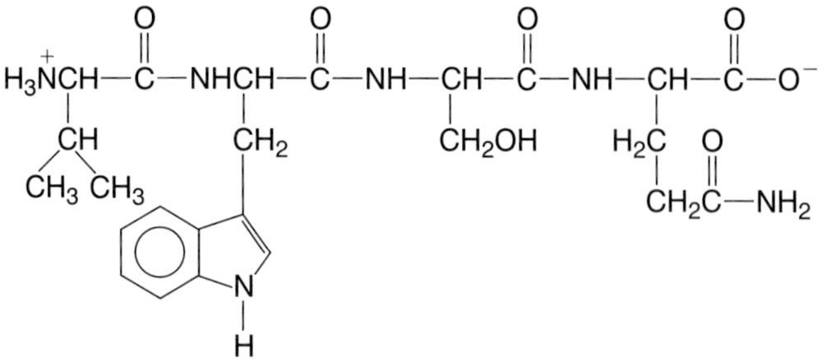

   b.   $Cl^-$ $\overset{+}{H_3N}CH_2\overset{O}{\overset{\|}{C}}-OH$

13.  $n! = 3 \times 2 \times \times 1 = 6$ However, since a glutamic acid residue could bond to cysteine with its $\alpha$ carboxyl or $\delta$ carboxyl group, 6 is doubled to get 12.

15.  a. A product yielded on the hydrolysis of a conjugated protein. Prosphetic groups may be either organic or inorganic substanes. Hydrolysis of a conjugated protein may yield more than one prosthetic group.
   b. The arrangement in space or the stereochemistry of a protein.
   c. Precipitation of a protein by treatment of the protein with acid, base, heat, ultraviolet light, high-molecular-weight metal ions, detergents, pressure, organic solvents, or other reagents.

  d. An enzyme that selectively attacks the peptide bond of the *C*-terminus.

  e. Compounds that react with both acid and base are said to be amphoteric.

  f. The experimental procedure used to separate mixtures of amino acids or proteins based on passing a current through the mixture placed on a filter paper or cellulose acetate gel (buffer pH 6) and determining the migration based on the net charge when an electric current is passed through the system.

17.    Val-Gly-Ser-Met-Cys-Gly

19.    Hydrogen bonding, disulfide bonds, salt bridges, London forces

21.    a.

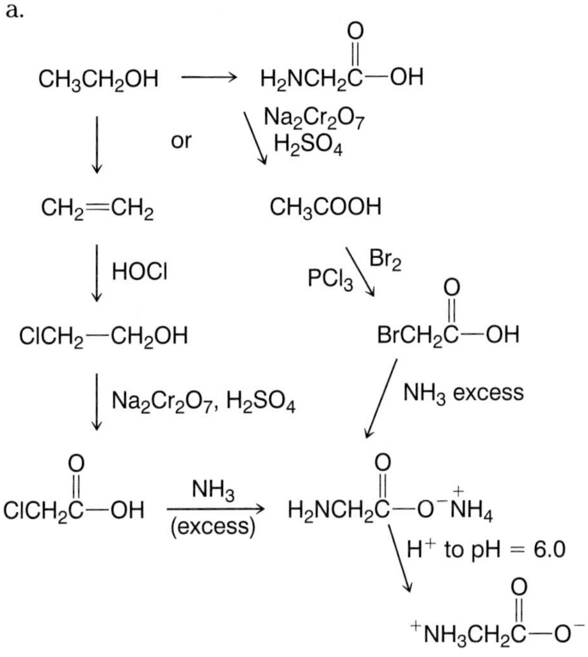

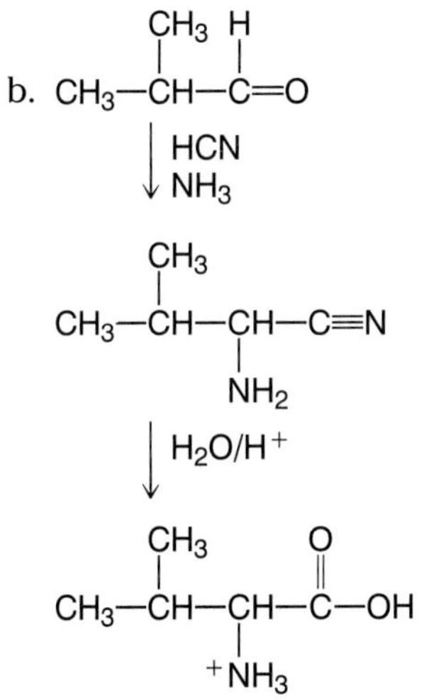

     b.

23.    a.                       b.

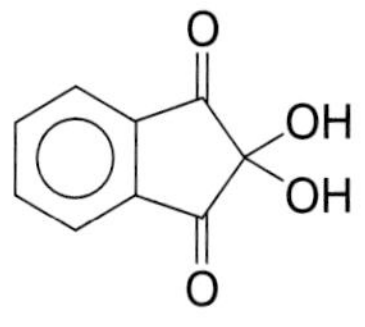

Ninhydrin

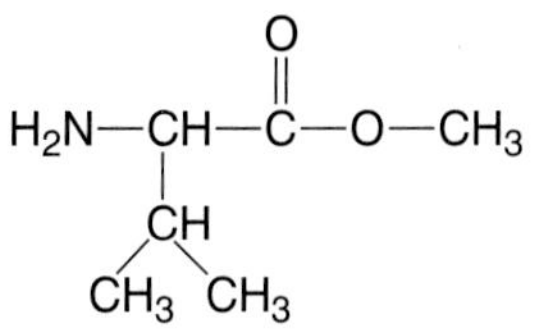

   c.

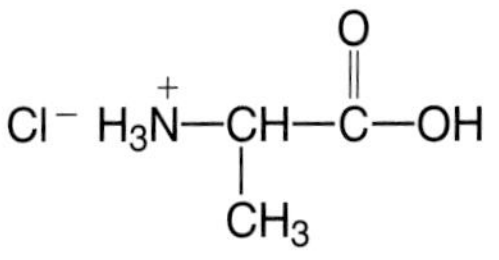

   d.

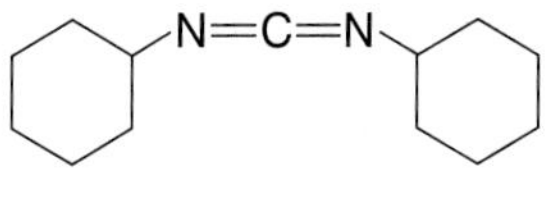

   e.

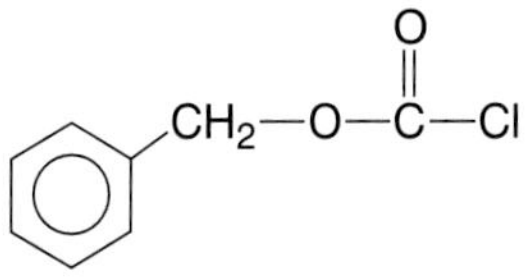

25.

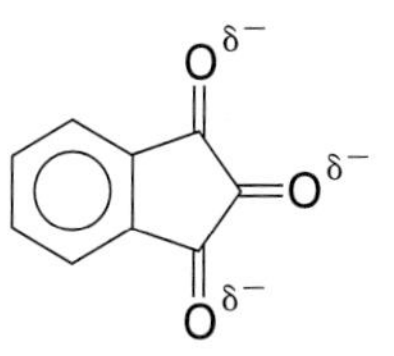

The triket form is instable due to repulsion of like $\delta^+$ charges on adjacent carbons—weakening the bonds.

## Chapter 29 Exercises

29.1.   $5.0 \times 10^{17}\ s^{-1}$

29.3.  

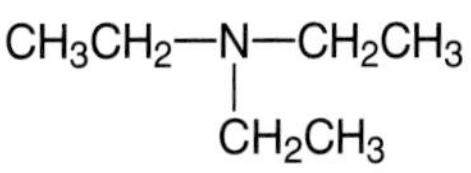

Triethylamine

29.5.   a.   7.3 pp (7.27 ppm)

        b.   727 Hz

29.7.   a.

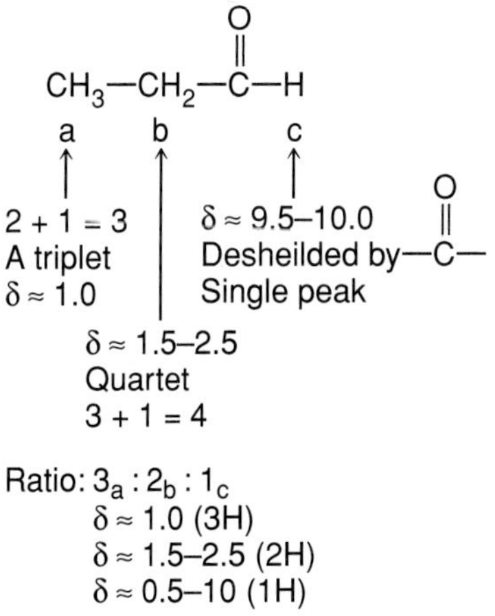

b.

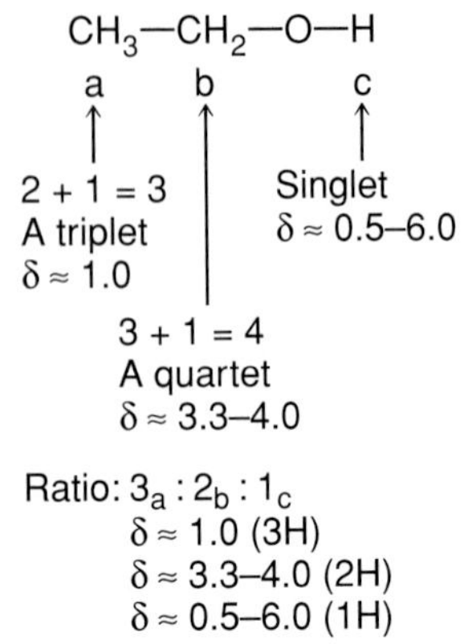

$CH_3-CH_2-O-H$

a        b        c

2 + 1 = 3        Singlet
A triplet        $\delta \approx 0.5\text{–}6.0$
$\delta \approx 1.0$

3 + 1 = 4
A quartet
$\delta \approx 3.3\text{–}4.0$

Ratio: $3_a : 2_b : 1_c$
$\delta \approx 1.0$ (3H)
$\delta \approx 3.3\text{–}4.0$ (2H)
$\delta \approx 0.5\text{–}6.0$ (1H)

c.

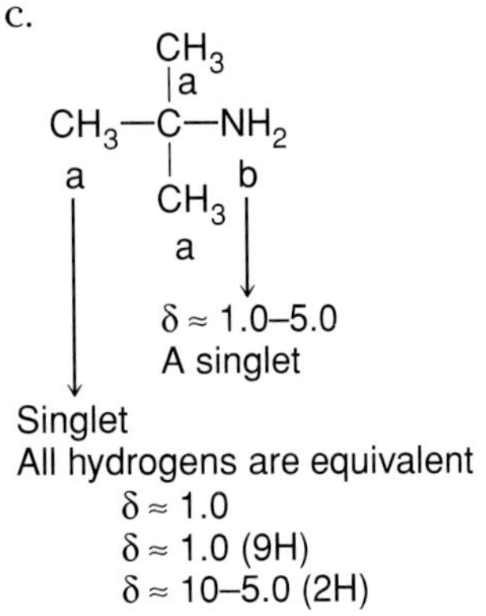

$CH_3$
|a
$CH_3-C-NH_2$
a        |        b
$CH_3$
a

$\delta \approx 1.0\text{–}5.0$
A singlet

Singlet
All hydrogens are equivalent
$\delta \approx 1.0$
$\delta \approx 1.0$ (9H)
$\delta \approx 10\text{–}5.0$ (2H)

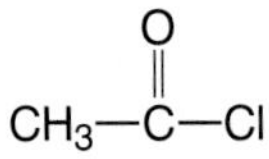

$CH_3-CH-CH_3$
29.9.        |
$Cl$

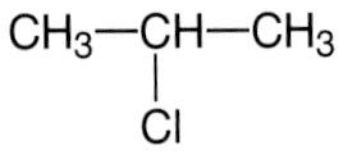

## Chapter 29 Problems

1.    a. $9.6 \times 10^{21}$ is more energetic; b. $3.3 \times 10^{15}$ Hz is more energetic.

3.    a. $1.709$ cm$^{-1}$; b. $35.71$ cm$^{-1}$

5.    a. $CH_3C\equiv N$
      b.

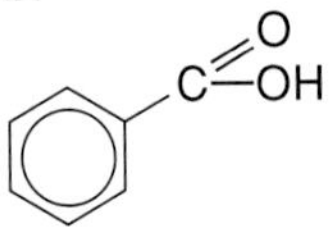

      c. $CH_3CH_2\overset{O}{\overset{\|}{C}}-H$   or   $CH_3\overset{O}{\overset{\|}{C}}-CH_3$

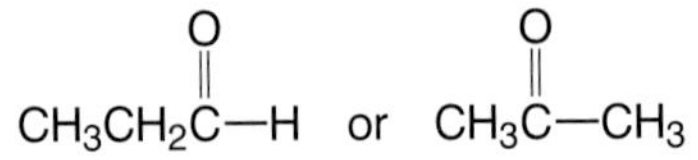

d.  $CH_3CH_2CH_2OH$  or  $CH_3-CH-CH_3$
                                                    |
                                                   OH

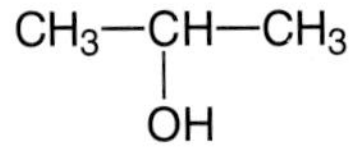

7.    $CH_3-\overset{O}{\overset{\|}{C}}-Cl$ and $(CH_3)_4$

9.    $CH_3CH_2\overset{O}{\overset{\|}{C}}-H$, $H-\overset{O}{\overset{\|}{C}}-OCH_2CH_3$, $CH_3CH_2CH_2NH_2$, and $CH_3CH_2I$

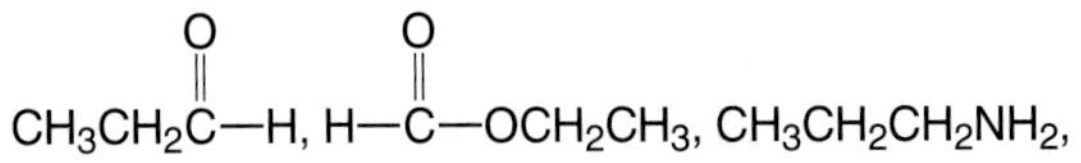

11.   a. Relatively high frequency in the NMR spectrum. Example: $\delta 1$ is upfield of $\delta 3$.

      b. Case when no electronegative groups are in the vicinity of the hydrogen to reduce the electron density of the hydrogen.

      c. The splitting pattern of the hydrogens on a given carbon can be predicted by adding the number of hydrogens on adjacent carbons and then adding one.

      d. The number of waves per unit time that pass through a given point when the wave is propogated through space.

      e. The distance from the peak (or valley) of one wave to the peak (or valley) of the next wave.

13.   The t-butyl alcohol sample was not pure. It was possibly contaminated with an aldehydic compound which is oxidized with $Na_2Cr_2O_7$ and would show an absorption between 1690 cm$^{-1}$ and 1740 CH$^{-1}$.

15.   a. $CH_3CBr_3$ Single peak, all hydrogens on methyl group are equivalent; b. 7; c. 13

17.   $CH_3-\overset{CH_3}{\underset{CH_3}{\overset{|}{\underset{|}{C}}}}-OH$   shows only two peaks

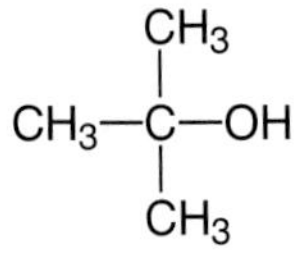

# Index

**Photos**

**Chapter 9**   9.4 Courtesy of IBM Corporation, Research Division, Almaden Research Center.

**Chapter 29**   29.4, 29.5, 29.6, 29.7, 29.8 Courtesy of Norman Schmidt 29.10, 29.11, 29.13, 29.14, 29.15 Courtesy of Dr. Jeffrey Orvis 29.16 Courtesy of Professor Erich C. Blossey, Rollins College. 29.17 © Science VU/Visuals Unlimited

**Line Art**

**Chapter 9**   9.3 Based on DD Ebbing. *General Chemistry.* Copyright © 1993. Boston: Houghton Mifflin, p. 276. 9.9 Based on J McMurry. *Organic Chemistry*, 3rd edition. Copyright © 1992. Pacific Grove, CA: Brooks/Cole Publishing, p. 144.

**Chapter 11**   11.8 Based on TWG Solomons. *Fundamentals of Organic Chemistry*, 4th edition. Copyright © 1994. New York: John Wiley & Sons, p. 182.

**Chapter 26**   26.14 Based on RJ Fessenden and JS Fessenden. *Organic Chemistry*, 4th edition. Copyright © 1990. Pacific Grove, CA: Brooks/Cole Publishing, p. 945.

**Chapter 28**   28.14B Based on H Hart. *Organic Chemistry: A Short Course*, 8th edition. Copyright © 1991. Boston: Houghton Mifflin, p. 485. 28.15 Based on AD Baker and R Engel. *Organic Chemistry.* Copyright © 1992. St. Paul, MN: West Publishing, p. 1035. 28.16 Based on TWG Solomons. *Fundamentals of Organic Chemistry*, 4th edition. Copyright © 1994. New York: John Wiley & Sons, p. 1014.

**Chapter 29**   29.9 Based on WW Linstromberg and HE Baumgarten. *Organic Chemistry*, 6th edition. Copyright © 1987. Lexington, MA: D. C. Heath, p. 479.

## Important Functional Groups

| Functional Group Family Name | Functional Group Structure | | Example |
|---|---|---|---|
| Alkane (paraffin) | 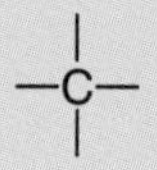 $-\overset{\vert}{\underset{\vert}{C}}-$ | 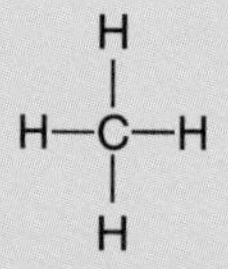 $H-\overset{\overset{\displaystyle H}{\vert}}{\underset{\underset{\displaystyle H}{\vert}}{C}}-H$ | Methane, major component of natural gas |
| Alkene (olefin) | 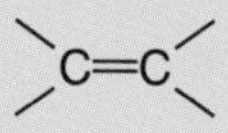 $\overset{}{\underset{}{>}}C=C\overset{}{\underset{}{<}}$ | 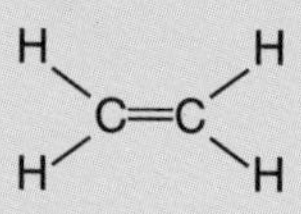 $\overset{H}{\underset{H}{>}}C=C\overset{H}{\underset{H}{<}}$ | Ethylene, raw material for the production of polyethylene |
| Alkyne | $-C\equiv C-$ | $H-C\equiv C-H$ | Acetylene, used in the oxyacetylene torch to generate a hot flame |
| Halide X = F, Cl, Br, and I | 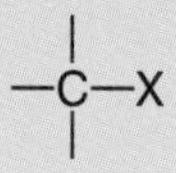 $-\overset{\vert}{\underset{\vert}{C}}-X$ | $CCl_4$ | Carbon tetrachloride, nonpolar solvent |
| Aromatic hydrocarbon | 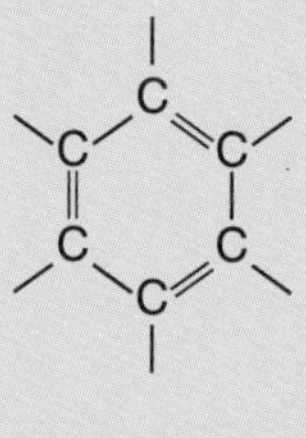 | 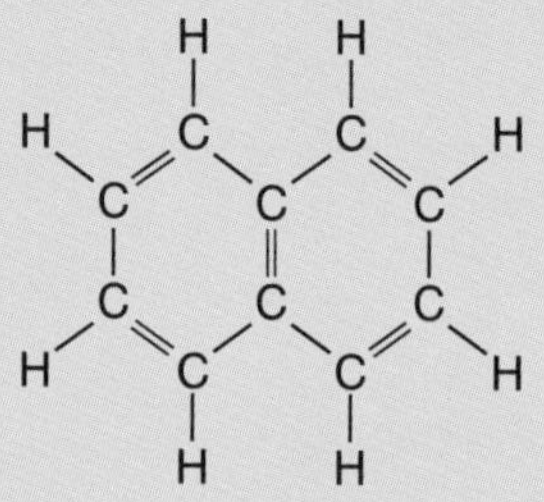 | Naphthalene, moth repellant |
| Alcohol | 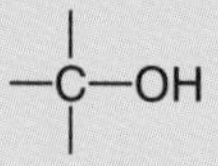 $-\overset{\vert}{\underset{\vert}{C}}-OH$ | 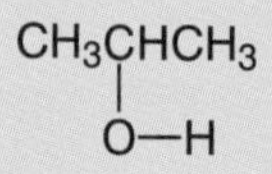 $CH_3\overset{}{\underset{\underset{\displaystyle O-H}{\vert}}{C}}HCH_3$ | Isopropyl (rubbing) alcohol, furnishes the "sting" of an after shave lotion |
| Mercaptan (thiol) | 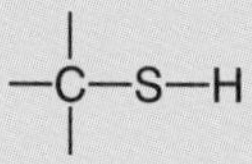 $-\overset{\vert}{\underset{\vert}{C}}-S-H$ | $(CH_3)_2CHCH_2CH_2SH$ | Isopentyl mercaptan, essence of skunk |
| Disulfide | 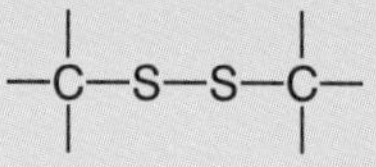 $-\overset{\vert}{\underset{\vert}{C}}-S-S-\overset{\vert}{\underset{\vert}{C}}-$ | $CH_3-S-S-CH_3$ | Dimethyl disulfide |